Proceedings of International Conference on

Electronic Components and Materials

November 7-10, 1989

Beijing, China

CIE

IEEE

ECS

CHMT

International Academic Publishers
A Pergamon-CNPIEC Joint Venture

Pergamon Press
Oxford · New York · Beijing · Frankfurt
São Paulo · Sydney · Tokyo · Toronto

IAP

DISTRIBUTORS:

U.K.	Pergamon Press plc. Headington Hill Hall, Oxford OX3 0BW. England
U.S.A.	Pergamon Press. Inc.. Maxwell House. Fairview Park. Elmsford. New York 10523. U.S.A.
PEOPLE'S REPUBLIC OF CHINA	International Academic Publishers. Xizhimenwai Dajie. Beijing Exhibition Centre. Beijing 100044. People's Republic of China
FEDERAL REPUBLIC OF GERMANY	Pergamon Press GmbH. Hammerweg 6. D-6242 Kronberg. Federal Republic of Germany
BRAZIL	Pergamon Editora Ltda. Rua Eça de Queiros. 346. CEP 04011. Paraiso. São Paulo. Brazil
AUSTRALIA	Pergamon Press Australia Pty Ltd.. P.O. Box 544. Potts Point. N.S.W. 2011. Australia
JAPAN	Pergamon Press. 5th Floor. Matsuoka Central Building. 1-7-1 Nishishinjuku. Shinjuku-ku. Tokyo 160. Japan
CANADA	Pergamon Press Canada Ltd.. Suite No. 271. 253 College Street. Toronto. Ontario. Canada M5T 1R5

First edition 1989

Library of Congress Cataloging-in-Publication Data

International Conference on Electronic Components and Materials (1989 : Beijing, China)
Proceedings of International Conference on Electronic Components and Materials : November 7-10, 1989, Beijing, China. — 1st ed.
p. cm.
Sponsored by CIE Electronic Components Society and others.
1. Electronic apparatus and appliances--Congresses.
2. Electronics—Materials—Congresses. I. CIE Electronic Components Society. II. Title. III. Title: Electronic components and materials.
TK7869.I57 1989
621.381—dc20 89-29226
CIP

British Library Cataloguing in Publication Data

International Conference on Electronic Components and Materials (1989: Beijing)
Proceedings of International Conference on Electronic Components and Materials, November 7-10, 1989, Beijing, China.
1. Electronic equipment
I. Title
621.381

ISBN 0-08-040175-9 Pergamon Press
ISBN 7-80003-087-3 International Academic Publishers

Printed in the People's Republic of China

Co-sponsors:

CIE Electronic Components Society

IEEE Components, Hybrids and Manufacturing Technology Society

IEEE Beijing Section

Cooperated with:

National Nature Science Foundation of China

Chinese Electronic Components Association

Conference Chairman:

Chen Kegong

INTERNATIONAL ADVISORY COMMITTEE

Chairmen:

Yin Zhiwen, China
Zhang Xiaowen, China
C.P. Wong, U.S.A.
Man F. Yan, U.S.A.
Hiroaki Yanagida, Japan
G. Arlt, F.R.G.
L.C. Burton, U.S.A.
C. G. Chen, U.S.A.
L.E. Cross, U.S.A.
D.K. Das-Gupta, U.K.
Makoto Kuwabara, Japan
S. Middelhoek, Netherlands
Qu Xixin, China
Don Smith, U.S.A.
R.W. Vest, U.S.A.
Yao Xi, China

PROGRAM COMMITTEE

Zhou Zhigang, China
Zhang Shiying, China
Kunihito Koumoto, Japan
F.W. Ainger, U.K.
Hisao Banno, Japan
Cai Qishan, China
Chen Guoguang, China
T.J. Coutts, U.K.
R.S. Duggan, U.S.A.
Lucien Eyraud, France
G.W. Farnell, Canada

PROGRAM COMMITTEE (Continued)

Gao Jinghao, China
Isemi Igarashi, Japan
Ji Hengkuan, China
John M. Kinn, U.S.A.
C.Y. Kuo, U.S.A.
J.H. Lau, U.S.A.
E.G.H. Lean, U.S.A.
Li Biaorong, China
Li Congzhou, China
Liu Hongyang, China
E. Lüder, F.R.G.
Ma Huaizu, China
Z.H. Maiksin, U.S.A.
Minoru Nagata, Japan
Yoshikatsu Namba, Japan
R.E. Newnham, U.S.A.
U.C. Paek, U.S.A.
Arnold Pfahnl, U.S.A.
Qian Zuqing, China
Qin Zikai, China
Akio Sasaki, Japan
W. Sansen, Belgium
J.M. Segelken, U.S.A.
J.M. Sipress, U.S.A.
William M. Speri, U.S.A.
Jozef Sroka, Poland
E. Suhir, U.S.A.
Sun Chengyon, China
Kiyoshi Takahashi, Japan
K.F. Teng, U.S.A.
Joe Unsworth, Australia
Wang Tingshu, China
Wu Boyuan, China
Wu Chongruo, China
Wu Yuansheng, China
Wu Yuxing, China
N. Yamazoe, Japan
Yang Linsheng, China
Zhang Jigao, China
Zhou Hengjin, China

ORGANIZING COMMITTEE

Sha Zong
Dong Yizhong
Bai Yuxian
Deng Lei
Duan Bingyi
Guo Jiucheng
Guo Yishu
Han Xizhen
Kou Yuqing
Li Fenglin
Li Jingyuan
Li Yuan
Song Yongxiang
Wang Wensheng
Wu Shuzhi
Yang Chenghua
Yin Qihua
Zhang Renren
Zhang Xi
Zhang Yaoxi
Zhao Guangyun
Zhao Pu
Zhou Mengqi
Zhuang Yan

Foreword

As the result of two years preparation, the International Conference on Electronic Components and Materials (ICECM) will be held in Beijing on November 7, 1989. Initiated by the Electronic Components Society of Chinese Institute of Electronics (CIE ECS), the conference is jointly sponsored by CIE ECS, IEEE CHMT Society and IEEE Beijing Section in cooperation with National Nature Science Foundation of China and Chinese Electronic Components Association. Dr. Leo Feinstein, Chairman of IEEE CHMT, has been in support and showing concern for ICECM from the very beginning. He even came in person to provide guidance for the preparation work for ICECM upon the founding of the CHMT Chapter of IEEE Beijing Section in December, 1988. Dr. C.P.Wong, Dr. C.G.Chen, Prof. Kunihito Koumoto and Dr. D.K.Das-Gupta have also given a great deal of help to ensure the publication of the Proceedings to be on schedule.

Electronic components and materials are building blocks of electronic industry and the achievements made in this respect will greatly facilitate the progress of electronic technology and the industry. This collection of Proceedings includes more than 150 papers dealing with various kinds of organic and inorganic materials, discrete components, hybrid microcircuits, fiber optics, sensors, superconducting materials, reliability, manufacturing technology and frontiers of packaging technology, etc. which represent the new development in research and production for today's electronic components and materials.

The authors involved include the science and technology personnel from over ten countries such as Australia, Canada, China, F.R.G., France, Japan, Netherland, U.K., U.S.A., U.S.S.R., etc., whose work comprising this issue of the Proceedings will further promote international exchange and cooperation in this field.

I wish to express my gratitude to the authors and all those who have worked for this publication.

Chen Kegong

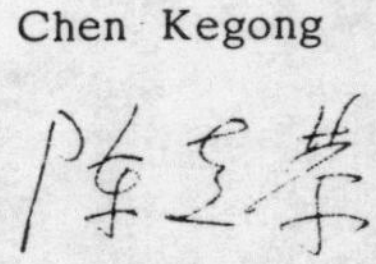

Chairman of ICECM
President of CIE ECS

Supporters:

China Zhen Hua Electronic Industry Corporation

Anshan Electronic Ceramic Company
Beijing Radio Components Factory No. 10
Beijing Third Radio Appliance Factory
China Electronic Components and Equipments Corporation
Fengkai Kaida Electrical Device Company
Hong Ming Electronic Components Factroy
Xiamen Overseas Chinese Electronic Co., LTD.

CONTENTS

Session XVII Frontiers of Packaging Technology

Session XVIII Crystal Materials and Devices

Session XIX Discrete Components

ELECTRONIC CERAMICS: CURRENT ISSUES AND FUTURE PROSPECTS

L. Eric Cross
Director, Materials Research Laboratory
Pennsylvania State University, University Park, PA 16802, USA.

The talk will discuss current and emerging new needs for electronic ceramic materials and composites with property combinations tailored for specific electronic and photonic fields of application. New needs in dielectrics for capacitors and electronic packages will be discussed emphasizing the problems and opportunities as applications move toward thinner films and lower processing temperatures. Thin ferroelectric films on silicon have sparked renewed interest in possible application for non-volatile memory but some problems with fatigue and retention must be resolved. Developments in piezoelectrics and electrostrictors for sensors and actuators suggest new possibilities for "agile" systems and for sensor:responder combinations which may engender a new generation of "smart" solids. In photonics the needs generated by fiber optic communications and optical computing are clamant, but the response is necessarily slower in view of the quality and homogeneity of crystals required. The status in electro-optic, non-linear optic, and photo-refractive crystals will be briefly reviewed and new thin film approaches explored. New "wild cards" in the electronic materials pack are the high T_c supeconductors and the recent emergence of techniques to generate diamond and diamond-like films by plasma assisted vapor deposition. Possible impacts of these developments will be discussed.

GRAIN BOUNDARY EFFECTS IN SEMICONDUCTOR CERAMICS

Yin Zhiwen, Wang Hong, Zhong Jipin
Shanghai Institute of Ceramics, Academia Sinica
Shanghai 200050, China

Abstract

A model of grain boundary Schottky barrier with an intergranular insulating layer was proposed and discussed for the grain boundary barrier layer (GBBL) capacitor (Fig. 1). With this model, the current-voltage relation and capacitance-voltage relation of GBBL capacitors were theoretically analyzed and derived. By means of these relations, the computerized imitation was performed for the experimental results. The reasonableness of above proposed barrier model was confirmed by the coincidence of the experimental data with the computerized results. When using this model for GBBL capacitors compared with the Schottky barrier model for ZnO varistor, two different breakdown mechanisms were found: the avalanche breakdown for the former and the tunnelling breakdown for the latter.

INTRODUCTION

It was well known that the electrical properties of semiconductor ceramics, such as GBBL capacitors, positive temperature coefficient resistors (PTCR) and ZnO varistors, are closely correlated with the characteristics of their grain boundaries. Several grain boundary barrier models have been proposed for the research theoretically. For instance, the grain boundary Schottky barrier model and the model of the grain boundary barrier with the intergranular layer have been used to explain the PTC effect[1,2] and the nonlinear characteristics of the ZnO varistors[3-10]. Park and Payne did discuss the grain boundary barrier of GBBL capacitors[11], however, only very few studies on the effect of grain boundary barrier on the properties of GBBL capacitors. In present paper, some results of further investigation on grain boundary barrier mainly for GBBL capacitors will be presented.

PROPOSED MODEL AND THEORETICAL DERIVATION

One important process in GBBL capacitor manufacture is through oxidation and diffusion to make the grain boundary to form an insulating layer. So, it can be considered that the grain boundary barrier in GBBL capacitors is the Schottky barrier with an intergranular layer. This intergranular layer is just the second phase insulating layer of the grain boundary. The existance of accepter states in the insulating layer leads to form the depletion layers lying within the grains between the intergranular layer and the n-type semiconductor grains so as to establish the Schottky barrier. The barrier model is described in Fig. 1. Based on this model, the current-voltage relation and capacitance-voltage relation can be derived.

(1) Current-Voltage Relation

The grain boundary barrier will be separated into three different regions when a voltage is applied as shown in Fig. 1:

Region I: Schottky barrier region with a forward field

Region II: Intergranular insulating layer

Region III: Schottky barrier region with a backward field

Due to the fact that the current flow through the three regions is identical, the current in region I can be used as the I-V relation of the barrier as a whole. The current change in region I is just the same as the forward I-V relation of a p-n junction:

$$I = A\exp(-\emptyset/kT)\left[\exp(bV/kT) - 1\right] \quad (1)$$

or simply

$$I = A_1\left[\exp(K_1 V) - 1\right] \quad (2)$$

where b is the distribution coefficient of the voltage on region 1, $\emptyset$ is the barrier height, A_1 and K_1 are constants only depend on temperature, V is the applied voltage. When V is increasing, the electron tunnelling will occur at the top of the barrier of region III where it becomes narrow[7]. As the current will be restricted by the high resisting region II, the voltage change on region III is not large and can be considered as a constant. Thus, from eq. (2),

$$I = A_2\left[\exp K_2(V-V_c-RI) - 1\right] \quad (3)$$

where V_c is a constant representing the voltage drop on region III, R is the resistance of region II and RI is the voltage drop on region II.

(2). Capacitance-Voltage Relation

The width of the depletion layer[10] is

$$X_{dI} = (2\varepsilon/qN_D)^{\frac{1}{2}}(\emptyset - bV)^{\frac{1}{2}} \quad (4)$$

$$X_{dIII} = (2\varepsilon/qN_D)^{\frac{1}{2}}(\emptyset + BV)^{\frac{1}{2}} \quad (5)$$

where N_D is the concentration of donor dopant, q represents the positive elementary charge, b and B are the distribution coefficient of the voltage on region I and III respectively, ε is the dielectric constant of the depletion layer. The total series capacitance of the barrier in Fig. 1 is given below:

$$(1/C) = (1/C_I) + (1/C_{II}) + (1/C_{III}) = (X_{dI}/Q\varepsilon) + (X_{dII}/Q\varepsilon') + (X_{dIII}/Q\varepsilon) \quad (6)$$

where Q is a constant correlated with the effective area and ε' is the dielectric constant of the intergranular layer. Based on eqs. (4), (5), (6) and with low voltage, i. e. $bV/\phi \ll 1$ or $BV/\phi \ll 1$, there is

$$(1/C) - (1/C_0) = (1/2Q)(1/qN_D\varepsilon\phi)^{\frac{1}{2}}(B - b)V \quad \text{------------------} \quad (7)$$

here C_0 is the capacitance as V = 0.

(3). Breakdown of the Grain Boundary Barrier

According to the prevailing theory, the breakdown of grain boundary is caused by the electron tunnelling effect of Schottky barrier (Fig. 2). The probability of the electrons tunnelling into the barrier could be described as follows[13]:

$$\Gamma = (\varepsilon a/\hbar)E\exp\left(-\frac{4(2m^*)^{\frac{1}{2}}}{2e\hbar}\frac{\phi^{3/2}}{E}\right) \quad \text{--------------------------} \quad (8)$$

and for breakdown current there is $I \propto \Gamma$,

where a is the cell parameter, m^* the effective electron mass and E the applied field. From Eq. (8), the breakdown current during tunnelling can be written as

$$J = KV\exp(-A/V) \quad \text{---} \quad (9)$$

here A and K are constants for definite material.

It shows that there is nearly no temperature dependence of the I-V curves in the highly nonohmic breakdown region, which is the case for ZnO varistors.

But for the grain boundary barrier with an intergranular layer as described in eq.(3), electron tunnelling will not lead to the breakdown of the grain boundary. In this case, it could be suggested that only the electron avalanche breakdown in intergranular layer taking place could the breakdown emerge.

The ionization coefficient of electron avalanche[13]

$$\alpha = (eE/2\hbar\omega_{op})\exp(-4\varepsilon_i\hbar\omega_{op}/e^2E^2l_0^2) \quad \text{---------------------} \quad (10)$$

and for breakdown current in this case, there is

$$I \propto \alpha$$

where $\hbar\omega_{op}$ is the energy of the optic phonon, l_0 is the average free path of an electron, ε_i is the threshold energy of ionization. Obviously the temperature will influence the thermoelectronic movement, i. e. the magnitude of ε_i, so the electron avalanche breakdown is temperature dependent.

EXPERIMENTAL RESULTS AND DISCUSSION

Fig. 3(a) demostrates the low-voltage behavior of capacitance of $SrTiO_3$-based GBBL capacitors which shows linear curves and coincides with eq.(7). In Fig.3(b) the current-voltage curves exhibit an exponential dependence which is correspondent with eq. (2). For computer imitation of the curves in Fig. 3(b), a factor V_0 is needed to introduce into the eq. (2), then

$$I = A_1 \left[\exp K_1 (V - V_0) - 1 \right] \quad \text{-------------------------------}(11)$$

where V_0 is a constant. The results for 4 specimens are as follows:

For specimen 1: $A_1 = 1.593 \times 10^{-3}$, $K_1 = 0.770$, $V_0 = 0.83$

For specimen 2: $A_1 = 1.420 \times 10^{-3}$, $K_1 = 0.742$, $V_0 = 2.05$

For specimen 3: $A_1 = 1.418 \times 10^{-3}$, $K_1 = 0.686$, $V_0 = 2.34$

For specimen 4: $A_1 = 0.892 \times 10^{-3}$, $K_1 = 0.201$, $V_0 = 20.08$

In Fig. 3(b), the curves formed by "+" are the results from the computer imitation which are coincident with the measured result. The V_0 denotes the effect of the intergranular layer.

From Fig. 4, it can be seen that there is a critical point of voltage at which the capacitance and current begin changing abruptly. It was recognized that the electron tunnelling in the region III is responsible for the phenomena. The computer imitation to the current curve in Fig. 4 was performed according to eq. 3. The result is coincident with the experimental data:

$$I = 7.272 \times 10^{-3} \left[\exp 0.828 \times (V - 8.23 - 1.314 \times 10^{5} I) - 1 \right]$$

As the electron tunnelling in region III will not lead to the breakdown of grain boundary, it is necessary with still higher voltage to cause the avalanche breakdown in region II which is temperature dependent. Fig. 5 and Fig. 6 are typical nonlinear characteristic of GBBL capacitor and ZnO varistor in breakdown region respectively. The grain boundary barrier of ZnO varistor is a Schottky barrier shown in Fig. 2. The intergranular layer hardly contributes resistance to electron flowing(4,5) and the electron tunnelling in the region of the reverse-biased Schottky barrier leads to the breakdown of grain boundary which is not temperature dependent(3-7, 12).

SUMMARY AND CONCLUSION

A grain boundary barrier model taking into account the intergranular layer sandwiched between Schottky barriers was discussed for GBBL capacitor. The intergranular layer plays an important role on the properties of the GBBL capacitors. When electron tunnelling happens in region III, the total capacitance tends to rising and the current turns to increasing more rapidly, however it does not lead to the breakdown of the grain boundary. Higher applied voltage will cause the avalanche breakdown of the intergranular layer which is temperature dependent. While for ZnO varistor, the electron tunnelling in reverse-biased Schottky barrier is responsible for the breakdown of grain boundary which is not temperature dependent.

References

1. W. Heywang, Solid State Electro., 3(1), 51, 1961.
2. W. Heywang, J. Am. Ceram. Soc., 47(10), 484, 1964.
3. L.M. Levinson, H.R. Philipp, J. Appl. Phys, 46(3), 1332, 1975.
4. J.D. Levine, CRC CRIT. Rev., Solid State Sci., 5, 597, 1975.
5. W.G. Morris, J. Vac. Sci., Technol., 13(4), 926, 1976.
6. K. Eda, J. Appl. Phys, 49(5), 2964, 1978.
7. G.D. Mahan, L.M. Levinson, H.R. Philipp, ibid, 50(4), 2799, 1979.
8. B. Hoffmann, Solid State Electr., 16(5), 617, 1973.
9. B. Hoffmann, Solid State Electr., 16(5), 623, 1973.
10. P.R. Emtage, J. Appl. Phys., 48(10, 4372, 1977.
11. H.D. Park, D.A. Payne, Adv. in Ceram., 1, 242, 1981.
12. K. Mukae, I. Nagasawa, ibid, 331, 1981.
13. 犬石嘉雄，浜川圭弘，Japan, «半导体物性 I—基础物性» «半导体物性 II—素子と物性», 1977, 朝倉書房

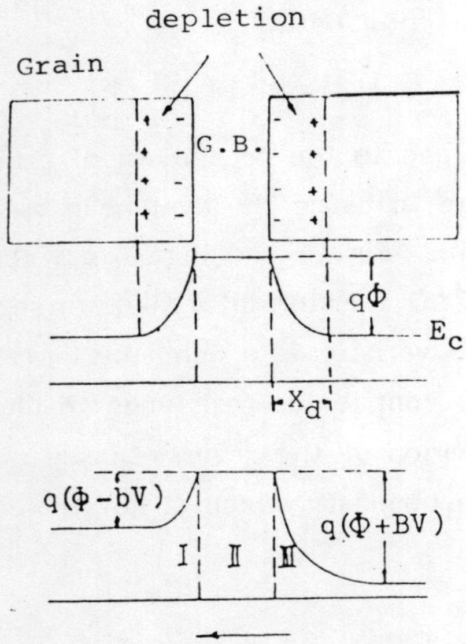

Fig.1 Grain Boundary Schottky Barrier with an Intergranular Layer.

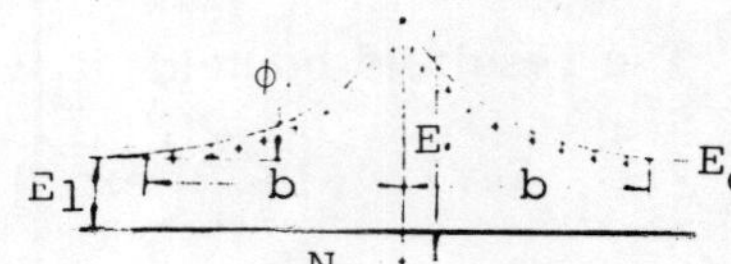

Fig.2 Grain boundary Schottky barrier.

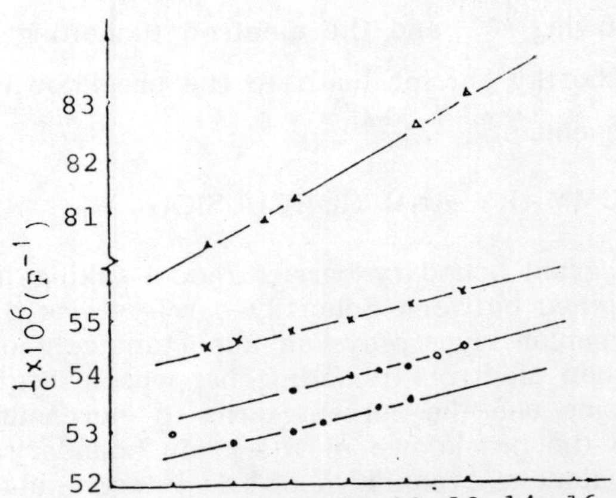

Fig.3 a)

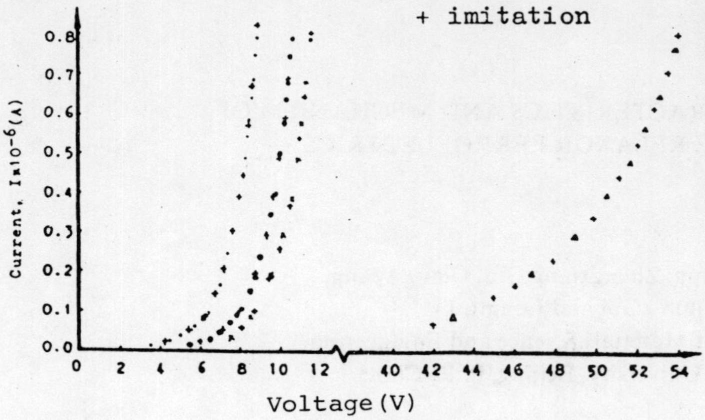

Fig.3 b)

Fig.3 $SrTiO_3$-based GBBL capacitors sintered in 1420°C for 4h. a) capacitance-voltage curves. b) current- voltage curves.

Quenched then oxidated :
. 800°C for 1h.
o 900°C for 1h.
×1000°C for 1h.
△1200°C for 1h.

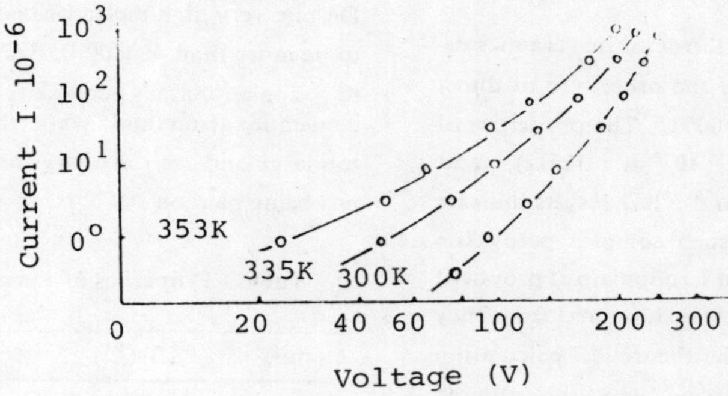

Fig. 5 Breakdown nonlinear current-voltage curves of GBEL capacitors at different temperatures.

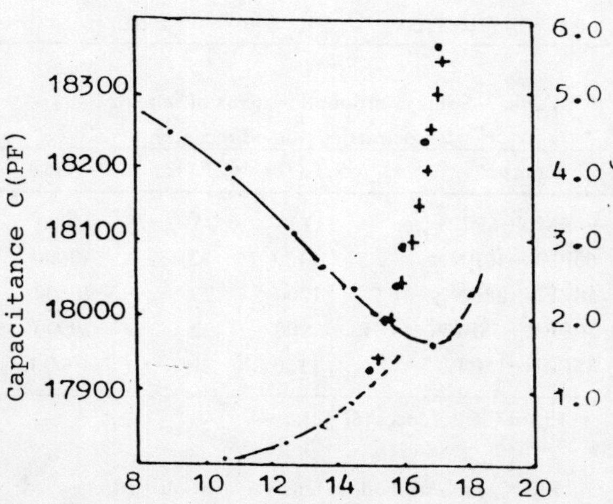

Fig. 4 Current-voltage curve and capacitance-voltage curve in the region of relatively high voltage.

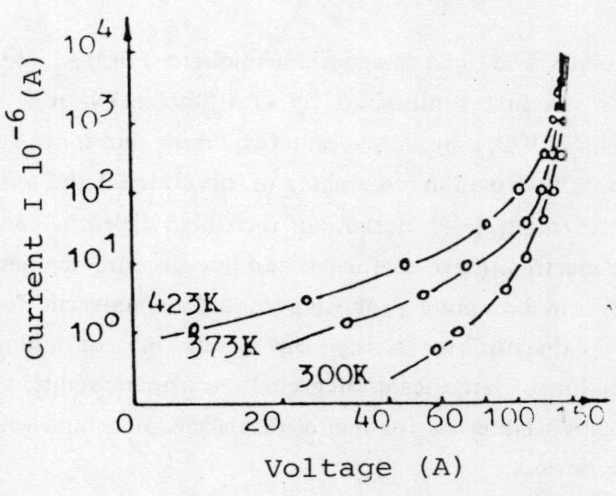

Fig. 6 Breakdown nonlinear current-voltage curves of ZnO varistor at different temperatures.

A STUDY OF CHARACTERISTICS AND MECHANISM OF PEROVSKITE RELAXOR FERROELECTRICS

Xiaowen Zhang, Zhengxiang Qiu, Qiang Wang,
Zhilun Gui and Longtu Li
Department of Materials Science and Engineering
Tsinghua University, Beijing, P. R. China

ABSTRACT

Lead magnesium niobate relaxor ferroelectric ceramics using chemical reagent raw materials by the procedure of direct formation could be sintered at 860~1000℃. The properties of $\varepsilon_{max}=16000-28000$, $\tan\delta<200\times10^{-4}$(at 1kHz) and $\rho>10^{12}\Omega\cdot cm$ were obtained. TEM and XRD results indicate that ordered microdomains exist in such complex perovskite compounds. A model of this ordered microdomain is proposed and used to explain the diffuse phase transition and frequency dispersion of relaxor ferroelectrics. The theoretical calculation by Kikuchi method is used to explain why the ordered microdomain of B′:B″=1:1 easily forms in the Pb(B′, B″)O_3 system no matter what valances of B′ and B″ are.

INTRODUCTION

The perovskite lead magnesium niobate Pb($Mg_{\frac{1}{3}}Nb_{\frac{2}{3}}$)$O_3$(PMN) was first synthesized by G.A.Smolensky in late 1950s[1]. Since 1970, especially recent years more and more attentions have been paid in researching of this material and other relaxor ceramics[2]–[4]. Because of their high dielectric constant, high electrostrictive coefficient and low sintering temperature, they are becoming promising candidate materials for multilayer capacitor to satisfy the needs of continuing miniaturization of large scale integrated electronic circuits, as well the disired materials for micro–displacement component in precious devices.

The complex compounds with perovskite structure have been extensively studied (see Table 1). It can be seen that near the temperature of ε_{max}, the dielectric constants of Pb($Zn_{\frac{1}{3}}Nb_{\frac{2}{3}}$)$O_3$(PZN), Pb($Fe_{\frac{1}{2}}Nb_{\frac{1}{2}}$)$O_3$(PFN), Pb($Fe_{\frac{2}{3}}W_{\frac{1}{3}}$)$O_3$ (PFW) and PMN can reach to more than 20000. The researches towards application of the materiacls based on the above perovskites have been got a great achivement. Table 2 is the properties of a part of the materials have been studied. Despite very high dielectric constant (ε_{max} was once reported to be more than 40000[5]), PZN is considered to be synthesized to single–phase perovskite form most difficultly by conventional method, while the dissipation factor of PFN is too large and the resistivity too small. So most attentions have just being paid on PMN these years.

Table 1.Properties of some typical relaxor ferroelectrics

Comp.	S.T.(℃)	C.T.(℃)	ε_{max}
PFN	1100	112	24000
PFW	850	−90	20000
PMN	1100	−10	18000
PZN	single crys.	150	30000
PT *	1100	490	10000

Table 2. Some compound systems of relaxor ferroelectrics studied recently *

Comp.	S.T.(℃)	C.T.(℃)	ε_{max}
94PMN−6PT	1150	25	24000
60PFN−40PNN	1150	12	17000
58PFN−38PNN−4PT	1000	29	20500
30PFW−58PFN−14PT	900	35	20000
85PZN−15PT	1150	30	10000

* From Conferences [6]–[8]

PMN was studied extensively in our country early in 1960s[9]–[10], but to produce a pyrochlor–free perovskite structure is one of the difficult problem all along. It has been reported that a procedure of prereacting of B–site refractory oxides to form the appropriate columbite or wolframite compounds and then followed by reacting with PbO is a effective method to reduce the amount of pyrochlore in Pb–based complex perovskite compounds[2]. We developed a direct synthesizing method of PMN and have got pleased results too.

In order to explain the excellent dielectric properties, and the characteristics of diffuse phase transition (DPT) and frequency dispersion of relaxor ferroelectrics, composition fluctuation[11], heterophase fluctuation[12] and structure fluctuation[13] have been proposed in the light of micro–inhomogenity due to more than two ions occupying a crystallographic site in these materials. These theories can explain DPT in certain extent but all of them have not been verified experimentally to now, and they are difficult to explain extraordinary high dielectric constant, very wide "Curie range" and frequency dispersion according to microscopic fluctuation of composition or heterophase or structure[13].

In 1980, N. Setter found that in $Pb(Sc_{\frac{1}{2}}Ta_{\frac{1}{2}})O_3$(PST) B–site cations Sc^{3+} and Ta^{5+} were partially ordered in the ratio of 1:1 and formed a $2a_o \times 2a_o \times 2a_o$FCC superlattice[14]. The order degree varied with annealing time and had great influences on DPT characteristics. The longer the annealing time, the larger the order degree and the narrower the "Curie range". So he believed that composition disorder was the main cause of DPT.

In 1986, J. Chen discovered that in PMN which had long been believed to be disordere completely there existed the same superlattice as that in PST[15]. In the superlattice Mg^{2+} and Nb^{5+} are arranged orderly in the ratio of 1:1 instead of 1:2 in the composition, which causes a deviation of micro–composition from the macro–composition and creats a local charge imblance which limits the ordered microdomain to grow. Because of the separated small areas the composition in the ordered microdomain could not be certified by using of electron probe, so this discovery have just been verified indirectly in experiment up to now.

After then, C. Randall et al. found that this superlattice appears in many Pb–based perovskite relaxor ferroelectrics and considerde that it is due to the order arrangement of vacancies of lead cations[16].

In this paper, the characteristics of PMN using direct synthesis method has been reported, a model of ordered microdomain in Pb–based perovskite relaxor ferroelectrics has been used to explain the DPT, frequency dispersion properties and so on, and formation mechanism of this ordered microdomain is elucidated theoretically.

A STUDY OF SYNTHESIS AND CHARACTERISTICS OF PMN

The columbite precursor method was considered to be successful in obtain PMN ceramics with perovskite structure[2]. In this method the columbite $MgNb_2O_6$was prereacted by two B–site oxides, MgO and Nb_2O_5, as follow:

$$MgO + Nb_2O_5 \rightarrow MgNb_2O_6 \qquad (1)$$

Then the $MgNb_2O_6$ and PbO were reacted sequentially to form the PMN:

$$MgNb_2O_6 + 3PbO \rightarrow 3Pb(Mg_{\frac{1}{3}}Nb_{\frac{2}{3}})O_3 \qquad (2)$$

A direct synthesis has been developed by using of conventional ceramics processing. The raw materials are C.P. oxides Pb_3O_4, Nb_2O_5, $Mg(OH)_2 \cdot 4MgCO_2 \cdot 6H_2O$ and TiO_2. Excess PbO and MgO and different amounts of $Pb(Cd_{\frac{1}{2}}W_{\frac{1}{2}})O_3$(PCW) were added. The composition of samples are listed in Table 3.

Table 3. The Composition of Samples

No.	Formula	x	y	z	u
1	PMN–xPT–yPCW+zmol%MgO+uwt%PbO	.08–.20	.01–.05	2–15	1
2	PMN–xPT–yPCW+uwt%PbO	.08–.20	.01–.05	/	1
3	PMN–xPT+zmol%MgO+uwt%PbO	.08–.20	/	2–15	1
4	PMN–xPT+zmol%MgO+uwt%PbO	.08–.20	/	2–15	1

The XRD patterns of sample 1 and 4 were shown in Figure 1. It can be seen that in both samples the pyrochlore–free PMN were synthesized for powder calcined at 800℃ and samples sintered at 860℃ and above 960℃ respectively.

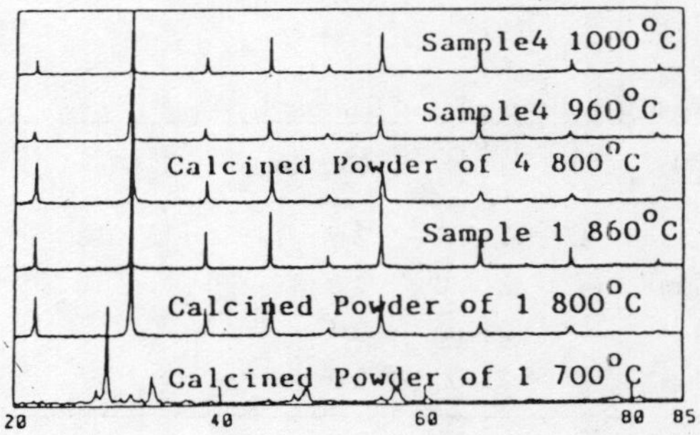

Fig.1. XRD Patterns of Sample 1 and 4

Table 4. The Dielectric Properties of Sample 1 and 2

No.	S.T. ℃	Tc ℃	ε_{max}	$\rho \times 10^{-12}$ Ω·cm	tanδ $\times 10^{-4}$
1	900 / 4hrs	40	18600	2	90
	860 / 4hrs	41	14000	2	86
2	920 / 4hrs	40	14200	2	82
	860 / 4hrs	40	11300	2	80
3	1200 / 4hrs	47	17156	2	81

The properties of sample 1 and 2 are listed in Table 4. The main different between sample 1 and 2 is with and without excess MgO. It seems that proper excess MgO could improve the PMN properties obviously.

Making a comparison between Samples 1 and 3, it was indicated that a small amount of PCW addition can lower the sintering temperature about 300℃ without degrading the properties.

For the same composition, directly synthesized Sample 1 (1A) has been compared with the columbite percursor method (1B) and the results are almost the same as shown in Table 5. So it is believed that the direct synthesis can be successful in preparing PMN.

Table 5. The Dielectric Properties of Sample 1 with Different methods of synthesis

No.	S.T. ℃	Tc ℃	ε_{max}	$\rho \times 10^{-12}$ Ω·cm	tanδ $\times 10^{-4}$
1A	900 / 4hrs	40	18600	2	90
1B	900 / 4hrs	40	18600	2.2	100

Content of excess PbO in Sample 4 is 2 wt% and the maxium dielectric constant $\varepsilon_{max} = 28000$ has been obtained, and the sintering temperature is 1000℃ as shown in Table 4. So a proper excess PbO could play a important role in this direct synthesis.

Table 6. The Properties of Sample 4

S.T.(℃)	Tc	ε_{max}	ρ(Ω · cm)	tanδ
960 / 4hrs	45	18500	2.5×10^{12}	130×10^{-4}
980 / 4hrs	47	21820	2.0	150×10^{-4}
1000 / 4hrs	47	28000	2.4	/
1020 / 4hrs	47	27000	2.0	/

A MODEL OF ORDERED MICRODOMAIN

Structure disorder upon a crystallographic site has long been considered to be responsible for the DPT, because only disorder can cause micro–inhomogenity. This idea can explain DPT in solid solution successfully, but for compounds which have defined composition it is not so suitable.

Many Pb–based complex perovskite compounds have been found with $2a_o \times 2a_o \times 2a_o$FCC superlattice. Some typical relaxor ferroelectrics with high dielectric constant inclduing PMN, PZN, PFN and PFW have been studied. It was found that all of them have superlattice structure but the range of ordered region is so small that we name it as ordered microdomain. Figure 2 and 3 show the existence of the microdomain from TEM diffraction pattern along <110> index and center dark field (CDF) image and XRD pattern of 0.9PZN–0.1BT. We think that the ordered microdomain instead of disorder is responsible for DPT and other characteristics of Pb–based complex perovskite compounds.

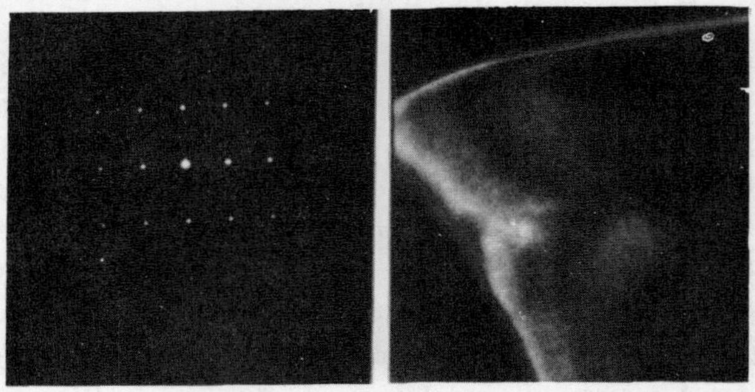

Fig.2. TEM diffraction pattern and CDF imange of 0.9PZN–0.1BT.

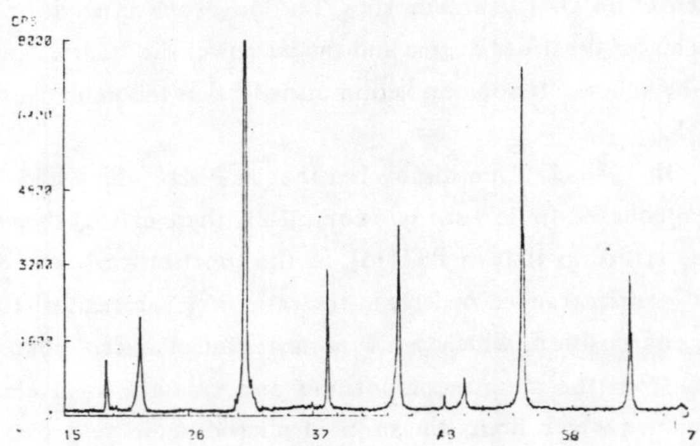

Fig.3. XRD pattern of 0.9PZN–0.1BT.

J.Chen et al. have verified that in the ordered microdomain of PMN the ratio of B–site cations is 1:1 which deviates from the composition of Mg:Nb = 1:2 in PMN[15]. It is proposed that this kind of superlattice which caused by order arrangement of B–site cations in the ratio of 1:1 would occur in other similar relaxor compounds in which the B–site cations have different valances. Figure 4 is the structure scheme of ordered and disordered $Pb(B',B'')O_3$. It is clear that B′ and B″ layers are arranged alternatively along <111> direction when the ratio of B′ to B″ is 1:1 in ordered structure (Figure 5a). Because the valances of B′ and B″ are different from the balance valance of center cation in oxygen octahedron +4, a effective dipolar moment should exist in ordered microdomain as shown in Figure 5b (suppose the valances of B′ and B″ are +3 and +5 respectively). This dipolar moment is related to the order–disorder transition which normally takes place in broad range of temperature because of high potential energy barrier of ion rearrangement, thus a net polarization will occur in a wide temperature range which causes diffuse phase transition.

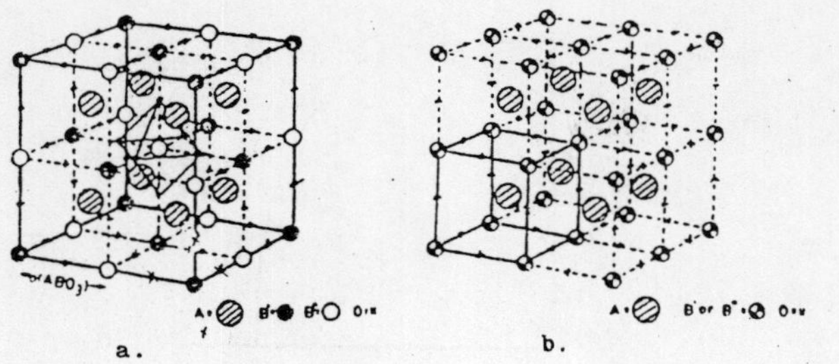

Fig.4. Ordered and disordered Pb $(B',B'')O_3$
(a) ordered (b) disordered

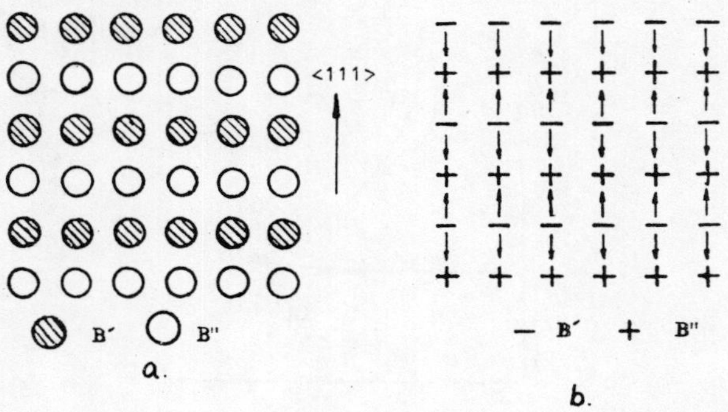

Fig.5. Arrangement of (a) B–site cations and (b) effective dipolar moment of $Pb(B',B'')O_3$ pervoskite along < 111 > direction

Almost all of this kind of compounds contain transition metallic cations, in which a large amount of electrons associated with the valance differences are frequently not fixed at one species ion site. In the same time, vaporization of PbO creats a lot of ionic defects. When a electric field is applied, these weak–bonded electrons and ionic defects will move in a limited area and cause a relax polarization which is sensitive to frequency within radio frequency range. This is why such compounds have frequency dispersion.

Besides the order polarization and relax polarizaton, there are ionic and electronic displacement polarization. Since the valances of B′ and B″ are different, anions of oxygen shift to the cation with higher valance B″, which will creat a displacement polarization under an applied field. All of these kinds of polarizations will induce high dielectric constants which is a macroparameter of polarizability of material of these Pb–based complex perovskite compounds.

$Pb(B'_{\frac{1}{3}}B''_{\frac{2}{3}})O_3$ is different from $Pb(B'_{\frac{1}{2}}B''_{\frac{1}{2}})O_3$ partially because of charge imbalance in the ordered microdomain. Detail theoretical calculations are under way. Valances of B–site cations, ionic radii and ionic polarizability have been considered.

THEORY OF THE FORMATION OF ORDERED DOMAIN

In perovskites $Pb(B',B'')O_3$, the lattice is of simple cubic when B′ and B″ are randomly arranged in B–site, while the lattice is of FCC with the lattice parameter $a = 2a_0$ if B′ and B″ are orderly arranged as shown in Figure 4(a) in which both B′ and B″ occupied in FCC sublattice α and β individually (Figure 6). If B′ occupies the α–site and B″ occupies the β–site completely or vice versa, it is the perfect order arrangement with order parameter $M = 1$. If the site propability of B′ (or B″) occupying α–sublattice is equal to that of B′ (or B″) occupying β–sublattice, it is the perfect disorder arrangement with $M = 0$. Between these two extereme situations partial order arrangement exists with $0 < M < 1$. The concept of order parameter can be used not only in the $A(B', B'')O_3$ system with nominal formulation B′:B″ = 1:1 but also in the system with nominal formation $B'_{2x}B''_{2(1-x)}$ in B–site.

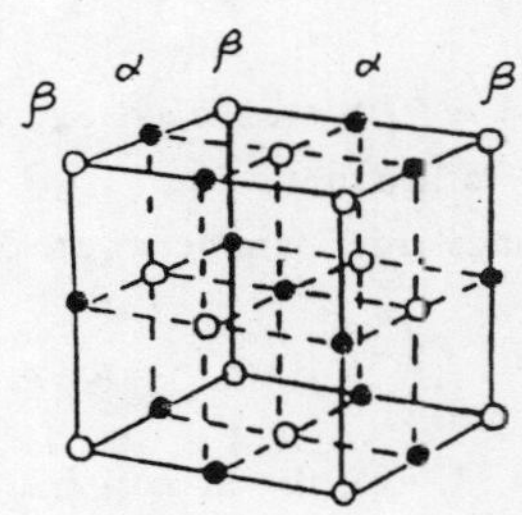

Fig.6. Simple cubic lattice composed of B–ions was divided into two interpenetrating sublattice α and β

The site probabilities on α–lattice and β–lattice are defined as P_i^α (i = 1 for B′ and 2 for B″) and P_j^β (j = 1 for B′ and 2 for B″) respectively. P_{ij} is the pair–probability of finding i–j species in neighbouring sites. The following relations are satisfied:

$$\sum_i P_i^\alpha = \sum_j P_j^\beta = 1$$

$$P_i^\alpha = \sum_j P_{ij}, \qquad P_j^\beta = \sum_i P_{ij} \tag{3}$$

$$P_1^\alpha + P_1^\beta = 2x$$

where 2x refers to the composition of B′ while the formula of B–site species was written as $B'_{2x}B''_{2(1-x)}$.

The effective interaction energy E_{ij} of neighbouring B–site species through A species and oxygen ions are chosen as follow:

$$E_{11} = yJ, \quad E_{22} = zJ, \quad E_{12} = E_{21} = \varepsilon J \tag{4}$$

where y,z > 0 and ε < 0.

The thermodynamics of the system is obtained through minimization of the free energy:

$$F = zN/2\sum_{ij} E_{ij}P_{ij} - TS \tag{5}$$

where z is the coordinate number, N is the number of sites in the lattice and S is the configurational entropy.

By using Kikuchi method[17], the pair-probability could be calculation:

$$P_{ij} = \frac{EXP(-E_{ij}/KT) * EXP(\mu \cdot a_{ij}/KT) * (P_i^\alpha \cdot P_j^\beta)^{\frac{5}{6}}}{\sum_{ij} EXP(-E_{ij}/KT) * EXP(\mu \cdot a_{ij}/KT) * (P_i^\alpha \cdot P_j^\beta)^{\frac{5}{6}}} \tag{6}$$

where K is Boltzmann constant, T is absolute temperture, μ is chemical potential, and a_{ij} is variable parameters, i.e., $a_{11} = 2$, $a_{12} = a_{21} = 1$, $a_{22} = 0$.

The long-range order parameter M can be defined as:

$$M = (|<P_1^\alpha> - <P_1^\beta>| + |<P_2^\alpha> - <P_2^\beta>|)/2 \tag{7}$$

where $<P_i^\alpha>$ and $<P_j^\beta>$ are the average value of P_i^α and P_j^β over all α- and β- lattice sites.

The short-range order parameters are defined as:

$$\delta_{ij} = P_{ij}/P_i^\alpha * P_j^\beta - 1 \tag{8}$$

According to (3)-(8), the M and δ_{ij} were computed at various reduced temperature KT / J. The results are shown in Figures 7 and 8. Furthermore, the order-disorder phase diagram as a function of composition vs temperature was obtained and is shown in Figure 9. It can be seen that the temperature range is maximum for order arrangement at ratio of 1:1 than at any other ratios in the $B'_{2x}B''_{2(1-x)}$ system. It means that 1:1 ratio is a perferred composition for the formation of ordered microdomain. It is also noticeable that besides the composition of 1:1, the ordering degree is nonzero, which means that there may be partial ordered states for different composition. We have seen from Figure 9 that the ordered microdomain could be formed in a wide range at low temperature and the range gradually becomes narrow with temperature increases, finally converges at a point. In conclusion that the ordered domain at a ratio of 1:1 is easily to be formed and could be formed at other composition range, even such formation would result in local charged state. For different species of B′ and B″, the effective interaction energy parameter are different, such as PMN, but the calculation showed that the charged microndomain has not important influence on the formation of 1:1 ordered microdomain.

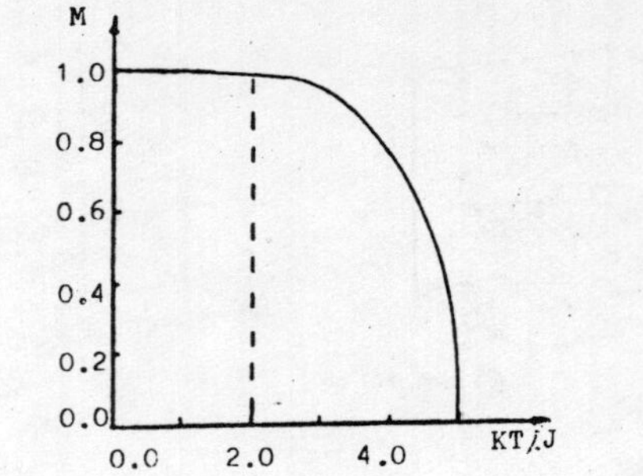

Fig.7. Plot of long-rang order parameter M vs reduced temp. KT / J at the composition point of B′:B″ = 1:1.

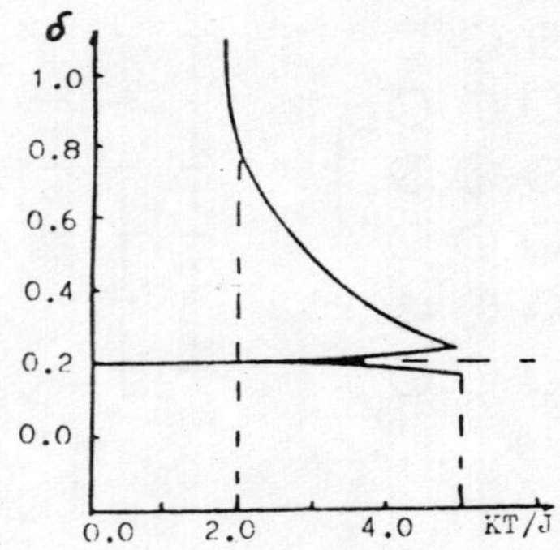

Fig.8. Plot of short-range order parameter vs KT / J at the composition point of B′:B″ = 1:1.

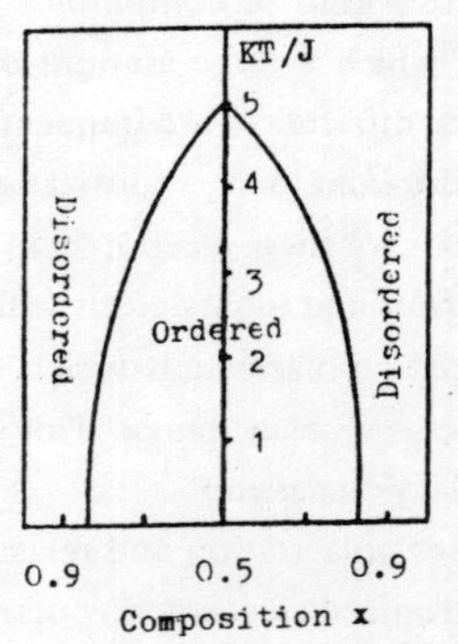

Fig.9. Order-disorder phase diagram of $B'_{2x}B''_{2(1-x)}$ system.

CONLUSION

PMN relaxor ferroelectric ceramics with good properties could be fabricated by direct synthesis method. A model of ordered microdomain could explain DPT and other characteristics of Pb-based complex perovskite compounds. Kikuchi method is an effective one to calculate the order-disorder phasediagram in $B'_{2x}B''_{2(1-x)}$ system.

REFERENCES

〔1〕. G.A.Smolensky and A.I.Agroanovskaya, Sov. Phys., Solid State, 1429-37 1 (1959)

〔2〕. S.L.Swartz and T.R.Shorut, Mat. Res. Bull., 1245-50 17 (1982)

〔3〕. T.R.Shrout and S.L.Swartz, Mat, Res. Bull., 663-67 18 (1983)

〔4〕. T.R.Shrout and A.V.Halliyal, Am. Ceram. Soc. Bull., 704-711 66[4] (1987)

〔5〕. Y.Yokomizo, et al., Jpn. J. Phys. Soc., 1278-84 28[5] (1970)

〔6〕. O.Tadashi, et al., Elec. Ceram. 10-16 7 (1987)

〔7〕. T.R.Shrout, S.L.Swartz and M.J.Haun, Am. Ceram. Soc. Bull., 808 63 [6](1984)

〔8〕. S.T.Fu and G.F.Chen, Am. Ceram. Soc. Bull., 1397 66[9] (1987)

〔9〕. No.1 Instiute of Electronic Ministry of China, Treatise of Low Temperature Sintering Multilayer Capacitor, July, 1977

〔10〕.Wang Yongzhuang, Electronic Ceramics of China, 1 (1979)

〔11〕.G.A.Smolensky, Jpn. J. Phys. Soc., 26-37 28[Suppl.] (1970)

〔12〕.Von W.Kanzig, Helv. Phys. Acta., 175 24 (1951)

〔13〕.Biaorong Li et al., Inorganic Medium Material, publicated by Science Press, (1984)

〔14〕.N.Setter and L.E.Cross, J. Appl. Phys., 4356-60 52[8] (1980)

〔15〕.J.Chen, H.M.Chan and M.P.Harmer, 89-E-87, Procceding of 89th Ann. Meet. of Am. Ceram. Soc., (1987)

〔16〕.C.Randall, D.Barber et al., Ferroelectrics, 277-82 76 (1987)

〔17〕.R.Kikuchi, J. Chem. Phys., 1071 60 (1974)

Interface Design of Functional Ceramics

Hiroaki Yanagida, Ph.D

Professor of Chemical Recognition Materials

Director of Research Center for Advanced Science and Technology

University of Tokyo, 4-6-1 Komaba, Meguro-ku, Tokyo, 153 Japan

Interesting phenomena arise from interfaces. To understand interfacial phenomena the present speaker proposes classification of interfaces with such terms as between same or different materials, open or close, and phenomenon taking place along or perpendicular the surface.

Interesting phenomena arise from nonlinear interactions between different materials with opposite properties. Examples of opposite properties are: acid/base, p-type/n-type semiconductors, red/ox, host/guest, etc.

Nonlinear interactions are designed at the interfaces. Most deeply discussed will be the cases where electric current across the open interface between p/n electrodes changes with ambient. The CuO/ZnO contact is effective to detect humidity without hysteresis and to distinguish carbon monoxide from hydrogen.

The open interfaces between different materials are origins of intelligent mechanisms such as self-recovery, stand-by, recognition, tuning ability from outside, etc.

SILICON: AN EXCELLENT MATERIAL FOR SENSORS

S. Middelhoek
Delft University of Technology, P.O. Box 5031
2600 GA Delft, The Netherlands

ABSTRACT

Worldwide many technologies for the fabrication of sensors are explored. Silicon, in particular, is quite promising, because it allows low-cost batch fabrication, shows many useful physical effects and permits the simultaneous integration of sensors and processing circuits on one chip. In this paper the suitability of silicon for the fabrication of sensors will be elucidated.

INTRODUCTION

The rapid development of microelectronics over the last decades, leading to a spectacular increase in the performance/price ratio of microelectronic components, has given an important impetus to what is customary called the "information society", one in which, along with materials and energy, information and information-processing systems play a major role.

Table 1 Generic technologies for sensors

Optical fibers
metal oxides
polymers
piezoelectric materials
Thin and thick films
Amorphous materials
III-V and II-VI semiconductors
Polycrystalline silicon
Monocrystalline silicon

A large-scale use of these VLSI components in information processing systems like measurement and control systems can also be expected. Though, nowadays, computers are widely used in this field, other microelectronic components have not yet found wide acceptance. Also the number of microelectronics-based innovative products is less than predicted. This is among other things induced by the non-availability of sensors and actuators, which have a performance/price ratio comparable to that of VLSI components.
As a consequence many researchers in academic and industrial laboratories are searching for new principles and technologies for sensors and actuators. The dominant technologies are shown in Table 1.

Based on these technologies ingenious sensors were designed for a multitude of physical and chemical parameters. Monocrystalline silicon plays a notable role, as it is speculated that only with silicon the performance/price barrier can be slashed.
There are three main reasons why silicon is considered to be highly suitable:
1. Batch production. The use of the well developed silicon planar technology permits the simultaneous fabrication of thousands of sensors on one silicon wafer. If large numbers are produced, design and mask costs will play a minor role.
2. Sensors are based on physical and chemical effects with which conversion from nonelectrical to electrical parameters can take place. A material is suited for sensors, if it is characterized by many large effects. Though unfortunately silicon is nonpiezoelectric and nonmagnetic, silicon is such a material.
3. Silicon is predominantly used for the fabrication of signal processing circuits. When silicon is also used for sensors, it becomes feasible to integrate sensors and circuits on one chip. Such so-called "smart sensors" can be made to have features such as a microprocessor-compatible output format, autocalibration or offset correction.

SILICON SENSOR TECHNOLOGY

Silicon planar technology is currently the principle processing method used in the fabrication of integrated circuits. In the last two decades all processing steps have been studied and optimised. Optical lithographical processes are capable of 1 micron resolution and chips with over 1 million components are on

the market. In most cases this complexity and sophistication is not required for the production of silicon sensors. However, in order to decrease defect density and increase yield, processing of sensors should be performed in clean rooms. Two major IC technologies exist, namely bipolar and (complementary) MOS. Nowadays most silicon sensors are produced with bipolar technology, but the number of MOS devices is steadily growing. In the future a combination of these processes called BICMOS might be the ultimate technology for the fabrication of silicon sensors and smart sensors.
The great advantage of using silicon technology for sensors, lies in the fact that a large number of wafers each containing hundreds of sensors can be simultaneously processed as one batch. When large numbers of devices with equal properties, e.g. Hall switches for computer keyboards, are required this has a very positive effect on the performance/price ratio of these devices.
Many silicon sensors, such as optical sensors, Hall plates and temperature sensors, can be produced by means of a sub-set of the standard processing sequence. To illustrate this, the cross-sections of a bipolar NPN transistor and a Hall plate are shown in Fig. 1

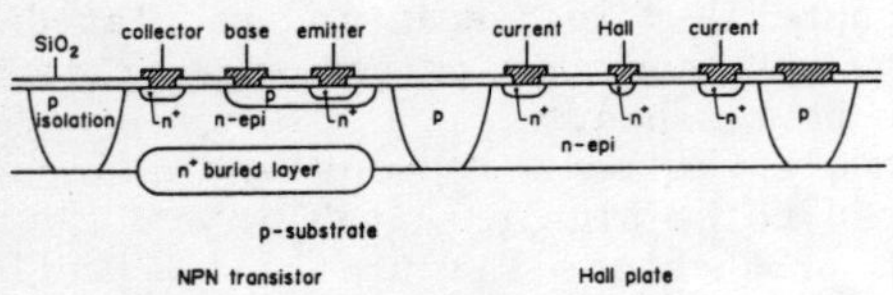

Fig. 1 NPN transistor and Hall plate.

Other devices such as the diffused diaphragm pressure sensor or the integrated thermopile require a special step called micromachining. For some sensors the wafer thickness of .5 mm is unsuitable and has to be reduced to 10 micron. Wet chemical and plasma dry etching techniques have been developed with which this thickness reduction can be achieved. As these processes are not compatible with regular IC processes, micromachining should always be the last step in a process sequence.
As we will discuss in the next section silicon is not piezoelectric and also not magnetic. When such properties are required it is possible to add these properties by depositing a ZnO piezoelectric layer or a NiFe magnetic layer on top of the silicon wafer. Also these layers should be added at the end of the processing sequence. In the literature these solutions are amply discussed.

PHYSICAL AND CHEMICAL EFFECTS

In a sensor a signal from a nonelectrical signal domain is converted into an electrical signal by means of a physical effect[1]. We can distinguish 5 nonelectrical signal domains:

Table 2 Nonelectrical signals

signal domain	examples
radiant signals	- light intensity
mechanical signals	- force, velocity
thermal signals	- temperature
magnetic signals	- magnetic field strength
chemical signals	- pH, concentration

For example, in a solar cell a light intensity is converted into a voltage by means of the photovoltaic effect. In a pressure sensor, a strain is converted into a small change of a resistance by means of the piezoresistive effect. Two kinds of effects can be distinguished:

self-generating effects
these effects permit a conversion without an auxiliary energy source. The photovoltaic effect, as mentioned above, is such an effect.

modulating effects
these effects can only be used when an auxiliary energy source is available. All resistive effects belong to this category. In a photoconductor the light induced change of the resistance can only be detected when a voltage or current source is connected.

A material is suited for the fabrication of sensors, when the material is characterized by many large effects, that allow conversion, if possible, from all 5 nonelectrical signal domains into the electrical domain.
As we have already discussed, from a batch-fabrication point of view silicon is an excellent material for sensors. However, without effects silicon would not be worth considering.
In the last three decades much research on silicon with respect to physical and chemical effects has been executed. The results are summarized in Table 3.
It shows, that silicon as a semiconductor shows many effects and that a number of them, e.g. the Seebeck effect and the piezoresistive effect even appear to be much larger than the same effects in metals. However, it was also found, that silicon misses on two rather useful effects. Silicon is not piezoelectric, because of its symmetrical crystal structure, and silicon is also not magnetic. As the mobility of charge carriers in silicon is rather small, magnetoresistivity in silicon is hardly detectable.
When yet sensors based on these effects are required, it is necessary to deposit piezoelectric or magnetic layers on top of the silicon substrate.

Table 3 Effects in silicon

Signal domain	self-generating effects	modulating effects
Rad.	photovoltaic e. Dember e.	photoconductivity photoelectric e. photocapacitive e.
Mech.	acoustoelectric e.	piezoresistivity piezojunction e.
Therm.	Seebeck e.	thermoresistivity thermojunction e.
Magn.	-	Hall e. Suhl e.
Chem.	Galvanoelectric e.	chemofield e.

SILICON SENSORS

Silicon is only a useful material for sensors, if, based on one of the effects, structures, that indeed are sensitive to the physical parameter of interest, can be obtained with standard process sequences. For instance, it is conceivable that the resulting structures, in order to show the desired sensitivity, have unrealistic dimensions. It is also possible that the temperature range in which the silicon sensor operates is so small that practical use of the silicon sensor is rather improbable. In this section a brief summary of the most important silicon sensors will be presented. It appears that a multitude of devices is feasible[2].

1. Silicon sensors for radiant signals.

Silicon can be used to construct devices that are sensitive for a wide range of electromagnetic radiation wavelengths, from gamma-rays to infrared[3]. Photoconductors, photodiodes and phototransistors are mass produced and have found a wide application field. A MOS capacitor is also sensitive to light. Based on this device, nowadays, one tries to design solid-state colour video cameras with high resolution. Mass production of these devices can soon be expected. For robot vision photodiodes have been developed of which the colour sensitivity can be electrically adjusted[4].
High-resistivity silicon is also successfully used in high energy-physics and for the detection of soft X-rays[5].
Radiation detection forms the most successful application field for silicon sensors.

2. Silicon sensors for mechanical signals.

The most successful device in this domain is the so-called diffused silicon strain-gauge diaphragm. The device was conceived twenty years ago, went through a rather difficult period, but is now mass-produced for the car industry. Most silicon pressure sensors are based on a Wheatstone bridge diffused into a thin diaphragm, which is fabricated using micromachining techniques[6]. In the R & D laboratories many devices for measuring strain, force, position, velocity, acceleration, tilt, and flow have been studied, but only a few of these devices have already found their market.

3. Silicon sensors for thermal signals.

The characteristics of diodes and transistors are temperature dependent. In most cases this is a undesirable effect, but for sensors it is quite useful. Especially dual transistor structures are of interest as with these devices temperature sensors can be made which have an output signal which is proportional to the absolute temperature[7].
Also the Seebeck effect in silicon is very large. Integrated thermopiles can easily be fabricated with standard diffusion steps and micromachining (Fig. 2). Temperature differences down to a few micro K can be easily detected on a chip.

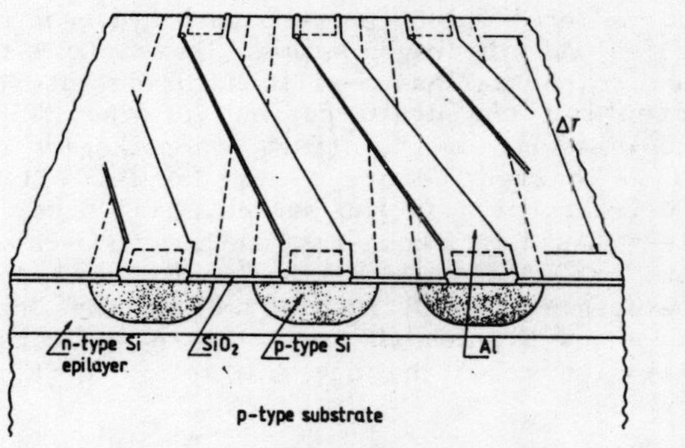

Fig. 2 Integrated Si/Al thermopile

Based on thermopiles sensors for vacuum, air velocity, infrared and chemical quantities have been conceived and presently are finding their markets[8].

4. Silicon sensors for magnetic signals.

Nowadays, many millions of silicon Hall plates are applied in computer keyboards and brushless electromotors. The Hall effect in silicon is not very large, but for above applications sufficient. The low sensitivity is caused by the rather moderate mobility of electrons and holes in silicon. Another inconvenient feature of Hall plates is the large offset that requires calibration after fabrication. Because of the unsatisfactory characteristics of Hall plates, many other silicon devices were designed and examined. Most of the devices are based on two collector or drain transistor structures. In the ideal case the emitter or source current flows in two equal parts to the collectors or drains. An in-plane magnetic field can upset this balance, the current difference is then a measure for the magnetic field. Based on this principle 2D and 3D magnetic field sensors for special applications have been built[9]. With respect to sensitivity and offset these new structures have no advantages compared to Hall plates.

Silicon sensors for chemical signals

In view of the increased interest in our environment, the need for in-line control of gas mixtures in industry and the medically attractive monitoring of oxygen in blood during surgery, many R & D laboratories were motivated to study low-cost silicon sensors. Up to now two kinds of devices have dominated this field. The first device is called an Ion Sensitive Field Effect Transistor (ISFET) and in it simplest form is a MOSFET without a gate contact[10]. When the ISFET is immersed in an electrolyte, the drain current depends on the ion concentration.
The other device is called a Pd gate MOSFET and, as the name suggests, consists of a MOSFET in which the gate material is palladium[11]. Palladium is a highly catalytically active metal which can decompose hydrogen molecules into hydrogen atoms. These atoms diffuse to the palladium-silicondioxide interface affecting the drain current of the MOSFET. Problems with drift, offset and packaging have to be solved before these devices will be widely accepted in the market. Many other devices based on chemo-resistors, chemo-capacitors, chemdiodes, miniature Clark cells, and chemosensors based on surface acoustic delay lines and silicon thermopiles are subjects of intensive research, but not yet commercially available.

SMART SENSOR

Most silicon sensors are based on elements that generate a small change in current, voltage, resistance, or capacitance in response to one or more measurands. In addition the output signal will be affected by other parameters like temperature or pressure in an undesirable way. Also a sensor will be plagued by offset, drift, nonlinearity and sensitivity changes with time. To apply the sensors in a system an intimate knowledge of the working of the sensor element is always necessary. In general, the systems engineer will be insufficiently trained for the task of applying sensors and will tend to avoid using innovative sensors. A possible way out is to make sensors more "smart". Sensors should be designed in such a way, that its application is about as difficult (or easy) as the use of a regular micro-electronic component[12]. The sensor should have a standard output and unwanted cross-sensitivities should be compensated. In addition offset, drift, etc. should be minimised. Also some kind of self-testing or calibration should periodically occur. The way to realise this concept is to combine a sensor device with a number of micro-electronic components into a single sensor package, often referred to as a "smart sensor". Though hybrid smart sensors certainly are of interest, a more elegant solution is obtained, when the sensor and the signal processing circuit are made in the same material, such that both components can be integrated on one and the same substrate. Because signal processing circuits can only be integrated in silicon and GaAs, these are the only materials available for smart sensor fabrication. As we have seen in the preceding section already many silicon sensors are available. In view of this situation it is timely to proceed further with the smart sensor concept. In the literature several ventures in designing and fabricating integrated smart sensors are described.
The readout of CCDs is integrated on the CCD chip doubling its size. Pressure sensors with temperature compensation and frequency output have been designed. PTAT temperature sensors with band-gap references integrated on the same chip have been presented in the literature. Smart Hall plates are nowadays widely used. Only in the chemical sensor field the development of smart sensors is less pronounced.
A lot of work has still to be done on electronic interfaces, in order to improve yield and performance of smart sensors. Several principles are available such as bridge-to-frequency converters, resonators, electrical oscillator based sensors, ring oscillators, frequency ratio based interfaces, flip-flop sensors and other A/D converters.

CONCLUSIONS

Though the difficulties most often connected with the development and packaging of silicon sensors and smart sensors have been grossly underestimated, silicon again and again shows to be an excellent material for sensors. Silicon sensors can be obtained using standard silicon processes. Micromachining and other non-standard processing steps, though, should be performed at the end of the processing sequence. Batch production improves yield and reduces cost. Silicon shows many physical and chemical effects, which often are much larger than those in metals. A large family of silicon sensors for almost any measurand is already available and resulted from worldwide efforts in this field. Several attempts to introduce smart sensors for measuring light intensity, pressure, temperature and magnetic field have been successful and more is to

come.
It can be expected that in the next decade smart sensors with bus-compatible output will reach the market.

REFERENCES

1. S. Middelhoek and A.C. Hoogerwerf, Sensors and Actuators, 10(1986), 1.
2. S. Middelhoek and S.A. Audet, Silicon Sensors, Academic Press, London (1989).
3. V. Härtel, Optoelectronics, McGraw-Hill, New York (1978).
4. R.F. Wolffenbuttel, IEEE J. Solid-State Circ. SC-22(1987), 350.
5. S.A. Audet and W.R.Th. ten Kate, Proc. Transducers'87, Tokyo (1987), 267.
6. S. Sugiyama, M. Takigawa and I. Igarashi, Sensors and Actuators, 4(1983), 113.
7. G.C.M. Meijer, Sensors and Actuators 10(1986), 103.
8. A.W. van Herwaarden and P.M. Sarro, Sensors and Actuators 10(1986), 321.
9. S. Kordic, Sensors and Actuators 10(1986), 347.
10. P. Bergveld, Sensors and Actuators 8(1985), 109.
11. I. Lundstrom and C. Svensson, In: Solid State Chemical Sensors, Academic Press, New York (1985), 1.
12. S. Middelhoek et al., Sensors and Actuators, 15(1988), 119.

Ultra-fine Structure Manipulation and Fine Composite Materials

YAO XI,
Department of Electronic Engineering,
Xian Jiaotong University, Xian, 710049, China.

During the past ten years, the advances in materials characterization are very rapid. It is now able to characterize the structure and composition on fine scale down to nanometer and atomic cluster. It has now become apparent that most ceramic materials are structurally and compositionally inhomogeneous on the nanometer scale. Very clear evidence of the influence of the nano-inhomogeneity on the electrical and optical behaviors of relaxor ferroelectric materials is presented.

It is very interesting to see if it is possible to manipulate the ultra-fine structure of materials deliberately to modify the electric and optic behaviors of the materials. Examples of electric field-induced, temperature-induced and doping-induced ultra-fine structure manipulation are discussed. Structure property relationship of the materials discussed is given.

A further step to the structure manipulation of given materials is to compose new composite materials with given structure connectivity and symmetry on very fine scale. 0-3 fine composite materials with crystalline polar ultra-fine powder as active components and liquid or transparent polymer as matrices is presented. Another easy way to compose fine composite materials is to induce very small oriented polar crystallite out of amorphous thin films to form 1-3 composite connectivity. Processing and property of the fine composite materials are discussed. The major application of the fine composite materials is in the field of integrated optics and nonlinear optics.

THE GLOBAL DEVELOPMENT OF IECQ

Li Mengwei
Vice President
Quality Certification Committee
for Electronic Components of China

It is my pleasure to offer to you, experts and specialists from all over the world, some information on the IECQ system on behalf of this organization under the IEC, on behalf of the Management Committee of IECQ, the CMC. Since I work for the Quality Certification Committee for Electronic Components of China, I would also like to take this opportunity to present to you what has happened in China in the field of certification of electronic components.

In today's world, electronic products have made themselves indispensable in global economy and in all aspects of life; and electronic components are the basic elements that form those products. The mission of IECQ is to assure the quality of components. The IECQ-the quality assessment system for electronic components - established by and operating under IEC, aims at facilitating the international trade in electronic components of assessed quality, as governed by the principles of reciprocity stipulated by the Rules of the System [1].

An international system, the IECQ achieves its purpose by the supply of electronic components of assessed quality through approved manufacturers and distributors in its member countries. The products thus certified can be accepted by users without being further tested. In this way, the trade in electronic components can be effectively promoted within a country and throughout the world.

The standards and specifications against which components are certified are an integral part of the IEC standards system. There are at present about 150 specifications covering 28 product families. To accelerate the certification process, the Rules of the System include a procedure which allows the adoption by the System of the documents prepared by its member countries in the absence of IEC Publications created by Technical Committees. These documents are made provisional specifications having the same structures as those of the IEC, consisting of generic, sectional and blank detail specifications. Up to the present day, the number of detail specifications supplied by participating countries has surpassed 130.

With eight years of experience, the Certification Management Committee (CMC) now has 24 participating countries, each of which has set up its own National Authorized Institution (NAI) and National Standards Organization (NSO), operating under IECQ within national boundaries. So far, 13(+2) of them,including China, have had their National Supervising Inspectorates (NSI) approved by the System's Inspectorate Coordination Committee (ICC), thus becoming full-member certifying countries.

Within the IECQ System, there are more than 150 types of certified products from approved manufacturers and distributors of 10 countries. They include cathode-ray tubes, magnetic oxide cores, capacitors resistors, PTC thermistors, PE ceramic filters, PE ceramic resonators, transistors and switches and co-axial r.f. and data cables. Users of these components are assured of good performance, high reliability and quality, and timely delivery. I feel happy to report to the conference that IECQ is undergoing rapid development. For instance, in the past two years, the approvals of components each year have increased by 50% annually. U.S. and Japan take the lead in this progress, with 72% of all the components approvals.

China is a developing country; it takes much interest in the concept of quality assessment for electronic components and is doing its best to accelerate the development of the system for that purpose. Since 1981 when Quality Certification Committee for Electronic Components of China (QCCECC) was founded, certification activities have been going on throughout China, and 1987 saw China obtain a status of certifying country (full member of the ICC as I already mentioned) of the IECQ, which marked the enhancement of quality assurance effort. Up to now, it has approved within the country manufacturers and types of components against national, industry and company standards recognized by the Chinese NSO. Among them, 8 firms and 10 products are on the IECQ's Qualified Products List. China's undertaking of quality certification intends to maintain the levels of quality and reliablity of products in order that the producers and users of electronic equipment and systems can use the certified components without re-testing them. On the other hand, China would like to see certification process provide more trade opportunities among different countries.

IECQ is also characterized by its stringent certification requirements. An approved manufacturer will see his approval suspended or withdrawn if he fails to certify products continuously. This has been demonstrated by the fact that several cases of this nature have emerged, though the cause is mainly on the part of specification developments.

Through its comprehensive surveillance techniques and its embodiment of the relevant principles of ISO Standards Series ISO 9000, a unified quality assurance system, the IECQ enjoys considerable advantages and competitiveness in terms of product quality, reliability, just-in-time delivery and pricing. Moreover, the appearance of IECQ-approved distributors shows a great potential for injecting this wonderful system into various countries. industrialized or developing.

I am sure that within the next ten years, the continuing development of the IECQ will make a permanent contribution to the components industries of the countries concerned. The QCCECC will certainly join with the efforts of other NAI's to speed up the growth of the System.

Finaly, I would like to quote a remark made by Mr. J.R.Isken, retiring Chairman of the CMC, that "For anyone looking for quality assessed electronic components, it is my firm belief that IECQ is the logical choice."[2]

REFERENCE

1. Clause 2, Basic Rules of the IECQ, QC 001001
2. Report from the Chairman of the Certification Management Committee (CMC) of the IECQ, July 1988 IEC Document 01(Central Office)774.

FERROELECTRIC PROPERTIES OF TUNGSTEN BRONZE MORPHOTROPIC PHASE BOUNDARY SYSTEMS

L. Eric Cross
Pennsylvania State University
University Park, PA 16802 USA

and

Ratnakar R. Neurgaonkar
Rockwell International Science Center
Thousand Oaks, CA 91360 USA

ABSTRACT

Morphotropic phase boundary (MPB) ferroelectric materials are well known to possess exceptionally large ferroelectric properties, such as their electro-optic, pyroelectric, polarization and dielectric characteristics, because of the proximity of two alternate ferroelectric states at the phase boundary. In the tungsten bronze structural family, the best known MPB system to date has been $Pb_{1-x}Ba_xNb_2O_6$, which has an MPB at x = 0.37. Because of the wide variety tungsten bronze compositions with differing (4mm or mm2) structural phases available in this family, the potential exists for numerous other MPB systems composed from the binary compositions of bronze end member phases such as $PbNb_2O_6$, $Ba_2NaNb_5O_{15}$, $Sr_2NaNb_5O_{15}$, $Pb_2KNb_5O_{15}$ etc. MPB compositions have been established in many of these systems, and ferroelectric and optical properties have been determined using hot-pressed ceramics, bulk single crystals and thin films. Exceptionally large ferroelectric and electro-optic effects are present in the vicinity of MPB regions; however, the interesting feature in these systems is the presence of large and nearly equal transverse and longitudinal ferroelectric and electro-optic coefficients in certain orthorhombic (mm2) MPB systems. Potential applications of these MPB materials are discussed, along with the problems associated with their bulk crystal and thin film growth.

Large Piezoelectric Effect Induced by Direct Current Bias in PMN:PT Ferroelectric Ceramics

W.Y.Pan, W.Y.Gu, D.J.Taylor and L.E.Cross
Materials Research Laboratory
The Pennsylvania State University
University Park, PA, 16802

The longitudinal piezoelectric effect induced by a DC bias field was invesitigated in the PMN:PT family of relaxor ferroelectric ceramics in the frequyency range from 0.1 Hz to 50 kHz. The piezoelectric d_{33} coefficient was studied by measuring the strain induced by the applied electric field using an AC ultra-dilatometer. Large Piezoelectric strain coefficients giving d_{33} values up to 800pC/N are obtained at 10 kHz. The piezoelectric strain coefficient derived from measured dielectric and electrostrictive parameters were also studied under direct current bias and the d_{33} coefficient was calculated from the equation $d_{33}=2Q_{11}P_3\varepsilon_{33}$. The calculated values are in good agreement with the directly measured ones. The dispersion of the piezoelectric effect appears to aririse mainly from dielectric dispersion.

ELECTRICAL PROPERTIES OF ZnSe LANGMUIR BLODGETT FILM MIS MICROELECTRONIC DEVICES

D.K. Das-Gupta & P. Townshend,
School of Electronic Engineering Science,
University College of North Wales,
Bangor, Gwynedd, LL57 1UT, U.K.

and

J.O. Williams, F.R. Myers, S.Y. Luk and N. Maung,
Department of Chemistry, UMIST,
Manchester, M60 1QD, U.K.

Abstract

Current-voltage characteristics and capacitance - voltage behaviour of MIS structure of In-GaAs-ZnSe-Mesoporphyrin dimethyl ester Indium Chloride (InMPDME-Cl) - Al are consistent with the conventional tunnel injection mechanism.

Introduction

In many electroluminescent (EL) devices minority carriers are injected into the luminescent region using a forward bias to a p-n homojunction semiconductor. In such cases the minority carrier injection ration γ needs to be quite high. With amphoteric semiconductors it is difficult to prepare both n- and p- type forms although these materials may have excellent EL properties. This is particularly true, for example of Zn-Se (Group II-VI) which is a direct bandgap material with an energy gap, 2.64eV which corresponds to the blue wavelength region of the visible spectrum. ZnSe can only be made an n-type in low resistivity form. An alternative EL structure is a metal-semiconductor (Schottky barrier) device which has unfortunately, a very low minority carrier injection ratio ($\gamma \sim 10^{-4}$), thus providing a low EL efficiency. However, a significant improvement in the γ value may be achieved by the incorporation of a thin insulating layer between the semiconductor and the metal electrode (i.e. MIS structure). The thickness and the uniformity of the insulating layer are extremely important parameters in achieving the desired level of the minority carrier injection which is a strong exponential function of the majority and minority carrier barrier height. Now the Langmuir-Blodgett (LB) technique[1,2] provides an excellent method for depositing uniform and thin insulating organic films which may be used to study the nature of electroluminescence in MIS devices. Recently, an MIS structure of ZnSe and silicon phthalocyanine LB films has been observed to produce blue electro-luminescence[3].

We have undertaken a comprehensive programme to investigate the electrical properties of such MIS devices. The present work reports the current-voltage (I-V) and the capacitance-voltage (C-V) characteristics of a sandwich structure of In-GaAs-ZnSe-Mesoporphyrin dimethyl ester IndiumChloride (InMPDME-CL)-Al.

Experimental

Epitaxial n-type ZnSe films of thickness in the range 0.96 to 9.6µm were grown on single crystals of silicon doped GaAs substrates with (100) orientation by the metal organic chemical vapour deposition (MOCVD)[4]. GaAs substrates were obtained from MCP Electronics Ltd. The carrier density N, of the conductive GaAs substrate was $1.6 \times 10^{24}/m^3$. In order to provide an ohmic contact for the back electrode to the GaAs substrate the surface was etched with 5% HF and 100% HCl. Indium electrode[3] was then evaporated on one surface of the GaAs in an inert atmosphere at 220°C. Prior to the deposition of the LB film (InMPDME-CL) ZnSe was cleaned using Br/MeoH (1%, 30s), followed by CS_2 rinse and millipore water rinse toprovide a hydrophobic sample. Layers[2] of LB films were then deposited on ZnSe by dipping it at a rate of 5×10^{-3} m s^{-1} at a surface pressure >45 mN m^{-1}. Finally aluminium electrode of 500 nm thickness were vacuum deposited on the LB film. Samples were then mounted on a veroboard and contacts made at the top and bottom electrodes using silver loaded epoxy and cu/sn wire (see figure 1).

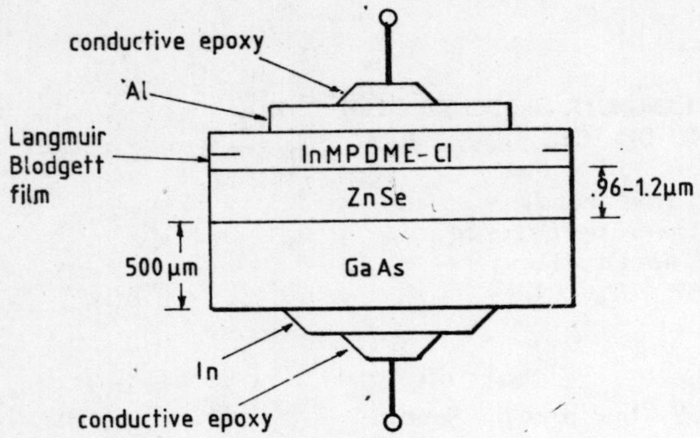

Figure 1. M.I.S. device.

All electrical measurements were made in a stainless steel chamber at a reduced pressure ($<10^{-5}$ torr) at a controlled temperature. Current-voltage characteristics were measured using two Keithley electrometers, type 602 and type 616 respectively. A high stability voltage generator was used for the externally impressed field. These measurements were automated with an IBM PC through the analog output of the current meter. All further manipulation and presentation of the data were done using custom software, written in Turbo C. Capacitance- voltage measurement data were obtained with a Wayne-Kerr 4225 automatic LCR bridge with suitable modification.

Results and Discussion

The In MPDME-Cl (figure 2) LB films used in this work, contain a metal-metal σ bond which may offer possibilities in modifying the light absorption-emission properties of such films[5].

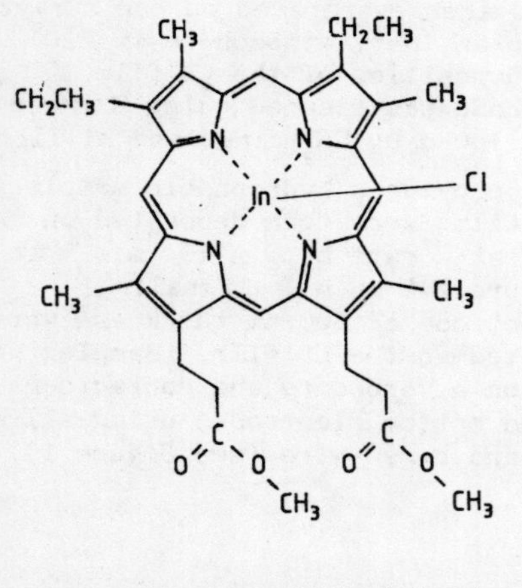

Figure 2. Mesoporphyrin dimethyl ester Indium Chloride.

The introduction of divalent metal ions has the advantage that they can replace the H^+ ions in the polar group of two acid molecules and form metal salts which improve the stability and prevent the slow collapse of the LB film due to dissolution of the film into sub-phase.

Generally the principle of the ZnSe-LB film MIS device operation is as follows. The LB film acts as an insulating layer which allows the energy bands in ZnSe to move with respect to the energy bands in the metal. With forward bias the valence band edge of the ZnSe moves towards the Fermi level in the metal, thus aligning the filled states in the ZnSe with empty sates in the metal. If the film is thin enough, minority carrier current should increase significantly with the bias voltage.

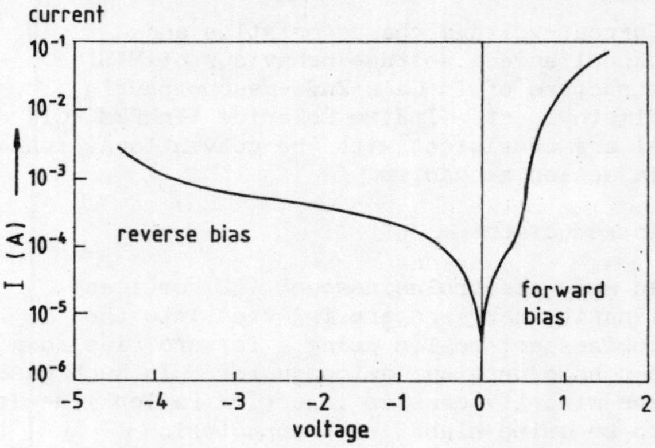

Figure 3. MIS device with 1.2 µm ZnSe and 2 layers of InMPDME-Cl.

Figure 3 shows the forward and reverse bias characteristics of our MIS device with 1.2µm thick ZnSe and two layers of In MPDE-C/ LB film Figures 4 & 5 show similar characteristics for the MIS device with 9.6µm thick ZnSe and one and two layers of the In MPDE-Cl LB films respectively.

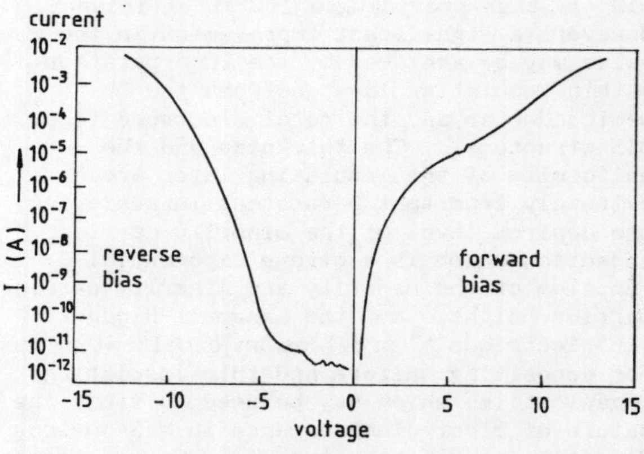

Figure 4. MIS device with 9.6 µm ZnSe and 1 layer of InMPDME-Cl.

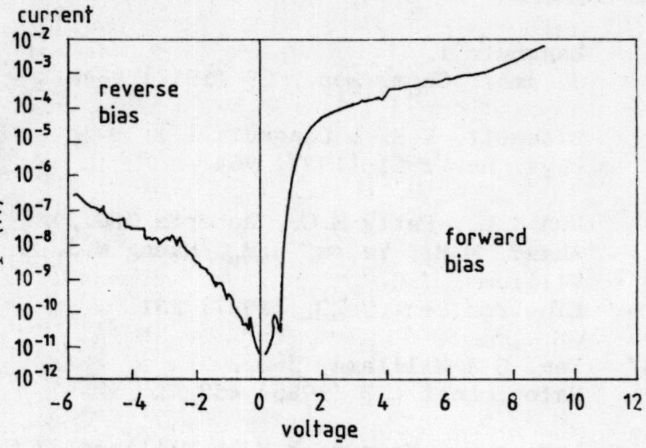

Figure 5. 9.6 μm ZnSe /2 layers InMPDME-Cl.

These results appear to be similar to d.c. tunneling characteristics of (Metal-Oxide-Silicon) structure with SiO_2 layers less than 50Å thick.[7] Esaki and Stiles have considered the phenomenon of electron tunneling between a metal and a degenerate semiconductor through a thin insulating layer. If there is a presence of surface states of the order of 10^{10} - 10^{12} states/cm^2/eV at semiconductor-insulator interface, tunneling characteristics will be significantly affected as these states may act either as available energy states in the energy gap or as recombination centres for tunneling of electrons from the metal[6]. The I-V characteristics depend critically on the insulator thickness. If the insulator layer is sufficiently thick, then the carrier transport through the insulator layer may become negligible and the MIS diode in the latter case will represent a conventional MIS capacitor. If on the other hand the insulator layer is extremely thin, the structure will represent a Schottky barrier diode and there will be little impediment for the carrier transport between the metal and the semiconductor. MIS diodes with degenerate semiconductors may exhibit negative resistance characteristics.

However, for the MIS tunnel diode with a very thin insulating layer on a non degenerate semiconductor negative resistance may not be expected. For such a case (as is ours) the current density is given by[8,9]

$$J=A^{*}T^{2}e^{-\alpha_T \phi_T^{1/2} d}\, e^{-q\phi_b/kt}\left(e^{qv/kt}-1\right) \quad \ldots(1)$$

where A is the effective Richardson constant α_T the tunneling probability for a rectangular barrier with an effective barrier height ϕ and width d. The other symbols have their usual meaning. Equation 1 is identical to the thermionic emission equation for Schottky barriers except for the term

$\exp\left(-\alpha_T \phi_T^{1/2} d\right)$, which represents the tunneling probability. It may be observed from equation 1 that for $\phi_T \sim 1$V and d> 50 Å the tunneling probability is extremely small and the current would be negligible. As ϕ_T and d increase the current increases rapidly for the thermionic emission current level. Present results are in agreement with the above model.

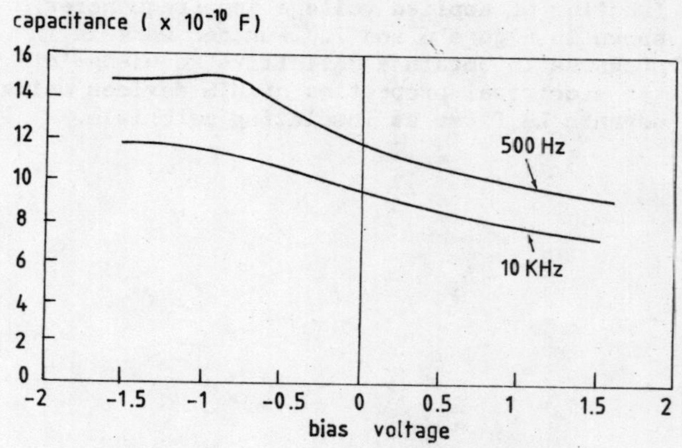

Figure 6. C-V at 500 Hz & 10KHz for MIS device with 9.6 μm ZnSe & 1 layer InMPDME-Cl.

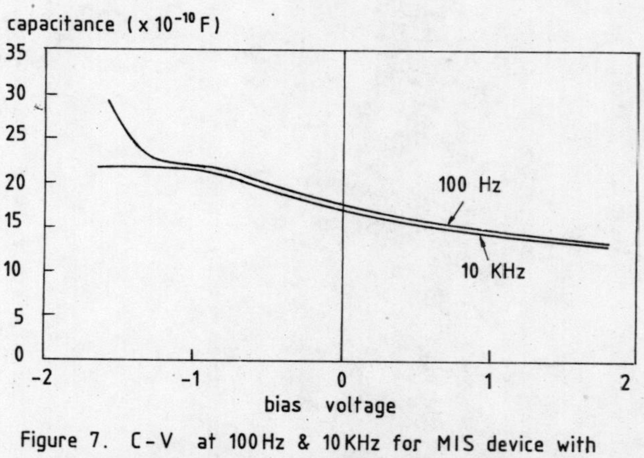

Figure 7. C-V at 100Hz & 10KHz for MIS device with 9.6 μm ZnSe & 2 layers InMPDME-Cl.

Figure 6 and 7 show the capacitance - voltage characteristics of two MIS devices at two different frequencies with 1 layer and 2 layers of LB film respectively deposited on 9.6μm thick ZnSe. The frequency dependence of tunneling in MOS devices has been investigated by Gray[10] for nondegenerate silicon. He introduced a transition probability[11] per unit time w_a, of an electron from metal into the interface states. In this model w_a, is linearly proportional to the density of interface states and it varies exponentially on the tunneling exponent. For low frequencies, the capacitance $C = \frac{dQ}{dV}$ and is proportional to the density of interface states. With increasing frequency C decreases as less charge is transferred to the interface states. Above observations are in agreement with the measured values of capacitances as a function of applied voltage and frequencies, shown in Figure 6 and 7. Further work is in progress to obtain a definitive knowledge of the electrical properties of MIS devices using organic LB films as insulating materials.

REFERENCES

1 Langmuir I.,
J. Amer. Chem. Soc., 39 (1917) 1848

2 Blodgett, K.B. & Langmuir I.,
Phys. Rev., 51 (1937) 964

3 Hua Y.L., Petty M.C., Roberts G.G., Ahmad, M.M., Yates, H.M., Maung N.J. & Williams, J.O.
Electron Lett., 23, (1987) 231.

4 Fan, G & Williams, J.O.,
Mater. Lett., 3 (1985) 453

5 Luk, Y.S., Mayers, R.M. & Williams, J.O.,
J.Chem.Soc., Chem. Commun, 1314 (1987) 215

6 Dahlke, W.E., & Sze, S.M.,
Solid State Electron, 10 (1967) 865

7 Esaki, L. & Stiles, P.J.,
Phys. Rev. Lett., 16 (1966) 1108

8 Kumar, V. & Dahlke E.,
Solid State Electron., 20 (1977) 143

9 Card, H.C. & Rhoderick, E.H.,
J. Phys. D. Appl. Phys. 4 (1971) 1589

10 Gray, P.V.,
Phys. Rev., 140, (1965) 179

11 Bardeen J.,
Phys. Rev. Lett. 6 (1961) 57.

A NEW TECHNIQUE FOR EXAMINATION OF POLARIZATION DISTRIBUTIONS IN THE SURFACE REGIONS OF FERROELECTRIC CERAMICS

Sidney B. Lang
Department of Chemical Engineering
Ben-Gurion University of the Negev
84105 Beer Sheva, Israel

ABSTRACT

Recently, a method for measuring the spatial distribution of either polarization or space charge in various ferroelectric materials has been developed.[1,2] The technique, called the Laser Intensity Modulation Method (LIMM), is based upon measurement of the pyroelectric current generated by the material when it is heated with a laser beam which is intensity modulated over a range of frequencies. A special analytical technique is used to transform the current-frequency measurements to spatial distributions.

A modification of this technique has now been developed which permits detailed profiling of the polarization distributions in a layer from the surface to a depth of 20-30 micrometers in poled ferroelectric ceramic samples. Because of the analogy to LIMM, the technique has been named the Surface Laser Intensity Modulation Method (SLIMM). In this technique, the laser modulation frequencies are much higher than in LIMM, typically from 1 to 100 kHz. The determination of the polarization distributions requires the solution of a Fredholm integral equation of the 1st kind using experimental data . Stable and accurate solutions have been obtained using the regularization method.

Experiments have been carried out on a number of different types of lead zirconate titanate ceramics. Effects of surface finish and electrode materials will be discussed.

REFERENCES

1. S.B. Lang and D.K. Das-Gupta, J. Appl. Phys. 59 (1986), 2151.
2. S.B. Lang and Yin Q.-R., Ferroelectrics 74 (1987), 357.

HIGH PERFORMANCE PIEZOELECTRIC CERAMICS WITH LOW TEMPERATURE SINTERING

Zhilun Gui, Longtu Li, Hongqing Lin, Qingjian Jiang, Ziaowen Zhang

Department of Materials Science & Engineering
Tsinghua University Beijing China

ABSTRACT

A new quarternary system of high performance piezoelectric ceramics with low temperature sintering has been investigated. Its composition is lead magnesium nickel niobate zirconate titanate (PMN-PNN-PZT). This ceramics with proper additives could be sintered at about 900°C and exhibited excellent piezoelectric properties Kp=0.60-0.65, $\varepsilon^T_{33}/\varepsilon_o$=2000-2800, d_{33}=(400-500)x10^{-12}C/N. The microstructure and sintering mechanicism of the piezoelectric ceramics has been analyzed by Scanning Electron Microscope (SEM), TEM and Positron Annihilation Technique(PAT). XRD of the Samples show a complete perovskite structure had been obtained.

INTRODUCTION

Normally for conventional PZT ceramics the sintering temperature is about 1200-1300°C, the volatility of PbO at that temperature is quite a serious problem both for compositional control and envirmental pollution, In addition the high firing temperature requires expensive metals internal electrode in multilayer piezoelectric devices. So the development of low temperature sintering piezoelectric ceramics will therefore reduce the energy consumption and environmental pollution. In general the sintering temperature could be lowered by doping with low-melting oxides, But previous some researches showed that the microstructure and properties of low temperature sintering piezoelectric ceramics are degraded.[1-2] One of the reasons is the excess low melting oxides as a nonpiezoelectric phase precipitated in the grain boundaries. The characteristics and goal of our investigation are to reduce the sintering temperature substantially without degrading the properties of the ceramics and even better properties obtained. So we must precisely control the composition of the materials and additives, equivalence and unequivalence ions substitute and addition, donors and acceptors compensate and form an eutectic at a lower temperature. The lower melting additions as an initial phase can promote the sintering in the early and middle stages of the sintering and as a final phase can enter to the lattice and modify the properties. Double benefit of aid firing and modify the properties is the key to realize low temperature sintering and high performance ceramics. We have been succeeded in studying the binary system PZT, ternary system PNN-PZT and quarternary system PMN-PNN-PZT ceramics.[3-5] The satisfactory piezoelectric properties at low sintering temperature had been obtained. In this paper we emphasize on high performance quarternary system PMN-PNN-PZT ceramics.

EXPERIMENTAL PROCEDURE

The chemical grade raw materials were used. Samples were prepared by conventional processing. Disc samples were sintered at 860°C-1000°C for 2-4hrs. The synthesis of the raw materials and sintering effect of the samples were analyzed by means of XRD. The piezoelectric properties of the samples were measured after poling at 4 KV/mm at 120°C in silicone oil bath. The microstructure and sintering mechanism were studied by means of SEM, TEM and PAT. The piezoelectric properties were measured by resonance method. The dielectric constant, dielectric loss were measured at 1 KHz.

The general formula of composition are given as follow:

$$Pb_{1-x}A_x(Ni_{1/3}Nb_{2/3})_y\ (Mg_{1/3}Nb_{2/3})_z\ Zr_u\ Ti_wO_3 + g_1 wt\%MnO_2 + g_2 wt\%SiO_2$$

where A-Sr, Cd, Zn.

$0<X\leq 0.08$, $0\leq y\leq 0.35$, $0\leq z\leq 0.40$
$u=0.25-0.35$, $w=0.35-0.45$
$0<g_1\leq 0.50$ $0.05\leq g_2\leq 0.50$

RESULTS AND DISCUSSION

Table 1 shows the dependence of the piezoelectric properties and different sintering temperature for PMN-PNN-PZT substitute with Cd ceramics. It can be seen in Table 1 that quite good properties could be obtained at 840°C-1000°C. This means PMN-PNN-PZT ceramics has the advantages to the processing control for their wide sintering range. It can be also seen that the

Table 1 Kp, $\varepsilon_{33}^{T}/\varepsilon_o$, d_{33} of the Samples at Different Sintering Temperatures for PMN-PNN-PZT substitute with Cd Ceramics

S.T. (°C)	840	860	880	900	920	940	960	980	1000
Kp	0.58	0.59	0.61	0.65	0.65	0.64	0.63	0.62	0.62
$\varepsilon_{33}^{T}/\varepsilon_o$	2300	2400	2605	2810	2980	2870	2750	2850	2900
d_{33} (10^{-12} C/N)	395	435	445	465	475	465	440	435	435

properties are better than that of low temperature sintering PNN-PZT and PZT ceramics.[3-5] In fact their properties are even better than that of conventional PZT at high sintering temperature.

Figure 1 a and b show the SEM micrograph which sintered at 920°C. It is indicated that the dense and uniform microstructure had been obtained at low sintering temperature. Figure 1b shows the fracture section of the sample. Most of the grains show transgranular fracture. This means the materials has better mechanical properties. The XRD of the raw materials and samples had been analyzed. It is showed in Figure 2. It is confirmed that the raw materials being used had been well synthesized into a perovskite structure after calcining and sintering at low temperature.

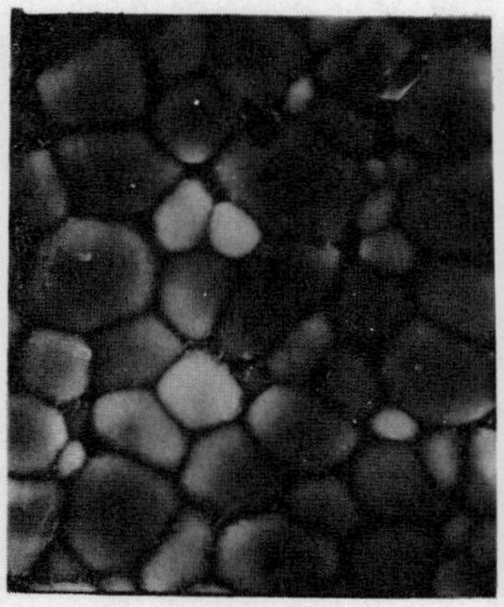

1a

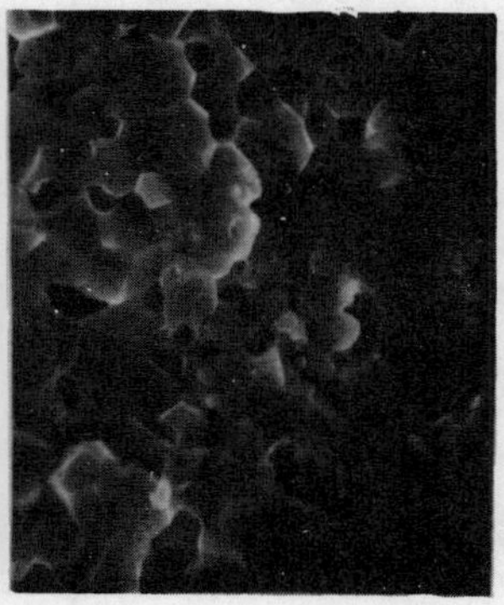

1b

Fig.1 SEM Micrographs of PMN-PNN-PZT Substitute With Cd Ceramics at 920°C (2-4μm)

Table 2 shows the properties of PMN-PNN-PZT ceramics with different substitute additives in A site, the Sr, Zn, Cd were used. It is indicated that the ceramics substitute with Cd^{2+} has the best properties and can be sintered at lower temperature (900-920°C), the ceramics substitute with Zn^{2+} can have better properties but the suitable temperature should be up to 980°C-1000°C, the properties of the ceramics substitute with Sr^{2+} is less better.

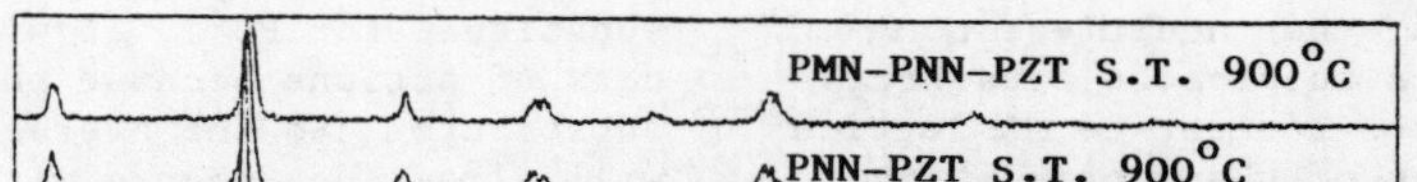

Fig.2 XRD of Raw Materials and Samples

Table 2 Properties of PMN-PNN-PZT ceramics with Different substitute Additives

S.T. (°C)	Substitute Additive	Content (mol)	Kp	d_{33} (10^{-12} C/N)	$\varepsilon^T_{33}/\varepsilon_o$	$\tan\delta$ (10^{-4})
900	Sr	0.02	0.46	288	1645	56
	Cd	0.02	0.60	414	2455	58
	Zn	0.02	0.50	384	1446	66
920	Sr	0.02	0.49	328	1816	51
	Cd	0.02	0.62	428	2702	46
	Zn	0.02	0.44	320	1661	107
940	Sr	0.02	0.55	402	1870	45
	Cd	0.02	0.65	430	2652	47
	Zn	0.02	0.60	399	2075	45
980	Sr	0.02	0.51	337	2044	55
	Cd	0.02	0.63	457	2841	45
	Zn	0.02	0.61	427	2213	42
1000	Sr	0.02	0.56	405	2070	46
	Cd	0.02	0.63	487	2908	49
	Zn	0.02	0.62	453	237	42

This result may be discussed as follow, according to their ionic radius of Sr^{2+}=1.44Å and Cd^{2+}=1.31Å. It is suggested that they can substitute for Pb^{2+}=1.49Å to occupy A site, Cd^{2+} belongs to IIB group element with 18 electron configuration. It has high polarization (polarizing power) than Sr^{2+}, IIA group element with 8 electron configuration. It seems the Cd^{2+} could combine with other additives to from an eutectic liquid phase to promote densification during sintering and reenter to the lattice to modify the properties of the matrix in the later stage of sintering. In the same IIB group Zn^{2+} can also lower the sintering temperature but didn't play the role as good as Cd^{2+} did. It is assumed that the ionic radius of Zn^{2+} is too small to enter A site. This result and suggestion had been also demonstrated by Positron Annihilation Technique (PAT).

The Positron Annihilation Technique (PAT) is an effective method to study lattice defect of piezoelectric ceramics and the mechanisum of the role of additives.

Positron is an antiparticle of electron. It is an electron with positive charge. When the cation vacancies occur in the matrix, they correspond to equivalent negative charge center and the positrons can be trapped. The average lifetime (τm) increases with increasing of the vacancies concentration of cations. Because the positron trapped rate increases with increasing of cations vacancies. However, in our experiment II group element Sr^{2+}, Zn^{2+}, Cd^{2+} were used as substitute additives. If we suggested that they all can enter A site of the ABO_3 structure to substitute for Pb^{2+}, It will not produce vacancies of cations because they are equivalent substitute. So the average lifetime of positron will almost not change with the content of substitute additives. We used La^{3+}, it is unequivalent substitute, as a contrast, La^{3+} will substitute for Pb^{2+} and produce a lead vacancy Vpb", the concentration of the vacancy increases with increasing of the La^{3+} content. τm also increases.

Figure 3 shows the average lifetime as a function of concentrations of substitute additives, it can be seen from Figure 3, Sr^{2+} did enter A site to equivalent substitute for Pb^{2+}, because the τm did not change with Sr^{2+} content. The τm of the samples did not change with Cd^{2+} content greatly until $X > 4$-5 mol%, but the sample substituted with Zn^{2+}, τm increased with increasing of Zn^{2+} content. This result indicated that Cd content<4-5 mol%, they mainly enter A

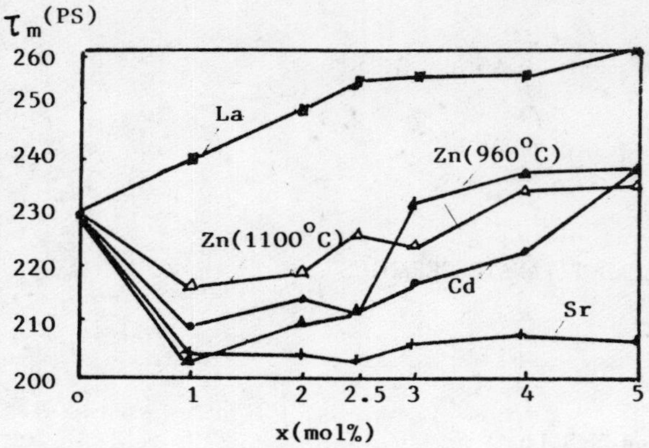

Fig.3 Average Life Time as a Function of Different Substitute Additive X(mol%)

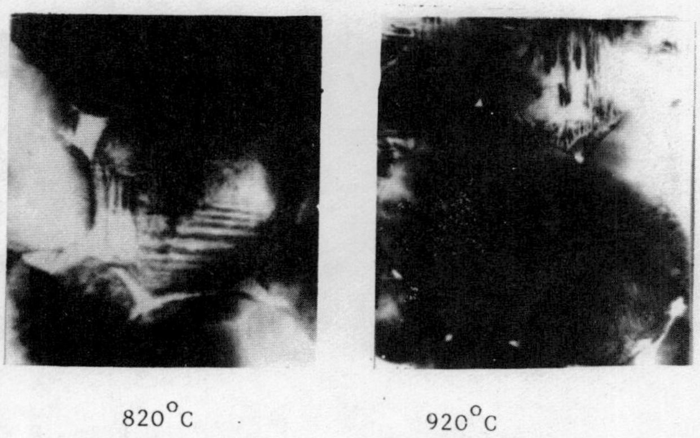

Fig.4 TEM Micrographs of PMN-PNN-PZT Substitute with Cd^{2-} Ceramics.

site to equivalent substitute for Pb^{2+}, but Cd content>4-5 mol% and substitute with Zn, τ_m inereased. This means nonequivalence substitute occurred. It is suggested that the Cd^{2+} and Zn^{2+} could enter B site in ABO_3 structure. Due to the nonequivalent substitute, the oxygen vacancy could occur in PMN-PNN-PZT ceramics as dopants are more than 4-5 mol% Cd^{2+} or Zn^{2+}. The oxygen vacancies are equivalent to positive charge center. Oxygen vacancies could be seen as quasi-F centers, which can attract two electrons shared with nearby Ti^{4+} ions. The positron can be trapped by oxygen vacancies attracted electrons to form third positron state. The lifetime of positron became long and τ_m increased. From our experimental result it is very instructive that oxygen vacancy may be the main trapping center in the samples.

The PMN-PNN-PZT substitute with Cd^{2+} ceramics can have excellent properties with low temperature sintering perhaps due to it can enter A site when the Cd^{2+} content<4 mol%, it produced the lattice distortion and improve the properties, when substituted with nigher Cd^{2+} concentration, partly of them entered B site to produce Vo", it can not promote sintering during the oxygen atmosphere, so when the content of Cd^{2+} increases, the sintering temperature will be increased. The samples substitute with Zn^{2+}, it entered B site and produced Vo", so the suitable sintering temperature raise up to 980°C-1000°C.

The Pb^{2+} replaced by Cd^{2+} etc. could form eutectic with SiO_2, the Teut=720°C-760°C [6] and we have also observed the liquid phase at about 700°C through the high temperature microscope.

The TEM micrographs of Figure 4 show that the sample with sintering temperature at 820°C shows a wider boundary, but at 920°C the sample shows almost all of the remained liquid phase concentrated in the triple intersections and the most liquid phase reabsorbed into the grains. The sintering mechanism seems to be controlled by solution-diffusion-precipatation.

CONCLUSION

1. PMN-PNN-PZT substitute with Cd^{2+} ceramics could be sintered at about 900°C and exhibited excellent piezoelectric properties. It is about 300°C lower than that of normal PZT ceramics.
2. Positron Annihilation Technique (PAT) is an effective method to study lattice defect of piezoelectric ceramics and the mechanisum of the role of additives.
3. The sintering mechanism seems to be controlled by solution-diffusion-precipitation through a transient liquid phase sintering.

REFERENCE

1. S.Takahashi, Jpn. J. Appl. Phys. 19. (1980), 771.
2. D.E.Wittmer et al., J.Am. Ceram. Soc., 64, 8. (1981), 485.
3. Longtu Li et al., Chinese Patent 85 100051. (1987).
4. Zhi-lun Gui et al., Chinese Patent 86108741. (1988).
5. Zhilun Gui et al., J. Am. Ceram. Soc., 72, 3. (1990)487.
6. E. M. Levine et al., Phase Diagrams for Ceramists (1964), Fig. 284.

DOMAIN STRUCTURE OF MODIFIED LEAD TITANATE CERAMICS

W.R.Xue,K.P.Liu and Y.N.Lu
Nanjing Institute of Chemical Technology
Nanjing,China

ABSTRACT

Lead titanate ceramics modified with samarium have been investigated.The materials exhibit large anisotropy in electromechanical coupling factors.As a possible reason for the large anisotropy in coupling factors,the 90° domain rotation was suggested.In the present paper,the domain structure of lead titanate ceramics modified with samarium were examined by electron microscopy.It is found that there are 90° and 180°domain in all the samples poling at 20-50 kv/cm,although the planar electromechanical coupling coefficient have shown the remarkable difference for the samples poling at different electric field.A kind of new microdomain was observed in the specimen.However,there are not ample evidences to prove the 90° domain rotation.

INTRODUCTION

Modified lead titanate ceramics are of increasing importance because of their relatively small dielectric constant,low loss tangent and large anisotropy in electromechanical coupling factors.Yamashita etal.[1]reported that the lead titanate ceramics modified with alkaline-earth and Takeuchi etal.[2]published the electromechanical properties of lead titanate ceramics modified with rare-earth.The ceramics in these two systems all exhibit large anisotropy in electromechanical coupling factors.Application of modified lead titanate ceramics for transducer arrays[3]and SAW devices[4]have been reported. The mechanism for the electromechanical anisotropy in rare-earth modified lead titanate was suggested by certain ionic radii of samarium[2] and in alkaline-earth modified lead titanate was supposed by the 90° domain rotation.[1] Xue etal.[5]examined the temperature behavior of the coupling factors for samarium modified lead titanate ceramics.The results indicated that the temperature has almost no influences on thickness coupling factors in the range of -100°C to +80°C.But,a minimum in the planar coupling factors can be achieved in the range near room temperature.The large anisotropy in electromechanical coupling factors was caused by very small Kp.Damjanovic etal.[6,7]published the temperature and poling electric field behavior of the coupling factors for modified lead titanate. A minimum in the planar coupling factor was explained by a change of sign of piezoelectric coefficient,d_{31}.In the present paper,the dependence of large anisotropy in electromechanical coupling factors on poling field has been investigated in order to examine the mechanisms for Sm-modified lead titanate. At the same time, the domain structure of the samples poled at different electric fields were observed by TEM. The relationship between large anisotropy in electromechanical coupling factor and domain structure is discussed.

EXPERIMENTAL

The composition of the modified lead titanate is $(Pb_{0.88}Sm_{0.10})(Ti_{0.98}Mn_{0.02})O_3$. After milling, the raw materials were calcined at 900°C to 920°C for 1 to 2 h and then pressed disks were fired at 1230°C for 2 to 3 h . The sintered ceramics were shaped in the form of thin disks and silver electrodes were made on the major faces. Samples were poled at the electric field of 20-60 kv/cm and then measured the electromechanical properties. At the same time, the thin foil TEM specimens were prepared by polishing down to 30μm and argon-ion-beam thinning. Using 4-5 kv argon incident to the foil plane, the thin foil specimens were observed by TEM.

RESULTS AND DISCUSSIONS

Fig.1 is the dependence of d_{33} of Sm-modified $PbTiO_3$ ceramic on poling field and the dependence of electromechanical coupling factors on poling electric field is shown in Fig.2. The dependences of d_{33} and k_t on poling field are similar. They increase with poling field increasing and tend to saturation when poling field reach 50kv/cm. d_{33} is about $60x10^{-12}c/N$ and k_t goes up to 0.50. kp, however, decrease with the poling field increasing and then drop down to 0.01 at the poling field of 50kv/cm. kt/kp is more than 50.

The ratio of kt/kp at different poling electric field is illustrated in Fig.3. When the poling field was varied from 20 to 50 kv/cm the value of kt/kp increased from 10 to 50. It is clear that the poling electric field plays an important role in causing large anisotropy in electromechanical coupling factors of modified lead titanate.

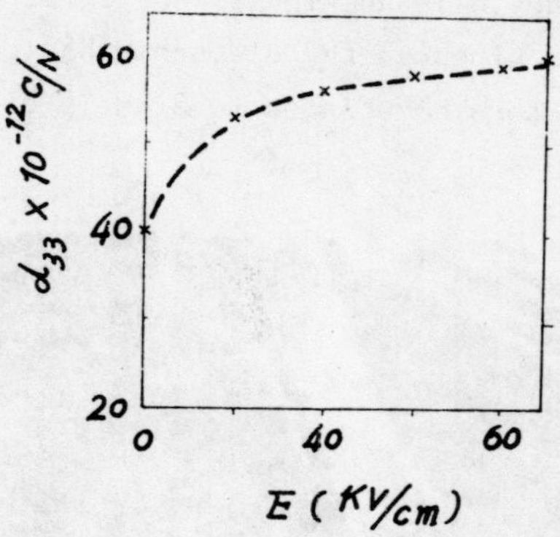

Fig.1 Dependence of d_{33} on poling field

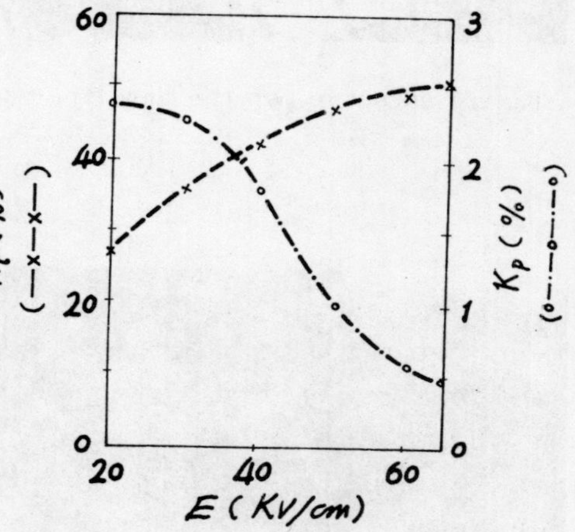

Fig.2 Dependence of kt, kp on poling field

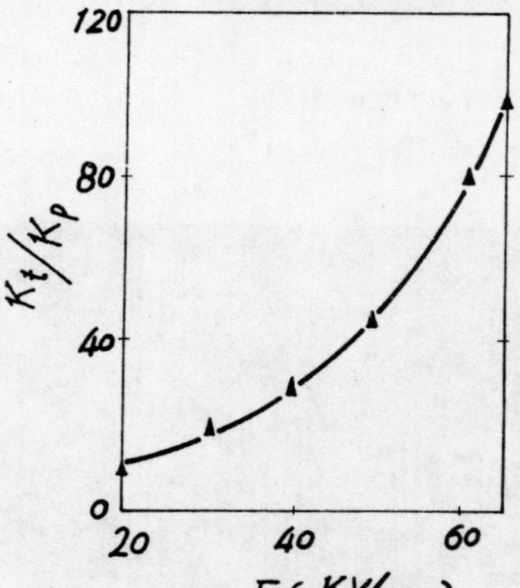

Fig.3 Dependence of kt/kp on poling field

Fig.4-6 show the domain structure of modified

lead titanate poled at different electric field. 90°and 180°domains were observed in all of the specimens.They have no significant difference in 90°domain although the electromechanical coupling factors have shown remarkable differences.

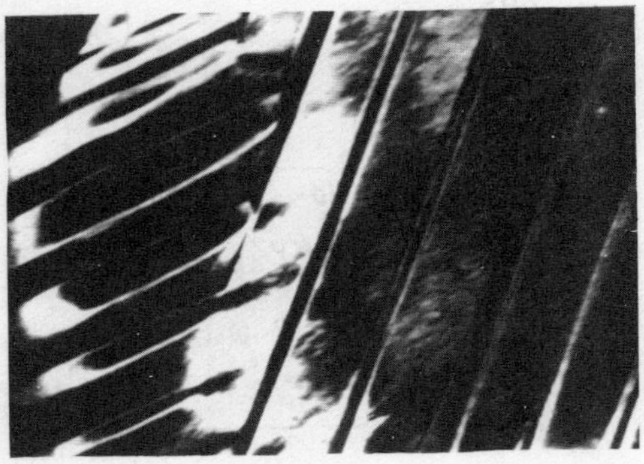

Fig.4 Domain structure of the unpoling sample

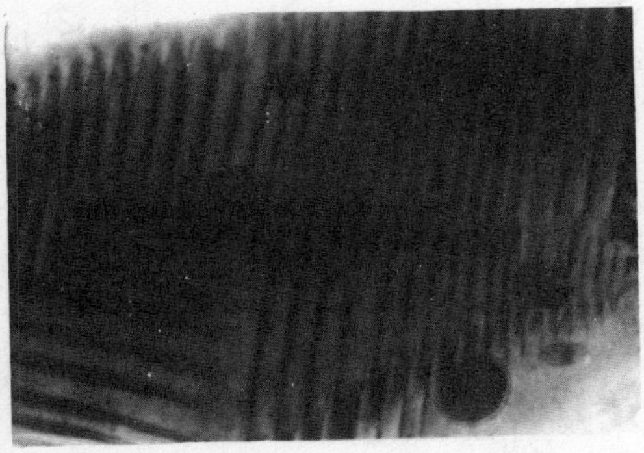

Fig.5 Domain structure of the poling sample at 20kv/cm electric field.

Fig.6 Domain structure of the poling sample at 50kv/cm electric field.

But,a kind of new microdomain was found in the specimens which were poled more than 50kv/cm (Fig.7).The width of the microdomain is about 300 to 400 Nm.There is a certain angle between microdomain and 180° or 90° domain.The microdomain could not be found clearly in the specimens which were poled below the electric field of 50kv/cm.

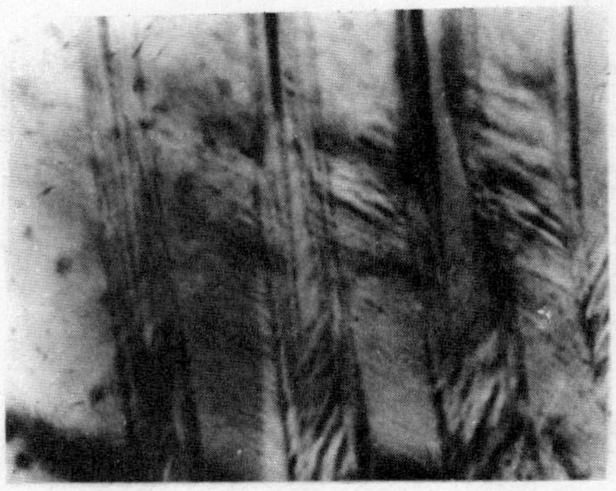

Fig.7 Microdomain of the poling sample at 50kv/cm electric field.

Large anisotropy in electromechanical coupling factors and appearance of microdomain in the samples seem to be caused by the same poling electric field.It is supposed that the appearance of microdomain might be the possible mechanism for the electromechanical anisotropy in Sm-modified lead titanate ceramics.The further research work about the influences of microdomain on electromechanical properties will be done in detail.From the observation of domain structure in Sm-modified lead titanate,there are not ample evidences to prove that the large anisotropy in electromechanical coupling factors were caused by 90°domain rotation.

CONCLUSIONS

The poling electric field plays an important role in causing large anisotropy in electromechanical coupling factors of Sm-modified lead titanate ceramics. The critical poling field for Sm-modified lead titanate ceramics seems to be 50kv/cm. Observation of domain structure could not prove the 90°domain rotation. A new kind of microdomain was found in the specimens poled more than the electric field of 50kv/cm. The appearance of microdomain may be a possible mechanism for electromechanical anisotropy in Sm-modified lead titanate ceramics.

REFERENCES

1. Y.Yamashita etal., Jpn.J.phys.20 Supplement 20-4 (1981), 183.
2. H.Takeuchi etal., J.Acoust.Soc.Am., 72 (4) (1982), 114.
3. H.Honda etal., proc.1982 IEEE Ultrasonics Symposium, (1982), 845.
4. Y.Ito etal., J.Appl.Phys.52(5)(1981), 3223.
5. W.R.Xue etal., Jpn.J.Appl.phys.24, Supplement 24-2, (1985), 718.
6. D.Damjanovic etal., Mater.Lett., 4 10 (1986), 414.
7. D.Damjanovic etal., Am.Ceram.Soc.Bull., 66 4 (1987), 699.

STUDY OF LARGE OUTPUT POWER PIEZOELECTRIC CERAMIC TRANSFORMER

Kuang Anxiang, Zhou Taosheng,
He changxin and Chai Liying
Department of Physics, Hubei University
Wuhan, China

ABSTRACT

The piezoelectric ceramic transformer has been more and more widely applied in electronic devices of high voltage and small current due to its novel features in the high voltage field: very high voltage step-up ratio, small in size, light in weight and non-metal material in use; it can't be burnt while the load circuit is shorted, and broken down in high voltage field; it doesn't suffer from humility of the air and electromagnetic disturbance, etc. Along with the development of the exploitation and application of the piezoelectric ceramic transformer, an investigation has been made on a piezoelectric ceramic transformer with larger output power and high time-temperature stability.

It has been shown that it is difficult to fabricate one-piece piezoelectric ceramic transformer with output power larger than 40 watts[1] by using piezoelectric materials from modified PZT binary system or PCM[2] and PMS[3] ternary system ceramics. In the present work, a study of the quaternary system piezoelectric ceramics consisting of $Pb(Mg_{1/3}Nb_{2/3})_A(Mn_{1/3}Nb_{2/3})_BTi_CZr_DO_3$ (called PMMN) has been thoroughly studied. And a transition position similar to those of binary and ternary system ceramics has been found in the tetragonal phase near the rhombohedral-tetragonal morphotropic phase boundary. But it is referred to as cross-transition position because the position not only depends on the first and second, but also on the third and fourth components of the composition. The material in this position has such an advantage as time-temperature stability being exellent, electromechanical coupling coefficient K_p and the mechanical strength high[4]. By using this kind of material, a plate piezoelectric ceramic transformer with 65 watts output has been fabricated with the size of 100x25x3(mm^3). It can continuously operate stably with 50 watts output in practice.

EXPERIMENTAL

I. The Material and its Basic Parameters

The basic composition of the PMMN quaternary system material used to fabricate the piezoelectric ceramic transformer is:

$$Pb(Mg_{1/3}Nb_{2/3})_A(Mn_{1/3}Nb_{2/3})_BZr_CTi_DO_3$$

where A=0.059, B=0.066, C=0.459, D=0.416. This composition is modified by adding an appropriate amoune of CeO_2 and by partly substituting Sr for Pb, and the characteristics of temperature of the material are thus improved. For example, the maximum relative deviation of the resonant frequency $|(\delta f_r)_m|$ and of the mechanical quality factor $|(\delta Q_m)_m|$ can be reduced by about 50%. The parameters of the material in this composition are shown in Table 1.

Table 1. The Material Parameters of the PMMN-1 Piezoelectric Ceramics

k_p	k_{31}	k_{33}	Q_m	$\varepsilon^T_{33}/\varepsilon_{33}$	tgδ
0.64	0.31	0.73	3500	1280	0.002

ρ	d_{33}	d_{31}	g_{33}	g_{31}	s^E_{11}	s^D_{11}
g/cm^3	10^{-12} C/N		10^{-3} V·m/N		10^{-12} m^2/N	
7.8	311	−110	30.1	−10.7	11.4	10.3

The PMMN quaternary system piezoelectric ceramic and the plate piezoelectric ceramic transformer with size of 100x25x3.6 (mm^3) are made by the conventional mixed oxides processing. The no-load alternating voltage step-up ratio of the piezoelectric transformer is 780 and its resonant frequency in full-wave mode is 30.58 KHz.

Results and Discussions

The main performances of the piezoelectric ceramic

transformer made by using the PMMN piezoelectric material have systematically been measured in order to decide the range of application. The results are obtained by conventional measuring circuits and methods.

1. Output Power Versus Load Impedance Characteristics

Fig.1 shows the dependence of the output power of the piezoelectric transformer on load resistance. The input voltage V_i=275V, and the room temperature T_o=25°C. When load resistance is R_L=9MΩ, the maximum output power is 65 watts, decreasing or increasing the load resistance results in the reduction of output power. The relations between output power and load impedance change slightly for different input voltage, because the input impedane and voltage step-up ratio of the piezoelectric ceramic transformer change as the load impedance changes.

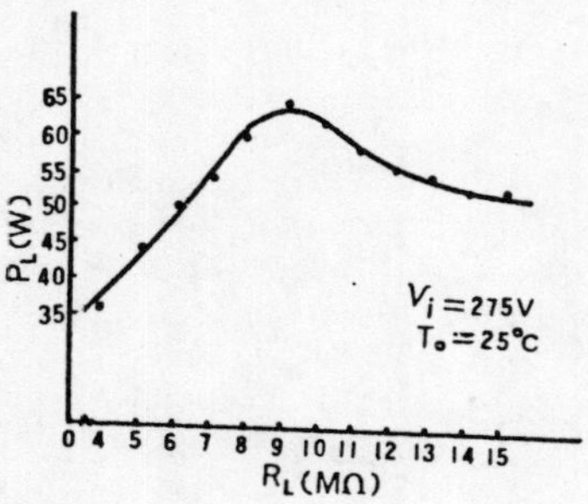

Fig.1 Output power P_L versus impedane R_L characteristics

2. Efficiency Versus Load Characteristics

The dependence of the efficiency on output power of the piezoelectric transformer is shown in Fig.2. This curve is determined after the piezoelectric transformer has reached its thermal equilibrium. The load resistance is R_L=8MΩ and the room temperature T_o=25°C. Under the condition of large power output, each parameter of the piezoelectric transformer changes slightly before thermal equilibrium, and after thermal equilibrium the parameters remain unchanged. Because the output power is measured on load resistance, the total loss of power supply includes the loss in the piezoelectric transformer and the rectifying circuit. Therefore, the actual efficiency of the ceramic transformer is higher than what is shown by the curve in Fig.2. Experiments show that, within 10 watts output power, the efficiency of the piezoelectric transformer can reach up to and in excess of 86%. When the output power is 40 watts, the efficiency reaches 70%, even if the transformer is operated with 50 watts, the efficiency of the piezoelectric transformer is still 63%.

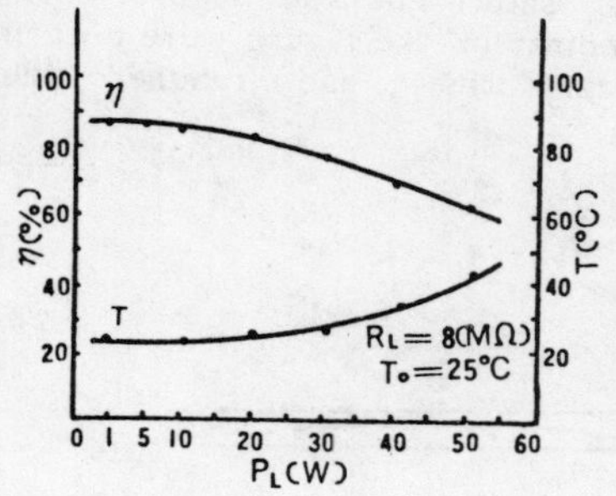

Fig.2 Efficiency η and wave node temperature T versus load P_L characteristics

3. Wave Node Temperature Versus Load Characteristics

As the stress at the wave node of the piezoelectric transformer has its maximum value, the temperature at the nodal points in the piezoelectric transformer is also the highest. For example, in the case of full-wave resonant mode, the temperature at 1/4 of transformer length reaches maximum; in other places, the temperature is lower. However, the temperature is the lowest at both ends of the piezoelectric transformer. The relations between the maximum temperatures at the nodal points and the output power are shown in Fig.2, where the load resistance R_L=8MΩ, the room temperature T_o=25°C. The experimental points in Fig.2 were all measured after the piezoelectric transformer had been operated for one hour and stayed completely in thermal equilibrium. It can be seen from Fig. 2 that when the output power of the piezoelectric transformer is 40 watts, the temperature at the nodal points is 34°C; when the output power is 50 watts, the highest temperature in piezoelectric transformer is 47°C; whereas the temperature of other parts of transformer becomes lower.

4. Resonant Frequency Versus Environmental Temperature Characteristics

Generally speaking, the rsonant frequency of the piezoelectric material shifts as a result of the heat of the piezolectric transformer itself in operation and the changes in the environmental temperature. Consequently, the temperature characteristic of the resonant frequency of the piezoelectric transformer is an important mark for determining the performance of the transformer. Particularly, regarding to the piezoelectric transformer operating with large output power, the study of the temperature characteristics of the resonant frequency becomes more important. Fig.3 shows how the resonant frequency in full-wave mode depends on the environmental temperature. The measuring conditions are shown in the figure. All the experimental points in the figure were determined after the transformer had reached thermal equilibrium.

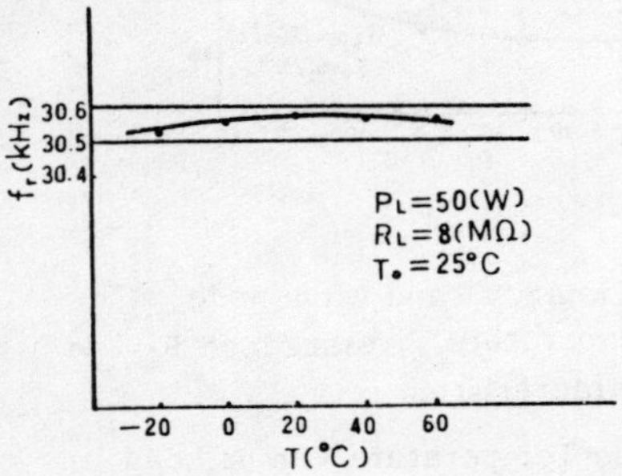

Fig.3 Resonant frequency f_r versus environmental temperature T characteristics

Fig.3 shows that when temperature rises from -20°C to 60°C, the largest deviation of the resonant frequency is 58 Hz; when the room temperature is 25°C, f_r= 30584 Hz, the highest relative deviation $|(\delta f_r)m|$ is not more than 0.2%. The high voltage power supply using this kind of transformer provides a high stable voltage output. The efficiency of these power supplies reaches 70%, and it can work stably in a broad temperature range[5].

CONCLUSIONS

The piezoelectric ceramic material of the PMMN quaternary system shows large mechanical quality factor Qm, large electro-mechanical coupling coefficient Kp and high as 3500. These are the favourable features which help us to fabricate large output power transformer with high voltage step-up ratio. Besides, the basic parameters of the material can be easily controlled by changing the proportions of each component of the PMMN ceramic material. The temperature characteristic can be improved appropriately by adding other materials and partly substituting for Pb, although the performance of the material reduces slightly. Therefore, it is necessary to add an appropriate amount of CeO and partly substitute Sr for Pb in the composition of PMMN.

A thorough examination has been carried out for the plate piezoelectric transformer with the size of 100x25x3.6(mm^3) by using the PMMN quaternary system ceramics. With regard to the voltage step-up ratio, time-temperature stability and efficiency, the largest output power of the piezoelectric transformer is 65 watts. it can continuously work with 40-50 watts for long period.

The large-power PMMN quaternary system piezoelectric transformer will be of advantage to the popularization and application of the piezoelectric transformer. Ceramic transformers with much larger output power and corresponding to high voltage power supply have become the subjects for further research.

REFERENCES

1. 川田雄彦, Electronic Futures (Japan), 7(1970), 5:67.
2. 朱风珠 许煜寰, Physics (China), 2(1973), 4:183-190.
3. Li Long Tu, et al, Ferroelectrics, 28(1980), 403-406.
4. A. X. Kuang, Journal of Hubei University, 8(1986), 2:51-61.
5. A.X.Kuang, et al., IEEE. Proc. Inter. Symp. Appl. ferroe., 1986, 689-692.

New Aspects of Aging Phenomena in Ferroelectric Ceramics

G. Arlt, H. Neumann, GAO Xihan*

Institut für Werkstoffe der Elektrotechnik
Aachen University of Technology
Fed. Rep. Germany

I. INTRODUCTION

The aging of material properties e.g. of the dielectric constant or of the elastic constant is an effect which is technically unacceptable. Yet for dielectric and piezoelectric applications ferroelectric ceramic materials with very low aging rates have been developped by systematic search.

The physical mechanisms which are the origin of the aging process, however, are still poorly understood. The aging has been attributed in particular to domain wall effects either to clamping or pinning of the walls or to their reorientation or to a reduction of the dielectrically active domain wall area. The driving forces towards the thermodynamic equilibrium were seen in elastic and/or in electric field energies which could be reduced by proper domain structure formation. Most of the aging processes proceed in time periods with different empirical logarithmic time laws.

We have observed aging effects in particular in acceptor-doped $BaTiO_3$ and PZT. In these ceramics there is a certain density of oxygen vacancies. These vacancies in contrast to all other ions in the lattice have at room temperature a mobility which can not be ignored. They can jump by diffusion to adjacent oxygen sites with activation energies between 0.8 and 2.0 eV. Thus they can be the origin of effects with long relaxation times if they find sites which lower the energy of the system. Wether this ordering leads to a variation of the dielectric and elastic constant is a matter of investigation.

90°-Domain walls contribute to the dielectric and elastic constants. A change of their mobility in time or of their density can also be a source of aging.

Finally mobile electric charges which compensate electric fields which may appear mainly at the grain boundaries where div P ≠ O can redistribute in dielectric relaxation times, in trapping times or in charge transfer times.

The three time dependent effects described above, which proceed in times comparable with aging in addition can be interdependent. Thus aging in ferroelectric ceramics certainly is the result of several mechanisms.

II. INTERNAL BIAS AND AGING

The internal bias is the result of the orientation of dipolar associates parallel to the spontaneous polarization. The associates consist of acceptor ions at the B-site and oxygen vacancies. In every domain these dipoles will orient by jumping of the oxygen vacancy such that there is a large amount of them (dependent on temperature) parallel to the polarization. This lowers the energy and establishes the internal fields in all domains.[1] It is known from Colemanite and TGS single crystals, that internal bias also modifies the intrinsic dielectric constant. We are studying wether this intrinsic effect also exists in Perovskite-ceramics.

The internal bias in the domains also stabilizes this region by the energy reduction : i.e. a 90°-domain wall which moves into the adjacent domain needs more energy then without an internal bias. The wall is trapped in an energy well the depth of which is strongly dependent on the internal bias field. This may strongly reduce its reversibel mobility. The contribution of the domain wall to the dielectric constant ε and to the elastic and piezoelectric constant then is reduced indirectly by the internal bias. This idea is strongly supported by a simple experiment: After aging (1 weak or shorter) of polarized Cr-doped $BaTiO_3$ the sample was submitted to one or two full hysteresis loops which affects strongly the whole domain system in the sample. After these loops the dielectric constant was the same as before the loops; aging continued as if there were no loops. In contrast, after half a hysteresis loop aging starts almost like in the first aging period . A possible explanation is the following : after the full loops the domains had exactly the same configuration as before, the domain walls were returned to their formerly formed energy wells. The idea of pinning the domain walls by point defects thus can no longer be sustained for the acceptor doped ceramics for aging during the first weak.

III. THERMAL ACTIVATION

On the lg t-scale during dielectric aging of Ni-doped and of Cr-doped $BaTiO_3$ time intervals can be observed in which ε' and ε" decrease approximately with lg t. In Cr-BT this interval is strongly dependent on temperature Fig. 1 . After

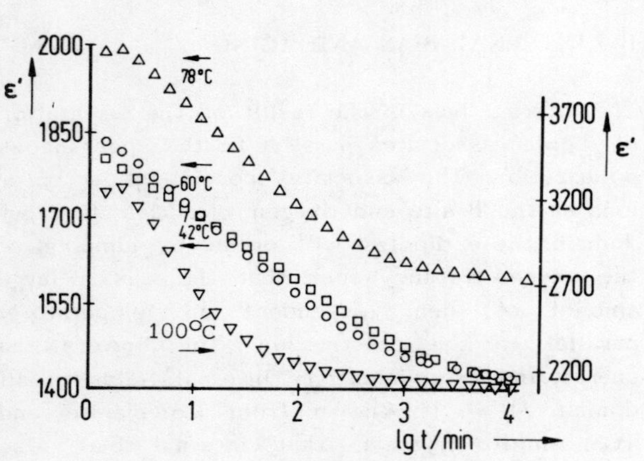

Fig. 1 Aging curves of unpolarized Cr-doped BT (0.5 mol%Cr) in the time period of one weak. Measuring frequency 10 KHz. Before aging the sample was thermally deaged for 4h at 150°C.

the logarithmic slope which ends at $t = \tau_o \exp(W/kT)$ with $\tau_o \approx 4 \cdot 10^{-7}$ s, $W \approx 0.71$ eV) a kind of saturation of ε' occurs. In Ni-BT (Fig. 2)

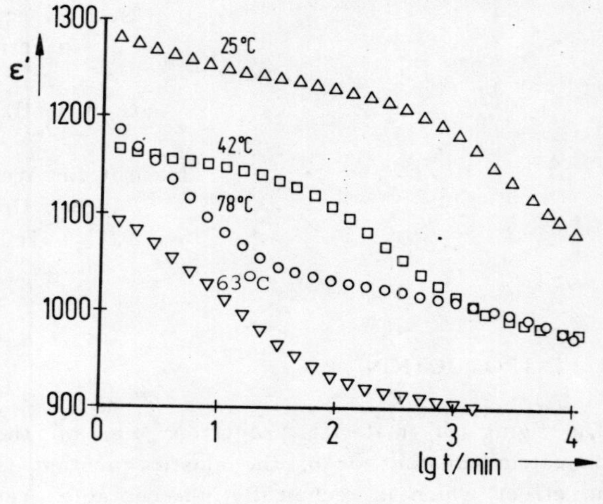

Fig. 2 Aging curves of unpolarized Ni-doped BT (0.3 mol% Ni). The other conditions as in Fig. 1.

the temperature dependence is more pronounced. The knee in the curves also follows the Arrhenius law with $\tau_o \approx 1.6 \cdot 10^{-15}$ s and $W \approx 1.26$ eV, with no following saturation. Figs. 1 and 2 represent unpolarized ceramics.

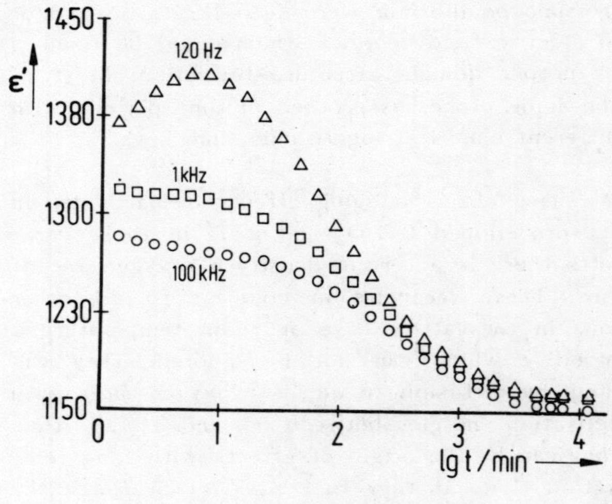

Fig. 3 Aging curves of polarized Ni-doped BT (0.3 mol% Ni) at 42°C. Before aging the sample was deaged by hysteresis cycling for several hours.

If the samples are polarized before aging, the curves of Cr-BT have a logarithmic and a saturation intervall the transition, however, being independent of temperature.

More drastic is the change of Ni-BT (Fig. 3) with poling before aging. At low measuring frequencies there is first an increase of ε' and ε" (more pronounced in ε") before saturation occurs. The origin of this effect is not clear. Its amplitude can not be correlated to the Ni-content, its representative time is shorter then the other aging effects. Therefore possibly the effect is due to electric defects at the grain boundaries.

IV. DOMAIN REORIENTATION

This kind of aging as most mechanisms was supposed already by Mason in 1955[2] and by Plessner [3]. Ikegami and Ueda [4] observed domain reorientation by light microscopy. Our experience with domain observation at the surface of ceramics shows, that grains at the surface are certainly not representative. The domain configuration is strongly dependent on the clamping condition, which is different for grains inside the ceramic and for those at the surface of the ceramic. In addition the polishing and etching conditions were not well developed at that time. Polishing in addition to the polish produces triboelectricity which introduces many 180° domain walls near the surface. Adding graphite to the polishing material reduces this effect considerably. The aging investigations by light microscopy have to be repeated under proper conditions before a statement about this effect can be given.

1. G. Arlt, H. Neumann, Internal Bias in Ferroelectric Ceramics: Origin and Time Dependence, Ferroelectrics 87, 109 (1988)
2. W. P. Mason, J. Acoust. Soc. Am. 27, 73 (1955)
3. K. W. Plessner, Proc. Phys. Soc. B69, 1261 (1956)
4. S. Ikegami, J. Ueda, J. Phys. Soc. Jap. 22, 725 (1967)

+ Present adress: Xidian University, Xian, P. R. China

AGING OF WEAK FIELD DIELECTRIC RESPONSE

IN POLYCRYSTALLINE $BaTiO_3$

Kaiwen Wu and Walter A. Schulze
NYS College of Ceramics at Alfred University
Alfred, NY 14802

Abstract

$BaTiO_3$ ceramics with grain sizes of 0.6 to 60 μm and relative densities of 89-92% were prepared by hot-forging and conventional sintering from extremely pure and fine Ba,Ti-oxalate derived powder. The aging of both the dielectric constant and the dielectric loss was examined with respect to grain size, temperature and frequency. It is suggested that the main aging mechanism in pure ceramic ferroelectric $BaTiO_3$ is the aging of domain wall motion. The dielectric aging is a thermally activated process and shows two aging rates when observed at temperature of 80 °C.

INTRODUCTION

The time-dependent changes of the dielectric properties in barium titanate, a theoretically and technically important ferroelectric material, were realized in the early capacitor applications in the late forties.[1] These include two different phenomena: aging and degradation. Aging describes the changes that may be troublesome in the maintenance of a device tolerance, and is a reversible process. It progresses with no external stimulus. Degradation is usually an irreversible failure of a property driven by an external force. The dielectric aging in $BaTiO_3$ has since then been studied for more than 40 years. The following aging models have been proposed by different authors to describe the time-dependent changes, but not necessarily indicate which possible contribution to the dielectric constant is affected.

1) Domains move to lower energy states; the driving force is to lower the mechanical and electrical stresses.[1]

2) Domain wall motion is reduced by:

a) Domain walls moving to lower energy states (physical positions); the driving force is to lower the high potential energy of domain walls.[1]

b) Diffusion of defects to domain walls thus lowering their energies and pinning their motions; the driving force is to lower the electrical energy and/or to relieve the internal stresses.[2-5]

3) Domain structure is split to produce more domains:

a) 180-splitting (or 180° domain walls increasing); the driving force is to lower the high electrostatic energy due to the spontaneous polarization and/or space charge polarization, or to neutralize the depoling field for poled ceramics.[4]

b) 90-splitting (or 90° domain walls increasing); the driving force is to lower the mechanical residual stresses built up at the ferroelectric transition.[6,7]

4) Ordering of defects in bulk domains; the driving force is to lower the dipolar interaction energy by the local spontaneous polarization.[8,9]

These models are quite different. A linear change of the dielectric properties with logarithmic time has been addressed in all the studies, which means that all the processes could be thermally activated with a broad distribution of activation energies. The domain structure changes were observed on extremely thin slab samples which could have quite different electric and mechanical boundary conditions and thus different domain structure from a bulk sample. The relation of the space charge polarization with aging is discarded because aging is a much longer time process than the electrical conduction which dissipates only in the order of one day.

Understanding of the dielectric aging in $BaTiO_3$ has to be related to the understanding of the dielectric contributions in the material. Many papers have addressed the dielectric contributions in ceramic $BaTiO_3$,[10-13] however a complete picture is still not available. In coarse grained ceramics and single crystals it seems reasonable to state that the main contribution to the dielectric constant (real part) is intrinsic, i.e. ionic and electronic displacement in the structure.[12] One of the minor contributions, the domain wall contribution is also important.[12] For the dielectric loss, the main contribution is not the intrinsic part, but the domain wall motion in the ferroelectric phase[13]. Domain wall motion can be hysteretic (out of phase) or non-hysteretic (in phase). The non-hysteretic part, since it follows the applied frequency, contributes little to the dielectric loss. The losses come from the hysteretic domain wall motion and/or the damping of domain wall motion.[14]

In ceramic $BaTiO_3$, the dielectric response is very much dependent on grain size, temperature, applied AC field level and frequency. At grain size of about 1 μm or less, the dielectric constant in the ferroelectric phase is 2-3 times higher than that of a coarse grained sample. This anomalous behavior has been attributed to the differences in the domain structures.[18,19] At moderate frequency, domain wall motion should show a much stronger frequency dependence than the intrinsic displacement.

Even though great effort was devoted in understanding the aging phenomenon in $BaTiO_3$ in the past, most studies mainly concentrated on the aging of the dielectric constant. The aging of the dielectric loss, which is as important as that of the dielectric constant, has been ignored for the reason that it is much more sensitive to the sample preparation conditions and the measurement conditions, therefore more difficult to be accurately measured. The linear decrease of the dielectric constant and the dielectric loss with logarithmic time, so-called the logarithmic aging law, was established a long time before automatic bridges were available to allow large scale, multiple frequency data collection. Most aging studies on $BaTiO_3$ did not report on the accuracy of the aging temperature and the purity of the materials used. All these factors add up to make it very difficult to compare the aging literature.

In this paper we report the aging studies of both the dielectric constant and the dielectric loss based on a extremely pure Ba,Ti-oxalate derived powder.

EXPERIMENTAL

1. Sample Preparation

Ba,Ti-oxalate derived powder (TAM HPB grade[1]) was used as the starting material. The powder was ball-milled for about 15 hrs in a 1 liter Nalgene bottle with YTZ medium[2] in purified methanol. The ball-milled powder had an average particle size of less than 1 μm. Chemical analysis was carried out on both the as-received powder and the ball-milled powders. For Ba and Ti, gravimetric analysis was used; for impurity elements, atomic absorption spectroscopy was used.

A PVA binder was added in the amount of 1% by weight to facilitate green shaping. The sieved powder was uniaxially pressed into 1.27 x 1.27 cm^2 square plates at a pressure of 1,000 kg/cm^2. Contamination from the die on the edges of green samples was removed by grinding the edges on #600 diamond paper.

Hot-forging was used to obtain fine grained samples. After pre-sintering the sample at the hot-forging temperature for one hour, a pressure (150 kg/cm^2) was applied and the sample was hot-forged for one more hour. After cutting the sample into the desired shape, it was again annealed at 1100 °C in air for one hour to relieve stresses. Conventional air-sintering was used to obtained coarse grained samples. A holding time of 2 hrs was used. Special attention was paid to maintain a constant density among the samples with various grain sizes. Grain sizes were determined from the SEM micrographs. Densities were measured using the water displacement method. A Cr/Au film was evaporated on both surfaces as the electrodes.

2. Dielectric Aging Measurement

A HP4192A impedance analyzer, controlled by a DEC Pro 350 computer, was used for the measurement of weak field dielectric response. In the dielectric response-temperature measurement, the sample was first brought to a temperature above 170 °C for about half an hour, then cooled down at 2 °C/min. In the case of the aging measurement, a HP3488A (an automatic switch scanner) was also used for multisample testing. An isothermal chamber was used to maintain a constant temperature by vigorously circulating the dielectric oil, i.e. dried kerosene oil. The RTD temperature detector was put next to the samples. For the 25°C aging runs, the temperature fluctuation was less than 0.2 °C, and for the 80 °C aging runs, the temperature fluctuation was less than 0.15 °C. Before starting aging, the sample was deaged at a temperature above 170 °C for at least half an hour. Then it was

[1]This powder was a commercial product but without ball-milling at TAM Ceramics Inc. It was calcined at 1150 °C for 4 hrs.

[2]TOSOH USA, Inc., 1700 Water Ptace, Suite 204 Atlanta, Georgia

transferred in 5 seconds to the isothermal chamber with the constant aging temperature, and time counting was started. The dielectric measurements and time counting were fully controlled by the DEC Pro 350 computer.

RESULTS AND DISCUSSIONS

1. Chemical and Physical Analyses

Table 1 lists the chemical analysis results of the as-received powder and the ball-milled powders. Ti was slightly excessive. The only impurity introduced into the powder during ball-milling was Zr. Zr is isovalent to Ti, and is expected to have little effect on the $BaTiO_3$ structure. There were almost no changes in other impurities after ball-milling. The powder was extremely pure, as shown.

Table 2 lists the grain sizes and densities of the samples. As shown, the densities were kept almost constant for all the samples.

2. Weak Field Dielectric Response vs Temperature

Fig. 1 shows the real dielectric constant (or dielectric constant) (A) and the dielectric loss (B) at weak field (less than 10 V/cm) as a function of temperature for samples with various grain sizes. A strong grain size dependence is observed, as described by other authors.[15-19] For three large grained samples, the dielectric constant is almost independent of temperature in the tetragonal phase until the Curie point is approached; while the dielectric loss increases steeply with increasing temperature up to the Curie point. After Tc, the dielectric loss suddenly drops down to a very low value. This has been attributed to the disappearance of the domain structure.[13] The increased dielectric loss with increasing temperature in the ferroelectric phase is believed to be largely due to the thermal activation of domain wall motion. However, this increased domain wall motion is very weakly evident in the dielectric constant because the domain wall contribution to the change in polarization is thermally nucleated and hysteretic with electric field giving principally an out-of-phase response.

For the two fine grained samples with grain sizes of 0.6 and 0.9 μm, the dielectric constant is increased by 2-3 times in the tetragonal phase compared to coarse grained samples. The Curie peak is strongly depressed and slightly shifted to lower temperature. In the paraelectric phase where domain structure no longer exists, the dielectric constant should be the same at all grain sizes. But, as shown, two fine grained samples have lower dielectric constant. This is probably due to the series effect of titania rich grain boundary phases found in fine grained ceramics with slightly excess Ti.[20]

The increased dielectric constant and the other phenomena at fine grain sizes are not fully understood. The internal stress model[18], which has been accepted since 1966 as the explanation, is currently put into question by a study by Arlt, et al[19]. The domain twinning structure, which was considered to be absent at the fine grain sizes, was shown to still exist and even increase. Therefore, domain wall motion maybe still exists at these fine grain sizes. But, the decreased loss suggests that domain wall motion may be mostly non-hysteretic at these fine grain sizes and therefore does not contribute much to the loss. The differences in the dielectric response between fine grain sizes and coarse grain sizes could be attributed to the differences in the domain structures.[18,19]

3. Aging at 25 °C

Fig. 2 shows the aging of both the dielectric constant and the dielectric loss for a 60 μm grain size sample (A) and a 0.6 μm grain size sample (B) at 25 °C. The 0.9 μm grain size sample behaves just like the 0.6 μm grain size sample; while the 20 μm grain size sample behaves like the 60 μm grain size sample. As shown, all the aging behaviors follow the logarithmic law:[1]

$$P_t = P_i - A_P \cdot log(t) \qquad (1)$$

where P_t can be the dielectric constant or the dielectric loss after aging time t, P_i is the initial value at aging time equal to one min. A_P is the absolute decrease in one decade, called the absolute aging rate. The relative aging rate is expressed as:[1]

$$R_P = \frac{A_P}{P_i} \times 100 \qquad (\%/decade) \qquad (2)$$

Both the absolute aging rate and the relative aging rate are used to characterize the aging behavior. The absolute aging rate can be more informative if the dielectric response is made up of components that may or may not be involved with aging.

The complex dielectric constant is expressed as:

$$\varepsilon = \varepsilon' - i\varepsilon" \quad (3)$$

where ε' is the real dielectric constant; $\varepsilon"$ the imaginary dielectric constant. As observed, both the aging of the dielectric constant and the aging of the dielectric loss follow the logarithmic law, that is,

$$-\frac{d\varepsilon'}{dlog(t)} = A_{\varepsilon'}; \quad -\frac{dtan\delta}{dlog(t)} = A_{tan\delta} \quad (4)$$

Therefore,

$$-\frac{d\varepsilon"}{dlog(t)} = A_{\varepsilon'}tan\delta + A_{tan\delta}\varepsilon' \quad (5)$$

Substituting the initial values into (5), then

$$-\frac{d\varepsilon"}{dlog(t)} = A_{\varepsilon'}tan\delta_i + A_{tan\delta}\varepsilon'_i - 2A_{\varepsilon'}A_{tan\delta}log(t) \quad (6)$$

At low absolute aging rates and short aging times as in this study, $A_{\varepsilon'}A_{tan\delta}\log(t) \ll A_{tan\delta}\varepsilon'_i$. Therefore,

$$-\frac{d\varepsilon"}{dlog(t)} \approx constant \quad (7)$$

This means that if the aging of the dielectric constant and the aging of the dielectric loss follow the logarithmic law, the aging of the imaginary dielectric constant will also follow the logarithmic law. Therefore, the aging of the dielectric constant and the dielectric loss can represent the aging of the complex dielectric constant.

Table 3 lists both the absolute aging rates and the relative aging rates for samples with various grain sizes at 25 °C. The aging rates are much higher at large grain sizes than at fine grain sizes for both the dielectric constant and the dielectric loss. For three coarse grained samples, the aging of the dielectric loss is very much frequency dependent (see the absolute aging rates); while the aging of the dielectric constant is much less frequency dependent, agreeing with earlier studies that indicated aging as frequency independent[6]. However, frequency dependence in the aging of the dielectric loss for the two fine grained samples is not observed.

The decreased aging rate and the lack of frequency dependence in the aging of the dielectric loss for fine grained samples could be explained by the lack of hysteretic domain wall motion and/or damping of domain wall motion. If aging is an internal stress relief, a higher aging rate should be expected for these fine grained samples since they would be more heavily internally stressed. This suggests, in fine grained ceramics, a different aging mechanism or the same mechanism operating on an additional contribution to the dielectric constant.

The bimodal grain size (6, 1 μm) sample behaves a little peculiar. For the loss at 1 KHz, two aging rates are observed, and both rates are higher than those for other samples. As observed by Arlt, et al [19], domain size starts to decrease at grain size less than 10 μm. Therefore, the higher aging rate at this bimodal grain size sample could be explained by the increased area of domain walls that can still move hysteretically, compared to other coarse grained or fine grained samples.

4. Aging at 80 °C

Fig. 3 shows the aging of both the dielectric constant (A) and the dielectric loss (B) at 80 °C for a 60 μm grained sample. Fig. 4 shows the aging of both the dielectric constant and the dielectric loss for a 0.6 μm grain size sample. The other two coarse grained samples behave the same way as the 60 μm grained sample; while the 0.9 μm grained sample behaves like the 0.6 μm grained sample.

For the coarse grained samples, two aging rates are clearly observed for both the dielectric constant and the dielectric loss at both frequencies even though frequency dependence is stronger for the dielectric loss. A similar aging behavior has been reported in a relaxor material.[21] Aging clearly begins to slow down after about 100 min. In the aging studies at 25 °C, it has been suggested that domain wall motion is very important and is probably stabilized by the ordering of defects with the local polarization. Increasing the aging temperature from 25 °C to 80 °C increases the dielectric contributions from the domain wall motion and ionic reorientation. The accelerated aging rate in the intial stage of aging is probably due to the accelerated removal of hysteretic domain wall motion. That is to say, aging of domain wall motion and ordering of defects are thermally activated and fast processes. This accelerated aging behavior appears to go to a completion after about 100 min, possibly due to the purity of the material. After about 100 min, the dominant aging process of domain wall motion becomes exhausted, and some other aging mechanism becomes dominant, and is seen to have a much lower aging rate. This could be a change in the intrinsic part due to the relief of internal stress or a stabilization process that slowly affects non-hysteretic domain wall motion. Compared to the hysteretic domain wall motion they are just minor contributions to the aging rate.

However, for the fine grained samples, even though aging rate is slightly accelerated by increasing the aging temperature, two aging rates do not show up except for the aging of the dielectric loss at 1 KHz. This difference again should be attributed to the restricted hysteretic domain wall motion or restricted damping of domain wall motion at these fine grain sizes, and this suggests a different aging mechanism in fine grained ceramics, which is probably related to the second aging mechanism in coarse grained ceramics.

5. Effect of Aging on the Dielectric Dispersion

The experimental evidences have supported the idea that the aging of domain wall motion is the dominant mechanism in the dielectric aging of coarse grained ceramic $BaTiO_3$ for dielectric loss at 25 °C and complex dielectric constant at 80 °C. This part of contribution should decrease with increasing frequency. Fig. 5 shows, as an example of coarse grained samples, this effect and the effect of aging at 80 °C. The decrease of K and D with frequency is due to the increasing clamping of hysteretic domain wall motion. At lower frequency, aging is stronger because of higher contribution of domain wall motion.

CONCLUSION

1) Dielectric aging in coarse grained ceramic $BaTiO_3$ mainly comes from the aging of domain wall motion. Domain wall motion can be hysteretic or non-hysteretic. Aging mainly involves the hysteretic domain wall motion or damping of domain wall motion.

2) Dielectric aging in fine grained ceramics is much slower, much less frequency and temperature sensitive than coarse grained ceramics.

3) Dielectric aging in ceramic $BaTiO_3$ involves more than one aging mechanism, or the effect of one mechanism on more than one dielectric contribution.

4) Domain wall motion and ordering of defects with local spontaneous polarization, which may cause aging, are thermally activated processes. These appear to combine to make the whole aging process in coarse grained ceramics much more dramatic at elevated temperature.

5) The logarithmic aging law does not necessarily hold for long-term aging when more than one contribution is involved as demonstrated in coarse grained ceramics.

Acknowledgement: The authors wish to thank TAM Ceramics, Inc. (Mr. Greg Bieniecki) for supplying the material. Funding for this research was from the National Science Foundation.

REFERENCES

[1]B. H. Marks, "Ceramic Dielectric Materials", *Electronics*, **21** [8], 116-20 (1948)

[2]Z. Pajak and J. Stankowski, "Polarization Changes During the Process of Aging in Ferroelectrics of the $BaTiO_3$ Type", *Proc. Phys. Soc.*, **72**, 1144-6 (1958)

[3]J. Stankowska and J. Stankowski, "Aging Process in Triglycine Sulfate", *ibid*, **25**, 455-6 (1960)

[4]J. Stankowski, *Bull. Soc. Amis. Sci. Lett. Pozn.*, **16**, Series B, 27 (1960-61)

[5]R. M. Gruver, W. R. Buessem, C. W. Dickey and J. W. Anderson, "State-of-Art Review on Ferroelectric Ceramic Materials", *Tech. Rept.*, **AFML-TR-66-164**, May (1966)

[6]S. Ikegami and I. Ueda, "Mechanisms of Aging in polycrystalline $BaTiO_3$", *J. Phys. Soc. Jap.*, **22** [3], 725-34 (1967)

[7]R. C. Bradt and G. R. Ansell, "Aging in Tetragonal Ferroelectric Barium Titanate", *J. Am. Ceram. Soc.*, **54** [4], 192-199 (1969)

[8]P. V. Lambeck and G. H. Jonker, "Ferroelectric Domain Stabilization in $BaTiO_3$ by Bulk Ordering of Defects", *Ferroelectrics*, **22**, 729-31 (1978)

[9]G. H. Jonker and P. V. Lambeck, "On the Origin of the Electrooptic Effects in Pyroelectric Crystals", *ibid*, **21**, 641-3 (1978)

[10]A. R. von Hipple, **Dielectrics and Waves**, John Wiley and Waves, Inc., New York, 1954

[11]W. A. Schulze and K. Ogino, "Review of Literature on Aging of Dielectric", *ibid*, **87**, 361-77 (1988)

[12]G. Arlt and H. Peusens, "The Dielectric Constant of Coarse Grained $BaTiO_3$ Ceramics", *Ferroelectrics*, **48**, 213-24 (1983)

[13]K. H. Hardtl, "Electrical and Mechanical Losses in Ferroelectric Ceramics", *Ceramics International*, 8 [4], 121-7 (1982)

[14]G. Arlt and H. Dederichs, "Complex Elastic, Dielectric and Piezoelectric Constants by Domain Wall Damping in Ferroelectric Ceramics", *Ferroelectric*, **29**, 47-50 (1980)

[15]L. Egerton and S. E. Koonce, "Effect of Firing Cycle on the Structure and Some Dielectric and Piezoelectric Properties in Barium Titanate Ceramics", *J. Am. Ceram. Soc.*, **38** [11], 412 (1955)

[16]G. Jonker and Noorlander, Science of Ceramics, Vol. I, pp. 255, edited by G. H. Stewart, Academic Press, New York, 1962

[17]E. A. Little, "Dynamic Behavior of Domain Walls in Barium Titanate", *Phys. Rev.*, **98** [4], 978-97 (1967)

[18]W. R. Buessem, L. E. Cross and A. K. Goswami, "Phenomenological Theory of High Permittivity in Fine-Grained Barium Titanate", *J. Am. Ceram. Soc.*, **49** [1], 33-6 (1966)

[19]G. Arlt, D. Hennings and G. De With, "Dielectric Properties in Fine-Grained Barium Titanate Ceramics", *J. Appl. Phys.*, **58** [4], 1619-25 (1985)

[20]K. Wu and W. A. Schulze, "Effects of Grain Sizes on the Dielectric Response in Polycrystalline $BaTiO_3$", to be submitted.

[21]W. Y. Pan, Q. Y. Jiang and L. E. Cross, "Isothermal Aging of the Complex Dielectric Permittivity in Mn Doped Lead Magnesium Niobate: Lead Titanate Relaxor Ferroelectric Ceramics", *Ferroelectrics*, **82**, 111-7 (1988)

TABLE AND FIGURE CAPTIONS

Table 1. Grain Dizes and Densities of the Samples

Table 2. Chemical Analyses of the Ba,Ti-oxalate Powders

Table 3. Aging Rates at Weak Field, 25 °C for Samples With Various Grain Sizes.

Figure 1. Dielectric Response at $E_{AC} < 10$ V/cm as a Function of Temperature for Samples With Various Grain Sizes: (A) Dielectric Constant; (B) Dielectric Loss.

Figure 2. Aging at 25 °C of Both the Dielectric Constant and the Dielectric Loss for (A) a 60 μm Grain Size Sample, (B) a 0.6 μm Grain Size Sample.

Figure 3. Aging at 80 °C of (A) the Dielectric Constant, (B) the Dielectric Loss for a 60 μm Grain Size Sample.

Figure 4. Aging at 80 °C of the Dielectric Constant and the Dielectric Loss for a 0.6 μm Grain Size Sample.

Figure 5. Effect of Aging at 80 °C on the Dielectric Dispersion for a 60 μm Grain Size Sample.

Table 1 Grain Sizes and Densities of the Sintered Samples

Temperature (°C)	Sintering Technique	Holding Time	Grain Size (μm)	Density (%)
1150	Hot-forging	1 hr Pressure off 1 hr Pressure on*	0.6	90-91
1200	Hot-forging	1 hr Pressure off 1 hr Pressure on*	0.9	90-92
1250	Hot-forging	1 hr Pressure off 15 min Pressure on*	6, 1**	89
1350	Air Sintering	2 hrs	20	90-91
1430	Air Sintering	2 hrs	60	91

*The pressure was 150 kg/cm². ** Bimodal grain size sample has large areas of 6 μm and small areas of 1 μm grains.

Table 2 Chemical Analyses of the $BaTiO_3$ Powders

Elements	As-received Powder (moles)	After 10 hrs Ball-milling (moles)	After 40 hrs Ball-milling (moles)
Ba	99.3	the same	the same
Ti	100	..	..
Na	0.008	..	..
K	0.0001	..	..
Ca	0.001	..	..
Sr	0.0037	..	..
Mg	0.0004	..	..
Al	∠ 0.0002	..	..
Y	∠ 0.0002	..	..
Fe	0.0001	..	..
Si	∠ 0.00003	..	..
Zr	0.0377	0.0617	0.247

Table 3 Aging Rates at Low Signal Field, 25 °C for Samples With Various Grain Sizes

d (μm)	Dielectric Constant				Dielectric Loss			
	A_K (per decade)		R_K (% per decade)		$A_D \times 10^3$ (per decade)		R_D (% per decade)	
	1 KHz	100 KHz	1 KHz	100 KHz	1 KHz	100 KHz	1 KHz	100 KHz
0.6	30	22	0.98	0.78	0.75	0.90	3.75	3.27
0.9	32	27	1.02	0.89	0.75	0.8	6.55	4.91
6,1	58	42	2.51	1.88	2 rates[1]	0.8	2 rates	10.53
20	40	35	2.42	2.18	1.30	0.3	6.57	5.08
60	40	33	2.58	2.22	1.78	0.37	7.31	4.63

[1]Two aging rates of the dielectric loss at 1 KHz were observed for this bimodal grain size sample. The initial stage of aging was slow, $A_D = 1.70 \times 10^{-3}$, $R_D = 8.72$; while for the second stage after about 100 min, $A_D = 3.12 \times 10^{-3}$, $R_D = 16.0$.

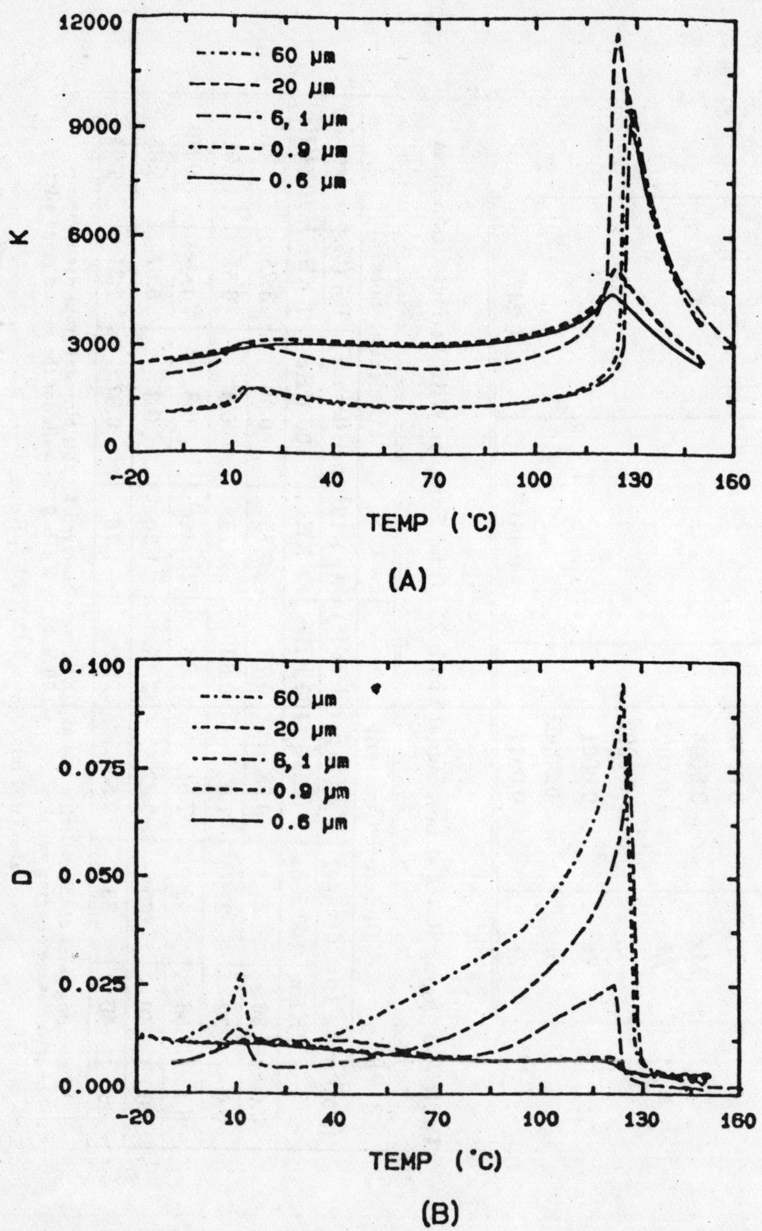
60 μm
20 μm
6, 1 μm
0.9 μm
0.6 μm
K
TEMP (°C)
(A)
D
TEMP (°C)
(B)

Fig. 1

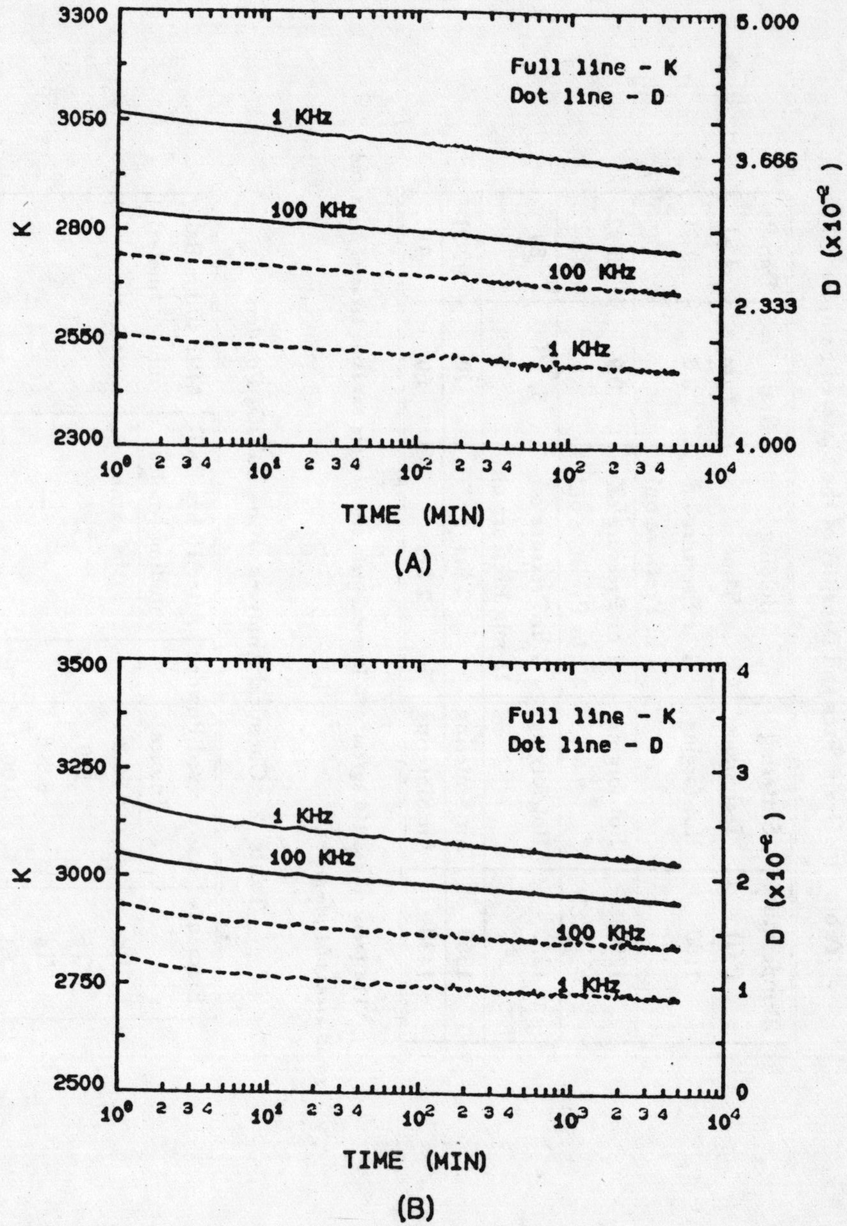
Full line - K
Dot line - D
1 KHz
100 KHz
K
D (x10^{-2})
TIME (MIN)
(A)
(B)

Fig. 2

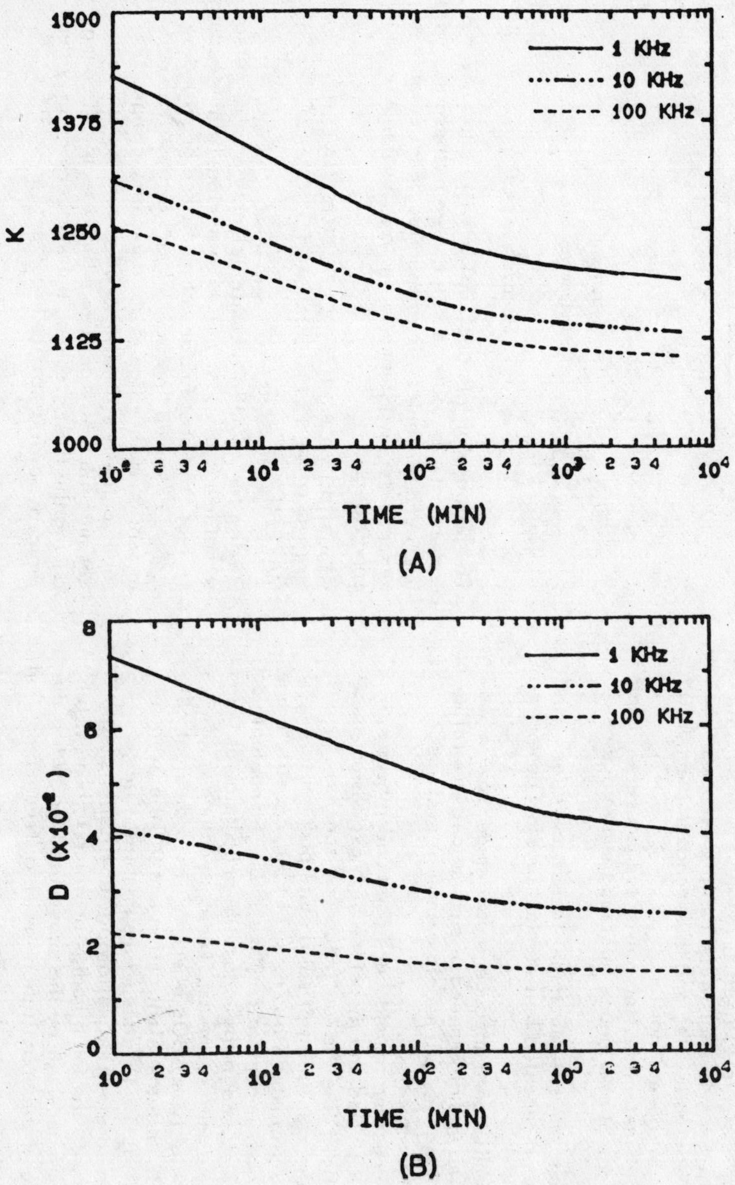

Fig. 3

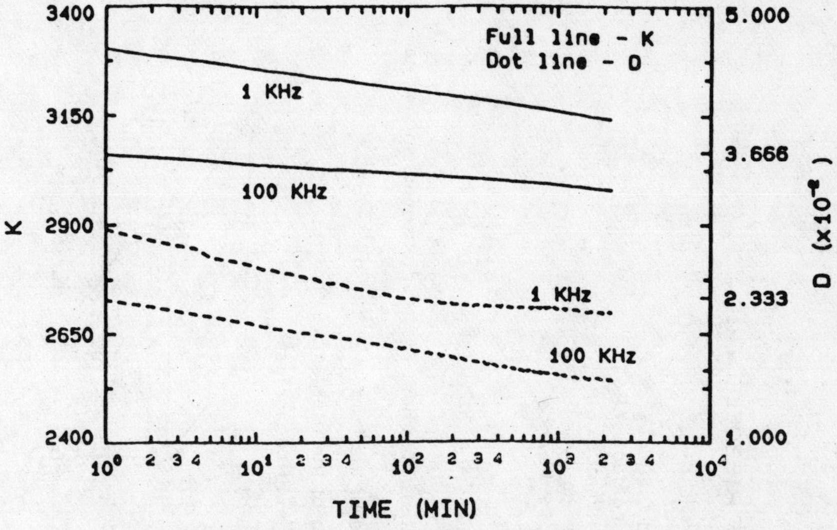

AGING OF FINE-GRAINED SAMPLE AT 80 C

Fig. 4

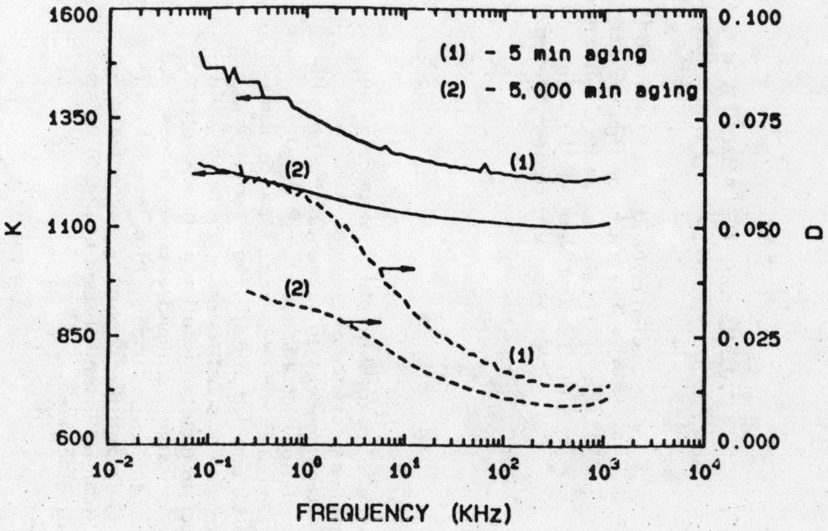

Fig. 5

CAUSES OF INSTABILITY AND AGEING OF PIEZOELECTRIC POWER CERAMICS

Lucien Eyraud, Paul Gonnard (*), Bernard Claudel (**)
(*) Laboratoire de Génie Electrique et Ferroélectricité
(**) Laboratoire de Cinétique et Génie Chimiques
Institut National des Sciences Appliquées de Lyon
20, Ave. A. Einstein, F-69621 Villeurbanne France

ABSTRACT

A simple model of perovskite of PZT type involving the simultaneous presence of electron donor and acceptor centers renders an account of the possibility of cancellation of the space charge density $\rho = -\operatorname{div} \vec{P}$ in polarized ceramics. A cause of instability can be avoided in realizing a convenient double heterovalent substitution in the cationic lattice in order to lock the location of these centers. In the case of materials aimed at power ultrasonic transducers, the losses under a high level of excitation are minimized by creation of oxygen vacancies. The origin of instability and ageing in these materials is discussed.

INTRODUCTION

The properties of piezoelectric ceramics lead to divide them into two main groups :
- "soft" materials, used at low levels of solicitation, the behaviour of which rapidly becomes non linear at high levels, and thus renders them unfit for use in these conditions,
- "hard" materials, for power transducers. Their conditions of use induce to class them into three types :
1) the mechanical-electrical transducers for ignition under repeated fast strokes or slow compression,
2) the emission transducers used under low acoustical load and high speed of mechanical strain like ink jet devices, ultrasonic cleaning systems, nebulizers etc.
3) the emission transducers used under heavy acoustical load and therefore high excitation electrical field like the low frequency sonars.

From the standpoint of solid state properties, lead zircono-titanates $Pb(Zr_{1-x}Ti_x)O_3$ constitute one of the most important families of ferroelectric materials used in the industrial production of transducer ceramic elements. Their structure is of perovskite type ABO_3 where A^{2+} is a divalent cation, B^{4+} a tetravalent cation and O^{2-} the oxygen anion (fig.1). This structure enables a large variety in composition, allowing substitutions with other ions, the ionic radius of which is compatible with the space left by the substituted ion. When the valence of the substitution ion is different from that of the substituted host ion, various mechanisms can occur :

a. change of the lead valence $Pb^{2+} \rightarrow Pb^{4+}$

b. change of the titanium valence $Ti^{4+} \rightarrow Ti^{3+}$

c. creation of vacancies.

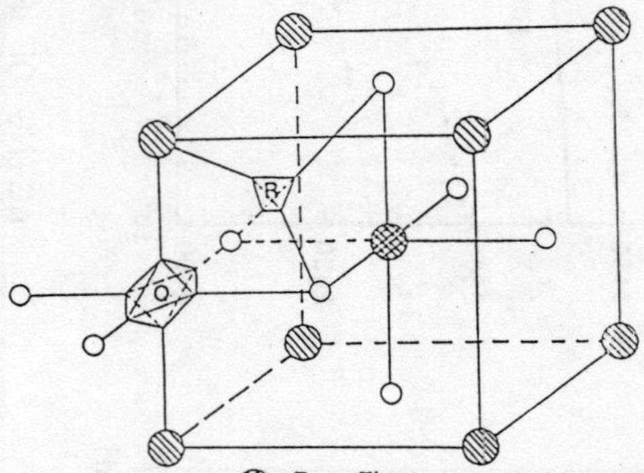

Fig. 1 Perovskite cell

The purpose of this publication will be to show how convenient substitutions can confer to these materials improved ferroelectric properties, particularly as regards the resistance to ageing in various working conditions.

STABILITY OF FERROELECTRIC MATERIALS UNDER REPETITIVE IMPACTS. CASE OF SOFT MATERIALS

Basing the choice of a material only on a high factor of merit (F.O.M.)[1] can be seriously misleading. In fact, the best factors of merit are provided by lightly coloured "soft ceramics" generally used for reception devices, and of no vacancy composition (or eventually with some lead vacancies). These ceramics actually deliver a high electrical energy during the first strokes, but are very quickly depolarized, owing to an important ageing [2],

(see Table 1)

Table 1 Comparaison between non-stabilized and stabilized ceramics. W_{max} at T_{max} = 600 bars, ignition duration 20 μ s, with sample of diameter 6,35 mm and height 16 mm.

Type	Components	Energy W yielded by matched resistive load (10^{-3} J)	F.O.M. 10^{-12} m^2/N	$\frac{d_{33} - d^{\circ}_{33}}{d_{33}}$
non-stabilized soft a	$PbSrBa(ZrTiNb^{5+})O_3$	4,7	16,8	- 2,5
non-stabilized soft b	$PbLa^{3+}(ZrTiNb^{5+})O_3$	4,2	13,5	- 8,8
stabilized soft c	$PbM^{+1}(ZrTiNb^{5+})O_3$	5	14,9	+ 0,6
stabilized hard d	$PbZrTiNi^{2+}O_3$	2,8	5,2	0

We shall briefly recall how we have interpreted the stability of the material called "stabilized soft c" in table 1, that we prepared by the " wet process[3]. Our interpretation rests upon the existence, in substituted perovskites of " sites " wich can act either as electron donors or acceptors according to mechanical or electrical excitations. These " sites " may be actual crystallographic sites, as we shall see later, or "multiplet sites" (denoted A_m or B_m) when they involve the ionic environment of an ion A or B. For instance, if we imagine a structure with A^{2+} and B^{4+} ions we shall have, as an "equilibrium", or paraelectric (planar) configuration

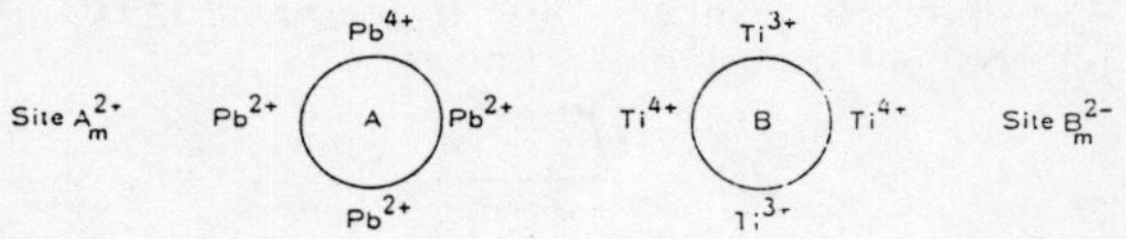

The commutation, in either direction, of 2 electrons, makes us pass to the other configurations

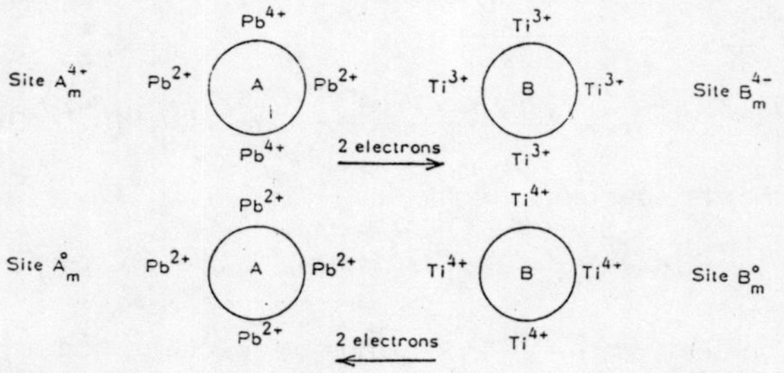

The possibility of "freezing" these configurations is offered by a double substitution : of monovalent K^+ ions on some crystallographic A sites and of pentavalent Nb^{5+} ions on some crystallographic B sites. The general formuia of these phases is

$$Pb^{2+}_{1-u-v}\ Pb^{4}_{u}\ K^{+}_{v}\ (Zr^{4+}_{1-x-w}\ Ti^{4+}_{x-2u+v-w}\ Ti^{3+}_{2u-v+w}\ Nb^{5+}_{w})\ O_3$$

(if it admitted that there is no oxygen vacancy)

The above configurations become :

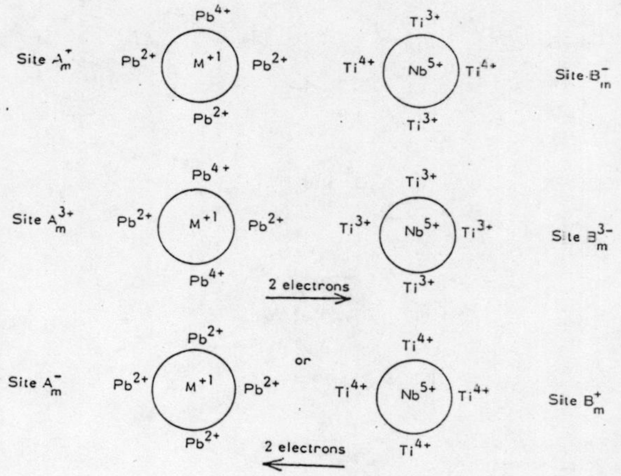

It can be shown[4,5] that with atomic fractions of M^{+1} on crystallographic A sites equal to $v = 5 \times 10^{-3}$ and of Nb^{5+} on crystallographic B sites equal to $w = 2 \times 10^{-2}$, the dipole moment induced by commutation of two electrons between the associated multiplet sites is of the order of magnitude required for cancellation of charge distribution, i.e. $P = 3\ \mu C\ cm^{-2}$. This value is the maximum reached when all the sites are used for commutation.

STABILIZATION OF "HARD" MATERIALS

These materials which contain Ni, Fe, or Mn (at atomic fractions ranging from 5×10^{-3} to 10×10^{-3}) are used in medium power acoustical emitters (a few watts / cm^2 of active face in stationary regime). These dark coloured ceramics are characterized, besides by low dielectric losses by

- low permittivity (ε_r < 1100),
- low coupling factor,
- high coercitive fields,
- high surtension coefficient Q , and low tan δ

and an important side phenomenon : a high internal field E_i , evidenced during the first hysteresis cycles by a marked dissymmetry. Indeed, a high internal field can give rise to two unwanted effects:

- an increase of internal losses due to a slight relaxation of internal field, at high level of solicitation,
- a variation in time of electromechanical properties during functioning and storage.

In order to explain the stabilization of hard ceramics by foreign elements, like Ni, Fe or Mn, it is logical to assume that the corresponding cations normally located on B sites and being normally at a valence less than 4, induce the formation of oxygen vacancies V_c [6,7,8,9]. It is then still possible to use the preceding model, with a like commutation between B_m and V_{om} sites.

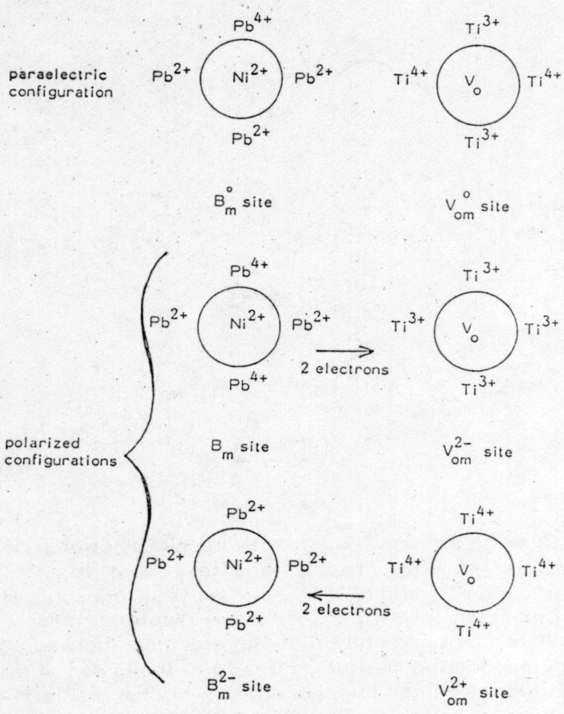

Nevertheless, it seems more appropriate to make transition elements play their usual role of electron exchangers, eventually climbing to higher valencies than 2 : 3 and even 4, the latter being attested in the solid state, even for Ni [10] and Fe [11]. Therefore the formula of a nickel substituted PZT can be written, in assuming Ni^{3+} on sites B, and singly charged oxygen vacancies

$$Pb^{2+}(Zr^{4+}_{1-x-z}\ Ti^{4+}_{x}\ Ni^{3+}_{z})\ O^{2}_{3(1-\frac{z}{3})}\ V^{-}_{oz}$$

This formula offers the possibility of charge commutation depicted by the following scheme

paraelectric configuration	Ni^{3+}		V_o^-
other configurations in polar state	Ni^{4+}	1 electron →	$V_o^{=}$
	Ni^{2+}	← 1 electron	V_o

Such a model involves only, contrarily to the preceding one, only " singlet " crystallographic sites A_s and B_s. It does not imply neither Ti^{3+} nor Pb^{4+}, which is probably an over-simplification, at least in the latter case.

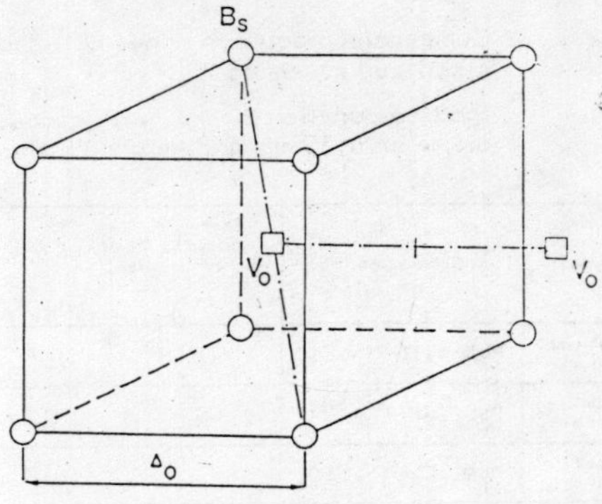

Fig. 2 Configuration of $B_s - V_o$ sites

If we assume that the solubility limit of oxygen vacancies V_o is of about 5/1000 in atomic percentage, the $V_o - B_s^o$ couples will constitute a set of dipoles, the moment of wich can be calculated by volume unit. Supposing a random distribution of these two types of defects, we can assume that they occupy the lattice points of two imbricated cubic lattices (fig.2). The dimension of the unit cell will thus be (with $a_o = 4\ \mathring{A}$) :

$$\Delta_o = \frac{a_o}{\sqrt[3]{5.10^{-3}}} = 23\ \mathring{A} = 23.10^{-10}\ m$$

The distance from a V_o site to the nearest B_s site in this configurations will be :

$$\ell = \Delta_o \frac{\sqrt{3}}{2}$$

Assuming that the commuting electron can be used to cancel the $\rho = -\operatorname{div}\vec{P}$ distribution, the commutation dipolar moment would be :

$$\mu_1 = e\ell = \frac{e\Delta_o\sqrt{3}}{2}$$

$$= \frac{1.732 \times 37.10^{-29}}{2}\ Cm = 32 \times 10^{-29}\ Cm$$

that is, per unit volume :

$$P_i = \frac{\mu_1}{(\Delta_o)^3} = 26.10^{-3}\ C\,m/m^3 = 2.6\ \mu\ C\ cm^{-2}$$

The calculation of P_i depends of course on the hypothesis made on Δ_o. Instead of the largest possible distance, the smallest distance can be taken into account. The latter can be evaluated in considering the perovskite cell of fig.1, $\Delta_{o\,min} = 2 \times 10^{-10}$ m. In this other extreme case commutation of an electron between a site B_s and a site V_o leads to a dipole of moment :

$$\mu_2 = e\,\Delta_{o\,min} = 2 \times 10^{-10} \times 1.6 \times 10^{-19}$$

$$= 3.2 \times 10^{-29}\ C\ m$$

The cell volume being $v = 64 \times 10^{-30} m^3$, the number of cells per m^3 is $N = \frac{1}{v} = 1.56 \times 10^{28}$ and therefore, the dipole number per m^3

$$n = 5 \times 10^{-3} \ N = 7.8 \times 10^{25}$$

It follows that the induced polarization is

$$(P_i)_{min} = \mu_2 = 7.8 \times 10^{25} \times 3.2 \times 10^{-29}$$

$$= 25 \times 10^{-4}\ C\ m^{-2} = 0.25\ \mu\ C\ cm^{-2}$$

The range of P_i variation is finally

$$0.25 < P_i < 2.6\ \mu\ C\ cm^{-2}$$

This induced polarization by charge commutation between B_s and V_o sites is sufficient to cancel the distribution of bound charges $\rho = \text{div}\ \vec{P}$ in this second type of material "hard ceramic", as in the first type "soft stabilized ceramic". This material is also stable when submitted to the stroke test, and can be used as igniter, with a lower efficiency but a high stability, ("stabilized hard d" in Table 1). As noted before, an important difference between these two types of ceramics ("soft" and "hard") is the existence of an internal field E_i wich appears during the first hysteresis cycles in the polarized elements. Several studies have already been devoted to the origin of the internal field. All agree with the existence of couples of oppositely charged defects [7,8,12,13,4]. Figure 3 shows the performances reached with such a material (denoted P762 of "Quartz & Silice"). In order to bring further decrease to the losses at high level of solicitation, it is possible to introduce increments of foreign elements exceeding the amount of oxygen vacancies that the host lattice can admittedly tolerate [15]. The internal field concomittantly rises, the losses decrease, but so do the other electromechanical characteristics and the F.O.M. (materials denoted P189 from "Quartz & Silice", X5105 from "Pons", PZT 8 from "Vernitron" in the figure 3).

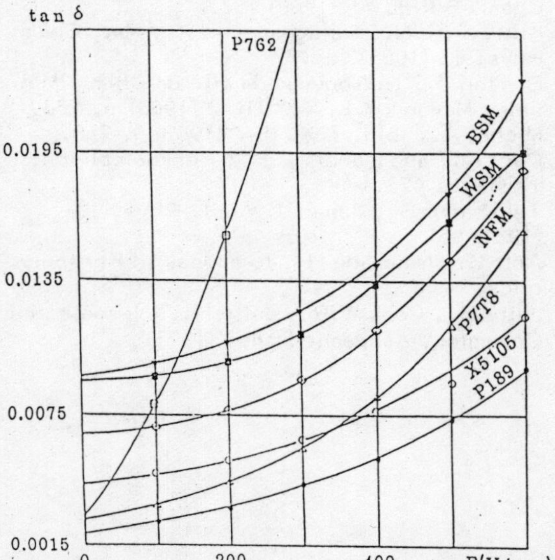

Fig. 3 Loss tangent versus the applied electric field (1 kHz)

NEW INVESTIGATED COMPOSITIONS

1. Principle of the stabilization by valency compensation.

These compositions satisfy two main criteria :

- stabilization of a basic "soft" PZT formula by a double multivalent substitution on B sites. The electronic configuration of each ion is such as to cancel the space charges without generating oxygen vacancies. For example :

$$Bi^{3+,\ 4+,\ 5+} / Sb^{5+,\ 4+,\ 3+}$$

$$W^{4+,\ 6+} / Sn^{4+,\ 2+}$$

$$Nb^{5+} / Fe^{2+,\ 3+,\ 4+}$$

- stabilization at high level of solicitation. In materials containing for instance Mn, the charge of oxygen vacancies can be partially avoided by including other ions, which give rise to redox mechanisms.

Bi, Sb, Mn substitutions (materials denoted " BSM ")

$$Bi^{3+} + Sb^{5+} + Mn^{2+} \rightarrow Bi^{3+} + Sb^{4+} + Mn^{3+}$$

$$Bi^{3+} + Sb^{5+} + 2\ Mn^{2+} \rightarrow Bi^{3+} + Sb^{3+} + 2\ Mn^{3+}$$

W, Sn, Mn substitutions (materials denoted " WSM ")

$$W^{6+} + Sn^{4+} + 2\ Mn^{2+} \rightarrow W^{6+} + Sn^{2+} + 2\ Mn^{3+}$$

$$W^{6+} + Sn^{4+} + 2\ Mn^{2+} \rightarrow W^{4+} + Sn^{4+} + 2\ Mn^{3+}$$

$$W^{4+} + Sn^{4+} + 2\ Mn^{2+} \rightarrow W^{4+} + Sn^{2+} + 2\ Mn^{3+}$$

Nb, Fe, Mn substitutions (materials denoted " NFM")

$$Nb^{5+} + Fe^{3+}\ Mn^{2+} \rightarrow Nb^{5+} + Fe^{2+} + Mn^{3+}$$

The results compiled in Table 2 show that the multivalent materials present some differences with respect to the industrial ones :

- increased permittivity and factor of merit
- a lowered ratio E / E_c
- an increased dissipation factor at low level (1 kHz)

It can be seen in figure 3, except for P762, the loss angle tangent at 1 kHz increases moderately and with the same slope as a function of the applied electric field. Figure 4 shows identical mechanical losses at high levels with a higher factor of merit.

Table 2 Electro-mechanical constants of the various materials

Specimen	ϵ_r	tgδ % 1 V. 1 kHz	Q	E_i/E_c	k_{33}	d_{33} pCN^{-1}	s^E_{33} $10^{-12} m^2 N^{-1}$	M_o $C^2 m^{-4}$
P189	1150	0.2	1400	0.6	0.66	250	13.6	330
X51.05	1200	0.4	1100	0.85	0.60	225	13.2	285
P762	1250	0.2	700	0.4	0.67	270	16	340
PZT8	1180	0.25	900	0.65	0.60	225	13.6	275
NFM	1470	0.7	1000	0.45	0.63	265	13.4	386
BSM	1640	0.9	1000	0.3	0.60	260	12.3	405
WSM	1670	0.9	1000	0.3	0.62	280	13.5	420

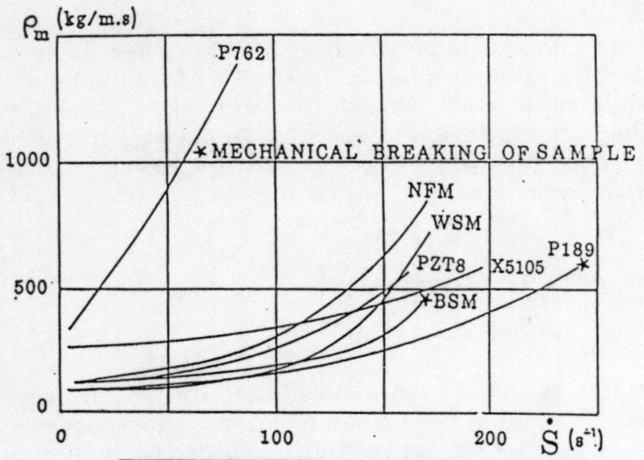

TYPE	PZT8	X5105	P189	P762	NFM	WSM	BSM
M_o	275	285	330	340	386	407	420

Fig. 4 Mechanical loss resistivity versus speed of relative displacement of active face

CONCLUSION

Four types of piezo electrical materials can be distinguished :

1 - "soft" materials, with no oxygen vacancies, which cannot be used as power transducers, but withstand the strokes,
2 - " hard " materials with low amounts of foreign cations on sites B exactly compensated by oxygen vacancies. They withstand the strokes, but undergo important losses at high level of solicitation,
3 - " hard " materials, "overloaded " with lower valency cations. They withstand the high levels of solicitations, but with a low F.O.M,
4 - " hard " materials, with multiple foreign cations, which undergo electronic exchanges. They withstand high levels of solicitation, with higher F.O.M.

The choice of a piezoelectric material well suited for a given power transducer is always very difficult. It can be noted that the standard specification, given by Naval Research Laboratory, clearly specifies the maximum values of the dielectric losses under high electric drive. However the values of mechanical losses at high level are not precisely stated and the strong dependence of the dielectric losses versus the driving frequency is not taken into account. In most cases a compromise between the highest factor of merit and the lowest dielectric and / or mechanical losses must be found. Among the tested materials, the industrial P189 and X51.05 specimens present the lowest dielectric losses at 1 kHz and the laboratory compositions are well suited for the high frequency moderately loaded transducers. This different behaviour may be due to the substitution on the B site of two multivalent ions inhibiting the charge of the oxygen vacancies.

NOTATIONS

d_{33} piezoelectric coefficient

k_{33} coupling coefficient

$$M_o = \left(\frac{d_{33}}{s^E_{33}} \right)^2 \text{, factor of merit (F.O.M.)}$$

ρ polarization

Q surtension coefficient

$\dot{S}$ speed of relative displacement of active face

s^E_{33} compliance at constant electric field

δ loss angle

ε_r relative permittivity

μ dipole moment

ρ_m mechanical loss resistivity

REFERENCES

1. Gonnard P., Eyraud L., and Troccaz M., in High Tech Ceramics, Elsevier, (1987) p.1645.
2. Ohanessian H., Gonnard P., Troccaz M., Eyraud L., Eyraud P., Rev. Phys. Appl. 18 (1983) 479.
3. Eyraud L., Eyraud P., Bauer F., Adv.Ceram. Mat, 1 (1986) 223.
4. Eyraud L., Eyraud P., Mathieu J.C., Claudel B., Ferroelectrics 50 (1983) 103.
5. Eyraud L., Eyraud P., Claudel B., J. Solid. State Chem. 53 (1984) 266.
6. Jonker G.H., and Lambeck P.V., Ferroelectrics 21 (1978) 641.
7. Lambeck P.V., and Jonker G.H., Ferroelectrics 22 (1978) 729.
8. Lambeck P.V., and Jonker G.H., J. Phys. Chem. Solids 47 (1986) 453.
9. Kutty T.R.N., Balachandran R., Mat. Chem. Phys. 13 (1985) 467.
10. Besson J., in Nouveau Traité de Chimie Minérale, Masson,Pub. T XVII-2 (1963) p. 753.
11. Michel A., ibid, T XVII-1 (1967) p. 758.
12. Carl K., and Hardtl K.H., Ferroelectrics, 17 (1978) 473.
13. Takahashi S., Japan J. Appl. Phys. 20 (1981) 95.
14. Arlt G., Neumann H., to appear in Ferroelectrics.
15. Jaffe B., Cook W.R., Jaffe H., Piezoelectric Ceramics, Academic Press (1971).

Electric Field Induced Antiferroelectric-ferroelectric Phase change in $(Pb,La)(Ti,Zr,Sn)O_3$ Family of Antiferroelectric Ceramics for Displacement Transducer Applications

W.Y. Pan, C. Dam, Q.M. Zhang and L.E. Cross
Materials Research Laboratory
The Pennsylvania State University
University Park, PA, 16802

The electric field induced antiferroelectric -ferroelectric phase changes in more than ten $(Pb,La)(Ti,Zr,Sn)O_3$ ceramic compositions near the antiferroelectric-ferroelectric phase boundary are studied for ultra-high field induced strain displacement transducer applications. The longitudinal field induced strain is in the range of 0.5-1%. The kinetics of the phase transition is studied both under single shot and continuous driving. Both the forward switching time and backward switching time are strongly dependent on the over electric field, but can be accomplished within 1μs under high over field. For continuous driving, the switching is affected by hysteretic heating. The degradation of switched charges and of associated strain depends on the surface smoothness, electrode character and applied field level.

DIELECTRIC CHARACTERISTICS OF SOME $MeTiO_3$ CERAMICS UNDER A.C. BIAS

Wu Yugong, Zhou Zhigang
Dep. Material Science and Engineering,
Tsinghua Univ., Beijing China

Jiang Zutong
Anshan Electronic Ceramics Corp. 177
Xingsheng Road, Tiexi District,
Anshan, Liaoning, China

INTRODUCTION

ABO_3-type materials and their solid solutions display many kinds of functional effects and can be used economically as components or materials with high performances.[1] These ensure them great importance in theoretical research and high-tech engineering.

However, most of the studies and applications on these materials were confined in weak electrical field regions. In resent years, of interest is their behavior under strong electrical field, especially A.C. ones, though systematic discussion is still in want.[2-5]

In this paper, we studied dielectric behavior under A.C. bias of some $MeTiO_3$-type ceramic materials where Me represents alkaline earth metal ions Mg, Ca, Sr and Ba. Special attention is payed for the discussion of relationship between breakdown strength of these materials and their crystalline chemistry parameters.

EXPERIMENTAL PROCEDURE

The specimens were prepared with conventional ceramic technology with sintering conditions shown in Table 1. X-Ray diffraction experiment indicated that the specimens were of monophase.

Table 1 Samples and their properties

Sample	$MgTiO_3$	$CaTiO_3$	$SrTiO_3$	$BaTiO_3$
sintering condition	1400°C x2hrs	1370°C xwhrs	1370°C x2hrs	1350°C x2hr
ε_0	18.4	187	320	1430
$tg\delta_0 (10^{-4})$	20	48	56	150

Data of breakdown strength were obtained through breakdown tests with Q4-r type electrostatic voltmeter. Dependences of dielectric constants and losses on electrical bias nere determined with a Schering Bridge at 50HZ.

EXPERIMENTAL RESULTS

1. Dependences of ε and $tg\delta$ on A.C. Bias

Dependences of dielectric constants and losses on A.C. bias were measured at 50HZ, shown in Fig.1-2.

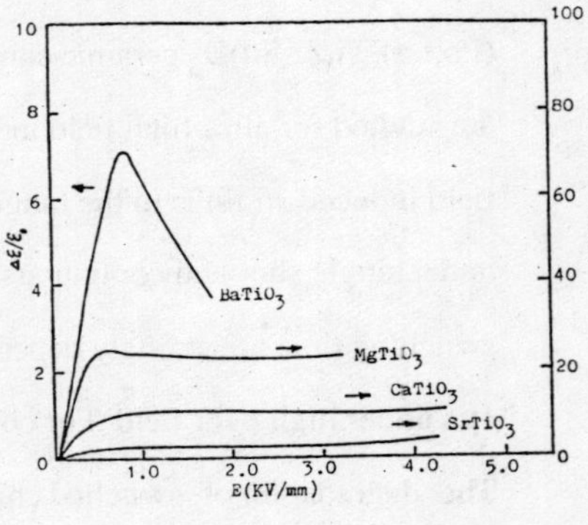

Fig.1 Dependence of dielectric constants on A.C. bias (f=50HZ)

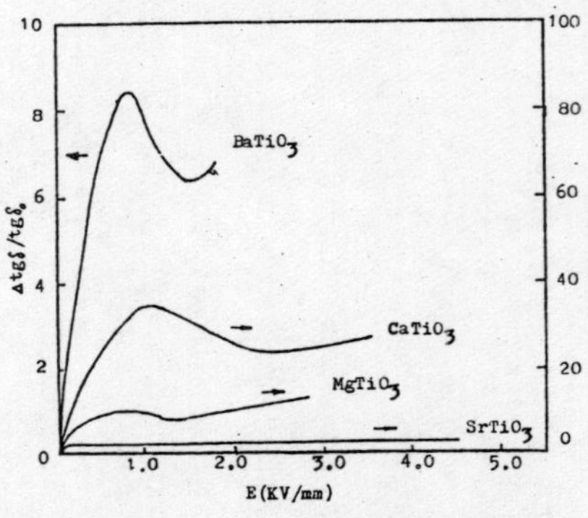

Fig.2 Dependence of dielectric losses on A.C. Bias (f=50HZ)

According to expression of energy density of electrical field in dielectrics[6]

$$W_e = \frac{1}{2}\varepsilon_o \varepsilon E^2$$

where ε_o=8.85x10-12 ($C^3/N \cdot m^2$), changes of energy density in the specimens with different bias were calculated (Fig.3).

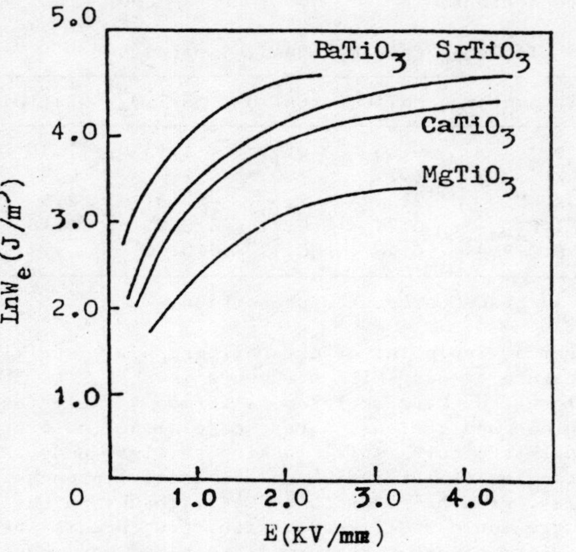

Fig.3 Dependence of energy density on A.C. bias.

2. Breakdown Characteristics

Breakdown strengthes under A.C. field as well as D.C. one were measured using specimens with thickness of 0.6mm. The results were shown in Fig.4, together with data obtained by D.M.Kazalnovski for a compareson.[5]

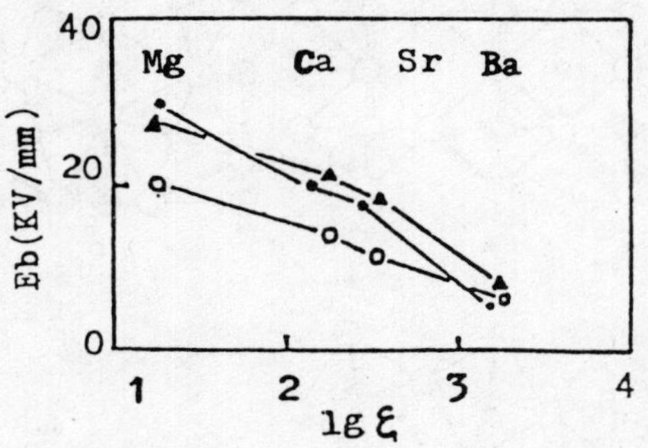

Fig.4 Breakdown strengthes Eb ploted against dielectric constant. ▲-under D.C.field; o-under A.C. field; ● -under A.C. field, from Ref.5

Variation of breakdown strengthes of ceramic specimens with their thickness is evidence. Generally, breakdown strength decreases with the increase of the specimen thickness. This may be partially contributed by the fact that breakdown phenomenon are of weakest-link theory with data in Weibull distribution[7,8]. We measured breakdown strengthes of specimens with thickness L in the range of 1.0mm to 8.0mm. Experimental data are combined with the formula

$$E = AL^{-n}$$

According to previous theoretical and experimental studies, breakdown occures in a purely electrical manner when n=0[7]; thermal breakdown happens with n=0.5 when temperature remains uniform throughout the specimen[9], or with n=1.0 while a temperature gradient exists[7], so that thermal breakdown most likely originates from the center of the specimen.

Agreement of data from our experiment with the formula mentioned above is shown in Table 2. Values of n were between 0.3 to 0.6, which indicated that breakdown mechanism may be a combination of those mentioned above. Further experimental and theoretical investigation into breakdown mechanism will be present else where.

Table 2 Dependence of Eb on thickness[a)]

	Sample	$MgTiO_3$	$CaTiO_3$	$SrTiO_3$	$BaTiO_3$
A.C.field	A[b)]	10.5	9.22	8.35	4.16
	n	0.62	0.36	0.50	0.33
	r[c)]	0.89	0.96	0.92	0.94
D.C.field	A[b)]	18.0	13.9	11.8	5.21
	n	0.66	0.55	0.60	0.40
	r[c)]	0.81	0.88	0.89	0.88

a) Data are combined into linear expression LnEb = LnA - LnL.
b) In KV/mm
c) Relative coefficient for the linear regression.

DISCUSSIONS

1. Dielectric Characteristics and Design of Ceramic Dielectric Materials

Advanced technology requires ceramic capacitors with higher performance.[3] Some of them are to work at electrical field quite high, making knowledge about the dependence of their dielectric constants and losses on A.C. bias important.

Some electrical field energy is stored in the ceramic body when the capacitor works. Experiment revealed that energy density increase continually as the increase of electrical field strength till the occurance of breakdown no matter whether the dielectric constants increase or not (Fig.3). Design of high voltage capacitor must take into account both its endurance to electrical field and dielectric constant. Energy density We combined electrical field strength and dielectric constant together and may serve as a parameter for the selection of ceramic capacitor material.

$BaTiO_3$ can acquire easily a relatively high energy density at low voltage due to its high dielectric constant, though its energy density increases smoothly at high voltage because of the decrease of its dielectric constant.However, the fact that dielectric constant of $SrTiO_3$ changes slightly with bias, together with a higher breakdown strength, ensures $SrTiO_3$ a energy density even higher than that of $BaTiO_3$ at high voltage. This may be one reason that energy is mainly concentrated on materials based on $SrTiO_3$ in the development of ceramic capacitors with high dielectric constants used in high voltage.

2. Covalent Bonding and Breakdown Strength

ABO_3-type oxides form an important class of materials of great scientific and technical value[1]. Such oxides are known to exist with a wide range of A and B ions, as well as solid solutions between them. Most of them have the structures of perovskite, ilmenite or pyrochlore.

There has been considerable amount of studies on the relationship between their properties and structures. Most of the work dealed with BO_6 octahedron[10].

Breakdown strengthes of $MeTiO_3$ ceramics (Me=Ba, Sr, Ca, Mg) increase progressively (Fig.4), which suggests the existence of a crystalline structure and chemical bonding basis.

Typical ionic crystal usually adopts a certain kind of closest packing structure for its anions to form the structure skeleton while cation, usually smaller than anions, occupy interstices in the anionic packing[11].

Adoption of closest packing is a character of ionic crystal where ionic bonding is omnidirectional, while there exists no closest packing in covalent crystal due to the directional covalent bond, resulting in a relatively loose arrangement of the ions.

However, ionic and covalent bond are merely two ideal cases, while chemical bond in real crystal is usually a mixing up of them with the covalent proportion estimated, to some accuracy, by Pauling's expression[12].

$$1-p=\exp[-\frac{1}{4}(X_A - X_B)^2]$$

where X_A, X_B are the electronegativities of the two bonding atoms.

Forthermore, we suggest the use of a parameter, space occupation proportion (S.O.P.) for inspecting covalent proportion of the chemical bond

$$\text{S.O.P.}=\frac{\text{sum of the volume of ions in a cell}}{\text{the volume of the cell}}$$

Each ion can be treated as a sphere having Pauling's radius. Different chemical bonding parameters were listed in Table 3, showing clearly progressive change of covalent bonding proportions.

Table 3 Chemical bonding parameter

Parameter	$MgTiO_3$	$CaTiO_3$	$SrTiO_3$	$BaTiO_3$
$r_{A^{2+}}$	0.65	0.99	1.13	1.35
X_A	1.2	1.0	1.0	0.9
X_A-X_o	2.3	2.5	2.5	2.6
$1-P_{A-o}$	0.27	0.21	0.21	0.18
S.O.P.a)	0.48	0.52	0.70	0.72

a) Space occupation proportion.

From a view point of crystallography, a striking feature of provskite structure is that its closest packing skeleton is formed jointly by anions and cations, a rare case among ionic crystal structures still lacking an clear understanding of its influence on material properties. From Ba^{2+}, Sr^{2+}, to Ca^{2+}, thanks to the large ionic radii, perovskite structure is able to form though structure distortions get more serious as the ion become smaller (Fig.5)resulting in more stronger bonding for A site ion with some certain oxide ions than others. In other words, directional covalent bonding emerges.

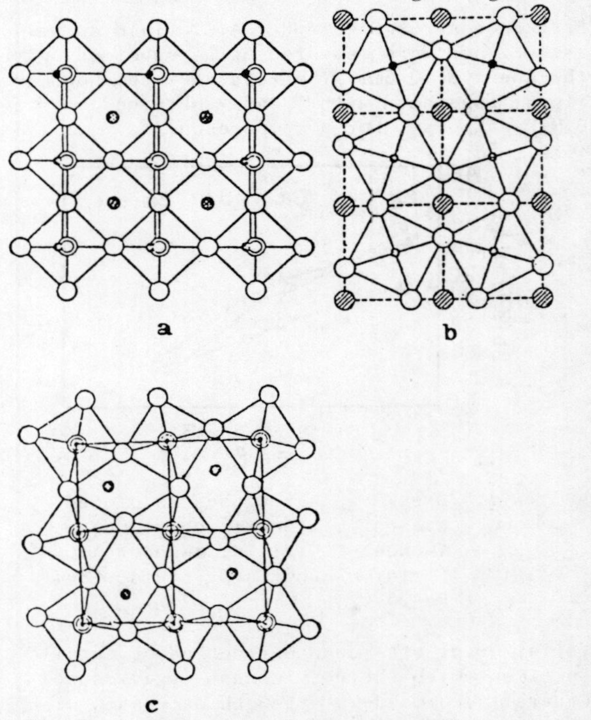

Fig.5 Arrangement of ions in a: $BaTiO_3$, b: $SrTiO_3$ and c: $CaTiO_3$.

In covalent bond, electrons is firmly localized between the two atom bonding, so that it has a strong resistance against electrical field to remain its bonding state. From $BaTiO_3$ to $CaTiO_3$, the increase of breakdown strength has a close relation with the increase of covalent bonding proportions (Fig.6).

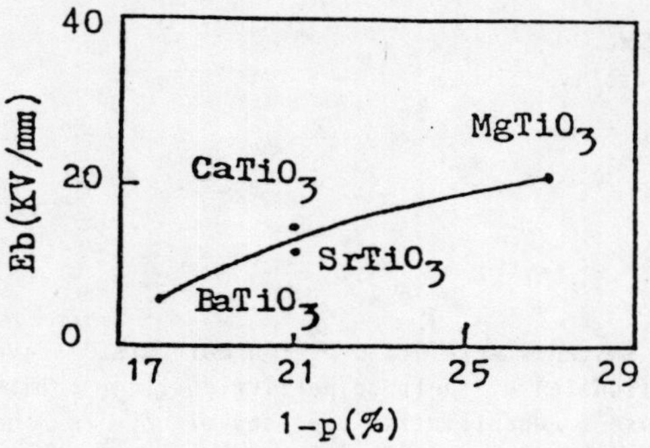

Fig.6 Breakdown strength Eb ploted against covalent bonding proportion.

As concerned with $MgTiO_3$ in ilmenite structure, higher proportion of covalent bonding, as well as the adoption of the structure with normal anionic closest packing, may contribute to its higher breakdown strength.

Discussions above may serve as a reference in selection of composition in developing ceramic material with high dielectric constant and breakclown strength.

CONCLUSION

Dielectric characteristics of some alkaline earth metal titanate under A.C. Bias were studied in this paper, together with a discussion on possible relationship between breakdown strength and covalent bonding in the structure. These may serve as references in development of high voltage ceramic capacitor.

ACKNOWLEDGMENT

The authors would like to thank Pro. Zhu Deheng and Li Jing, an enjineer in Tsinghua University for their experimental assistance.

REFERENCE

1. N.Ramadass, Mat. Sci. and Eng., V.36(1978) P231.
2. J.V.Bigger, et al, Ceramic Bulletin V.51, No.8(1972).
3. Y.Fujiwara, J.Electronic Ceram.(in Japaness), V.9(1975).
4. T.R.Gururaja, et al, Ferroelectrics, V. 27 (1980), P277.
5. D.M.Kazalnovski, Seignetokeramishski Kondensatorei, Gos enelgoizdat (in Russian), 1956.
6. B.Tareev, Physics of Dielectric Materials, Mir Publisher, Moscow, 1979.
7. Chen Jidan, et al, Physics of Dielectrics (in Chinese), Mechanical Industry Publisher, 1982.
8. Wu Yugong, Thesis, Tsinghua University, China (1988).
9. R.Coelho, Physics of Dielectrics for the Engineer, Elsevier Scientific Publishing Compony, 1979.
10. J.P.Goodenough and J.M.Longo, Landolt-Bornstein Tabellen, New Series, III/4a, Springer Verlag, Berlin, 1970.
11. R.C.Evans, An Introduction to Crystal Chemistry, second edition, Cambridge University Press, London, 1964.
12. L.Pauling, Nature of the Chemical Bond, Third edition, Cornell University Press, Itaca, N.Y., 1960.

FERROELECTRIC LEAD TITANATE ZIRCONATE PREPARED BY SOL-GEL TECHNIQUE

Q.F.Zhou, A.X.Kuang and L.S.Wu
Physics Department, Hubei University
Wuhan, China

ABSTRACT

Ferroelectric PZT ($P_bZr_xTi_{1-x}O_3$) powders were prepared by sol-gel technique from metal alkoxides. The particle size range was about 150 to 300Å , the purity was about 99.7%, the major contaminant is silicon from the glass apparatus. DTA and X-ray analyses indicated that powders of PZT fired at 650℃ for two hours and had the phase of perovskite PZT. PZT ceramics were sintered by hot-pressing. These ceramics have high density, low dielectric dissipation factor and high piezoelectric constant. Variations of temperatures, pressure and hot-pressing time during prepared process were also studied.

INTRODUCTION

Piezoelectric lead zirconate titanate (PZT) ceramics have been used in a wide range of applications since the 1950'S[1]. However, the most of them were prepared by the conventional ball milling method. Because the synthesized temperature was very high (850℃), it was very difficult to control the problem of PbO volatilization and the inhomogeneous mixing. In the present work, PZT powders were prepared successfully by sol-gel technique from metal alkoxides with high purity, high homogeneity, fine particle size and low processing temperature. Subsequently, these powders were used for the fabrication of PZT ceramics by hotpressing technique. The properties of ceramics prepared by sol-gel processing were measured and the optimum hot-pressing conditions were discussed.

EXPERIMENTAL PROCEDURE

1. Synthesis of metal alkoxides

As-received spectrograde anhydrous 99.95% $ZrCl_4$ were used in preparing the Zr alkoxide. The zirconium alkoxide was synthesized by the method of Bradley et al[2]

$$ZrCl_4 + C_2H_5OH + NH_3 \xrightarrow{C_6H_6} Zr(OC_2H_5)_4 + NH_4Cl \downarrow$$

All the solvents were dried over a molecular sieve and fractionated at their respective boiling points before use. Quantitative analyses of Zr in the Zirconate alkohol was performed by chemistry method.

2. PZT powder perparation

Figure 1 shows a typical sol-gel preparation process in a metal solvent. Metal alkoxides $Zr(OC_2H_5)_4$, liquid $Ti(OC_4H_9)_4$ and anhydrous lead acetate $[Pb(C_2H_3O_2)_2]$ were used for the preparation $Pb(Zr_xTi_{1-x})O_3$ powders. These materials were mixed to form a homogeneous solution at first, the solution was then hydrolized by exposure to air and polymerized to form a transparent gel. This gel was ground to powder and calcined at 750℃ for two

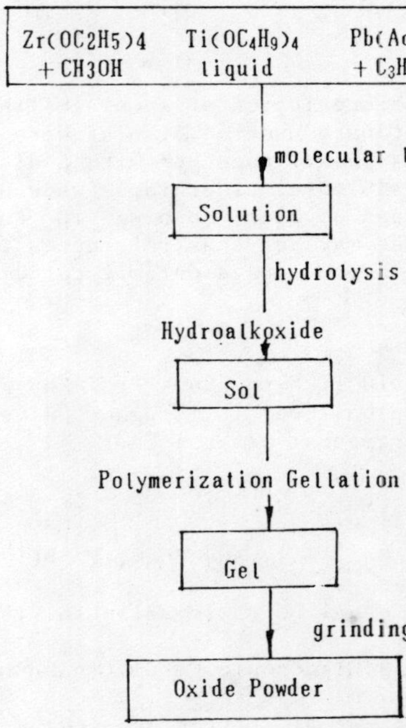

Fig.1 The Preparation Process of PZT Powder by Sol-Gel Technique

hours to achieve cystalline PZT powders. Powders were cold-pressed in a stainless steel die at 30000 psi to form discs 10 mm in diameter and 2 mm in thick. These discs were then put into an Al_2O_3 die and surrounded by coarse Al_2O_3 powers for hot-pressing. The following hot-pressing conditions were used to prepare ceramic samples. (1) temperature from 1180℃ to 1230℃, (2) pressure from 1200 to 3500 psi, and (3) time form 1 to 5 hours. In the present experiment, the composition was selected near protopic phase boundary where zirconate-titanate molar ratio was 51/49 with two to four mole percent excess lead oxide.

RESULTS AND DISCUSSIONS

1. Morphology of the Gel

The particle size of as-prepared gel was determined by TEM of a dispersion of the gel in either water or ethyl alcohol sprayed onto a Carbon substrate on a Copper mech screen. The results at x 30000 are shown in Fig. 2. It could be observed that average particle size was about 300Å . Figure 3 was an electron micrograph of the same gel at x40000 after five day in air, it is very evident that three dimension mesh structure of gel would be formed by hydrolized and polymerized reaction.

Fig. 2 Electron micrograph of PZT dispersion gel

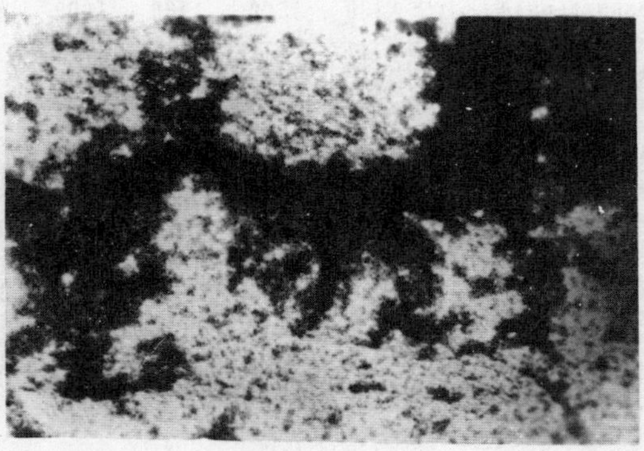

Fig. 3 Electron micrograph of PZT gel

2. Emission Spectrographic Analysis

Table 1 shows the results of the emission spectrographic impurity analysis for the PZT powders after calcination at 750℃ for 2 hours. PZT powders averaged 99.70% purity, the major impurities were Si, Fe, Na and Ca in the range of the 300 to 1000 ppm each, according to analyses of other starting materials. Most of the impurity levels are similar to those of the starting materials with the exception of Si. However, the higher purity material can be synthesized either by use of spectrograde starting materials or by controlling the synthesis steps.

Table 1 Emission Spectrographic Analysis of Alkoxy Derived PZT

element	Si	Fe	Na	Ca	Al	B
PZT(ppm)	1000	800	500	300	100	< 50

Bi	Nb	Mo	Cd	Y	Cu	Ag
50	100	<5	<100	100	<100	<10

3. Differential Thermal Analysis

Figure 4 shows the differential thermal analysis (DTA) of $PbZr_{0.51}Ti_{0.49}O_3$ gel.

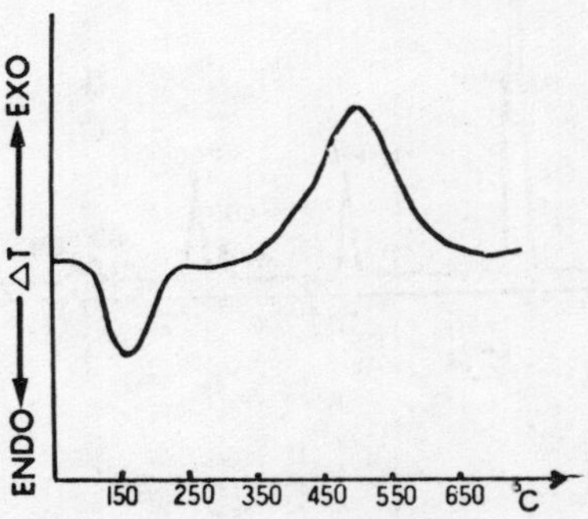

Fig. 4 The DAT curves of $PbZr_{0.51}Ti_{0.49}O_3$ gel

At 160°C an endothermal peak on the curve was found, which might result from H_2O and ROH volatilization, at 550℃, exothermic peak on the curve was found, at 650℃ solid-state reaction of $PbZr_{0.51}Ti_{0.49}O_3$ was end. It was very different from that of conversional method which had three stages[3], their synthesis reaction rules by sol-gel method have also been investigated by means of infrared absorption spectrometry[4].

4. X-Ray Diffraction studies

The X-ary Diffraction patterns using CuK α radiation were shown in Fig.5 after calcining at different temperature from 650℃ to 800℃ for two hours. X-ary diffraction patterns showed that both perovskite PZT and lead oxide were present between 650℃ to 700℃. However, after sintering lead oxide diffraction peaks could not be detected, indicating that the excess leadoxide was volatilized during sintering. It was only existence of the tetragonal phase. It was a agreement with result of DTA analysis.

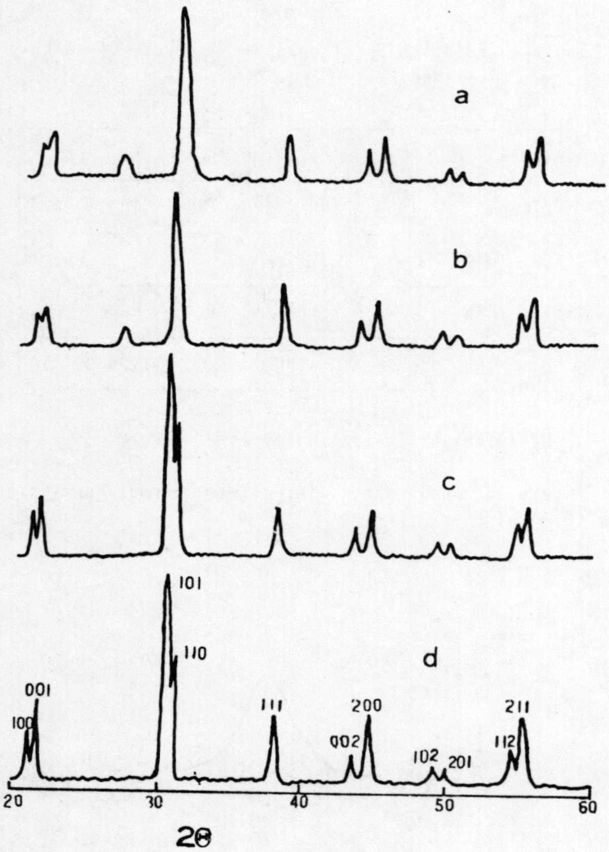

Fig.5 X-Ray Diffraction patterns of $PbZr_{0.51}Ti_{0.49}O_3$

a, 650℃, 2hr b, 700℃, 2hr
c, 750℃, 2hr d, 800℃, 2hr

5. Hot-Pressing Sintering and Electric Property

Different hot-pressing parameters such as temperature, pressure and time were investigated in order to improve the ferroelectric and piezoelectric properties of PZT ceramics in O_2. Table 2 shows the effect of temperature, where $PbZr_{0.51}Ti_{0.49}O_3$ ceramics were hot-pressed at constant pressure of 1700 psi for 2 hours, but at different temperature from 1190℃ to 1230℃. It is found that density and piezoelectric constant increased with temperature up to 1200℃

Table 2 Effect of Temperature on Hot-pressed $PbZr_{0.51}Ti_{0.49}O_3$

Specimen	A_1	A_2	A_3	A_4	A_5
Temperature(℃)	1190	1200	1210	1220	1230
Pressure(psi)	1700	1700	1700	1700	1700
Time(hour)	2	2	2	2	2
Density(g/cm^3)	7.38	7.61	7.49	7.32	7.10
ε(at 25℃)	858	886	802	712	705
tgδ(%)	0.20	0.10	0.20	0.50	0.60
$d_{33}(\times10^{-12}C/N)$	167	193	160	135	110

Table 3 Effect of pressure on hot-pressing $PbZr_{0.51}Ti_{0.49}O_3$

Specimen	A_1	A_2	A_3	A_4	A_5
Temperature(℃)	1200	1200	1200	1200	1200
Pressure(psi)	1300	1500	1700	1900	2100
Time(hour)	2	2	2	2	2
Density(g/cm^3)	7.52	7.58	7.62	7.56	7.50
ε(at 25℃)	645	688	870	670	613
tgδ(%)	0.40	0.20	0.10	0.30	0.40
$d_{33}(\times10^{-12}C/N)$	153	183	197	163	150

After that the volatilization of PbO would occur and density and piezoelectric constant decreased with the temperature increase.

Table 3 shows the effect of pressure, where hot-pressure was at constant temperature 1200℃ for two hours. It can be seen from table 3 that the best electric properties were obtained at 1700 psi. Table 4 shows the effect of hot-pressing time, where the hot-pressing temperature was 1200℃ and pressure was 1700 psi, but the time from 1 to 5 hours. It is shown that the best properties sere obtained at 2 hours sintering.

Table 4 Effect of Sintering Time on Hot-Pressed $PbZr_{0.51}Ti_{0.49}O_3$

Specimen	B_1	B_2	B_3	B_4	B_5
Temperature(℃)	1200	1200	1200	1200	1200
Pressure(psi)	1700	1700	1700	1700	1700
Time(hour)	1	2	3	4	5
Density(g/cm^3)	7.58	7.63	7.56	7.50	7.47
ε(at 25℃)	590	889	567	543	520
tgδ(%)	0.16	0.10	0.18	0.40	0.60
$d_{33}(\times 10^{-12}C/N)$	186	197	165	157	147

On the basis of above results, the optimum condition would be considered that the pressure and temperature and sintering time were 1700 psi, 1200℃ and 2 hours, respectively. The PZT ceramics with high density (up to $7.63g/cm^3$), high piezoelectric constant (197×10^{-12} C/N) and low dissipation factor (0.001) were achieved.

CONCLUSION

PZT powders have heen successfully prepared by sol-gel technique. These powders processes very high purity (99.7%) and ultrafine particle size (300A). Their ultimate homogeneity makes their calcining temperatures about 200℃ lower than that of the convensional method. The optimum hot-pressing condition of 1200℃, 1700 psi and 2hr have heen found to fabricate nice PZT ceramics from these powders derived from sol-gel method. The experiments suggest that the alkoxybased preparation of oxides may provide a broad base for improving the quality and reproducibility of electric and structure ceramics.

REFERENCES

1. J.B.Jaffe, W.R.Cook and H. Jaffe, Piezoelectric ceramic, London, Academic Perss, 1971.

2. D. C.Bradley etal J.Chem, Sol. London, 1952, PP2027-32.

3. Danal, Hankey etal, Communication of Amer ceramic Society, 1981, C-172.

4. Q.F. Zhou, Master Thesis, Hubei University, 1989.

THE SURFACE ACOUSTIC WAVE FILTER FOR SATELLITE COMMUNICATIONS EARTH STATIONS

SHENG ZHONGDE YU ZIZHEN
First Research Institute of Ministry of P & T
No. 48, Ping-jiang Road, SHANGHAI, CHINA

Abstract

This paper described the designs of both wideband and narrowband 70 MHz SAW filters series used in satellite communications earth stations. The filters designed by these methods are all under the requirement of INTELSAT's specifications and have been applied to satellite communications earth stations in China successfully. This paper discusses the design specifications and test results.

Introduce

The technical specifications of 70 MHz filters now commonly used in satellite communications earth stations have been strictly specified by INTELSAT. Generally, these filters are fabricated with LC components at abroad. The shortcomings of LC filters are of large size and poor stability. Due to the not high Q values of inductors and capacitors, the spuareness factor of the filter is difficult to be made small, especially when to meet the group delay characteristics requirement specified by INTELSAT, this kind of filters still must adopt up to 5-6 orders phase equilizer to make phase correction. Thus, it will put the design and adjustment to inconvenience.

SAW filter is a new component developed in recent years. It is a filter to realize frequency selection by activating and detecting an elastic wave transmitted on the surface of a piezoelectric substrate. It can be made with semi-conductor planar technology on the piezoelectric crystal, therefore it is of small size, light weight, good consistency and reproducibility, and good stability to temperature, being easy to be fabricated in small amount. Due to using the linear phase response technology, the filter has good group delay characteristics. It is an ideal component to replace LC filters used in satellite communications earth stations.

Structure and working principle of the SAW filter

A simple SAW filter's structure is shown in Fig. 1. It is formed by two interdigital transducers and a metal shielding bar by photoetching on a piezoelectrical substrate, with little acoustic absorptive paste at both ends of the bar. The function of the transmit interdigital transducer is to convert the input electric signal to surface acoustic wave with reverse piezoelectric effect and to make the SAW propagate from the surface of the piezoelectric substrate to both ends outside of the interdigital transducer. To prevent the reflection when leftward SAW reaches the end of the substrate, which is unwanted, and thus having effect on the performance of the component, we use acoustic absorptive paste to absorb it. When the rightward SAW propagates to the bottom of the electrode of the receive interdigital transducer, the SAW will be converted to electric signal due to the piezoelectric effect and will be outputed by the receive interdigital transducer after detecting. The frequency response of the output signal depends on weighting to the interdigital transducers. If the good squareness factor of frequency characteristics are to be got, one must put weighting to both receive/transmit interdigital transducers. And now the total frequencies response of the filter should equal the product of both receive/transmit interdigital transducers' frequencies response. Namely

$$H(\omega)=H_1(\omega)*H_2(\omega) \qquad (1)$$

Digitals length unequal overlapping weighting is a widely used weighting method. Yet, as beams are ununiform, the frequencies response of the filter, which are formed by digitals length unequal overlapping weighting to interdigital transducers, are no longer their products. Multistrip coupler must be introduced so long as design is concerned. The multistrip coupler is made of a series of metal strips. It can be divided into A and B channel, see Fig. 2. The function are: channel A receives the spatial ununiform SAW caused by transmit interdigital transducer, then the received ununiform SAW will be converted to the spatial uniform SAW by the multistrip coupler, the spatial uniform SAW will be transmitted to the receive interdigital transducer by channel B. Furthermore, the multistrip coupler can greatly improves the performance of the SAW filter by effectively suppressing second-order effects such as bulk wave, three times transitional signal. Therefore, the structure of a technically good SAW is rather complicated, as shown in Fig. 3.

Design of SAW filter

Specified by INTELSAT, the frequencies domain characteristics and time domain characteristics of 70 MHz IF filter used in satellite communications earth stations must all agree with Fig.4.

Take a filter of 25 MHz bandwidth, the required ripple of its frequency characteristics at frequency band A/3 must be <0.3dB, and the squareness factor between B and C must be <1.144. As for its time domain characteristics, the required group delay ripple at frequency band A/3 must be <3ns, and at frequency band A must be <5ns. From the single specifications above, we can see that these technical specifications are very difficult to be realized, the difficulty is even more so when these high technical specifications to be realized in one filter simultaneously.

The design of SAW filter is mainly the design of interdigital transducer. We know that the pulse response of the interdigital transducer is correspondent to its geometric structure. That is, if the pulse response of the interdigital transducer is known, the structure parametres can be fixed. From filter's theory we also know that there is a pair of Fourier transforms between pulse response H(t) and frequency response H(ω), i.e.

$$H(\omega)=\int_{-\infty}^{\infty} H(t)\exp(j\omega t)dt \qquad (2)$$

$$H(t)=\int_{-\infty}^{\infty} H(\omega)\exp(j\omega t)d\omega \qquad (3)$$

In theory, the Fourier transform can be used to help design the SAW filter if the filter's frequency response specifications are known. Yet it must be pointed out that contradiction exists between mathematic processing and physical feasibility. As the length of the piezoelectric substrate is limited, the length of the interdigital transducer pulse reaponse can not be limitless, we must try to cut it off. We also know that when it is a continuous integral procedure in mathematics, it is just a discrete sampling procedure in physics. And there is a great difference between the frequency responses we actually get and we hope to get. Therefore, when in designing, a time limited function is needed to replace the rectangular aperture function, thus, to change the limitless pulse response into limited one. "Kaiser Aperture" function is proved to meet the requirement through analysis. Its expression is:

$$\omega_{(t)} = \begin{cases} \dfrac{I_0(\omega\sqrt{\tau^2 - t^2})}{I_0(\omega\tau)} & \text{when } |t|<\tau \\ 0 & \text{when } |t|>\tau \end{cases} \qquad (4)$$

In the formula, I_0 is first class of zero order modified Bessel function, while ω, τ are parameteres of the function, and need to be selected after optimizing both bandwidth and sidelobe suppression. The width of SAW filter's passband relays on electrode digital pairs number of the interdigital transducer, i.e.

$$B_0 = f_0 / N \qquad (5)$$

in which B_0 is 3 dB bandwidth, f_0 is central frequency of the filter, N is digital pairs number of the interdigital transducer. From formula (5), we know that with the percentage increase of filter bandwidth, electrode digital pairs number of the interdigital transducer will decrease correspondently. When designing wideband filter, every increase or decrease of 0.5 pair (1 electrode) will result in several MHz difference in filter's bandwidth. For example, if 3 pairs of digitals in 25 MHz bandwidth filter are adopted, then 3 dB bandwidth is only 23.3 MHz, that's too narrow; if 2.5 pairs are adopted, then 3 dB bandwidth will be 28 MHz, too wide. The technical specifications of the filter designed in this way will never meet what have been shown in Fig.4. To solve the problem, we adopt the design method of central frequencies' high and low ends overlapping. In designing, these requirements are needed:

1. The passband width of single interdigital transducer must be designed

wider than the required specification.
2. To make the central frequencies of both interdigital transducers overlapping separately with high and low ends of filter's central frequencies. How many part is to be overlapping depends on that the high end frequency response of one interdigital transducer can meet the requirement of high end frequency response of the filter and the low end frequency response of another interdigital transducer can meet the requirement of low end frequency response of the filter.
3. The height of both interdigital transducers' central frequencies depends on controlling width of interdigital transducer's electrode digital bars. Different width of the digital bars are formed by carving figure before making the mask. Thus, to complete once-imaging by using the same minifying factor lens when mask-making.

With the methods described above, various bandwidth filters can be made.

From formula 5, we also know that the electrode digital pairs number of the interdigital transducer will be much more when the requirement to the passband of the filter becomes very narrow. As capacitance exists between electrodes, when the capacitance volume becomes high enough, the filter can not be realized. Ganerally, small electromechanical coupling factor materials like piezoelectric crystal are often adopted to solve the problem. Yet due to the too small electromechanical coupling factor of the piezoelectric crystal, it is impossible to introduce multistrip coupler in designing and this makes the small squareness factor's frequency response impossible to be realized. To solve the problem, we introduced series capacitance coupling weighting method in designing narrow-band filter. Series capacitance coupling weighting is indicated in Fig.5.

Weighting method is to cut each pair interdigital electrode of the interdigital transducer to several with each other serial connected interdigital electrode according to weighting requirement. Take a bisecting series coupling interdigital transducer, see Fig. 6, as the length and distance of a floated electrode in the middle is the same as that of two up and down electrodes connecting with bus bar, the middle floated electrode is at intermediate potential. The two serial interdigital pairs' electric field intensity and directions are the same of their own. Yet the electric field strength between electrodes is only the half to that between the unseparated interdigital pair electrodes. The capacitance between electrodes will get greatly reduced due to the result of series connection, and thus makes it possible to realize the narrow-band SAW filter of small squareness factor.

With the methods described above, we developed 70 MHz central frequency's filters with bandwidth 2.5-25 MHz. The specifications all meet requirement of INTELSAT. Many of them have succesfully replaced LC filters. Long term using in Shanghai and Beijing satellite communications earth stations prove the reliable and stable performance. Fig.7 and Fig.8 are the frequency responses and time domain characteristics of 25 MHz bandsidth filter. The passband ripple of the filter is <0.2dB.

References

[1] C.F.Vasile, IEEE Trans., SU-21, p.7-11, (1974)

[2] T.L.Szabo, IEEE Trans., su-21, p.114-119, (1974)

RFbandwidth (MHz)	A (MHz)	B (MHz)	C (MHz)	D (MHz)	a (dB)	b (dB)	c (dB)	d (dB)	e (dB)
2.5	9.8	2.25	2.75	8.0	0.7	1.5	2.5	25	0
5	8.6	4.5	5.25	16.0	0.5	2.0	3.0	25	0
7.5	5.4	6.75	7.75	17.0	0.4	2.5	4.0	25	0
10	7.2	9.0	10.25	19.0	0.3	2.5	5.0	25	0.1
15	10.8	13.5	15.5	16.0	0.3	2.5	5.5	25	0.1
17.5	12.6	15.75	16.00	26.5	0.3	2.5	6.5	25	0.1
20	36.6	10.0	20.5	28.0	0.3	2.5	7.5	25	0.1
25	50.0	22.5	25.75	34.0	0.3	2.5	8.0	25	0.2
36.0	28.8	36.0	45.25	40.0	0.4	2.5	10.0	25	0.3
TV	21.0	30.0	"		0.5	2.5	"	"	0.3

RFbandwidth (MHz)	Λ (MHz)	H (MHz)	f (nS)	k (nS)	Λ (nS)
2.5	2.0	2.1	16.0	16.0	20
5	3.0	4.1	12.0	12.0	20
7.5	5.4	6.2	12.0	12.0	20
10	7.2	8.3	9.0	9.0	18
16	10.0	12.4	6.0	6.0	16
17.5	12.6	14.2	6.0	6.0	15
10	14.4	16.6	4.0	5.0	15
25	18.0	20.7	3.0	5.0	15
36.0	20.0	22.1	3.0	5.0	15
TV	24.0	30.0	5.0	5.0	15

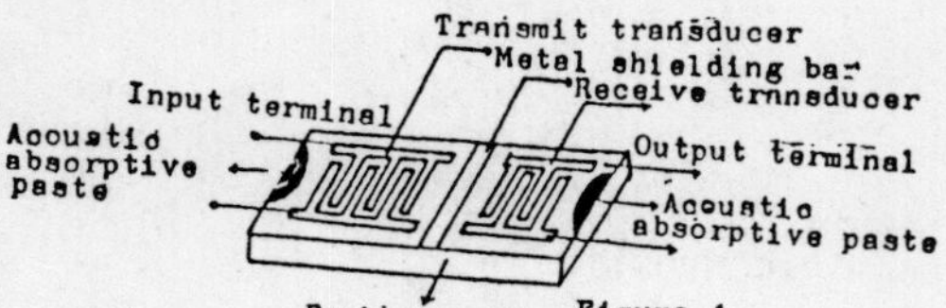

Figure 1

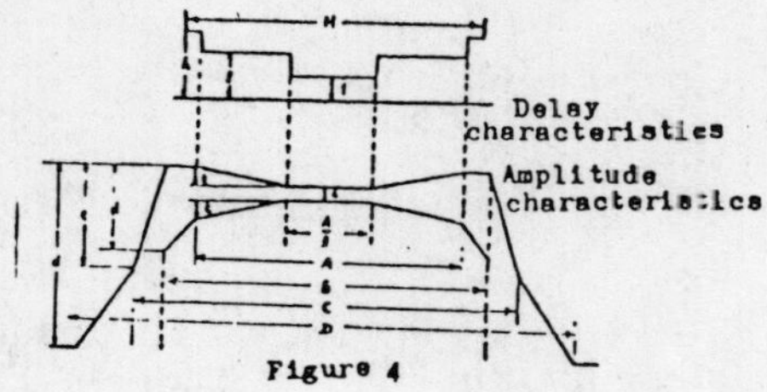

Figure 4

Figure 2

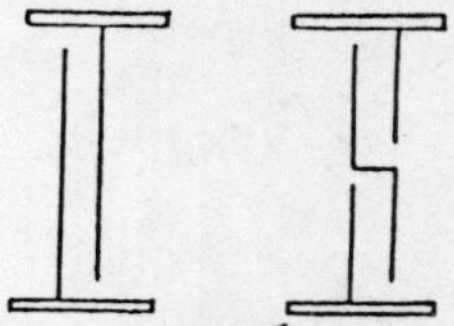
Figure 6

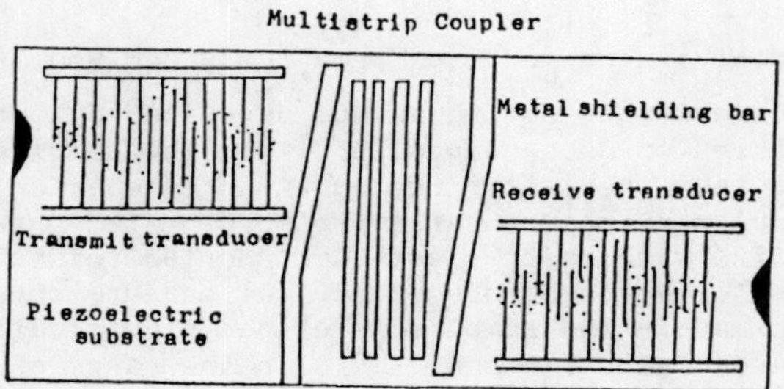

Figure 3

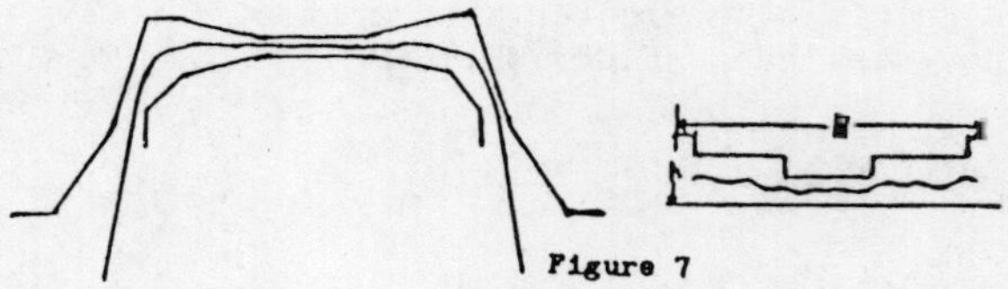
Figure 7

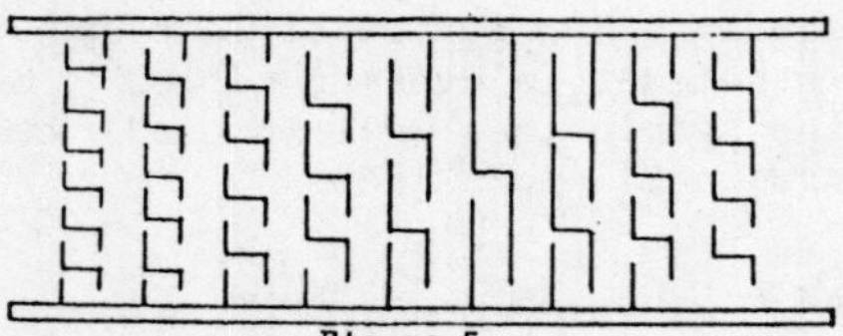
Figure 5

PYROELECTRIC PROPERTIES OF A POLYMER-CERAMIC COMPOSITE

D.K. Das-Gupta and M.J. Abdullah
School of Electronic Engineering Science
University College of North Wales
Bangor, Gwynedd LL57 1UT, UK

Abstract

Composite films of lead zirconate titante (PZT5)/Vinylidene fluoride- trifluorethylene (VDF-TRFE) appears to be an attractive electro-active material with a pyroelectric coefficient 1.4 x 10^{-4} ($m^{-2}K^{-1}$) which is significantly greater than that of polyvinylidene fluoride (PVDF).

Introduction

Composite films containing ferroelectric ceramic inclusions in the matrix of suitable polymers may possess high dielectric constants, with good electrical breakdown strength and enhanced piezo- and pyro-electric properties. Such composite materials will have commercial applications as piezo- and pyro-electric sensors, including medical uses. It has been our objective to prepare and study the electro-active properties of such composite films with suitable ceramics such as lead zirconate titanate, dispersed in the matrix of the copolymers of vinylidene fluoride and tetra-fluoroethylene (VDF-TRFE).

PZT 5 powder was obtained from the Unilator Division of Morgan Matroc. This material is donor doped with niobium, compensation being achieved by electrons or cation vacancies. Due to a significant domain wall mobility, the ceramic polarises rapidly with a very large proportion of its activity being due to 180° domains. Pellets of the copolymer VDF-TRFE were kindly provided by the Laporte Industries of Luton U.K.

Sample Preparation

Copolymer-ceramic composite samples were prepared by mixing PZT5 powder (average grain size 20 microns) with the copolymer using a rolling mill operated at 433K. The composite hide was then pressed in a temperature controlled hydraulic press to thickness in the range of 100-800 μm. The samples were then thoroughly cleaned with ethyl alcohol to remove the surface contaminants.

Experimental

Aluminium electrodes of area 2cm x 2cm and thickness 1000 Å were vacuum deposited on both surfaces of each composite film. The samples were then thermally treated in an evacuated measurement chamber at a pressure of 10^{-6} torr at 373K with their electrodes shorted for at least 12 hours before electrical measurements were made. The samples were then electrically stressed with external field in the range of 1.75 x 10^{5} Vm^{-1} to 8 x 10^{6} Vm^{-1} in the temperature range of 293K - 3373K at each field (thermal poling) at a pressure of <10^{-5}torr.

The pyroelectric responses were obtained with a direct method[1] by applying a linear heating rate of approximately 1^{0}c per minute to the prepoled sample. The pyroelectric responses were also studied by a dynamic method[2,3] by exposing the sample to a well-focused radiation from a tungsten filament lamp.

Results and Discussion

The pyroelectric current Ip, is related to the pyroelectric coefficient p(T), thus:

$$I_p = A\ p\ (T) \frac{dT}{dt} \qquad \text{..... (1)}$$

where A is the elctroded sample area and dT/dt the rate of change of temperature. Thus the reversible pyroelectric coefficient can be directly determined from a thermally stimulated discharge current (TSDC) experiment provided the trapped space charges (irreversible component) have been eliminated.

A typical pattern of TSDC spectra of the composite PZT 5/VDF-TrFE (with 50:500 Vol.%) is shown in figure 1.

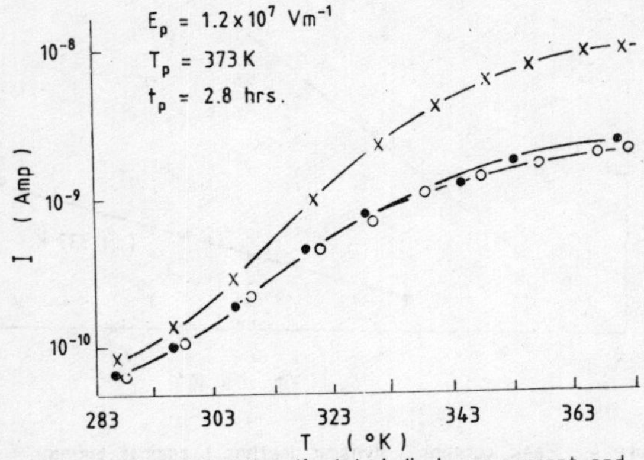

Figure 1. Typical thermally stimulated discharge current and pyroelectric current in PZT5/VDF-TrFE (50:50 Vol. %) composite.

The sample was poled at a field of 1.2×10^7 Vm^{-1} at 373K for 2.8 hours and then cooled to room temperature in the presence of the field. The electrodes of the sample were then shorted and it was heated at a uniform rate of 1^0c/minute from 285K to 373K. Similar absence of TSDC peak had also been observed[4] for a composite of PZT/polyethylene at temperatures not exceeding 373K. It may be observed from figure 1 that the second TSDC run shows a large reduction in current. However, the third subsequent heating shows no appreciable current reduction, thus establishing a reversible pyroelectric current. The magnitude of the pyroelectric coefficient p(T) at 343K, calculated with equation 1 from these data, was found to be 2×10^{-4} (m^{-2}K^{-1} which is significantly higher than that of the well-poled Form 1 (polar) PVDF ($\simeq 1\times10^{-5}$ (m^{-2}K^{-1}) at the same temperature[5]. Although the pyroelectric coefficient of the present composite material is lower than that of PZT ($p(T) \sim 5 \times 10^{-4}$ Cm^{-2}K^{-1}) the pyroelectric figure of merit ($^{p(T)}/\varepsilon^1(T)$) of the former material is approximately five times greater than that of the latter at 343K. This is due to the fact that the real part of the dielectric constant (ε^1) of the composite material is significantly lower than that of PZT. The pyroelectric coefficient as a function of temperature, obtained by the direct method[1] is shown in figure 2.

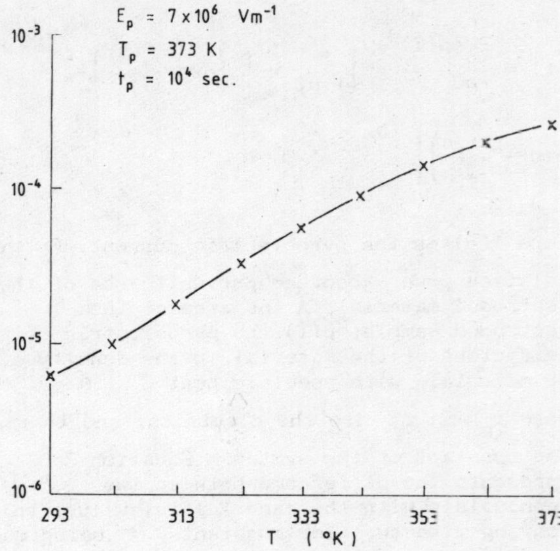

Figure 2. Pyroelectric coefficient against temperature for PZT5/VDF-TrFE (50:50 Vol. %) composite.

The dynamic method[2,3] provides an alternative way of measuring the pyroelectric coefficient of a material.

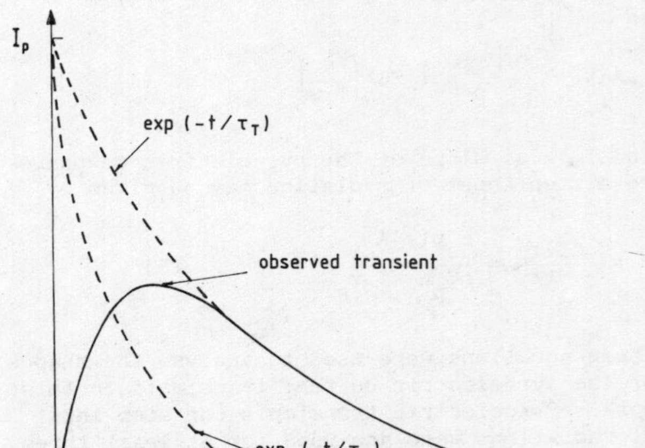

Figure 3. Schematic representation of dynamic pyroelectrical response seperated into thermal and electrical components.

Figure 3 shows a schematic pyroelectric current transient response arising from a material with its electrodes shorted through an electrometer following a step input of thermal radiations incident on the front electrode. An analysis of this pyroelectric current transient shows[2].

$$I(t) = \frac{F_o Ap(T)}{\rho C_p L} = \left(\frac{1}{1-\theta}\right) \times \left[\left(\exp - \frac{t}{\tau_T}\right) - \left(\exp - \frac{t}{\tau_E}\right) \right] \quad \ldots (2)$$

where I(t) is the pyroelectric current, F_o the radiation power absorbed per unit area of the electroded material, A the area of the electroded sample, p(T) the pyroelectric coefficient of the material, ρ the density of the material, with specific heat C_p. $\theta = \tau_E/\tau_T$ where τ_E and τ_T are the electrical and thermal time constant of the system. Equation 2 represents the difference between two exponentials with the same initial value, but decaying with two time constants, τ_E being much less than τ_T (see figure 3). The peak value of the current I_p and the time t_p for the pyroelectric current to reach the value are given by[2],

$$I_p = \frac{F_o p(T)A}{\rho C_p L} \theta^{\frac{\theta}{1-\theta}} \quad \ldots (3)$$

and

$$t_p = \tau_E \ell n \left(\frac{\theta}{1-\theta}\right) \quad \ldots (4)$$

The initial slope of the pyroelectric response to a step input of radiation may be given by[2]:

$$M = \frac{dI(t)}{dt} = \frac{F_o p(T)A}{\rho C_p L} \frac{1}{\tau_E} \quad \ldots (5)$$

These equations were used to analyse the shapes of the pyroelectric current transients in this work. Pyroelectric transients for step input of radiations were analysed for at least three samples for each poling field after the completion of the third TSDC run in order to establish the true reversibility of the pyroelectric response. Figure 4 shows the dynamic method as a function of the poling field at 333K and 373K. It may be observed that the data show on approximately linear relationship at the respective temperatures and within the range of the poling fields employed in this work.

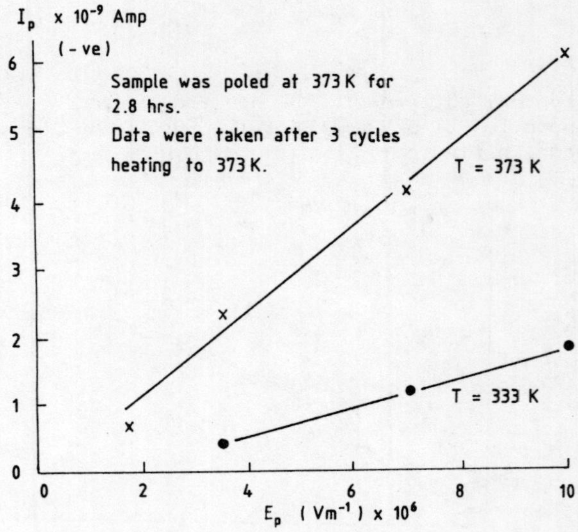

Figure 4. Peak current (dynamic method) against poling field for PZT5/VDF-TrFE (50:50 Vol. %) composite.

Now I_p, measured with the dynamic method, must be related to the pyroelectric coefficient p(T), obtained by the direct method.

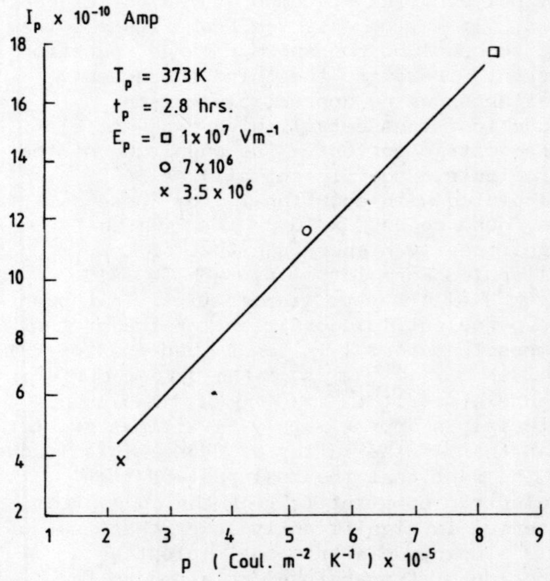

Figure 5. Peak current dynamic method against pyroelectric coefficient at 333 K for PZT5/VDF-TrFE (50 : 50 Vol. %) composite.

Figure 5 shows a plot of I_p (dynamic method) /p(T) (direct method) for the composite film at 333K, thus relating the two methods. This relationship appears to be linear, as expected, and is expressed thus:

$$I_p = Gp(T) \qquad \dots (6)$$

where $G = 2.1 \times 10^{-5}\ m^2 s^{-1} K$ which may be considered as a constant for the same material with the same electrode geometry for the same incident power of radiation. The thermal and electrical time constant of the system, of course remain unchanged.

An efficient poling is obviously achieved at high temperature t_p, fields E_p, and at long poling time T_p. Care should, however, be exercised that the material does not suffer an electrical breakdown. The results of a study of these three poling parameters have been summarised in figure 6 for this composite material.

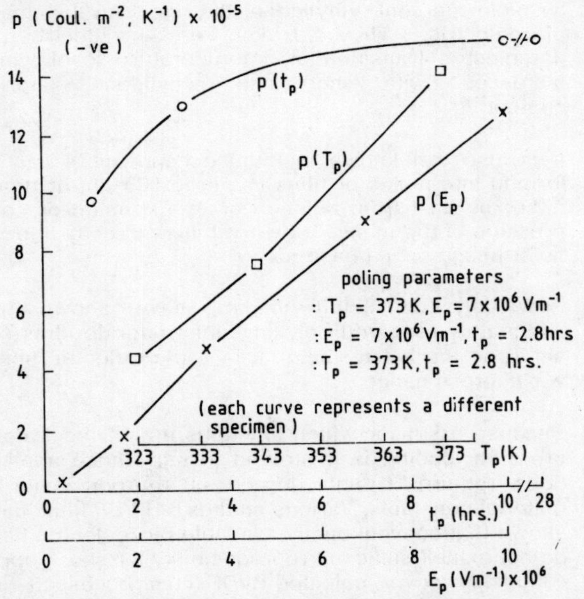

Figure 6. The dependence of pyroelectric coefficient on poling parameters at 343 K for PZT5/VDF-TrFE (50:50 Vol. %) composite.

It may be observed that a pyroelectric coefficient of $1.4 \times 10^{-4} (m^{-2} K^{-1})$ can be obtained for the PZT5/VDF-TrFE (50:50 volume %) composite with $E_p = 1 \times 10^7\ Vm^{-1}$, $T_p = 373K$ and $t_p = 8$ hours with thermal poling.

Acknowledgement

This work was financed by a research contract from the European Research Office of the U.S. Army. One of the authors (MJA) is grateful to the Public Service Department of Malaysia for a scholarship and to the University Sains Malaysia for granting a study leave and a maintenance allowance.

References

1. Byer, R.L. and Roundy, C.B., IEEE Trans Sonic & Ultrasonics, SU-19 (1972), 333.

2. Simhony, M. and Shaulov, A., J. Appl. Phys., 42 (1971), 3741.

3. Das-Gupta, D.K. Ferroelectrics, 33, (1981) 75.

4. Shakhtakhtinskii, M.G., Guseinov, B.A., Kurbanov, M.A., Gazaryan, Y.N. and Guliev, A.O.: Sov. Phys. Solid State, 25, (1983) 2145.

5. Inoui, N., Takada, T., Sakai, T., and Nakamara, K., Jap. J. Appl. Phys., 21 (1982) 706.

FERROELECTRIC PROPERTIES AND HIGH PRESSURE DEPENDENCIES OF PVDF POLYMER AND VF_2/VF_3 COPOLYMERS APPLICATION TO SHOCK GAUGES

François BAUER
ISL, 12 rue de l'Industrie, P.O. Box 301
68301 SAINT-LOUIS, France

ABSTRACT

Ferroelectric properties of PVDF polymers and VF_2/C_2F_3H copolymers are presented. The polarization process of these organic materials influences the piezoelectric properties and their behavior under dynamic compression and shock loading. Using the (PWP) pressure wave propagation, it can be observed that the polarization homogeneity is very good over the whole thickness of the samples.

Piezoelectric and dielectric properties of uniform poled PVDF and VF_2/VF_3 copolymers are reported. The intrinsic characterization of the d_{33}, k_T and d_h is presented.

The electrical response of PVDF ISL shock gauges fabricated from a high quality, biaxially stretched film, and electrically poled to high values of remanent polarization (9.5 $\mu C/cm^2$) according to the ISL process was studied under shock loading (< 30 GPa). The observed response shows good fidelity until 30 GPa. Shock calibration data of standardized gauges are examined in the case of impact loading.

The response of PVDF ISL gauges subjected to complex loading and unloading are compared to VISAR measurements.

The frequency, temperature and hydrostatic pressure (< 10 kbar) dependencies of the dielectric properties, molecular relaxations and phase transition in PVDF and a copolymer with a 30% trifluorethylene are reported. For the copolymer, both the ferroelectric transition and dynamic melting temperatures exhibit large increases with pressure. The dynamic melting temperature (T_m) of PVDF shows also a large increase with pressure. This suggests an explanation why it is possible to use these polymers as shock wave gauges at relatively high shock pressure.

A characteristic example of PVDF shock pressure measurement is shown.

INTRODUCTION

The polar polymer polyvinylidene fluoride PVF_2 or PVDF exhibits unique piezoelectric, pyroelectric and ferroelectric properties which have made themselves useful in a variety of conventional transducer and detector applications[1, 2, 3].

Vinylidene fluoride and trifluorethylene copolymers P(VDF-TrFE) or VF_2/C_2F_3H are also known to be of ferroelectric nature. These copolymers with more than 50 mole percent vinylidene fluoride can be made piezoelectric. They exhibit a ferroelectric to paraelectric transition at a temperature level which increases with increasing vinylidene fluoride content[3, 6].

It is also well-known that these copolymers can be formed into plates or films (1 μm to 500 μm or more) either by cast from solution or by extrusion or compression of the melts. They crystallize directly into the polar phase or β phase.

Previously conducted studies have shown that adequately prepared polyvinylidene fluoride films can serve as the active material in a piezoelectric shock wave stress gauge[1, 2].

Because a sensor which functions under shock compression loading is destroyed in use, these applications require a high degree of reproducibility in material constants. Various studies of PVDF films have shown that conventionally available piezoelectric films do not exhibit sufficient reproducibility for shock applications. However, selected PVDF films processed with our electrical poling method seem to achieve the desired level of reproducibility[3].

Over the past few years, a cooperative effort between "Institut de Recherches de Saint-Louis (France)", Sandia National Laboratories (U.S.A.) and Ktech Corporation (U.S.A.) has worked to develop an improved standardized PVDF gauge based on the ISL electrical poling process[5] which could be fabricated in large quantities and produced commercially. Such a standardized gauge element is now commercially available. It is the purpose of the present article to give the essential elements of the gauge standardization process as well as their shock wave response at high pressure and to report the results of hydrostatic pressure studies of PVDF and of one VF_2/VF_3 copolymer as well as some dielectric and piezoelectric properties.

HOMOGENEITY OF THE POLARIZATION INDUCED INTO PVDF AND VF_2/C_2F_3H

In the preparation of piezoelectric polymers and copolymers it is in general necessary to apply a high poling electric field to an essentially insulating material. During this operation three basic processes can occur. The dipoles tend to rotate and to orient themselves, ions migrate under the internal electric field and space charges can be injected at the interfaces. The relative importance of these processes depends on a number of parameters such as poling temperature and the art of the electric field applied.

The polar material (biaxially stretched or transverse stretched PVDF films or VF_2/C_2F_3H pressed in thin plates after melting) was poled using the Bauer method. This also permits to run hysteresis loops at room temperature and to measure directly the polarization of the material under investigation, Fig.1[5]. The hysteresis loop was traversed many times at room temperature in order to drain the ions and local space

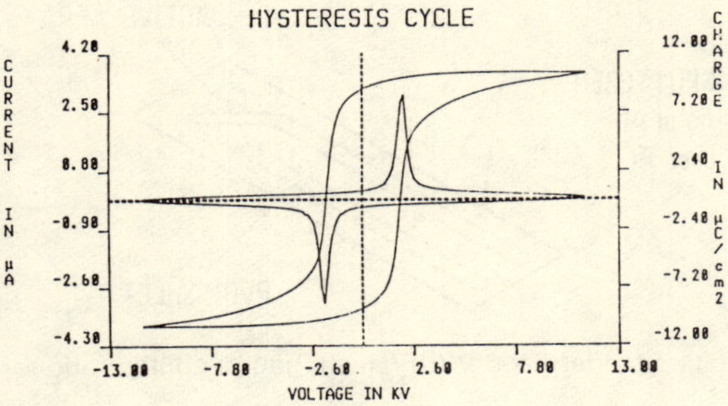

Fig. 1 Hysteresis curve of a PVDF gauge

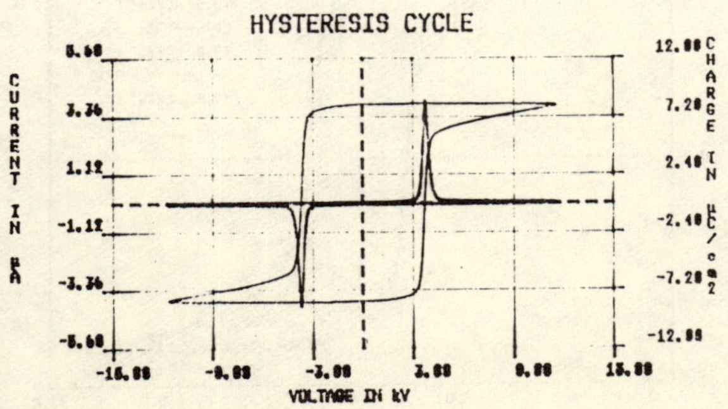

Fig. 2 Hysteresis curve of a VF_2/VF_3 (70/30) copolymer

charges. PVDF showing this hysteresis loop is "ferroelectric"[4]. The remanent polarization attains 10 $\mu C/cm^2$ which is the highest achieved on PVDF. Furthermore, various samples have demonstrated a high degree of reproducibility.

The polarization of the material is stable between -40°C and +80°C. It is verified that the coefficient d_{33} (measured with a Berlincourt piezometer) is proportional to the remanent polarization. Reproducible d_{33} values of PVDF samples can be obtained with this poling process. Maximum values of d_{33} are obtained for electric fields just below the value of dielectric breakdown.

Using the pressure wave propagation (pulse pressure of few nanoseconds duration (PWP)) C. ALQUIÉ and J. LEWINER[7] have performed some tests on our poled PVDF materials in order to examine the polarization distribution obtained by the "cycling hysteresis method". Results obtained show that our PVDF poled in such a manner has a good spatial polarization homogeneity.

1. Ferroelectric Polarization of VF_2/C_2F_3H Copolymers

Measurements on copolymers VF_2/C_2F_3H pressed at ISL (resin available from ATOCHEM in France) could be achieved thanks to the kind assistance given by Mr. Kappler. Copolymers with 60 to 80 mole% VF_2 were studied and polarized. These copolymers with such VF_2 content reveal ferroelectricity. They give square-shaped P-E hysteresis curves (Fig.2) with large remanent polarization (6 to 9 $\mu C/cm^2$, Table I) and lower coercive fields than those observed in PVDF. For VF_2/C_2F_3H with 60 mole% VF_2, d_{33} values (measured with a d_{33} Berlincourt piezometer) attain 35 pC/N.

Copolymer	Mole VF_2 %	Remanent polarization
VF_2/C_2F_3H	60/40	P_r = 7.0 $\mu C/cm^2$
VF_2/C_2F_3H	70/30	P_r = 8.8 $\mu C/cm^2$
VF_2/C_2F_3H	75/25	P_r = 9.0 $\mu C/cm^2$
VF_2/C_2F_3H	80/20	P_r = 8.5 $\mu C/cm^2$

TABLE I

Direct evaluation of intrinsic thickness-mode characteristics is impossible in very lossy piezoelectric VF_2/C_2F_3H copolymers. A characterization method based on the previous work of C. MEQUIO and R.H. COURSANT has been developed by R. HEINTZ at ISL. In particular, he has studied an intrinsic characterization of thickness-mode samples made from VF_2/C_2F_3H piezoelectric copolymers, based on an entire automatic fitting of the simulated and measured electrical impedance frequency curves. For the thickness mode of vibration the electrical impedance Z_T in the frequency domain is given by:

$$Z_T(j\omega) = \frac{1}{jC_0\omega(1 - jtg\delta_e)} \cdot \left[1 - k^2 \frac{(1 - jtg\delta_e)}{(1 + jtg\delta_m)} \cdot \frac{tg\left[\frac{\omega\tau}{2\sqrt{1 + jtg\delta_m}}\right]}{\frac{\omega\tau}{2\sqrt{1 + jtg\delta_m}}} \right]$$

where C_0 is the clamped capacitor value, k is the electromechanical coupling coefficient in the thickness mode, τ is the time of flight of the longitudinal ultrasonic wave through the piezoelectric plate, $tg\delta_e$ and $tg\delta_m$ are respectively the dielectrical and mechanical losses.

The experimental impedance curves are obtained from a Hewlett-Packard 4193 vector impedance meter. The data are transferred on a IBM 3090 VF, VM HPO which is necessary for the computations of all parameters. All parameters are chosen as non-linear function of frequency in the automatic developed fitting process.

The following Table II summarizes the main intrinsic characteristics of this material.

Composition	Copolymer (70/30)
- density ρ (10^3 kg m^{-3})	1,90
- sound velocity c^D_{33} (m/s)	2286,2
- acoustical impedance Z_t (10^6 kg m^{-2} • s^{-1})	4,355
- coupling factor k_T(%)	22,8
At resonance frequency	
- relative permittivity $\varepsilon^s_{33}/\varepsilon_0$	5,19
- dielectric loss factor tanδ	0,094
- electrical impedance z [kΩ]	1,235
Geometrical characteristics	
- thickness (μm)	457
- diameter (cm)	2,26
Resonance frequency F_r (MHz)	2,389

TABLE II

We notice that the value of the electromechanical coupling factor is of the order of 20% for a non-annealed VF_2/C_2F_3H (75/25). But we observe that the permittivity is remarkably stable in the frequency range. This is true also for the losses. It should be pointed out that the copolymer sample was poled by using the process described in the precedent section.

THE STANDARDIZED PVDF GAUGE

A shock gauge based on physical properties of a gauge element must be based on a source of highly reproducible material. Shock gauges are typically destroyed in use and cannot be calibrated for an arbitrary stress pulse shape over the many year duration required to characterize the response under shock compression. Once the gauge response is characterized, an experimentalist wishing to use the gauge must be assured that the gauge element has the same characteristics as those used in the controlled shock-response. Thus, the first requirement is for a reproducible gauge element which is widely available over an extended period of time.

The material reproducibility problem is particularly critical for a polymeric material which is generally regarded as having properties which are not highly reproducible.

Our standardization program is based on the following elements[8]:

- high quality, biaxially-stretched PVDF film,
- the Bauer electrical poling process,
- sputtered gold over platinum electrodes/leads,
- remanent polarization of 9.2 μC/cm^2, and
- quantitative specification of physical and electr cal characteristics.

The PVDF film used for the improved polymer stress gauges is a 25 μm thick, biaxially-stretched film produced by the Rhône-Poulenc Company in France. The standardized stress gauge electroding process was developed after evaluating a number of metallization techniques and mask designs. Magnetron sputtering of Au over Pt produced electrodes with very good adhesion. The electrode and sensing area design utilized for the standard gauge is shown in figure 3. The poling process applied to the gauges, incorporates the essential features as described in the French and U.S. patents and in the open literature[5]. This process has been proven by shock compression tests to condition the PVDF for accurate reproducible output[3, 9]. The remanent polarization of PVDF shock gauges, produced in France at ISL and Metravib RDS, 64 Chemin des Mouillis, 69 ECULLY (France) is the same as measured at Ktech Corporation, Alburquerque (U.S.A.), (Fig. 4), nominally 9.2 + 0.2 μC/cm^2.

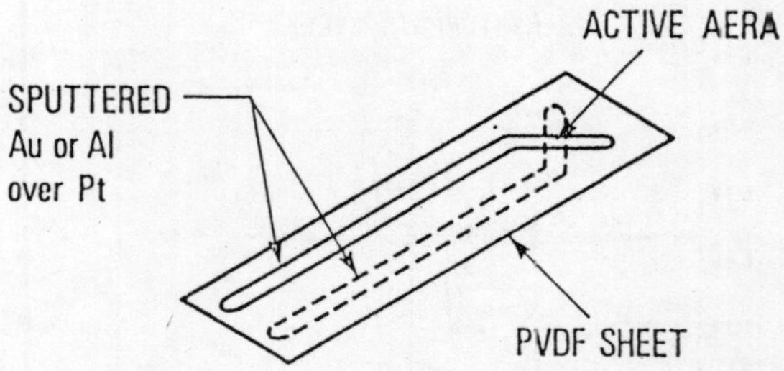

Fig. 3 Standardized PVDF shock gauge configuration.

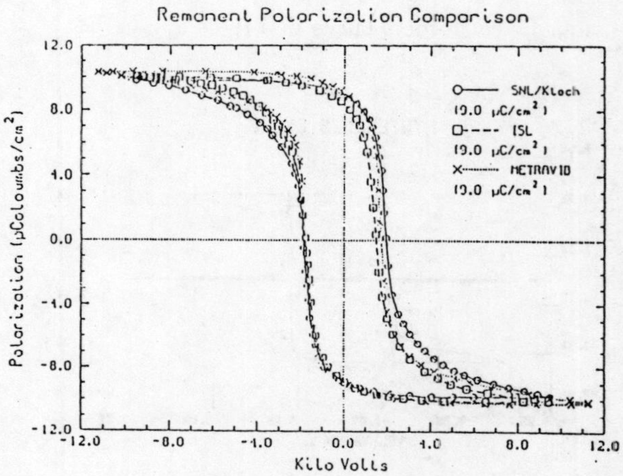

Fig. 4 Hysteresis loop comparison for SNLA/Ktech, ISL and Metravib standardized PVDF shock gauges.

SHOCK COMPRESSION RESPONSE[3, 9, 15]

Controlled shock compression is used to characterize the piezoelectric response of the standardized PVDF gauge elements. The method of loading is that used in our prior studies[3]. The symmetrical impact of an impactor and target material provides the loading. The gauge element is placed on the impact surface of the target material. In this configuration, the initial stress produced by impact is that typical of the impact of the impactor on the polymer material. This wave then reverberates between the impactor and target until an equilibrium stress is reached equal to that for the standard impactor and target. Time to establish equilibrium through the gauge element is about 60 nsec for the case of standard z-cut quartz and z-cut sapphire. The electrical signal from the shock--compressed PVDF is recorded in the "current-mode".

In this mode, the current is proportional to the stress difference across the electrodes. Thus, for most stress pulses with finite rise times, the gauge provides a signal dependent on the stress rate. Such a stress-rate measurement provides an unusually sensitive measure of wave profiles. Upon integration of the current pulse the stress versus time relation is obtained (Fig. 5). For reliable signals it is necessary to propagate the shock pulse from the positive electrical electrode to the negative lead with the positive lead connected to the ground shell of the current-viewing resistor; however, it was demonstrated that the polarization is homogeneous through the thickness of the sample[7].

Standardized PVDF gauges ISL		Shock pressure GPa	Charge released $\mu C/cm^2$
active area	1 mm^2	4.65	2.45
active area	1 mm^2	6.25	2.93
active area	1 mm^2	7.32	3.68
active area	1 mm^2	9.5	4.02
active area	1 mm^2	12.5	4.7
active area	1 mm^2	15.2	5.12

TABLE III.

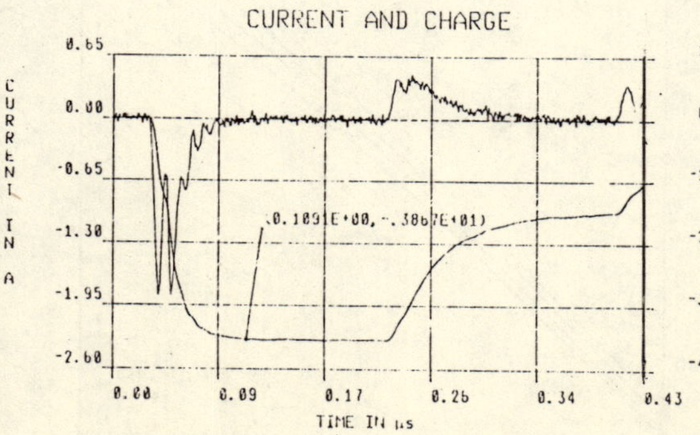

Fig. 5 PVDF shock stress gauge calibration by symmetric projectile impact.

A summary of the experiments conducted to a maximum of 16 GPa is shown in Table III. Higher stress data have been obtained but are not yet available for publication until the shock impedances of the tungsten carbide impactor/target can be better defined. These data show that the gauge has a useful output to pressures of about 46 GPa[10].

Electrical responses of two PVDF gauges were compared at the same shock level 20 GPa[8] and in the same experimental conditions. Figure 6 shows the results. The ISL gauge follows with fidelity the induced stress. The commercial gauge poled in a conventional manner does not follow the induced stress: this behavior is disturbed by the existing ions and space charges which contribute to increase the divergence of the polarization as well as it is true that div $\vec{P} = \rho$ (ρ electrical charge density present in the material).

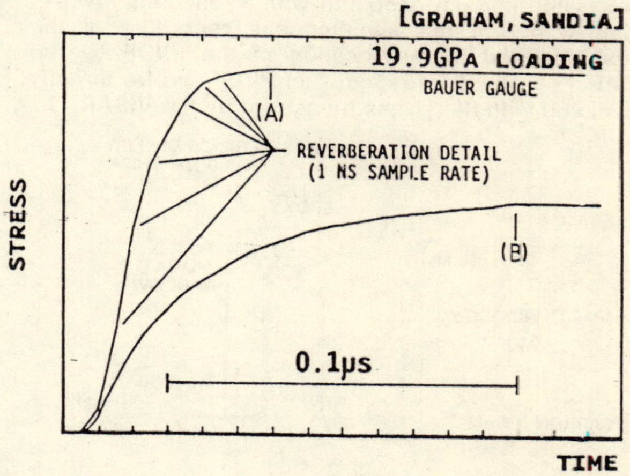

Fig. 6 Impact induced shock compression via PVDF gauges by symmetric projectile impact:
(A) PVDF poled with ISL method
(B) PVDF poled in conventional manner (R.A. Graham, SNL).

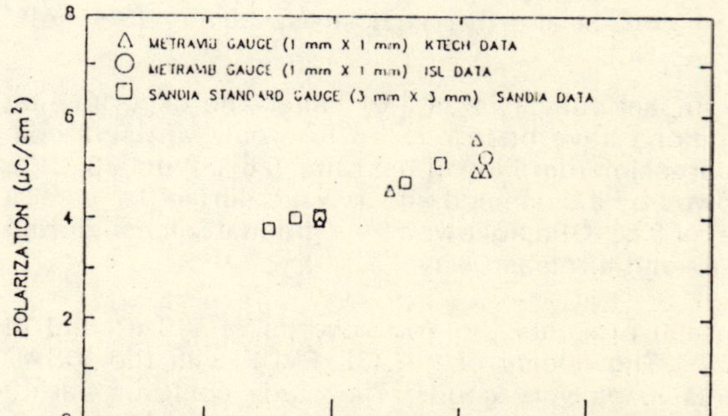

Fig. 7 PVDF gauge charge as a function a shock stress (preliminary data, detailed data analysis in process) (L.M. Lee, Ktech; R.A. Graham, SNL).

A comparison of results from different gauges shock loaded using the controlled impact technique are shown in figure 7. The shock compression data shown in figure 7 were generated at SNL and ISL. The results indicate minimal scatter, with similar responses being obtained for the gauges produced in France and the United States. The ability to produce repeatable PVDF shock transducers in the different laboratories to exacting specifications now provides the availability for general application of the shock transducers for unique measuring requirements.

1. ISL PVDF response under complex shock structure[11]

The reported study on the ISL gauge was realized at Sandia National Laboratories. The experimental assembly is given in figure 8. The impactor is in fused silica. The interface is coated with a reflecting layer in order to permit the simultaneous recording of the stress induced at the location of the PVDF by the VISAR. The delivered charge of PVDF can be directly correlated with the stress measured by the VISAR.

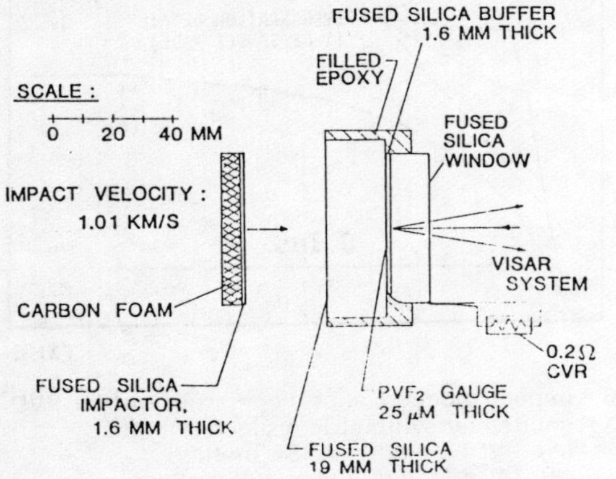

Fig. 8 Experimental configuration used to simultaneously measure the PVDF gauge current and the axial stress history (Setchell, SNL).

The impact velocity was 1.01 km/s. The experimental conditions have been chosen to create an isentropic compression (duration of the ramp 0.6 μs) until 3 GPa, followed by a sustained shock wave during 0.4 μs at a level of 5.86 GPa, followed by a gradual release during 0.2 μs and a release wave[11].

Figure 9 presents the two records of VISAR and of PVDF. The fidelity of the ISL PVDF with the VISAR seems to be very good. The study confirms earlier results[3]. The small delay observed under shock compression and released wave corresponds to the time of the equilibrium of the gauge.

HYDROSTATIC PRESSURE STUDIES OF POLYVINYLIDENE FLUORIDE (PVDF) AND FOR A VF_2/VF_3 COPOLYMER[12]

The pressure dependencies of T_m (melting temperature) of PVDF and T_c (normal ferroelectric transition) and T_m of a VF_2/VF_3 (70/30) copolymer were determined from complex dielectric constant and differential thermal analysis (DTA) measurements. The complex dielectric constant $\varepsilon = \varepsilon' - i\varepsilon''$ was measured as a function of frequency (10^2 - 10^6 Hz), temperature (80 - 500 K) and hydrostatic pressure (0 - 10 kbar).

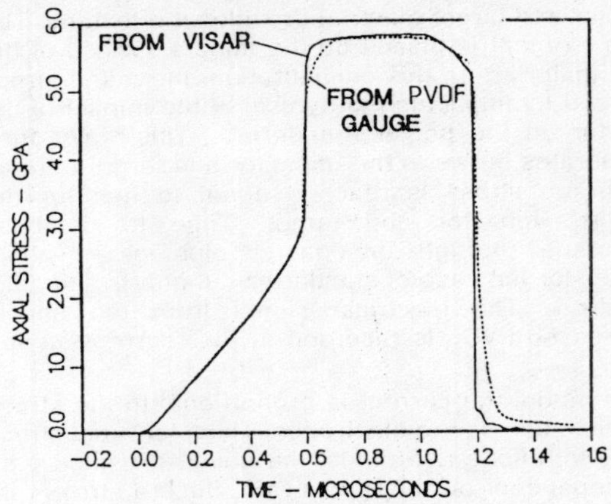

Fig. 9 Comparison between the stress history obtained from VISAR and from PVDF gauge (Setchell, SNL).

1. Homopolymer PVDF

The dielectric constant of PVDF exhibits strong pressure and temperature dependencies and frequency dispersion in the frequency range investigated 10^2 - 10^6 Hz[12]. Above 320 K, with increasing T, ε' exhibits a broad maximum and then decreases fairly rapidly up to the melting point ($\approx$ 440 K) above which temperature it rises very sharply. The magnitude of the ε' peak depends on the degree of poling of the sample. The pressure dependence of T_m is shown in figure 10 which also shows that the frequency dispersion of T_m increases with pressure. The slope dT_m/dP is frequency-dependent and is of the order of 30 K/kbar.

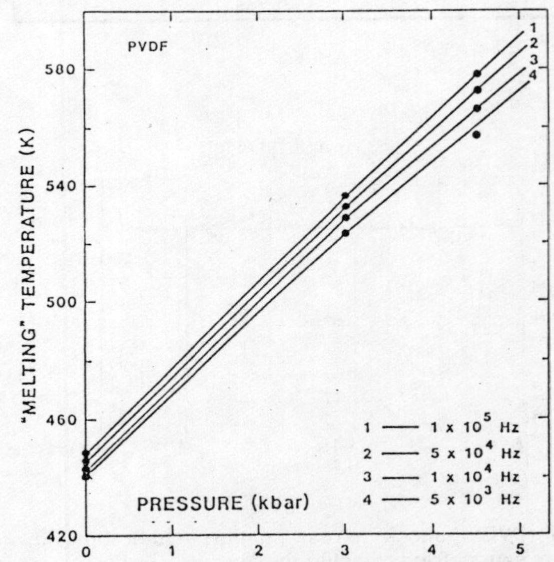

Fig. 10 Pressure dependence of T_m of PVDF at different frequencies.

2. Copolymer VF_2/VF_3 70/30

Figure 11 shows the pressure dependencies of both T_c and T_m of VF_2/VF_3 70/30 copolymer. For T_c we show two lines, one corresponding to the onset of the FE/PE transition (solid line) and one corresponding to the peak in $\varepsilon'(T)$ (dashed line). The slope dT_c/dP is 30 K/kbar below 2 kbar. The initial slope is dT_m/dP = 53 K/kbar at 10^5 Hz.

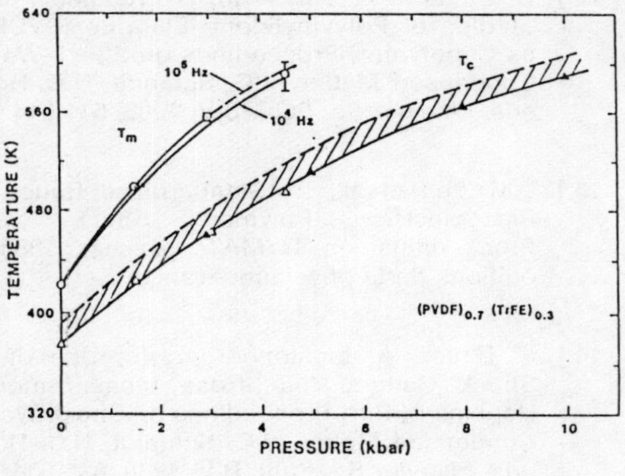

Fig. 11 Pressure dependencies of T_m and T_c of a copolymer of PVDF with 30 mole% TrFE.

CHARACTERISTIC EXAMPLE OF PVDF APPLICATIONS

1. Laser impact[13]

PVDF finds naturally applications for measuring dynamic stress induced in materials. Measurements which are difficult to realize with conventional gauges in detonics, pyrotechnics and in dynamic material behavior are made possible with piezoelectric polymeric gauges.

The application presented here, voluntary chosen, is in itself a first conclusion of the state of the art of the PVDF gauge. Figure 12 shows two stress versus time curves. These two records have been realized by L.M. LEE [13] on two PVDF gauges: one from ISL, the other characterized by the fact that the gauge was poled in a conventional manner. The experience consists of a laser impact on a 100 microns thick aluminum target: the gauge was placed on a PMMA support placed behind the target. The laser pulse has the following characteristics: 0.25 μm wavelength, 15 ns duration, 80 J/cm² energy.

Only the ISL PVDF gauge acts as a true piezoelectric gauge.

Some other applications have been studied in following domains:

- internal ballistics,
- explosive shock loading[3],
- shock pressure measurements in the vicinity of small size detonators[3],
- accelerometers,
- Hopkinson bar studies[14].

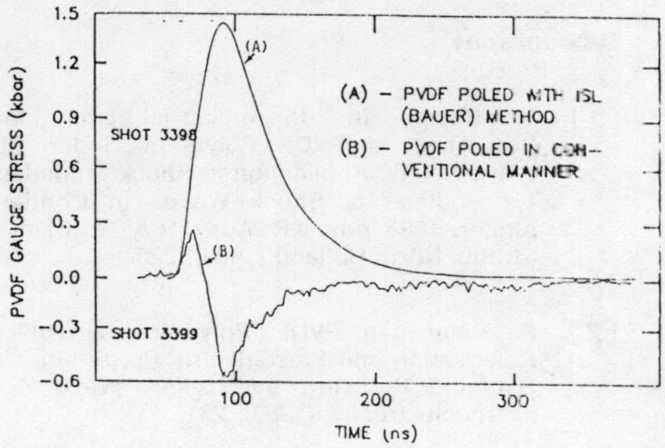

Fig. 12 Laser-induced shock compression via PVDF gauges
(A) PVDF poled with ISL method
(B) PVDF poled in conventional manner (L.M. Lee, Ktech).

CONCLUSION

The research work presented here, shows the progress realized in the studies of the ferroelectric and piezoelectric properties of PVDF. The shock compression data shown here, in addition to data obtained at the ISL at Saint-Louis, Sandia National Laboratories and Ktech Corporation have demonstrated that the shock response of standardized PVDF has a high degree of reproducibility. When gauge elements are prepared with the Bauer process and a standard remanent polarization of 9.2 μC/cm² is achieved, they provide the basis for a highly reliable, unusually sensitive shock gauge. If the gauge element is used in a standard, impedance-matched host, such as Kel F, the maximum operating limit of 46 GPa in the gauge corresponds to pressures of hundreds of GPa in a metal target.

The results obtained under hydrostatic pressure have shown that the melting temperature T_m is frequency-dependent and a strong increasing function of pressure. We can believe that PVDF remains ferroelectric and could undergo pressure- or temperature-induced depoling on time scale of shock experiment. A similar behavior is observed on VF_2/VF_3 copolymer but its potential under shock pressure remains unexplored.

Because of its ease of use, unobtrusive character, excellent time resolution, sensitivity of stress rate, large output signal and stress range of application, the standardized PVDF piezoelectric polymer gauge gives us a new, revolutionary capability to monitor and study shock compression events[15].

But the scientist, the experimentalist must be attentive. There are different PVDF, but it seems clear today that there is only one piezoelectric PVDF which can be used for shock gauge transducers.

Bibliography

[1] F. Bauer, In: Piezoelectric and Electric Properties of PVDF Polymers Under Shock Wave Action: Application to Shock Transducers, Proceedings of Shock Waves in Condensed Matter, 1983, eds. J.R. Asay, R.A. Graham, G.K. Straub, North Holland (1984), 225.

[2] F. Bauer, In: PVDF Polymers: Ferroelectric Polarization and Piezoelectric Properties Under Dynamic Pressure and Shock Wave Action, Ferroeclectric 49, (1983), 231.

[3] F. Bauer, In: Properties and High Pressure Loading Response of Poled Ferroelectric PVDF Polymer Gauges, Proceedings of Techniques and Theory of Stress Measurements for Shock Wave Applications, eds: R.B. Stout, F.R. Norwood, M.E. Fourney (ASME, New York, 1987), 19-28.

[4] R.C. Kepler, In: Saturation Remanent Polarization of Polyvinylidene Fluoride, Org. Coatings Plastic Chem., 38, 278 (1978).

[5] F. Bauer, In: Method and Apparatus for Polarizing Ferroelectric Materials, French Patent 8221025, U.S. Patents 4611260 and 4684337.

[6] G.T. Davis, M.C. Broadhurst, A.J. Lovinger, T. Furukawa, In: Hysteresis in Copolymers of Vinylidene Fluoride and Trifluorethylene, Ferroelectrics, 57, 73, 1984.

[7] C. Alquié, J. Lewiner, In: A New Method for Studying Piezoelectric Materials, Rev. Phys. Appliquée, Vol. 20, p. 131, 1985.

[8] R.P. Reed, In: Recent Developments in Pizoelectric Polymer Stress Gauges, Proceedings of The Range Commanders Council, Instrumentation/Transducer Committee, 14th Transducer Workshop, ed. S.F. Kuehn, Colorado Springs, CO, June 15-18, 1987.

[9] L.M. Lee, W.D. Williams, R.A. Graham, F. Bauer, In: Studies of Bauer Piezoelectric Polymer Gauges (PVDF) Under Impact Loading, Proceedings on Shock Waves in Condensed Matter, Y.M. Gupta, ed., Plenum Press, New York, (1986), 497.

[10] R.A. Graham, Private Communication.

[11] R.E. Stechell, In: Response of Polyvinylidene Fluoride (PVDF) Gauges to Structured Waveforms, Proceedings on Shock Waves in Condensed Matter, S.C. Schmidt, N.C. Holmes, eds. Elsevier Sc. Publ. B.V. 1988, 623-626.

[12] G. Samara, F. Bauer, In: Hydrostatic Pressure Studies of Polyvinylidene Fluoride (PVDF) and its Copolymer, Proceedings on Shock Waves in Condensed Matter, S.C. Schmidt, N.C. Holmes, eds. Elsevier Sc. Publ. B.V. 1988, 611-614.

[13] L.M. Lee et al., In: Standardized Bauer PVDF Piezoelectric Polymer Shock Gauge, Proceedings on DYMAT, Ajaccio 1988, Les éditions de la physique, France.

[14] F. Bauer, A. Lichtenberger, In: Use of PVDF Shock Gauges for Stress Measurements in Hopkinson Bar, Proceedings on Shock Waves in Condensed Matter, S.C. Schmidt, N.C. Holmes, eds. Elsevier Sc. Publ. B.V. 1988, 631-634.

[15] R.A. Graham, L.M. Lee, F. Bauer, In: Response of Bauer Piezoelectric Polymer Stress Gauges (PVDF) to Shock Loading, Proceedings on Shock Waves in Condensed Matter, 1987, Elsevier Science Publishers B.V. (1988), 619.

DIELECTRIC PROPERTIES OF PFN-PFEW-PMN TERNARY SYSTEM RELAXOR FERROELECTRICS

Qiu Zhengxiang, Li Longtu, Wang Jiang, Zhang Xiaowen
Department of Materials Science and Engineering
Tsinghua University, Beijing, P. R. China

ABSTRACT

The $Pb(Fe_{1/2}Nb_{1/2})O_3$ - $Pb(Fe_{2/3}W_{1/3})O_3$ - $Pb(Mg_{1/3}Nb_{2/3})O_3$ (PFN-PFW-PMN) ternary system relaxor ferroelectric ceramics were synthesized by precalcined $FeNbO_4$, Fe_2WO_6 and $MgNb_2O_6$ method. The influences of the amount of PMN, $PbTiO_3$ (PT) and sintering temperature (S.T.) on the dielectric properties of the system were studied. The composition of 0.53 PFN-0.30PFW-0.17PMN +5at%PT +0.2wt%MnO_2 was chosen as the material for low sintering temperature multilayer capacitor with complete silver electrods, which has good dielectric properties of $\varepsilon(25°C) = 17400$, D.F. $<150\times10^{-4}$ and $\Delta C/C=-50\sim+8\%$ between +10 to +85 °C.

INTRODUCTION

High quality ceramic materials for multilayer capacitor with high dielectric constant, small temperature coefficient and cost reduction(relating to electrode materials) are focused on in recent years[1]-[3]. The ceramic sintering temperature should be lowed when Ag electrodes were used, for the melting point of Ag is 960°C. The system consisting of combination of complex perovskite compounds containing Pb are good candidate materials to get the aim.

PFN, PFW and PMN all have good dielectric properties[4](Table 1) and have been studied before[5]-[6]. But each of them has dissipation factor (D. F.) more than 5% and cannot be used to capacitors. In order to utilize the high dielectric constant of them and low sintering temperature of PFW, the ternary system of PFN-PFW-PMN is studied in this investigation.

EXPERIMENT

It has been reported that the perovsike PFN, PFW and PMN are difficult to be fabricate as single-phase ceramics due to appearance of stable pyrochlore phase on calcination and a process of prereacting the B-site refractory oxides to form the appriate columbite or wolframite compounds and then followed by reacting with PbO to reduce the amount of pyrochlore phase was

Table 1. Properties of PFN, PFW and PMN

Comp.	C.T.(°C)	S.T.(°C)	ε max	D F.*
PFN	112	1100	24000	6.5%
PFW	-90	850	20000	7.0%
PMN	-10	1100	18000	5.5%

* From our experiment results.

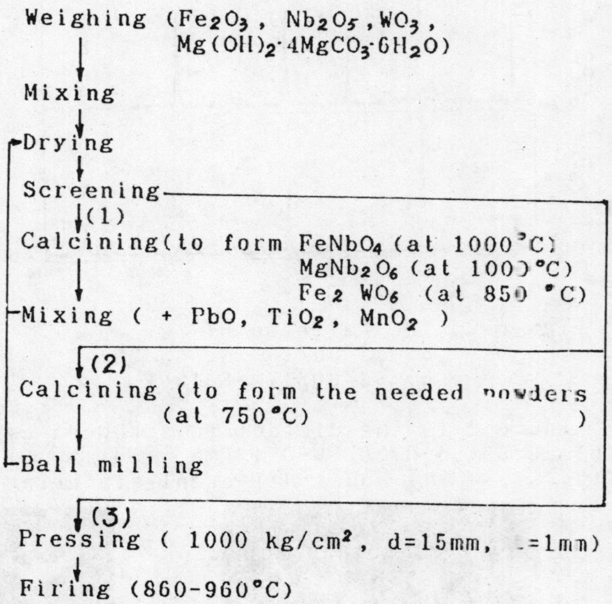

Figure 1. Sample preparation process for disk capacitors

developed[7]. In this investigation the precalcining method was used and the proparation processes are shown in Figure 1. The purities of all the raw materials used are larger than 99%.

RESULTS AND DISCUSSION

From the XRD patterns of the samples (Figure 2) we can see that the samples studied are almost single-phase perovskite form and the pyrochlore phase are undetectable under the experiment conditions. It means that the preparation of B-site refractory oxides is very effective to reduce the amount of pyrochlore phase in the system studied and heance increase the dielectric properties.

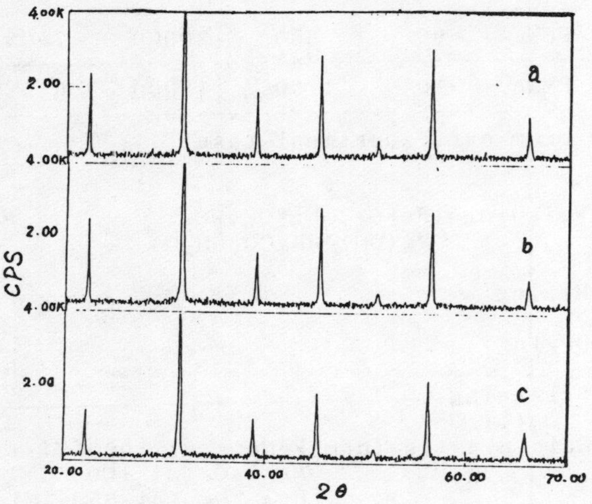

Figure 2. XRD Patterns of
a. 63/36/1 b. 53/30/17
c. 53/30/17 +5at%PT

Table 2 is the dielectric properties of 0.53PFN-0.30PFW-0.17PMN +5at.% PT +0.2wt.% MnO_2 under different sintering temperature. It shows that the sintering temperature has great effects on the dielectric properties. This is related to the density, grain size and porosity of the samples (see Figure 3).

It can be see from Table 3 that certain amount of PMN is very important to the properties of this system. When the amount is less than 17at.% , PMN can increase the dielectric constant greatly with other properties unvaried. But when the amount is larger than 17at.% it decreases the resistivity and permittivity as well increases dissipation factor.

Table 2. Properties of .53PFN-.30PFW-.17PMN +5at%PT +0.2wt%MnO_2*

S.T. °C		ε max	$\rho\times10^{-11}$ ohm·cm	D.F. %	ΔC/C 10~85°C %
860	8320	8920	1.0	1.3	-44~+7
880	11180	11970	1.0	1.5	-48~+7
900	13970	15320	1.4	1.6	-51~+10
920	14810	15810	1.4	1.5	-54~+8
940	16500	18290	1.5	1.4	-56~+11
960	16090	18240	1.0	0.9	-55~+13

* The Curie Temperatures of all the sample are 35°C.

In order to get small temperature coefficient and make it satisfy the demands of Z5U, the Curie temperature of about 35°C is best, so we add a certain amount of $PbTiO_3$ to increase Curie point. The result are listed in Table 4.

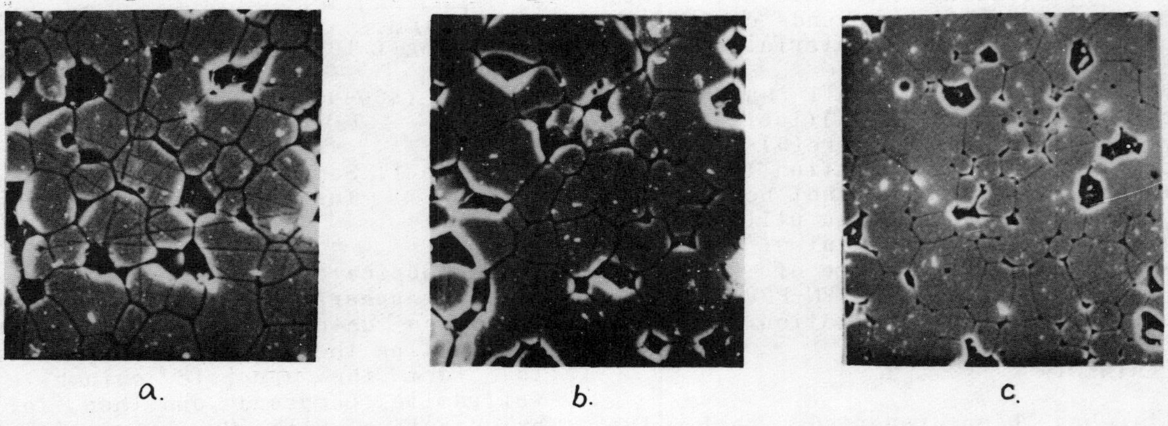

Figure 3. SEM Micrographes of 53/30/17 +5at%PT +.2wt%MnO at
a. 880 b.900 c.920°C for 4 hours

Table 3. The effects of the amount of PMN on the properties in PFN-PFW-PMN ternary system

Comp.*	ε 25°C	ε max **	$\rho X10^{-11}$ ohm·cm	D.F. %
63/36/1	9610	10460	1.8	0.6
61/34/5	11820	12290	2.3	0.8
58/33/9	13220	14920	5.0	1.0
56/31/13	16950	17290	2.4	1.0
53/30/17	16740	17440	4.0	0.7
49/28/23	16020	17060	0.1	1.7

* In all the compositions, the ratio of PFN to PFW is 64/36 and remains unchanged.

** The Curie points of all the compositions are between 15~20°C.

Table 4. properties of 53/30/17 +x at% PT +0.2wt.% MnO

x	C.T. °C	ε 25°C	ε max	$\rho X10^{-11}$ ohm·cm	D.F. %	$\Delta C/C$(%) 10~85°C
0.0	15	16740	17440	4.0	.68	-70~+4
5.0	35	14970	16500	1.4	.80	-51~+10
5.5	43	13700	15800	1.2	1.1	-51~+15
7.0	45	11950	15000	1.1	1.7	-38~+26

Table 4 shows that PbTiO3 is offective to increase the Curie temperature, meanswhile it decreases dielectric constant and resistivity and increases dissipation factor. So suitable amount is important.

We use the composition of 0.53PFN-0.30PFW-0.17PMN +5at.% PT+0.2 wt% MnO_2 to fabricate a five-layer monolithic capacitor using complete silver electrodes with dimensions of 12X12mm. The properties are: S.T.=880°C, C.T.=35°C, ε(25°C)=17600, εmax=19200, D.F.=110X10^{-4}, and $\Delta C/C$=-50~+8% between +10 to +85°C.

CONCLUSION

The PFN-PFW-PMN ternary system appears to be a suitable one for low sintering temperature multilayer capacitors with high dielectric constant and small temperature coefficient which can satisfy the demands of Z5U standard and silver electrodes can be used completely. The experiment results indicate that certain amount of PMN and $PbTiO_3$ are important to obtain desirable properties.

REFERENCES

[1]. W.R.Buessem and T.I.Prokopowicz, Ferroelectrics, 225-230 10(1976)

[2]. I.Burn and G.Maher, J. Mater. Sci., 633-640 10 (1975)

[3]. Y.Sakabe,et al., Jpn. J. Appl. Phys.147-150 20[Suppl.20-4](1976)

[4]. Y.Yohati, et al.,Electronic Ceramics, 22-26 7(1987)

[5]. T.R.Shrout,et al.,Am. Ceram. Soc. Bull., 808-810 63[6](1986)

[6] T.Tawfik and M. I. Abd El Ati, Jpn. J. Appl. Phys., 1168-1170 24[9](1985)

[7]. S.L.Swartz and T.R.Shrout, Mater. Sci. Bull., 1245-1250 17(1982)

Dielectric and Electromecanical Properties of Soft PZT-family Ceramics Containing WO_3 Component and Their Applications to Electromecanical Transducers

Hisao BANNO

NTK Technical Ceramics Div., NGK Spark Plug Co., Ltd.

14-18, Takatsuji-cho, Mizuho-ku, Nagoya, 467, Japan.

[ABSTRACT]

1. Materials

Dielectric and electromechanical properties of $Pb(Zr,Ti)O_3$ ceramics modified with WO_3 additives (PZWT) and $PbZrO_3$-$PbTiO_3$-$Pb(W_{1/2}Ni_{1/2})O_3$ were investigated at compositions in vicinity of the morphotropic phase bouodary (MPB) in the phase diagram.

A behavior of the MPB in PZWNT system resembles to that of $PbZrO_3$-$PbTiO_3$-$Pb(Nb_{2/3}Mg_{1/3})O_3$ or $PbZrO_3$-$PbTiO_3$-$Pb(Nb_{2/3}Ni_{1/3})O_3$. The dielectric and electromecanical properties of PZWNT were enhanced by a partial substitution of Pb with Sr and an addition of Ta_2O_5 or Bi_2O_3.

Typical properties of the above-mentioned materials are shown in Table 1.

2. Their Applications to Elecromecanical Transducers

2-1. Piezoelectric Gas Ignitor

An application of soft PZT-family ceramics to gas ignition is discussed in view of durability of the ignitor and energy of spark compared to the hard PZT ceramic material.

2-2. Ceramic Discriminater for FM

Temperature dependences of resonant frequency in the counter-extentional mode and thickness mode of PZWT ceramic plate were investigated at compositions in the tetragonal crystal structure region. 455 kHz and 10.7 MHz discriminators are designed using this PZWT material.

2-3 Piezoelectric Actuators

Mutilayer and bimorph type piezoelectric actuators are described in view of displacement characteristics.

2-4 Low Frequency Ladder Type Ceramic Filter

A 173 kHz ladder type ceramic filter were desiged using PZWT ceramics. This ceramic filter are available for automobile traffic control in Japan.

2-5 Ultrasonic Air Transducer

An ultrasonic air transducer using soft PZT material is discussed compared to the hard PZT material.

Table 1. Typical Properties of Piezoelectric ceramic material A~E of PZWT and PZWNT.

	Unit	A	B	C	D	E
k_p	%	63	60	67	63	62
k_{31}	%	34	32	35	34	33
$\varepsilon_{33}^T / \varepsilon_0$		1800	2000	2400	3800	5800
g_{31}	$\times 10^{-3}$ Vm/N	11.5	10.7	10.0	8.0	6.2
d_{31}	$\times 10^{-2}$ m/V	180	180	210	260	310
Y_{11}^E	$\times 10^{10}$ N/m^2	5.7	6.3	5.7	6.0	5.9
s_{11}^E	$\times 10^{-10}$ m^2/N	17.5	16.0	17.5	16.8	17.0
Q_m		70	80	75	65	45
ρ	$\times 10^3$ kg/m^3	7.7	7.5	7.8	7.7	7.8
T_c	℃	350	270	250	220	110

GRAIN CORE-GRAIN SHELL STRUCTURE AND CHARACTERISTICS IN MODIFIED-BT CERAMICS

Zhang Weilong, Zhou Zhigang

Dept. of Materials Science and Engineering,
Tsinghua University, Beijing 100084, China

Sun Jigung, You Liping
HVEM Research,
General Research Institute for Non-ferrous Metals,
Beijing 100088, China

ABSTRACT

During the past few years, grain core-grain shell (GC-GS) structure in modified-BT ceramics has attained increasing importance as materials for capacitors. In this paper, we discuss the characteristics and mechanism of GC-GS structure of several modified-BT ceramics. The GC-GS structural model will be a structural basis and of potential applications in the area of electronic ceramics.

INTRODUCTION

An interesting grain structure, GC-GS structure, has aroused considerable attention recently because it shows some promise of being candidate for ceramic capacitors with high performance. Such GC-GS structure is composed of two-phase of ferroelectric and non-ferroelectric. The grain core (ferroelectric) was found to be relatively pure $BaTiO_3$, and the grain shell (non-ferroelectric) contained significant amounts of additives, such as Zr, Ca, Sn, Bi, and so on. The GC is embedded in and surrounded by a GS with little grain growth and limited diffusion.

As seen by the relationship between property-structure-technology, manufacturing techniques play a major role in the high performance ceramics. In this work, therefore, attention was focused on GC-GS structural characteristics, growth conditions and mechansim in an attempt to explore a method of controlling GC-GS structure effectively.

EXPERIMENTAL

The specimens were prepared with conventional ceramic technology using reagent grade oxide powers as raw materials. After batching, mixing, drying, forming and finally sintered in air at 1100-1200°C for 2-4h.

The crystalline phases present in the calcined powers and samples were identified by XRD at room temperature using Cu radiation. Dielectric data were obtained with bridges from the specimens electroded with silver. The microstructures were examined by electron microscopes (JEM-1000, JEM-2000FX), the line-analysis of element distributions were performed using an EDAX, after the thin sections of specimens obtained by ion beam thinning technique.

RESULTS AND DISCUSSION

On the basis of our investigation, the characteristics, mechanism and behaviour of modified-BT ceramics with GC-GS structure are discussed as follows.

Characteristics

As mentioned above, GC-GS structure may result from heterogeneous additive distribution in a grain, which is comfirmed by the experiment of this work. The typical GC-GS structure of grain in the present samples is given in Fig.1, showing that ferroelectric domain patterns existed only in the center of a grain (GS). The EDAX analysis revealed that compostional differences could be responsible for this structural difference in the same grain (see Table I). Compositions, phase and structure of modified-BT with GC-GS structure are shown in Table I.

The phase transition of GC-GS structure can be seen in Fig.2 for $BaTiO_3$-$PbTiO_3$-$Bi_2(SnO_3)_3$ ceramics, the domain pattern of a grain core would disappear rapidly as the temperature reached 130°C, whereas no domain patterns of a grain shell emerge even when temperature down as low as minus 90°C.

It stands to reason that the Bi^{3+}, Pb^{2+} substitute for Ba ion and Sn^{4+} for Ti^{4+}. This fact supplies a reason for nonferroelectric phase formation.

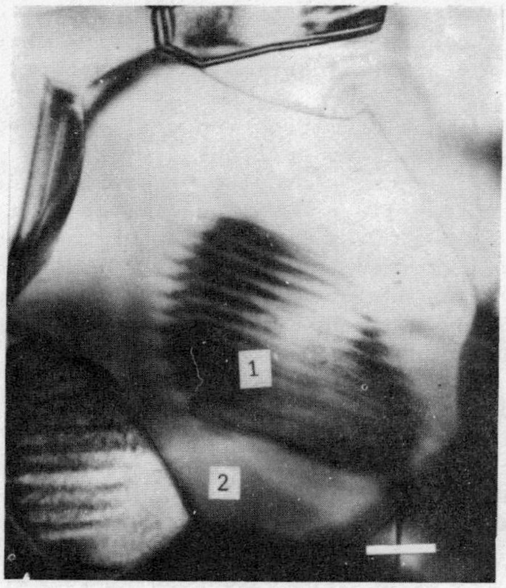

Fig.1 TEM photomicrograph of GC-GS structure of a grain in the present samples
1-grain core, 2-grain shell
(a bar=200nm)

Table 1 Compositions, phase and structure of modified-BT with GC-GS structure

NO.	Compositions	Phase and Structure GC	GS	Ref.
1	$BaTiO_3$ $Bi_4Ti_3O_{12}$ Nb_2O_5	T., F. perov.	Bi, Nb C., PF perov.	[3]
2	$BaTiO_3$ $CaSnO_3$ $CaTiSiO_5$	T., F. perov.	Zr, Sn, Ca -, NF. perov.	[1-2]
3	$BaTiO_3$ $CaSnO_3$ $CaTiSiO_5$ $PbTiO_3$	T., F. perov.	Zr, Sn, Ca, Pb -, NF perov.	[4]
4	$BaTiO_3$ $PbTiO_3$ $Bi_2(SnO_3)_3$	T., F. perov.	Bi, Pb, Sn -, NF. perov.	This work
5	$BaTiO_3$ $SrTiO_3$ $CaTiSiO_5$	T., F. perov.	Sr, Zr, Ca, Mg -, NF. perov.	[5]

Note: T-Tetragonal, C-Cubic, F-Ferroelectric
NF-Non-ferroelectric, PF-Paraelectric
Perov.-Perovskite

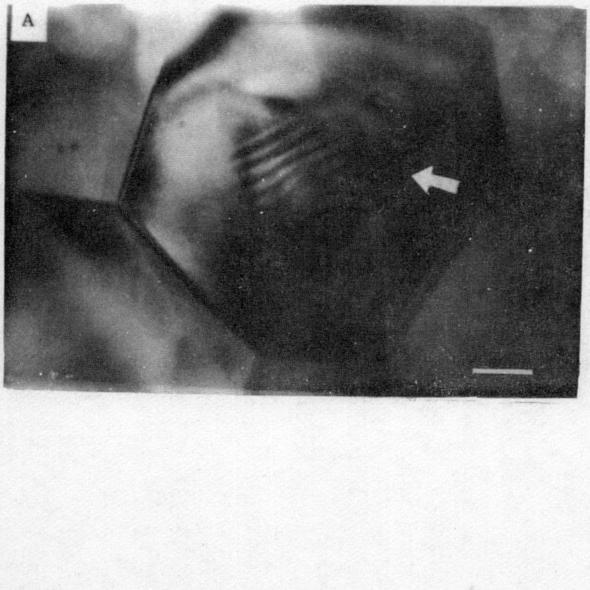

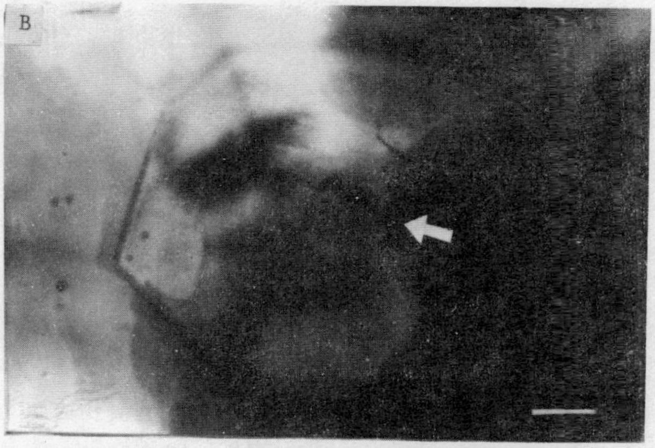

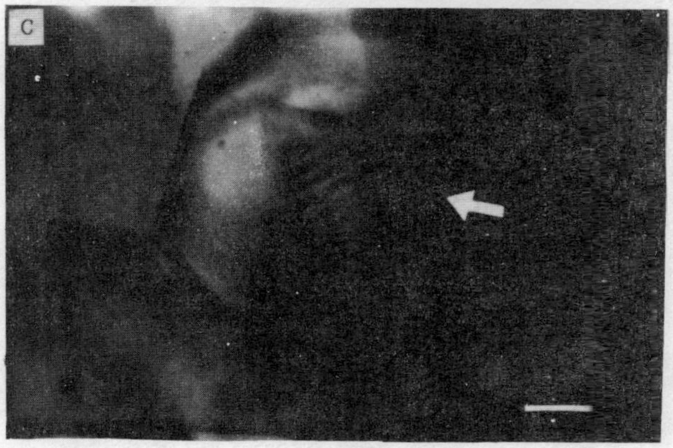

Fig.2 GC-GS structure changes with surrounding temperature at
A-room temperature, B-130°C, C--90°C
(a bar=200nm)

Model and Mechanism

The two-phase structural models frequently occured in ceramics are shown in Fig.3, together with GC-GS model developed in this work. The GC-GS structure, as we will see later, has many advantages as comparied to other models.

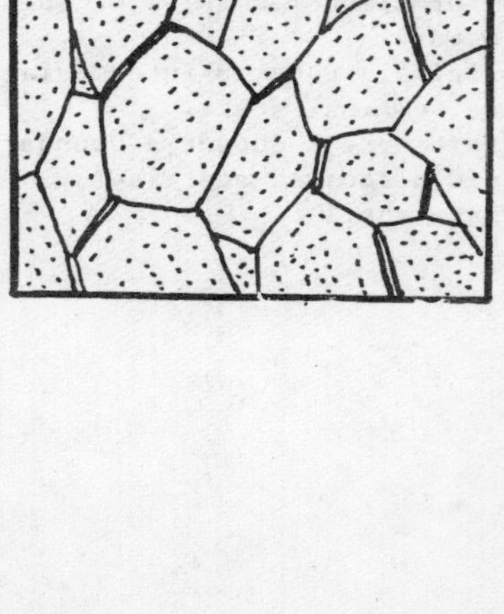

(a)

(b)

(c)

Fig.3 Three kinds of models for two-phase structure in ceramics
(a) Mix-up, (b) Solid solution,
(c) GC-GS structure

The formation of GC-GS structure has a direct relationship with manufacturing process. one of mechanisms for GC-GS growth process is represented schematically in Fig.4.

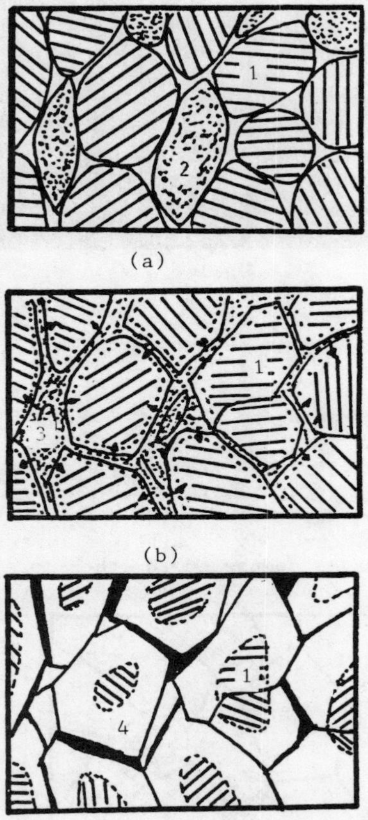

Fig.4 The formation mechanism of GC-GS structure.
a-original state, b-intermediate. state, c-final state.
1-grain core, 2-dopant, 3-boundary liquid dissolved additives,
4-grain shell

As seen in Fig.4 that the growth process of GC-GS structure can be described in three states: original, intermediate, and final state.

The experiment of microstructure indicated that the GC-GS structure develops from $BaTiO_3$ as nucleation as the sintering process proceeds. With little grain growth and limited diffusion, the resluting GC-GS structureis is remained. However, prior to that, compositions, additives and power technology must be considered to ensure the development of GC-GS structure during sintering.

Behavior and Control

As discussed above, of particular interest is GC-GS model among two-phase models. Fig.5 shows the dielectric constant temperature behavior of $BaTiO_3$-$CaTiO_3$, $BaTiO_3$-$BaSnO_3$ and $BaTiO_3$-$PbTiO_3$-$Bi_2(SnO_3)_3$ ceramics, corresponding to Mix-up, Solid solution and GC-GS model, respectively.

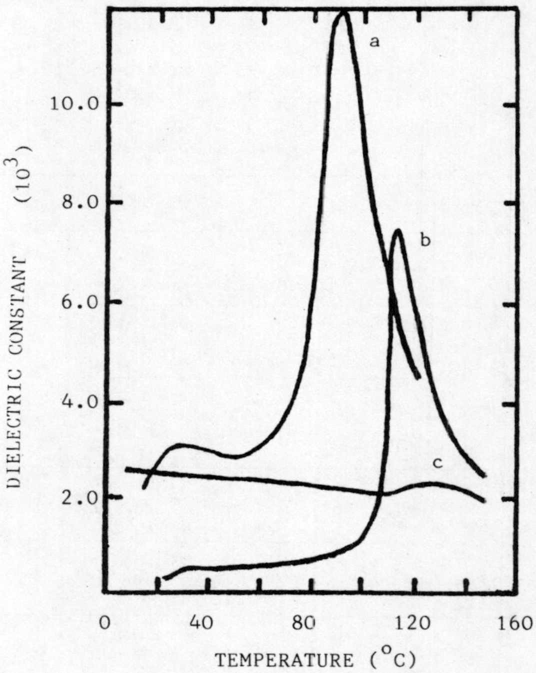

Fig.5 Dielectric constant vs temperature for a:76 $BaTiO_3$-24$CaTiO_3$[6], b:98 $BaTiO_3$-2$BaSnO_3$[7] and c: $BaTiO_3$-$PbTiO_3$-$Bi_2(SnO_3)_3$

Both $BaTiO_3$-$CaTiO_3$ and $BaTiO_3$-$BaSnO_3$ ceramics exhibit a remarkable peak in their dielectric constant-temperature curves, which is undesirable for capacitors, whereas $BaTiO_3$-$PbTiO_3$-$Bi_2(SnO_3)_3$ ceramics with GC-GS structure shows a slight peak.

It shoud be noted that dielectric constant temperature behavior depends not on the presence of GC-GS structure but also on the volume fractions of grain cores. The samples fired at 1150°C for 2h to 10h exhibit a slight differences in their dielectric constant-temperature curves shown in Fig.6. Increasing in holding time decreases the magnitudes of dielectric constant at Curie point, attributed to the decreasing propertions of grain cores volume. With certain assumptions, the volume fractions of grain core can be estimated from the magnitudes of dielectric constant at Curie point according to logarithmic mixing rule:

$$\lg \varepsilon_{peak} = V_1 \lg \varepsilon_1 + V_2 \lg \varepsilon_2 [8]$$

where ε_1=7000[5], dielectric constant of grain core at Curie point, ε_2=1900, dielectric constant of grain shell at Curie point, V_1, V_2 the volume fractions of grain core, grain shell,

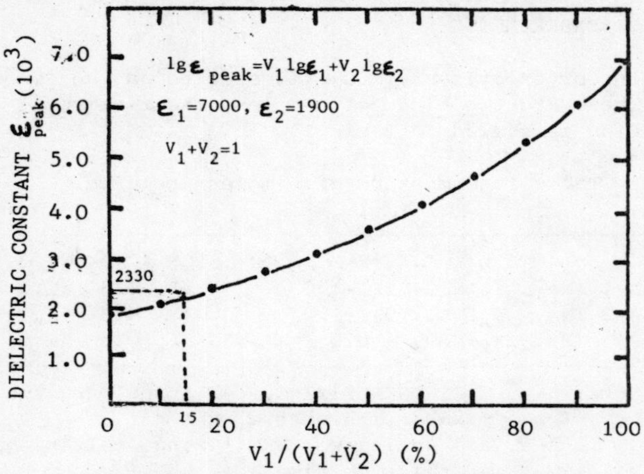

Fig.7 Dielectric constant at Curie point ploted against the volume fractions of grain cores.

respectively. The calculated grain core volume fractions for samples fired at 1150°C for 2h to 10h range from 10% to 15% as shown in Fig.7. Further work is in progress to correlate GC-GS model with behavior and to explore its potential applications in the area of electronic ceramics.

CONCLUSIONS

1. Modified-$BaTiO_3$ with various additives show GC-GS structure where the GC is embedded in and surrounded by a GS when fired in air. Such GC-GS structure consists of two-phase of ferroelectric and nonferroelectric.
2. GC-GS structure grows from $BaTiO_3$ as nucleation by diffusion-controlled process, in which original, intermediate and final state are developed.
3. Compositions, additives and power technology are important factors in the manufacture of modified-BT ceramics with GC-GS structure.
4. The ratios of the volume of grain cores to that of grain shells, which have a effect on dielectric behaviour, can be programmed by a controlled firing-history.
5. GC-GS model will be a new microstructure as a basis and of potential applications in the area of electronic ceramics.

ACKNOWLEDGEMENT

The authors would like to acknowledge the National Nature Science Foundation of China, Ministry of Electronic Industry, and National Bureau of Building Materials for their financial assistance.

REFERENCES

1. Zhou Zhigang, et al, Proc. of the 2nd International Conference on Properties and Applications of Dielectric Materials, P144-147. (sept. 12-16, 1988, Beijing, China)
2. Zhou Zhigang, et al, Ferroelectrics,(to be published)
3. B.S.Rawal, et al, Advance in Ceramics Vol.1 P.172-188 (Am. Ceram. Soc., 1981)
4. Zhou Zhigang, Zhang Weilong, Proc. of the 39 the Electronic Component Conference, P.33-36 (May 22-24, 1989, Houston, U.S.A.)
5. Zhang Weilong, Thesis, T singhua Univ.(1989)
6. T.Mitsui, et al, Phys. Rev.124, P.1354(1961)
7. G.H.Jonker, Philips Tech. Rev., 17, P.129(1955)
8. W.Niesel, Ann.d. Physik, 6[10] P.336 (1952)

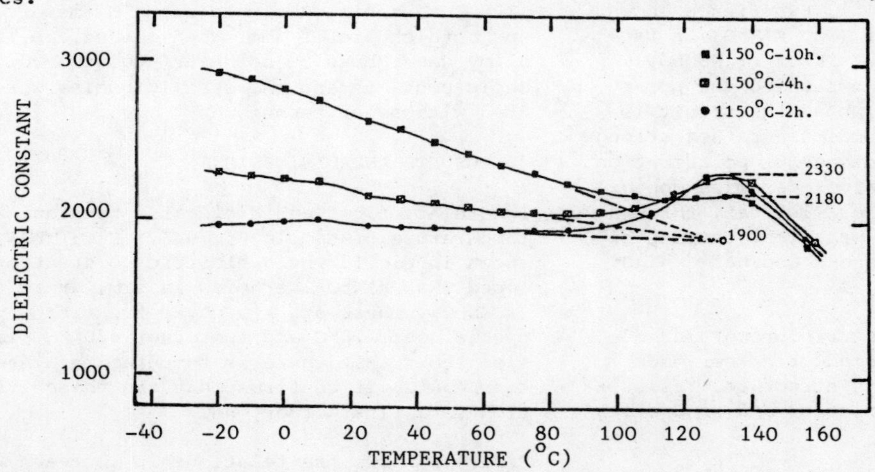

Fig.6 Dielectric constant vs temperature of $BaTiO_3$-$PbTiO_3$-$Bi_2(SnO_3)_3$ ceramics with GC-GS structure.

PREPARATION AND STUDY OF Mo-DOPED (Na, K)NbO_3 FERROELECTRIC SEMICONDUCTING CERAMICS

Bai Tiecheng, Wang Peiying and Li Xingjiao
Huazhong University of Science and Technology,
Wuhan, China

ABSTRACT

The preparing technique of Mo-doped (Na, K)NbO_3 ferroelectric semiconducting ceramics is reperted. By studying the dielectric properties of the ceramics, we have discussed the variation of its resistivity at room temperature with the various doping of Mo. The effect of Mo-doping on the structure and property of the ceramics is also discussed. On $\rho(T)$ curves, near 200°C, a change in resistivity of three orders is found. The conducting mechanism of the semiconducting ceramics is studied primarily by using Heywang model with the presence of intergranular layers of glassy phase.

1. INTRODUCTION

It is well known that $KNbO_3$ is a ferroelectrics with the perovskit type structure. It has a similar phase transition to $BaTiO_3$ [1]. But the practical usage of $KNbO_3$ ceramics is limited because it is difficalt to prepare compact $KNbO_3$ ceramics by using traditional techniques.

Reseaches show that $NaNbO_3$ and $KNbO_3$ can be mixed, at any proportion, to form continuous (Na, K)NbO_3 solid solution, which has the similar properties to $KNbO_3$ [2]. Experience shows the preparing technique of (Na, K)NbO_3 is far more developed than that of $KNbO_3$ or $NaNbO_3$. Dungan and Galding [3] have studied the polar properties of ($Na_{0.5}$ $K_{0.5}$)NbO_3, the result is that the solid solution remains a rather strong ferroelectrics, with a spontanecus polarization of 21μC/cm^2. Similar to $BaTiO_3$ ceramics doped with donors, (Na, K)NbO_3 ceramics can, through suitable doping, become a practical thermistor material which is nonpoisonous and has a high transition temperature.

Through studying the dielectric properties of Mo-doped (Na, K)NbO_3 semiconducting ceramics, the paper has discussed its structure, resistivity-temperature characteristics and conducting mechanism.

2. EXPERIMENTS

In our experiment, MoO_3 is selected as the donor. There are two kinds of samples, their composition is showed in Table 1.

Table 1. Composion of samples prepared. (mol. %)

		MoO_3	GeO_2	SiO_2	Al_2O_3	K_2CO_3
A-1	($Na_{1-x}K_x$)NbO_3	0.1-1.0		2.0	0.3	1.5
A-2	($Na_{1-x}K_x$)NbO_3	0.1-1.0		2.0	0.3	1.5
B	($Na_{1-x}K_x$)NbO_3	0.5	6.0	2.0	0.3	1.5

Note: A-1, MoO_3 + NKN mixing, calcining, ang then mixing with other.
A-2, all additives + NKN mixing, calcining at the same time.
x = 0.1-0.9.

After being dryed up, analytically pure Na_2CO_3 and K_2CO_3 are mixed with Nb_2O_5, MoO_3 and other additives according to chemical dosage. Then the mixture is added with absolute alcohol and ball ground 48 hs. The ceramics is prepared using the traditional process, We set the calcining temperature of (Na, K)NbO_3 semiconducting ceramics to be 780°C. The sintering temperature is 1050-1120°C. A finished sample has a diameter of 14mm and a thickness of 3mm. When needed, an In-Ga alloy was rubbed on both surfaces to provide ohmic contacts and the ceramic bodies were treated with ather techniques.

3. RESULTS AND DISCUSSIONS

The relation between dielectric constant and temperature of sample B, ($Na_{0.3}$ $K_{0.7}$)NbO_3, is shown in Fig.1. The dielectric constant of Mo-doped (Na, K)NbO_3 ceramics is similar to that of $KNbO_3$. There are two phase transition points, one is near 220°C and the other 430°C. Figure 1 also shows that there is temperature hysteresis near 430°C, it confirms that the phase transition is a first order one.

Figure 2 shows the relationship butween room temperature resistivity of sample A-1, (Na...

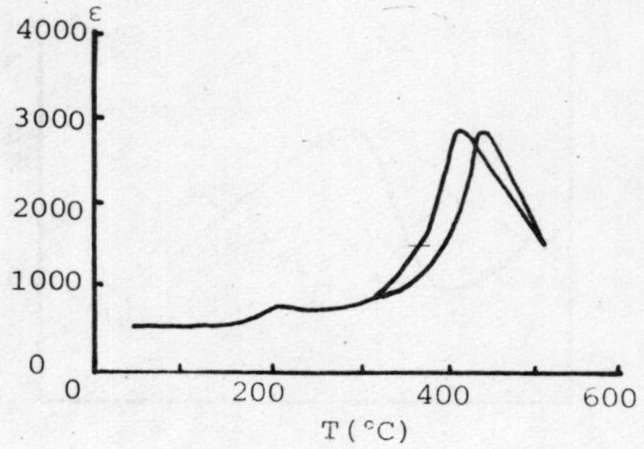

Fig. 1. The variation of the dielectric constant of sample B with temperature.

$K_{0.7})NbO_3$, and the amount of doped MoO_3. The dependence of resistivity on the amount of MoO_3 has a shape of 'U'. This behavior is similar to that of Nb-doped $BaTiO_3$ semiconducting ceramics[4].

An X-ray diffraction experiment has been done on sample A-1. From the indexed X-ray diffra-

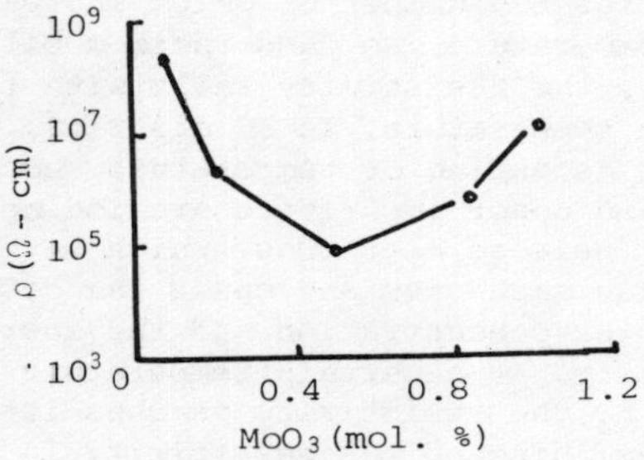

Fig. 2. Resistivity of sample A-1 at room temperature as a function of added amount of MoO_3.

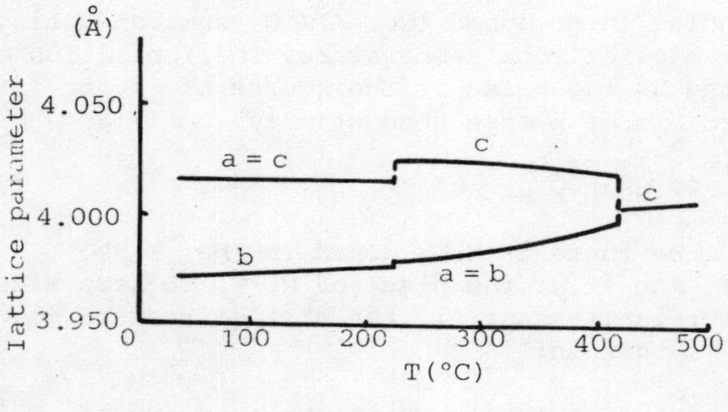

Fig. 3. Temperature dependence of lattice parameter of sample A-1.

ction spectrum lines, we have calculated the change of the lattice parameter with the temperature. Figure 3 shown the variation of the lattice parameter with the temperature. It is apprent that there are structure phase transitions at 220°C and 430°C. This behavior is almost the same as the undoped $(Na_{0.3}K_{0.7})NbO_3$ solid solution. It shows that the doping of Mo does not change phase transition temperature and Curie temperature.

Figure 4 is a scanning electronic microscope picture of the natural surface of sample B (x = 0.7). It shows that the ceramic grains have a shape of a cuboid, very different from those of the ferroelectrics $BaTiO_3$ and $PbTiO_3$.

Fig. 4. The scanning electronic microscope of the natural surface of sample B.

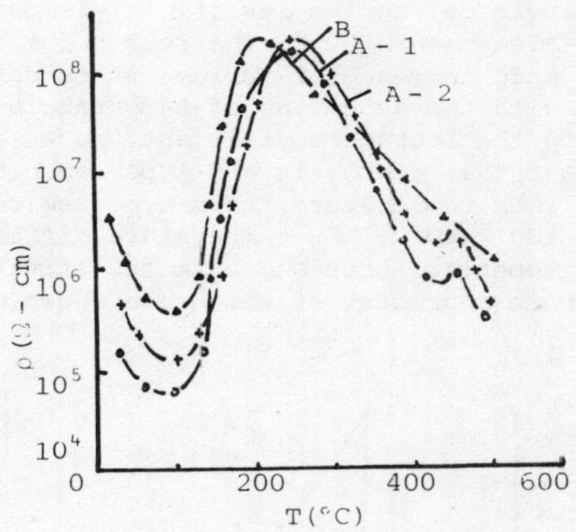

Fig. 5. The resitivity-temperature characteristics of samples examined.

As to Mo-doped (Na, K)NbO_3 ferroelectric semiconducting ceramics, the most interesting thing to us is its resistivity-temperature characteristic. Figure 5 shows this character of the samples (x= 0.7, MoO_3 doping 0.5mol%). Compared with the resistivity-temperature characteristics of doner doped $BaTiO_3$, the former has a smaller resistivity change near the Curie Point (about 430°C) but has a resistivity leap of more than three orders near 200°C. At room temperature, the resistivity has a negative temperature coefficient, it reaches its minimum in 90∿100°C, then the resistivity increases rapidly with the rising of the temperatire, after reaching its maximum, it declines and has a small leap near the Curie point.

The reference (5) has deduced the surface barrier expresion for grain boundaries of polycrystal ferroelectric semiconductors with the presence of intergranular phase:

$$\phi_{\mp} = 1/2\left(\frac{en_s \pm P_s}{\sqrt{\varepsilon_o\varepsilon_g N_g} + \sqrt{\varepsilon_o\varepsilon_p N_p}}\right) = \frac{e^2 N_g l_{g\mp}^2}{2\varepsilon_o\varepsilon_g} = \frac{e^2 N_p l_{p\mp}^2}{2\varepsilon_o\varepsilon_p} \quad (1)$$

$$n_s = N_s\{1 + \exp[(E_F - E_s + \phi)/kT]^{-1}\}^{-1} \quad (2)$$

Where E is surface energy levels, E_F Fermi level, N_s the surface state concentration and ε_g, ε_p, N_g, N_p, l_g, l_p are dielectric constant, donor concentration and barrier width of grains and intergranular phase.

Using this modified Heywang model, we can explain the resistivity-temperature characteristic of Mo-doped (Na, K)NbO_3 ferroelectric semiconducting ceramics.

In equations (1) and (2), let $N_s \sim 4\times10^{14}cm^{-2}$, $N_g \sim 10^{17}cm^{-3}$, $\varepsilon_p N_p \sim 10^{23}cm^{-3}$, $(E_s - E_f) \sim 0.6eV$[5], we can work out the relationship between exp (ϕ/kT) and T (shown in Fig.6). It is apparent that there is a big peak near 200°C. It means there should be, on the resistivity-temperature curve, a leap near 200°C. The reason for this is the rapid increase of surface state concentration with the ascending of temperature. By measuring the Seeback coefficient, we have found that (Na, K)NbO_3 is a p-type semiconductor at room temperature, an n-type semiconductor in 120∿140°C, the temperature differential electromotive force and bulk resistivity-temperature character of which are shown in

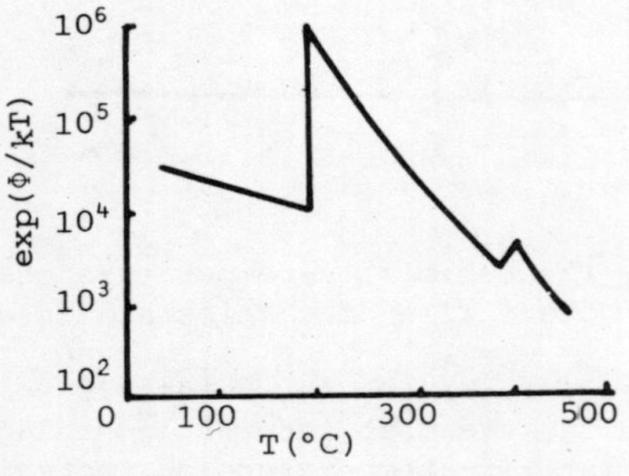

Fig. 6. The relation between surface barrier and temperature calculated from eqs. (1) and (2).

Figure 7. It is well-known in semiconductor physics that the resistivity-temperature character of doped semiconductors is rather complicated.

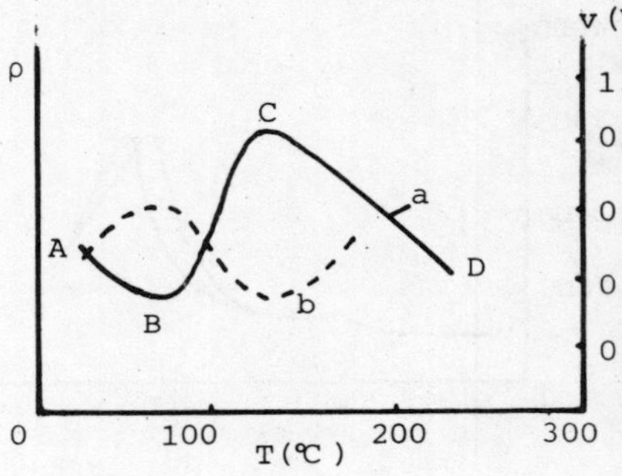

Fig. 7. The relation between (a) p-type impurity semiconductor a resistivity (b) temperature differential electromotive force and temperature.

This is partly because the coexistence of ionized impurities and intrinsic carriers, and partly because the coexistence of ionized impurity scattering and lattice scattering. As shown in Figure 7, from A to B is the low temperature district, where intrinsic excitation can be omitted. So the carriers are mainly ionized vacancies the number of which increases with the temperature, so does their mobility. As a result, the resistivity falls with the rising of the temperature. In BC district, with the further ascending of temperature, both the vacancies and donor impurities are ionized, the latter will release electrons, which may recombine with the vacancies and cause the decrease of the carrier concentration and the increase of resistivity. At a certain temperature (120∿140°C), p = n, the resistivity reaches its maximum. The declining of the resistivity in CD is because the further ionization of donor impurities.

In the positron annihilation experiment, it is proved that there are potassium and sodium vacancies in Mo-doped (Nb, K)NbO_3 semiconducting ceramics at room temperature. The ionization of K and Na vacancies is the source of vacancy carriers of p-type conductivity:

$$V^{\times}_{K,Na} = V'_{K,Na} + h' \quad (3)$$

Because there is MoO_3 doped in (Na, K)NbO_3, Mo^{6+} ion enter the place of Nb^{5+}, To keep elect electrically neutral, the Nb^{5+} is reduced to Nb^{4+}, that is:

$$(Nb_{1-x}K_x)NbO_3 + yMo^{6+} \rightarrow Na_{1-x}K_x(Nb^{5\pm}_{1-2y}Mo^{6+}_y Nb^{4+}_y)O_3 + yNb^{5+} \quad (4)$$

where Nb^{4+} represents Nb^{5+}. e. This weakly binded electron is the source of free electrons.

Based on analysis above, we can see that Mo-doped (Na, K)NbO_3 ferroelectric semiconducting ceramics has not only surface barrier effect on the grain boundary, but also the changing process from p-type to n-type conductivity. As a result, the leaping point of resistivity on the ρ(T) curve shifts to a lower temperature position.

As to the very small change of resistivity near 430°C on the ρ(T) curve, maybe it is because the high temperature has made the ceramics in the state of intrinsic excitation. When heated, the trapped electrons in grain boundaries have been excited, which has caused the fall of surface barrier. Though there are dielectric anomaly and the disappearance of spontaneous polarization near the Curie point, the barrier leap caused by them is not very big, so the resistivity change is not great either.

REFERENCES

1. Lcastet-Mejean, J. Phys. C: Solid State Phys. 19 (1986), 1637.
2. G. Shirane, R. Newnham and R. Pepinsky, phys, Rev., 96 (1954), 581.
3. R. Hdungan and R. D. Golding, J. Am. Ceram. Soc., 48 (1965), 601.
4. G. H. Jonker, Solid-State Electronics Pergamon Press., 7 (1964), 895.
5. O. I. Prokopalo et al, Ferroelectrics Letters, 2 (1984), 1.

THE EFFECTS OF GLASS VISCOSITY AND RuO_2 PARTICLE SIZE ON THICK FILM RESISTOR PROPERTIES*

R.W. Vest, B.S. Lee and S. Vasudevan
Purdue University
W. Lafayette, IN, USA

ABSTRACT

The system of three different particle size RuO_2 powders and lead borosilicate glass with varying amounts of Al_2O_3 was investigated. Addition of alumina to the glass increased the viscosity from 3.5 to 52.1 Pa·s at 850°C. With a fixed particle size RuO_2, the room temperature sheet resistivity went through a maximum as the 850°C glass viscosity was increased for low RuO_2 contents (high value resistors). At higher RuO_2 contents, the room temperature sheet resistivity only decreased as the viscosity of glass increased, and the magnitude of the effect decreased with increasing volume fraction of RuO_2. The conductive particle size effect on the magnitude and the temperature dependence of resistivity for RuO_2 thick film resistors was studied with a fixed glass composition. A model was developed to correlate the dependence of sheet resistivity on glass viscosity, particle size and the firing conditions. This model is based on the microstructure development processes of chain formation, conductive sintering, and chain rearrangement. One conclusion which resulted from modeling the particle size effects was that percolation theory is applicable only to very high value resistors. The temperautre dependence of resistance was modeled using a multi-path approach to simulate the chain microstructure. The naturally occurring distribution of sintered and non-sintered contacts of conductive particles in the chains is shown to be the primary contributor to the small temperature dependence characteristic of thick film resistors.

* This work was supported in part by the National Science Foundation under grant DMR-8507908 and by the Basil S. Turner Foundation.

INTRODUCTION

During the firing of RuO_2 - based resistors, a microstructure develops in which the conductive phase rearranges itself into three dimensional chains, and some of them may be continuous throughout the matrix. The conditions at the peak firing temperature are suitable for the development of a highly connected three dimensional network which meanders through the insulating matrix. Since the electrical conduction takes place through this conducting network, the observed electrical properties are due to the microstructure. Vest [1] proposed that the major processes in microstructure development during thick film resistor (TFR) processing are glass sintering, glass spreading and infiltration, microrearrangement, glass densification and liquid phase sintering of conductives. Over the past decade, studies [2-7] on the contributions of each process to the microstructure development have been conducted. These studies developed the kinetic equations pertinent to the above listed processes and determined the dominant prarameters for these important stages in microstructure development. It was established that the glass viscosity affects kinetics of each step in the microstructure development. It was also established [8] that the small RuO_2 particles or agglomerates segregate to the intersticies between the larger glass particles prior to firing, and that the degree of segregation varies with the volume fraction of RuO_2 and the particle size of RuO_2. Therefore, this study considered the effects of glass viscosity, volume fraction and RuO_2 particle size on the electrical properties of TFR's.

EXPERIMENTAL

Materials and Processing

In order to see the effects of glass viscosity, volume fraction and particle size of RuO_2, all other processing parameters were kept constant. Glasses of similar composition with different viscosities were obtained by addition of different amounts of alumina to a standard lead borosilicate glass. It is impossible to get glasses with the same composition and different viscosities, but by making changes in the amounts of a major constituent it is possible to get glasses with similar chemistry and different viscosity. Because all the major constituents were maintained at the same ratio except for alumina, any effects due to glass chemistry were minimized. The materials system used was as follows:

- Conductive: RuO_2(SSA = 30, 50 and 85 m^2/g)
- Glasses: 63% PbO - 25% B_2O_3 - 12% SiO_2 modified with varying amount of Al_2O_3
- Screening agent: 10% ethyl cellulose (N-300) dissolved in diethylene glycol monobutyl ether (butyl carbitol)
- Substrates: 96% Al_2O_3 (AlSiMag 614)
- Terminations: Engelhard A-3770, DuPont 7712

The glass frit was prepared by mixing appropriate amounts of oxides, melting them at a high temperature to get a homogeneous liquid, and friting it into deionized water. The glass powder was prepared by grinding the frit and ball milling for about 8 hrs followed by sieving to -325 mesh. The average glass particle size was 4.7 μm. The appropriate amounts of glass and RuO_2 were initially mixed until a uniform grey color was obtained. Screening agent was then added and the ink was blended on a glass sheathed roll mill for 30 min. The resistor inks were screen printed using a 200-mesh stainless steel screen, and the conductors were printed and fired on the substrate prior to the printing of the resistor inks. The prints were allowed to level at room temperature for 30 min. and then dried at 150°C for about 20 min. and 350°C for 30 min. The resistors were heated at 40°C/min. from room temperature to 850°C and cooled to room temperature at the same rate after holding at the peak firing temperatures for 12 min. The thickness of each resistor was measured so that the experimentally determined sheet resistance could be normalized to a thickness of 25 μm.

Viscosity Measurement

The high temperature viscosity of the glasses as a function of temperature was determined by the sphere method [9]. The velocity (v) of an inert sphere in a liquid due to an applied load (L) is related to the viscosity (η) of the liquid by

$$\eta = \frac{LgF_a}{3v\pi d} \quad (1)$$

in which g is the gravitational acceleration, F_a is the Faxen correction factor equal to $[1 - 2.104\,(d/D) + 2.09\,(d/D)^3 - 0.95\,(d/D)^5]$, d is the diameter of the sphere, and D is the diameter of the crucible. A gold sphere was used in this work due to its inertness towards the glasses and air.

The gold sphere, 9 mm in diameter, was suspended from the sample pan of an automatic recording microbalance by means of a platinum wire. Beam displacements were sensed by an LVDT, whose signal was proportional to the beam displacement, and the amplified signal drove a chart recorder. The glass contained in a platinum crucible was placed in a furnace which could be positioned vertically by means of a jack. The glass temperature was measured by a Pt/Pt + 10% Rh thermocouple with the hot junction immersed in the glass.

Temperature Dependence of Resistance Measurements

A computer controlled experimental system was used to collect resistance data as a function of temperature. The system consisted of a cryostat, a sample holder assembly, a temperautre controller, a computer, and a data acquisition system. The sample holder assembly was completely immersed into liquid nitrogen in a cryostat, cooled to the lowest possible temperature, and then slowly heated. The temperature controller used two different platinum sensors for controlling and measuring the temperature of the sample. The data acquisition system measured the resistance at each preset temperature, and the computer controlled all the devices connected to the system through an IEEE 488 bus. The system was capable of measuring resistance from 78 K to 500 K, and was programmed to reach an equilibrium temperature before the resistance values were measured. The four terminal technique was used in resistance measurements to compensate for lead wire resistance. The sample chamber was kept in a helium atmosphere to avoid any moisture condensation on the sample. At each set temperature, the resistance value stored on the disc was an average of 20 readings.

RESULTS AND DISCUSSION

Glass Viscosity Effects

The measured viscosities of glasses with different amounts of Al_2O_3 at 850°C are shown in Fig. 1. The logarithm of the viscosity at 850°C changes linearly with the amount of alumina added, and a least square fit showed that the relation was $\log \eta = 0.54 + 0.084$ P, where the unit of viscosity (η) is Pa.s and P is the amount of alumina dissolved in weight percent. The fitted viscosity data at 850°C are shown in Table 1.

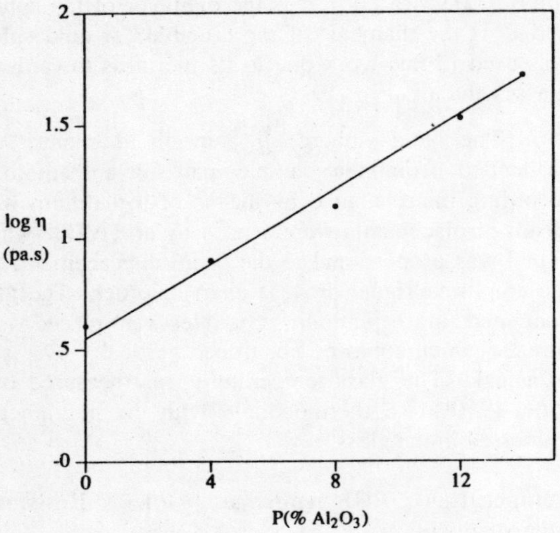

Figure 1. Variation of glass viscosity with alumina content at 850°C.

Table 1. Change of glass viscosity with alumina content (P) at 850°C.

P	0	2	4	6	8	10	12	14
η(Pa.s)	3.5	5.1	7.5	11.1	16.3	24.0	35.3	52.0

At the peak firing temperature of 850°C the viscosity changed from 3.5 Pa.s for the standard glass to 52.0 Pa.s for the glass containing 14 weight percent Al_2O_3. This change in η by a factor of 15 is large enough to show the effect of viscosity on microstructure development and the electrical properties of fired films.

The sheet resistivities for resistors with different glasses and with different volume percents of RuO_2 are shown in Fig. 2 as a function of glass viscosity. The sheet resistivity data are the average of three different processing runs. The general trend of the data shown in Fig. 2 is a decreasing sheet resistivity with increasing viscosity of the glass, and the change in resistivity is more prominent at the lower volume percents. At 3 volume percent RuO_2 there was a maximum of sheet resistivity at lower glass viscosities that was significantly greater that the experimental scatter of the data; Fig. 3 shows an expanded view of this maximum. At higher glass viscosities for the 3 v/o resistors, ρ_s decreased as η increased while the opposite behavior was observed at lower viscosities. The general features of the results shown in Figs. 2 and 3 can be explained by considering the relationship between glass viscosity and the various microstructure development processes in TFR's.

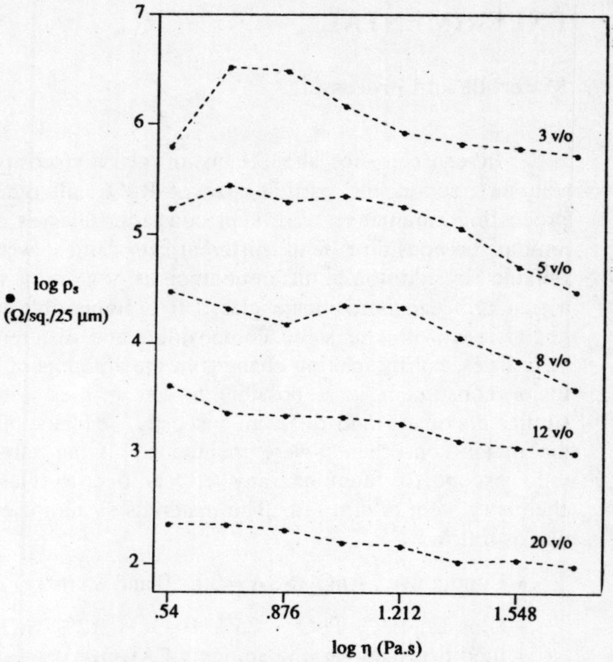

Figure 2. Glass viscosity dependence for ρ_s for different volume percents of RuO_2.

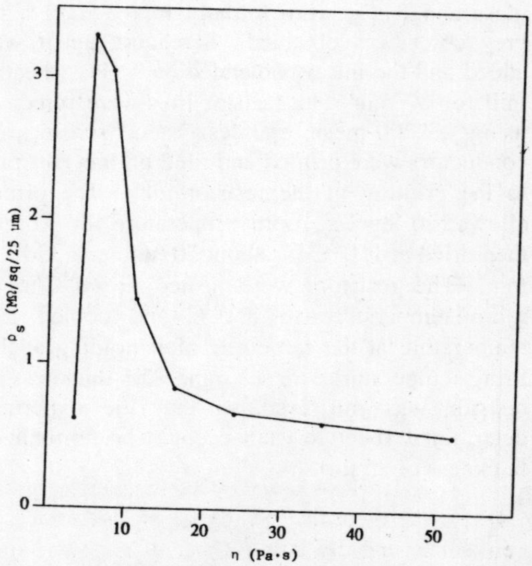

Figure 3. Glass viscosity dependence of sheet resistivity for resistors with 3 v/o RuO_2.

It was initially believed that a higher glass viscosity at the processing temperature would retard the microstructure development kinetics and increase the electrical resistivity of the TFR's because it was reported [7] that the time required for equivalent RuO_2 neck growth during initial stage sintering in the glass of higher alumina content was longer than that in the glass with lower alumina content. The initial assumption is true for the sintering of conductive particles, but the overall microstructure development and conductive network formation, which control the electrical properties, may vary with viscosity of the glass in a different way.

At temperatures above the glass sintering stage, the final resistivity is most strongly influenced by the motion of individual conductive particles (or particle agglomerates) to and from the conductive chains, and the sintering of conductive particles within the chains. It is believed that Brownian motion is the primary driving force for the motion of individual particles or agglomerates in the glass. The distance (x) that a particle of radius r will move during time t in a liquid of viscosity η at temperature T is given by [10]:

$$x = \left[\frac{kT}{3\pi\eta r}\right]^{1/2} t^{1/2} \qquad (2)$$

If it is assumed that the particles are RuO_2 agglomerates of radius 0.1 μm, which is in agreement with SEM and TEM observations using 30 m^2/g RuO_2, and taking the firing temperature of 850°C, Eq. 2 gives the Brownian motion velocities of 69 nm/s for the SO glass and 18 nm/s for the S14 glass. Considering the firing time of 12 minutes at 850°C, Brownian motion is certainly a significant contributor in moving the particles to and from the chains. Whether or not a particle with leave a chain depends on the size of neck that has developed by sintering to an adjacent particle.

These microstructural development considerations lead to opposing predictions as to the dependence of resistivity on glass viscosity. A higher glass viscosity retards the sintering [7] and reduces the rate at which individual particles join the chians, both of which will increase the resistivity. However, a higher glass viscosity will reduce the rate at which individual particles leave the chains, which will decrease the resistivity. The balance of these effects leads to the behavior shown in Figs. 2 and 3.

RuO_2 Particle Size Effects

It has been shown by various investigators [11-14] that the electrical properties of TFRs such as sheet resistance and TCR not only change with volume fraction of the conductive (V_c), but also with particle size of both the conductive (R_c) and glass (R_g), and with processing conditions such as time and temperature of firing. A study [8] of microstructure development during firing a series of RuO_2 conducting phase and lead borosilicate glass thick film reistors using different particle sizes and volume fractions of RuO_2 identified five important features: 1) agglomeration of high surface area RuO_2 powders; 2) varying degrees of coverage of the large glass particles by the small RuO_2 particles or agglomerates; 3) segregation of the small RuO_2 particles or agglomerates to the intersticies between the larger glass particles; 4) formation of glass islands, 3-5 times the mean diameter of the initial glass frit; and 5) formation of rings of RuO_2 in the glass matrix. These features of microstructure development were analyzed in terms of a coverage factor, which was defined as the ratio of projected area of conductive phase particles to surface area of glass. The coverage factor (F) was written in terms of the particle size of glass, the particle size of conductive phase particles or agglomerates, and the volume fraction of the conductive phase particles or agglomerates to give

$$F = \frac{R_g}{4R_c}\left(\frac{V_c}{1-V_c}\right) \qquad (3)$$

It was determined that a ring like microstructure of RuO_2 conducting phase was developed for values of F greater than 0.24, and further that the distribution of RuO_2 ring sizes was a near perfect replica of the initial particle size distribution of the glass used in the ink. A second critical value of F was found to be 0.083, because F values below this gave unstable and mostly discontinuous resistors. Resistors with F values between these two critical values had changing microstructure with the distribution of RuO_2 ring sizes greater than the initial glass particle size distribution.

Simple percolation theory [15] predicts that the sheet resistivity should be given by

$$\rho_s = \rho_o(p-p_c)^{\mu} \qquad (4)$$

where ρ_o is a normalizing constant, p_c is the percolation threshold and equals 0.24, and μ is the critical exponent and equals 1.7. For a system containing equal size particles of conducting and insulating phases, $p = V_c$, but for differing particle sizes is given by [16]:

$$p = \frac{V_c}{V_c + (1-V_c)\frac{R_c}{R_g}} \qquad (5)$$

Combining Eqs. 3 and 5 gives

$$p = \frac{4F}{4F+1} \qquad (6)$$

Equation 4 is valid only for the range $p_c < p < p_c + 0.2$ [15]. Using the theoretical value of $p_c = 0.24$ [15] and Eq. 6, leads to the conclusion that percolation theory is valid only for the range $0.079 < F < 0.196$. The experimental observation of unstable microstructure for F values less than 0.083 is in good agreement with the predicted percolation threshold ($F = 0.079$), and the experimental value of $F = 0.24$ to produce resistors with the same microstructure of the conducting chains compares favorable with the upper limit of applicability of percolation theory ($F = 0.196$). Ruthenium dioxide based resistors with F values less than 0.2 have sheet resistivities greater than 10^6 ohm/sq/25 μm, which means that almost all thick film resistors are operating well above the percolation threshold in the region where percolation theory does not apply.

A decrease in particle size of RuO_2 decreased ρ_s drastically at small V_c values, but the effect was negligable for $V_c = 0.2$. The effect of RuO_2 particle size on the sheet resistivity of resistors made with the P = 14 glass is shown in Fig. 4 for $V_c = 0.046$. The ρ_s data are plotted versus the reciprocal of the RuO_2 specific surface area (SSA), which is proportional to particle size. The coverage factor (Eq. 3) increases as R_c decreases and as V_c increases, and ρ_s is most sensitive at low F values. Therefore, ρ_s should be most sensitive to changes in R_c at low values of V_c, which was the behavior observed.

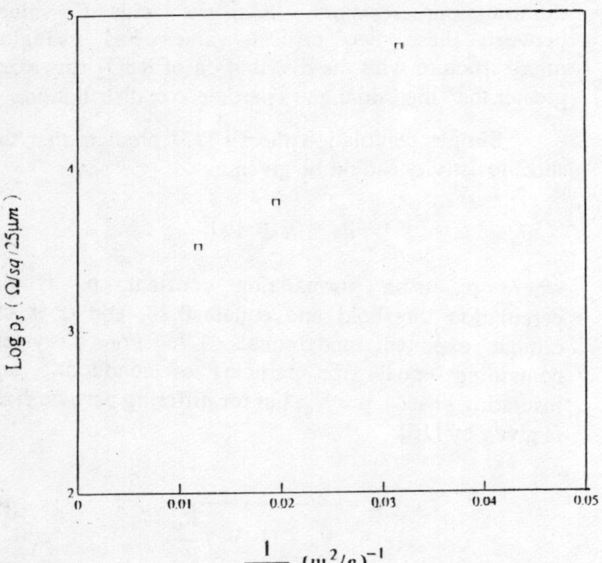

Figure 4. The variation of log ρ_s with inverse of specific surface area for resistors with 4.6 v/o RuO_2.

A typical normalized resistance ($R/R_{25°C}$) versus temperature (T) curve showed a shallow minimin. The temperature of the resistance minimum (T_m), the sheet resistivity at T_m, and the TCR values were functions of the RuO_2 content and the particle size of RuO_2 at fixed glass composition. With an increase in the amount of conductive, the $R/R_{25°C}$ versus T behavior changed in such a way that T_m moved towards lower temperatures and the TCR became more positive. With an increase in volume fraction of RuO_2, the density of the conductive on the surface of the glass particles prior to firing increases and the interstices between the glass particles contain more RuO_2 particles. As a result, the available number of conductive particles to form conducting chains early in the firing is higher, and more sintered contacts can be developed. This results in a decrease of T_m and also an increase in TCR. For larger particle size RuO_2 and low volume fraction (0.03) of the conductive, T_m was well above 500 K giving more negative TCR values, while for a small particle size and high volume fraction ($V_c \geq 0.062$) of the conductive, T_m was very low giving a high positive TCR values. To compare particles size effects on temperature dependent behavior, 0.046 volume fraction of the conductive was selected.

The effect of conductive particle size on $R/R_{25°C}$ versus T behavior is shown in Fig. 5 for resistors containing 0.046 volume fraction of RuO_2 and the

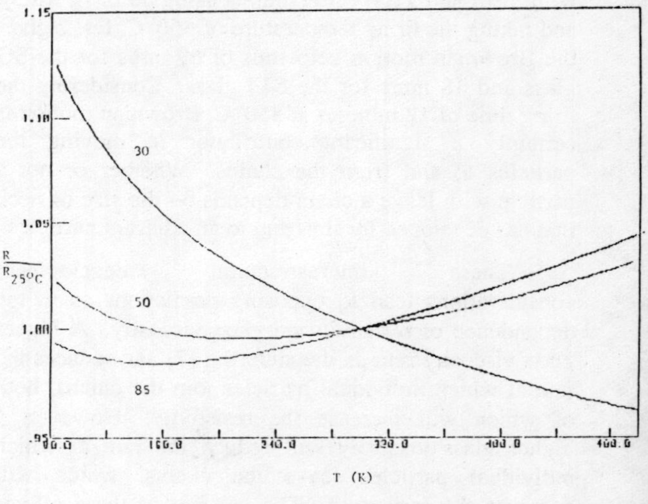

Figure 5. The effect of conductive particle size on temperature dependence of resistance for resistors with 4.6 v/o RuO_2.

P=12 glass. With a decrease in particle size, T_m decreased and the TCR became more positive. For the same volume fraction and processing conditions, the number of available conductive particles to form conducting chains increases with a decrease in particle size. In addition, the density of the conducting particles on the surface of the glass particles prior to firing is higher, which means that the interparticle distance between two particles which can form a sintered contact is lower. Hence, the number of conducting paths that can be developed in the glass matrix is much higher for smaller particles. As a result, T_m decreases and TCR becomes more positive. For all the volume fractions and processing conditions studied, T_m was lower and hence the TCR was always higher for smaller particles than for larger particles.

The data for the 30 m^2/g RuO_2 resistors in Fig. 5 can be fit reasonably well to an exponential temperature dependence. Such a fitting yields an activation energy of 18 meV, which is typical of values reported by other investigators [17] at temperatures below T_m. It was these small activation energies that convinced the researchers to discard mechanisms such as Schottky emission, impurity conduction and space charge limited flow for electron transport through the thin glass region in non-sintered contacts between conducting particles. The microstructure of TFRs show complex networks of sintered and non-sintered contacts in the insulating matrix [4]. Vest, et.al. [18] proposed a multi path model for TFRs in which the dominant current carrying paths can be viewed as a set of parallel resistors. Each parallel path can be considered to consist of resistors with different numbers of sintered and non-sintered contacts. The equivalent resistance of these paths and its temperature dependence is a function of number of sintered and non-sintered contacts and also the geometry of the contacts and the charge transport mechanisms. The complex network of resistors present in a TFR was simplified by considering a set of most dominant conduction paths, with each conduction path containing different numbers of metallic and insulating units modeled as shown in Fig. 6. The charge transport through the RuO_2 particle and the sintered contact is metallic, and through the non-sintered contact it may be any of a number of mechanisms. For a simple network of resistors, the temperature dependence of resistance was calculated using seven parallel conductances, each of which contained 10^6 sintered contacts. As a simple case, the charge transport through the non-sintered contact was assumed to be impurity conduction with an activation energy of 0.455 eV. The number of non-sintered contacts was increased by an order of magnitude for each parallel conductance so that the first had one non-sintered contact for each 10^6 sintered contacts and the seventh had an equal number of sintered and non-sintered contacts. The results of this calculation over a 70°C interval gave a good fit to an exponential temperature dependence with an activation energy of 17 meV, which is in good agreement with the experimental value for the 30 m^2/g RuO_2 resistors in Fig. 6. This calculation shows that a charge transport process with a high activation energy (455 meV) can result in a very small apparent activation energy (17 meV) when the interconnected microstructure is considered. Hence, charge transport mechanisms such as Frenkel-Poole emission, Schottky emission, impurity conduction, and space charge limited flow near the trap filled limit may be important for the non-sintered contacts.

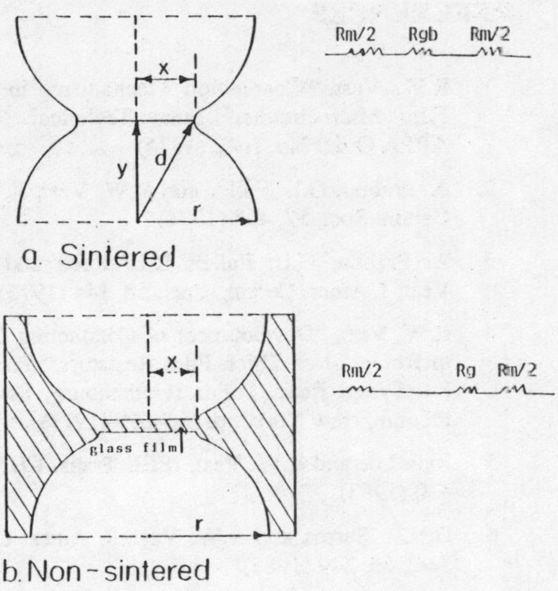

Figure 6. The possible electrical contacts.

SUMMARY

It is well known that the physics and chemistry of the glass used play an important role in thick film resistor processing. The viscosty of the glass is the physical property which influences every step of microstructure development of thick film resistors. Even though the glass viscosity retards the sintering kinetics of RuO_2 conductives, the overall microstructure development and network formation are favored for higher of glass viscosities.

All the changes in electrical properties observed with different particle sizes of RuO_2 were correlated with a proposed microstructure development model. A multi-path conduction mechanism model was invoked to correlate the temperature dependence of resistance at temperautres below the minimum in the R versus T plots, and it was shown that charge transport mechanisms with high activation energies can give a much lower measured activation energy.

REFERENCES

1. R.W. Vest, "Conduction Mechanisms in Thick Film Microcircuits," Final Technical Report, ARPA Order No. 1642 (1975).

2. A. Prabhu, G.L. Fuller and R.W. Vest, J. Amer. Ceram. Soc., 57, 408 (1974).

3. A. Prabhu, G.L. Fuller, R.L. Reed and R.W. Vest, J. Amer. Ceram. Soc., 58, 144 (1975).

4. R.W. Vest, "Development of Conducting Chains in RuO_2 Glass Thick Film Resistors," Proc. 8th Int. Symp. Reac. Solids (Gothenburg, Sweden), Plenum, New York, pp. 695-699 (1977).

5. Joon Lee and R.W. Vest, IEEE Trans. CHMT, 6, 430 (1983).

6. D.H.R. Sarma and R.W. Vest, J. Amer. Ceram. Soc., 68, 249 (1985).

7. P. Palanisamy, D.H.R. Sarma and R.W. Vest, J. Am. Ceram. Soc., 68, C215 (1985).

8. S.M. Chitale and R.W. Vest, IEEE Trans. CHMT, 11, 604 (1988).

9. T. Shartis and S. Spinner, J. Res. NBS., 46, 176 (1951).

10. E.A. Moelwyn-Hughes, "Physical Chemistry," 2nd Ed., Pergamon Press, Oxford (1961).

11. T. Inokuma, Y. Taketa and M. Haradome, IEEE Trans. CHMT, 7, 166 (1984).

12. T. Inokuma, Y. Taketa and M. Haradome, IEEE Trans. CHMT, 8, 372 (1985).

13. P.F. Carcia, A. Ferretti and A. Suna, J. Appl. Phys., 53, 5282 (1982).

14. C. Schaffer and J. Sergent, "The Effect of Particle Size Distribution on the Electrical Properties of RuO_2 Thick Film Resistors," Proc. Int. Microelectronic Conf., pp. 60-64 (1977).

15. S. Kirkpatrick, Rev. Mod. Phys., 45, 574 (1973).

16. D.P.H. Smith and J.C. Anderson, Phil. Mag. B, 43, 497 (1981).

17. G.E. Pike and Seager, J. Appl. Phys., 48, 5152 (1977).

18. R.W. Vest, S.M. Chitage and A.K. Kollipara, "The Dependence of Charge Transport on Microstructure of Thick Film Resistors," Proc. 5th European Hybrid Microelectronics Conference, Stressa, Italy, May 22-24, pp. 406-412 (1985).

THERMOELECTRIC PROPERTIES OF AMORPHOUS SILICON ALLOYS

S. YUGO, H. WATANABE, J. YAMAKAWA AND T. KIMURA
UNIVERSITY OF ELECTRO-COMMUNICATIONS
1-5-1 CHOFUGAOKA, CHOFU-SHI, TOKYO 182, JAPAN

ABSTRACT

Electrical conductivity and thermoelectric power were investigated for amorphous silicon alloys with high contents of boron and phosphorus (Si-B, Si-P and Si-B-P) prepared by chemical vapor deposition (CVD). Si-B and Si-P showed semiconductor-like behaviours in the temperature dependence of conductivity between room temperature and 800 °C. The thermoelectric power was positive for Si-B and negative for Si-P. With increasing temperature, their absolute values increased in the low temperature region, reached maxima around 400 °C and then decreased sharply above that temperature. Si-B-P alloys which were prepared by plasma CVD also showed semiconductor-like behaviours in the temperature dependence of resistance with a negative sign of thermoelectric powers, whereas those prepared by thermal CVD showed metallic behaviours with positive thermoelectric powers. In spite of different behaviours in conduction, both Si-B-P alloys were characteristic of steadily increasing thermoelectric powers up to 800 °C in contrast to Si-B or Si-P.

1. INTRODUCTION

Thermoelectricity has been investigated on various kinds of materials, but high efficiency sufficient for practical use have not yet been obtained for any materials. Some reports on peculiar electrical behaviours of a-Si and its compounds [1] have been stimulating studies on the thermoelectricity of these amorphous materials. These peculiar behaviours may be due to their crystalline as well as structural complexities and anisotropies which are highly dependent on preparation conditions and subsequent annealing. They may be also affected strongly by impurities which can be incorporated into host materials without any solubility limit to make compounds (alloys). For instance, amorphous Si-Ge has been known to give a relatively high figure of merit (about 0.9×10^{-3}/K [2]) in the temperature range of several hundreds centigrade. This figure of merit was further increased to 1×10^{-3}/K by adding GaP into Si-Ge [3]. In a previous paper, we reported on a high thermoelectric power on boron phosphide refractory semiconductors prepared by CVD [4]. Its large bandgap energy (about 2 eV) gave rise to a high transition temperature between the impurity- and intrinsic-regions, resulting in a high thermoelectirc power up to high temperatures. Based on the above background, this study concernes with the thermoelectric properties of a-Si alloys with high contents of B and P prepared by CVD.

2. EXPERIMENTAL PROCEDURE AND MEASUREMENTS

Amorphous silicon alloys composed of different B and P contents were deposited on alumina glass substrates by rf plasma decomposition of a gaseous mixture of 10% SiH_4/H_2 with various quantities of added dopants (5% B_2H_6/H_2 and 5% PH_3/H_2). An rf power of about 200 W was applied to the gaseous mixture held at a pressure of $10^{-1} \sim 10^{-2}$ Torr. The substrate tempeature was considered to be about 100 °C during the deposition. After the deposition, the samples were heat-treated at 900 °C for 15 min. in argon atmosphere. For comparison, Si-B-P alloys were also prepared by thermal CVD at a substrate temperature of 900 °C. In Table 1 are shown the specimens used in this experiment. The samples were provided with two gold electrodes, two silver contacts and alumel-chromel thermocouples. A schematic diagram of our thermoelectricity measurement apparatus is shown in Fig.1. Samples were placed in a vacuum chamber and heated with infrared light which was introduced from outside the chamber via a quartz rod. The temperature difference for the measurement of thermoelectricity was obtained by sliding a shield plate set infront of the samples, resulting in a nonuniform irradiation. The sample mid-temperature was changed from room temperature up to 800 °C and the temperature difference was kept constant at 5 °C. By this arrangement, electrical noise or potential fluctuation usually encounterd in case of an electrical heater was avoided and a relatively small thermoelectric power was able to measure with a good accuracy even at high temperatures. In this experiment, the voltage fluctuation at 800 °C was suppresed below ± 10 μV.

Table 1 The gaseous impurity ratio used to prepare the samples.

Sample	SiH_4/H_2 10%(SCCM)	B_2H_6/H_2 5%(SCCM)	PH_3/H_2 5%(SCCM)
1 (plasmaCVD)	100	2	
2 (plasma CVD)	100		4
3 (plasma CVD)	100	5	1
4 (thermal CVD)	100	5	4

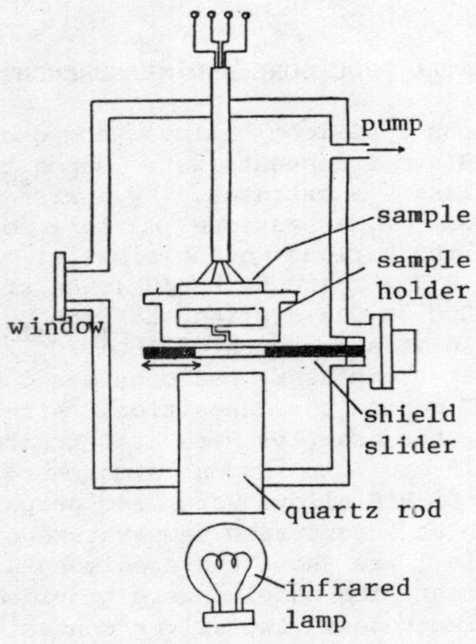

Fig. 1 Apparatus for measuring Seebeck coefficients

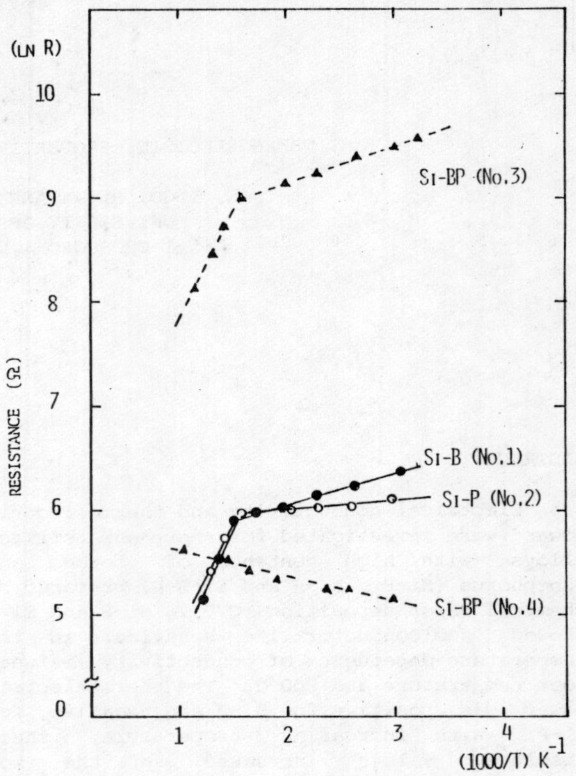

Fig. 2 The logarithmic variation of electrical resistance as a function of reciprocal temperature for Si-alloy.

3. RESULTS AND DISCUSSION

Figure 2 shows Arrhenius plots of the electrical resistance of four representative samples shown in Table 1. No.1(Si-B) and No.2(Si-P) samples showed resistances which, with increasing temperature, decreased slowly from room temperature to about 400°C and then sharply above 400°C, indicating semiconductor-like behaviours. The activation energies were 0.02-0.03 eV and 0.4 eV below and above the kink point, respectively. No.3 (Si-B-P) samples, though very high resistive, showed also semiconductor-like behaviours with almost the same activation energies as those of Si-P and Si-B. No.4 (Si-B-P) samples which were prepared by thermal CVD showed, however, metallic behaviours.

The temperature dependence of the thermoelectirc power of these samples were shown in Fig.3. No.1 (Si-B) samples showed positive thermoelectric powers, whereas No.2 (Si-P) samples showed negative values. With increasing temperature, their absolute values

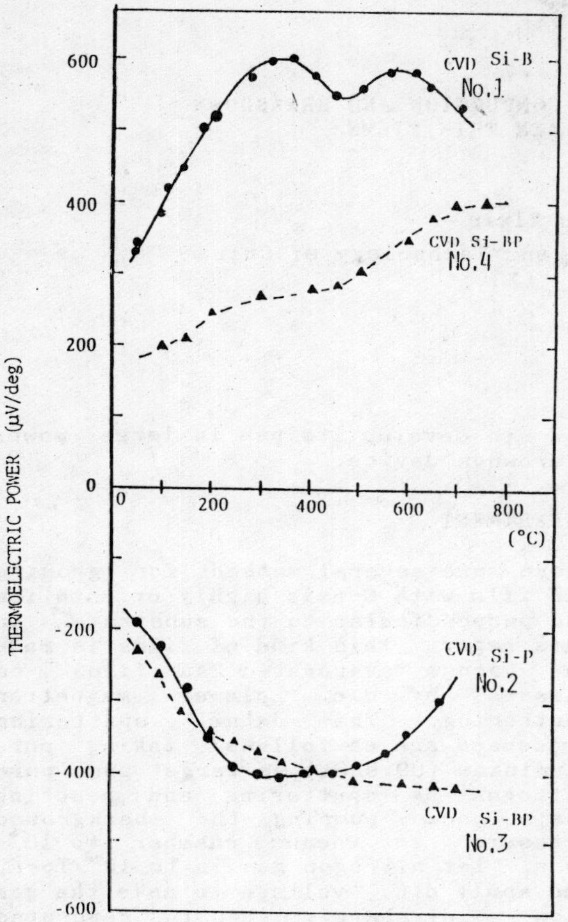

Fig. 3 The thermoelectric power as a function of temperature for Si-alloy.

increased first, showed maxima and then decreased. These results may be understood such that the increase in the low temperature region would be caused by the increase in the ionized impurity scattering relative to the neutral impurity scattering and that the sharp decrease in the high temperature region would be due to the ambipolar contribution in the intrinsic region. However, the temperature at the maximum did not coincide with the kink poits in the resistance vs. temperature curves. Moreover, their detail behaviours such as a presence of two maxima can not be at present fully understood. The temperature dependences of thermoelectric powers of No.3 and No.4 (Si-B-P) samples were quite different from those of Si-P and Si-B. Though the sign was differnet between No.3 and No.4, the absolute values of their thermoelectric powers increased steadily up to 800 °C without no hints of decrease. The above inexplicable results might suggest a presence of unknown conduction mechanisms peculiar to these amorphous alloys. In conclusion, the results of Si-B-P alloys are very suggestive in developing amorphous alloys of high thermoelectric powers at high temperatures.

REFERENCES

[1] For instance,
W.E.Spear and P.G.Le Comber, Phil. Mag. 33 (1976) 935, D.I.Jones, P.G.Le Comber and W.E.Spear, Phil. Mag. 36 (1977) 541

[2] J.P.Dismukes, L.Ekstrom, E.F.Steigmeier. I.Kudman and D.S.Beers, J. Appl.Phys. 35 (1964) 2899

[3] J.W.Vandersand, C.Wood and S.Draper, Mat. Res. Soc. Symp. Proc. vol.97 (1987) 347

[4] S.Yugo and T.Kimura, phys. stat. solidi(a) 59 (1980) 363

INVESTIGATION OF D. C. ELECTRIC CONDUCTION AND BREAKDOWN OF PIEZOELECTRIC AlN THIN FILMS

Liufu De*, and Qu Xixin
University of Electronic Science and Technology of China
Chengdu, China

ABSTRACT

The purpose of this report is to introduce the d. c. electric conduction and breakdown properties of piezoelectric AlN thin films grown by d.c. planar magnetron sputtering. The experimental results show that the d. c. electric conduction mechanism of the films under strong electric field is that of Poole-Frenkel type. The breakdown electric field strength of 3.4-3.9MV/cm are obtained from the statistic mean on the experimental data for the film thickness ranging from 0.6 -1.6um. Furthermore, the breakdown strength decreases with the increasing film thickness.

INTRODUCTION

Piezoelectric AlN thin film, with its large electromechanical coupling coefficient, stable chemical properties and good insulating characteristics, is a kind of high quality electronic meterial. Because the longitudinal acoustic velocity of AlN doubles that of piezoelectric ZnO, it is a promising material as large power microwave transducers in microwave acoustics. So far, too little of breakdown electric field strength (Eb) of C-axis oriented AlN films is known. Because of the structural difference between amorphous and C-axis oriented crystallite films, one should not use the amorphous Eb to design large power piezoelectric devices. Thus, this paper intends to investigate the d.c. conduction and breakdown properties of the C-axis oriented AlN film so as to determine the conducting mechanism under strong electric field and breakdown strength, and to develop its use in large power microwave device.

* Present address:
Box 8898, Xian Jiaotong University
Xian, China

EXPERIMENT

There are several methods for growing AlN film with C-axis highly orientation and perpendicular to the substrate[1-7]. In this report, this kind of films is made on vacuum evaporated Au films on glasses by d.c. planar magnetron sputtering. The detail sputtering processes are as follows: taking pure aluminium (99.999%) as target and pure nitrogen as sputtering and reacting gas; once pumping the background pressure in vacuum chamber to 10^{-6} Torr, let nitrogen gas in to 10^{-2} Torr, and apply d.c. voltage to make the gas glow discharge; then the generated high energy particles bombarding the target, and gradually growing AlN films on substrates. Prior to film deposition, the substrates were cleaned with organic solvents and deionized water. Aluminium is an active metal and easily oxidizes in usual conditions, which make the film contain impurity and colour. Hence, sputtering the target for half an hour prior to grow films would greatly improve the quality of the growing films. With small power sputtering, the deposition rate is too low and the film does not form with C-axis orientation. But with too large power sputtering, the grown films contain defects resulted from the high energy particles bombarding the formed film surface and damaging the film. From experiment, the optimum conditions are obtained for growing C-axis highly oriented AlN films as follows:

Substrate temperature	150-250°C
Distance from target to substrate	35-45mm
Gas pressure	3×10^{-3} to 3×10^{-2} Torr
D.c. sputtering voltage	320-400V

Sputtering current 350-500mA
Deposition rate 0.3-0.6μm/h

The film thickness is determined by the ellipsometry. X-ray diffraction is used to analyze the structure of the films. The currents are measured by electrometer. And the breakdown voltages are obtained from the indicator which functions when the leak currents of the sample are larger than that of the detecting limit by increasing the voltage continuously.

MEASUREMENT AND DISCUSSION

1. Structure of The Films

The structure of the films is determined from the X-ray diffraction spectrum using Cu-Kα radiation. Fig.1 shows the X-ray diffraction pattern of 1.67μm-thick AlN film deposited on vacuum evaporating Au film on glass. It proves that the film is C-axis highly oriented.

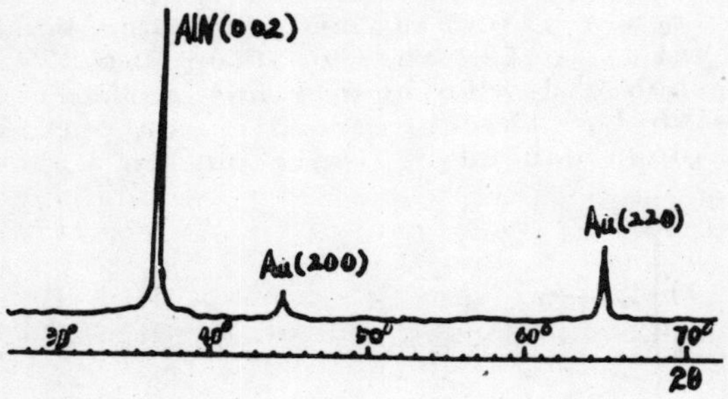

Fig.1 X-ray diffraction pattern of AlN/Au/Glass structure

2. D.C. Conduction Behaviors

For the 986Å thick film, the measured current-voltage are shown in Fig.2. It is demonstrated that in low electric field the currents are proportional to the voltages, i.e. behaving ohmic properties as in the region I; then the currents increase exponentially with the applied voltages, corresponding to the region II in the Fig.; with the further increasing of the voltages, the currents increase steeply, reaching to the critical breakdown state, as in the region III in Fig.2. Fig.3 draws out the current-field relationship from Fig.2, which exhibites that the logI is proportional to $E^{1/2}$ in the region II indicating that the electrical conduction mechanism may belong to either Schottky or Poole-Frenkel type. For Schottky type, the current density Js is expressed as

$$J_s = AT^2 \exp[(\beta_s E^{0.5} - \phi_o)/kT] \qquad (1)$$

where $\beta_s = (e^3/4\pi\varepsilon_o\varepsilon_i)^{0.5}$, A is the Richardson constant, T is the absolute temperature, E is the applied electric field, ϕ_o is the barrier height at the metal-insulator interface in the absence of field, e is the electronic charge, k is the Boltzmann constant, ε_o is the permitivity of free space, ε_i is the dielectric constant in high frequency, for AlN, $\varepsilon_i = 8.7$[8]. For Poole-Frenkel type, the current density Jf is expressed as

$$J_f = A'E^{0.5} \exp[(\beta_f E^{0.5} - \phi_i)]/kT \qquad (2)$$

where $\beta_f = (e^3/\pi\varepsilon_o\varepsilon_i)^{0.5}$. The experimental value obtained from Fig.3 is 4.18×10^{-24} which is nearly equal to the calculated β_f value of 4.11×10^{-24}. It is proved the eletric conduction mechanism of piezoelectric AlN films is Poole-Frenkel type. The carriers might probably come from impurities and/or nonstoichiometry.

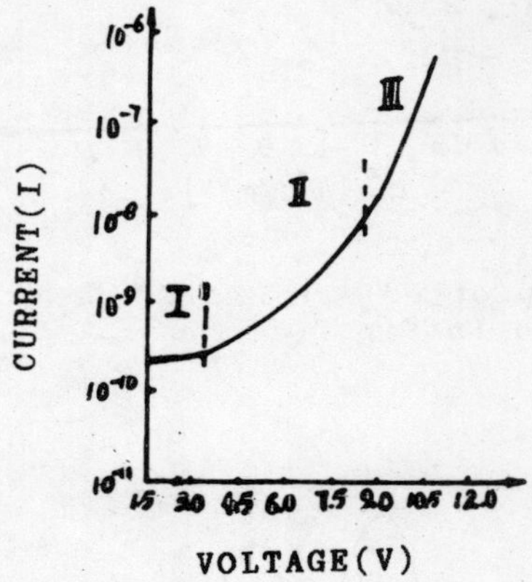

Fig.2 The curve of d.c. current against voltage for piezoelectric AlN film with the thickness of 986Å (Temperature 20°C)

3. Electric Breakdown Strength

The breakdown voltage of the Au/AlN/Au/Glass structure are measured by increasing voltage continuously. Fig.4 draws out the breakdown strength probability distribution of the specimens of a, b, c, and d. It is seen that the curves behave Gaussian distribution and can be analyzed with statistic method. The mean breakdown strength $\bar{E}b$ and the standard deviation δ

may be obtained from the expression as follows:

$$\overline{E}_b = \sum_{i=1}^{N} E_{bi}/N \qquad (3)$$

and

$$\delta = \left[\sum_{i=1}^{N} (E_{bi} - \overline{E}_b)^2/N\right]^{1/2} \qquad (4)$$

where Ebi is the value tested in each time, and N is the total numbers on testing. Table 1 lists the mean values of the breakdown strength and the standard deviation of the specimens, from which one sees that Eb is in the range of 3.44-3.91 MV/cm.

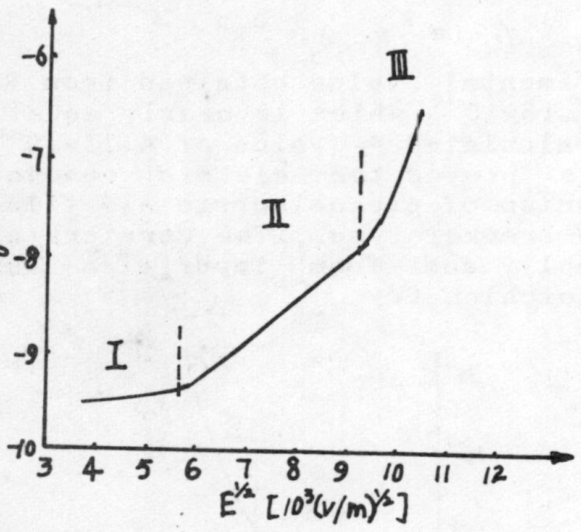

Fig.3 logI plotted against $E^{1/2}$ from the curve in Fig.2

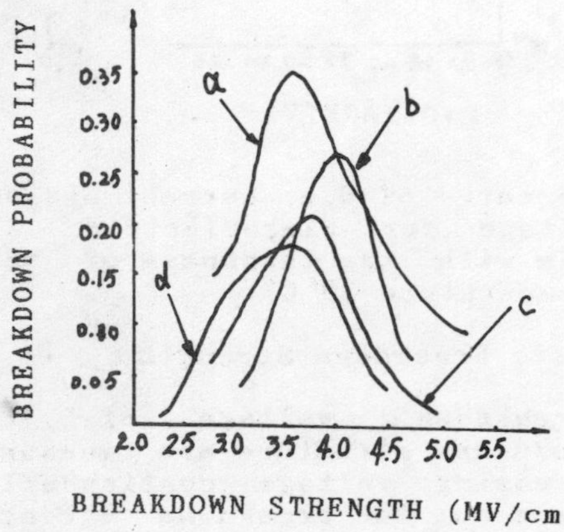

Fig.4 The breakdown probability distribution

Table 1 The statistic results on the measuring values of the breakdown strength

Specimen	Average $\overline{E}b$ (MV/cm)	Standard deviation (MV/cm)	Thickness (μm)
a	3.84	0.626	0.63
b	3.91	0.367	0.83
c	3.67	0.479	1.08
d	3.44	0.490	1.57

It is also seen from table 1 that the breakdown strength has something to do with thickness as shown in Fig.5. When the thickness (d) of the piezoelectric AlN films is large than d_D=0.83μm, the Eb decreases with the increase of d which corresponds to the breakdown theory proposed by Forlani et al[9] who pointed out that for large band gap dielectric, the breakdown is caused by electron avalanching and the breakdown strength meets the relation $Eb \propto d^{-1/2}$. But when $d<d_D$, Eb nearly has nothing to do with the thickness, which suggests a breakdown caused by ionic motion.

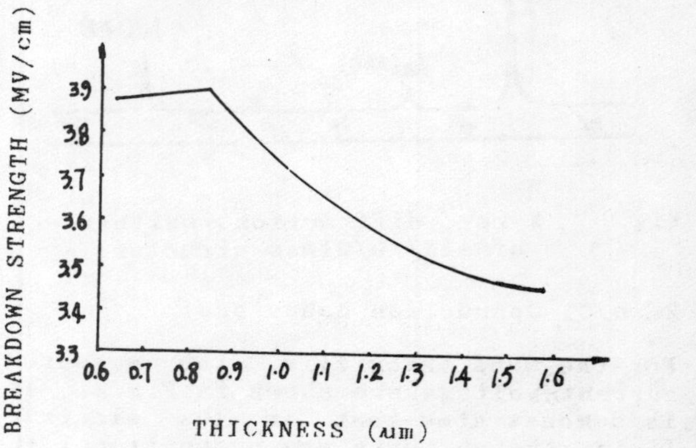

Fig.5 Variation of breakdown strength as a function of thickness of piezoelectric AlN films

CONCLUSION

This report investigates the properties of d.c. electric conduction and breakdown of piezoelectric AlN films constructed with the structure of Au/AlN/Au/Glass. The AlN films are grown by d.c. planar magnetron sputtering. It is proved that the d.c. electric conducting mechanism of the piezoelectric films is Poole-Frenkel type under strong electric field, in which the carriers might be generated by impurities and/or nonstoichiometry. The breakdown electric field strength of 3.4–3.9MV/cm are obtained from the statistic mean on the experimental data for the film thickness ranging from 0.6 to 1.6um. Compared with the breakdown strength of 4.5MV/cm of the amorphous AlN film(0.3um-thick)[10], the result of the C-axis highly oriented film is lower, which lies in the structural difference of the films as the boundaries of C-axis oriented crystallite films are more convenient for carriers to move. Furthermore, the breakdown strength decreases clearly with the increasing thickness for the range of $d>d_b=0.83um$ which implies that the breakdown is caused by impact ionization. But when $d<d_b$, the breakdown strength nearly has nothing to do with the thickness which suggests a breakdown caused by ionic motion.

The author expects to thanks associate professor Yang Bangchao and senior engineer Zhou Hongren for their helps in experiments.

REFERENCES

1. S.Yoshida, J. Vac. Sci. Technol., 16(1979), 990.
2. Y.Uemura, Thin Solid Films, 20(1974), 11.
3. Y.Murayama, J. Vac. Sci. technol., 17(1980). 796.
4. C.R.Aita, J. Appl. Phys., 53(1982), 1807.
5. L.G.Perce, Appl. Phys. Lett., 39(1981), 878.
6. Y.Pauleau, J. Electrochem. Soc., 127(1980), 1532.
7. M.Morita, Japan J. Appl. Phys., 20(1981), 17.
8. K.M.Taylor and C.Lenie, J. Electrochem. Soc., 107(1969). 308.
9. F.Forlani and N.Minnaja, Phys. Stat. Sol., 4(1964), 331.
10. S.Bhat, Journal of Electronic Materials, 14(1985), 405.

THE STUDY OF NEW $Ga_xAl_{1-x}PO_4$ PIEZOELECTRIC CRYSTALS

Han Jian-ru Chen Chen Zhang Lei
Yuan Wu Xu Bin Qin Zi-kai
(Shandong University, P.R.China)

ABSTRACT

The new non-ferroelectric piezoelectric crystals $Ga_xAl_{1-x}PO_4$ have been firstly successfully developed. It has been approved that these crystals belong to class 32, form binary system, and have larger piezoelectric constant than $AlPO_4$ and $GaPO_4$. Also the cell parameters and some electro-elastic constants of $Ga_xAl_{1-x}PO_4$ have been given.

I. INTRODUCTION

In the course of ferroelectric development, the study of ferroelectric solid solution has accelerated the development of binary and triple systems ferroelectric, improved their qualities, and enhanced the varieties. Therefore, ferroelectric materials have been applied to many fields of practical applications. For non-ferroelectric piezoelectric crystals their piezoelectric properties were considered maybe to be decreased through dopant substitution. So up to now, there only have been the study of single system piezoelectric crystal growth and properties (such as SiO_2, $AlPO_4$[1,2,3] etc), but there are not any study of the binary system piezoelectric crystals. Considering that Ga^{3+} and Al^{3+} have approximate radii, same chemical equivalent, and are capable of reacting with H_3PO_4 to form $AlPO_4$ and $GaPO_4$ crystals, we can expect that it is completely possible to form $Ga_xAl_{1-x}PO_4$ solid solution piezoelectric crystals (where $0 < x < 1$). The success on non-ferroelectric binary system piezoelectric crystal growth will not only increase the piezoelectric crystal study fields but also carve out a new way to improve piezoelectric properties.

The present paper deals with the study of $Ga_xAl_{1-x}PO_4$ crystal growth using heating method, the solubility of $Ga_xAl_{1-x}PO_4$ at different composition x. The structures, cell parameters, and defects were carried out by means of x-ray diffraction. Also some electro-elastic constants of these crystals have been determined.[4,5,6]

II. CRYSTAL SYNTHESIS

The starting materials are Al, Ga and the phosphoric acid (H_3PO_4) solvent. Growth experiments were conducted in autoclaves with glass liners using slow heating method. The compositions of the growth solutions are:

$$(GaPO_4)_x + (AlPO_4)_{1-x} + H_3PO_4$$

where x being a series of different values. The autoclave temperature was controlled by means of DWT-702 auto-temperature controller. The growth solutions were presaturated at the initial temperature, then the crystals were obtained using slow heating method (Heating rates ranging from 0.5-1 °c/day). In the majority of the experiments, the percent fill ranged from 40%-70%, the experimental temperature ranged from 150 -200°c.

In order to understand solubility variation at different compositions x, we measured solubility versus x at 40% fill and 191 °c, as shown in Table 1. It can be seen that s decreases while x increases, crystallization varies while x is different.

We have successfully developed 36 mm length $Ga_{0.1}Al_{0.9}PO_4$ crystals which having perfect morphology.

Table 1 The solubility and crystallisation versus x of $Ga_xAl_{1-x}PO_4$ crystals

x	s(wt%)	H_3PO_4 (mol)	Crystallization(grains)
0.59	4.49	6.96	uniform, quite transparent
0.39	6.84	6.79	quite large, untransparent
0.21	8.54	6.45	small, quite transparent
0.14	8.81	6.28	small, transparent
0.10	9.43	6.12	small, transparent
0.065	10.26	5.94	small, transparent
0.04	14.34	5.78	white powder
0.02	15.15	5.61	white powder

The compositions of $Ga_xAl_{1-x}PO_4$ crystals were determined by means of S-520 scanning electron microscope. The structures and cell parameters of these crystals were measured using D/max - rA x-ray powder diffractometer. It was observed that $Ga_xAl_{1-x}PO_4$ crystals belong to class 32 and form binary system piezoelectric crystal series. The list of cell parameters at temperature 18 °c are given in Table 2.

Table 2 Cell parameters of $Ga_xAl_{1-x}PO_4$ crystals

x	a(Å)	c(Å)
0	4.9417	10.9585
0.10	4.9385	10.9667
0.254	4.9307	10.9640
0.475	4.9235	10.9870
0.52	4.9161	11.0000
1	4.9030	11.0507

III. MEASUREMENT

1. Theory Analyses

The matrixes of dielectric, piezoelectric stress and elastic stiffness constants of piezoelectric crystal in class 32 are given below:

$$\begin{bmatrix} \varepsilon_{11} & 0 & 0 \\ 0 & \varepsilon_{11} & 0 \\ 0 & 0 & \varepsilon_{33} \end{bmatrix} \quad \begin{bmatrix} e_{11} & -e_{11} & 0 & e_{14} & 0 & 0 \\ 0 & 0 & 0 & 0 & -e_{14} & -e_{11} \\ 0 & 0 & 0 & 0 & 0 & 0 \end{bmatrix}$$

$$\begin{bmatrix} c^E_{11} & c^E_{12} & c^E_{13} & c^E_{14} & 0 & 0 \\ c^E_{12} & c^E_{11} & c^E_{13} & -c^E_{14} & 0 & 0 \\ c^E_{13} & c^E_{13} & c^E_{33} & 0 & 0 & 0 \\ c^E_{14} & -c^E_{14} & 0 & c^E_{44} & 0 & 0 \\ 0 & 0 & 0 & 0 & c^E_{44} & c^E_{14} \\ 0 & 0 & 0 & 0 & c^E_{14} & c^E_{66} \end{bmatrix} \quad c^E_{66} = \tfrac{1}{2}(c^E_{11} - c^E_{12})$$

The piezoelectric plate may be considered as an infinite plate if its thickness is much smaller than its other dimensions. In this case, the particle displacement, strain, stress and electric displacement corresponding to standing wave vibration are only the functions of thickness coordinate, and the analyses about thickness vibrations of piezoelectric plate is much simplified.

Through piezoelectric constitutive relations and dynamical equations of acoustics, one gets the determinant equation.

A. 0° Y-cut plate

The determinant Eq. is

$$\begin{vmatrix} \bar{c}_{66} - c_y & 0 & 0 \\ 0 & c^E_{11} - c_y & -c^E_{14} \\ 0 & -c^E_{14} & c^E_{44} - c_y \end{vmatrix} = 0$$

one root of the determinant Eq. is

$$c_{y1} = \bar{c}_{66} = c^E_{66}(1 + k^2_{26}) \tag{1}$$

where

$$k^2_{26} = e^2_{11} / c^E_{66}\varepsilon^S_{11} \tag{2}$$

is the thickness shear electromechanical coupling coefficient.

The sum and product of another two roots of the determinant Eq. are respectively

$$\Sigma_y = c_{y2} + c_{y3} = c^E_{11} + c^E_{44} \tag{3}$$

$$\Pi_y = c_{y2}c_{y3} = c^E_{11}c^E_{44} - (c^E_{14})^2 \tag{4}$$

By way of the free boundary conditions, $T_2 = T_4 = T_6 = 0$, one gets

$$c_{y1} = 4\rho(f^1_{pn}t/n)^2 \tag{5}$$

$$c_{y2} = 4\rho(f^2_{sn}t/n)^2 \tag{6}$$

$$c_{y3} = 4\rho(f^3_{sn}t/n)^2 \tag{7}$$

where ρ is the mass density, t is the thickness of the piezoelectric plate, n is the overtone order, f_p is the parallel resonant frequency, and f_s is the series resonant frequency.

B. (YXL)ϕ-cut plate

The determinant Eq. is

$$\begin{vmatrix} \bar{c}'_{66} - c'_y & 0 & 0 \\ 0 & c'_{22} - c'_y & c'_{24} \\ 0 & c'_{24} & c'_{44} - c'_y \end{vmatrix} = 0$$

in which one root is

$$c'_{y1} = \bar{c}'_{66} = c'_{66}(1 + k'^2_{26}) \tag{8}$$

where

$$c'_{66} = c^E_{44}n^2 + c^E_{66}m^2 + 2c_{14}mn, \tag{9}$$

$$k'^2_{26} = e'^2_{26}/c'_{66}\varepsilon^{S'}_{22} \tag{10}$$

$$e'_{26} = -e_{11}m^2 - e_{14}mn \tag{11}$$

$$\varepsilon^{S'}_{22} = \varepsilon^S_{11}m^2 + \varepsilon^S_{33}n^2 \tag{12}$$

$$m = \cos\phi, \quad n = \sin\phi$$

The sum and product of another two roots

are respectively

$$\Sigma_y' = c_{y2}' + c_{y3}' = c_{22}' + c_{44}' \qquad (13)$$

$$\Pi_y' = c_{y2}' c_{y3}' = c_{22}' c_{44}' - (c_{24}')^2 \qquad (14)$$

where

$$c_{22}' = c_{11}^E m^4 + c_{33}^E n^4 + (4c_{44}^E + 2c_{13}^E) m^2 n^2 - 4c_{14}^E m^3 n \ ,$$

$$c_{44}' = (c_{11}^E + c_{33}^E - 2c_{13}^E) m^2 n^2 + 2c_{14}^E mn(m^2 - n^2) + c_{44}^E (m^2 - n^2)^2 \ ,$$

$$c_{24}' = (c_{33}^E n^2 - c_{11}^E m^2) mn + (2c_{44}^E + c_{13}^E)(m^2 - n^2) mn - c_{14}^E m^2 (m^2 - 3n^2) \ .$$

The relations between effective elastic constant and resonant frequency can be got by means of free boundary conditions. The relation between c_{y1}' and parallel frequency is analogous to Eq. (5), and the relations between c_{y2}', c_{y3}' and series resonant frequencies are, respectively, similar to Eqs.(6) and (7).

2. Measurement Results

By means of the theory on thickness vibration of piezoelectric plates, using 0° Y-cut, (YXl) 28°-cut, and (YXl) 31°-cut plates at perpendicular field, we can determine some electro-elastic constants of $Ga_{0.1}Al_{0.9}PO_4$ crystal. In testing, HP-4191A, HP-4192A impendance analyzer and HP-3585A spectrum analyzer had been used.

a. Measuring the capacitances of test specimens at 1 KHz and 30 MHz respectively, we can get the free dielectric constant ε_{11}^T and the clamped dielectric constant ε_{11}^S, calculate out ε_{33}^T and ε_{33}^S through the measured capacitances by Eq.(12).

b. Measuring overtone frequencies f_{s1}, f_{s2}, f_{s3}, ···, we can get piezoelectric and some elastic constants of $Ga_{0.1}Al_{0.9}PO_4$ crystal.

Table 3 Mass density, Electro-elastic constants of $Ga_{0.1}Al_{0.9}PO_4$

Constants	Unit	Ours	$AlPO_4$[3]
ρ	Kg/m^3	2760	2618
ε_{11}^T	relative unit	6.18	5.47
ε_{33}^T		5.75	5.37
ε_{11}^S		5.34	4.60
ε_{33}^S		4.99	4.48
e_{11}	C/m^2	-0.29	-0.223
e_{14}		0.17	0.15
c_{44}^E	$10^9 N/m^2$	34.38	43.15
c_{66}^E		33.29	30.56
c_{14}^E		-12.92	-12.14

In Table 3 the density, dielectric, piezoelectric and some elastic constants have been listed.

IV. CONCLUSIONS

1. Non-ferroelectric binary system piezoelectric crystal $Ga_{0.1}Al_{0.9}PO_4$ with larger size has been firstly successfully produced by means of hydrothermal growth. Experimental results have shown that this crystal has better growth quality and larger piezoelectric constant e_{11} than $AlPO_4$ and $GaPO_4$. Therefore, for non-ferroelectric piezoelectric crystals applicable dopant substitutions may not decrease but enhance and improve piezoelectric properties. By further study of non-ferroelectric piezoelectric crystals, It can be expected to obtain a series of large size binary system piezoelectric crystals and to apply these materials to many practical fields.
2. X-ray powder diffraction results have indicated that $Ga_xAl_{1-x}PO_4$ crystals have the same diffraction peak position, belong to class 32. The cell parameters a decreases and c increases while the composition x increases.
3. It was demonstrated that the dielectric constants of $Ga_{0.1}Al_{0.9}PO_4$ are slightly larger than $AlPO_4$ and change little with frequency.

REFERENCES

1. E.D.Kolb, J.C.Grenier and R.A.Laudise, J. Crystal Growth, 51 (1981), 178.
2. K.Barappa, etal, J. Mater. Sci., 21 (1986), 2202.
3. H.Wang, B.Xu, X.L.Liu, J.R.Han, and S.X.Shang, Acta Physica Sinica, 34 (1985), 1634.
4. Z.P.Chang and G.R.Barsch, IEEE Trans. Sonics and Ultrasonics, SU-23, (1976) 127.
5. W.P.Mason, Piezoelectric Crystal and Their Application to Ultrasonics, Van Nostrand, New York, (1950), 205.
6. Z.Y.Zhao, X.C.Guo, Z.J.Shi, Acta Physica Sinica, 11 (1986), 1473.
7. M.Onoe, H.F.Tiersten and A.H.Meitzler, J.Acoust.Soc.Am., 35 (1963) 36.

TUNABLE BANDPASS FILTERS USING METAL-DIELECTRIC WAVEGUIDES

Vladimir D. Voronkov, Boris M. Leved, Eugeney S. Redko
USSR

The basic tendency in the field of microwave technics is to raise the upper level of the frequency range. It relates to bandpass ferrite tuned filters as well. The metal-dielectric waveguide provides a sufficient effective coupling with ferrite resonator in mm-wave. The problem studied in the paper is the determination of the requirements of magnetic materials used in above microwave devices.

1. Ferrite resonator coupling with metal-dielectric waveguide.

a) Uniform precession.

The uniform precession approximation is true to the condition of a small size of ferrite resonator in comparison with wavelength and biased $\bar{H}_o$ and exciting $\bar{H}$ field uniformity around ferrite.

The filter characteristic analysis should be carried out by the eguivalent circuit approach. In this case the coupling of the ferrite resonator with waveguide field is defined by the coupling ratio of the ferrite with the waveguide.

$$\beta = j2\omega V \frac{H_x \chi_x - H_y \chi_y}{2N}, \text{ where} \quad (1)$$

N - normalized nondisturbed wave factor according to ref. [1]

H_x, H_y - components of the wave field

X_x, X_y - components of the susceptibility tensor.

Having put the calculated values N, H_x and Hy (metal-dielectric waveguide) into (1), we should get a formula of the coupling ratio in resonance β if the ferrite resonator is positioned inside of the dielectric rod. (figure 1):

$$\beta_o = A\left(1+\frac{\sin 2k_\partial t}{2k_\partial t}\right)^{-1} \cos^2 k_\partial\left(\frac{a}{2}-x\right)\left(1+\Im_H^2\right) \quad (z)$$

and outside of the dielectric rod

$$\beta_o = A\frac{2t}{a}\left\{\left(1-\frac{2t}{a}\right)\left(\frac{\cos^2 k_\partial t}{\sin k_o\left(\frac{a}{2}-t\right)}\right)^2\left[1-\frac{\sin 2k_o\left(\frac{a}{2}-t\right)}{2k_o\left(\frac{a}{2}-t\right)}\right]\right\}\sin^2 k_o x\left(1+\Im_H^2\right) \quad (3)$$

into (2) and (3) $A = \frac{2\pi}{\lambda}\cdot\frac{m_\partial V}{\mu_o 2tb}\cdot\frac{M_o}{2\Delta H}$

λ - wavelength in free space,

m_∂ = - slowing-down factor of a waveguide, αt - width of a dielectric rod; ℓ - height of waveguide; M_o - saturation magnetisation, $2\Delta H$ - ferromagnetic resonance linewidth; $\mathcal{E}_o$ - dielectric constant of a rod, γ_o - propagation constant in free space,

$k_o = (\omega^2\mu_o\varepsilon_o-\gamma_o^2)^{1/2}$

$k_\partial = (\omega^2\mu_o\varepsilon_\partial\varepsilon_o-\gamma_o^2)^{1/2}$

The coupling ratio of the ferrite resonator, positioned in the centre of waveguide as a function of waveguide filling factor of the dielectric 2+/a forthree dielectric permeability values of the rod $\mathcal{E}_\partial$ = and the above mentioned wavelength a/λ = 0,5 is presented in the figure 1

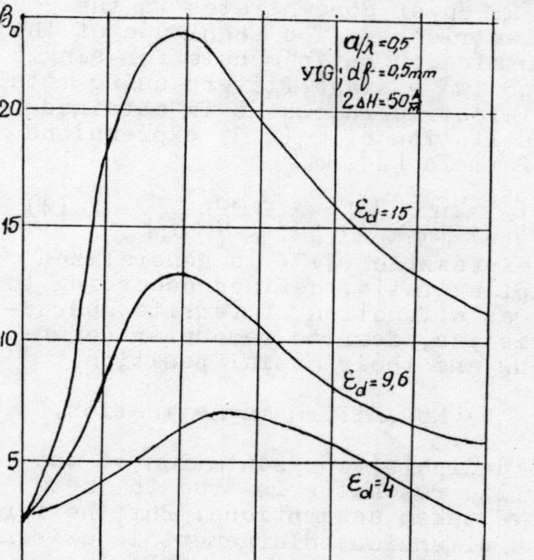

The initial growth of the coupling ratio takes place due to the slowing-down factor increase and to the energy concentration in the rod. With further increase of the filling factor the RF field concentration around ferrite position decreases, that results in general decrease of the coupling ratio β.

Having removed the ferrite resonator in the intersection plane of the waveguide, we localize the coupling ratio about the air-dielectric interface (figure 2).

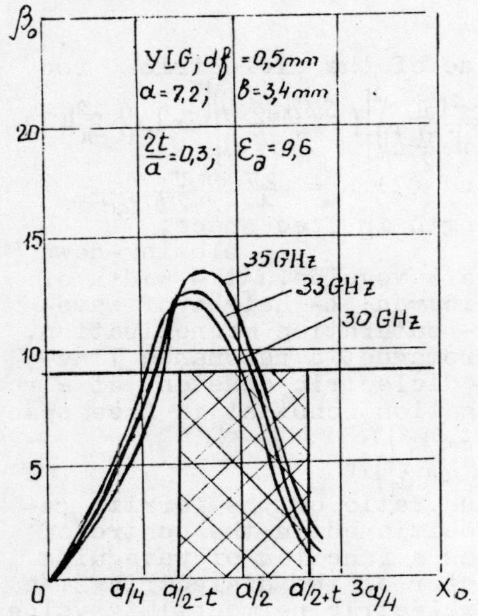

The circular polarization plane is removed to waveguide sidewall if freguency increases. At the same time the RF power concentrates in the dielectric rod. The magnitude of the transmission ratio module for single - and two - stage filters using metal - dielectric wavequide is obtained by substitution of β (2,3) expressions in formula [2]

$$|S'_{21}| = \frac{2\beta_o}{1+2\beta_o}; \quad |S''_{21}| = \frac{2\beta_o \sigma_o}{(1+\beta_o)(1+\sigma_o^2)} \qquad (4)$$

In expression (4) σ_o - is generalized coupling ratio, defined according to [3] as a function of ferrite characteristics, ferrite resonator demensions and their mutial position.

b) Nonuniform magnetisation.

Field dipole representation of the ferrite resonator is true to tne above taken assumptions. But the ferrite dimension enlargement is arisen from the opportunity to take into account the uniformity of magnetisation in ferrite volume.

The generation of various magnetostatic modes is a function of the presence of the suitable spherical harmonic in the exciting field expansion. The magnetostatic mode amplitude is defined by the ratio of the suitable harmonic.

According to ref. [1] waveguide fields generated by ferrite resonators are defined by the expression:

$$\vec{H} = \sum_n C_{\pm n} \vec{H}_n; \quad \vec{E} = \sum_n C_{\pm n} \vec{E}_{\pm n} \qquad (5)$$

In this case the expansion is carried out by the intrinsic waves of the nondisturbed waveguide, but the disturbance factor Cn is generally defined by the formula:

$$C_{\pm n} = \frac{j\gamma_o}{N_n} \int_V \vec{H}_{\mp n} \vec{M}\, dV \qquad (6)$$

The resulted expressions of $C_{\pm n}$ don't cite because of their complexity. The analysis of the exciting factor Cn as a function of the waveguide filling by a dielectric shows its general-purpose character in all magnetostatic modes, the uniform precession included.

The calculations of the frequency dependence of the exciting factors are shown in fig 3.

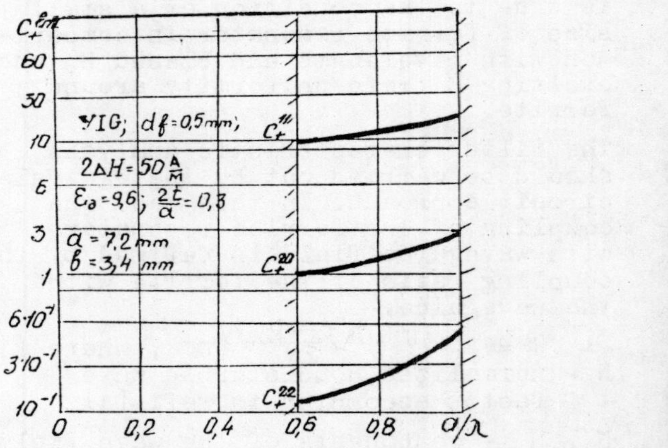

In the highest modes this dependence has a more uncommon character, due to the microwave power increase in the rod.

2. Magnetic materials used as a gyromagnetic resonators in mm-waves.

High guality single crystal garnet resonators are commonly used in band-pass filters with narrow bandwidth in freguency range up to 53 GHz.

Table 1. Relations of hexaferrite resonators characteristics and frequency ranges,

Frequency range, GHz	Composition	Anisotropy field, kOe	ΔH, Oe	$Q_0 = \frac{w}{\int 2 \Delta H}$
37,5 + 53,67	$BaJe_{0,3}Fe_{11,7}O_{19}$	10,0	35	350
53,6 + 78,3	$BaFe_{12}O_{19}$	17,0	25	600
78,3 + 118	$BaAl_1Fe_{11}O_{19}$	26,0	70	400
118 + 150	$BaAl_3Fe\ O_{19}$	40,0	120	300

We use $BaFe_{12}O_{19}$ hexaferrite. The Sc^{3+} and Al^{3+} substitution allowes to synthesize series of materials with required anisotropy fields, but at the same time it reduces the quality of magnetic materials. The additional feature that complicate hexaferrite application is the great resonance response discontinuity due to the magnetostatie mode excitation. The magnetostatic mode intensity is increased with substitutions in the body of the hexaferrite.

3. The experimental results of bandpass filters.

The basic feature of the design of the bandpass filter is a orthogonal intersection of metal-dielectric waveguides by a coupling aperture diaphragm. In two-stage filter (80 GHz) every resonator is positioned on the ceramic holder in the waveguide just against the coupling aperture. In the case of a single-stage filter (up to 120 GHz) the ferrite resonator is adhered in the coupling aperture. The frequency dependence of insertion loss of single and two-stage filters with various ferrite resonator are shown in figure 4. The resonator materials data are also presented in figure 4.
In this frequency range in some in cases the single crystal ferrite spinels such as Lithium and Lithium Zinc Jron spinels with $4_\pi Ms$ = 5500G and 4500G are used.

The application of magnetic materials with small anisotropy is limited (50+60GHz) by their saturation induction in magnetic cires of contorl systems. There it is advisable over 40 GHz to use ferrite with great anisotropy fields.

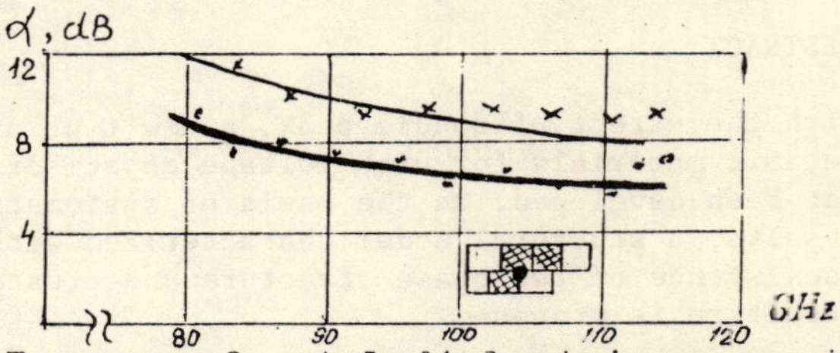

In case of metal-dielectric waveguide a dielectric with a permeability $\mathcal{E}$ = 9,6 (Al_2O_3 ceramic) and waveguide filling factor 2 t/a 0,3 are studied.

The experimental results are in good agreement with theoretical calculations. The increasing discrepancy in the top region of the tuning range can be partly explained by the fact that the frequency dependence of the ferromagnetic resonance linewidth and the dielectric loss in the rod was not accounted.

The spurious resonance suppression in single-stage filters is no worse than 20dB, but in two-stage filters it increases up to 25-30 dB. The growth of the nonuniformity of resonance response of the filter with a frequency increase is due to the use of resonators made of magnefic materials with high substitution. According to the said above this nonuniformity is explained by the highest magnetostatic mode excitation.

References

1. Vainstein L.A. Electromagnetic fields and waves. M., Soviet Radio, 1957
2. Rogozin V.V., Churkin V.N. Ferrite filters and limiters. M., Soviet Radio, 1985
3. Mikaelyan L.A., Antonyantz, V.Ya. The mutial coupling phenomenon in magnetized ferrite system". "Radiotechnika and Electronica", 1962, t. V*1,

AN EFFECTIVE METHOD TO MINIMIZE THE TEMPERATURE COEFFICIENT OF PERMITTIVITY IN SPCT CERAMIC CAPACITORS

Xiong Zhaoxian
Department of Chemistry
Xiamen University
Xiamen, China

Chen Shoutian
Department of Electrical Engineering
Xi'an Jiaotong University
Xi'an, China

ABSTRACT

With the effect of double-peak, a new type of ceramic materials for high-voltage capacitors has been developed. On the basis of systematic results, a stuctural model characterized with coexistence of two phase structure and crustal structure is proposed.

INTRODUCTION

SPCT, i.e., $(Sr_xPb_yCa_{1-x-y})TiO_3.mBi_2O_3.nTiO_2$, is a new type of ceramic materials for high-voltage capacitors, used in color TV, laser sets, etc. However, it is difficult for SPCT ceramics to have not only high permittivity and low dissipation factor but also small temperature coefficient of permittivity, because its Curie point , Tc, is below the room temperature.[1]

In this paper, the difficulty has been overcome with the effect of double-peak, and, a structural model has been proposed. The effect of double-peak results from the recombination of two kinds of calcined composite powders with different Tc.

MODEL AND COMPUTATION

A structural model characterized with coexistence of two phase structure and crustal structure is proposed, as schematically Fig.1. The model can be described more typically, as Fig.2. In Fig.2(a), the ceramic body is consisted of two type of grains with different Tc; however, in Fig.2(b), there is only same kind of grains, but the structure in every grain appears crustal, and the different crust has different Tc.

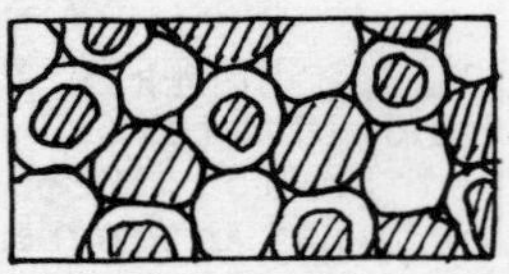

Fig.1 A structural model proposed schematically for SPCT ceramics, characterized with coexistence of two phase structure and crust structure.

(a) (b)

Fig.2 The model described more typically:
(a) The two phase structure
(b) The crustal structure

The temperature dependences of permittivity and dissipation factors are computerized, as Fig.3, based just on the model described before.

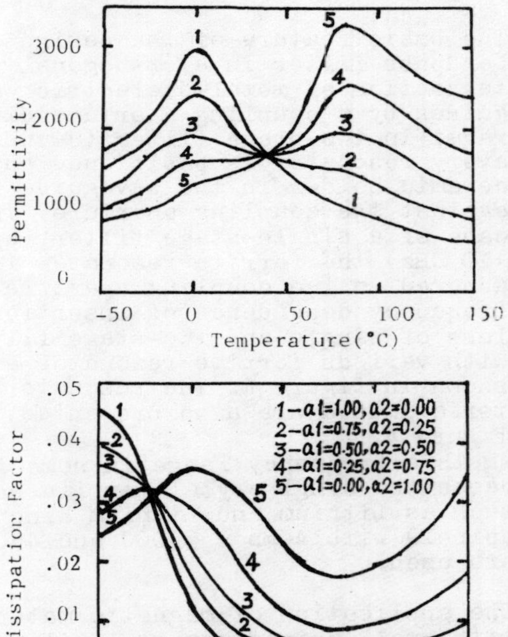

Fig.3 The temperature dependence of permittivity and dissipation factor computerized.

It is clear that the temperature coefficients of permittivity can be minimized if the grains or crusts with different Tc are taken at a certain volume ratio. During the computerizing, the Lichtenecker formula is taken, as follows,

$$\ln \varepsilon = a_1 (\ln\varepsilon_1) + a_2 (\ln\varepsilon_2)$$

$$\ln (\tan\delta) = a_1(\ln\tan \delta_1) + a_2(\ln\tan \delta_2)$$

and, a Series-parallel Formula is used. The details can be seen in another paper.[2]

EXPERIMENTAL AND RESULTS

The samples, ceramic capacitors, were prepared by ordinary ceramic techniques. Raw materials ($\geqslant$ 99% $SrCO_3$, PbO, $CaCO_3$, TiO_2, Bi_2O_3) were combined in the system, $(Sr_x Pb_y Ca_{1-x-y})TiO_3$, $mBi_2O_3.nTiO_2$, to produce two kinds of calcined composites with different values of x and y, respectively. The compositions were calcined at 1200 °C for 2h, and recombined at a certain volume ratio, then, ground for 12h, pressed at 200kg/cm2 in the form of dislike pellets, and fired at 1200 ± 50 °C for 1h.

The temperature dependences of permittivity and dissipation factor were measured by an automatic capacitance bridge, as Fig.4.

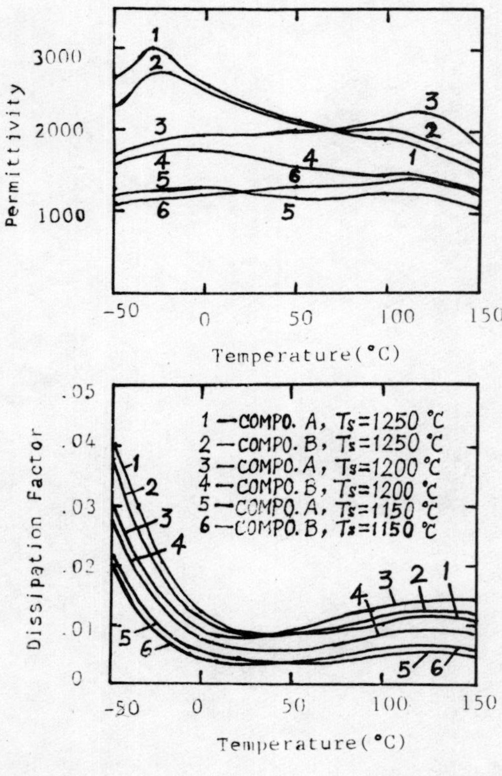

Fig.4 The temperature dependence of permittivity and dissipation factor measured.

It is sure that the temperature coefficients of permittivity is minimized in some samples, and, the values of permittivity and dissipation factor are also reasonable.

The microstructure observation and electron beam probe analyses show that there is a chemically inhomogeneous state in the ceramic body, as Fig.5.

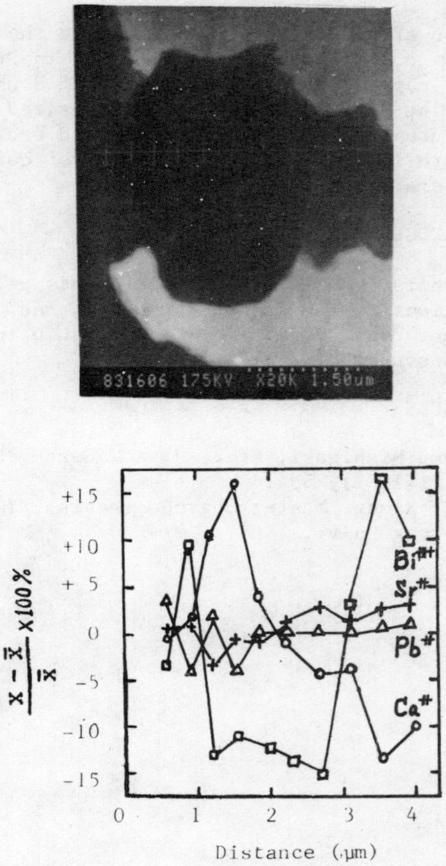

Fig.5 The microstructure observation and electron beam probe analyses:
(a) STEM and the positions selected for EDAX
(b) A chemically inhomogeneous state in SPCT ceramics with EDAX

It is just such chemically inhomogeneous state that results in the dielectric properties desired. And, this state results from the techniques controlled carefully.

DISCUSSION

The model proposed in this paper is used to described the chemically inhomogeneous state for SPCT ceramic body. However, there are derivations for the temperature dependences of permittivity and dissipation factor between computerized and measured. The derivations come majorly from two aspects. In one hand, the model proposed is simplified, but the actual ceramic body is more complicated; in another hand, the exact control of technique is difficult. Nevertheless, the effect of double-peak is obvious in the results either computerized or measured, and, the dielectric properties desired for the ceramic capacitors has been reached.

CONCLUSIONS

It is an effective method to minimize the temperature coefficient of permittivity in SPCT ceramic capacitors with the effect of double-peak. The structural model characterized with coexisture of two phase structure and crustal structure is suitable for such kind of ceramic capacitors.

ACKNOWLEDGEMENTS

The authors thank Prof. Yao Xi for his helpful suggestions on the structural model, and Assistant Prof. H. K. Xie and F. Y. Liu for their discussion.

REFERENCES

1. Susumu Nishigaki, etc., J. Am. Ceram. Soc., 65, 11(1982), 554.
2. Z. X. Xiong, Master Degreee thesis, Xi'an Jiaotong Univ., 1988.

THE EFFECTS OF THICKNESS ON GAS SENSING METAL OXIDE SEMICONDUCTOR THIN FILMS

Lü Hong lang, Li Jianming and Qiu Nanyuan
Physics Department of Shandong Polytechnic University
Jinan, China

ABSTRACT

ZnO and SnO_2 thin films were deposited by powder sputtering, their electric characteristics and gas sensing properties were tested. The experimental results indicate that the effects of thickness on gas sensing characteristics in ZnO thin films by powder sputtering are similar to those by reactive sputtering, their gas sensing characteristics are not affected by film thicknesses much remarkablely, but the effect in SnO_2 thin films are different from those in ZnO thin films. In addition, these results are discussed.

INTRODUCTION

We[1] have made investigations of gas sensing characteristics in reactive sputtered ZnO thin films with different thicknesses, the results make know that the gas sensing characteristics are not affected by their thicknesses much remarkably. So we also prepared ZnO and SnO2 thin films by powder sputtering, and made a further investigation of the effects of thickness on gas sensing characteristics.

EXPERIMENTAL RESULTS

Experimental equipment and methods in this article all described elswhere[2].

1. Gas Sensing Properties and Electric Characteristics in Powder Sputtered ZnO Thin Film

A. Relations between the gas sensing properties and film thicknesses

Fig.1 and Fig.2 illustrate that the conductance G of ZnO thin films with different thicknesses have power law relations with the concentration C of C_2H_5OH vapour, and the power law exponents P and sensitivity S (S is defined as the ratio of the conductance of a thin film in 380 ppm and 38 ppm C_2H_5OH vapour respectively) all have relations with film thicknesses, when film thickness d is about 400Å, P and S are all the largest.

B. Relations between the electric characteristics and film thicknesses

Fig. 3 indicates the relations between conductance of ZnO thin films both in the air and in 380 ppm C_2H_5OH vapour and their thickness

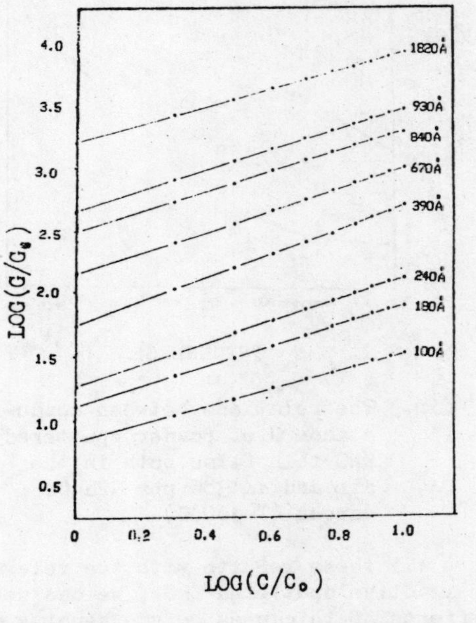

Fig.1 The response of conductance G of powder sputtered ZnO thin films with different thicknesses to C_2H_5OH vapour ($T=360°C$; $C_0=38ppm$; $G_0=1x10^{-7}ohm^{-1}$)

If these ZnO thin film with different thicknesses are treated with vacuum annealing, and then their electrical conductivity σ and activation energy Eact. are tested, we find the results illustrated by Fig,4 and Fig.5.

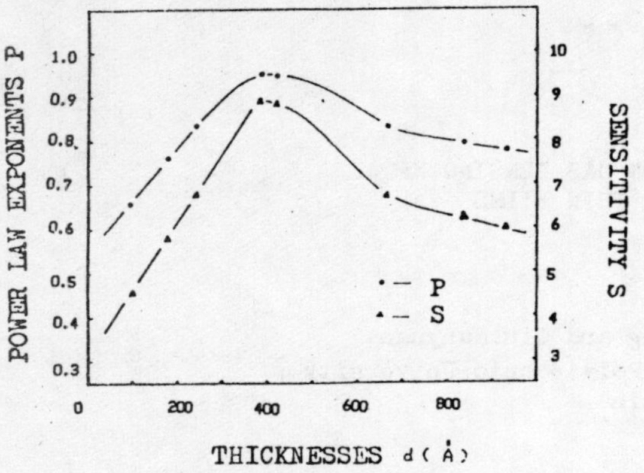

Fig. 2 The relation of power law exponents and sensitivity to film thicknesses (T=360°C)

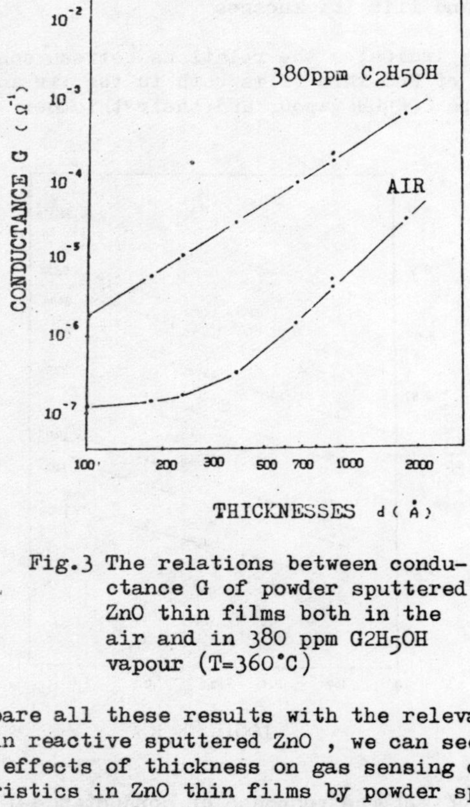

Fig.3 The relations between conductance G of powder sputtered ZnO thin films both in the air and in 380 ppm G2H5OH vapour (T=360°C)

Compare all these results with the relevant ones in reactive sputtered ZnO , we can see that the effects of thickness on gas sensing characteristics in ZnO thin films by powder sputtering are similar to those by reactive sputtering, and the relations between their gas sensing characteristics and film thicknesses are not very remarkable, the relations of electric conductivity σ and activation energy Eact. of the two kinds of ZnO thin films after vaccum annealing to their thicknesses are all identical with the results of theoretical calculation with monocrystal[3].

2. Electric characteristics and gas sensing properties in powder sputtered SnO_2 thin films.

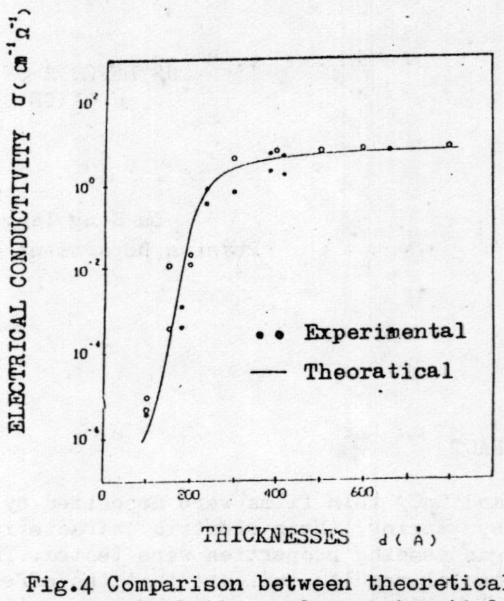

Fig.4 Comparison between theoretical calculation and experimential results of the relations between electrical conductivity of ZnO thin films and their thicknesses (• • respectively show experimental results of powder sputtered and reactive sputtered ZnO thin films treated with vacuum annealing)

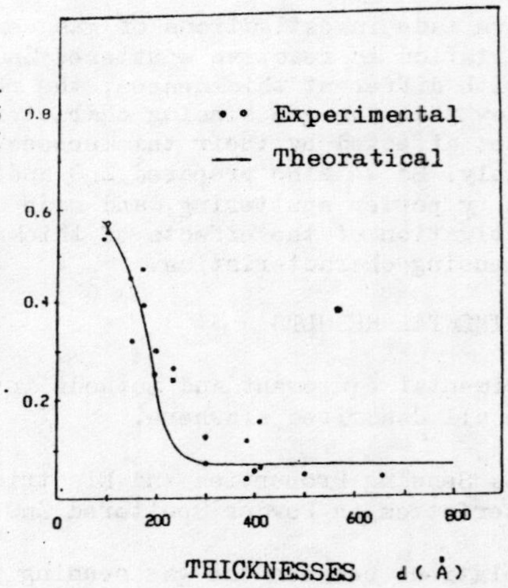

Fig.5 Comparison between theoretical calculation and experimental results of the relations between activation energy Eact. of ZnO thin films and their thicknesses (• • respectively show experimental results of powder sputtered and reactive sputtered ZnO thin films treated with vacuum annealing)

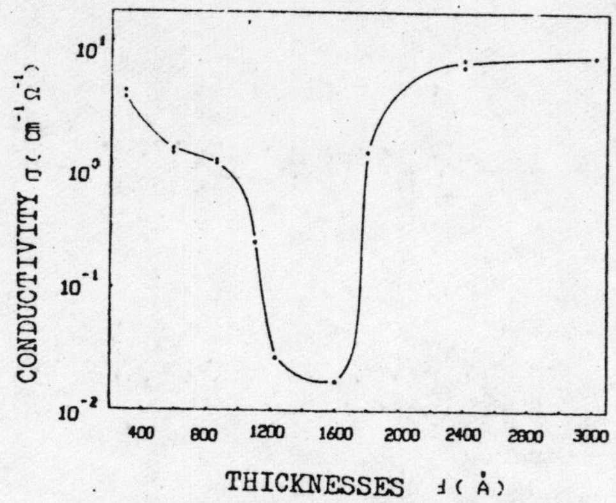

Fig.6 The relation between electrical conductivity σ of powder sputtered SnO_2 thin films in the air and their thicknesses (T=10°C)

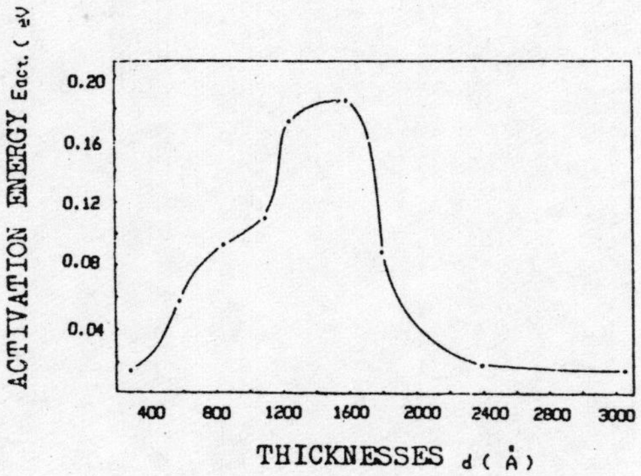

Fig.7 The relation between activation energy of powder sputtered SnO_2 thin films and their thicknesses

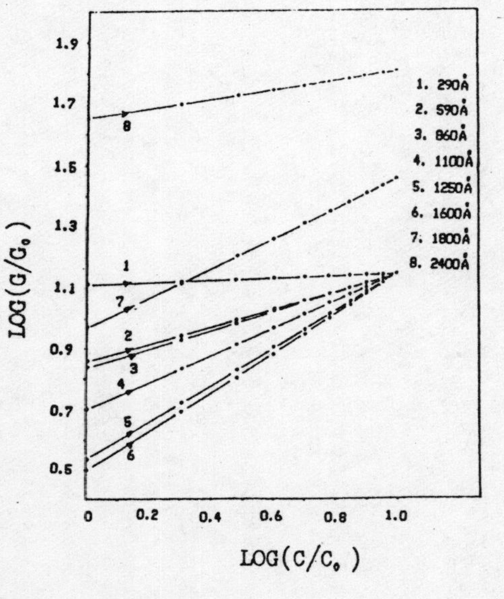

Fig.8 The response of conductance G of powder sputtered SnO_2 thin films with different thicknesses to C_2H_5OH vapour (T=260°C; C =38ppm; G = 1×10^{-5} ohm^{-1})

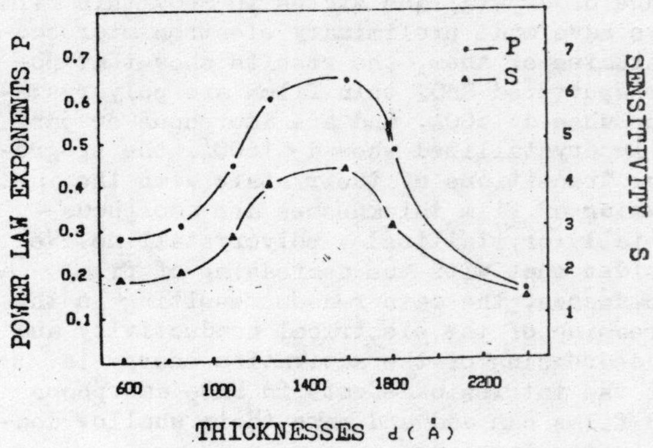

Fig.9 The relations of power law exponents P and sensitivity S to film thicknesses (T=260°C)

A. Relations between the electric characteristics and film thicknesses.

Fig.6 and Fig.7 show the relations of elctric conductivity σ and activation energy Eact. of SnO_2 thin films with different thicknesses before and after the vacuum annealing to their thicknesses.

B. Relations between the gas sensing properties and film thicknesses

Fig.8 and Fig.9 indicate that the conductance G of SnO2 thin films with different thicknesses also have power law relations with the concentration C of C_2H_5OH vapour, and their power law exponents P and sensitivity S all have remarkable relations with film thicknesses.

All of the results make clear that the electrical conductivity and activation energy of powder sputtered SnO_2 thin films all have relations with their thicknesses, and these bear no relation to whether those thin films have been treated with vacuum annealing or not; besides, their gas sensing characteristics have distinct relations with film thicknesses, these are all different from those of ZnO thin films.

In addition, we have also tested relations of gas sensing properties in powder sputtered SnO_2 thin films doped with a little amount of additive to their thicknesses, and find that the relations are basically similar to those in pure SnO_2 thin films.

DISCUSSIONS

1. Fig.6 and Fig.7 show: when the thicknesses

of powder sputtered SnO_2 thin films are less than some value (1600 Å), their electrical conductivity σ increase but their activation energy Eact. decrease with the decreasing of film thicknesses d. This may result from the change of crystalline states in SnO_2 thin film. So we have made preliminary electron microscopy studies of them, the results show that powder sputtered SnO_2 thin films are polycrystalline when $d > 1600$Å, and are amorphous or partially crystallized when $d < 1600$Å, the progressive transitions of their state with the increasing of film thicknesses are amorphous→ partially crystallized→ polycrystalline. We consider that with the decreasing of film thicknesses, the main reason resulting in the increasing of the electrical conductivity and the decreasing of the activation energy is that the intrinsic defects in SnO_2 amorphous thin films can add and make their shallow donor density increase.

2. We think that the differences in electric characteristics and gas sensing properies of between ZnO and SnO_2 polycrystalline thin film may mainly result from their internal structure and the adsorptive power of oxygen on grain boundaries. T.L. Tansley et al[4] have pointed out that ZnO thim films prepared by R.F. sputtering have columnar grain structure. Thus it can be considered that under a rather high gas sensing working temperature, adsorbed oxygen (O^-) can diffuse along the grain boundaries in ZnO thin film, get into and out with some freedom. So, although the sensing reaction take place on the surface of ZnO thin films, they can also influence the oxygen adsorbed on the grain boundaries remarkablely, such gas sensing reaction can be considered a kind of bulk effect, and have not very remarkable relations with film thicknesses. But in relation to SnO_2 thin film, owing to the characteristics of its grain structure and the relative strong adsorptive power of oxygen (O^-) adsorbed on thegrain boundaries, it can be thought that oxygen (O^-) adsorbed on the grain boundaries can not go out and come into the body of SnO_2 thin film with relative freedom, the effects of oxygen adsorbed on the grain boundaries are not very remarkable. So the gas sensing effects of SnO_2 polycrystalline thin films mainly show a kind of surface effects.

REFERENCES

1. Li Jianming et al, Thesis of This Meeting.
2. Lü Honglang, M.S. Thesis of Shandong Polytechnic University, (1989).
3. Li Jianming et al, Journal of Shandong Polytechnic University, 18(1) (1988) 82-89.
4. T.L. Tansley et al, Thin Solid Film, 117 (1983) 19-32.

MIDDLE TEMPERATURE SINTERED CERAMIC NPO–30 FOR MLC

Zhou Shiliang and Dai Fugui
Hong Ming Electronic Components Factory, P.O.Box. 82, Chengdu, China

ABSTRACT

Fine crystallite ceramic, having excellent frequency–temperature stability, is based on $4BaO \cdot 75TiO_2 \cdot 21Nd_2O_3$, (5—10) wt% of the Bao · $SiO_2 \cdot Pb_3O_4$ low–melting glass and just right amount SiO_2 clay is taken as additive This ceramic is sintered at 1135 ± 10℃, and suitable for making high quality MLC.

I INTRODUCTION

It is known that there is a lot of ceramic material systems with the dielectric permittivity $\varepsilon \approx 30$, such as: BaO–TiO_2(hereafter referred to as simply BT), ZrO_2–TiO_2(ZT), MgO–TiO_2(MT), BaO–TiO_2–Nd_2O_3(BTN) etc. But their sintering temperature are higher than 1300℃, however the sintering temperature of the middle temperature sintered ceramic is about 1150℃. The melting point of the BaO · SiO_2 · Pb_3O_4(BSP) glass is lower, tangent of dielectric loss angle $\tan\delta$ is smaller, and have a good compatibility with basic composition. When 10wt% of the BSP was added into it, sintering temperature of ceramic could be decreased by 200℃ or more and the properties would be unchanged. Of couse, the powder grain–size ($< 2\mu m$) is very important, the subject grain–size is $< 1.5\mu m$. The dielectric properties of the sintered ceramic are as follow:
$\varepsilon \approx 30$, temperature coefficient of capacitance of specimen:$\alpha_C \approx 0 \pm 30ppm$, $\tan\delta < 4 \times 10^{-4}$, $\rho_{v150℃} > 10^{13} \Omega.cm$. electric strength: 15KV / mm.
The technology of the forming film for the ceramic is good, the film substrate is uniform, so it is suitable for making high quility MLC.

II PREPARING OF CERAMIC MATERIAL SPECIMENS AND MEASURING INSTRUMENTS

1. Raw material: technical clay, $BaCO_3$, TiO_2, ZrO_2, Nd_2O_3, Pb_3O_4, ZnO and SiO_2.
2. Technological process for preparing the ceramic specimens is shown in fig.1.
3. Size of specimen: Φ 12x1.5 for measuring capacitance C, $\tan\delta$ and insulation resistance R, Φ 8.5x1.0 for measuring frequency dependence of ε and temperature dependence of the capacitance.

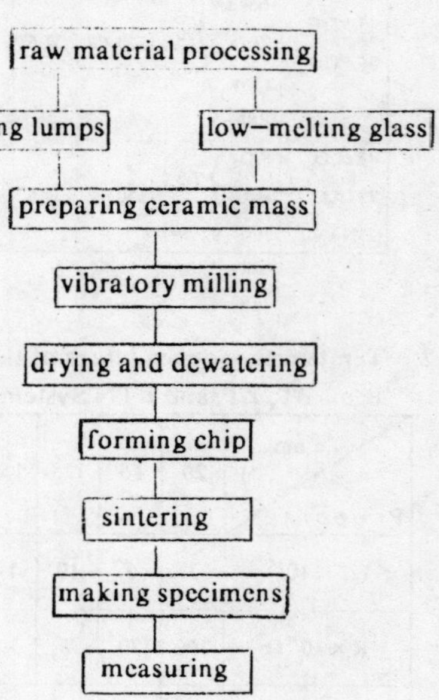

Fig.1 Technological process for preparing the ceramic powder and specimens.

4. Measuring instruments:
For measuring C: CCF–1 (made in China)
For $\tan\delta$: CJS–2B 〃 〃
For R: NF–2511 〃 〃
For frequency dependence of ε: HP4191A and HP4192A(made in HP Co. USA)
For temp. dependence of capacitance: HP4277 (made in HP Co. USA).

III EXPERIMENT RESULTS

1. Selecting material system
During manufacturing material, BSP is taken as flux(mineralizer), and BTN system has been selected after BT, ZT, BTN systems were compared.

The dielectric properties are shown in Table 1. The temperature dependences of the tanδ and R are shown in Table 2.

Table 1. Dielectric properties for typical composition of BT, ZT, BTN Systems.

Composition: Base (mol%)	Composition: Additive (wt%)	Firing temp. (℃) and Time(h)	ε	tanδ $\times 10^{-4}$	$\alpha_c \times 10^{-6}$ /℃
$19.2BaCO_3$ $80.8TiO_2$	ZrO_2, WO_3, Clay } 8.5; BSP–12	1150 / 2	31.5	3.5	−50
$43.5ZrO_2$ $56.5TiO_2$	SnO_2, $BaCO_3$, ZnO, Clay } 30; BSP–5	1170 / 2	32.6	2.7	−25
$4BaCO_3$ $73TiO_2$ $23Nd_2O_3$	SiO_2, B_2O_3 } 7.5; BSP–9	1150 / 2	34.6	2.1	−31

Table 2. Temperature dependences of the tanδ and R of BT, ZT, and BTN Systems

Propert. \ Temp. (℃)	20	85	125	155	matepial systems
tg$\delta \times 10^{-4}$	3	5	10	12	BT
$R \times 10^{10}\Omega$	200	70	7	3.9	BT
tg$\delta \times 10^{-4}$	2.7	6.	7	7.5	2T
$R \times 10^{10}\Omega$	900	225	50	10	2T
tg$\delta \times 10^{-4}$	2.3	2.1	2.7	2.5	BTN
$R \times 10^{10}\Omega$	6000	2900	115	40	BTN

2. Formulation

According to literature[1], three composition points are selected:20BaO • $56TiO_2$ • $24Nd_2O_3$, 7BaO • $64TiO_2$ • $29Nd_2O_3$and 4Bao • $75TiO_2$ • $21Nd_2O_3$. After comparison test, it was found that sintering temperature of ceramic 4BaO • $75TiO_2$ • $21Nd_2O_3$in which 10wt% BSP, 2wt% $SiO_2$and 1.0wt% clay were added is 1135± 10℃ and its dielectric properties are very good.

3.Effect of chemical composition of the material on its properties

A. Effect of chemical composition of basic material on properties of the ceramic. It is given in tab.3

Table 3. Effect of chemical composition of the basic material on properties of ceramic

Added base composition(wt%): $BaCO_3$	TiO_2	Nd_2O_3	ε	tanδ $\times 10^{-4}$	$\alpha_c \times 10^{-6}$ / ℃	Firing temp. (℃) and time (h)
/	/	/	30.1	2.0	−23	1135 / 1
1	/	/	36.6	2.5	−45	1135 / 1
4	/	/	39.8	2.5	−28	1135 / 1
6	/	/	42.4	3.0	−25	1135 / 1
/	3	/	34.6	2.5	−51	1135 / 1
/	6	/	33.5	3.0	−57	1135 / 1
/	9	/	32.4	3.2	−65	1135 / 1
/	/	3	31.1	2.5	−27	1135 / 1
/	/	6	31.5	3.0	−14	1135 / 1
/	/	10	30.7	3.2	+18	1135 / 1

B. Effect of clay content on properties of the ceramic. It is given in fig.2

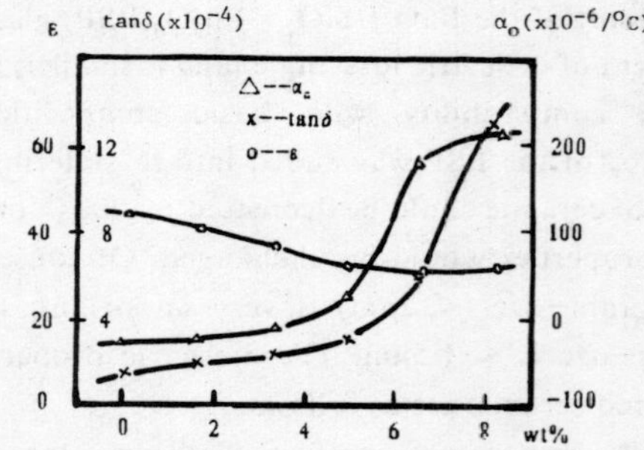

Fig.2 Effect of clay content on properties of the material

C. Effect of SiO_2content on properties of the ceramic. It is given in fig.3.

After their comparison it was confirmed that 4BaO • $75TiO_2$ • $21Nd_2O_3$with 10wt% BSP, 2wt% $SiO_2$and 1.0wt% clay is the NPO–30 ceramic.

4. Temperature–frequency dependence of dielectric properties of NPO–30 ceramic. The temperature dependences of the tanδ and R of the NPO–30 are shown in fig.4.

The temperature dependences of the α_c of the NPO–30 are shown in fig.5.
The frequence dependences of the ε of the NPO–30, BSP are shown in fig. 6.

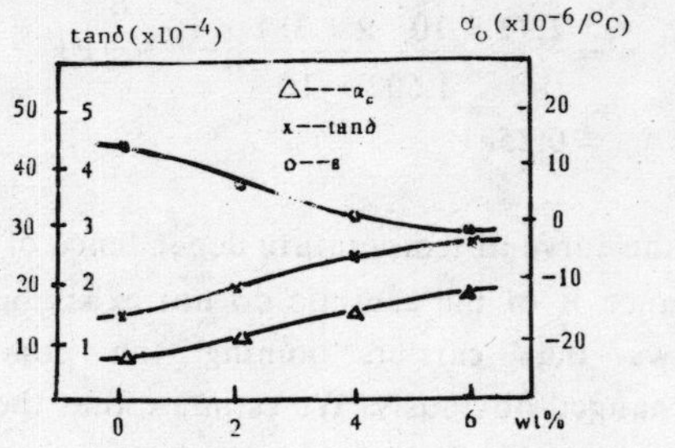

Fig.3 Effect of SiO_2content on properties of the material.

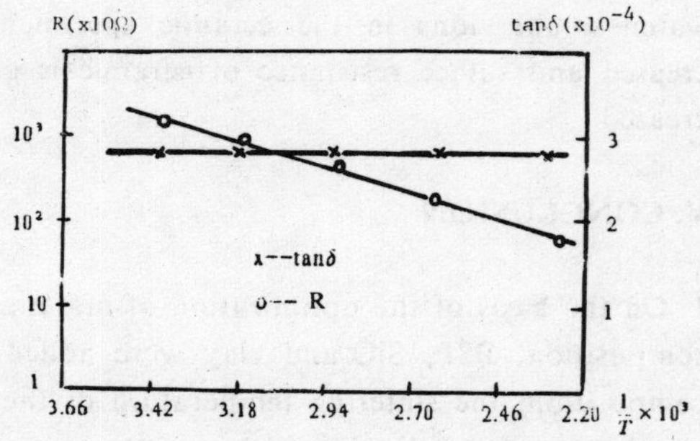

Fig.4 Tanδ and R of ceramic NPO–30 versus temperature T.

The data for ecah point in fig.4 are given as follw:

Temp. t(℃)	20	40	70	100	155
$\frac{1000}{T}=\frac{1000}{t+273}$	3.41	3.19	2.91	2.68	2.34
Rx10^{10}(Ω)	1400	900	600	300	60
lgR	3.14	2.95	2.77	2.47	1.77
Tanδx10^{-4}	2.8	2.9	2.8	2.8	2.8

IV DISCUSSION FOR THE TESTING RESULT

1. Sintering temperature of the ceramic. According to literature [1], the sintering temperature is 1400—1450℃ when the composition is changing along the line of $BaTiO_3$–$Nd_2Ti_2O_7$. The sintering temperature is 1300—1350℃ when the composition is changing along the line of $BaTiO_3$–TiO_2. The eutectic point in the system $BaTiO_3$–$BaTi_3O_7$–$BaNd_2Ti_5O_{14}$is 1240℃ . The sintering temperature 1420℃ / 1h was selected for 4BaO.75TiO_2.21Nd_2O_3.

According to the experiment, 10wt% of the BSP were respectively added in 13BaO • 57TiO_2 • 30$Nd_2O_3$and 20BaO • 56TiO_2 • 24Nd_2O_3, sintering temperatures de–

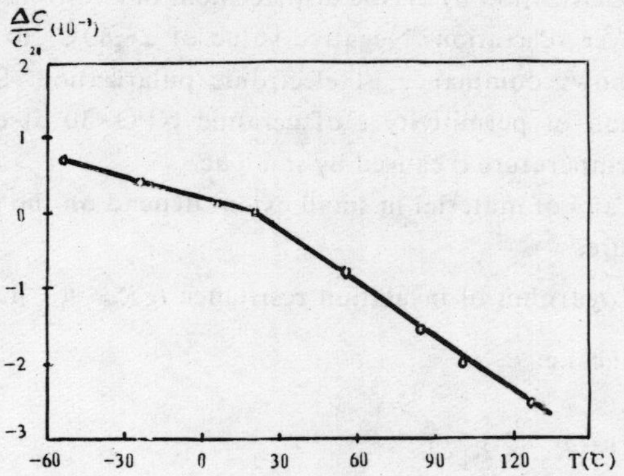

Fig.5 Capacitance C of specimen of Ceramic NPO–30 versus temp. t(℃)

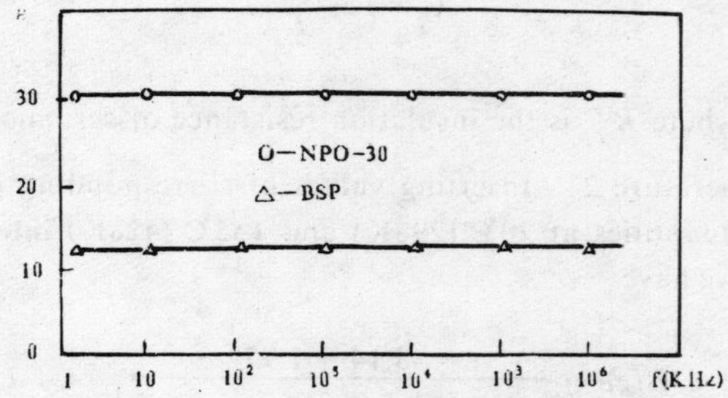

Fig.6 Permitivity ε of NPO–30 and BSP versus frequence

creases from 1580℃ / 1h and 1550℃ / 1h to 1350℃ / 2h and 1320℃ / 2h correspondently, i.e. the sintering temperature has been decreased by 230℃. It is evident that if 10wt% of the BSP will be added in 4BaO • 75TiO_2 • 21Nd_2O_3, the sintering temperature will be decreased to 1150℃, but $\varepsilon \approx 45$. BSP play the role of flux(mineralizer) and directly take forming glass phase and bonding the grains and them forming ceramic, so the sintering temperature was decreased. The dielectric permitivity of the ceramic was further decreased by adding $SiO_2$with low permitivity ε. But during forming ceramic on its surface appeared crystalization which has influence on its surface smoothness. If $SiO_2$and clay were simultaneously added in it, the crystalization not only was restrained but also the desired dielectric properties of the ceramic were obtained, i.e. $\varepsilon \approx 30$, $\alpha_c \approx 0 \pm 30$ppm.

2. Dielectric properties of the material

Independence of permitivity ε of the ceramic on field frequences at 1 to 10^6KHz shows large dominance of polarization by elastic displacement of electrons and ions over relaxation. Negative value of α_c−60℃ to +125℃ shows dominance of electronic polarization. Stabilization of permitivity ε of ceramic NPO−30 at changing temperature is caused by small αc.

Tanδ of material in small extent depend on the temperatures.

Logarithm of insulation resistance $lgR = f(\frac{1}{T})$is straight line, i.e.

$$R_T = R_o e^{\frac{B}{T}}$$

or

$$lgR_T = lgR_o + (Blge)\frac{1}{T}$$

$$\therefore \quad Blge = \frac{lgR_{T_2} - lgR_{T_1}}{(\frac{1}{T_2}) - (\frac{1}{T_1})} \qquad (1)$$

where R_{T_i} is the insulation resistance of ceramic at temperature T_i. Inserting values of corresponding physical quantities at 20℃ (293K) and 155℃ (428K) into eq.(1), we have

$$Blge = \frac{3.14 - 1.77}{(3.41 \times 10^{-3}) - (2.34 \times 10^{-3})} = 1.27 \times 10^3 (K)$$

$$B = \frac{1.27 \times 10^3}{lge} = 2.3 \times 1.27 \times 10^3 = 2.92 \times 10^3 (K)$$

Supposed, that the conductivity is caused by migration by ions over potential barrier, we have

$$B = \frac{W}{k}$$

where w is the hight of the potential barrier, k is the Boltzmann constant. From (2) we find

$$\omega = Bk = 2.92 \times 10^3 \times 1.381 \times 10^{-23} J$$
$$= \frac{2.92 \times 10^3 \times 1.381 \times 10^{-23}}{1.602 \times 10^{-19}} eV$$
$$= 0.25eV$$

At the curve of temperature dependence of insulation resistance R of the ceramic do not exists bend. This fact shows that carriers joining the conductivity are unchanged obviously. We estimate that the conductivity of ceramic NPO−30 is caused by migration of ions of impurity in it. It is found else in our experiment that the insulation resistance R of the ceramic NPO−30 is increased in great extent after boiling it in water, because water−soluble ions in the ceramic specimen are decreased and suface resistance of ceramic is greatly increased.

V. CONCLUSION

1. On the basis of the optimization of material systems composition, BSP, $SiO_2$and clay were added in BTN composition, the sintering temperature of the material was decreased and the electrical properties are very good.
2. The ceramic material is fined by traditional vibratory milling (average grain size: $<1.5\ \mu m$, specific surface: up to $2.99 m^2/g$) which is helpful to improve its technological properties.
3. It seems that the material palarization is caused by elastic displacement of electrons and ions and the conductivity is caused by migration of impurity ions in ceramic.

REFERENCE

1. Ber.Dt. Keram. Ges. 55(1978)Nr7 p. 346−−349.

LOW FREQUENCY DISPERSION CHARACTERISTICS OF ELECTRONIC MATERIALS

Larry C. Burton
Departments of Electrical and Materials Engineering
Virginia Polytechnic Institute & State University
Blacksburg, VA 24061 U.S.A.

ABSTRACT

Low frequency (f < 10kHz) electrical measurements can be very useful in the modelling of a variety of electronic materials and devices. Measurements are described, and models discussed, for GaAs Schottky barrier diodes, multilayer ceramic capacitors, and carbon black filled rubber.

INTRODUCTION

In this paper I would like to discuss the usefulness of low frequency electrical measurements for studying certain characteristics of electronic materials. The discussion is divided into three main parts, pertaining to three general materials types that we have studied: semiconductors, ceramics, and filled polymers. Strong dispersion in the electrical parameters (dielectric constant, loss and resistivity) at frequencies below about 1kHz are seen for all of these materials. A major objective is to relate low frequency measurements to electrical models of the materials, in order to learn more about the material, and its electrical contacts, which may also be of importance. As we will see, contacts can play a principal role in low frequency dispersion; such a dominant role may be good or bad, depending on the device under test. In other instances, such dispersion is dominated by the bulk of the material.

Low frequency measurements and modelling for three specific materials types or devices that we have measured are discussed: gallium arsenide (GaAs) Schottky diodes (semiconductor); multilayer ceramic (MLC) capacitors (ceramic); carbon black filled natural and styrene-butadeine rubber (polymer). These considerations could also pertain to many other materials from these and other materials families.

Low frequency dispersion of electrical parameters (dielectric constant, loss, resistivity) can be attributed to various electric dipole mechanisms with long relaxation times. Two general categories can be defined:

a) Dipoles on an atomic or molecular level are distributed throughout the solid. An example is electron hopping between adjacent traps, in response to an applied alternating field. If a single relaxation time is involved (which is usually not the case), a Debye-type response results.

b) Heterogeneous dielectrics, including interfacial polarization. Cases here include solids composed of two or more materials of different dielectric constant and resistivity, and contacts. The polarization that arises in such solids is often referred to as the Maxwell-Wagner type.

Other types of polarization, such as those due to displacement of electron cloud from nucleus, ionic vibration, and molecular rotation, do not pertain at low frequencies to the electronic solids discussed here, although they may be important for higher frequencies or other materials.

EXPERIMENTAL

Measurements were performed using two impedance analyzers, an HP 4192A (down to 5Hz), and a Solartron 1260 (down to 10^{-6}Hz). Most of the measurements were done at zero voltage bias; if bias voltage was used, it is so indicated. Measurement temperatures are also indicated.

Since three different materials types were studied, each with different processing, sample preparation is discussed separately below for each case.

RESULTS AND DISCUSSION

1. GaAs Schottky Barrier Diodes

The capacitance of such a diode does not typically depend to any substantial degree on frequency below at least several hundred MHz, if the interface is prepared properly, on "good" GaAs. However, some interesting low frequency characteristics appeared for diodes made on ion bombarded material. Such results helped us to understand the ion bombardment damage.

Details of diode preparation have been discussed elsewhere [1]. The main things of interest to us in this paper are the effects of low energy Ar^+ ion bombardment at energies of 1keV and 3keV on certain electrical properties of the diodes, and the modelling of such effects.

Capacitance dispersion is shown in Figure 1 for the non-bombarded (virgin) and ion-bombarded (1keV, 3keV) cases. Similar results were obtained for GaAs wafers obtained from two vendors. Substantial low frequency dispersion ($f < 10$kHz) was seen for ion-damaged GaAs; none for the virgin. Such low frequency dispersion also showed strong temperature and voltage bias dependences (Figure 2). Such dispersion is eliminated by reducing temperature, or applying weak reverse bias.

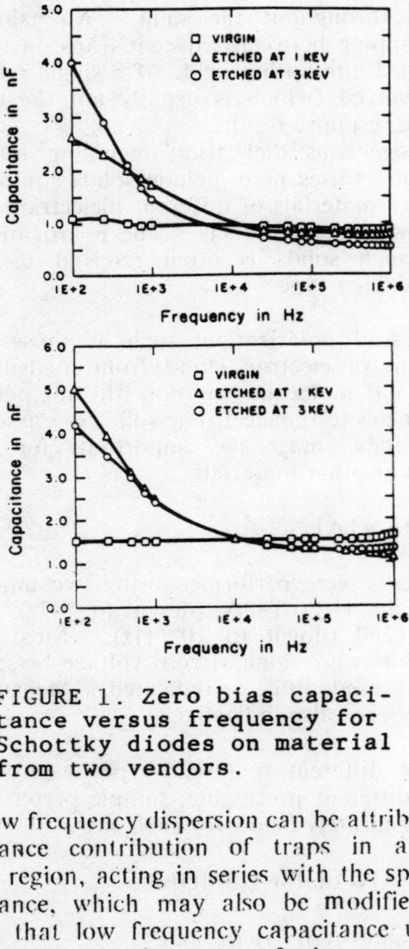

FIGURE 1. Zero bias capacitance versus frequency for Schottky diodes on material from two vendors.

Such low frequency dispersion can be attributed to the capacitance contribution of traps in a damaged surface region, acting in series with the space charge capacitance, which may also be modified. If we assume that low frequency capacitance under zero bias is due to the sum of deep and shallow contributions (i.e. deep traps due to ion damage, and shallow Si donors), then total capacitance can be expressed as [2]

$$C = C(f) + C_D$$

where $$C(f) = \left[\frac{2\varepsilon q^2}{kT} N_t \frac{f_1}{f} \right]^{1/2}$$

with $f_1 = A \exp(-qE_t/kT)$; A = constant
Here N_t is ion damage trap density, f is frequency, and E_t is trap energy.

In our measurements, low frequency capacitance C(f) varies roughly as $f^{-0.6}$, compared to the theoretical $f^{-0.5}$. Runs at different temperatures have yielded a trap energy of about 0.5eV and a density of over $10^{17}cm^{-3}$.

In part from such low frequency measurements, the models shown in Figure 3 were constructed [1]. The lumped components represent the regions shown in the cross sectional representation. All of the salient capacitive features we have measured (including deep level transient spectroscopy - DLTS) can be accounted for by this model.

The damaged and partly amorphous layer near the top surface has a larger effective electron concentration at frequencies below f_1, and controls capacitance. This large electron concentration clamps the depletion layer thickness as long as frequency is low enough and temperature high enough such that the traps can respond. Under these conditions, C_{diode} can be neglected because the space charge is confined to the top (partly amorphous) layer. Under reverse bias, low temperature or high frequency conditions, the space charge extends into the lower, less damaged region, and C_{diode} ultimately predominates. The high frequency capacitance is less for ion-bombarded material due to damage in the bulk material which partly compensates the background donors, resulting in a larger space charge width.

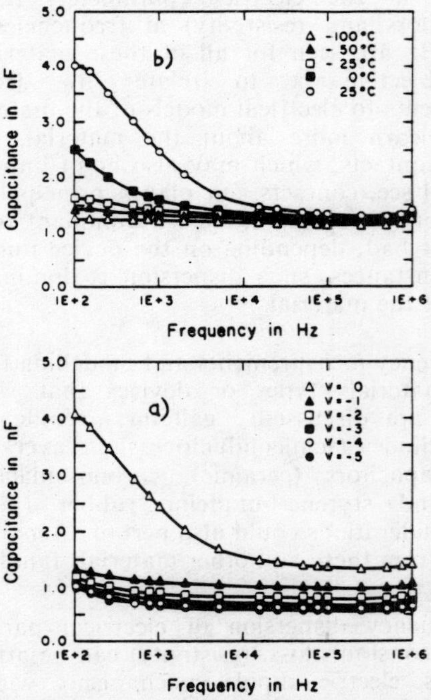

FIGURE 2. Capacitance versus frequency at different a) reverse bias, b) temperature.

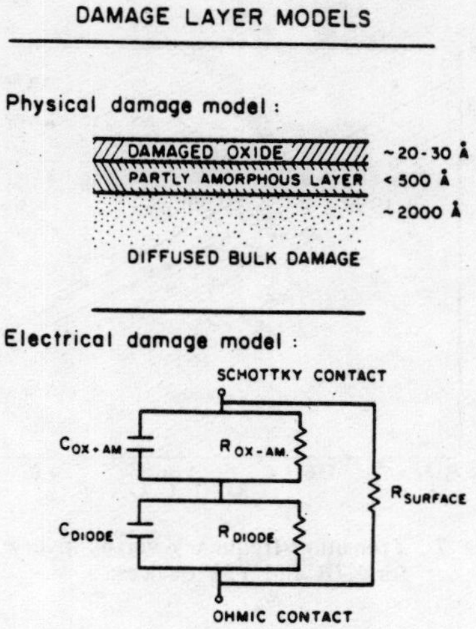

FIGURE 3. Cross sectional model for IBE GaAs, and an electrical equivalent circuit.

Low frequency dispersion in this case is thus attributed to deep electron traps caused by ion damage in a shallow surface region. Such low frequency characteristics don't appear for virgin GaAs material, and can therefore not be attributed to properties of the metal-semiconductor interface or the bulk material.

2. MLC Capacitors

Another device which shows some interesting low frequency characteristics is the MLC capacitor. This is (at room temperature) a very high resistance device ($R > 10^{12}\Omega$), unlike the Schottky diode and filled polymer, discussed elsewhere in this paper. Due to the very high resistance, DC current I relaxes slowly following application of voltage; such an I(t) can be related directly to the frequency dependence of complex dielectric function, $\varepsilon^*(f)$, which determines capacitance and loss. [3]

We have discussed the time dependence of leakage current for MLC capacitors in some detail, and the correlation with $\varepsilon^*(f)$. One model is based on charge trapping at interface states [4]. The interface can be the metal-ceramic, or grain boundary. Both cases can produce similar results, since each can be viewed as different forms of Schottky diodes. Bulk traps in the defect-filled grains could also contribute [5].

We have shown that charge trapping results in current of the form $I(t) = I_o t^{-n}$ where $0 \leq n \leq 1$. As pointed out by Lewis [6], if I(t) is known, the frequency dependence of $\varepsilon^* = \varepsilon' - j\varepsilon''$ can be deduced; ε' should change as f^{n-1} and ε'' as $\sigma/2\pi f$, where conductivity σ varies as f^n.

The accuracy of this model was checked by measuring time dependence of current and capacitance, and frequency dependence of capacitance and conductance, for commercial MLC capacitors. Decay of current and capacitance with time is shown in Figure 4. Both types of decay are consistent with the model. The power factor n remains at near unity until temperature increases above 70°C, then decreasing (Figure 5). It was found that the temperature and voltage dependences of n followed the model quite well [4].

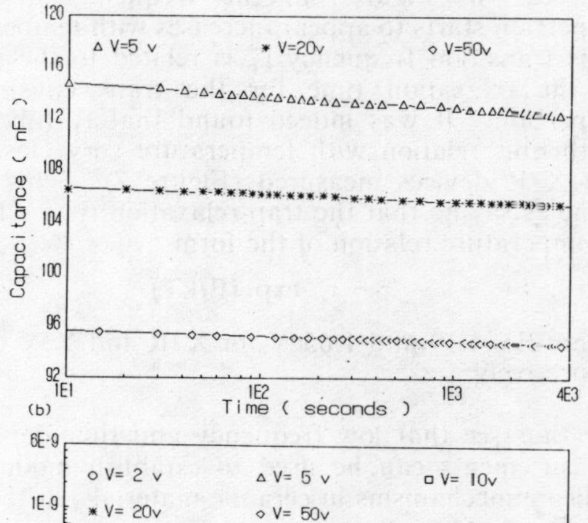

Figure 4. Time decay of current (a) and capacitance (b), for X7R device at 25°C, at several bias voltages.

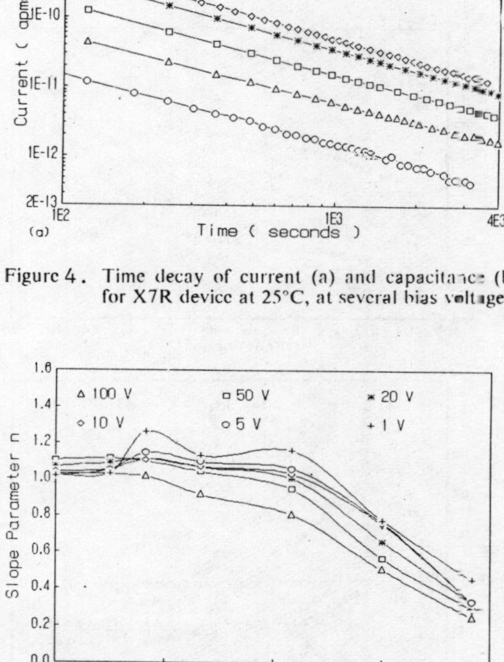

Figure 5. Temperature dependence of current exponent n, expressed as $1/n - 1$, at 6 bias voltages (X7R, 25°C).

Low frequency dispersion curves for capacitance and conductance are shown in Figure 6. At room temperature there is no capacitive dispersion evident, to frequencies as low as 10^{-4}Hz. Conductance increases approximately linearly with frequency at room temperature.

At elevated temperatures, n decreases toward zero (Figure 5), and this is also evident in the dispersion curves of Figure 6. Low frequency conductance becomes flatter ($n \to 0$), and capacitance becomes dispersive ($n - 1 \to -1$).

As seen in Figure 6a, the frequency at which dispersion starts to appear increases with temperature. This transition frequency f_T is related to the inverse of the relaxation time for the traps causing the dispersion. It was indeed found that f_T follows an Arrhenius relation with temperature very closely for the X7R devices measured (Figure 7). This is the same as saying that the trap relaxation time τ follows a temperature relation of the form

$$\tau = \tau_o \exp\ (E/kT)$$

where $E \simeq 1.19$ and 1.65eV for X7R and Y5V devices respectively.

We thus see that low frequency and time-dependent measurements can be used to establish models and indicate mechanisms in ceramic materials, in this case MLC capacitors.

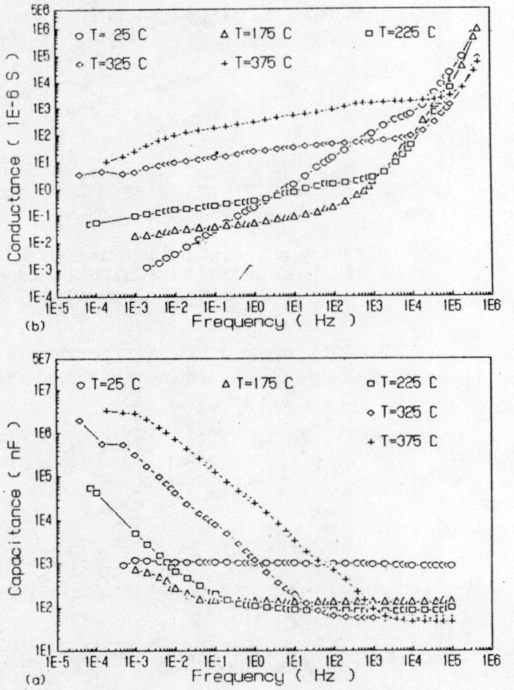

Figure 6. Frequency dependence of capacitance (a) and conductance (b) for Y5V device at several temperatures.

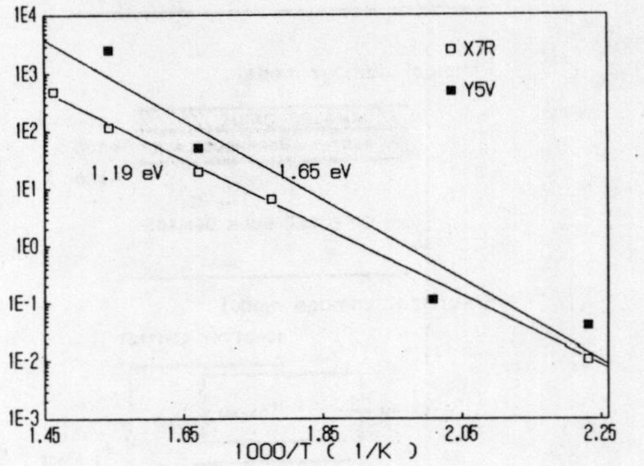

Figure 7. Transition frequency versus inverse temperature for X7R and Y5V devices.

3. Carbon Black Filled Rubber

It is well known that carbon black plays a dominant role in the electrical and mechanical properties of filled polymers [6]. We have studied electrical properties of such systems in some detail [7].

The samples discussed below are based on styrene butadeine rubber (SBR 1500), with 0-70 parts-per-hundred-rubber (phr) carbon black filler. Sample preparation was described in detail elsewhere [8].

The carbon black filler is a relatively good conductor, and such a filled system is expected to show strong Maxwell-Wagner type dispersion at low frequencies. As seen below, this is indeed the case. In addition, the choice of a good low resistance electrode material presented somewhat of a problem, and shows another area where low frequency measurements may be useful.

Low frequency dispersion characteristics for capacitance and resistance of a 30 phr sample are seen in Figure 8, for four types of electrical contact. It is seen that the contact dominates low frequency resistance, since it is in series with the bulk. Contact capacitance has a smaller effect, since the contact capacitance is large, and it is in series with the smaller bulk capacitance. Many contact methods were tried; silver paint was the best and metal plates applied using mechanical pressure the worst. Samples discussed below had silver paint contacts.

Dispersion characteristics of dielectric constant and resistivity are shown in Figure 9, for a 50 phr sample. Both of these electrical parameters, for moderate carbon black loading, follow fairly closely the dispersion model outlined by Kawamoto [9]. Such dispersive behavior of bulk material can be modelled electrically as shown in Figure 10. Polymer gaps separating the carbon black structures are what determine both R and C, there being a larger number of narrower gaps with increased loading, thus decreasing effective resistivity and increasing effective dielectric constant. From Fig. 10, at low frequencies the net resistance equals $R_A + R_C$, at high frequencies it equals R_A. The effective dielectric constant is flat at low frequencies, falling at higher frequencies, the transition frequency depending on amount of loading.

A broader range of dispersion characteristics accompanies a range of carbon black compositions, as shown in Figure 11. Dielectric constant and loss are plotted versus frequency for 0-70 phr loadings.

The increase in dielectric constant with loading in the low and mid phr ranges is due to an average reduction in distance between carbon black aggregates. With further loading (> 60 phr), increased leakage between aggregates reduces the charge storage ability of the matrix, and a slight decrease in dielectric constant results.

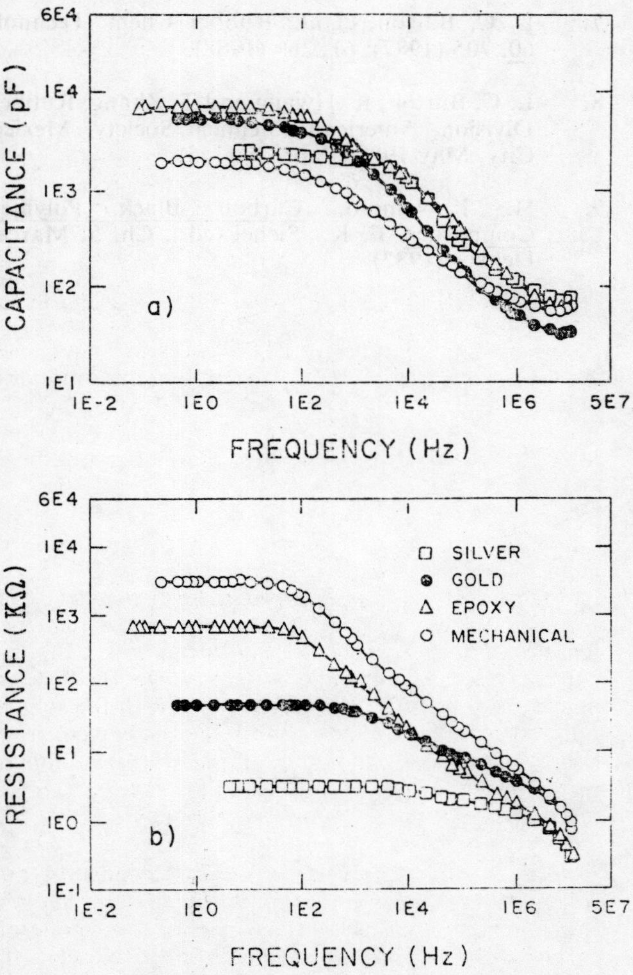

Figure 8. Capacitance and resistance versus frequency for four contacts.

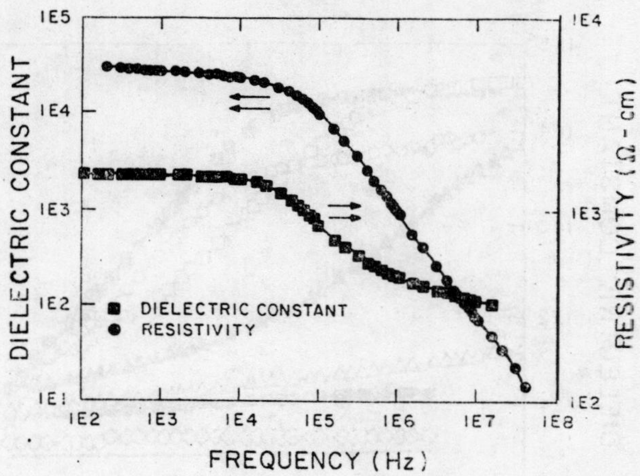

Figure 9. Dispersion characteristics for 50 phr sample.

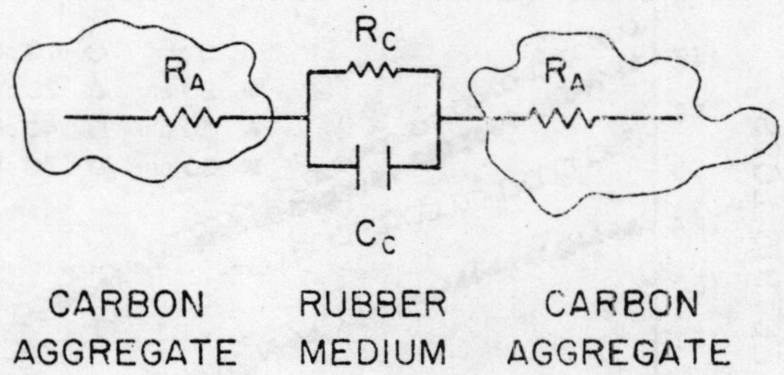

Figure 10. Lumped electrical model for carbon black-polymer composite.

CONCLUSION

We have tried to indicate in this paper that low frequency dispersion measurements can help a great deal in the modelling of vastly different types of electronic materials. In the case of ion-bombarded GaAs, a damaged surface layer containing traps dominates low frequency capacitance. For MLC capacitance, elevated temperatures are required in order to see such dispersion, due to the much longer room temperature relaxation times of the traps, which are deeper in the band gap. And for the filled polymer, a simple lumped model suffices to describe dispersion characteristics over certain loading ranges, as long as contact impedance is not a problem.

ACKNOWLEDGEMENTS

Texas Instruments, the Office of Naval Research, the U.S. Army-TACOM and Virginia's Center for Innovative Technology are acknowledged for supporting various ports of the work reported here. The following graduate students are acknowledged for their major roles in these and related studies: E. Cole, S. Sen, A. Vaseashta, K. Hwang, P. Johnson, T. Zhang, I. K. Yoo, J. Epp.

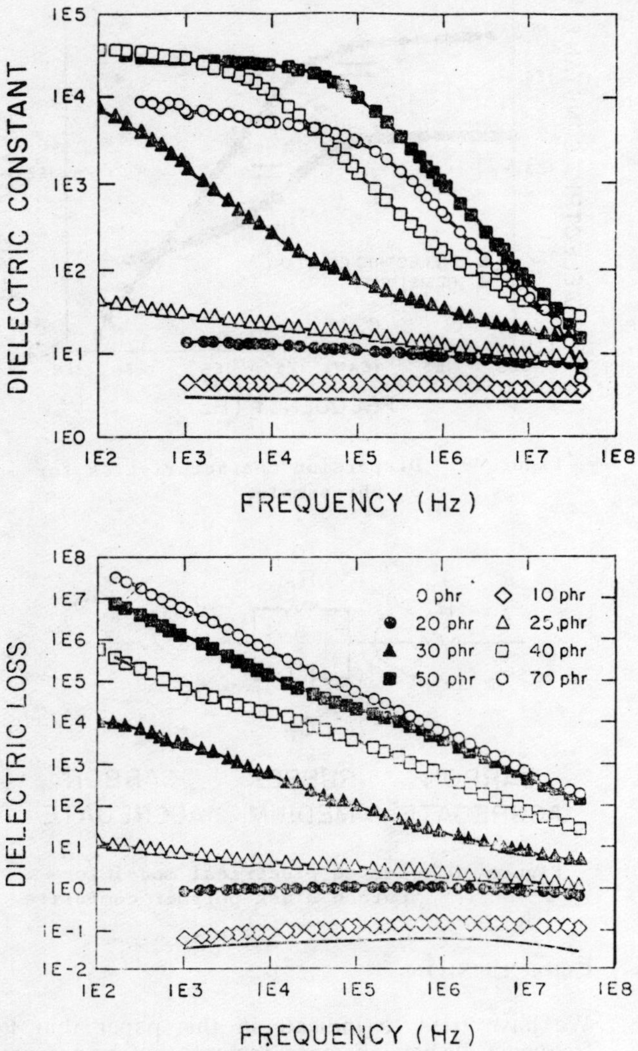

Figure 11. Dispersion characteristics for eight carbon black loadings.

REFERENCES

1. E. D. Cole, S. Sen and L. C. Burton, Jour. Elec. Mater. 18, 527 (1989).

2. E. Schibli and A. G. Milnes, Sol. St. Elec. 11, 323 (1968).

3. A. K. Jonscher, Dielectric Relaxation in Solids, Chelsea Dielectrics Press (1983).

4. Z. Zhang, I. K. Yoo and L. C. Burton, ECC Conference Proceedings (Houston, 1989). To be published in IEEE Transactions-CHMT.

5. M. A. Lampert and P. Mark, Current Injection in Solids, Academic Press (1970).

6. A. I. Medalia, Rubber Chem. Technol. 59, 432 (1986).

7. L. C. Burton, et al., Rubber Chem. Technol. 60, 705 (1987); 61, 269 (1988).

8. L. C. Burton, K. Hwang and T. Zhang, Rubber Division, American Chemical Society, Mexico City, May 1989 (Paper No. 45).

9. H. Kawamoto, Carbon Black Polymer Composites, E. K. Sichel (ed.), Ch. 5. Marcel Dekker (1982).

ELECTRICAL BEHAVIOUR AND MÖSSBAUER SPECTROSCOPIC STUDIES ON $Ba(Sn,Sb)O_3$ SYSTEM CERAMIC RESISTANCE MATERIALS

Zhuang Yan, Yang Zunde
Guangzhou Communications Institute, The Ministry of Machinery and Electronic Industry, China
Jin Mingzhi
Department of Physics, Jilin University, Changchun, China
Yu Zhengfang and Zheng Yufang
Department of Physics, Zhongzhan University, Guangzhou, China

ABSTRACT

The influences of the silicate sintering agent, the content of BaO and Sb_2O_3 on the electrical conductivity of $Ba(Sn,Sb)O_3$ system ceramic resistance materials were studied. It was found that the conductivity depends on the substitution of Sb^{5+} for Sn^{4+} and the phase constituents, low-resistance phases $BaSnO_3$, SnO_2, high-resistance phases $Ba_3Sn_2O_7$, Ba_2SnO_4 and insulating phases $BaSb_2O_6$, $Ba_2Si_3O_8$.

INTRODUCTION

$BaSnO_3$ system resistance materials have found their practical applications in the fixed resistors because there are no phase transitions within a wide range of temperature, and because of their stable properties, small TCR and high working temperature. Russian(1-3) and Frence(4) scholars had studied the electrical conductivity of $BaSnO_3$ ceramics and $Ba(Sn_{1-x}Sb_x)O_3$ powders respectively, and the latter pointed out that antimony dissolves in the stannic sites of stannate perovskite up to 0.17 at. in pentavalent state(Sb^{5+}), above this limit, the formation of the insulating phase antimonate with pyrochlore structure $Ba_2Sb_2O_7$ causes the decreasing of the electrical conductivity. Meanwhile, the electrical behaviour of $BaSnO_3$ with different Sb contents was explained. In this work, by means of x-ray diffraction and Mossbauer spectroscopy, the relationships between the electrical conductivity and the contents of silicate agent, BaO and Sb_2O_3 in $BaO-SnO_2-Sb_2O_3$ system are studied.

EXPERIMENTAL

All the samples were with industrial purity $BaCO_3$, SnO_2 and chemical purity Sb_2O_3 and silicates. The raw materials were carefully weighed out, mixed by viberation ball milling, these mixtures were precalcined at 1220-1240℃, and pulverized by viberation ball milling, and then, pressed into disk, all of green compacts were sintered at 1350-1400℃ in air, with a sintered size of 12.0×1.5mm. Silver electrode was used and fired at 800℃. Electrical resistance and TCR were measured by a micro-computer controlled measuring system involving a FLUKE 8520A digital multimeter and a BMA low/high temperature test chamber. Cu target and graphite filter were used on a Riguka D/maxⅡA x-ray diffractometer for analysis. Mossbauer spectrometry measuring system had been described in reference(5), $Ca^{119m}SnO_3$ was used as a radioactive source. Calibration of velocity was performed with a MVC 450 laser absolute velocity calibrator. All the Mossbauer spectra were computer-fitted.

RESULTS AND DISCUSSION

1. Influence of Silicate Sintering Agent

$BaSnO_3$ is difficult to be sintered owing to its high sintering temperature. In order to obtain dense resistance materials, $MgSiO_3$ was added as sintering agent as reported in reference(6). In this work, various silicates were adopted as sintering agents, and it was found that the suitable amount of added silicate was 13-18wt%. The relationships between the resistivity and the silicate content in the systems of $2BaO-0.90SnO_2-0.05Sb_2O_3$ and $BaO-0.94SnO_2-0.03Sb_2O_3$ are

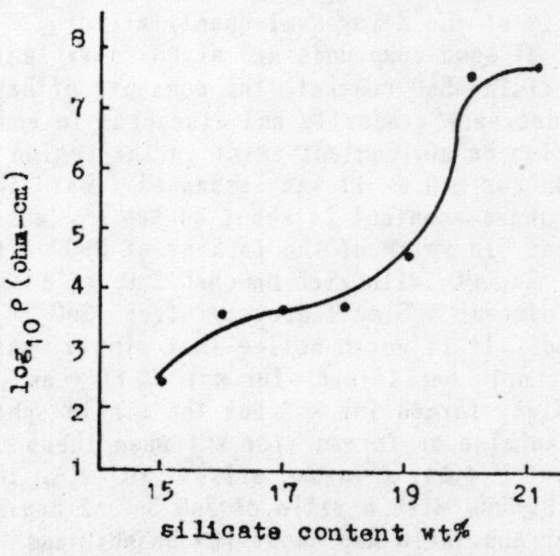

Fig. 1 Semilog plot of ρ as a function of silicate for x=2.

illustrated in Fig.1 and Fig. 2 respectively. All the materials discussed as follows are all with an addition of 15wt% of silicate as sintering agent, i.e., all the materials can be formulated as $xBaO-(1-y)SnO_2-y/2Sb_2O_3$-15wt% silicate.

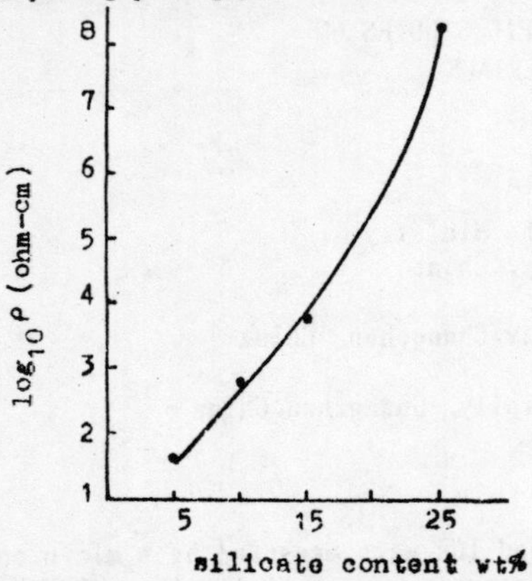

Fig. 2 Semilog plot of ρ as a function of silicate for x=1.

The substitution of heterocations for Sn ions causes the change of the electrical field gradient at Sn nuclear site, which leads to quadrupole splitting of the Mössbauer spectra (Fig. 6). Mössbauer spectrum analysis also shows that when the content of sintering agent varies from 0 to 30wt%, the non-symmetry of two doublet intensity of Mössbauer spectrum increases, which indicates the orientation of the main axis of electrical field gradient varies with the silicate content, so as to cause the variation of the resistivity of the materials. The X-ray diffraction semi- quantilative analyses show that with the rising silicate content, in the phase constituents the high-resistance phases $Ba_3Sn_2O_7$ and Ba_2SnO_4 increase, while the low-resistance phases $BaSnO_3$ and SnO_2 decrease, furthermore, as the silicate content comes up to 25wt%, lots of insulating phase $Ba_2Si_3O_8$ appear.

2. Influence of BaO Content

The results of the X-ray semi-quantilative analyses of some compounds are given in table 1. With the rising BaO content, the contents of $Ba_3Sn_2O_7$ and SnO_2 decrease gradually and disappear in end, and the high $BaSnO_3$ content exist in the region of x=2.0-3.0. For x=0.5, it was assessed that the $Ba_3Sn_2O_7$ phase content is about 40-50%, which shows that in spite of the lackage of BaO, the compound $Ba_3Sn_2O_7$ with rich BaO(BaO/SnO_2 =1.5) is still produced. Simultaneously free SnO_2 is engendered. It is worth notice that single phase $BaSnO_3$ cannot be formed for x=1, it can be approximately formed for x=2, but the single phase $BaSnO_3$ can also be formed for x=1 when there is no silicate. As x value arises to 2.5, the compound Ba_2SnO_4 with a ratio of BaO/SnO_2=2 begins to appear. And there are about 70% $BaSnO_3$ and 30% Ba_2SnO_4 when x value increases to 3.5. It is clear that part of BaO does not participate in forming stannates within the x range from 1.5 to 3.5. The relation between resistivity of the samples and BaO contents is shown in Fig.3.

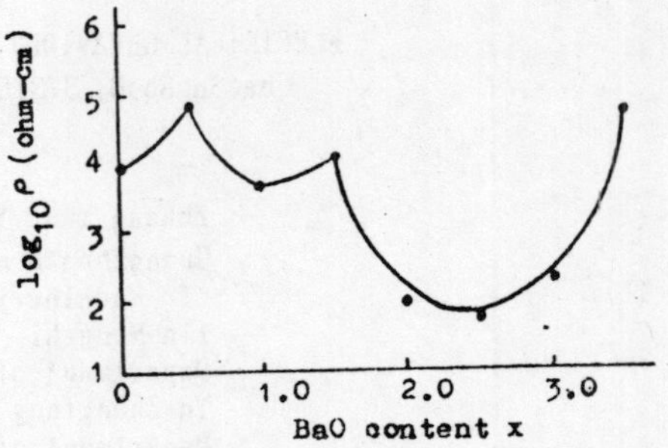

Fig.3 Plot of ρ against the content of BaO.

Table 1. Effects of BaO content on the phase constituent determind by X-ray analysis.

content (mole)			phase constituent (%)				
BaO x	SnO 1-y	Sb O y/2	BS	B_3S_2	B_2S	SO_2	BS_2 *
0	0.94	0.03				100	
1.0	1.0	0#	100				
0.5	0.94	0.03	15	45		30	
1.0	0.94	0.03	60	40			
1.5	0.94	0.03	70	30			
2.0	0.94	0.03	97	3			
2.5	0.94	0.03	85		15		
3.0	0.94	0.03	80		20		
3.5	0.94	0.03	70		30		

* BS-$BaSnO_3$, B_3S_2-$Ba_3Sn_2O_7$, B_2S-Ba_2SnO_4, SO_2-SnO_2, BS_2-$BaSb_2O_6$.
# without silicate

$Ba_3Sn_2O_7$ is the interim compound between $BaSnO_3$ and $BaSnO_4$, with a tetragonal structure. The investigations on $Ba_3Sn_2O_7$ and Ba_2SnO_4, and their electrical behaviour have seldom reported. However, in the case of this work, the electrical conductivity of these formed compounds could be estimated. It could be thought that $BaSnO_3$ and SnO_2 are low-resistance phases, and $Ba_3Sn_2O_7$ as well as Ba_2SnO_4 are high-resistance phases. As shown in Fig.3, the resistivity of the doped SnO_2 is less than $10^4\,\Omega$.cm. Approximate signle phase $BaSnO_3$ comes to existance for x=2, and the resistivity of the material is less than $10^2\,\Omega$cm. It was reported that $BaSnO_3$ is the most liable to become semiconductive among all the stannates and perovskites containing Ba. By comparing the proportions of $BaSnO_3$ and $Ba_3Sn_2O_7$ as well as the resistivity of the samples with x=1.5 and 3.5, we can see that $BaSnO_3$:$Ba_3Sn_2O_7$=7:3 for x=1.5, and $BaSnO_3$: Ba_2SnO_4=7: 3 for x=3.5, i. e. both ratios are approximate, however the resistivity of the latter is greater than that of the former. In the same way by comparing the resistivity of the samples with x=0.5 and 3.5, it can be seen that the resistivity values of both samples are similar, but the content of the high-resistance phase

Ba_2SnO_4 for x=3.5(about 30%) is less than that of high-resistance phase $Ba_3Sn_2O_7$ for x=0.5 (about 40-50%). In consequence, it cound be evaluated that the electrical resistivity of Ba_2SnO_4 is greater than that of $Ba_3Sn_2O_7$. Therefore it could be considered that these compounds in order of the increasing resistivity are as follows: $BaSnO_3$ < SnO_2 < $Ba_3Sn_2O_7$ < Ba_2SnO_4. The correlation between the added amount of BaO and the resistivity shown in Fig.3 has basically represented the content proportion regularity of the low and high-resistance phases in the materials.

3. Influence of Sb_2O_3 Content

According to the composition formula $xBaO-(1-y)SnO_2-(y/2)Sb_2O_3$, for x=1 and x=2, the electrical resistivity variations of the materials with different y values are given in Fig. 4 and Fig.5 respectively. Table 2 shows the results of

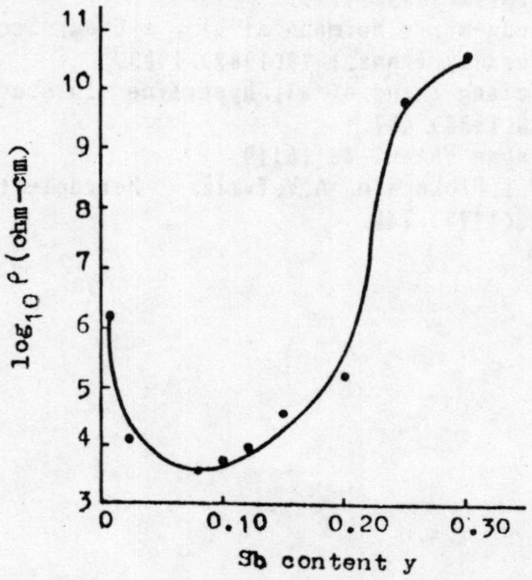

Fig.4 Variations of ρ as a function of Sb content for x=1.

the semi-quantalative analysis. The minimum value of ρ for x=2 is less than that for x=1, both around y=0.10. and the electrical resistivity with different y values for x=2 is also much less than that with the corresponding y values for x=1.

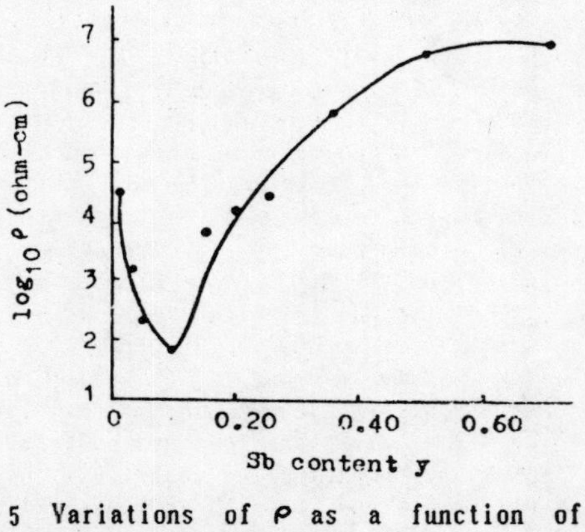

Fig. 5 Variations of ρ as a function of Sb content for x=2.

Table 2. Effect of Sb_2O_3 content on the phase constituent determined by X-ray analysis

content (mole) BaO x	SnO 1-y	Sb O y/2	phase constituent(%) BS	B_3S_2	B_2S	SO_2	BS_2 *
1.0	0.994	0.003	60	25		15	
1.0	0.950	0.025	55	35		10	
1.0	0.940	0.030	60	40			
1.0	0.920	0.040	50	50			
1.0	0.750	0.125	45	55			
1.0	0.700	0.150	40	60			
2.0	0.99	0.005	90				
2.0	0.90	0.050	90				
2.0	0.85	0.075	80	10			
2.0	0.65	0.175	75	25			
2.0	0.50	0.250	65	20	10		5
2.0	0.30	0.350	65				35

* the signs are the same as in table 1.

The curve in Fig.4 is similar to the ρ-Sb_2O_3 curve of $BaSnO_3$ with added Sb_2O_3 reported in Ref.6. Their minimum valuves of resistivity are both around by 0.1 at. added Sb content, but the resistivity in Fig.4 is about one order of magnitude greater than that reported in Ref.6. which may be ascribed to the more amount of high-resistance phase contents. Ref.4 studied the electrical conductivity of $Ba(Sn_{1-x}Sb_x)O_3$ powders, and found that the conductivity has an abrupt decreasing for the further increasing of Sb content for x > 0.5, owing to the occurrence of the insulating phase $Ba_2Sb_2O_7$. The limit of solid solubility of Sb in $BaSnO_3$ is 0.17 at. according to Ref.4, the electrical conduction mainly comes from the doping effect of Sb. In fact, the addition of Sb also gives rise to the variations of phase constituents(Table 2). For example, in the region of y=0.006-0.1 and x=1, with the rising Sb content, $BaSnO_3$ content keeps about 60% , but the SnO_2 content decreases gradually from about 15% to zero, and the high-resistance phase $Ba_3Sb_2O_7$ increases progressively. Since the doping effect of Sb^{5+} in stannate and tin oxide markedly predominates, the conductivity increases drastically with increasing y values. The content of $Ba_3Sn_2O_7$ phase gradually increases for y≥0.1. When the y value is above the solid solubility limit, the doping effect of Sb^{5+} does not increase any more, but the high-resistance phase grows progressively, accompanied with the appearance of quite a number of kaolinita insulating phase, which leads to the rise of electrical resistivity of the materials. In the system for x=2, the insulating phase $BaSb_2O_6$ even comes to existance for the materials with high Sb^{5+} dopant.

The Mössbauer spectra of the samples with an excess of BaO are illustrated in Fig.6. All the spectra consist of two doublet subspectra with quadrupole splitting. With the rising Sb content, Sn content decreases gradually, and consequently the relative transmission decreases too. Pure $BaSnO_3$ has a non-distorted cubic cell, due to the addition of Sb atoms, its lattice structure will produce different extents of distortions. The substitution of Sn atoms by Sb and other impurity atoms will change the electrical field gradient of Sn nuclear site, leading to the quadrupole splitting of the Mössbauer spectra.

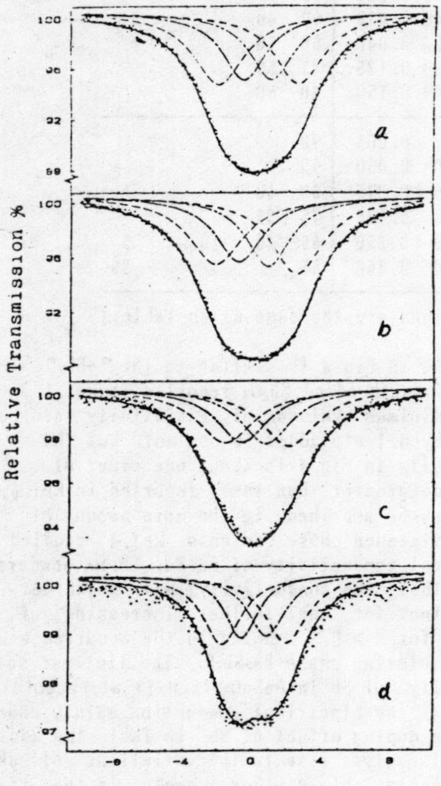

Fig.6 Mössbauer spectra of $xBaO-(1-y)SnO_2-Y/2Sb_2O_3$ at RT.
(1) y=0.006; (2) y=0.026; (3) y=0.115; (4) y=0.209.

CONCLUSIONS

1. $BaO-SnO_2-Sb_2O_3$ system is a kind of resistance materials with practical values whose electrical resistivity could be adjusted in the range of 10^2-10^4 Ω cm by varying the contents of BaO, Sb_2O_3 as well as sintering agent. This kind of materials have a TCR of 1000±200 ppm/°C and stable electrical properties after aging.
2. The substitution of Sb^{5+} for Sn^{4+} to form the donor centres is the essential cause for the electrical conduction in this system. Different contents of silicate, BaO and Sb_2O_3 will change the phase constituents of the materials. Since there exist low-resistance phases of $BaSnO_3$, SnO_2 and high-resistance phases $Ba_3Sn_2O_7$, Ba_2SnO_4 as well as the insulating phases $BaSb_2O_6$ and $Ba_2Si_3O_8$ and other silicates in this system, the phase constituents will influence directly the electrical resistivity of the materials.
3. The addition of sintering agent and Sb_2O_3 gives rise to produce the doublet subspectra with quadrupole splitting in ^{119}Sn Mossbauer spectra, which indcates that different extents of distortions in the lattice structure have occured.
4. The optimum amount of added sintering agent is 15wt% . Excessive BaO can generally improve sintering, and the proper excess of BaO is advantageous to the formation of low-resistance phase $BaSnO_3$. Consequently, excessive BaO is available for low electrical resistivity materials.

REFERENCES

1. V. N. Pogatina, Crystallization and the Characteristics of Crystals, 1, (1974), 124.
2. VINITI, 2886(1979).
3. VINITI, 2545(1979).
4. Jean-marie Hermann, et al., J. Chem. Soc., Faraday Trans. 1, 78(1982), 1983.
5. Yufang Zheng et al., Hyperfine Interactions, 40(1988), 467.
6. Japan Patent 48-16119.
7. O. I. Prokopalo, A. V. Tvrik, Ferroelectrics, 22(1978), 749.

COMPLEX ELECTRODE MATERIALS OF Cu^+ SYSTEM POTENTIAL MEMORIODE

Chen Ai *Wang youjun* *Deng Hong*
University of Electronic Science and Technology of China
Chengdu. S.C. 610054 P.R.C.

ABSTRACT

Cu^+system potential memoriode is a new model of inexpensive electrochemical integral devices, which takes fast–ionic conductor, $Rb_4Cu_{16}I_7Cl_{13}$,as the solid state electorlyte. It has been found from this paper that the perfect principal electric characteristics, especially high keeping potential E_k,and wide range of charging and discharging capacity C_{ch},C_{dc}, etc, depend not only on the composition and the amount of electrolyte but also that of electrode system. Mixing α–phase $Cu_{2-x}Se$, γ–phase $Cu_{2-x}S$ and single phase area $Cu_xMo_6S_{8-y}$with suitable ratio electrolye separately to do the complex electrode and make it become a three electrode system memoriode . Its changing capacity and keeping potential are as follows respectively 190 μA. hr , 180mv for Cu $_{2-x}$Se, 200 μA, hr. 318 mv for $Cu_{2-x}S$, and ~2000 μA.hr, 380 mv for $Cu_xMo_6S_{8-y}$.This improvement will widen the application area of this devices.

INTRODUCTION

Cu^+ionic conductor material $Rb_4Cu_{16}I_7Cl_{13}$has been the focus of worldwide attention because of its high ionic conductivity $\sigma_i = \approx 0.34.\Omega^{-1}cm^{-1}$.The authers developed the potential memoriode which is make as a timer or turner using $Rb_4Cu_{16}I_7Cl_{13}$ for solid electrolyte. It has been found through research that the principal characteristics of this devices, such as keeping potential E_k, charging and discharging capacity C_{CD} etc. not only depend on the fast ionic conductor itself but the electrode material ,including the working, counter and reference electrode decisively .

Many experiments have been performed on electrode system $Cu_{2-x}Se$, $Cu_{2-x}S$ and $Cu_xMo_6S_{8-y}$in this paper. The structure of potential memoriode is

PM+E / E / PM+E / E / PM+E

Where E is solid electrolyte Rb_4Cu_{16} I_7 Cl_{13}, PM is electrode material.

EXPERIMENTAL RESULTS

1.Electrode system $Cu_{2-x}Se$

The electrode material of Potential memoriode normally requires a mixing conductor which has high electronic conductivity (σ_e),ionic conductivity (σ_i),and high activity of ion Cu^+but is chemical inertia. If some other chemical / electro–chemical reaction takes place in the electrode, the charging and discharging curve will deviate from linearity somewhat ,and the keeping potential will be unstable. It is thus clear from Cu–Se System phase diagram (Fig.1) that the two low temperature phase areas α– and β– are present near Cu_2Se [1] . Its electrical conductivity is shown in table 1.

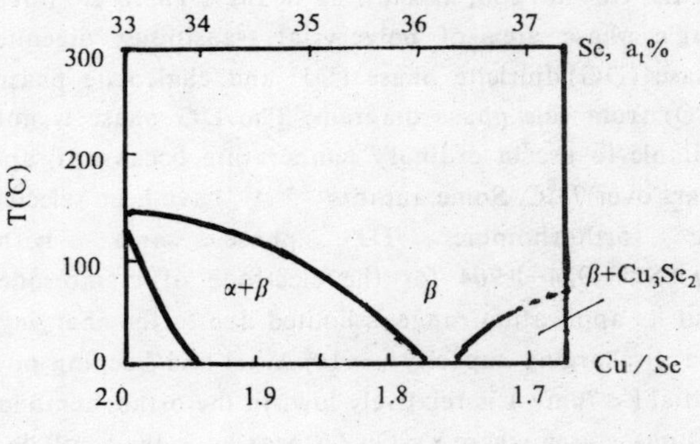

Fig.1 Phase diagram for the Cu–Se system in the range between 33.5 to 3.7 at .% Se

Some researchers [2] [3] selected $Cu_{1.75}Se$ for the electrode material of potential memoriode in virtue of more high σ_i and σ_e of β–phase, but the deficiencies are that the keeping potential $E_k$is on the low side and the potential attenuation is rather poor. As a result of the activity coefficient K of α– phase is higher than that of β– phase , the charging area of memoriode useing electrode α–$Cu_{2-x}Se$ is wider, so it will raise E_k. In the other hand, the Lower $\sigma_i$of pure $\alpha-Cu_{2-x}Se$ also causes the unstability of E_k, so it is necessasy to mix some solid electrolyte. If the mixing electrolyte is overweight ,the over–potential Eov will increase. In adddition, an over thick working electrode leads a greater concertration gradient on electrode and a larger transport of ions, it also bring about some high over–potential of the device. Takeing $x<0.04$,right amount of mixing solid electrolyte and suitable thickness of electrode, the E_k of memoriolde is almost 180mv within the linear rang. But the deviation of keeping potential $\triangle E_k$is difficult to less than 0.3mv after 48 hr. so far.

Table 1 The electrical conductivity of $\alpha-$, $\beta-$ $Cu_{2-x}Se$

phase area	σ_i $(\Omega^{-1}\cdot,cm^{-1})$	σ_e $(\Omega^{-1}\cdot,cm^{-1})$
α–phase	5×10^{-2}	~1000
β–phase	15×10^{-2}	~2300

2. Electrode system $Cu_{2-x}S$

Cupper(I) sulfide exists a numbe distinc forms depending on the Cu / S ratio, as showing in Fig.2.There are three single phase area of polycrystal transitions, digenite phase (DG),djurleite phase (DJ) and chalcocite phase (CO) from this phase diagram. The DG phase is not suitable to use in ordinary temperature because it, appears over 73°C. Some authors [4] [5] have been selected the orthorhombic DJ phase area with Cu / S = 1.934–1.964 for the electrode of memoriode .But its application range is limited due to the charging and discharging capacity (~ 10μA.hr) and keeping potential (< 70mv) is relatively low. In the orthorhombiic CO phase area where x = Cu / S near by 2, the small deviation of copper content can leads to remarkable change of σ_i.There are three polymophic phases ,namely α–phase (>470℃), β–phase (470℃ –110℃) and α–phase(< 110℃) appear in this phase earea. In the α–phase ,it behaves as a P–type semiconductor, and the total conductivity increase exponentially (up to $10^3\Omega^{-1}.Cm^{-1}$) with temperature, but the ionic conductivity is rather poor, its ratio $x(=\sigma_i/\sigma_e)$ is only 10^{-5}.
In the β–Phase ,the rather remarkable ionic current together with the sharp decrease of electronic current ,the x is almost $\gg 1$. In the α–phase, the ionic current disappears rather abruptly, even though the electronic current is not too low, it also can't be used at ordinary temperature. In addition, the transition energy from r–phase to β–phase (~ 4.8Cal / gr.) is almost 10 times larger than that from β–phase to α–phase. It means that the anomalous thermo–electromotive force is caused by the heterogeneous distribution of mouable copper ions, so it is a kind of stable modified electrode material.
Synthesize the opinions of above mentions that the composition of this electrode system is adopted at $x<0.004$. Moreover, to raise the keeping potential and reduce the over–potential, we mingled suitable proportional solid electrolyte powder with $Cu_{2-x}S$ and mixed it well. It can made the non–uniformity of charging in the surface and interface of the material powder as small as possible .The charging and discharging curve of memoriode is shown in curve 2 of Fig.6, $E_k$is 318mv and $C_{CD}$is 200 μA.hr.

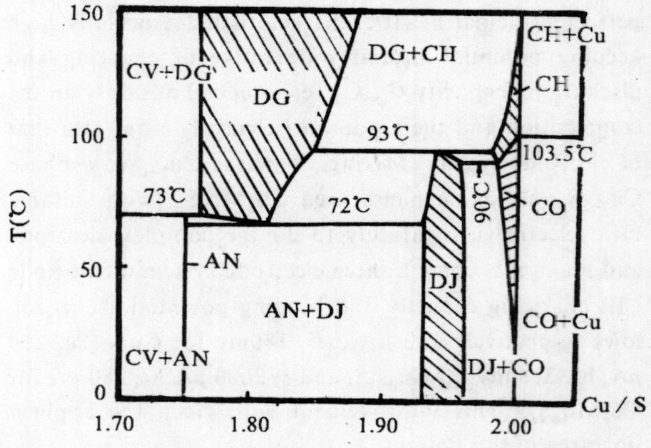

Fig.2 phase diagram for the Cu–S system below 150℃ (part.)

3.Electrode Systems $Cu_xMo_6S_{8-y}$

In order to get the eleclrode material which has greater activity of Cu^+, thus farther raises the value of C_{CD}, $E_k$and stability of E_k, we have also done some researches on system $Cu_xMo_6S_{8-y}$.When the ionic vacancies is within the range of $0<y<0.4$, the deviation of x for single phase chevrel is shown as Fig.3 [5] [6] In our case, it is indicated that there is wider single phase range at y = 0.15~ 0:30 when $x\approx 2.1$. If the composition is $Cu_2Mo_6S_8$(that is x = 2 y = 0), then some free phase MoS_2 can be found. If an anneal processing is going on at some suitable temperature for a long period ,the content of free phase can be reducced or even disappear at all. It means that the heat treatment can extend single phase area of Fig.3.

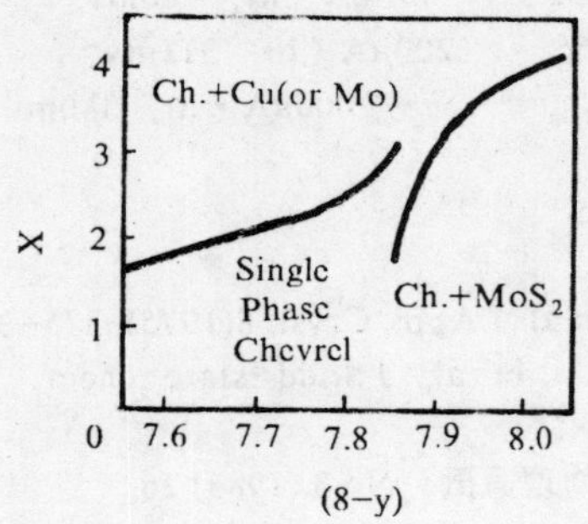

Fig.3 Single phase range of $Cu_xMo_6S_{8-y}$ and the relationship between x and y within this range.
(Sintering temp. is 1000°C)

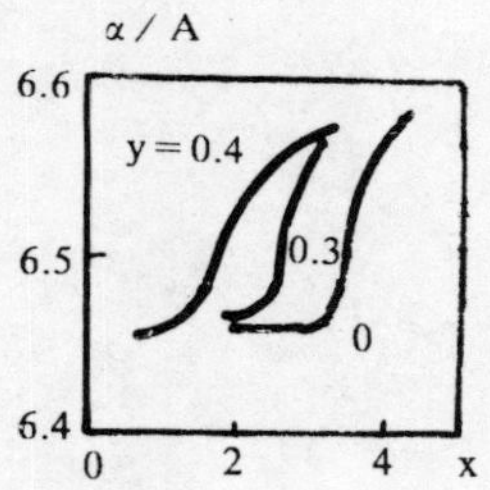

Fig.4 Rhombohedral lattice Parameters, α,A of $Cu_xMo_6S_{8-y}$ as a function of x and y.

On the other hand, the change of x,y can cause the transfer of crystal parameter α / A of this triangle crystal system fast ionic conductor (Fig.4).The horizontal line segment in Fig.4 is a condition for keeping crystalline parameter constant .Ion Cu^+ can be distributed over different positions I or II near 8 group centers of Mo_6S_{8-y}, and position I is more stable than II [7] .Greater part of Cu^+ is located in position I when y=0.4, x<1.8. some of them will transfer to position II from I when x>1.8 .The lattice parameters are beginning to change when y=0.4 and x≈3.0 when x>3.0 , 1.4 out of 6 positions I are occupied by Cu^+,and position II by others. It is thus obvious from Fig.4 that x decreases as y grows up. It indicates the preventive effect of vacancy of S^{2-} on the entering of ion Cu^+ from postition I to II . We consider that the stoichimetric deviation on Mo_6S_8 is resulted from the interstitial of M_0^+ which is located at position I instead of vacancy of S^{2-}.For the System $Cu_xMo_6S_{8-y}$, the relationship among activity α_{cu} of ion C^+u, keeping potential E_k and x,y is shown in Fig.5.
In view of analysis above ,three characteristics compositions $Cu_2Mo_6S_8$, and $Cu_4Mo_6S_8$. have been experimented .The results indicate that $Cu_2Mo_6S_{7.7}$ has the optical properties ,$\sigma_i \approx 0.02\Omega^{-1}.cm^{-1}$,$\sigma_c \approx 97\ \Omega^{-1}.cm^{-1}$. The charge and discharge curves of the potential memoriode which is made from the complex electrode using $Cu_2Mo_6S_{7.7}$ mixed with suitable propertional fast ionic conductor is shown in Fig.6, all E_k(~ 380m−)and charging, discharging capactity ($C_{CD} \approx 200\mu A.hr$) increase greatly.

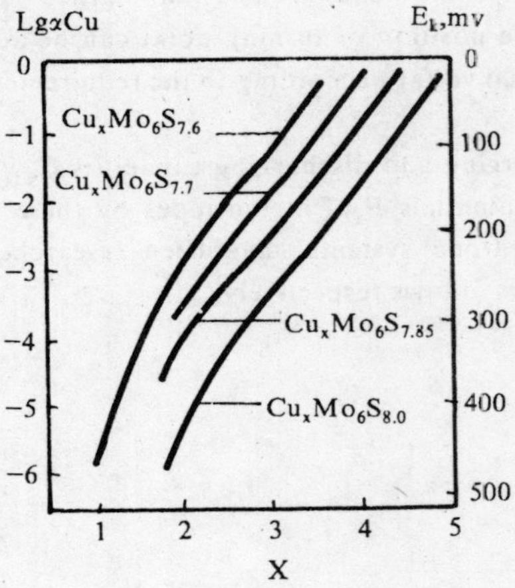

Fig.5 α_{Cu} adn E_k of system $Cu_xMo_6S_{8-y}$ as a function of x, and y.

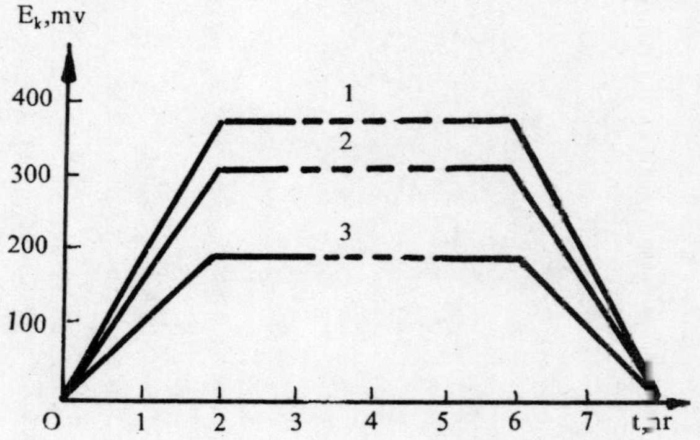

Fig.6 charging and discharging curves of the memoriodes in different electrodes
1.$Cu_2Mo_6S_{7.7}$, I_{ch}= 1mA
2.$Cu_{1.996}S$, I_{ch}=0.1mA
3.$Cu_{1.95}Se$, I_{ch}=0.1mA

CONCLUSIONS

1.The effect of electrode system on electrical characteristics of potential memoriode is very seriously. Except the composition of electrode material ,must pay attention to its amount ,particle size and the mixing ratio with the solid electrolyte (fast ionic conductor).

2.The electrode system $Cu_{2-x}Se$ is selected in α−phase area of Cu −Se phase diagram, $Cu_{2-x}S$ electrode is selected in α−phase area and $Cu_xMo_6S_{8-y}$electrode is selected in the single phase area showing in Fig.3. There is a turnning point appears in the curve 1 of Fig.6 which is caused by the changement of lattice parameter α/ A. The position of turning point can be adjested by pre−applied voltage according to the requirement.

3.The charging and discharging capacities C_{CD} and the keeping potentials $E_k$of memoriodes by these three different electrode systems have been researched in the work are as follows respectively.

$Cu_{1.95-2.00}Se$: 190μA · hr, 180mv.
$Cu_{1.99-2.00}S$: 200μA · hr, 318mv.
$Cu_2Mo_6S_{7.70}$: ~2000μA · hr, 380mv

REFERENCES

1. A.Toneje et al .J.Appl. Cryst. 8(1975), 375−379.
2. T.Takahashi et al, J.Solid state chem. 16(1976) 35−39.
3. 肖超亮.《物理通讯》No 3.(1983) 26
4. A.D.Jonath et al ,J.Vac. Sci. Tech. 16(2) (1979)200−202.
5. R.Chevrel et al ,J.Solid state chem. 3(1971) 515−519
6. J.Mizosaki et al ,Solid state Ionics, 11(1984)293−297.
7. K.Yuon et al, Acta Cryst. B 33(1977) 3066−3072.

A NEW APPLICATION OF SILVER-CADMIUM OXIDE COMPOSITE MATERIALS-THE ELECTRODE PASTE OF MONOLITHIC CERAMIC CAPACITORS

Yu Shougeng
Institute of Precious Metals
Kunming, China

ABSTRACT

A Ag(31)CdO(69) electrode paste can be sintered in the atmosphere at a higher temperature than melting point of silver. The coalescence of liquid phases during sintering can be reduced by the control of pretreating condition of CdO and the electrodes keep characteristics of lower electric resistance (less than 0.07Ω/sq). Cermet electrodes with close texture can be formed by the aid of the diffusion (solution) / reprecipitation with a rise in temperature during liquid-phase sintering. Perfectly destructive temperature of the electrode is 1046°C. When matching with sintered dielectric composed as $PbMg_{1/3}Nb_{2/3}O_3$-0.1 $PbTiO_3$-0.55$PbCd_{1/2}W_{1/2}O_3$, measured tgδ-t and ε-t curves, as compared with measured both from silver electrode are about the same. It is used as internal electrode and matched with former dielectric to make monolithic ceramic capacitors. When rised sintering temperature from 880 to 950°C, the advantage descending loss apparently is obtained.

INTRODUCTION

The AgCdO composite material originally has been sintered material for electrical contacts in power engineering. The CdO adding improves certain properties of pure Ag such as increasing antiwelding ability etc[1]. The CdO content is 8~15%. Here report on another AgCdO composite material, electrode paste. It is used as internal electrode when making monolithic ceramic capacitors.

Monolithic ceramic capacitors are superior in wider range of capacity, small bodies and nonpolarity etc. They extensively have applied to electronic packages which are of higher density, smaller and thinner models. Making kinds of capacitors has needed sintering at high temperature generally, and using Pt and Pd as internal electrodes. When bigger capacity is required and areas of electrode must expand thereby, the cost of electrode material can rise apparently, so that the commercialization is difficult. At present monolithic capacitors whose internal electrodes consist of Pd with high Ag content have been applied in practice. With electrode materials are replaced gradually by pure Ag or Ni low cost may be hoped to reach. However questions resolved necessary are that the dielectric material can be sintered at lower temperature or in reducing atmosphere[2]. Moreover to set up equipment sintering in reducing atmosphere increases the cost, too. It is reported that Zn paste instead of Ag had became an electrode after sintering in the atmosphere at 460-900°C[3]. This study has pursued a low cost electrode paste sintered in the atmosphere at a higher temperature than melting point of Ag. It will be advantageous both to study dielectric and to reduce producing cost of monolithic ceramic capacitors.

Composite material has been selected as a new material to tackle this kind of problems. Many of this kind of material important to technology have been made by liquid-phase sintering process. But some studies have showed that the liquid-phase components which favoured the compact densification during sintering just were the sources destroying this kind of material applied at a high temperature, especially in the presence of stress[4,5]. Fortunately the imperfection can be avoided when this kind of material is used as an electrode paste of monolithic capacitors. It is obviously important to densify the electrode by liquid-phase sintering because the past itself can not be compacted. The pretreating conditions for CdO have been studied so that internal electrodes do not sweat liquid phase during sintering, hence remain whole patterns. Its conductance and influence on electrical properties of the dielectric material have been examined.

EXPERIMENTAL

Finely-ground powders were formulated to several kinds of past with solution of ethyl cellulose in terpineol. They were printed on 96% alumina substrates and sintered. According to the technology producing capacitors, drafted sintering condition was: In the atmosphere samples heated gradually from ambient temperature to 980°C for about 14 hs and held for 2 hs, then cooled down in the oven. The results were as follows.

A. The resistance of $Ag(60)SnO_2(40)$ was less than 0.05 Ω/sq, its pattern was whole, but the texture was not close. Then it as electrode paste was applied to unsintered dielectric chips composed as $PbMg_{1/3}Nb_{2/3}O$ -0.1$PbTiO_3$-0.55$PbCd_{1/2}W_{1/2}O_3$, heated to 880°C and held for 2 hs in the atmosphere. Results showed that electrodes adhered to chips weakly and sheded easy. Loss tangent values of samples were more than 56×10^{-2}.

B. In AgCdO system Ag powder consisted of spherical particles of 0.2-0.5 μm in diameter in dominance and a few cubical particles of ~1 μm in length. Through same conditional test as former different results showed. When CdO pretreating temperature was at the low side, the resistance increased. Added quantity of CdO decreased, the liquid phase sweated during sintering. When CdO pretreating temperature was at the high, the resistance, too, increased and molten drops might show. The resistance of Ag(31)CdO(69) formulated with CdO pretreated properly was less than 0.07Ω/sq. Its pattern was whole and the texture was close. Fig.1 is its metallograph.

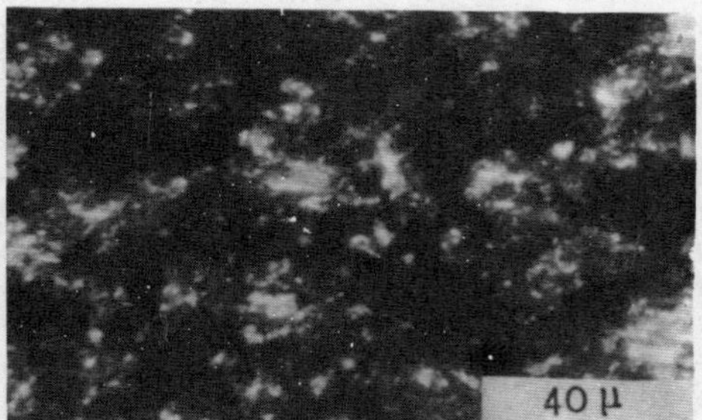

Fig.1 The metallograph of Ag(31)CdO(69) formulated with CdO pretreated properly.

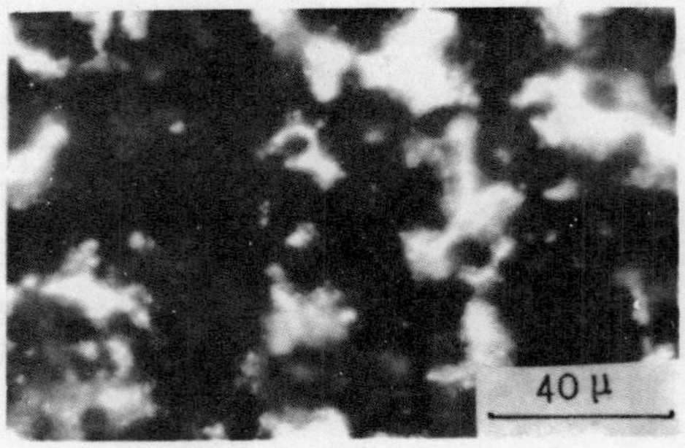

Fig.2 The metallograph of Ag(31)CdO(69) formulated with CdO pretreated at the high side of temperature.

Fig.2 is provided for comparison with Fig.1. The sample was formulated with CdO pretreated temperature at the high, but liquid phase did not sweat yet during sintering. Fig.3 shows DTA curves of Ag(31)CdO(69) solid powder. There is a wide endothermic peak with tip temperature 864°C, in the range 713 to 954 °C. There are 4 sharp endothermic peaks respective at 958, 979, 1026 and 1046°C.

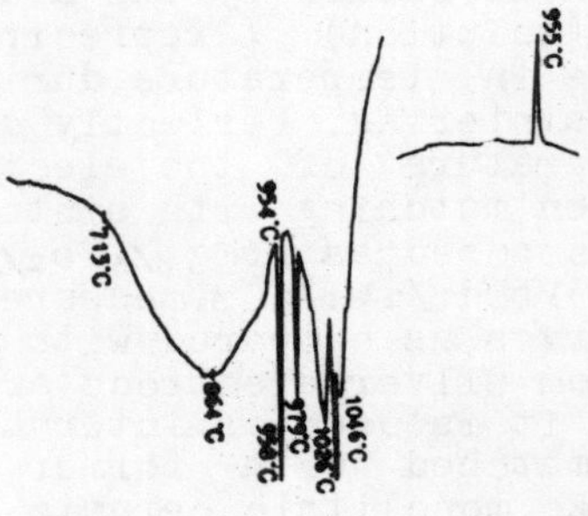

Fig.3 DTA curves of Ag(31)CdO(69)

Among them, the last tip value was very low so that it exceeded ±10μV, the limits of sensitivity regulated primarily. The tip is not observed through ±25,50, until the third regulation ±100μV. When cooling there is an exothermic peak at 955°C only. It as electrode paste was applied to chips and sintered as the former. Results showed that electrodes adhered to chips strongly. Loss tangent values of samples, $tg\delta=(0.53\text{-}0.60)\times10^{-2}$. Then applied electrode paste to sintered chips, made same sintering test. Both changes of dielectric constant, ε, and loss tangent, tgδ, respectively with temperature, which are shown in Fig.4. Both curves obtained when used Ag electrode are shown as well for comparison. Losses resulted from both electrodes are same principally; dielectric constant, ε, brought by Ag(31)CdO(69) shows higher slightly.

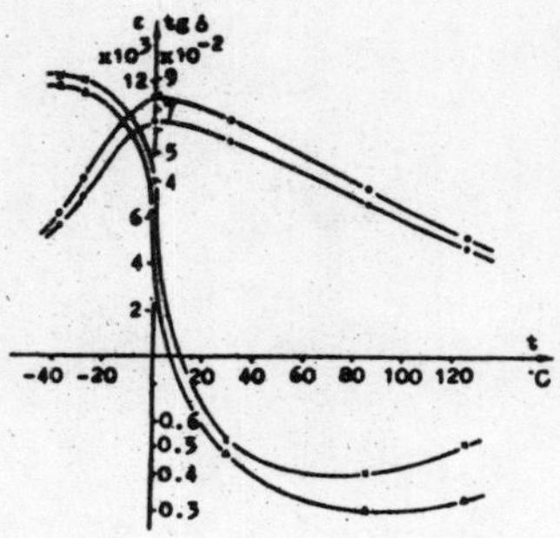

Fig.4 ε-t, tgδ-t curves of both $Ag(31)CdO(69)$, Ag electrodes.
o——o ε-t curve of $Ag(31)CdO(69)$;
•——• ε-t curve of Ag.
x——x tgδ-t curve of $Ag(31)CdO(69)$;
▲——▲ tgδ-t curve of Ag.

$Ag(31)CdO(69)$ paste, as internal electrode, was printed on single side of unsintered dielectric films. After over lapping the films we took 2 sintering schedules respectively at the highest temperature 880 or 950°C. Measured capacity and loss tangent values on monolithic capacitors showed c= 13800-20500 pf, $tg\delta=(0.6-1.0)\times10^{-2}$ for sintered at 880°C; c=13100-21100pf, $tg\delta=(0.25-0.46)\times10^{-2}$ for 950°C. For comparison silver electrode paste was printed on 2 sides of films. Measured resultants on monolithic capacitors sintered at 880°C showed c=14200-25800 pf, $tg\delta=(0.5-0.9)\times10^{-2}$. When sintered at 950°C the silver electrode melted down.

DISCUSSION

The sintering in the atmosphere for longer period and the holding at higher temperature, 980°C, than melting point of Ag for 2 hs, both $Ag(60)SnO_2(40)$ and $Ag(31)CdO(69)$ could withstand such a severe test. But the texture of former sintered film was not close as compared with that of the latter. Furthermore the adhesion to dielectric material was weak. When matching with dielectric as an electrode very big loss was shown. Obviously $Ag(60)SnO_2(40)$ here could not be used as electrode material.

$Ag(31)CdO(69)$ was subjected to a liquid-phase sintering in the process from the paste to forming the electrode. Surface energy of CdO is lower because the solid CdO contains cations with incomplete electron shells, Cd^{2+}, having lower polarizing power. But liquid Ag has a high surface tension. Consequently it is a nonwetting solid-liquid system. M. N. Rahaman et al. have reported that CdO compact could be sintered to higher density in the temperature range 700 to 850°C[6]. Therefore CdO obtained had different surface area according to pretreating condition. It would affect structure and properties of the cermet electrode formed by sintering. The microstructure showed in Fig.1 is rather favorable, the dispersion of phases is rather uniform. The electrode has characteristic of lower electric resistance. On the other hand, Fig.2 shows that Ag coalesces more. The resistance of electrode is higher. This is because pretreating temperature was at the high side, total area of CdO decreased. Increasing Ag coalescence would decrease liquid-solid area whose energy is high, and increase solid-vapor area whose energy is low, thus total free surface energy of the system would decrease. When further rising pretreating temperature Ag coalescence would be aggravate correspondingly and cause liquid phase to sweat during sintering. Decreasing pretreating temperature for CdO caused total area of CdO to increase. The resistance showed increase. Decreasing added quantity caused total area of CdO to decrease and could make liquid phase sweat during sintering.

DTA result of $Ag(31)CdO(69)$ powder (see Fig.3) shows that there are a series of diffusion (solution) / reprecipitation processes in a wider temperature region. Solid phase diffuses into liquid phase (solution), then the precipitate separates out; it is remelted with a rise in temperature, hence new liquid phase forms. Diffusion of solid phase (solution), precipitation, time and again, such transport of matter taking place with a rise in temperature changes composition of the liquid phase, decreases the solid-liquid interfacial energy, improves the wetting and makes the texture of obtained $Ag(31)CdO(69)$ electrode close. Very low tip value of peak at 1046°C shows the greatest part of the electrode was solid yet before it, although the melted phase was presented with a rise in temperature, hence original pattern of electrode kept well.

Fig.4 presents 2 groups of curves concerning changes in the dielectric constant, ε, and the loss tangent, tgδ, with the temperature respectively. In each group 2 curves have the same tendency of change and near each other in position. It shows $Ag(31)CdO(69)$ electrode does not reflect on dielectric properties. When matched with unsintered dielectric to make monolithic capacitors, sintering temperature could rise from 880 to 950°C. Results showed the losses descended although capacity values unchanged essentially. Here, it should be stated that the object of discussion in this paper is a Ag(31)CdO(69) composite electrode paste. Selected dielectric has been only used for a reference. It is reported that Ag(70)Pd(30) as internal electrode matched with mentioned dielectric. The dielectric constant, ε, could increase to 14300 after sintering temperature rised to 1100-1140°C[7]. To this temperature range neither pure Ag nor Ag(31)CdO(69) electrode withstands.

CONCLUSION

For preparing Ag(31)CdO(69) composite electrode paste which can be sintered in the atmosphere at a higher temperature than melting point of Ag, the pretreating condition for CdO needs to set up by experiments. A series of diffusion (solution) / reprecipitation processes take place with a rise in sintering temperature. Such transport of matter makes paste be sintered an electrode with a close texture. The temperature destroying the electrode perfectly is 1046°C. Its resistance is less than 0.07 Ω/sq. When matching with selected dielectric the monolithic ceramic capacitors made at a higher sintering temperature than that of silver electrode are characterized by an advantage descending losses.

ACKNOWLEDGEMENTS

The author is indebted to Mr. Lu Banghong, Ms. Yang Yulu for assistance in analyses and to Mr. Li Tingliang, Ms.Li Qiuzheng for assistance in experiments.

REFERENCES

1. H. Schreiner, Powder Metallurgy International, 12(1)(1980), 16.
2. Gen Itakura, Takashi Iguchi, Takayuki Kuroda, National Technical Report, 31(3)(1985), 145.
3. Journal of Electronic Engineering, 18(178)(1981), 32.
4. J.E.Marion, A.G.Evans, M.D.Drory, D.R.Clarke, Acta Metall., 31(10)(1983), 1445.
5. M.D. Thouless, A.G.Evans, J. Am. Ceram. Soc., 67(11)(1984), 721.
6. M.N.Rahaman, L.C. De Jonghe, J. Am. Ceram. Soc., 67(10)(1984), c-205.
7. Gao Hongqing, Electronic Components and Materials, 5(2)(1986), 52.

COPPER-BASED CONDUCTIVE PASTES

Sun Wentong, Zhao Lin, Tan Fubin
Institute of Precious Metals
Kunming, China

ABSTRACT

Interest in thick film made of copper-based materials for electrodes uses has increased considerably during the past few years. A series of copper(coated with silver) containing compositions were prepared in this paper. F-01 resole resin was used as adhesive. These pastes have good performances for conductive connector of carbon-film resistors and other conductive-adhesive uses. Scanning electron microscope and other devices were used to analyze conductive and adhesive mechanism and other performances.

INTRODUCTION

This paper reveals a series of copper-based conductive pastes to meet the requirement of the electrode of carbon resistance and carbon rheostat. Using copper as a substitute for most parts of silver can greatly reduce the cost of the raw material. Therefore, this has a latent applying power.

EXPERIMENTS AND RESULTS

The mixture of Ag-coated Cu powder and silver flake were taken as conductive phase. A resole resine, which contained linear and branched compound and some network chains, was regarded as an adhesive phase. The pastes also contained a few additions.

Conductive pastes were screen-printed onto the nonconductive phenolic plastics substratum using a laboratory screen printer and a 200-mesh synthetic fabrics mask. The printed conductor consisted S-shaped pattern of 1.0x100 mm² used for measuring circuit resistance. These test parts were heated in the air blast electroheated furnace at a peak temperature 200 °C for 20-25 minutes after drying at 120 °C for 5-10 minutes. All resistance of test parts were measured by Automatic Component Bridge B605.

Thermal cycling tests were carried out by subjecting parts to 30 minutes at -30 °C and 30 minutes at 80°C, a total of 10 cycles were performed. The results are shown in Table 1. Aging tests that the test parts of these three samples were put at 120 °C for 500 hours were completed. The resistivity changes are shown in Table 1.

Humidfying tests at temperature 40±1 °C, relative humidity 95% for 200 hours, were finished in the Thermostatical Humidstat DL302 and results are shown in Table 1.

A square conductive pad 2.0mm in each side were printed onto the substratum. In order to solder easily for pulling test we make the test patterns coated with silver by electroless plating. Then a nailhead with the same pattern was soldered to the test pad perpendicularly. Adhesion was measured as pull strength using Y36-1 Pulling Tester at the rate of 1.5 millimeter per second. The results are shown in Table 1, too.

Table 1 contains a series of requirements of carbon-rheostat, which were suggested by many carbon-made factories. When we compared these results we can see CA30 sample can meet successfully the requirements of low resistance carbon conductor, CA20, CA10 had reached the qualities index of high resistance carbon conductor except aging test. The cross-section of CA30 sample pad was prepared by mechanical polishing and chemical corroding. Corroding can lessen the reflection of copper particle. Then a Universal Camera Microscope 'MeF₂' was used to take a photo for CA30 section sample, the result is shown in Fig.1. In this picture white flake is the symbol for silver, piebald for copper and grey for resol. We also used 'MeF₂' to analysis for the thickness changes from heat-

Table 1 Copper-based pastes performances

Sample	Composition Cu	Ag	F-01	Sq.resist. Ω/□	Adhe. N/m²
CA30	49.7	21.3	25	0.05	3×10^6
CA20	59.2	14.8	22	0.13	3×10^6
CA10	69.3	7.7	19	0.18	3×10^6
Low [b]	/	/	/	0.10	good
High [b]	/	/	/	0.50	good

Variance in resistance $\Delta R/R$ [a]

95% R·H test	therm.cycling	aging test
0.063	0.005	0.063
0.100	0.029	0.43
0.065	0.067	1.00
0.10	0.10	0.10
0.10	0.10	0.10

a) $\Delta R=R_{before}-R_{after}$ $R=R_{before}$

b) Low and High indicate the performances of electrode of low and high resistance-carbon rheostat.

time at 150°C. No changes were found except for the first 5 minutes.

The micrograph of the surface appearance of CA30 sample pad was taken by JEX-2000 Electronic Scanning Microscope. The result is shown in Fig.2. Because resol couldn't reflect electron we can not see the figure of resol.

Fig.3 shows that CA10 sample resistivity follows the changes with heat-time at peak temperature 150°C.The sample resistance changed from metal contg. with a curve, shown in Fig.4.

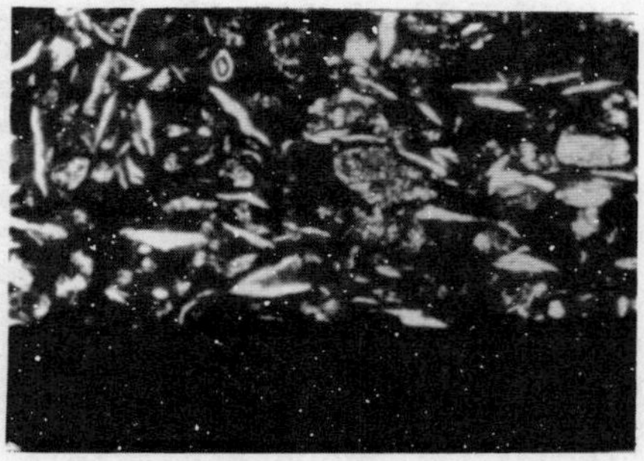

Fig.1 Opitical Micrograph of CA30 cross section(800x)

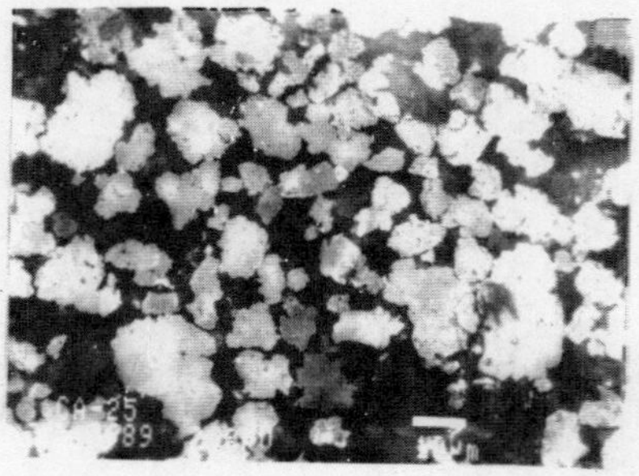

Fig. 2 Scanning electron micrograph of CA30 sample(800x)

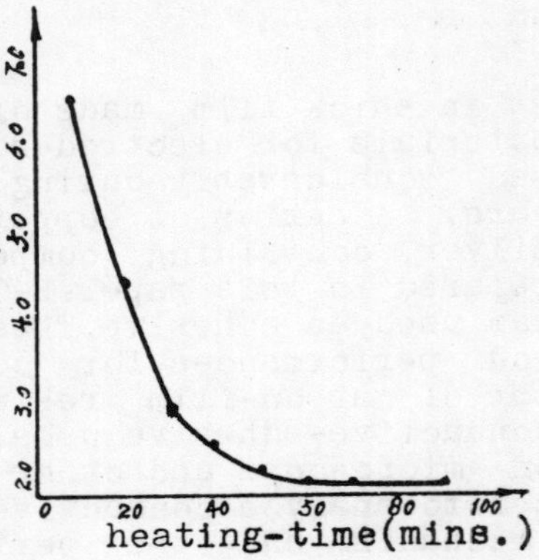

Fig.3

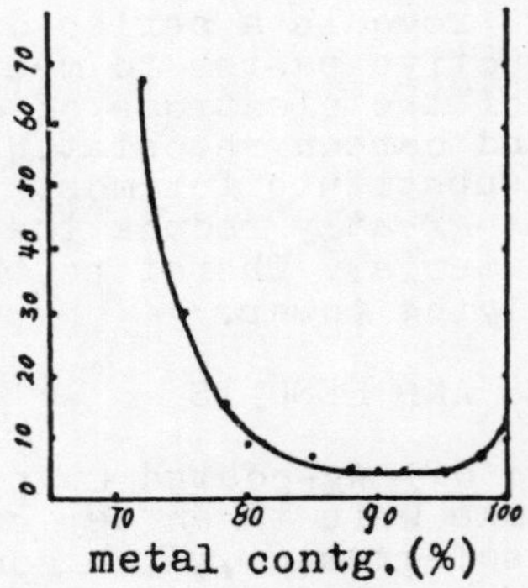

Fig.4

DISCUSSIONS

When silver-coated copper powder and silver flake were mixed with F-01 resol, the metal particles were enveloped by resol and they couldn't contact each other.This film had the resistivity near to the resol's.

When the copper-based pastes were heated,two processes happened:(1) Solvent vaporization and thermo-molecular motion made the metal particles rearrangement in order and the pad volume contracted. (2) Polymeadditive and condensation polymeric reactions took place in the resole phase. A large three-dimensional network compound was produced,which made the metal particles expose and oonnect to each other.

We may reasonably conclude that this process would proceed step by step with the heat-time. Therefore,CA10 sample resistance reduced with heat-time, see Fig.3.

All these results brought about increase of condensation of film and continuous metal particle chains performance, see Fig.1 and Fig.2. In these two pictures, silver flake and copper particles are well distributed.The flake silver spanning the copper particle would greatly increase the conductivity of the paste.

An excellent metal-to-resin ratioes should be obtained in each conductive paste. These ratioes relied on the qualities of resin and metal which were used.When we selected F-01 resol and Ag-coated copper and silver flake, we obtained the excellent ratioes: 77:19 for CA10, 74:22 for CA20, 71:25 for CA30.In these ranges,not only did the metal particles contact closely, but obtained excellent adhesive strength as well.See Table 1 and Fig.4.

Generally,the film resistance drops when increasing metal composition.However,it doesn't always increase the qualities of film.When the ratio exceeds to an excellent point it almost useless due to the lack of resin. Furthermore,adhesive would descend and paste would be difficult to print.The surface tension,which the resin acts on metal particles to rearrange in order and to make the film contract when heated, would not appear obvious.Therefore metal particles could not contact each other closely and the density of the film would reduced.Especially when the ratio was far from the excellent point the resistivity of film would rise again, see Fig.4.

Copper particle surface is so active that it always is covered with oxidic film, as a result that copper could not contact directly. In this paper, copper particle is coated with silver film by electroless plating, which can prevent copper from oxidation effectively.

In Table 1 the result of aging test are not excellent. These result are for two reasons: (1) when resol resin was kept at a high temperature for a long time, the resin molecular may be oxided and split and their bond structure would be loosened. Furthermore, copper ion,as a catalyst,would cause the resin to be oxided and split rapidly[1].

$$RR' + O_2 \longrightarrow R'O_2\cdot + R\cdot$$
$$R\cdot + O_2 \longrightarrow ROO\cdot$$
$$ROO\cdot + RH \longrightarrow ROOH + R\cdot$$
$$ROO\cdot + R'OO\cdot \longrightarrow ROOR' + O_2$$

(2) coating with silver is only a mechanical envelope.Defect usually exists in the coating layer,so that the oxidic velocity in the defect is more rapid due to the microelectrode corrosion. The standard potentials for copper and silver are shown below:

$$E^\circ_{Cu^{++}/Cu} = 0.337\ V$$
$$E^\circ_{Ag^+/Ag} = 0.7995\ V$$

When copper contacts silver a microcell would be composed. Among the two electrodes copper is anode being oxided and silver is cathode,O_2 and other compounds would be reduced around the cathode. In the aging test, we found that a slite grey appeared on the film surface.

Humidfying test results in Table 1 are not heart-stirring for CA10,CA20, even for CA30 due to the above reasons especially.But the oxidation was not severe enough to affect resistance notably.Generally speaking, all performances of CA30 can meet the requirements of low resistance carbon rheostat and CA20, CA10 can near to the high resistance-carbon requirements.

A good pull strength of samples CA10, CA20,CA30 in the pulling test are shown in Table 1.These resulted from mechanical diffusive and chemical adhesives.

In Fig.3,F-01 resol indicating with grey diffused all over the picture and joined up with phenolic substratum without a crackle. This result in the fact that F-01 resol can easily wet the surface of metal particles and substratum.When the resol which distributed all over the film was heated, a large three-dimension network formed.Metal particles could get in touch with each other and fixed firmly by the network, This mechanism was good enough for film adhesion. However, it is an essential prerequisite for metal particles to deprive some gas which attached to the surface of metal particle in order to absorb the resol. This process preoceeded slowly because of the high viscosity of F-01 resol. In Fig.3, some crackle which couldn't reflect light as photomicrographing existed around the copper particle, indicated with black. But this phenomenum was not too expensive to affect the performances of the film severely.

As described above, F-01 resol can wet the substratum surface. Especially at high temperature, surface tension overcame capillary impedence, thus the element chains, even full chains, can diffuse into the substratum[2]. These chains linked up with other resol chains and anchored in the substratum when heated. On the other hand, -OH and $O=\underset{|}{C}-H$ radicals in the resol and phenolic substratum could react to each other:

$$R\text{-}CHO + R'\text{-}OH \longrightarrow R\text{-}CH(OH)\text{-}OR'$$
$$R\text{-}CH_2\text{-}CHO + R\text{-}CHO \longrightarrow R\text{-}CH(CHO)\text{-}CH(OH)\text{-}R'$$

Thus, some new chemical bond existed in the interface between film and substratum, we call the interface a diffused layer. After the pulling test carried out this layer was often pulled off.

CONCLUSION

1. Ag-coated Cu particle can effectively be prevented from oxidation and has a good conductivity. Silver flake linking up with copper particle increases the conductivity obviously in the film.
2. CA30 can meet the requirements of conductive connector for low resistance carbon rheostat and CA20, CA10 near to the high resistance-carbon requirements. They have potential power for application, even in resistance network and capacitor sphere.
3. A good adhesive obtained among the metal and substratum, resol, which was induced by the action of diffusion and chemical adhesion.
4. Increasing the thickness of Ag-coated layer would increase all the samples performances.

REFERENCES

1. Yang Yikun et al., Synthetic Adhesive, Scientific Publishers (1980), 407.
2. Dyckerhoff, D.A., Sell, P.J., Anew. Makr. Chem., 21(1982), 169.

THE ANALYSIS OF THE STRUCTURE & PROPERTY OF LEAD RUTHENATES SERIES RESISTOR PASTE

Xu Hengsheng
Dept. of Electronic Engineering, Xi'an Jiaotong University
Xi'an, Shaanxi China

ABSTRACT

This paper, by x-ray diffraction analysis and combining with synthesis condition, determines that the material of conductor phase of Lead Ruthenates Series Resistor Paste is the substituted type of Lead Ruthenate, $Pb_2[Ru_{2-x}M_x]O_{6.5}$. And in term of its energy band structure of electrons, also analyses its metallic conductivity and the effect of substitution of different metal elements (M) on its conductivity. The resistivity of elementary material of conductor phase, $Pb_2[Ru_{1.9}Pb_{0.1}]O_{6.5}$, is 7×10^{-4} ohm-cm, and that of conductor phase material with high resistivity, $Pb_2[Ru_{1.6}Ta_{0.4}]O_{6.5}$, is 5×10^{-2} ohm-cm. The analysis by using an electron microscope confirms that the size of conductor phase grains is less than 1000 A, and that the grains are high dispersed in vitreous phase of resistor film. It is convenient to form a structure of tunnel type of conductive path which has lower temperature coefficient of resistance (TCR).

INTRODUCTION

Lead Ruthenate series resistor pastes have such characteristics as stable properties, lower costs and less TCR than RuO_2 series (TCR of sheet resistivity less than 1M is 100 PPM/°C). It is an important variety for making up relatively precise thick film resistor as well as metal glaze potentiometer. In our country, it was beginning to study and apply this kind of resistor paste since 1970s, and its properties are now equal to the level of DuPont Company of the U.S.A.

Material of conductor phase of this resistor paste is Lead Ruthenate Series which is produced by PbO & RuO_2 reacting at 650°C, or adding some oxides such as Ta_2O_5, etc. This material can be used for preparing resistor paste whose range of resistivity is wide (5 ohm/sq— 10Mohm/sq). However, Lead Ruthenate has various types while synthesis conditions are different. It is necessary to determine its crystal structure and chemical formula by X-ray diffraction analysis, and on the basis of which we can understand its metallic conductivity and the influence of adding Ta_2O_5 on its conductivity in term of the analysis of its energy band structure of electrons. In the meanwhile we study its conduction mechanism and the nature of having lower TCR by means of electron microscope analysis to material of conductor phase and resistor film. Thus, we may clarify the essential factors making its characteristics superior.

THE ANALYSIS OF CRYSTAL STRUCTURE

So far it has not been found that stoichiometric Lead Ruthenate($Pb_2Ru_2O_7$) was reported to have been produced. But defect type of $Pb_2Ru_2O_{7-y}$ had been produced by sintering PbO and RuO_2 with 1:1 molar ratio at 850 °C, by Longo et al.[1] . X-ray diffraction analysis indicates that it's a face-centered cubic structure of PYROCHLORE, its unit cell dimension is 10.253 Å. Next, Sleight, et al.[2] reported that $Pb_2Ru_2O_{6.5}$ was produced at 3000 atm. Then, Horowitz, et al.[3] reported that Lead-subsitituted type of Lead Ruthenate Series $Pb_2[Ru_{2-x}Pb_x]O_{6.5}$, was produced below 700 °C. X-ray diffraction analysis indicates that it is still PYROCHLORE structure. The lower the reaction temperature is, the larger the values of X. To the extent of $0 < X \leq 1$, the lattice parameter increases from 10.253 Å to 10.595 Å, and the values of X has approximately linear incresing relationship to lattice parameter. The quantities of oxygen in Lead Ruthenate in this series are wholly equal to 6.5, which have nothing to do with the values of X. That has been proved by neutron diffraction analysis.[4]

Elementary material of conductor phase in this kind of resistor paste in our country is produced by the reaction between PbO & RuO_2, with PbO : RuO_2=1 : 1 molar ratio and at 650 °C. Thereby, it should be Lead-substituted type Lead Ruthenate, a little quantity of Lead has substituted for Ruthenium.

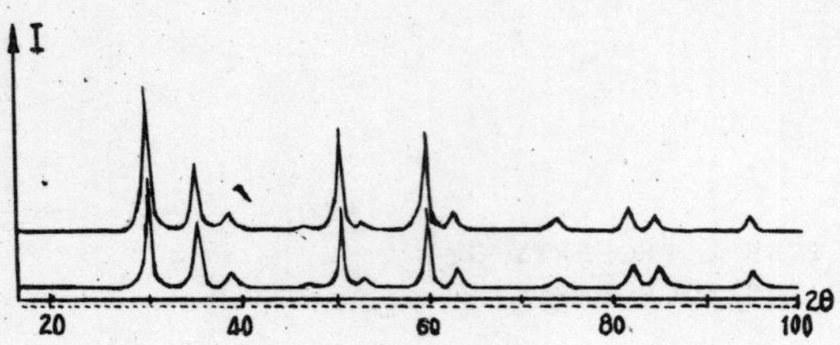

Fig.1. The Xray diffraction scheme of the conductor phase (the upper) and resistor film (the lower)

Here is shown in Fig.1, the X-ray diffraction scheme of this material and resistor film made of it. The diffraction analysis was carried out on the Japanese Rigaku diffractometer, and by using CuK_{α} radiation and Ni filter. The Bragg angle of diffraction ray was calibrated by the standard specimen of high purity Si powder.

In Fig.1, we can know that the two diffraction scheme resemble very much, and that sintering has no influence on the type of crystal structure of conductor phase. While their diffraction data was contrasted with standard diffraction data on two JCPDS Cards ($Pb_2Ru_2O_{6.5}$ 34-471, $Pb_2[Ru_{1.69}\ Pb_{0.31}]O_{6.5}$ 34-472), it was found that they were basically conformable. Therefore, it can be confirmed that their crystal structure is that of PYROCHLORE. Their lattice parameters can be obtained by extrapolation, which is 10.289 Å. According to the approximately linear relationships of lattice parameters to substituted coefficient X as previouly stated, it can be inferred that the value of X of elementary conductor phase is about 0.1, or $X \cong 0.1$, and corresponding reaction temperature is about 650 °C, which is basically in accordance with synthesis conditions. Hence, it can be determined that its molecular formula is $Pb_2[Ru_{1.9}Pb_{0.1}]O_{6.5}$ and there are about 5 percent of RuO_2 which haven't joined the reaction, it is in agreement with the observed result by electron microscope.

THE METALLIC CONDUCTION CHARACTERISTIC OF THE CONDUCTOR PHASE

The stoichiometric formula of this compound having PYROCHLORE structure is $A_2B_2O_7$, 1/4 of its unit cell of this compound is shown as in Fig.2. In each unit cell, there are respectively 16 cation sites of A or B and 64 anion sites of Oxygen. Oxygen lies in tetrahedral interstices of face-centered cubic lattices which is composed of A and B. There are totally three types of interstice sites i.e, 48(f) sites , 8(a) sites and 8(b) sites. And in all kinds of PYROCHLORE structure for $A_2B_2O_7$, 8(b) sites are always empty. In defect and substituted type of Lead Ruthenate, 8(a) sites are partially empty. The existance of 8(b) oxygen vacancies lead to oxygen at 48(f) sites moving to two B in near neighbouhood

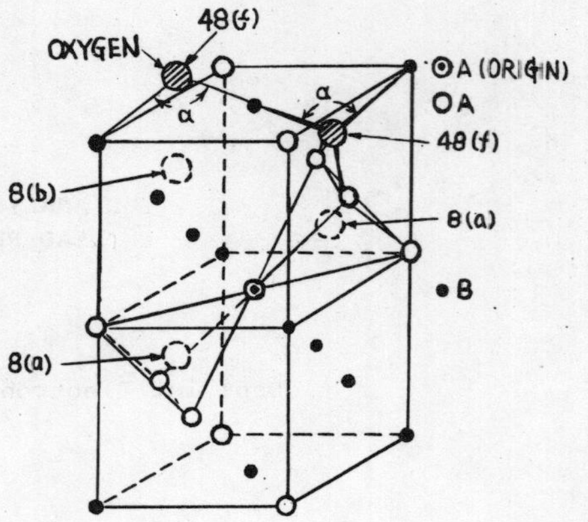

Fig.2 1/4 of unit cell of PYROCHLORE struture

from their tetrahedral center, which thereby will result in th full overlap of oxygens outer s.p electron shell with outer d electron shell of Ru at B sites and form stable covalent bonds. For the same reason, the existance of 8(a) oxygen vacancies will lead to the outmost electron shell ($6s^2 6p^2$) of Pb at A sites spreading out into the oxygen vancancies, getting a full overlap and forming stable band states. This is the intrinsic reason why the defect type PYROCHLORE structure easily occurs. Fig.3 shows a diagram of electron energy bands of defect type of Lead Ruthenate[1], electron distribution in the energy bands is closely related to that of vacancies. The outmost π^* band can be filled in twelve electrons, and there are 6+2y electrons in band of $Pb_2Ru_2O_{7-y}$. when y=0, there are no vacancies in 8(a) oxygen sites, and, Pb can't overlap by means of it, so the energy band can be only filled in forty-two electrons of seven O atoms($2s^2 2p^4$) and sixteen electrons of two Ru atoms ($4d^7 5s^1$). It can be totally filled in sixty-four electrons from [2]x[18] band to π^* band, π^* band can be only filled in six electrons, so it is half-full bands. And when y=1, it can be filled in sixty electrons of six O atoms and two Ru atoms as well as two Pb atoms (6^2 $6p^2$), so there are eight electrons in π^* band.

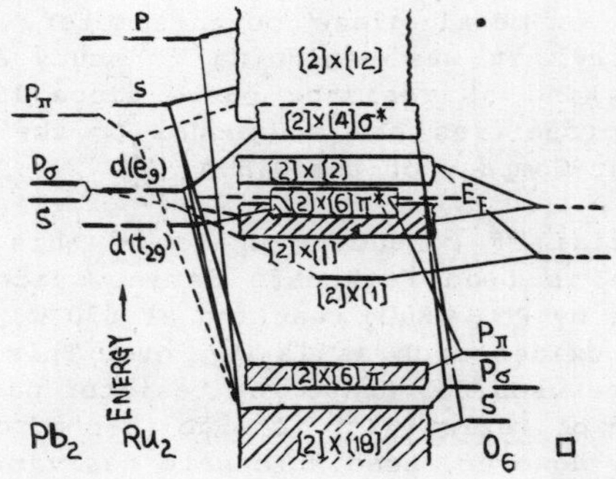

Fig.3 The structure of $Pb_2Ru_2O_{7-y}$ electron energy bands

It may be on the analogy of this under other circumstance. The outmost π^* bands in all kinds of Lead Ruthenate are non-full bands, for this reason, all of them are of metallic conductivity. The resistivity of $Pb_2Ru_2O_{6.5}$ is 2.7×10^{-4} ohm-cm.

Conductor phase material with high resistivity is gained by incorporating Ta_2O_5, in fact, partically substituting Ta for Ru , because there are five electrons in the outside shell of $Ta(5d^2 6s^2)$, instead of eight ones in Ru, the replacement of Ta must decrease numbers of electrons in π^* band and this leads to the increasement of resistivity. The resistivity of $Pb_2[Ru_{1.6}Ta_{0.4}]O_{6.5}$ is 5×10^{-2} ohm-cm. For the same reason, a little quantity of Pb substitute for Ru in elementary conductor phase, $Pb_2[Ru_{1.9}Pb_{0.1}]O_{6.5}$, which also leads to a little increasing of its resistivity, by, about 7×10^{-4} ohm-cm.

THE ANALYSISES OF THE CONDUCTOR PHASE AND RESISTOR FILM BY ELECTRON MICROSCOPE.

Because the grains of the conductor phase are very small and easy to adhere together, The key to electron microscope observation is how to prepare specimen. Fig 4 shows the photograph of conductor phase grain, which is gained by HU-12A TEM made in Japan. Most of grains have round and smooth shape, and their diameters are usually smaller than 1000Å.

It is very difficult to make the TEM specimen for observing resistor film, not only grinding away ceramtic substrate as much as possible by using Emery Grinding Device, but also making it thinner and punching it at one side of ceramic substrate by using Ar-ion beam thinning machine. Fig 5 shows the TEM photograph of resistor film. There are two elementary phase in the film. One is grey vitreous glass, the other is black conductor phase grain with different sizes. And the former adheres the latter together. Electron diffraction has proved that glass is amorphous and the electron differation data of conductor phase grains are very near to that of X-ray diffraction analysis. The sizes of conductor phase grain are usually smaller than 1000 Å, therefore it is often seen that they are seperated from each other, this is different from that the most of grains of conductor phase of RuO_2 series resistor contact with each other [5]. It shows the conduction mechanism of these two kinds of the thick-film resistor are very different, which is the essential reason why TCR of Lead Ruthenates Series is superior to that of RuO_2 series. For the stability of PYROCHLORE structure, TCR of Lead Ruthenate conductor phase itself (400 ppm/°C) is superior to that of RuO_2 (7800ppm/ °C). Furthermore, it has good interdiffusing capability in melten glass, [6] and can form tunnel type conductor path structure supposed by Pike et al[7]. However, RuO_2 has very bad interdiffusing capability in melten glass, so be pressed to form the structure of conductor path contacted with each other. So the size of glass powder has an obvious influence on its conduction performance. Fig 6. shows the model of conduction path structure of these two kinds of resistor. In ideal Lead Ruthenates resistor, the grains of conductor phase are uniformly distributed in vitreous phase, and thin glass layers between grains of conductor phase have tunnel conductivity, whose TCR can compensate that of the grains of conductor phase. Therefore the size of conductor phase grains and their distribution in vitreous phase have great in-

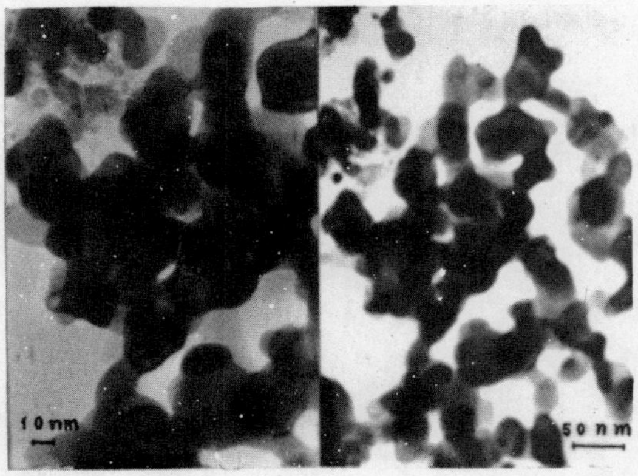

Fig.4. The TEM photograph of grains of conductor phase

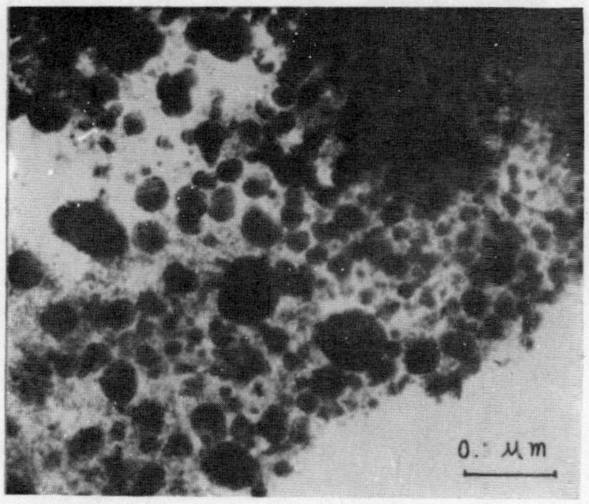

Fig.5. TEM photograph of resistor film

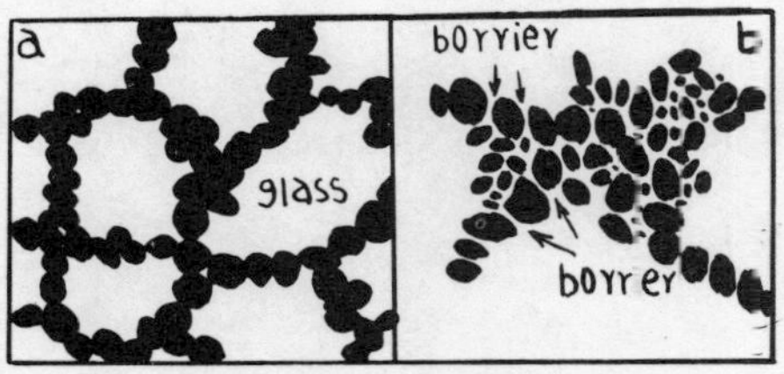

Fig.6. The conductive path model of (a) RuO_2 series resistor and (b) lead Ruthenate series resistors

fluence on TCR of this kind of resistor.

At last, we have observed the photograph of electron diffraction of RuO_2, in elementary material of conductor phase as well as in resistors made of them. This indicates that only $Pb_2[Ru_{1.9}Pb_{0.1}]O_{6.5}$ can occur at reaction temperature 650 C, and not $Pb_2Ru_2O_{6.5}$.

CONCLUSIONS

1. Elmentary conductor phase of domestic Lead Ruthenates Series resistor paste is Pb substituted type. Its molecule formula is $Pb_2[Ru_{1.9}\ Pb_{0.1}]O_{6.5}$ and its structure is that of face-centred cubic PYROCHLORE . Whose lattice parameter is 10.289 Å. There are fully convalent bonds in this structure to gain stable chemical performances and metallic conductivity. Its conductivity can be adjusted through incoporating other metal oxides in preparing ingredients so as to produce substituted type Lead Ruthenates to change the number of outer-shell electrons in non-full bands. The resistivity of elementary material of conductor phase, $Pb_2[Ru_{1.9}Pb_{0.1}]O_{6.5}$, is $7x10^{-4}$ ohm-cm, however, in material of conductor phase with high resistivity is $5x10^{-2}$ohm-cm.

2. Analysises by electron microscope have proved the grains of Lead Ruthenates conductor phase are very small, usually less than 1000 Å, having stronger adsorptivity. But they can be fully dispersed in glass, not to be pressed together to form the intercontacted structure of conductor path. In fact, they form tunnel type conductor path structure with thin glass layer in the middle. This primarily proves the conduction mechanism supposed by Pike et al.

3. The analyses of structure confirm that only Pb substituted type of Lead Ruthenate can occur under the synthesis condition of 650 °C, 1:1 molar ratio needs to be adjusted so as to save nobel metal.

Finally, I owe sincere thanks to senior engineer, ZHOU Jisheng working in 4310 factory, for his providing analysis specimen and warmly supporting and helping.

REFERENCE

1. J.M.Longo, et al., Mat.Res.Bull. 4, 191 (1969).

2. A.W.Sleight, Mat.Res.Bull. 6, 775 (1971).

3. H.S.Horowitz, et al., Mat.Res.Bull. 16, 489 (1981).

4. R.A.Beyerlein, et al., New York Meeting of The American Physical Society (March, 1980).

5. 猪熊敏夫，电气化学 (Japan)51, 721 (1980).

6. 武田义章，日本大学生产工学部报告, Vol 8, No.1 (1950).

7. G.E.Pike, et al., J.Appl.Phys. 48(12), 5152 (1977).

THE DEVELOPMENT OF SPECIAL FIRE-RETARDENT COATINGS FOR ELECTRONIC COMPONENTS

Zhang Yong Xiang, Bao Xing, Wang Xue Jun
Electronic Engineering Department, Tianjin University
Qilitai, Nankai District, Tianjin, P. R. China

INTRODUCTION

At present, the protective coatings of electronic components used in electronic industry and home electronic appliances are commonly high polymer. Its great weakness is inflammable. Especially, when the electronic components are commonly high polymer. Its great weakness is inflammable. Especially, when the electronic compnents are overloaded, the entire electronic instalation is easy to be burnt. Such as these cases are not rarely seen, there are thousand times happended in China and Japan every year. Therefore, it is very urgent to adopt insulating fire-retardent coatings to meet the demands of electronic components. Our country start lately in the field of the development and production of fire-retardent coatings for electronic components. We started to develop the coatings since 1986. This paper deals with the development of the fire-retardent coatings, including the choice of fire-retardent formula and the typical test ways and results of the coatings.

EXPERIMENTAL AND MECHANISM

1. Fire-retardent mechanism.

According as the organism can produce free radicals and bring about a series of chain reaction which can cause the components to be damaged, while the organism is burning. In order to prevent the organism from burning, it must be isoluted from contacting the burning body. The temperature of the burning body must be reduced, and the active radicals must be quenched suddenly to stop thermal decomposition raction. In accordance with the imagination of fire-retardent cources, two ways followed are described,

A. Fire-retardent in the gaseousness.

Under a certain temperature, fire-retardent can be decomposed to produce free radical where can quench high polymer in a thermal power or change radicals, the smaller active radicals substitute
the bigger ones and the thermal decompostion reaction can be controlled secondly, a kind of flame-proof gas (HX, H_2O) decomposed under high temperature can dilute the buring body, with absorbing a large amount of heat energy to reduce temperature, so the burning gets controlled. For instance, when epoxy resin bromid is heated, it releace hydrogen bromide (HBr) floating on the surface of the coating to cut off the supplies of air and heat energy which play a part in retarding fire. Hydrogen bromide (HBr) is capture reagent of OH. group, it can first react with the greatest active free radical,

$$HBr + \cdot OH \longrightarrow H_2O + Br\cdot \quad \cdots\cdots (1)$$

$$Br\cdot + RH \longrightarrow HBr + R\cdot \quad \cdots\cdots (2)$$

Thus, it can make free radicals profuced from burning occur brecding reaction, cut off the chain reaction on burning and form a barrier layer. In addition, a synergist can be used as utilizing the fire-retardent halogen-containing reagents, such as antimonous oxide. it happens halogen/antimony coordination effect, its equation is ,

$$Sb_2O_3 + 6Rx \longrightarrow 2SbX_3 + 3R_2O \quad \cdots (3)$$

Because of its big density, Antimony trihalogenated (SbX_3) gather on the surface of the burning body acting as isolating air and fire-source. Antimony trihalogenated(SbX_3) can also capture free radicals that is inhibiting thermal decomposition reaction, thus the fire can be quenched.

B. Fire-retardent in the form of coacervate

Fire-retardent reagents play its action in the form of caacervation phase. At high temperature, a protective layer which is difficult to flame and volatilize is formed on the surface of the burning body, which can separate from fire source. According to requirements, inorganic fire-retardent reagents may be utilized for making up the faults of corrosion, Smoking, poison ect. existed in organism. The actions of mainly inorganic matter have to absorb the heat, lose water, cool and dilute inflammable gases, in the fire-retardent reagent, so as to inhibit the burning. Its equation is,

$$Al(OH)_3 \xrightarrow{\text{High Temp.}} Al_2O_3 + H_2O - 71.6Kal \cdots\cdots (4)$$

This Aluminium compound serves as retarding fire and extinguishing smoke.

2. Choosing of fire-reardent reagents

We must also consider the following several aspects when choosing fire-retardent reagents.
A. The decomposition temperature of fire-retardent reagents must be higher than the curing temperature of Coatings.
B. It musn't release the poisonous and great corrosion gases.
C. Fire-retardent reagints with polyner mix well.
All fire-retardent used can be diveded into three forms, they are hatogen group, phosphorus family and inorganic fire-retardent agents.

RESULTS AND DISCUSSIONS

1. Formulary of fire-retardent coatings and testing digits
A. Formulary,

name	proportion	name	proportion (by weight)
bromic epoxy resin	100	deca-bromid-phenylether	20
E-51 epoxy resin	100	fire-retardent diluent	10
tributyl-phosphate	10	curing agent	5
Sb_2O_3	5	filler and pigments	200

The coatings consists of two components, Dominant agents and curing agents. A curing temperature is from 150 degree to 100 degree. and the curing time is from three minutes to five minutes (for resistant productive line).

2. Main testing digits,

Table 1. Main characteristics of coating

n	per. abso. of water	elect. charac.		resist. to resolv.	flam. bility
		per. of IR (Ω · cm)	bre. Volt. (V)		
1	0.24	4.0×10^{16}	Averages 3.71	chemi. resist. propet.	Not flam. under alcohol lamp
2	0.22	2.3×10^{16}			
3	0.36	2.1×10^{16}	SOUK. in wat. after 24 hour. 36.7		Not flam. load over six point two five time rate. power
4	0.42	1.0×10^{16}			

REFERENCES,

1. Modern plastics, 56(10A), 670
2. J.Appl. polym. Sci., 25, 451(1980) 1, 1986
3. Yuan Lu Bin, 《LNHG》 1988, No.4, p35
4. U.S.P., 4038248 (1977)
5. C.A.97.73446(1982)
6. C.A.87.85031(1977)

TG-TSSP METHOD STUDY THE POLARITY OF CARRIERS IN DIELECTRIC MATERIAL

Wang Li-heng Feng Yu-zhu
Xi'an Jiaotong University
Xi'an, China

ABSTRACT

TSC and TG-TSSP measurements are applied to the study of judging type and polarity of carrier in insulating protection film of high voltage silicon devices in this paper. Authors have judged the type and polarity of carries existed in the kind of 643 film by TSC and TG-TSSP measurement respectively. The experimental results indicate that carries in this insulating film include clipole and positive movable ions.

INTRODUCTION

A kind of insulating protection film is often coated on the sides of high voltage silicon devices. Such a treatment may make the tolerance of voltage and temperature of the devices deteriorated. According to our research work, it has been found that the voltage and temperature characteristics of the devices are related with the sort and the polarity of the carrier in the film. It is necessary to determine the features of the carriers in the insulating material. The protection film which is used in the research work is No. 643 insulating material made by Tao's Kangling Company in Hongkong. Though the composition and the structure of the material are not well known, it is regarded that the material belongs to an organic silicon insulating material when tested by means of the infrared spectrum analysis method.

In order to determine the polarity of the carriers in dielectrics we should know the kind of carriers beforehand. The thermally stimulated current method (TSC)[1] and the thermally stimulated surface potential method using a temperature gradient (TG-TSSP) are applied to determine the kind and polarity of carriers respectively. Some experimental results on the No. 643 insulating film are shown as follows.

SAMPLE PREPARATION

Samples are prepared in the form of MIM sandwich structure. The lower electrode is a circular Al foil (Ø50 mm) and the middle layer is a No. 643 insulating film uniform in thickness (50 um) which is achieved by adjusting the rotating speed of the spinner.

After being placed at the atmospher or room temperature about half an hour and being dried for 15-16 hours, the samples are ready for use.

MEASURING PRINCIPLES AND RESULTS

1. Determination of the kind of Carriers in the No. 643 Insulating Film

The kind of carriers can be determined by the TSC method. The depolarization current of carriers in dielectrics is measured as the temperature is changed. Fig. 1 (a) shows the principle of measurement to get the TSC spectra. Due to the difference in activation energy of carriers in the insulating film, the charged carriers will be separated when the sample temperature rises. It is not difficult to understand that carriers with lower activation energy release the YSC firstly, and then show greater ones in the proper order. The measuring procedure is shown in Fig. 1 (b). Under given conditions, there are two peaks,

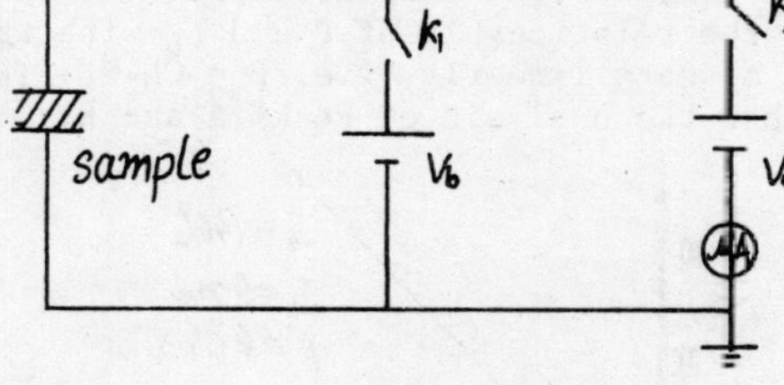

(a) The Measuring Principle

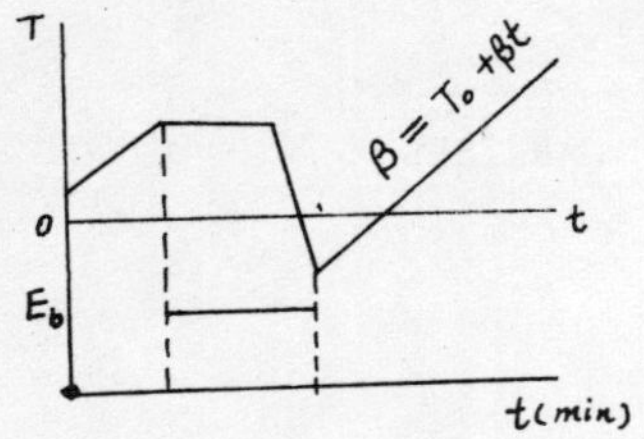

(b) The Measuring Procedure

Fig. 1 The Measuring Principle and Procedure of the TSC

(Peaks A and B) in the No. 643 film TSC spectra, as shown in Fig. 2. This is because two kinds of carriers are contained in the No. 643 dielectric film. It is known from the dielectric theory that the polarization strength of mobile ions represented by P (of course the same as depolarization strength) of the amount of the total charge related to the area surrounded by the TSC curve (Q_{TSC}) and the polar electric field Ep have the relation as:

$$P \quad (QTsc) = A \sinh (BE_b)$$

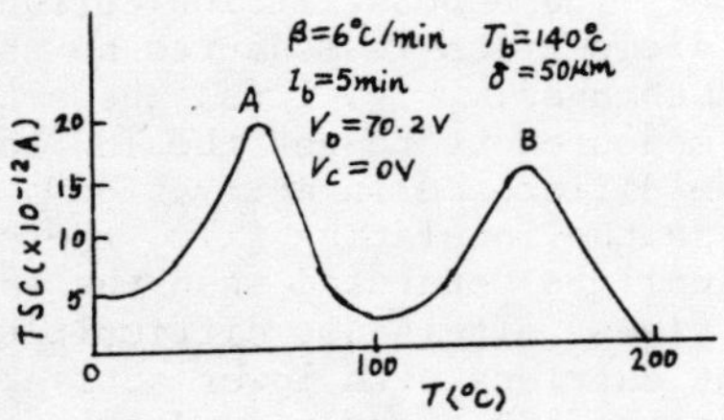

Fig. 2 The TSC Spectra of No. 643 Insulating Material

This means that in the P and E_b mobile ions there is a hyperbolic function relationship. However the relationship of P and E_b with the dipoles appears linearly, i.e. $P = CE_b^2$. Fig.3 and 4 show the analysis of Peaks A and B.

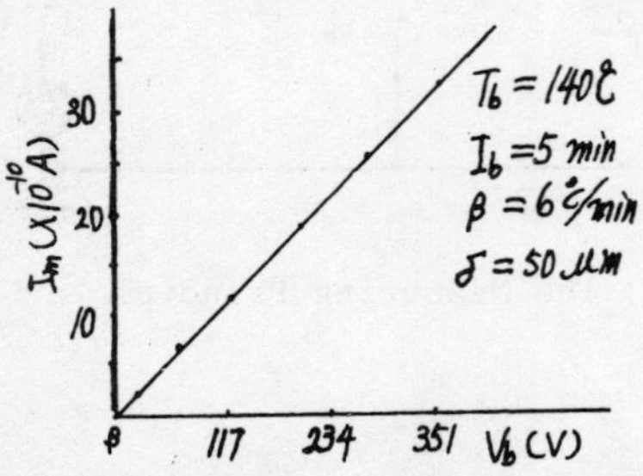

Fig. 3 The Relation between Imax and Vb of Peak A

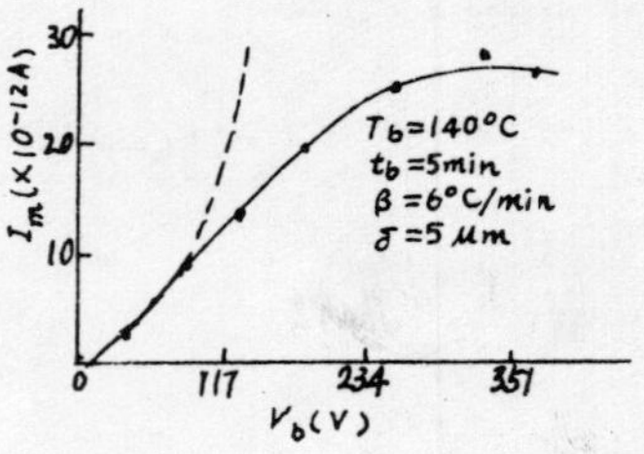

Fig. 4 The Relation between Imax and Vb of Peak B

Peak A is caused by the dipole which exists in the No. 643 film and Peak B is caused by mobile ions. To ensure the vision of Peak B, the author has carried out further determination by using the correct voltage Vc method. It is known from theory and experiments that another feature of mobile ions is that the TSC value is always positive corresponding to the reference zero current whether Vc is positive or negative[3]. The curves in Figs 5 (a) and (b) have confirmed that Peak B is caused by mobile ions indeed.

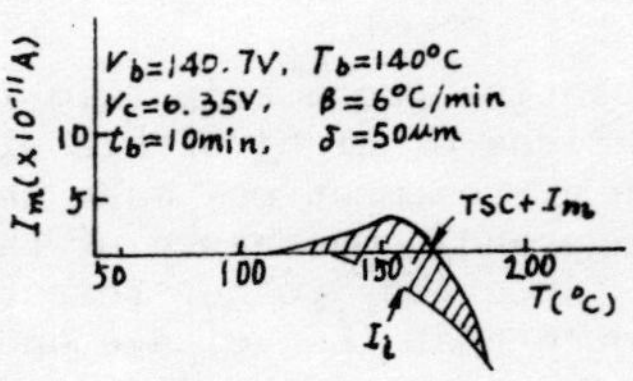

(a) The TSC below + Vc

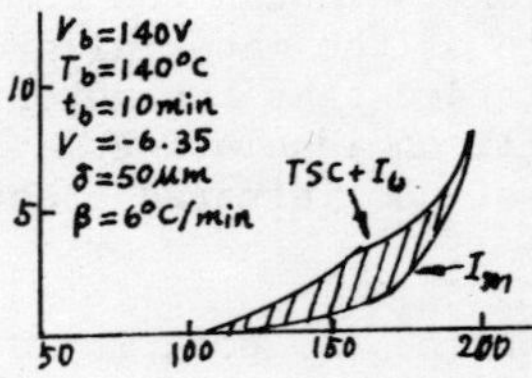

(b) The TSC below - Vc

Fig. 5 m of Peak B under ± Vc

From the above, it can be seen that under certain testing condition one can find some mobile ions as well as a kind of dipole. In order to determine whether the polarity of mobile ions is positive or negative the TG-TSSP method is used by the author.

2. The Polarity of Mobile Ions Determination

The principle schematic explanation of the TG-TSSP method is shown in Fig. 6. The apparatus is designed to measure the TSC when K_1 and K_3

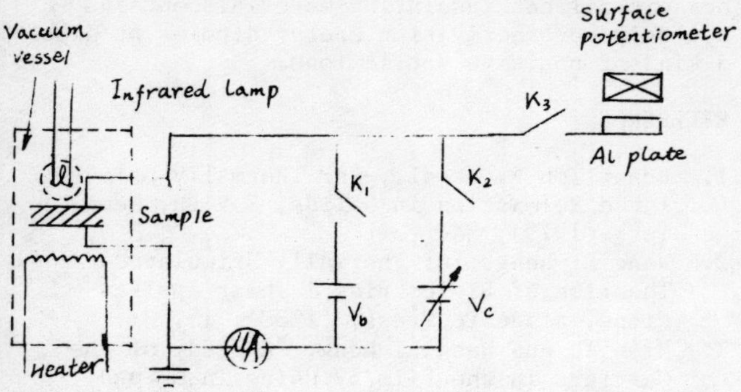

Fig. 6 The Measuring Principle of the TTG-TSSP

are opened but K_2 is closed and to measure the TSSP when K_1 and K_2 are opened but K_3 is closed. If there exists a temperature difference between two electrode plates, the TSSP obtained under the condition called the TG-TSSP results[4].

In other words, when the temperature of the upper electrode is T_1 and the lower electrode T_2, the TG-TSSP is the TSSP when $T_2 - T_1 = \Delta T \neq 0$. The temperature control process for measuring the TG-TSSP is shown in Fig. 7. If the Temperature difference between two electrodes is kept constant (i.e. $T_2 - T_1 = \Delta T =$ constant) and T_1 and T_2 rise with the same speed , the value of the TG-TSSP can be measured accurately. In order to ensure a constant, two microcomputer controllers are used to control the rise speed of T_1 and T_2. The analysis of Peak B is shown in Fig. 8 (from above 100°C, another

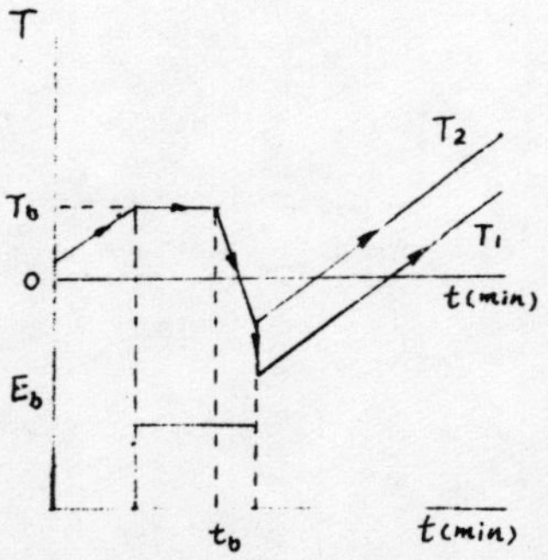

Fig. 7 The Temperature Control process when measuring TG-TSSP

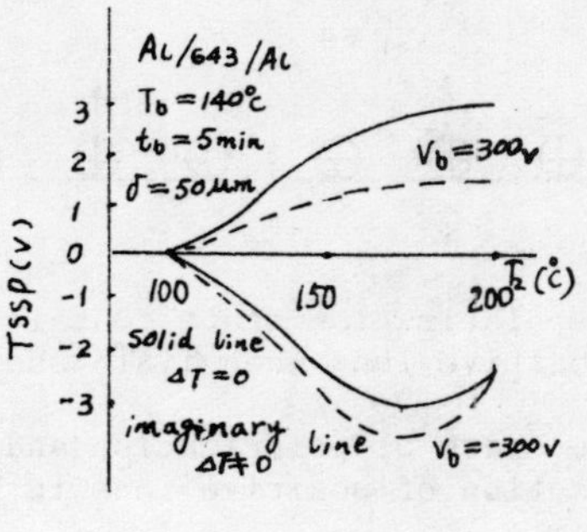

Fig. 8 The TSSP when $\Delta T=0$ and $\Delta T \neq 0$

method is used), in which the solid and imaginary lines correspond to $\Delta T = 0$ and $\Delta T \neq 0$ respectively.

From the testing results, it is clear that the TSSP value in the case of $\Delta T \neq 0$ tends to be greater in the negative direction than $\Delta T=0$. This indicates that the Peak B is caused by mobile positive ions. The conclusion can be explained as follows: Whether the TSSP or TSC varies with the temperature can be used as an indication of the depolarization of charged carriers (here referring to ions). If the same sample is used in the same measuring conditions, the depolarization speed in two states, i.e. $\Delta T = 0$ and $\Delta T \neq 0$, will be different. Therefore the TSSP we have obtained at the same temperature will be different under two ΔT conditions. Fig. 9 is shown by supposing that the No. 643 film is in the state of polarization and depolarization of positive carriers. In the cases of $\Delta T = 0$ and $\Delta T \neq 0$, the speed of the depolarization on of polarized ions is apparently distinct, or say, under the same T_2, we have $V_{\Delta T \neq 0} > V_{\Delta T=0}$. Hence the TSSP values in the two states appear as $TSSP_{grad.} > TSSP_{Equal}$ and vice versa when the bias voltage is reversed.

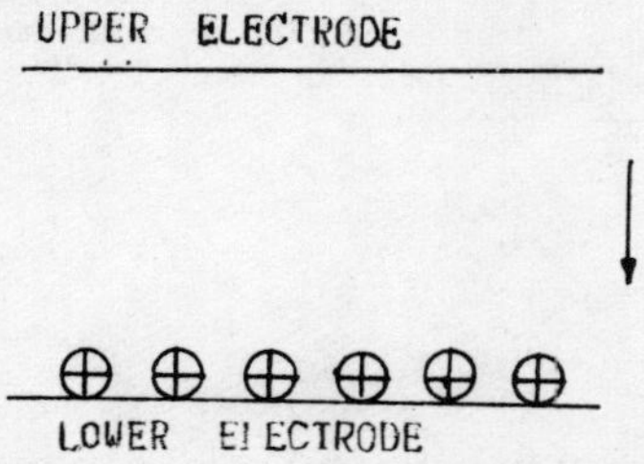

(a) Polarization state containing positive ions

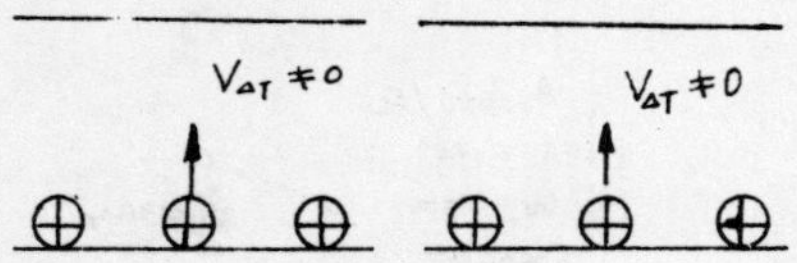

(b) Depolarization state containing positive ions when $T_1<T_2$ and $T_1=T_2$

Fig. 9 The state of polarization and depolarization of positive ions in No. 643 film.

If the polarity of ions obtained in the insulating film is negative, the measuring result will be opposite to the positive one. It is required to indicate that the above analysis on the experimental mental results are acquired only when the activation energy and mobility of each kind d of mobile ions are not equal, or the problems would tend to be more complex. Fortunately, there always exist some differences in the activation energy or mobility. As a result they can be easily displayed in the TSC spectra.

CONCLUSION

We can precisely determine the polarity of carriers in dielectric materials in a simple way by using the TG-TSSP method. The No. 643 organic silicon insulating material contains not only mono-activation energy dipoles bt also a kind of positive mobile ions.

REFERENCES

1. Braunlich P. et al., In: Thermally Stimulated Relexation in Solids, S-V BH, New York,(1979), 34.
2. Wang Li-heng, In: Thermally Stimulated Theories of Dielectrics & Their Applications, Academic Press, (1988), 15, 31.
3. Hino T. and Wang Li-heng, "A Study of the Carriers in the Film by Using Ta_2O_5 and their Conductivity"(in Japanese) , Technical Journal of the Nipon Denki Association (A),(1982), 465.
4. Hino T. "Thermally Stimulated Current Due to Dipolar Depolarization Under a Temperature Gradient", Japanese Journal of Applied Physics, Vol. 26, No. 1, (1987) 70-73.

CHEMISTRY AND APPLICATION OF UV-CURED SOLDER RESIST INK
I. INFLUENCE OF REACTIVE DILUENT

Pang Zhengzhi Ge Zhibang
Beijing Inst. of Chem. Technol. Dept. of Polym. Sci.
P.O.Box 105
Beijing, China

ABSTRACT

The acrylate monomer which is used as reactive diluent in the UV cured system was reviewed. UV cured rate of prepolymer/monomer system was observed by Fourier-transform infrared spectrometry and Soxhlet extraction. The conversionality of double bond and gel's content were calculated. Then, the effect of acrylate monomer on the cured rate was investigated. The adhesive property in UV system was discussed by the measurement on the shrinkage of cured volume of monomer and the internal stress. In general, when the functionality of monomer increased, the cured rate of system became faster, the shrinkage of cured volume and the internal stress were increased, then the adhesion force with substrates was changed bad, and there is important influence of steric effect and group's property of monomer on the cured rate of system.

INTRODUCTION

The solder resist inks are important organic chemical material for printed circuit plates in electronic industry. UV curing process has many advantages[1], by comparision with the traditional heat curing, such as high cured rate, lower use of energy, saved space, reduced pollution, without evaperation of solvent, and offer the possibility of coating substrates which are adversely affected by heat. Therefore, UV cured solder resist ink is widely used[2,3]. UV cured ink consists of pigment dispersed in a UV curable vehicle. The principal ingredients of the vehicle have reactive diluents, prepolymers and photoinitiators. The reactive diluents are important components in UV cured system. It requires that reactive diluents must be low toxicity, low odour and low volatility, good viscosity reduction and high cured rate. As for UV inks, the dispersed pigments absorbs UV irradiation that reduce the cured rate and the penetration depth of UV irradiation[4]. The most important advantage of UV inks is that it is so rapidly cured that it is beneficial to obtain efficient production, to save energy and to produce hard films. Therefore, it is necessary to investigate acrylate monomer with high cured rate. The adhesion force is the key parameter of solder resist ink, and has important influence on the solder resistance[5]. It is well known that the adhesionforce of cured film is low, if the internal stress is high. When the internal stress is greater than that adhesion force, it is readily for the coating film to peel off from the substrates. In the same way, lower internal stress increases adhesion force. The shrinkage of cured volume is one of the factor to affect the internal stress of cured film. Therefore, it is possible to find out the way too improve the adhesion force by observing the effect of acrylate monomer on shrinkage, and relation between shrinkage and internal stress.

EXPERIMENTAL

The acrylate monomers evaluated are (a) pentaerythritol triacrylate (PETA);(b) trimethylol propane triacrylate (TMPTA);(c) 1.6-hexane diol diacrylate (HDDA); (d) neopentyl glyco diacrylate (NPGDA); (e) cyclohexyl acrylate (CHA); (f) 2-hydroxyethyl methylacrylate (HEMA); (g) isobornyl acrylate (IBA). Among them (c), (d), (e) and (g) are synthesized according to the references.

1. Sample Preparation

acrylate resin prepolymer (made in Wuxi, brand

mark F-51) 3 wt.% of photoinitiator 2-EAQ (made in Jilin Dyestuff Factory) and 32.3 wt.% of various acrylate monomer used as reactive diluent (made in Beijing Dongfang Chemical Industry Factory, Shanghai Shanhu Chemical Industry Factory and etc. respectively). Each mixture is slightly heated, a little higher than room temperature and abraded to ensure homogeneously mixing used as the samples. The sample monitored by infrared spectroscopy are cast directly on KBr plates(35×20mm), with a thickness that allows at least 20% of conversionality in the infrared region (1000-4000cm^{-1})[6] before and after each UV irradiation. The mentioned samples used to determine internal stress are cast on the polyester plates. The samples are that acrylate monomers contained 3 wt.% photoinitiators 2-EAQ to determine shrinkage of cured volume. To cure the sample, high pressure mercury lamps (λ=365nm) is used as the irradiation source with 80 W/cm. lamp's distance is 145mm.

2. Assessment Methods

A. Infrared Spectroscopic Measurements. Infrared spectra of thin film of sample cast on KBr plates is measured by a Fourier-transform infrared spectrometer (Nicolet 5DXC made in USA). The characteristic C=C absorption at 1630cm^{-1} and the characteristic O-C=O absorption at 1721cm^{-1} that is used as the internal maker. The proportion of the height of characteristic peak C=C and O-C=O band represents the relative content of double bond.

B. Soxhlet Extraction. The cured samples are taken from the basic plate and placed in the Soxhlet extractor. Acetone is used as the solvent. The insoluble polymer material is dried under vacuum up to constant weight. Weighting and determining the gel's content.

C. Measurement of Shrinkage of Cured Volume and Internal Stress. The shrinkage of cured volume of monomers is determined by a specific gravity bottle and super constant temperature bath before and after irradiation. According to the formula, $\eta_v=(d_s-d_l)/d_s$, the shrinkage of cured volume is calculated respectively after the measuring out solid specific gravity of monomers d_s and liquid specific gravity of monomers d_l. The internal stress is measured by Deffuse Right Polariscope WZM-300 with the principle of photoelastic method.

RESULTS AND DISCUSSIONS

1. The Effect of Acrylate Monomers on UV Cured Rate.

The cured rate is expressed in terms of the changes of the gel's content and the conversionality of double bond in the system, that is determined by infrared spectroscopy. The content of double bond of relevant sample is measured under different irradiation time, the relevant conversionality of double bond is calculated. Fig.1 shows the effect of functionality of the monomer on conversionality of double bond.

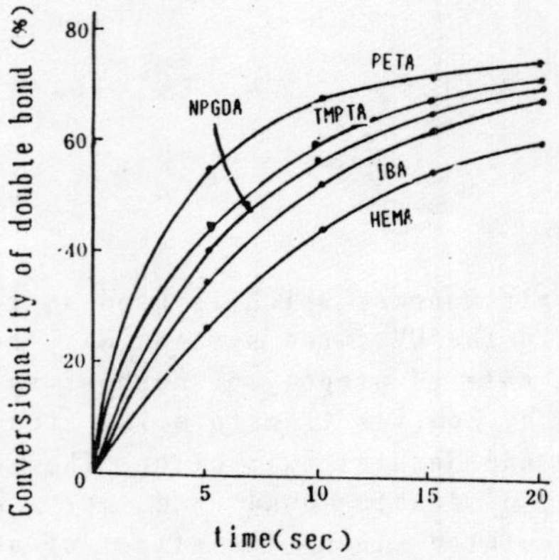

Fig.1 Effect of irradiation time on inversion rate of double bond

As shown in Fig.1, under the same irradiation time, the sample containing PETA monomer has the highest conversionality of double bond, the sample of NPGDA monomer ranks the middle, the sample of HEMA has the lowest. PETA monomer contains trifunctional groups, NPGDA contains difunctional groups and HEMA contains monofunctional group. When increasing functionality of monomer, the number of double bond taking part in the reaction increases, so does the producing free radical, and the photopolymerization reaction becomes faster, then the cured rate is quickened. Because monomer's molecule differs in structure, the cured rate of monomer having same functionality is difference, for example, the functionality of HEMA and IBA are both one, the cured rate of HEMA less than that of IBA, because of the steric effect of methyl group in HEMA, the functionality of PETA and TMPTA are both three, the cured rate of PETA more than that of TMPTA. It is due to that hydroxy group has active hydrogen atom to be initiated to form free radical in PETA. Therefore the photopolymerization reaction is reduced or increased, and the cured rate of them has change, From Fig. 2, it can be observed that for the sample contained TMPTA, when it is irradiated in the first five seconds, the conversionalityof double bond is 45.2%, but when in the second five seconds, it is changed to 14.2%, in the third five seconds, it is 7.8% and in the fourth five seconds, it is 4.4%. It can be considered that the cured rate is decreases with the time of irradiation. HEMA has the same tendency too.

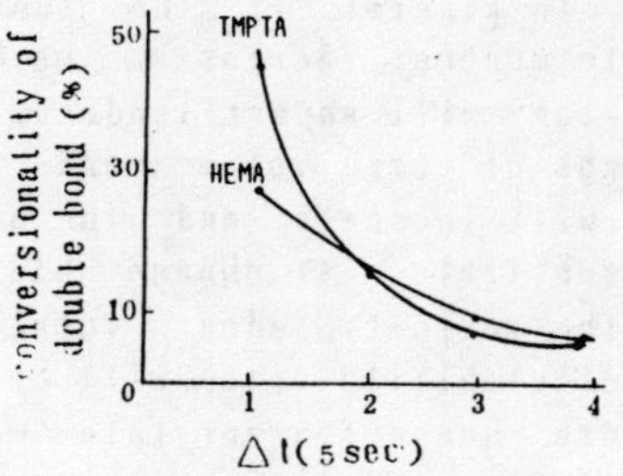

Fig. 2 Conversionality of double bond at differnt time

1, 1st 5 Second 2, 2nd 5 Second
3, 3rd 5 Second 4, 4th 5 Second

It can be ascribed that photosensitive group in polymer continuously disappears, i.e. its concentration continuously decreases with time of irradiation. At the beginning of irradiation, the concentration of photosensitive group is high, the curedrate is large; but with decreasing the concentration, the cured rate is decreased, i, e. concentration of photosensitive group in polymer influences the cured rate. At this time producing free radical which takes part in additional reaction under the existance of photoinitiator and gets crosslinked film of three demensional

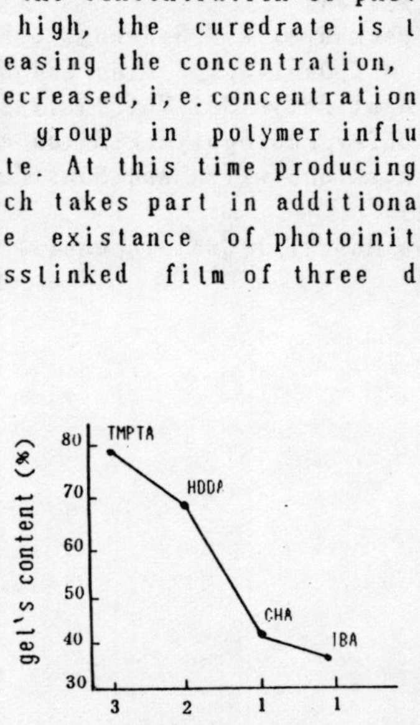

Fig. 3 Relationship between gel's content and functionality of monomer

network structure. The gel's percentage content of samples are determined under the same radiational time and the characteristic curve is drawn in Fig. 3.

It is showed that gel's percentage content and the crosslinked density of system are high for the poly-functionality's TMPTA. But IBA monomer of low-functionality gets contrary result. Of course, as Fig. 4 shows, the concentration of monomer HEMA is too high, the terminating rate of double molecule will rise, and monomer will precedently homopolymerize, at this time, gel's percentage content decreases. However, when the concentration of monomer is small, its gel's percentage content also falls. It can be ascribed that prepolymer occupies a dominant position in crosslink state.

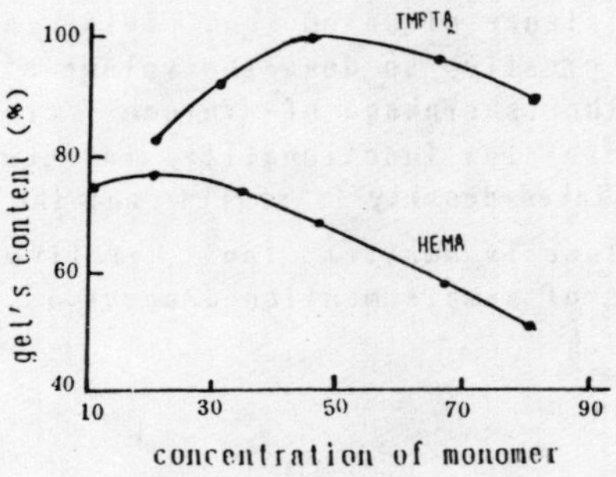

Fig. 4 Relationship between gel's content and concentration of monomer

2. The Shrinkage of the Cured Volume, Internal Stress and Adhesion Force

According to formular (1), the shrinkage of cured volume of the acrylate monomer can be obtained and characteristic curve is plotted in Fig 5.

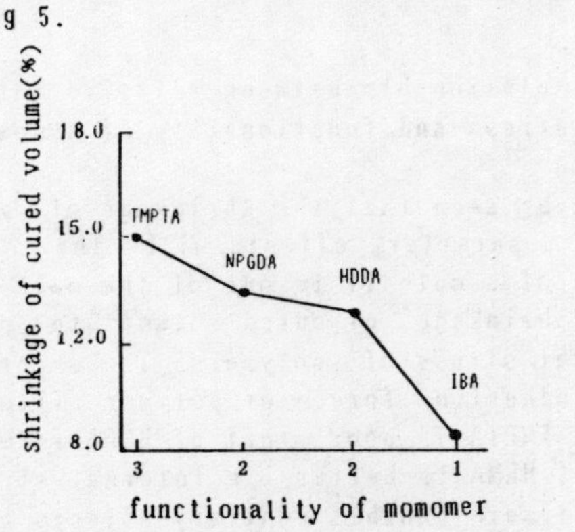

Fig. 5 Relationship between shrinkage of crued volume and functionality of monomer

$$\eta_v=(d_s-d_l)/d_l \quad (1)$$

where, d_s, the specific gravity of solid after the monomer cured, d_l, the specific gravity of liquid before the monomer cured. It shows that different acrylate monomers have different shrinkage of cured volume (η_v)[7]. The shrinkage of polyfunctional monomer (such as TMPTA) is higher than that of monofunctional monomer. The shrinkage of IBA is the lowest, η_v=8.3%. It can be ascribed that then forming to film, the dispersed single and free molecule of liquid monomer will instantaneously change into three demensional network or linear macromolecule of solid film, and the distance between moleculae decreases to atom-like distance. Hence the shrinkage of cure! volume occured. Under the same condition, if functionality of monomer is higher, the crosslinked density of polymer is greater, and after irradiation the relative position and distance of atom and molecule will reduces greatly. so does the volume of molecule, and the shrinkage of volume increases, but monomerof low-functionality, in general, the crosslinked density is smaller, and the shrinkage of volume is smaller too[8]. Relative internal stress of sample mentioned above is determined in Fig. 6.

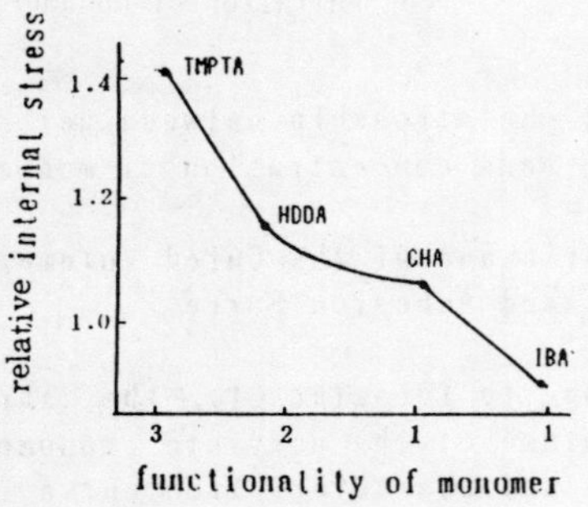

Fig. 6 Relationship between relative internal stress and functionality of monomer

It can be seen that the shrinkage of monomer η_v has paraller effects with the internal stress of sample. It is one of the main reasons that shrinkage of cured volume to produce internal stress of polymers. The experiments show; adhesion force of polymer film contained TMPTA is poor, that of HDDA is middle, that of HEMA is better. i.e. internal stress of the polymer exhibit contrary effects on adhesion force, so that the polyfunctional monomer has harmful influence on adhesion force of system, in general, the monomer of low-functionality is beneficial to increase adhesion force. It can be concluded that in order to design optimal recipe of UV cured solder resist ink to select acrylate monomer used as reactive dilute is quite important. The results show principal effect of the functionality, the steric effect and the group's property of monomer on the cured rate of system. In general, if the functionality of acrylate monomer increases, the UV cured rate of system will appear tendency to rise, the shrinkage of cured volame and the internal stress will increase, and the adhesion force with substrate will change bad. Apart from that, the toxicity, odour, volatility, viscosity, dilutability etc. will be considered. Therefore none of a acrylate monomer satisfies all the conditions, but it is possible to select a comprehensive recipe including requisite poly-and low-functional monomers.

REFERENCE

1. J.R. Youbger, J. Oil. Col. Chem. Assoc. 1976, 52 197-201.
2. JP. 60, 113, 237
3. JP. 60 240, 703
4. Yasifime Otsubo, Tadeshi Amari. and Doichiro Watanabe, J. of Appl. Polym. Sci. Vol. 31 1986, 209 9-2108.
5. US 4, 533, 445
6. T.A. Speckharo, K.K.S. Hwang, S.B. Lin, S.Y. Tsay, M.K. Oshiba, Y.S. Ding, and S.L. Cooper, J. of Appl. Polym. Sci. Vol. 30 1985, 647-666
7. C.G, Roofey, Photopolymerizaton of Surface coatings, John wilex and Sons Publishers, (1982)
8. Shimbo Masaki, J. Coat. Technol. 1984, 56(713)

ADSORPTION STUDY FOR CO SENSING BY SnO_2 BASED ELEMENTS

Noboru Yamazoe, Norio Miura, Jun Tamaki, and Chaonan Xu
Department of Materials Science and Technology, Graduate School of Engineering Sciences, Kyushu University, Kasuga, Fukuoka 816, Japan

ABSTRACT

Adsorption behavior of CO, NO and NO_2 on SnO_2 and Pd(2 wt%)-SnO_2 was investigated by means of a TPD technique. CO was captured preferentially by the adsorbed oxygen at room temperature to desorb as CO_2 on heating, while at 100 °C the lattice oxygen attained reactivity comparable with the adsorbed oxygen. NO and NO_2 were adsorbed in three states and desorbed as NO on heating. The addition of Pd to SnO_2 influenced markedly the amounts and adsorption strength of these adsorbates through the electronic interaction between Pd and SnO_2. The CO adsorption was strongly interfered by NO or NO_2 only on SnO_2 because β oxygen species, on which CO, NO, and NO_2 were competitively adsorbed, existed only on SnO_2 surface.

INTRODUCTION

Semiconductor gas sensor has been widely used as a detector or monitor of inflammable gases having leaked out in air. Many investigations have been accumulated on gas sensing characteristics[1], sensing mechanisms[2-5], and promoter actions[6,7], but basic problems still remain to be investigated for their new applications. For example, a combustion monitor sensor to be equipped on a gas boiler should respond to incomplete combustion products, typically CO, in the exhaust. However, the other exhaust components such as CO_2, O_2, H_2O, NO, NO_2 and H_2 may severely interfere with monitoring CO. In fact, the interference problem has never been cleared out yet, although there have been several technical efforts to increase the selectivity to CO by adding BaO[8], Y_2O_3[8] and Bi_2O_3[9] to the sensor element. To provide a sound base for a combustion monitor sensor, the present study aims at elucidating the adsorption-desorption behavior of CO, NO and NO_2 on the SnO_2 surface as well as the interference among them by using a temperature programmed desorption (TPD) technique. The effects of Pd on the gas-adsorptive properties of SnO_2 was also examined by using a Pd-loaded SnO_2 sample.

EXPERIMENTAL

The specimen powder of SnO_2 was prepared from an aqueous solution of $SnCl_4$ by hydrolyzing it with an ammonia solution. The resulting precipitate (stannic acid) was thoroughly washed with deionized water, dried at 100 °C, and calcined in air at 600 °C for 5 h. To prepare Pd-loaded SnO_2, the above powder was impregnated with a dilute sulfuric acid solution of $PdCl_2$ and calcined in air at 600 °C for 5 h after dryness. The Pd content was fixed at 2 wt% to SnO_2. The surface areas of SnO_2 and Pd-SnO_2 powder were 12.7 and 10.0 m^2/g, respectively.

For TPD measurements, these powder sample were pressed and ground to granules of 20-60 mesh, and 1 g of them was packed in a TPD apparatus which was essentially the same as one used previously[10]. Each TPD experiment was commenced with the pretreatment to produce the well specified surface. Exposure to O_2 (100 Torr) at 600 or 700 °C for 30 min followed by cooling to room temperature was undertaken to produce the oxidized surface. The partially reduced surface was produced by a specified amount of H_2 pulse at 450 °C. The degree of reduction was determined from the H_2 consumption and the surface area, and expressed in terms of the percentage of reduced surface tin ions (Sn^{2+}). For gas adsorption, CO, NO, NO_2, or their mixtures, all diluted with N_2, was kept in contact with the sample in the specimen chamber (ca. 150 cm^3) usually at room temperature for 30 min (static method). The initial partial pressure of each component was usually controlled to 1 Torr. In some occasions, gas adsorption was carried out as a pulse of a known amount (ca. 10 μmol) of specimen gases in an He flow (pulse method). Finally, TPD was carried out in an He flow (60 cm^3/min) at a heating rate of ca. 10 °C/min. The desorbed gases were monitored with a thermal conductivity detector (TCD) and a quadrupole mass spectrometer (MASS).

RESULTS AND DISCUSSION

1. Adsorption of CO

TPD chromatograms of oxygen from SnO_2 were reported previously[10]. In agreement with them, the oxidized SnO_2 sample in the present study was found to give oxygen desorption, β and γ, on heating as indicated by TPD chromatogram (a) in Fig. 1. It has been shown that β is the desorption of adsorbed oxygen in the form of O^- or O^{2-}, while γ is the desorption of lattice oxygen on the surface.

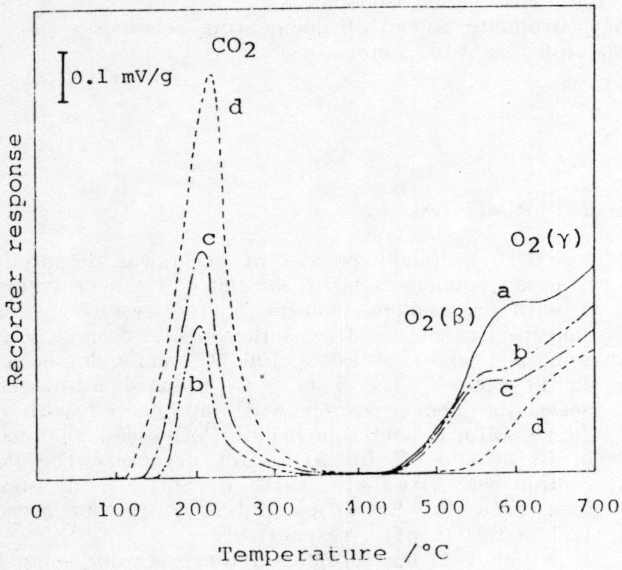

Fig. 1 TPD chromatograms from oxidized SnO_2 after exposure to various CO atmospheres at room temperature.
a: oxidized sueface (no CO exposure),
b: CO 100 Torr 60 min, c: CO 200 Torr 30 min,
d: CO 500 Torr 60 min.

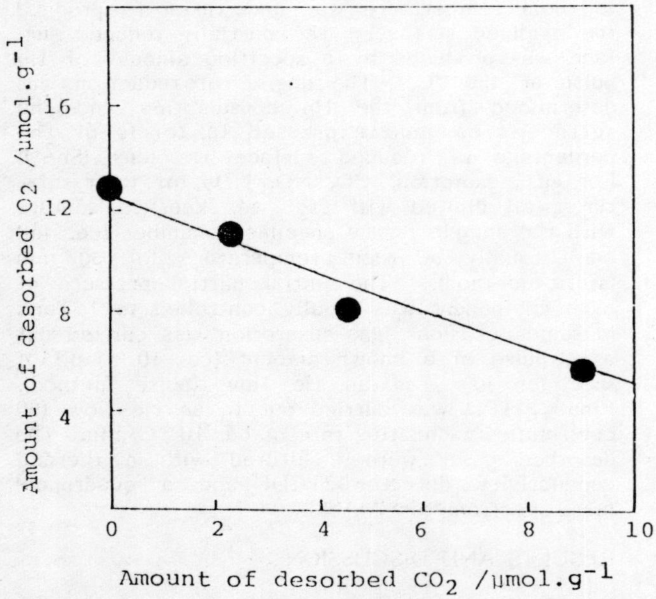

Fig. 2 Quantitative relationship between the desorbed CO_2 and the desorbed O_2 in the temperature range of 400-700 °C.

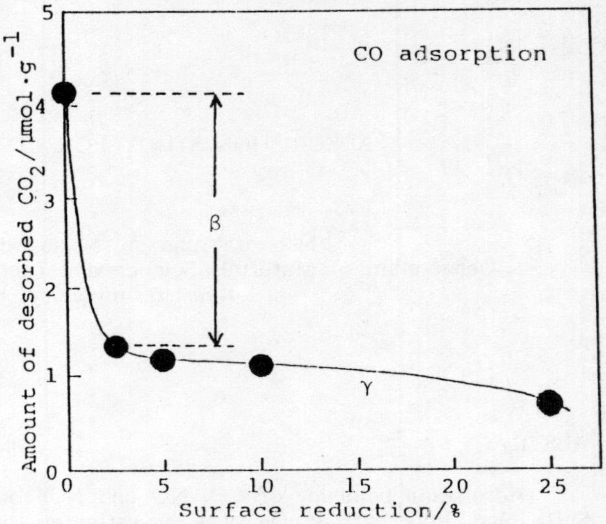

Fig. 3 Dependence of the amount of desorbed CO_2 on the degree of reduction of SnO_2 surface.

The same sample was exposed to CO at room temperature, and the resulting TPD chromatograms were also shown in Fig. 1(b, c and d). Desorption appeared around 220 °C and above 450 °C in each case. By coupling with MASS analyses, the desorption centered at ca. 220 °C was assigned to the desorption of CO_2 (M/e=44), and that above 450 °C was to O_2 (M/e=32), respectively. Notably CO was not detected even as a trace in the desorbed gas. This means that, during the gas adsorption treatment, CO was captured and oxidized by the surface oxygen to be desorbed as CO_2 on heating. As seen from Fig. 1, the CO_2 desorption increased in amount with increasing the initial pressure of CO. At the same time, these changes were accompanied by a decrease in oxygen desorption above 450 °C, especially β desorption. Figure 2 shows the quantitative relation between the desorbed CO_2 and the desorbed oxygen in the temperature range of 450-700 °C, as determined from the TPD chromatograms. The amount of desorbed oxygen linearly decreases with an increase of desorbed CO_2, and its slope (1/2) indicates that 1 mol CO_2 is produced as a cost of 0.5 mol O_2 or 1 mol surface oxygen. This assures that the surface reaction

$$CO + O_s \qquad CO_{2\,ad}$$

takes place on the SnO_2 surface, where O_s and $CO_{2\,ad}$ are the surface oxygen and adsorbed CO_2, respectively.

As already mentioned, O_s consists of β(O^- or O^{2-}) and γ(lattice oxygen), and β is estimated to be more reactive to CO at room temperature. To verify this estimation, similar TPD experiments were carried out by using a series of partially reduced SnO_2 samples. Figure 3 shows the amount of desorbed CO_2 after exposure to 1 Torr of CO for 30 min at room temperature as a function of the degree of reduction of SnO_2 surface. The CO_2 desorption decreased drastically even after very minor surface reduction, while further reduction resulted in only a slight decrease. The oxygen adsorbates β on the oxidized SnO_2 sample are known to amount to only a few percent of monolayer in surface coverage, and they are exhausted even with minor surface reduction. Therefore the CO_2 formation behavior in Fig. 3 can be understood as arising from the participation of two oxygen species, β and γ, in the reaction. That is, the part of CO_2 formation corresponding to the initial drastic decrease is attributed to the reaction with β, while the remaining part is to γ. These results lead to a picture that CO molecules would be captured much more rapidly (or with a much larger sticking coefficient) by the β oxygen sites, which are limited in number, than by the abundant surface lattice oxygen (γ). CO molecules are on these oxygen species to form carbonate species (CO_3^{2-}), as shown by Thornton and Harrison[11]. The relative reactivities of β and γ species to CO could be more clearly demonstrated by using a pulse technique. The SnO_2 samples, oxidized as above as well as evacuated at 700 °C for 30 min to eliminate β species, were exposed to a pulse of CO (ca. 10 μmol) at room temperature or 100 °C. The resulting chromatograms are compared in Fig. 4. After the CO pulse at room temperature, the oxidized sample gave 3.4 μmol CO_2 (a), while the evacuated sample gave only 0.4 μmol CO_2 (b). On the other hand, corresponding values after the CO pulse at 100 °C were 5.2 and 4.9 μmol CO_2, respectively. These results indicate that β oxygen is overwhelmingly more reactive than γ at room temperature, but both are of comparable reactivities at 100 °C.

2. Adsorption of NO and NO_2

The adsorption of NO and NO_2 on SnO_2, both oxidized and partially reduced, was investigated. Typical chromatograms after exposure to

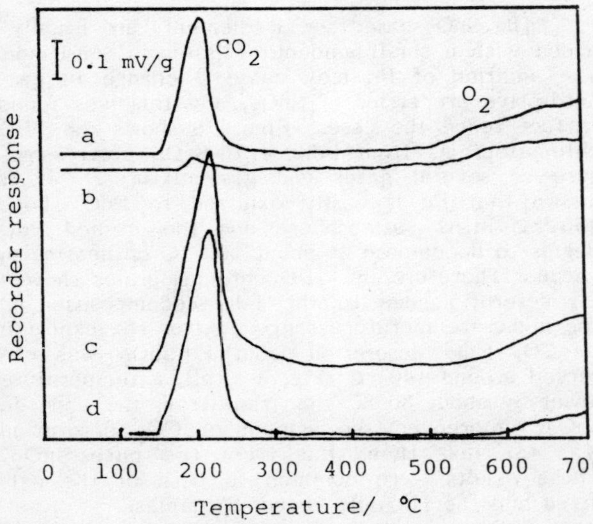

Fig. 4 TPD chromatograms from oxidized and evacuated SnO_2 after CO adsorption at R.T. or 100 °C by pulse method.
a: oxidized R.T., b: evacuated R.T., c: oxidized 100 °C, d: evacuated 100 °C.

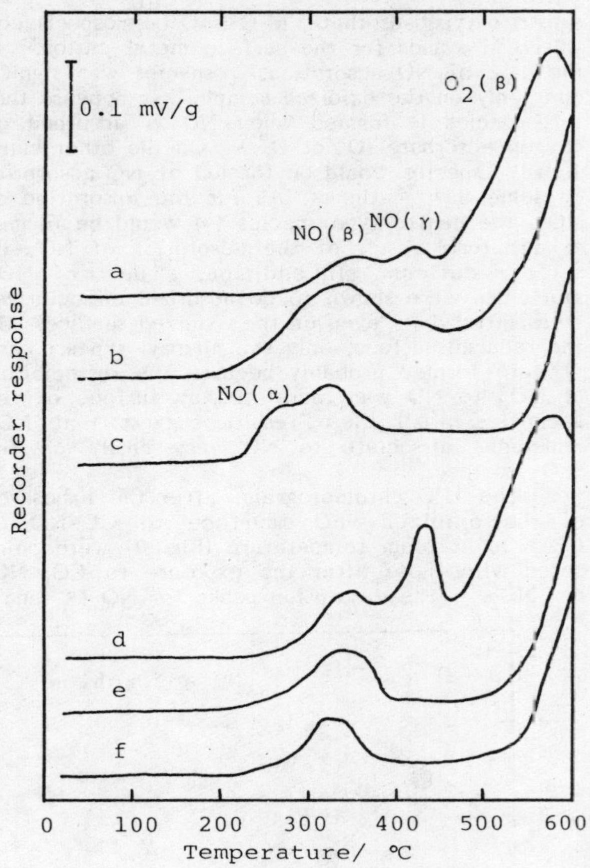

Fig. 5. TPD chromatograms from the SnO_2 surface after exposure to NO (a,b,c) or NO_2 (d,e,f).
a,d: oxidized surface, b,e: 5% reduced surface,
c,f: 10% reduced surface.

NO and NO_2 at room temperature are shown in Fig. 5. Several desorption peaks appeared below 450 °C, all of which were ascribed to the desorption of NO (M/e=30) in all cases. Even after exposure to NO_2, the desorption of NO_2 was not observed. Although the chromatograms changed in a rather complex manner depending on the oxidation state of SnO_2 samples and the exposed gas (NO or NO_2), it was possible to analyze all the TPD profiles of NO desorption as being composed of three component peaks centered at ca. 250 °C (α), ca. 350 °C(β) and ca. 420 °C(γ), respectively. It was notable that such decomvolution was possible commonly for the both cases of NO and NO_2 adsorption. This indicates that the same adsorbates are formed from NO and NO_2. However, the population of the three components was dependent on the degree of surface reduction in considerably different manners for the cases of NO and NO_2 adsorption, as shown in Fig. 6 (a) and (b).

It has been reported that NO is adsorbed on metal oxides in three states[12-14]. Based on the observed dependence just mentioned, the three component peaks were estimated to be two nitrosyl type adsorbates, M-NO^- (α) and M-NO^+ (β), and a nitrito type adsorbate, M-O-N=O (γ), respectively, where M stands for the surface metal cation. In the case of NO adsorption, species was significant only on the oxidized sample. It appears that this species is formed when NO is adsorbed on oxygen adsorbate (O^- or O^{2-}). On the other hand, α and β species would be formed by NO adsorption on some Sn^{4+} cations. As for the adsorption of NO_2, the nitrito type species (γ) would be formed as a natural result of the adsorption of NO_2 on surface cations. In addition, a part of NO_2 molecules were shown to be adsorbed dissociatively as nitrosyls(β) even on the oxidized surface. On the reduced surface, only the nitrosyl types (α and β) were formed probably because the dissociation of NO_2 to NO was facilitated by surface oxygen vacancies. All these results suggest that NO_2 molecules dissociate to NO very easily on the SnO_2 surface.

The TPD chromatograms after the exposure of the oxidized SnO_2 surface to CO+NO or CO+NO_2 at room temperature (Fig. 7) were compared with those after the exposure to CO, NO, or NO_2. The desorption peaks for NO (β and γ)

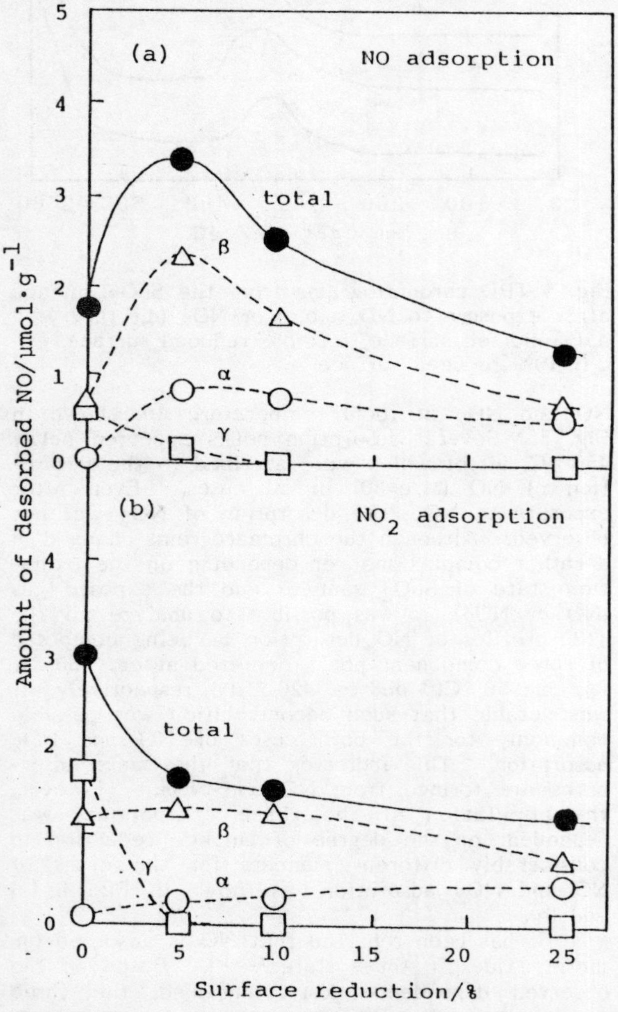

Fig. 6. The amounts of adsorbates from NO(a) and NO_2(b) versus the degree of reduction.

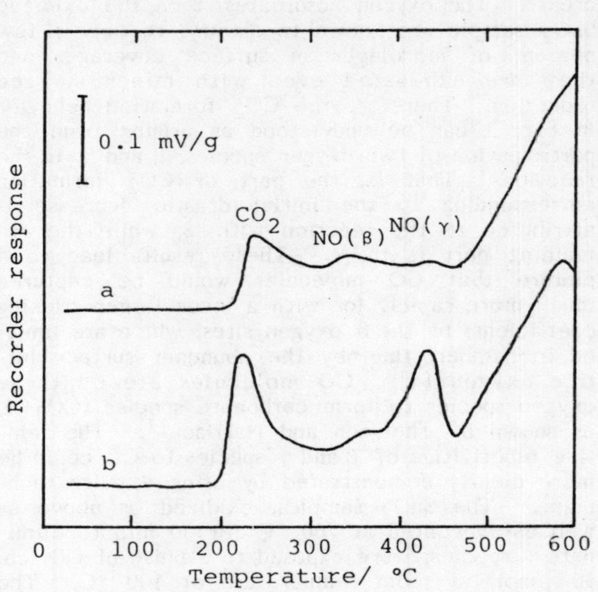

Fig. 7. TPD chromatograms from the oxidized SnO_2 surface after coadsorption of CO and NOx. a: CO+NO, b: CO+NO_2

were little affected by the presence of CO, but the desorption of CO_2 decreased to about one fourth under the copresence of NO or NO_2. Therefore the presence of NOx was found to hinder the adsorption of CO markedly. It is inferred that the strongly adsorbed NO species (the nitrito type, γ) preferentially occupy the surface oxygen (β) on which CO molecules also adsorb competitively and thus cause the significant decrease of CO adsorption.

3. Adsorption on Pd-SnO_2

The SnO_2-based sensor elements are usually added with a small amount of Pd as a sensitizer. The addition of Pd may cause a change in gas-adsorptive properties of SnO_2, and this was found in fact to be the case. Figure 8 shows the TPD chromatograms from Pd(2 wt%)-SnO_2 after exposure to several gases and gas mixtures. It is known that Pd is easily oxidized to PdO above 150 °C in O_2 atmosphere and the formed PdO begins to decompose at about 500 °C on heating in vacuo. Therefore the TPD chromatograms showed O_2 desorption due to the PdO decomposition in the higher temperature range. After the exposure to CO, CO_2 desorption from Pd-SnO_2 was observed around 140 °C (Fig. 8-a, a'), a temperature lower by about 80 °C than that from the undoped SnO_2. Moreover, the amount of CO_2 desorption was far less than that from the pure SnO_2. These results were common to both of the oxidized and the partially reduced samples.

TPD chromatograms after the exposure to NO or NO_2 showed NO around 350-400 °C (Fig. 8-b, b', c, and c'). The desorption changed its profile with a change in the degree of reduction of the sample, as was similarly observed on the pure SnO_2 sample. The peaks centered at 350 °C

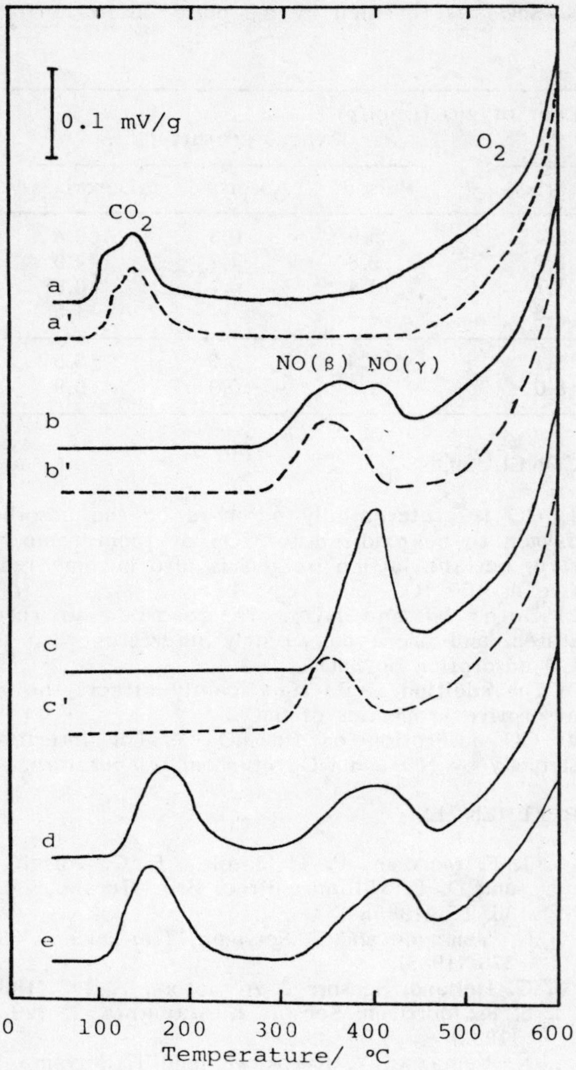

Fig. 8. TPD chromatograms after exposure of the oxidized Pd-SnO_2 surface (- -) and 2.5% reduced surface (---) to various gases.
a,a': CO, b,b': NO, c,c':NO_2, d: CO+NO, e: CO+NO_2

and 400 °C were assigned to NO(β) and NO(γ) species, i.e., M-NO^+ and M-O-N=O, respectively. M-NO^-(α) species was not observed even on the 50% reduced surface. Characteristically the adsorbed species after the exposure of the oxidized Pd-SnO_2 sample to NO_2 was exclusively NO(γ). The desorption temperature of NO(γ) was lower by 20 °C than that from the undoped SnO_2, though the desorbed NO amount was not different so much. A comment should be given here. In the above description, the degree of reduction was expressed as a nominal value on the assumption that the reduction with the H_2 pulse took place only on the SnO_2 surface. However, the Pd particles on it would also have taken part in the reduction. Therefore the actual degree of reduction of SnO_2 surface would be less than that for the Pd-SnO_2 sample.

In the chromatograms after exposure to CO+NO (d), the desorption of CO_2 and NO was not so much different from those after the single component adsorption of CO (a) and NO (b), respectively. This suggests that the interference between NO and CO is not so significant on the Pd-SnO_2 surface. With the coadsorption of CO+NO_2 (e), however, the desorption of the nitrito species (NO(γ)) decreased compared with the case of the single component adsorption of NO_2 (c).

4. Effects of Pd doping on gas adsorption

To make clearer the effects of Pd addition on the gas adsorptive properties, TPD results after the single component adsorption of CO, NO, and NO_2 are listed for the oxidized samples of SnO_2 and Pd(2 wt%)-SnO_2 in Table 1. Also listed together are our previous results on O_2 adsorption[15]. The addition of Pd is seen to cause the total disappearance of desorption of O_2, a marked decrease in CO_2 desorption after the CO adsorption, and the total disappearance of β desorption of NO after the NO_2 adsorption.

The above data were collected after gas adsorption by the static method. For more quantitative comparison, the pulse adsorption method coupled TPD was carried out for SnO_2 and Pd-SnO_2 and the results are listed in Table 2. The amount of adsorbed gas on a gas pulse was determined from the amount of the gas having passed through the sample, which was analyzed with a TCD detector. As seen from Table 2, the adsorbed amount thus determined coincided quite well with the desorbed amount determined from the subsequent TPD in each case. This confirms that each TPD chromatogram covers all the adsorbed species on the sample. Two interesting facts can be pointed out from Table 2.
1) On the oxidized samples, NO adsorption (at 100°C) is much larger in amount on Pd-SnO_2 than on SnO_2, while such a difference disappears on the evacuated samples.
2) The interaction of CO on the surface at 100 °C

Table 1 Gas adsorption behavior of oxidized SnO_2 and Pd(2 wt%)-SnO_2 (Static gas adsorption at room temperature)

Adsorbed gas	Desorbed amount(μmol/g) SnO_2	Desorbed amount(μmol/g) Pd-SnO_2	Desorbed gas
O_2	β 2.0(570)	β 0	O_2
CO	4.2(220)	0.85(140)	CO_2
NO	β 0.8(350) γ 0.6(420)	β 1.1(350) γ 1.0(400)	NO
NO_2	β 1.2(350) γ 1.8(420)	β trace γ 2.4(400)	NO

The parentheses show desorption temperatures (°C).

Table 2 The adsorption behavior of SnO_2 and Pd(2 wt%)-SnO_2 as revealed by a pulse adsorption-TPD technique

Sample	Adsorbed gas	Amount of gas (μmol/g) Oxidized surface Pulsed	Adsorbed	Desorbed	Evacuated surface Pulsed	Adsorbed	Desorbed
SnO_2	CO (R.T.)	9.5	3.1	3.4	8.9	0.3	0.4
	CO (100°C)	9.3	5.3	5.2	9.8	4.7	4.9
	NO (100°C)	10.2	1.6	1.7	10.4	0.7	0.6
	NO_2(100°C)	2.1	-	1.8	2.0	-	1.5
Pd-SnO_2	CO (100°C)	9.2	5.2	5.1	9.4	5.5	5.5
	NO (100°C)	9.2	7.9	8.0	9.2	0.9	0.9

is at large very similar irrespective of the presence of Pd and the oxidation state of the surface.

According to our XPS study[16], the Pd particles dispersed on SnO_2 interact strongly with the SnO_2 surface. In the oxidized Pd-SnO_2, the Pd particles (existing as PdO) act as an acceptor of electrons from SnO_2 and thus give rise to high electric resistance of SnO_2. The disappearance of adsorbed oxygen(β) on the Pd-SnO_2 is a result of such exhaustion of free (or conduction) electrons in the space charge region of SnO_2. In other words, the SnO_2 surface of oxidized Pd-SnO_2 can be assumed as being oxidized far more highly than that of oxidized SnO_2. The above decrease in CO_2 desorption from Pd-SnO_2 is considered to be a direct result of the lack of β adsorbates of oxygen. The disappearance of β adsorbates from NO_2 seems to reflect the lack of the reductive sites at which NO_2 molecules undergo dissociation into NO.

It is noted that the Pd addition also affects the desorption temperatures. The desorption temperature of CO_2 was lowered by as much as 80 °C. Slight lowering of desorption temperature was also seen for γ adsorbate resulting from NO and NO_2, while no such change was recognized for β adsorbate of NO. Presumably these phenomena are related with the electronic state of each adsorbate, i.e., whether the adsorbate donates or accepts electrons to or from SnO_2, respectively. It is speculated that electron accepting adsorption becomes less encouraged on Pd-SnO_2 because SnO_2 has already been depleted in electrons by the Pd particles. This would lead to the desorption of the adsorbate involved at a lower temperature.

Unlike SnO_2, Pd-SnO_2 exhibited no strong interference of NO or NO_2 on the CO adsorption at room temperature. This is probably because the adsorbed oxygen acting as the competitive adsorption sites for CO and NO on the pure SnO_2 surface can not be formed on Pd-SnO_2. On the other hand, NO_2 adsorption interfered considerably by CO. It seems that the copresence of CO produces some reductive sites on Pd-SnO_2 by which NO_2 dissociates into NO.

CONCLUSIONS

1) CO is preferentially adsorbed on the adsorbed oxygen to be oxidized to CO_2 at room temperature, but the lattice oxygen is also become reactive at 100 °C.
2) Both NO and NO_2 are adsorbed in three states, and one type strongly interferes with the CO adsorption on SnO_2.
3) The addition of Pd significantly affects the gas adsorptive properties of SnO_2.
4) CO adsorption on Pd-SnO_2 is not interfered strongly by NO and NO_2 at room temperature.

REFERENCES

1. J. F. McAleer, P. T. Moseley, B. C. Tofield and D. E. Williams, Proc. Brit. Ceram. Soc., 36, 89 (1985).
2. N. Yamazoe and T. Seiyama, Transducers '85, 376 (1985).
3. G. Heiland, Sensors & Actuators, 2, 343 (1982).
4. S. R. Morrison, Sensors & Actuators, 2, 329 (1982).
5. N. Yamazoe, Y. Kurokawa, and T. Seiyama, Proc. Int. Meeting on Chemical Sensors, Fukuoka (1983) p35.
6. M. Nitta, S. Kanefusa, Y. Takita, and M. Haradome, Appl. Phys. Lett., 32, 590 (1978).
7. N. Yamazoe, Y. Kurokawa and T. Seiyama, Sensors & Actuators, 4, 283 (1983).
8. K. Fukui and K. Komatsu, Proc. Symp. Chem. Sensors, (1987) p187.
9. G. S. V. Coles and K. J. Gallagher, Sensors & Actuators, 7, 89 (1985).
10. N. Yamazoe, J. Fuchigami, M. Kishikawa and T. Seiyama, Surf. Sci., 86, 335 (1979).
11. E. W. Thornton and P. G. Harrison, J. Chem. Soc., Faraday Trans. I, 71, 461 (1975).
12. H. Arai and H. Tominaga, J. Catal., 43, 131 (1976).
13. F. Solymosi and J. Kiss, J. Catal., 41, 202 (1976).
14. M. Niwa and T. Minami, J. Catal., 53, 198 (1978).
15. K. Wada, N. Yamazoe and T. Seiyama, Nippon Kagaku Kaishi, 10, 1600 (1980).
16. S. Matsushima, Y. Teraoka, N. Miura and N. Yamazoe, Jpn. J. Appl. Phys., 27, 1798 (1988).

PERFORMANCE OF SPINEL TYPE OXIDES AS AIR/FUEL RATIO OXYGEN SENSORS

Makoto Egashira, Yasuhiro Shimizu, Shuji Kusano, Hiroyuki Kuwayama, and Katsuyuki Tanaka*
Department of Materials Science and Engineering, Faculty of Engineering,
*Graduate School of Marine Science and Engineering, Nagasaki University,
1-14 Bunkyo-machi, Nagasaki 852, Japan

ABSTRACT

Temperature dependence of electrical resistance of several spinel type oxides has been measured in exhaust gases of air-methane combustion, to evaluate the performance of the oxides as an air/fuel stoichiometric oxygen sensor. It was found that the sensitivity and the response time were markedly affected by the point defect structures of the oxides. Among the oxides studied, the $ZnCr_2O_4 + Cr_2O_3$ (1.0 : 0.1 in molar ratio) exhibited the most promising properties as the oxygen sensor.

INTRODUCTION

To reduce pollutants in the exhaust gases from automobile engines, the air/fuel ratio (A/F) controlling systems have been put into practical use, by utilizing an A/F stoichiometric oxygen sensor combined with a three-way catalyst. To date, many efforts have been focused on developments and improvements of semiconducting A/F oxygen sensors. But, most of the studies have been directed to utilizing single metal oxides of n-type semiconductivity[1-3], and only a few results were reported about the potential of mixed oxides.[4]
The aim of the present work is to investigate suitable materials among the spinel type oxides for the A/F oxygen sensor. The results will be discussed in terms of sensitivity and response time at the stoichiometric A/F point, based on the point defect structures and the catalytic activities of the oxides.

EXPERIMENTAL

1. Preparation of Sensing Elements

Several spinel type oxides were prepared by two procedures. In the preliminary screening tests, the oxides were prepared by calcining a mixture of the constituent metal oxides in the desired proportions at given temperatures (see Table 1.) for 5h. In another procedure, metal nitrates were employed as starting materials. An aqueous solution of a desired mixture of the constituent nitrates was evaporated to dryness on a hot plate, followed by thermal decomposition at 500°C for 1 h. The resulting solid was ground and calcined at given temperatures for 2 h. Hereafter each specimen will be expressed by abbreviation in Table 1.
The powder of each spinel type oxide was tightly packed within a 2 × 2 × 1 mm cavity of an alumina substrate whose structure was essentially the same as the literature[3], and was then sintered at given temperatures (see Table 1) for 2 h. A pair of Pt wires as electrodes were embedded within the specimen compact. The distance between the two Pt wires could be controlled within a negligible error in one element to another by employing small leading holes in the substrate. Thus, differences in resistance of the elements correspond directly to the differences in conductivity of the specimens.

2. Measurement of Sensitivity, Response Time, and Catalytic Activity

Temperature dependence of the resistance of the specimen was measured in the range of 400°C to 800°C in air and in two exhaust gases using a flow apparatus ; one was the exhaust

Table 1 Preparation conditions of sensing elements

Key	Specimens	Starting materials	Calcination & sintering Tepm./°C
MF-O	$MgFe_2O_4$	oxides	1200
ZF-O	$ZnFe_2O_4$	oxides	1200
NF-O	$NiFe_2O_4$	oxides	1200
MC-O	$MgCr_2O_4$	oxides	1200
ZC-O	$ZnCr_2O_4$	oxides	1000
NC-O	$NiCr_2O_4$	oxides	1200
MC-N	$MgCr_2O_4$	nitrates	1000
ZC-N	$ZnCr_2O_4$	nitrates	900
ZCC-N	$ZnCr_2O_4 + Cr_2O_3$ (1.0 : 0.1 in molar ratio)	nitrates	900

of the air and methane mixture with $\lambda = 1.13$ (lean-burn condition) and the other $\lambda = 0.89$ (rich-burn condition). Where λ was defined as

$$\lambda = \frac{Air/CH_4}{(Air/CH_4)_{stoich.}} \quad (1)$$

The ratio of the resistance in $\lambda = 1.13$ to in $\lambda = 0.89$ for n-type semiconductive oxides, or the reverse for p-types, at the same operating temperature represents the sensitivity in detecting the stoichiometric combustion point. The resistance change was calculated from the voltage change across a load resistor connected in series with the specimen under the applied voltage of 5.0 V dc.

In measurement of response times, the exhaust gas of the mixture with $\lambda = 1.13$ flowing over the specimen was quickly changed to another exhaust gas with $\lambda = 0.89$ (denoted as "Red" response time), and vice versa ("Ox" response time). The time required for the output voltage to reach the 80% value of the total change was defined as the response time.

Granules of each specimen 0.2 g were placed in a fixed bed reactor in the same gas-flow apparatus. An air and methane mixture of $\lambda = 1.0$ was fed to the specimen bed at a flow rate of 1800 m^3h^{-1} (space velocity was about 14000 h^{-1}). The catalytic activities of various specimens were evaluated in terms of the conversion of methane.

RESULTS AND DISCUSSION

1. Sensing Properties

Figure 1 shows typical temperature dependence of resistance in three environments and response times of the ZF-O specimen. At every temperatures, the resistance decreased in the order of air > $\lambda = 1.13$ > $\lambda = 0.89$, suggesting that the specimen was n-type semiconductive over the whole oxygen partial pressure range between 2.1×10^4 Pa (in air) and about 10^{-14} Pa (in $\lambda = 0.89$)[5]. It is apparent that the

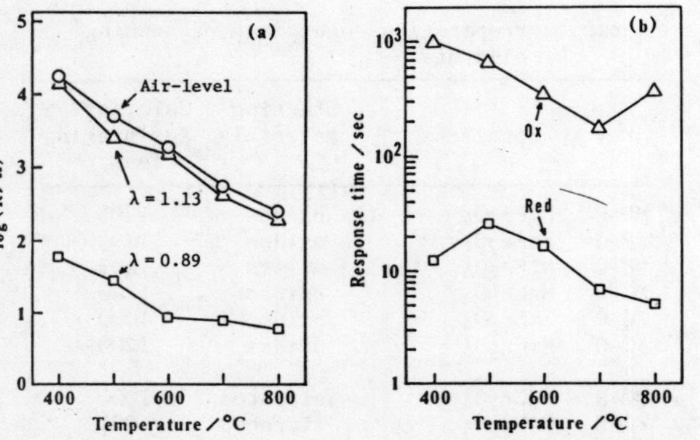

Fig. 1. Temperature dependence of (a) resistance in three environments and (b) response times of ZF-O specimen.

Table 2 Sensing properties at 600°C and catalytic activities of the specimens

Key	Semi-conductivity	Sensitivity $k^{a)}$	80% response time/sec Ox	Red	Surface area $m^2\ g^{-1}$	$T_{80\%}$$^{b)}$ °C
MF-O	n	35	117	15	0.16	785
ZF-O	n	170	354	16	0.27	785
NF-O	n	48	228	6	–	–
MC-O	p	1000	0.4	2	1.5	715
ZC-O	p	1800	0.4	1.8	2.1	670
NC-O	p	28	4	14	–	–
MC-O	p	750	0.4	0.8	4.7	625
ZC-N	p	3300	0.35	0.7	6.9	615
ZCC-N	p	3800	0.45	0.5	4.8	660

a) $k = R(\lambda = 1.13)/R(\lambda = 0.89)$ for n-types,
$k = R(\lambda = 0.89)/R(\lambda = 1.13)$ for p-types.
b) Temperature at which conversion level is 80 %.

sensitivity is in the magnitude of several hundreds at 400°C and it decreases as the temperature increases. From Fig. 1(b), the "Red" response time is much shorter than "Ox". But, even the "Red" times are too long from a practical point of view.

Sensing properties of the spinel type oxides prepared from the constituent metal oxides are summarized in Table 2. Broadly, an interesting relation is notable between the sensing properties and the semiconductivity; the sensitivity is much higher and the response times are much shorter for the p-types except for NC-O. As to the response times, another feature is noted. In the case of n-types, the values of "Ox" were considerably larger than "Red". In contrast, the values of "Ox" were shorter than "Red" in the p-types.

From the result on the specimens from the constituent oxides, it is apparent that the p-type semiconductive MC-O and ZC-O were more promising as oxygen sensors than the n-types. To improve the sensing properties of the p-types, the specimens were prepared from metal nitrates so as to reduce the calcination temperatures and to enhance the surface areas. Sensing properties of the two specimens are also summarized in Table 2. In the case of ZC-N, the sensitivity was enhanced significantly, whereas such improvement was not observed for MC-N. This suggests that the surface area is not important in determining the sensitivity.

Temperature dependence of resistance in the exhaust gases and response times of ZC-N are shown in Fig. 2. For comparison, the results on ZC-O are also plotted. On the whole, the response times were slightly improved at temperatures higher than 500°C by employing metal nitrates (see Fig. 2(b)). This may be attributed to the increased surface area (see Table 2). More interesting is the change in sensitivity. The resistance of ZC-N in $\lambda = 1.13$ decreased at every temperatures compared with ZC-O, and the magnitude of the decrease was comparable at every temperature. In $\lambda = 0.89$,

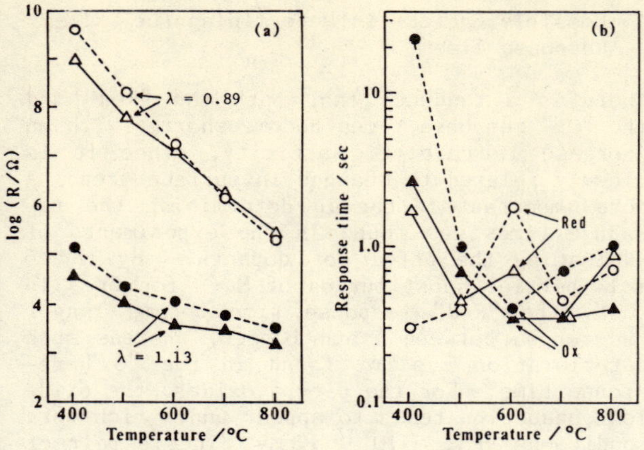

Fig. 2. Temperature dependence of (a) resistance in the exhaust gases with different λ values and (b) response times of ZC-O and ZC-N specimens. ●, ○ ; ZC-O, ▲, △ ; ZC-N

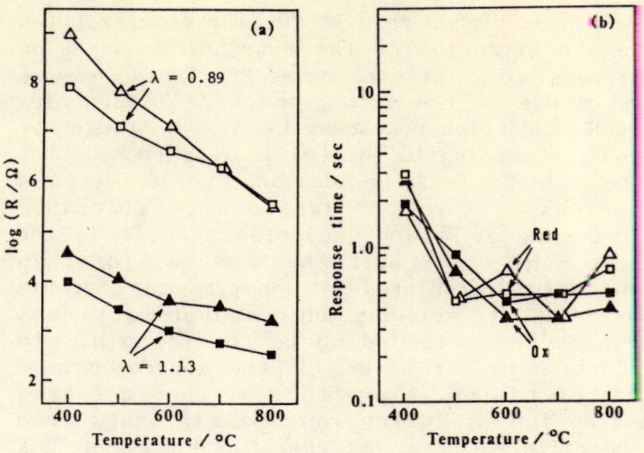

Fig. 3. Temperature dependence of (a) resistance in the exhaust gases with different λ values and (b) response times of ZCC-N and ZC-N specimens. ■, □ ; ZCC-N, ▲, △ ; ZC-N

on the other hand, the resistance slightly decreased at 400-600°C, while it increased a little above 700°C. As a result, the sensitivity was enhanced at temperatures higher than 600°C without any significant deterioration in response times. In addition, this resulted in an increase of the gap between the resistance in λ = 0.89 at 800°C and that in λ = 1.13 at 400°C, indicating that less compensation is needed for the effect of operating temperature on the resistance in its practical use.

2. Attempts of Improvement of Sensing Properties by Doping

Further attempts were made to improve the sensing properties of ZC-N, by 5 mol% partial substitution of Mg^{2+} ions for Cr^{3+} in ZC-N as well as by addition of 10 mol% ZnO or Cr_2O_3 to ZC-N. The three specimens were prepared from metal nitrates. These experiments were aimed also to evaluate the effect of charge defects in the ZC-N lattice on the sensing properties. Among the three specimens, the most significant improvement in sensitivity was found for ZCC-N (Fig. 3). The resistance in λ = 1.13 decreased over the whole operating temperature range by the addition of 10 mol% Cr_2O_3. Especially, the decrease in resistance was most pronounced at temperatures higher than 700°C. On the other hand, the resistance in λ = 0.89 remained unchanged at temperatures higher than 700°C, whereas it decreased at temperatures lower than 600°C. Thus, the addition of 10 mol% excess Cr_2O_3 resulted in the increase of the sensitivity at temperatures higher than 600°C without any marked changes in response times and also in the enlargement of the gap between the resistance in λ = 0.89 at 800°C and that in λ = 1.13 at 400°C. Therefore, it was revealed that ZCC-N was most suitable for the A/F stoichiometric oxygen sensor among the specimens studied.

In the case of ZCC-N, the following defect reaction is conceivable.

$$x\,Cr_2O_3 \xrightarrow{ZnCr_2O_4} x\,V_{Zn}'' + 2x\,Cr_{Cr}^{x} + 3x\,O_O^{x} + x\,V_O^{\cdot\cdot} \quad (2)$$

This is essentially the same as the following reaction.

$$x\,Cr_2^{3+}O_3 + Zn^{2+}Cr_2^{3+}O_4 + x/2\,O_2(g) \longrightarrow Zn^{2+}Cr_2^{3+}Cr_{2x}^{4+}O_{4(1+x)} \quad (3)$$

Therefore, the decrease in resistance by the addition of excess Cr_2O_3 was attributed to increased positive holes available for electrical conduction. However, the increase in the sensitivity could not be explained only by this mechanism.

In the formula representing the dependence of the electrical conductivity on oxygen partial pressure for the p-type semiconductive oxides, $\sigma \propto P_{O_2}{}^{1/m}$, the value of m can vary from 4 to 6 depending on whether the conduction process is extrinsic or intrinsic, that is, depending on the amount of impurities in the lattice. In addition, the m value increases with an increase in the transport number of oxide ions, especially at lower oxygen partial pressures. It is easy to imagine that the sensitivity at the stoichiometric point decreases with an increase in the m value. Thus, an important factor in determining the sensitivity is the oxygen partial pressure at which the conduction process is changed from extrinsic to intrinsic. If the transition point is higher than the oxygen partial pressure of 10^3 Pa, the sensitivity becomes large. Therefore, the more pronounced decrease in resistance of ZCC-N at λ = 1.13, which was observed at temperatures higher than 700°C, may be attributed to the extension of the extrinsic conduction process to a higher oxygen partial pressure range due to the formation of Zn^{2+} vacancies by the addition of excess Cr_2O_3. On the other hand, according to eq. (2) it is supposed that the effect of the oxide ion conductivity is serious in $\lambda = 0.89$ due to a considerable amount of oxide ion vacancies. If it is the case, the sensitivity becomes smaller than the value calculated by assuming the formula $\sigma \propto P_{O_2}{}^{1/4}$, because the conductivity does not decrease according to the formula. In addition to the effect of the increased hole concentration, the effect of the increased oxide ion vacancies was clearly found from the differences in the resistance between ZC-N and ZCC-N in $\lambda = 0.89$ at temperatures lower than 500°C. In contrast, the resistance of ZCC-N in $\lambda = 0.89$ was almost the same as ZC-N at temperatures higher than 700°C. Although this is one of the origins of the increased sensitivity, the reason for this phenomenon is not clear at present.

3. Effect of Catalytic Activity on Sensing Properties

In the exhaust gases from automobile engines, there is a certain amount of residual fuel even in the lean-burn region ($\lambda > 1.0$), and also a certain amount of free oxygen even in the rich-burn region ($\lambda < 1.0$)[1]. Thus, if the catalytic activity of the sensing element is poor, it leads to the longer response time and the smaller sensitivity at the stoichiometric point, because thermodynamic equilibration in the gaseous phase over the element cannot be achieved rapidly. In the case of the commercially produced A/F semiconducting oxygen sensors, the fine particle Pt electrodes and/or the Pt catalyst are employed to catalyze the reaction between the residual fuel and oxygen. Therefore, there is a possibility that the catalytic activity of the present specimens themselves may be an important factor to determine the sensitivity.
From the experiments of catalytic oxidation of methane over each specimen, $T_{80\%}$ temperatures at which the conversion of methane attained 80% were obtained and summarized in Table 2. It is noticed that the catalytic activities of the n-type specimens, ZF-O and MF-O, were negligibly small because the $T_{80\%}$ values were close to the value for the blank run without the specimen, while all the p-type specimens showed relatively high activities. However, it is suggested from comparison between ZC-N and ZCC-N that the sensitivity is not affected by the catalytic activity. Anyway, in a practical application to automobile engines the ZCC-N offers the potential that excellent performance can be realized without Pt catalyst owing to its high catalytic activity.

4. Possible Factors in Determining the Response Time

There is a tendency that both the "Red" and the "Ox" response times become shorter with an increase in catalytic activity, since it is closely related to changes in surface area. A more important factor in determining the response times was found in the experiments of evaluating the effect of dopants. By the 5 mol% partial substitution of Mg^{2+} for Cr^{3+} in ZC-N, the "Red" response time became longer and ranged between 4 and 10 sec, whereas such deterioration was not found in the "Ox" response time. For the p-type oxides, the oxide ion conduction tends to appear under rich-burn conditions ($P_{O_2} \simeq 10^{-14}$ Pa). Since a pair of Pt wire electrodes are not expected to catalyze the electron transfer between oxide ions and gaseous oxygens because of the extremely small surface area, oxide ions would be blocked at the Pt wire/semiconductor interface. This would lead to a relatively long response time when λ is changed from 1.13 to 0.89. Thus, it is suggested that the oxide ion conductivity affects the response time. This discussion also offers an answer to explain the fact that n-type oxides exhibited the longer "Ox" response time (see Table 2). It is easy to imagine that in the case of the n-type oxides the contribution of the oxide ion conductivity is dominant in the lean-burn region, if the concentration of oxide ion vacancies is high enough.

CONCLUSION

The most promising sensing properties could be realized with $ZnCr_2O_4 + Cr_2O_3$ (1.0 : 0.1 molar ratio) specimen. The role of the excess Cr_2O_3 was anticipated to extend the extrinsic conduction process to a higher oxygen partial pressure range especially at higher operating temperatures. It is concluded that the point defect structure is the most important factor in determining the sensitivity as well as the response times.

REFERENCES

1. T. Y. Tien, H. L. Stadler, E. F. Gibbons, and P. J. Zacmanidis, Ceram. Bull., 54 (1975), 280.
2. E. M. Logothetis and W. J. Kaiser, Sensors and Actuators, 4 (1983), 333.
3. K. Ihokura, K. Tanaka, and N. Murakami, Sensors and Actuators, 4 (1983), 607.
4. Y. Shimizu, Y. Fukuyama, H. Arai, and T. Seiyama, In : Fundamentals and Applications of Chemical Sensors, ACS Symp. Ser., 309 (1986), 84.
5. C. Yu, Y. Shimizu, and H. Arai, Chem. Lett., 1986, 563.

INFLUENCE OF ADDITIVES ON THE THERMAL STABILITY OF SnO_2-BASED GAS SENSORS

Chaonan Xu, Jun Tamaki, Norio Miura and Noboru Yamazoe
Department of Materials Science and Technology, Graduate School of Engineering Sciences, Kyushu University, Kasuga, Fukuoka 816, Japan

ABSTRACT

The thermal stability of tin oxide powder impregnated with various additives was investigated in the temperature range of 600-900 °C. The data on specific surface area (SA) as well as SnO_2 crystallite size (D) showed that some dopants such as Ba, P, Nb and Sm were effective to keep D below 100 Å at 900 °C. Non-metal dopants of P, S and B gave extremely large SA at 600 °C. Even larger stabilizing effects were observed with double doping such as P-Ba and P-V. Such doping hardly degraded the gas sensitivity and promoted it significantly in some cases.

INTRODUCTION

Tin oxide has long been exploited for use in semiconductor gas sensors. Recent investigations have shown that the addition of a small amount of certain metal oxides promotes selectivity, sensitivity and stability of SnO_2-based gas sensors[1-4]. However, the nature of the promoting effects is far from being well understood. In principle, the addition of a metal oxide may modify various properties of tin oxide, such as microstructure, resistance to thermal sintering, oxidation activity, acid-base property, gas adsorption property, electronic property and so on. Control of these properties of tin oxide by a foreign metal oxide appears to be very important for elaborate design of gas sensors. Especially the stability of the microstructure of polycrystalline elements is decisively important. Such situation prompted us to seek for those additives which promote the thermal stability of SnO_2-based gas sensors. In this paper, the effects of various additives on the thermal stability were investigated on the basis of specific surface area and SnO_2 crystallite size. The electrical as well as gas sensing properties of the doped samples were also investigated.

EXPRIMENTAL

The powder of SnO_2 was prepared by neutralizing a cold solution of $SnCl_4$(27%) with an aqueous ammonia solution (28%). The resulting precipitate was washed thoroughly with deionized water, dried at 100 °C and ground into granules of 1-10 μm. The sample was actually hydrous SnO_2 containing 13.8% water. To prepare a doped sample, the hydrous SnO_2 powder was impregnated with an aqueous solution of a salt (mainly acetate) of metal (M) and/or a polyoxy acid of non-metal (A). The amount of M or A was fixed to 5% in atomic composition M(orA)/(M(orA)+Sn). The resulting suspension was dried at 100 °C and ground to granules. The other metal salts used were nitrates (In, Ga and Bi), an oxalate (Nb), NH_4VO_3, $(NH_4)_6Mo_7O_{24}$ and $5(NH_4)_2O \cdot 12WO_3 \cdot 5H_2O$. The doped samples thus prepared were labeled in formulae of M-SnO_2, A-SnO_2 and A-M-SnO_2. The specific surface area was measured by the B.E.T. method using N_2 adsorption. The average crystallite size of SnO_2 was evaluated from the X-ray diffraction line (101) based on the Scherrer's equation.

To fabricate a sensor element, the powder sample was calcined at 600 °C for 1h in air and, after mixing with water, applied 1.0mm thick on an alumina tube attached with two Pt coil electrodes 1.5mm apart from each other (Fig. 1).

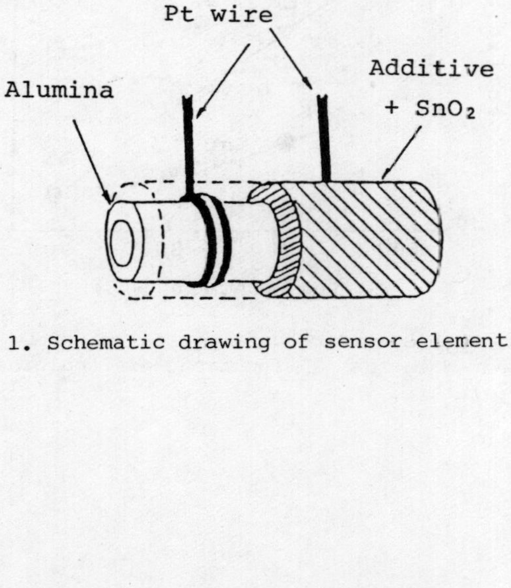

Fig. 1. Schematic drawing of sensor element.

RESULTS AND DISCUSSION

1. Specific Surface Area

The effect of additives on the thermal sintering of SnO_2 particles was first evaluated from the specific surface area (SA) of the doped samples after calcination in air at elevated temperatures for 1h. Fig. 2 shows SA of some typical samples as a function of calcination temperature. The undoped SnO_2 had SA as large as 154 m^2/g after calcination at 300 °C, but SA decreased rapidly with increasing calcination temperature to 24.4 and 6.4 m^2/g at 600 and 900 °C, respectively. On the other hand, all the doped samples kept larger SA than the undoped sample at 600 °C. This indicates that all the additives examined are effective to increase resistance to thermal sintering up to the medium temperature of 600 °C. At higher temperatures, however, the doped samples behaved differently depending on the kinds of additives. That is, some additives such as P and Ba were still effective to keep far larger SA than that of the undoped SnO_2 even at 900 °C, while some others such as V and S were not. Thus the additives can be classified into two types, i.e., one type which is effective to retard the sintering of SnO_2 particles in the whole temperature range examined (Type I) and the other effective only up to the medium temperature (Type II).

The specific surface area data for the prepared samples after calcination at 600 or 900 °C for 1h are listed in Table 1. It is seen that Type II dopants are composed of alkali metals, V, S, B and Tl. On the other hand, Type I dopants, which remain effective at 900 °C, consist of alkaline earth metals, some transition metals, rare earth metals and P. Among them P, Ba, Nb, W and Pb exhibited especially remarkable effects in this order at 900 °C. The effect became more pronounced when two dopants were added simultaneously. Thus the samples doubly doped with P-V and P-Ba showed the highest SA of ca. 40 m^2/g after calcination at 900 °C.

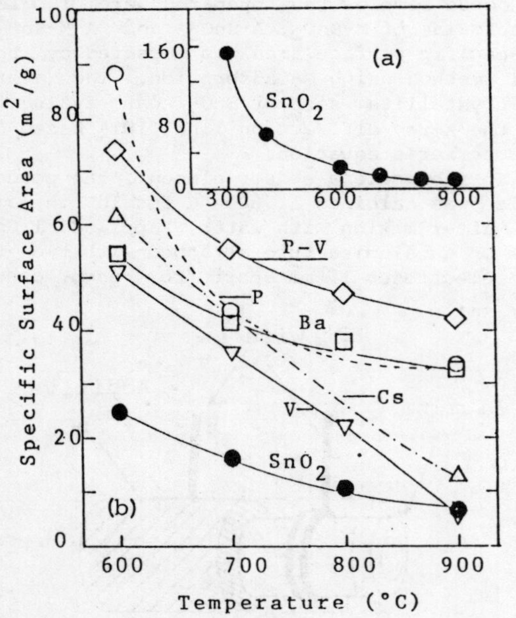

Fig. 2. Specific surface area of SnO_2 and doped SnO_2 as a function of calcination temperature.

Table 1 Specific surface area (SA) and SnO_2 crystallite size (D) after calcination at 600 and 900 °C for 1h

Dopant 5%	600 °C SA (m^2/g)	600 °C D (Å)	900 °C SA (m^2/g)	900 °C D (Å)
none	24	138	6	272
Li	38	117	4	-
Na	44	100	5	-
K	54	92	5	330
Rb	55	-	-	-
Cs	62	90	12	262
Mg	52	52	22	-
Ca	40	54	20	110
Ba	54	57	33	86
Sr	58	-	23	
V	51	67	6	382
Cr	41	71	-	99
Mn	37	-	-	-
Fe	39	-	20	-
Co	28	71	-	-
Ni	44	58	23	131
Cu	29	73	-	-
Zn	48	57	21	118
Ga	35	-	-	176
Nb	48	55	31	100
Mo	36	63	-	87
In	42	71	-	99
La	41	90	24	142
Ce	35	-	-	-
Pr	41	-	-	-
Nd	35	-	-	-
Sm	38	56	-	83
Gd	38	-	-	-
W	55	66	29	109
Tl	30	-	11	-
Pb	70	59	28	174
Bi	53	64	24	121
S	86	-	6	254
B	84	64	6	-
P	88	43	34	88
S-V	93	52	7	-
B-V	87	-	6	412
B-Ba	98	-	31	107
P-Ba	77	44	39	67
P-V	74	49	42	120

Among both type dopants, the non-metal dopants of S, B and P are seen to give particularly large SA exceeding $80m^2/g$ at 600 °C. These dopants are expected to be bonded to the SnO_2 surface as polyoxy cluster anions. Difficulty in surface diffusion of such cluster ions seems to be a reason for the very marked stabilizing effects of the non-metal dopants at the medium temperature (600 °C). A similar explanation may be possible to the less but still marked effects of high valent metal dopants such as V, Nb and W which are also expected to exist as polyoxy anions on the SnO_2 surface. The other dopants showing remarkable effects consist of Pb, Cs, Sr and Ba, all of which have large ionic radii. Therefore the presence of less mobile species seems to be the origin of the stabilizing effects also in these cases.

2. Crystallite Size of SnO_2

After calcination at various temperatures for 1h , the doped samples were subjected to X-ray diffraction (XRD) analyses. No significant diffraction peaks other than those of SnO_2 itself were observed, except for $Cr-SnO_2$ system for which free Cr_2O_3 was detected. The diffraction peaks of SnO_2 were rather broad in most cases, showing that the SnO_2 crystallites included were small in size. The average crystallite size (D) were determined from (101) diffraction for a number of samples after calcination at 600 and 900 °C, and listed in Table 1 along with the SA data. For comparison, some samples were also subjected to direct observation by a transmission electron microscope (TEM). The crystallite size thus obtained coincided well with the above XRD data. For $V-SnO_2$ and SnO_2(undoped) calcined at 600 °C, for example, TEM showed crystallites of 50-80 Å and 100-150 Å in diameter, respectively, while XRD showed average sizes of 67 Å and 138Å, respectively, as listed in Table 1.

As seen from Table 1, the D values after calcination at 600 °C were in the range of 40-70 Å for most of the doped samples, to be compared with that (138 Å) of the undoped sample. After calcination at 900 °C, it increased to about 80 Å or above, which were in many cases still less than the value (272 Å) of the undoped sample. These data show that most of the dopants are effective to retard the crystal growth of SnO_2 at the medium temperature (600 °C), whereas a more limited number of dopants remain effective at the high temperature (900 °C).

The stabilizing effects of the dopants as revealed from the D data are essentially in good agreement with those revealed from the SA data, but the D data would possibly provide more microscopic information. For example, alkali metals gave comparatively large crystallites at 600 °C despite giving rather larger SA, and at 900 °C they gave D values nearly equal to or even larger than D of the undoped samples (Type II dopants). Particularly interesting are the cases of K, V and B-V doped systems, which gave D values larger than the undoped sample at 900 °C. Probably these dopants accelerate the growth of SnO_2 crystallites at the high temperature. Based on the data of D and SA after calcination at 900 °C, it is possible to classify the stabilizing effects of dopants as follows:

A; very marked (D≤100 Å):
P-Ba, Sm, Ba, P, Mo, In, Cr

B; marked (100 Å<D≤130 Å):
Nb, B-Ba, W, Ca, Zn, P-V, Bi

C; noticeable (130 Å<D≤180 Å):
Ni, La, Pb, Ga

D; little or non.(180 Å<D):
S, B, V, alkali metals

Groups A, B, and C belong to Type I and group D belongs to Type II, respectively, according to the previous classification.

It is seen that Type II dopants (group D) are subject to either melting or sublimation at 700-900 °C. Disappearance of these dopants from the surface of SnO_2 crystallites by diffusion or sublimation seems to be a main cause of the losses of stabilizing effects at high temperature. In the case of $V-SnO_2$, a part of V ions were confirmed to form a solid solution with SnO_2 by means of XRD analysis. On the other hand, Type I dopants (groups A, B and C) appear to stay more firmly on the surface of SnO_2 crystallites. It is surprising that P, Mo and W are included in group A or B, because their oxides are known to be volatile at high temperatures. It follows that these dopants stay on the surface as particularly stable polyoxy anions. At the same time, these results suggest that the actual stabilizing effect of a given dopant may often be totally different from what is expected from the properties of its bulk compounds.

3. Electric and Gas Sensing Properties

The electric properties as well as gas sensing properties of doped samples were investigated in the form of gas sensor elements (Fig. 1). The electric resistance of a sensor element can be influenced not only by the microstructure of the polycrystalline element but also by the temperature of measurement as well as the chemisorption of gases such as O_2 and water. To finger out the effect of additives on the microstructure, it is imperative to compare the electric resistance data of the elements with a specified thermal history. The data collected at 600 °C for some elements sintered (calcined) at 600 °C for 1h are listed in Table 2. Although the undoped element had a small resistance (0.5 kΩ), the doped elements showed widely dispersed values up to 240kΩ ($La-SnO_2$). Such a variety suggests that the state of additives and their effects on electric resistance are by no means simple.

It was further found that the state of additives was often subject to changes with an increase in sintering (calcination) temperature (Ts). Fig. 3 shows relative changes (δ) in electric resistance (measured at 500 °C in air) of some typical elements with a change in Ts, where $\delta = \log(R(Ts)/R(600))$, and R(Ts) and R(600) are the resistance values of each element sintered at Ts and 600 °C, respectively. For the undoped element, increased monotonously with increasing Ts, reflecting simply that SnO_2 crystallites grew larger at higher Ts to result in a decrease in the number of contacts between SnO_2 crystallites or their aggregates. For the doped elements, however, δ behaved very differently and even a minimum (B-SnO_2, V-SnO_2) or a monotonous decrease (Nb-SnO_2) was observed.

As seen from Fig. 3, three types were recognized in δ-Ts behavior, i.e., monotonous increase (a), decrease followed by increase (b), and monotonous decrease (c). The behavior type of each doped element was shown in Table 2. Such behavior would be associated with changes in microstructure of the elements, such as growth of SnO_2 crystallites, segregation or disappearance of additives between SnO_2 crystallites, and formation of new compounds or solid solutions between additives and SnO_2. Among these, formation of solid solutions seems to be important for the dopants of V, P, Nb and W, the cationic sizes of which are smaller than that (0.83 Å) of $SnO_2{}^{4+}$, as indicated in Table 2. If these cations are dissolved into SnO_2 lattice to increasing extents with increasing Ts, type c or b behavior would result. For the basic dopants of Ca, Bi, La, Pb and Ba, on the other hand, formation of surface componds would be more important. However, more studies are necessary for elucidating the actual microstructure of doped elements.

Table 2 Electric properties and gas sensitivities of doped elements

Dopant	Radius (Å)	R (kΩ)	Type*)	Sensitivity H_2	Sensitivity $i\text{-}C_4H_{10}$
none	-	0.5	a	36	3
V^{5+}	0.68	46	b		21
B^{3+}	(0.25)	0.4	b	83	
P^{5+}	0.31	5	b	110	
Nb^{5+}	0.78	74	c	57	7
W^{6+}	0.74	7	c		22
Zn^{2+}	0.88	63	a	10	
Ca^{2+}	1.14	120	b	36	3
Bi^{3+}	1.17	25	a		
La^{3+}	1.17	240	a	1000	
Pb^{3+}	1.33	16	a	457	
Ba^{2+}	1.49	30	c		7
P-V		17	b		26

*) Type in δ-Ts behavior (see in text).

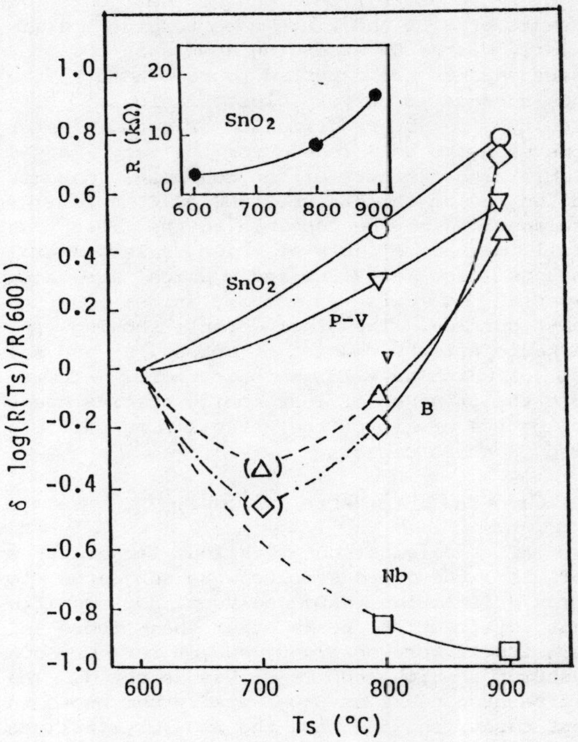

Fig. 3. Relative change in electric resistance with a change of sintering temperature.

Finally, the sensitivities of doped elements to 800 ppm H_2 and 1000 ppm $i\text{-}C_4H_{10}$, both diluted in air, were measured at 300 °C. The sensitivity (K) was defined as $K = R_{air}/R_g$ where R_{air} and R_g stand for the electric resistances in air and in the diluted gas, respectively. The obtained data were also listed in Table 2. It is seen that most of the doped elements exhibit higher sensitivities than the undoped element. This confirms that the additives can promote the thermal stability of SnO_2 without degrading its sensing properties. Moreover, some doped elements showed very high sensitivities to H_2 or $i\text{-}C_4H_{10}$, compared with those of undoped element. This would indicate even the possibility of obtaining a highly sensitive and selective element by such a doping method.

REFERENCES

1. K. Takahata, "Chemical Sensors Technology", edited by T. Seiyama, Kodansha/Elsevier, 1, (1988) 39.
2. Y. Nakamura, S.Yasunaga, N. Yamazoe and T. Seiyama, Proc. 2nd Int. Meeting on Chem. Sensors, Bordeaux, 1986, p.163.
3. K. Fukui and K. Komatsu , Digest of Technical papers of Transiducer '87, Tokyo, 1987, p.614.
4. Y. Matsuura, N. Murakami and K.Ikokura, Proc. Int. Meeting on Chem. Sensors, Fukuoka, 1983, p.18.

GAS SENSING PROPERTIES OF TIN OXIDE ULTRAFINE PARTICLE FILMS

Dai Guorui, Kang Chanhe, Zhang Qingiong
Department of Electronic Science, Jilin University
79 Jie Fang Da Lu, P.O. Box 130021, Changchun, China

ABSTRACT

Gas sensing properties of tin oxide ultrafine particle films have been studied in the paper. The ultrafine particle films having porous columnar structure are found to have high sensitivity and high selectivity in detecting combustible gases. The use of the structure of Fe_2O_3-SnO_2 multilayer thin films greatly improves long-term stability of gas thin film sensors.

INTRODUCTION

Tin dioxide, SnO_2, in its pure state is an n-type metal oxide semiconductor. Its unique properties make the material useful for many application[1,2] such as resistor, NESA glass, solar cell, and so on. Recently, SnO_2 has attracted particular interest as a gas sensor material. The properties of SnO_2 in the form of sintered polycrystals[3], thin films[4,5], thick films[6] as gas sensor have been reported. The sintered type sensor has been put into practical applications for sensing combustible gases. However, the sintered sensor has some disadvantages, for example, being troublesome to manufacture, brittle and having sensitivities that vary greatly between sensor elements. Accordingly, some improvements in the sensor elements are required. Widespread use of thin films in mordem techmology has prompted the development of new deposition techniques. We have tried to develop a new type of sensor with the aid of semiconductor technologies and to manufacture thin film sensors. In this article, the tin oxide and ferric oxide ultrafine particle films have fabricated by ultilizing plasma enhanced chemical vapor deposition (PECVD). The porours columnar structure of the films has been constructed from the ultrafine particles having the median particle size of several hundreds of angstroms. Gas sensing properties of the ultrafine particle films have been investigated and The thin film sensors of high sensitivity in detecting combustible gases have been fabricated.

EXPERIMENTAL

The ultrafine particle films of SnO_2 and Fe_2O_3 onto the silicon or ceramic substrates have been fabricated by PECVD. Figure 1 shows the plasma deposition system utilized in this study.

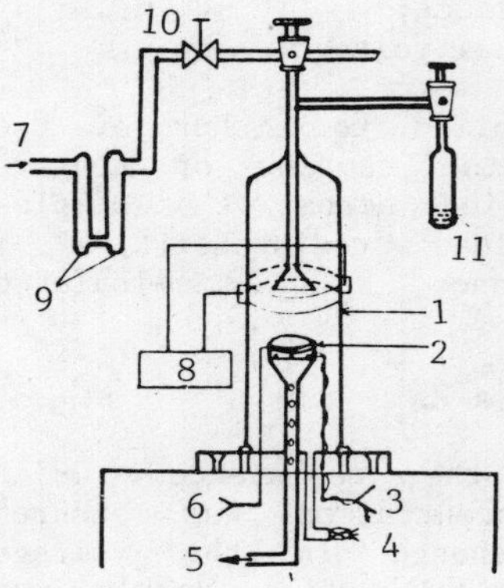

Fig. 1 Plasma ehhanced chemical vapor deposition system

The present work describes the structure of SnO_2 and Fe_2O_3 films which were deposited from $SnCl_4$, $Fe(CO)_5$ and O_2 in an r.f. glow discharge at 200°C. In general, flow rates were set at the desired values, at a flow ratios of 5:1 O_2 to $SnCl_4$ or $Fe(CO)_5$, the deposition rate ranged between 0.005 um/min and 0.01 um/min.

Figure 2 shows the sensing thin film sample and sensor structure.

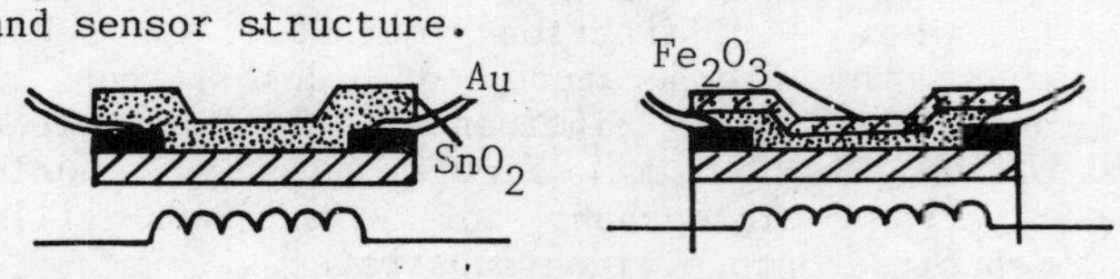

(a) Monolayer SnO_2 samples (b) Multilayer thin film sensors

Fig. 2 Schematic drawing of the sensing thin film samples and sensors

The sensing materials, the ultrafine particles films of SnO_2 and $SnO_2 - Fe_2O_3$, were deposited between the two gold-coated interdigital electrodes. Electrical contacts were made with 0.05 mm gold wires attached to the gold electrodes. The contact resitance between ultrafine particle films and gold electrodes was measured at the operating temperature.

Figure 3 shows the electrical measuring circuit.

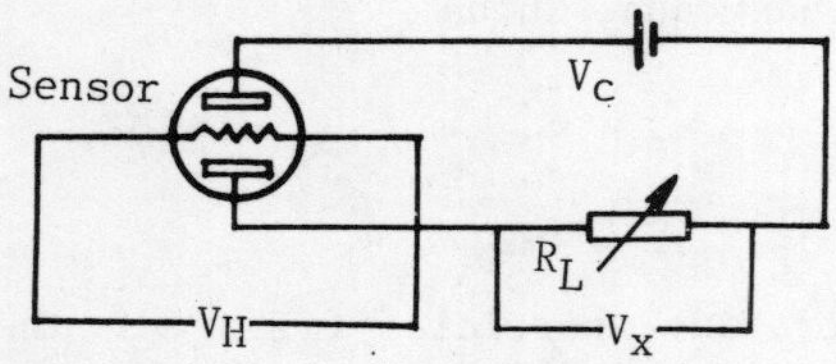

Fig.3 Electrical measuring circuit
V_H: Heating voltage,
V_c: Circuit voltage,
R_L: Load resistance,
V_x: Output signal voltage

Approx. 5V was supplied to either of the coils for heating the samples or sensors and 10V was supplied across the samples or sensors and load resistance(R_L=2 K). Here, the resistance, R_x, was calculated as follows

$$R_x = \left(\frac{V_c}{V_x} - 1 \right) \times R_L$$

The variation in the resistance, which changed with gas concentration, was measured indirectly as a change in the voltage across R_L. As the SnO_2 film sample was an n-type semiconductor, the resistance in room air, R_0, became larger than the resistance in the reductive gas atmosphere, R_g. The gas sensitivity, S, was defined as the value of R_0/R_g.

The chemical components, surface morphologies and the other properties of the SnO_2 films were investigated by the modern analysis instruments[8]. The in-depth profiles of compositions in the SnO_2 films were investigated by means of Auger electron spectroscopy and infrared spectra techniques. The structure of the SnO_2 films has been determined by x-ray diffraction method. It shows that the temperature of deposition and annealing may influence the structure of the SnO_2 films. Figure 4 shows scanning electron micrographs of the SnO_2 films deposited onto silicon substrates.

The mean particle sizes of the SnO_2 films increased as the annealing temperature increased. When the SnO_2 films grew along any direction to the furface of the substrate, they became columnar and very much more porous structure like spongy in appearance.

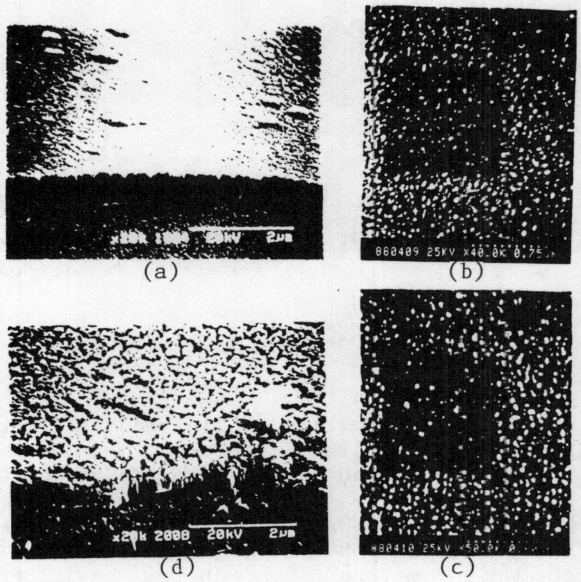

Fig. 4 Scanning electron micrographs of SnO_2 deposited onto silicon substrates at temperature of 170°C(a), 230°C(b), at annealing temperature of 300°C(c), 400°C(d)

RESULTS AND DESCUSSIONS

Gas sensing properties of SnO_2 ultrafine particle films operated as sensors at elevated temperatures and exposed to gases were measured. Figure 5 shows sensitivities of ultrafine particle films deposited onto silicon substrates, as a function of the operating temperature.

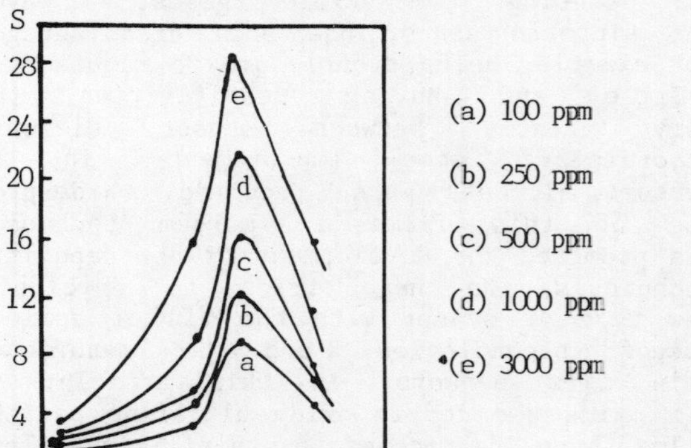

Fig.5 Sensitivities of SnO_2 ultrafine particle films as a function of the operating temperature for detecting C_2H_5OH

It shows that the ultrafine particle of SnO_2 films had high sensitivity to C_2H_5OH gas in a gas containing different amounts of fresh air. Therefore, the detection of C_2H_5OH gas could be realized by operating the sensor at the temperature of 230°C.

Figure 6 shows the sensitivities of SnO_2 ultrafine particle films as a function of the concentration of C_2H_5OH gas at variable temperature.

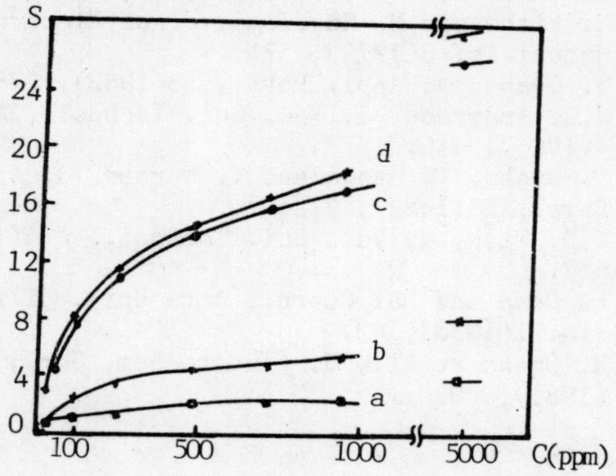

Fig.6 Sensitivities of SnO_2 ultrafine films as a function of C_2H_5OH gas concentration at temperature (a) 110°C, (b) 145°C, (c) 200°C (d) 230°C

The sensitivity increased as the C_2H_5OH concentration increased. The liner relationship exteds down to 5 ppm of C_2H_5OH concentration, indicating a very low limit of detection.

Figure 7 shows sensitivities of SnO_2 ultrafine particle films as a function of the sort of combustible gases, such as ethanol, hydrogen, cool gas, butagas.

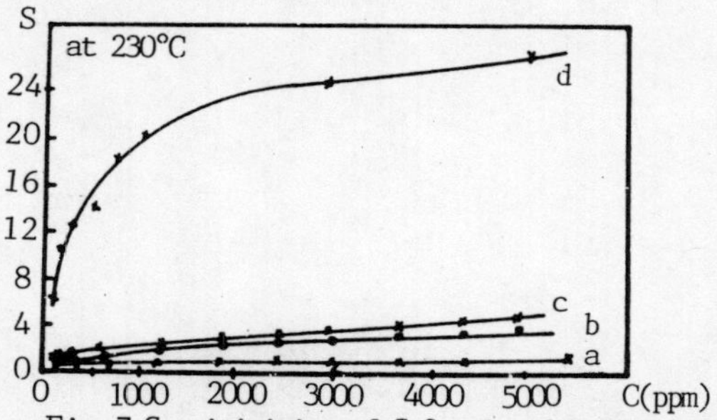

Fig.7 Sensitivities of SnO_2 ultrafine particle films as a function of the sort of combustible gases. (a) Butagas, (b) Cool gas, (c) Hydrogen, (d) Ethanol vapour

This experimental results indicate that the SnO_2 thin film is very sensitive to C_2H_5OH gas, but not sensitive to butagas. High gas selectivity for detecting C_2H_5OH gas has been obtained. As mentioned above, the SnO_2 films have a large specific surface area and became the porous columnar structure constructed from the ultrafine particles. When the effects of the SnO_2 films are noticeable, the conductance modulation by gas adsorption is indicated to have unique characteristics. The thin film thicknesses as sensors is less than 2000Å that make it easy for gas molecules to enter into the film and to exchange electrons, so the conductance medulated response is fast. In this work, the response and the resume time of the SnO_2 films is less than two seconds.

Truly thim film gas sensors, deposited by conventional thim film methods, have generally shown poor gas selectivity and stability. We have devoloped multilayer structure which overcome these problems. This structure used are shown in Fig. 2(b). The film thicknesses were 1000 Å for the first layer of ferric oxide and 800 Å for the second layer of tin oxide. Figure 8 shows sensitivities of the multilayer thin film gas sensors as a function of heating current in 1000 ppm of the test gases.

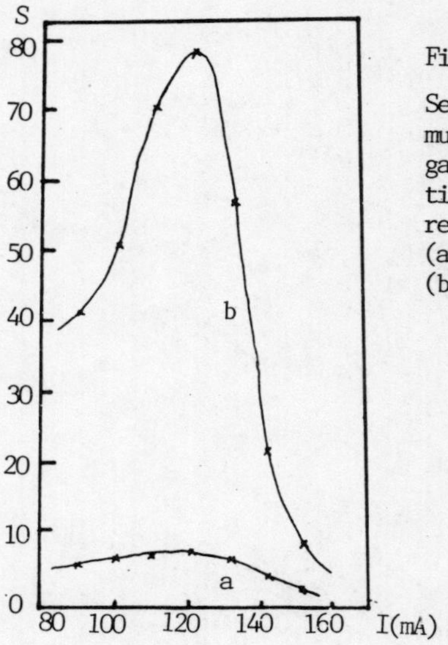

Fig. 8

Sensitivities of the multilayer thin film gas sensors as a function of heating current
(a) Gasoline gas,
(b) Ethanol vapour

The discrimination between ethancl gas and gasoline was defined by the separate coefficient, C , was calculated as follows

$$C = \frac{V_1 - V_0}{V_2 - V_0}$$

Where V_1 was the voltage appied across the device in ethanol Vapour, V_2 was the voltage applied across device in gasoline gas. This separate coefficient was 12 in the experimental result.

The authors claim that the use of SnO_2 - Fe_2O_3 multilayer structure greatly improves the reversibility and long-term stability, compared with that achieved with SnO_2 film alone. We do not find any poisoning effects by other gases.

CONCLUSION

Tin oxide ultrafine particle films have been successfully fabricated by using plasma enhanced chemical vapor deposition and the porous columnar and spongy structure of films are observed. The high sensitivity of the ultrafine particle films is considered to be caused by a large specific surface area and the porous film structure which makes it easy for gas molecules to enter into the film and to be adsorbed at the surface of SnO_2 film and to exchange its electron. The experimental results indicate that use of the structure of SnO_2 - Fe_2O_3 multilayer thin films greatly improves the selectivity and long - term stability of thin film sensors, without adding of any catalyzers such as Pt and Pd.

REFERENCES

1. J.C. Manifacior, Thin Solid Films, 90(1982), 297.
2. T.R. Nash and R.L. Anderson, IEEE Trans., Electron Devices, 24(1977), 127.
3. M. Nitta and M. Haradome, J. of Electronic Materials, 8(1979), 571.
4. T. Oyabu, J. Appl. Phys., 53(1982), 2785.
5. J.C. Anderson, J. Vac. Sci. Technol., A 4(1986), 610.
6. T. Oyabu, T. Osawa and T. Kurobe, J. Appl. Phys., 53(1982), 7125.
7. K.D. Wise, J. Vac. Sci. Technol., A 4(1986), 617.
8. Lu Dean and Dai Guorui, Acta Univ. Jilinensis, 1(1988), 43.
9. H. Ogawa et al., J. Electrochem. Soc., 128 (1981), 2020.

WORKING MECHANISMS OF SIC/ZNO HETERO-CONTACT CHLORINE GAS SENSORS

R. Koyama, Y. Nakamura, M. Miyayama and H. Yanagida
Research Center for Advanced Science and Technology, The University of Tokyo.
4-6-1, Komaba, Meguro-ku, Tokyo 153, Japan

ABSTRACT

Chlorine gas sensing properties and conduction mechanism were investigated for the hetero-contact of silicon carbide and zinc oxide. The current-voltage characteristics of the contact showed a rectifying character in which the forward bias is positive on SiC. The current under forward bias decreased by the introduction of chlorine gas. The logarithmic ratio of the current with chlorine gas to the current without chlorine gas at 2.0V forward bias had a linear relationship to the square of chlorine gas concentration. This phenomenon was observed only in wet air. The current at the forward bias linearly increased with increasing humidity. The following mechanism is supposed for the chlorine detection of this sensor; The current transfer of the interface at the contact is governed by the barrier height of ZnO surface. By the reaction between chlorine and water, ZnO surface is oxidized. As a result, the barrier height of ZnO surface increase and the current under forward bias decrease.

INTRODUCTION

Presently, several types of chlorine gas sensor have been proposed; potentiometric-type sensors using chloride ion conductors such as $SrCl_2$-KCl[1] and $PbCl_2$-KCl[2)3)], and coulometric-type sensors using bromine aqueous solution[4]. However, those sensors have complicated cell structures and selectivity for chlorine gas are not high enough. Heating above about 150°C is also needed for operation of the potentiometric chlorine gas sensor.

In this study, new type chlorine gas sensor using ceramic hetero-contact was investigated. The hetero-contact is two ceramics contacted by mechanical pressing. Usual semiconductor gas sensors obtain the information of atmosphere by measuring its electrical resistance. For the case of the hetero-contact sensors, current-voltage characteristics of the contacts vary according to atmosphere. We can get a lot of information by changing applied voltage to the contact. The hetero-contact type sensors such as Li-doped NiO/ZnO and CuO/ZnO have been found to show good humidity- or CO gas-sensing characteristics[5)-7)]. In this paper, the properties of hetero-contact chlorine gas sensor is reported and working mechanisms are discussed.

EXPERIMENTAL METHOD

1 Sample Preparation

Beta-SiC powder (99% pure, Central Glass Co., Ltd.) was isostatically pressed into a pellet under 36MPa and sintered at 2000°C for 3 h in N_2 gas . Relative density of SiC sintered body was about 60% of theoretical. The electrical conductivity at 20°C was about 10 $\Omega^{-1}cm^{-1}$ and an n-type semiconduction was confirmed by thermoelectric power measurement. Zinc oxide powder (99.99% pure, Ko-jundo Kagaku, Co., Ltd.) was isostatically pressed into a pellet under 20MPa and sintered at 900°C for 3 h in air. Relative density of ZnO sintered body was about 94% of theoretical. The electrical conductivity at 20°C was about $10^{-2}\,\Omega^{-1}cm^{-1}$. They were cut into a rectangular plates like Fig. 1. Their surfaces were polished with SiC abrasive paper (#400), and ohmic electrode (Indium metal) was applied to

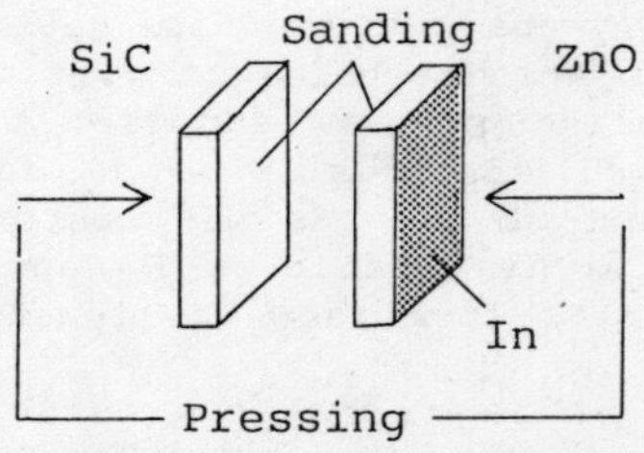

Fig. 1 Diagram of SiC/ZnO hetero-contact. Each plate's size is 5mm x 5mm x 1mm.

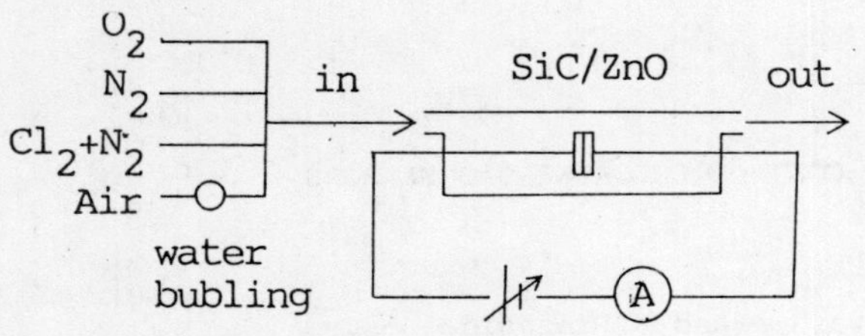

Fig. 2 Diagram of System for electrical measurement of hetero-contact under controlled atmosphere.

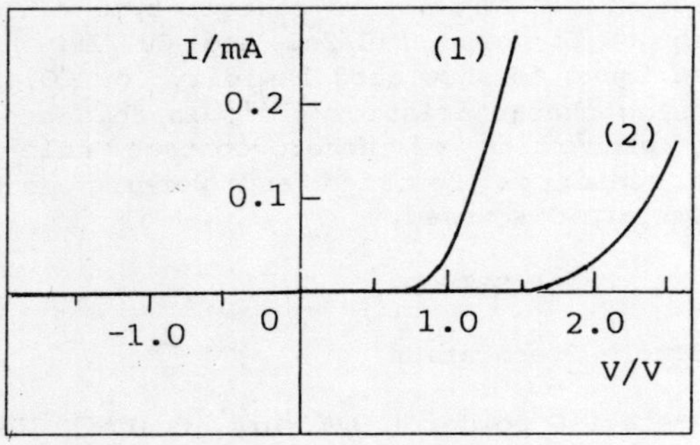

Fig. 3 Current-voltage curves of SiC/ZnO hetero-contact in air (relative humidity 35%) with and without Cl_2 gas (1600ppm) at 20°C. Indicated voltage is of the SiC.
(1) in air (2) in Cl_2/air (1600ppm)

one side of the plate. The hetero-contact was made by mechanical pressing under about 1.5 MPa (shown in Fig. 1).

2 Electrical Measurement

The system for measuring the properties of the hetero-contact chlorine gas sensor is illustrated in Fig. 2. Relative humidity was controlled by changing the flow ratio of dry and wet air which bubbled through water at 20°C. Chlorine gas was supplied by a gas cylinder (chlorine is 1% and N_2 is 99%). Concentration of Cl_2 gas was controlled by changing the flow ratio of N_2 and Cl_2+N_2 gas. The total flow rate was fixed at 250 cm^3/min.

Current-voltage characteristics was measured using a programmable DC power supply (Hioki Co., Ltd. #7005) and an electrometer (Takeda-Riken Co., Ltd. TR8651). Applied voltage was swept at a speed of 100 mV/sec up and down to ±2.0V.

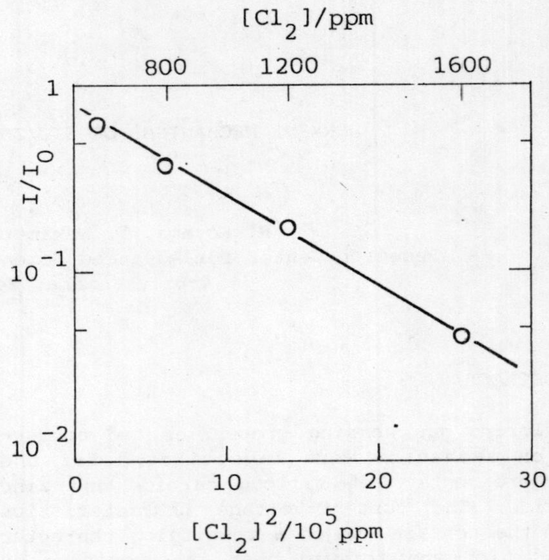

Fig. 4 Relative current I/I_0 vs. Cl_2 gas concentration. I and I_0 are current with and without Cl_2 gas.

RESULTS AND DISCUSSION

The typical behavior in current-voltage curve of SiC/ZnO hetero-contact is shown in Fig. 3. In air, current-voltage curve showed a rectifying character in which the current increase exponentially only when the SiC is positively biased. The current at the forward bias was found to be decreased by an introduction of chlorine gas. The voltage at which current starts to increase, shifted to a higher potential. Such a current change by chlorine gas was not observed in dry air.

Figure 4 shows the current changes by chlorine gas. The vertical scale of Fig. 4 is a logarithm of relative current I/I_0, where I and I_0 are the current values with and without chlorine gas at 2.0V forward bias, respectively. The I/I_0 decreased with increasing chlorine gas concentration $[Cl_2]$ and the relation between I/I_0 and chlorine gas concentration was found to be expressed by,

$$1-\log(I/I_0) \propto -[Cl_2]^2 \qquad (1)$$

There is some difference between "contact" and "junction". However, we substitute the hetero-junction model[8], because of no useful model for the hetero-contact. The band gap is 2.2eV for β-SiC[9] and 3.2eV for ZnO[10]. Since the density of surface level must be very large due to polishing, Fermi level at the surface is assumed to be fixed at the position of one third of the band gap

(bardeen limit). Accordingly, when SiC and ZnO are contacted, the barrier height of the surface of ZnO will be higher than that of SiC. As a result, electron transfer across the interface of the SiC/ZnO hetero-contact is influenced mainly by the barrier height at the ZnO surface. From the hetero-junction model[8], the current density J is given by

$$J = A'N_2\exp(-eV_{D2}/kT)[\exp(eV_2/kT)-\exp(-eV_1/kT)] \simeq A'N_2\exp(-eV_{D2}/kT)\exp(eV/ikT) \quad (2)$$

where, V_1 and V_2 are the partial voltages applied to SiC and to ZnO, respectively. N_2 is the donor density of ZnO, V_{D2} the barrier height at the surface of ZnO, and A' is the constant. The surface potential V_{D2} is given as the solution of the Poisson equation.

$$V_{D2} = eN_2d^2/2\varepsilon_0\varepsilon_s = Q_{ss}^2/2\varepsilon_0\varepsilon_s N_2 \quad (3)$$

Q_{ss} is the charge in surface states ($=eN_2d$), and d is the barrier width. If Q_{ss} is proportional to chlorine gas concentration, eq. (1) can be explained by eqs. (2) and (3).

Figure 5 shows the relation between current at 2.0 V forward bias and relative humidity in air. The current increased linearly with increasing humidity.

As the hetero-contact was not able to detect chlorine gas without water vapor, it seems that the product by the reaction between chlorine and water affects the electrical properties of the hetero-contact. Chlorine reacts with water and HClO is produced.

$$Cl_2(g) = Cl_2(aq) \quad (4)$$
$$Cl_2(aq) + H_2O = H^+ + Cl^- + HClO \quad (5)$$

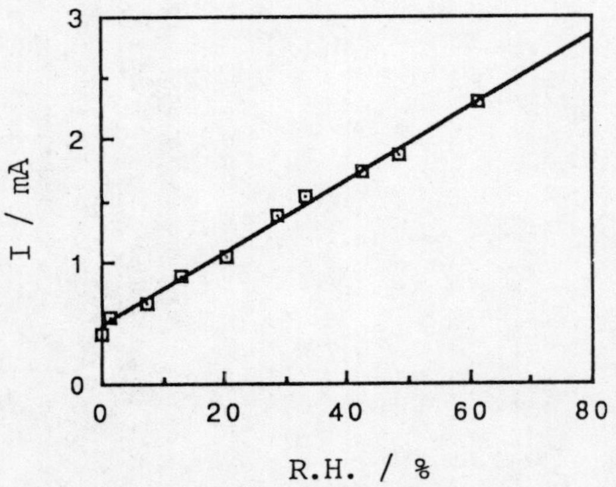

Fig. 5 The relation between current at 2.0V forward bias and relative humidity in air.

HClO shows very strong oxidizing power by the reaction as,

$$HClO + 2e^- = HCl + O^{2-} \quad (6)$$

If oxygen produced by the reaction (6) adsorb the ZnO surface, Q_{ss} is in proportion to $[Cl_2]$ and resultantly, log J should be proportional to the square of $[Cl_2]$. This model are in agreement with the expelimental data.

The contacted surface of the hetero-contact is supposed like Fig. 6. Since their surfaces are polished with rough sand paper, the contacted area is small and the large space is present between SiC and ZnO. In wet air, physisorbed water will exist around the contact points. When voltages are applied to this hetero-contact, next two conduction passes are expected.

(A) electron transfer through the contacted surface.

(B) through adsorbed water and two interfaces between semiconductors and adsorbed water, i.e. electrochemical reaction of water molecules.

Above discussion is applied to the pass (A). But, it is difficult to explain the humidity effect. The result shown in Fig. 5 suggests the contribution of the pass (B). To discuss about the pass (B), more information is required.

CONCLUSION

In this paper, the following results were reported.

(1) The current-voltage curve of SiC/ZnO hetero-contact showed a rectifying character

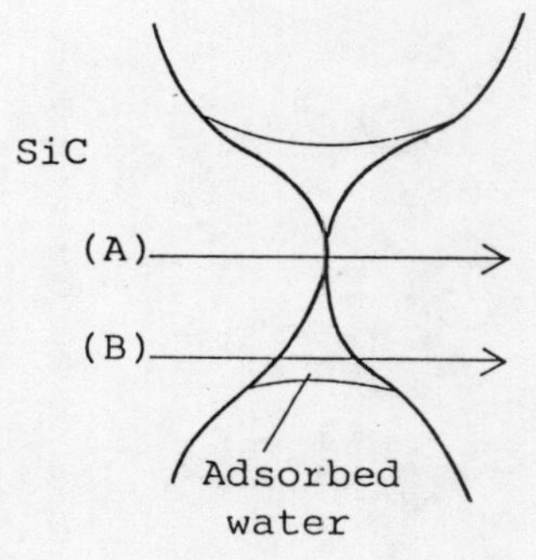

Fig. 6 diagram of the interface of the hetero-contact and conduction passes.

in air. This curve varied by chlorine gas.
(2) The logarithmic ratio of the current with chlorine gas to the current without chlorine gas at 2.0 V forward bias had a linear relationship to the square of chlorine gas concentration.
(3) This sensor also has a sensitivity for humidity in air. The current under 2.0 V forward bias increased with increasing humidity. In dry air, this sensor is no sensitive for chlorine gas.

REFERENCES

1) A Pelloux, P. Fabry and P. Durante, Sensors and Actuators, 7, 245-52 (1985).
2) Y. Niizeki, S. Itabashi, Y. Onodera and O. Takagi, Denki Kagaku, 54, 948-53 (1986).
3) Y. Niizeki, O. Takagi and S. Toshima, Denki Kagaku, 54, 961-65 (1986).
4) K. Takahashi, "Introduction of Sensor Technology", Kogyo-Chosa-Kai, Tokyo, 1978, pp.239, 241.
5)Y. Toyoshima, M. Miyayama, H. Yanagida and K. Koumoto, Jpn. J. Appl. Phys., 22, 1933 (1983).
6) Y. Nakamura, M. Ikejiri, M. Miyayama, K. Koumoto and H. Yanagida, Nippon-Kagakukai-Shi, 1985, 1154-59, (1985).
7) Y. Nakamura, A. Ando, T. Tsurutani, O. Okada, M. Miyayama, K. Koumoto and H. Yanagida, Chem. Lett., 1986, 413-16 (1986).
8) K. Takahashi, "Semiconductor technology", Morikita Press, Tokyo, 1975, pp.157,173.
9) G.A. Lomakina and Yu.A. Vodakov, Sov. Phys. Solid State, 15, 83 (1973).
10) K. Hauffe, Angew. Chem., 72 730 (1960).

The Alcohol Gas Sensor Made by ZnO Thin Film

Chang Ying   *Qu Xi–Xing*   *Yan Kang–Nin*
University of Electronic Science and Technology of China

ABSTRACT

In this paper, an alcohol gas sensor made by ZnO thin film is reported. With a new structure of the film and doped by catalytic additives, the sensor has many advantages, such as high sensitivity, good selectivity and long stability.

INTRODUCTION

In recent years, many gas sensors with various type have been developed and is or will be used in various fields (1–4). There have been some reports about alcohol gas sensor and its use in traffic control. [5,6] But these gas sensors still have some deficiencies, they always have bad selectivety and can't tell alcohol from gasoline gas and cigarette smoke, so they often give wrong signal. And they are not able to work in long time becausse their stability are not good enough. These deficiencies made these sensors difficult to use in more and more important fields.

In this paper, an alcohol gas sensor made by ZnO thin film is reported. With a new structure of the film and doped by some catalytic additives , the sensor has many advantages, such as high sensitivity, good selectivity and long time stability. It can be easily used in traffic control and environnment protection fields.

EXPERIMENTAL

The alcohol gas sensor was made by ZnO thin film as its detective part and a resistor heater. It's technical processes is shown in Fig.1.

ZnO thin film which was sensitive to alcohol gas was made by D.C. magnetron sputtering. The target was pure zinc plate(99.99% pure, Φ120× 5 mm). Use pure oxygen as the sputtering gas. The common sputtering parameters were shown in Table.1.

The film was deposited on AL_2O_3 ceramic substrate, with thickness of 5000 A. The structure of as sputtered film was porous and nonoriented polycrystal, has large surface to volume ratios and a lot of surface defects. According to X–ray deffraction, the average crystal size was about 700 A.

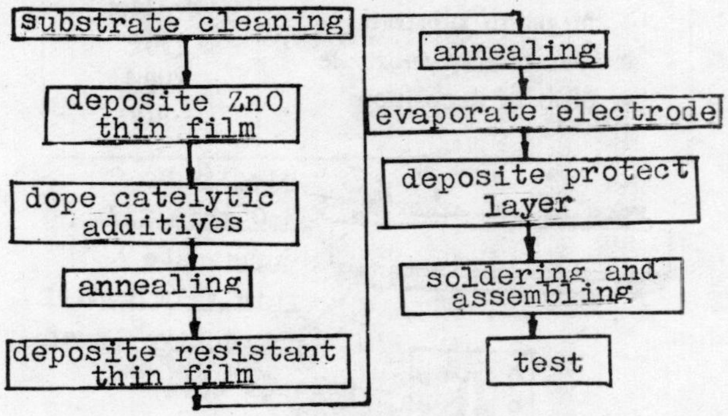

Fig.1. The technical processes of Zno thin film gas sensor

The selectivity of pure ZnO thin film was bad, and the sensitivity was not high enough to be used in practice. In order to approve its property, some additives should be doped into the film. As for alcohol gas sensor, it is usually dope some rare–earth elements into the film to increase the redox reaction speed on the surface. After doping, the film should be annealing under high temperature for some time. In this process, the additives could be deffused to all the film and the defects number in film may decrease, so the property of the film may be more stable.

The gas sensors always work under high temperature, because the surface reaction only takes places on high temperature condition. The sensor made by ZnO thin film usually works on 300℃ ,so it needs a heater to meet this condition. We made the heater to meet this condition. We made the heater by resistive film or Nr–Cr wire wounded resistor, resistive film may chose Ta–AL–Si thin films or W–Si thin films because of their good stability under high temperature, these film can be deposited on the back of the substrate which have deposited ZnO thin film on its other side already. The resistive film was about 2000 A thick, its resistance per square is 20Ω / □. The resistive film is always covered by a AlN thin film as a protective layer so to increase its long time stability. The working power of the heater was about 0.75W. These heater can be well combined to the gas sensitive thin film and easy to make. The structure and form of the sensor is shown in Fig.2.

Table.1.The sputter condition of Zno thin film.

sputtering parameter	condition
pre—vaccum pressure	2×10^{-5}Torr
sputtering pressure	4.5×10^{-2}Torr
D.C. Voltage	400V
Current	400mA
the distance from target to substrate	4~5cm
temperature of substrate	120℃
ratio of deposition	200Å/min

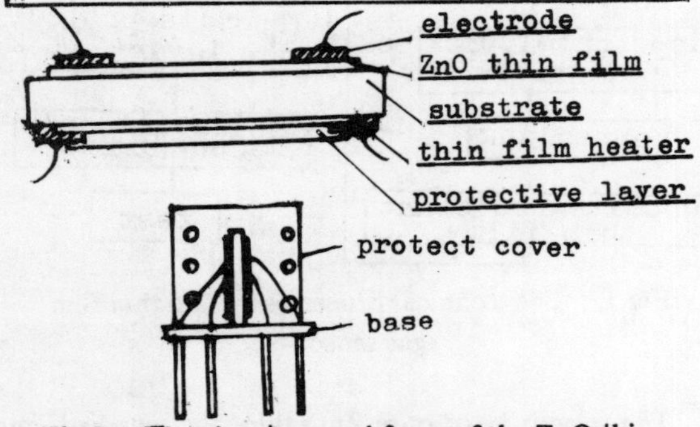

Fig.2. The structure and form of the ZnO thin film sensor

RESULTS AND DISCUSSIONS

1.The crystal structure of the film.

The Zno thin films made by different conditions always have different structure, composition, density and different number of surface defects, and these parameters have large influence to the surface properties of the thin film. In order to improve the gas sensitivity, it needs a large surface to volume ratios, and this demands small crystal size and more porous in thin film. To meet these conditions, we select a high deposition ratio (200 A/ min) and a low substrate temperature. Under this condition, the deposited partical was difficult to move on the surface when process going on, so the film may be more porous and have more surface defect,,its density was low. In this way we may get larger ratio of surface to volume and better property of the film.

Another factor which influemce the film properties is the substrate surface. Since the thin film is deposited on the surface of substrate, its crystal structure and size would be influenced by the crystal structure of the substrate. We have studied several substrate materials, such as glass, glass with microcrystal structure, AL_2O_3ceramic with polished surface or with nonpolished surface. The properties of thin film on the AL_2O_3ceramic substrate with non—polished surface was much better than those on other substrates, its sensitivity was very high, and had good stability. This may owe to the porous property of the ceramic surface and the high temperature stability of AL_2O_3.

2.The catalytic additives doping

Pure ZnO thin film was sensitive to many reductive gases, and their sensitivities were not high enough to be used in practice. If one wants to use this film to defect one type of gas, he must increase the sensitivity to this gas so to tell this gas from other gases. Through adding some catalytic additives into the film, we can arrive this goal. From our study, we found that sensitivity of ZnO film to alcohol gas may be changed to very high by doping some rare—earth elements into the film, and at the same time its sensitivity to other reductive gases (such as hydrogen, methane, carbon monoxide and gasoline gas) remains very low. So we may improve both the selectivity and the sensitivity of the film.

The doped elements always have variable valence, such as Ce^{3+}/Ce^{4+}, $Pr^{3+}/Pr^{4+}$and Sm^{2+}/Sm^{3+}. They are both good oxidizers and good reductors, and their oxidization and reduction properties are very active in surface absorption and electron exchanges. So their doping may increase the speed of the surface chemic reaction. The results of doping in sensitivity and selectivity are shown in Fig.3.

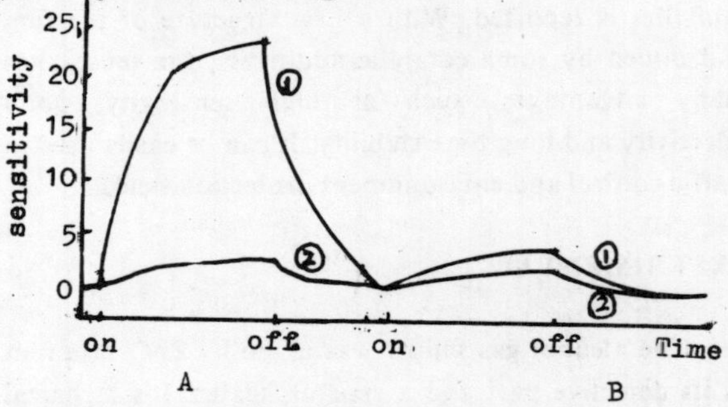

Fig.3. The results of doping. Test gases are: A: 300 ppm alcohol gas. B: 300 ppm Gasoline gas. Sample (1) Zno thin film doped by rare—earth elements. Sample (2) is pure Zno thin film.

When worked under 300℃ for long time, the additive element on the surface of the thin film may be volatilized away or may be polluted, then the sensitivity of the sensor may decrease. In order to improve the long time stability of the thin film, we use a new doping structure —— sandwich structure, which means place a catalytic layer between two layers of ZnO thin film (Shown in Fig.4). Through annealing, the additive element could get to top surface of upper layer by thermic diffusion. In long time working, the additive atoms could move gradually from catalytic layer to the top surface of the film so to compensate the losses in long time use. This kind of structure enhanced the long time stability of ZnO thin film gas sensor.

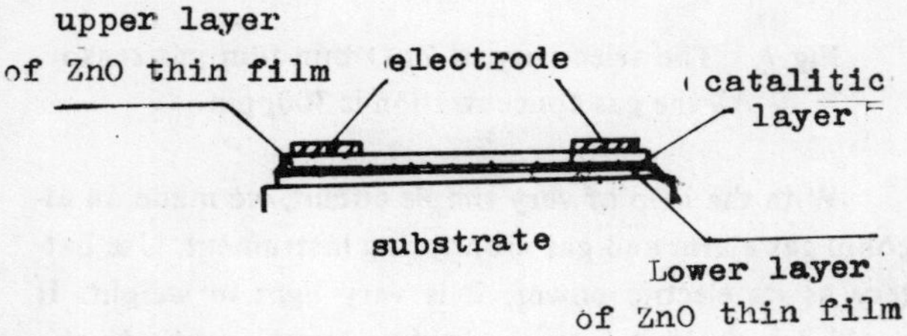

Fig.4 The doping structure of the film.

PROPERTY OF THE ZnO ALCOHOL GAS SENSOR

1.The first restoration property.

Put the electric power on a new gas sensor or a sensor which have not been used for a long time, as the temperature of the sensor increase, its resistivity will first decrease rapidly, and then increase gradually, finally get a stable value. All these changes would take place in several minites, as shown in Fig.5. This property of the gas sensor in called the first restoration property.

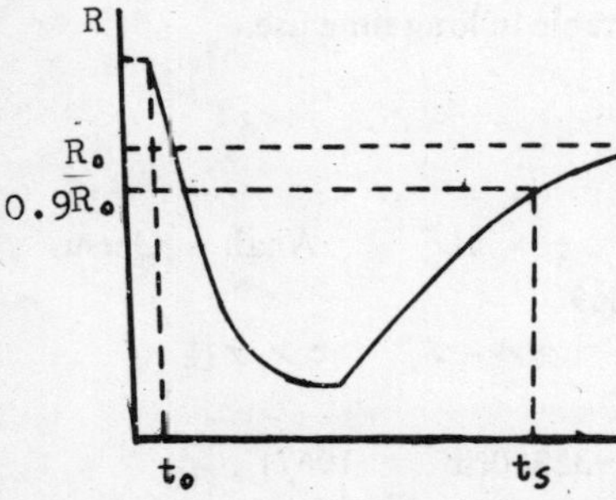

Fig.5. The first restoration property of gas sensor. $t_o$is the time putting on electric power. $t_s$is the time the resistance of the sensor getting t_o90% of final stable value. t_s−$t_o$is always called as first restoration time.

The first restoration property is mainly caused by the change of thermic carries in the film and oxygen absorption on the film surface. The ZnO thin film belongs to semiconductive materials, when temperature increase, its thermic carries increase very fast, so its resistance decrease. But when temperature of the surface become high enough, the oxygen absorbed on the surface may changed from physical absorption to chemical absorption. As the temperature getting stable (above 250℃), the adsorbed oxygen are mainly O^-ion. When oxygen changed from atom to ion, it would grasp the free electron from the film , so in this process the carries in film decrease and the resistance of the film increase. Usually the time from putting electric power to the resistance of the sensor getting to 90% of stable value is called the first restoration time. For common gas sensor, this time is about 5~15min, but for our film sensor, the time is only 2~4 min.

2.The response and restoration property.

Set the gas sensor in a test box with clean atmosphere, its resistance is a stable value. When put some reductive gas into the box, the resistance of the sensor will rapidly changed to another value (always decrease). The time from putting in reductive gas to the resistance of the sensor getting to 63% of the new stable value is called the response time. When put off these reductive gas from the test box, restore the clean atmosphere in the box, the resistance of the sensor will restore to its first value, the time needed by this process is called the restoration time. The property of response and restoration of ZnO thin film gas sensor is shown in Fig.3.

These property is mainly caused by surface reaction on the film , when reductive gas get to the surface of the film, it may be reacted with the oxygen ion which was absorbed on the surface.

The time of response and restoration of ZnO thin film gas sensor is 5 second and 10 second respectively, that is much faster than other gas sensor made by sintering process. This is the results of its large ratio of surface to volume. In surface reaction process, the oxygen in the material may diffuse to the surface, and it will take some time to get equilibrium. As the film is very thin compared with other sintering material, it need only very short time. So the reaction speed of ZnO thin film sensor is very fast.

3.Sensitivity

The ratio of resistance of sensor under reductive atmosphere with given concentration to its resistance under clean atmosphere is defined as sensitivity of given concentration, noted by S:

$$S = \frac{\textit{resistance under clean atmosphere}}{\textit{resistance under reductive atmosphere}}$$

Under given temperature, the sensitivity of the ZnO thin film gas sensor is changed as the concentration of reductive gas changes. Its relation is shown in Fig.6.

From 10 to 3000ppm, the sensitivity is in direct proportion to the concentration of alcohol gas, when concentration is very high, the sensitivity will gradually get to saturated, its change with concentration will become small.

The principal advantage of thin film gas sensor is its high sensitivity, especially under very low concentration. when concentration of alcohol gas is only 50ppm, the sensitivity of ZnO thin film sensor is about 10, which is much higher than other type of sensor. So this sensor can be easily used to detect alcohol gas with very low concentration.

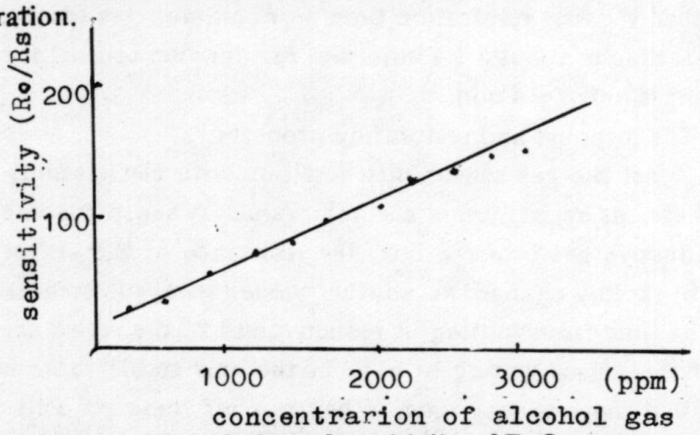

Fig.6. The relation of sensitivity of ZnO gas sensor to consentration of alcohol gas. The temperature is 300℃.

4.Selectivity and stability

Selectivity is one of the important parameters of the gas sensor. It is the ability the sensor can tell one type of gas from many other type of gas. So far, there are few gas sensor which is sensitive to one type of gas and nonsensitive to other type of gases. So, when the resistance of sensor changes, it is always difficult to tell which gas the sensor discover. After doping rare–earth elements into the film, the ZnO thin film is very sensitive to alcohol, at the same time its sensitivity to gasoline gas, methane, hydrogen, carbon monoxide and smoke remains low, as it shown in Fig.7. Under same concentration, the sensitivity to alcohol is about 5 times that to other gases. The detector made by this sensor may be easily used in traffic control to detect wether the driver have drunk. It need not complex circuit and will not be interfered by other reductive gas.

As the gas sensor always works under high temperature, its stability is very important. Compared with other semiconductive materials, ZnO thin film have stable physical and chemical property. As we made the sensor in new structure and annealing it properly, we get very good stability. In two years use, its resistivity and sensitivity changes were less than 10%.

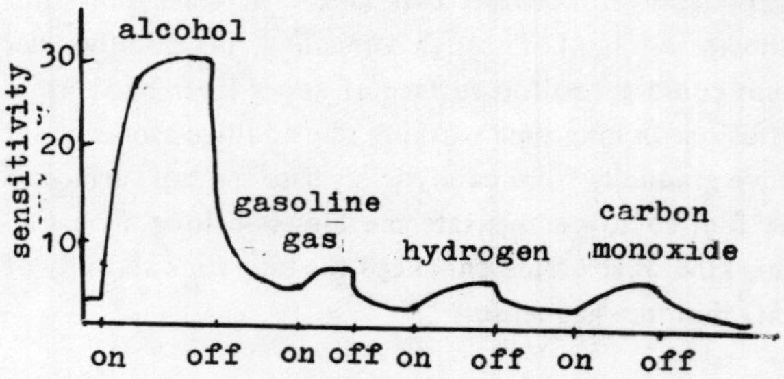

Fig.7. The selectivity of ZnO thin film gas sensor. All the gas concentration is 300ppm.

With the help of very simple circuit, we made an alcohol gas alarm and gas monitoring instrument. Use battery as its electric power, it is very light in weight. It could detect alcohol gas qualitatively and quantitatively, and gave light and phono alarm signal if the concentration is higher than a given value. The set have been used by traffic control department and got good reputation.

CONCLUSION

An alcohol gas sensor is made by ZnO thin film. Doping some rare–earth elements and using new structure of thin film, the sensor have high sensitivity, selectivity and stability. The sensor can be used for fast detect of alcohol gas with low concentration, the lowest detect concentration may be 10 ppm. Because of its good selectivity, the detect results will not be interfered by other reductive gases. It is stable in long time use.

REFERENCE

1. Tetsuro seiyama et al. , Anal. Chem. Vol.38(No.8)(1966) P.1069
2.マー クテイング・ グループ, センサ技 Vol.4(No.6)(1984) P.86
3.田口尚 ,特公昭, 45–38200(出 1962)
4.一ノ, 东芝レビエ–Vol. 30. (昭 50) P.585
5.桂正树, エレタトロニケ・セラミス夏号,(1984) P.85
6.池上昭, 子材料, No.5.(1983)P.39

RESEARCH ON SENSOR MATERIALS OF ALCOHOL VAPOUR

X.L.Yan, Y.L.Xu†, Q.X.Cao, X.H.Zhou, P.Li and J.X.Gao
Department of Technical Physics, Xidian University
Xi'an, China

ABSTRACT

The sensor materials of alcohol vapour are studied. The addition effects of a kind of nitrate on the alcohol selectivity of these sensing materials have been prominently investigated. Using these materials we have prepared many sensors of partial sintering type. The sensors, made of $Sn(OH)_4$ mixed with α-Al_2O_3 which was soaked in nitrate solution, without precious metal, have high sensitivity, excellent selectivity, low power consumption and quick response.

Key words: Alcohol-Sensitive Gas Sensor, Partial Sintering Ceramics

I. INTRODUCTION

There has been a great interest shown in alcohol vapour sensors in recent years [1-7]. There are two reasons for that: (1) to avoid the false alarm problem of LPG sensors or town gas sensors, caused by alcohol vapour, use alcohol vapour sensors as compensation elements; (2) to use alcohol vapour sensors as breathalyser devices to test drivers. However alcohol vapour sensors still have many disadvantages so far, such as high power consumption, slow response and poor selectivity. Comparing oxides ZnO, Fe_2O_3 and ABO_3 type composite oxides with SnO_2, it is found that SnO_2 is the best sensor material of alcohol vapour. Using SnO_2 mixed with some content α-Al_2O_3, soaked in nitrate solution, we got some alcohol vapour sensors with excellent performance.

II. EXPERIMENTAL

1. Preparation of $Sn(OH)_4$ or SnO_2

Using wet method we prepar $Sn(OH)_4$ or SnO_2 from $SnCl_4$ as Figaro[8]. We modified the manufacturing engineering of Figaro: $Sn(OH)_4$ or $SnO_2 \cdot H_2O$ were mixed with α-Al_2O_3 without calcining.

2. Preparation of the additive

Commercial α-Al_2O_3 was soaked in different concentration nitrate solution for 24 hrs. Then the powder was fired at 850-950°C for 2 hours in air, and was ground into finer powder.

3. Preparation of the paste

$Sn(OH)_4$ with 80wt % and α-Al_2O_3 with 20 wt %, which was soaked in different concentration nitrate solution, were mixed. Then the mixed powder with some H_2O, organic binder and strengthening agent was ground into pastes. These pastes were painted on an alumina pipe, and then were fired at 650-750°C for 2 hours in air to get partial sintering elements. The element construction is shown schematically in Fig.1.

4. The performance testing of sensors

The experimental system and circuit are shown in Fig.2. A pickup resistor is connected in series to the sensor and a dc voltage is applied across the circuit. The resistance

† Senior member of IEEE

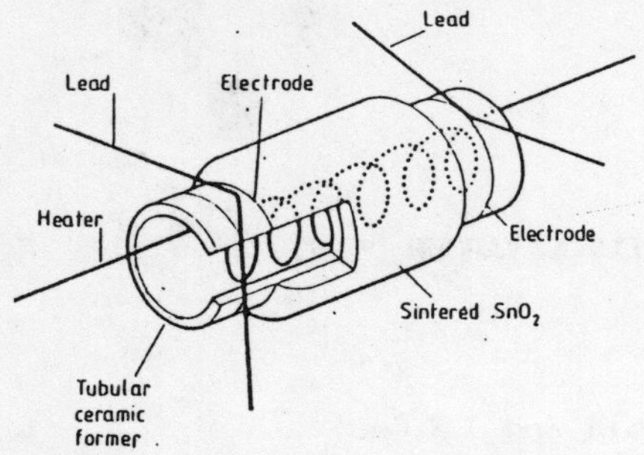

Fig.1 Configuration of the semiconductor gas sensor element.

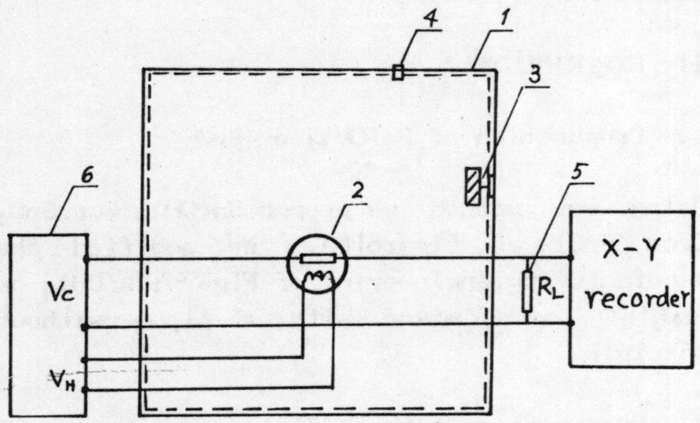

Fig.2 The experimental system
1. chamber 2. Sensor 3. fan 4. gas valve 5. pickup resistor 6. power source

variation of the sensor element, which changes with gas concentration, is measured indirectly as a change in the voltage across the pickup resistor R_L. The sensitivity Kv of the sensors has been calculated from the change in the detection voltage across the pickup resistor. The test gas is introduced into the chamber and is stirred by a fan to dilute the test gas with air in the chamber uniformly. The sensor operating temperature is controlled by changing the applied voltage across the heater.

III.RESULTS

Fig.3 shows the relation between the concentration of a kind of nitrate solution, in which the α-Al_2O_3 powder was soaked, and the sensitivities of alcohol, benzine, CO,CH_4

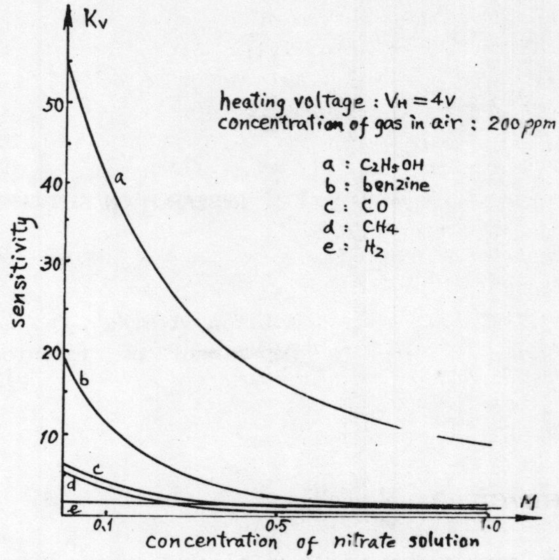

Fig.3 Sensitivity Kv as function of the concentration of a kind of nitrate solution.

and H_2. From this figure it is clear that the sensitivities of alcohol, benzine, CO, CH_4, and $H_2$are all restrained. But the restrainning function for alcohol is different from restrainning function for benzine, CO, $CH_4$and H_2 . In order to further illustrate this fact we have Fig.4. Fig.4 shows the relation between the ratio of the sensitivity of

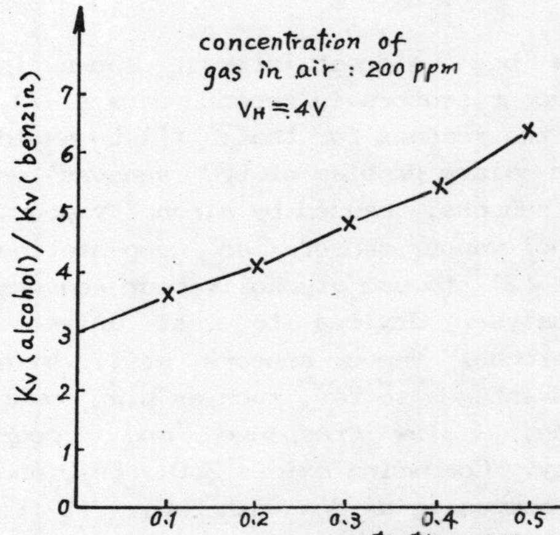

Fig.4 The relation between the concentration of a kind of nitrate solution and the sensitivity ratio.

C_2H_5OH (200ppm) to that of benzine (200ppm.) and the concentration of a kind of nitrate solution. The selectivity is indicated by sensitivity ratios. A large ratio means the more excellent alcohol selectivity. From Fig.4 it is clear that a certain concentration of nitrate gives the most prominent selectivity of alcohol to benzine, CH_4, CO and hydrogen.

Fig.5, Fig.6 and Fig.7 show respectively the temperature (heating voltage) dependence of sensitivity, the concentration dependence of sensitivity and the response of sensors. The concentration of nitrate solution, in which the α-Al_2O_3 powder was soaked, is 0.1 M for all sensors. The resistance of the heater is about 33 Ω. Fig.5 and Fig.6 exhibit that the heating power consumption of sensors is less than 0.3 W. Fig.6 shows the minimum gas concentration, which can be detected, is less than 10 ppm.. The response time of the sensor is dependent on temperature of the sensor operation. The typical response curve is shown in Fig.7. From Fig.7, we have found that the response time is less than one second. The response time is defined as the time it takes for the output voltage to reach 67 percent of the saturation. Fig.7 also shows that the recovery time is one second. The definition of the recovery time is similar to that of the response time.

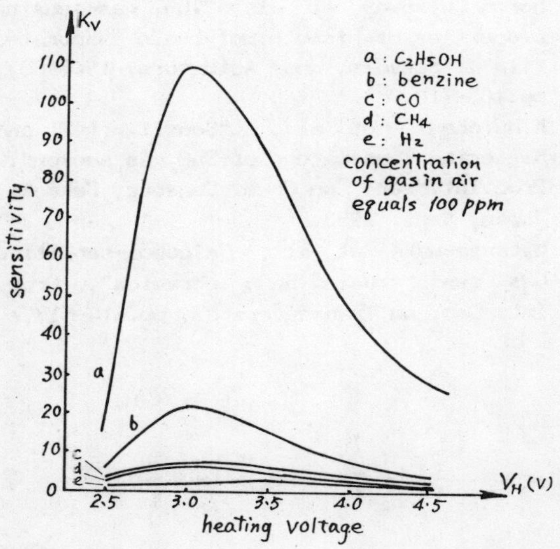

Fig.5 Sensitivity Kv as a function of heating voltage V_H.

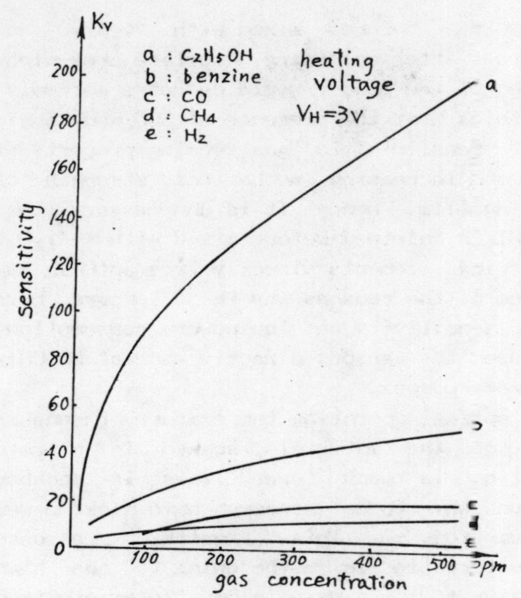

Fig.6 Sensitivity Kv as a function of gas concentration

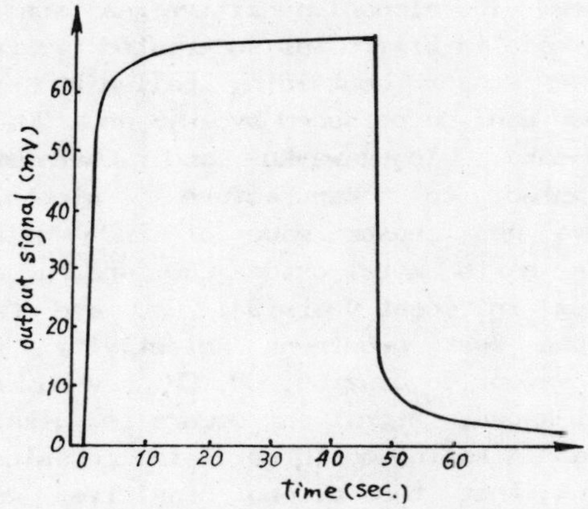

Fig.7 Response of the sensor. Concentration of gas in air equals 2000ppm. V_H =4v.

IV. DISCUSSION

An analysis by x-ray powder diffractometer and by scanning electron microscope shows that the main composition of sensors made of $Sn(OH)_4$ and that made of SnO_2 are both SnO_2 crystals. But the former crystallizes poorly, size of its grains is small and its surface area-weight ratio is very large. In order to aviod the presence of SnO and Sn in the sensor based on SnO_2, normally the $Sn(OH)_4$ are calcined into SnO_2 before mixed with α-Al_2O_3 as Figaro. After calcining, surface area-wight ratio of the powder would decrease markedly. We think that the presence of SnO and Sn is maybe beneficial to gas sensing property of SnO_2 and increasing mechanical strength of sensing film. Hence it is unnecessary that $Sn(OH)_4$ is calcined before mixed with α-Al_2O_3. Sintering elements directly from $Sn(OH)_4$ may be one of the reasons why these sensors have high sensitivity and low power consumption, because the sensors directly made of $Sn(OH)_4$ are more porous.

The optimal operating temperature of sensors made of the materials soaked in nitrate solution is much lower. This is another reason why these sensors have low power consumption. This quick response characteristic is maybe owing to the high activity of the sensing element characterized by a very large surface area-weight ratio as mentioned above. Usually to improve the selectivity of alcohol, some precious metals such as palladium and platainum are used as additives. The alcohol-sensitive gas sensor TGS812 made in Figaro was so treated . But the gas sensor containing palladium and platinum can be poisoned by H_2S and NO_x. Mitsuya-naka, Yodogawa-ku and Osaka-shi investigated to manufacture alcohol-sensitive gas sensor made of SnO_2 using alkaline earth metal oxides CaO and MgO as additives in recent years[9]. CaO and MgO give the most prominent selectivity of ethanol vapour to benzine, CO, CH_4 and H_2. As is well-known, migration occurs in alkali metal and alkaline earth metal in ceramics. We think that the alcohol-sensitive gas sensor using SnO_2 ceramics investigsted by Mitsuya-naka et al. maybe has a poor long-term stability. The present paper presents a new additive in which precious metal palladium and platinum do not exist, neither do alkali metal and alkaline earth metal. The alcohol vapour sensor using SnO_2 and this new additive maybe has a more excellent selectivity and better long-term stability.

ACKNOWLEDGEMENTS

We wish to thank Madam X.L.Zhu for her help to manufacture materials. We also acknowledge the financial support from the Former Ministry of Electronic Industry.

REFERENCES

1. X.H.Wu et al., Proceeding of 1st countryside conference on gas sensors in China. Nanjing, China, Nov. 1988, pp.166-172(in Chinese).
2. B.G.Mu, H.Y.Ling, Proceeding of China-Japan Conference on electronic sensing technology, Harbin, China, 1986, pp.419-425(in Chinese).
3. J.K.Wu, L.Y.Sun et al., Jounal of Transduction Technology, Vol.1, No.1(1988), pp.1-7, (in Chinese).
4. Japanese patent 6843-2G, "Gas sensors", July 8, 1986.
5. Japanese patent J-6843-2G, "Alcohol gas sensors", December 13, 1986.
6. Japanese patent C-6843-2G, "Gas sensors with high selectivity for alcohol vapour", January 6, 1987.
7. Norma Hykaway et al, "The gas-sensing properties of bismuth molybdate evaporated film", Sensors and Actuators, 15(1988), pp.105-118.
8. K.Murakami et al., "Sensitivity and Sintering Temperature of SnO_2 gas sensor". Proc.Int.Meet. on Chem. Sensors, Fukuoka, Japan, Sept. 1983.
9. Mitsuya-naka et al., "Alcohol-sensitive Gas sensors using SnO_2 ceramics", Proc. Int. Con. on Transducers 87, pp.614-617.

AN INVESTIGATION OF THE GAS SENSING MECHANISM AND PROPERTIES OF R.F. SPUTTERED ZnO THIN FILMS

Li Jianming, Lu Honglong and Qiu Nanyuan
Physics Department of Shandong Polytechnic University
Jinan, China

ABSTRACT

The role of the grain boundaries in the response of the conductance and the effect of the film thickness on the gas sensitivity of R.F. sputtered ZnO thin films are investigated with theoretical calculation and experiment. The results show that the grain boundaries in the polycrystalline ZnO thin films play an important role in the gas response of the film conductance and the film thickness has no significant effect on the gas sensitivity.

INTRODUCTION

In recent years, a lot of papers on metal oxide semiconductor gas sensor has been published and the mechanism of the sensor is widely discussed. In case of thin film type gas sensors, it is usually supposed that the gas adsorption and reaction, which cause a conductance change in the film, is limited at the film surface. Consequently, the film thickness is considered to be an important parameter to the gas sensitivity. But few systematic investigation on the relation between gas sensitivity and film thickness, to the author's knowledge, is present. On other side, the semiconductor thin films used as gas sensors usually possess a polycrystalline nature. The conductance of the film may be remarkably affected by the oxygen adsorption at the grain boundaries. In the gas response of the film conductance, the grain boundaries might also play an important role. However, the effect of the grain boundaries is usually ignored in the papers analyzing the gas response of thin film gas sensor. The correctness of such an approximation is not verified satisfactorily.

In this paper, theoretical calculation results are compared with experiments with R.F. sputtered ZnO thin films as samples. The role of grain boundaries in the gas response of film conductance and the effect of film thickness on the gas sensitivity are discussed.

THEORETICAL CONSIDERATIONS

It has been certain that for most widely used gas sensing semiconductors such as ZnO and SnO_2, oxygen adsorption play an important role in the gas response of the semiconductors' conductance: The charge carrier in the semiconductor is depleted by the ionic adsorped oxygen and the conductance is decreased remarkably. Reducing gas can react with the adsorped oxygen, and decrease its density thereby cause an increase in the conductance. Thus, for a good gas sensing material, it is necessary that the conductance of the material should be able to be affected remarkably by oxygen adsorption. As the thickness of the deplete layer caused by oxygen adsorption exists a limit (Weisz limit), the depth of the subsurface layer of the semiconductor, which conductance is able to be notably affected by reducing gas, is also limited. As a result, the texture of the semiconductor become an important factor to the gas sensitivity.

Theoretical calculations[1] show that, if the oxygen adsorption is limited at the film surface, the effect of film thickness is significant. (Fig.1) is the theoretical result for ZnO thin films. Where N is the average density of charge carrier in the conduction band, Nss and Ea are the density of oxygen adsorption state and the energy level of the state calculated from the bottom of the conduction band at the surface respectively, d is the film thickness. The donor density is taken to be $Nd = 10^{17}/cm^3$. It is obvious that only the thiner films' conductance is affected remarkably by oxygen adsorption. As the film thickness is increased, the affecting of oxygen adsorption on the film conductance decreases. For the films thicker than a critical thickness, oxygen adsorption has no notable affecting on the film conductance. Based on this result, it should be expected that if the oxygen adsorption is limited at the film surface, only the thiner films possess high gas sensitivity. Futher calculations with the same method show that, in case of the oxygen adsorption being limited at the film surface, both the conductivity and its activation energy will be thickness dependent. (Fig.2) and (Fig.3) give these results, which can be compared with experiments to determine if the oxygen adsorption is limited at the film surface.

Other than surface adsorption, oxygen adsorption may take place at the grain boundaries in a polycrystalline thin film. In this case, there should be no notable relation between the film thickness and the conductivity, activation energy and gas sensitivity as well.

EXPERIMETAL RESULTS AND DISCUSSION

Various ZnO thin films with different thicknesses are deposited with R.F. reactive sputtering and the conductance and the gas sensitivity are measured. The detail of the procedure is discribed else where[2]. The ZnO thin films possess high sensitivity to alcohol vaper. The sensitive temperature ranges from 300°C to beyond 450°C. For a high sensitivity, fast respose and good stablity, the working temperature is chosen at 380°C.

(Fig.4) is the conductance of ZnO films with different thicknesses respons to alcohol vapor in the ambient of air. The film thickness has no much effect on the gas senditivity, and the thicker films are still sensitive. Conductivity measurement shows that all the film posess low conductivity and there exists no obvious relation between the film thickness and the conductivity and its activation energy. These results imply that the oxygen adsorption is not limited at the film surface. This is reasonable for the films are deposited in the ambient of O_2/Ar mixture gas.

The oxygen adsorped at grain boundaries may be cleared away by annealing the films in high vacuum up to a temperature of 400°C[3]. After annealing, either the conductivity or its activation energy of the films become thickness dependent, and the relation is in good agreement with the theoratical calculations for the case of oxygen adsorption being limited on the film surface. (Fig.5) and (Fig.6) give these comparison.

In case of air, the vacuum annealed films are stable only as the temperature is below 180°C. As the temperature is beyond 180°C, the conductivity of the film will gradually decreased to the value before vacuum annealing in a short time. The higher the temperature, the faster the decreasing. Considering that the ionic adsorption state of oxygen on ZnO begins to convert form O_2^- to O^- as the temperature is beyond 180°C[4] the author proposes that under high temperature, the oxygen adsorped as O^- be able to diffuse along the grain boundaries with a relatively high speed, and permeate ZnO film again. For the working temperature of ZnO thin film gas sensor is usually beyond 300°C, it should be certain that the oxygen adsorption takes place at the grain boundaries

To investigate further the role of the grain boundaries in the gas response of conductance, the relation between the film thickness and the conductance of the film in certain concentration of reducing gas is examined. (Fig.7) gives the result. Either in air or in reducing gas, the conductance of the films increases as the film thickness is increased, implying that the conductivity throughout the thickness range of the film is gas sensitive. This could only be explained by supposing that the oxygen adsorbed at grain boundaries is also affected by the reducing gas. In other words, the oxygen adsorped atgrain boudaries takes part in the gas response. Otherwise, because the area of grain boundaries is much larger than the area of the film surface, the vary of the oxygen adsorped at the film surface itself can not cause a conductivity change throughout the thickness range of the film. As the mechanism, the author proposed that at high temperature, the oxygen adsorped as O^- at the grain boundaries is in equilibrium with the oxygen adsorped on the film surface by diffusing along the boundaries, and as the surface oxygen is disturbed by reducing gas, the boundary oxygen is also changed. Another experimental result that support the proposition is that the gas response time of the conductance increases as the film thickness is increased. (Fig.8) gives this result.

The R.F. sputtered ZnO thin films possess high sensitivity and good selectivity to alcohol vapor, and it is seen from (Fig.4) that the relation between the film conductance and the concentration of alcohol vapor in ambient obeys power law well. The exponent of the power law vary with the working temperature, and is thickness dependent. (Fig.9) indicates the relation between the exponent P, sensitivity S and the film thickness, where S is defined as the ratio of the conductances of the film in 800 ppm and 80 ppm alcohol vapor respectively. This relation is supposed to have some connection with that the grain boundaries take part in the gas response, but has not been explained well. As (Fig.9) is compared with (Fig.5), it is seen that the thickness that P and S get a maximum value is almost the thickness that the conductance of the vacuum annealed films begin to decrease sharply as the thickness is decreased. This is probably not a coincidence. On other side, although the film thickness has some effects on the gas sensitivity, the effects are not significant. This is the very result of grain boundaries' taking part in the gas response.

SUMMARY

The role of the grain boundaries in the gas response of the conductance and the effect of the film thickness on the gas sensitivity of R.F. sputtered polycrystalline ZnO thin films are investigated with theoretical calculations and experiments. The results suggest that the grain boundaries in the polycrystalline ZnO thin films play an important role in the gas reponse of the film conductance. At high temperature, the oxygen adsorption, which can affect the film conductance remarkably, not only takes place on the film surface, but also take place at the grain coundaries. As the oxygen adsorbed at the grain boundaries is able to diffuse along the boundaries with a relatively high speed, an equilibrium might build up quickly between the boundary oxygen and the surface oxygen, and as the surface oxygen is disturbed by the reducing gases in the ambient, the amount of the boundary oxygen is changed. As a result, the conductivity throughout the thickness range of the film is gas sensitive and the film thickness has no important effect on the gas sensitivity.

The R.F. sputtered ZnO thin films possess high sensitivity and good selectivity to alcohol vapor. The response obeys power law well, and the exponent of the power law is affected by the working temperature and the thickness of the film.

REFERENCES

1. Li Jianming et al, Journal of Shandong Polytechnic University, 18 (1) (1988), 82.
2. Li Jianming, M.S. Thesis of Shandong Polytechnic University, (1988)
3. S. Major et al, Thin Solid Films, 122 (1984), 31.
4. Chon et al, J. Catalysis, 14 (1969), 257.

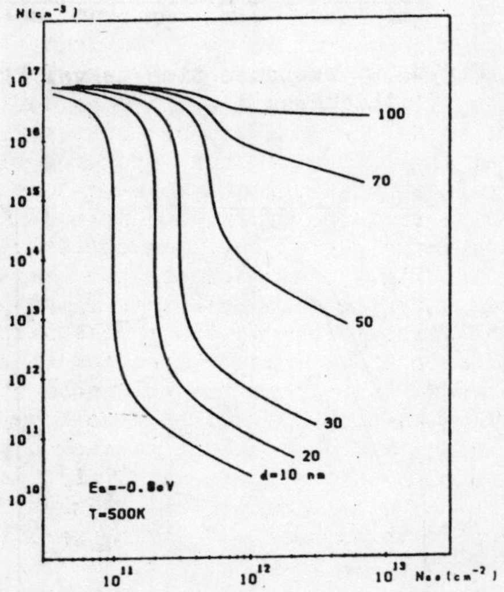

Fig.1 Average charge carrier concentration N in ZnO films with thickness d vs. density of oxygen adsorption state Nss (computer solution)

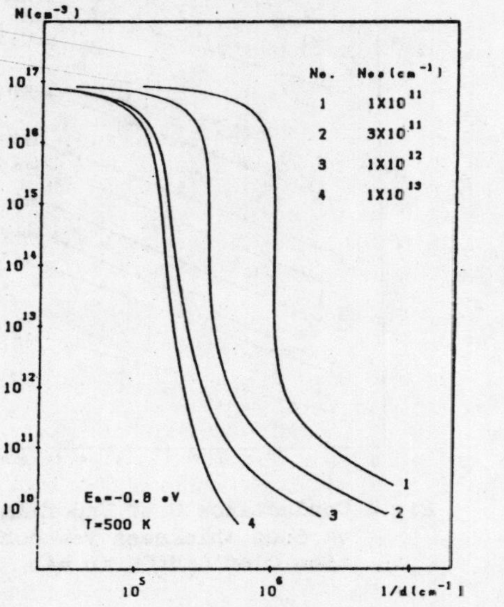

Fig.2 Average charge carrier concentration N in ZnO film vs film thickness d (computer solution)

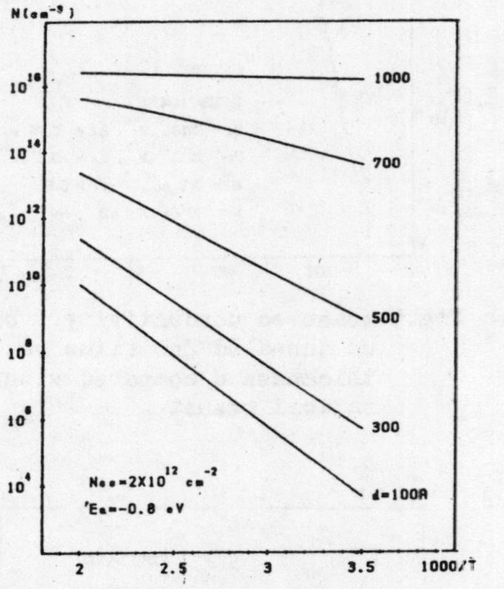

Fig.3 Average charge carrier concentration N in ZnO film with thickness d vs. absolute temperature T. The density of oxygen adsorption state Nss is supposed to be a constant (computer solution)

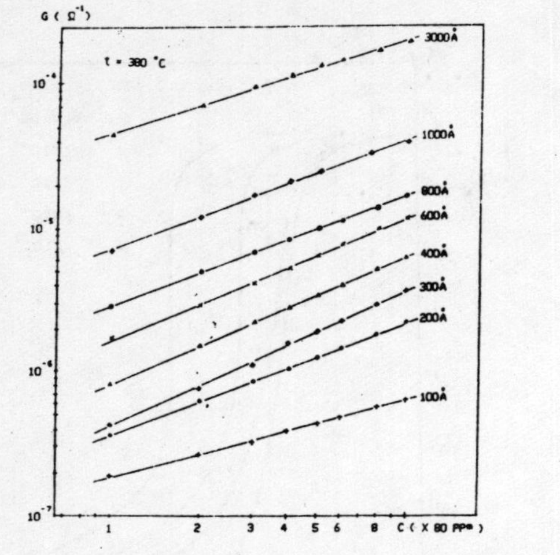

Fig.4 Conductance G of ZnO films with various thickness vs. concentration C of C_2H_5OH in air

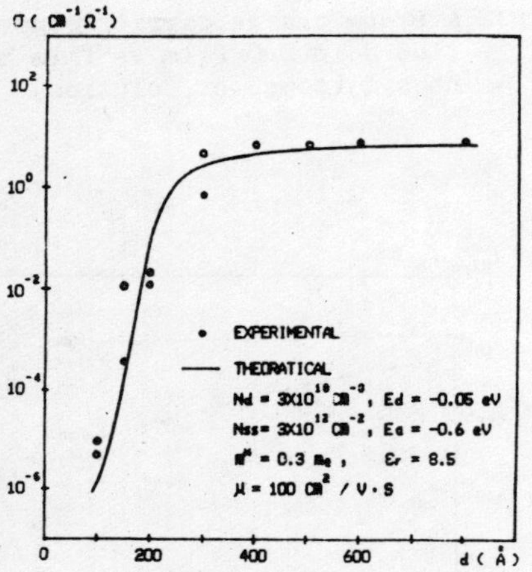

Fig.5 Measured conductivity of vacuum annealed ZnO films vs. film thickness d compared with theoratical result

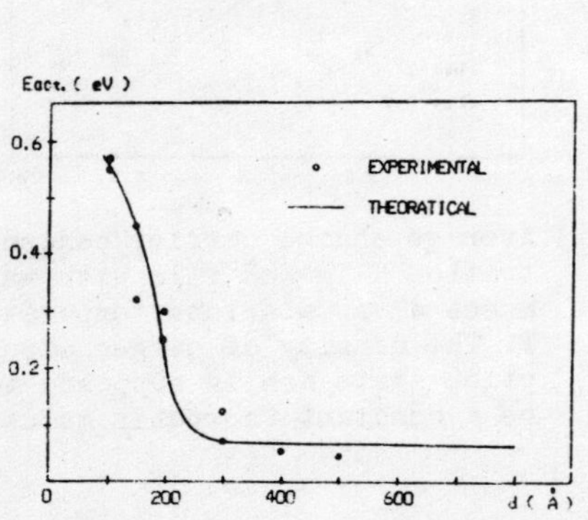

Fig.6 Measured activation energy of conductivity Eact. of vacuum annealed ZnO films vs. film thickness d compared with theoratical result

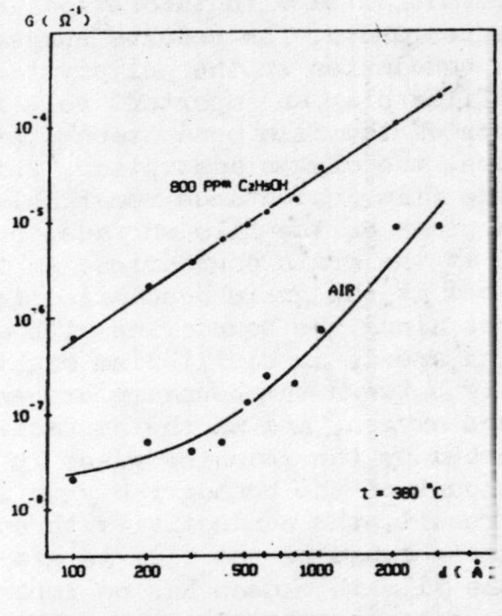

Fig.7 Conductance G of ZnO films in reducing gas and in air vs. film thickness d

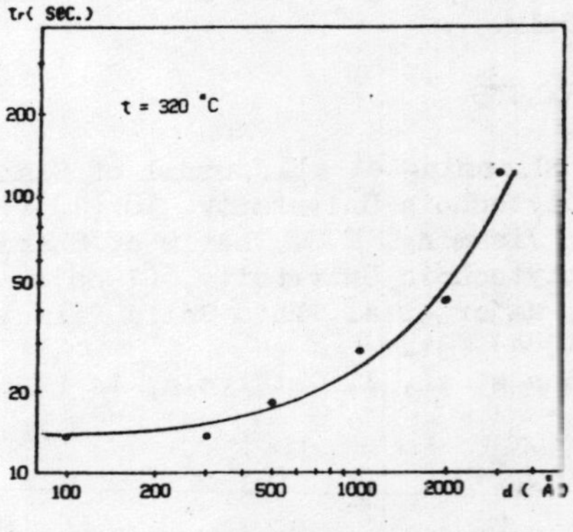

Fig.8 90% response time tr vs. film thickness d

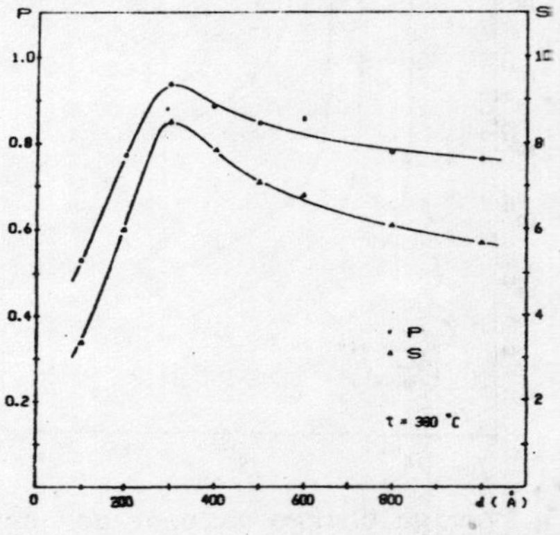

Fig.9 Exponent P of power law and sensitivity S vs. film thickness d

QUASI-ULTRAFINE PARTICLE THIN FILM NO_x SENSOR AND ITS STABILITY

Liu Kuo Wei, Fang Zheng, Mao Jing Ying,
Jin Geng Chi, Zhou Qun, Hu Ming, Xu Gu Hua*
Hangzhou Institute of Electronic Engineering
P.O.Box 55 Hangzhou China No.116 Factory of Jin Hua. China

ABSTRACT

First of all, this paper is focus on the fabrication of the NO_x sensor and the measurements of electronic-chemical properties. Base on planar-technology, in the hybrid or monolithic integrated we made heater resistor using the p-n junction isolation diffusion, p-n junction temperature sensor and quasi-ultrafine particle sensitive film by means of plasma chemical deposition. There are complement between planar-technology and quasi-ultrafine particle film.

The measurements indicate that sensor's optimum operating temperature lies in about 180°C, it is the lower value, the limit of detected NO_x gas reaches 0.4 ppm, it is a good value.

At last, we take two methods to decrease any measured error. Taking four methods, the stability of this sensor is improved. Thus, we made this NO_x sensor and its second instrument for measure and alarm.

PREPARATION

Resently, Ultrafine particle film of SnO_2 has been reported used for humidity, C_2H_5OH and so on [1 2 3]. This paper makes research on NO_x sensor of quasi-ultrafine particle film of SnO_2, the NO_x gas is a mixture of NO_2 and NO.

The quasi-ultrafine particle SnO_2 film is made by means of plasma chemical deposition. The vacuum deposition chamber is used in this study. The specific features arranged in the chamber were (a) glass planes is guiding the ultrafine particles from an evaporator boat to a substrate, and (b) the plasma and glow discharge of the oxygen in the vacuum chamber were produced by the RF coil.

The electron-microscope photo for this film indicates that the average particle size is about 1000A°. Its bigger than ultrafine particle, strictly, this is a guasi-ultrafine particle film.

The heater in this NO_x sensor is a p-n junction isolation diffusion resistor. Used planar technology, the chip of this sensor is small, about 3x4 mm^2.

Adopted constant source of little current, I = constant, V_{eb} is linear to tamperature T from-50 to +150C. This is a very good tamperature sensor. Using I = 250 A, keeps the thermalization it-

---

* Xu gu Hua is the member of No.116 Factory of Jin Hua

self smaller and forward characteristics of the p-n junction better.

CHEMICAL-ELECTRONIC CHARACTERISTICS

1. SnO_2 quasi-ultrafine particle thin film

This film resistance has good repetition of resistor value also shown in Fig.1. When the temperature T increases, This repetition is pretty well which is expected.

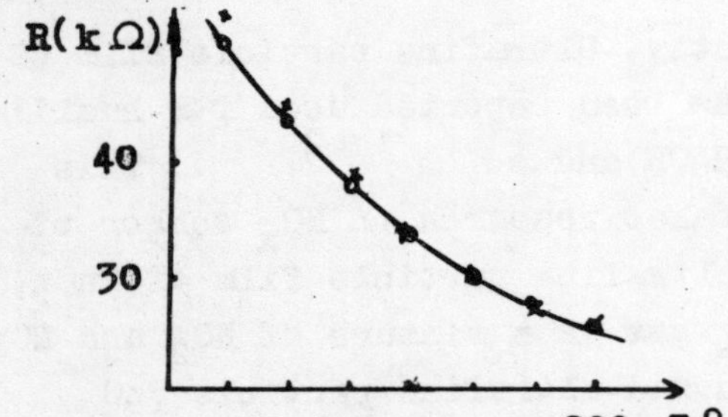

Fig.1 Film resistor as a function of temperature

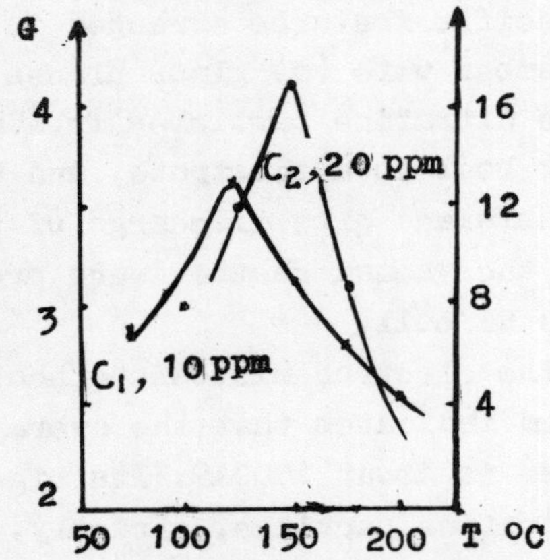

Fig.2 Optimum operating temperature of this NO_x sensor

2. Optimum operating temperature of sensor

It is well know, no matter it belongs to physical absorb or chemical absorb, NO_x gas act with SnO_x film is dependent on the operating temperature. Therefore, there is an optimum operating temperature. Define the sensitivity of sensor is G

$$G = \frac{R_g - R_a}{R_a}$$

where, Ra, the film resistance in clear air; R_g, the film resistance in NO_x gas injuncted.

The results is shown in Fig.2. much data indicate that optimum operating temperature of this film is about 170°C, it is decreases apparently for 300 °C, 500 °C to ZnO_2 and SnO_2 ceremic sensor.

3. Sensor sensitivity for NO_x gas concentration

Fig.3 and Fig.4 show that it is a linear in low consentration and it is a saturat in high concentration. This is expected because of the progress of absorb is mono-atomic sheet. It is clear that large the film area, enhance width the linear area.

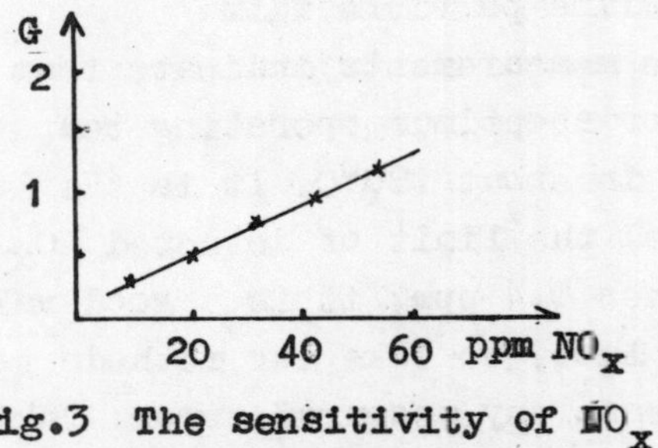

Fig.3 The sensitivity of NO_x sensor as a function of NO_x gas concentration in low concentration

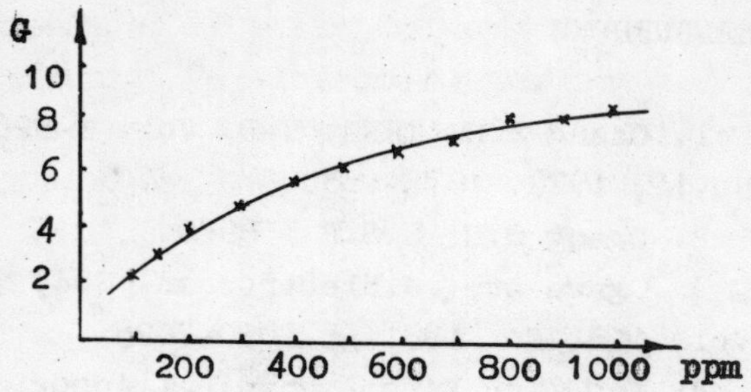

Fig.4 The sensitivity of NO_x sensor as a function of NO_x gas in high concentration

CALIBRATION OF GAS CONCENTRATION

It is difficult for us to give gas calibration at about 1 ppm now. During the measurement, the great error cames from absorb and deposition of this rare gas, so, three methods are used:

1. Based on law of part pressure of gases, measurement of the NO_x concentration is from pressure difference of height in the Hg pole.

2. The standard NO_2 gas source is given from Peking.

3. The NO_x concentration is measured by a spectrophotometer.

The three methods have be study and considerable attention should be paid to absorb and deposition of NO_x rare gas, the absorb of NO_x rare gas is shown in Fig.5.

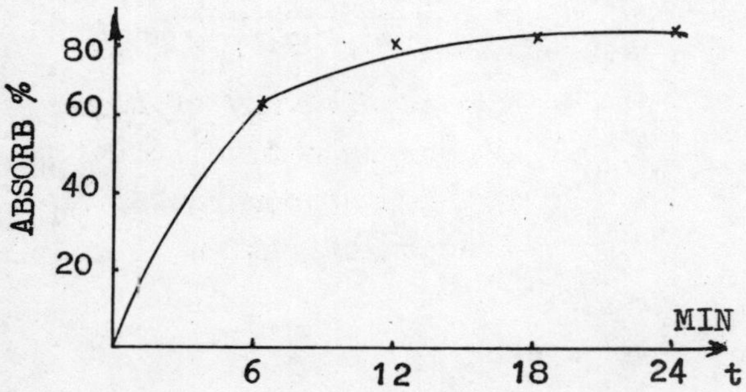

Fig.5 The absorb on the surface of glass chamber

Thus measured error can decrease to 8%, otherwise, it can reach a few tenths.

Later, we'll publish the rare gas calibration and second instrument on the other papers.[4] [5]

STABILITY IMPROVEMENT

Primary, the stability of this NO_x sensor is not so good. Thus, we have taken the following four measures:

1. When measure the sensor, to reach a finite temperature is from the higher temperature down to, for example, to reach the 180 °C is from 200 °C down to 180 °C, not from 0 °C up to 180 °C.

2. Digital electrode of this sensor changes from Al to Al-Ni-Au which is a sandwich, about it, we had a lot of research: Al-Au, Ni-Au, Ti-Au, Al-Ni-Au, the best is Al-Ni-Au.

3. Making the sensor ageing: ageing of this sensor is important, usually this sensore ageing is made at 180°C and 24 hr. Before every measure, the sensor ageing is made at 180°C and 20 min.

4. Adherence of another p-n junction temperature sensor: The construction of this sensor is shown in Fig.6 schematicly. When sensor is contact with second

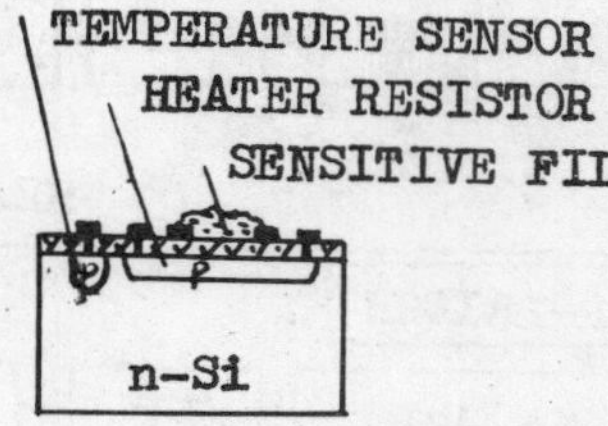

Fig.6 The construction of this sensor schematicly

instrument, the p-n junction leak is increased because of the higher operating temperature and it is also became the sensor is contact with the air directly.

It influences circuit of temperature control. Thus, we made another p-n junction temperature sensor placing on the surface of SiO_2 to replace origin temperature sensor. By this means, to avoid p-n junction leak influences circuit of temperature control, we call it hybrid IC.

By the four methods, stability of this sensor is improved, of course, we hope the stability of this sensor can be improved more and more.

NO_x GAS MONITOR FOR MEASURE AND ALARM

NO_x gas sensor is contact with the second instrument which makes the NO_x gas monitor used for measure and alarm.

This monitor consists of

1. Temperature control unit.
2. Measurement unit.
3. Alarm unit.

This schematic diagrams is shown Fig. 7, it can monitor NO_x gas for measurement and alarm in factory and Army because of NO_x gas 2.6 ppm is the limit for protestion of labour health.

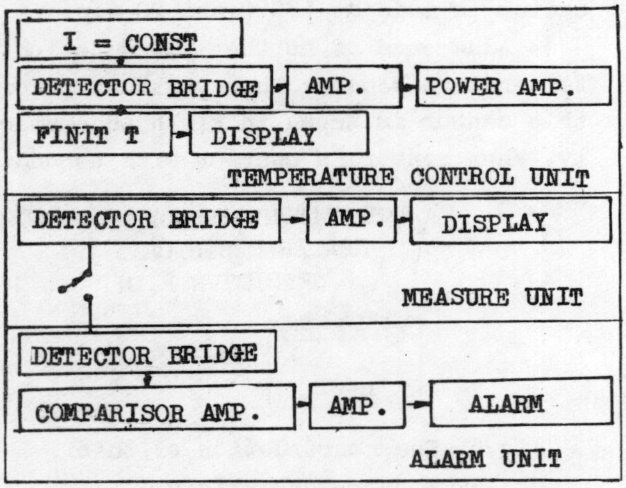

Fig.7 The schematic diagrams of the NO_x gas monitor

REFERENCES

1. Chang S.C. IEEE, E-D, Vol. E-D26, No.12, 1979, 1875-1880.
2. Chang S.C. U.S.P 358950.
3. Ogawa etc. J.Electrochem. Soc, Vol. 128, No. 1981, P.2020-2025.
4. Mao Zing Ying J.Chemical Ansor, 4(1986)32-35.
5. Jin Geng Chi J.Sensor Technology 3(1988)31-35.

THE PERCOLATION MODEL AND POWER LAW OF POROUS SEMICONDUCTOR CERAMICS GAS SENSORS

Zhao Chao, Li Jianming* and Qiu Nanyuan*
Dep. of Silicate, Shandong Light Industrial College,
Physics Dep. of Shandong Polytechnic University
Jinan, China

ABSTRACT

A new model, which we termed percolation model, is proposed, in which a porous ceramics gas sensor is dealt with as a 3-dimensional bond percolation network, and a more reasonable power law, $R \sim P^{-\alpha\cdot\beta}$, is obtained, where β is a simple integer ratio depending on the surface gas reaction and α is not a simple integer ration and related to the bulk properties of porous ceramics and operating temperature. Above results are in good agreement with the measurement of porous ceramics gas sensors.

INTRODUCTION

Two mechanisms of gas sensitivity of porous ceramics, the grain boundary barrier model[1] and AET model[2], have been described. The first model can describe the conductance of a pressed pellet sensor well[3], and the last one deals with sintered sensors. AET model states that the conductance of a porous n-type semiconductor ceramics which consists of pores, grains and necks between grains depends on that of the necks in which the thickness of depletion layer resulted from the chemisorption of oxygen is comparable with neck radius, or a severe depletion has established. Introducing gas, the density of chemisorption of oxygen is decreased by the oxidation of gas, the severe depletion is destryed, and the conductance of ceramics increases. Such a model is too simple to concern with the effect of microstructure of ceramics. In this paper, a more reasonable model is proposed.

THEORY

1. Percolation Model

A gas sensor can be simplified as a network of resistive necks in which the effective concentration of free charges depends on the surface oxidation of gas. The dependence is so tightly on neck radius as the neck radius and the cratical radius of severe depletion are comparable[4] that the practical distribution of neck radii is much wider than cractical zone of AET. This means the adsorption effects of necks possessing different radii are not synchronous.

Two approximations, the binery distribution of neck conductance in which two values σ_1 and σ_2 occur with $\bar{r}_n > L_c$ and $\bar{r}_n < L_c$, and the normal distribution of neck radii

$$f(r_n) = \frac{1}{\sqrt{2\pi}\,\sigma} \exp\left[\frac{-(r_n - \bar{r}_n)^2}{2\sigma^2}\right] \tag{1}$$

are considered, where $\sigma_1 \gg \sigma_2$, $L_c = 2N_s/N_o$, (N_s, the surface charge density, and N_o, the bulk carrier concentration), and $\sigma = \bar{r}_n/3$. These give rise to a 3-dimensional bond percolation network in which bonds corresponds necks and probability P is related to Lc,

$$P = 1 - \int_0^{L_c} f(r_n)\, dr_n \tag{2}$$

According to Kirkpatrick[5], the conductance of such a network is

$$G(P) = \left\{\left(\tfrac{1}{2}ZP-1\right)\sigma_1 + \left(\tfrac{1}{2}Z(1-P)-1\right)\sigma_2\right\}/(Z-2) + \left\{\left\{\left(\tfrac{1}{2}ZP-1\right)\sigma_1 + \left(\tfrac{1}{2}Z(1-P)-1\right)\sigma_2\right\}^2 + 2(Z-2)\sigma_1\sigma_2\right\}^{\frac{1}{2}}/(Z-2) \tag{3}$$

where Z is the number of bonds at each node of network.

The dependence of resistance of a sensor on Lc (Fig.1) is given by above equations and $R = R_o/G(P)$, where R_o is the bulk resistance. At the start, Lc is maximum (Lcm) and R is Ra. Introducing gas, Lc and Ns decrease.

$$\Delta L_c = L_{cm} - L_c \tag{4}$$

$$\Delta N_s = \frac{1}{2} N_o \cdot \Delta L_c \tag{5}$$

So, a $R \sim \Delta N_s$ curve can be made from $R \sim L_c$ plot.

2. Power Law

A $R \sim \Delta N_s$ curve in which the parameters are determined from an experiment[6] by Kaker's model[7] is shown in log-log plot (Fig.2) in which the curve is nearly linear, that is

$$R \sim \Delta Ns^{-\alpha} \qquad (6)$$

(the dash line are the experimental results).

Some authors[8,9,10] have obtained a power relation from mass action law in which the exponent is a simple integer ratio, but such a relation is completely different from eq.(6) and should be

$$\Delta Ns \sim P^{\beta} \qquad (7)$$

where P is the partial pressure of gas. So, the power law of porous ceramics gas sensor are

$$R \sim P^{-\alpha \cdot \beta} \qquad (8)$$

where the exponent of power is a product of two terms in which β is the simple integer ratio from mass action law and α is not a integer ratio and results directly from percolation model. This is in agreement with Clifford's observation.[11,12]

In the view of percolation, the figure of R~Lc line which depends on the distribution of neck radii and the relative positions of Lcm and the threshold on R~Lc curve affect the R~ΔNs curve, the exponent of power law of a porous ceramics gas sensor bears a relation to A. the category of gas which determined β, B. the surface properties of semiconductor, C. the microstructure of ceramics, D. the bulk carrier density, and E. the operating temperature which Nsm is relative to.

EXPERIMENTAL

Zinc oxide powders (reagent grade) were ground for 5 h in an agate ball mill containing agate pebbles and pure alcohol, and then calcined in air at 400°C for 1 h. After calcination, the powders were pressed into plates (15 mmϕ and 4 mm thick), then the plates were sintered at the temperature shown in table 1.

The porisitys of samples were obtained by Archimedes method using mercury. The microstructures were examined by SEM and the mean of grain size were determined from micrograph by numerating 50 grains. Silver electrodes were applied on both surface of plates by sputtering, and the resistances of samples were measured on a heating holder in a sealing container. CO, C_4H_{10} and CH_4 (99.99% pure) were injected into the measuring room by injectors and temperatures were measured with a thermcouple placed near samples.

RESULTS AND DISCUSSIONS

The sintering conditions and microstructural parameters of samples are shown in table 1. in which the data in 5th column are determined by Kaker's model.

Table 1 Sintering condition and microstructural parameters

No	Sintering condition	Average grain radius (μm)	Porosity (%)	$\bar{r}_n$ (μm)
1	650 C, 1h	0.133	36.6	0.081
2	750 C, 3h	0.145	35.1	0.089
3	850 C, 1h	0.147	32.0	0.094
4	900 C, 1h	0.218	30.1	0.168
5	1100 C, 1h	0.428	18.6	0.307

In electric measurements of samples in air, a great decrease of resistance occurs as temperature rise in No.1-4 samples, and such a decrease disappears in No.5. This is expected because Lc in No.5 which possesses a much biger $\bar{r}_n$ than others is all along in low resistance zone far from threshold of R~Lc curve.

The response curve in C_4H_{10}, CO and CH_4 (Fig.4) show a good power law at higher concentrations of gases and the exponent of power is dependent tightly with the category of gas and in some degree with the microstructure of ceramics. Considering the proposal about exponent and let β be 3/4, 1/2 and 1/4 in C_4H_{10}, CO and CH_4, nearly same values of α for each samples are obtained from the curve in three kinds of gases. The values of Ra of samples possessing different $\bar{r}_n$ are distinct from each others as well as α.

It is interesting that these curves are bent in some degree at lower concentrations. Such a bending can not be explained by mass action law, and is probably a speciality of porous ceramics sensors, at minimum for anything we know the bending has never occured in thin films. We hold it come from the R-ΔNs curves which are usually bent. Different operating temperature, as expected, also result in the distinctions of Ra and α even in same sample. To unify Ra of different samples, a distinction of α is found as an evidence of effect of micro-structure to R-Lc line. Bulk carrier density No which Lc is inverse proportion to is another effective factor to Ra and α. The proof has been found[13] in R-P curves of porous ZnO ceramics undoped and doped with Li_2O by soaking method which make the microstructure of ceramics invariable. In doped sample a greater Ra and an obvious Biger α appeared.

REFERENCES

1. S. Roy Morrison, Sensors and Actuators, 11 (1987), 283.
2. H. Mitsudo, Ceramics (Jap.), 15 (1980),339.
3. S. Roy Morrison, Sensors and Actuator, 12 (1987), 485.
4. D. Bednarczyk, J. Bednarczyk and A. Wegrzyn, Thin Solid Films, 36 (1976), 165
5. S. Kirkpatrick, Phys. Rev. Lett., 27(1971), 1722.
6. S. Satio, M. Miyayama, K. Koumoto and H.

Yanagida, J. Am. Ceram. Soc., 68(1985), 40.
7. S. Ri, K. Hamano and Z. Nakagawa, J. Ceram. Soc. Jap., 93(1985), 231.
8. A. Jones, T.A. Jones, B. Mann and J.G. Firth, Sensors and Actuators, 5(1984), 75.
9. S. Sträsler and A.Reis, Sensors and Actuators, 4(1983), 465.
10. H. Windischmann and P. Mark, J. Electrochem Soc., 126(1979), 1813.
11. P.K. Clifford, Proc. Int. Meet. on Chemical Sensors, (1983), 135.
12. P.K. Clifford and D.T. Tuma, Sensors and Actuators, 3(1982/83), 233.
13. M. Miyayama and H. Yanagida, Electrochem. (Jap), 50(1985), 231.

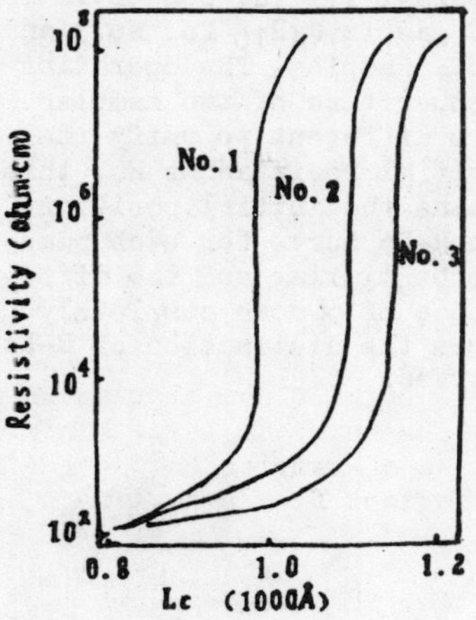

Fig.1 Resistivity vs. cratical radius of severe depletion Lc for porous ceramics percolation network. Values of $\bar{r}_n$ for each sample are show in table 1. $\sigma_1=1$, $\sigma_2=10^7$ $R_o = 10^2$ (ohm·cm)

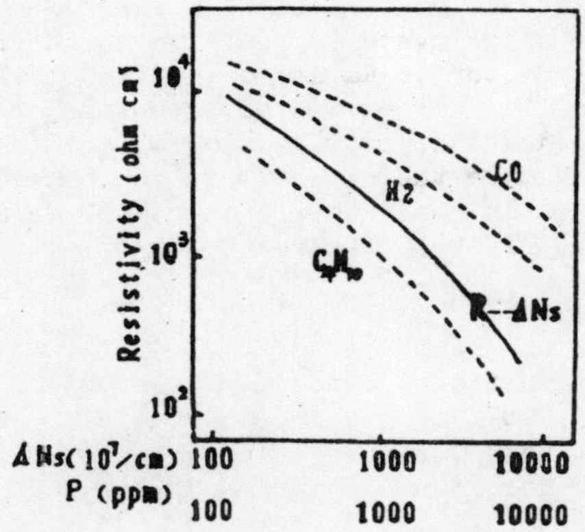

Fig.2
Solid Line: Resistivity of a porous ceramics percolation network vs. decrease of surface charge density Ns. Values of $\bar{r}_n$ is determined as 0.12 m by Kaker's model from Saito's experimental data.[10] $\sigma_2=10^{-6}$, Z=6

Dash Line: Experimental result of porous ZnO ceramics samples, R-P, given by Saito.[10] Relative density of sintered bodies was 52-55%, grain size was 0.1-0.6 μm, and R (The resistivity in N_2) =40 (ohm.cm), Ra(The resistivity in carrier gas) = 2×10^4(ohm.cm), carrier gas is $N_2 + O_2$ (1:3), measurement temperature is at 400°C.

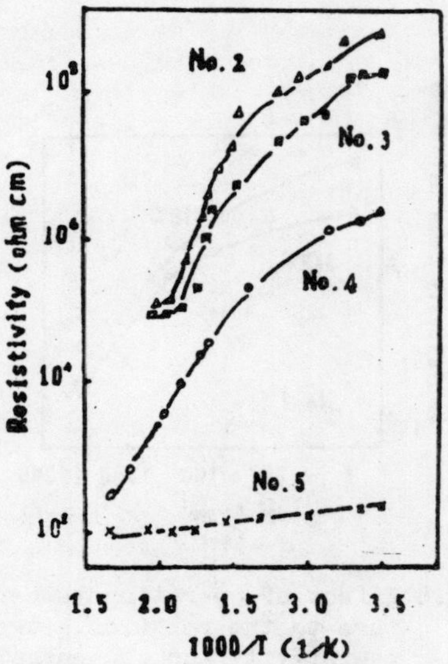

Fig.3 Dependence of resistivity of porous ceramics gas sensors on temperature. Microstructural parameters of each sample are shown in table 1.

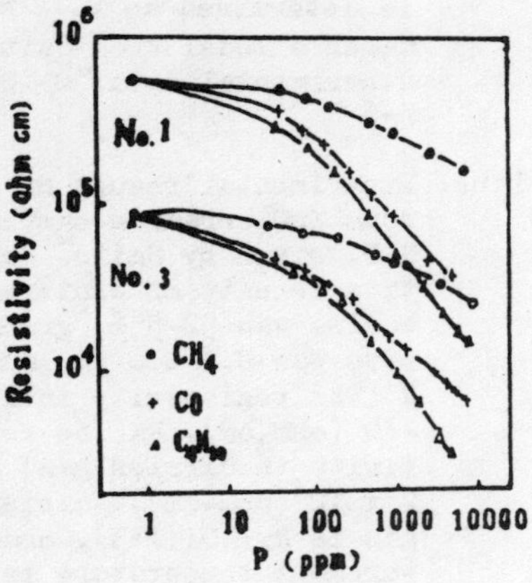

Fig.4 Resistivity vs. concentration of reducing gas at 430°C for porous ZnO ceramics samples.

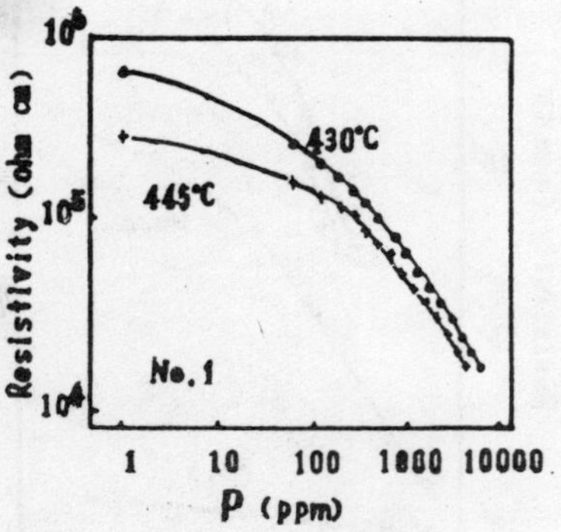

Fig.5 Effect of operating temperature to the relation between resistivity and concentration of gas in C_4H_{10} at different temperature for No.1 sample. Different values of Ra and α are obtained at these temperature.

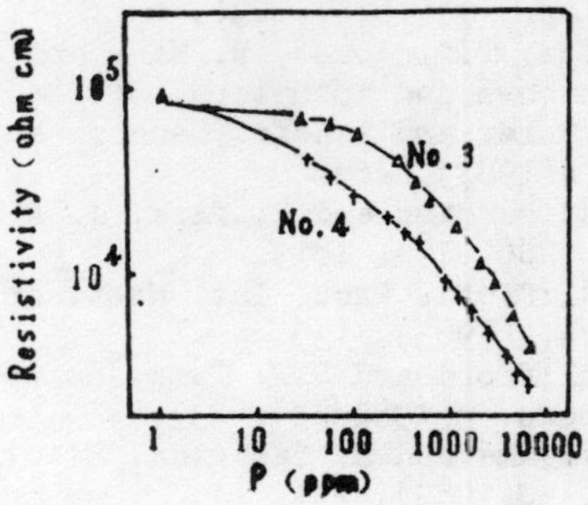

Fig.6 Resistivity vs. concentration of gas in C_4H_{10} for No.3 and No.4 samples. The operating temperature of two samples are different to unify the initial resistances Ra, this means the initial position on R-Lc curve for each sample be similar and the difference of α come completely from the distinction of R-Lc curves.

Modification of the Microstructure of (Ba,Pb)TiO_3 Ceramics by Doping with Specific Impurity Elements and Their PTCR Characteristics

Makoto Kuwabara and Yuji Kato

Department of Applied Chemistry, Kyushu Institute of Technology,
Sensui-cho, Tobata, Kitakyushu, 804 Japan

Abstract

The effect of the addition of specific impurity elements such as B and Bi on the microstructure of semiconducting barium-lead titanate ceramics has been investigated in an attempt to make clear the correlation between the microstructure of the materials and the PTCR (positive temperature coefficient of resistivity) effect observed. The composition of the materials used in this study is described by the formula $Ba_{0.398}M_{0.002}Pb_{0.600}TiO_3$ (M=Sb or La) with the Curie point around 360°C, where Sb or La was used to make the materials semiconductive.

The impurity elements, B and Bi, were added in the materials in the forms of BN and Bi_2O_3 respectively, the concentrations of these elements examined in the present study ranging from 0 to 1.5 mol% for BN and from 0 to 0.2 mole% for Bi_2O_3. The co-doping with B and Bi elements at their specific concentrations different was found to be quite effective to obtain the materials with stable, high-performance PTCR characteristics. This can be understood as the result of a successful control of the microstructure of the materials to have distinct grain morphologies consisting of small (1-2 μm), uniform grains.

An intensive investigation of the microstructural changes of the mateials with the amount of Bi_2O_3 added has further revealed that the addition of Bi_2O_3 within a certain concentration range of it yielded the materials with microstructures having a double grain structure featured by large grains exaggeratedly grown surrounded by smaller grains insufficiently grown. Although such a double grain structure as was observed in these materials is normally considered to be undesirable one to get better electronic properties in the materials, they were found to unexpectedly exhibit improved PTCR characteristics with large PTCR effects and simultaneously low room temperature resistivities.

The improvement on the PTCR characteristic realized in these materials will be interpreted in terms of its microstructural dependence, and the role of the impurity elements doped in the modification of the microstructure of the materials will be discussed.

RESEARCH ON SAW TEMPERATURE SENSOR

Wu Yili, Hu Lanfong and Liu Jiaben
Radio-Electronics Department, Sichuan University
Chengdu, China

ABSTRACT

The principle and design of SAW temperature sensor are presented. A test sample is built with JCL-cut quartz substrate and comb SAW delay line. It is shown by measuring and calculating that the sensitivity and resolution of this test sample is 1.388 KHz/°c and 0.08 °c respectively. The non-linear coefficient is 1%.

INTRODUCTION

Companying with the popularity and development of imformation process technique based on microcomputer, various sensor having excellent performance are demanded. The SAW Sensor is a new type sensor developed in 1980s. Up to now, many SAW Sensor, such as temperature sensor, pressure sensor, velocity sensor, accelerate sensor, voltage sensor and gas sensor have been investigated [1]-[5]. The main advantage of SAW Sensor are:

(1) The output quantity of SAW Sensor is frequency. This benefits high accurately measuring, digital display and connecting with computer to realize intelligent control.
(2) High sensitivity and excellent linearity.
(3) It is apt to realize remote control and measuring.
(4) The plane and solid constructure, simple package.

In this paper, we present the researching work of ours.

THE PRINCIPLE OF SAW TEMPERATURE SENSOR

The sensing temperature unite of SAW Temperature Sensor is a particular SAW Oscillator (SAWO). The principle of SAW Temperature Sensor is based on frequency-temperature characteristic of SAWO. The output frequency of SAWO is the function of environmental temperature. It can be expressed by following equation:

$$f(T)=f(T)[a_0(T-T_0)+a_1(T-T_0)^2+a_2(T-T_0)^3+---] \quad (1)$$

where a_0, a_1, and a_2 is one, two and three rank temperature coefficient respectively. Their value are decided by the kind of material and its cuts. The SAWO consists of a SAW Delay Line or a SAW Resonator and a low noise electronic amplifier. It's basic constructure is as shown in Fig.1 The

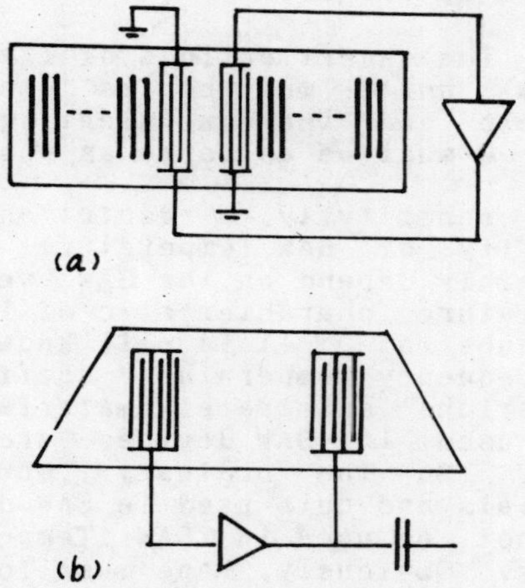

Fig.1 SAW Oscillator
(a) SAW Delay Line Oscillator
(b) SAW Resonate Oscillator

oscillation frequency of SAWO is decided by SAW propagation velocity Vs on the surface of substrate, and the value of Vs is depend upon the envirorment temperature. So the changes of environment temperature can be sensed by the changes of output frequency of SAWO.

DESIGN CONSIDERATION OF SAW TEMPERATURE SENSOR

Making use of the temperature-frequency characteristic of SAWO, the SAWO used for sensing unite must have be high stability of frequency, especially the short-term stability. The short-term stability of a SAWO can be expressed by using single-sideband noise power in 1 Hz bandwidth [6]:

$$P_{ssb} = 10\mathrm{Log}\,\frac{4KT(NF)G^2}{QPo}\left(\frac{f_o}{\Delta f}\right)^2 \qquad (2)$$

where, G -- amplifier gain.
NF-- amplifier noise figure.
Po-- oscillator power at amplifier output.
Q -- equivalent quality factor of SAW Delay Line or SAWR.
KT-- thermal noise.
fo-- oscillation frequency of SAWO.
Δf-- offset or modulation frequency.

It can be seen from equation (2) that in order to obtain good short-term stability, the following condition must be satisfied:

(1) Amplifier must be low noise and Po must be larger.

(2) The insertion loss of frequency control unite must be as small as possible and the equivalent quality factor Q must be as large as possible.

The sensitivity, resolution and linearity of SAW Temperature Sensor are mainly depend on the SAW velocity-temperature characteristic of loading wave substrate. It is well known that the frequency temperature coefficient of various substrates material and cuts used in SAW devices are very small. So the ordinary substrate materials and cuts used in SAW devices can not be used in SAW Temperature Sensor. Obviously, one must look for some of new substrate materials and cuts. Based on the demands of sensitivity and linearity proposed to SAW Temperature Sensor, the new substrate materials and cuts must have large first order frequency temperature coefficient and smaller second order frequency temperature coefficient. Reference [7] presented two new quartz cuts--JCL cut and LST cut. They can provide very excellent frequency-temperature characteristic. So they are very adapt to make SAW Temperature Sensor.

Table 1 [7] shows theoretical values of propagation velocity, electro-mechanical coupling factor and temperature coefficients of first and second order of two quartz new cuts called JCL cut and LST cut.

Table 1. Theoretical value of JCL-cut and LST-cut quartz

cuts	cut angle ϕ	cut angle θ	ψ	v m/s	k $\times 10^{-2}$	a $\times 10^{-6}/K$	b $\times 10^{-9}/K$
JCL	0	42.1	39	3275	0.21	22	-.06
LST	11.4	59.4	35	3360	0.11	23	-.04

Experimental frequency-temperature characteristics of JCL cut and LST cut are shown on Fig.2. In the first step, we adapt JCL-cut for substrate of SAW Temperature Sensor.

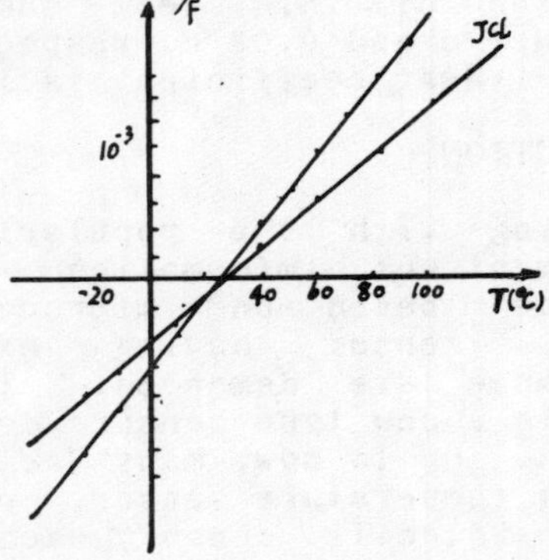

Fig.2 Frequency shift versus temperature for JCL-cut and LST-cut quartz.

SAW Delay Line and SAWR could be employed as frequency control unite of SAWO. The disadvantage of employing SAW Delay Line is poor short-term stability and excessive sensitivity to circuit perturbation[8], and employing SAW Resonlator, the high Q, lower insertion loss, and non-sensitivity to circuit perturbations could be obtained. In the first step, we adapt SAWDL. In order to increase equivalent quality factor and reduce the second order effects, we adapt comb SAW Delay Line as frequency control unite, as shown in Fig.3.

The important constituent of SAW Temperature Sensor is electronic amplifier. It may be tune amplifier or wideband IF amplifier. The former consists of transistor and tune circuit, and possess function of amplifying and selecting frequency. It's shortcoming is lake of stability and troublesome in setting. The later

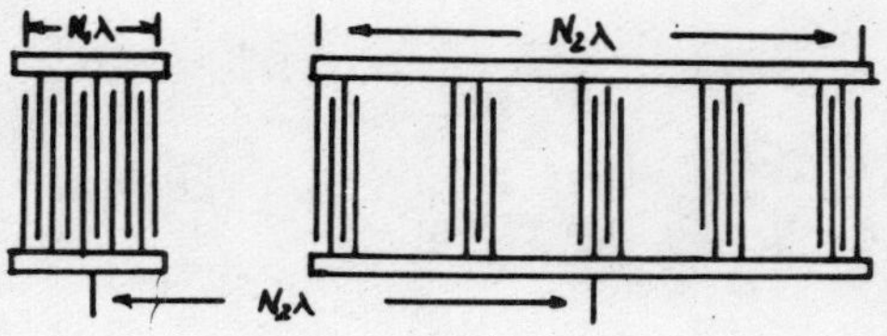

Fig.3 SAW Comb Delay Line

consists of transistor and RC circuit or wideband transfomer. Its advantage is wideband, high gain, excellent stability and apts to be integrated. We adapt tune transistor amplifier as shown in Fig.4. It includes three stage circuit. The common-emitter transistor amplifier circuit is used as the first stage. It is helpful to consider noise figure and gain.

A common-base transistor amplifier circuit is used as the second stage. Because of it's lower input impedence and good stability, the function of the third stage circuit is to provide higher output power. The gain of amplifier circuit shown in Fig.4 is larger than 30dB.

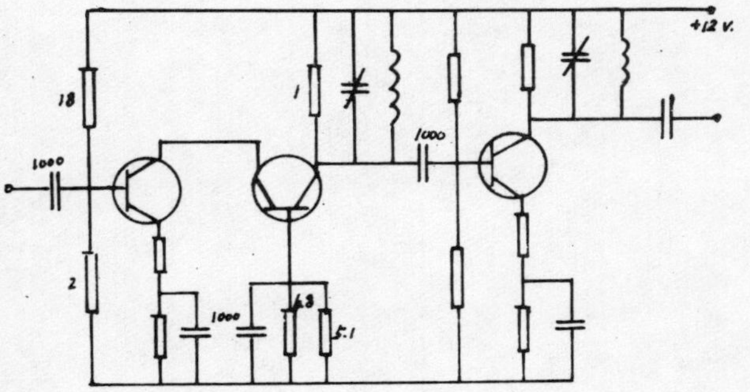

Fig.4 Tuned Transister amplifier

RESULTS AND EXPERIMENT

Using JCL-cut quartz as londing SAW substrate and comb SAW delay line as frequency control unite. We have made a primary test sample of SAW temperature sensor by planar fabrication technique. It's operative frequency is 73.3MHz, with the confination of measuring condition. The frequency-temperature characteristics was measured in two section of temperature. One is -20°c-+20°c, and the other is +20°c-100°c. The result was shown in Fig.5. The influence of power supply voltage to output frequency of SAW Temperature Sensor was shown in Fig.6. By measuring and calculateing ,the basic performance of the primary test sample are:

Sensitivity of measuring temperature: 1.387KHz/°c
Result of temperature: 0.08°c
Non-linearity coefficent: <1%

This research work was completed with the help of Scientific Fund of Sichuan Province. The authors hope to express their thanks.

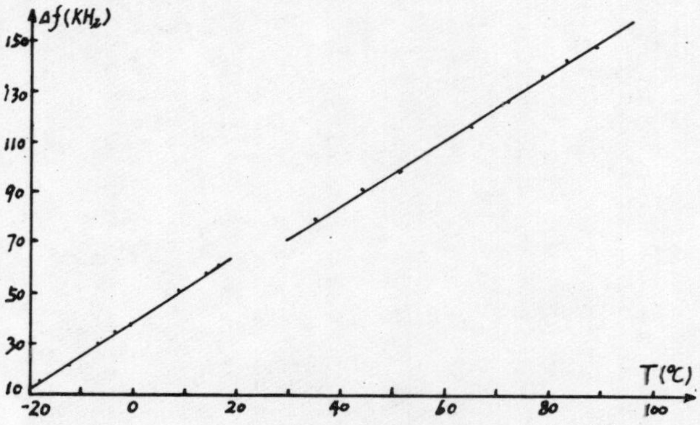

Fig.5 Frequency vs temperature characteristic of SAW Temperature Sensor

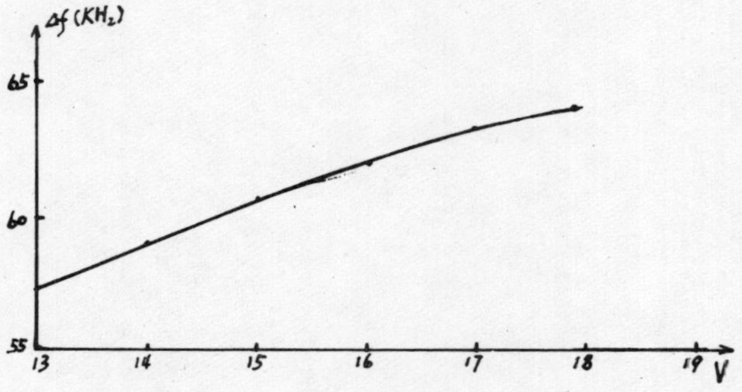

Fig.6 Frequency vs power supply voltage

REFERENCE

1. D.Hauden,IEEE Trans LFFC-34, No.2 (1987), 253-258
2. M.R.Risch, Sensor and Actuators, 6 (1984), 127-133
3. H.F.Tiersten,1980 Ultrasonic Symposium Proceedings,692
4. E.Gatti, Sensors and Actuator,

4 (1983),45-54
5. A.Bryant, M.Poirier ect., Sensors and Actuator, 4 (1983),105-111
6. A.A.Oliner, Acoustic Surface Wave, Springer(1978)
7. D.Hauden ect., Proc. of 36-th A.F.C.S., (1982), 284-289
8. D.E.Cullen and G.K.Montress, 1980 Ultrsonics Symposium Proceedings, 696-701
9. M.F.Lewis, 1973 Ultrasonic Symp. proceeding, 344-347

TEMPERATURE-HUMIDITY SENSING CERAMICS AND ITS SENSOR

Lin Jingxian Hu Zhongxu Jiang Zhihua
Wu Dacan Cen Xiaoliang Jin Xige
University of Electronic Science and Technology of China

ABSTRACT

In present work, $BaTiO_3-SrTiO_3$ (BST) temperature–humidity sensing porous ceramics has been investigated experimentally. Undoped $Ba_{0.5}Sr_{0.5}TiO_3$ and $Ba_{0.4}Sr_{0.6}TiO_3$ porous ceramics obtained from high–purity, ultra–fine barium oxalate and strontium oxalate powders have good porosity and three–dimensionally linked pores.

A double alumina substrate assembly type multifunctional temperature–humidity sensor has been made from the porous ceramics. Temperature measuring range of the sensor is −20(or−40) to +100℃ (10kHz, 1V), and its humidity measuring range is 10 to 100% RH(5Hz, 3V). Response time of humidity measuring is less than 15sec. Heat cleaning time is 60sec (400℃).

INTRODUCTION

Along with the progress of automation and application of microcomputer, the realization of temperature, humidity or gas concentration measuring of atmosphere with single multifunctional sensor has been an important trend. In 1980, Jiro Terada et. al. reported a multifunctional temperature–humidity sensor made from ferroelectric $Ba_{0.5}Sr_{0.5}TiO_3$ which was obtained from solid–state reaction of $BaCO_3$, $SrCO_3$ and TiO_2 and doped with small amount of $MgCr_2O_4$ to form porous structure. Its Curie–Weiss law property of dielectric constant vs. temperature above the Curie point was used to measure temperature, and resistance variation due to humidity absorption of porous structure was used to measure humidity respectively by a frequency separation technique [1].

In present work, the influences of different synthetic technologies of BST fundamental material, dopants and doping amount and sintering process on the porous microstructure and temperature–humidity sensing properties have been investigated experimentally. a compact double substrate assembly type multifunctional sensor which includes a thick film heat cleaner has been made whith the porous ceramic element.

EXPERIMENTAL METMODS

Specimen preparing. Preparing ceramic specimens with different raw materials and ceramic synthetic technologies and ordinary doping methed. Formation pressure is 50MPa. Sintering in early densification period(1000℃ to 1300℃, 0.5 to 1hr) to obtain three dimensionally neck–linked network type porous structure.

Property measuring. After firing RuO_2 porous thick film electrodes on both sides of BSTspecimens, adhesing silver leads and heat treating, measure the temparature–humidity sensing properties respectively. Resistance vs. humidity property is measured by voltage dividing method using saturated salt solution vessel series as humidity sources. Measuring signal is 5 to 10Hz, 3V. Humidity range is 10 to 100%RH. Measuring signal for capacitance vs. temperature is 10KHz, 1V. Temperature range is −40 to +100℃. Measuring porosity and pore–volume distribution by displacement and mercury porosimetry methods. Observing appearance of grains and pores with SEM. Analyzing phase composition and structure of the ceramics by XRD and EMP.

RESULTS AND DISSCUSION

1. Synthetic technologies of BST fundamental materials. Four technologies were used. (a) Synthesizing from AR $BaCO_3$, $SrCO_3$ and TiO_2 by solid–phase reaction.

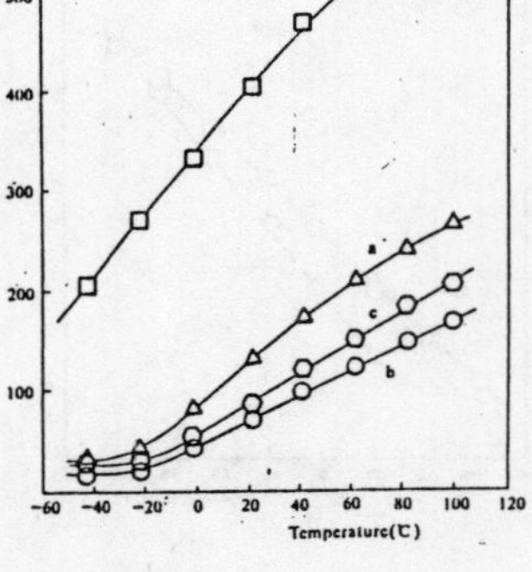

Fig. 1 Influence of synthetic technologies of fundamental materials on temperature sensing property of BST ceramics.

(b)Sintering from high–pure, ultra–fine liquid phase coprecipitated $BaTiO_3$and $SrTiO_3$ powders. (c) Sintering directly from high–pure ultra–fine liquid phase coprecipetated barium oxalate and strotium oxalate powders [2] . (d) Synthesizing from $BaCl_2$, $SrCl_2$ and $TiCl_4$ by coprecipetation in one procedure. The measuring results of specimens prepared from above methods are shown in Fig. 1,2 and SEM Photo 1 a, b, c and d. Among them the ceramics from method c has relatively homogeneous microstructure, nearly linear temperature–humidity sensing characteristic curves and high sensitivity, and repeatability.

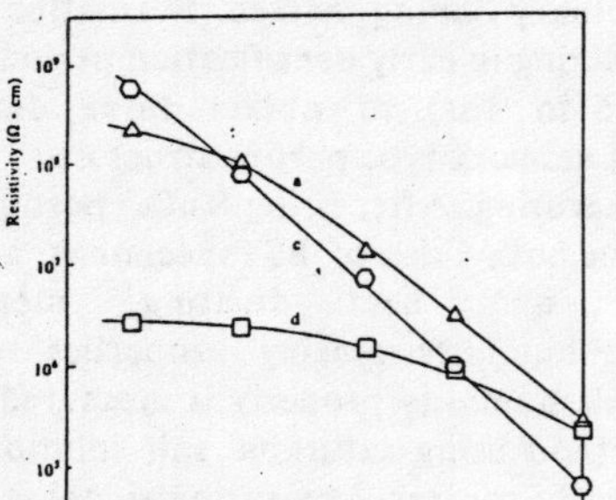

Fig. 2 Influence of synthetic tecchnologies of fundamental materials on hummidity sensing property of BST ceramics.

2. Sr / Ba molar ratio. Temperature–humidity sensing properties of $Ba_{0.5}$ $Sr_{0.5}$ TiO_3 and $Ba_{0.4}Sr_{0.6}TiO_3$ ceramics are shown in Fig. 3 and 4. Curie temperature of the latter is about −70℃, and sensing property has better lincarity in range from −40 to+150℃. The linear sensing range of temperature can be adjusted by controlling the Sr / Ba molar ratio.

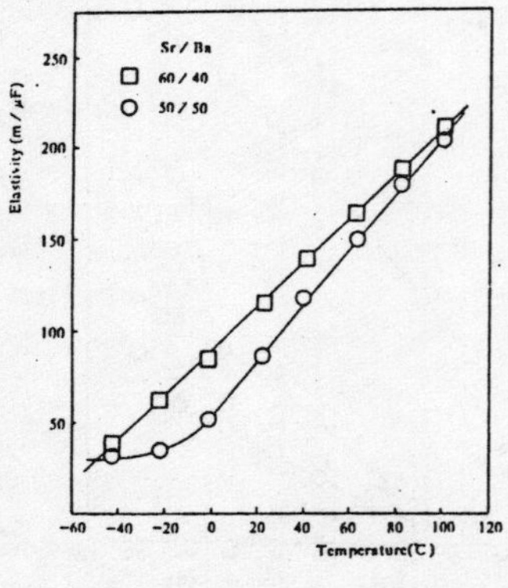

Fig. 3 Temperature sensing property of undoped BST ceramics.

3. Dopants. Temperature–humidity sensing properties of $Ba_{0.5}$ $Sr_{0.5}$ TiO_3 ceramics are shown in Fig. 5 and 6.

From the temerature and humidity sensing properties it can be seen, doping causes the decreasing of ε and ρ. The results of crystal structure, microstructure and micro–region composition analyses show that, the doped and undoped BST ceramics all have single pervoskite structure, the introducing of dopant (less than 5 mol%) has very little effect on the lattice constants, and the dopant (c.g.Cr^{3+}) is mainly concentrated at the grain boundaries and benificial to the formation of porous structure. Thus, the doping of Cr_2O_3, $MgCr_2O_4$, $ZnCr_2O_4$ and SnO_2 is helpful to improve humidity sensing property.

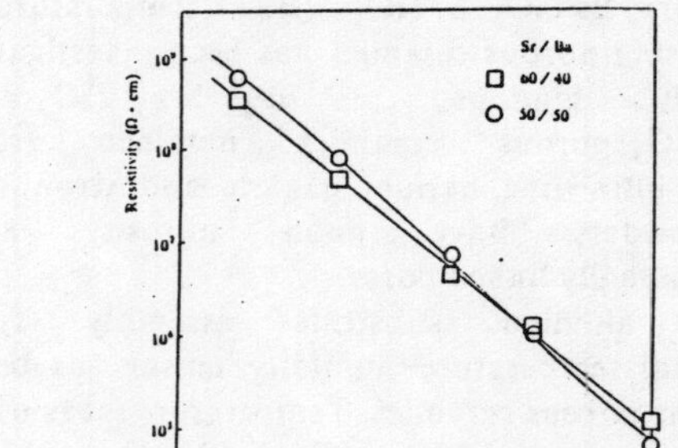

Fig. 4· Humidity sensing property of undoped BST ceramics.

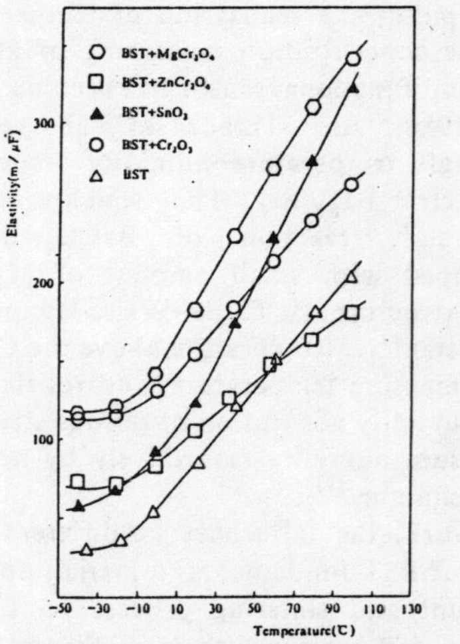

Fig. 5 Influence of doping on temperature sensing property of BST ceramics.

4. Doping amount and sintering technology. The influence of doping amount on humidity sensing properties of $Ba_{0.5}$ $Sr_{0.5}$ TiO_3 porous ceramics are shown in Fig. 7.Technological experimental results show that, between improvement of doping amount on temperature–humid-

ity propertices and sintering processes there exist some mutual adjusting and restricting relations. Usually, sintering at higher temperature, doping about 2 mol% Cr_2O_3(or $MgCr_2O_4$) can effectively imncrease porsity, improve pore structure and temperature–humidity properties, while sintering at lower temperature in early densification period, doping less than 2 mol% or no doping is required. SEM appearance of BST porous ceramics is shown in Photo. 2 e and f. It presents a typical three–dimensionally neck–linked network type porous structure. The porosity is 35 to 40%.

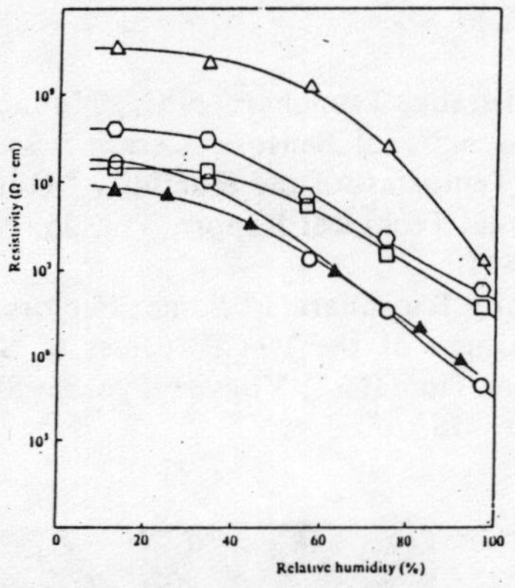

Fig. 6 Influence of doping on humidity sensing property of BST ceramics.

5. The constructional diagram of the temperature–humidity sensor of present work is shown in Fig. 8. The porous ceramic element with porous RuO_2 thick film electrodes is connected to the RuO_2 thick film conductor pattern on the upper alumina substrate. On the lower alumina subustrate, there is a Pd–Ag thick film conducter pattern, and a part of it with more narrow width forms a heating cleaner. Temperature measuring range

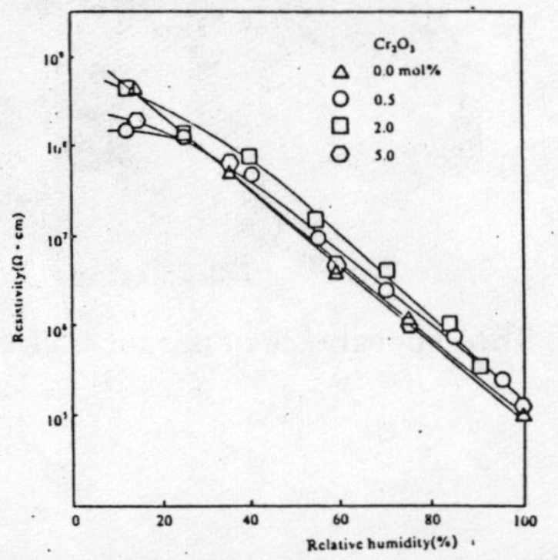

Fig. 7 Influence of doping amount on humidity sensing property of BST ceramics.

of the sensor is −20 (or −40) to +100℃ (10kHz, 1V) and humidity measuring range is 10–100% RH(5Hz, 3V). An example of humidity absorption and deabsorption response property is shown in Fig. 9. The heat cleaning property of the sensor is shown in Fig. 10. Heating temperture is about 400℃ and heating time is 1 min. The time needed from the beginning of heat cleaning to the recovery of sensor resistance (to 1.36 times of its original value)is about 200 sec.

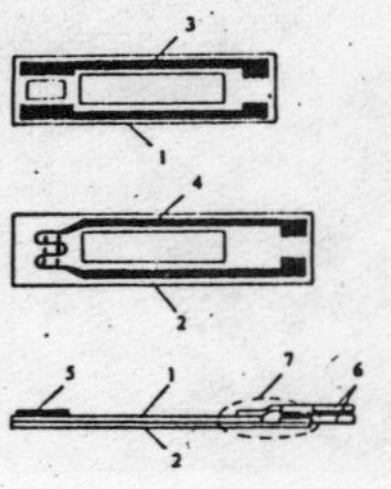

Fig.8 Construction of temperature–humidity sensor. 1—upper substrate(Al_2O_3); 2—lower substrate(Al_2O_3); 3—RuO_2 thick film; 4—Pd–Ag thick film; 5—temp.–hum. sensing element; 6—leads; 7—encapsulation

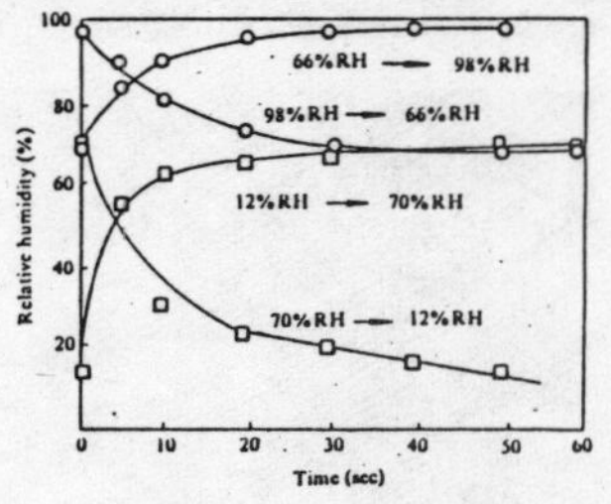

Fig.9 Humidity response of temperature–humidity sensor.

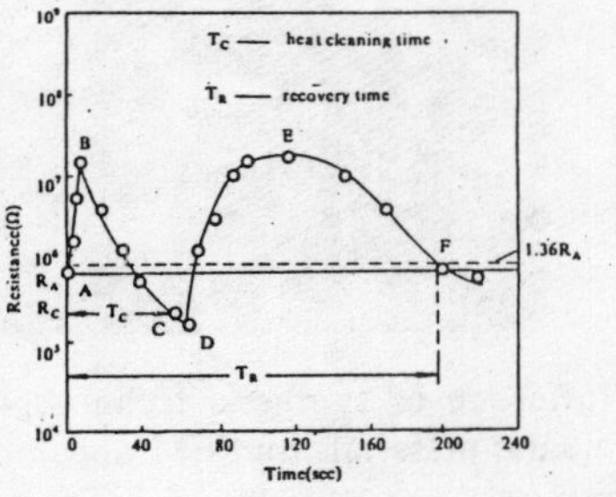

Fig.10 Heat cleaning property of temperature–humidity sensor

CONCLUSION

1. BST temperature–humidity sensing porous ceramics which has good repeatability can be obtained from high–pure, ultra–fine liquid phase coprecipitated barium oxalate and strontium oxalate powders whithout any dopant or whith small amount of dopant, and ordinary ceramic technology (Very similar property has been re-

peated whith industrial high–pure, ultra–fine powder products). The resistivity varies 3 orders of magnitade in whole humidity range. The humidity sensing property is nearly linear and response time is less than 15 sec. The temperature sensing property presents linearity above the temperature which is 30℃ higher than the Curie temperature, and the sensing range can be adjusted by varying the Sr / Ba molar ratio.

2. The construction of temperature–humidity sensor in present work is compact, miniature and solid. With BST porous ceramic element of good sensing property, temperature (–20 or –40 to +100℃) and humidity (10 to 100%RH) can be measured by frequency separation technique respectively.

REFERENCES

1. Jiro Terada, Tsuncharu Nitta, Nobubaru Tsukiji, " Multifunctional Sensor– Ceramic Sensor to Detect Both Temperature and Humidity ' Humicerm–II ' ", National Technical Report, Vol. 26, No. 3 (1980), p. 433–441.
2. Makoto Kuwabara, " Some Factors Affecting the Magnitude of the PTCR Effect in Semiconducting Barium Titanates", Yogyo–Kyokai–Shi, 90(8), 1982, p. 470–475.

a

b

c

d

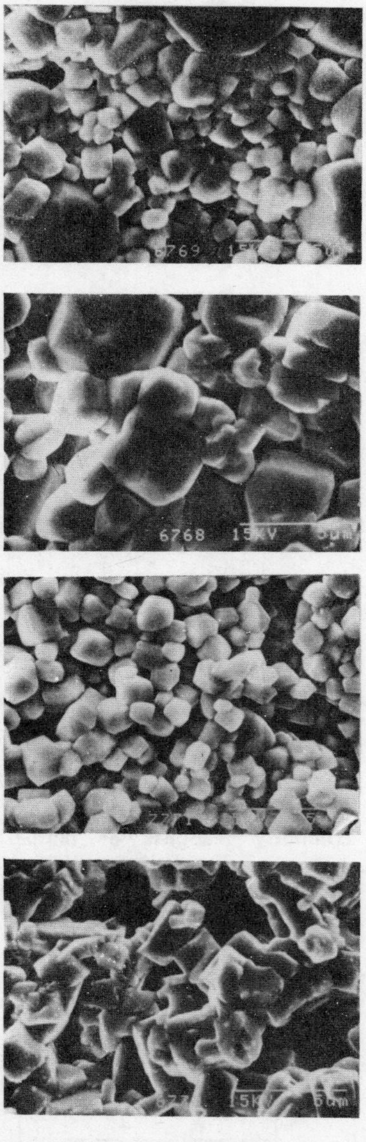

Photo.1 Influence of synthetic technologies of fundamental materials on SEM apperace of BST ceramics.

e

f

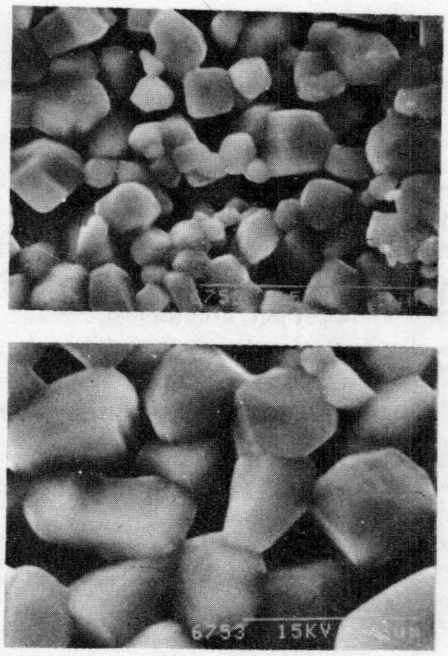

Photo.2 SEM appearance of porous BST ceramics.

A HIGH PERFORMANCE INTEGRATED TEMPERATURE SENSOR

Huang Jin-Biao, Tong Qin-Yi and Lu Yong
Microelectronics Center, South-east University
Nanjing 210018, P.R. China

ABSTRACT

A CMOS integrated temperature sensor including eletro-thermal feedback network is introduced. It is heated and operates at a constant temperature with the help of the thermal feedback on the silicon sensor substrate. Its work is based on the heat transfer between the sensor chip and the ambient, and the heating power maintaining constant chip temperature is used as the measure of the ambient temperature.

INTRODUCTION

In general, integrated temperature sensors operate at such a temperature balanced with the ambient that the behaviour of the signal processing circuits on the sensor chip is greatly affected by the ambient temperature [1]. A novel CMOS integrated temperature sensor described in this paper operates at a constant chip temperature (CCT) which is higher than ambient temperature. The sensor has the advantage of high stability, fast response,and wide range of working temperature, etc. with respect to traditional temperature sensors. Moreover, a lot of signal processing circuits can be integrated on the sensor chip without performance degeneration of these circuits.

The solid line part of Fig. 1 is a typical configuration of traditional integrated temperature sensors, which is a open loop configuration. Its transfer function (Laplace transform form) can be described as follows:

$$F(S)=A/(1+TS) \qquad (1)$$

where A is the static sensitivity (including the gain of the amplifier) of the sensor, T=RtCt, is the time constant, Rt and Ct are thermal resistance (between the sensor chip and ambient) and thermal capacitance of the sensor, respectively.

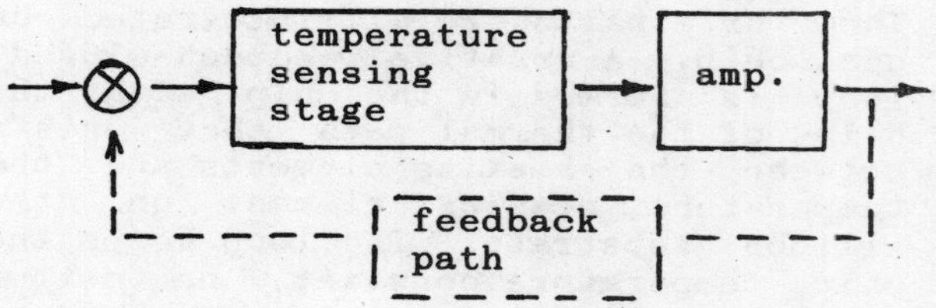

Fig.1 Configurations of open loop (solid line part) and closed loop (the whole) integrated temperature sensor.

It is well known that the great improvement in the performance of an electronical system can be reached when the negative feedback technology is introduced [4]. In fact, the feedback is also attractive in the measuring system. The Fig. 1 including both solid line part and dashed line part is the configuration of the integrated temperature sensor having the feedback, which is a closed-loop configuration. Its transfer function is

$$G(S)=F(S)/(1+F(S)B) \qquad (2)$$

where B is the feedback coefficient of the sensor. Substituting eqn. (1) into eqn. (2), we have

$$G(S)=(A/(1+TS))/(1+AB/(1+TS))$$
$$=(A/(1+AB))/(1+TS/(1+AB))$$

$$= A'/(1+T'S) \qquad (3)$$

where A' is the closed-loop static sensitivity, T' is the closed-loop time constant. It is obvious that $T' \ll T$, i.e., the closed-loop system has much faster response and wider work frequency than the open loop configuration. In particular, if $|AB| \gg 1$, then

$$A'=A/(1+AB) \doteq 1/B \qquad (4)$$

and the sensitivity may be made to depend entirely on the feedback network. Therefore, the improvement in accuracy and stability may indeed be pronounced.

CONFIGURATION AND OPERATION

A CMOS integrated temperature sensor using above feedback technology was designed. It consists of the temperature sensing stage(diode D and constant current source I1) having positive temperature coefficient, a CMOS op-amp used for reverse phase amplification, and a heating branch circuit(Q and R)(as shown in Fig. 2). The three parts are all integrated on one chip. A negative feedback closed-loop is formed in the chip with the help of the thermal path that exists between the heating elements and the temperature sensing element on the silicon substrate. The loop keeps the chip temperature constant, no matter how the ambient temperature changes. This is similar to the TSS (temperature-stabilized substrate) technology used in some low drift circuits. The detail circuit configuration is similar to those of constant-temperature flow and vacuum sensors developed by us [2,3].

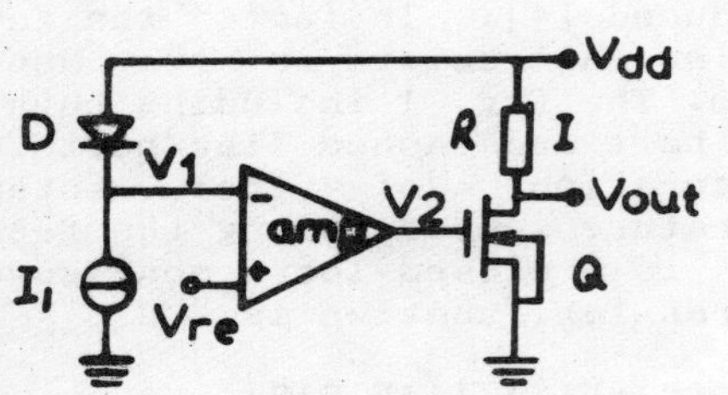

Fig. 2 Circuit configuration of the sensor

When the ambient temperature decreases, the heat transfer between the chip and the ambient will increase and consequently the chip temperature has a decrease trend. As a result of the negative feedback loop, the heating current will be increased. It is expected that if the loop gain is high enough, the chip temperature will recover to the initial value after the steady state is achieved. Therefore, the change of the heating current to maintain constant chip temperature can be used as a measurement of the ambient temperature. The energy balance equation of the sensor can be depicted as follows:

$$IVdd = (Tc-Ta)/Rt \qquad (5)$$

where IVdd is the heating power of the chip (in which I and Vdd are, respectively, the heating current and power supply voltage), Tc is the chip temperature, Ta is the ambient temperature, Rt is the thermal resistance of the heat transfer between the sensor chip and the ambient. When the sensor is placed in a motionless fluid of changeable temperature, the heat transfer is mainly the natural convective and Rt is only the function of temperature Ta. If a certain velocity exists in the fluid, the sensor is also applicable because proper package design(e.g., a hollow package) can make the flow velocity at the chip surface be close to zero. When the sensor is used to the solid body case, the Rt is only the heat conductance thermal resistance and the function of Ta. Therefore, the equation (5) shows that the current I is a function of the ambient temperature Ta if Tc is kept constant.

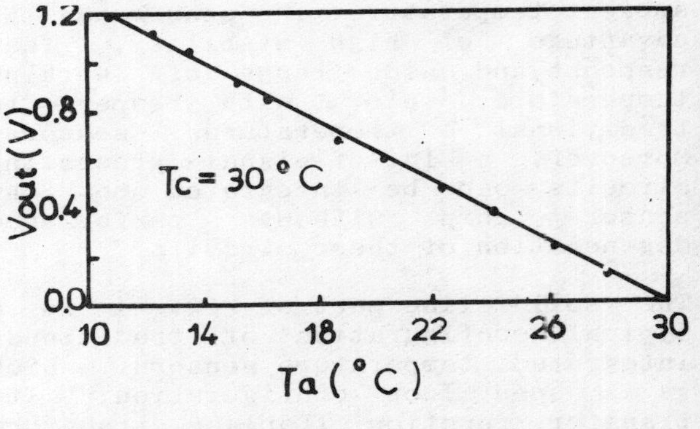

Fig.3 Experimental results

EXPERIMENTS AND CONCLUSIONS

The sensor has been fabricated by a conventional CMOS process and its temperature measurements were carried out between 10°C and 30°C at Tc=30 ° C. The output voltage of the sensor, caused by the heating current across a sampling resistor, shows a temperature sensitivity of about -64 mV/°C.

In conclusion, a feedback-type CMOS integrated temperature sensor superior to traditional integrated temperature sensors has been developed. Because the sensor works at a certain constant temperature, it is suitable in a quite wide temperature range, even the very low temperature which previous temperature sensors and general silicon devices can't work. Moreover, much more signal processing circuits can be integrated on the sensor chip to promote further the sensor performance and the output interface flexibility. Although only lower ambient temperatures than the chip temperature can be measured by the sensor, the measurement of higher ambient temperatures can also be realized by designing a cooling element instead of the heating element on the sensor chip.

ACKNOWLEDGMENT

This work was in part supported by the National Natural Science Foundation (6866009). The authors are grateful for the many fruitful discussions with their colleagues at the Microelectronics Center, and Mr. Xie Yu-Li at Nanjing No.24 Factory.

REFERENCES

1. Hosticka, B.J.: 'Integrated monolithic temperature sensor for acquisition and regulation', Sensors and Actuators, 1984, 6, pp.191-200

2. Tong, Q.-Y., and Huang, J.-B.: 'A novel CMOS flow sensor with constant chip temperature(CCT) operation', Sensors and Actuators, 1987,12,pp.9-21

3 Huang, J.-B., and Tong, Q.-Y.: 'Constant-temperature integrated vacuum sensor', Electron. Lett., 1988, 24, pp.1429-1430

4 Millman, J.: 'Microelectronics',(McGraw-Hill Book Co., New York, 1979)

5 Meijer, G.C.M.: 'Thermal sensors based on transistors', Sensors and Actuators, 10(1986), pp.103-125

COMPOSITE MATERIALS WITH A PTC EFFECT

Wu Baiyuan, Zou Xiang, Liu Xiaoyang and Y.L.Chen
South China University of Technology
Guangzhou, China

ABSTRACT

Vanadium oxides fillers were incorporated into polymer matrix to produce composites that exhibit very low room temperature resistivity and large PTC effect (up to 10 orders of magnitude). The state of carbon was found to be changed in the interface between the two phases. To explain the PTC effect a three dimensional network model and electron tunneling was proposed.

INTRODUCTION

Composites consisting of transition metal oxide with polymer has been developed since 1973.[1] K.A. Hu indicated that 50 volume percent TiO-epoxy composite provides excellent thermistor properties: low room temperature resistivity (about 1-5 Ω.cm) and large PTC effect (about 10 orders of magnitude).[2] The resistivity vs temperature curve for V_2O_3-polymer resembles a square well with three regines in which the composite behaves like a semiconductor, a metal and an insulator. To understand the PTC effect of these composites, the concept of percolation was proposed.[3]

In this paper we report our work on composites with vanadium oxides and various polymers. Composite exhibiting excellent thermistor properties was prepared and a new PTC mechanism with three dimensional network model and electron tunneling was proposed.

EXPERIMENTAL

V_2O_3 and V_2O_5 were used as conductive fillers in this study. Three commercial polymer were used: high density polyethylene (HDPE), low density polyethylene (LDPE) and polypropylene (PP). Composites was fabricated by roll-milling filler and polymer at suitable temperature for 10 min. Ga-In solution electrodes were used. The D.C. resistivity of the composites was measured as a function of temperature (with 1°c/min heating and colding speed) with a microammeter and a low resistivity box bridge. All concentrations stated in this paper refer to volume percent of vanadium oxides.

RESULTS AND DISCUSSIONS

The resistivity vs temperature curves for several V_2O_3 composites with HDPE are shown in Fig.1.

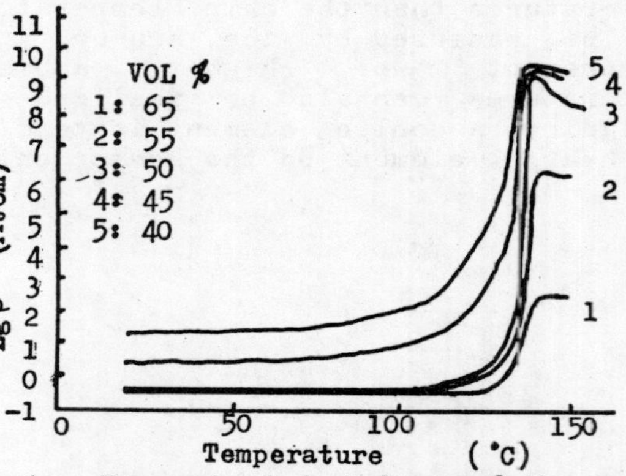

Fig.1. Temperature dependence of the resistivity of V_2O_3-HDPE composites prepared with different volume concentration of V_2O_3.

For the 50% V_2O_3 composite, room temperature resistivity is very low (only about 0.4 Ω.cm) and almost constant in the temperature range of 20 to 120°C. Near the peak melting point of polyethylene (Tm=132°C), resistivity increase sharply by approximately 10 orders of magnitude. As illustrated in Fig.2, increasing V_2O_3 room temperature resistivity decrease rapidly and the hight of PTC effect increase when concentration is lower than 50%, and there is a little decreace in room temperature resistivity but great decreace of PTC hight when concentration is great than 50%. Similar behaviou was observed for V_2O_5-HDPE composite with much greater resistivity at room temperature and large PTC anomaly (about 5 orders of magnitude) (Fig.3). The NTC effect below 120°C is due to the resistivity decreace of V_2O_5 crystal.[4,5] Compared with V_2O_3-HDPE, V_2O_3-PP composite also displays low resistivity at room temperature and strong PTC effect (about 8 orders of magnitude) over a wide range of about 80°C which presumably due to wide melting range (127-173°C) of PP.

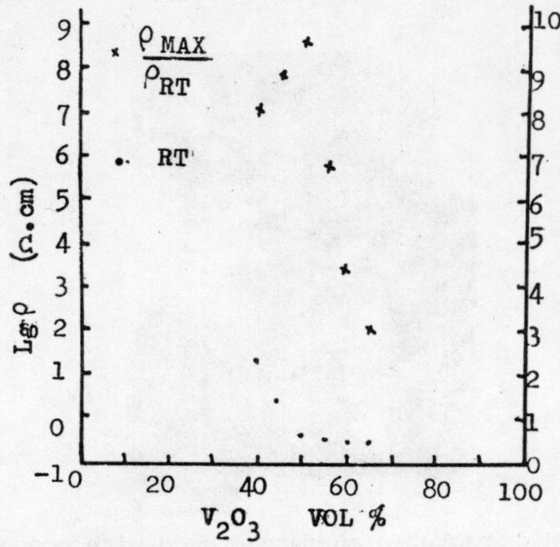

Fig.2. Room temperature resistivity (ρ_{RT}) and hight of PTC effect (ρ_{MAX}/ρ_{RT}) of V_2O_3-HDPE composites with various concentrations of V_2O_3.

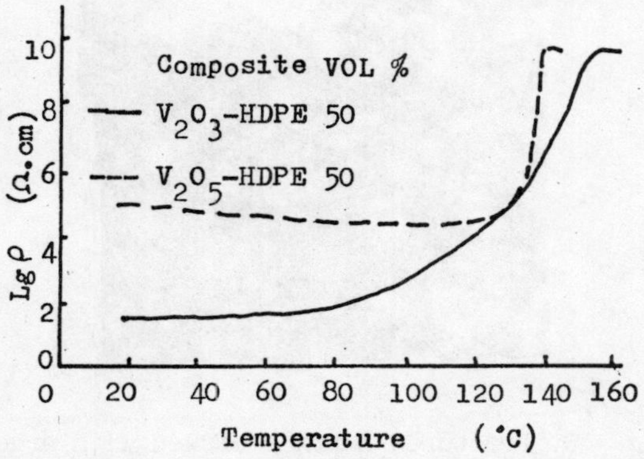

Fig.3. Temperature dependence of resistivity of composites.

The effect of polymer crystallinity and sizes of V_2O_3 particles are shown in Fig.4. Low crystallinity polymer (LDPE) made the room temperature resistivity much greater compared with the high one (HDPE). The PTC effect was somewhat diminished when the size of particles was smaller.

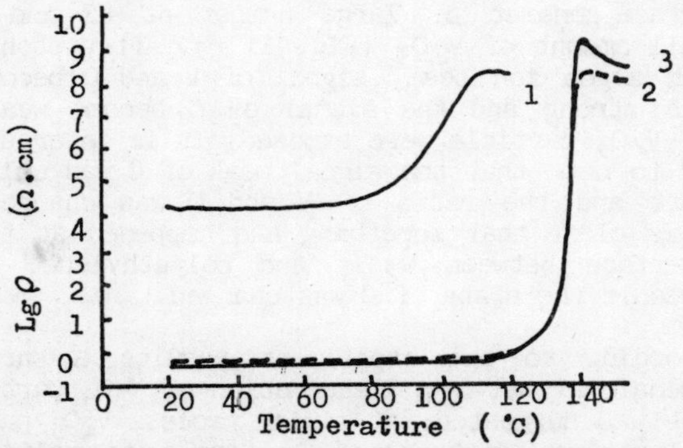

Fig.4. Effect of polymer crystallinity and sizes of V_2O_3 particles on PTC curve.
1: V_2O_3-LDPE; 2: V_2O_3-HDPE with narrow distribution of V_2O_3 particles sizes; 3: V_2O_3-HDPE with wide distribution of sizes.

It is interesting that the resistivity of these composites with 50% vanadium oxides decrease and tend towards the order of about 10 .cm above the PTC effect.

1. Reproducibility

Fig.5 shows reproducibility of the PTC effect in a V_2O_3-HDPE composite containing 50% V_2O_3.

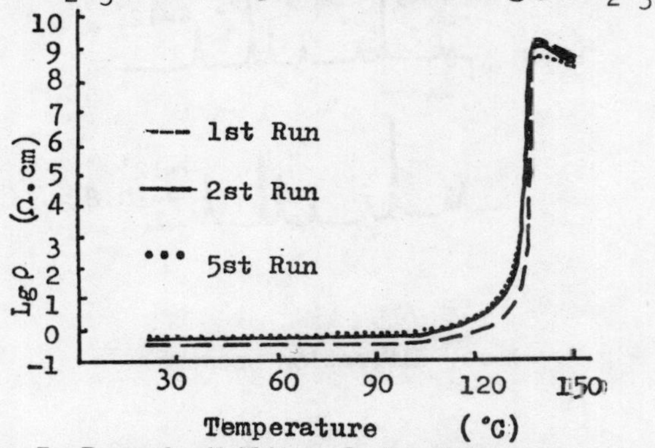

Fig.5. Reproducibility of the PTC anomaly in a V_2O_3-HDPE composite containing 50% V_2O_3.

Compared curve 1 and 3, the room temperature resistivity increase by a small amount (from 0.4 to 0.6Ω.cm) from the first run to the fifth and the hight of PTC effect was slightly diminished. The position of the PTC phenomenon moves to slightly lower temperature upon cycling

2. I—V behaviour

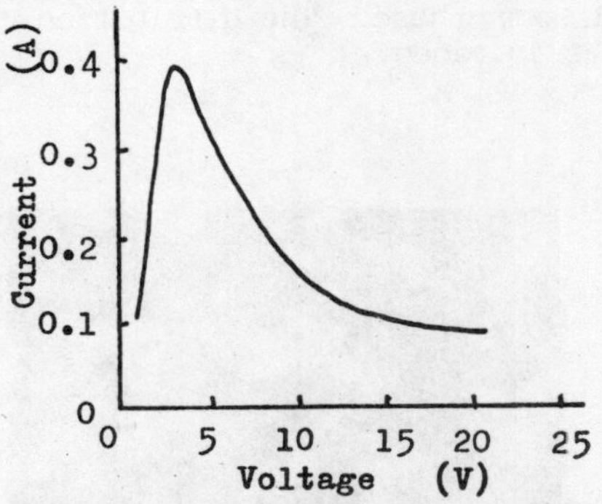

Fig.6. Current-voltage relationship for V_2O_3-HDPE composite with 50% V_2O_3.

As illustrated in Fig. 6, strong nonliner relationship between current and voltage indicates switching PTC effect in V_2O_3-HDPE composite.

3. Microstructure

X ray diffration spectra for V_2O_3, HDPE and V_2O_3-HDPE composite were given in Fig.7.

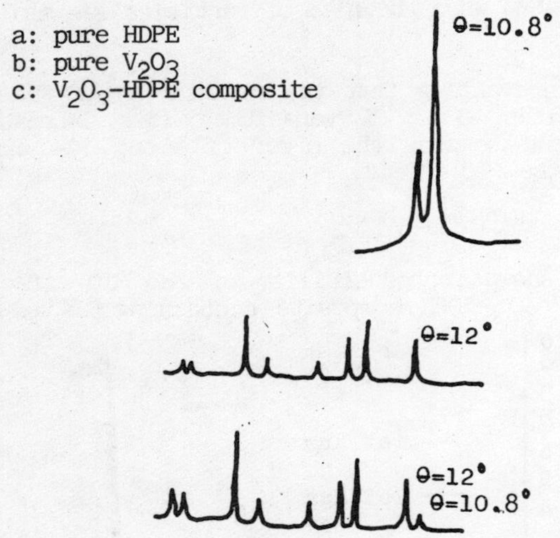

Fig.7. X ray diffration spectra.

It is apparent that there is crystalline PE in composite, as evidence by the diffracting peak of 10.8°.

Fig. 8 to Fig. 10 illustrate the microstruction of V_2O_3-HDPE composite observed under an electronmicroscope: fractured surface untreated, fractured surface etched with concentrated nitric acid and argon ion beam. It is clear that PE exist in composite with crystalline (like film, tree, ect) and amorphous form, and form a continuous phase. The ditribution of V_2O_3 particles is random.

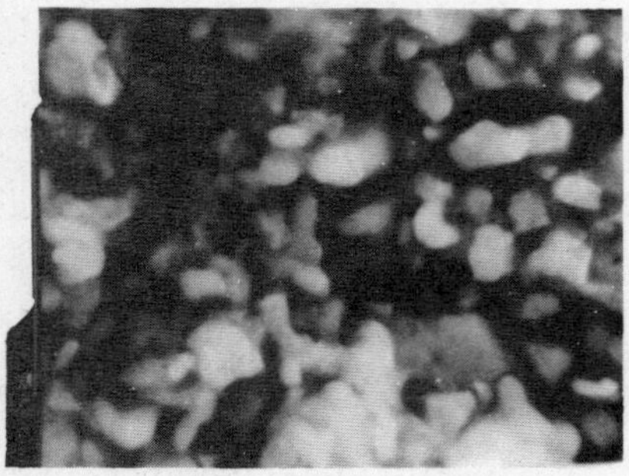

Fig. 8 Fractured surface untreated of V_2O_3-HDPE composite.

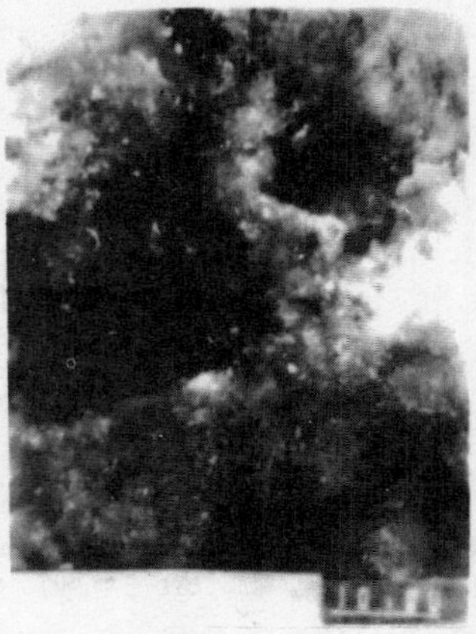

Fig. 9. fractured surface etched with concentrated nitric acid.

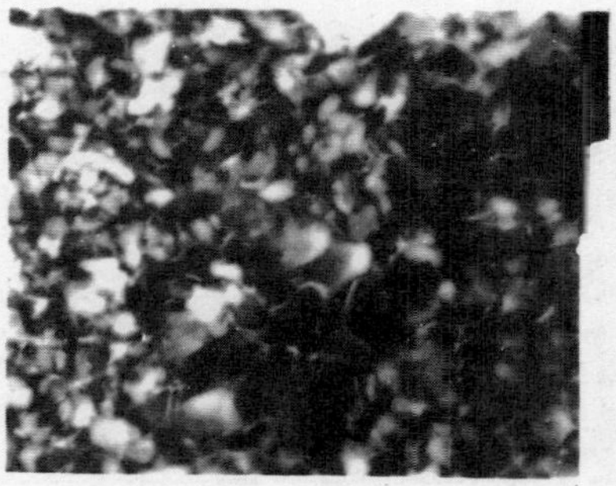

Fig. 10. fractured surface etched with argon ion beam.

The auger electron spectroscopy of pure V_2O_3, pure HDPE and V_2O_3-HDPE composite are shown in Fig. 11. The strong signal of carbon and the weak one of vanadium indicate that fractured surface consist of large number of PE and a small amount of V_2O_3 (Fig. 11 c). After etched with argon ion beam, signal of V and O become much strong and the signal of C become weak, the V_2O_3 particle were exposed. It is interesting to note that the signal peak of C was split apart and the ratio of V and O was changed. It is clear that something had happened at the interface between V_2O_3 and polyethylene, as a result the state of C was changed.

According to the experiment results a three dimensional network distribution of V_2O_3 particles was suggested: at mixing process V_2O_3 particles were distributed uniformly in melting PE, then were swept into amorphous forming a three dimensional network because of polymer

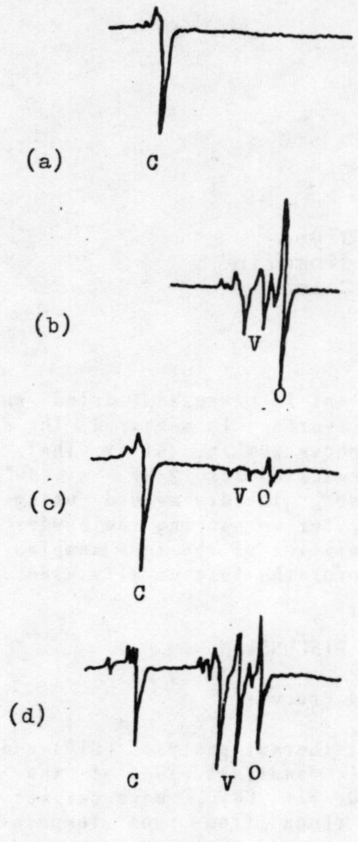

Fig. 11. Auger electron spectroscopy.
a: polyethylene; b: V_2O_3; c: composite and d: composite etched with argon ion beam.

crystalline at colding. There is an unknown interface layer between V_2O_3 particle and PE. If composite was heated to polymer melting temperature, the distribution of V_2O_3 filler would become comparatively uniform again. This view is supported by the same result in carbon black filled polymer composite.[6] Using electron tunnel[7] theory a PTC mechanism is given. At low temperature the minimum intergrap gap between V_2O_3 particles is small enough to allow sinificant tunneling to occur, resulting low resistivity. Near polymer melting piont the ditribution of V_2O_3 particles become uniform and there is no more three dimensional network, so that the minimum gap width become too large to allow sizable electron tunneling to occur with a resultant great rise in resistivity, though the average gap width is not greatly changed. The NTC effect follows the PTC may be caused by the two sides: at first the resistivity of polymer decrease with temperature rise, and the other hand though the minimum gap width become large the maximum gap width is smaller for more electron tunneling to occur.

According to this mechanism the crystallinity of polymer and the size of conductive particles would affect thermistor properties by affecting the distribution of particles and the intergrap gaps. If the polymer melt completely, composite with the same conductive filler concentration and various polymer would display resistivity of the same order of magnitude because of uniform distribution of particles. It is corresponding with experiment results (see Fig. 3 and Fig. 4).

CONCLUTIONS

A. Composites consisting of vanadium oxides and polymer exhibit excellent thermistor properties: very low resistivity at room temperature and large PTC effect.
B. Various PTC materials can be prepared by adjusting composition of composite.
C. A new PTC mechanism with three dimensional network model and electron tunnel theory was proposed.

REFERENCES

1. F. Buech, J. Appl. Phys., 54(1973), 532.
2. K. A. Hu, J. Runt, A. Safari, and R. Newnham, Phase, Trans., 7(1986), 1.
3. K. A. Hu, Dawns Moffatt, J. Runt, A. Safari, and R. Newnham. J. Am. Ceram. Soc., 70(1987), 583.
4. T. Allersma, R. Hakim, T. N. Kennedy, and J. D. Mackenzie, J. Cham. Phys., 76(1967), 154.
5. J. Bullot, O. Gallais, M. Gauthier and J. Livage, Appl. Phys. Lett., 36(1980), 986.
6. J. Meyer, Polym. Eng. Sci., 14(1974), 706.
7. K. Ohe and Y. Natto, J. Appl. Phys., 10(1971), 99.

CERAMIC THERMISTORS OF THE CdO-SnO_2-WO_3 SYSTEM

Xu Tinxian, Liang Hui and Shen Jiyao
Dept. of Materials Science and Engineering
Tianjin University, Tianjin, China

ABSTRACT

Sintering, microstructure and electrical conductibility of the CdO-SnO_2-WO_3 system have been studied. It shows that vapor-phase-material transfer process is existed throughout the sintering, and the liquid phase can improve the sintering characteristics. The system contains two kinds of the crystal phases. It is found that they are $CdSnO_3$ and $CdWO_4$. Measurements with carefully prepared samples over a wide temperature range (from -50 to 200°C) show that the relation between the resistance and temperature is linear.

INTRODUCTION

It is known that the electrical resistivity of most NTC thermistors is exponential function with respect to temperature. Although their sensitivity is high, they cannot be used for accurate measurement. Therefore, linear thermistors have received more attention in recent years as they can be used in digital instrument together with microcomputers. One of such materials for mamufacture of this kind of thermistors is so called CdO-SnO_2-WO_3 system ceramics. But the material is so difficult to sintering that it is almost impossible to its production and application. So, in this paper, the study of the mechanisms of sintering and electric conduction of such material is reported.

EXPERIMENTAL METHODS

The powders of $CdSnO_3$ and $CdWO_4$ were prepared by means of chemical coprecipitation process so as to lower the sintering temperature of the system CdO-SnO_2-WO_3. The wet method was chosen to produce the samples used in this research because it offers better control of stoichiometric ratios than the dry method, especially with those compounds involving volatile oxides, such as CdO. In the wet method, $CdSnO_3$ and $CdWO_4$ are obtained from the solution of $Cd(NO_3)_2$, Na_2SnO_3 and K_2WO_4 according to the reaction:

$$Cd(NO_3)_2 + Na_2SnO_3 \rightarrow CdSnO_3 \downarrow + 2NaNO_3$$
$$Cd(NO_3)_2 + K_2WO_4 \rightarrow CdWO_4 \downarrow + 2KNO_3$$

The precipitant is cleansed, dried and then its specific surface is measured. The specific surface is above $20 m^2/g$. After that the material is calcined for 2 hr. at 850°C. Also for comparison, the dry method using $CdCO_3$, SnO_2 and WO_3 for preparing the powder is utilized. Preparation of the test samples is carried out according to generally used ceramic techniques.

RESULTS AND DISCUSSION

1. Sintering Process

Differential thermal analyses (DTA) and thermogravimetric analyses (TGA) of the powders mixed $CdSnO_3$ with $CdWO_4$ were carried out to cover the range from room temperature to 1200°C. The thermal analysis curves are shown in Fig.1. Figure 1 (a) is the thermal analysis curves for the powder of the wet method, it shows a broad exothermic peak at 975°C; (b) is that of the dry method, it shows a exothermic peak at 1080°C, that is, the temperature of the dry method is about 100°C higher than that of the wet method. Meanwhile, the TG curves show that the slight weight loss started 1070°C and the heavy weight loss is at 1200°C. Obviously, this is caused by the volatility of CdO at high temperature.

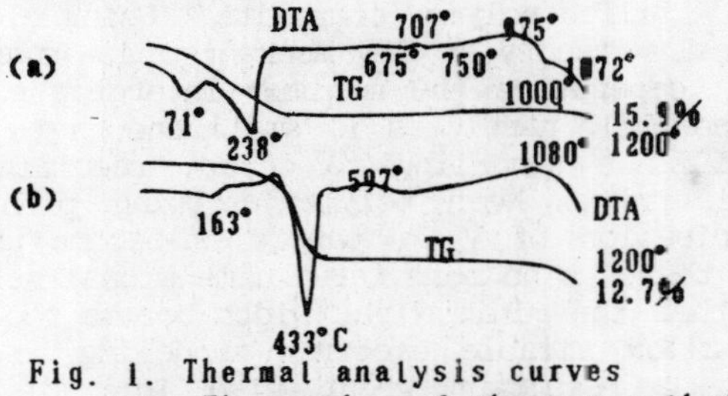

Fig. 1. Thermal analysis curves
(a). The powder of the wet method
(b). The powder of the dry method

The X-ray diffraction curve is shown in Fig.2, for the sintered samples for 2 hr. at 1100°C, using the powder of the wet method. It is similar to that for 2hr. at 1180°C, using the powder of the dry method. It is enough to prove that one of the reasons of the difficult-

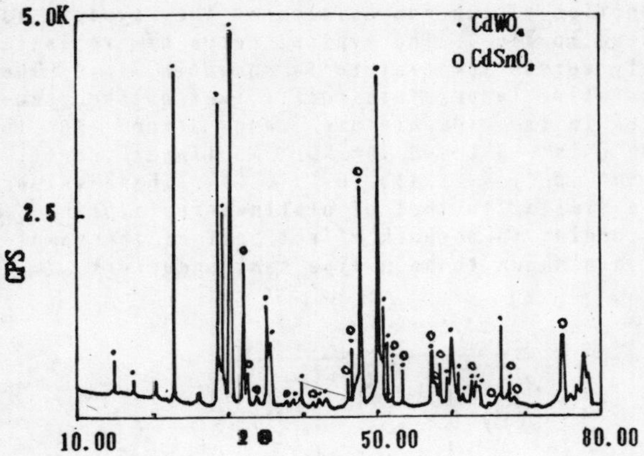

Fig. 2. X-ray diffraction curves

ty to make sintering for the dry method is that a great deal of CdO is vaporized at its sintering temperature (1180°C). As contrast with that, the samples of the wet method are sintered at lower temperature (1100°C), so the vaporization of CdO is prevented. The sintering is then much easier.

As it is seen from X-ray diffraction analyses, two kinds of the crystal phases are existed in the sinters of the CdO-SnO_2-WO_3 system. One of them is $CdWO_4$ which is classified as monoclinic crystal system. The lattice constants for $CdWO_4$ is found to be a=5.01545, b=5.83984, c=5.06577Å, β=91.28°. Another is $CdSnO_3$ which is orthgrhombic with a=5.45325, b=7.86044, c=5.57168Å. Fig.3 shows SEM patterns from surface of samples. The two kinds of the crystal phases are completely different from each other in the crystalline behaviours. The grains of $CdWO_4$ are column, 10-30 um long and 3-10 um wide, and $CdSnO_3$ is spheroid, its diameter is 1-6 μm.

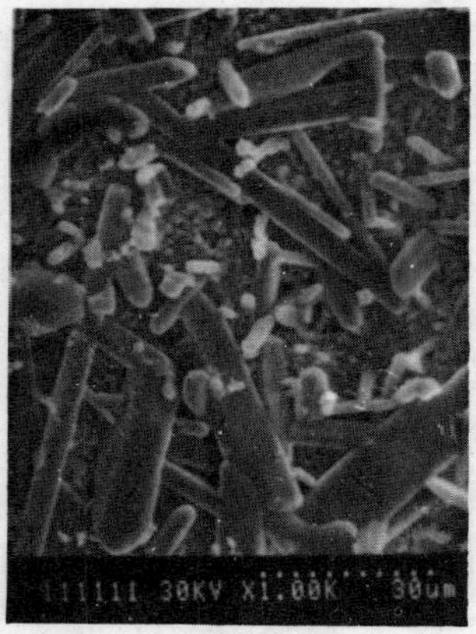

Fig.3. SEM pattern
(The wet method, sintered at 1100°C)

As it is seen from Fig.4 and Table 1 that, the solid solubility of the two types of the crystal phases from each other is smaller. Sn

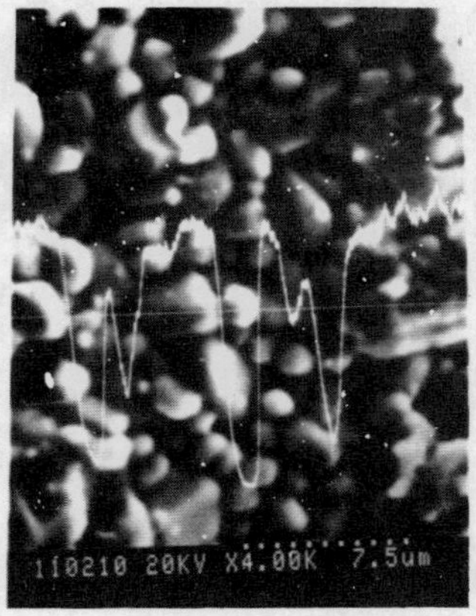

Fig.4. Linear scanning electron spectrum for W atom

Table 1. Data of electron spectra

Samples, No.		W-0#			D-6#		
Dimension		%wt	%AT	%S.E.	%wt	%AT	%S.E.
Smaller grains	Cd	50.59	52.43	0.62	49.02	52.01	1.00
	Sn	46.72	45.86	0.78	41.86	42.07	1.32
	W	2.69	1.70	10.22	9.12	5.92	6.00
Larger grains	Cd	36.48	48.44	1.01	42.68	54.91	1.17
	Sn	0.00	0.00	0.00	0.00	0.00	0.00
	W	63.52	51.56	1.19	57.32	45.09	1.62

atoms aren't found in $CdWO_4$. The solid solubility of the atoms W in $CdSnO_3$ is below 1-6%, therefore the molecular formula of $CdSnO_3$ should be

$$Cd(Sn_{1-x}W_{2x/3})O_3$$

where X=0-0.09.

From the facts mentioned above, the longer and larger grains of $CdWO_4$ set up a "shelf". Some empty holes of this "shalf" are filled with smaller grains of $CdSnO_3$, but they aren't full, that is, a lot of the holes are still empty. So there remain a great number of the pores. This phenomenon may be called "shelf" effect. this is another important reason as regards the difficulty of the sintering of the system.

The influence of the "shelf" effect on the sintering can be improved as the proportions of $CdSnO_3$ and $CdWO_4$ is varied. Fig.5 shows the open porosity of the sinters as function of the mole rates $CdSnO_3$ and $CdWO_4$. The porosity is the smallest when the rates is 1,1.

The influence of the "shelf" effect on the sintering can be lessened when some liquid phases are introduced. Fig.6 is the SEM photo of the quenched sample. The sample is doped 1%wt dopants to produce some liquid phases at high temperature. Obviously, From Figure 6 we see that there are a small amount of liquid

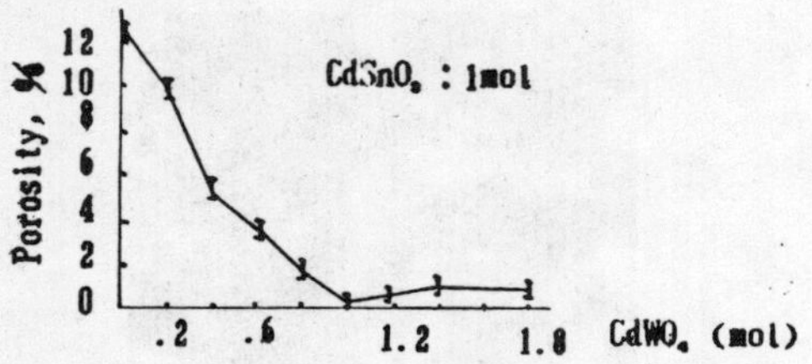

Fig.5 The porosity of the sinters as a function of incorporated $CdWO_4$, $CdSnO_3$=1mol.

phases around the grains, at the same time, there is about 11% linear contraction in the sample, that is, there are liquid phases cer-

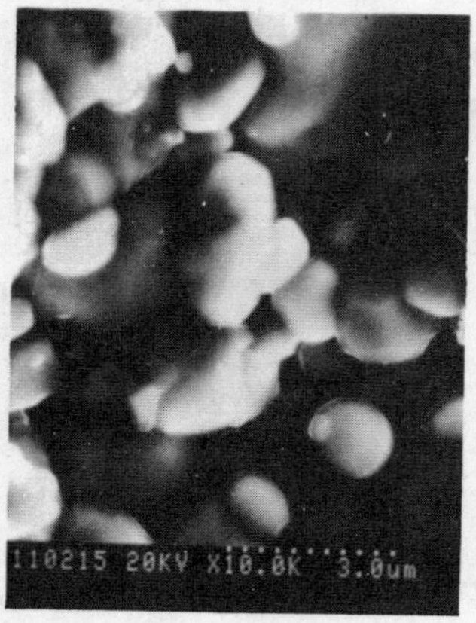

Fig.6 SEM photo of the quenched sample.

tainly during the process of the sintering. In addition, theions of W^{6+} are soluted in $CdSnO_3$, these ions replace Sn^{4+} and form the vacancies of Sn^{4+} or Cd^{2+}, also, it is of great advantage to the diffusion of the crystal lattices. Hence, it is an advancement to the sintering of the system.

2. Elcetrical Conductibility

The measuring circuits for electrical resistivity and resistance—temperature (R-T) curves of the samples are shown in Fig.7. The samples are introduced into the tester of controlled temperature from -100 to 300°C. The resistance of the samples is calculated by

$$R_x = \frac{R_1}{\frac{U_0}{U_x} - 1}$$

here the meanings of these signs are illustrated by Fig.7. Resistivity of the samples is Calculated according to the geometrical sizes of samples and the known formula. Resistivity of the samples at room temperature (25°C) as a function of the composition of the system is shown in Fig.8. The typical curve of resistivity versus temperature is shown in Fig.9. The resistive temperature coefficient of the samples in the temperature range from -50 to 200°C is (-0.5~-0.7)%/°C. Nonlinear coefficient is $(1.5\sim5.5)\times10^{-5}/(°C)^2$. These values are similar to that of platinum resistor. According to Seebeck effect, all of the samples are shown to be n-type semiconductors.

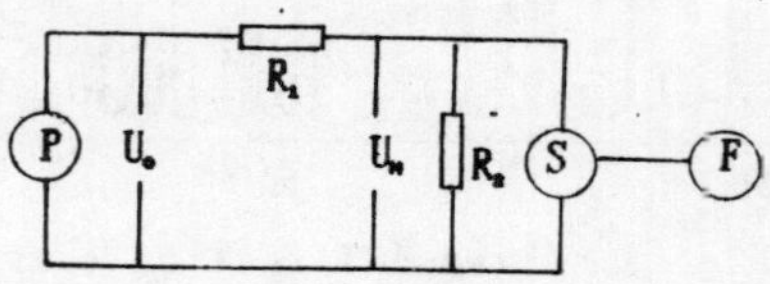

Fig.7. Measuring circuit, P-DC source, S-electrostatic voltmeter, F—X—Y recorder.

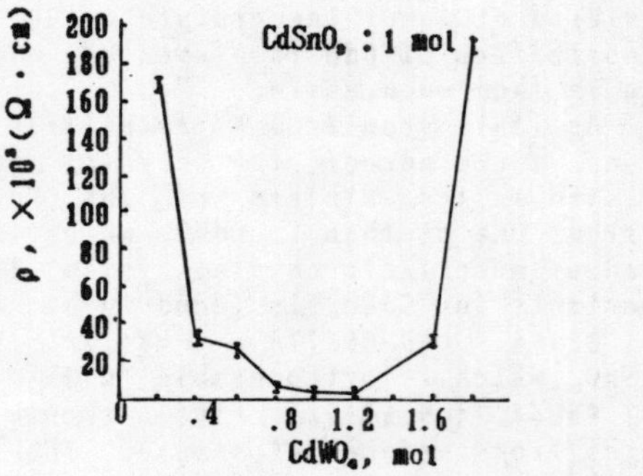

Fig.8 Resistivity at room temperature as a function of the composition of the system.

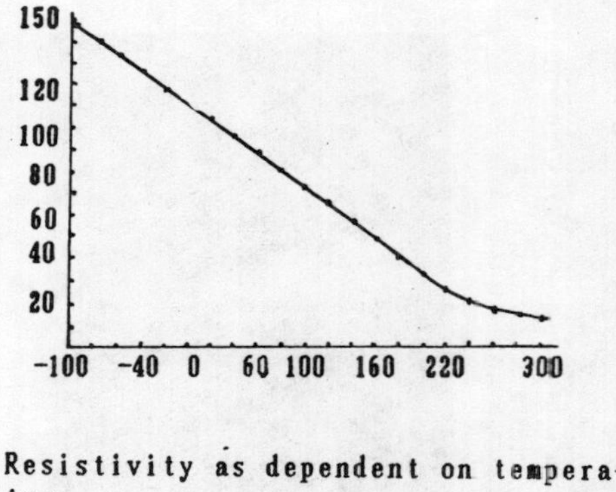

Fig.9 Resistivity as dependent on temperature.

The samples are treated in O_2, N_2 or air for 10 hr. at 300°C, their resistances at room temperature are then measured and the data are shown in Table 2. The resistances before and after heat treatment are increased. Let T represents the resisitance ratio after and before the treatment, i.e. T=R(after)/R(before), we find

$T(\text{in } O_2) > T(\text{in air}) > T(\text{in } N_2)$.

Hence, it is supposed that the oxygen ion vacancies ($V_\bullet$) can cause a quasi-free electrons (e′) in the conduction band having a negative charge. This is given by the following quasi-chemical reactions,

$$O_\bullet \rightleftharpoons V_\bullet + \frac{1}{2} O_2 \uparrow$$
$$V_\bullet \rightleftharpoons V_\bullet^{\cdot} + e'$$
$$V_\bullet^{\cdot} \rightleftharpoons V_\bullet^{\cdot\cdot} + e'$$

Table.2 Resistance at room temperature before and after heat treatment(300°C, 10hr.)

conditions	before heat treatment	after heat treatment in O_2	in air	in N_2
R (25°C)	58Ω	160Ω	103Ω	94Ω
R(after)/R(before)	1	2.759	1.776	1.621

The reactions can reach the state of thermodynamic equilibrium under the condition of certain temperature and partial pressure of oxygen. If the O_2 partial pressure is increased, the equilibrium moves towards the left-hand-side of the reactions. The concentrations of the defects are then lessened and the resistance values of the samples are increased. The samples are processed by heat treatment at 200°C. During this process, the resistance values of the samples become lower rapidly at the begining, and then change slowly and stabilize gradually, as shown in Fig.10. The curve isn't related to the atmospheres, such as O_2, N_2 or air of heat treatment obviously. Hence, the stabilized samples may be obtained by using this method.

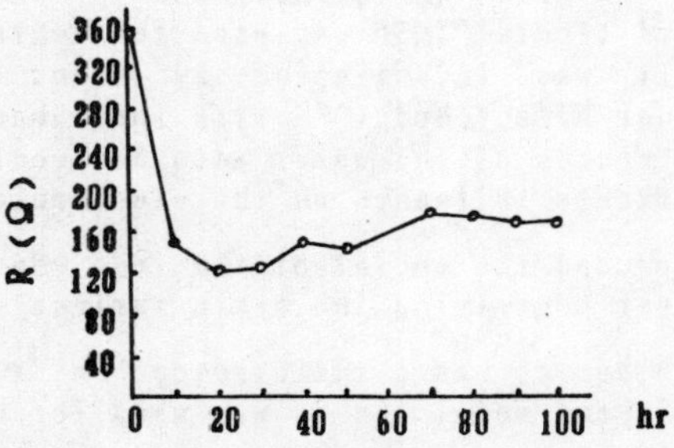

Fig.10 Heat treatment curve at 200°C

Fig.11 is Mössbauer spectroscopy of heat treated samples at room temperature. Radioactive source is ^{119}Sn. The velocity scale is calibrated by α-Sn. The standard isomer shift (IS) of Sn^{+} ion is -1.998 ± 0.0015 mm/sec, measuring value is -2.067 mm/sec. It is proved that the maximun is a special peak of stable Sn^{4+}. The outer electron configurations of Sn^{4+} is $4S^2 4P^6 4d^{10}$. According to energy band structure of $CdSnO_3$ of perovskite type, the valence electrons of Sn^{4+} ion should fill σ^* (eg) band. σ^* band becomes the conduction band[1]. Since σ^* band is wider than σ^* (eg) band, $CdSnO_3$ should be a broad-band semiconductor[2]. Large-polaron conduction may be expected in the compound[3]. For this reason we have carried out the calculation as follows. The conductivity (σ) of semiconductor material can be calculated by

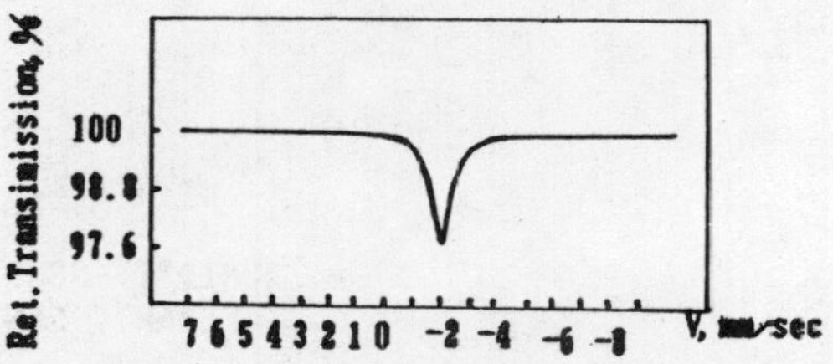

Fig.11 Mössbauer spectrum of the system $CdO\text{-}SnO_2\text{-}WO_3$ at room temperature.

$$\sigma = AT^{C} \exp\left(-\frac{\Delta E}{RT}\right) \qquad (1)$$

Where ΔE is effective activation energy of carrier, C is the value related to mobility, A is a constant. Generally, $C \approx 0$, $\Delta E \approx 0.5 \sim 2.5$ e.v., the conductivity of NTC thermistors is therefore an exponential function of temperature. For the semiconductor of $CdO\text{-}SnO_2\text{-}WO_3$, Table 3 shows A, C and ΔE of this material. ΔE is $10^{-2} \sim 10^{-1}$ e.v. in the range of $-100 \sim 300$ °C and it isn't a constant. In this case, ΔE can not be assumed to be an ionization energy or donor level. It should be the activation energy of a large polaron. So we can conclude that the R-T linearity of this semiconductor is caused by the effective movement of the large polaron under control of a point defect.

Table 3 Data of A, C and ΔE in equation(1)

Temperature range(°C)	A	C	ΔE(ev)
$-100 \sim 0$	7.859×10^{-4}	-1/2	0.0165
$0 \sim 100$	3.058×10^{-3}	-1/2	0.0437
$100 \sim 200$	2.372×10^{-2}	-1/2	0.100
$200 \sim 300$	2.228×10^{-1}	-1/2	0.179

In summary, the conductive mechanism of the material is the result of an effective movement of a large polaron. The feature of the material is that the resistance vs. temperature is linear over a wide temperature region (from -50 to 200°C)

REFERENCES

[1] J.B. Goodenough, Prog. Solid State Chem., 5, (1971), 145.

[2] J.M.Honig, Electrodes of Conductive Metallic Oxides, Part A, Elesvier, (1980), 1~96

[3] Z.M.Jarzebski, Oxide Semiconductors, Pergamon press, (1973), 130~144.

HUMIDITY-SENSITIVE MECHANISM FOR SEMICONDUCTIVE CERAMIC SENSOR OF PEROVSKITE TYPE STRUCTURE

Zhao Gang
Jiangsu Radio Science Research Institute, P.O. Box 110, Wuxi, China
Zhou Zhi-gang
Dept. of Material Science And Eng., Tsinghua University, Beijing, China

ABSTRACT

In the paper, the humidity-sensitive mechanism of the semiconductive ceramic sensor BTS of ABO_3-type structure has been investigated in detail by means of the "Nuclear Magnetic Resonance Effect" and the "Mossbauer Effect". The results indicate that there is nearly no "Positron-conductive Mechanism" range from low humidity to middle one at least, and the "Electron-conductive Mechanism" dominates the humidity-sensitive mechanism in the whole humidity.

INTRODUCTION

At present, various new - type ceramic sensors have been applied more and more widely in many fields. But there are a lot of problems needing to be solved such as uniformity, reproducibility and aging resistance of sensor's properties; as well as linearity, hysteresis of humidity-resistance relation for humidity sensors and so on, which promote further deep investigation of sensors for sensitive mechanism such as the humidity-sensitive mechanism.

In the past, many theories such as "Positron-conductive Mechanism", "Electron-conductive Mechanism" and so on, had proposed in order to explain sensitive behaviors of humidity sensors. But it was only qualitative. So far, no informiation for directly experimental identification on these theories has been reported.[1]

The humidity - sensitive mechanism has been researched in detail in this paper, and which would try to be on the basis of experiments to establish some proper theories quantitatively describing the humidity-sensitive behaviors, to seek after the associated mathematical model making the theory conform with the practice.

EXPERIMENTAL METHOD [2],[3]

A new-type multifunctional sensitive ceramic $BaTiO_3$-$BaSnO_3$ (for short BTS), a P-type metal oxide porous ceramic semiconductor of ABO_3-type structure was used as a specimen studied, which is capable of detecting temperature-humidity-gas with a single sensing element. The standard humidity source was created by saturated salt solution range from 12%RH to 92%RH. The type of semiconduction for the ceramic sensor BTS was determined by the characteristics of conductivity as a function of oxygen pressure at high temperature. The resistance and capacitance of sensor BTS were measured by Model B905 automatic precision bridge instrument under the conditions of testing voltage 3 (V), frequency 1 kHz, temperature 25℃ and pressure 1 atm.

On the other hand, in process of sensing humidity for the sensor BTS range from humidity 12%RH to 92%RH, the electronic situation around the hydrogen nuclei in the water (H_2O) adsorbed on the grain surface was researched by means of the Nuclear Magnetic Resonance (for short NMR) effect (Model FX-90 instrument), where the standard liquid "TMS" was used for contrast. And then it was investigated by means of the Mossbauer Effect(Model FM-1918 instrument) that the molecules of the water adsorbed could exert a remarkable influence on the electronic situation around the nuclei of tin ions (Sn^{4+}) in the layer underlying the grain surface, where radiation source was a radioisotope Sn^{119} (intensity 10mC) and metal tin Sn was used for contrast.

RESULTS AND DISCUSSIONS [4]

1. NMR Effect on Hydrogens For Adsorbed Water

Theoretic analysis shows that the situations of the Energy levels and the electrons around hydrogen nuclei are different under the different chemical environments, which are determined by the ratio of the NMR frequencies for the hydrogen nuclei in the specimen and the standard liquid TMS, named as the chemical displacement δ However, the larger is the average value of equivalent electron density $|\psi(H)|^2$ around the hydrogen nuclei, the greater is the shield effect(corresponding to shield coefficient) on the hydrogen nuclei by electrons, i.e., the less is the value δ. The values δ of the hydrogen nuclei in the water adsorbed on the BTS grain's surface and the solutions of various PH values, which are prepared by sulphuric acid (H_2SO_4), were measured under the same conditions. According to the theory, it is known that the function of displacement δ vs. the PH value is linear. So the displacement δ_{PH7} was obtained mathematically by the least-square fit when the PH value was equal to seven. The experimental results of the NMR effect are shown in Table 1.,

Table 1. Parameters of NMR Spectrum for sensor

Test No.	PH Value	Humidity (%RH)	Displacement Sign	δ Value
in the acid solution				
1	1.5	—	δ 1	4.814
2	3.0	—	δ 2	4.804
3	5.0	—	δ 3	4.799
4	6.4	—	δ 4	4.794
5	7.0	—	δ_{PH7} a)	4.791
in the water adsorbed				
6	—	12	$\delta_{12\%RH}$	4.083
7	—	92	$\delta_{92\%RH}$	4.367

a) The value be estimated at least-square fit as PH = 7.0

which indicate clearly that the displacement $\delta_{12\%RH}$ and $\delta_{92\%RH}$ are quite less than the displacement δ_{PH7}. That is to say, in the range of low humidity or even middle one at least, the average value of the equivalent electron density $|\psi(H)|^2_{ADS}$ around the hydrogen nuclei in the water adsorbed is larger than the average value $|\psi(H)|^2_{SOL}$ of one in the acid solution, whose electronic situations, in other words, would hardly change toward one around nuclei of positron H^+, conversely, but one around the nuclei of hydrogen atoms H.

So it was concluded that the water molecules adsorbed reacting with the surface states on the grain surface each other offer some acceptors at surface level on the sensor BTS in the form of chemisorption. In fact, the past studies indicated that there is a positive electric field near to the hydrogen nuclei in a polar water molecule, which can accept electron directly not only from the oxygen ions O^{2-} or O^- adsorbed on semiconductive ceramic surface, but also from semicinductor's valence band, and make those traped at the region where the water molecules were adsorbed.

On the other hand, in the range of high humidity, the water moleculesare adsorbed on the grain surface of BTS ceramics in the form of multi-layer structure, and the outer molecules are adsorbed physically upon theinner molecules by Hydrogen Bond and Van Der Waals' Attraction, which would reduce the shield effect on the hydrogen nuclei in outer molecules. Hence, the electron density $|\psi(H)|^2_{ADS}$ of outer hydrogen nuclei trends to decrease. Because of the average effect of magnetic fast-exchange, the displacement δ of the hydrogen nuclei in water adsorbed shifts much towards one of Positron H^+ at high humidity than that does it at low one. The fact has been proved by the experimental result that $\delta_{92\%RH} > \delta_{12\%RH}$ (Tab. 1). Based on the analyses above, it would be summarized that, in process of sensing humidity of the sensor BTS, there is nearly no "Positron-conductive Mechanism" range from low humidity to middle one at least, but it might be valid to some extent in the range of high. one.

2. Energy Band Structure For Semiconductor BTS

Experiments of electrical conductivity at various temperature supplied informations about that the sensor BTS is a P-type semiconductor under the normal surrounding atmosphere. The gas-sensitive effect proved also that. Besides The sensor BTS is a ceramics composed of many grain and many crystal phases. Interface potential barrier is formed in the crystal boundary phases due to the interface states deduced by dislocations, defects and distortion in crystal lattice,[5] which makes the resistivity of the crystal boundary region become very great and have no concern with the ambient humidity.

Surface physics' theories on semiconductor[5],[6] had shown that the periodicity of the crystal potential is disturbed or lost at the crystal surface; further, the surface atoms are usually displaced from their ideal lattice positions, resulting in a surface structure that deviates electrically from the bulk structure. These disturbances lead to localized electronic states at the surface referred to as surface states or interface states mentioned above, named usually as Shockley Energy Level and Tamm Energy Level.

At the surface of a semiconductor crystal, a net positive charge is found because its valence bounds end abruptly there, with electrons missing. These positive charges may result in a depletion or inversion of an underlying P-type surface or in an accumulation of an underlying n-type surface. In practice, the real semiconductor surface in air is usually covered by a extremely thin oxide layer insulator. The ions adsorbed on the surface outside the oxide layer keep eletrically in equilibrium with the space charges inside the semiconductor(Fig. 1-a). There are fast and slow surface states, i.e., localized election energy levels, which can exchange charges rapidly or slowly with the semiconductor space charge region, respectively. The slow states are at the outer surface of insulator, the fast states are at semiconductor-insulator interface. The energy band structure of the grain surface of P-type ceramic BTS is illustrated by the

Fig. 1-b. It is obviously analogous to the ideal MIS system under the depletion or inversion conditions if only the Fig. 1. compared with the Fig. 2. In fact, the water molecules adsorbed on the grain surface in the form of the positive ions provide it with the slow surface states (acceptors), and make a depletion or inversion of an underlying P-type surface formed, which is analogous to what the ideal MIS system will behave as the bias(V_G) is positive. On the other hand, the general capacitance of the sensor BTS as function of the relative humidity and measuring frequency(Fig. 1-c.) is also analogous quite to normalized theoretical MIS capacitance (C/C_0) as function of gate voltage (V_G) and measuring frequency (Fig, 2-c). Hence, which all make sure that studies on the humidity-sensitive mechanism for the sensor BTS could be imitated by research on the electric properties for the ideal MIS system. According to the MIS structure's theory, the surface potential ϕ_s in the P-type semiconductor increases with the bias V_G, but the capacitance C_{MIS} decreases with V_G increases under the conditions of the surface accumulation or the surface depletion, and increases with the bias V_G under the surface inversion. Analogously, the density of the slow surface states as well as the general capacitance of the sensor BTS increases with the relative humidity (corresponding to V_G). Based on the analogy above, it is acceptable that the water molecules adsorbed can make the grain surface of sensor BTS form inversion layers.

Supposing that the physic model on the inversion layer is correct to describe humidity sensitive mechanism, then it could be expected that,

1) the electronic situation of the ions tin Sn^{4+} in quite wide grain surface layer will be affected by the water molecules adsorbed;

2) the mathematical model established on basis of the physical model can correctly describe the humidity-sensitive properties of sensors.

3. Mossbauer Effect on The Sensor BTS

Because the negative space charges in the inversion layers accumulate into the quite thin grain surface layer, the crystal layer in the quite deep region from the surface must carry some positive charges compared with the bulk, i.e., the equivalent electron density $|\psi(Sn^{4+})|_s^2$ around the nuclei of the ions tin Sn^{4+} in this crystal layer is larger than density $|\psi(Sn^{4+})|_b^2$ in the bulk of sensor BTS.

The theory on the Mossbauer Effect had shown that the change of the electron density around the nuclei of tin ions or atoms would result in a displacement of nuclear energy level, referred to as isomerism-dispalcement (for short DI), it is used for studies on the behavior of sensor It is necessary that the displacement DI_b of the ions tin Sn^{4+} in the grain bulk is forced to keep constant because the density $|\psi(Sn^{4+})|_b^2$ is not affected by the water molecules adsorbed, and the bi-peak fit is made for the Mossbauer spectrum of the sensor BTS. The results are shown as Table 2. Here,

$$DI \propto [|\psi(Sn)|^2_{SAMPLE} - |\psi(Sn)|^2_{SOURCE}], \quad (1)$$

and the displacement $DI[P_s^{(i)}] < DI[P_b^{(i)}]$, where = 1 or 2, so the density $|\psi(Sn^{4+})|_s^2 < |\psi(Sn^{4+})|_b^2$. Furthermore, the displacement $DI[P_s^{(2)}] < DI[P_s^{(1)}]$, so the density $|\psi(Sn^{4+})|_s^2$ decreases as the ambient humidity increases. These experimental results have indicated that, the electrical properties of the grain surface layer and the bare necks of the crystal boundary are remarkably affected by the water molecules adsorbed on them.

Table 2. Parameters of Mossbauer Spectrum For Sensor BTS

Test No.	peak No. %RH	Displac. DI [a] mm/sec	FWHM GA mm/sec	Intens I %	square error χ^2
	Model I, at the ambient humidity 12%RH				
1	$P_b^{(1)}$	-1.96	0.54	80	
	$P_s^{(1)}$	-2.28	0.38	20	205
	Model II, at the ambient humidity 92%RH				
2	$P_b^{(2)}$	-1.96	0.54	82	
	$P_s^{(2)}$	-2.32	0.39	18	168

a) $DI(Sn^{4+}) = DI(Sn^{4+})_{BTS} - DI(Sn^{4+})_{Metal}$

4. Humidity-sensitive Mechanism And Mathematic Model

As mentioned above, obviously, it is the law that dominates the humidity-sensitive behaviors of the sensor BTS, which describes the relations for the concentration and the behaviors of the charge-carriers in those inversion layers to change with the ambient relative humidity, it is, furthermore, what the humidity-sensitive mechanism was said. Now, the mathematical expressions can be deduced out on the physic model as below,

$$R(H) = \frac{Ro \times K_H}{H \times e^{(\alpha \times H)} + K_H} \quad (2)$$

$$\alpha_H(H) = -(\alpha + \frac{1}{H}) \times [\frac{H}{H + K_H \times e^{(-\alpha \times H)}}], \quad (3)$$

where,
- H = relative humidity;
- K_H = linear factor, a constant;
- α = sensitve factor, a constant;
- Ro = intrinsic resistance, a constant
- $R(H)$ = general resistance of sensor;
- $\alpha_H(H)$ = sensitivity for humidity.

From now on, the humidity-sensitive behaviors of many sensors can be expressed mathematically by the essential formula (2) and formula (3) . For example, in the whole humidity, fitting the R(H) data for the sensor BTS, BST and MCT with H at least-square method, the results are shown as Table 3. Where the sensor BST ($Ba_{1-x}Sr_xTiO_3$) and MCT ($MgCr_2O_4-TiO_2$) are bi-functional ceramic sensors of ABO_3-type structure and other type structure, respectively.[1],[2] It is clear that the correlation coefficients are very high, which means that these essential formulas can give an excellent description for the humidity-sensitive properties of sensors of various type structure. Besides, those can also explain the behaviors of heat-cleaning cycle at high temperature. In fact, as temperature $T \rightarrow +\infty$, the resistance $R(H) \rightarrow Ro$ from the formulas, i.e., the heat-cleaning can eliminate various slow surface states on the sensor's grain surfaces, and revivify the sensors. In practice, as only the temperature $T > 450$℃, the heat-cleaning cycle can be finished.

CONCLUSION

In process of sensing humidity fot ceramic sensor, there is nearly no "Positron-conductive Mechanism" range from low relative humidity to middle one at least, and " Electron-conductive Mechanism" dominates the humidity-sensitive mechanism in the whole humidity. The essential formulas from it can express the main features of the humidity-sensitive behaviors for semiconductive ceramic sensors of many type structures.

Table 3. characteristics factors of sensors

Sample	Ro (M Ω)	α	K_H	Cor a) (%)
BTS	200	0.076	11.88	99.99
BST	400	0.044	0.81	99.99
MCT	900	0.037	0.40	99.98

a) "Cor" is a correlation coefficient

REFERRNCES

1. T.Nitta, F. Fukushima, J. Terada et al., National Technical Report, Vol. 26, No. 3, June (1980)433-441, 450-456; IEEE, Vol. ED-29, NO.1, JAN. (1982) 95-101.
2. Zhou Zhi-gang, Zhang Zhong-tai and Zhao Gang, A New Multifunctional Sensitive Ceramics BTS, Proceedings of 36th Electronic Components Conference, (1986).
3. Zhou Zhi-gang, Zhao Gang and Zhang Zhong-tai, Temperature-Humidity-Gas Multifunctional Sensitive Ceramic BTS, China-Japan International Scientific Symposium On Sensor Harbin'86.
4. Zhao Gang, The Studies On The Multifunctional Sensitive Ceramics BTS , Master's thesis, Unpublished.
5. C.S. Smith, Grain, Phases and Interfaces, An Interpretation of Microstructure, Trans. AIMA, Vol. 175 (1948) 15-51.
6. Helmut Wolf, Semiconductors, Published by Pergamon Press, Oxford, 1969.
7. S.R. Morrison, Semiconductor Surface, Sol. St. Chem. P. 203, (1976).

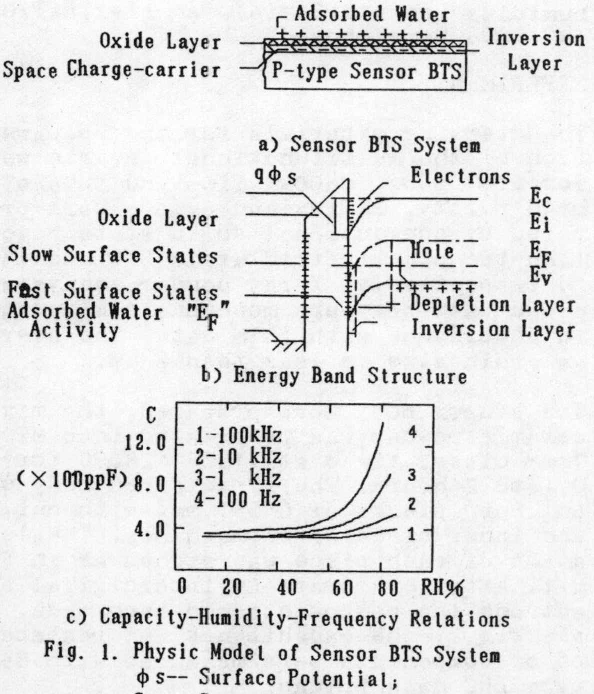

c) Capacity-Humidity-Frequency Relations

Fig. 1. Physic Model of Sensor BTS System
φs-- Surface Potential;
C -- General Capacitance of Sensor BTS.

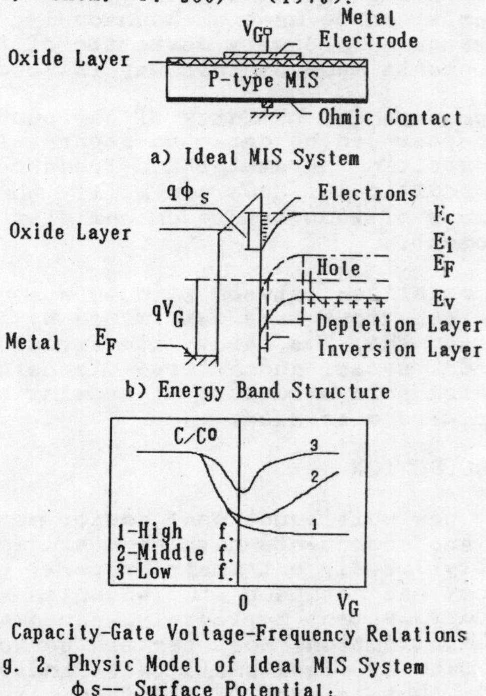

c) Capacity-Gate Voltage-Frequency Relations

Fig. 2. Physic Model of Ideal MIS System
φs-- Surface Potential;
C/Co- Normal General Capacitance.

A WAY OF ADVANCED PROPERTY OF MULTIFUNCTIONAL CERAMIC SENSOR

Huang Songtao, Dept. 301
Beijing General Research
Institute for Non-ferrous Metals
100088 Beijing China

Zhou Zhigang
Dept. Mat. Sci. & Eng.
Tsinghua University
100084 Beijing China

ABSTRACT

In order to obtain fine sensing property and good consistency of multifunctional ceramic sensor, the method to combine conventional ceramic process with semiconductor technique is used. The ceramic sensor is based on $(1-x)BaTiO_3+xCaSnO_3$ $(x=0,0.3)$. At first, it has been produced by ceramic process: The mixed powder is prepressed into Ø50*7mm^3 discs, fired at 1180--1280°C for 0.1 to 2 hours; and then using semiconductor technique to cut the fired body into pieces and evaporate Au interdigital electrode on it. Components made by this way maintain the fired ceramic charactor, the body and surface electronic property of the components can be used simutaneously, and the sensing property deviation of the components made by this way is lower.

Temprature and humidity of surrounding atmospher can be detected separatelly by detecting parameter and frequency. The addition of SnO_2 can modify the influence of temperature on humidity measurement.

The relations between structure and sensing behaviour were determined by XRD, SEM and PMA. The perovskite structure of BCST sensor and a three-dimensional network pore structure on sensing behavior were discussed.

INTRODUCTION

Many new multifunctional sensor materials and components have been studied recently(1,2,3). but their property consistency has not been studied,which affect its development seriousely, especially its application. Most ceramic sensors are made by fired small bases individually, that is very difficult to maintain the consistency of process for all bases. Semiconductor technique overcomes this shortcoming. To cut large body into pieces may keep not only the component size exactly, but also the process consistency, therefore the consistency of product can be improved.

$BaTiO_3$-$CaSnO_3$ can form monogenic perovskite solid solution, which has ferroelectric property, so we can use its capacitance to detect temperature(1,3). The impedance will change when water is absorbed by the pore structure, therefore it could detect temperature and humidity simutanneoussly by means of a single component. Ca, Sn element can increase the humidity sensitivity(4), so $(1-x)BaTiO_3$ $+xCaSnO_3$ were chosed.

EXPERIMENT

The starting materials for the preparation of the multifunctional ceramic sensor are $BaCO_3$, $CaCO_3$, TiO_2 and SnO_2 of high purity. $BaTiO_3$ and $CaSnO_3$ were prepared by conventional solid state reaction technique: fired at 1250°C and 1350 °C respectively. X-ray powder analyses prove that they are monophasic structure in accordance with ASTM data. The average grain size is less than 2 µm.

The sensor BCST were prepared, the mixed ceramic powder was prepressed into Ø50* 7mm^3 discs, fired at 1180--1280°C for 0.1 to 2 hours. The large fired body was cut into pieces of 6*5*1 mm^3 with outer and inner circular cutmashines. The location of each piece was signed as in Fig.1. After evaporate Au interdigital electrode the components had been made, see Fig.2. the capacitance and resistence of components were measured with 8500 type LRC data bridge.

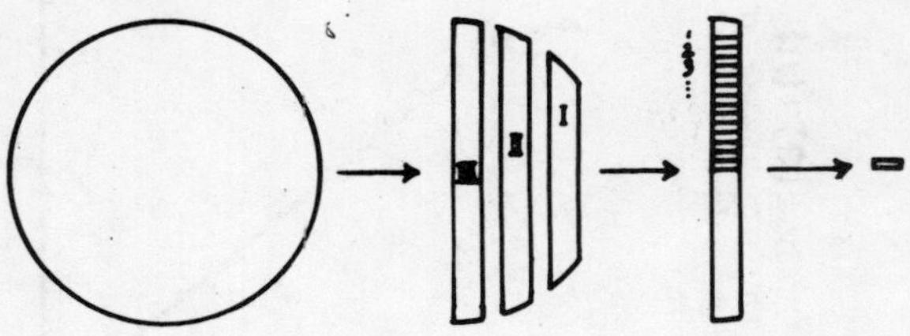

Fig.1 A scheme of cuting large body into pieces

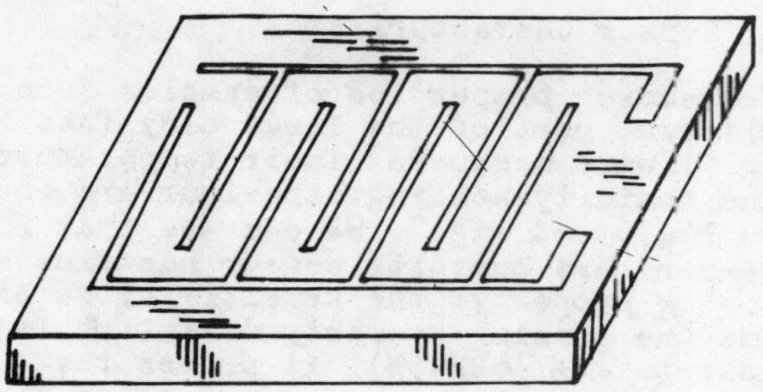

Fig.2 A scheme of BCST sensor

RESULTS AND DISCUSSION

A. Detection Mechanism

The measurement of temperature and humidity by one component can be done by detecting parameter and frequency(1). our experiment shows that when detecting frequency is over 10^5Hz, the influence of environment humidity on capacitence can be negligible, see Fig.3. Therefore 1 MHz and 100 Hz for detecting temperature and humidity respectively have been chosed.

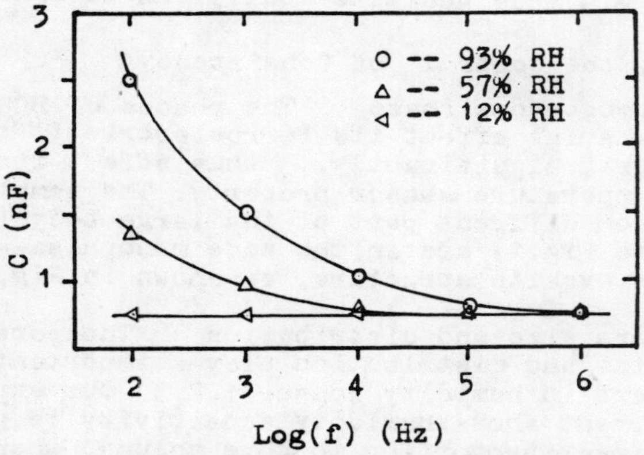

Fig.3 Capacitance frequence dispertion

The BCST material is monophase of perovskite by X-ray diffraction analysis see Fig.4. It has ferroelectric property and obeys Curie-Wass's law, the reciprocal of capacitance (1/C) is in direct proportion to temperature (T). When 20mol% $CaSnO_3$ was added, the Tc of $BaTiO_3$ was shift from 125°C to -50 ^{o}C, so BCST sensor has wide temperature detecting range and good sensitivity, see Fig.6.

Porous structure and distribution of the components are very important in humidity sensing property (3). Since stannates are generally difficult to sinter to high densities, the sintered compact tends to exhibit a typical porrous structure. The fractured section

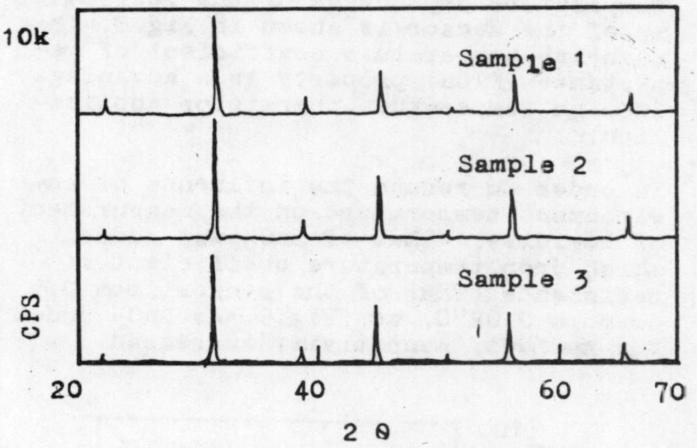

Fig.4 X-ray diffraction

Fig.5 SEM of typical section fracture of poros ceramic

exhibits intergranular poros in combination with neck parts, as shown in Fig.5. It forms a three-dimensional network pore structure, the pore structure consists of open proes that form a kind of capillary tubes. The porous ceramic characterized by this pore structure can easily exhibit absorption and condensation of water vaper, which changes the impedance of the components.

B. Sensor Characteristics

The sensor properties of samples from diffrent part of the large body (see Fig.1).were measured. Their temperature and humidity sensing behaviour are shown in Fig.6 and Fig.7. We can see that BCST temperature-humidity sensor has fine sensing property: the sensitivity is high and the sensing pyoperty deviation is lower($\Delta T < 4\%$, $\Delta RH < 5\%$). It proves that this preparation proces maintains fine sensing consistency.

Temperature dependence of the resistance of the sensor is shown in Fig.8. This negative temperature coefficient of resistance (NTCR) property is a advantage in high temperature thermistor application.

In order to reduce the influence of environment temperature on the measurement of humidity, 1-5%wt of SnO_2 was added, which drop temperature coefficient of resistance (TKR) of the sensor from 0.5 down to 0.02°C, see Fig.9. As SnO_2 added the humidity sensitivity increased.

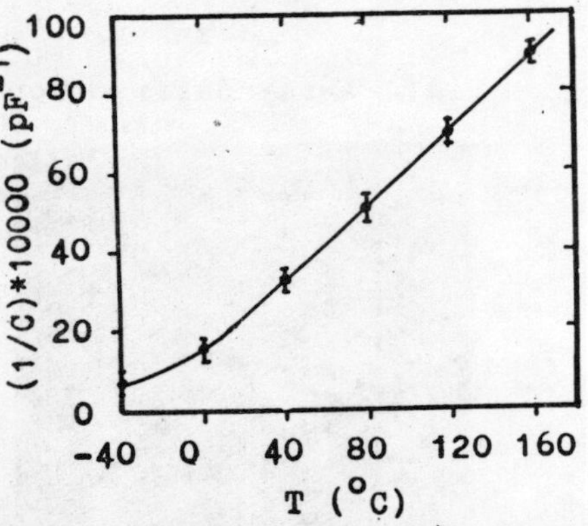

Fig.6 Temperature sensor property

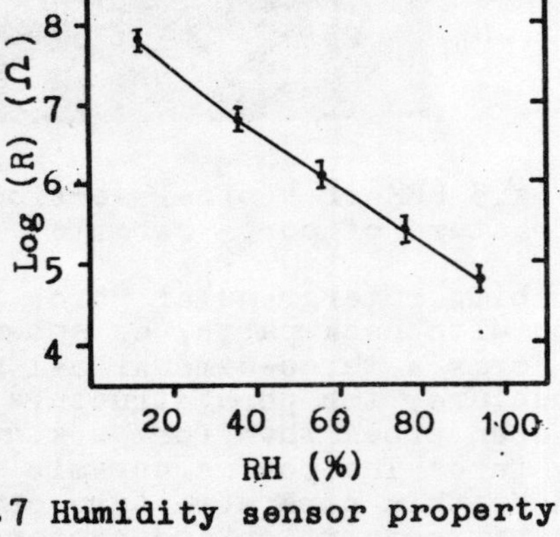

Fig.7 Humidity sensor property

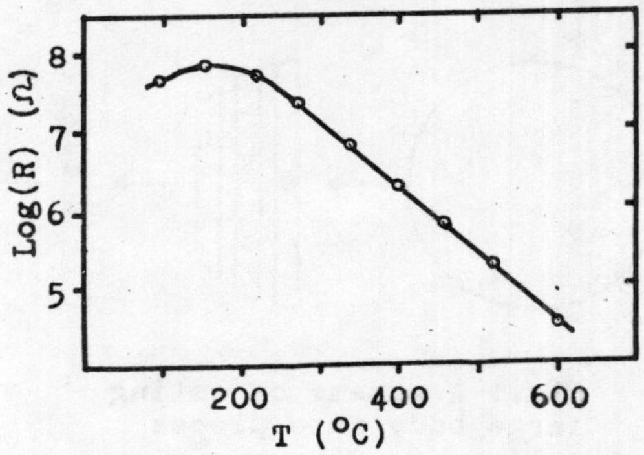

Fig.8 NTCR property at high temperature

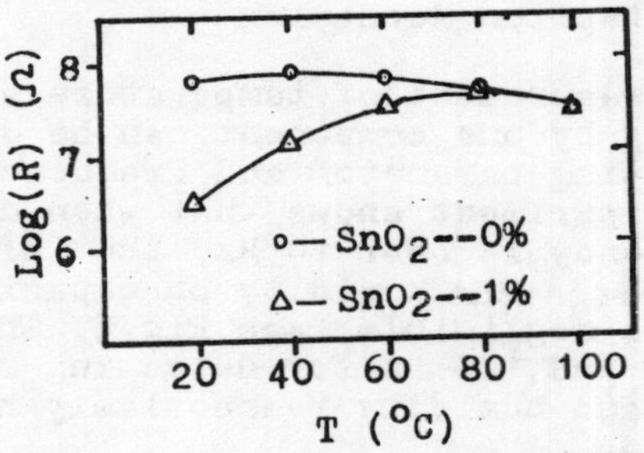

Fig.9 SnO_2 additive modify TKR of BCST

C. the Control of Consistency

Homogeneous Phase   The phases of BCST material affect its ferroelectric property significantly, thus affect its temperature sensor property. The samples from diffrent part of the large body (see Fig.1) are in the same monophase--perovskite structure, as shown in Fig.4.

Pore size and distribution   The pore size and distribution play a important part in humidity sensor(1,2,3).our experment shows:humidity sensitivity is in direct proportion to pore volume, when it ranges from 0.04 to 0.08cm^3/g. Aside from SEM analysis, the pore size and distribution has been studied by press mercury analysis (PMA),see Fig.10. The pore sizes are mainly distributed from 0.1 to 1.0μm, this range is the optimum size distribution for humidity sensor (1,3). From Fig.5 and Fig.10, we can see our sensors maintain uniform in their microstructures.

The sizes and electrodes of components are also very important in the control of sensing property consistency. Semiconductor technique is able to keep the components sizes exactly and interdigital electrodes uniform.

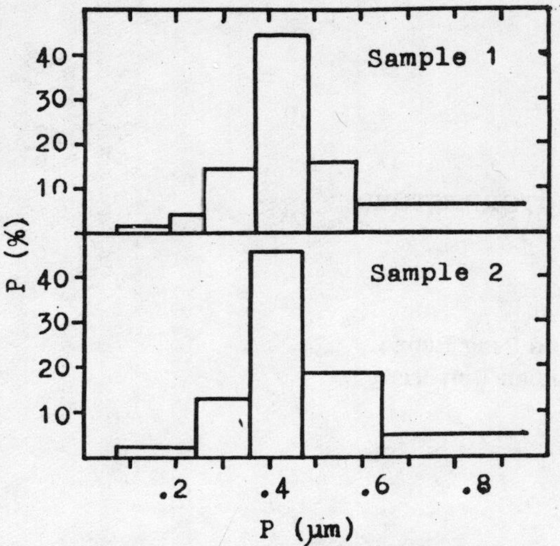

Fig.10 PMA of pore distribution

CONCLUSION

1. The BCST solid solution has fine temperature-humidity senso property.

2. Good property consistency of temperature-humidity sensor can be obtained by combining traditional ceramic process with semiconductor technique.

3. Homogeneous phase, well-distributed pore structure, electrode structure and exact base size are the guarantee of the property consistency of sensor components.

ACKNONLEDGMENT

The authors wish to thank Dr. Zhang zhongtai, proffessor of Material Science and Engneering Deptment, Qinhua University, for his helpful diacussions and suggestions.

REFERENCES

1. J. Terada et al, National Technical Report, vol.26, no.3,1980,pp.433-441
2. F. Fukushima et al, National Technical Report, vol.26,no.3,1980,pp.450-456.
3. Zhou zhigang et al, paper presented at China-Japan Science Symposium on Sensor meeting, 1986,p. 13.
4. H.Arai et al, Chemical Sensors, Analytical Chemical Symposia Series, vol.17, pp.393-398.

SCREEN PRINTED THICK FILM PHOTORESISTORS

Zhang Xianzhi, Wang Xusheng, and Fang Hubao
Technical Physics Department, Sidian Univeristy
Xi'an, China

ABSTRACT

Thick film $Cd_{1-x}Zn_xTe$ semiconductive photoresistors have been prepared on the Al_2O_3 substrate by screen printing and sintering. The $Cd_{1-x}Zn_xTe$ semiconductive resistors have good photosensitivities, its spectral response is wide and fast, the peak spectral response and resistivity can be shifted by changing the ratio of Zn to Cd.

INTRODUCTION

Electronic components produced by screen printing and sintering methods have excellent characteristics of resistance to oxidation and thermal shocks, and of high stability and reliability under radiation. Including their technological process and low cost, they are more and more atractive. Nowadays, the screen printing and sintering technology has been used not only in making conventional resistor, capacitor and conductor, but also in making various sensors. In this paper, we will present screen printed thick film $Cd_{1-x}Zn_xTe$ photoresistors. They have large photosensitivity and wide spectral response (from near ultraviolet to near infrared). Their resistivity and the peak of spectral response can be shifted by changing the ratio of Zn to Cd. These photoresistors can find great use in some photocontrolled devices.The $Cd_{1-x}Zn_xTe$ film can also be used in composing heterojunction solar cells with $Cd_{1-x}Zn_xS$ materials.

SAMPLE PREPARATION

1. Paste

Making up the pure (99.999%) powders of Te, Zn and Cd according to the composition of $Cd_{1-x}Zn_xTe$ resistor,we mix the powders, grind and press them into wafers, then sinter them at 300°C for 2 hours in N_2 atomsphere. The $Cd_{1-x}Zn_xTe$ compound are synthesized from these powders, then smash, grind and sieve the presintered $Cd_{1-x}Zn_xTe$. 5Wt% $CdCl_2$ and 7-8Wt% Propylene glycol are added as flux and shaping agent respectively.After mixing and grinding,the paste is available for printing.

2. Printing and Sintering

The prepared $Cd_{1-x}Zn_xTe$ paste is doubled on the Al_2O_3 or microcrystal glass substrate. After infrared dryiny, the printed films are sintered at 680 °C for 1 hour in the N_2 atomsphere.

The thick film resistors of different x (=0.1,0.2,0.4,0.6,0.8,1.0) and sintering temperatures are produced,their experimental characteristics of resistivity vs sintering temperature are given in Fig.1.The resistivity of $Cd_{1-x}Zn_xTe$ has a minimum value for sintering temperature of 680 °C.It is shown that the structure of crystalline grains of the film is more compact and has less holes.The SEM measurement result given in Fig.2 also shows the same results.Therefor we can conclude that the optimal sintering temperature is about 680°C.

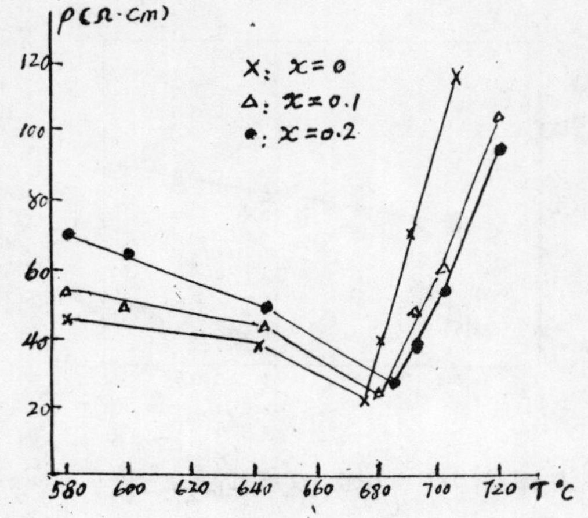

Fig.1 The resistivity of $Cd_{1-x}Zn_xTe$ vs sintering temperature

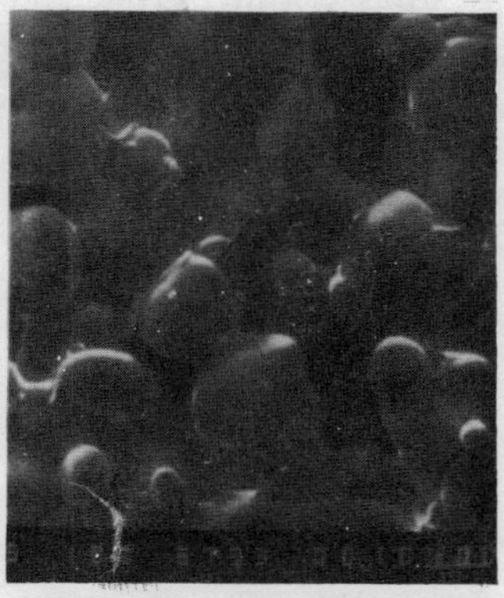

(a) 650 C

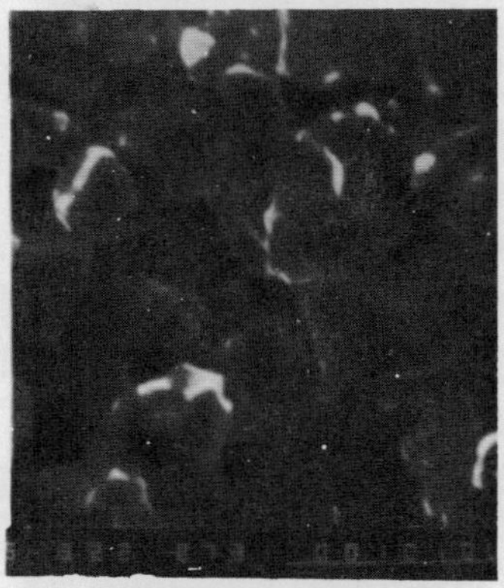

(b) 680 C

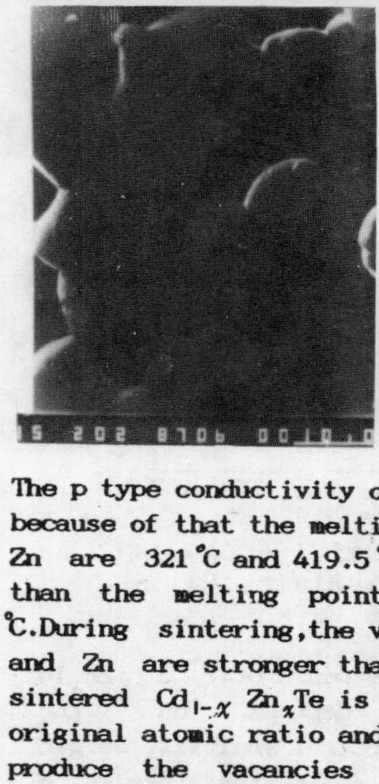

(c) 720 C

Fig.2

SEM image of $Cd_{1-x}Zn_xTe$ thick film resistors in different sintering temperature

MEASUREMENTS AND ANALYSES OF THE CHARACTERISTICS

1. Electric feature

From measurements for $Cd_{1-x}Zn_xTe$ resistor of different x,we find that this kind of thick films are p type semiconductor resistors.The resistivity ρ is decreased as x increase.The resistivity may vary from several hundred kΩ·cm to several Ω·cm of CdTe for x=0,Fig.3 gives the resistivity ρ of $Cd_{1-x}Zn_xTe$ for different x.

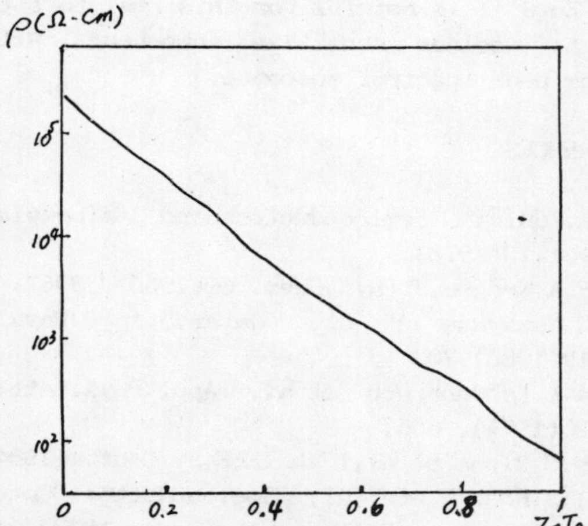

Fig.3 The resistivity of $Cd_{1-x}Zn_xTe$ vs x

The p type conductivity of these resistors is because of that the melting points of Cd and Zn are 321 °C and 419.5 °C respectively,lower than the melting point of Te,i.e. 449.8 °C.During sintering,the volalizations of Cd and Zn are stronger than that of Te,so the sintered $Cd_{1-x}Zn_xTe$ is deviated from the original atomic ratio and is riched in Te.It produce the vacancies of Cd and Zn in $Cd_{1-x}Zn_xTe$, therefore the conductivity of p type appeares, this also can be shown from the electron prob analysis.
CdTe and ZnTe are belong to cubic crystal and

have zincblende structure, the lattice constants are 0.648 Å and 0.610 Å; bond distances are 0.28 Å and 0.274 Å respectively. Because their similarities of structure, they are unlimitedly and mutually soluable and can be combined to form ternary compound $Cd_{1-x}Zn_xTe$ semiconductor. The dissociated Zn, Cd, Te and CdTe, ZnTe are very little, it can be seen that the semiconductor resistors are fully composed, the electric features must be related to the ratio of Zn to Cd.

2. Photoelectric feature

The $Cd_{1-x}Zn_xTe$ semiconductive resistors have good photosensitivity, which can be shifted by changing the composition. We have measured the photosensitivities of printed and sintered $Cd_{1-x}Zn_xTe$ resistors with incandescent lamp illumination having the power of 35.8 mw/cm^2. When x=0, the dark resistivity $\rho_D = 2.64 \times 10^5\ \Omega \cdot cm$, light resistivity $\rho_L = 8.85 \times 10^4\ \Omega \cdot cm$, the ratio $r = \rho_D/\rho_L$ =3; when x=0.1, the r=1-2; x=0.2, the r=0.5-0.8 and x=0.4 the r=0.1-0.2. The light resistivities vs. x are shown in Fig.4. From the above measurements, it can be shown

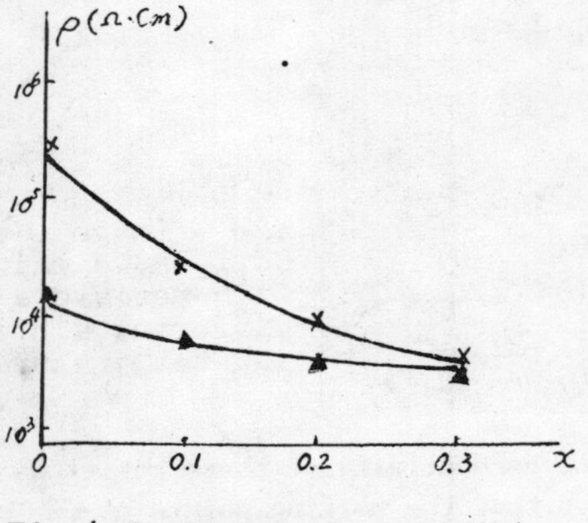

Fig.4 Dark resistivity (X) and light resistivity (▲) of $Cd_{1-x}Zn_xTe$ vs x

that the spectral response of $Cd_{1-x}Zn_xTe$ ternary compounds depends on the compositions. With spectral analysis, we get the bandgap of $Cd_{1-x}Zn_xTe$ resistors which varies with x (see Fig.5), the $Cd_{1-x}Zn_xTe$ resistors have response from near ultraviolet to near infrared. There are stronger infrared radiations for the incandescent lamp, and the

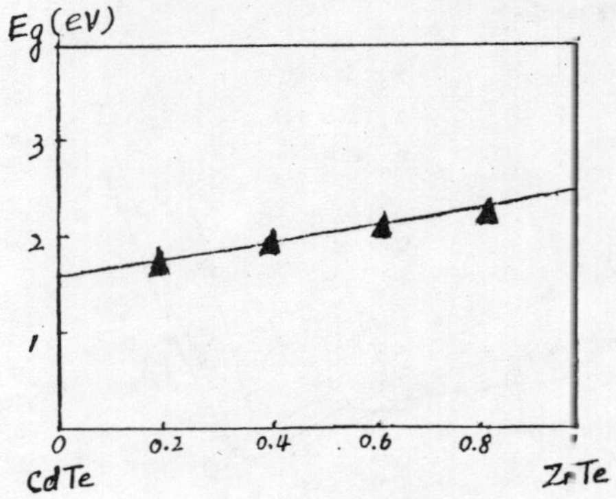

Fig.5 Bandgap of $Cd_{1-x}Zn_xTe$ vs x

bandgap of $Cd_{1-x}Zn_xTe$ for x=0 is 1.4ev, the response peak is at near infrared region, so the resisstor of $Cd_{1-x}Zn_xTe$ for x=0 has larger photosensitivity.

SUMMARY

The good photosensitivity of printed $Cd_{1-x}Zn_xTe$ thick film semiconductive resistors are produced. The resistors are stable, low cost and simple technological process, its spectral response is wide and fast, the peak spectral response and resistivity can be shifted by changing the ratio of Zn to Cd. This kind of resistors can be used in photocontrolled devices and composing heterojunction solar cells with $Cd_{1-x}Zn_xS$. It is hopeful for this resistors to use as colour sensitive components with narrow peak spectral response.

REFERENCES

1. K.Zanio, Semiconductors and Semimetals, vol.13(1978).
2. F.A.Krögen,J.Phys.Chem. 69(1965), 3367.
3. N.Nakayama et al, Japan.J.Appl.Phys., 19(1980),703.
4. A.L.Fahrenbruch et al, Appl.Phys.Lett., 25(1974), 605.
5. W.H.Bloss et al,17th IEEE PV Conf.,(1984)
6. T.S.Moss et al, Semiconductor Opto-Electronic, (Butterworth,London,1973),36.
7. G.J.Korsh et al, Solid-State Electron., 21(1978),1045.
8. B.Jensen, IEEE J.Quantum Electron., QE-18 (1982),1361.

PTCR EFFECT IN $(Ba_{1-x}Pb_x)(Ti_{1-y}Nb_y)O_3$ PREPARED BY MOLTEN SALT SYNTHESIS METHOD

Ki Hyun Yoon and Man Haw Lee
Department of Ceramic Engineering, Yonsei University,
Seoul, Korea

ABSTRACT

The PTCR behavior and the variation in lattice constants in barium-lead titanate have been investigated as a function of the Nb_2O_5 dopant and the Pb content. As the amount of Pb increases in the $BaTiO_3$ lattice, the length of the c-axis increases but that of the a-axis decreases. And a difference in lattice parameter was observed between two methods of preparation, i.e., molten salt synthesis and calcining of mixed oxides. In Nb doped $(Ba_{0.8}Pb_{0.2})TiO_3$ ceramics, the density and the grain size decrease above 0.4 mol% Nb ions and room temperature resistivity increases above 0.6 mol%. These results can be explained by the solubility limit of Nb ions and charge compensation.

INTRODUCTION

There have been many studies on the PTCR effect since 1955 (see Haayman et al.[1]). Among the theoretical descriptions of this effect, the works of Heywang[2] and Jonker[3] are considered most suitable. The PTCR effect in $BaTiO_3$ ceramics is strongly dependent on microstructure and in our previous reports [4,5,6] ceramics produced by the molten salt synthesis (MSS) method showed better microstructure and larger PTCR effect than ceramics produced by the calcining of mixed oxide method. High Curie temperature $BaTiO_3$ based ceramic thermistor can be obtained by replacing the Ba ions with Pb ions[7].

In this study we observed structural differences in barium-lead titanate due to the synthesis method. Also, for $(Ba_{0.8}Pb_{0.2})TiO_3$ ceramics, a widely used composition for application, the effects of Nb dopant were investigated in terms of the change in density, microstructure, and PTCR characteristics.

EXPERIMENTAL PROCEDURE

Barium-lead titanate powders, $(Ba_{1-x}Pb_x)TiO_3$ with x in the range of 0.1 to 0.5 mole were prepared by the molten salt synthesis (MSS) method and the conventional calcining of mixed oxides (CMO) method. To prepare barium-lead titanate powders, $BaCO_3$, PbO, TiO_2, and KCl as a molten salt were used. The purity of each of the starting materials was above 99%. In the MSS method, the starting materials and the salt (with a weight ratio of KCl to starting materials of 0.8) were weighed and milled for 20 hrs with stabilized zirconia balls in ethyl alcohol, and the mixture were pan-dried with stirring. The dried powders were then heated to 800°C using an alumina crucible in an electrical furnace and allowed to react for 2 hrs. After the reaction the salt was washed out with deionized water until no free Cl ions were detected using $AgNO_3$ solution[4]. In the CMO method, the oxide mixture was calcined at 900°C for 2 hrs.

The synthesized powders were granulated with 2 wt% PVA and pressed into discs 1 cm in diameter by 0.2 mm thick under a pressure of 900 kg/cm². The discs were sintered at 1300°C for 30 min. in an electrical furnace on a Pt plate. The powders and also the sintered specimens were examined by X-ray diffraction analysis using Cu $K\alpha$ radiation. Accurate observation of the change in 2θ values above 90° were made[8] and all data were evaluated by the least squares method.

Powders with composition of $(Ba_{0.8}Pb_{0.2})(Ti_{1-y}Nb_y)O_3$ (y=0.002 - 0.010) were prepared by the MSS method. In the same way as above, the mixtures were allowed to react at 800°C for 6 hrs and washed with distilled water and then pressed into discs. The pressed discs were sintered in an electrical furnace at 1320°C for 30 min.

In order to measure the resistivity, Ni electrodes were coated on opposite surfaces of the polished specimens by the electroless plating method[9]. The density of the specimens were determined by the ASTM test method C372-72, and the temperature-resistivity characteristics were measured by the voltage drop method[4]. Micro-structures in the specimens were observed using a metallurgical microscope (Nikon AFX Type 104, Japan) after etching with an 85% H_3PO_4 solution at 800°C for 15-30 sec.

RESULTS AND DISCUSSION

1. Lattice Constants

Fig.1 shows lattice constants for various compositions of the specimens prepared by the MSS and CMO methods. As the content of Pb in the lattice increases, the length of the a-axis decreases but that of the c-axis increases. The dotted lines represent Vegard's law, and this figure shows negative deviation for the c-axis. The results shown in Fig.1 are comparable with those of Shevanov et al.[10] and Shirane et al.[11]. However a slight difference between MSS and CMO specimens can be seen in the lattice constants. This difference may be due to vaporization loss of the Pb ions during the calcining step in the CMO method because in this case powders were calcined at 900°C which is over the melting temperature of PbO (880°C).

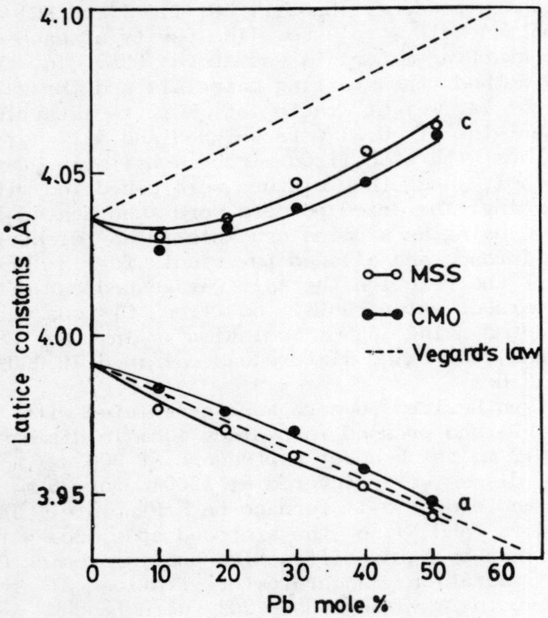

Fig. 1 Lattice constants of MSS and CMO specimens as a function of Pb contents in $(Ba_{1-x}Pb_x)TiO_3$.

2. Effects of Nb^{+5} Ion Dopant on $(Ba_{0.8}Pb_{0.2})TiO_3$

Fig.2 shows the change in density of $(Ba_{0.8}Pb_{0.2})TiO_3$ prepared by the MSS method as a function of Nb^{+5} ion content. As the amount of Nb^{+5} increases, the density of the specimens increase up to 0.4 mol% and then decrease again. According to the results of Swilam et al.[12], in the sintering stage for $BaTiO_3$ the determining factor is the diffusion of Ba ions and the sinterability is improved by the Ba vacancies created by substituting trivalent ions such as La^{+3} at Ba sites. And Chan et al.[13] report that the substitution of Nb ions for Ti ions in the $Ba(Ti_{1-y}Nb_y)O_3$ lattice generates conduction electrons which then leads to the formation of Ba vacancies as in the following equations,

$$Nb_2O_5 = 2Nb^{\bullet}_{Ti} + 2e' + 4O_O + 1/2O_2 \quad (1)$$

$$2e' + 1/2O_2 = V''_{Ba} + O_O \quad (2)$$

It can be considered that the substitution of Nb ions at Ti sites may improve the sinterability due to Ba vacancies and thus the density of the specimen increases. This result agrees well with our previous report[6].

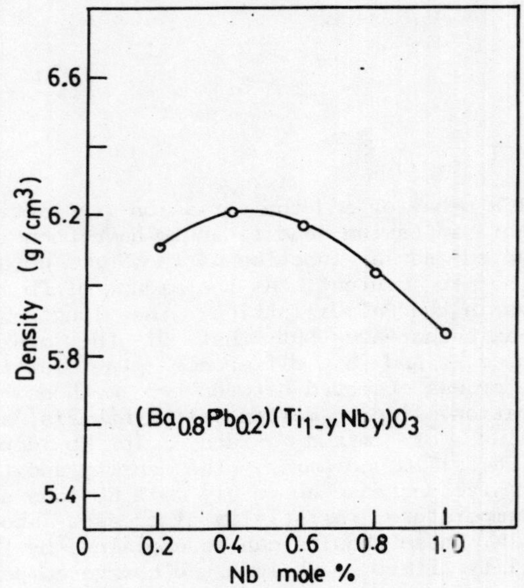

Fig. 2 The apparent density vs. Nb mole% for $(Ba_{0.8}Pb_{0.2})(Ti_{1-y}Nb_y)O_3$ specimens prepared by MSS with a KCl weight ratio of 0.8.

However, when a dopant exceeds the solubility limit, the excess dopant segregates by formation of an oxide phase at the grain boundaries and thus the grain growth of the specimen is inhibited due to the decrease in mobility at the grain boundary[14]. As shown in Fig.2, the densities decreased above 0.6 mol% of Nb ions. It can be considered that the excess Nb ions exist at the grain boundaries and inhibit the grain growth so that the density decreases. This result corresponds well with observations on the microstructures of the specimens given in Fig.3.

As shown in Fig.3, the grain size of the specimens increases up to 0.4 mole% of Nb ions, and above 0.6 mole% grain size decreases but the porosity increases. Kahn[15] reported that $BaTiO_3$ with less than 0.35 at% Nb ions did not grow grains above 1 μm in size. The solubility limit of Nb ions in $BaTiO_3$ was about 0.3 at% in our previous report[6] and that in $PbTiO_3$ was about 2-5 at%[16]. From the above results, we can see that the solubility limit of a dopant strongly influences the grain growth and density of specimens.

Fig.4 shows the resistivity - temperature characteristics with variation of the Nb content where the weight ratio of KCl to raw materials is 0.8. As the amount of Nb ions increases the room temperature resistivity decreases up to 0.6 mole% and then slightly increases again. As mentioned above, the

a)

b)

c)

d)

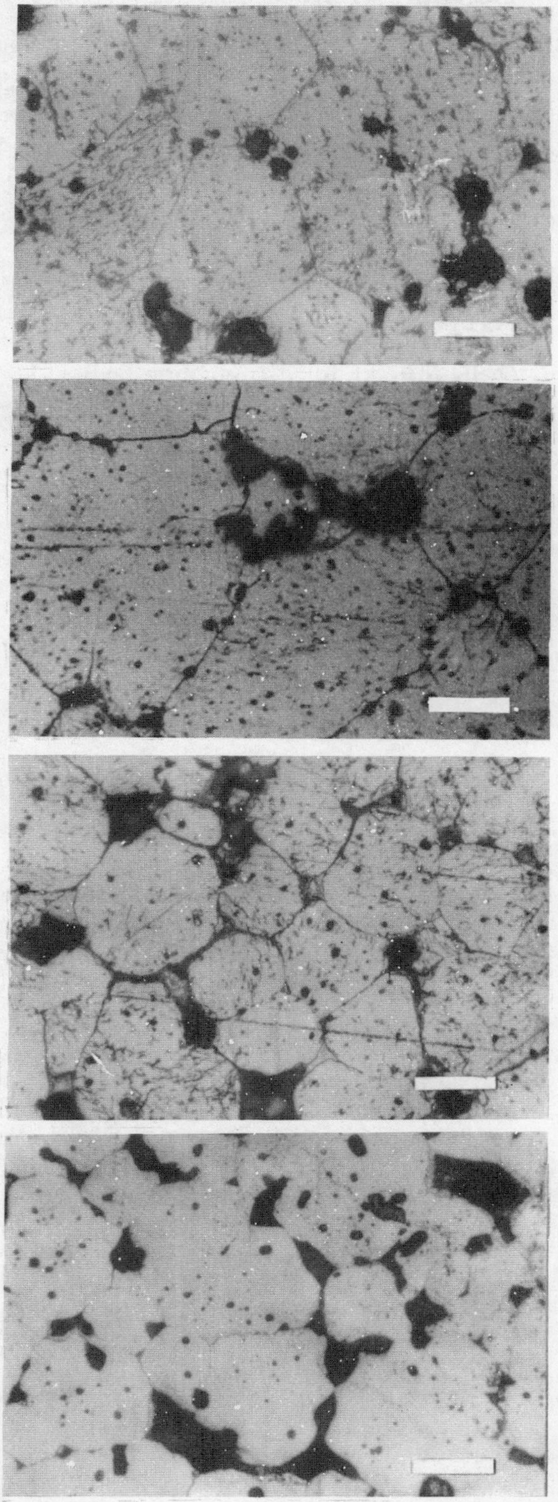

e)

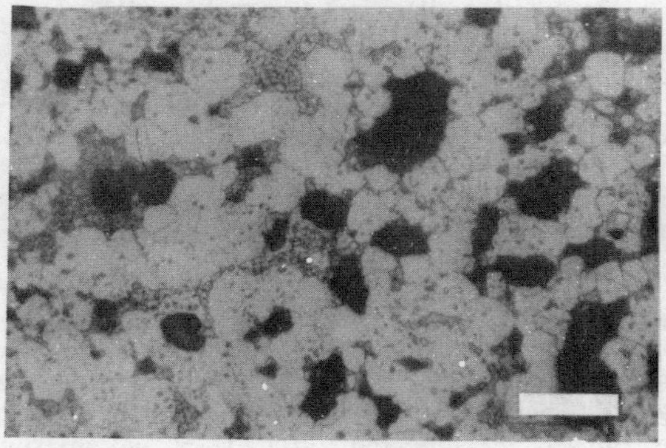

Fig. 3 Microstructures of $(Ba_{0.8}Pb_{0.2})TiO_3$ with various Nb content for a KCl weight ratio of 0.8 ; a)0.2, b)0.4, c)0.6, d)0.8, e)1.0 mole% Nb (bar=10 μm).

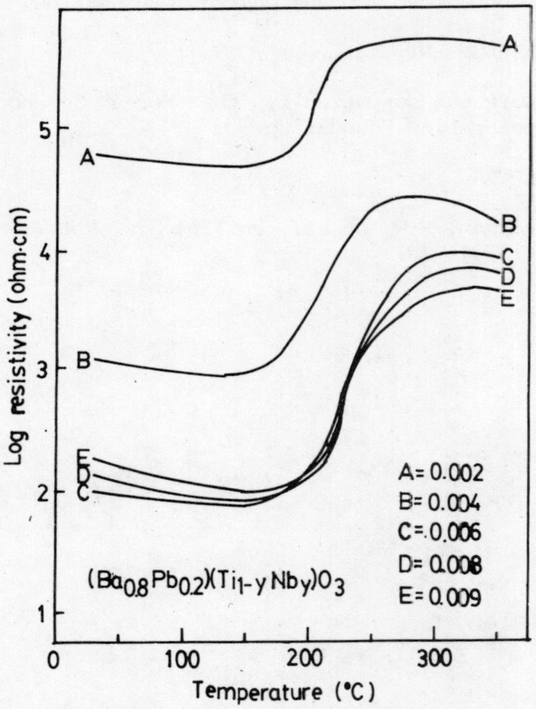

Fig. 4 Resistivity-temperature characteristics with variation in Nb content for a KCl weight ratio of 0.8.

explanation for this is that as the Nb ions are substituted at Ti ion sites within the solubility limit, the density of the conduction electrons increase according to eq.(1). Therefore lowest room temperature resistivity should appear in the 0.4 at% Nb doped specimen according to the boundary layer model. However, we observe a small change in the relation between microstructures and room temperature resistivity. A conjectured reason is that the Pb vacancy is created by vaporization of PbO during the sintering and this generates a hole as in eq.(3) and thus a compensation effect occurs with the conduction electrons.

$$PbO = V_{Pb}'' + 2h^{\cdot} + Pb(g) + O_O. \quad (3)$$

Above the solubility limit, the excess Nb ions form resistive layers at grain boundaries resulting in a slight increase in room temperature resistivity and a decrease in the PTCR effect.

CONCLUSIONS

1. $(Ba_{1-x}Pb_x)TiO_3$ ($0<x<0.5$) can be prepared by the MSS method at 800°C and this method is effective for reducing PbO loss according to results of the lattice constant.
2. The density and the grain size of specimens decrease about 0.4 mole% and the room temperature resistivity increases above 0.6 mole% in Nb doped $(Ba_{0.8}Pb_{0.2})TiO_3$ ceramics. These result can be explained by the solubility limit of the dopant Nb ion and charge compensation.

ACKNOWLEDGEMENT

This work was supported by the Korea Science and Engineering Foundation.

REFERENCES

1. Haayman, P.W. et al., Philips Res. Rpts., 5(1970), 73.
2. Heywang, W., J.Am.Ceram.Soc., 47(1964), 484.
3. Jonker,G.H., Solid State Electron., 7 (1964), 895.
4. Yoon, K.H. et al., Mat.Res Bull., 21(1986), 1429.
5. Yoon, K.H. et al., In: High Tech Ceramics, Elsevier Sci. Pub., Amsterdam, (1987), 1873.
6. Yoon, K.H. et al., Mat.res.bull., 23(1988), 1527.
7. Saburi, O., J.Am.Ceram.Soc., 44(1961), 54.
8. Dungan, R.H. et al., J.Am.Ceram.Soc., 35 (1952), 318.
9. Turner, D.R. et al., J.Electrochem.Soc., 107(1960), 250.
10. Shevanov, L.A. et al., Sov.Phys.Solid State, 23(1981), 373.
11. Shirane, G. et al., J.Appl.Phys.Jpn., 6 (1951), 274.
12. Swilam, M.N. et al., J.Brit Ceram.Soc., 74 (1975), 165.
13. Chan, N.H. et al., J.Am.Ceram.Soc., 67 (1984), 285.
14. Kingery, W.D. et al., In: Introduction to Ceramics, John Wiley and Sons Inc., New York, (1976), 452.
15. Kahn, M.J. et al., J.Am.Ceram.Soc., 54 (1971), 452.
16. Jaffee, B. et al., In: Piezoelectric Ceramics, Academic Press Inc., London, (1971), 122.

LOW TEMPERATURE THERMISTORS FOR 20--200K

Zhuang Shong Chang, Tuo Wan Lu, Wang Shou Lian and Kwladan
Xinjiang Institute of Physics of the Academia Sinica. Wulumuqi, China

ABSTRACT

The present paper reports the configuration design and main electrical properties of two type of thermistors used in low temperature. Have been put into use for more then 10 years, result shows it is reliable in configuration, stable in specific properties and can be used as a kind of low temperature sensor with low cost.

INTRODUCTION

Low thermistor has been successfully used in crygenic engineering especially space technique[1], because it has such features as small demension, fast response, high sensitivity, excellent stability, not needing to compensate for zero point and convenience for remorte measurement. To improve reliability, we investigated the material composition, structure design and major characteristics of low temperature thermistors.

1. COMPOSITION OF MATERIAL FOR THE THERMISTORS

In general, the resistivity of seiconducting material for NTC thermistor can be expressed by following equation

$$\rho = \rho_0 \exp(B/T) \qquad (1)$$

where B is material constand. In the case of material that can be used in a temperature range of 20-77K, a suitable value of B should be in a range of 150-300K. For material in a region 77-200K, B should be 500-1000K. The focue point to develope low-temperature thermistors is to prepare a kind of composition of material which exhibits good stability and producibility as well as a suitable value of B. Experimental results indecate that Mn-Ni-Cu-La-O system oxide semiconducting ceramic with La $(Ni, Mn)O_3$ as its main crystalline phase is good for use in the region 20-77K and Mn-Ni-Cu-Fe-O system with main phase $(Cu, Ni)(Mn, Fe)_2O_4$ is suitable use in the range of 77-200K.

2. CONFIGURATION OF THE THERMISTORS

Three kinds of structures were tested:
A. Chip-type glass seal thermistor with Pt wires electrodes as shown in Fig. 1 a.
B. Chip-type resin seal themistor with Pt electrodes coated on its both sides (Fig.1b)
C. Chip-type glass seal thermistor with electrodes coated on its both sides (Fig.1c).

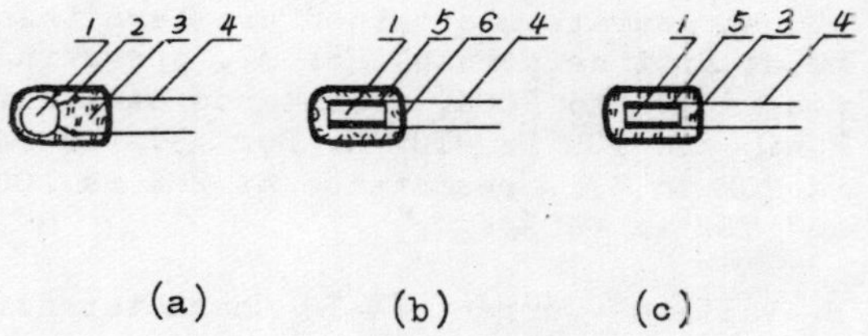

1-ceramic body
2-electrode wire (Pt)
3-glass
4-dumet wire
5-electrode coat
6-resin

Fig. 1 Configuration of low-Temp. thermistors

For all three types of configuration, thermal cycling test between 20K (LH_2) and room temperature was carried out for 100 times. The repeatabilities measured are (a) 10 mk, (b) 60 mk and (c) 20mk. Configuration (a) is prior to others because the contact of electrode wire (Pt) with ceramic is better after sintering at high temperature and glass is better than resin for seal.

3. SPECIFIC PROPERTIES OF THE THERMISTORS

A. Resistance-Temperature (R-T) Characteristic

In operating temperature range, the R-T relationship basically written as

$$R=A \exp (B/T) \qquad (2)$$

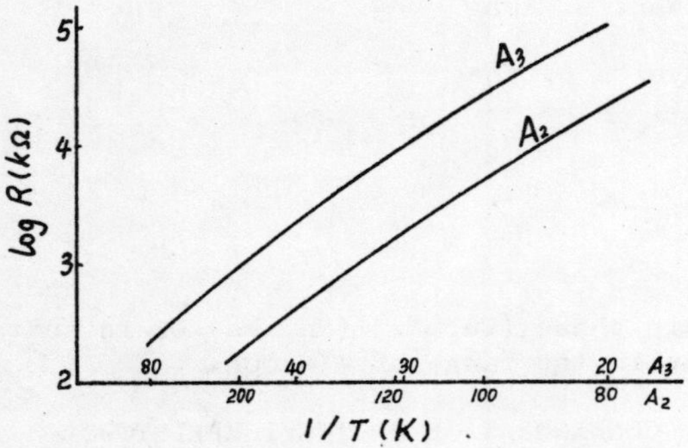

Fig.2 R-T Characteristic thermistors.

Fig.2 shows typical log R-1/T curves of type A2 and A3 thermistors. It can be seen that at higher temperature region, the curves obviously appear to be nonlinear. Therefor, it will give rise to a great error to calculate on the basis of eq.(2). With Chebyshev fifth order polynomial to fit in R-T data, which are accurately measured, the residual deviation obtained are less than 1 mk at fitting points[2]. For A2, operating range is 77K to 200K, resistance at 77K is 20KΩ and TCR is -10%/K. For A3, the range is 20K to 77K, resistance at 20K is 100KΩ and TCR is -40%/K.

B. Voltage--Current (V-I) Characteristic

Fig.3 demostrates the V-I curves of above mentioned thermistors, which were obtained by measuring the thermistors in LN_2 and LH_2 respectively. It is know by computting from measurements dissipation constant, H, is 5 mw/k for A2 in LN_2 and 15 mw/k for A3 in LH_2.

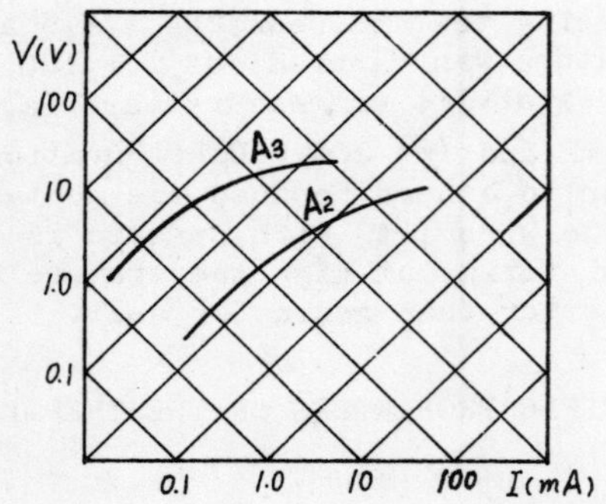

Fig.3 V-I Characteristic thermistors.

C. Repeatability

The repeatabilities of the thermistors were measured[3]. The resistance changes for A2 at 77K and for A3 at 20K were measured after each thermal cycling between 77K (LN_2) and room temperature and between 20K (LH_2) and room temperature respectively, and the cycling was taken for 8 times, and then the resistance changes measured were converted into temperature changes ΔT_{ji}. Short-term repeatability is represented with standard deviation

$$\delta i=\pm\left(\sum_{j=1}^{m}\Delta T_{ji}/(m-1)\right)^{\frac{1}{2}} \qquad (3)$$

The results thus found show that for type A2 and type A3, the repeatability is 5 mk and 10 mk respectively.

D. Long-term stability

The stability of thermistor A3 was measured during operating for 10 monthes[4]. The resurt is that the temperature derivation is less than 40 mk.

4. CONCLUSIVE REMARKES

Low temperature thermistors type A2 and A3 have some features in material composition, configuration and preparing technology. They are suitable for use in low-temperature range of 20K to 200K. Have been used in crygenic engineering like Rocket Chang Zhen for more than 10 years. The results proved it is reliable in contiguration, stable in characteristics and really a kind of practical low temperature sensor with low cost.

Duing the product developing, we were given much cooperation and support by Mr. Dai WenLu, He WenXion, Bi Qiang, Lu Yuo Yong. (Aeronautical Industry Departmant), Zhou GuiEn (Chinese Uni. of Sci. Techn.). Here, a special thanks is extended to all with togather.

REFERENCES

1. B. Sachse, Temperature in mass and control in Science Industry, Vol 3, part 2, 347-353, (1962)
2. Wen Zheng Guo, Data fitting for low-temperature thermistor (to publish).
3. Bi Qiang et al., Acuracy analysis on and applications of low-temperature thermistors. (to publish).
4. Dai WenLu et al., Repeatability of low-temperature thermistor (to publish).

THE RESEARCH OF THIN PLATINUM FILM USED FOR TEMPERATURE SENSOR

Wu Yunzhong, Chu Weibing and Wang Weiyuan
Laboratory of Transducer Technology
Shanghai Institute of Metallurgy, Academia Sinica

ABSTRACT

Platinum thin film temperature sensors have been paid great attention due to its unique properties and extremely high performance-cost ratio. The key to the settlement of the question of such kind of sensors lies in rising its temperature coefficient of resistance(TCR) to the internationally fixed value. In this paper, the effects of impurities, defects and surface scattering on conductivity of platinum films have been studied and the methods of increasing the TCR have been discussed. We have successfully developed the thin platinum film temperature sensors with the performances accorded with standard IEC751 as follows:
TCR=3850±10 ppm/℃
R_0=100±0.1 ohm
Temperature range: -200c— +600℃

INTRODUCTION

Platinum resistors are the most widely used temperature sensors. The thin film platinum resistors, new type temperature sensors, due to its unique properties and extremely high performance-cost ratio, have been extensively valued by many engineers in the world and are being developed competitively. The key to success to develop such kind of sensors lies in increasing the temperature coefficient of resistance(α) of thin platinum films to the value of 3850ppm/℃, which is influenced by material textures and structures. It is well known that the α of bulk materials is much higher than that of thin films. Up to now, the researches on the electrical properties of thin platinum films are very few. This paper explores the factors affecting the α of thin platinum films used to be the references to increase the α and fabricate the thin film platinum resistance temperature sensors successfully.

EXPERIMENTAL

The thin platinum films were deposited by high frequence sputtering on the polished ceramic substrates. The film sheet-resistance($R_\square$) and thickness(d) was measured by accurate four-probe method and Taylor-Hobson step instrument respectively, and the film resistivity(ρ) was calculated from $\rho = R_\square d$. The resistors, each of 100 ohm, were fabricated by planar technique, and its resistance at 0℃ and 100℃, R_0 and R_{100} , were measured by using of the QJ18 temperature-measuring bridge with grade of 0.02 and temperature-controlling precision of 0.1℃. Then the α was calculated from the equation $\alpha = (R_{100} - R_0)/100 \cdot R_0$.

RESULTS AND DISCUSSIONS

The α is defined by following equation

$$\alpha = \frac{1}{\rho}\frac{d\rho}{dT} \qquad (1)$$

where ρ is the material resistivity and T the temperature. Generally, there is a relation $\rho \propto T$ for metals over the Debye temperature. Thus we have

$$\alpha\rho = k \qquad (2)$$

here k is a constant. So the only way to increase the α is to decrease the ρ. According to Matthiessens rule, the resistivity of material is composed of two parts

$$\rho = \rho_T + \rho_r \qquad (3)$$

where ρ_T is the part contributed by the lattice thermal vibration, the physical basis of the resistance type temperature sensors, and ρ_r is the part contributed by the other scattering sources, such as the inner stress, structure defects, impurities and surface etc. The purpose of this paper is that how to decrease ρ_r, increase α.

1. Effect of Annealing Temperature

A set of platinum films with the same thickness were annealed at different temperatures in atmosphere for half an hour and then cooled down to room temperature togehter with the annealing furnace. The measured ρ-T and α-T relationship were displayed in Fig.1 and Fig.2 respectively. The ρ-T relation can obviously be divided into three parts: at temperature between 300 and 900c, the ρ decreases drastically with rising of T, below or above the temperature range, the ρ decreases slowly with rising of T. The α-T relation has a similar characteristic, but when the annealing temperature is extremely high, for example T>1150°c, the α decreases slightly with rising of T.

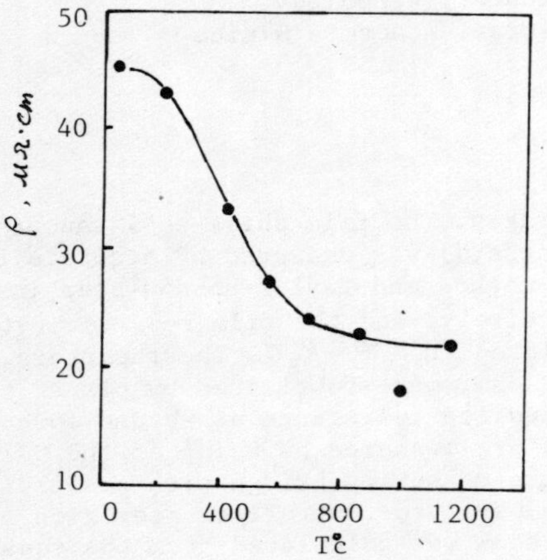

Fig.1 Resistivity vs. annealing temperature

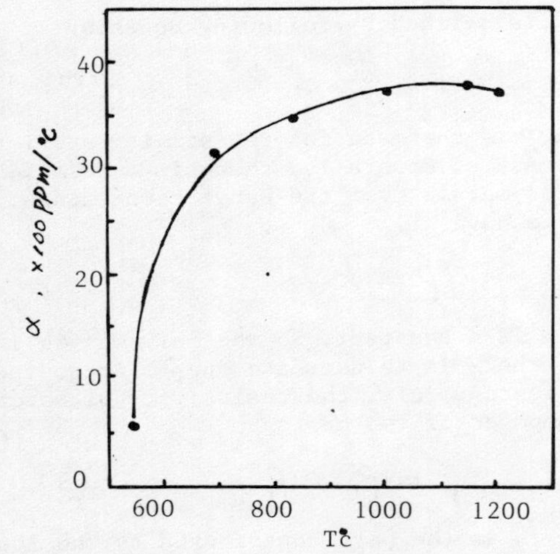

Fig.2 TCR vs. annealing temperature

2.Effect of Impurities, Textures and Structures

The sputtered platinum film surface is contaminated by oxygen and carbon as determined by Auger energy spectrum analysis. The samples were ion-etched for fifteen minutes before analysis to eliminate the impurity adsorption on surface, because the carbon relative intensity adsorped on the platinum film surface is as high as 54.3(the platinum peak is taken to be 100). The oxygen and carbon relative intensities in platinum films thermally annealed under different temperatures are given in Table 1. It shows that the oxygen content in annealed platinum films decreases below 1000°c while increases above 1000°c with rising of T and is higher than that in sputtered state. The carbon content in annealed films has the same variation trends but is much lower than that in sputtered films at temperatures above 1000c.

Table 1 The oxygen, carbon relative intensity in platinum films thermally annealed under different temperatures

sample	Tc	Pt	C	O
10	as-sputtered	100	10.2	0.64
13	550	100	22.8	6.58
14	690	100	10.0	4.09
16	1000	100	2.03	0.76
17	1150	100	3.34	1.78
11	1200	100	4.00	2.67

The variation of oxygen or carbon content caused by thermal activation can be described as follows

$$C \propto \exp(-E/kT) \qquad (4)$$

where c is the oxygen or carbon content, T the temperature and E the process activation energy. By substituting the data of Table 1 in expression (4), the process activation energy of oxygen of 0.98 or 0.44ev can be obtained at temperature above or below 1000c respectively, which corresponds to the diffusion activation energy of oxygen in inner-grain or grain boundary in thin platinum films.[1]

The measured electron diffraction results by us show that the sputtered platinum films are amorphous structure with micro-crystal but the films after been annealed at temperature above 900c are bcc structure with macro-crystal, in coincidence with the results of two different activation energies. Similarly, the process activation energy of carbon of 0.52ev at temperatures below or above 1000c is calculated, which fits to the diffusion activation energy of carbon in platinum.[2] Based on the above discussion, we can conclude that the variation of the oxygen or carbon content in platinum films is limited by the diffusion process.

The ρ-T relation can be summarized as follows:
(1) at temperature between 300 and 900c, with the increasing of annealing temperature, the inner stress in films eliminates, sputtering defects decrease and grains are growing bigger ; on the other hand, at temperature above 600c , the impurities in films such as oxygen and carbon also decrease continually. The whole effect stated above leads to the drastic decreasing of ρ with the increasing of T.
(2) at temperature between room temperature and 300c, accompaning with the annealing treatment, the inner stress and point defects of low formation energy in films partly eliminate. Due to the effect of point defects on ρ is

rather small, the ρ decreases slowly with T. (3) at temperatures above 900c, with the rising of T on the one hand, the grains are growing bigger and the resistivity contributed by grain boundaries also decreases continually; on the other hand, however, the oxygen and carbon content increase slowly, which leads to increasing of ρ slowly.

3. Effect of Film Thickness

Surface is a scattering source for conducting electrons. In bulk materials, the surface scattering can be neglected. When the material thickness is in the order of the electron mean free path, especially when the collision between electrons and surface is nonspecular reflection, the resistivity contributed by surface scattering becomes very large and can not be neglected.
By solving the Boltzmann transport equation, we can get [3]

$$\frac{\rho_0}{\rho} = 1 - \frac{3}{8K}(1-P) \qquad K \gg 1 \quad (5)$$

$$\frac{\rho_0}{\rho} = \frac{3}{4}\left(\frac{1+P}{1-P}\right) K \ln\frac{1}{K} \qquad K \ll 1 \quad (6)$$

where ρ_0 is the bulk material resistivity, ρ the thin film resistivity, λ the electron mean free path, d the film thickness, $k=d/\lambda$ and p the coefficient of specular reflection. From equation (5), it can be seen that when k>10, the relation $\rho \approx \rho_0$ can be obtained.
The platinum films with different thicknesses were annealed at 1000c for half an hour and the experimental relationship α - T is given in Fig.3. The electron mean free path of platinum at room temperature is about 100A. Fig.3 shows that α does not depend on d only when k>90, much higher than the data obtained from equation (5). The reasons are as follows.
(1) As stated above, a large amount of carbon adsorped on the surface of the sputtered thin platinum films diffuse into the films during annealing. The more thinner the films are, the more higher the carbon density in the films is, which leads to decreasing of α.
(2) The thinner the films are, the more obvious the effect of surface roughness is, which results in decreasing of α more rapidly with the decreasing of d.

CONCLUSIONS

Due to the different textures and structures between thin films and bulk materials, and also the effect of surface, the α of thin films is much lower than that of bulk materials. By proper annealing the thin films, the inner stress can be eliminated , the textures and structures can be improved, and the impurities such as oxygen and carbon can be decreased. Controlling the film thickness over a certain value, the effect of surface can be eliminated. Using the methods outlined above, the ρ of platinum films can be decreased and the α raised to the value of 3850ppm/c. The performances of the developed thin film resistors accord with standard IEC751 and are shown in Table 2.

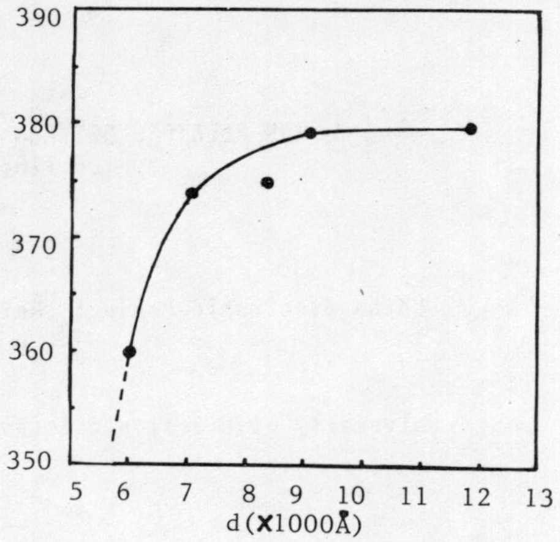

Fig.3 TCR vs. film thickness

Table 2 The reached technical levels

measured temperature range	-200-- +600°c
resistance at 0°c	100 0.1ohm
TCR(ppm/°c)	3850 10
responsing time(s)	0.35
self-heating(°c)	0.04
reliability (R_s) (under extreme condition)	0.99

REFERENCES

1. F.K.Moyhadam, D.A.Stevenson, J. Electrochem. Soc., 1335(1986)1329.
2. P.Ferguson, K.H.Westmacott, R.M.Fisher, and Dahmen, Materials Sci. and Tech., 1(1)(1985)53.
3. A.H.Wilson, The Theory of Metals,University Press, Cambridge, 1954.

THE RELATONS BETWEEN HUMIDITY CHARACTERISTICS AND MICROSTRUCTURE OF TiO_2-V_2O_5 HUMIDITY SENSOR

Chen Yi
China Electronic Product Reliability And environmental Testing Research Institute
P.O.BOX 1501-5 Guangzhou, China
Yun Zhengzhong
University of Electronic Science And Technology of China. P.O.BOX 83 Chengdu, China

ABSTRACT

The paper describes the process of making TiO_2-V_2O_5 humidity sensors. SEM, X-ray diffraction, Auger energy spectroscopy, X-ray flourescence and Mercury intrusion techniques have been used to study the microstructure of sensors and the characteristics of materials have been discussed. The results reveal that the content of impurity elements on the surface of failed sensors is higher than that of good ones. Porous structure of surface adsorbs ambient vapor. Electrons and protons take part in electrical conduction of elements

INTRODUCTION

The TiO_2-V_2O_5 humidity sensor was first developed by Chibu cement Co. ltd of Japan at the end of 1970's. The sensor has many advantages such as high sensitivity, good linearity, fast speed of response and wide measurement range, etc. But, on the other hand, the stability of TiO_2-V_2O_5 humidity sensor is not ideal enough to satisfy all needs. The microstructure has played an important role in determining the characteristics of sensors. Hence it is important to study the relations between humidity characteristics and microstructure of sensors.

SAMPLE PREPARATION

The fabrication of TiO_2-V_2O_5 humidity sensor follows standard ceramic techniques. The simplified flow diagram is, weighing materials→ball milling→pressing→sintering→cleaning→printing electrode→sintering→assembling→testing.
The raw materials used in preparing samples are analytically pure. Pellets are sintered at a high temperature(typically 1100～1350℃), which are electroded with a fired ruthenium oxide or silver, then leads are attached to fired RuO_2 or Ag electrode. Measurement humidity is produced by saturated aqueous salt which can vary from 11%RH to 98%RH.
Specifications of sensors are, response time(absoption; desorption), ＜20 secs; operating humidity, 10～100%RH; humidity hystersis, ＜4%RH; temperature coefficient, ＜0.3%RH/℃; range of resistance, 10^4～10^7 Ω.
The typical humidity-resistance curve is shown in Fig.1.

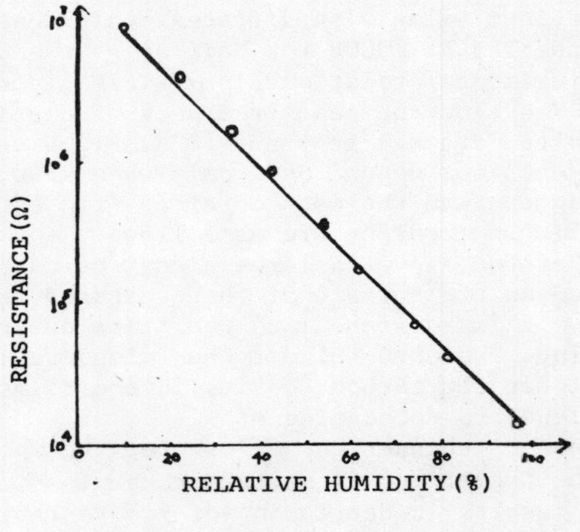

Fig.1 Humidity-resistance curve (measured at AC 150Hz, 8V)

MICOSTRUCTURE CHARACTERISTICS OF MATERIALS

Titania is an ionic compound. Because the outer electronic structure of titanium atom is $3d^2 4s^2$, its valence can be +2, +3 Or +4, and +4 is a stable valence. Ti^{4+} is easy to grasp external electrons and that makes part of Ti^{4+} ions become Ti^{3+}(Ti^{4+}·e) ions. Reaction process can be expressed as,
$Ti^{4+}O_2 - xO \rightarrow Ti^{4+}{}_{1-2x}(Ti^{4+}\cdot e)_{2x}O^{2-}{}_{2-x}$.
The electron grasped by Ti^{4+} ion is at a high energy state. It is weak binding and easy to escape. As the electric field is applied, weak binding elec-

trons can move in the direction opposite to the electric field and thus form an electric current. According to the theory of energy band, they are equivalent to electrons of conduction band. As a result, the more Ti^{3+} ions, the higher electrical conduction of materials. The reasons of causing lower valence titanium ions are as follows[1];

A. The influence of sintering atmosphere and high sintering temperature, In the sintering process, when reduced atmosphere(CO , H_2) or high sintering temperature are exerted on TiO_2 material, oxygen vancancy is produced in crystal lattice, resulting in the formation of Ti^{3+} ions. i.e.

$$TiO_2+xH_2 \rightarrow (Ti^{4+}{}_{1-2x}Ti^{3+}{}_{2x})O^{2-}{}_{2-x}VO_x+xH_2O\uparrow$$

$$TiO_2+xCO \rightarrow (Ti^{4+}{}_{1-2x}Ti^{3+}{}_{2x})O^{2-}{}_{2-x}VO_x+xCO_2\uparrow$$

$$TiO_2 \rightarrow (Ti^{4+}{}_{1-2x}Ti^{3+}{}_{2x})O^{2-}{}_{2-x}VO_x+x/2O_2\uparrow$$

B. The influence of impurity, When TiO_2 material contains high valence metal oxide such as V_2O_5, Ta_2O_5, Nb_2O_5 etc, the radius of five valence metal ions approaches that of Ti^{4+} ion's, hence it is possible that M^{5+} ion take place of Ti^{4+} ion. In this reaction process, the fifth valence electron of impurity atom is not accepted by oxygen ion in Ti-O octahedral structure. To keep electrical neutrality, the unnecessary valence electron is given to neighboring Ti^{4+} ion, which reduces a number of Ti^{4+} ions to Ti^{3+} ions.

C. The influence of environment, When TiO_2 material works at high temperature, high humidity and direct current electric field for a long time, because of the influence of environment, a few of oxygen ions on the surface, grain boundary and other defects can escape from crystal lattice sites. This phenomenon leads to the formation of Ti^{3+} ions.

In TiO_2-V_2O_5 material, the radius of Ti^{4+} ion is 0.68Å , and V^{5+} ion is 0.59Å [2]. As there is little difference between the radius of V^{5+} and Ti^{4+} ions, as mentioned above, V^{5+} ions are easy to replace Ti^{4+} ions at lattice sites, and the fifth electron of vanadium atom is given to Ti^{4+} ion in the vicinity, i.e,

$$(1-x)TiO_2+x/2V_2O_5 \rightarrow (Ti^{4+}{}_{1-2x}Ti^{3+}{}_xV^{5+}{}_x)O_2+x/4O_2$$

In the end, the TiO_2-V_2O_5 system exhibits a single phase solid solution with a pure rutile titania structure. With the mixture of V_2O_5, the electrical conduction of the material increase substantially. This means that V_2O_5 material has played an important role in semiconducting behaviour of TiO_2-V_2O_5 system.

From Fig. 2, we can see that all peaks are caused by rutile TiO_2. In addition, X-ray fluorescence method analysis reveals that there is V component in the material. This shows that substitute solid solution has been formed.

According to the results of scanning electron micrograph (SEM) on surface of sensors, the structure of ceramic surface is porous, and so is the surface of electrode. The porous surface has played an important role in determining humidity sensitive characteristics of sensors. In other words, the shape and parameters of micropores are very important.

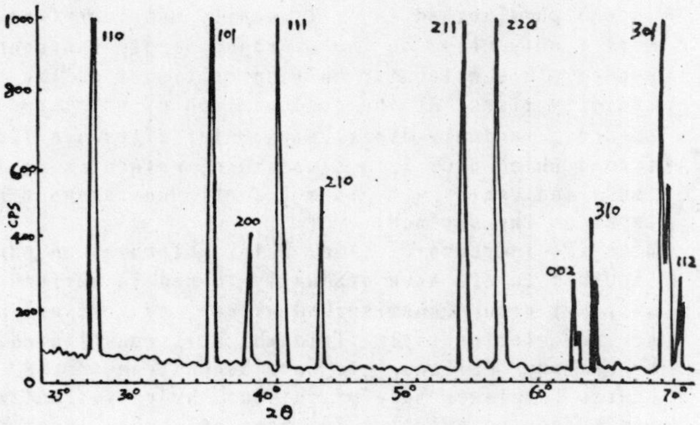

Fig. 2 X-ray diffraction spectroscopy of TiO_2-V_2O_5 material(peaks are all caused by rutile TiO_2)

Specifications of micropores of sensors measured by Mercury intrusion techniques are, total intrusion volume, 0.0714 cm^3/g; total pore area, 1.9428 m^2/g median pore radius(volume), 0.6999 um; median pore radius(area), 0.0036 um; average pore radius, 0.0735 um; porosity, 27.4%.

It is generally good when pore radius of ceramic humidity sensor is between 50Å ~ 5000Å and porosity is between 25 ~ 45 %. From the above measured data, we have known that the pore radius and porosity of TiO_2-V_2O_5 sensor are mostly within the range. It indicates that specifications of micropores are somewhat suitable. But in order to obtain the best effect, some improved methods must be applied.

Because rutile TiO_2 has powerful crystallization ability, it is easy to grow up big and rough crystal grains in sintering process. On this occasion, grains only relat each other by neck part. Hence a large number of open pores have been produced. In addition, part of V_2O_5 material probably evaporates to ambient in the sintering process. As a result, the porous surface structure has been formed. According to the adsorption isotherm, it is possible that ink bottle form micropores have probably taker the dominant place.

DISCUSSION

1. Surface Adsorption and Electrical Conduction[3,4,5,6]

Water vapor has two kinds of adsorption on metal oxide surface, the physical and chemical adsorption. In the adsorption process, when water vapor contacts the porous surface of sensors, the monomolecular water layer has been chemisorbed on the surface. The chemisorbed process happens with the decomposition of water molecules which results in the formation of surface hydroxide called hydroxy layer. After the chemisorbed layer has been finished, more water molecules are adsorped physically above chemisorbed water molecules layer to form phsisorbed multimolecular water layers. In the first physisorbed layer, each water molecule is adsorbed on two hydroxyl groups through hydrogen bonding. The second physisorbed layer is mobile and therefore more loosly held, on the average, singly hydrogen bonded. This molecular buildup continues as the humidity rises. As the condensation of moisture commences, randomly distributed water sites are first formed which grow into clusters of molecules or islands and coalesce to form a continuous aqueous adlayer on the surface.

With the increase of time, water molecules on physisorbed layers have gradually turned to surface hydroxyl groups(chemisorbed water). As a result, surface electrostatic field which is constituted by hydroxy, cations, etc have been strengthened. Charge complexes made of cations, hydroxyls and water molecules existing in the chemisorbed layer of the sensor support localized electrostatic field which not only attracts water molecules for physisoption but also dissociates them.

Both protons and electrons have contributed to electrical conduction of sensors. Physisobed water molecules are dissociated as OH^- and $H_3O^+(H_2O+H^+)$ ions, i.e $2H_2O=H_3O^++OH^-$. the charge carrier of protons is hydroxonium ion(H_3O^+) which moves by Grotthus chain reaction. The model of Grotthus chain reaction is; A proton adheres to one water molecule and they form a hydroxonium. The hydroxonium releases one proton to the second water molecule and they form a hydroxonium again. In this way, hydroxoiums move through continuous water film.

At low humidity, electrons have played an important role in conduction. Its rough model is; The adsorbed water has a donor action. Thus the adsorbed water on surface can transfer electrons to it This transference of electrons causes the appearance of holes in the water molecules and a positive charge is produced on the surface. At low humidity, water molecules can only form the sub-monolayer, on which the conduction pathway is not continuous. To overcome the discontinuous pathway, an electron can tunnel from a water molecule to a hole in another water molecule. In all probability, these sites are all not at the same energy. Hence, the transition is always accompanied by the emission or absorption of a phonon. In summary, the model is of a phonon-assisted electron tunneling from a donor water site to another donor water site.

In short, the low humidity and high humidity response characteristics of $TiO_2-V_2O_5$ sensor depend upon different surface conduction mechanisms, on which electrons and protons take mainly part in conduction respectively.

2. Theoretic Explanation on Stability of Sensors

Fig.3 is the experiment curve of long term stability of sensors at low and high humidity. The simple explanation are given as below.

When sensors have been at low humidity for a long time, most of water molecules chmisorbed on surface and few of physisorbed. With the theory mentioned above, physisorbed water molecules have gradually turned to chemisorbed water molecules. for this reason, hydroxoniums have decreased. Only electrons take part in conduction, hence resistance increases. At high humidity, there are more physisorbed water molecules on the surface. At first, the electrostatic field formed by charge complexes is not strong enough, which only makes part of the physisorbed water molecules to be decomposed as hydroxoniums. With the increase of time, because of change of adsoption form of water molecules, more and more hydroxyl groups have been formed. Hence, the electrostatic field has become stronger and stronger, which results in the incease of dissociation ratio of physisorbed water. But the formation speed of hydroxyl groups becomes slower and slower, and gradually tents to saturate. It means that the number of prontons on surface is constant. In the end, electrostatic field and conductivity of the sensor become comparatively stable.

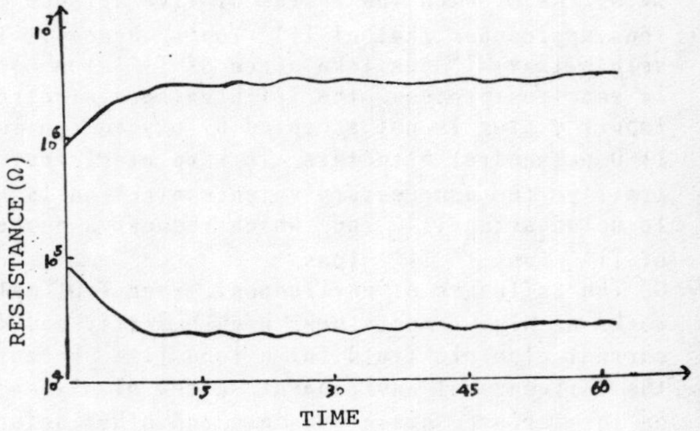

Fig.3 The variation of resistance of sensors with time

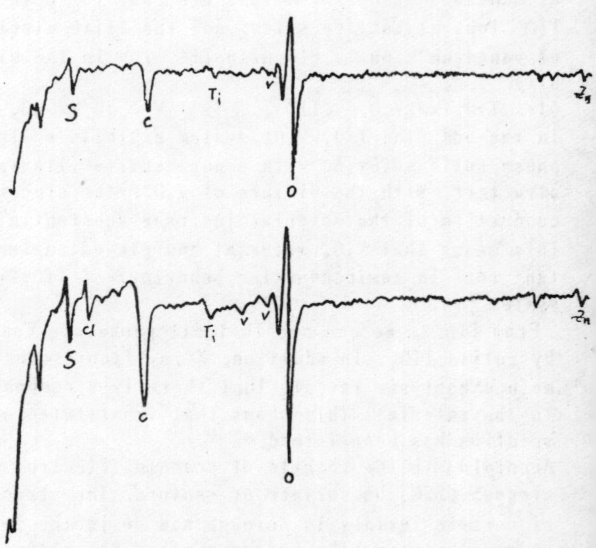

Fig.4 Auger energy spectroscopy of $TiO_2-V_2O_5$ senso (above is of a good device, below is of a failed one

Because the sensor is exposed to an atmosphere containing dust, oil gas and other kinds of gases, air pollution greatly affects the characteristics of sensors. i.e continued exposure of the sensor produces a progressive increase in resistance at constant humidity. The resistance change is accompanied by changes in the composition of the surface.

Auger energy spectroscopy analyses find that the content of S and O elements on the surface of a failed sensor is more than that in a good one (Fig.4) and it was found that the content of C and O elements decreases sharaply with the increase of depth. Meanwhile, Cl element was also discovered. They indicate that impurity elements compound with surface cations. And gases such as CO_2, SO_2 which are easily soluble to water enter into the surface and not being desorbed. All these have an important effect on normal conduction of sensors, which makes devices lose their sensitivity.

When the TiO_2-V_2O_5 sensor is heated above 400°C which is called heat cleaning', impurity, oil pollution and chemisorbed water molecules on the surface could be partly expelled. But experiments show that this method can not perfectly recover the characteristics of sensors. Perhaps this is attributed to the impurity ions entering the body of disk. Meanwhile, some compounds have been formed by the combination of materials and impurity ions. For this reason, the physical and chemical behavior of sensors have changed.

CONCLUSION

1. TiO_2-V_2O_5 is a more promising material for making humidity sensors, which can obtain good, well-distributed porous structure after suitable firing process, and the electrode is important in determining humidity behaviour.
2. Both electrons and protons take part in conduction, electrons mainly at low humidity and protons mainly at high humidity.
3. Because physical and chemical change take place on the surface and in the body of devices, the resistance of sensors varies with time. Heat cleaning can not recover humidity characteristics perfectly.

In the future, we must do more research work on the stability of sensors. The specifications and shape of micropores have played an important role in determining humidity characteristics. Many mensures must be taken to control them.

The authors are pleased to acknowledge helpful discussions with Qiu Zudong senior engineer.

REFERANCE

1. The Third Department of Chengdu Institute of Radio Engineering, Electronic Ceramic(textbook), 1983
2. K.Katayama, T.Akiba.Rutile humidity sensors. International Chemical Sensors Symp,1983
3. Victor E.henrich,Ultraviolet photoemission studies of molecular adsorption on oxide surfaces. Pro Surf Sci, 1979;9;143-164
4. F.Uchikawa et al, Time variability of surface ionic conduction on humidity-sensitive SiO_2 films, Am Ceram Soc Bull. 1985;64(8);1137-1141
5. V.K.Khanna,R.K.Nahar, Surface conduction mechanisms and the electrical properites of Al_2O_3 humidity sensor, Applied Surface Science.1987; 28;247-264
6. T,Yamamoto,Some consideration on stability of electrical resistance of the TiO_2/SnO_2 ceramic moisture sensor, IEEE Trans.Components,Hybrids and Manufacturing Technology, 1979;CHMT-5(2); 238-241
7. T.Nitta et al, Ceramic humidity sensors, IEEE Trans.Components,Hybrids and Manufacturing Technology.1980;CHMT-3(2);237-243

AN INVESTIGATION ON THE CONDUCTING AND AGEING MECHANISM OF CERAMIC HUMIDITY SENSITIVE COMPONENTS

B.R. Li, X.G. Pan and J.L. Zhang
Department of Solid State Electronics
Huazhong University of Science and Technology, Wuhan, P.R. China

ABSTRACT

The expermental investigation on the charge-discharge characteristics under DC voltage and the ageing characteristics under different conditions of $ZnCr_2O_4$, $Ba_{1-x}Sr_xTiO_3$ etc. was carried out. The accumulative ion charge at different relative humidity was calculated using numerical integral method. The conducting and ageing mechanism were analyzed and discussed.

INTRODUCTION

Ceramic moisture-sensitive components are considered to be physically, chemically and thermally stable compared to the electrolytic or organic materials. However, most of them are lacking in regard to considerably high resistance and long term resistance stability, which results in quite troublesome for practical use.

Recently, two kinds of conduction mechanism, namely electronic and ionic conduction theories, have been developed[1,2]. But neither of them can explain the charge carriers and moisture sensitive properties adequately. On the other hand, the ageing mechanism of the materials still need further investigation although some work has been done[3,4].

The present study was conducted to investigate the conducting and ageing properties of $ZnCr_2O_4$, $Ba_{1-x}Sr_xTiO_3$ etc. ceramic moisture-sensitive elements and to clarify their conducting and ageing mechanism.

EXPERIMENTAL

1. Measurement of Charge-discharge Characteristics

The measurement of charge-discharge characteristics of different samples at various humidity was carried out using a microcomputer system. Fig. 1 is the charge-discharge curve of $ZnCr_2O_4$ -- $LiZnV_2O_4$ element.

2. The Life-tests Under Different Environment

The results of the life-tests of the components are shown in Fig. 2, which indicates the

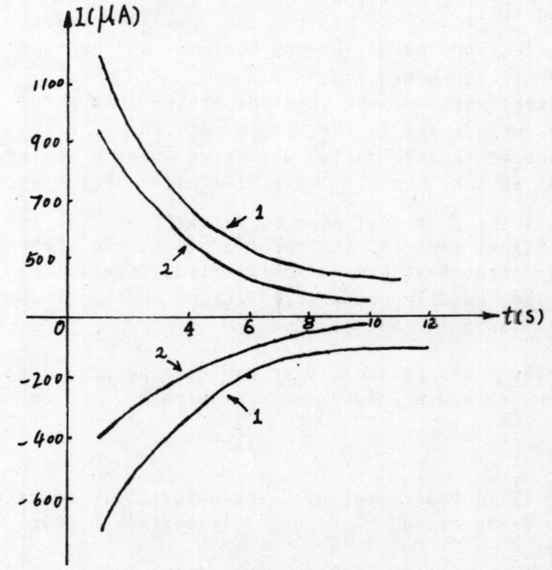

Fig. 1(A) Charge-discharge characteristics
curve 1 : 100 RH%   curve 2 : 96.9 RH%

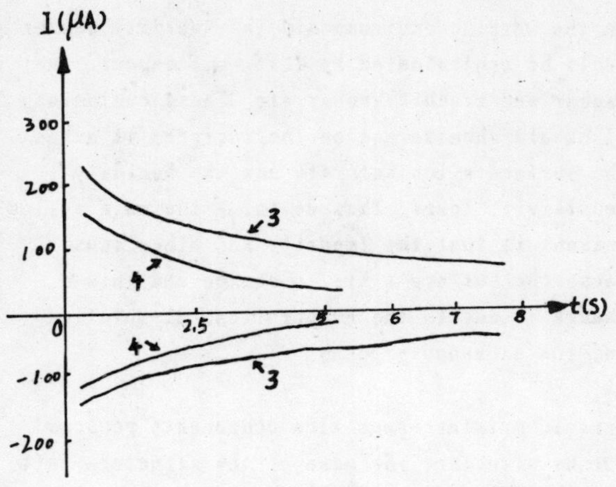

Fig. 1(B) Charge-discharge characteristics
curve 3 : 80 RH% curve 4 : 75.8 RH%

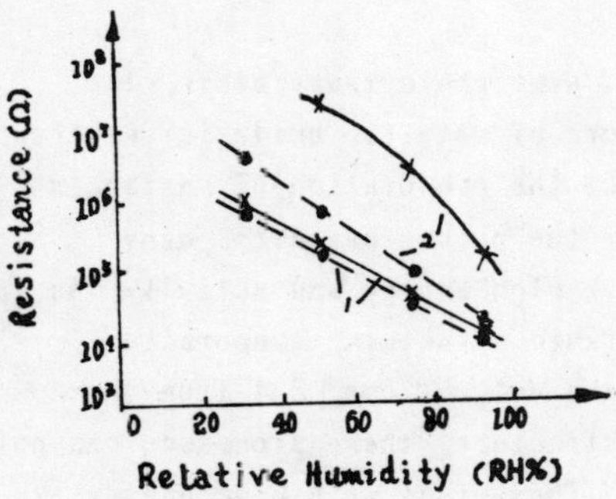

Fig. 2 Typical results of Life-tests

comparatively stable operation of the elements in the cleaner environment. In Fig. 2, curve 1 shows the characteristics of the resistance versus relative humidity of the tested device which was measured just after the fabrication. Curve 2 shows the results after the tested device had been left for two months in the atmosphere(real line) and in a clean container (dotted line) respectively.

3. Ageing Test Under Different Conditions

Fig. 3 shows the typical results of speeding age test. Curve 1 and 2 are the characteristics of electrical resistance versus relative humidity before and after sample is heated in water at 80℃ for 72hr.

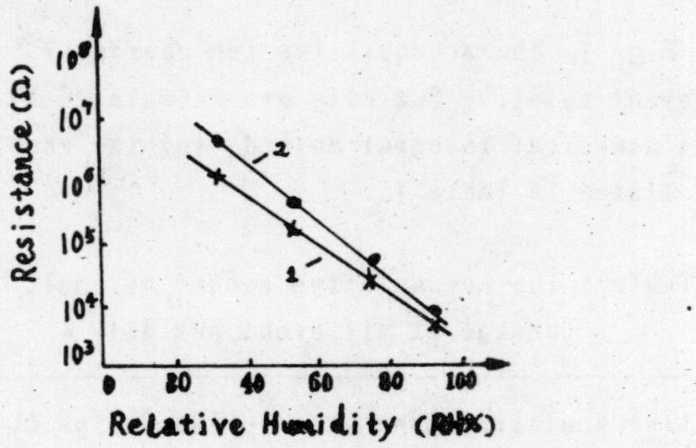

Fig. 3 Speeding-age curve

Fig. 4 shows the ageing curve of resistance versus time. The dust and gas adsorped in the sample was removed by heating the element intermittently up to 450℃ for a short time. The results indicate that variation of resistance decreases with the time while the resistance increases with the time.

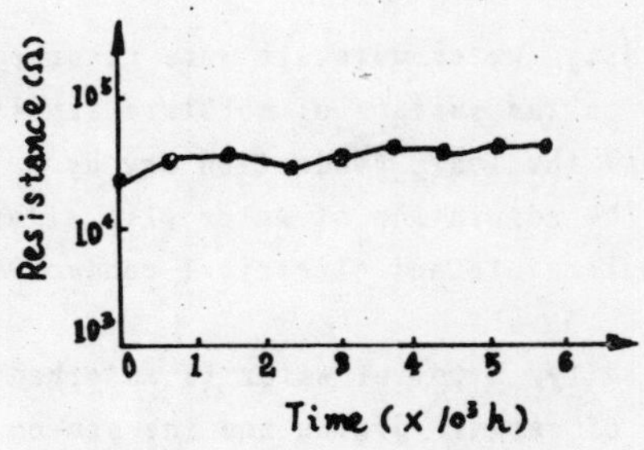

Fig. 4 Resistance variation curve

RESULTS AND DISCUSSIONS

In our opinion, both electronic and ionic conduction are considered to make contribution to the electrical conduction of humidity sensitive materials, and the electronic and ionic conduction are dominant at low and high humidity respectively. To verify this, it is essential to distinguish the ionic and electronic conduction in the electrical conduction process. Since the ionic conduction under DC voltage results in the accumulation of ionic charge and the charge-discharge properties. We can get some useful informationm on conduction mechanism from the charge-discharge characteristics.

From Fig. 1, the accumulative ion charge at different relative humidity was calculated by using numerical integral method, and the results were listed in Table 1.

Table 1 The accumulative amount of ionic charge at different humidity

Relative Humidity (RH%)	Ion Charge (C)
100	2.518×10^{-3}
96.9	2.023×10^{-3}
80.0	4.710×10^{-3}
75.8	1.940×10^{-3}

It can be seen from Table 1 that the proportion of ionic conduction as well as the accumulative amount of ionic charge increase with the relative humidity.

At low humidity, water molecule were adsorbed continually on the surface of moisture sensitive elements, and the ionic conduction may be neglected, The adsorption of water will alter the electronic state and electrical conduction.

At high humidity, a lot of water is adsorbed on the surface of ceramic grains and the proton in water is the main carriers. In the first stage of water adsorption, a few water molecules chemisorb on the surface of the ceramics to form hydroxyl group, then a proton may be transfered from hydroxyl groups to water molecules to form H_3O^+. In the second stage more water is adsorbed physically on the hydroxide surface to form the hydroxyl multilayer.

From Fig. 2 -- Fig. 4, Three features on the ageing phenomena are:

A. Regardless of the circumstance, the resistance of the sensors always increases with the time except for that the ageing rate alters with the ambient condition.

B. At the same relative humidity, the sensor's ageing rate in the atmosphere is greater than that in an enclosed clean container.

C. The ageing rate of the humidity sensors decreases with the time.

In the working environment, the humidity sensor would be contaminated by dust, oil vapour, fuel vapour and organic vapour etc. These contaminations are chemisorbed on the surface. It makes the surface's ion saturate and the humidity sensitivity lower. Thus we think the main ageing reasons is that the impurity and other ions cause the surface state to change and this change is due to the evaporation, dissolution and ion exchange process.

Ceramic moisture-sensitive components possess porous structure. Because of the structure full of bumps and holes the ceramic body is sensitive to moisture. The ionic density in water membrane on the convex surface is greater than that on the concavity, so the ion in water shifts from land to concavity.

At the normal atmosphere temperature, the vapour pressure of metallic oxide is very low. We may neglect the evaporation of surface's particle. For the porous ceramics, many particles have high energy and activity. As for the whole ceramic structure, evaporation, congealment are not obvious. But from the microscopic structure, these processes can not be neglected. Those ions with high energy are very easy to dissolve and evaporate. It is the main factor that the surface is sensitive to the moisture. Even if the amount of the ions decreases a little bit, the activity will change obviously.

On the other hand, the ions in surface arrange in loosen and confusion, and ionically bond with opposite ions near surface, but the ionic binding power in surface is weak than that near the surface. Thus, the ions in surface(mostly in the land) have exchanging membrane. When the surface were contaminated, a few ions may be dissolved in water in membrane. When the ions in surface are attracted by opposite ions and repelled by the same ions in water membrane, they may leave primary site to water membrane, and the same impurity ions will replace them. It is only those ions which strongly band with ions near surface after they exchanged can exchange. The impurity ions become a part of the ceramic grains after they exchange with highly active ions in the surface of ceramic grains. So the impurity ions in water membrane decreases, which spurs the impurity to dissolve and the ions to exchange. The exchanged ions in surface dissolve in water membrane. After they become saturated, they will be deposited in the surface(mostly in he low holes). Although the process above goes on slowly, it revises the active surface state of the ceramic grains, lowers the activity and the moisture sensitivity, and increases the resistance of the sensors.

CONCLUSIONS

We may summarize the moisture-sensitive mechanism as follows, adsorption mechanism, conducting and ageing mechanism. The adsorption of water molecule usually transforms physical adsorption into chemisorb. Both electronic and ionic conduction make contribution to the electrical conduction of the humidity sensors. The electronic conduction is dominant at low relative humidity while the ion conduction is dominant at high relative humidity, and the ion conduction increases with the relative humidity.

For the ceramic moisture-sensitive components, the surface's activity and attraction force decreases because of the dissolution, evaporation in land and resolving crystal, congealment in holes. The activity decreases with time, so the resistance increases with time. Some ions, such as aluminum, iron, calcium and sodium dissolve in the surface of ceramic grains, which will lower the surface activity and increase the resistance. The high active ions are easy to dissolve and ervaporate, which makes the newly emerging sensors age fast at the beginning and the ageing rate decrease while the percentage of active ions decrease.

REFERENCES

1. Fleming William, J. SEA Prepr n 810432 (1981)
2. Tsuneharu Nitta, J. of the Amer. Cream. Soc., 63 (1980), 295.
3. Tatsuo Yamamoto, IEEE-CHMT, CHMT-5(1982), 238.
4. Tsumeharu Nitta, IEEE-ED, ED-29(1982), 95.

MICROSTRUCTURE AND PROPERTIES OF BSTZ-LZV MULTIFUNCTIONAL SENSOR

Zhang Xiaolin and Meng Zhongyan
Department of Electronic Engineering
Xi'an Jiaotong University
Xi'an, Shaanxi, P.R. China

ABSTRACT

$(Ba,Sr)(Ti,Zr)O_3$ (BSTZ) based ceramics with impure phase of Li_2O-ZnO-V_2O_5 (LZV) have been investigated. The sensors fabricated with these ceramics are independently sensitive to both temperature and humidity. X-ray diffraction patterns and electron microscopy analysis (TEM,SEM) illustrated that the multiphase compound ceramics are composed of BSTZ grains forming a porous ceramic structure, and of $Zn_3V_3O_8$ and $Ba_3(VO_4)_2$ situating on the surface of BSTZ grains. Such distribution of the impurities is beneficial to the sensing characteristics. Finally, a large body of experimental data is presented to indirectly validate the proposed conduction mechanism.

INTRODUCTION

Recent developments in automated systems have made increasing demands for detecting and controling both of temperature and humidity. several kinds temperature-humidity multifunctional sensors have been reported[5,6]. The recent trend toward such sensors is based on the development of metal-oxide ceramic humidity sensors which appear high thermal stability. Temperature-humidity sensors reported in this paper was developed for application to air conditioning systems. The sensing materials developed in this work are multiphase compound ceramics which are composed of main crystallin phase $(Ba,Sr)(Ti,Zr)O_3$ and impure phase Li_2O-ZnO-V_2O_5. The microstructure and the impure phase distribution in the ceramics were examined. The conduction mechanisms for humidity sensitivity are discussed.

EXPERIMENTAL PROCEDURE

Samples of different compositions were prepared by mixing reagent grade chemicals according to the following formula,

$$(1-x)BaCO_3+xSrCO_3+(1-y)TiO_2+yZrO_2 \text{ --- } (Ba_{1-x},Sr_x)(Ti_{1-y},Zr_y)O_3$$

where x=0-0.6, y=0-0.1
The mixture was calcined at 1250 °C for 2 h in air. After which the synthesis was mixed with small amounts of reagent grade ZnO,Li_2CO_3 and V_2O_5 used as impuritis, and with 15wt% powdered carbon used to form porosity. The resulting mixture was pressed into disks about 9 mm in diameter and 2 mm thick at a pressure of 900 Pa. The pressed disks were sintered at 1150 °C for 0.5 h in air. The sintered samples were polished to 0.25 mm thickness. Porous Ru electrodes were fired for electrical measurements, followed by ageing for 7 days under high relative humidity (97%RH) at 90 °C. Samples for TEM were mechanical ground to 40 um, followed by Ar^+ ion milling to perforation.

Crystalline phases were identified by X-ray diffraction (XRD), microstructures were examined by scanning electron microscopy (SEM) and transmission electron microscopy (TEM) and the phases present was characterized by electron diffraction. The sensing characteristics and electrical properties were measured using Hp-4274A, Hp-4275A and Hp-3585A meters.

RESULTS AND DISCUSSIONS

1. Microstructures

A SEM photograph of a fractured section is shown in Fig. 1, which reveals a typical porous structure, these pores with different radii

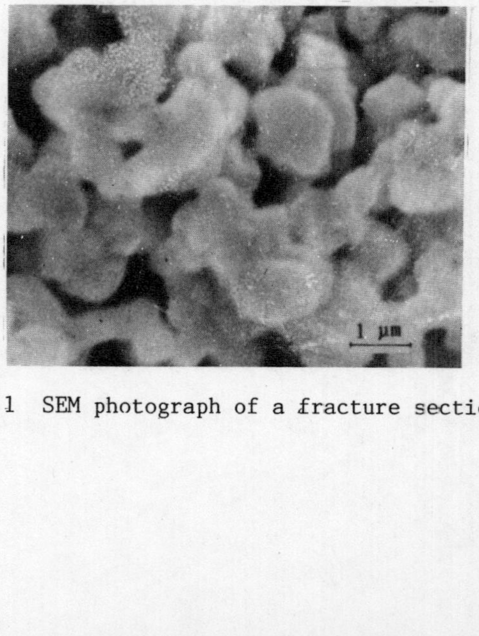

Fig. 1 SEM photograph of a fracture section.

interconnect each other and make up a three dimensional network. Forthermore adjacent grains also interconnect each other and form necks. X-ray diffraction pattens show that the sensing ceramics contains three crystallin phases of BSTZ perovskite, $Zn_3V_3O_8$ spinel[4] and $Ba_3(VO_4)_2$ hexagonal[3], as shown in Fig. 2.

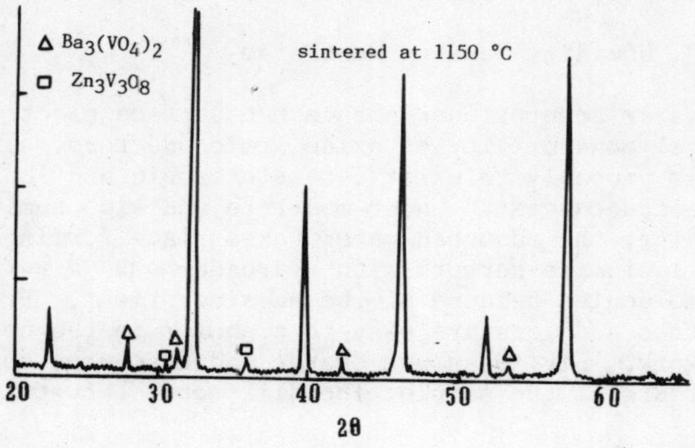

Fig. 2 X-ray diffraction pattens

A bright-field image of a triple-grain junction and electron diffraction pattens are shown in Fig. 3. The grains were determined to be main

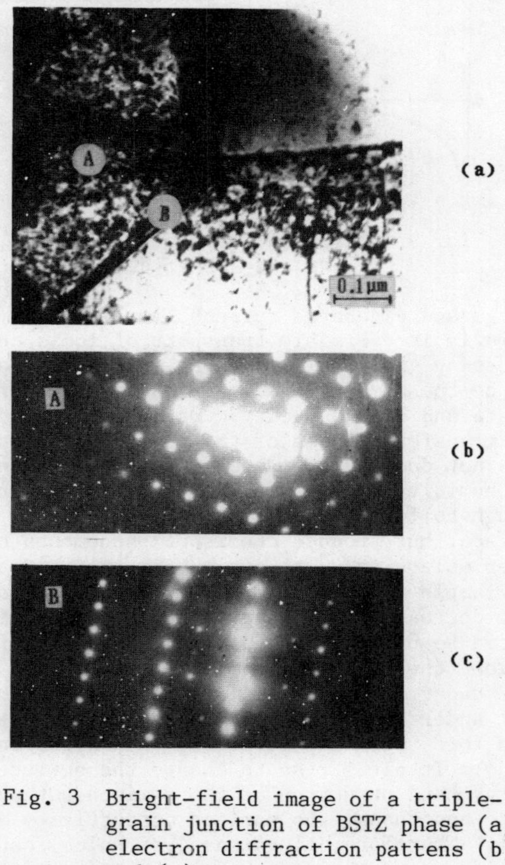

Fig. 3 Bright-field image of a triple-grain junction of BSTZ phase (a), electron diffraction pattens (b) and (c).

crystalline phase BSTZ, and the impure crystalline phase was not found at the grain boundaries by electron diffraction. Fig. 4 shows a dark image of $Zn_3V_3O_8$ spinel. It is indicated that $Zn_3V_3O_8$ is situated on the neck and surface of BSTZ grains.

Fig. 4 Dark-field image of $Zn_3V_3O_8$ spinel

From the results just mantioned, it can be recognized that BSTZ grains form a porous ceramic structure and humidity sensing impurities are situated on the surface of BSTZ grains. However, there is no impure crystlline phase found at BSTZ grain boundary. A schematic microstructure is shown in Fig. 5. Obviously, the distribution of impurities is beneficial to the sensing characteristics of the sample.

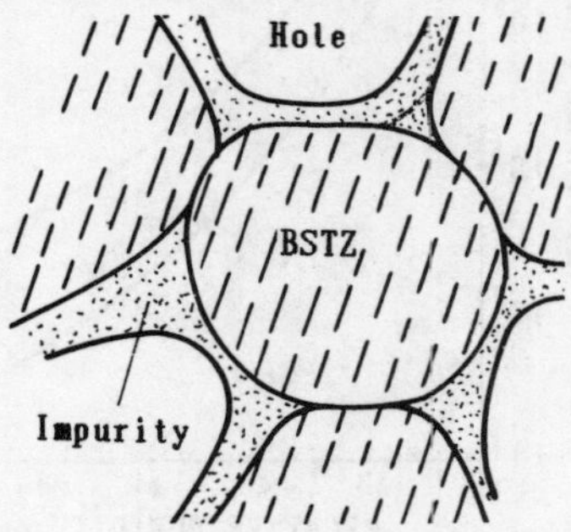

Fig. 5 Scheme microstructure

2. Sensor Characteristics

A temperature detection is based on the dielectric constant obeying the Curie-Weise law above the Curie point of the BSTZ ferroelectric phase and a relative humidity-sensitivity based on the impedance or resistance of the porous ceramics containing LZV additives.

Fig. 6 shows temperature dependence of dielectric constant. It can be seen that the dielectric constant decreasing is over one order with the temperature increasing from -10 °C to 200 °C, and that the temperature dependence of the reciprocal dielectric constant reveals good linearity. Moreover there is no obvious influence of ambient humidity on the capacitance of the sample at high frequency, as shown in Fig. 10 (a). Fig. 7 shows the logarithmic resistance versus relative humidity. It appears

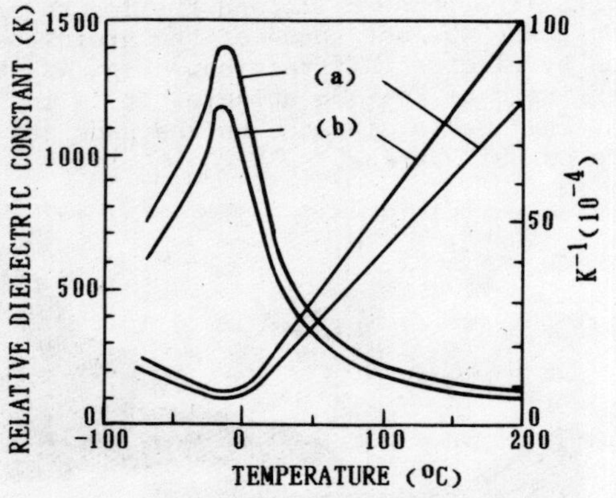

Fig. 6 Temperature dependence of dielectric constant of the sample (a) add 5wt% LZV and (b) add 3wt% LZV.

good linearity in the humidity range from 10% RH to 80% RH. The temperature coefficient for

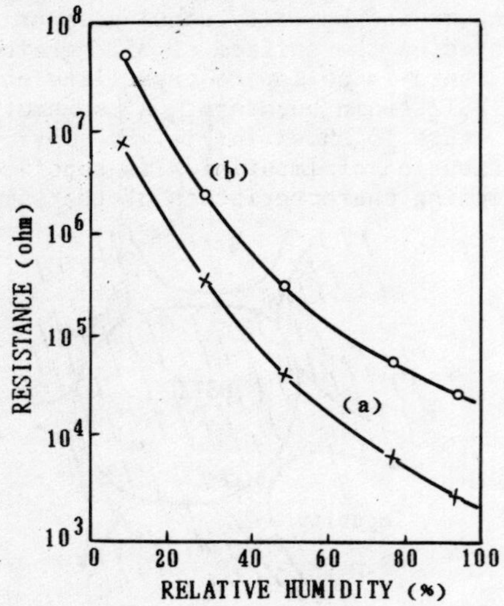

Fig. 7 Resistance-humidity characteristics of the sample (a) and (b)

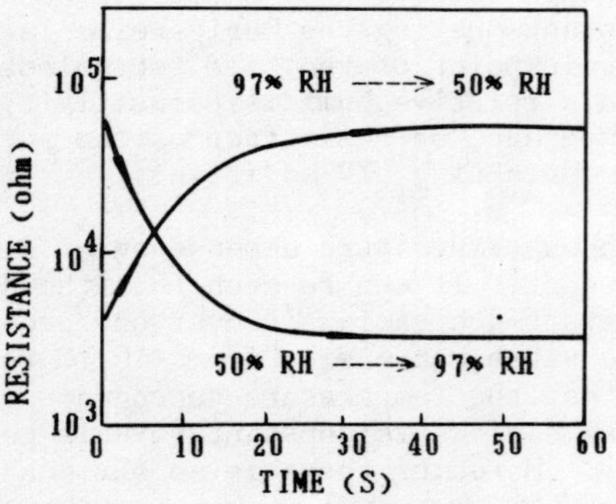

Fig. 8 Humidity response characteristics.

humidity measurement is around 0.3% RH/°C between 10 °C and 40 °C. From the ageing results, it is found that the sensors appear high stability. In addition, there is no hysteresis observed for both temperature and humidity detectings. The response time is less than 40 seconds in both adsorption and desorption processes, as shown in Fig. 8.

3. Humidity Sensing Mechanism

Water adsorptions enhance the surface electrical conductivity of oxide semiconductors. It is probably to exist both electronic and ionic conductances[1]. Under moderate and high humidity, the adsorbed water takes place forming a liquidlike network with hydrogen-bounded water molecules because of the physisorption[1]. Protons and ions are easy to transfer in the network. Fig. 9 shows charge and discharge currents of the sample. The difference between

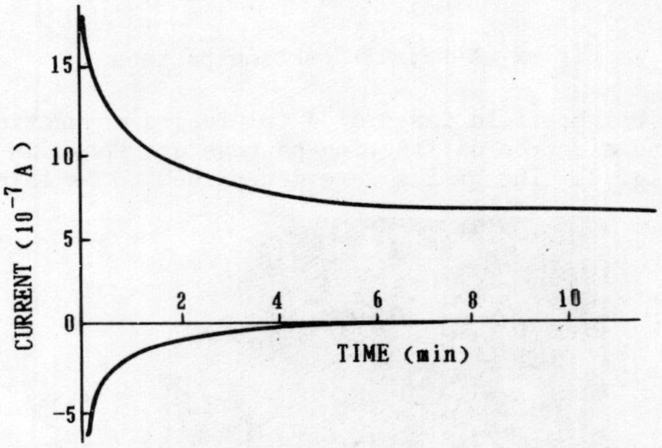

Fig. 9 Charge and discharge curve at 15 °C 57% RH and 1.0 V (dc)

charge and discharge currents implies that there is irreversible transport of ionic charge acrossing the sample, protons (H_3O^+) and ions may be the dominating charge carriers under moderate and high humidity. On the other hand, the transfer process of the protons and ions does not dominate the conducting process under low humidity, since the adsorbed water is not enough to form a liquidlike network on the grain surface, but produce chemisorb phenomenon of water molecules. The dielectric relaxation of the sample is shown in Fig. 10. The relaxation time decreases with increasing ambient humidity. It can be understood that the interaction force between the adsorbed water and the ceramic surface under low humidity is more strong than that under high humidity, and that the interaction force may be more strong than hydrogen bond[2]. It gives rise to change the surface energy band structure of the semiconducting oxide and change the surface conductivity as well. Therefore, electrons (or holes) can be considered as the dominating charge carriers under low humidity.

It has be pointed above that the distribution

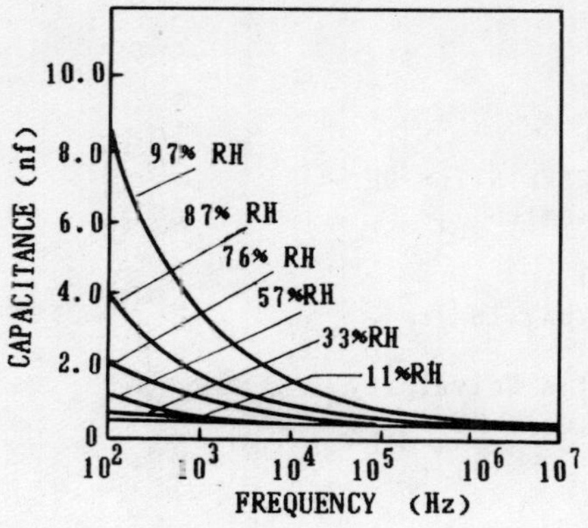

(a) Capacitance-frequency characteristics.

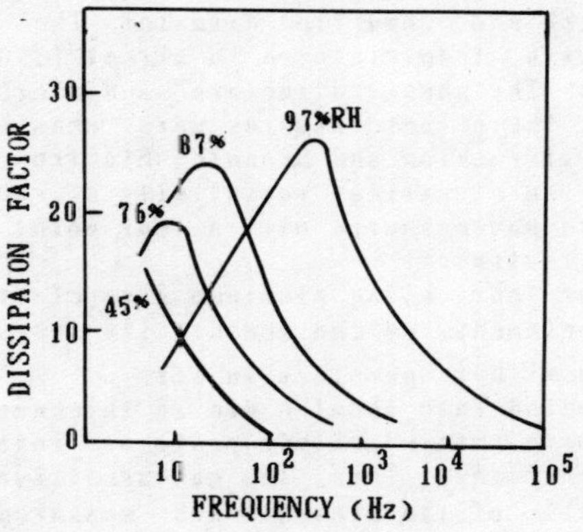

(b) Dissipaion factor-frequency characteristics.

Fig. 10 The dielectric relaxation of the sample

of impurities is beneficial to humidity sensitivity. It may be interpreted that the $Zn_3V_3O_8$ in the ceramic is composed of a normal spinel, $Zn^{2+}(V_2{}^{3+})O_4$, and a inverse spinel, $Zn^{2+}(Zn^{2+}V^{4+})O_4$. Each of V^{3+} and V^{4+} ions occupy the octahedral sites. The quasi-chemical reaction of $V^{3+} + V^{4+} = V^{4+} + V^{3+}$ occurs very easily. The electrical behavior of the $Zn_3V_3O_8$ is similar with that of a p-type semiconductor, since the number of V^{3+} ion is twice of V^{4+} ion. The water absorbing on the ceramic grain surface modifies the concentration of V^{4+} ions on the surface in consequent the resistance of sensing element changes.

CONCLUSIONS

1. The BSTZ-LZV multiphase compound ceramics consist of main crystalline phase BSTZ perovskite and impure phase $Zn_3V_3O_8$ spinel and $Ba_3(VO_4)_2$ hexagonal grains. The BSTZ grains form a porous ceramic structure and the impurities are situated on the surface of the BSTZ grains.

2. The sensor is sensitive to both temperature and humidity with high sensitivity and reliability, especially it can be continusly operated without heat-cleanings.

3. The main conducting carriers are holes under low humidity and ions under high humidity.

ACKNOWLEDGEMENT

The work was supported by the National Science Foundation of China.

REFERENCES

1. J.H.Anderson and G.A.Parks, J. Phys. Chem., 72[10] 3662-68(1968).
2. E. Mccafferty and A.C.Zettlemoyer, Discuss Faraday Soc., 239(1971).
3. JCPDS Powder Diffraction File 29-211 (1971).
4. JCPDS Powder Diffraction File 31-1477(1981).
5. T. Nitta, J. Terada and F. Fukushima, IEEE Trans. on Electron Device ED-29, No.1 99 (1982).
6. Li Fang and Meng Zhongyan, Proceeding of The Second National Transducer Conference (in Chinese). Vol.2, 510

THE STUDY OF THE SENSITIVE CHARACTERISTICS OF SnO_2-ZnO SOLID SOLUTION CERAMICS

Bao Fengqi, Niu Mengnian, Liu Yongxin
and Fu Zhiqiang
Department of Physics, Inner Mongolia University
Huhhot, P.R.China

ABSTRACT

The purpose of this paper is to investigate the effect of different proportions of ZnO or SnO_2 content on the structure and sensitive characteristics of SnO_2-ZnO ceramics. With changing the proportions of ZnO and SnO_2 in the SnO_2-ZnO solid solution ceramics, we can obtain a material sensitive to heat, gas and humidity, which satisfies the need of designs.

INTRODUCTION

The N-type semiconductive ceramics SnO_2-ZnO is a new kind of multifunction sensitive ceramics which we first developed in 1978. The ceramics has better electrical and optical characteristics, and the characteristics of being sensitive to heat, gas and humidity are better, too. It has practically been used as the conductive ceramics and transparent conductive thin film at home and abroad[1,2,3]. The sensitive material has been studied in[4], Zn_2SnO_4 ceramics with anti-spinel type structure has been studied in [5,6].

In this paper, we study the effect of different proportions of ZnO or SnO_2 content on the structure and sensitive characteristics of SnO_2-ZnO ceramics. With changing the proportion of ZnO and SnO_2 in raw materials, we can obtain the material sensitive to heat, which satisfies the need of designs. The material's resistivity at room temperature is from 10^{-3} to 10^{-8} Ω-cm, its B value (K) is between 200 and 22,000, and its operating temperature is from 77 to 1,200K. We can also see that changing the proportion of ZnO or SnO_2 in the SnO_2-ZnO solid solution ceramics and doping the rare-earth oxide into the single phase Zn_2SnO_4 ceramics have similar sensitive action for ethanol (C_6H_{10}) vapour. Being sensitive to humidity, the ceramics has the advantage of less hysteresis effect.

EXPERIMENTAL

A. In a common ceramic procedure the mixture of the ZnO and SnO_2 (chemical--pure on the market) in different proportions was compressed into discs which were approximately 2mm thick and 20mm in diameter. The green discs were then sintered in air at 1350°C for 8 hours. The phase structure and microstructure of the ceramic samples were examined by X-Ray Diffraction and Scanning Electron Microscope. The electrical resistivity at room temperature was measured with a four point resistivity instrument.

B. After fabricating platinum wire electrodes on the element we can measure its resistance varying with temperature in air.

C. Grinding into about 0.6mm in thickness, the discs were smeared silver paste as interdigital electrodes. Then, the gas sensitive characteristic of the element was measured. The discs of the single phase Zn_2SnO_4 ceramics, which structure was determined by XRD, were grinded into powder, then milled with rare-earth oxide (about 1wt%) in an agate mortar for 30 minutes. The sintered gas sensor, then, were sintered in air at 1350°C for an hour. We used heating voltage (V_H) to express operating temperature of the element. Gas sensitivity was defined as the ratio of the electrical resistance of element in air (R_a) to that in a sample gas (R_g), namely R_a/R_g.

D. The milled powder of the SnO_2-ZnO ceramics mixed with an amount of binder was smeared onto an Al_2O_3 substrate. After sintering in air at 800°C for 2 hours and smearing silver paste as interdigital electrodes, we can obtain a humidity sensor.

RESULTS AND DISCUSSION

1. The Ceramics' Structure

The XRD analyses are shown in Fig.1-1. The specimens are the single phase solid solution SnO_2 and ZnO when ZnO content is less than 5wt% and over 95wt%, respectively. If the ZnO content is just 40wt%, the specimen is the single phase solid solution Zn_2SnO_4 ceramics with anti-spinel type structure. The specimen is the mixture of SnO_2 phase and Zn_2SnO_4 phase when the ZnO content is less than 40wt%, and is the mixture of ZnO phase and Zn_2SnO_4 phase when ZnO content is over 40wt%.

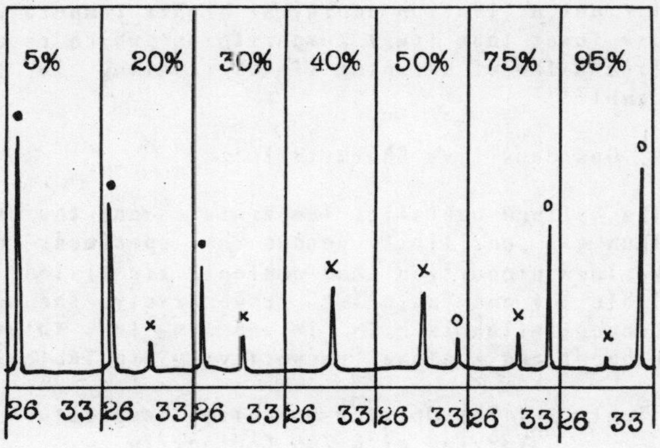

Fig.1-1 The XRD Spectrum of the Specimens of Different Proportions
•-SnO_2 phase; ×-Zn_2SnO_4 phase; ∘-ZnO phase

As shown in Fig.1-2, the microstructure examined by SEM indicate that the single phase solid solution is the porous ceramics with more homogeneous grains, and the solid solution is the aggregate of many different grains in which the gas holes are small and less. The physical characteristic of the specimen strongly depends on the microstructure.

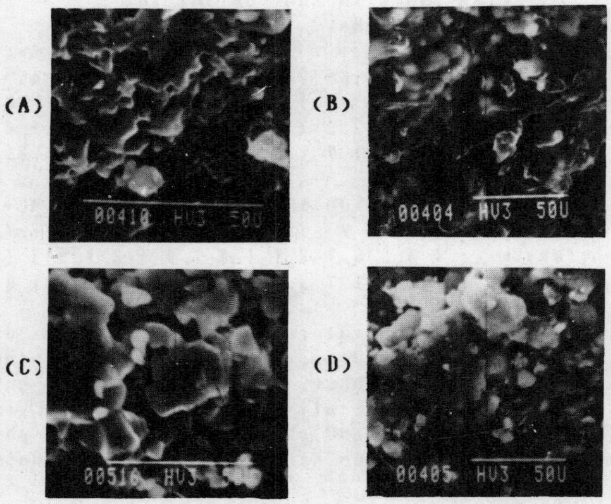

Fig.1-2 The Photographes Taken by SEM ZnO Content of 5%, 20%, 40% and 70% in Samples are Shown in A, B, C and D, Respectively.

The specimen's resistivity at room temperature varying with the ZnO content is shown in Fig.1-3. There is a minimum of resistivity when the ZnO content is about 20wt%. The minimum is just at the middle of pure SnO_2 phase and Zn_2SnO phase (ZnO content is 40wt%). The specimen of which ZnO content is 70wt% has the similar feature (See Table 2-1). The sensitive characteristic of the two kinds of specimens is very different from others.

2. Temperature Sensitive Characteristic

As we can see in Fig.2-1, the specimen's resistance might be expressed as a formular in a certain temperature range:

$$R=R_0\exp(E_c/kT)$$

where E_c is efficient activation energy. The Table 2-1 shows the specimen's resistance range and B value in a certain temperature range for various proportion of ZnO content. According to Table 2-1, we can observe the similar extreme value laws

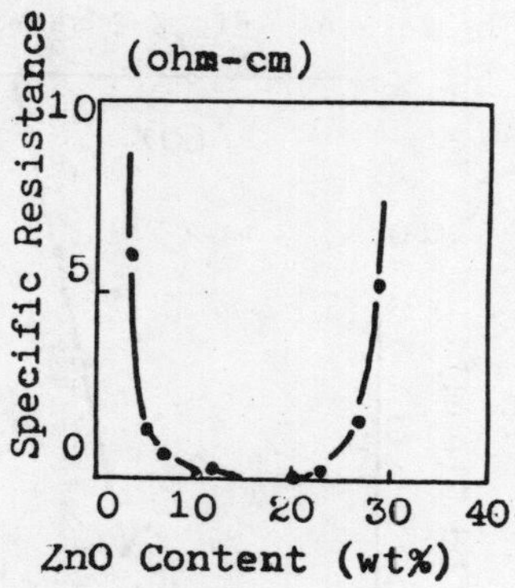

Fig.1-3 Relationship between Resistivity and ZnO Content

Table 2-1 Relation between Temperature Sensitive Characteristic and Proportion of ZnO and SnO_2

ZnO(%) Content	Temperature Range(°C)	Resistance Range(Ω)	B Value (K)
5	40--278	10.3k-0.5k	2100
5	278--553	0.5k-35	5333
10	30--127	81 -51	550
10	127--393	51 -20	1000
15	130--352	27 -7.8	1600
15	352--477	7.8 -3.9	2666
20	30--227	16.8 -13	200
25	30--202	1.2k-0.5k	667
25	202--394	0.5k-51	5000
30	150--355	1934k-21.5k	8500
30	415--900	4.5k-25	12500
35	290--393	191k-3k	13000
35	523--900	2k-63	12000
40	240--315	1579k-46.9k	13300
40	525--900	2.9k-115	13300
45	270--393	1331k-13k	18000
45	523--900	6k-63	13000
50	310--394	1815k-24.2k	22000
50	524--900	10.2k-65	14000
55	50--127	1265k-430k	1846
55	171--394	202k-0.44k	10000
60	270--510	1618k-0.88k	10000
60	640--900	0.85k-57	10400
65	90--330	20.6k-1.6k	2400
65	330--727	1.6k-0.3k	8667
70	20--800	3.75k-17.9	2000
75	171--350	1.6k-54.7	6000
75	350--700	54.7 -15.2	2400
80	30--60	11.5k-2.45k	5300
80	60--510	2.45k-30	2400
90	30--127	9.9k-2.1k	2000
90	127--298	2.1k-11.3	8500
95	40--77	2.25k-0.62k	3700
95	77--315	0.62k-17.1	3000

of heat sensitive characteristic in which ZnO content is 20wt% and 70wt%. The specimen's resistance versus temperature curve which ZnO content is 70wt% is a straight line from 290K to 1,200K as shown in Fig. 2-1. As Fig. 2-2 shows, the lower tempera-

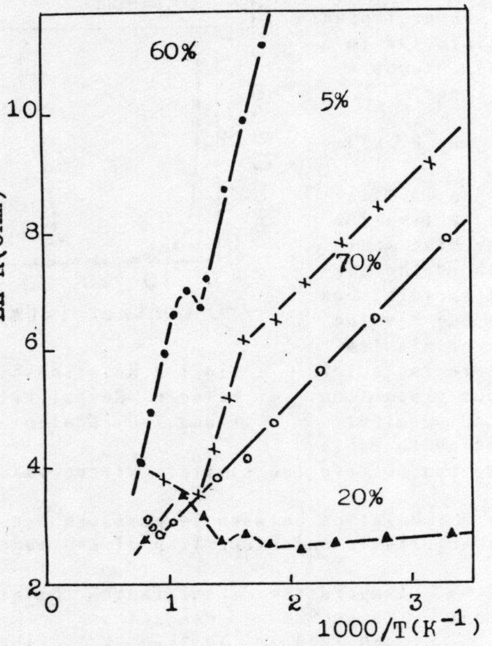

Fig. 2-1 The Specimen's Resistance Versus Temperature Curves which ZnO Content is 5wt%, 20wt%, 60wt%, and 70wt%, Respectively

ture experiment of the specimen which ZnO content is 20wt% indicates that the resistance versus temperature curve from 77K to room temperature is nearly liniar, which is drawn by an autorecorder used for studing superconductor, its temperature coefficient is about 5 times than platinum wire's. All these results show that we can obtain the material sensitive to heat from 77K to 1,200K, which satisfies the need of designs, with changing the

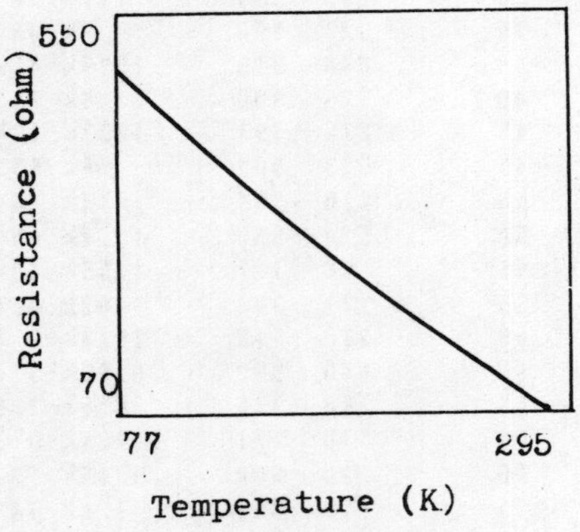

Fig. 2-2 Relationship between Specimen's Resistance and Temperature in Lower Temperature Range

proportion of ZnO and SnO_2. Most curves of resistance versus temperature have an undulation at about 900K. It may be caused by the change of states of oxygen absorption. Some sample's efficient activation energy at higher temperatures are lower than lower temperatures', which results from a few of acceptor states existing in the sample[7].

3. Gas Sensitive Characteristic

The optimum operating temperatures and the maximum gas sensitivity versus the specimens of various proportion (ZnO content) are listed in Table 3-1 and Table 3-2, respectively. The gas concentration is 0.5%, 1% and 1% for toluene, ethanol and acetone, respectively. In Table 3-1,

Table 3-1 The Optimum Operating Temperatures(°C) Varying with ZnO Content in Different Gas Environments

	5%	10%	15%	20%	25%	30%	35%	40%	45%
toluene	110	190	410	80	320	120	?	?	80
ethanol	100	280	410	230	290	200	270	160	120
acetone	80	110	110	140	260	140	230	?	80

	50%	55%	60%	65%	70%	75%	80%	85%	90%
toluene	170	?	?	390	140	280	200	140	50
ethanol	160	260	50	310	170	110	340	140	110
acetone	140	250	90	380	300	250	250	150	110

Table 3-2 The Maximum Gas Sensitivity (R_a/R_g) Varying with ZnO Content in Different Gas Environments

	5%	10%	15%	20%	25%	30%	35%	40%	45%
toluene	7.2	4.2	3.1	1.1	3.5	1.8	?	?	28
ethanol	7.1	8.0	3.0	2.1	6.5	13	49	7.4	1.4
acetone	11	3.1	1.9	1.2	6.9	5.1	57	?	25

	50%	55%	60%	65%	70%	75%	80%	85%	90%
toluene	10	?	?	5.7	5.2	2.0	2.5	1.3	1.7
ethanol	1.4	4.5	6.7	2.7	1.4	1.9	4.2	1.5	2.7
acetone	1.1	2.3	1.7	3.7	2.1	1.6	3.8	1.8	1.8

the operating temperature may be sharply decreased if there is less Zn_2SnO_4 phase in the ZnO phase or SnO_2 phase. In Table 3-2, the gas sensitive will be greatly enhanced if there is less SnO_2 phase or ZnO phase in the Zn_2SnO_4 phase with anti-spinel type structure. With changing

Table 3-3 The Maximum Gas Sensitivity (R_a/R_g) and the Optimum Operating Temperatures Varying with Doped Impurities (about 1wt%)

	Zn_2SnO_4	Pd	Fe_2O_3	RuO_2	Gd_2O_3
V_H(v)	1.6	1.6	1.4	1.1	1.4
R_a/R_g	12.9	18.6	24.3	20.9	30.6

	Eu_2O_3	Sm_2O_3	Y_2O_3	Nd_2O_3	CeO_2	La_2O_3
V_H(v)	1.1	1.3	1.2	1.0	1.2	1.4
R_a/R_g	54.0	57.0	43.9	50.0	50.9	21.9

the proportion of raw materials (ZnO and SnO_2),

we can obtain the gas sensitive material with lower operating temperature and higher sensitivity without adding any other catalysis. For the sake of contrast, we doped a dozen impurities into the single phase Zn_2SnO_4 ceramics. The optimum operating temperatures and the maximum of gas sensitivity versus the doped impurities are listed in Table 3-3 under 1% of ethanol in air. Fig.3-1 shows the relation between gas concentration and sensitivity. The sample gas is also ethanol vapour. The experiment results indicate that whether adding catalysis or changing the proportion of raw materials (ZnO and SnO_2) the gas sensitive characteristic has similar effects.

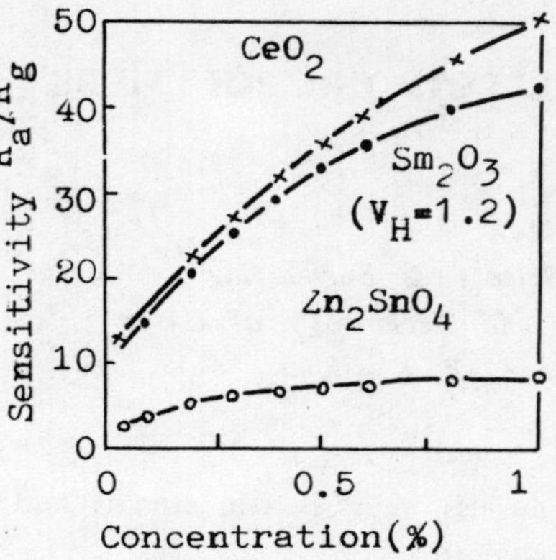

Fig.3-1 Relationship between Gas Concentration and Sensitivity (R_a/R_g)

4. Humidity Sensitive Characteristic

With changing the processing, we can fabricate a humidity sensor for all-humidity and higher humidity. The hysteresis curve of higher humidity is shown in Fig.4-1, which hysteresis effect is very small. The response time is less than 4 seconds and 10 seconds for the moisture absorption and desorption, respectively.

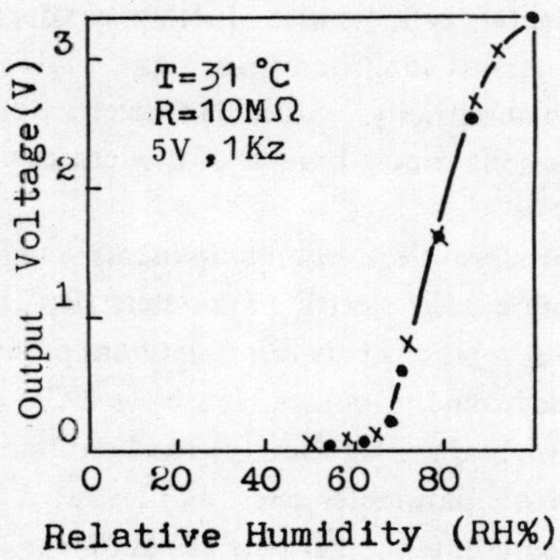

Fig.4-1 The Hysteresis Curve of Humidity
×--absorption; •--desorption

CONCLUSION

With changing the proportion of the raw materials ZnO and SnO_2, we can obtain the material sensitive to heat, gas and humidity, which satisfies the need of designs. Although the sensitive mechanism of multiphase solid solution ceramics becomes more complex, designing of sensitive material gets more flexibity. Therefore, the study of multiphase solid solution ceramics may have more practical significance than the single phase's.

REFERENCES

[1] Bao Fengqi et al., WULI(Physics, Bimonthly), vol.8, No.1,(1979) 56-58 (in chinese)

[2] Ji Binghuo et al., Bulletin of Materials for Energy System, vol.3, No.1,(1985) 26-30 (in chinese)

[3] Schubert C.C. et al., Jap. J. Appl. Phys., vol.20, No.6, L437(1982)

[4] Bao Fengqi et al., "TRANS" (The Trans of Papers Presented at the National Conference on Materials, Devices and Application of Sensors, Chongqing, China, Oct.,1985), 130-133

[5] Zang Zhongtai et al., The Collection of Papers ('86 Harbin China--Japan International Scientific Symposium on Sensor), 35-45

[6] Cheng Zuyao et al., Journal of China University of Science and Technology, vol.17, No.3, (1987) 343-351 (in chinese)

[7] Huang Kun, and Xie Xide, Physics of Semiconductors, Scientific Press, 1985, p.77 (in chinese)

HUMIDITY COMPONENTS OF ELECTRONIC POLYMERS

Yang Daben
Department of Materials Science & Engineering
University of Electronic Science and Technology of China
610054 Chengdu, Sichuan, P. R. China

ABSTRACT

This paper illustrates the significance of electronic polymer functional components and devices ,especially the characteristics of linear and nonlinear polymer humidity components .As an example of linear humidity components endurable to high temperature ,we studied the properties of PI humidity capacitive component ,namely its linear functioning of capacity over the whole range of humidity (0–100%RH) with response time $<1 \rightarrow 5$ seconds and work temperature$>100 \rightarrow 200$℃ .A dew switch type PC composite polymer resistive component with a nonlinear functioning in humidity was also investigated in the high humidity range(80→ 100%); the humidity swittch functioning has reached the order of $10^2 \rightarrow 10^3$. From the electronic polymer microstructure analysis of these two kinds of humidity response mechanics, it provides a basis for further design and research of high–functional humidity components of electronic polymers.

INTRODUCTION

The development in the science of modern electronic materials,components and devices has two remarkable characters: first, the development of modern electronic materials, symbolized by functional metal, electronic ceramics and electronic polymers (EP) has promoted the exploration of a series of new electronic material with high qualities and multiple functions; secondly, the modern electronic components have now in the transition from the conventional electronic basic components (e.g.sourcefree resistor R, capacitor C, inductor L, resonator f, sourcefull transistor TG, integrated circuit IC,etc.) to a variety of new electronic functional components and devices (e.g.sensors and transducers, piezoelectric and ferroelectric transducers, electret storage devices, photoelectric devices, conductors and power devices, microelectric circuits and molecular circuits, etc.).

EP is a new concept developed in the 1980s analogous to electronic ceramics. It describes those electronic functional polymers and their composite materials which, based on their electron states and function transformation effects, have great applications in the field of electronics, With its large variety of special electronic functional types, such as diclectric type (insulating, dielectric, microwave), sensitive type (physical, chemical, biological), piezoelectric type (piezoelectric, ferroelectric, acoustoelectric), pyroelectric type (thermoelectric, pyroelectric), electret (electret), conducting type (semiconducting, conducting, superconducting, fast ion conducting) and biologically electric type (bioelectricity, molecular electricity). EP is now becoming a new branch of modern electronic material science.

Modern electronic components and devices (ECD) use some basic electric parameters (R,C,L,f) to express various relations between functional parameters, i.e.different electronic responses.Hence the ECD can be divided into two main categories: (a) fixed parameter ones and (b) variable parameter ones; as illustrated in Fig.1. The latter represents the variable effect (linear or nonlinear) under all kinds of physical, chemical and biological stimulations:

$$E = f(P,C,B) \tag{1}$$

Here E indicates the electric parameter; $E = r,c,l,\rho,v,\varphi,i$,etc, and P,C,B are physical, chemical and biological stimulations, respectively; $P,C,B = T,f,RH,t,\tau$,etc.

It is obvious that this kind of functional parameter components and devices, characterized by functional relations (linear or nonlinear) of electric parameters instead of electric parameters themselves is really one of the trends in the development of modern ECD. That is, it is the transition of ECD from conventional fixed parameter components to electrosensitive components, sensors, and other electronic functional components and devices (EFCD).

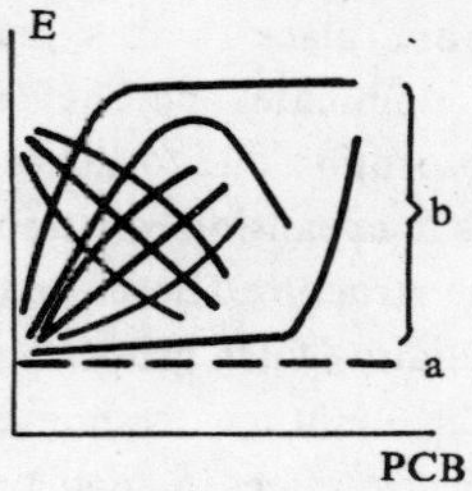

Fig.1

Among the EFCD using a chemical parameter of humidity, those which are based on EP, both linear and nonlinear, are very remarkable compared to inorganic nonmetal components, and hence are attracting great interests all over the world in recent years. The types of electronic polymer humidity components (EPHC) which have been explored are shown in table 1.

LINEAR HUMIDITY COMPONENTS

The first step in the exploration of EPHC is linear humidity components.Nowadays, the research of various kinds of resistive and capacitive polymer components which have linear functioning over the whole humidity range (0→ 100%RH) has already achieved the level of practical usage. Compared to inorganic nonmetal humidity components, EPHC has the advantages of having good linear characteristics, short response time, no need of heaters, fine structures, high stability, easy to be used in sliceformed mixed IC, and simple techniques. The combination of excellent function and economical advantage has led to a rapid improvement in the EPHC very recently.

However, most of the linear EPHC now available requires a very low work temperature ($T_w < 100℃$). The research of EPHC endurable to high temperature (i.e.$T_w > 100℃$) is therefore very attractive. We will discuss below the possibility of using PI thin film as high temperature humidity components based on our investigations of the dielectric spectroscopy of the polyimide—H films, which is remarkable for its high temperature endurance (thermoresolution temperature 550℃).

Our study shows that the functional relation between the PI humidity components and the environmental relative humidity is caused by the PI films' diffinity for water due to its molecular structure of strong polarity cluster, and the electro—negative factor at different positions of the molecular chain due to the water absorption by the hydrogen bonds. see Fig.2. According to Looyenga equation, the composite dielectric coefficient ε_Σ of the polymer films after water absorption depends on the composition of two parts:

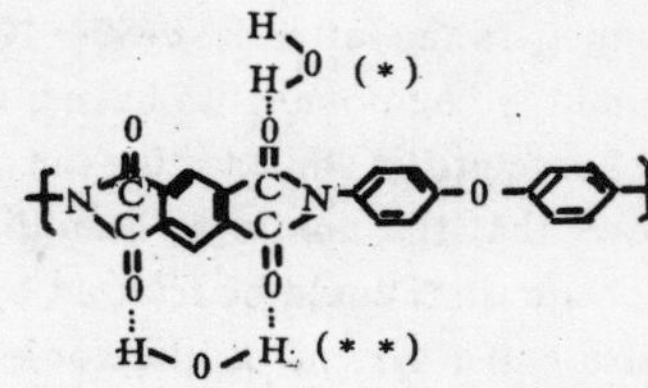

Fig.2

$$\varepsilon_\Sigma = [V_H(\varepsilon_H^{1/3} - \varepsilon_p^{1/3}) + \varepsilon_p^{1/3}]^3 \qquad (2)$$

where ε_p and ε_H are the dielectric coefficients of polymer and water, respectively. V_H is the partial volume of the absorbed water.

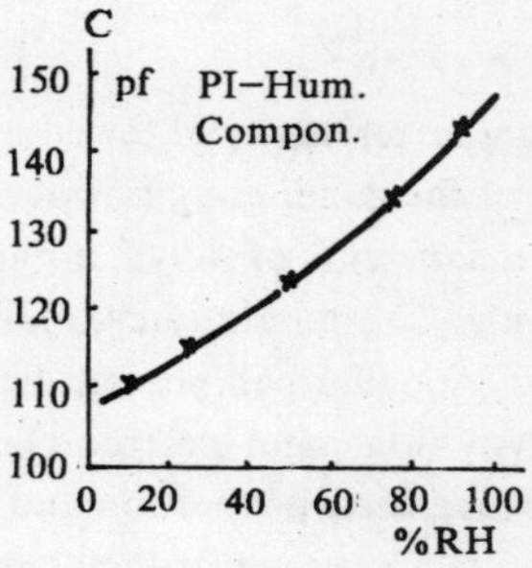

Fig.3

Our experimental results of the standard humidity characteristic curve of the PI humidity components are shown in Fig.3. It has a good approximate linearity, corresponding to the Looyenga equation. When we use the substrate capactive—type penetrated electrode structure, with the PI film thikness around 1μm, we achieved very good humidity characteristics: linearity of the capacitive components over the whole range of humidity with capacitance from 100 to 300pf; response time less than 1 to 5 seconds and the temperature coefficient less than ± 0.1%RH / ℃ . In particular, the H films' outstanding endurance to high temperature makes it possible to have a work temperature higher that 100 to 200℃ for PI humidity sensitive components.

It is obvious that the use of polyimide electronic polymers for linear humidity components paves a new path for EPHC endurable to high temperature. This is very likely to greatly enhance the wide application of EPHC in the techniques of humidity testing and controlling.

NONLINEAR HUMIDITY COMPONENTS

The present difficulty in technologies of humidity testing and controlling focuses on both the extremely low and high humidity ranges. In the latter case (80→100%RH), it requires the humidity components to have a nonlinear humidity switch characteristic while testing the dew state. Our analysis shows that the nonlinear humidity switch property of composite films could be realized by the molecular designs using the system of electronic polymer composite films. This is a new kind of humidity components.

After analyzing the composite modes of various humidity-absorbed polymers and conducting particles, we chose the PC resistive nonlinear humidity component structure optimizely composited by the cellulose polymer MC and acetylene carbon black. The resistivity of its conducting system satisfies the Bulgin D. equation:

$$\rho_v = exp(a / cp) \qquad (3)$$

where ρ_v is the volume resistivity of the composite film, c is the proportion of the conducting powder and a and p are the correlation constants of polymer and conducting powder, respectively. The humidity characteristic of the PC composite polymer film component depends on the intrinsic conductivity (ionic and electronic) of the polymer and carbon powder, their proportion and the structure of the composite film.The result of the characteristics of the PC nonlinear humidity component of our experimentally optimized prescription is shown in Fig.4.

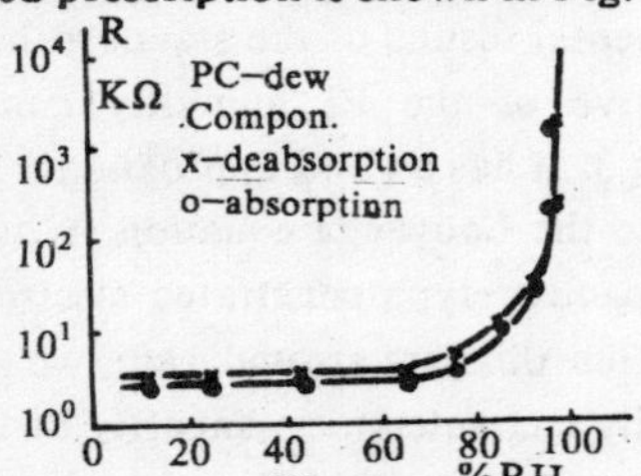

Fig.4

Our experiments demonstrate that the nonlinear resistance functioning in high humidity has reached the order of $10^2$to $10^3$with a very high switch sensitivity (i.e.the slope of the curve).This completely meets the requirements of humidity switch property in the dew state in high humidity.

With repeated microstructure analyses of the PC composite films, we observed two characteristics in the structure factors from the SEM images (see Fig.5): one is a three dimensional net-like distribution(partially oriented) of the cellulose molecule beam including carbon blacks;the other is the exitence of empty holes in the molecule texture of the composite films.Obviously ,the former is helpful to the electronic conductivity determined by the carbon black (with a graphite-like structure) and is the dominate one.The latter,however,can help the water absorption due to the cellulose molecular polarity and leads to expansion-contraction size effect of the composite film structure.Therefore,the mechanism of the nonlinear humidity switch property of composite film comes from the mutual response of electronic conductivety and size effect.In low humidity range,the cellulose resin absorbs little moisture and the distance between conducting particles is small. Thus the electrons posses a high conductivity and the film resistance is kept in a low state. On the other hand, the cellulose resin will absorb much moisture in the high humidity range and expands so that the distance between conducting particles increases quickly. When the humidity passes a certain threshold, the electronic conductivity will drop dramatically. Furthermore, this effect will overplay the increase of ionic conductivity due to the high humidity, and the film resestance will then increase very quickly and exhibit an abrupt switching.

Fig.5

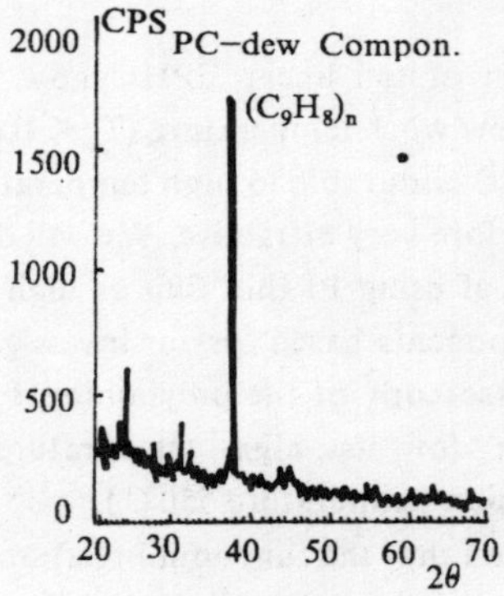

Fig.6

Further x-ray diffraction analysis, as shown in Fig.6. indicates that the PC composite film, compared with the original cellulose resin, has created a new phase—$(C_9H_8)_n$ at the diffraction angle $2\theta = 38°$. This a highly oriented composite polymer polymethylphenylacethylene.Its molecular formula might be:

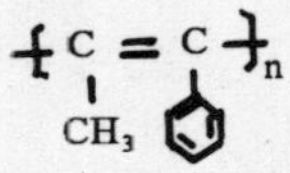

(4)

Obviously, $(C_9H_8)_n$ belongs to polyacetylene–type polymer.It has conjugate π electron double bonds which enhances the channels of electronic conductivity. In the meantime,the solubility due to its asymmetric polarity structure helps the size effect of the molecular structures after absorbing moisture. Thus, the humidity component PC composite film in this study has a very similar switch characteristic to the diodes.

Our comparative experiments have demonstrated that the combination of optimized PC composite film structure design, prescription, and microstructure analysis is very important in the improving the switch characteristic, sensitivity and repeatability of the nonlinear humidity components of EP.

SUMMARY

EP has now opened a new area for development of modern EFCD The great research efforts devoted to the EPHC will open a new area for the applications not only in linear humidity testing,but also in the techniques of modern video electronics,automatic controlling and various kinds of hi–tech nonlinear humidity processing and controlling.

REFERENCES

1. J.Mort,G.Peister, Electronic Properties of Polymers, John Willey and Sons,1982.
2. D.Yang ,J.Melcher,G.Arlt,Watersensitive,Dielectric Relaxation Process in Polyimide in a Temperature Range from 90K to 350K, ICPADM, 85–CH 2115–4,IEEE,1985.
3. M.M.Dubinin,Fundamentals of the Theory of physical Absorption of Gases and Vapors in Micropores,Academic,1972.
4. V.E.Kerrey,Research and application of conductive polymers,Japenese,1970.
5. YANG Daben,Investigation of the Dielectric Spectroscopy of Polymer Film KAPTON, Joural of chengdu–Institute of radio Engineering,Vol.16,1987.

Table 1

Aspects	Types	Materials	Functions	Properties
mechanism	water–affined polymers (structure effecct)	CA.PA etc	capacitive	Change in dielectric const. $C=f(\varepsilon_\Sigma)$
		SA–type	resistive	ionic conductivity $R=f(\rho_\Sigma)$
		PA–type	resonant	resonant property
	expansion–contraction polymers (size effect)	$\frac{PAA}{HEC}$+con.powder	conducting	electron conducting
humidity	relative humidity type(%RH); absolute humidity (water content ppm)			
test	full humidity type (0 ➝100%RH,linear);dew type (high RH switch, nonlinear)			
signal	crossbetasten type(capacitive, resistive, conducting);resonant–type (resonant); FET type (current).			

PREPARATION AND CHARACTERISTIC OF THE COLOUR SENSORS

Gao Qi-an, Wan Wen-hui and Zhang Wei
Depart. of Applied Physics, Hefei University
of Technology, 230009 , Hefei, China

ABSTRACT

The colour sensors prepared by us possessed total spectral response in visible light range, higher photo-electronic sensitivity and photo-electronic conversion quantum efficiency. The devices may be used to detect weak light and to distinguish different colours and so it may be used for the robot to recognize different colours. The photo-sensitivity of common silicon photo-diodes to short wavelength is quite low. Through using the technologies of shallow junction, of high-low junction by ion implantation, of antireflection film and of integrated filter, we prepared the silicon colour sensors which possess the spectral response to the wave length between 320-1100nm with the peak sensitivity of 0.49A/W. The sensitivities of the device to 3 basic colours, i.e. red, green and blue, expressed as peak wave lengths are 700, 550 and 450nm respectively.

INTRODUCTION

Most informations are gathered by human beings through sense organs and among these informations, colour information, as perceived by visual organ plays most important part. So there is no doubt about the fact that the colour sensor technology and colour sensitive devices are very important.[1]
The colour sensor is mainly used for detecting weak light, for distinguishing colours, for furnishing the robot with a colour recognition system and for colour measuring.[2] It can also be used in the place of photomultiplier tube in some occasions and has wide application in automachanism for sorting out products, detecting grains and food, distinguishing colours of dyes for printing and reading the colour bar codes etc.

BASIC PRINCIPLE AND THE DESIGN ENVISAGED

When extracircuit is open, that is I = 0, we have

$$V_{oc} = \frac{KT}{q} \ln(1 + \frac{I_L}{I_o})$$

and $V \propto A$, A is the logarithm of the incident light intensity.
Light absorption spectrum of silicon material is closely related to the wavelength. For high energy photons with short wavelength, silicon has a very big absorption coefficient. The blue and violet parts of the light which shines on the silicon will be absorbed within a thin layer under the surface. For instance, for the violet light of $\lambda = 0.4\mu m$, when the surface reflection is neglected, the absorbing depth can be estimated as $< 0.25um$. Therefore in order to raise the efficiency of collecting the photogenerated carriers caused by short wavelength photon, the surface recombination should be reduced and p-n junction ought to be as near the surface as possible. And since the p-n junction is a shallow one, we must consider the electromagnetic field caused by the ionized impurity. Supposing that the mobility in ion implantation is a constant, we can get (according to the equation of current density)

$$E = -\frac{KT}{q} \cdot \frac{1}{N(x)} \cdot \frac{dN(x)}{dx}$$

If the exponent distribution is used instead of Boltzman distribution, we have

$$E = \frac{KT}{q} \frac{1}{x_j} \ln \frac{N_s}{N_j}$$

From above it is clear that the doping concentration and distribution of p area effect greatly the response of short wavelength, because it involves the life of carriers and the intensity and direction of drifting field.
To reduce the doping concentration of surface can decrease defect and recombination centre caused by impurity, and the barrier potential field of p-n junction can thus move to the surface, which is of advantage to raise the colle-

cting efficiency of photo-generating carriers. The distribution of the steep concentration of impurity may generate rather powerful drift-field, so as to speed up the movement of the photo-generating carrier and so to help raise the collecting efficiency and the response time. So it is necessary for us to select suitable impurity concentration.

Besides, we utilized the structure of high-low junction, i.e. we implanted low dosage of boron ion of high energy to make a deep junction of low concentration of surface and then implanted high dosage of boron ion with low energy to make a shallow junction of higher surface concentration. This high-low junction not only keeps a powerful drift field on the surface, but also easily forms even junction. Thus photo-generating carriers are collected more efficiently and the quantum efficiency and the sensitivity of photo-electronic conversion are raised.

Moreover, the loss from light-reflex from the surface of silicon material is rather serious. It is usually > 30%. Its reflex is related to the wave-length of light. The shorter the wavelength is, the greater the loss will be. Therefore, anti-reflection measures must be taken to raise the absorption efficiency of light.

From above, it is obvious that in order to raise the sensitivity of the device to short wavelength, beside ordinary requirements for making diode, the following three conditions must be satisfied:

1. The X_j must be shallow (the X_j of high-low junction is not small).
2. The impurity concentration of the surface should not be too high and must be suitable.
3. There must be anti-reflection film on the surface.

According to this, the parameters and the design of the structure of devices we choose are as follows:

1. The depth X_j of p-n junction is $< 0.2\mu m$ (for shallow junction).
2. The impurity concentration of surface is controlled at about $10^{18} cm^{-3}$.
3. A SiO_2 thin-film is made on the surface of device. The thickness of which is about 700-800 Å. The SiO_2 thin-film as above is calculated according to the thin-film optical principle and is used to resist reflex and to raise transmission. This SiO_2 film is technologically used as a shelter. Impurity get into the silicon through SiO_2 film. This can not only climinate unfavourable factors such as surface recombination etc. caused by impurity, but also help get the required low doping concentration in surface and the shallow junction. In addition, it plays a role in clearing up the back-field caused by ion implantation. And also the SiO_2 film can be used as a passivation film which has the function of protection. Because the film is of many uses it simplifies the working processes. The SiO_2 film is very important.
4. Since the doping concentration on the device's surface is low, it is harmful to ohm contact. Thus, we have designed the area of deep-shallow junction so as to decrease ohm contact resistance; it may also form protecting ring in the deep-junction area to prevent shallow junction from being brokendown. It is advantageous to improve the device's performance.
5. The filter system: Since this device is used to distinguish colours, each colour-sensor set consists of three elements which are responsive to red, green and blue light respectively. The red elements is made up of the semiconductor chip and the red filter. And so do the green and the blue ones.

THE STRUCTURE AND TECHNOLOGY OF THE DEVICE

From the above considerations, the design of the structure of the device chosen is shown in Fig. 1.

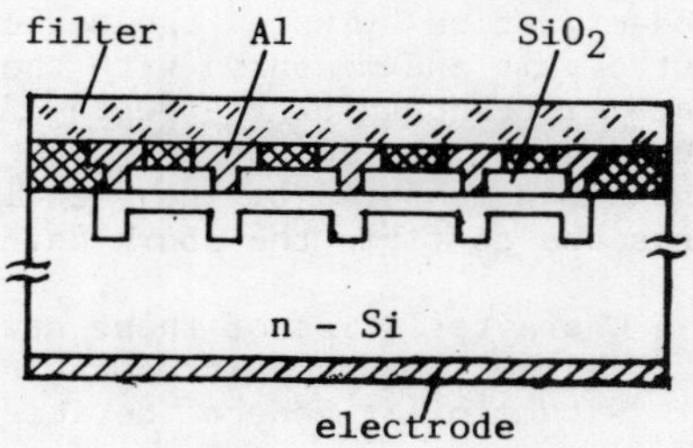

a) Section view of the device

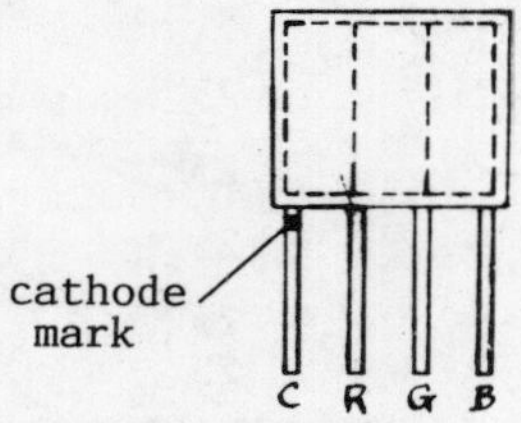

b) Parttern of the device

c) the integrated device for color recognition

Fig.1 Structure of the color sensor

To use technology of ion implantation can control separately the junction depth and the distribution of impurity concentration,[3] which is easier to reach the requirement of raising sensitivity of short wavelength than the diffusion method. The experiment has shown that the more suitable technological parameters of boron ion implantation should be as follow: the implanting energy is 15-20 keV and the implanting do-

sage is 10^{13}-$10^{14}cm^{-2}$ to make shallow junction. Then it is heat treated for about 10' at 1,000-1,100°C. Such devices made in this way have wide spectral response, high photoelectronic sensitivity and high quantum efficiency.
We choosed three basic-colour filters (red, green and blue ones) to make the device able to distinguish colours.[4] When the device is used to measure colour, we will plate a thin film of optics medium, so that they can correct spectrum respond and fulfil the requirement of the standard stimulant valve.

THE RESULTS OF MEASURING

Characteristics of these new devices in comparison with those of similar devices of old model as measured in several items are shown in Tab.1.
We have also measured some other parameters such as back-breakdown and noise equivalent power and so on, beside spectral characteristics, and got the similar parameters as those got from the devices made abroad.
We did anod-oxidation by using the method of water-electrolysis and measured with the 6JA interference microscope for the depth of junction of the device, the result measured is ⩽ 0.2μm. The datum measured by the scanning electron-microscope confirms the above datum.

Table 1 Characteristics of these new devices measured in comparison with similar general devices

Type	Photosensitive area(mm^2)	Spectral response range(nm)	Spectral response peak wavelength λp(nm)
S876-606 BR	29	320-1054	800
VTB8441B	8	330-720	720
2CR-21	25	400-1050	830
PCS-R	25	400-1100	700
PCS-G	25	400-900	550
PCS-B	25	320-600	450
PCS-O	25	320-1100	700

radiant sensitivity Re(A/W) $\lambda=\lambda p$	Dark current I_d (pA)	Dark current V_R (V)	Co. in which the device was made
0.4	200	10^{-2}	HAMAMATSU
0.28	100	2	Vactec
0.16-0.3	100	2	China BPD
0.49	100	10^{-2}	Our Lab
0.26	100	10^{-2}	Our Lab
0.19	100	10^{-2}	Our Lab
0.49	100	10^{-2}	Our Lab

REFERENCES

1. Abc, C. et al, Electronic Parts and Materials. (Japanese), 21(1982), 4, 73.
2. Omine, E. et al., Electronic Parts and Materials. (Japanese) 22(1983), 4, 83.
3. Young, R.T. et al., IEEE Trans., ED-29 (1980), 4, 807.
4. Gayoshi, N. Electronic Parts and Materials (Japanese), 21(1982), 2, 79.

PIEZOELECTRIC QUARTZ LINEAR ACCELERATION SENSOR

Zheng Yongxiang, Hou Zhengjun and Kou Zhengyi
Sichuan Institute of Piezoelectric and Acoustooptic Techniques
Sichuan, China

ABSTRACT

Linear accelerometer is one of the most important sensors used in inertial control and navigation. It's principles and structures are of many types. In this paper we present a newly developed bicrystal plate piezoelectric linear accelerometer. The experimental results show that its resolution is better than 10^{-3}G, nonlinearity is less than 1% and the measurement superior limit can reach 100G.

INTRODUCTION

Acceleration sensors are mainly used to measure the longitudinal, transverse and normal acceleration of the vehicle in the space and aviation field. It is the information source of the inertial navigation and control. By sensing the acceleration of the missle and feeding them to the computer for caculating and processing , the accuracy of the missle hit the target can then be obtained. Since 1940 when the pulse integrating pendulous accelerometer (PIGA) was first utilized by German, many countries in 1950's have been involved in the research of PIGA. Even up to today these accelerometers still have wide applications. These traditional accelerometers can obtaine high accuracy but they are of large size, high cost and difficult for manufacture. With the development of the mordern science and technology, now, various been developed. Among them, piezoelectric linear acceleration sensor is one of these types. It is characterized by frequency output, high accuracy, good stability, simple structure, high reliability, small size, light weight, low cost and it can also stand the overload. So in recent years, many countries such as America, Japan and Britain have done much research in this area and have already achieved some progress. The piezoelectric crystal linear accelerometers are of variouse types, including turning-forks, cantilever beams as well as disks etc... In this paper however, we only discuss the piezoelectric quartz bicrystal cantilever beam including their principles and structures in detail. The specifications of our newly developed products are also presented.

OPERATIONAL PRINCIPLES

The basic principles of the piezoelectric linear acceleration sensor is to make the quartz bicrystal plates and the mass into cantilever beam structure and then use the self-exiting oscillating circuit to produce resonance. According to Newton's second law, when there is an input acceleration there will be an inertial force on the mass causing the two plates bending, one is stretched, the pther is compressed. Due to the force frequency properties of the crystal the resonance frequency rises and falls respectively. After the output signal of the two quartz oscillators is mixed and the differential frequency signal is obtained, the frequency variation of the differential frequency signal then will reflect the acceleration measurement. As it is shown in Fig.1, two piezoelectric quartz plates deposited with ring electrodes are matched to each other, one end is fixed to the base, the other is connected with the mass as the free end. The two plates are combined separately with their oscillation circuits forming a self-oscillator to produce resonance on the fundamental frequency of the thick shear mode. When the sensor inputs an acceleration in its sensing direction, due to its inertial action of the mass, it then receives a force which keeps its position unchanged and which is opposite from the direction of the acceleration. Hence, the two plates are caused to deform by the strain, one is stretched, the other is suppressed. Asuming that the unloaded resonant frequency are f1 and f2 respectively, when there input an acceleration it then becomes as $f_1 \pm \Delta f_1$ and $f_2 \mp \Delta f_2$, where Δf_1 and Δf_2 are due to the acceleration and are proportional to the strain (linear acceleration), given by

$$\Delta f = \frac{f \cdot F}{2C_{ij}} \cdot \frac{dC_{ij}}{dF} \quad (1)$$

$$f = \frac{1}{2t} \sqrt{\frac{C_{ij}}{\rho}} \quad (2)$$

where Δf is the variation of the resonant frequency; f is the unloaded resonant frequency; F is the load strain acted on the piezoelectric plates; t is the thickness of the plate; C_{ij} is the elastic constant of the quartz crystal; ρ is the density of the quartz crystal.

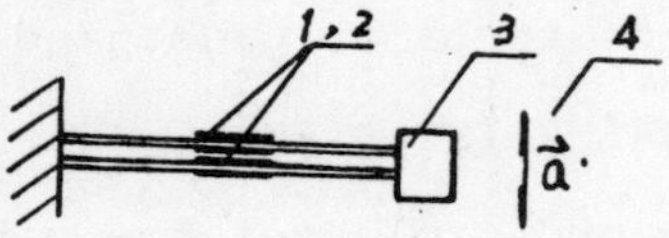

Fig. 1 The principle structure of piezoelectric sensor.

1,2: piezoelectric plates
3: mass
4: input acceleration

The differential signal is obtained by mixing the two output signal and filtering them through the low pass filter. It frequency are $f_1-f_2 \pm (\Delta f_1 + f_2)$, where f_1-f_2 is the bias, i,e, the output of the zero acceleration; $(\Delta f_1+\Delta f_2)$ is proportional to the acceleration. It can also be transfered into DC voltage by frequency-voltage converter.

If $f_1=f_2$, then $\Delta f_1=\Delta f_2$. Thus, the output signal frequency will be $\pm$ 2Δf. Due to the reason that the frequency can not be nagative, during the design, we must let $f_1 \neq f_2$ and f_1-f_2 must be greater than the frequency variation $(\Delta f_1+\Delta f_2)_{max}$ due to the maximum input acceleration, i,e, $(f_1-f_2)>(\Delta f_1+\Delta f_2)_{max}$. When $f_1 \neq f_2$, Δf_1 and Δf_2 may not be equal $(\Delta f_1 \neq \Delta f_2)$. It has been demonstrated that when f_1 and f_2 are limited to a certain range, there is only a slight difference between Δf_1 and Δf_2.

THE STRUCTURE AND ELECTRICAL PRINCIPLE OF BICRYSTAL PIEZOELECTRIC LINEAR ACCELEROMETER

The accelerometer, as shown in Fig. 3, comprises a rectangular quartz plate, a mass, a damper, and a housing with nitrogen(N_2) filled in its cavity. The demension of the plate is 15x10x0.15. Its thickness depends on the resonant frequency and its cut is AT birotation with 40° azimuth which is determined in the frequency temperature characteristics, the measurements and the stability of the force frequency sensitivity of the plates. The theoritical analysis show that the relation between the normalized force sensing coefficient K_f of Y-rotated quartz crystal and the azimuth ψ(angle between action and x axis) is given by

$$K_f(\psi) = P + Q\cos(24) \quad (3)$$

where P and Q depend on the cut parameters. The temperature performance of P and Q can then determine the dependence of K_f on temperature. The temperature performance of AT cut K_f with different azimuth is shown in Fig. 2. when the azimuth is at 40°, the K_f temperature performance of AT cut appears to be good.

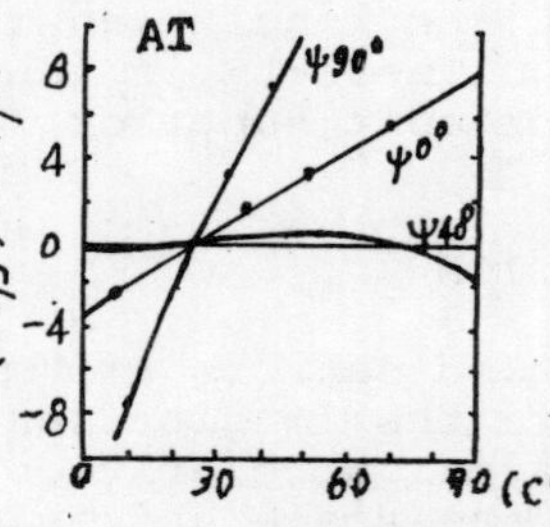

Fig. 2 Temperature dependence of AT cut K_f.

The mass, damper and housing are made of special alloy material with its thermal expansion coefficients corresponding with the plate. The ratio factor is determined by the mass size which can be properly adjusted according to the requirements of the users. The damper can not only be served as damper but also can reduce the transverse sensitivity.

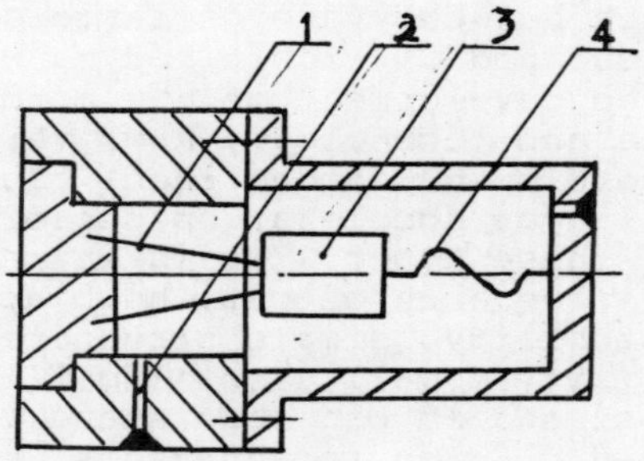

Fig. 3 The structure diagram of the accelerometer. 1): piezoelectric plate; 2): airing hole; 3): mass; 4): damper;

The electrical principle diagram of the piezoelectric linear accelerometer is shown in Fig. 4.

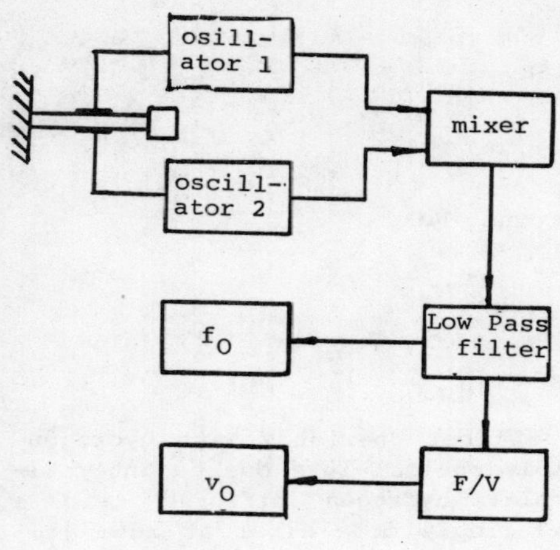

Fig. 4 Electrical principle of the piezoelectric accelerometer

THE SPECIFICATIONS OF THE ACCELERATION SENSOR

Table 1 Lists the specifications of our newly developed sample.

Parameters	88-02#	88-04#	88-05#
Superior limit of the measurement	100	100	100
sensitivity (Hg/G)	190	145	60
resolution (G)	$1x10^{-4}$	$2.1x10^{-3}$	$1.1x10^{-3}$
nonlinearity	1%	1%	1%
transverse sensitivity (%)	2	2.5	2
operation temperature (°C)	-30~50	-30~50	-30~50
frequency response (Hz)	0~200	0~200	0~200
weight (g)	50	45	30
volume (mm^3)	φ30x50	φ30x50	φ30x50
power consumption (mW)	200	200	200

Among them, the output signal can be: 1) frequency output having the forms of sine wave, square wave or unidirectional rectangular pulse wave; 2) voltage output. Its bias V_O and sensitivity mV/G can be abitrarily adjusted with the requirements of the customers.

REFERENCE

1. Chang Fuxie, "Piezoelectric Crystal Force and Acceleration Sensors" A Book Published by Sichuan Scientific Press.

HYDROGEN GENERATION IN SL CABLE DUE TO MANUFACTURE AND SERVICE LIFE

Chung-Shin Ma
AT&T Bell Laboratories, Crawfords Corner Road
Holmdel, New Jersey, USA

ABSTRACT

In optical fiber communications systems, the presence of excess hydrogen in a cable may increase optical loss and thereby impair transmission. This study focuses on the mechanisms of hydrogen generation during the manufacture of an undersea optical fiber cable and the amount of hydrogen accumulated in a cable during its service life. Hydrogen can be introduced into the fiber environment by a variety of mechanisms: thermal degradation of polymers, outgassing of other constituent materials, and aqueous attack of cable materials by moisture. To assess the amount of hydrogen generated due to polymer degradation and outgassing during manufacture, gas samples extracted from cables were analyzed for hydrogen concentration. Also, the amount of moisture entrapped in a cable during manufacture was studied. The likelihood of water condensation in a cable at the lowest operational cable temperature was analyzed and determined to be very low. For cable storage, hydrogen generation was studied by aging 18 inch long cable samples at different storage temperatures. The overall conclusion is that the hydrogen concentration in a cable during its service life without cable damage will be about 1500 parts per million (ppm).

INTRODUCTION

For an optical fiber cable, the cable structure encloses optical fibers in an environment that may have a variable amount of hydrogen. The presence of hydrogen in a cable is not desirable because it can diffuse into the optical fibers resulting in increased optical attenuation. Subsequent to the work by Stone et al. on the absorption Raman spectrum of hydrogen diffused optical fiber[1], many studies[2][3][4] have shown that the transmission capability of optical fibers is degraded if they are immersed in a high hydrogen concentration environment. The increase in optical loss due to hydrogen is caused by two mechanisms: interstitial molecular hydrogen in the silica fiber and reactions of the fiber chemistry with hydrogen. The reversible optical loss due to interstitial molecular hydrogen diffused into a silica-based single mode fiber at room temperature can be 0.2 to 0.3 dB/km/atm at 1.3 μm,[4][5][6] and about two and half times these values at 1.55 μm[4]. The permanent optical loss increase caused by interaction of the fiber constituent chemistry with hydrogen is also wavelength dependent. Owing to these loss effects, the presence of hydrogen in a cable, especially in an impermeable cable structure, plays an important role in the prediction of aged cable optical loss and is crucial to the loss budget of an optical fiber transmission system.

The amount of hydrogen generated in a cable depends on the materials and the structure of the cable. Different cable designs will present different hydrogen environments. Limited data is available for the hydrogen levels in optical fiber cables. For land based systems, the work of Ashpole et al.[7] has shown a hydrogen environment of less than 1000 ppm in a cable. However, for undersea optical cable systems, due to the mechanical cable structure that is necessary to fulfill the more stringent requirements of a harsh environment, optical fibers are usually protected within a hermetically closed cable structure, and the hydrogen generated in the cable does not easily escape. The hydrogen is trapped and accumulates within the cable.

The structure of AT&T's SL Undersea Lightguide Cable, studied in this paper, is shown in Fig. 1. Generally, this submarine cable can be divided into two parts: the center cable core inside the nylon (including nylon) and the external structure of the cable. The core includes optical fibers in addition to the center kingwire, Hytrel® elastomer and nylon layers. Outside the core, steel wires are helically wrapped around the core with waterblocking material filling the intersti-

® Hytrel elastomer is a registered Trademark of E. I. duPont de Nemours & Co..

tial voids between the wires. This entire subassembly is held together by a copper sheath formed and welded from copper tape. Outside the copper sheath is a layer of polyethylene, that acts as an insulation and jacketing layer. The copper sheath associated with this type of cable construction serves as an almost impermeable barrier in terms of radial hydrogen diffusion.

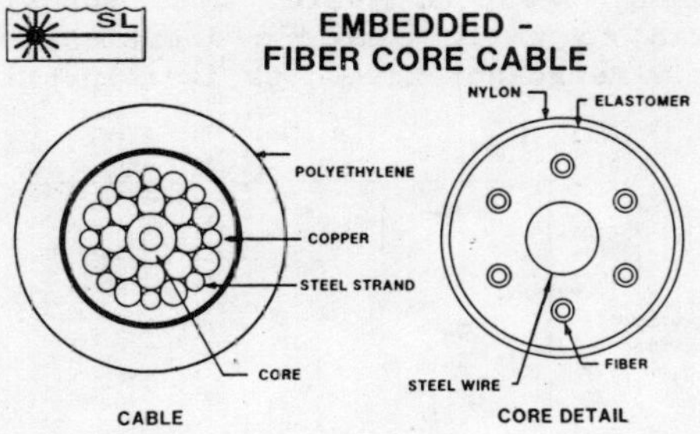

Fig. 1 Structure of SL Undersea Lightguide Cable.

This paper presents the results of a study of the amount of hydrogen generated in the AT&T's SL cables during the manufacturing processes, storage and service. The amount of moisture trapped in a cable during manufacture has also been estimated and the likelihood of condensation and the amount of condensed water are studied. The total results yield the amount of hydrogen in an undersea cable from manufacturing throughout service.

EXPERIMENTAL METHOD OF CABLE HYDROGEN SAMPLING

In this study, the amount of hydrogen in a cable was measured by extracting gas samples from specified locations in a cable. The gas samples were analyzed by a hydrogen detector (RGD2 Reduction Gas Detector) with GC1 gas chromatography from Trace Analytical Co., USA. The minimum amount of gas sample needed for analysis is about 1.5 to 2 cubic centimers (cc) for the GC used. The sensitivity of this unit can detect hydrogen in the range of 20 parts per billion. The principle of this detector is to measure the light intensity change due to the amount of generated mercury vapor in a heated bed of mercuric oxide, which reacts with hydrogen or any other reducing gas to release mercury vapor. The detector block is shown in Fig. 2.

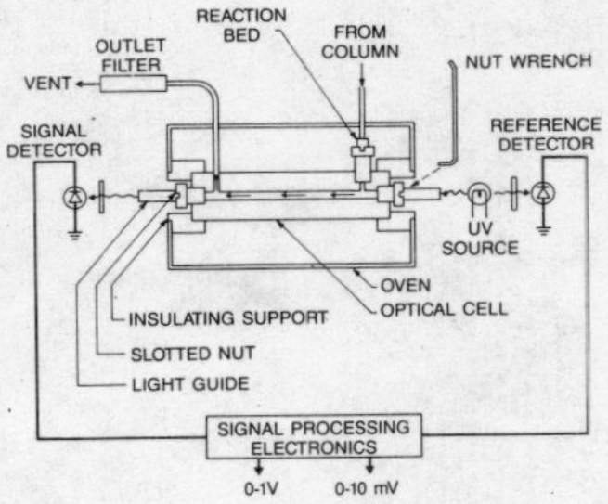

Fig. 2 Detector block of the RGD2 Reduction Gas Detector (by courtesy of Trace Analytical Co.).

To obtain gas samples from long production cables, a piercing valve (Fig. 3) is needed to penetrate the copper sheath of a cable. Before the piercer penetrates the copper, the gas in the reservoir of the piecing valve is removed by a leak-tight gas syringe through a septum-blocked sampling port. A vacuum gauge is used to ensure that the reservoir is evacuated.

Fig. 3 Gas sampling through piercing valves on cable.

After piercing, the gas from the cable fills the reservoir and is taken away by a gas syringe and purged into the GC and detector system. This technique can be used to obtain gas samples from cables of any length. Long length cables were tested in the as-manufactured and aged conditions. For aging tests, the cable samples are 18 inches long. Both ends of the cable samples are sealed by Swagelok® tube fittings. A gas syringe is used to obtain gas samples through the septum-block sampling port (Fig. 4).

® Swagelok is a registered Trademark of Crawford Fitting Company.

Fig. 4 Gas sampling through Swagelok tube fitting by syringe.

HYDROGEN GENERATED IN THE CABLE DUE TO CABLE MANUFACTURE AND NORMAL STORAGE

The amount of hydrogen generated during manufacture and normal storage is determined via measurement and calculation. Hydrogen concentration measurements are performed on recently manufactured cables and cables which were stored normally for nearly three years. These measurements characterize the hydrogen generated by polymer thermal degradation and materials outgassing. Calculations describing the likelihood of water condensation which could allow aqueous degradation of cable materials are performed.

1. Measured Hydrogen Generation In Manufactured Cable

During cable manufacturing, once the copper sheath is welded and reduced to the nominal size, the fibers are housed in an environment from which it is difficult for hydrogen to escape. The subsequent insulation extrusion process heats the cable core with the fibers in an almost closed system. Any hydrogen generated during this process is trapped within the copper sheath. The only path for the hydrogen to escape is to diffuse along the cable. However, from a simple model described in Carslaw and Jaeger's book,[8] the outward diffusion of hydrogen with a constant concentration in a long hollow tube was determined to be small; about a 10% reduction in hydrogen concentration in the tube at a location of 100 meters from the tube end after one year. The SL cable contains waterblocking material filling the interstitial spaces of the steel wires which slows down the diffusion of hydrogen much more. Even considering this slow rate of diffusion, all the gas samples were taken at least 70 meters away from the cable ends to ensure an accurate measurement.

Due to the destructive nature of the sampling method which uses the piercing valves, a limited number of system production cables were examined. Various cables were measured, beginning with one immediately after manufacture and continuing to a piece of cable which is about two and half years after its manufacture. The results show the influence of manufacturing and also the general storage processes on the cables. The hydrogen concentrations of several cables after their manufacture were measured. For each cable, concentrations were measured at a minimum of two locations. It was found that the concentrations measured were not functionally related to their sampling locations in the cable. Representative data is illustrated in Fig. 5.

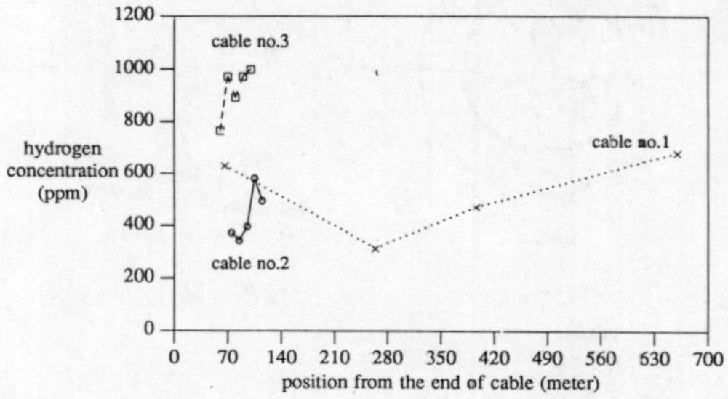

Fig. 5 Hydrogen concentration measured in SL cables with their measured locations. (total length of cable no.1=27.6 km, no.2=2.83 km, no3.=44 km)

This data confirms that the gas generated in the cable does not readily move from one location to another and that the trapped hydrogen does not quickly diffuse from the cable end. The variation in hydrogen concentrations among the different locations is most likely due to the non-homogeneous presence of the waterblocking material and the residual wire drawing compounds which coat the steel wires. Thus we conclude that the gas sampling location of a cable is not an important factor in the measurement of hydrogen in a cable, unless the sample is extracted at a location very close to the cable ends.

To determine the amount of hydrogen in manufactured cables, the hydrogen concentrations of six production cables were measured after their manufacture. Because of the different time periods between measurement and manufacture, the results represent the generation of hydrogen due to manufacture and due to aging during normal cable storage. In each measurement, the gas pressure in the cable was found to be about the same as atmospheric pressure. The maximum, average and minimum hydrogen concentrations measured on each cable are summarized in Table 1 and plotted in Fig. 6 versus the time of test after manufacture. From Fig. 6, it can be seen that while the maximum value measured in a cable seems to increase as the time of test after manufacture increases, the average value of a measured cable does not exhibit this behavior.

Table 1 The level of hydrogen concentration in production SL cables after manufacture and during normal storage

cable	no. of measurements	H_2 concentration (ppm) ave.	max.	min.	days between manuf. and test
A	4	523	680	313	218
B	2	295	390	200	66
C	5	438	581	344	6
D	2	741	877	605	181
E	5	913	1074	690	307
F	5	545	1200	280	900
average		576	800	405	

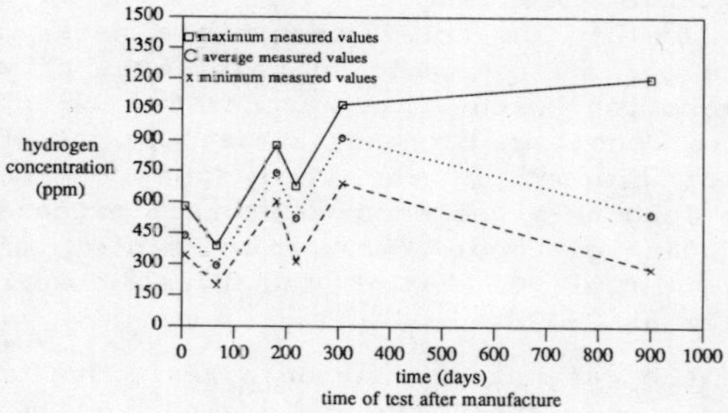

Fig. 6 Statistical data of hydrogen measured with different time after manufacture.

Also, the uneven distribution of hydrogen in a cable could be attributed to the uneven distribution of the waterblocking material and the residual wire drawing compounds. Therefore, in terms of optical transmission loss, the maximum value measured does not represent the hydrogen concentration in an entire span of cable. The average hydrogen concentration more appropriately reflects the hydrogen level for loss budget purposes. However, to be conservative, we estimate a hydrogen concentration of 1200 ppm for the manufactured cable span.

2. Hydrogen Generation Due To Aqueous Degradation Of Cable Materials By Moisture

The measurement of hydrogen levels in the above manufactured cables is random in terms of weather and humidity and therefore the measurements represent hydrogen levels in cables manufactured under nominal weather and humidity conditions. To determine the likelihood of moisture condensation in a cable, the amount of moisture present under more severe manufacturing conditions is studied.

During the manufacture of an SL cable, there are three processes in which moisture can be trapped in the cable: the manufacture of a cable core, the long-term storage of the core on a reel, and the copper conductor sheath manufacturing process. The copper sheath manufacture involves steel wire stranding, waterblocking material application, and the copper sheath welding process.

In the cable core manufacturing process, the pre-extruded Hytrel elastomer and nylon materials are kept extremely dry; there is about 0.002% moisture in the materials by weight after the drying process. Therefore, the amount of moisture contributed by these materials immediately after extrusion is considered to be negligible in this study. However, immediately after the nylon extrusion of the core outer surface, the core passes through cooling troughs containing running water at room temperature. This step allows the core to be vulnerable to moisture introduction because water directly contacts the core. When the core exits the cooling troughs, it is wound on a core reel. During the period of core take-up onto a reel, the single core is exposed to ambient factory air for about one hour. This step also provides the potential for moisture introduction into the core.

The next process that is vulnerable to moisture introduction is the core storage process. After core manufacture and take-up on a reel made of steel, the core reel is placed into storage under ambient factory conditions. Because the cable core reel may be in storage for an extended time, ambient moisture can enter the outermost layers and diffuse radially inward.

In the process of copper conductor manufacture, which consists of paying-out the core, laying down steel wires, introducing waterblocking material around the steel wires, and applying a welded copper jacket, the core, under normal conditions, is exposed for a short time period to the ambient factory environment. During this time, the core that has been paid-off of the reel, but that is not yet enclosed by the steel wires, as well as the top most layer of core on the core reel, is exposed to the factory environment for about three hours. In addition to the potential for water absorption during core pay-out, water can also be introduced internally to the cable by water dissolved in the waterblocking material. However, the applied waterblocking material is relatively dry; the moisture content is less than 0.05% by weight, or 11% relatively to the saturation solubility of the material at 24°C.

From the above discussion, the subset of potential water absorption steps identified for modeling to estimate the amount of water absorption in the cable during manufacture is listed as follows:

1. water cooling of molten core,
2. take-up of cooled core onto reel and pay-out of core during conductor manufacture,
3. long term storage of core on reels.

For item 1, water cooling of the molten core, a two layer composite cylinder is used to model moisture diffusion into the cable core. An approximate solution considering the diffusion coefficients as a function of temperature is found by using the methods from Jaeger[9] and Mulholland and Cobble[10]. The result of the moisture distribution in the core after 18 seconds in the water trough is shown in Table 2.

Table 2 The approximate distribution of relative moisture concentration in a cable core after leaving the core water cooling troughs

core location	radial distance from core center (mm)	relative moisture concentration
elastomer kingwire interface	0.36	0
within elastomer layer	0.69	0
within elastomer layer	1.04	0
elastomer-nylon interface	1.39	0
middle of nylon layer	1.44	0.115
outer nylon surface	1.485	1

Assuming the concentrations between each location in Table 2 are linearly distributed and the nylon outer surface is saturated to 0.0238 by volume or 0.02224 by weight (at 25°C), the total amount of water in the nylon is 6.69×10^{-5} grams per centimeter length of cable core.

The actual amount of water absorbed in the core during submersion in the water trough is measured by weight difference of a core sample before and after drying the core sample to remove the moisture. The moisture content is determined to be 0.18 grams for about 30 meters of total core length. This compares well with the above calculated amount of water for a core 30 meters long of 0.20 gms.

To model the second moisture absorption step, diffusion into a single core during the take-up of the core onto the reel and during pay-out of the core during copper conductor manufacture, a model of a two-layer composite using the conduction problem from Jaeger[9] is used. Extreme conditions of 100% relative humidity at 25°C are considered for conservative estimates. The maximum time which a single core is exposed to the factory environment is about 4 hours. During this period, moisture diffuses into the cable core and remains within the cable structure. The relative concentration distribution of moisture at the end of the 4th hour of core exposure and at 25°C is calculated, and the results are listed in Table 3.

Table 3 Relative moisture concentration in a cable core after a four hour exposure to ambient air at 25°C

core location	radial distance from core center (mm)	relative moisture concentration
elastomer kingwire interface	0.36	0.002
within elastomer layer	0.69	0.004
within elastomer layer	1.04	0.009
elastomer-nylon interface	1.39	0.017
middle of nylon layer	1.44	0.418
outer nylon surface	1.485	1

If the worst condition is considered, that is, the relative humidity of the factory air is assumed to be 100%, then the water trapped at the fourth hour is calculated using a similar method as the water cooling process. From Table 3, the total moisture content is determined to be 0.00011 gms per centimeter of cable core length. This amount of water is 18% of the total absorbable water in a core at 25°C. Combining this level of water absorption with that associated with the water cooling process at 25°C, the total water trapped in the core (not considering the long term reeled core storage process) is 0.00016 gms per centimeter cable length. This is about 27% of the total absorbable water at 25°C.

For the case of storage on a reel, due to the impermeable steel flanges and hub of the core reel, a model of 5 plain multi-layer composites of nylon and Hytrel elastomer is used to simulate the layers of cable core on a reel. The solution of this model also uses the method from Mulholland and Cobble[10]. The distributions of normalized moisture concentration (with the moisture concentration of the outermost nylon surface as 1, fully saturated) in the center of each Hytrel elastomer layer after 10-day and 60-day exposure periods are shown in Table 4.

Table 4 Hydrogen concentration in the middle of each Hytrel layer simulating 5 layers of core on a reel

exposure time (days)	layer number from top of reel 1	2	3	4	5
10	0.29	0.022	0.0008	0.00003	0.00001
60	0.71	0.29	0.089	0.021	0.005

From the results, we find that only the outermost layers are affected by the ambient humid air of the storage environment and the amount of moisture in these layers increases with time. Fortunately, the core reel is usually stored for about 10 days prior to the next copper sheath process. This waiting period minimally affects the outermost layer of the core. For long term storage of a cable core on a reel, moisture absorption is a potential problem even though there are 60 to 70 layers on a core reel. Therefore, a cover sheet was designed to cover the reel for long-term storage to minimize radial moisture diffusion.

During normal processing of the undersea cable, the total water absorbed in the cable relative to saturation is found to be low. This is summarized in Table 5.

Table 5 Water absorption in normal SL cable manufacturing process

process	total water absorbed (gm/cm)	conditions
1. core cooling in water trough	$6.x10^{-5}$	from experimental data
2. copper sheath manufacture	$1.64x10^{-5}$	moisture in water-blocking material <0.05% by wt. (25°C)
3. exposure of cable core to ambient air	$16.x10^{-5}$	25°C, 100% R. H. 4 hr exposure
summation	$23.64x10^{-5}$	

The total absorbable amount of water of the undersea cable at 3°C is calculated from the solubility in the waterblocking material, Hytrel elastomer and nylon. It is found to be $53.55x10^{-5}$ gm/cm. Comparing this value with the amount of water absorbed in an undersea cable during processing which is tabulated in Table 5, it is determined that the moisture contained in a cable is about 44% of the total absorbable water (at 3°C) in an undersea cable. If the normal exposure of a core reel is also involved, the total water absorbed in an undersea cable should not be over 50% of the total absorbable water in a cable. Thus, the occurrence of condensation in a normal undersea cable is highly unlikely.

HYDROGEN GROWTH DURING CABLE STORAGE AND SERVICE AT DIFFERENT TEMPERATURES

To find the growth of hydrogen in undersea cables after manufacture, 18 inch cable samples with one capped end and a bellows valve attached at the other end were subjected to two types of aging tests. These tests simulated aging during storage and service.

For the storage aging experiment, 18 inch cable samples were flushed to reduce the initial hydrogen concentration in the cable to only a few ppm. The generation and accumulation of hydrogen in an SL cable with time was found for aging at 30, 40 and 50°C. The results are plotted in Fig. 7.
The data appears to show the tendency of hydrogen saturation in the cable with time at different temperatures. For the service aging test, the 18 inch manufactured cable samples with 1.6 amperes direct current running

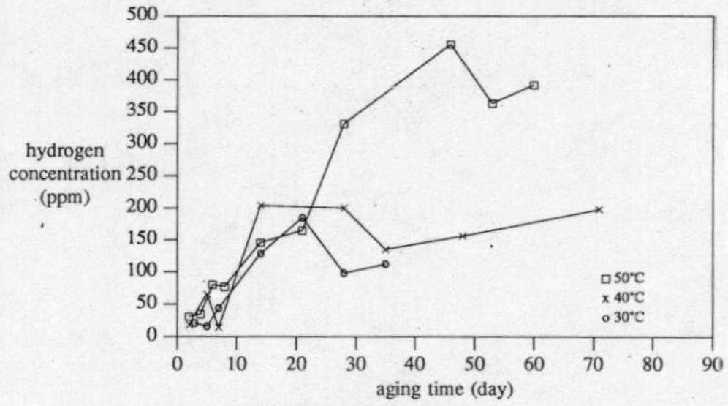

Fig. 7 Hydrogen generation in a dry, nitrogen flushed undersea cable during simulated storage aging at different temperatures.

through them were tested at 3, 10, and 20°C. The results are shown in Fig. 8.

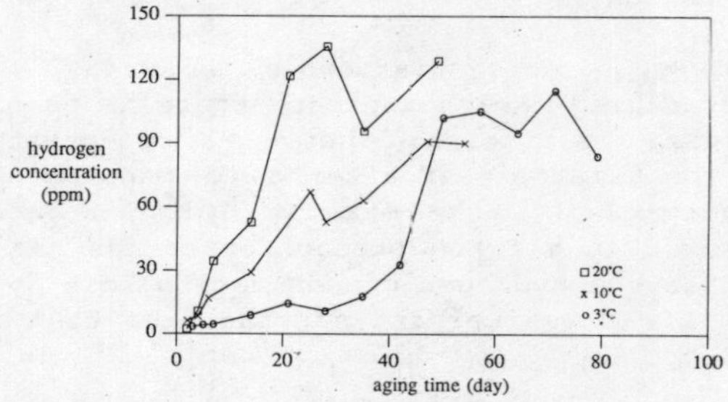

Fig. 8 Hydrogen generation in a dry flushed undersea cable during simulated service aging at different temperatures with 1.6 amp direct current.

All the dry, as-manufactured cable samples in the storage and service aging tests exhibited similar tendency of hydrogen saturation. Since the unevenly distributed waterblocking material and wire drawing compounds in the cable caused significant hydrogen concentration fluctuations, the hydrogen saturation level is calculated as the average of the last fluctuating data points in the aging tests. The relationship of these saturated hydrogen levels with temperature is shown in Fig. 9 and summarized in Table 6. Due to the temperature of cable storage and service is rarely beyond 40°C, a reasonable hydrogen concentration estimate for this period is about 300 ppm. This level is relatively low compared with the hydrogen estimate for cable manufacture. No significant evidence of the impact of electrical current on hydrogen generation is observed.

Table 6 Saturated hydrogen concentrations under various aging condition

temp (°C)	simulation	maximum measured concentration (ppm)	average saturated concentration (ppm)
3	service*	115	100.2
10	service	91	90.5
20	service	136	120.7
30	storage	185	130.9
40	storage	204	178.6
50	storage	456	403.7

* only service samples with 1.6 amp direct current

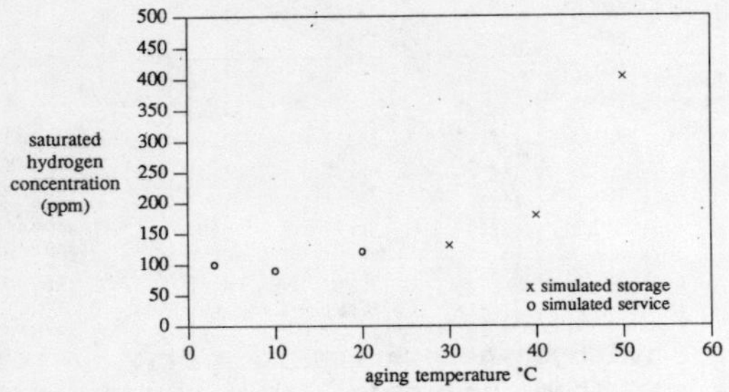

Fig. 9 Saturated hydrogen concentrations in undersea cables aging at different temperatures (simulated storage and simulated service).

CONCLUSION

Hydrogen concentrations measured in manufactured cables indicate there is no significant relationship between concentration and the time of test after cable manufacture. Conservatively, the maximum hydrogen amount measured in all the production cables can be used as the hydrogen concentration in a typical SL cable after its manufacture; that is, 1200 ppm.

The growth of hydrogen during the storage and service period of a dry, as-manufactured cable increases with time but tends to slow to a saturated state. Thus, the amount of hydrogen generated from normal storage and service in a cable should be less than 300 ppm. In addition, the likelihood for cable to be wet from condensation is very low. Therefore, the total hydrogen amount in an SL cable from its manufacture throughout its 25-year service life is estimated conservatively to be about 1500 ppm or less. The increase in optical loss resulting from this level of hydrogen in an SL cable is on the order of few thousandths of a dB/km at 1.3 µm and is accounted for in SL loss budgets. In addition, the fiber and cable manufacturing processes are rigidly controlled to ensure product uniformity so that the allocated hydrogen loss in the system loss budget is sufficient for all manufactured product.

ACKNOWLEDGEMENTS

I would like to give thanks to Ray Tuminaro from AT&T Bell Laboratories for his helpful suggestions and the personnel from Simplex Wire and Cable Company for their support.

REFERENCES

1. Stone J., Chraplyvy A. R. , and Burrus C. A.: Optics Letter, 7(1982), 296-299.

2. Mochizuchi K., H., Namihira Y., and Yamamoto H.: Elec. Lett., 19 (1983), 743-745.

3. Uesugi N., Murakami Y., Tanaka C., Ishida Y., Mitsunaga Y., Neigishi Y., and Uchida N.: Elec. Lett., 19(1983), 762-764.

4. Mies E. W., Philen D. L., and Reents W. D., Jr.: Proc. OFC 1984, New Orleans.

5. Rush J. D., Beales K. J., Cooper D. M. and Duncan W. J.: IEE Colloquium, June(1984), paper 3.

6. Pitt N. J. and Marshall A.: ibid, paper 6.

7. Ashpole R. S., Lewis D. L., Powell R. J. W., and Hinchliffe J. D. S.: IEE Proceedings, 132(1985),162-168.

8. Carslaw H. S. and Jaeger J. C.: Conduction of Heat in Solids, Clarendon Press(1986), 59.

9. Jaeger J. G.:Phil. Mag., 32(1941), 324-335.

10. Mulholland G. P. and Cobble M. H.: Int. J. Heat and Mass Transfer, 15(1972), 147-160.

RAPID RESTORATION SYSTEM USING NEW CONNECTORS AND CONNECTORIZED SINGLE-MODE FIBER CABLE

H. Katayose and H. Yokosuka
Fujikura Ltd. 1440 Mutsuzaki Sakura
Chiba Japan.

Abstract

With the wide use of single-mode optical fiber cables in many fields, development of a rapid restoration system for sudden cable breakdown has been demanded.

The components and performance of a rapid restoration system for a single-mode optical fiber are described in this paper.

This system consists of a connectorized cable which has compact closures at both ends, rapid-restoration connectors, and a small tool kit for connection use.

The connectorized cable, approximately 10 m in length, is a single mode two-fiber cable. Damaged fibers are connected to the connectorized cable, using the restoration connectors in the small closures at the both ends.

Each closure is 280 mm long and 32 mm in diameter , and weighs 500g. The closures are firmly fixed to the cable, giving the connectorized cable unit an allowable tensile strength of approximately 200 kgf. The restoration connector uses a single-mode FC type ferrule and a push-on type connector housing.

The average coupling loss of the connector is 0.38 dB when an index matching agent is used. Loss fluctuation remains within 0.3 dB, as demonstrated in temperature cycle testing between - 20°C and 60°C.

Emergency restoration of a two-fiber cable can be carried out in about 20 minutes, approximately 10 minutes for connection of each end of the connectorized cable.

1. Introduction

The rapid restoration system is used for emergency restoration of a broken fiber optic cable, both for outdoor and indoor fiber optic transmission systems. The restoration system has been designed to be extremely lightweight and compact, with consideration of ease of restoration work and ease of portability of the system.

This paper presents the components, restoration procedures, and describes various characteristics of the system.

2. Rapid Restoration System

Fig. 1 shows the restoration method for an optical fiber cable when the new rapid restoration system is used.

The connectorized restoration cable is used as a replacement for the damaged section of an optical fiber cable. The restoration procedure is shown below.

Cut the optical fiber cable on both sides of the damaged point, attach the rapid-restoration connectors to the cable fibers, then couple these connectors with the

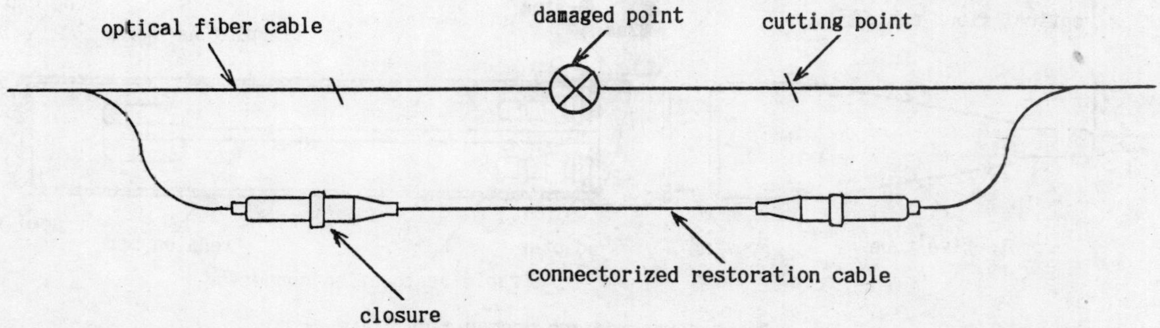

Fig. 1 RAPID RESTORATION SYSTEM

connector adapters of the connectorized restoration cable.

The requirements for this system are as follows.

1) The system must be sufficiently compact and lightweight to be carried by a single worker.
2) Handling and operation of the system must be easy.
3) It must be possible to do cable restoration in a short period of time.
4) The system must have low coupling loss.
5) The system must be re-usable.
6) The connected section must have high optical and mechanical reliability.
7) The system must be water-resistant.
8) The system must have satisfactory stability in outdoor environment.

2.1 Connectorized Restoration Cable and Closure

Fig. 2 shows the connectorized restoration cable and its closure. The specification of the connectorized cable are shown in Table 1.

The cable is made of non-metallic materials and has two single mode optical fibers. External diameter of the cable is 5 mm, and its length is 10 m.
Each end of the cable has two FC ferrules in two adapter units, attached to the housing. The damaged cable is attached to the clamping hardware at the end of the housing on each end of the cable. Maximum external diameter of the housing is 32 mm. The housing is closed with a closure by fastening a nut, and becomes water-tight.

Each closure consists of a rapid-restoration connector, waterproof rings (an O-ring and a seal ring), and an end cap. The closure, shown in Fig. 2, is connected to the connectorized cable housing by the nut. The joint between the closure and the housing is kept waterproof by an O-ring. The damaged cable is inserted into the hole at the end of the closure, and the seal ring is compressed by the end cap to keep the joint between the closure and the cable section waterproof. The rapid restoration connector installed on the damaged fibers are connected to the connector adapters of the housing, then covered by the closure. The closure functions to protect the fiber and the connector when a broken cable has been connected with the connectorized restoration cable.
Emergency restoration work is made easier because of the closure design in which all necessary parts including the rapid restoration connectors are housed inside the closure. The housing with the closure is 280 mm in length, and weighs 500 g, due to ts extremely light-weight, compact components.
The total weight of all components used for restoration is only 1,200 g, very light in weight and easy to carry.

2.2 Rapid Restoration Connector

Fig. 3 shows the structure of the rapid restoration connector. The connector was developed based on the FC type ferrule and a push-on type coupling method. The back of the ferrule is changed to have a fiber clamp which has four jaws and tapered fastening nut. The fiber clamp is designed mainly for a nylon coated fiber having a diameter of 0.4mm~0.9mm. When the fastening nut is fully clamped its clamping force is strong enough to still perform well even in a harsh environment.

Table. 1 SPECIFICATION OF CONNECTORIZED CABLE

ITEMS			SPECIFICATION
OPTICAL FIBER	TYPE core/clad	 (um)	SM 10/125
CABLE	diameter weight length	(mm) (g/m) (m)	5 20 10
ALLOWABLE TENTION BENDING RADIUS		(kgf) (mm)	100< 60>

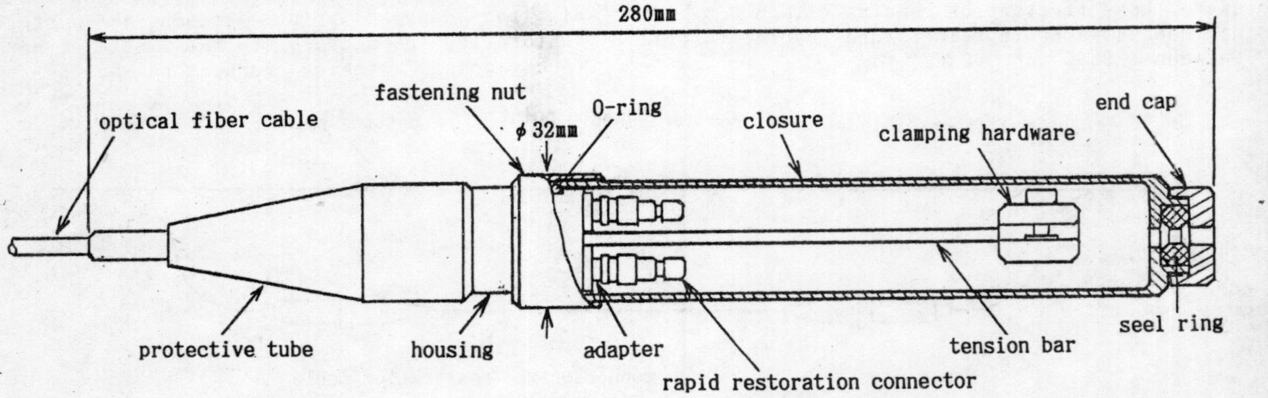

Fig. 2 CONNECTORIZED RESTORATION CABLE

2.3 Tool kit for the Rapid Restoration System

A tool kit is used for the rapid restoration. The cable cutter, fiber stripper, fiber cutter, screwdriver, index matching agent, and loupe, are all provided, in a tool kit case. The rapid restoration connectorized cables can also be stored in the case, so everything necessary for restoration can be carried in a single container. The loupe is used to determine the condition of the optical fiber at the end surface of the ferrule, after the optical fiber has been attached to the rapid restoration connector.

2.4 Restoration Procedure

The rapid restoration procedure is shown below.
(1) Bring a connectorized restoration cable to the point where there is a break in the cable.
(2) Cut off the damaged section of the cable. Remove the closure and the end cap of the connectorized restoration cable, and insert the trimmed end of the cable into the closure section.
(3) Remove the sheath from the cable to be restored, and take out the optical fibers.
(4) Remove the nylon coating and buffer layer from the optical fiber, and cut the fiber using the fiber cutter, to ensure satisfactory connection of the fiber.

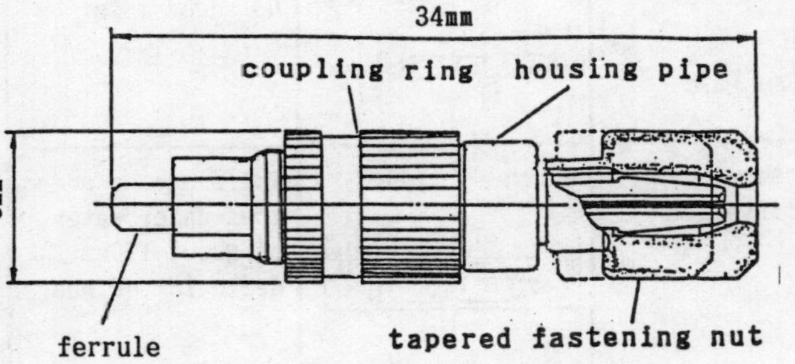

Fig.3 RAPID RESTORATION CONNECTOR

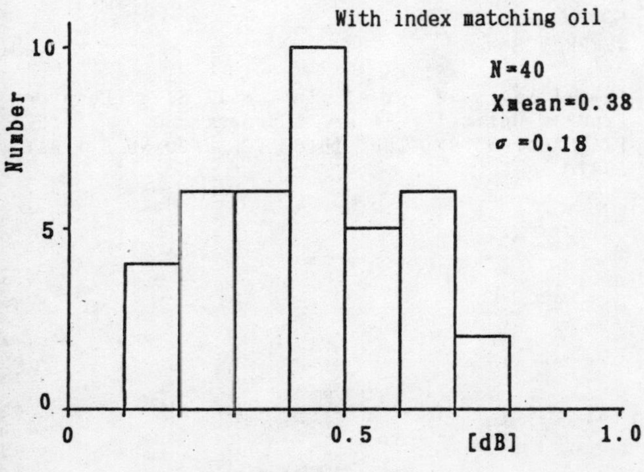

Fig.4 COUPLING LOSS

(5) Insert the cut fiber into the rapi restoration connector, and match the end surface of the fiber with the end surface of the ferrule of the connector. The fiber is fixed in place using the fixing nut of the rapid restoration connector. Use the loupe to check the end surface of the fiber whether or not there is any chipping.
(6) Apply refractive index matching oil to the end surface of the ferrule of the rapid restoration connector and connect to the adapter.
(7) Attach the cable to the cable clamping hardware section in the closure.
(8) Move the closure so it can be attached to the housing, and close it tightly for satisfactory waterproofing.

The above work procedures, from (1) to (8) are to be done at both ends of the rapid restoration connectorized cable.

3. Test and Result

3.1 Coupling loss and Restoration time

Fig. 4 shows coupling loss of the rapid restoration connector. Measurement of the coupling loss was done for the connection point, using the matching agent. The average coupling loss for single-mode optical fiber was 0.38 dB. This result means that the average coupling loss for the system can be determined as 0.76 dB.

Fig. 5 shows the results of measurement of the time needed for connection of a closure and rapid restoration connector to the end of the rapid restoration connectorized cable.
The time needed to assemble one end of the connectorized cable is an average of 10.6 minutes, and the average total for restoration time using the rapid restoration system is 21.2 minutes.

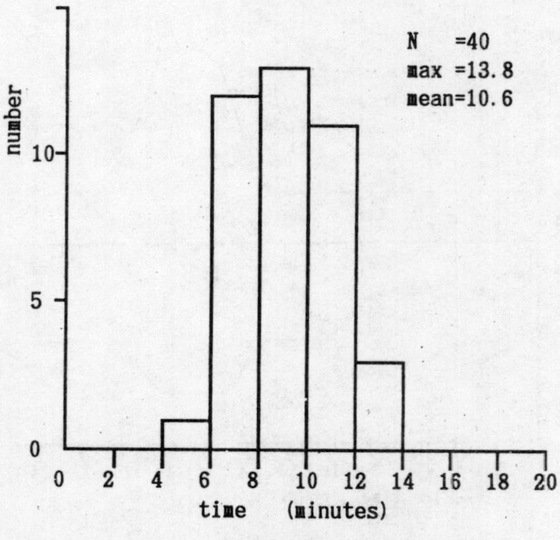

Fig.5 RESTORATION TIME

3.2 Thermal Characteristics

Fig. 6 shows the increase of coupling loss of the thermal test when both ends of the rapid restoration connectorized cable have been connected. The temperature range was from -20 ℃ to 60 ℃. The results showed a maximum loss increase of 0.28 dB at -20℃.

3.3 Mechanical characteristics

The mechanical testing methods and their results are shown in Table 2. The results of the tensile strength test and vibration test of the connectorized cable proved its excellent properties. The cable clamping hardware for a damaged cable has sufficient strength.

3.4 Water Invasion

The closure section of the rapid restoration cable was immersed in water at a depth of 100 mm for 48 hours, but no water invasion into the inside of the closure was noted.

4. Conclusion

A rapid restoration system using connectorized single-mode optical fiber cable has been developed. The system is sufficiently compact and light in weight for one person to carry, and the work of restoration using the system is easy.

Restoration time for a single worker is approximately 20 minutes. Coupling loss for the system, which uses single-mode optical fiber, averages 0.76 dB. The increase of coupling loss due to temperature change, from -20℃ to 60℃ is 0.3 dB or less.
In mechanical testing, the increase of coupling loss was 0.1 dB or less. The closure section is suitably waterproof.

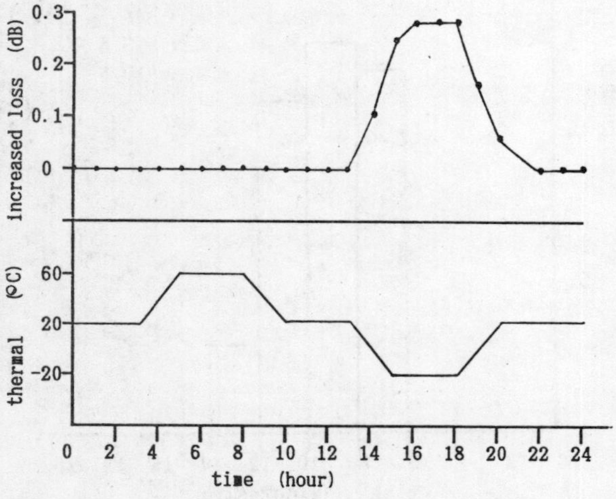

Fig.6 THERMAL TEST

Table.2 MECHANICAL TEST AND RESULTS

ITEM	TEST CONDITION	RESULT
Tensile strength	tensile speed :10mm/min 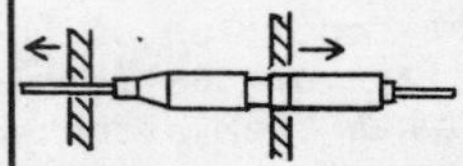	Breakdown tensile strength was more than 200kgf.
Vibration	amplitude:2mm frequency:10Hz 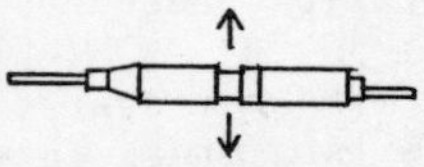	The loss change was less than 0.05dB.
Compression	weight :100kgf 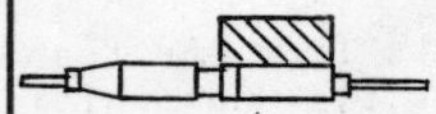	There was no damaged. The added loss was less than 0.05dB.
Cable retention	tensile strength :50kgf 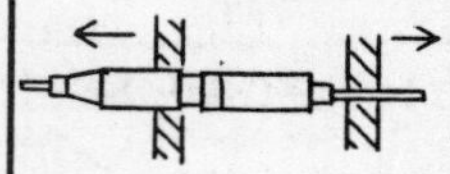	There was no damaged. There was no added loss.
Water invasion	depth :100mm 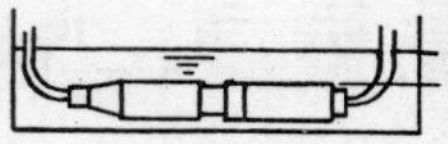	There are no added loss under water of about 100mm depth for 48 hours.

REFERENCE

Sinji Nagasawa, et al., In: Small-Size Push-on Type Optical Fiber Connector , THE TRANSACTION OF THE IEICE, VOL. EA70, NO. 5 MAY 1987.

COUPLING PROPERTIES OF SINGLE-MODE OPTICAL FIBER COUPLER

Zhang Xiang, Zhang Yi and Xu Jisen
Shanghai Transmission Lines Research Institute
Shanghai, China

ABSTRACT

A new method to analyze the coupling properties in fused biconical taper coupler has been reported in this paper. The coupling power of the single-mode coupler can be easy calculated using these matrixes with the personal computer by combining the coupled-mode equations with amplitude transmission matrix. The theoretical results are compared with the experimental data, good agreement is found.

INTRODUCTION

Many fiber optic systems in operation today such as LANs, ISDN, single-mode fiber communication systems and single-mode fiber sensors require low loss single-mode couplers to passively split or combine light between two or more single-mode fibers. Fused biconical taper couplers are employed because of their easy of fabrication and very stable performance over a wide range of operating conditions.

In recent years, different models and analysis methods to describe the coupling properties of coupler have been extensively reported in the literature.[1,2,3,4] The common way is regarding the fused fiber region as rectangle fiber and neglecting the fiber core, fields in this rectangle fiber are employed instead of fields in two fiber.[5,6] It's simple but the theoretical results show difference compare with experimental results. Some different way[8] has been calculated by considering the core as perturbation. It's very complicated and difficult to calculated but with similar result. In this paper, we combine the coupled-mode equations between two fibers with amplitude transmission matrixes in one taper fiber to describe properties of coupler. This method is available in calculating the coupling between N fibers to study star couplers.

THEORY

For a taper single-mode fiber, the core reduce and fields diffuse into cladding. It is considerable that optic fields transmit in new core of original cladding index surrounding with new cladding of air index in the coupling region with neglecting the original core. In this case, it is convenient to adopting scalar approximation.

1. Fields

The normalize fields of fundamental mode in a fiber are

$$\psi_o = \begin{cases} \dfrac{u\,K_o(w)}{aVK_1(w)\sqrt{\pi}} \dfrac{J_o(ur/a)}{J_o(u)} e^{jm\theta} & r<a \\[2ex] \dfrac{u\,K_o(w)}{aVK_1(w)\sqrt{\pi}} \dfrac{K_o(wr/a)}{K_o(w)} e^{jm\theta} & r>a \end{cases} \tag{1}$$

Eigenvalue equation

$$\frac{u\,J_1(u)}{J_o(u)} = \frac{w\,K_1(w)}{K_o(w)} \tag{2}$$

Where a is radius of core, J , Kare Bessel and modify Bessel function, u , w are eigenvalue

$$\begin{aligned} u^2 &= a^2 (4\pi^2/\lambda^2 * N_1^2 - \beta^2) \\ w^2 &= a^2 (\beta^2 - 4\pi^2/\lambda^2 * N_2^2) \\ V^2 &= u^2 + w^2 \end{aligned} \tag{3}$$

2. Analysis of taper fiber

For a taper fiber shown in Fig. 1. Its radius along longitudinal Z are not equal everywhere. It is difficult and troublesome to solute the fields as a function of Z. Instead, we divide the long taper fiber into several sections as shown in Fig. 1 and assume that every section is equal-radius, then connect them using boundary conditions to obtain final results.

The principle of division is that the radius difference of two nearby fiber is much less than the radius of individual section and length of every section is much longer than the wavelength.

3. Fields transmit between two sections

Fig. 2 shows two sections. We assume that ψ_1 ψ_2 are incident field and reflected field in section 1, ψ_3, ψ_4 are transmission field and opposite incident field in section 2. ψ_3 in section 2 can be expressed with ψ_1 in section 1

$$\psi_3 = \psi_1 + \frac{d\psi_1}{da}\Delta a + \frac{1}{2}\frac{d^2\psi_1}{da}(\Delta a)^2 + \dots \tag{4}$$

Multiplication of Equation(4) by ψ_1^*, then integration over S. Neglecting high order terms and with help of Equation(3), Equation(2), we obtain

$$\int \psi_3 * \psi_1^* \, ds = 1 - P \tag{5}$$

Where P=P(a,u,w, a) is a function of radius, eigenvalue, step height and Bessel function.

Tangentail electromagnetic fields are continuous at surface S

$$A\psi_1 + a_r\psi_2 + \int q(\rho)\psi_2(\rho)d\rho = c_t\psi_3 + B\psi_4 + \int g(\rho)\psi_3(\rho)d\rho \tag{6}$$

Where A is amplitude of incident field, ar is amplitude of reflected field, $q(\rho)$ is amplitude of reflected radiation, c_t is amplitude of transmission fields, B is amplitude of opposite incident field, $g(\rho)$

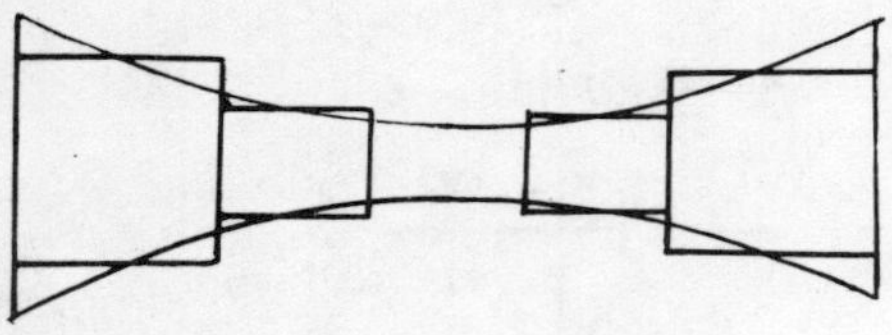

Fig. 1 Geometry and division of a taper fiber

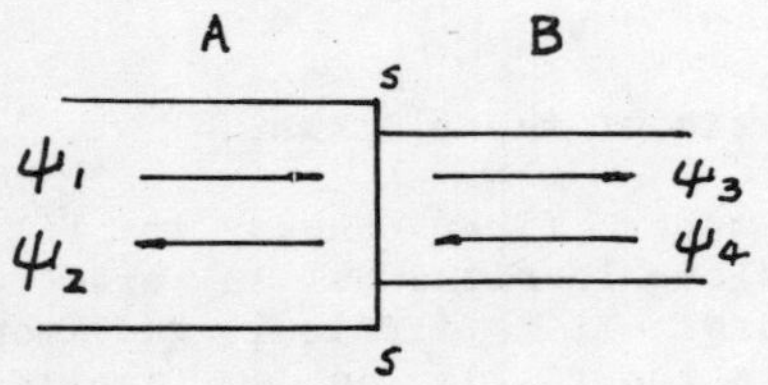

Fig. 2 Fields in two sections

is amplitude of transmission radiation field, ψ $\psi(\rho)$ are guide mode and radiation mode fields respectively.

The modes of section 1 are not orthogonal to the modes of section 2 because of the different size of fiber. To simplify calculation, we assume that they are orthogonal[7].

For a taper coupler, its reflected field is more smaller than incident field. So first we achieve the c_t by neglecting ψ_2,

$$A\psi_1 = c_t\psi_3 + \int g(\rho)\psi_3(\rho)d\rho \tag{7}$$

Multiplication Equation(7) by ψ_3^* and integration over S. using orthogonality condition and Equation(5)

$$c_t = A\int \psi_1 * \psi_3^* ds = A\,(1-P) \tag{8}$$

Substitute Equation(8) into Equation(6) for further calculation and use orthogonal condition we obtain

$$a_r = -P(2-P)A + (1-P)B \tag{9}$$

The law of conservation of energy shall be followed at surface S[9]

$$A^2 + B^2 = c_t^2 + a_r^2 \tag{10}$$

Amplitude B of opposite incident field is small than amplitude A of incident field. Express it and neglecting high order terms

$$c_t = \sqrt{A^2 + B^2 - a_r^2} = \sqrt{1-P^2(2-P)^2}\,A + \frac{P(2-P)(1-P)}{\sqrt{1-P^2(2-P)^2}}\,B \tag{11}$$

The matrix about optical field transmit through a homogeneous fiber

$$\begin{bmatrix} e^{-j\beta\Delta Z} & 0 \\ 0 & e^{j\beta\Delta Z} \end{bmatrix} \tag{12}$$

Combine all of these expressions, we obtain a transmission matrix about optical field transmit through any individual section fiber

$$\begin{bmatrix} c_t \\ B \end{bmatrix} = \begin{bmatrix} e^{-j\beta\Delta Z} & 0 \\ 0 & e^{j\beta\Delta Z} \end{bmatrix}\begin{bmatrix} H_{11} & H_{12} \\ H_{21} & H_{22} \end{bmatrix}\begin{bmatrix} A \\ a_r \end{bmatrix} \tag{13}$$

Where

$$H_{11} = \frac{1}{\sqrt{1-P^2(2-P)^2}} \qquad H_{12} = \frac{P(2-P)}{\sqrt{1-P^2(2-P)^2}}$$
$$H_{21} = \frac{P(2-P)}{1-P} \qquad H_{22} = \frac{1}{1-P} \tag{14}$$

4. Coupling between two fiber

Fig. 3 shows two adjacent fibers. We assume fields in each fibers are $A_1\psi_1$ and $A_5\psi_5$. Coupling Equation:[10]

$$dA_1/dZ = -j\beta_1 A_1 + jA_5 k_{51}$$
$$dA_5/dZ = -j\beta_5 A_5 + jA_1 k_{15} \qquad (15)$$

k_{15}, k_{51} are coupling coefficient. For two identical fibers $\beta_1=\beta_2=\beta$, $k_{15}=k_{51}=k$.

For step index fiber, k can be simplified[11]

$$k = \frac{u^2 K_0(wd/a)\sqrt{(N_1^2 - N_2^2)}}{N_1\, a\, V^3 K_1^2(w)} \qquad (16)$$

a is radius of fiber, d is distance between two fiber centers, here d=2a. Solute Equation(15) with condition $A_1=A_1(0)$, $A_5=A_5(0)$ at Z=0. We obtain

$$\begin{bmatrix} A_1(Z) \\ A_5(Z) \end{bmatrix} = \frac{1}{2}\begin{bmatrix} M_{11} & M_{12} \\ M_{21} & M_{22} \end{bmatrix}\begin{bmatrix} A_1(0) \\ A_5(0) \end{bmatrix} \qquad (17)$$

Where

$$M_{11} = M_{22} = e^{-j(\beta-k)Z} + e^{-j(\beta+k)Z}$$
$$M_{12} = M_{21} = e^{-j(\beta-k)Z} - e^{-j(\beta+k)Z} \qquad (18)$$

We neglect coupling between opposite transmission fields.

5. Optical fields transmit in taper coupler

Fig. 4 shows a certain section of fiber. The direction of arrow is the direction of field

Fig. 3 Coupling between two fiber

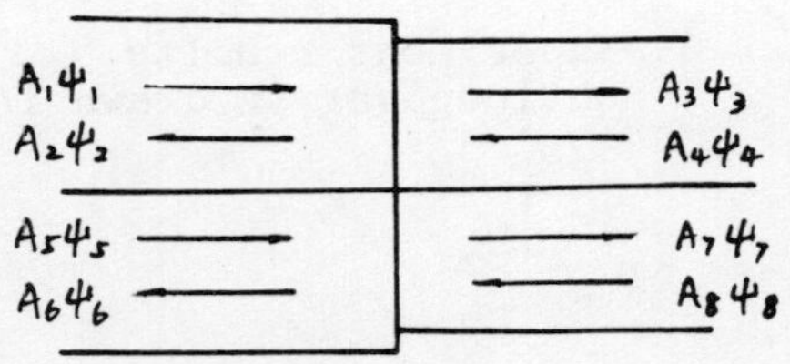

Fig. 4 Transmision and coupling between two taper fibers

transmission. A_1, A_2, A_3, A_4 are amplitudes of fields in one fiber which meet Equation(13). A_5, A_6, A_7, A_8 are amplitudes of fields in another fiber which meet Equation(13). A_1, A_3, A_5, A_7 meet Equation(17), and A_2, A_4, A_6, A_8 also meet Equation(17). By combining these relations, we obtain a comprehensive matrix which describes the optic fields transmitting in a taper coupler.

$$\begin{bmatrix} A_3 \\ A_4 \\ A_7 \\ A_8 \end{bmatrix} = \frac{1}{2}\begin{bmatrix} M_{11} & 0 & M_{12} & 0 \\ 0 & M_{11}^* & 0 & M_{12}^* \\ M_{21} & 0 & M_{22} & 0 \\ 0 & M_{21}^* & 0 & M_{22}^* \end{bmatrix}$$

$$\begin{bmatrix} H_{11} & H_{12} & 0 & 0 \\ H_{21} & H_{22} & 0 & 0 \\ 0 & 0 & H_{11} & H_{12} \\ 0 & 0 & H_{21} & H_{22} \end{bmatrix}\begin{bmatrix} A_1 \\ A_2 \\ A_5 \\ A_6 \end{bmatrix} \qquad (19)$$

M matrix is described by Equation(18), M^* are thier complex conjugate, H matrix is described by Equation(14).

5. Calculation

different speed of drawing fiber lead to different shape of taper. In our calculation we found small difference of final results among different variations of radius we adopt. We adopt function in following term

$$a = L1 - L2 * Z \qquad (20)$$

a is radius of fiber, Z is length of fiber, L1, L2 are parameters of taper shape. We divide the fibers into N sections, every radius difference between two sections $\Delta a_n = 1/m\, a_{n-1}$. Every section length of Z :$\Delta Z_n = a_n*(L1-1)/L2$. We assume $A_1(0)=1$ at Z=0, then obtain final output power A_3^2 and A_7^2 .

Fig. 5 shows three kinds of calculation results, they have equal length of taper but with different L1, L2 parameters.

We have parameters for
Fig. 5a : m=20, L1=135, L2=1, N=94
Fig. 5b : m=20, L1=77, L2=1/2, N=98
Fig. 5c : m=27, L1=42, L2=1/3, N=140

Fig. 5c have more sections because of sharply changing in the curve.

Consider the fact that coupler have very small excess loss, so we only calculate the fundamental mode. For reference of other mode, we will discuss later.

EXPERIMENT

Experiment are conducted on the fiber fusion station. A short segment of fiber coating are removed in the middle of two lengths of fiber, then these regions must be carefully cleaned using chemical solution. Load the fibers into the vacuum clamp on fusion station. Light (LD) source are injected and detectors are connected to monitor the optical power output. Computer is employed to monitor the stability of optical source and store output data. We can obtain different taper shape by changing the draw speed. Fig. 6 shows three curves of experiment.

Compare Fig. 5 and Fig. 6, a good agreement is found. The approximation used in our theory are acceptable and adoptable.

CONCLUSION

In this paper, we report a new way to calculate the properties of optical fields transmit in fused biconical taper single-mode fiber coupler. This method is available to extend to calculate coupling of N fibers.

REFERENCES

1. J.F.Digonnet and H.J.Shaw, IEEE J. Quantum Electron, QE-18(1982), 746
2. S.K.Sheem and T.G.Giallorenzi, Optics Letters, 4(1979), 29
3. B.S.Kawasaki, K.O.Hill and R.G.Lamont, Optics Letters, 6(1981), 327
4. G.Georgiou and A.C.Boucouvalas, IEE. Proc. J. Optoelectron, 132(1985), 297
5. F.P.Payne, C.D.Hussey and M.S.Yataki, Electron Letter, 21(1985), 461
6. J.D.Love and M.Hall, Electron Letter, 21(1985), 519
7. D.Marcuse, Light Transmission Optics, Van Nostrand reinhold company, (1982), 389
8. K.S.Chiang, Electron Letter, 23(1987) 717
9. D.Marcuse, B.S.T.J., 49(1970), 1665
10. D.Marcuse, B.S.T.J., 52(1973), 817
11. P.D.Mcintyre and A.W.Snyder, J. Opt. Soc. Am. , 64(1974), 285

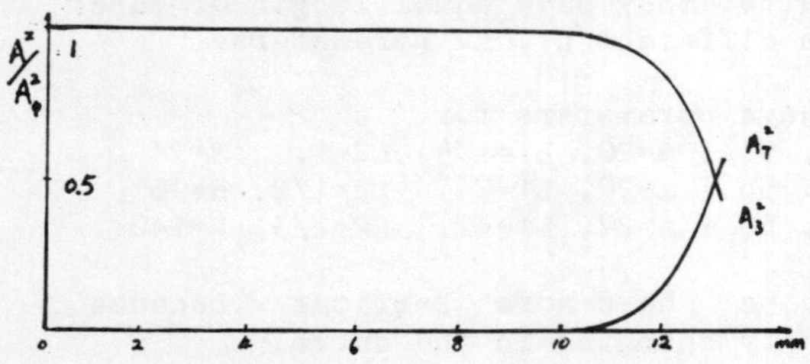

Fig. 5a

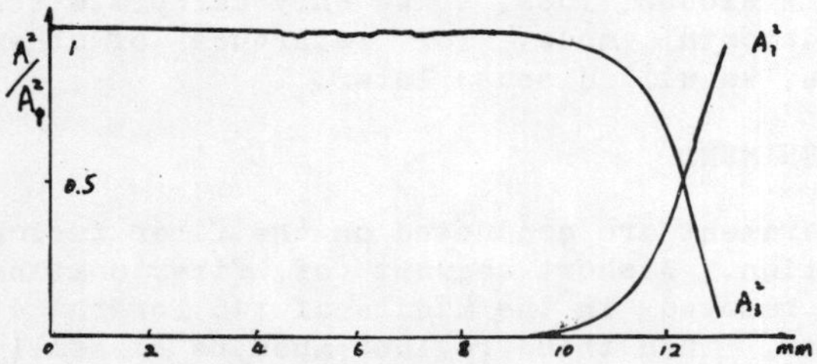

Fig. 5b

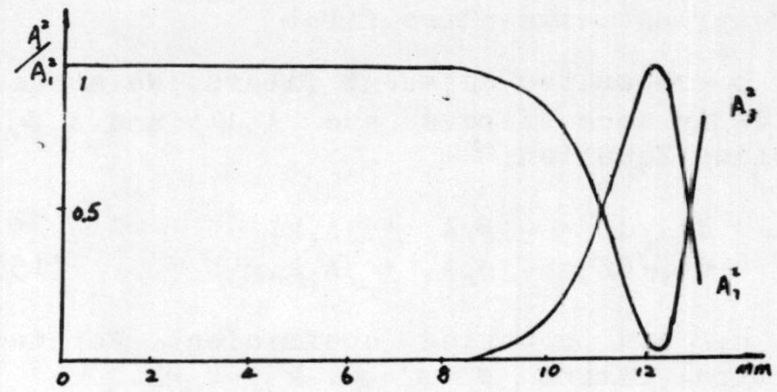

Fig. 5c

Fig. 5 Theoretical calculation Optic power vs drawn length

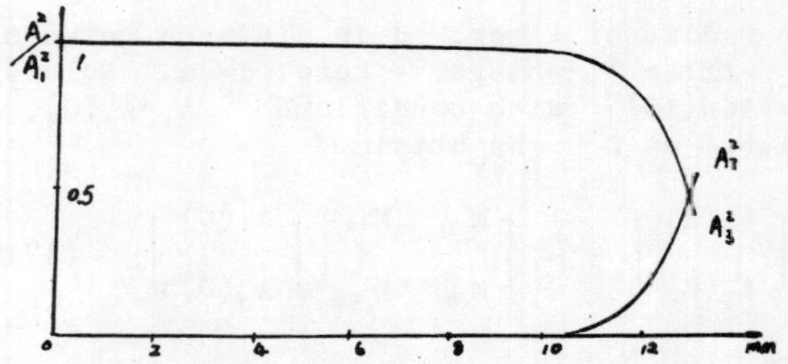

Fig. 6a

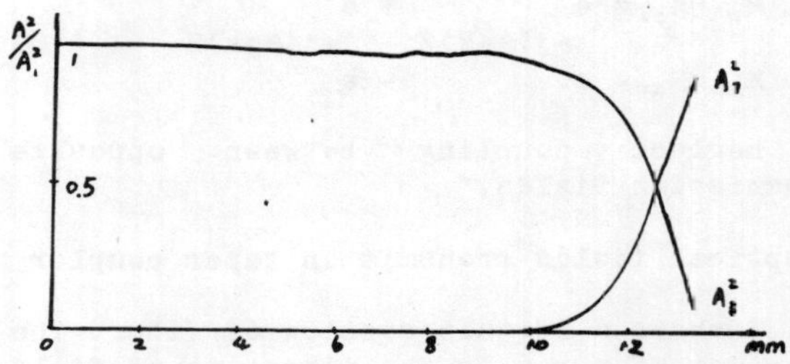

Fig. 6b

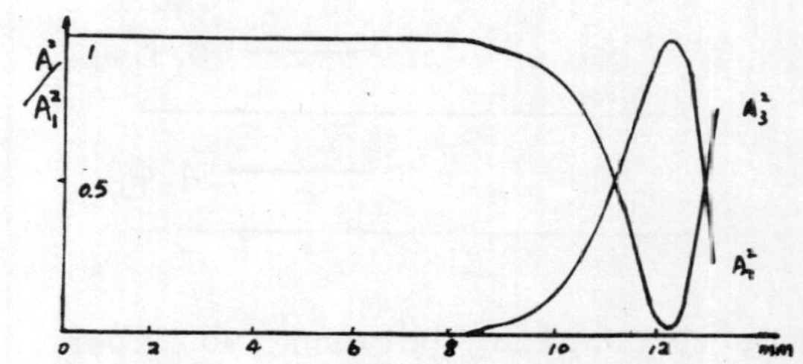

Fig. 6c

Fig. 6 Experiment results Optic power vs drawn length

HIGH SILICA WAVEGUIDE TECHNOLOGY FOR INTEGRATED OPTICAL CIRCUIT FABRICATION

Shin Sumida and Tadashi Miyashita
Photonic Integration Research, Inc.
1375 Perry Street, Columbus, Ohio
43201 U.S.A.

Abstract

Silica glass waveguide circuits with low loss and well controlled configuration are fabricated on silicon substrates by a combination of flame hydrolysis deposition, photolithographic and reactive ion etching processes. Based on this silica glass waveguide technology, integrated optical circuits such as 1 x 8 splitter, wavelength division multi/demultiplexing(WDM) module, modulator/switch, and ring resonator have been fabricated and tested.

1. Introduction

Various kinds of passive optical components are required in optical systems such as communication and sensing. Regarding fabrication methods, these components are classified roughly into three groups:Bulk, fiber and waveguide types. The waveguide type[1-4] is very attractive from the viewpoints of mass productivity, low loss, low cost, and compactness. These points are of importance in order to introduce optical systems into practical fields. The formation of optical waveguides in various materials such as polymers[4] and glasses[1-3,5] has been researched and several kinds of devices have been already developed based on these material related technologies. An optical fiber compatible silica glass waveguide technology based on a combination of flame hydrolysis deposition(FHD)[6], photolithographic and reactive ion etching processes has been developed which provides high performance and cost effective waveguide circuits.

In the following sections, first, silica glass waveguide technology is described. Second, construction and performance of 1x8 splitter, WDM module, modulator/switch, and ring resonator are reported comprehensively.

2. Waveguide Fabrication

Figure 1 shows fabrication steps of silica waveguides. Silica glass guided-wave optical circuits are fabricated by a combination of flame hydrolysis deposition and reactive ion etching. First, planar waveguides are fabricated on silicon substrates. Fine glass particles synthesized by flame hydrolysis of $SiCl_4$ /$TiCl_4$ or $SiCl_4$/$GeCl_4$ are deposited on the substrates.

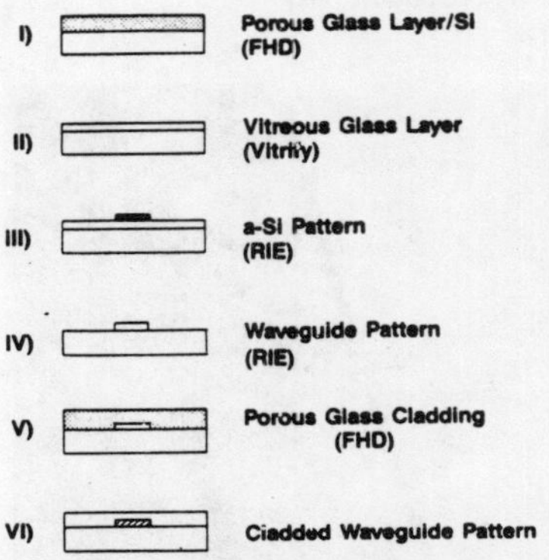

Fig. 1. Silica waveguide fabrication process

After deposition , the substrates are heated to high temperature for conslidation. The refractive index difference can be controlled precisely by changing the flow rate of row material gases. The thickness of the glass layer is also controlled in the range of 10 to 200 um within a 2% thickness variation. Figuure 2 shows the structure of a planar waveguide fabricated on a silicon substrate.

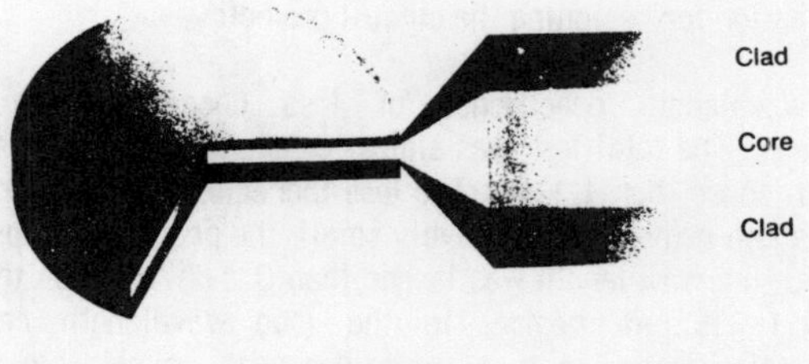

Fig. 2 Structure of silica planar waveguide fabricated on a silicon substrate.

Ridge waveguide is formed by a combination of photolithographic patterning recognition and reactive ion etching. Amorphous silicon is used as a mask material for etching silica glass. Thanks to the photolithographical patterning, any desired circuit configuration, including circular patterns, can be easily fabricated and identical circuits can be simultaneously processed on a single substrate and repeatedly duplicated by the same photomask. Ridge waveguides with vertical and smooth side walls are obtained by selecting suitable etching conditions. Scanning electron micrograph of ridge waveguides is shown in Fig. 3. The ridge single-mode waveguides are ordinarily embedded again by the FHD process. The stability of silicon and silica ensures high performance of the circuits under a wide variety of enviromental conditions.

To sumarize this section, the present silica waveguide fabrication techonlogy has the advanteges of high precision waveguide fabrication, arbitrary circuit geometry, productivity, and reproducibility.

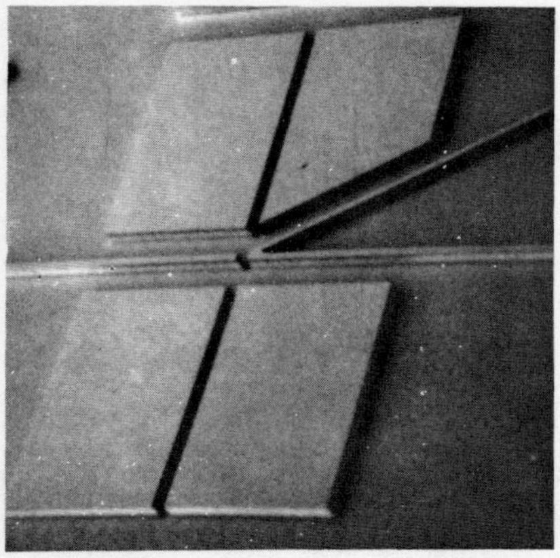

Fig. 3 Scanning electron micrograph of ridge waveguides (Branching section)

3. Guided-wave optical circuits

Fundamental waveguide circuit elements such as straight and bending waveguides are briefly related here. The characteristics of these basic elements are very useful information for designing the circuit geometry.

The wavelength dependence of loss characteristics was measured. The total loss was almost constant in the wavelength region longer than 1.1 um. The loss increment in the shorter-wavelength region was relatively small: the propagation loss at the 0.6-um wavelength was larger than 0.2 dB/cm than that in the 1.1-1.6 um region. In the long wavelength region, propagation losses in the waveguides with a relative index difference of 0.25% and 0.75% are as small as 0.1dB/cm and 0.2 dB/cm, respectively. The measured coupling loss between the waveguide and single mode optical fiber was smaller than 0.05 dB. This value is small enough to allow practical use of the waveguides.

The bending loss dependence of 90 degree arc waveguides on the radius of curvature R at 1.55 um wavelength for a low index(delta=0.25%) and a high index(delta=0.75%) were measured. The bending loss was strongly dependent on the refractive index difference between the core and cladding. No loss increase was observed in the curvature range of R> 4mm for a high delta and of R>15 mm for a low delta. With regard to the total loss and the device size, a higher delta waveguide is preferable to a lower delta waveguide, even if the propagation loss for a higher delta is higher than for a lower delta.

3-1. 1 x 8 splitter

Two of the most important elements for splitting circuits are a Y coupler and a directional coupler. The typical insertion loss for both types are smaller than 0.5 dB, including input/ output optical fiber coupling loss. In the directional coupler, spacing of of a few microns between the two waveguide cores can be controlled precisely by the silica glass waveguide fabrication technique presented in this paper. The directional coupler shows a wavelength dependent coupling efficiency, which can be used not only for power splitter but also for wavelength filter. For example, directional coupler acts as a filter for 1.3 um /1.55 um with the loss of 0.5 dB and the extinction ratio of as high as 25 dB or more.

A 8 port splitter is realized with a cascade of directional couplers. Figure 4 shows the schematic diagram of 1 x 8 splitter using optical directional couplers. The circuit size is 25 mm x 2 mm x 1 mm. The single directional coupler has a 2.5 um waveguide spacing in the interaction region and a 5 mm interaction length . The refractive index difference is 0.25%. The waveguides at the 8-port end are separated by 250 um. The excess loss of the coupler is <1 dB at the wavelength of 1.3 um. In Figure 8, the coupler consists of cascaded directional couplers but it is also possible to use Y couplers instead. The characteristics of the Y coupler type 1 x 8 splitter are the same as that of directional coupler type 1 x 8 splitter. A Y-coupler splitter shows the wavelength independent characteristics unlike a directional-coupler splitter.

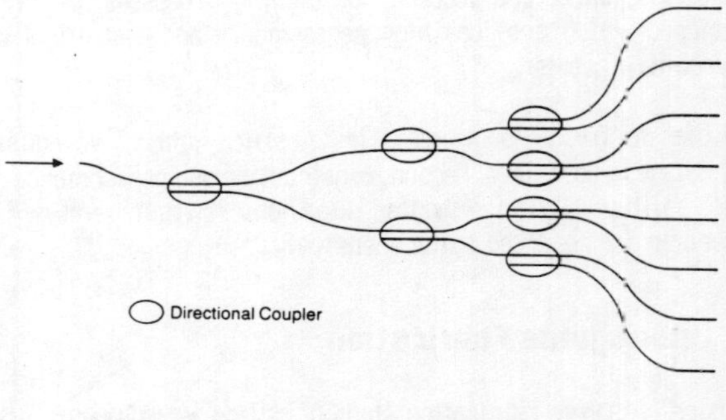

Fig. 4 1 x 8 splitter(directional coupler type)
Circuit size:25 mm x 2 mm x 1 mm

3-2. WDM module

The optical directional coupler shows wavelength dependent coupling efficiency and can be used as a WDM with a broad wavelength bandwidth of 100-200 nm. Figure 5 shows a schematic of the WDM module with a metalic package. Pigtailed package size is 30 mm x 12 mm x 50 mm. The operating wavelengths are 1.3 and 1.55 um. The module is composed of a directional coupler, a laser diode, and PIN photo-diode(PD) The output power at the pigtail end is 0 dBm. The coupling loss between the pigtail end and PIN PD is 1.4 dB. The laser diode is placed in the isolated metalic package from PIN photodiode for reduction of electrical interference between two diodes. The crosstalk is <-30 dB.

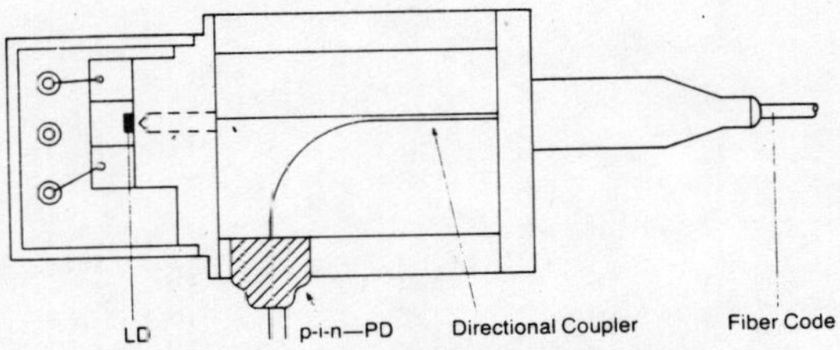

Fig. 5 1.3 um/1.55 um WDM module.
Package size:30 mm x 50 mm x 12 mm

3-3. Modulator/switch

In glass waveguides, the optical modulator/switch can be fabricated by using the thermooptic effect instead of the electrooptic effect for an LiNbO3 waveguide. Figure 6 shows the optical switch/modulator fabricated. A thin film heater on one of two waveguide arms can act as a phase shifter because the light path length of the heated waveguide arm changes due to the temperature dependence of the refractive index. For silica glass, the change of refractive index with temperature ($\Delta n/\Delta T$) is 1×10^{-5}; for example 5-mm-long waveguide is heated by 30 C increments, the light path length changes by 1.5 um and thus the phase shift is 2π for a 1.5-um lightwave.

The modulator/switch were formed using low delta waveguides. The separation between the two waveguide arms was 250 um. The chromium thin film heater, 300 nm thick, 5 mm long, and 12 um wide, was formed just above one waveguide arm through a 20-um cladding layer by electron beam evaporating and wet etching. The resistance of the heater was 240 ohms.

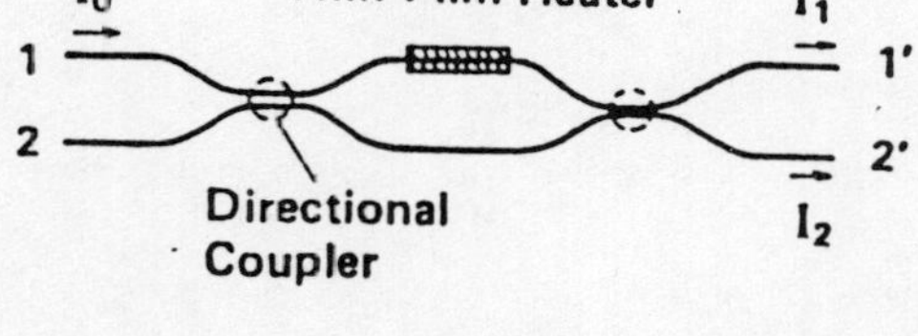

Fig. 6 Thermooptic modulator/switch
Circuit size: 10 mm x 40 mm x 1 mm

The transmission of the 1.3 um LD light was measured while the voltage was applied to the heater. The extinction ratio was 17 dB and the excess loss was 0.8 dB, including input/output coupling loss. The response time of the circuit was 1 ms.

3-4. Ring resonator

Figure 7 shows the schematic diagram of the ring resonator. It consists of an input waveguide, a resonator ring, and output waveguide and two directional couplers. The ring resonator radius was designed to be 6.5 mm for a free spectrum range of 5 GHz. A high delta(0.75%) waveguide was used for circuit compactness purpose. Chromium thin film heaters were formed onto the ring as a thermooptic phase shifter. Resonance characteristics were measured by sweeping a DFB laser diode frequency with incident current.

The finess was estimated to be 15 and the insertion loss including input/output fiber coupling loss was 16 dB at the resonant frequency. Such ring resonator has many applications such as optical frequency selective filters, LD frequency stabilizers and rotation sensors.

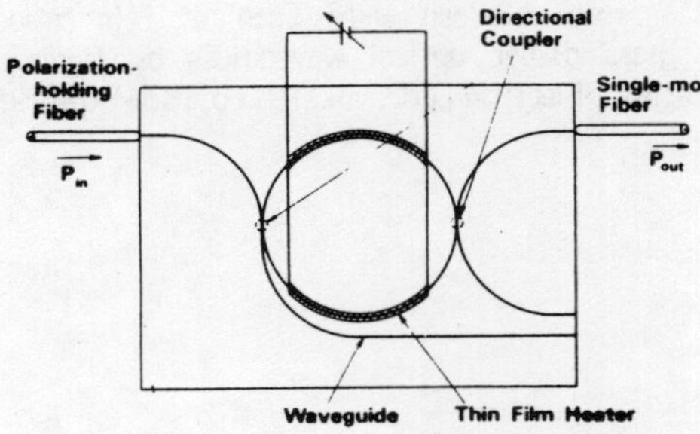

Fig. 7 Ring Resonator
Circuit size: 18 mm 28 mm x 1 mm

4. Conclusion

Guided-wave devices employing silica-based waveguides have excellent features, thanks to low propagation loss, compatibility with conventional single-mode optical fibers and highly reproducible waveguide parameters. The combination of basic waveguide circuit elements such as curved waveguides, directional couplers, and Y couplers makes it possible to fabricate the desired waveguide circuits for application.

References

1. N. Takato, K. Jinguji, M. Yasu, H. Toba, and M. Kawachi, "Silica-based single-mode waveguides on silicon and their application to guided-wave optical interferometers", J. of Lightwave Tech. vol. 6. pp.1003-1010(1988).

2. T. Miyashita, S. Sumida, and S. Sakaguchi, "Integrated optical devices bsed on silica waveguide technologies", SPIE Vol.933 Integrated Optcical Circuit Engineering VI pp.288-294(1988).

3. N. Takato, M. Yasu, and M. Kawachi, "Low-loss high-silica embedded channel waveguides", Electron. Lett., vol. 22, pp.321-322(1986).

4. Charles T. Sullivan, "Optical waveguide circuits for printed wire-board interconnections", SPIE, vol.994 Optelectronic Materials, Devices, Packaging and Interconnects II, pp.92-100.

5. M. Seki, R. Sugawara, H. Hashizume, and E. Okada, "Purely-thermal ion exchange technique for sinbgle-mode waveguide devices in glass", OFC '88 Postdeadline papaers, PD4-1.

6. M. Kawachi, M. Yasu, and T. Edahiro, "Fabrication of SiO2-TiO2 glass planar optical waveguides by flame hydrolysis deposition", Electron Lett., vol. 19, pp.583-584(1983).

SPLICING, JOINTING AND INSTALLATION OF THE OPTICAL FIBER CABLES

Hiroshi Murata
The Furukawa Electric Co., Ltd.
2-6-1, Marunouchi, Chiyoda-ku, Tokyo 100, Japan
Fax +81-3-286-3708 Phone +81-3-286-3416 Telex 2222365 FECTOK J

Abstract

Mass fusion splicing method with high splicing-efficiency and simple glass sleeve splicing method with high-economy were developed for splicing optical fibers on site. The single-mode fiber splicing loss of the former is about 0.05 dB on average and that of the latter is about 0.2 dB on average. Small sized plastic connectors were developed. And the convenient closure with light weight for cable joint was developed.

1. Fiber splicing 1)-10)

Fusion splice by electric arc discharge is normally used to splice SM fibers. Three types of splicing machine shown in Table 1.1 are currently used for splicing of SM fibers.

Table 1.1 Splicing machine 1)-8)

Splicing machine	Local injection Local detection	Core monitoring	No control*
Alignment	Automatic	Automatic	Setting on V-groove (no control)
Operation time	2 min.	2 min.	1 min.
Splicing loss (SM fibers)	0.05 dB	0.05 dB	0.2 dB*
Machine size	Middle	Large	Small

* Eccentricity < 0.5 μm

The splicing machine by core monitoring has been used widely in Japan. This machine has a feature which enables the splicing condition to observed directly through TV camera. However, its size is a little larger than other machines and also, attention should be paid for its maintenance.

A mult-fiber fusion splicing machine was developed for 2-8 SM fiber ribbon. To splice ribbons, a fiber setter is used to set fibers to the splicing machine by arranging fibers in parallel. Also, if fibers to be spliced are stuffed by about 30 μm in the axial direction at discharging process, it reduces splicing loss. Fig. 1.1 shows the experimental data. This process is carried out automatically by the splicing machine. The splicing loss, in this case, is about 0.05 dB (2-8 fiber ribbons).

Application of glass sleeves to splice SM pair-fiber ribbon has been put to a practical use. Fiber splicing using glass sleeve was studied and used in the early stage of development, and applied to pair-fiber ribbons again (Fig. 1.2). Fibers to be spliced are inserted and butted in the glass sleeve, and matching compound is injected from a small hole located in the center of the sleeve. UV curable resin is applied on the base and cover plates, and these plates, fibers and glass sleeves are fixed by irradiating ultraviolet ray. The splicing loss of the SM fiber by this method is about 0.2 dB (average). The properties of spliced fiber are shown in Table 1.2. Because this method does not need any special splicing machine, it is easy to be used on site for the cable with small number of fibers. This operating time required for the glass sleeve splicing is about 1/5 of that required for fusion splicing of single fiber, and the operating time is almost equal to that of mass fusion splicing.

2. Plastic connector 1)9)

Small sized multi fiber plastic connector was developed. Its size is 4x4x30mm for pair fiber ribbon. Its splicing loss stands at about 0.3 dB on average. This connector can easily be equipped on site. The characteristics of the connector are shown in Table 1.2. As the connector is small sized, it allows to ship the cables by setting the connectors at the ends of cable beforehand in the plant. The splicing time for cable with connectors is about 1/3 of that of mass fusion splice.

Table 1.2 Properties of spliced fibers by glass sleeve method (pair-fiber ribbon) and plastic connector (pair, 4-6 fiber ribbon) 1)9)10)

Item	Condition	Loss change	
		Glass sleeve	Connector
Thermal test	-25°C - +70°C/6h 4 cycle/day 20 cycles	< 0.1 dB	< 0.2 dB
Humidity	60°C x 95% RH 100 h	< 0.06 dB	< 0.2 dB
High temperature	80°C 100 h	--	< 0.2 dB
Low temperature	-30°C 100 h	--	< 0.2 dB
Vibration	Frequency: 10Hz - 55Hz Amplitude: 1.5 mm Direction: Longitudinal & radius Time: 2 h	< 0.04 dB	< 0.2 dB

3. Plastic closure 1)9)

A plastic closure was developed for cable joint. Its dimensions is 120mm dia. x 700mm long, and can be used with up to 200-fiber cable. The closure can accommodate fibers spliced by fusion or glass sleeve and has a construction for easy cable branching. It is light and thin and easy to assemble/disassemble.

4. Cable installation

Long cable installation technology is significant to construction of optical fiber cable network for reduction of installation and splicing costs. The automatic controlled caterpillar-puller is in actual use for midway pulling for the installation of the long cable in the duct. The process of installing the cable in the manner to form an 8-figure in the midway is called a back-feed method and is also used for the installation of the long fiber cable.

5. Conclusion

Mass fusion splicing method and glass sleeve splicing method which are easily applicable on site were developed.

Small sized plastic connector being able to be set on site and plastic closure being convenient to handle were developed.

All these are currently in practical use worldwide.

Reference

1) F. Nihei, Y. Yamamoto and N. Kojima: Optical subscriber cable technologies in Japan, J. Lightwave Technol. LT-5, 6, June, 1987, pp. 809-821.

2) M. Hirai and N. Uchida: Melt splice of multimode optical fibre with an electric arc, Electron. Lett. 13, 5, March 1977, pp. 123-125.

3) N. Kashima and F. Nihei: Optical fiber fusion splice using high frequency discharge with high voltage trigger, Trans. IECE J. E64, 8, August 1981, pp. 529-535.

4) Y. Kato, S. Seikai and M. Tateda: Arc-fusion splicing of single-mode fibers, 1: Optimum splice conditions, Appl. Opt. 21, 7, April 1982, pp. 1332-1336.

5) M. Tachikura and N. Kashima: Fusion mass-splices for optical fibers using high-frequency discharge, Jour. Lightwave Technology LT-2, 1, Feb. 1984, pp. 25-31.

6) O. Kawata, K. Hoshino, Y. Miyajima, M. Ohnishi and K. Ishihara: A splicing and inspection technique for single-mode fibers using direct core monitoring, Jour. Lightwave Technology LT-2, 2, April 1984, pp. 185-191.

7) M. Matsumoto, M. Tachikura, T. Katagiri, M. Sato, A. Ide and M. Miyauchi: Fully automatic, high speed fusion splice machine for single-mode optical fiber ribbon, 13th ECOC (Helsinki), Sept. 1987, Vol. 1, pp. 155-158.

8) M. Kawase, M. Tachikura, F. Nihei and H. Murata: Mass fusion splices for high density optical fibre units, 8th ECOC (Cannes) Sept. 1982, pp. 348-353.

9) Y. Katsuyama, M. Miyauchi and N. Kashima: Single-mode fiber cable and jointing technologies for broadband networks, GLOBECOM '87, 33-8 (Nov. 1987) pp. 1310-1314.

10) M. Miyazaki, M. Mizutani and S. Tachigami: Development of mechanical splice of pair-fiber ribbon, Study G. Telecom, IEE J, CMN-88-27 (July 1988) pp. 61-65.

Fig. 1.1 Mass-Fusion Splicing (5-Fiber SM)[9]

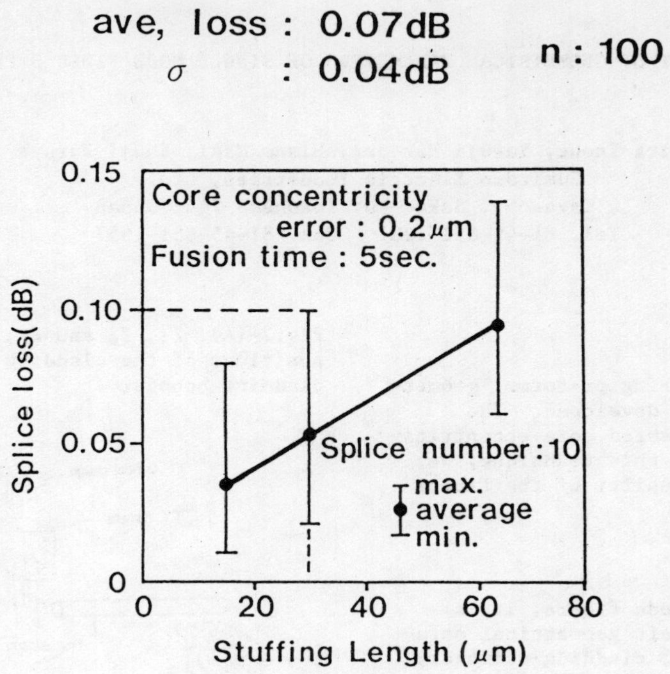

Fig. 1.2 Glass-Sleeve Splicing of Pair-Fiber Ribbon[10]

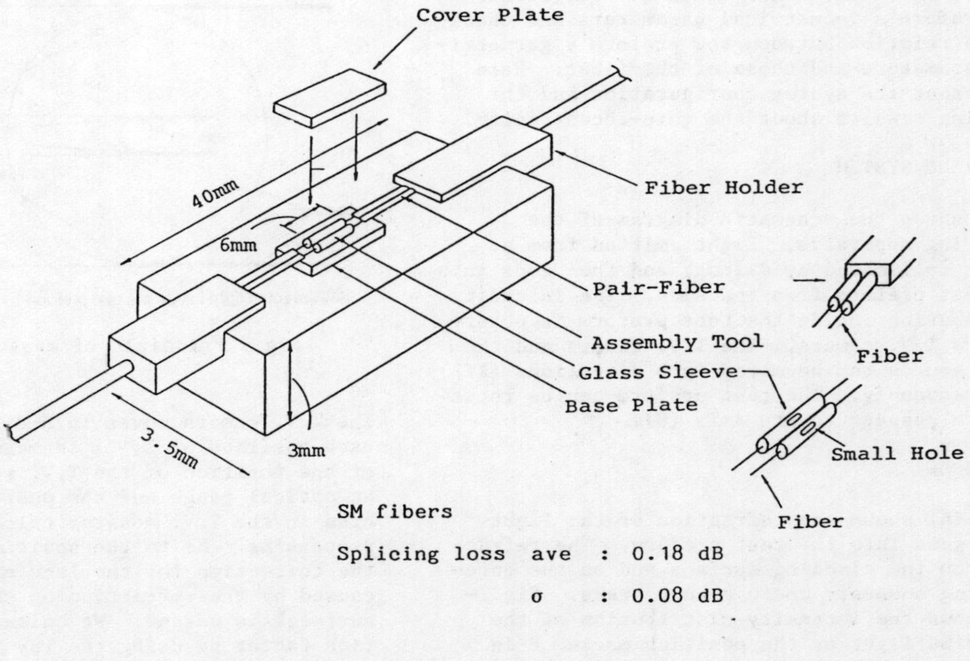

MEASUREMENT OF GEOMETRICAL PARAMETERS OF SINGLE MODE FIBER'S PREFORM

Akira Inoue, Yasuji Hattori, Hisao Maki, Shuji Furuya
Sumitomo Electric Industries, Ltd.
1, Taya-cho, Sakae-ku, Yokohama 244, Japan
Tel. 81-45-851-1281 Fax. 81-45-851-1557

ABSTRACT

A new technique for measuring preforms' geometrical parameters has been developed. The standard deviation of measured core-eccentricity is only 0.005%. Using this technique, we can accurately know the quality of the fibers before drawing process.

INTRODUCTION

In manufacturing single mode fibers, it is necessary to stabilize their geometrical parameters (core-eccentricity, cladding-diameter, cladding-non-circularity). To control these parameters, it is important to estimate these parameters at the preform stage. If a high relationship is guaranteed between the preform's geometrical parameters and the fiber's geometrical parameters, we can accurately know the quality of the fibers before drawing process.
We have developed a new technique for measuring the preform's geometrical parameters and checked the relation between the preform's geometrical parameters and those of the fiber. Here we present the system configuration and the checking results about the core-eccentricity.

MEASURING SYSTEM

Fig.1 shows the schematic diagram of the measuring apparatus. Light emitted from an LED is collimated by a lens, and then goes into the test preform from the side. The intensity distribution inside the test preform is observed by a T.V. camera. The T.V. camera and the light source can be moved in 3 directions (XYZ) simultaneously. The test preform can be rotated with respect to its axis (θ).

PRINCIPLE

Fig.2-(A) shows the refraction of the light which goes into the test preform. The refractions on the cladding surface and on the core-cladding boundary cause shadow areas. Fig.2-(B) shows the intensity distribution of the refracted light at the position marked P in Fig.2-(A). Z_1, Z_4 and Z_2, Z_3 correspond to the positions of the cladding surface and the core-cladding boundary.

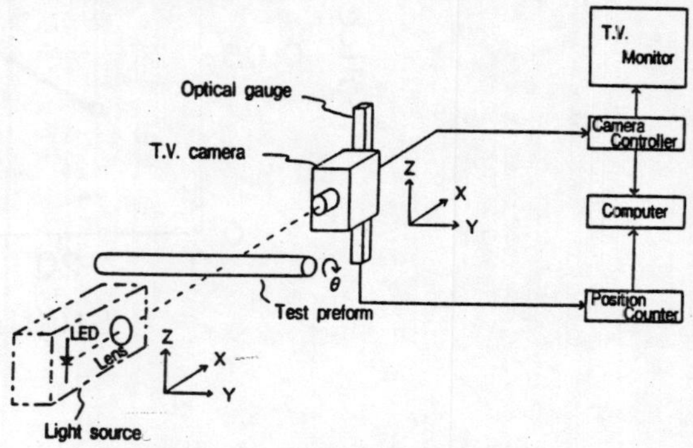

Fig.1 Schematic diagram of the preform measuring system

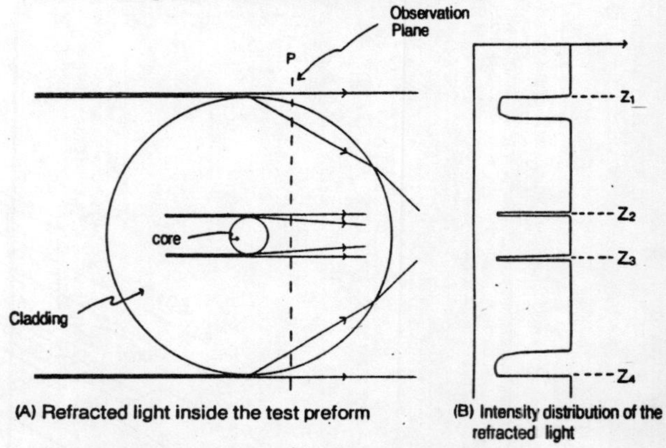

Fig.2 Principle of measurement

The T.V. camera moves in Z-direction and the each position of $Z_1 \sim Z_4$ is measured as the sum of the position of the T.V. camera measured by an optical gauge and the position of the shadow area in the T.V. monitor calculated by image processing. As to the position of the core, the correction for the lens effect, which is caused by the refraction on the cladding surface, is needed. We calculated the correction factor by using the ray tracing method.

The test preform is rotated and the core-eccentricity is obtained by fitting the sinusoidal function to the measured core-eccentricities at each rotation angle. The amplitude of the sinusoidal function corresponds to the amount of the core-eccentricity, and the phase of it corresponds to the direction of the core-eccentricity.

RESULTS

Diameters of Preforms are 200~600 times as large as those of fibers. Therefore we can measure the core-eccentricity far more accurately at the preform stage than at the fiber stage. The standard deviation of the measured core-eccentricity is 0.005%, which is as small as 1/20 of the standard deviation in the fiber measurement. Fig.3 shows the cross-checking result between the preform's core-eccentricities and the fiber's core-eccentricities. A large preform (ϕ68mm) which had been manufactured by VAD method was measured. The preform's core-eccentricities coincide with the fiber's core-eccentricities.

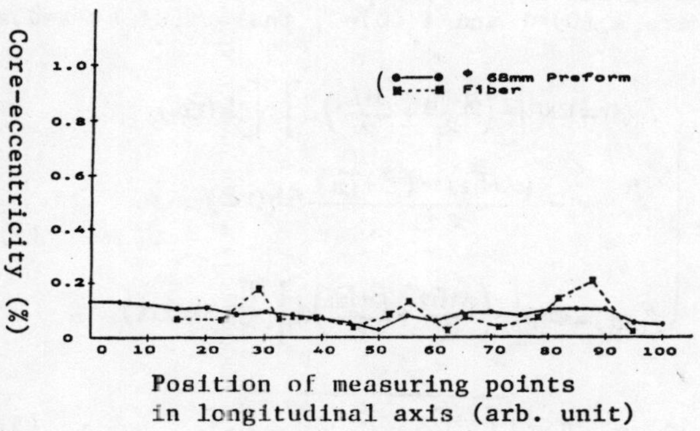

Fig.3 Cross checking result between the preform's core-eccentricity and the fiber's core-eccentricity

CONCLUSION

We have developed a new technique for measuring preforms' geometrical parameters. Using this technique, we can accurately know the quality of the fiber before drawing process. Therefore this technique is a most effective one in manufacturing fibers.

REFERENCE

M. Horiguchi, H. Itoh, "Nondestructive Measurement of Core-Eccentricity for Single-Mode Fiber Preforms", IEICE (Mar. 1988)

PHENOMENOLOGICAL THEORY FOR PLASTIC OPTICAL FIBER COUPLER

Xu Jisen
Shanghai Transmission Lines Research Institute
25 Yixian Road, Shanghai, China

ABSTRACT

A simplified model for plastic optical fiber coupler and its phenomenological theory are presented that it is possible to accurately predict the coupling power, coupling ratio and excess loss of the fused plastic fiber coupler with the parameters of manufacturing process. The experiment results are compared with the theory and good agreement is found.

INTRODUCTION

A new coupler structure and a special method of fabricating the plastic optical fiber couplers both for symmetric and asymmetric couplers have been developed recently[1-3]. In depth studies on the coupling phenomena in fused multimode couplers have been widely reported[4-7]. But a different theoretical approach is reported here for this kind of large number of modes plastic optical fiber coupler with the coupled-power equations. The present approach is different from previously reported efforts by using physical parameters rather than complex modal analysis. This phenomenological theory is compared with the experiment results and this measured data is fitted with the theory by parameter optimization, good agreement is found.

THEORY

A 2x2 coupler, as a simplest case, is shown in Fig.1. Two plastic optical fibers, with the plastic cladding of the segment removed, will be pressed and fused together during the heating process. The deformation appears for both of the fibers at the same time and optical power coupled from one fiber to the other fiber. It is no difficult to derive the well-know coupled-power equations, that is

$$\frac{d}{dx}\begin{bmatrix}A_1(x)\\A_2(x)\end{bmatrix}=\begin{bmatrix}-\alpha(x)-\Gamma_{12}(x), & \Gamma_{21}(x)\\ \Gamma_{12}(x) & ,-\beta(x)-\Gamma_{21}(x)\end{bmatrix}\cdot\begin{bmatrix}A_1(x)\\A_2(x)\end{bmatrix} \quad (1)$$

where $A_1(x)$ and $A_2(x)$ are optical power in fiber 1 and fiber 2, $\alpha(x)$ and $\beta(x)$ are some kind of loss coefficient about deformation of fibers during the process (the fibers are always supposed lossless here), $\Gamma_{12}(x)$ and $\Gamma_{21}(x)$ are some kind of coupling coefficients from fiber 1 to fiber 2 and from fiber 2 to fiber 1 respectively. This equations, in general, can not be solved. For approximation, suppose that $\alpha(x)$, $\beta(x)$, $\Gamma_{12}(x)$ and $\Gamma_{21}(x)$ are constant (we will find late that the results from this approximation for our manufacturing process agree with experiment results). If the coupling region, for above coupler, is from x=0 to x=d, the input power in fiber 1 and fiber 2 are $A_1(0)=1$ and $A_2(0)=0$, the output at x=d, are

$$\begin{cases} A_1(d)=\exp\left[-\left(\frac{\alpha+\Gamma_{12}}{2}+\frac{\beta+\Gamma_{21}}{2}\right)d\right]\cdot\left[ch(\tau d)-\right.\\ \qquad \left.-\frac{(\alpha+\Gamma_{12})-(\beta+\Gamma_{21})}{2\tau}sh(\tau d)\right] \\ A_2(d)=\exp\left[-\left(\frac{\alpha+\Gamma_{12}}{2}+\frac{\beta+\Gamma_{21}}{2}\right)d\right]\cdot\frac{\Gamma_{12}}{\tau}sh(\tau d)\end{cases} \quad (2)$$

where $$\tau=\sqrt{\left(\frac{\alpha+\Gamma_{12}}{2}-\frac{\beta+\Gamma_{21}}{2}\right)^2+\Gamma_{12}\Gamma_{21}} \quad (3)$$

if the input in fibers are $A_2(0)=1$ and $A_1(0)=0$ the output at x=d are

$$\begin{cases} A_1(d)=\exp\left[-\left(\frac{\alpha+\Gamma_{12}}{2}+\frac{\beta+\Gamma_{21}}{2}\right)d\right]\cdot\frac{\Gamma_{21}}{\tau}sh(\tau d) \\ A_2(d)=\exp\left[-\left(\frac{\alpha+\Gamma_{12}}{2}+\frac{\beta+\Gamma_{21}}{2}\right)d\right]\cdot\left[ch(\tau d)+\right.\\ \qquad \left.+\frac{(\alpha+\Gamma_{12})-(\beta+\Gamma_{21})}{2\tau}\cdot sh(\tau d)\right]\end{cases} \quad (4)$$

For couplers, coupling ratio CR and excess loss EL are the most important parameters. If the subscript 1 and 2 are for fiber 1 and 2 input, ther

$$CR_1=\frac{\tau ch(\tau d)-[(\alpha+\Gamma_{12})-(\beta+\Gamma_{21})]\,sh(\tau d)/2}{\Gamma_{12}sh(\tau d)} \quad (5\text{-}1)$$

$$CR_2=\frac{\Gamma_{21}\, sh(\tau d)}{\tau ch(\tau d)+[(\alpha-\Gamma_{12})-(\beta+\Gamma_{21})]\, sh(\tau d)/2} \qquad (5-2)$$

$$EL_1=1-[A_1(d)+A_2(d)]_1$$
$$=1-\exp\left[-\left(\frac{\alpha+\Gamma_{12}}{2}+\frac{\beta+\Gamma_{21}}{2}\right)d\right]\cdot\left[ch(\tau d)+\frac{(\Gamma_{12}+\Gamma_{21})-(\alpha-\beta)}{2\tau}\cdot sh(\tau d)\right] \qquad (6-1)$$

$$EL_2=1-\exp\left[-\left(\frac{\alpha+\Gamma_{12}}{2}+\frac{\beta+\Gamma_{21}}{2}\right)d\right]\left[ch(\tau d)+\frac{(\Gamma_{12}+\Gamma_{21})-(\alpha-\beta)}{2\tau}sh(\tau d)\right] \qquad (6-2)$$

It is also very useful to understand that how the excess loss changes with the coupling ratio CR_1 and CR_2 for different fiber input during manufacture process, then

$$\begin{cases} EL_1=1-\dfrac{(1+t_1)\Gamma_{12}}{[(t_1\Gamma_{12}+A-B)^2-\tau^2]}\left[\dfrac{(t_1\Gamma_{12}+A-B)-\tau}{(t_1\Gamma_{12}+A-B)+\tau}\right]^{\frac{(A+B)}{2\tau}} \\ EL_2=1-\dfrac{(1+t_2)\Gamma_{21}/t_2}{[(\Gamma_{21}/t_2-A+B)^2-\tau^2]}\left[\dfrac{(\Gamma_{21}/t_2-A+B)-\tau}{(\Gamma_{21}/t_2-A+B)+\tau}\right]^{\frac{(A+B)}{2\tau}} \end{cases} \qquad (7)$$

where $t_1=CR_1$, $t_2=CR_2$, $A=\frac{(\alpha+\Gamma_{12})}{2}$ and $B=\frac{(\beta+\Gamma_{21})}{2}$

As to now, the important results for 2x2 coupler are shown above. From equation (2) and (4), one can understand that how the power transfer from one fiber to the other during the manufacturing process, from (5) and (6), the coupling ratio and excess loss chang during the process. The equations (7) is a special relation about the excess loss with couling ratio during the process.

DISCUSSION

1. Generalization

Thereare several ways to manufacture the coupler with coupling region from x=0 to x=d and a constant coupling coefficient (as supposed above). For example, the heater moves from x=0 to x=d during the process, the coupling coefficient is kept as a constant, or the coupling region is fixed from x=0 to x=d but the coupling coefficient changes from 0 to the desired value. It can be generalized, supposed $\tau d=Q$ is a parameter about coupling capacity, the formula (2) is then

$$\begin{cases} A_1(Q)=\exp\left[-\left(\dfrac{\alpha+\Gamma_{12}}{2\tau}+\dfrac{\beta+\Gamma_{21}}{2\tau}\right)Q\right]\cdot\left[ch(Q)-\dfrac{(\alpha+\Gamma_{12})-(\beta+\Gamma_{21})}{2\tau}sh(Q)\right] \\ A_2(Q)=\exp\left[-\left(\dfrac{\alpha+\Gamma_{12}}{2\tau}+\dfrac{\beta+\Gamma_{21}}{2\tau}\right)Q\right]\cdot\dfrac{\Gamma_{12}}{\tau}sh(Q) \end{cases} \qquad (8)$$

where Q is coupling capacity and formula (4), (5) and (6) can do the same substitute. Q can be a function of time or length of coupling region or somthing else. The results by this substitute is compared with experiment result and good agreement is found.

2. Coupling Coefficient

The coupling coefficient, in some case, can be calculated by using a geometric model or a modal analogy[8]. It is no difficienty to get a general result for star coupler,

$$\Gamma_{ij}=C*(N.A._j)^2*S_j \qquad (9)$$

where C is a constant, S_j is the cross area of jth fiber, the coupling coefficient Γ_{ij} are now related with the fiber physical parameter directly.

3. Coupling Ratio

By equation (5), the coupler with any desired coupling ratio can be made without any difficulty by stopping the process somewhere. It shows that the coupling ratio, from one fiber input, is mono-decreased tending to its limit and the coupling ratio, from the another fiber input, is mono-increased tending to the same limit. For the ideal lossless coupler, $\alpha=\beta=0$, then it can relate them with fiber parameters only now.

$$CR_{id}=\lim_{Q\to\infty} CR_1=\lim_{Q\to\infty} CR_2=\Gamma_{21}/\Gamma_{12} \qquad (10)$$

this is the theoretical limit of the coupling ratio for the lossless coupler.

4. Parameter Optimization and Process Controlling

From equation (7), it shows that the excess loss of the coupler will increase as the coupling goes strong and increase rapidly as the coupling ratio tend to its limit. The very important thing is that it is impossible theoretically to manufacture a coupler with coupling ratio equal to its theoretical limit. A coupler can only be made with low excess loss but with coupling ratio close enough to its theoretical limit. Now after to fix the experimental data with the above theoretical results by parameter optimization, the manufacture process can be easy controlled.

5. Experimental results

Fig.2 shows the excess loss changes with coupling ratio and power changes with coupling capacity for 2x2 coupler with two same kind of fibers. Fig.3 shows the same relation but for asymmetric coupler with two different kind of fibers. Similar results have been got for star couplers. More details will be published in other papers.

CONCLUSION

A simplified model and its phenomenological theory are presented. It is possible to accurately predict the coupling power, coupling ratio and excess loss of fused plastic star coupler with the fusion process, and the dependence of coupling ratio with excess loss. It shows good agreement between theory and experiment. All these couplers manufactured with our method exhibit low excess loss and desired coupling ratio.

Part of the experiment was done when the author was with Aster Corporation, Milford, Massachusetts, U.S.A.

REFERENCES

1. J. Xu et al, "Direct Bonding Yields Low Cost Plastic Star Couplers" Laser Focus, April 1988.
2. J. Xu et al, "Low Cost High Quality Plastic Fiber Couplers" SPIE' Fiber Optics, Optoelectronics and Laser Applications. Boston, Sept. 1988.
3. J. Xu et al, "Plastic Fiber Couplers for Local Area Networks". ibid.
4. K. Ogawa, Bell Sys. Tech. J. Vol. 56, 729 (1977).
5. K. Ogawa, A.R. McCormick, Appl. Opt. Vol. 17, 2077 (1978).
6. F. Szarka et al, Fiber & Integrated Opt. Vol. 3, 285 (1980).
7. Y. Li, J.W.Y. Lit, J. Opt. Soc. Am. Vol. 2, 1301 (1985).
8. J.D. Muhs, "Coupling Limitations of Asymmetric Multimode Couplers". SPIE Vol. 839, Components for Fiber Optic Applications (1987).

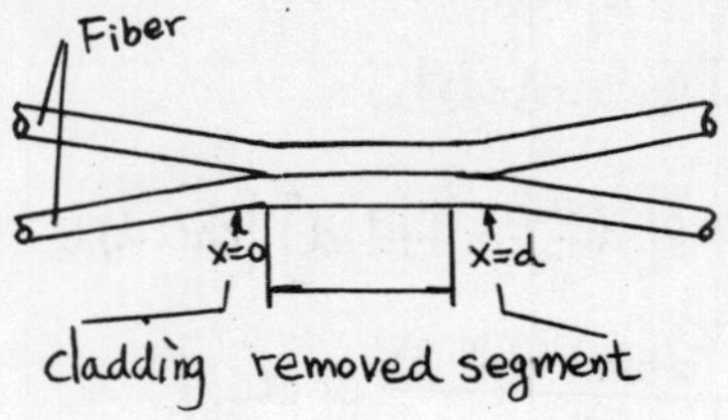

Fig.1 Structure of 2x2 coupler

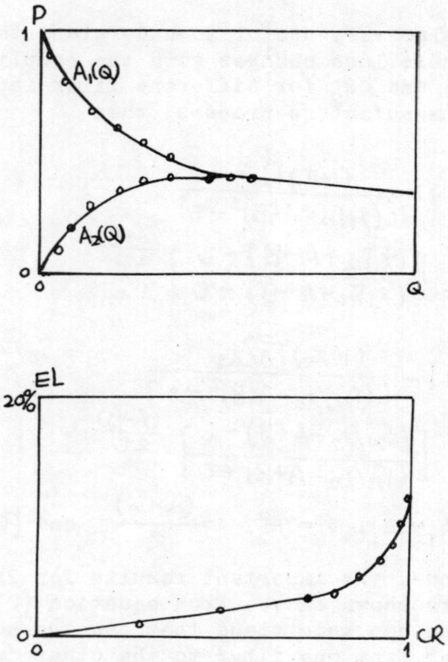

Fig.2 Experiment compare with theory for 2x2 symmetric coupler

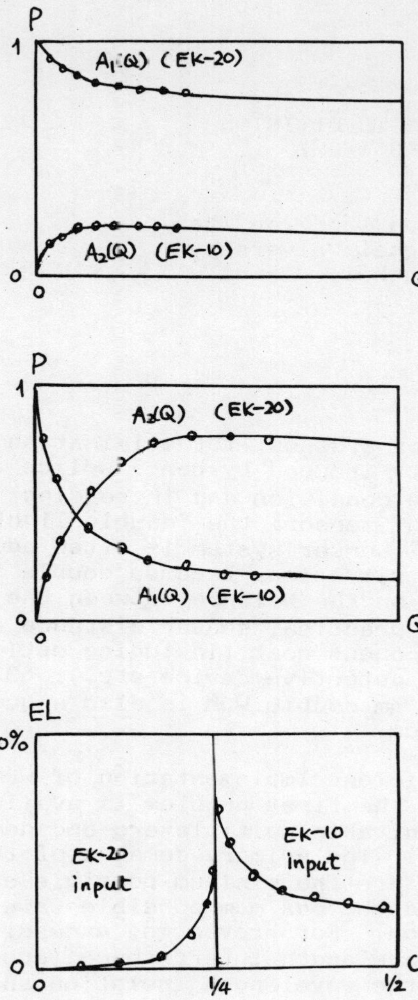

Fig.3 Experiment compare with theory for 2x2 asymmetric coupler

SINGLE MODE FUSED TAPER WAVELENGTH DIVISION MULTIPLEXERS/DEMULTIPLEXERS (WDM)

Wang Daogang, Qin Guanhua, Chen Hua and Yan Zen
Wave Science Laboratory, Shanghai University of Science and Technology, Shanghai, China

ABSTRACT

Single mode wavelength division multiplexers/demultiplexers (WDM) is used both for single mode optical fiber communication systems and for advanced single mode optical fiber sensor systems. This paper describes the principle and technology of manufacturing single mode fused taper wavelength division multiplexers/demultiplexers.

INTRODUCTION

In optical fiber communication, besides time division multiplexing (TDM) and frequency division multiplexing (FDM), we can also increase the channel capacity by wavelength division multiplexing (WDM). Because information capacity of single mode fiber is more than multimode fiber and 1300 nm and 1550 nm are two major windows of low loss, 1300 nm and 1550 nm single mode double wavelength division multiplexing is a very good selection. However, the 1300/1550 nm double channels WDM is only appropriate for the single mode optical fiber communication system at 140 Mbit/s bit rate. The most important reason is a too high chromatic dispersion (15-20 ps/km/nm) of conventional single mode fiber at 1550 nm. Although, the special fibers that chromatic dispersion from 1300 to 1550 nm is flat and comes near zero have been researched by various methods, the technology is too complex and the cost is too expensive. Therefore, the interests have been concentrated on the multichannels WDM of narrow wavelength gap about 1300 nm. For example, four channels 560 Mbit/s laboratory WDM system at 1200, 1240, 1280 and 1320 nm has been realized. 1/

In sensor systems, for eliminating unstability induced by bend, splice and ambience condition and increasing precision of sensor, the "double light sources" sensor system is often adopted. In this situation, because double WDM depends on the balance between the demand of practical sensor distance and the component cost (including optical source, detective device etc.), 633 nm and 850 nm double WDM is also a good selection.

For efficient implementation of WDM, of course, the first problem is availability of suitable multiplexers and demultiplexers. The primary demands of these devices are the minimum possible excess loss and the maximum possible wavelength isolation. For preventing excessive inter-wavelength interference (cross talk), the wavelength isolation should be the most important parameter. This paper is just to discuss the manufacturing principle and technology of fused taper single mode double wavelength division multiplexers/demultiplexers.

MANUFACTURING PRINCIPLE

In seventies, two basic phenomena were used for manufacturing wavelength division multiplexers/demultiplexers:

1. The wavelength dependence of the angle of reflection or transmission in angularly dispersive devices, such as prisms or diffraction grating. 2/

2. The wavelength dependence of transmission and reflection coefficients of multilayer interference filters. 3/

Recently, for manufacturing single mode WDM, fused taper technology has been adopted. 4/ It has a series of advantage, for example, low excess loss, high wavelength isolation, low cost and high reliability.

Fused taper single mode WDM practically is a special single mode coupler. When two fibers that are twisted together are burned and drawed, the relatively optical powers of injecting arm and coupling arm show a period variation as showed by figure 1.

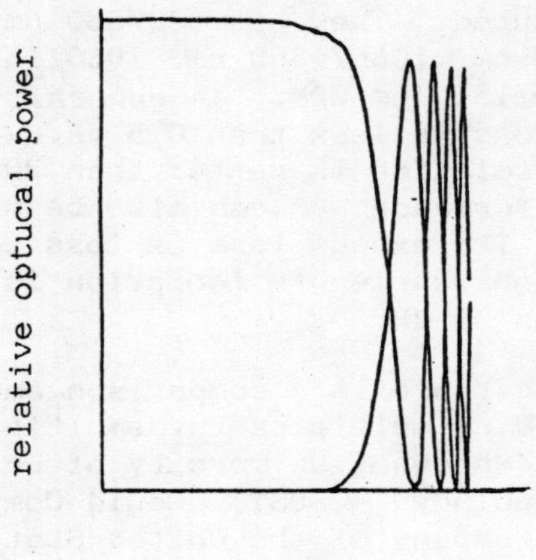

Figure 1.

When the drawing stops in the extreme value point of splitting ratio, optical signal of injecting wavelength always concentrates to one output arm. This device always has a sinusoidal spectrum as showed by figure 2.

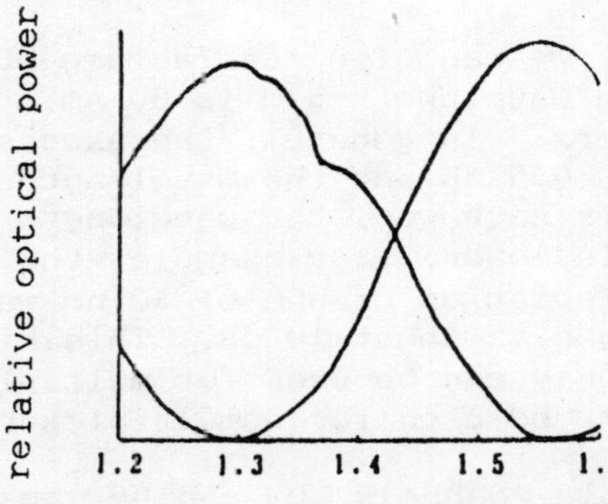

Figure 2.

There is another wavelength that is shorter or longer than injecting wavelength and splitting ratio also has an extreme value on this wavelength. Therefore, this device can be considered as a double wavelength (injecting wavelength and above another wavelength) division multiplexer. The another wavelength depends on the number of the extreme value point of splitting ratio which passed in drawing process. The "number" is larger, the "another wavelength" is nearer the injecting wavelength. In other words, the "number" is larger, the wavelength gap of WDM is narrower.

TECHNOLOGY

Fused taper single mode WDM has similar technology as general fused taper single mode coupler. Low excess loss is very important. For this, various factors that should be regard are the same as that of general fused taper single mode coupler. The wavelength isolation is the most important parameter for WDM. For increasing wavelength isolation, the stability of the flame is very important. If the flame is unstable, wavelength isolation will decrease for wave guide is heterogeneous.

Our research has showed that three factors should be considered at least for manufacturing an excellent wavelength division multiplexers/demultiplexers.

1. Drawing method

Horizontal and symmetrical drawing method is better than other drawing methods for the symmetry of character of WDM.

2. Heat softening and drawing rate

Two fibers that are partly removed organic coating and twisted together are burned by an oxy-butane flame. The "fused" means that two fibers receive an appropriate heat softening. The "heat softening" depends on both temperature field and burned time. Drawing rate and heat softening are relative factors each other. "Synchronism" between drawing rate and heat softening is very important for obtaining low excess loss WDM.

3. Packing technology

The bare devices must be protected by the packing technology. The principle is that the bare devices don't suffer the excess stress from difference of expand coefficent among packing materials. The excellent elasticity of coating, adhesive and sealing materials is very useful for relaxing the excess stress.

After packing procedure, it is inevitable that the absolute value of wavelength of WDM will shift and the gap of wavelength of WDM will expand. Fortunately, above change is regular. Therefore, this change can be corrected by some deviation in advance.

Figure 3 shows the spectrum of a bare WDM. The wavelength gap of this WDM is about 65 nm and wavelength isolation is about 15 dB.

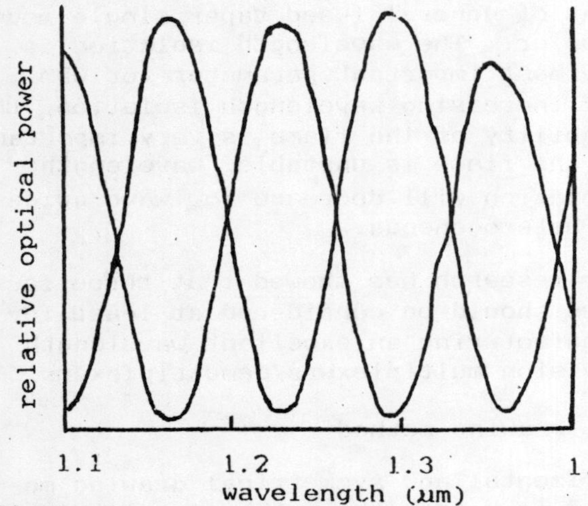

Figure 3.

Figure 4 shows the spectrum of above WDM after packing procedure. Its wavelength gap and isolation have became 80 nm and better than 20 dB respectively. This WDM can be regard as a 1240/1320 nm double channels WDM.

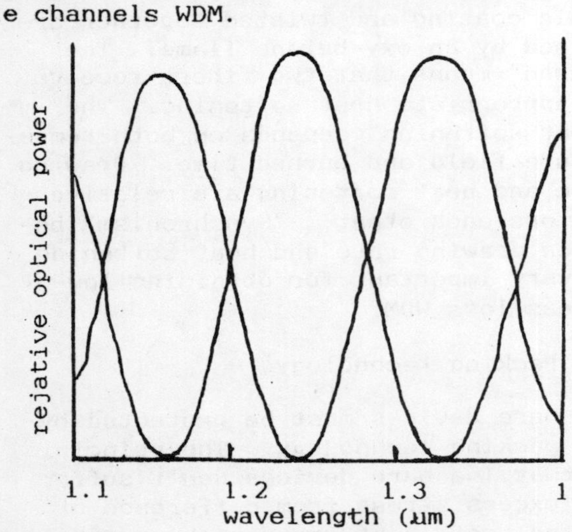

Figure 4.

We are surprised at finding that the wavelength isolation of WDM is obviously increased by the packing procedure. This phenomenon is specially obvious when WDM of narrow wavelength gap is packed. Of course, this is a useful phenomenon.

RESULTS AND DISCUSSIONS

In our laboratory, various fused taper single mode double wavelength division multiplexers/demultiplexers have been manufactured. They are 633/850 nm, 850/1300 nm, 1060/1300 nm, 1060/1550 nm and 1300/1550 nm WDM. In general, the excess loss is less than 0.5 dB, wavelength isolation is better than 20 dB. High performance WDM can also be manufactured. Its excess loss is less than 0.1 dB and wavelength isolation is better than 30 dB.

Following table is a comparison among 1300/1550 nm multiplexers/demultiplexers of ours (Shanghai University of Science and Technology -- SUST), Gould Company and ADC Company of the United States of America.

	Excess Loss	Wavelength Isolation	Package Size
SUST	0.5 dB	20 dB	D3X30 mm
Gould	0.5 dB	16 dB	D4X40 mm
ADC	1 dB	17 dB	D8X75 mm

Rcently, we can also manufacture WDM that the wavelength gap is 80 nm, 40 nm and others. In general, the excess loss is about 0.5 dB and the wavelength gap of WDM is narrower, the wavelength isolation is lower. For example, the wavelength isolation of WDM of 40 nm wavelength gap is about 10 dB. This kind of device only can be used for multiplexers, but can't be used for demultiplexer.

The double channels WDM can be considered as a base. Multichannels WDM can be constructed by cascade of double channels WDM. For example, 1060/1300/1550 nm three channels WDM can be constructed by cascade of 1060/1550 nm WDM and 1060/1300 nm (or 1300/1550 nm) WDM. 1200/1240/1280/1320 nm four channels WDM can be constructed by cascade of 1200/1280 nm, 1240/1320 nm and 1200/1240 nm (or 1280/1320 nm) WDM.

Figure 5 shows the constructing of 1060/1300/1550 nm WDM. Because the wavelength isolation of 1060/1550 nm WDM and 1060/1300 nm (or 1300/1550 nm) WDM are both better than 20 dB, 1060/1300/1550 nm threechannels WDM can be used both for multiplexer and for demultiplexer.

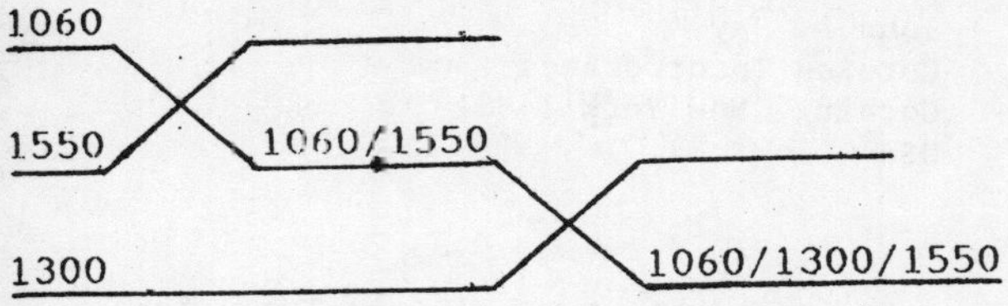

Fig. 5 1060/1300/1550 multiplexer/demultiplexer

Figure 6 shows the constructing of 1200/1240/1280/1320 nm WDM. Up to the present day, 1200/1240/1280/1320 nm fused taper single mode WDM only can be used for multiplexer, but can't be used for demultiplexer. The reason is that the wavelength isolation of 1200/1240 nm (or 1280/1320 nm) double channels WDM only is about 10 dB.

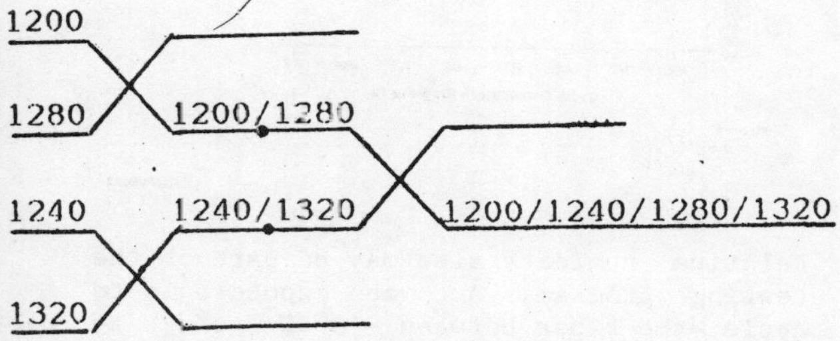

Fig. 6 1200/1240/1280/1320 multiplexer

For increasing the wavelength isolation of WDM, Japanese has adopted the method of cascade between fused taper single mode WDM and the film filter. 5/

In comparison with above hybrid WDM, all-fiber WDM with high wavelength isolation should have many advantages. Therefore, we suggest following two possible ways:

1. A fused taper single mode WDM can be regarded as a filter. Therefore, the cascade of fused taper single mode WDMs can be regarded as a way of increasing the wavelength isolation.

2. Biconical taper coaxial coupler filter 6/ that manufactured by depressed single mode fiber is a very good all-fiber filter. The cascade between all-fiber WDM and all-fiber filter will construct all-fiber WDM with high wavelength isolation.

REFRENCES

1. K. Fussgaenger, etal., ECOC (1986);
2. H. Ishio, Proc. 5th. ECOC (1979);
3. H. Hashimoto, etal., Proc. 5th. ECOC (1979);
4. G. Georgiou, etal., Proc. Sino-British Joint Meeting on OFC (1986);
5. H. Yanagawa, Electro. Lett. 24(1988);
6. A. C. Boucouvalas, etal., Electro. Lett. 21(1985).

ENVIRONMENTAL AND MECHANICAL TESTING ON OPTICAL FIBER TO ENSURE IMMEDIATE AND LONG TERM SYSTEM PERFORMANCE

John F. Spencer
Optical Waveguides Australia
Noble Park, Victoria 3174
Australia

John A. Jay
Corning Incorporated
Corning, New York 14831
USA

INTRODUCTION

Mechanical and environmental testing of optical fiber by the manufacturer is an accepted approach to addressing technical concerns regarding both immediate performance and long-term operating expectations in the field. Exposure testing has to cover both physical and optical changes that could influence system operation throughout its lifetime. The tests are designed either to study immediate usage conditions or accelerate effects on both the glass and coating which can degrade the fiber performance over time.

COATING EFFECTS

Physical property mismatch between the UV cured acrylate typically used as optical fiber coatings today and the glass fiber itself initially posed a technical challenge. With proper design and application the coating can be functionally compatible with the glass. A well designed environmental testing program further ensures their compatibility in the field.

The tests described below are typical for this assessment. For each exposure, the induced attenuation, intrinsic strength, microbending resistance, and coating strip force may be measured depending on the effects under study.

Temperature cycling is the test most frequently performed. Conditions depend on the application, but -60° to $+85^{o}$ C is common for terrestrial applications, though only -20° may be required in moderate climates. Submarine systems typically specify more moderate temperature ranges, but require attenuation changes almost an order of magnitude less than terrestrial systems. The number of temperature cycles can vary depending on the application. Typical results for single-mode fibers with 250 micron acrylate coating made by the Outside Vapor Deposition process are presented in figure 1.

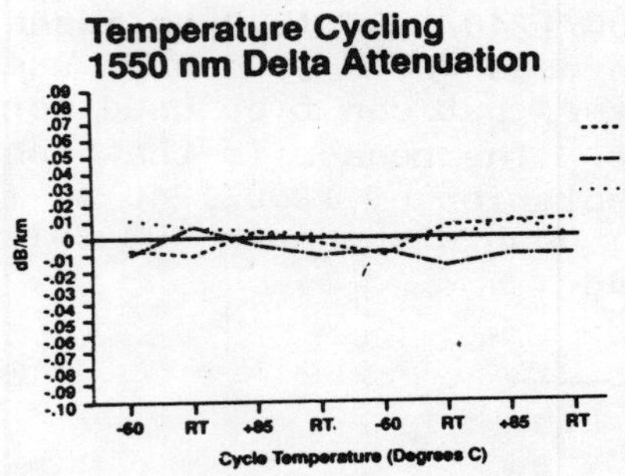

Figure 1

CORNING

Relative humidity also may be part of the testing program. A common exposure is to cycle the fiber between -10° C @ 4% RH to $+65^{o}$ C @ 98% RH. The cycle usually takes several hours, and the fiber is sampled on a logarithmic basis (day 0, 1, 3, 7, 14, 30, etc.).

Various aqueous soaks often are part of the testing program. While de-ionized water is used most frequently, other media include sea water, basic, and acidic solutions.

Thermal aging of the coating is conducted at the maximum coating rated temperature, often near $+85^{o}$ C. Attenuation measurements are made to characterize any microbending due to dehydration of the acrylate and evaporation of non-matrix coating constituents.

GLASS EFFECTS

Evidence of the general material stability of glass is on display at any museum;

however, some sub-molecular effects have been reported that can impair the transmission properties of optical fibers over time. Appropriate environmental testing can measure susceptibility to these effects.

Several optical effects have been associated with vibrational absorptions of different hydrogen species. Unreacted interstitial H_2 which has diffused in to the glass has an overtone at 1.24 micron. Hydroxyls can be formed with Si, Ge, and P which absorb at 1.39, 1.41, and 1.6 micron respectively[1]. A transient hydride with Si absorbing at 1.52 micron has also been observed[2].

Agreement on appropriate kinetic models has been difficult to achieve, so a variety of test conditions have been proposed to predict long term results. These generally assume operating conditions of hydrogen partial pressures under 1000 ppm and ambient temperatures. Lifetimes are accelerated by increasing H_2 partial pressure (usually 1-2 atm) and temperature (anywhere from 70° - 150° C) for up to 1000 h.

Exposure to gamma radiation causes a sharp increase in attenuation which decays afterwards. The rate of decay and eventual attenuation relative to original can vary depending upon composition and materials. Proposed applications include nuclear fallout secure communications, tactical military communications, use in an accelerometer, and submarine optical cables. Thus no one set of test conditions is widely accepted as a proper predictor for long term service.

MECHANICAL AND RELIABILITY TESTING

Parameters defining the mechanical system life are minimum fiber strength (defined by proof test) and the fatigue characteristics of the glass to estimate what sub-critical loads can be applied without detrimentally weakening the fiber.

A proof test is imposed on the fiber to establish a minimum strength. Though the intrinsic fiber strength may be upward of 600 kpsi, the proof test typically ranges from 50 kpsi (terrestrial applications) to 200 kpsi (submarine) for telecommunications applications. This is due to the statistical nature the distribution of strength limiting flaws along the fiber surface. The proof test typically is applied by a loaded pulley during a winding process. This is done either at the draw or on a dedicated machine subsequent in the process.

The strength distribution is characterized by dynamic fracture testing. Fibers are fractured by increasing tensile load, using an Instron or rotating capstans, and the distribution of fracture strength is plotted. The distribution will have width, defined by the Weibull statistic M, consisting of one or more modes, each describing a particular regime of flaw sizes.

The results are sensitive to test conditions. Longer gauge lengths will effect the distribution among the failure modes, as the probability of observing a low stress flaw increases with length. Decreasing the strain rate moves the distribution to a lower strength as the failure mechanism includes a larger component of sub-critical crack growth. Similar effects can be seen by increasing temperature or relative humidity.

Once the proof test establishes a minimum fiber strength, an understanding of the fatigue characteristics is necessary to properly design usage conditions without undesired sub-critical flaw growth.

For static loads that growth is described from a ratio of the applied stress to the proof stress by the following equation:

$$(\sigma_1/\sigma_2)^n = t_2/t^1 \qquad (1)$$

Where σ_i and t_i are static stresses and times to failure and n is the fatigue resistance parameter.

One method to determine n is to apply sub-critical loads to the fiber and characterize the flaw velocity by the time to failure. This is known as the static fatigue test. However, such testing is not trivial. The amounts of sample, facility and time required usually are prohibitive.

Dynamic fatigue testing by fracturing at different stress rates substantially reduces the resources necessary for fatigue characterization. It is accepted that the flaw growth has a similar dependence on the stress rate as it does on the load.

$$(\sigma_1/\sigma_2) = (\dot{\sigma}_1/\dot{\sigma}_2)^{1/(n+1)} \qquad (2)$$

Thus fracture distribution differences for different stress rates can be used to calculate n. A slope of the log of the stress rate and the log of the failure stress equals 1/(n+1) (see figure 2). In fact, dynamic fatigue results tend to be conservative compared to static fatigue.

Dynamic Fatigue Plot

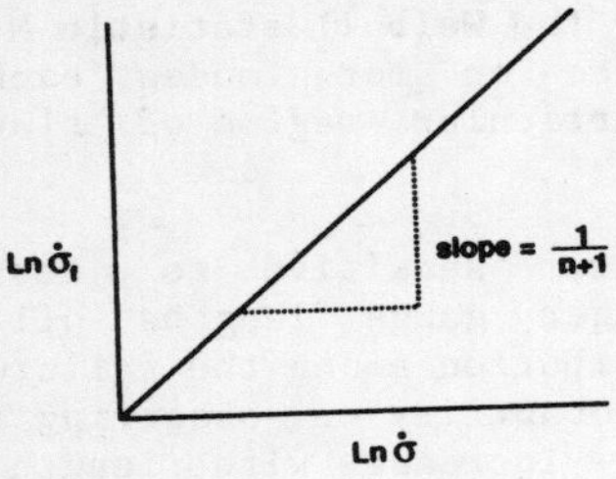

Figure 2

CORNING

MACROBENDING TESTING

The attenuation stability when wound in loops is an important characteristic of optical fibers. Coils are the typical method of storing excess fiber length in the field, such as in a splice enclosure. The same properties determine how well attenuation is maintained during a low frequency/high amplitude perturbation, such as a pinch.

Loss associated with these deployments is called macrobending. This is characterized by winding the fibers on mandrels of known diameter and measuring the excess loss, which increases with mode field diameter and decreasing bend diameter. For single-mode fibers the loss appears as a wavelength dependent edge moving down from longer wavelengths as the bend diameter decreases. The loss increase is wavelength independent for multimode fibers. The increase is linear with length under bend. Figure 3 shows typical results for matched-clad OVD fibers.

Attenuation Loss Vs Mandrel Diameter

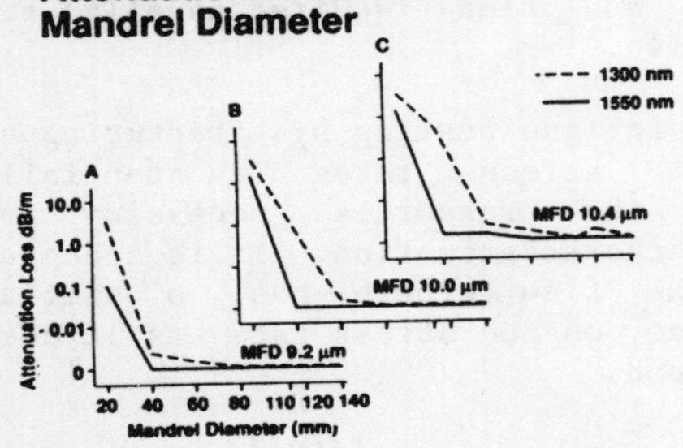

Figure 3

CORNING

Several precautions must be taken during the test. Smooth mandrel surfaces and consistent, low tension winds are essential for measurement repeatability. In addition, power coupled back into the core in phase at the radiation caustic, known as "whispering gallery modes", can cause noise in the measurement[3].

MICROBENDING TESTING

The actual spectrum of perturbations imposed by the cable on the fibers typically is thought to be better represented as a high frequency, low amplitude distribution, referred to as microbending. Unlike macrobending, the responses of single-mode and multimode fibers to microbending are taken to be similar, a wavelength independent increase in the spectral attenuation baseline.

Much work has been done to simulate the microbending spectrum of a cable on a fiber. In theory, such a test would be used to screen new fiber and coating designs, thus avoiding expensive test cable manufacture.

The two historic tests have been the lateral load and pin array test. Each attempts to impose a controlled microbending spectrum on the fiber/coatings systems. The lateral load test uses increasing lateral force against a controlled surface such as specified commercial grade sand paper (see figure 4 for typical results for matched-clad OVD fiber) or sieve sizes. The pin array test controls the microbending spectrum by spacing pins in a controlled manner and looking for attenuation increases as the fiber is wound through the array.

Lateral Load: Loss vs. Weight At 1550 nm

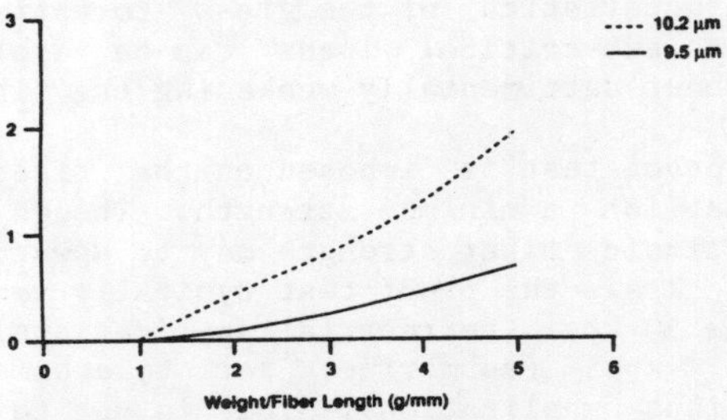

Figure 4
Dixon, et al. "Proceedings Of The Optical Fiber Communication Conference", 1987 p. 41

CORNING

The success of these tests in relating to specific cable design has been mixed. Two new approaches have shown more promise. One approach uses a surface analyzer to measure the actual spectrum of perturbations the fiber will contact[4]. By characterizing this spectrum as a Lorenzian function, the parameters can be used in a Petermann analysis to predict fiber attenuation response to cable loading.

The second approach wraps tubed fiber around a drum, and then splits the top off the tube. The fiber is stressed and the attenuation increase measured[5].

CONCLUSION

A series of environmental tests can be designed to simulate field conditions and accelerated lifetimes. Testing optical fibers under these conditions can anticipate issues for field operability and system lifetimes.

BIBLIOGRAPHY

1. Uesugi, N. et al. Appl. Phys. Lett. 43 (4) 15 August 1983. pp 327-8.

2. Ogai, M. et al. Optical Fibers Communication 1987 Proceedings. Optical Society of America.

3. Harris & Castle, J. Lightwave Tech. LT-4 p. 34. Jan 1986.

4. Grasso, G., Meli, F. European Conference on Optical Communication 1988 Proceedings

5. Bjerkland, L., et al. Technical Digest, Symposium on Optical Fiber Measurements, 1988. pp89-91

LOW LOSS 1×2 SINGLE-MODE OPTICAL SWITCH

H. Yanagawa, H. Miyazawa, M. Ogai and K. Okubo
Furukawa Electric Co., Ltd.
No. 6, Yawata-Kaigandori, Ichihara, Chiba 290 Japan

ABSTRACT

A low loss and compact 1×2 single-mode optical switch was fabricated by employing a pair of multiple-fiber connectors with oblong alignment pin holes. Two switching states were held mechanically with a see-saw mechanism, so a driving power was necessary only when the states were changed. Its insertion loss was less than 1dB, and the switching time was 2 - 3 ms.

INTRODUCTION

Because of its economy and development of joining technology, single-mode fibers are replacing multimode fibers even for short distance applications, where multimode fibers have been dominantly used so far. Although microoptics-based optical switches[1] are well developed, they are not always suitable for single-mode systems, because their assembly is involved in many complicated high precision alignments.

On the other hand, a novel fiber-based approach[2] was proposed recently and successfully applied to various optical passive components such as filters and wavelength division multi/demultiplexers. The approach was featured by a low insertion loss value and easy fabrication even with single-mode fibers.

In this paper, another fiber-based approach[3-5] is proposed, where high precision multiple-fiber connectors[6] are used, and it is demonstrated that an optical switch with this approach exhibits excellent performances.

SWITCH CONFIGURATION

1. Alignment Mechanism

The lower insertion loss limit of an ideal optical switch should be equal to its value of an optical connector which doesn't need a moving mechanism. This is the reason why an optical-connector-based design was adopted here. Furthermore, a multiple-fiber switch can be easily constructed by using a multiple-fiber connector.

An alignment mechanism of a new optical switch is illustrated in Fig.1.

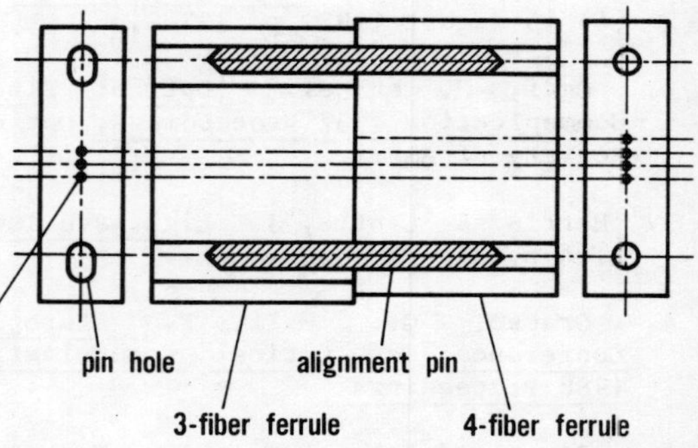

Fig.1 Schematic diagram of alignment mechanism of new switch

A pair of multiple-fiber connector ferrules was butted to each other, and a three-fiber and a four-fiber connector ferrules were actually used in the fabrication to form a two-fiber 1×2 optical switch as shown in Fig.2 (a). In this mode of switching, two states were possible, and they were (A-B, A'-B') and (A-C, A'-C'). On the other hand, by looping back as shown in Fig.2(b), this switch could be used as a single-fiber 2×2 optical switch. In this mode of switching, two states were (A-D, B-C) and (A-C, B-d).

As easily inferred, this switch configuration can be extended to that of a N-fiber switch by using a pair of 2N-fiber connectors. Since a twelve-fiber single-mode connector is already established, a six-fiber switch can be easily constructed at present.

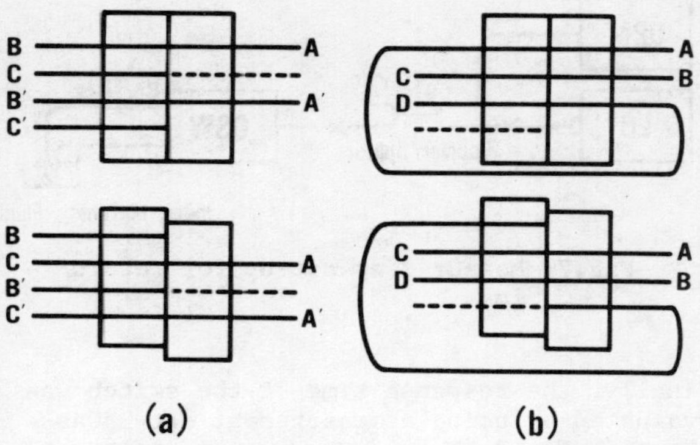

Fig.2. Conceptual diagram of switching modes. (a): two-fiber 1×2 and (b): single-fiber 2×2.

Although the three-fiber connector ferrule used in the fabrication was a conventional one , the four-fiber ferrule was specifically designed for the purpose of switching, that is, it had a pair of oblong alignment pin holes to accommodate two cylindrical alignment pins. The width along the longer axis of the holes was designed to be P_f+dp, where P_f is one fiber pitch of the connector and dp is the diameter of the alignment pins. Consequently, the fibers were aligned without a transverse offset, when the alignment pins were in contact with the inner walls of the oblong pin holes at two extreme positions. Therefore, the insertion loss of the switch could be very low, because it is basically the same as that of well-developed multiple-fiber connectors.

2. Self-Holding Mechanism

As an optical switch for outside plant use, a self-holding property is one of the most important requirements because of the difficulty in power feeding in the field, so the two switching states illustrated in Fig.2 must be held mechanically without supplying external electric power. As a result of investigation of a self-holding mechanism, a see-saw mechanism which has been widely used in an

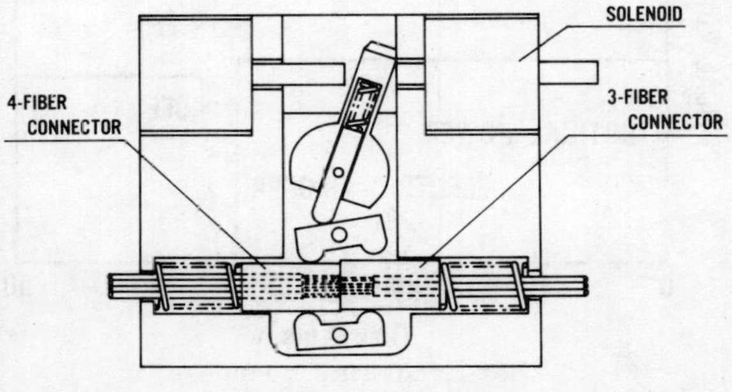

Fig.3 Self-holding mechanism

electric switch was considered to be preferable also for an optical switch. Consequently, a see-saw mechanism which is similar to the one used in an electric toggle switch, was employed to hold two switching states stably without supplying external holding power as shown in Fig.3. Therefore, a driving power was needed only to change the states, that is, to move the three-fiber connector ferrule transversely by one fiber pitch. This push-pull motion was activated by a pair of solenoids.

ASSEMBLY

As is easily understood in the previous explanation on the alignment mechanism shown in Fig.1, the assembly was performed as follows.

A. A self-holding mechanism with see-saws and a pair of solenoids were fixed on a package base,
B. A pair of multiple-fiber connector ferrules was mated in the way similar to that of an optical connector,
C. The completed alignment mechanism was inserted between the see-saws,
and
D. A cover was fixed on the package base.

Clearly the assembly was not involved in adjustable alignments and performed very easily. The fiber pitch of the fabricated switch was chosen to be 0.25mm in such a way that the switch might be compatible with standard fiber ribbons, and a standard four-fiber single-mode ribbon was used. The single-mode fiber had a mode field diameter of 9.4 µm at the wavelength of 1.3 µm, cut-off wavelength of 1.20 µm, and core eccentricity of 0.4 µm.

The fabricated switch was very compact, and it measured 85 mm in width, 35 mm in length and 16 mm in height as shown in Fig.4.

Fig.4 Photograph of fabricated prototype switch.

PERFORMANCE EVALUATION

The insertion loss of the fabricated switch was evaluated at the wavelength of 1.3 µm in the 1×2 switching mode by using an light-emitting-diode and an optical powermeter. The loss values for connection A-B, A'-B', A-C and A'-C' in Fig.2(a) were 0.55dB, 0.71dB, 0.83dB, and 0.64dB, respectively. These were very low and less than 1dB even with single-mode fibers. The insertion loss was confirmed to be similar to that of a multiple-fiber connector used in the fabrication.

The insertion loss, isolation and return loss of the switch were evaluated at the wavelength of 1.3 μm during a 1000-cycle repeated switching test by using a stabilized laser diode and an optical powermeter.

The experimental set-up for insertion loss and isolation measurements is shown in Fig.5.

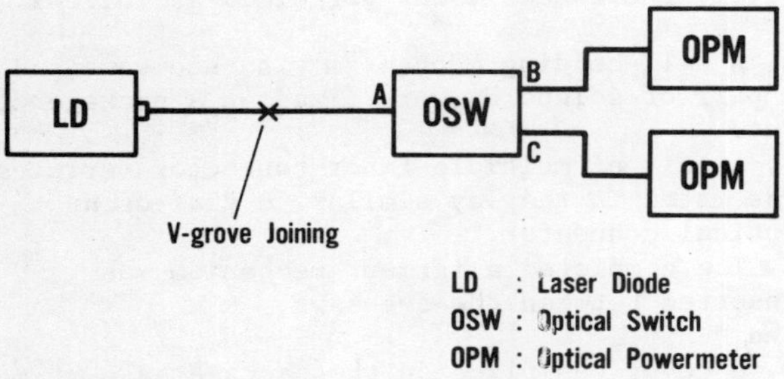

Fig.5 Measurement set-up for insertion loss and isolation

The insertion loss variation during the test was very small, and it was less than 0.1dB as plotted in Fig.6. The isolation between two adjacent ports was always more than 50dB, which corresponded to the detection limit of the measurement set-up.

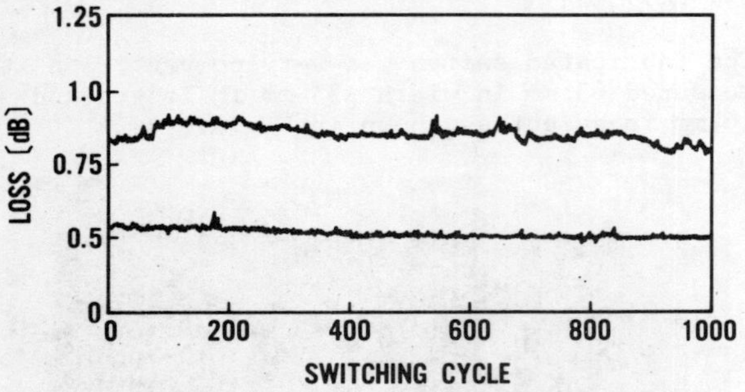

Fig.6 Insertion loss variation during repeated switching test.

The return loss measurement set-up is illustrated in Fig.7, and a fused biconical taper coupler was used as a coupler/splitter.

The measured return loss was 50dB±5dB, and this was almost equal to be 55dB which was the return loss of the coupler/splitter. The return loss was considered to be sufficiently larger than the required value of more than 30dB.

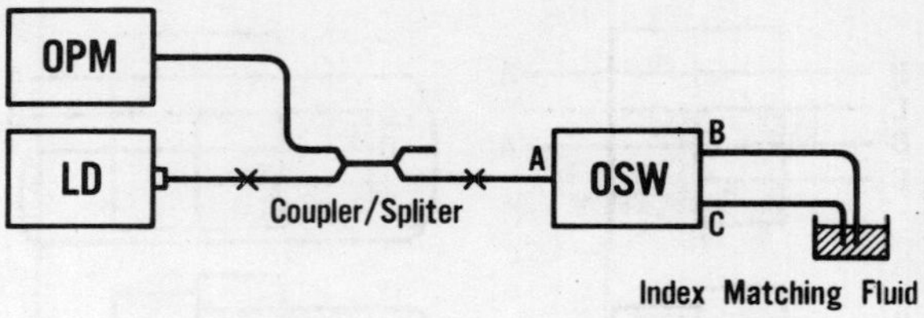

Fig.7 Measurement set-up for return loss

Finally, the response time of the switch was evaluated by using a measurement set-up as shown in Fig.8. The response time of the measurement set-up itself was less than 0.1ms.

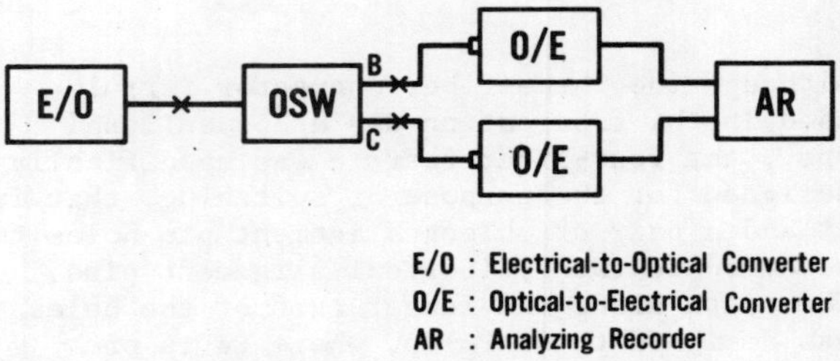

Fig.8 Measurement set-up for response time

The output of the analyzing recorder is illustrated in Fig.9. The delay was observed before a change in optical power occurred, and it was around 24ms after an electric trigger was given. The switching time which was defined to be the time duration of the optical signal suspension, was measured to be 2.9ms and its standard deviation was 0.2ms.

All the previous measurements were performed at a driving power of 15V×1A, and the results are summarized in Table 1.

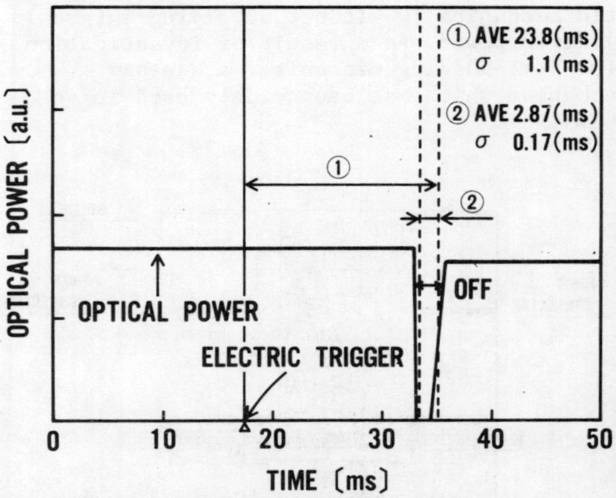

Fig.9 Measured time response of fabricated switch

Table 1 Measured characteristics of prototype switch

Item	Measured Results
Insertion Loss	less than 1dB
Isolation	more than 50dB
Return Loss	50±5dB
Switching Time	2～3ms
Driving Power	15V×1A

CONCLUSION

A low loss and compact two-fiber 1×2 self-holding single-mode optical switch was fabricated by adopting the following design lines.

A. As an alignment mechanism, a pair of multiple-fiber connectors with alignment pin holes was employed. Therefore, a low insertion loss value comparable with that of multiple-fiber connectors was obtained, and it was less than 1dB even for single-mode fibers. Furthermore, the assembly of the mechanism was not involved in adjustable alignments, and thus easily performed. It was shown that this design was easily extended to a N-fiber switch by using 2N-fiber connectors.
B. As a self-holding mechanism, a see-saw was used, so a driving power was necessary only when the states were changed.

The switching time was also evaluated, and it was 2 - 3ms.

A further study is going on to optimize the design and the fabrication.

ACKNOWLEDGMENT

The authors would like to thank Mr. Ochiai and Mr. Raijo for their help in the fabrication.

REFERENCES

1. (For example) M. Nunoshita, Y. Nomura, T. Matsui and T. Nakayama : "Optical Switch for Multimode Optical Switch Systems", Optics Letters, vol. 4, No. 1, 1979, pp.27-27-28.
2. H. Yanagawa, T. Ochiai, H. Hayakawa and H. Miyazawa : "Filter-Embedded Design and Its Applications to Passive Components", to be published in J. Lightwave Technol.
3. S. Nagasawa, H. Furukawa, and N. Kashima : "A New Type of Optical Switch with a Plastic-Molded Ferrule", Trans IEICE Japan, vol. E70, No. 8, 1987, pp.696-698.
4. T. Satake, S. Nagasawa, and N. Kashima : "Single-Mode 1×50 Switch for 10-Fiber Ribbon", Trans. IEICE Japan, vol. E70, No. 7, 1987, pp.623-624.
5. H. Yanagawa, T. Ochiai, H. Raijo, and H. Miyazawa : "Very Large-Scale Single-Mode 1×N Optical Switch", Trans. IEICE Japan, vol. E72, No. 4, 1989, pp.269-271.
6. T. Satake, S. Nagasawa, and N. Kashima : "SM Multi-Optical-Fiber Connector for High Density Cable (in Japanese)", Natl. Conf. Rec. on Commun. of IEICE Japan, 1986, p.278.

γ-RAY IRRADIATION INDUCED LOSS OF PURE SiO_2 CORE SINGLE-MODE FIBER WITH LOW DOSE RATE

M.Ooe, Y.Chigusa, M.Kyoto, M.Watanabe, S.Okamoto and T.Yamamoto
Sumitomo Electric Industries, Ltd.
1, Taya-cho, Sakae-ku, Yokohama, Japan
Radiation center of Osaka Prefecture
704, Shinke-cho, Sakai-shi, Osaka, Japan

ABSTRACT

By resolving the recovery curves, the radiation hardening characteristics of the optical fiber were investigated. This investigation showed that a defect with long decay time causes the induced loss of low dose rate and long term irradiation. It was also shown that the induced loss with low dose rate and long term irradiation can be estimated by extracting the induced loss of this defect.

INTRODUCTION

Fiber optic signal transmission systems are now being applied to radiation environments and the research on radiation sensitivity of optical fibers has intensified. Most radiation induced optical loss studies have been concerned with the characterization of the fiber behavior of high dose rate irradiation. In order to introduce optical fibers with high reliability in communication systems, it is important to investigate the radiation resistant characteristics with low dose rate over extended periods of time [1]. However, there is a problem that the induced loss depends not only on total dose but also on dose rate (Fig.1). This paper describes the behavior of defects by resolving the recovery curves and attempts to clarify the differences radiation hardening characteristics between high and low dose rate irradiation. Furthermore, radiation induced loss of single-mode fiber with low dose rate over extended periods of time was estimated and the reliability of this estimation method was investigated.

EXPERIMENTAL

2 types of single-mode fiber were investigated. One with a pure SiO_2 core and F-SiO_2 cladding (Si-SM). The other with a GeO_2 core and pure SiO_2 cladding (Ge-SM). They were exposed to γ-ray from a ^{60}Co source in γ-cell of in pool type γ-ray irradiation facilities. The measurement wavelength was 1.30 μm.

RESULTS AND DISCUSSION

1. Resolution of recovery curve

Fig.2 shows the increase in attenuation at 1.30 μm wavelength under a relatively high dose

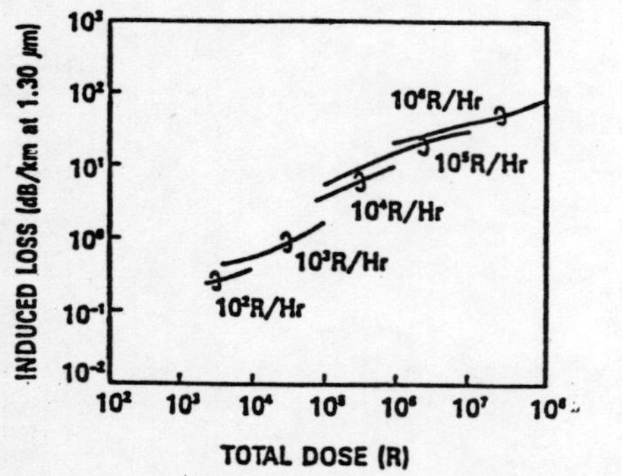

Fig.1 γ-ray irradiation characteristics (Si-SM)

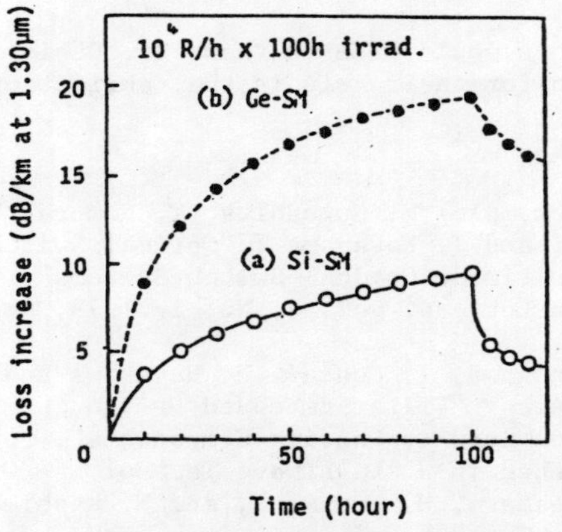

Fig.2 Radiation hardening characteristics of single-mode fibers at 1.30 μm

rate (10^4 R/h) and recovery after irradiation has ceased. This figure indicates that attenuation of Si-SM increased approximately 10 dB/km after exposure to 10^4 R/h for 100 hours and recovered to less than 4 dB/km in another 15 hours. This increase is approximately one half that of Ge-SM and the recovered attenuation is approximately one third as low [2].

In general, the loss increase and recovery model from exposure to nuclear radiation can be written as follows.

(Model)

precursor i $\rightleftharpoons$ defect i

When optical fibers are exposed to γ-ray irradiation, formation of defect centers and recovery from defects to precursor occur at the same time. The irradiation induced loss is generally reversible so when γ-ray irradiation stops, some recovery from damage is observed.

Because there are many defects (precursors) in the glass, induced loss after irradiation has ceased is represented by the sum of the induced loss of each defect i.

$$\Delta\beta_R(t) = \Sigma\Delta\beta_{R(i)}(t) = \Sigma\ell_i \exp(-t/t_i)$$

$\Delta\beta_R(t)$: Induced Loss.

$\Delta\beta_{R(i)}(t)$: Induced Loss of each defect (i)

ℓ_i : Induced Loss of each defect (i) immediately after irradiation is ceased.

τ_i : Decay time.

t : Time after irradiation is ceased.

The behavior of defects in the optical fiber can be investigated by resolving the recovery curve.

The value of ℓ_i and τ_i of Si-SM and Ge-SM were calculated so as to fit recovery curves. Table 1 shows the τ_i value of both fibers.

Table 1 Decay time of defect τ_i

Defect	Decay time/Hr	
	Si-SM	Ge-SM
1	97.4	118.4
2	2.59	2.54
3	0.15	0.15
4	0.02	0.03

Recovery curves were both resolved to 4 terms. Decay time τ_i of both fibers were almost the same. Table 2 shows the value of ℓ_i and τ_i of both fibers after being exposed with 10^5 R/Hr for 25 hours and 100 hours. The value of ℓ_i and τ_i of Ge-SM are obviously larger than that of Si-SM. These results indicate that the difference in radiation hardening characteristics between Si-SM and Ge-SM do not depend on the kind of defect (There are 4 kinds of defect centers (τ_i) in Si-SM and Ge-SM), but on the amount of defect.

Except for ℓ_0, the value of other ℓ_i(i=1~3) had already reached the saturation point after 25 hours irradiation (Table 2). For example, the value of ℓ_i of Si-SM after 100 hours irradiation is 12.9 dB/km, which is not so different from 10.1 dB/km, the value after 25 hours irradiation. On the other hand, the value of ℓ_0 had not reached the saturation point after 100 hours irradiation. That is to say, γ-ray irradiation over a long period of time causes a large ratio of induced loss of defect i=0 to that of all defects. From these facts it is important to investigate the behavior of the defect center i=0 which has a long decay time, in the case of long term irradiation (high total dose).

Table 2 Resolution of recovery curves (irradiation time dependence)

$$\Delta\beta_R(t) = \Sigma\Delta\beta_{R(i)}(t) = \Sigma\ell_i \exp(-t/t_i)$$

(a) Si-SM

ℓ_i(dB/km) / τ_i (hour) / irrad. time	ℓ_0	ℓ_1	ℓ_2	ℓ_3
	(−)	2.59	0.15	0.02
25 hours	15.7	10.1	10.0	15.2
100 hours	43.2	12.9	12.5	14.6

(b) Ge-SM

ℓ_i(dB/km) / τ_i (hour) / irrad. time	ℓ_0	ℓ_1	ℓ_2	ℓ_3
	(−)	2.54	0.16	0.03
25 hours	59.2	26.3	5.9	12.9
100 hours	75.9	26.9	14.0	8.0

Fig.3 presents the transmission loss spectra for 2 types of single mode fibers after 2 weeks from γ-ray irradiation (dose rate 10^6 R/Hr, irradiation time 25 hours, a: Si-SM b: Ge-SM). The absorption tails from the shorter and the longer wavelength region were shown. Compared with the absorption in short wavelength region, the absorption in long wavelength is broad and has a strong influence on the wide wavelength region. This absorption tail from the longer wavelength for Ge-SM is

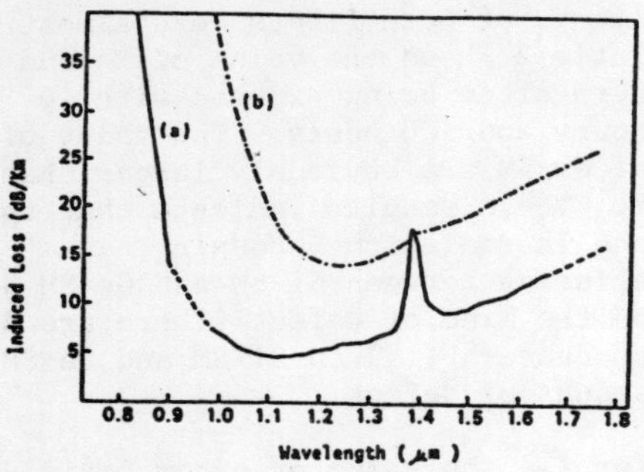

Fig.3 Loss increase due to γ-ray irradiation. (10^4 R/Hr x 25 hours, 2 weeks after irradiation)
(a) Si-SM (b) Ge-SM

much larger than that for Si-SM.

This result indicates that the defect center i=0 which has a long decay time causes irradiation induced optical absorption in the long wavelength region.

2. Estimation of induced loss

In the case of low dose rate and long term irradiation, the formation of defect centers and recovery from defects to precursor occur at the same time. Only defect center which causes the optical absorption in long wavelength region hardly recovers to its precursor, because it has long decay time.

(Model - in the case of low dose rate irradiation)

precursor ⟶ defect
||
Optical absorption in long wavelength region

Therefore, it is considered that the value of ℓo corresponds to the induced loss with low dose rate over extended periods of time.

Fig.4 shows the induced losses of Si-SM and Ge-SM obtained by experimental method and ℓo. The dose rate was 2.12 R/h and 259 days have passed since irradiation was started. Those experimental data agree well with the curve of ℓo. This result indicates that the the induced loss of low dose rate and long term irradiation can be estimated from the resolution of recovery curve of high dose rate irradiation.

Fig.4 also indicates that the induced loss of Si-SM is lower than that of Ge-SM that was exposed to γ-ray irradiation with low dose rate for extended periods.

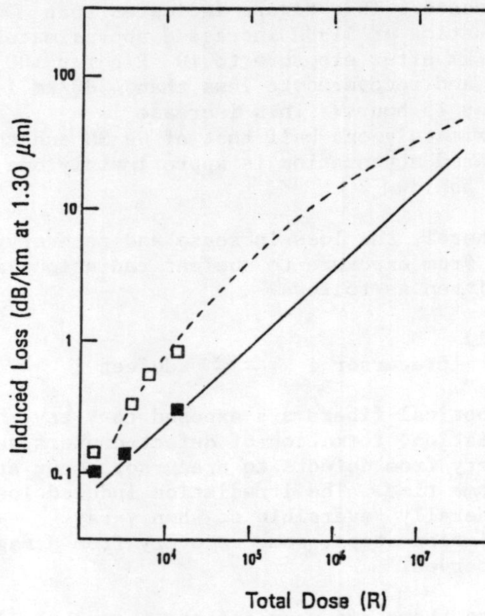

Fig.4 Induced loss of optical fibers after 2.12 R/h at 1.30 μm
Experimental (■ Si-SM, □ Ge-SM)
ℓo (—— Si-SM, ------ Ge-SM)

SUMMARY

By resolving the recovery curves, the radiation hardening characteristics of the optical fiber were clarified as follows.

1) The difference in radiation hardening characteristics between Si-SM and Ge-SM do not depend on the kind of defect, but the amount of defect.
2) There are 4 kinds of defects in these fibers.
3) In the case of long term irradiation (high total dose) it is important to investigate the behavior of the defect center i=0 which has a long decay time.
4) The value of ℓo corresponds to the induced loss of low dose rate and long term irradiation. Therefore, the induced loss can be estimated using the value of ℓo.
5) The induced loss of Si-SM is lower than that of Ge-SM that was exposed to γ-ray irradiation with low dose rate for extended periods.

REFERENCE

1) E. J. Friebele et al., J. Appl. Phys., 45 (8), 3424 (1974).
2) I. Yoshida et al., OFC 1986 Tech, Digest, TU12;
Y. Chigusa et al., IEEE Trans. Nuc. Sci, 35, 894 (1988)

STABILITY OF POLARIZATION MAINTAINING FIBER IN HIGH TEMPERATURE

Eisuke Sasaoka, Hiroshi Suganuma, Sigeru Tanaka
SUMITOMO ELECTRIC INDUSTRIES, Ltd.
1, Taya-cho, Sakae-ku, Yokohama 244, JAPAN

ABSTRACT

Stability of polarization maintainint fiber in high temperature was confirmed by measuring the birefringence up to 300°C by phase difference measurement between two orthogonal polarization modes. Zero birefringence temperature was extrapolated as 1000°C which is consistent with softening temperature of glass.

INTRODUCTION

Polarization maintaining fibers (PMFs) are widely used for many kinds of optical fiber sensors. In some applications, PMFs will be used in high temperature. Hence, it is important to study the stability of PMFs in high temperature. This paper clarifies the birefringence behavior of PMF in high temperature.

ANALYSIS OF MEASUREMENT METHOD

Schematic diagram of a new birefringence measurement system is shown in Fig.1.

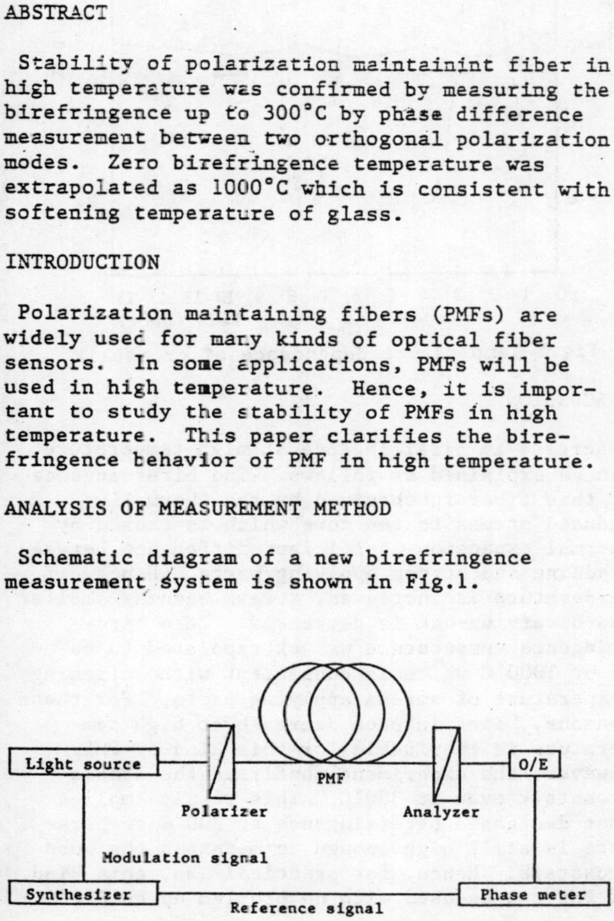

Fig.1 Birefringence measurement system

In this measurement system, amplitude modulated and linearly polarized light is injected into PMF under measurement. Group delay is different between two orthogonal polarization (X-polarization and Y-polarization) modes because of birefringence. By measuring the phase difference between two polarization modes using phase meter, group delay difference is detected. This measuring method can be applied to a relatively long fiber, which enables changing the environment of the fiber under measurement. Relation between birefringence B and measured phase difference ϕ_0 is given by Eq.1.

$$\phi_0=2\Pi f LB/c \qquad \text{(Eq.1)}$$

In Eq.1, f is modulation frequency of input light, L is fiber length and c is light velocity in vacuum.

In this measurement method, if there is certain amount of crosstalk between two polarization modes, the measured phase difference will be affected by the polarization mixing, which results in a possible measurement error. Hence, we first estimate birefringence measurement error caused by crosstalk. When the crosstalk is assumed to be caused by random mode coupling along the fiber length, measured phase difference ϕ_1 is expressed as Eq.2.

$$\phi_1=\mathrm{Arctan}\left[\frac{\sin\phi_0+2q(1-\cos\phi_0)}{\cos\phi_0+2q\sin\phi_0+2q^2(1-\cos\phi_0)}\right]$$

$$q=ch/(2\Pi f B) \qquad \text{(Eq.2)}$$

ϕ_0 denotes phase difference without crosstalk which is given by Eq.1, and h denotes mode coupling coefficient in the fiber. Using Eq.1 and 2, birefringence is precisely calculated based on the measured phase difference. However, when mode coupling coefficient h is unknown, only Eq.1 is available, instead of Eq.2. The measurement error induced by Eq.1 in presence of crosstalk was analyzed. Generally, measurement error decreases with increasing modulation frequency f and fiber length L. Fig.2 shows an example of calculation for the 50 m fiber with birefringence of 5.0×10^{-4} under modulation frequency of 800MHz as a function of crosstalk. We can see from this figure that measurement error is less than 2%, if crosstalk is lower than -17 dB.

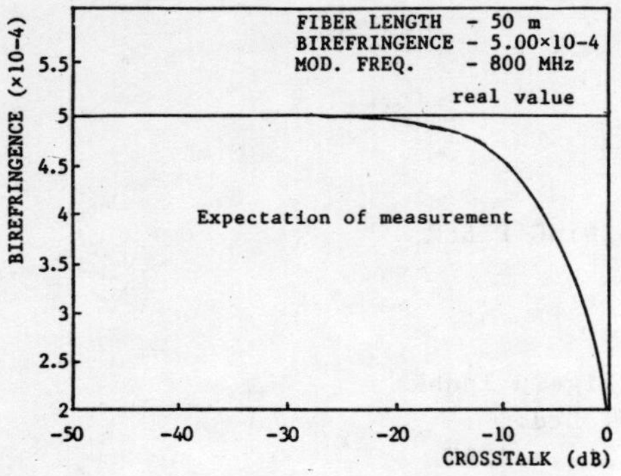

Fig.2 Birefringence error by crosstalk

III. EXPERIMENT

Birefringence behavior in high temperature was measured about PMF whose cross section and structural parameters are shown in Fig.3.

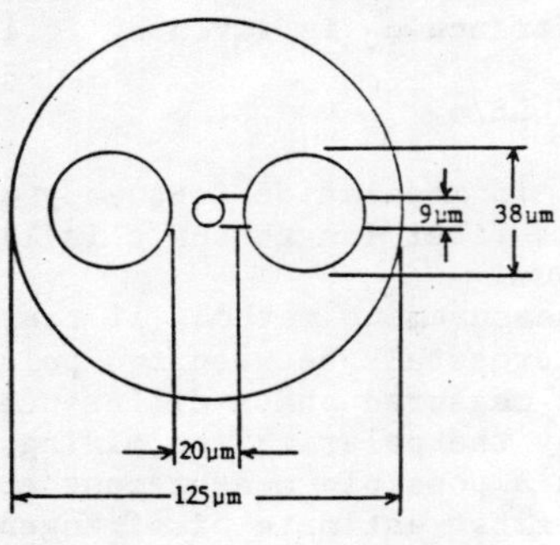

Fig.3 Cross section of measured PMF

Core, cladding and stress applying parts are composed of GeO_2 doped SiO_2, pure SiO_2 and B_2O_3 doped SiO_2, respectively. Refractive index difference between core and cladding is 0.3% and cut-off wavelength is 0.90 μm. The fiber is coated with polyimide which is more resistant to high temperature than conventional silicone and UV resin. Fiber length was 50 m and loosely wound. The light source was 1.3 μm wavelength LD. The result of measurement is shown in Fig.4.

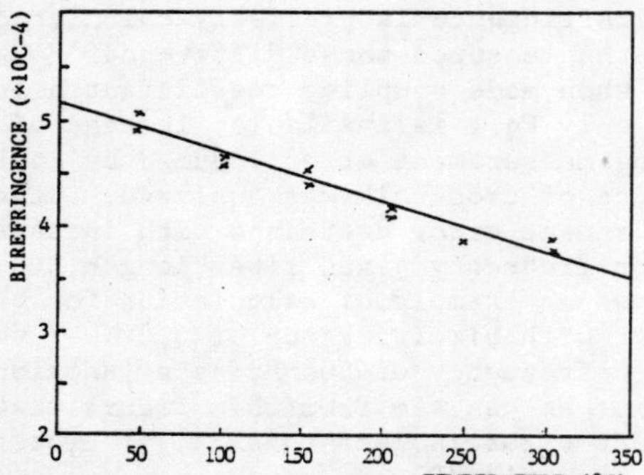

Fig.4 Temperature dependence of birefringence

Birefringence of the fiber decreases linearly with temperature increase, but the fiber still had high birefringence of 3.7×10^{-4} even at 300°C temperature. Temperature dependence of crosstalk was also measured for the same PMF. The result in Fig.5 shows that crosstalk does not degrade against temperature change, although the birefringence itself decreases high temperature. This result proves that birefringence of 3.7×10^{-4} is high enough to maintain the good crosstalk characteristics.

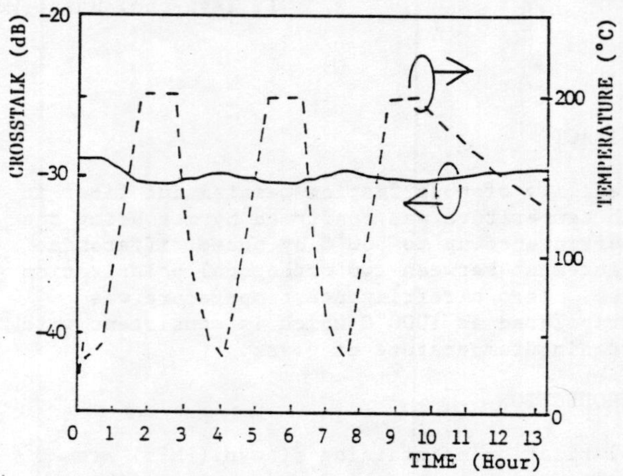

Fig.5 Temperature dependence of crosstalk

DISCUSSION

Decrease in birefringence at high temperature can be explained as follows. The birefringence of this fiber is obtained by the thermally induced stress to the core which is caused by thermal expansion coefficient difference between cladding and stress applying parts. Hence, if temperature is increased, stress becomes smaller and birefringence is decreased. Zero birefringence temperature was extrapolated to be about 1000°C which is consistent with softening temperature of stress applying parts. For these reasons, birefringence decrease at high temperature is inevitable for this kind of PMF. However, the experiment confirmed the stable crosstalk even at 300°C. This result implies that decreased birefringence at 300°C temperature is still high enough to maintain the good crosstalk. Hence, for practical use, this kind of PMF can be used with no problem up to 300°C.

CONCLUSION

By measuring phase difference between two orthogonal polarization modes, birefringence of PMF was measured. Linear decrease in birefringence with temperature rise was observed and zero birefringence temperature was extrapolated to be 1000°C which is consistent with softening temperature of stress applying parts. However, the fiber was proved to have good polarization maintaining ability up to 300°C temperature, which promises the application of this fiber to the optical fiber sensors for high temperature operation.

REFERENCES

(1) A. Ormazd et. al. Applied optics Vol.22, No.15 pp 2374-2379, 1983

(2) Malcolm P. Varnham et.al. J. lightwave tech. Vol. LT-1, No.2 pp 332-339, 1983

Elimination of the Errors Caused by Spectral Width of Light Source in Optical Fiber Loss Measurements

Li Luo
Institute of Optical Fiber Technology, Shanghai Jiao Tong University,
Shanghai 200052, P.R. CHINA

ABSTRACT

In optical fiber loss measurements the errors will inevitably be produced in using the light source with wide spectral linewidth espcially at long wavelengths, In this paper, we indicate the major factor producing the errors and calculated the errors by formula. A typical modification curve is given for LED at 1300nm wavelength and by using the curve we can easily eliminate the great part of the errors.

INTRODUCTION

In general, operators use a light source with narrow spectral linewidth in fiber loss measurements. But it is difficult for LED which is very usful source in measurements. In order to eliminate the errors caused by spectral linewidth of source, we start with the basic principle to calculnate the loss and discuss the major factor causing the errors. Later a typical modification curve is given which is available for LED at 1300 nm wavelength.

The Principle Methods for Calculnating The Errors

No matter what methods we used in fiber loss measurements, the basic principle consists of measuring the power in both ends of input and output of fiber. Let p_1 stands for the power launching into fiber and p_2 for emitting out from it. The fiber loss α could be expressed in dB by:

$$\alpha = 10\log(P_1/P_2) \qquad (1)$$

From basic physics concept, α could be further expressed as:[1]

$$\alpha = 10\log\frac{\int_{\lambda_1}^{\lambda_2} S(\lambda)R(\lambda)d\lambda}{\int_{\lambda_1}^{\lambda_2} S(\lambda)K(\lambda)R(\lambda)d\lambda} \qquad (2)$$

Where $S(\lambda)$ stands for the spectral distribution of light sourec, $R(\lambda)$ for spectral responsibility of detector and $K(\lambda)$ for loss spectrum of fiber. $\lambda_2-\lambda_1$ is full linewidth of spectrum of source. From the expression (2), we could see that α is relevant to the loss spectrum of fiber closely. The other two factors $S(\lambda)$ and $R(\lambda)$ are relatively less important. Let us briefly analysis the feature of the loss spectrum of fiber(fig.1):

At the short wavelengths the curve of loss spectrum of fiber decreases rather smoothly by the increase of wavelength. So the error α_s caused by spectral linewidth of light source will become slight due to the effect of integration. At the long wavelengths, however, the curve of the loss spectrum of fiber nearby 1300 nm and 1550 nm is just at the lowest points between the absorb peaks respectively. So, due to the effect of integration, the error α_s in long wavelength is large and value measured is always bigger than actual value owing to the action of high loss peaks. The wider the width of source is, the bigger the error α_s will be. But the flatter the loss spectrum of fiber is, the smaller error α_s will be. Figure 1 shows the loss spectrum K_1 and K_2 which indicate fiber 1 and fiber 2 respectively. We could see that the curve of K_1 is flatter than K_2 nearby the region of 1300nm. The error α_{s1} which stands for fiber 1 is smaller than the error α_{s2} which stands for fiber 2. Table 1 shows the error α_{s1} and α_{s2} in dB/km at 1300 nm.

Figure 2 shows the error α_2 at both 850 nm and 1300 nm wavelength[2].

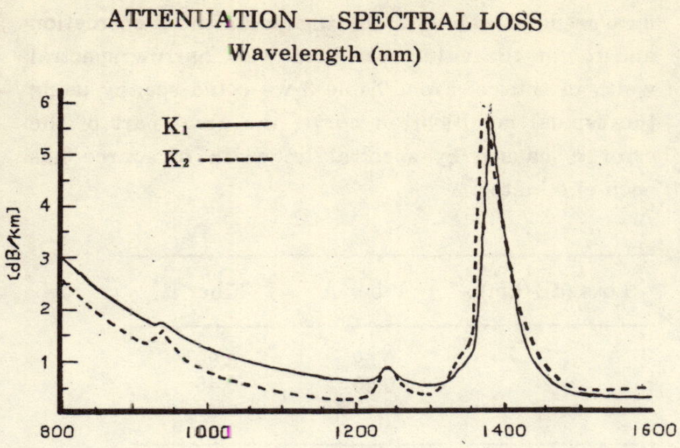

Figure 1. The loss spectrum K_1 of fiber 1 and the loss spectrum K_2 of fiber 2.

$\Delta\lambda$(nm)	k_1(fiber 1)	k_2(fiber 2)
60	0.10	0.16
80	0.18	0.24
100	0.27	0.31

Table 1. The error α_{s1} and α_{s2} from two different loss spectrum of fibers (fiber 1 and fiber 2).

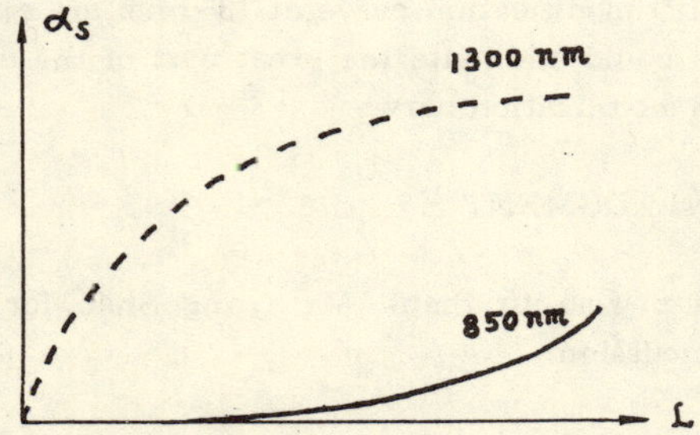

Figure 2. A typical relation of the error α_s with fiber length at 850 nm and 1300nm wavelengths.

Now let us see, through the comparion of the results of calculation, how the K (spectrum loss of fiber) or R (responsibility of detector) effects the error α_s. Figure 3 shows the curves of spectral responsibility of two detectors made by Ge and InGaAs. The error α_{s1} from the two detectors with the same fiber is shown in Table 2 respectively.

$\Delta\lambda$(nm)	Ge	InGaAs
60	0.16	0.17
80	0.23	0.25
100	0.30	0.32

Table 2. The error α_s from two different detectors (Ge and In Ga As) with same fiber in dB/km.

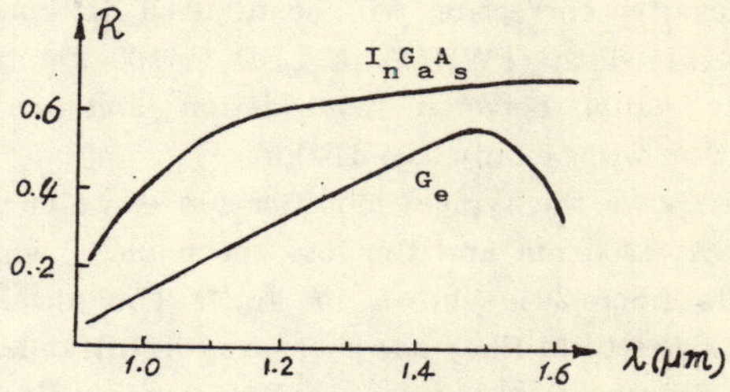

Figure 3. The spectral responsibility of two detectors made of Ge and In Ga As.

Comparing figure 1 with fiber 3 we could find that the difference of the two curves (Ge and InGaAs) in figure 3 is obviously bigger than the difference of the two curves (K_1 and K_2) in figure 1. On the contrary, however, the change of the error α_s corresponding figure 3 (shows in Table 2) is smaller than the change of the error α_s corresponding figure 1 (shows in Table 1). That is to say, the characterestic of loss spectrum of fiber is the major factor producing the error α_s caused by wide spectral linewidth of source.

To Eliminate The Error By Typical Modification Curve

If we know the loss spectrum of fiber in advance, we could eliminate the error α_s through formula (3), but in commen case, we couldn't know the loss spectrum of fiber before measurements, how could we eliminate the error α_s? Here we give an error modification curve of a typical fiber which is choosen from four standard fibers. Although the precision of the measurement value after correction by using typical modification curve is not superior to the one corrected by using loss spectrum of the actual fiber being measured, it could eliminate the great part of the error α_s. It is improtant that the difference between the loss spectrum k_m of the fiber being measured and the loss spectrum k_t of the typical fiber being given is not very wide.

Fortunately, this difference, as a general rule, is not very wide in the same class of fiber. For instance, in Fig.2 this difference between k_1 representing fiber 1 and k_2 representing fiber 2 is the largest set among four fibers choosen from three different manufactures respective.[2]

If we calculnate the error α_{s2} with formula by using k_1 instead of k_2, the different value between modification and actual calculnation will less than 0.06 dB/km (see Table 1). That is to say, suppose the loss of fiber 2 is α=0.50dB/km, the spectral width λ (FWHM) of LED is 80 nm. The measurement value is α_m=0.74 dB/km, and the value after correction will be α_c=0.56 dB/km. If the spectral width (FWHM) of LED is 100 nm, the difference value between modification and actual calculnation will be only 0.04 dB/km.

Figure 4 shows the typical modification error currve for LED at 1300 nm and the loss spectrum of which represents fiber 2 is shown in Fig.1. The abscissa represents the total fiber loss measured in dB and the ordinate represents the error α_s. For example, if the measurement value is 2 dB, and the linewidth of source $\Delta\lambda$ (FWHM) is 80 nm, we could find, from the ordinate, the modification error value on the curve corresponding to 2 dB on abscissa is 0.4 dB. As the result, the fiber loss measured by narrow width of source is 2−0.4=1.6 dB in total fiber length.

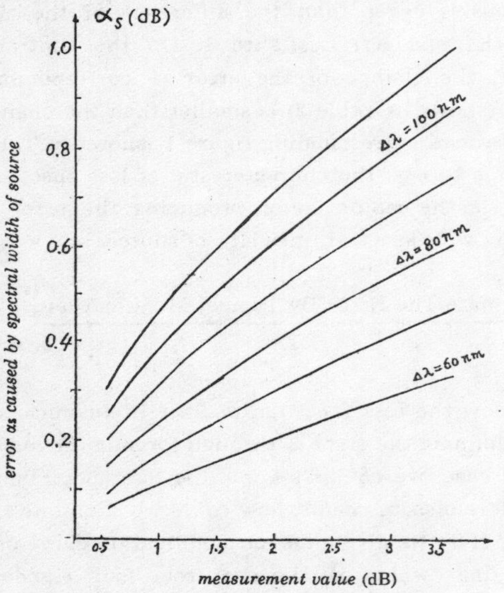

Figure 4. The typical modification error curve for LED at 1300 nm wavelength.

In recent comparison of single mode optical fiber loss measurements[2], one participant has measured the two fibers using LED at 1300 nm and the (FWHM) $\Delta\lambda$ of which is 80nm. We corrected the measurement value by the typical modification curve. Table 3 is the results of the correction in dB/km. α_m stands for the measurement value, α_c for the value after corrcetion and α_n for the value measured with narrow spectral width of source. From Table 3 we could see, by using the typical modification curve, the great part of the error α_s caused by spectral linewidth of source has been eliminated.

Loss (dB/km)	Fiber A	Fiber B
α_m	0.59	0.76
α_c	0.50	.0.46
α_n	0.47	0.47

Table 3. The results of correcting the error α_s in measurements. α_m stands for the value with LED, α_c for with LED after correction by the typical modification curve and α_n for with narrow spectral width of source.

Conclusion

The error in fiber loss measurements caused by spectral linewidth of source is very large at long wavelengths and relies closely on the loss spectrum of fiber. The methods of modifying the errors and a typical LED modification curve at 1300 nm are given. Operators could eliminate the great part of the error by typical modification curve.

ACKNOWLEDGMENT

The author wish to thank Mr. Yang Shao for his helpful discussion.

References

(1) Yang Shao and L.Luo,'Effect of Spectral linewidth of source on optical fiber attenuation measurements,'Optical Fiber & Electric Cable (China). Vol. 5, pp.26-29, Oct.1987.

(2) Li Luo and J. Tougcun,'A Recent nationwide comparison of single mode optical fiber attenuation measurements,'IEEE Trans. Instrum. Meas., vol. IM-37, no.4, pp.655-656, Dec.1988.

THE USEFUL CROSS-SECTION & FOURIER TRANSFORM OF SELFOC ROD LENS

Chen Gaunghei, Shanghai Transmission Lines Research Institute, P.O.Box 7833
Shanghai, China

ABSTRACT

An important property, useful cross-section, of SELFOC rod lens will be put forward in this paper, and a proximate radius of the useful cross-section will be given. Besides, the paper tries to discuss the Fourier transform of SELFOC rod lens within its useful cross-section.

INTRODUCTION

SELFOC rod lens is one of tiny optical lenses which are often used in optical fiber sensors and optical fiber components. In China, the researches on its manufacture, applications and on the measurment of its properties have been developed for several years, but the effect of its applications is not as good as expected.Some people try in vain to calculate an optimum length of the rod lens to get the best effect of its application,basing on Eq.(5).It was proved that a main reason is most users may ignore existance of the useful cross-section in a SELFOC rod lens fabricated by ion-exchange method. In order to provide the users with a helpful reference, this paper first discusses the existance of the property and gives approximately the radius of the useful cross-section. Meanwhile to provide a theoretical basis for the use of SELFOC rod lens in optical information processing, this paper tries to discuss Fourier transform of SELFOC rod lens within its useful cross-section.

REFRACTIVE INDEX PROFILE OF SELFOC ROD

The refractive index profile of domestic SELFOC rod is fabricated via ion-exchange method[1], which process can be described by diffusion eqution:

$$\begin{cases} \dfrac{\partial U}{\partial t} = D\nabla^2 U \\ U(r,\theta)\big|_{t=0} = U_0 \\ U(a,t)\big|_{r=a} = 0 \end{cases} \qquad 0 \leq r \leq a \qquad (1)$$

where U=distribution of the ion concentration at a distance r from the axis of mother glass rod; D=diffusion coefficient; a = radius of mother glass rod.

Assuming the mother glass rod is homogeneous in both axial and angular directions, we got the distribution of the ion concentration:

$$U(r,t)=U_0\frac{2}{K}\cdot\frac{J_0(Kr/a)}{J_1(K)}\cdot e^{-K^2Dt/a^2} \qquad (2)$$

where K is the first root of Bessel function J_0.

Based on the proximately proportional relation between the difference of refractive index $N(r,t)-N_a$ and the ion concentration $U(r,t)$, the refractive index profile of ion-exchanged glass rod can be obtained:

$$N(r)=N_a+N_0\cdot A(a/K)^2J_0(K\cdot r/a) \qquad (3)$$

where

$$A=2K(N_0-N_a)/a^2\cdot\frac{e^{-K^2Dt_0/a^2}}{N_0J_1(K)} \qquad (4)$$

is a constance, N_a, N_0 are respectively the refractive index in side surface and axis of the glass rod, t_0 is diffusion time when the diffusion arrives at axis of the mother glass.

USEFUL CROSS-SECTION & ITS RADIUS

If the refractive index profile of a circular glass rod can be described by a parabolic function[2]:

$$N(r)=N_0(1-Ar^2/2) \qquad (5)$$

it has self-focusing characteristics.Comparing Eq.(3) with Eq.(5), it was found that the Eq.(5) is a satisfactly approximate form of Eq.(3) when r is a less value.Therefor,the ion-exchanged glass rod mentioned above is often refered to SELFOC rod. But with r increases, both difference becomes gradually larger, especialy when the first inflection point of Eq.(3) appears. In other words, the profile will deviate from parabola when r becomes large enough, the part of lens (where r is a large value)loses the optical characteristics of square-law medium. A a result, we think, there is an useful cross-section in SELFOC rod lens, difine it by

$$S_e=\pi R_e^2 \qquad (6)$$

where R_e is the distance from the axis to the first inflection point of Eq.(3), called useful radius. Let the second derivative of Eq.(3) equal zero:

$$(K\cdot r/a)\cdot J_0(K\cdot r/a)-J_1(K\cdot r/a)=0 \qquad (7)$$

Then the useful radius R_e can be obtained by micro-computer as

$$R_e=0.765a \qquad (8)$$

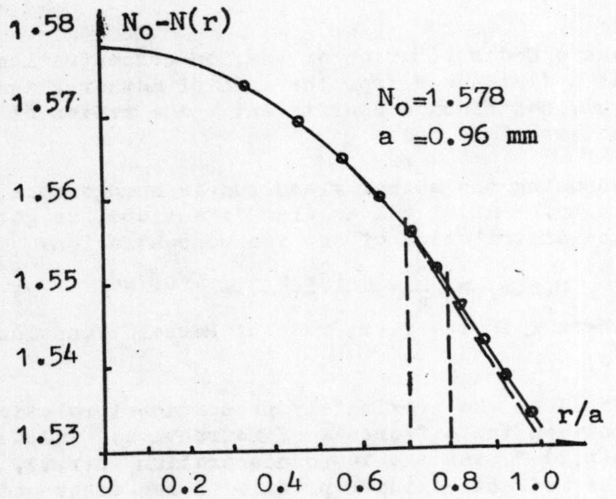

Fig.1 Refractive index profile of a domestic SELFOC rod

Fig.1 shows a refractive index profile curve of a domestic SELFOC rod fabricated by ion-exchange method. It is not difficault to find an inflection point in the curve between r/a =0.7 and r/a=0.8.

FOURIER TRANSFORM OF SELFOC ROD LENS

Within the useful cross-section (where $r \leq 0.765a$), the SELFOC rod lens mentioned previously can be regared as a circular square-law medium, which refractive index profile is shown in Eq.(5). Therefor, the Kernel[3], Eq.(9), of a circular square-law medium can be quoted to discribe optical wave propagation in a SELFOC rod lens.

$$T(x,y,x';y';z)=\frac{\exp[jWNo/C+j(\xi^2+\eta^2)ctg(Z\sqrt{A})/2]}{j2\pi\sin(Z\sqrt{A})}$$

$$\cdot\exp[\frac{-j(\xi'\xi+\eta'\eta)}{\sin(Z\sqrt{A})}-\frac{jctg(\xi'^2+\eta'^2)}{2}]\cdot\alpha^{-2} \qquad (9)$$

where

$$\xi=(W^2N_0{}^2A/C^2)\cdot x \qquad \eta=(W^2N_0{}^2A/C)\cdot y$$

$$\xi\xi'=2\pi N_0A\cdot xx'/\lambda \qquad \eta\eta'=2\pi N_0A\cdot yy'/\lambda$$

$$\alpha^2=(2\pi N_0\sqrt{A}/\lambda)^{-1}$$

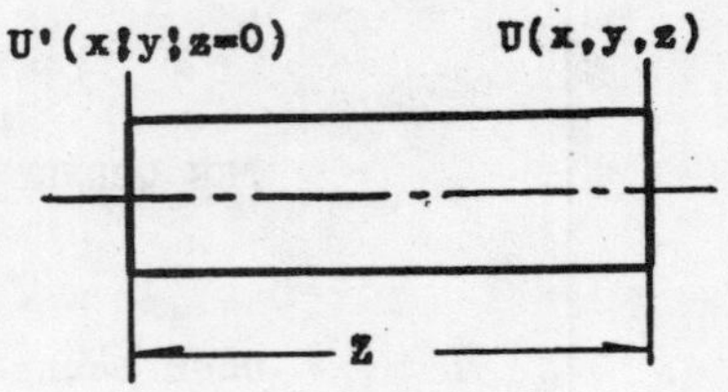

Fig. 2

If U'(x';y';z=0) represents the light wave field pattern in the front surface of a SELFOC rod lens, shown in Fig.2, then the field distribution in the back surface would be given by

$$U(x,y,z)=\iint_{-\infty}^{+\infty}T(x';y';x,y,z)U'(x';y';z=0)dx'dy' \qquad (10)$$

To simplfy Eq.(9), set a constance $M=\sin^2(Z\sqrt{A})$ for a SELFOC rod lens and quote its focus $F=\frac{1}{N_0\sqrt{A}\cdot\sin(Z\sqrt{A})}$, Eq.(9) can be written as

$$T(x,y,x';y';z)=\frac{e^{j\varphi(x,y,z)}}{j\lambda F_sM}\exp[\frac{-j2\pi}{\lambda F_sM}(xx'+yy')]$$

$$\cdot\exp[\frac{j2\pi\cos(Z\sqrt{A})}{\lambda F_sM}(x'^2+y'^2)] \qquad (11)$$

where

$$\varphi(x,y,z)=\frac{2\pi N_0}{\lambda}+\frac{2\pi N_0\sqrt{A}}{2\lambda}ctg(Z\sqrt{A})\cdot(x^2+y^2)$$

In fact, people are insteristed in the distribution of light intensity. Because $e^{j\varphi(x,y,z)}$ is only a shape factor in Eq.(11) that is unrelational to the integral in Eq.(10) and the light intensity distribution, it can be ignored here. Under "incoherent image formation", an optical system of image formation is liner to light intensity[4], besides, prectically the length of SELFOC rod lens is slightly less than (2n+1)P/4 (n=1,2, 3,···, $P=2\pi/\sqrt{A}$). Based on these, the factor $(x'^2+y'^2)$ needn't be taken into consideration here. However, the Kernel of SELFOC rod lens can be approximately written as

$$T(x';y';x,y,z)=\frac{1}{j\lambda F_sM}\exp[\frac{-j2\pi}{\lambda F_sM}(xx'+yy')] \qquad (12)$$

then Eq.(10) can be re-written as

$$U(x,y)=e^{j\varnothing(x,y,z)}\cdot\frac{1}{j\lambda F_sM}\cdot\mathcal{F}\left[U'(\frac{x'}{\lambda F_sM},\frac{y'}{\lambda F_sM})\right] \qquad (13)$$

Eq.(13) shows the field distribution of light wave in the back surface of a SELFOC rod lens is a product of a phase factor and the Fourier transform of the field pattern in its front surface.

EXPERIMENT

Fig.3 shows a typical optical system formed with two SELFOC rod lenses in optical fiber components. Lens A is used to collimate light from Fiber a, Lens B to focus light beam into Fiber b. Because of the uncompleteness of refractive index profile of SELFOC rod lens,it is impossible the light beam coming out from Lens A be completly parallel. The beam has a minium scatter angle θ_{min}. In addition Lens B has an useful cross-section, the gap d between both lenses will be limited within a certain range. In order to get as more as possible light power at the end of Fiber b,

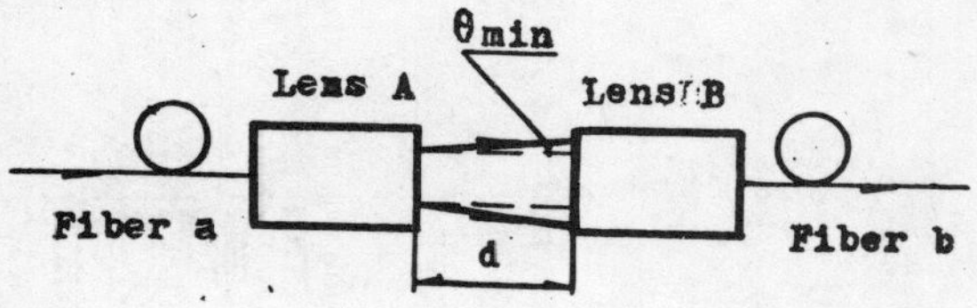

Fig.3 Typicl optical system in optical fiber components

the gap d must be estimated according to Eq.(8) and the minium scatter angle θ_{min}. Table 1 gives the data of our experiment with two domestic SELFOC rod lenses, which radius a=1 mm; useful radius R_e=0.765 mm.

Table 1 Experiment data

d mm	P_{in} dBm	P_{out} dBm	system loss dB
1.0	13.2	12.3	0.9
6.0	13.2	12.3	0.9
15.0	13.2	12.3	0.9
20.0	13.2	11.7	1.5
30.0	13.2	9.8	3.4

It is obvious that the gap d should be less than 20 mm, and the system loss is relatively less.

CONCLUSION

The useful cross-section of SELFOC rod lens and Fig.(8) will be a practical reference to the users of SELFOC rod lens. The Fourier transform of SELFOC rod lens discussed here is merely a theoretical proximation, which may give a clue to the use of SELFOC rod lens in optical imformation processing. But it must be pointed out that the Fourier trans-form should need further study.

REFERENCES

1. M.S.Sodha & A.K.Ghatak, INHOMOGENEOUS OPTICAL WAVEGUIDES, 1977, 239
2. James.D.Ress, Appl.Opt.Vol.21, No.6 1982, 1009
3. M.S.Sodha & A.K.Ghatak, INHOMOGENEOUS OPTICAL WAVEGUIDES, 1977, 85
4. J.W.Goodman, INTRODUCTION TO FOURIER OPTICS, Mc Graw-Hill, 1968

LOW-LOSS WAVELENGTH DIVISION MULTI/DEMULTIPLEXER AT THREE LONG WAVELENGTHS

Ye Ai Lun, Liang You Jian and Yang Tian Sheng
Institute of Fiber Optic Technology, Shanghai
Jiao Tong University, Shanghai, P. R. C.

Recently, wavelength division multiplexing (WDM) technology has become an interesting subject and has found a practical application in fiber optic communication systems. A variety of multi/demultiplexer products has been developed. However, most of available WDM devices with 3 channels or over 3 channels operates in a hybrid-way, this means a part of channels at a short wavelengths and others at a long wavelengths, in general. The hybrid type device results in an unbalance of loss budget in WDM communication systems because the loss of optical fiberat short wavelengths is larger than that one at long wavelengths in 1-2 dB/km, so that WDM communication systems are limited.

This paper presents an all long-wavelength 3-channel multi/demultiplexer having the advantages of low loss, wide passband and low crosstalk. The operating wavelengths are 1.15, 1.3 and 1.5 µm. The insertion loss of multiplexer and demultiplexer developed are less than 2dB and 2.5 dB, respectively. The far-end crosstalk of the demultiplexer is above 35 dB and the passband is more than 80 nm. In Fig. 1, 2 and 3, we show transmission spectral characteristics of the demultiplexer which are measured by using an optical spectrum analyzer.

The WDM devices we report have a conventional structure as shown in Fig. 4. But, in order to get a high-quality characteristics of the devices we pay a great attention to a designing of dielectric thin film filters and to a carefully handling of other key details in the structures. We briefly discuss the important points of the practice in a realizing high quality and reproducibility of the devices in the following:

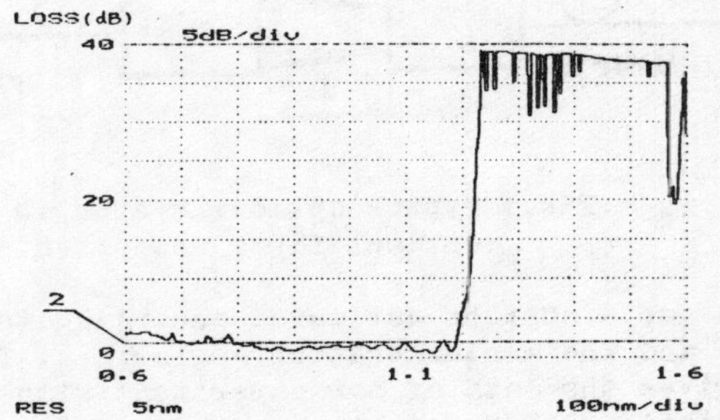

Fig. 1 Spectral characteristics of 1.15 µm channel of demultiplexer

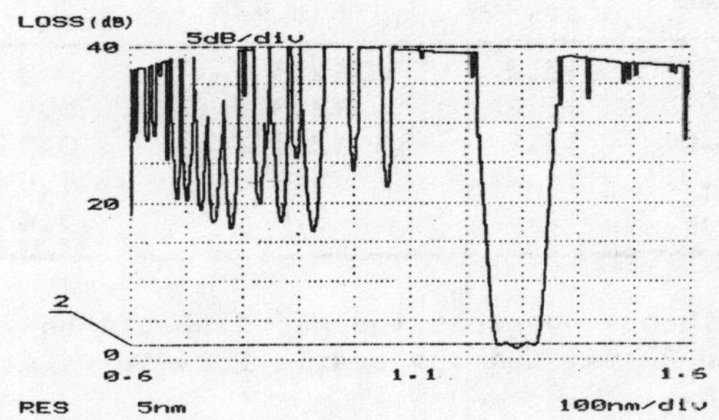

Fig. 2 Spectral characteristics of 1.3 µm channel of demultiplexer

1. By developing a dielectric thin film used as an optic filter with passband loss less than 0.6

dB, rejection-band attenuation more than 30 dB and passband more than 80 nm, one can reduce both insertion loss and cross-talk, and the devices are endbowed with the compatibility for WDM communication systems with LED or LD as light source.

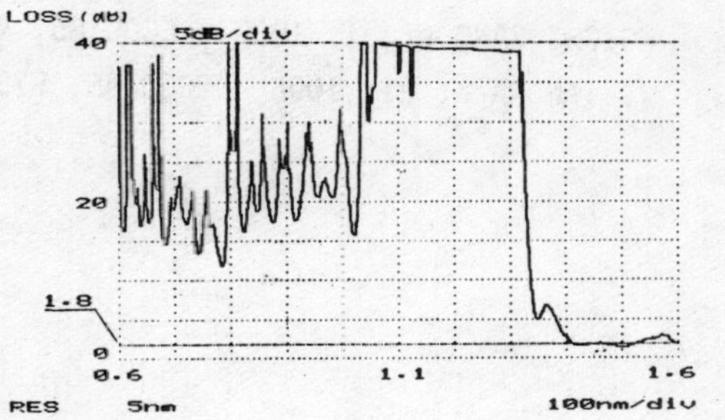

Fig. 3 Spectral characteristics of 1.5 μm channel of demultiplexer

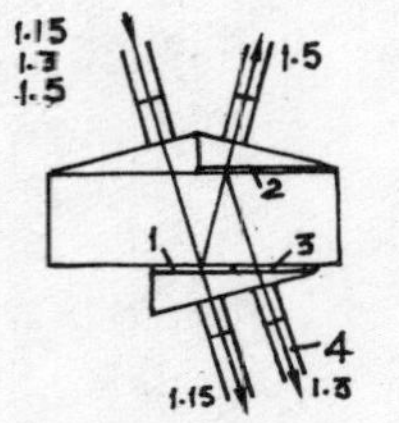

Fig. 4 1-optically filtering film of 1.15 μm short wavelength pass;
2-optically filtering film of 1.5 μm long wavelength pass;
3-optically filtering film of 1.3 μm passband;
4-fiber optic collimator.

2. There is a fact that the separation of central wavelengths on the passband in 3-channel WDM devices is too small, for example, the separation of central wavelengths for 1.15 μm and 1.3 μm merely is 150 nm, it requires a thin film filter with passband width more than 100 nm and steep stopband edge. The difficulty may occur in fabrication of above-mentioned filters. In order to realize a steep edges of the stopband, we have adopted a measures for a increase of the repeating symmetric period number in the film, but meanwhile the ripples of the passband tend towards a severeness. Obviously, the ripples increase insertion loss and effect astability of spectral characteristics of the WDM devices when the central wavelength of a light source drifts. Thus, we must adopt a effective measures, for example, the compensating method for rejection of the passband ripples. A spectral characteristics of the fabricated filters are shown in Fig. 5,6 and 7.

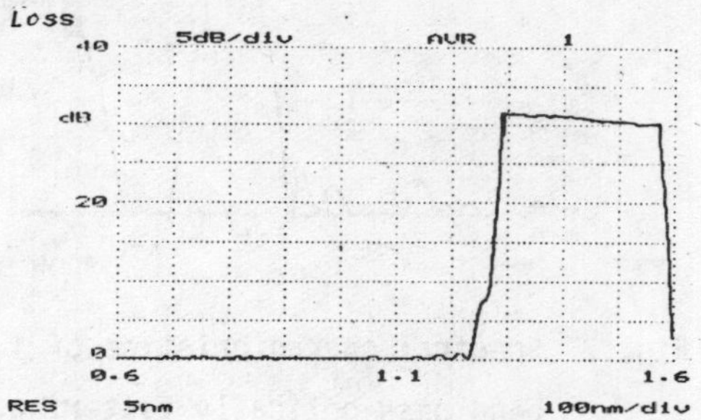

Fig. 5 Spectral characteristics of 1.15 μm short wavelength pass optically filtering film

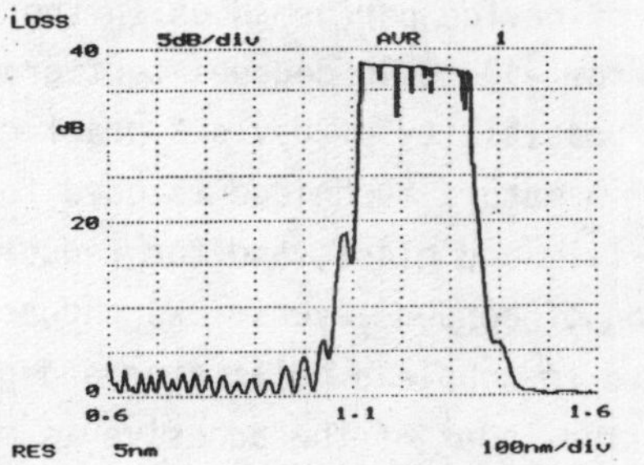

Fig. 6 Spectral characteristics of 1.5 μm long wavelength pass optically filtering film

3. Development of the fiber optic collimators with low loss at collimating distance of 15-30 mm is necessary for WDM devices. In general, a collimator which consists of a optical fiber with core diameter of 50 μm and a Selfoc rod lens has a large loss, while its collimating distance is larger than 20 mm. As mentioned as above, the collimating distance from the sum-end to the split-end may be 30 mm, nearly. In order to make the collimator a low loss one, we firstly prepare the sum-end collimator by choosing a appropriate length Selfoc,then make the

split-end collimator with suitable lengths Selfoc in accordance with variable collimating distances. We have reached a coupling loss less than 0.6 to 0.8 dB at wavelength of 1.15 to 1.5 μm and the collimating distance of 15 to 30 mm.

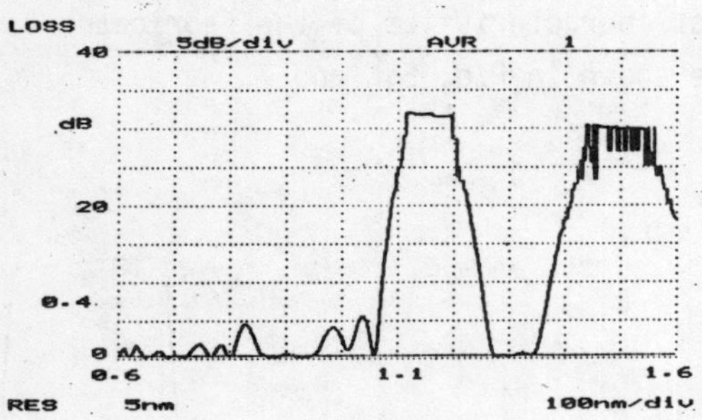

Fig. 7 Spectral characteristics of 1.3 μm band pass optically filtering film

4. Structure design of main optic path and optic bonding. The principle of main optic path is to minimize incidence angle to reduce the polarization effect and to ensure the stability of device performances in the temperature range from -10 to 40 degrees centigrade to provide a possibility of devices practical applications. A mature technique is used for bonding of the film and glass, and for reducing of absorption of bonded layers. We choose the epoxy adhesive to couple a collimator and glass body. Refractive index of the adhesive is matched to one of the collimator optics and the glass, so that Fresnel reflectance loss can be minimized and the characteristics of main optic path can be stable in a great extent.

5. Development of a set of apparatus for collimating, positioning and adjusting. We have worked out a set of apparaturs, including the collimating jig, special positioner and adjustor, it is helpful for a fabrication of high-quality WDM devices.

In conclusion, we again emphasize that due to low loss of optical fiber operating at long wavelengths of 1.15-1.50 μm 3-channel WDM devices we develop can be successful for use in a long distance fiber optic digital communication systems or analog TV signal transmission systems.

REFERENCES

1. HIDEKI ISHIO et al. JOURNAL OF LIGHTWAVE TECHNOLOGY, VOL. LT-2, NO. 4, 1984, 448.
2. KOICHI SANO et al. IOOC _ ECOC'85, 449.
3. Y. TAMURA et al. IOOC _ ECOC'85, 579.

A NEW TAPE OF OPTICAL CIRCULATOR

Wu Guofeng
The 34th Research Institute of Mechanical &
Electronic Ministry P.O. Box 5 Guilin China

ABSTRACT

A new tape of optical circulator by using of the birefringence effect of calcite crystal is decribed. By adopting of Selfoc. rod lens and monomode fiber, the insertion loss of the device is less than 1.5 dB, and the extinction ratio is more than 25 dB at 1.3 μm wavelength.

INTRODUCTION

The optical circulator is suitable for biredirection transmition, high bit speed optical communication, coherent optical communication, interferometry, fiber sensor, etc. At present, the insertion loss of circulator which uses Glan-Taylor prosms is hgih because of that only one polarization beam is used. In this paper, the two polarization beam of the calcite which is polished into special shape can be used effectually, so its insertion loss is low.

PRINCIPLE

As shown in Fig. 1, the natural beam cast perpendicularly on the surface of calcite, the optical axis of which is parallel AB cross-section. When the natural beam transmit to the air interval A'A" (thick 20μm), the odinary beam is reflected completely into calcite 5; the extraordinary beam transmit directly into calcite 3; then transmit into YIG Faraday rotator 1. In the crystal 5, ordinary beam is reflected completely at the AK' interface into YIG Faraday rotator 1'. Ordinary and extraordinary beam are rotated at 45° by YIG Faraday rotator, in the same direction respectively, then transmit into quartz natural rotatory compensator 2, to be rotated at 45° reversely respectively. Now, the direction of polarization of ordinary and extraordinary beam exchange each other. Ordinary and extraordinary beam of before rotation become into extraordinary and ordinary beam of after roatation, respectively. The two polarization beams transmit into the second calcite (it has the same shape with the first one, but the position is rotated at 180°). The extraordinary beam of after rotation transmit directly into position (2); in the calcite 4 and 6, ordinary beam of after rotation is reflected completely at the SH' and S'S" interface respectively, then transmit into position (2). The beam routes are not inversely, in other words, the beam cannot go through from position (2) to position (1). The whole beam routes are as following:

(1) The routes that beam can go through:

(1)—(2), (2)—(3), (3)—(4), (4)—(1);

(2) The routes that beam cannot go through:

(2)—(1), (3)—(2), (4)—(3), (1)—(4), (3)—(1), (1)—(3), (2)—(4), (4)—(2).

So the beam routes have the characters of optical circulator.

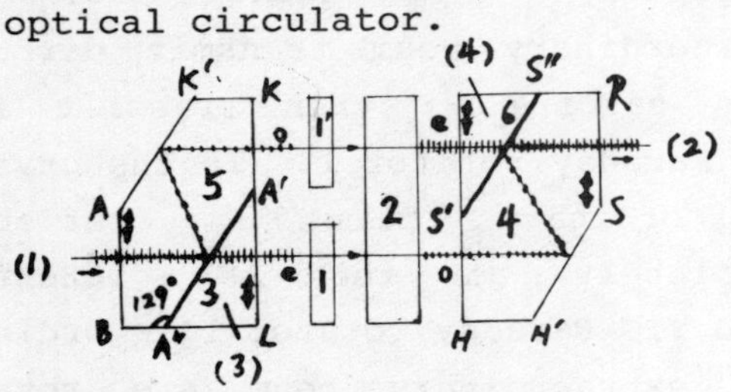

Fig. 1 The principle of circulator
1'1: YIG 45° Faraday rotator;
2: quartz 45° natural rotatory compensator;
3,4,5,6: Calcite;
: ordinary beam;
: extraordinary beam.

SELECTION OF MATERIAL

(A) In Fig.1, __BA"A'=__RS"S'=129°. AK'//A"A', H'S//S'S"", __A"A'L=__S"S'G=39°. At the AK', A"A', S'S"" and H'S interface, the incident angles of extraordinary and ordinary beam are 39°, respectively. For the calcite material, n_o=1.63767, n_e=1.47831, at 1.3 μm wavelength. The critical angles of ordinary and extraordinary beam which is reflected completely are 37°24' and 43°6', respectively. So ordinary beam can be reflected completely, but extraordinary beam cannot be. Actually, there is about 3% extraordinary beam to be reflected. The loss of optical circulator by using of calcite is about 0.2 dB, and there is no absorbing loss for calcite at 1.3 μm wavelength.

(B) We select the crystal of left rotatory quartz for 45° rotatory compensator, its thick is 11 mm, and its optical axis is perpendicular to its cross-section.

(C) We use Sm-Co(Sm_2Co_{17}) perpetual magnet to produce saturated magnetic field. Its size is 20 X 10 X 3, there are two holes on it, the diameters of the holes are 2 mm, respectively. Two magnets which have the same size are used.

(D) YIG single crystal has low absorbance (Fig.2), high Verdet constan (Table 1), and temperature characters stabilization (Fig.3). It can be suit for 10° - 80°C range. Its parameters is listed on Table 1. The thick of YIG slice is 2.1 mm at 1.3 μm wavelength, its absorbing loss is 0.273 dB. The crystal axis (111) of YIG is perpendicular to its cross-sections.

Table 1 Parameters of YIG crystal

Parameter λ	1.3 μm	1.55 μm
θ (degree·mm⁻¹)	21.5	14.9
absorbing loss (dB·mm⁻¹)	0.13	0.11
θ/α' (degree)	717	596
H_{min} (Oe)	1300	
dθ/dT (degree·mm⁻¹·℃⁻¹)	-1.3×10^{-2}	

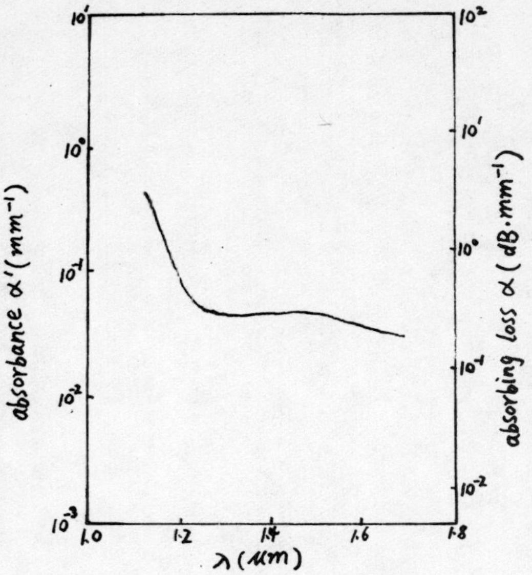

Fig. 2 The absorbance and absobing loss of YIG in 1.1—1.7 μm wavelength

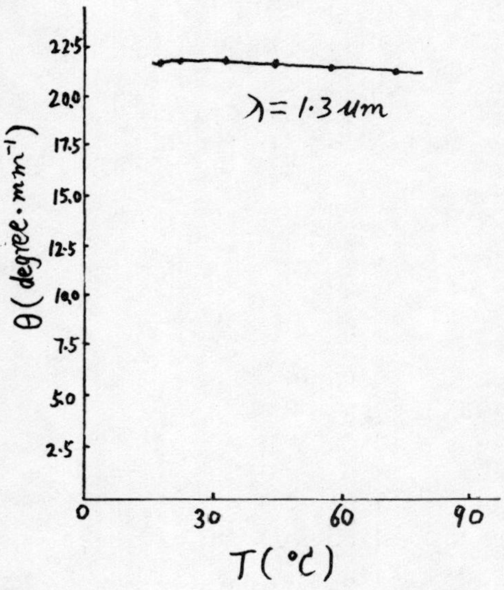

Fig. 3 Relation of Faraday rotation angle with temperature, and H=90 O_e.

ESTIMATION OF ERROR

In practical production, since the restriction of precision by metal procession and beam collimation, the beam that come from optical fiber and is collimated by Selfoc. rod lens is not completely perpendicularly to cast on the surface of calcite (there is a angle α, as shown in Fig. 4).

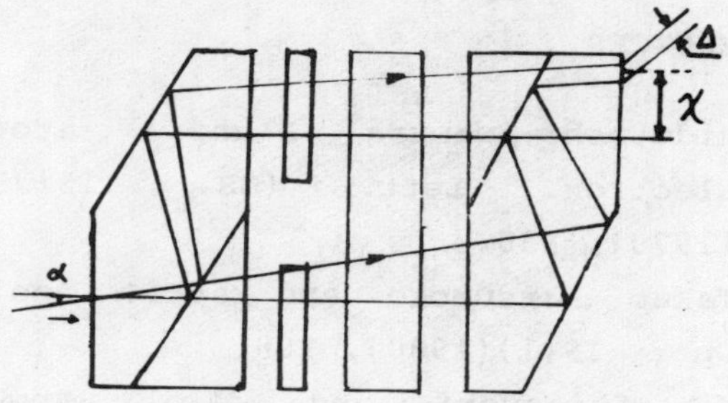

Fig. 4 Estimation of error

When the beam come through out of the optical circulator, the beam become into a beam bundle of wide Δ, and its centre deviate distance X from the ideal state.

If α=15', such Δ =50 μm, X=160 μm at 1.3 μm wavelength.

In this paper, the beam wide that cast on the cross-section of calcite is 1.4 mm, when the tilt angle α is 15', there is 0.4 dB tilt loss.

Besides, it is should be consider that the position tolerance of calcite, intervals of components. The components are plated films on their surfaces to reduce the power loss of reflection. It is very important to adjust the beam road in a certain degree.

CONCLUSION

In this paper, we give out of a new tape of optical circulator. When coupling by Selfoc. rod lens and monomode fiber, the insertion loss is less than 1.5 dB, and the extinction ratio is more than 25 dB at 1.3 μm wavelength. The whole size is 39 X 12 X 53. The optical circulator has widely application in fiber optical communication, optical fiber sensor, interferometry, etc.

REFERENCES

1. Hidetoshi-Iwamura and etc., Electron. Lett. (GB), 15(25) (1979), 830.
2. Takao Matsumoto and etc., Appl. Opt., 19(1)(1980), 108.
3. M. Shirasaki and etc., Appl. Opt., 20 (15)(1981), 2683.
4. Booth R.C. et al., 8th European Conference on optical communication, (1982), 238.
5. Atsushi Shibukawa et al., Appl. Opt., 18(21)(1979), 3700.

A NEW PRACTICAL OPTICAL FIBER SENSOR
FOR DETECTION OF TWO LAYER INTERFACES

Guan Tie Liang, Zhou Chen, Zou Jing, Li Fei and Liu Tian Xiong
Tianjin Electronic Materials Research Institute. P.O. Box 412
Tianjin, China

ABSTRACT

A novel optical fiber sensor has been successfully applied to detect two layer interfaces formed by air, non-absorption liquid and absorption liquid. The device consists of sensor head, optical fiber and measuring instrumentation. The theoretical analysis based on optical transfer matrix was carried out and compared with experimental results.A measuring accuracy of ±1mm was obtained.

INTRODUCTION

In the technological process of manufacturing colour film, it is necessary to wash the emulsion with distilled water. After mixing water into emulsion and precipitation of the emulsion, it is needed to draw off the water and remain the emulsion. Therefore, the position of the tow interfaces formed by air/water and water/emulsion should be detected exactly. All above procedure must be done in a darkoom.

In recent years there are new developments of optical fiber sensor for measuring liquid level.[1,2] But no one now is available to do present work competently.We designed an optical fiber sensor which can be used to measure the tow interfaces continually with the same sensor head.

BASIC CONSIDERATION

A. The operation principles of this sensor is based on comparison of absorption and refraction of different liquid.
B. In order to aviode exposure of the emulsion, the LED at 0.85µm wavalength is used as the light source.
C. Considering the higher viscosity of the emulsion, a wider sensed space is required. So the expanded beam technique[3] was be chosen.

EXPERIMENTAL ARRANGEMENT

The sensor proposed in this paper is shown schematically in Fig.1 (a). Light from LED is coupled into the multimode input fiber F1, and transferred to the sensor head. The end of output fiber F3 is positioned in the focal plane of the lens which collimate the emanant light to a nearly parallel beam. Then the beam passing through the sensed volume in which air , water or emulsion maybe filled with is reflected back to F3. The fiber F3 is served to transfer all optical signals to and from the sensor head. the reflected light is measured through a fiber coupler by a photodiode and followed by an amplifier.

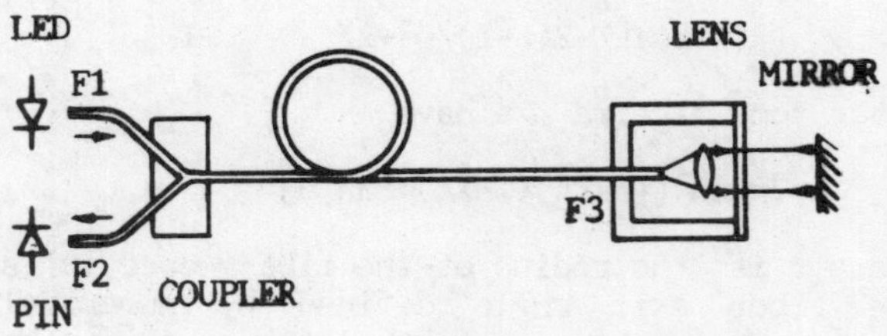

(a)

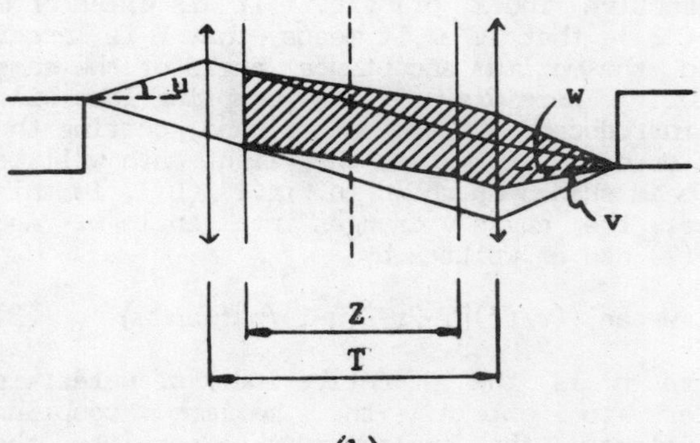

(b)

Fig. 1 Scheme of liquid level sensor
(a) arrangement (b) light path

PRINCIPLES

The main aim of the design of this new optical fiber sensor is to find the two interfaces successively using the same sensor. It is easy to detect the water/ emulsion interface by comparing absorption of the two medium. Because the emulsion is composed of AgBr↓, KNO_3 and gelatin which exhibit strong absorption at near infrared band. In order to find the air/water interface using the same sensor , we noted that when you put the sensor head from air into water , the output power from fiber F2 is increased obviously. The reason is as follows. Firstly,it is well known that the absorption of water at 0.85μm wavelength can be neglected.[4] Secondly, the Fresnel reflection loss on the surface of window attached to sensor head was decreased greatly, because the refractive index of the window is very close to that of the water. Finally , in order to make it clearly, the image of fiber F3 and the lens are placed on the right side of the dotted line at which the mirror is located as shown in Fig.1 (b). Let the distance T between the two lenses be larger than 2f (f is the focal length). When the sensed volume Z was filled with air, the light path was indicated as an unshaded beam. The optical transfer matrix of this system is given by

$$\begin{vmatrix} -1 & 0 \\ (1/f^2)[T-Z(n-1)/n]-2/f & -1 \end{vmatrix} \tag{1}$$

after some algebra, we have

$$V=\tan^{-1}[(r/f^2)(T-2f)+\tan u] \tag{2}$$

Where r is the radius of the fiber core, u is the fiber exit angle defined by numerical aperture (N. A.) of the fiber. V is the convergence angle given by lens. And n=1 is the rafractive index of air. It is clear from Eq.(2) that v>u. It means that V is larger than the maximum acceptance angle of the same fiber. Therefore certain coupling loss will be introduced. On the other hand, putting the sensor head into water, the light path will be a beam shaded as shown in Fig.1 (b). In this case, the angle v changes into angle w. And Eq.(2) can be written by

$$w=\tan^{-1}\{(r/f^2)[T-2f-Z(n-1)/n]+\tan u\} \tag{3}$$

Where n is the refractive index of water. In order to obtain the maximum coupling efficiency, the angle w must be equal to the angle u. Comparing Eq.(2) and Eq.(3), we have

$$T=2f+Z(n-1)/n \tag{4}$$

Choosing parameters T , f and Z to satisfy Eq. (4), it is possible to obtain the maximum coupling.

As mentioned above, when the sensed volume is filled with water instead of air, the light output intensity passing through the sensor head will be increased enough to detect the air /water interface.

RESULT AND DISCUSSION

The intensity ratio R between the water and air filled in the sensed volume (i.e. The intensity ratio R between the two cases where the sensed volume is filled with water and air respectively) depends on some parameters. Experiments show that when T= 11mm , Z=8mm , N.A.=0.25 and r=40μm , we have R= 1.4. It is big enough to distinguish between higher and lower intensity level, because ratio R is not changed with the fluctuation of the light intensity in transmission fiber. in fact, R increases as the r increases and the f decreases. So selsction of larger core fiber and shorter focal length will contribute to increment of ratio R i.e. improvement of reliability.

In summary, the method presented in this paper is applicable for detection of tow interfaces formed by air, non-absorption liquid and absorption liquid.

REFERENCES

1. K.Spenner et al., Proceedings of the First International Conference on Optical Fiber Sensor, (1983), 96.
2. M.K.Ravishanker, Appl. Opt., 25(1986),480.
3. A.Nicia, Appl. Opt., 20(1981), 3136.
4. J. P.Dakin et al.,Proceedings of the First International Conference on Optical Fiber Sensor, (1983), 91.

THE INVESTIGATION OF SENSITIVITY FOR
SINGLE MODE FIBER REFRACTOMETRIC SENSOR

Shen Pengnian, Xue Zheng
Institute National D'optique, 369, rue Franquet
C.P. 9970, Sainte-Fay, Canada, GIV 4C5

Wang Dening, Wang Kefeng
Shanghai Institute of Metallurgy, Academia Sinica
865 Chang Ning Road, Shanghai 200050, China

ABSTRACT

In this paper, the sensitivity of single mode fiber refractometric sensor is investigated. Based on the attenuation theory of the polish exposed single mode optical fiber with the imersing medium, at the different polishing depths d, the effect of the difference between medium refractive index n_{31} and cladding material n_{21} on attenuation is analysed. We can find the sensitivity of sensor is higher when $n_{31}<n_{21}$, it is used in measuring the solution concentration, but $n_{31}>n_{21}$, the sensitivity of sensor is less. The dependences of temperature and the wavelength of optical source on sensitivity are also investigated. The results show that the sensor can be used as a temperature sensor and one can use different wavelengths to increase the temperature range. It is very important to develop and apply the sensors used in temperature or concentration measurement.

INTRODUCTION

The optical fiber refractometric sensor based on the attenuation of the optical power travelling through a section of optical fiber surrounded by refractive index changing medium is a distributed type sensor. Both multimode fiber and single mode fiber can be used. The single mode fiber approach has the advantage that it don't depend on the model power distribution in the fiber core.

With the increasingly use of single mode fiber in optical fiber communication system and sensor applications, now the single mode fiber approach is realistic both in economical and technical aspects.

In this paper, we will analyse the sensitivity of the single mode fiber refractometric sensor to the refractive index change of the surround medium and also the temperature response of the sensor. We will also analyse the dependence of this type of sensor on the wavelengths of the optical source used.

There are many techniques which can make one to get access to the fiber core of a single mode fiber. The polishing technique is the most popular one. Directly polishing a section of optical fiber and getting several mm of optical fiber core exposed are possible[1]. Or, one can use a half coupler block with a curved groove to adapt the single mode fiber with epoxy and then get the fiber polished[2].

ANALYTICAL PRINCIPLE OF SENSITIVITY

First of all, consider a polished section of single mode optical fiber with a length of Z mm and immersed in a homogeneous medium with refractive index of n_{31} (at λ_1) (Fig.1). In the figure, a is the radius of the fiber core, n_{11} is the effective refractive index of the core at the source wavelength λ_1, n_{21} is the cladding refractive index at source wavelength λ_1.

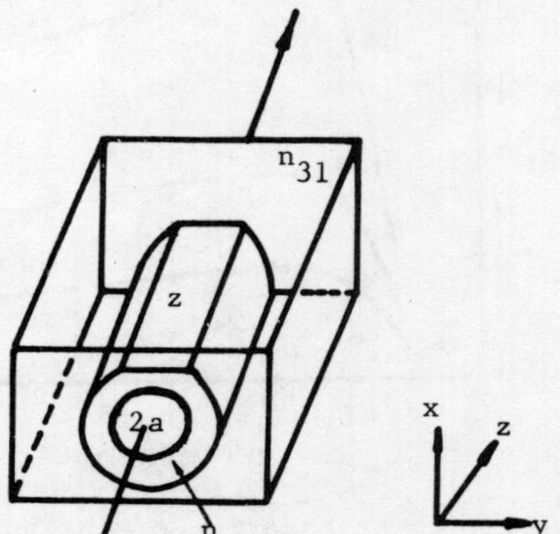

Fig. 1 A polished single mode fiber section immersed in a medium with refractive index of n_{31}.

According to the coupling theory between the fundamental mode of the single mode fiber and the LP mode of the medium which may be considered as a mode single[3,4], the power attenuation coefficient is:

$$r = \frac{4\pi\beta_0}{n_2^2 k_0^2} N_0^2 [1-(\omega/V_3)^2] \int_0^1 \sqrt{1-x^2} \cdot$$

$$\exp[-2\frac{d}{a}\sqrt{(V_3^2-\omega^2)x^2+\omega^2}]\,dx \qquad (1)$$

$$\omega = 2\pi \cdot \frac{a}{\lambda}\sqrt{n_{11}^2 - n_{21}^2} \qquad (1a)$$

$$V_3 = 2\pi \cdot \frac{a}{\lambda}\sqrt{n_{31}^2 - n_{21}^2} \qquad (1b)$$

The overall attenuation of the whole section of the single mode fiber exposed is then:

$$r_{TOT} = \int_o^z r\,dz \qquad (2)$$

And the loss in dB can be readily obtained from r_{TOT}. For the structures where the d value is not the same at every z position, as in the case of half coupler, one can use d(z) as a substitute for d.

RESULTS AND DISCUSSION

Fig.2 is the relationship of the attenuation coefficient of the polish exposed single mode optical fiber with the imersing medium refractive index Δ_{31} at different polishing depths.

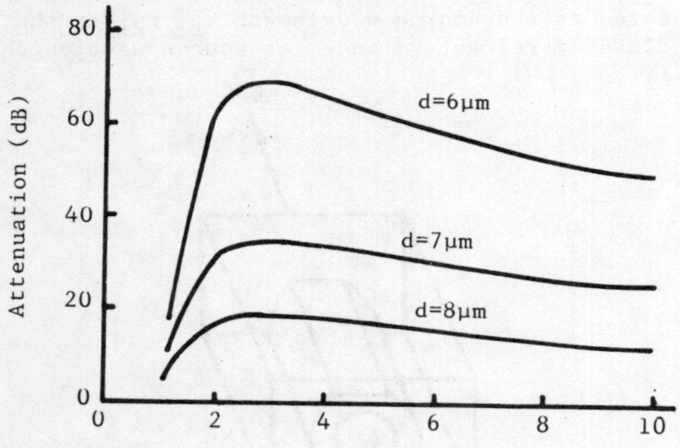

Fig. 2 Attenuation of the polish exposed single mode optical fiber
$a=4\mu m$ $n_{21}=1.444$
$\lambda_1=15000\text{Å}$ $n_e=n_{11}=1.446$

It is found that at $\Delta_{31}=0.002\sim0.004$, there is a peak for the attenuation coefficient and at right or the left side of the peak, the rise and the fall of the attenuation coefficient is not the same. So, we differenciate the r with respect to the index differences of the medium. The results is a new set of curve, the refractive index sensitivity curve of the polish exposed single mode fiber refractometric sensor (Fig.3).

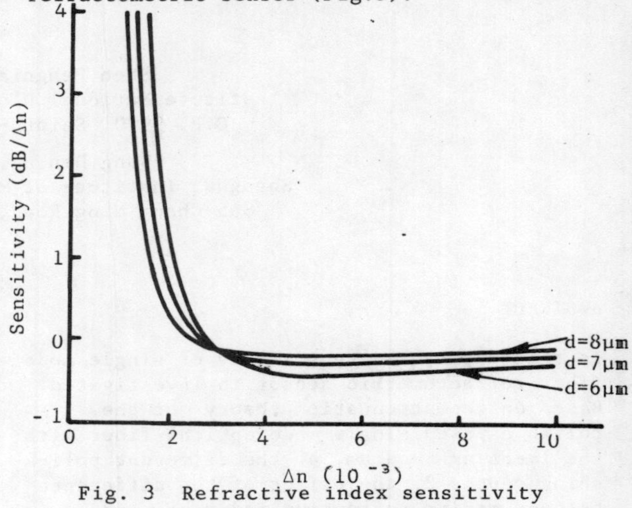

Fig. 3 Refractive index sensitivity of the single mode fiber refractometric sensor.

It is concluded that because this type of sensor is based on the coupling between the fundamental mode of the single mode fiber core and the LP mode of the imersing medium served as a mode sink, the loss will begin at some refractive index value higher than that of the cladding material (n_{21}). This type of sensor may conveniently used for medium whose refractive index is higher than that of SiO_2. Though these will be also loss when the medium whose refractive index is lower than n_{21} is used and there exist at the same time some sort of absorbing materials in the medium. But, in such cases, the sensitivity is rather low[5].

To use exposed core single mode fiber as refractometric sensor for medium with low refractive index (as some water solutions), one can use the curved fiber with very small radius of curvature R. For multimode optical fiber, this type of sensor is already demonstrated to be sensitive for water solution measurement[6], and for its single mode companion to realize, it is j[illegible] a technical problem.

Now, we analyse the temperature sensitivity of the sensor. When temperature changes, both the refractive index of the fiber and medium will change. If the temperature range is not great and the refractive index change of the medium is larger compared to that of the fiber, one can simply transform Fig.2 with the scale in abscissa changed from Δn to temperature. Fig.4 is the temperature and loss diagram calculated for two liquids whose refractive index change in the temperature span is 0.0001/C^{o}, but whose refractive index values are located at the left and right side of the peak of curves in Fig.2 respectively. All other parameters of the optical fiber keep invaried as the fiber in Fig.2.

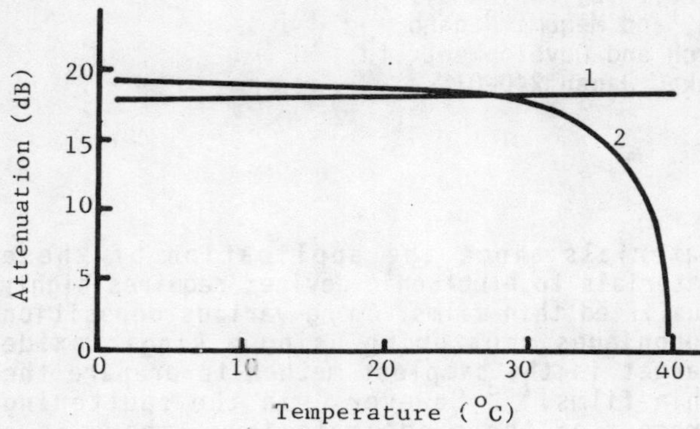

Fig. 4 Temperature dependence of the attenuation of two medium with same refractive index change coefficient to temperature but different room temperature refractive index.
1. $n_{31}(20^{o}C)=1.446$ ($\lambda=15000Å$)
2. $n_{31}(20^{o}C)=1.458$ ($\lambda=15000Å$)

Sensitivities are not same for this two liquids.

Now, we consider the influence of this type of sensor. When we change the wavelength of the source, the n_1, n_2, n_3 all change. We use the method in[7] to calculate the refractive index change of the silica at different wavelengths. Then we can compare the different wavelengths. The results are depicted in Fig.5 for the same fiber at different wavelengths. So, for a temperature sensor based on the refractive index change of the medium, one can use different wavelengths to increase the temperature range of the sensor. With some optimization and selection of the source wavelengths, one wavelength can be used as the carrier for communication, the other wavelength can be used to some physical parameter, such as refractive index and some refractive index dependent parameters: temperature and solution concentration.

With the advent of the wavelengths, tunable semiconductor laser and the wavelengths selectable, it is possible to use the wavelengths dependent properties of this type of structure to determine the refractive index dependent parameters.

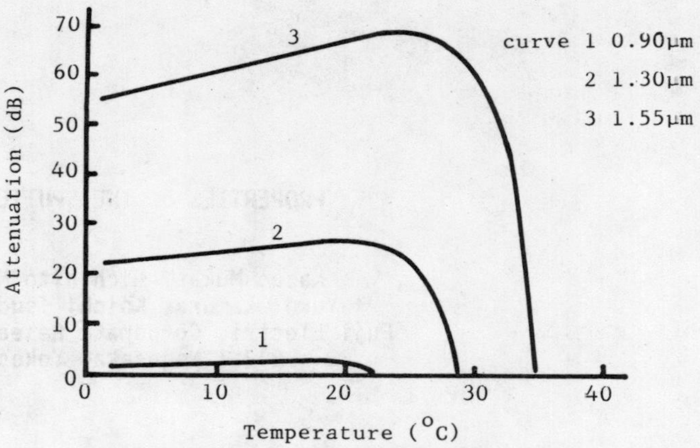

Fig. 5 Wavelength dependence of a refractive index change based temperature sensor.

The principle and analysis carried out in this paper would also applicable to the integrated optical single mode waveguide type distributed refractometric sensors.

REFERENCES

1. C.D. Hussey, and J.D. Minelly, "Optical fiber polishing with a motor-driven polishing wheel", Electron. Lett., 24, pp.805-807 (1988).
2. M.J.F. Digonnet and H.J. Shaw, "Analysis of tunable single mode optical fiber coupler". IEEE J. Quantum Electron., Vol.QE-18, pp.746-754, Apr. (1982).
3. A.W. Snyder and J.D. Love, Optical Waveguide Theory, London, V.K. Chapman and Hall, (1983).
4. O.G. Leminger and R. Zengerle, "Determination of single mode fiber coupler design parameters from loss measurements". IEEE J. Lightwave Technol. LT-3, 864-867 (1985).
5. L. Falco, H. Berthou, F. Cochet, B.Scheja, O. Parriaux, "Temperature sensor using single mode fiber evanescent field absorption", SPIE, Vol.586, Fiber optic sensors, pp.114-119 (1985).
6. T. Takao, Japanese J. Appl. Phys., 21(10) p.1509-1512 (1982).
7. J. Gowar "Optical communication systems" Prentice/Hall International, p.44, (1983).

PROPERTIES OF THE SPUTTERED YBCO THIN FILMS

Kazuo Mukae, Michihito Muroi, Yuji Koinuma,
Yuko Okamura, Koichi Tsuda, and Megumi Nagano
Fuji Electric Corporate Research and Development Ltd.
2-2-1 Nagasaka, Yokosuka, Japan 240-01

ABSTRACT

YBCO thin films were prepared by rf-planar magnetron sputtering with a single target on the MgO(100) substrates at elevated temperature. Various configurations of the substrates and the target were examined to investigate the effects of bombardment of high energy particles on YBCO films. The contents of Cu and Ba decreased when the position of the substrates were above the erosion area of the target(position-a) because of the resputtering phenomenon by bombardment of high energy particles. On the other hand, the films deposited at above the center of the target(position-b) had similar contents to that of the target. Higher Tc's were obtained at position-b than position-a.

INTRODUCTION

The discovery of the high temperature superconductors have inspired many investigators to synthesize thin films of these materials since the application of these materials to electronic devices requires highly qualified thin films. Among various deposition techniques sputtering using a single oxide target is the simplest method to prepare the thin films.[1-3] However, in the sputtering process of the oxide materials, there is a serious problem of the resputtering phenomenon which is caused by bombardment of high energy particles toward the deposited films.[4-5] This phenomenon generates not only the deviation of the composition from the target but also the crystal damage of the deposited films. Because of this problem high annealing temperature is required to obtain higher Tc. In this paper a method to reduce the resputtering phenomenon is discussed by comparing the effect of the configuration of the target and the substrate.
In addition a new technique is described in which a metal mask is used to overcome this problem.

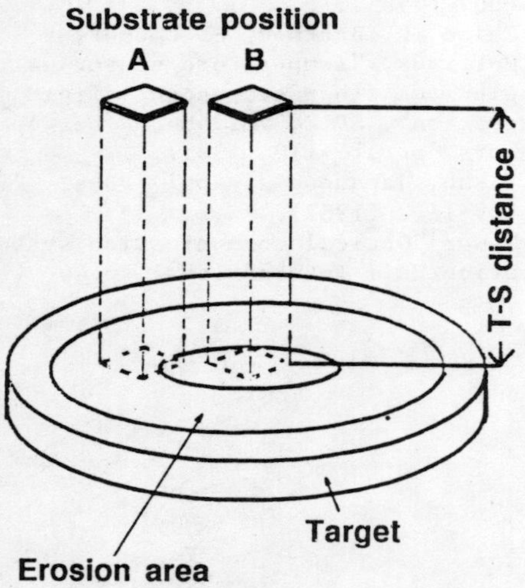

Fig.1 Position of the substrate.
A:above the erosion area
B:above the center of the target

EXPERIMENTAL

YBCO films are prepared using a rf-planar magnetron sputtering system. The targets were sintered YBCO discs with various compositions. The diameter of the targets was 150mm. MgO(100) crystals were used as substrates. The sputtering gas pressure and the distance between the target and the substrate(T-S distance) were varied 2-10Pa and 28-67mm respectively. Resistance versus temperature characteristics were measured using conventional dc four-probe method with gold electrodes evaporated at intervals of 4mm. Chemical composition was analyzed by Inductively Coupled Plasma Emission Spectroscopy(ICP).

RESULTS AND DISCUSSIONS

1. Deposition at position-A

At first the relationship between the sputtering conditions and the composition of the films was examined. The substrate position

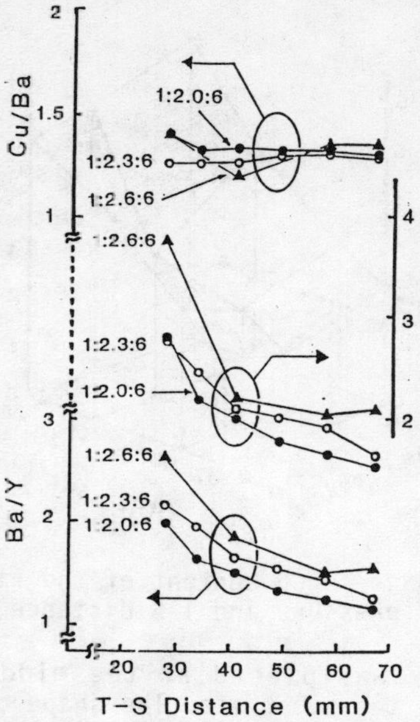

Fig.2 Change in film composition with T-S distance.

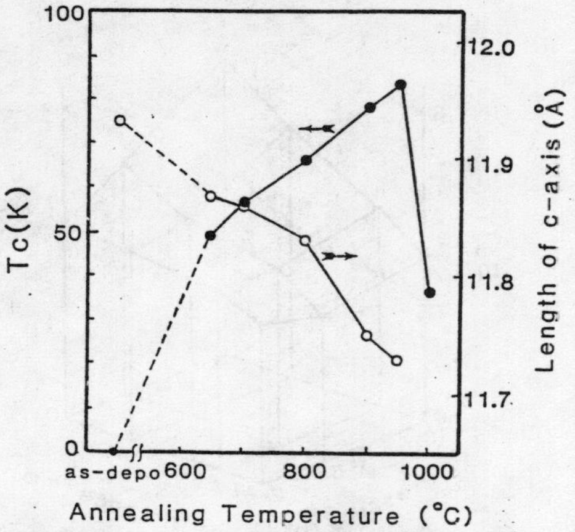

Fig.3 Changes in Tc and c_o of films deposited at position-A by annealing temperature.

was chosen above the erosion area of the target(position-A in Fig.1). Figure 2 shows the change in composition with T-S distance. The sputtering gas pressure and the substrate temperature(Ts) were 10Pa and 650°C, respectively. The composition is expressed in the atomic ratios of metals, that is, Ba/Y, Cu/Y, and Cu/Ba. It can be seen that contents of Ba and Cu decrease with the increase in T-S distance. This is explained by the preferential resputtering of Ba and Cu from the growing films. The deposition rate decreases with T-S distance due to the scattering of sputtered atoms by the sputtering gas. On the other hand, the number of oxygen ions or neutralized atoms arriving at the substrate seems to be less dependent on T-S distance than that of the other neutral atoms because they are always accelerated away from the target as long as they have negative charges. This means that the composition of the films is more altered by the preferential resputtering as T-S distance becomes longer. A steeper increase in the contents of Ba and Cu is observed when the T-S distance is shortened to 28mm. This is probably because the substrate enters the plasma region where an insulator tends to be charged negative due to the difference of the mobilities between electrons and positive ions. Even if some negative charges leak out through a metallic substrate holder, resultant negative bias of the substrate must have reduced the sputtering effect. It is noted that Cu/Y of the film is dependent on target composition in spite of the same Cu/Y ratio in each target and that Cu/Ba is almost constant at 1.3 throughout the samples. This fact suggests the strong attraction between Ba and Cu, which must have prevented excessive resputtering of Cu to keep Cu/Ba constant.

According to the X-ray diffraction(XRD), the as-deposited films with stoichiometric composition had the (123)-structure with perfect c-axis orientation. They showed, however, no superconducting transition above 20K. Figure 3 shows the change in critical temperature(Tc), as well as that in the lattice constant of c-axis(c_o), with annealing temperature. Annealing was done in flowing oxygen atmosphere for one hour followed by a cooling to room temperature with a rate of 5°C/min. It can be seen that as the annealing temperature is raised higher up to 950°C, Tc becomes higher with gradual shortening of c_o. These changes in Tc and c_o should be attributed not to the supply of oxygen but to the reduction of crystal defects[6] which have been created by the bombardment of high-energy particles because Tc and c_o depended only on the annealing temperature and they were not improved by the longer annealing.

2. Deposition at position-B

Next the substrate position just above the center of the target(position-B) was tried to reduce the bombardment of high-energy particles. Figure 4 and Fig.5 show the changes in Ba/Y and Cu/Y when 1-2-6-target was used. For comparison the composition change of the films deposited at position-A is also indicated by the dotted line in these figures. The films deposited at position-B seemed to have less different composition from the target than that at position-A, and showed no rapid decrease of Cu/Y with the decrease of gas pressure. These facts suggests that the bombardment of the films by high-energy particles and resultant preferential resputtering of Ba and Cu were reduced by locating the substrate just above the center of the target.

Nearly stoichiometric films were obtained with T-S distance of 41mm and gas pressure of 5Pa. These films had smooth surfaces and rather high Tc of 80K without annealing. XRD measurement revealed that they had the crystal structure of

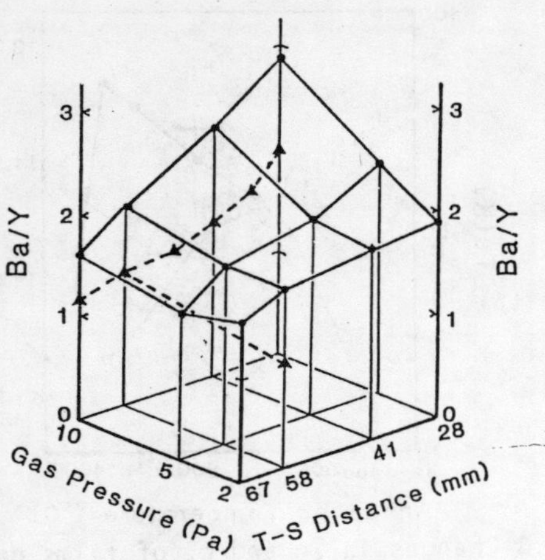

Fig.4 Change in Ba content of the films with gas pressure and T-S distance.

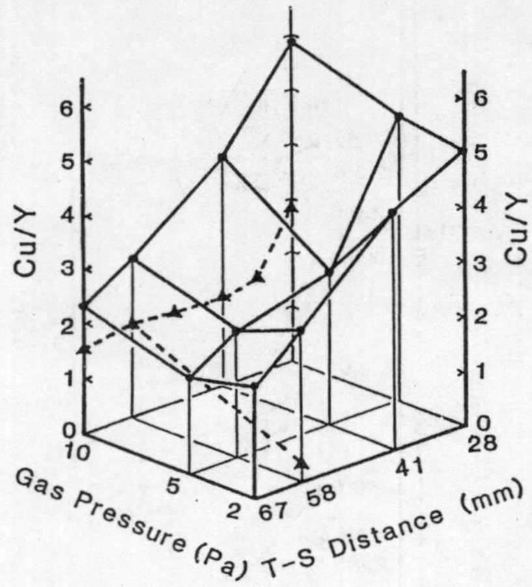

Fig.5 Change in Cu content of the films with gas pressure and T-S distance.

the (123)-phase with perfect c-axis orientation and shorter c_o of 11.80Å. Figure 6 shows the rocking curves of the (005)-diffraction lines of this film(film-B) and the as-grown film deposited at position-A(film-A). Smaller full width at half maximum(FWHM) for film-B indicates that it has more perfect c-axis orientation than film-A.

3. Deposition with a mask

In the previous section, it was shown that the deposition above the center of the target improved the crystallinity of the films. However, the decrease of Ba and Cu and the enlargement of the lattice constant c_o at the lower sputtering gas pressure were not completely suppressed. This is because the effect of high-energy particles still remain at the lower gas pressure. In order to prevent from such bombardment at lower gas pressure, a metal mask was placed at the middle of the target and the substrate. The shape of the mask was a disc with the diameter of 60mm which would intercept the direct bombardment of high-energy particles. The mask was connected to the chamber with a metal wire to be grounded. Figure 7 shows the variation of the deposition rates with the sputtering gas. The deposition without the mask was also indicated in Fig.7. The deposition rate without the mask increased with increasing gas pressure below 1.0Pa due to the increase in Ar ions. The deposition above the gas pressure of 1.0Pa decreased with the gas pressure because of the scattering of sputtered particles by the gas. In this region the deposition should occur mainly by the diffusion process. On the other hand, when a mask was used, films were formed only above the gas pressure of 0.5Pa. In this case the high-energy particles were intercepted by the mask resulting in the reduction of the bombardment with the growing film. XRD revealed that the all films deposited with the gas pressure between 2 and 10Pa had the (123)-structure with c-axis orientation whether

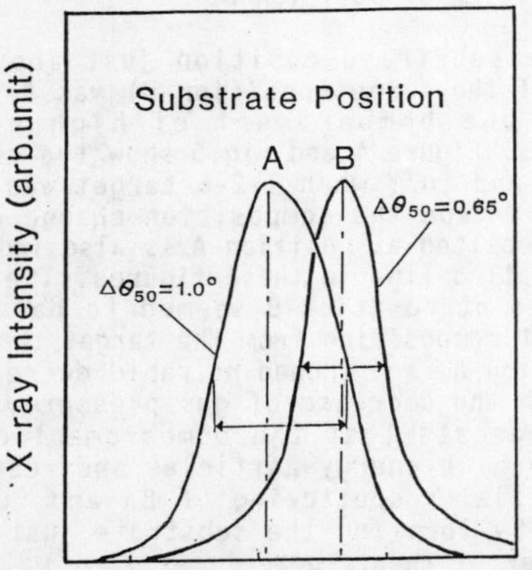

Fig.6 Rocking curves of films deposited at position-A and position-B.

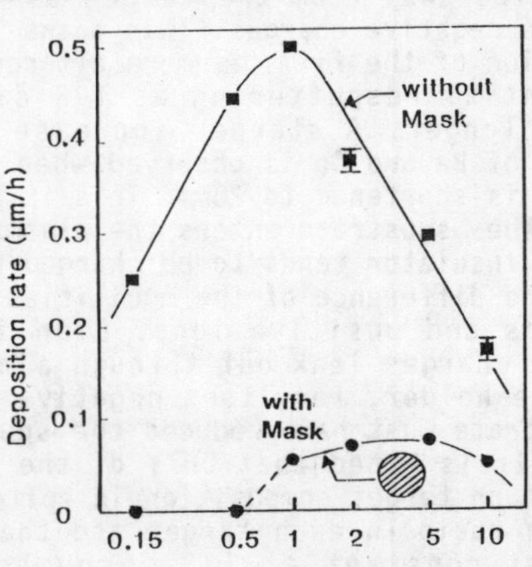

Fig.7 Gas pressure dependence of the deposition rate with and without a mask.

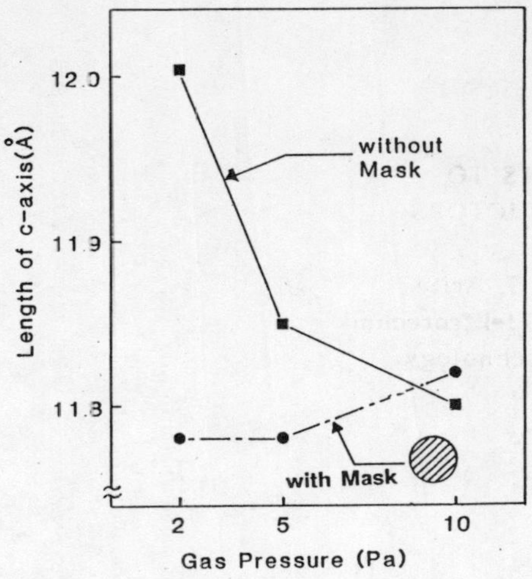

Fig.8 Dependence of c_0 on gas pressure.

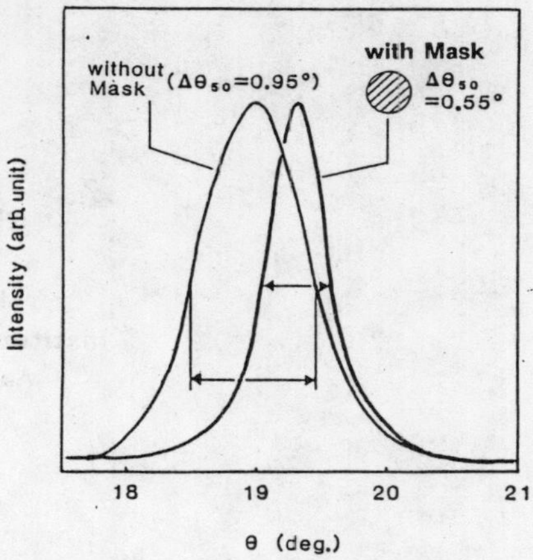

Fig.9 Rocking curves of (005)-line for films deposited with and without a mask.

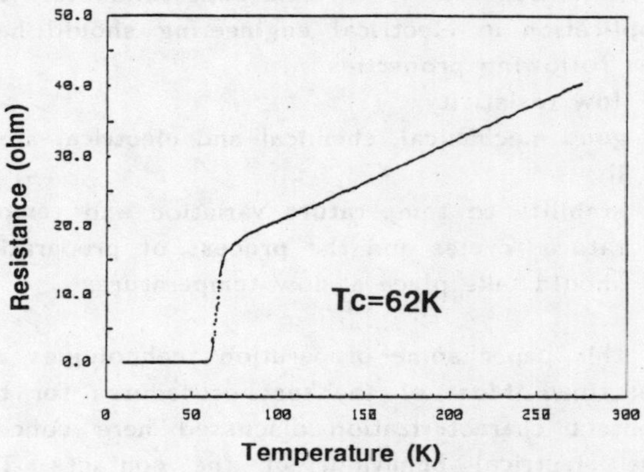

Fig.10 R-T characteristc of the film deposited with a mask.

the mask was used or not. Figure 8 shows the variation of c_0 as a function of the gas pressure. The deposition with the mask kept the c_0 almost constant at 11.8Å regardless of the gas pressure in contrast to the deposition without the mask which caused the c_0 to enlarge with the decrease in the gas pressure due to the stronger bombardment at the lower gas pressure. The rocking curves of (005)-lines for these films showed the higher crystallinity of the film deposited with a mask as illustrated in Fig.9. Full width at half maximum was 0.55° for the film deposited with the mask, while it was 0.95° for that without the mask.

The R-T characteristics of the films deposited with the mask were almost the same regardless of the sputtering conditions. The Tc's were 60-65K, though the resistance linearly decreased with decreasing temperature above the Tc accompanied with a sharp superconducting transition as can be seen in Fig.10. The moderate Tc's should probably be attributed to the reaction between the film and the substrate and/or to the larger mismatch of the lattice constant between YBCO and MgO than $SrTiO_3$.

CONCLUSION

(1)Two different substrate positions were examined to reveal that the position above the center of the target reduced the bombardment by the high-energy particles which caused the preferential resputtering and crystal damage of the films.

(2)Tc's above 80K were obtained without any annealing for the films deposited at this position.

(3)A mask between the target and the substrate was tried to prevent from the direct bombardment of the high-energy particles. As a result, further improvement of the crystallinity was achieved even at lower gas pressure below 5Pa.

REFERENCES

1. M. Suzuki et al., Jpn. J. Appl. Phys., 26(1987), L524
2. S.H. Liou et al., Appl.Phys.Lett., 52(1988), 1735
3. M. Muroi et al., IEEE Trans. Mag., 25(1989), 2451.
4. K. Tominaga et al., Jpn. J. Appl. Phys., 21(1982), 688.
5. N. Terada et al., Jpn. J. Appl. Phys., 27(1988), L639
6. Y. Matsui et al., Jpn. J. Appl. Phys., 27(1988), L350

METALLIC CONTACTS TO CERAMIC SUPERCONDUCTORS

H. Kliem, A. Weyers, G. Arlt
Institut für Werkstoffe der Elektrotechnik
Aachen University of Technology
5100 Aachen
Fed. Rep. Germany

1.) INTRODUCTION

Metallic contacts to ceramic superconductors for application in electrical engineering should have the following properties:

a) low resistivity
b) good mechanical, chemical and electrical stability
c) stability to temperature variation and temperature cycles and the process of preparation should take place at low temperatures

In this paper some preparation technologies are described. Most of the test procedures for the contact characterization discussed here concern the electrical behaviour of the contacts. The absolute values of the contact resistivities as well as their change with temperature, current density (non-linearity) have been measured. Most of the experiments and tests have been carried out using evaporated metal electrodes.

2.) MEASUREMENT TECHNIQUE

All contacts have an area of about 0.4....0.5 cm^2. A current I (Fig. 1) is fed through the contact and the voltage U across the junction metal-superconductor is measured in a four-probe arrangement. The current I is either a dc-current or it is a dc-current superimposed with a small ac-current : $I = I_{dc} + I_{ac}$. In the first case we measure the curve U(I) and in the second case the differential contact resistivity $\partial U/\partial I$ (U) or $\partial U/\partial I$ (I) is measured. The frequency of the ac-current lies between 28 Hz and 580 Hz. The measured voltage U consists of the main contribution arsing from the junction itself and another contribution due to the resistivity of the

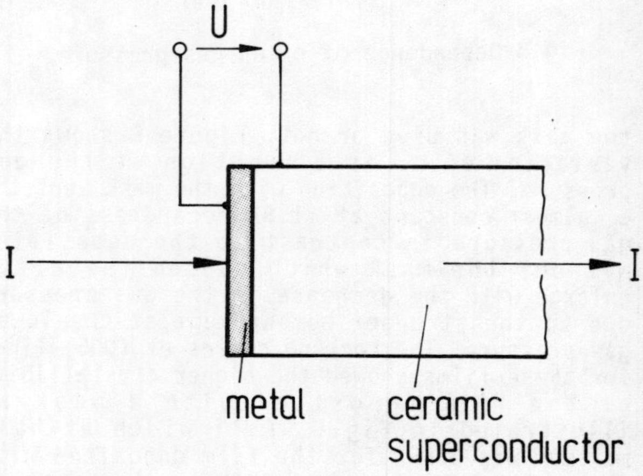

Fig. 1: The measurement technique

bulk (above T_C). This second contribution can be calculated and is subtracted. In most measurements, however, this part is negligibly small.

3.) SURVEY OF SOME PREPARATION TECHNOLOGIES

All ceramics were produced by mixed oxide preparation and a final sintering process in O_2 for 16 hours at 950°C and a cooling rate of 50 K/h.

a.) FIREABLE COMPOSITIONS

A fireable composition of silver particles and organic binder which burns out above 400°C was painted to the ceramic and fired at 600°C in air. This technology yields a contact resistance of $R = 2.5 \cdot 10^{-2}\ \Omega cm^2$ at 273 K which increases to $5 \cdot 10^{-2}\ \Omega cm^2$ at 77 K. Burn in temperature of 800°C (tetragonal phase) yields poor contacts with low stability and high resistivity.

b.) ULTRASONIC BONDING

Ultrasonic bonding of In (melting point 151°C) results in contact resistances of 0.5....2.5 $\cdot$ 10^{-2} Ωcm^2 at 295 K and similar values at 77 K. These contacts have a non-linear U(I) dependence comparable to contacts made by metal evaporation (see below)

c.) WIRES SINTERED INTO THE CERAMIC

During the firing process of the ceramics Au-wires were sintered into the material [1]. Contact resistivities as low as 5 $\cdot$ 10^{-5} Ωcm^2 at room temperature and 5 $\cdot$ 10^{-7} Ωcm^2 at 77 K are obtained.

d.) EVAPORATION OF METAL ELECTRODES

Al, Au, Ag or Cu was thermally evaporated to the ceramics at a pressure of 10^{-5} mbar. Thicknesses of the electrodes are between 10 000 Å and 20 000 Å. In order to prevent oxidation from the air Al electrodes were covered with a 1000 Å Au film.

4.) PROPERTIES OF EVAPORATED METAL ELECTRODES

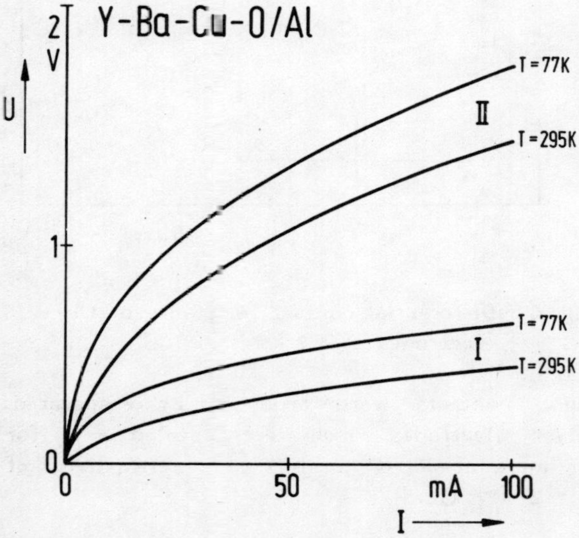

Fig. 2: Non-linear current voltage characteristic of an evaporated Al-contact.

Fig.2 shows the current-voltage characteristic of two Al-YBaCuO contacts at 295 K and at 77 K. The U(I)- behaviour is non-linear. Its temperature dependence is stronger in the low current (low voltage) region than in the higher current (higher voltage) region. Such a behaviour can be interpreted as a tunneling process of charge carriers through a thin insulating layer . Possibly an AlO_x layer is formed between Al and YBaCuO.

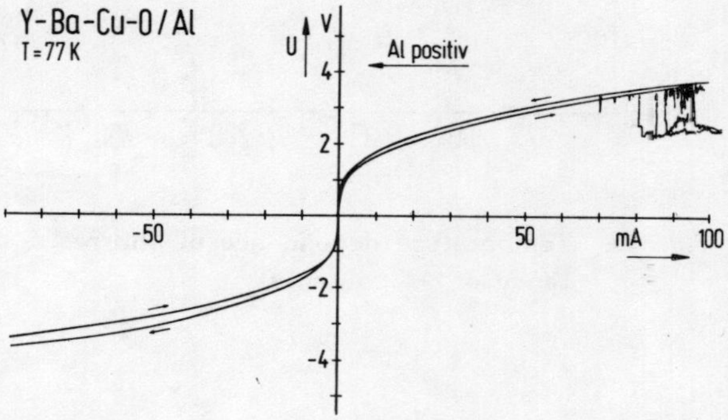

Fig. 3: The U(I) hysteretic characteristic of an evaporated Al-contact.

These contacts have a slight hysteresis of their U(I) characteristic (Fig. 3). During current flow O^--ions can be transported to or from the interface and enhance or diminuish the effective layer thickness.

Furthermore electrical breakdown events can be found at about U ≈ 3 V. For breakdown field strength of 2...5 MV cm^{-1} the thickness d of the insulating layer can be calculated to be d_1 ≈ 60....150Å. From high frequency measurements (10^6 Hz) the capacitance of the junction is known and a thickness of $d_2 \approx 10 \cdot \varepsilon_r$ is calculated. Oxides often have permittivities ε_r in the range from 3....10. With these two estimations for d it seems to be not to unrealistic to assume that the layer has a thickness between 50 Å and 100 Å.

The temperature dependence of the differential resistance $R_K = \partial U/\partial I$ at $I \rightarrow O$ is given with Fig. 4. R_K has a peak at T_c and decreases for $T > T_c$. The contact resistances lie between R = 5 $\cdot$ 10^{-2} Ωcm^2 and 25 Ωcm^2 at room temperature and R = 0.15 Ωcm^2 and 120 Ωcm^2 at 77 K.

The differential resistance at 77 K (below T_c) and 97 K (above T_c) is plotted in Fig. 5.

Two anomalies appear : (a) The maximum of the differential resistance is not centered at U = O V.

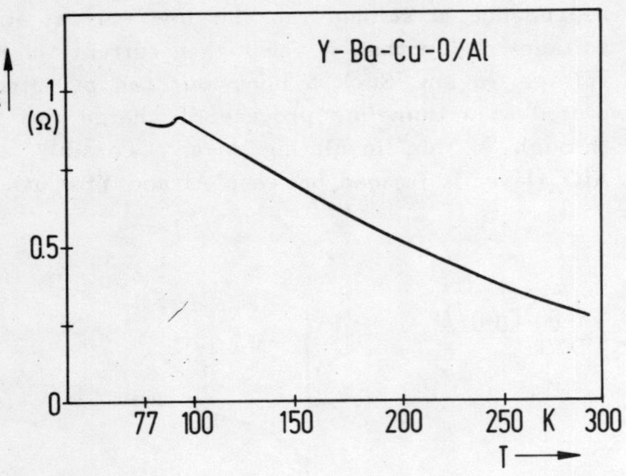

Fig. 4 : Temperature dependence of the resistance of Al contacts.

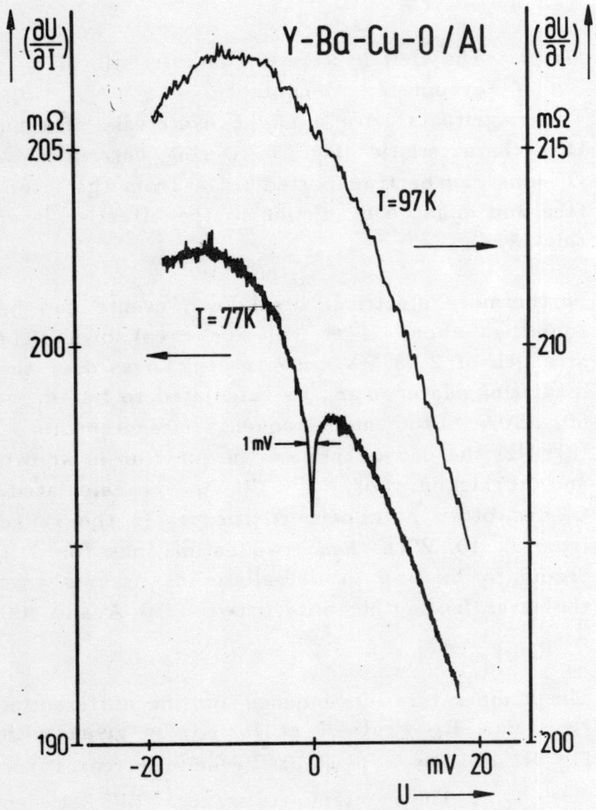

Fig. 5 : Differential contact resistance above and below T_C

(b) Below T_C a zero bias anomaly is found. The resistance has a sharp minimum which is associated to the superconducting state.

The zero bias anomaly can even be found using Cu, Ag or Au electrodes. Fig. 6 shows the differential resistance of evaporated Cu-electrodes. Onto Cu-electrodes of thickness d > 1 μm wires can be soldered. The contacts resist against mechanical tensions up to 150 N cm^{-2}.

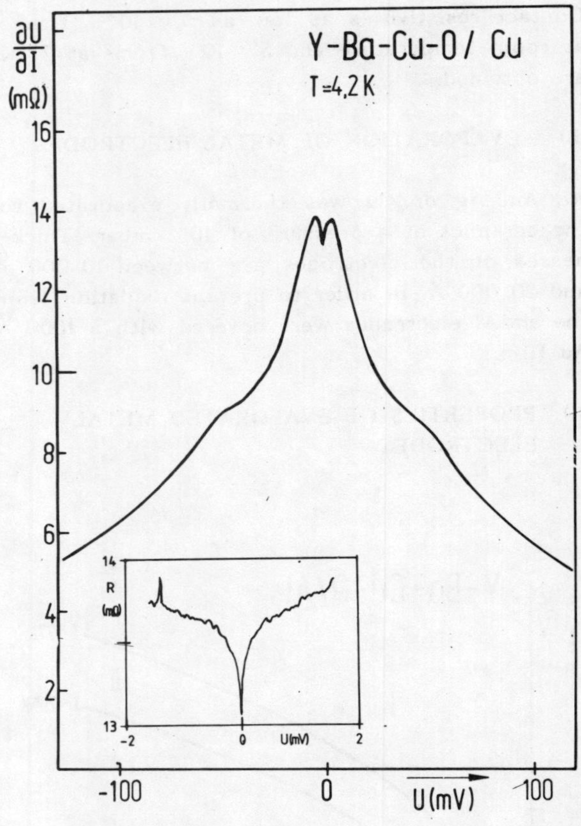

Fig. 6 : Differential contact resistance of Cu-electrodes at 4.2 K.

Good contacts were produced by evaporated silver electrodes which were annealed in O_2 for 10 hours at 500°C. In this case a resistivity of R = $2 \cdot 10^{-6}$ Ωcm² was achieved.

Ekin [2] reports on contact resistivities in the range of 10^{-10} Ωcm² by sputtered Ag-electrodes with a subsequent oxygen anneal.

5.) SYNOPSIS OF CONTACT RESISTIVITIES

In the following table the contact properties resulting from the different technologies and materials are summarized.

	R(293 K) $\Omega\,cm^2$	R(77 K) $\Omega\,cm^2$
Ag contacts by fireable compostiones	$2.5 \cdot 10^{-2}$	$5 \cdot 10^{-2}$
In-ultrasonic bonding	$5 \cdot 10^{-3} ... 2.5 \cdot 10^{-2}$	$2.5 \cdot 10^{-2}$
Au-wires sintered into the ceramic		$5 \cdot 10^{-7}$
evaporated metals:		
Al	$1.5 \cdot 10^{-2} ... 25$	$1.5 \cdot 10^{-1} ... 1.2 \cdot 10^{2}$
Cu	$10^{-4} ... 5 \cdot 10^{-3}$	$2 \cdot 10^{-4} ... 10^{-2}$
Au	10^{-3}	$10^{-5} ... 6 \cdot 10^{-5}$
Ag	10^{-3}	$4 \cdot 10^{-6} ... 4 \cdot 10^{-5}$
Ag annealed in O_2	10^{-3}	$2 \cdot 10^{-6}$

A dependence between the metal work function and the contact resistance as reported in [3] could not be found since the resistance depends strongly on the processing procedure.

[1] A. D Wieck, Appl. Phys. Lett 52, 1017 (1988)

[2] J. W. Ekin et al., Appl. Phys. Lett. 52, 1819 (1988)

[3] M. Suzuki et al., Jap. J. Appl. Phys. 27, 2003 (1988)

IMPROVEMENT OF CRITICAL CURRENT DENSITY IN CERAMIC $YBa_2Cu_3O_x$ SUPERCONDUCTOR

Xu-guang Zheng, Hisao Kuriyaki and Kazuyoshi Hirakawa
Department of Electronics, Faculty of Engineering, 36
Kyushu University, Higashi-ku, Fukuoka,812 Japan

ABSTRACT

Highly oriented $YBa_2Cu_3O_x$ ceramic samples were prepared by a mechanical aligning method which involved of large crystal growing and vibration aligning. A maximum critical current density of 4200 A/cm^2 (77K) was obtained. It was also found that the J_c value varied according to the resintering heat treatment.

INTRODUCTION

Critical current density in conventionally prepared ceramic samples of the high T_c superconductor $YBa_2Cu_3O_x$ has been low (generally a few hundred A/cm^{2}at 77 K).[1,2] However, much higher J_c is obtainable by enhancing the crystal orientation and packing density, since that value of the single crystals is very high. Based on this thinking, we have developed a simple and convenient process for preparation of high J_c ceramic samples which involves large crystal growth and aligning vibration as being described in our previous letter.[3] In the following text, further experimental results will be reported.

SAMPLE PREPARATION

Powders of Y_2O_3 (purity:99.9%), $BaCO_3$ (99%), and CuO(95%) were mixed by the ratio 1:2:3 and heated at 900℃ for 12 hours, then pelletized and heated again at 900℃ for 6 hours. All heat treatments here and below were carried out in flowing oxygen gas. The pellets were ground, repelletized, heated at 1000℃ for 60 hours, and finally ground into individual grains. A major part of these crystal grains were large rectangular plates, with grain size L ranging from $\sim$ 10 μm up to $\sim$ 100 μm (L is defined as the length of the long plate). The large grains were collected by a 20 μm sieve and then were poured into a rectangular cell, to which a mechanical vibration with regulated intensity and frequency was applied.[3] Large crystals are employed to take advantage of their anisotropic morphology. The vibration is intended to help align these crystals to achieve better orientation and denser packing. L<20 μm grains were also treated by the same process for investigation. After this process, the grains were pressed at $\sim$ 15 ton/cm^2 and heat treated at 950℃ for 5 hours. After this resintering, the temperature was decreased at a rate of 50℃/hour to 600℃, then cooled slowly ($\sim$0.3℃/minute) to room temperature. The samples obtained were $\sim$ 0.2x1x8mm^3 in dimensions.

EXPERIMENTAL RESULTS AND DISCUSSIONS

1. Asessment of Present Method

Experiments on J_c, X-ray diffraction intensity pattern and SEM images were conducted. Beside samples of the present preparation, conventional sample (sintered at 950℃ for 5 hours after several heat treatments) has been also investigated for comparison. For convenience, these samples are denoted as sample A (the conventional sample), B (sample prepared from the L<20 μm grains) and C (sample prepared from the L>20 μm large grains) respectively. J_c values (77K, at 1 μV/cm with steady current) for sample A, B and C were determined to be 250, 600 and 4200 A/cm^2 respectively. X-ray diffraction intensity patterns revealed highier crystal orientation for samples B and C.[3] As an indicator of the orientation degree, σ has been defined by $\sigma=(P'-P)/(1-P)$ (2θ:20-60 degree). Here P' is the proportion of (00ℓ) intensities to the overall intensity for the sample on measure-

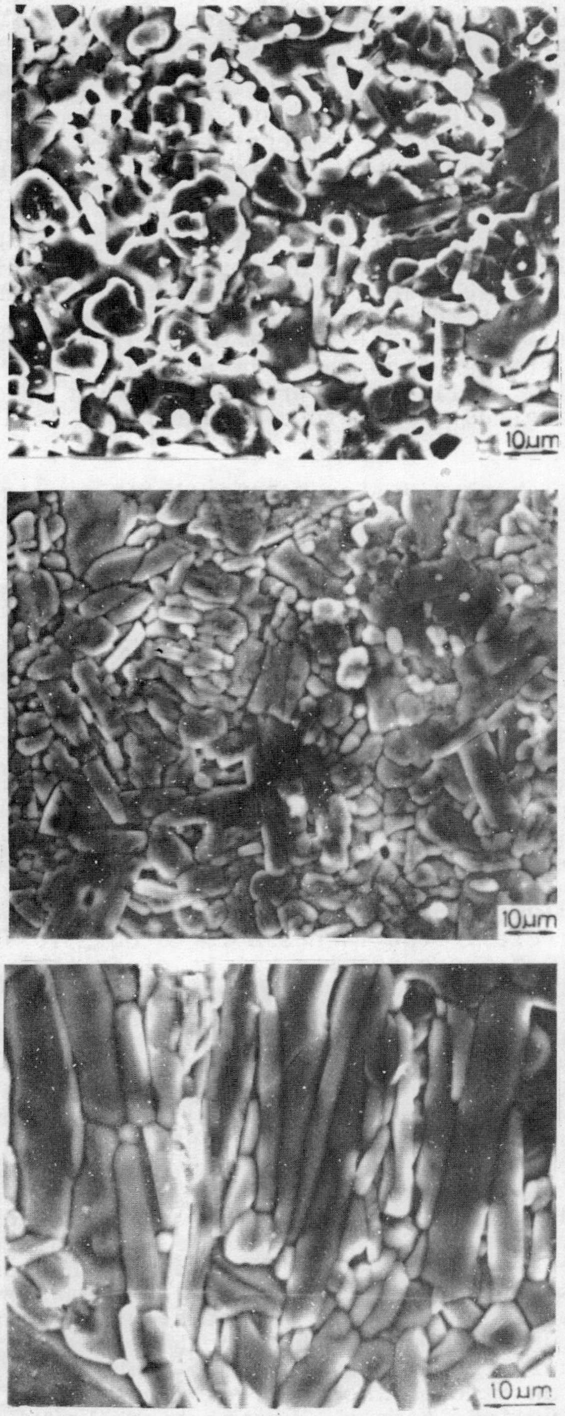

Fig. 1 Surface SEM images of samples A, B, and C (from top downward).

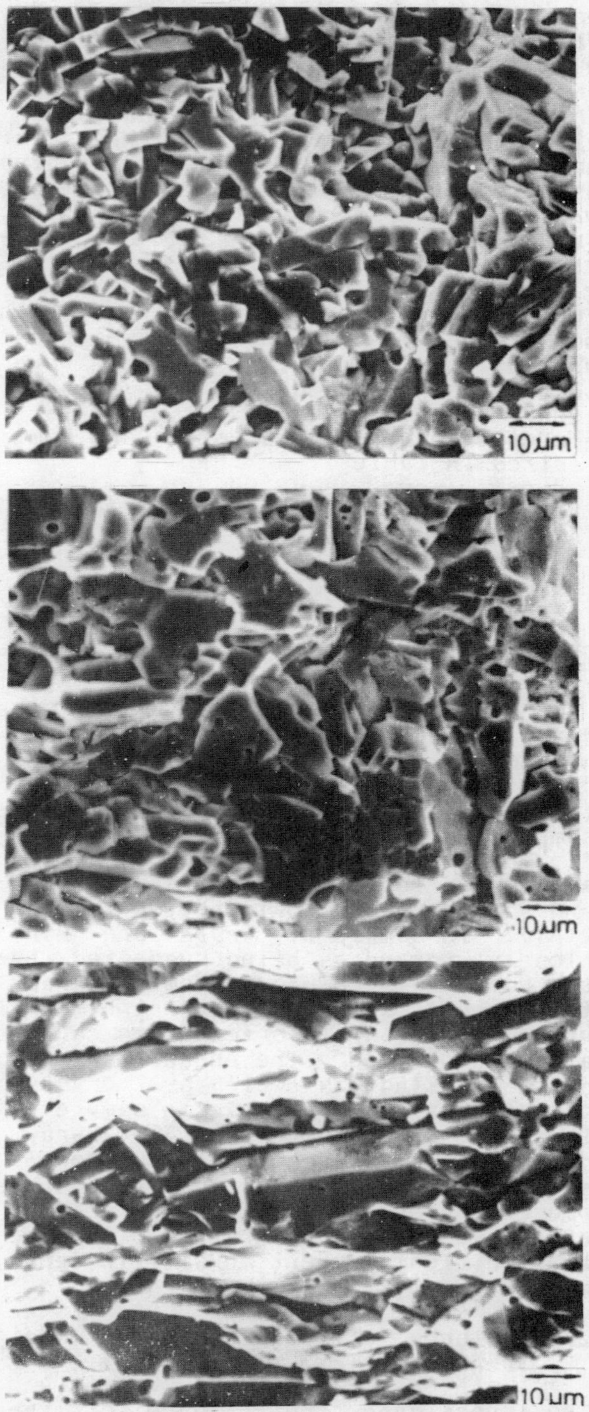

Fig. 2 SEM images of the fractured cross sections of samples A, B, and C (from top downward).

ment, which can be expressed as the sum of (00l) intensities versus the sum of all peak intensities, while P is that of a totally randomly oriented sample which was calculated to be 0.14 in the same range of 2θ from the powder X-ray spectrum.[4] According to this criterion, the orientation degree σ of sample A, B, and C was found to be 30%, 59%, and 85% respectively. This also seems to support our assumption that the larger crystals are more suitable for obtaining highly oriented samples because of their more pronounced anisotropy.

Bulk density ρ was also measured for these samples. It was found to be 5.20 g/cm^3, 5.68 g/cm^3, and 5.97 g/cm^3 for A, B, and C respectively. It is noted that the packing densities, as well as the orientation degree, have been also enhanced by the present technique.

The experimental measurements are summarized in Table 1.

Table 1 Experimental measurements for samples A, B and C

sample	J_c(A/cm^2)	σ(%)	ρ(g/cm^3)
A	250	30	5.20
B	600	59	5.68
C	4200	85	5.97

To give a visual image of the grain alignment in samples A, B and C, SEM photographs of the surfaces as well as fractured cross sections of these samples were shown in Fig. 1 and Fig. 2. The conventionally prepared sample A shows a loose packing (Fig. 1(A)) and no obvious orientation (Fig. 2(A)). In sample B, grains are more closely packed as being suggested by the surface microimage. From the fractured cross section photograph, improvement of grain orientation can be also recognized. The enhancement of dense packing as well as crystal orientation is remarkable in sample C. Therefore the results from J_c, X-ray, density and SEM experiments are consistent to prove the practicality of the present process. It is especially effective with large grains to improve the critical current density in ceramic $YBa_2Cu_3O_x$.

2. Influnce of the Resintering Heat Treatment

To investigate the influnce of the resintering heat treatment in the present process of sample preparation for exploring suitable conditions for high J_c sample, the last heat treatment in this process was tried in varied conditions. Beside the 950℃ (5 hour) resintering, samples were also prepared by heat treatment at 900 ℃ (48 hour), and 920℃ (12 hour). The cooling rate is the same as the 950℃ sintering. Since it is generally reported that the suitable sintering temperature for $YBa_2Cu_3O_x$ is in the range of 900℃ to 960℃, sintering at those three temperatures should be representative. Sintering time of 48 hours and 12 hours was choosed for 900℃ and 920℃ heat treatment respectively refering to references 5 and 6, in which these sintering time were reported most suitable for the sintering temperature. In addition, a short (1 hour) heat treatment at 1000℃ (a littele below the melting point which is a little over 1000℃[7]) was also applied. J_c values is shown in Table 2.

Table 2 J_c for samples heat treated at varied temperatures

heat treatment	J_c (A/cm^2)
900℃48hrs	1700-2100
920℃12hrs	1300-2000
1000℃1hr	350-440

Samples heat treated at 900℃ and 920℃ did not produce as high value as the 950℃ sample, and J_c of the sample heat treated at the high temperature (1000℃) was greatly reduced. It is also observed that in the J_c measurement of the 900℃ heat treated samples, the first sweeping cycle always gives lower J_c than the following cycles (the increase is about 10%). From these results, it is clear that J_c is very sensitive to the conditions of the last heat treatment. SEM images of the fractured cross sections of these samples, as shown in Fig. 3, may help to explain the result of J_c measurement. As can be seen from these photographs, grains in the relatively low temperature (900℃) treated sample are not well inter-connected, which is not desirable because of the weak path for current flowing. This is perhaps the reason of the lower J_c and the increasement of the value in J_c measurement, where the first high density current flowing through the sample may help the adjoining grains to be better connected by some effect. In the 920℃ treated sample, connection between adjoining grains seem strong neither, and besides, the grains are porous. On the contrast, in 1000℃ treated sam-

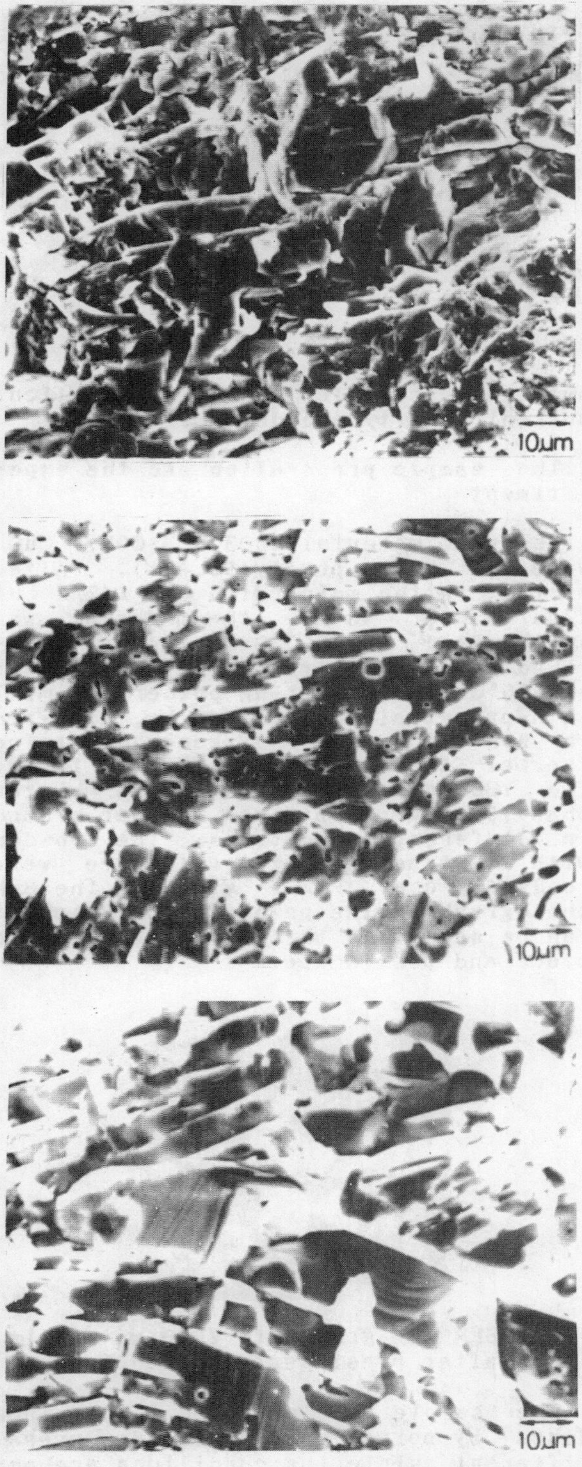

Fig. 3 SEM images of the fractured cross sections of 900℃, 920℃, and 1000℃ resintered samples (from top downward).

ple, grains seem strongly connected to each other, and the pores existing in large quantities in the 920℃ treated sample nearly disappeared completely. However, the grains are ill aligned due to the over regrowth in the heat treatment. This is considered the main reason for the much reduced J_c . The porouslessness is also perhaps undesirable since oxygen absorbing in the sample during cooling may be insufficient.
The conclusion that J_c is influnced by the resintering heat treatment can be drawn from the above analysis. With finer adjustment of the conditions (temperature as well as sintering time) better samples should be available.

3. Conclusions

Samples prepared by present technique were assessed by experiments of X-ray diffraction, density and SEM beside J_c measurement. From analysis of these results, it was clear that critical current density was improved by the present preparation process as the result of close packing and orderly alignment of crystal grains. As demonstrated by the investigated effect of the resintering temperature, definite conditions in the preparation process are important. By working out best conditions, more satisfying results are expected.

REFERENCES

1. R. J. Cava, B. Batlogg, R. B. van Lover, D. W. Murphy, S. Sunshine, T. Siegrist, J. P. Remeika, E. A. Reitman, S. Zahurak and G. P. Espinosa, Phys. Rev. Lett. 58 (1987), 1676.
2. E. M. Gyorgy, G. S. Grader, D. W. Johnson, Jr., L. C. Feldman, D. W. Murphy and W. W. Rhodes, Appl. Phys. Lett. 52(1988), 328.
3. X. G. Zheng, H. Kuriyaki and K. Hirakawa, Jpn. J. Appl. Phys. 28(1989), L52.
4. E. Takayama-Muromachi, Y. Uchida, K. Yukino, T. Tanaka and K. Kato, Jpn. J. Appl. Phys. 26(1987), L665.
5. D. Pavuna etc., Physica C 153-155(1988), 1339.
6. T. Ikemachi etc., presented at 1988 Nagoya (Japan) International Conference on Superconducting Materials.
7. K. G. Frase, E. G. Liniger and D. R. Clarke, J. Am. Ceram. Soc. 70(1987), C-204.

FABRICATION OF YBCO SUPERCONDUCTOR BULK MATERIAL WITH C-AXIS TEXTURE

H.P.Chou   Y.Wang   M.L. Liu*
Dept. of Materials Sci. and Eng., *Dept. of Phys.,
Tsinghua University, Beijing, P. R. China

ABSTRACT

In this paper, the preferred orientation dependence on the composition, average grain size, forming pressure, sintering temperature and sintering time are studied for the YBCO samples formed by the dry press process. The results show that with the forming pressure of higher than 5ton/cm^2 and sintering temperature below 990° C, the materials with notable preferred orientation are obtained in the pure YBCO composition with the grain size of 10 µm or so. The critical current density of the bulk material is higher than 10^3 A/cm^2.

1. Introduction

Since the discovery of YBCO superconductors, great attention has been paid to the applicational prospect of the material. Thus it becomes a very important researching aera to improve the behavior of the new superconductor material in strong electrical field. With the melt-texture method, Jin[1]-[3] of Bell Laboratory prepared the YBCO bulk material with the critical current density of 10^4 A/cm^2. Many factors[4-6] have been known to dominate the current density of superconductors, among which the most important is, the material must be in high purity of YBCO superconductor phase, clear in grain boundary and with high bulk density. Besides, high degree of preferred orientation along c-axis is another way to improve the behavior of current transportation due to the anisotropy of the material. In this paper, the dependence of the preferred orientation on the powder composition, the grain size and shape of the particals, the forming pressure, the sintering temperature and time are studied. Also, the relation between orientation degree and critical current density are determined.

2. The sample preparation and the experiment

The raw material, Y_2O_3, $BaCO_3$, CuO were in high purity(99.99%). After mixing in the mole ratio of 1/2/3 of Y/Ba/Cu for 48 hr, the material was synthesized in 960°C for 24 hr, followed by thermal treatment in fluid air of 610°C. The resulting powder was mill in n-propyl alcohol for 48 hr, sintered in 995°C/60 hr in oxygen and twice thermal treatment in 610°C for 12 hr. The YBCO grains with the size bigger than 70 µm were obtained. After ground in different conditions, the powder with various average grain size were ready for use. Fig.1 gives us the SEM photograph of the powder, from which it can be seen that these partical have a piece and post shape.

Fig.1 SEM of grains in YBCO materials after presintering (×300)

The sample were prepared into 15mm discs by normal process of dry press. Different sintering conditions are employed to study the dependence of preferred orientation degree on the process. The orientation degree was detected and analysed quanlitatively by XRD and defined by the intensity ratio of different peaks (006) to (200), but it can be neglected in quanlitative analysis because when (001) preferred orientation presented obviously, the diffraction intensity of (020) and (006) are changed in different tendency and the increase of diffraction intensity of (006) is more contribative compared with the decrease of (020) diffraction intensity. Besides, the SEM and other microanalysis technique were used to study the microstructure of the sample with preferred orientation.

3.Results and discussion

3-1.Preferred orientation of the green body

The green powder with the same value of average grain size was pressed into discs in different pressure. The XRD method was used to analyse the plane perpendicular to the press direction. The results showed that the intensity ratio of I(006)/I(200) became higher with the increase of forming pressure, which indicate that there existed [001] preferred orientation perpendicular to the surface of the disc. Fig.2 gives us the result for the average grain size of 15 μm and 50 μm. When the forming pressure is 1×10^4 lb, the diffraction intensity ratio I(006)/I(200) for the samples with average grain size 15 μm and 50 μm are 1.2 and 3.3 times higher respectively than the samples without preferred orientation. This may be due to the preferred arrangement of piece and post crystals and YBCO layer structure. In Fig.3, we have the SEM photograph of the cross section of green body for the sample with average grain size 20 μm, forming pressure 5×10^4 lb, which show us the preferred orientation of the sheet cryatal.

When the forming pressure maintained unchange, the average grain size of green powder played a very important role in affecting the preferred orientation degree of the green body. Within the experimental range, the increase of preferred orientation degree was almost proportional to the increase of average grain size. This is because of the good fluidity of the material with small grain size.

We have seen from Fig.2 that there are two regions existed where the preferred orientation degree increase linearly with the forming pressure.When the forming pressure is lower than 3×10^4 lb, the linear coefficient is equal to 28.8 and 61.6 for the materials with grain size of 15 μm and 50 μm respectively. When

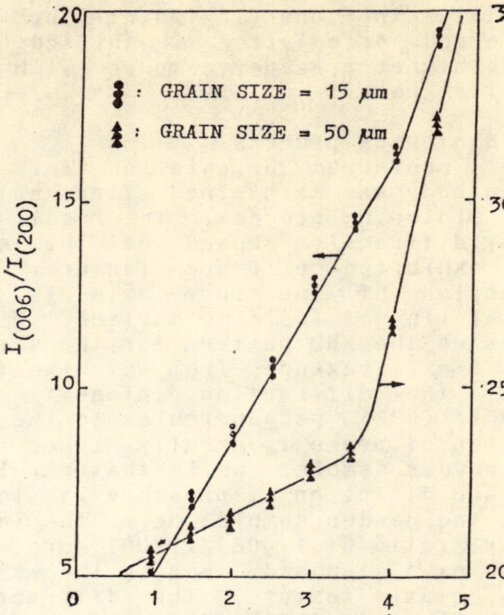

Fig.2 Dependence of preferred orientation degree on forming pressure of green body

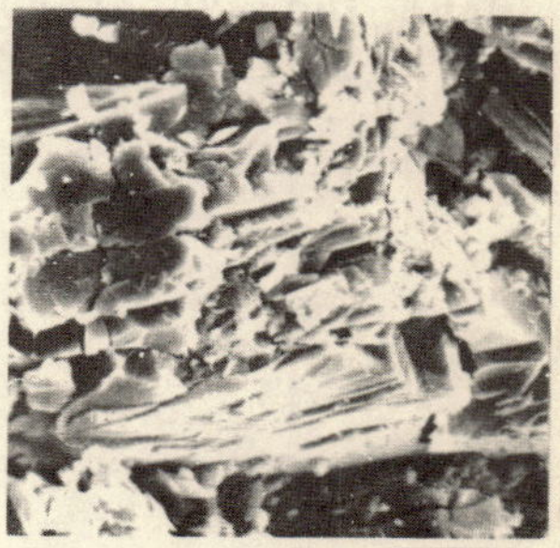

Fig.3 SEM of cross section parallel to the pressure direction (×3000)

the pressure is higher than 4×10^4 lb, the coefficient is about 89.9. The turning point of two regions for the material with larger average grain size is in higher pressure region compared with that with smaller grain size. The relationship of the discontinuous change can be discussed as following. When the forming pressure is low, the main contribution to the preferred orientation of the green body is the oriented arrangement of piece and post crystal. While in the region of high pressure, the forming pressure is high enough to overcome the movement resistance among grain, this results in the slip of the grain in the material and improve the preferred orientation degree. For the resistance of movement among big cryatal is higher than that among small crystal, the changing point for the preferred orientation is shifted towards higher pressuer compared with the smaller grain.

3-2. Sintering process

The preferred orientation for the green body was maintained after suitable sintering process. The results of X-ray diffraction showed that the sample exhibited a fibre texture. The direction of the fibre axis is the normal line of the disc surface. Fig.4 gives us the XRD pattern for the sample with c-axis texture, from which we find that the diffraction intensity of I(006)/I(002) perpendicular to the direction of press is greatly higher than the powder sample, while that parallel to the direction of pressure is lower than the powder sample. Here, the intensity ratio of I(006)/I(200) are not used as a standard. For in the sample with c-axis texture, the diffraction intensity of (006) for the surface parallel to the direction of press is so weak that it is submerged by the diffraction intensity of (020). The diffraction intensity change of (005) is then used as a standard of comparison. Because of the synchronous change

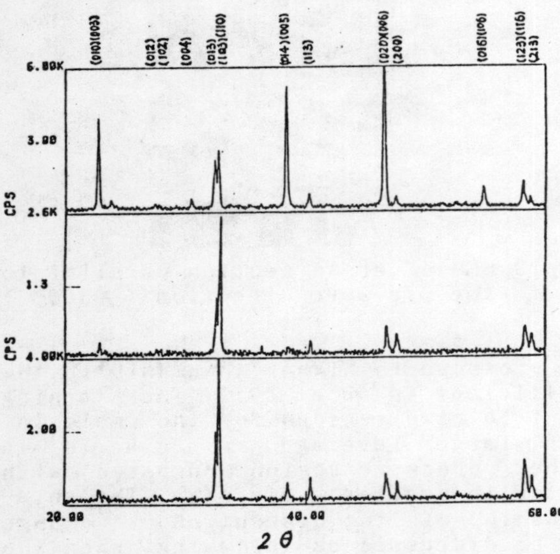

Fig.4 XRD patterns for YBCO sample with c-axis texture
a. powder
b. parallel to the pressure direction
c. perpendicular to the pressure direction

for the intensity of two diffraction peaks, (005) and (014), there are no submerged effect between the two peaks.

The preferred orientation of the green body was maintained after sintering. The orientation degree was affected by the sintering process. The experimental data for the samples sintered for 6 hr in different temperature are listed in Tab.1, wherein the green body with the average grain size of 20 μm was formed at the pressure of 6.5 ton/cm². The X-ray diffraction plane was perpendicular to the direction of the pressure. The table shows that the higher the sintering temperature, the more obvious the detexture. That is, the higher sintering temperature makes the degree of preferred orientation decreased. The observation of microstructure for the sintering samples by SEM are shown in Fig.5. With the increase

Tab.1 Preferred orientation degree in different sintering temperature

No.	Sint.Temp.	Sint.Time	I(006)/I(200)
1	R.Temp.	/	17.69
2	980°C	6 hr.	15.71
3	990°C	6 hr.	6.00
4	1000°C	6 hr.	4.78

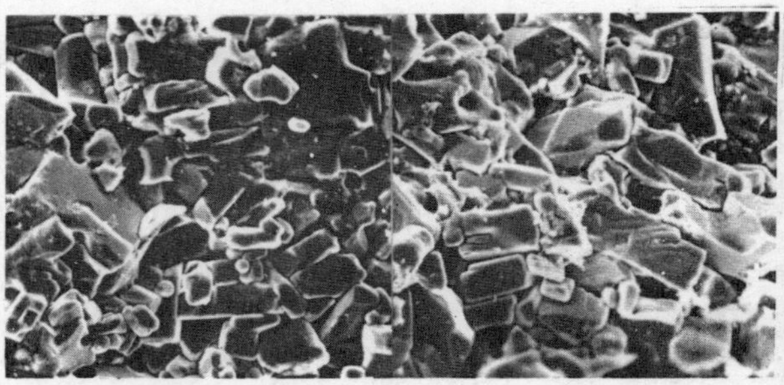

a. 970°C, 6hr. b. 990°C 6hr.

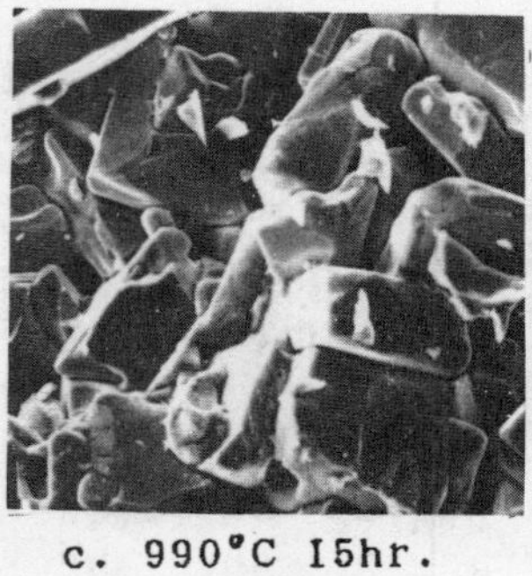

c. 990°C 15hr.

Fig.5 SEM of YBCO samples sintered by different precess (×1000)

of temperature, the former arrangement in the green body was rearranged and detextured. The results also show that the effect of detexture become notable with the decrease of the grain size.

With the lengthening of the sintering time for the green body with preferred orientation, the detexture is occourred. The decreasing tendency of preferred orientation degree is stopped when it changed to a certain level. The dependence of sintering time on the detexture is much smaller than on the sintering temperature.

3-3.Degree of preferred orientation dependence on the composition

The comparison are made between pure YBCO composition and YBCO phase with a little among of other phases (BaCuO2, Y2Cu2O5 etc.). The average grain size of YBCO phase in two materials are 30 um. After formed at the pressure of 15 ton/cm², the discs were sintered at 980°C for 6 hr. The results of XRD and SEM show, the preferred orientation of the material with pure YBCO phase was well maintained, whereas, the material with other phases was not. With the increase of other phases, the degree of preferred orientation decreased and resulted in the completely disordered arrangement. This is because of the existence of chemical reaction and rearrangement of the grain during sintering.

3-4.Current density dependence on preferred orientation

The measurement results by four-probe method showed that the zero-resistance temperature for the sample are 91-91 K. The critical current density at zero magnetic field for the sample without preferred orientation are in the order of 10^2 A/cm². While for the sample with preferred orientation, the current density is about 10^3 A/cm². Fig.4 gives us the XRD results for the sample with the size of 0.3×0.7×0.9mm, its current density is 1200A/cm². Also , Fig.6 show the X-ray diffraction pattern for the sample with the grain size of 20 μm, forming pressure of 14 ton/cm², sintered at 980°C in air for 6hr. Fig.7 tells us

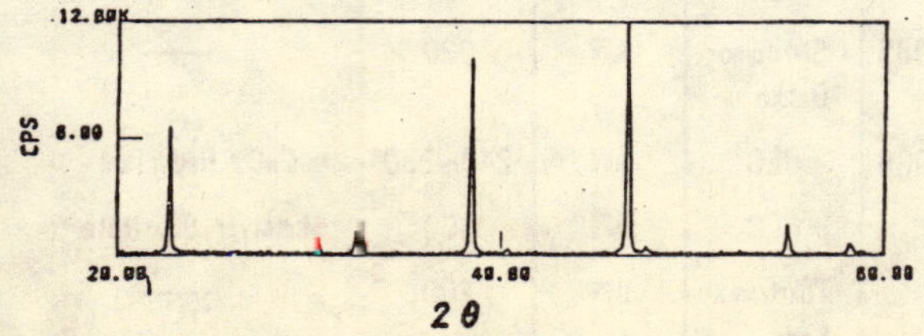

Fig.6 XRD pattern of the sample with high orientation degree

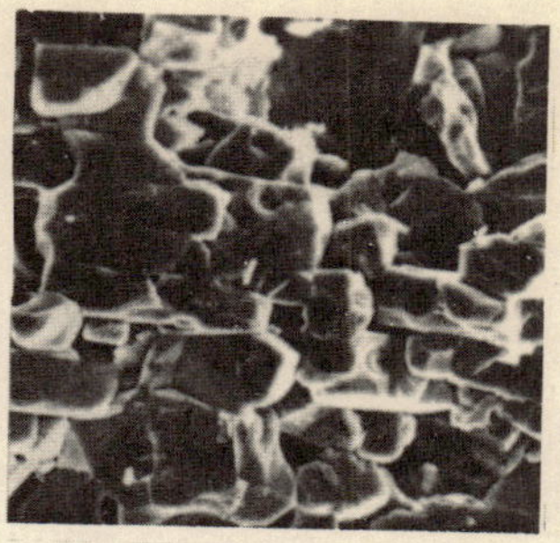

Fig.7 SEM of the sample with high orientation dedree (×1500)

this sample exhibits obvious oriented arrangement of the grain. When the size of the sample is 0.3×0.7×0.9mm(length × high × width), the critical current density reaches as high as 2600A/cm².

4.Conclusion

We can conclude from above discussions,

1).The grain size of starting material influences the preferred orientation degree of the green body. When the grain size is bigger than 10 μm, it is easy to obtain the c-axis texture. The pressure makes higher degree of preferred orientation.

2).The green body with c-axis texture can be detextured with the long time sintering. The sintering temperature dependence is greater than that of sintering time.

3).The preferred orientation degree of the green body with pure YBCO phase is well maintained after sintering.

4).It is favourable for the improvement of critical current density to increase the degree of preferred orientation along the c-axis.

References

[1].S.Jin, T.H.Tiefel,R.C.Sherwood,et al., Appl. Phys. Letters, 51(12), pp.943(1987).

[2].M.M.Waldrop, Science, 238(4834), pp.1655 (1987).

[3].S.Jin, R.C.Sherwood, E.M.Gyorgy,et al., Appl. Phys. Letters,54(6),pp. 584 (1989).

[4].J.E.Bleudell, et al., Adv. Cera. Mat., 2, pp.512 (1987).

[5].J.W.Ekin Adv. Cera.Mat.,2, pp.582 (1987).

[6].H.Watanabe, Y.Kasai, T.Mochiku,et al., Jpn.J. Appl. Phys., 26(5),pp. 657 (1987).

THE THERMAL AND MECHANICAL PROPERTIES IN THE ALN-BN COMPOSITES

Noboru Ichinose, Yasuharu Miyauchi and Kazuhiro Ikuina
Department of Materials Science and Engineering
School of Science and Engineering
Waseda University
3-4-1 Ohkubo, Shinjuku-ku, Tokyo 169, Japan

Abstract

High thermal conductive AlN-BN composites added sintering agents such as CaO, Y_2O_3 and B_2O_3 were obtained by hot-pressing at the temperature of 2000°C and in the reduced pressure of 10^{-3}-10^{-2} Torr. The relative density of the sintered body is dependent on the content of BN and decreases with the increasing amount of BN. The CaO and Y_2O_3 dopants are effective for highly relative density. The thermal conductivities in the AlN-BN composites are anisotropic. At the amount of 10wt.% BN, they are about 150W/m·K and 80W/m·K for the direction perpendicular and parallel to axis of hot pressing, respectively. They are decreasing with the increasing amount of BN. Furthermore, they have good machinable properties.

INTRODUCTION

Ceramic substrates based on AlN ceramics have excellent characteristics such as: high thermal conductivity, low permittivity and high mechanical strength. Furthermore, it is readily sintered at normal pressure with some dopants such as CaO[1] and CaC_2[2], and can be used in fabricating packaged substrates consisting of green sheet laminations.

The thermal conductivity of these ceramics is influenced by impurities such oxygen and transition metal ions that exist in AlN. The highly thermal conductive AlN ceramics have been obtained by using high purity AlN powder and sintering aids such as yttrium and calcium compounds as shown in table 1.

Here, we report the composition of reactants and its correlation to the thermal conductivity of the AlN-BN composites modified by CaO, Y_2O_3 or B_2O_3.

The mechanical properties in the AlN-BN composites are also reported.

EXPERIMENTAL

1. Sample preparation

AlN powder was prepared by the reduction process of Al_2O_3 with carbon in nitrogen atmosphere. This powder is commercially available from Tokuyama-Soda in Japan[3]. Chemical analysis of AlN powder is shown in table 2.

AlN-BN composites were fabricated by the following procedure as shown in figure 1. AlN, BN and sintering agent powders were mixed in butyl alcohol. After drying, the mixed powder was isostatic-pressed at 1ton/cm^2.

The AlN-BN composites with dopants were obtained by hot-pressing at the temperature of 2000°C and in the reduced pressure of 10^{-3}-10^{-2} Torr.

Table 1 Recent trend in the development of highly thermal conductive substrates.

Year	Company	Material	Thermal Conductivity	Topics
1982	Hitachi	SiC	270w/mK	——
1983	Toshiba	AlN	60	Thyristor for Train
1984	Hitachi	SiC	270	ECL LSI Package
	Toshiba	AlN	100	Y_2O_3 Additive
	Tokuyama Soda	AlN	110	High Purity AlN Powder Translucent AlN
	Heraeus	AlN	140~170	——
	NEC	AlN	160	Reducing Sintering
	NEC	AlN	160	Transistor, LED, Laser Diode
1985	Sumitomo Denko	AlN	220	——
1986	NEC	AlN	240~260	CaC_2 Reductant
	NEC	AlN	160	Multilayer Substrate
	Tokuyama Soda	AlN	260	——
1987	Toshiba	AlN	260	Reducing Atmosphere
	Toshiba	AlN	70	Multilayer PGA

Table 2 Chemical analysis of AlN powder.

element	chemical analysis
Al	65.4%
N	33.5%
O	1%
C	200ppm
Ca	103ppm
Si	35ppm
Fe	<10ppm
Surface specific area	$3.9m^2/g$

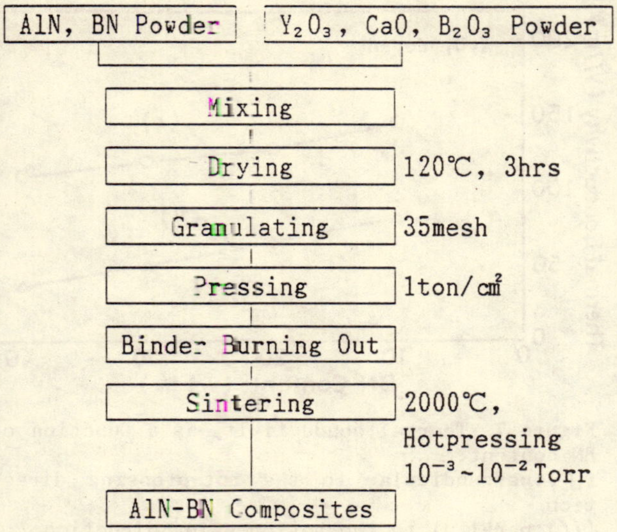

Figure 1 Flowchart for AlN-BN composites.

2. Thermal conductivity measurement

Thermal diffusivity of samples were measured by the laser-flash method using a ruby laser (=0.6943μm) and the liquid-nitrogen-cooled InSb infrared detector[4] at room temperature. Figure 2 shows the principle of laser flash method for thermal measurement. The specimen discs of 10 mm in diameter were cut from the sintered bodies which have various kinds of thickness. The specimen surface, on which the laser flash was incident, was coated with gold and both surfaces were coated with carbon to assure complete absorption of the flash.

The way of determing the thermal diffusivity α has been deduced from the following equation[5]

$$\alpha = 0.1388 \cdot L^2/t_{1/2} \qquad (1)$$

where $t_{1/2}$(ms) is the time required for the back surface to reach half of the maximum temperature rise and L is thickness of sample in (mm), as shown in figure 2.

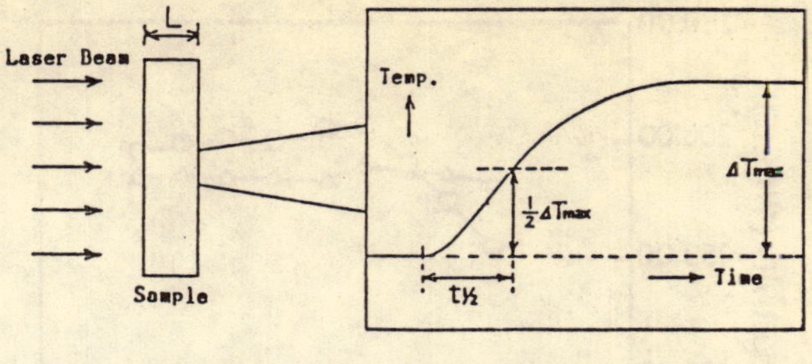

Figure 2 Measurement of thermal diffusivity by laser flash method.

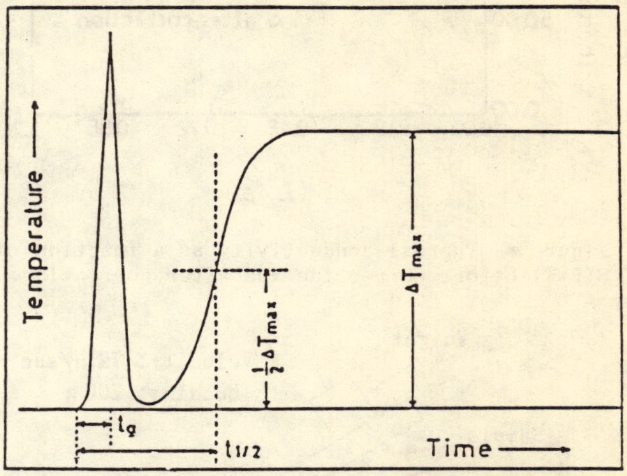

Figure 3 Experimental chart in case of measurement of thermal diffusivity for corrction of $t_{1/2}$.

Therefore, the thermal conductivity k(W/m·K) is found from the relationship

$$k = \alpha d c \qquad (2)$$

where d is the density in (g/cm^3) and c is the heat capacity in (cal/g·K).

The foregoing treatment has neglected the variation of thermal diffusivity with temperature. Although the method produces an effective value of diffusivity for the sample, an effective value of the corresponding temperature is yet to be determined. This problem is common to all types of diffusivity measurements and is usually minimized by the fact that the range of temperatures in a single measurement is kept as small as possible. Clearly, the time of transit of heat pulse will depend upon the range of temperature encountered en route.

Therefore, it is necessary to attempt a rigorous analysis in order to determine the thermal diffusivity. Taking account of factors such as shape of laser pulse, heat losses and distribution function of luminance, the rigorously checked equation has been given by T. Azumi[6], in the following:

$$\alpha = 0.1388 \cdot Y_b \cdot Y_c \cdot L^2/(t_{1/2} - t_g) \qquad (3)$$

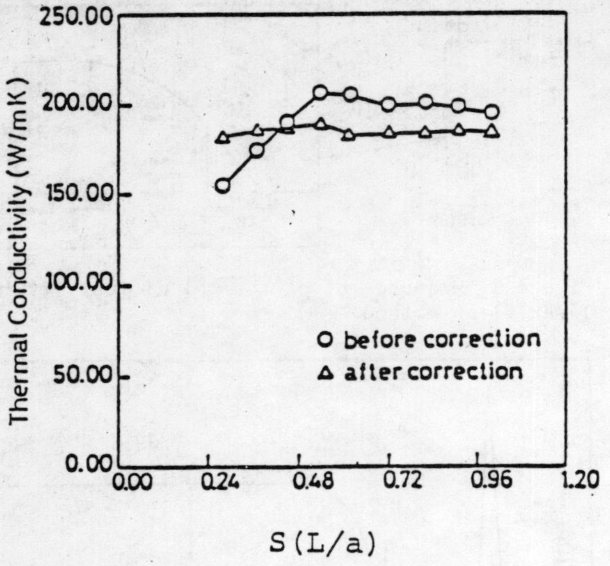

Figure 4 Thermal conductivity as a function of S(L=a) before correction and after correction.

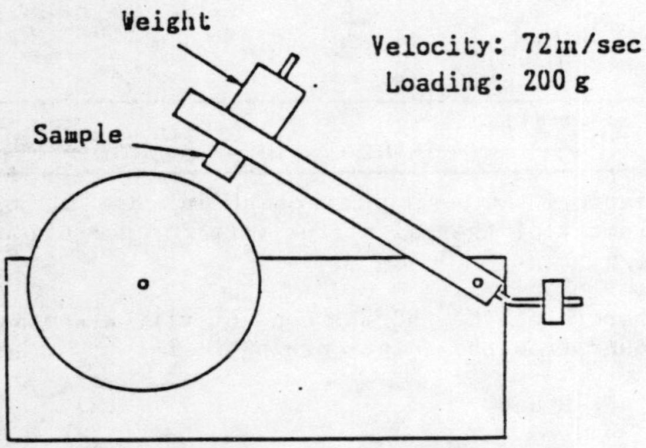

Figure 5 Machinable test for composites.

where Yb is compensatory factor for heat losses and Yc is for distribution function of luminance. Furthermore, tg is the time shift of pulse as shown in figure 3.

Figure 4 shows the thermal conductivity of Y_2O_3 (5wt.%) modified AlN as a function of S(L/a) where a is the diameter of sample. It is found in figure 4 that their thermal conductivities deduced from equations 1 and 2 are dependent of the thickness of samples. However, its dependency can not be found after correction using equation 3. So, we are using these corrected values of thermal conductivity in the following section.

3. Mechanical property measurement

Mechanical properties of AlN-BN composites were measured by the conventional method. Furthermore, the machinable properties of these materials were measured using the test apparatus as shown in figure 5.

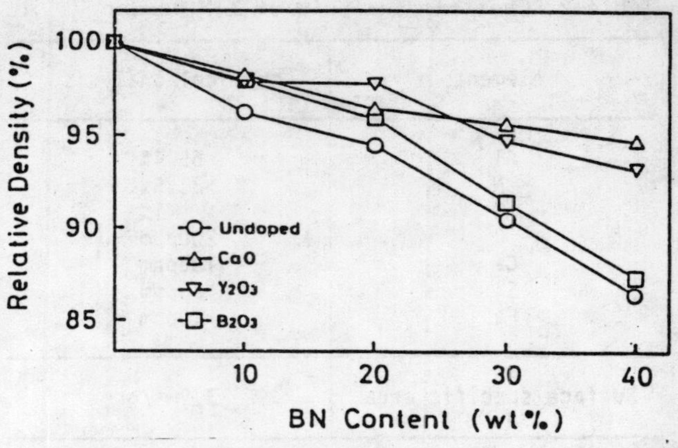

Figure 6 Relative density as a function of BN content.

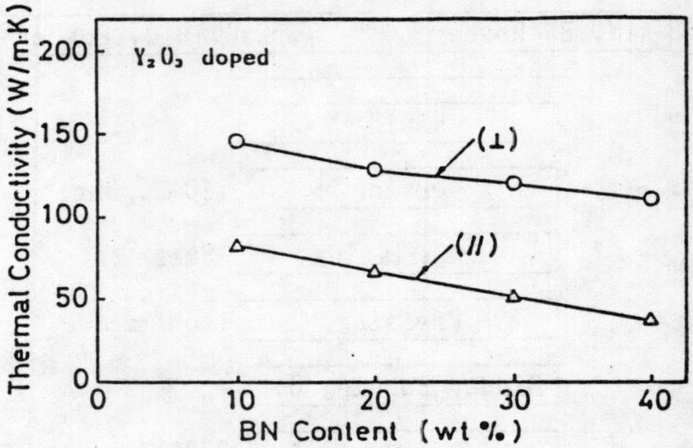

Figure 7 Thermal conductivity as a function of BN content.
(⊥):perpendicular to the hot-pressing direction
(//):parallel to the hot-pressing direction

4. X-ray measurement

The powder X-ray diffraction data were obtained by a diffractometer using the graphite-monochromated CuKα radiation. Microstructural analysis was carried out using a scanning electron microscope and an energy dispersive X-ray (EDX).

RESULTS AND DISCUSSION

1. Densitification

The relative density of the sintered body in the AlN-BN system is dependent of the content of BN as shown in figure 6. The CaO and Y_2O_3 dopants are effective for highly relative density.

2. Thermal conductivity

The thermal conductivities in the AlN-BN composites are anisotropic as shown in figure 7. At the amount of 10wt.% BN, they are about 150W/m·K and 80W/m·K for the direction perpen-

Model I

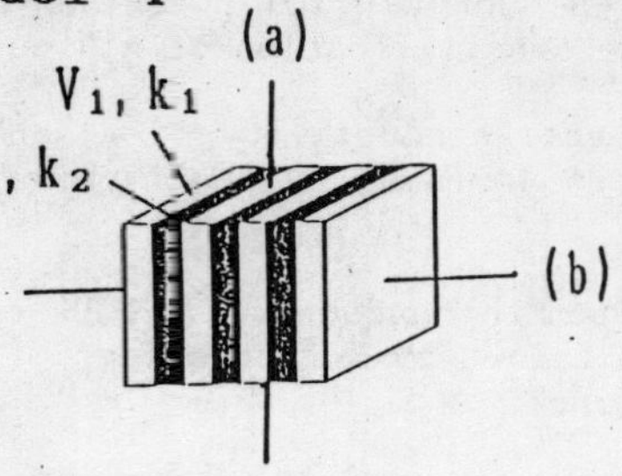

Model II

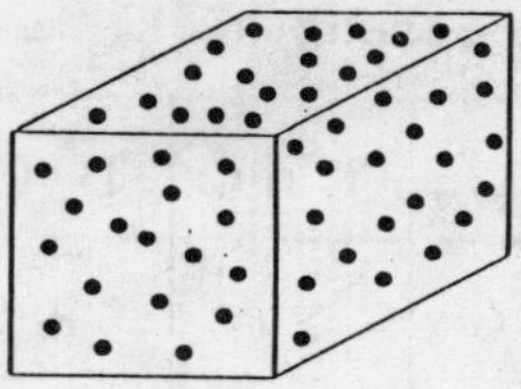

Maxwell's Equation

(a) direction

$$k_c = k_1 V_1 + k_2 V_2$$

(b) direction

$$k_c = \frac{k_1 k_2}{k_1 V_2 + k_2 V_1}$$

$$k_c = k_m \times \frac{k_d + 2k_m - 2V_d (k_m - k_d)}{k_d + 2k_m + V_d (k_m - k_d)}$$

k_d: Thermal conductivity for dispersed phase

k_m: Thermal conductivity for matrix

V_d: Volum fraction of dispersed phase

Figure 8 Structural model for composite materials in thermal conductivity calculation.

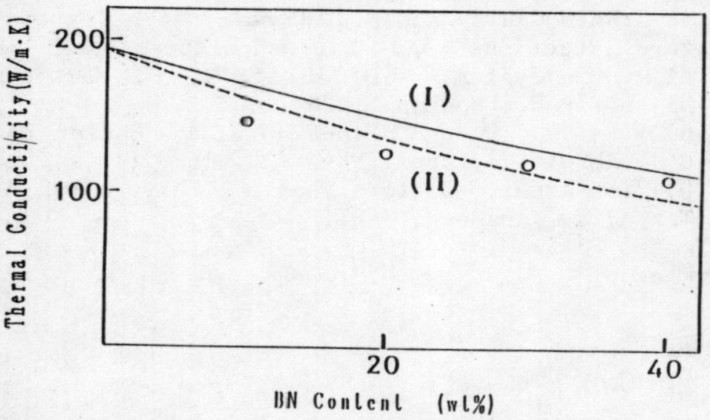

Figure 9 Thermal conductivity vs. BN content; Model calculation.
(Perpendicular to the hot-pressing direction)

dicular and parallel to axis of hot-pressing, respectively. They are explained by considering the anisotropic properties in BN.

The value of the thermal conductivity for the direction perpendicular to axis of hot-pressing was calculated using the model for composite materials as shown in figure 8. Figure 9 shows the result for model calculation. It is found in this figure that model calculation (II) fairly agree with the experimental value.

3. Microstructure

X-ray diffraction patterns show the phases of AlN and BN are found separately as shown in figure 10 for AlN-BN composites.

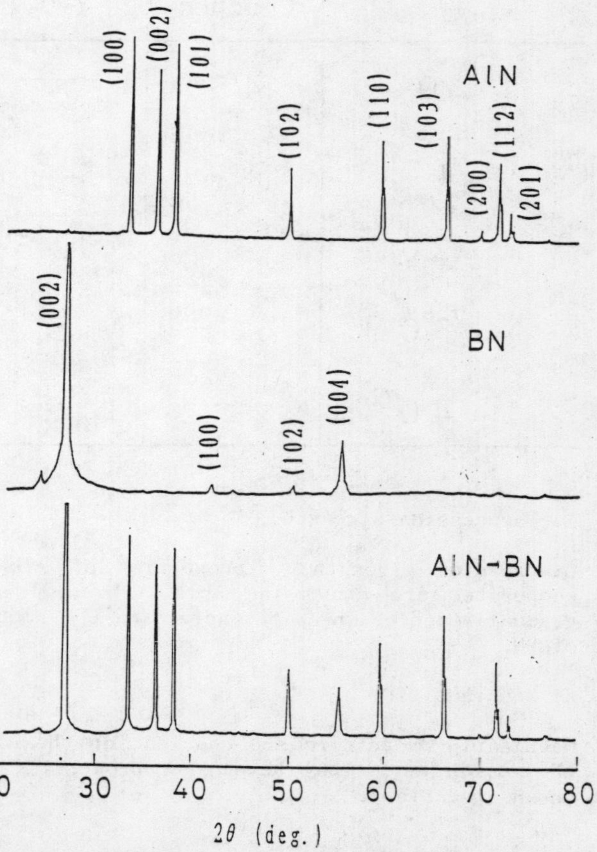

Figure 10 X-ray diffraction patterns for various kinds of compounds.

Table 3 Mechanical properties of ALN-BN composites.

BN (wt%)	Compression Strength (kg/mm^2)			
	Undoped	Y_2O_3	CaO	B_2O_3
10	24	26	26	26
20	14	14	14	14
30	8.0	13	11	8.3
40	5.9	6.4	6.7	4.9

Table 4 Machinable properties of AlN-BN composites.
(sample: 10mmx10mm)

BN Content(%)	Machinable Time(Min)	
	Undoped	Y_2O_3
0	48	——
10	15	15
20	9	9.5
30	4.5	5
40	——	2

4. Compression Strength

Compression strength properties of AlN-BN composites are shown in table 3. They are gradually decreasing with increasing the amount of BN.

5. Machinability

Machinable properties are checked for the AlN-BN composites. They have good properties as shown in table 4.

CONCLUSION

We summarized our results in the following;

(1) Additives such as Y_2O_3, CaO and B_2O_3 have promoted the densitification of AlN ceramics.

(2) The thermal conductivity of AlN-BN composites has been improved by addition of Y_2O_3 or CaO.

(3) High thermal conductive AlN-BN composites have been obtained by hot pressing. (2000°C, 10^{-3}-10^{-2} Torr)

(4) The thermal conductivity of machinable AlN-BN composites has 150W/m·K at the amount of 10wt.% BN.

(5) The thermal conductivity of AlN-BN composite system has been explained by the model calculation (II).

(6) The AlN-BN composites have good machinable properties.

REFERENCES

1) N. Kuramoto and H. Taniguchi, J. Mat. Sci. Lett. 3, 471 (1984).
2) T. Kurokawa, H. Hamaguchi, Y. Shimada, K. Utsumi, ISHM Proc. 654 (1987).
3) Tokuyama-Soda Catalogue.
4) J.W. McCauley and N.D. Corbin, High Temperature Reactions and Microstructures in the Al_2O_3-AlN System, NATO ASI Ser., E, Martinus Nijhoff Pub., Boston, 1983.
5) W.J. Parker, R.J. Jenkins, C.P. Butler and G.J. Abbott, J. Appl. Phys. 32, 1679 (1961).
6) T. Azumi, Doctor Thesis (University of Tokyo) in Japanese, 1981.

HEAT TREATMENT OF $SrTiO_3$ BASED CAPACITOR-VARISTOR

REN LIN, MENG ZHONG-YAN, WU MING-TANG
Department of Electronic Engineering
Xi'an Jiaotong University, China

ABSTRACT

The diffusion process of sodium oxides is associated with an appearence of liquid phase during heat treatment. Such process has significant influence upon the electrical properties. There were no obvious intergranular phase observed. A double diffusion layer model has been proposed to interprete the origination of the $SrTiO_3$ BL capacitor-varistor behaviors.

INTRODUCTION

During the secondary heat-treatment process, the painted metal oxide diffuse into the grain boundaries of the $SrTiO_3$ based semiconducting ceramics to form a second phase layer. The resultant BL capacitor materials reveal a high apparent permittivity, low dissipation factor, high resistivity and the small temperature dependence of their dielectric properties over a wide temperature range[1]. It is generally believed that the internal-boundary-layer capacitors are achieved by the appearence of insulating second phase layers between n-type semiconducting $SrTiO_3$ grains[2-4]. R.Wernicke[2] has proposed a selective melting reaction mechanism to explain the formation of second phase layer.

Recently, a new type of $SrTiO_3$ based BL capacitors which also show nonohmic electrical characteristics have been reported[5,6]. The preperation of these capacitors are similar to that of their predecessors. The primary difference is that they are sodium duffused in the secondary heat treatment. Fujimoto et al.[7] have studied the diffused layer profile in such capacitor materials, using HRTEM and STEM. No intergranular phases have been observed, but a narrow ($\sim$10nm) sodium diffusion layer was observed. Most of adjacent $SrTiO_3$ grains are in direct contact. It was proposed that alkali ions enter solution on strontium sites in $SrTiO_3$ as acceptors, which upon segegation to grain boundary give rise to varistor electrical barriers.

This paper deals with the electrical properties of the sodium diffused $SrTiO_3$ based BL capacitor-varistor under different heat treatment. The microstructure was examined by SEM and SIMS. A double diffused layer model has been proposed to interprete the capacitor-varistor behaviors.

EXPERIMENTAL PROCEDURES

Starting materials $SrCO_3$,TiO_2 were used to synthesis stoichiometric $SrTiO_3$ by calcining at 1200°C for 2h. The substance after calcining was verified by X-ray powder diffraction to be pervoskite $SrTiO_3$. Then it was blended with 0.5mol% Nb_2O_5, 0.5mol% TiO_2, 0.02wt% Al_2O_3, and 0.04wt% SiO_2 of chemical reagent-grade purity by wet milling for 16h. The mixture was mixed with an organic binder and pressed into disks of Φ10mmX3mm at a pressure of 1.9 X 10^9Pa. The pressed disks were sintered at 1420°C for 4h in a reducing atmosphere (90%N_2+10%H_2 cylinder gas with a flow rate of 10cm³/min) and furnace cooled to room temperature. The sintered specimens were polished to 1mm in thickness and painted NaF paste. Subsequent heat-treatments were carried out at a series of different temperature for 1h in air, afterward furnace cooled (250°C/hr) or rapid cooled (700°C/hr) to room temperature to obtain two kinds of samples. Silver electrodes were applied for electrical measurements. The apparent dielectric constant, dissipation factor were measured with HP4274A LCR meter. The nonohmic characteristics and thermal stimulated current were measured using a HP4140B pA meter/dc voltage source. The nonlinearity value α was determined according to the empirical formula

$$I = KV^{\alpha} \quad (1)$$

Where K is a constant.

RESULTS AND DISCUSSIONS

The relations of dielectric and varistor properties with annealing temperature were shown in Fig.1. It can be seen that the apparent dielectric constant (εr) and dissipation factor

($\tan\delta$) drop down, and the DC resistivity (ρ) goes high when annealing temperature increases. An obvious change in ε_r, $\tan\delta$ and ρ was revealed in the temperature range 940°C to 1000°C. Upon at 1100°C annealing for 1h, the $SrTiO_3$ based BL capacitors with $\varepsilon_r \sim 27{,}000$, $\tan\delta \sim 0.5\%$ and $\rho \sim 1.35 \times 10^9 \Omega$.cm at room temperature were achieved. These samples also have nonohmic characteristics. With annealing temperature increasing, α and threshhold voltage V_{1mA} decrease. In particular, both curves drop sharply in the temperature range just mentioned. The results drawing from Fig.1 predict that the status and profile of grain boundary relatively alter more obviously in annealing temperature range 940°C to 1000°C.

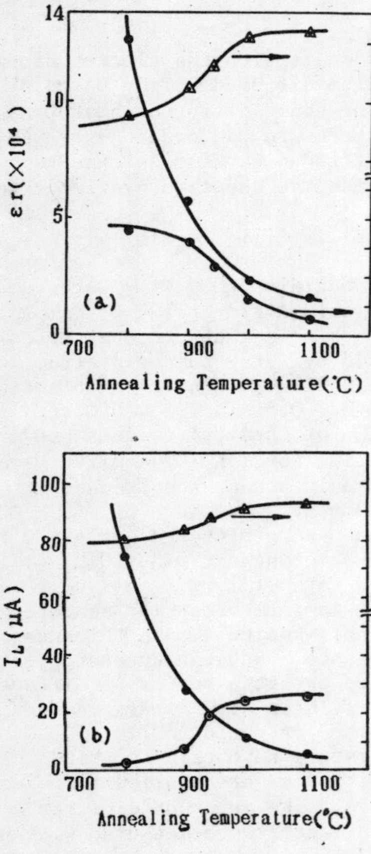

Fig. 1 The relations of (a) dielectric and (b) varistor behaviors with annealing temperature

Fig.2 shows the temperature dependence of dielectric constant, dissipation factor of typical furnaced-cooled samples annealed at a series of temperature. The frequency dispersions of dielectric constant over temperature range from -100°C to 200°C were minimized as annealing temperature going up. Upon annealing at 1100°C for 1h, the temperature coeffient of dielectric constant (at 1Kz) is lower than 0.4%/°C over a wide measurement temperature range. Futhermore, a typical relaxation peak appears in the temperature dependence of loss tangent for the samples annealed above 940°C. The activation energies of the appearing polarization were determined by loss tangent peak shifting under different frequencies to be .61eV, .65eV and .71eV for 940°C, 1000°C and 1100°C annealing samples repectively. It is implied that the peak revealing is caused by the sodium diffusion process. In direct examinations of painted sodium oxide under the secondary firing, following sequence of steps occuring were observed. The colour of painted oxide was slightly changed at 780°C, the oxides melted and began to contract at 900°C, up to 940°C the contracting process accelarated, and above 1000 °C the molten oxide dissapeared rapidly. According to the phase diagram Na_2O-TiO_2[8], there is a eutectic point at 986°C which will be lowerd by the appearence of SiO_2 and Al_2O_3 ad-

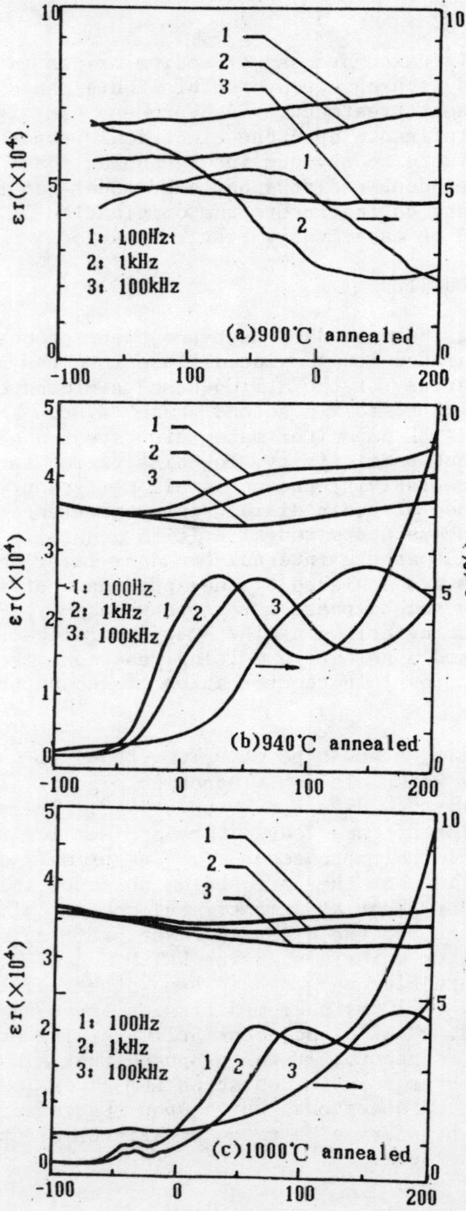

dictives at grain boundary. Upon wetting the $SrTiO_3$ grains, the melting sodium oxide react with excess TiO_2 and form a liquid phase rapid penetrating along grain boundary[2]. This kind of process has been strongly supported by the results in Fig. 2. It can be understood that deffused Na^+ in solution in the $SrTiO_3$ are more likely substitute for Sr^{2+} as acceptors than for Ti^{4+} because of better matching of ionic radius and valence, and consequently the sodium acceptors upon segregation to the grain boundary cause the negative space charge to pile up and exhibit typical relaxation phenomina in Fig. 2.

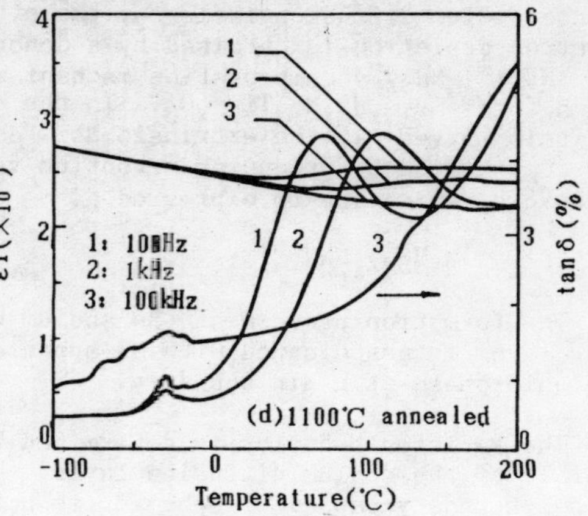

Fig. 2 The temperature dependences of dielectric properties

On the other hand, the posttreatment cooling rate having a significant effect upon the electrical properties of $SrTiO_3$ based BL capacitor-varistors can be expected. Fig. 3 shows a logarithmic plot of the nonohmic current-voltage traces for two different samples. The rapidly cooled sample exhibits a bad varistor behavior as compared to the furnace-cooled sample and reveals a smaller value of α and V_{1mA}, larger dissipation factor and leakage current. But there is no significant difference in their dielectric constant at room temperature. Above measurements indicate that the electrostatic barriers existing at grain boundary are substantially influenced by the posttreatment cooling rate.

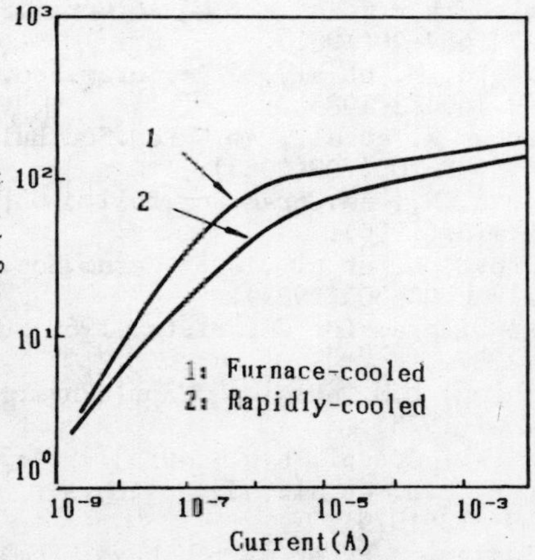

Fig. 3 Current-voltage traces of the two different samples

The thermal stimulated depolarization current (TSDC) spectra of these two different samples are illustrated in Fig.4. By comparing, the peaks in Fig. 4 are clearly related to the sodium ions segregation at grain boundary, which give rise to acceptor electron traps. The rear-part of current peak in curve 2 is broader than that in curve 1, which means the rapidly cooled sample has a broader energy distribution of the electron traps. It implys that the solution and segregation of sodium ions in $SrTiO_3$ lattice at grain boundary are strongly influenced by the posttreatment cooling rate.

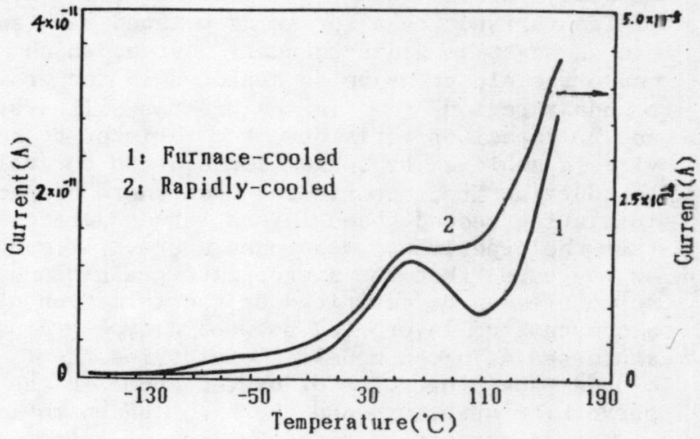

Fig. 4 TSDC spectra of the samples with different cooling rate

The microstructure of a sample annealed at 1000 °C for 1h is shown in Fig. 5. As can be seen, most of the grains directly contact and there is no obvious intergranular phase (>.1um) observed. Some small grains ($\sim$1um) enriched in Ti (confirmed by EDAX analysis) dispersed in the $SrTiO_3$ matrix.

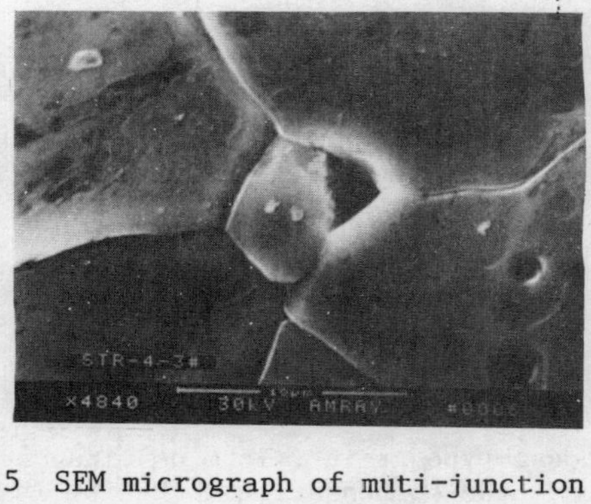

Fig. 5 SEM micrograph of muti-junction

The surface and intersection of a sample were analysed by SIMS with argon ion etching (Table. 1). Because the intersection contains more areas of grain boundary region, the results in Table. 1 demonstrate that most of sodium impurities locate at grain boundaries and grain-boundary region.

Table 1 Composition changement with Ar^{+} etching

Na/Sr \ Time	0mins	5mins	10mins
surface	0.30	0.20	0.16
intersection	0.74	0.36	0.14

Previous studies have found that the grain-boundary barrier in n-type semiconductors which yields varistor behavior is associated with an excess negative grain-boundary charge, which cause the electrons being depleted in the grain-boundary region[9-11]. In the present $SrTiO_3$ based BL capacitor-varistors, the varistor behavior is achieved by sodium segregation to grain boundary[7]. It is considered that there are no insulating second phase layers between grains from the experiments mentioned above. It can be suggested that the appropriate grain-boundary structure can be described by the formation of a compensated layer (c) between n-type grains expressed as n-c-n model. Considering the closed-pack properties of oxygen atoms in the pervoskite structure and the high concentration of oxygen vacancy produced by loss of oxygen from the material in reducing atmosphere, high rates of oxygen diffusivity at elevated temperatures can be expected[12]. Therefore the oxygen diffusion layer must be thicker than sodium diffusion layer. A double diffusion layer ($n-c_1-c_2-c_1-n$) model as shown in Fig. 6 can substantially describe the grain-boundary structure of $SrTiO_3$ BL capacitor-varistors as follows:

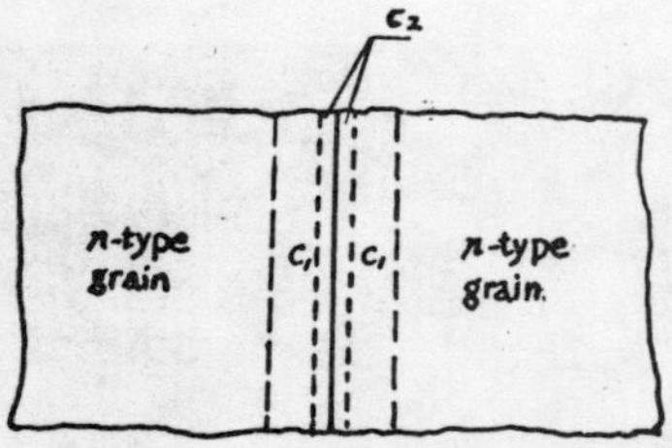

Fig. 6 Diagram for $n-c_2-c_1-c_2-n$ model

1. Good n-type grain (<1ohm.cm) induced by reduction, the compensation mechanism is electron compensation by the scheme $SrNb_x^{5+}Ti_{1-x}O_{3-y}$. In the sodium diffusion layer c_2, the effective electron depletion is obtained by a donor-acceptor ($Nb_{Ti}^{+}-Na_{Sr}^{-}$) compensation mechanism, which forms $Sr_{1-x}Na_xNb_xTi_{1-x}O_3$. In the oxygen diffusion layer c_1, the extrinsic Nb^{5+} donors are compensated by producing strontium vacancies V_{Sr}^{2-}, which can be expressed as

$$Sr_{1-x/2}(V_{Sr})_{x/2}Nb_x^{5+}Ti_{1-x}O_3.$$

2. The formation process of the sodium diffusing layer is associoated with an appearence of a liquid phase at grain boundary.

3. The varistor behavior is determined by the profile of the sodium diffusion layer. The internal-boundary-capacitor behavior is achieved by both the oxygen and sodium diffusion layers.

CONCLUSIONS

In the heat treatment process, diffused sodium oxide react with excess TiO_2, SiO_2 and Al_2O_3 existed at grain boundary and form a liquid phase which rapidly penetrates along grain boundary, followed by cooling, the sodium ions diffuse into $SrTiO_3$ lattice to form a diffudion layer. Most of adjacent $SrTiO_3$ grains are in direct contact, there is no obvious intergranular phase observed. A double diffusion layer model has been proposed to describe the grain-boundary of $SrTiO_3$ BL capacitor-varistor.

ACKNOWLEDGEMENT

The work was supported by the National Science Foundation of China.

REFERENCES

1. Takaashi, Y. et al., U.S.Pat. No. 3933688, 1976.
2. Wernicke, R., pp.261-81 in Advances in Ceramics, Vol.1. Edited by Levinson,L.M. American Ceramic Society, Columbus, OH. 1981
3. Franken, P.E.C. et al., J.Am.Ceram.Soc., 64[12] 687-90(1981).
4. Fujimoto, M. et al., J.Am.Ceram.Soc., 68[4] 169-73(1985).
5. Yamaoka, N. et al., Am.Ceram.Soc.Bull., 62[6] 698-700,703(1983).
6. Yamaoka, N., Am.Ceram.Soc.Bull., 65[8] 1149-1152(1986).
7. Fujinoto, M. et al., J.Am.Ceram.Soc., 68[11] C300-303(1985).
8. Phase Diagram for Ceramists, 1969 Supp. Fig.2287.
9. Levinson, L.M. et al., J.Appl.Phys., 46[3] 1332-41(1975).
10. Eda, K., J.Appl.Phys., 49[5] 2964-72(1978).
11. Mahan, G.D. et al., J.Appl.Phys., 50[4] 2799-812(1979).
12. Paladino, A.E. et al., J.Phys.Chem.Solids 26, 391(1965).

A COMPUTER AIDED SIMULATION STUDY FOR THE MACROSCOPIC PERFORMANCE OF ZINC OXIDE VARISTOR

Song Run-Sheng and Cao Ze-Chun
Shanghai University of Science and Technology
Shanghai, 201800, China

ABSTRACT

In this paper, effort is paid to simulate the performance of zinc oxide varistor. The computer aided simulation is conducted in two different scales; simulation the performance of component from property of local material, using microscopic parameters to predict property of small macroscopic region. In the simulation, a series-parallel equivalent circuit is adopted. The result is in good agreement with the experimental data.

INTRODUCTION -- NONUNIFORMITY

The outstanding I-V characteristics of zinc oxide varistor (ZNR) comes from the barrier effect of grain boundary (GB) between numerous grains constituting this material. The GB network distributes through out the whole material but its volume percentage is small, so that ZNR ceramics is a material with microscopic nonuniformity. On the other hand, the feature of ZNR in composition is its multiple component and minor additives. Some authors consider that Bi_2O_3 and other minor additives concentrated mainly at GB.[1] Hence it make up nonuniformity of ZNR in composition. These foregoing nonuniformities make the performance of the device very sensitive to its composition and processing. Therefore it is meaningful to study the influence of nonuniformities on the electrical properties of ZNR. Furthermore this subject can be studied in two aspects or two levels: that is influence of the nonuniformity of local material on the performance of device and relationship between microscopic nonuniformity and macroscopic property. Although this is a subject of great importance both in theory and practice and has been paying attention for a long time, but it relates how to treat various factors and numerous data. So that little or no progress has been made on the macroscopic system of GB network yet.[2]

INFLUENCE OF LOCAL CHARACTERISTICS ON THE PERFORMANCE OF ZNR DEVICE

1. Experimental determination of I-V curve

Generally, the processing is under critical control but the nonuniformity unavoidably influences quality of the device. In order to examine the subject, samples of lightening arrester components are available from Shanghai Electric Porcelain Works. The size of components are 50(dia)x9.5(thick)m.m. After I-V curves of the components are measured (the data and α-value derived are listed in Tab.I), it is sliced into small thin wafers with thickness of 0.94m.m. A pair of ohmic contacts of 8m.m. in diameter is

Table I Experimental I-V data of components

No.	I(mA) 0.1	0.2	0.4	0.6	0.8	1.0	– value
	V(kV)						
2	1.54	1.59	1.61	1.62	1.34	1.64	36.0
3	1.30	1.35	1.40	1.43	1.45	1.46	19.8
4	1.43	1.45	1.49	1.51	1.53	1.53	34.0
5	1.24	1.30	1.37	1.41	1.43	1.44	15.3

made on both surfaces of each wafer, then I-V curves of the wafers are determined and nonlinear coefficients, α, are calculated (shown in

Table II α-value of wafers sliced from components

No.	α-value					
2	39.10	37.55	36.49	36.19		
	18.35	20.55	18.52	25.42	24.81	7.473
3	27.34	22.62	12.57	16.38	3.934	22.07
	22.01	17.30	20.77	19.49		
4	31.20	29.96	25.07	30.44	34.00	31.72
	28.37					
	16.74	17.71	17.02	5.320	12.65	15.83
5	18.05	18.60	12.38	17.94	15.61	9.60
	15.78					

Table II). From Table I and II, it can be seen that in components No.2 and No.4, the α-value is larger, corresponding α-value in wafers cut from these components are also larger and concentrative. On the other hand, as the performance of No.3 and No.5 are no good, α-value of its wafers are rather small and dispersive.

2. Computer simulation for α-value of ZNR component

In breakdown region of ZNR, I-V curve may be expressed as follows:

$$\ln(I) = \alpha \times \ln(V) + A \qquad (1)$$

where A is a constant. This equation fits the experimental I-V curve by computer aided least square method, and α-value of each wafer is determined. For the simulation of electrical performance of component from the data of wafers, the component can be equivalent to a configuration consists of 10 layers with 40 wafers in each layer (10 columns and 40 lines) according to ratio of the area of the component to that of ohmic contacts of the wafer. Its equivalent circuit is shown in Fig.1. Then I-V curve of the component may be simulated by series-parallel connection of a number of wafers from this model and α-value is calculated from these curves. The flowchart of simulation is shown in Fig.2, and it is explained in brief as follows: From M wafers of a component, labeled from 1 to M, M characters obtained experimentally constitute a data base. Then 40X10 random numbers are extracted from the base and endowed with random array N(J,K), which makes up a component with 10 columns (K=10) and 40 lines (J=40). When I-V characteristic of the component is simulated,

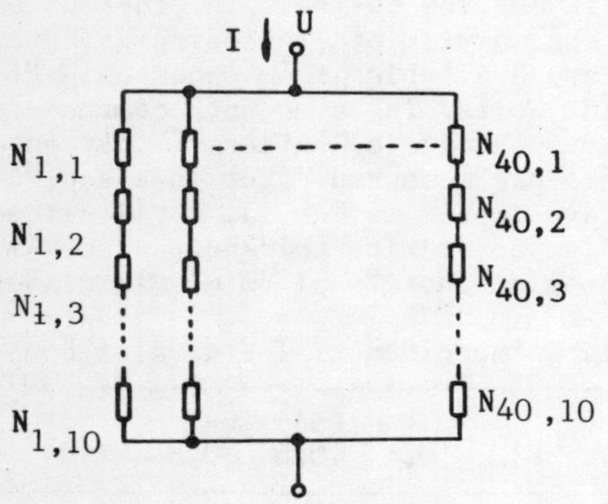

Fig.1. Equivalent circuit of component

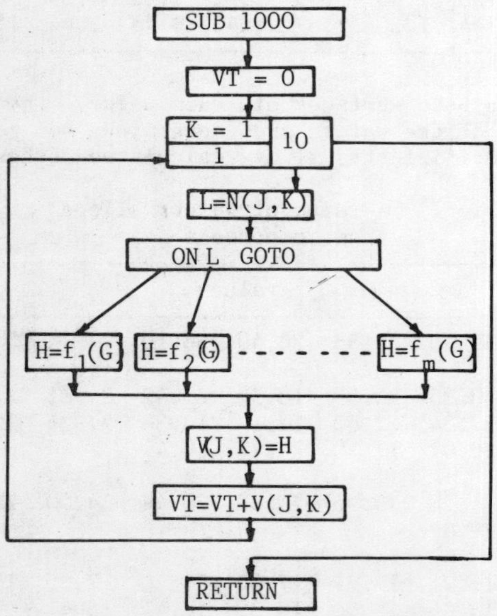

Fig.2(a) Flowchart of the subroutin 1000

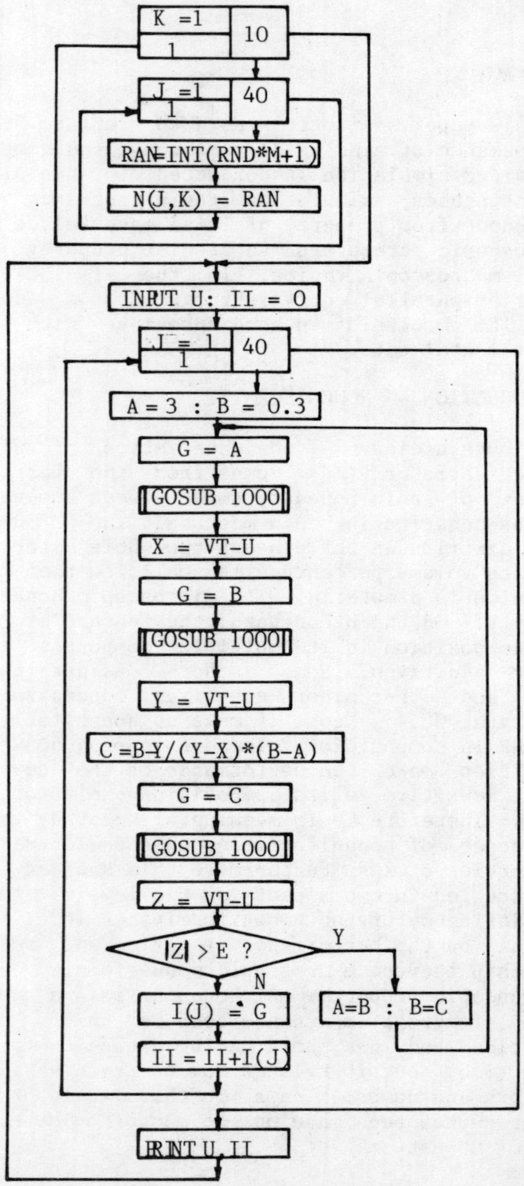

Fig.2(b) Flowchart of simulation

current,I, in each of the 40 parallal branches under certain voltage is determined by iterative computing method.
Subroutine 1000 is designed for the calculation of voltage drop across each of the branches. Since I-V characteristic curves for each of the 10 wafers, which constitute a branch, are endowed with random array N(J,K), when program begins,relationship between voltage,H, and current,G,of the wafer is read with ON-GOTO statement, then the voltage drops across every branch in a wafer are added together. After the currents of 40 branches are calculated separately and summed up together,the total current of the simulated component under certain voltage, V, is determined. When input different value of V, the corresponding total current ,I, can be determined. Thus the whole I-V characteristic of the component is obtained.
According to the program shown in above flowchart, α-value of the components are calculated, and list in Table III.In the table,case 1 and 2

Table III Comparison of determined and simulated α-value

No.	2	3	4	5
	35.0	19.8	34.0	15.3
case 1	37.3	16.3	30.0	13.8
case 2	37.3	16.6	30.1	13.4

are characters of components, which are simulated result with two different random array, and compared with the result derived from experimental curves. According to this table, the attempt of simulation seems successful.

3. Influence of local property on the performance of the component

From the data listed in Table II, it is noticed that α -values of individual wafers are very low in component 3 and 5. For example, for No. 3-6 and No.3-11, it is 7.47 and 3.93 respectively. Influence of relative amount of the poor wafer on the α-value is listed in Table IV.
Using the same method, when No.3-11 wafer is

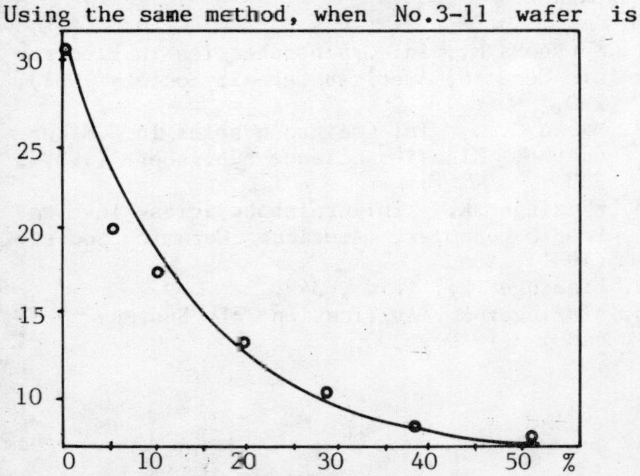

Fig.3 Influence of No.3-11

Table IV Influence of No.3-6 and No.3-11 on the α-value of component 3

percentage of No.3-6 and No.3-11	12.25	6.75	4.5	3.5	3.0
α -value	16.30	17.45	17.77	18,50	18.57

added randomly to component 4, whose property is better, with different amount. The α-value are calculated and results are shown in Fig.3. It can be seen that the properties of local material have significant influence upon the performance of the whole component.

GB MODEL AND MICRO I-V CHARACTERISTIC CURVE

1. GB model

It is common accepted now that the feature of ZNR GB is its double Schottky barrier. However further research shows that a part of GBs are dominanted by Bi_2O_3 segregation layer with its α-value between 1 and 2.[3] In addition, there may exist another kind of GBs, which have neither Bi_2O_3 segregation nor barrier, its resistivity is comparable with that of grains. In the above three kinds of GBs, the first one is the source that leads to nonlinear characteristic. So it may be called "effective GB". As for the percentage of effective GBs, it can be estimated as follows: For the wafers given in Table II, the average size of grain is about 10μ , hence there are about one hundred grains along the direction of wafer thickness, and it must be 100 GBs at least in series for each branch in Fig.1.
If there were Schottky barrier in every GB, the breakdown voltage of wafer should be 340 v. But the experimental value is only 140 to 150v. Therefore the percentage of effective GBs is only about 40%.
The equivalent circuit of grain (include GB) is shown in Fig.4.

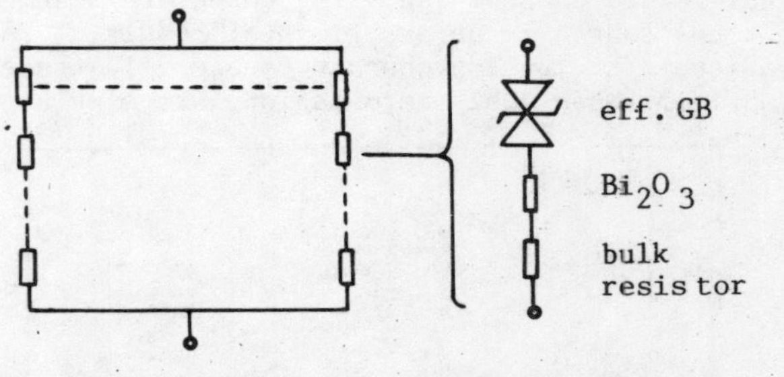

Fig.4 Equivalent circuit of grain

Besides the resistor which represents the grain there are three kinds of element representing GB emerging randomly and excluding each other. According to the above-mentioned estimation from experimental result, probability for the effective GBs is 40% and that for the Bi_2O_3

segregation is 15%.

2. The microscopic I-V characteristic

The I-V charateristic of effective GBs can be formulated as follows:

$$I = V^{\alpha_1}/C \qquad V < Vb$$
$$I = (V - V_1)^{\alpha_2}/C \qquad V \geq Vb \qquad (2)$$

where $V_1 = Vb - (Vb^{\alpha_1})^{-\alpha_2}$
C is a constant, $\alpha_2 = 100$, $\alpha_1 = 2\text{-}4$.[3] α_1,C are derived from experimental I-V curves, and these are seven sets of data available. The I-V character of Bi_2O_3 segregation layer is also formulated from the same source.[3]

$$I = V^{1.39} / e^{14.13} \qquad (3)$$

THE SIMULATION OF MACROSCOPIC PROPERTY FROM MICRO-PARAMETERS OF GBs

1. Model

In this section the simulate method used is similar to section 2.2,it is based on the equivalent circuit illustrated in Fig.1 and the computer program sketched in Fig.2.The simulate model is a parallel circuit with 10 branches, each consists of 100 series microscopic units. They make up a small macroscopic region (SMR).The area of SMR depends on the numbers of branches, i.e.the value of K.Simulate computation shows that α-value is nearly independent of the number of parallel branches and also of the area of SMR. Therefore a model of 10 parallel branches is enough for the simulation, and a circuit with 10 branches is adopted.

2. The influence of GB breakdown voltage

It is accepted that the breakdown voltage of GB Schottky barrier is 3.4V.[4] As a matter of fact the breakdown voltage have its own distribution.
In order to study the influence of the voltage distribution on α-value, four distribution characters are chosen. The first three are assumed and the fourth is an experimental result.[5] The various distribution characters are illustrated in Fig.5. When α of micro-regions are simulated with the various kinds of distributions the following expression is used:

$$\alpha = \lg (I_2/I_1) / \lg (V_2/V_1) \qquad (4)$$

the result is listed in table V. So that α-value becomes larger as the breakdown voltage is more concentric. The above result is obtained when bulk resistivity of grain is 0.1 Ω-cm,[5] and the ratio of effective GB is 40%.

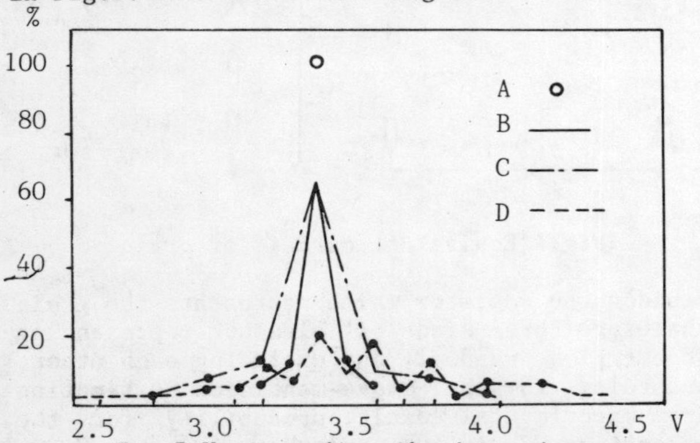

Fig.5 Various distributions characters

Table V The influence of the distribution of the breakdown voltage

distribution of the breakdown voltage	A	B	C	D
α-value	79.2	46.5	40.5	38.1

DISCUSSION AND CONCLUSION

In this papers α-value, the most important parameter, has been examined in two levels. Stress is paid to the influence of imhomogeneity of local material on the whole performance. In the first level,the performance of component is simulated by that of wafers. The calculated result is in good agreement with the measured data. It indicates that the model and calculative program designed are correct. It also lays the foundation for this kind of simulation.
In the study of second level, within the microscpic region, various factors which influence the α-value are revealed more deeply. The factors listed are: ratio of effective GB and its distribution, ratio of Bi_2O_3 segregation,distribution of breakdown voltage and bulk resistivity of grain. But up to now a complete set of experimental data of micro-parameters didn't available yet. Therefore a more complete simulation should wait for the further accumulation of experimental data.

REFERENCE

1. Matsuoka M., In: Grainboundaries in Electronic Ceramic, American Ceramic Society(1981), 290.
2. Mahan C.D., In: Grainboundaries in Semiconductor, Elsevier Science Publishers (1982), 333.
3. Einzinger R., In: Grainboundaries in Electronic Ceramies, American Ceramic Society (1981), 356.
4. Einzinger R., ibid., 343.
5. Einzinger R., Application of Surface Science, 3(1979), 390.

A RESEARCH IN ZnO VARISTOR PREPARING PROCESS

Wang Peiying, Wang Shiliang and Rao Wenhua
Huazhong University of Science and Technology,
WUHAN, P. R. CHINA

ABSTRACT

ZnO and additives in powder form of fine particles with high activity but free from pollution can be obtained by liquid phase reaction methods. In the ZnO varistor, the additives of solid phase oxides of Bi, Sb, and Mn are usually large particles, so that they can not be easily crushed, and evenly mixed by the conventional method. Since they play an important role in the electric properties of the varistor, we have adopted the solution method for powder preparation, and rather satisfactory results have been obtained.

INTRODUCTION

In ZnO varistors the ZnO powder is used as the base composition and several other metal oxides are added to it. The ZnO varistor prepared by the conventional ceramic process is of high compactness, small coefficient of volume expansion and accurate size. However, the solid phase reaction among different kinds of powder is closely related to the particle size. The average diameter of ZnO powder is somewhere between 1 and 2um, and that of Sb_2O_3 powder and MnO_2 powder is greater than 10um, then in the process of mixing and grinding, not only other impurities are easily introduced into them but also the reaction takes place mainly among large particles of the powder oxides. As for the trace dopant, since the number of its particles is much less than that of ZnO particles, an even contact can not be established among the particles of the base composition and those of the trace dopant. Therefore it is impossible to obtain an ideal homogeneous mixture.

Liquid phase reaction methods have been adopted to improve the homogeneity of the mixtrue in the process of the powder preparation. There are many kinds of liquid phase reaction methods, one of which is the salt separation method. By this method tiny salt particles are separated from salt solution of high solubility, and then they are decomposed into ultrafine oxide particles by heating.

The second method of liquid phase reaction is the coprecipitation method. By this method acid or alkali is added in the solution of solvable metal salts. After they react, insoluble solution is produced and the precipitate is developed in the solution. The sol-precipitation method is an example of the coprecipitation method.

The third method is the Sol-Gel mehod. By this method various metal alkoxides are dissolved in some organic solvent and they form a homogeneous solution. The size of the sol particles is somewhere between 1 nm and 0.1μm. This solution is easy to become a very homogeneous and transparent gel. By calcining the gel ultrafine and active powder the ceramic products can be obtained. This method has many advantages. For example, reactions can proceed among molecules or atoms of various metal alkoxides, making the reaction quite full and the homogeneity of the products satisfactory.

The method of powder preparation must be selected according to the electronic ceramic product to be produced. As for ZnO varistors, there are a lot of dopants in it, all the corresponding metal alkoxides must be prepared, and they must be able to dissolve in the same solvent. In a word, the conditions of the powder preparation of ZnO varistors are very strict, and the cost is high. Therefore, the sol-gel method is very difficult to be adopted.

This paper discusses the sol-precipitation method and the inorganic salt solution method. By these two methods ultrafine and active powder can be produced after being calcined at lower temperatures. The electrical properties of ZnO varistors thus prepared have been improved.

EXPERIMENTAL METHODS

The composition shown in Table 1 remains the same

for all the following samples.

Table 1 Composition of ZnO varistor(mol. %)

ZnO	Bi_2O_3	MnO_2	Sb_2O_3	Co_2O_3
98	0.5	0.5	0.5	0.5

preparation of sample A, by conventional method.

ZnO powder was mixed with various metal oxide dopants according to the directions given in Table 1. Then the mixture was ballmilled, calcined, shaped and sintered.

preparation of sample B, by aslt separation method[1].

According to the directions given in Table 1, various metal salts listed in Table 2 were dissolved thoroughly in different solvents, forming solutions with definite concentration. Then they were mixed with ZnO powder. The mixture was calcined, shaped and sintered. Therefore, the processes of this method were the same as those of the conventional one except that the dopants were inorganic salts.

Table 2. Inorganic salts used in the solution reaction method.

salts	$Co(NO_3)_2$	$Bi(OH)_9(NO_3)_4$	$Mn(NO_3)_2$	$SbCl_3$
solvents	H_2O	HNO_3	H_2O	HCL

Preparation of sample C, by sol-precipitation method[2]

Some metal inorganic salts were dissolved in different solvents to form solutions. Then these solutions were reacted with NH_4OH solution in sequence, forming some hydroxide colloidal precipitates. In the experiment, NH_4OH was added till all the colloidal solution was precipitated. Therfore, the PH value of the colloidal solution must be controlled. Otherwise, there would be stoichiometry deviation. The salts and solvents used to form a colloidal solution and other data are listed in Table 3.

Table 3. Materials for preparaing sample c and other data.

salts	$Zn(NO_3)_2$	$Co(NO_3)_2$	$Bi(OH)_9(NO_3)_4$	$Mn(NO_3)_2$	$SbCl_3$
solvents	H_2O	H_2O	NHO_3	H_2O	HCL
washing agent	H_2O	H_2O	H_2O	H_2O	NH_4OH H_2O
PH	7.5	9.3	8.6	11.0	8.0
color	white	green	white	brown	white

To prevent $SbCl_3$ from hydrolyzing, $SbCl_3$ was first dissolved into HCl colution. Then NH_4OH was added in it, forming a white colloid hydroxide precipitate. To remove all the trace chloride ions in the solution, it was thoroughly washed with dilute NH_4OH, tested by the standard silver nitrate reagent, and rewashed with water. Other hydroxide precipitates were washed and rewashed with water to remove surplus dielectrics in the solutions until mist precipitates were observed. These washed precipitates were then put in to a bottle. At least three times water was added in the bottle. The wet precipitate mixture was stirred in a highspeed stirrer, causing further dispersion of the mixed oxide sol.

In sample C(1) the mixed oxide sol contains the base composition Zn, while in sample C (2) the mixed oxide sol contains only additives. ZnO powder was added after it had been stirred for 24 hours, and then they were stirred with water for another 24 hours. Sample C (1) and sample C (2) were heated at 100℃. The mixed oxide sol was subsequently evaporated and the gel was obtained. Other processes of this method were the same as those of the conventional method.

RESULTS AND DISCUSSION

1. Density and activity

Table 4. Density of samples*

sample	A	B	C(1)	C(2)
density g/cm³	5.33	5.32	4.51	5.37

It is well known that the smaller the size of particles, the higher the dispersity, and the greater the specific surface, the higher the surface free energy. Furthermore, the higher the system activity, the easier the physical chemistry reaction required will take place and therefore, the more helpful to sintering it is. The average size of oxide powder particles prepared by sol precipitation is less than 0.5 μm, which are fine particles and of high activity. We can find from Table 4 that the density of sample C(1) is the smallest, only 80% of the theoretical density of 5.6g/cm³. If the hot-pressing method is not used, the density can not be increased. This is because there are residual hydroxyl · in the body after the sol is decomposed, and because the ceramic becomes porous when sintering and removing of gases proceed at the same time. Fig. 1 shows the microphotographs of the free surface of samples A, B, C(1) and C(2). C(1) is porous.

Sample B and sample C(2) were prepared by improved processes. The base composition used was still ZnO powder but solution additives were added instead of solid powder. On the one hand, the additives were mixed between molecules, making the mixture

* All samples presintered at 700℃ and sintered at 1150℃.

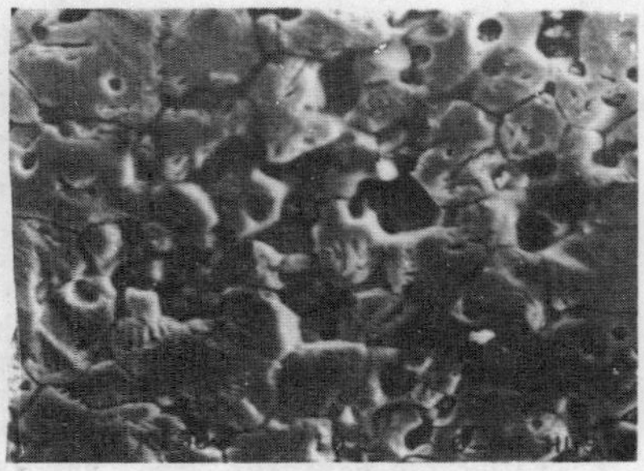

Sample C (1)

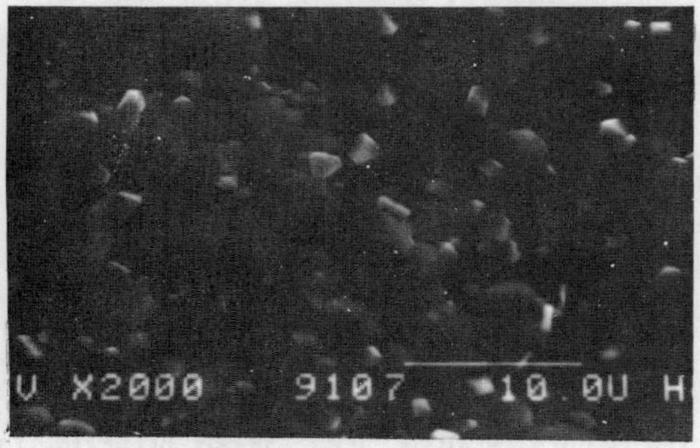

Sample C (2)

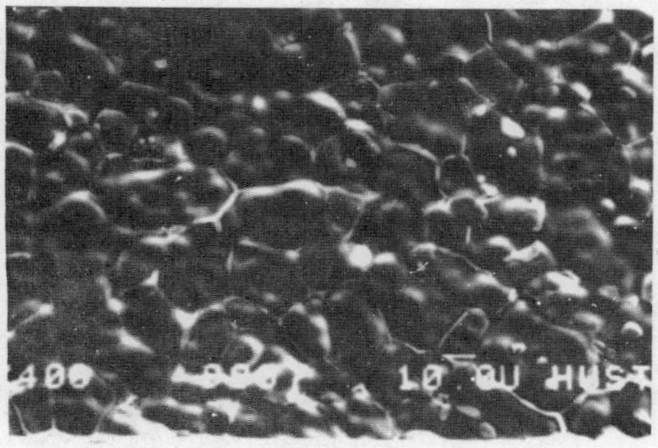

Sample A

Sample B

Fig. 1 micrephotographs of the free surface of sample A, B, C(1), and C(2).

homogeneous, and the activity of particles was great. Therefore, after the base composition ZnO was added, the particles of it can be wrapped by the solution, achieving sufficient contact. On the other hand, the amount of the hydroxyl and acid radicals in the raw semifinished product was much less than in sample C(1), so that in sample B and sample C(2) the amount of gases to be removed was much less. Therefore their respective density was increased.

2. Electrical properties

All the samples were in the shape of a disc with diameter about 1 cm and thickness 1-2mm, and their electrodes were made of silver. To make comparisons all the samples were sintered at 1200°C. Their electrical properties were measured and listed in Table 5.

Table 5. Electrical properties of samples

sample	$\frac{V_{1mA}}{mm}$	nonlinear exponent α	leakage curreut (μA)	100A $\frac{\Delta V}{V}$(%)	V - T coeff. $\times 10^{-4}$
A	90.2	48.7	1.6	5.3	-6.3
B	91.5	51.19	0.1	3.8	-2.6
C(1)	153.8	40.7	3.3	3.3	-3.0
C(2)	84.9	50.42	1.0	3.5	-3.2

Note that the expression of the nonlinear exponent α is $\alpha = 1/Log(V_{1mA}/V_{0.1mA})$. Table 5 shows that there are great differences in the values of V_{1mA}/mm of the samples prepared by different methods. The value of V_{1mA}/mm of sample C(1) is tens of volts greater. Furthermore, sintering and calcining temperatures have great influence on the value of V_{1mA}/mm, for example, the value of V_{1mA}/mm may be increased to somewhere 300 if the sintering temperature is 1150°C. Fig. 2 shows the relation of samples C(2) and B between V_{1mA}/mm and temperature of sintering and presintering. Table 5 shows that the voltage temperature coefficient and the rate of voltage change of sample A at 100 A 2ms current shock are greater than those of samples B, C(1) and C(2). All these show the advartages of the processes.

The microstructure and the homogeneity of samples are tested by 2ms large electric current shocking. The temperature at the grain boundary of the ZnO varistor will rise after large electric current shocking. If the ZnO varistor is of good homogeneity and of high heat transfer rate, heat will be transported from the grain boundary to the inner part of the grain, and heat will evenly distribute in the bulk. Otherwise, heat at the boundary area can not be promptly transported to the inner part, so that the temperature at some area becomes so high as to cause heat breakdown. The experimental result shows that the homogeneity of samples B, C(1), and C(2) is better than that of sample A.

The V-I characteristics of samples A, B, and C(2) are shown in Fig. 3. The nonlinear exponent α of varistors prepared by the salt separation method is greater than that by other processes.

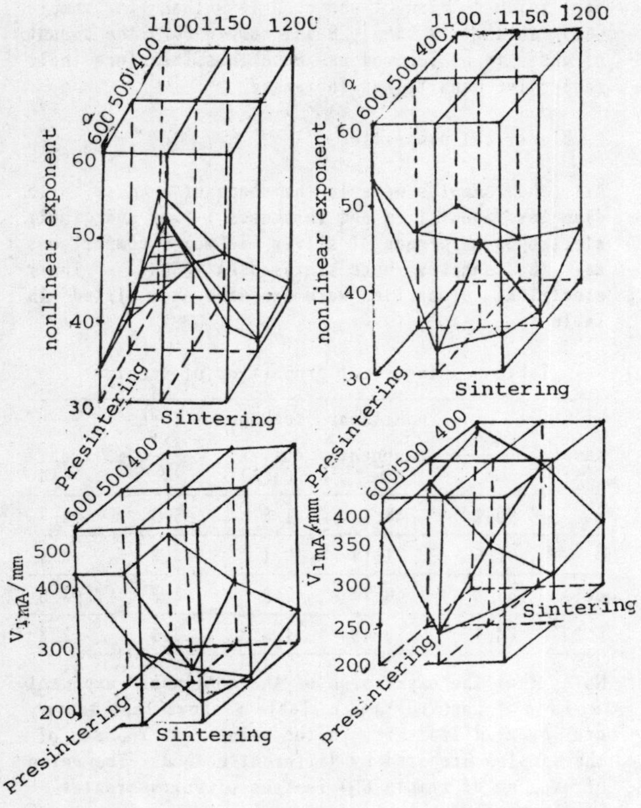

Sample B   Sample C(2)

Fig. 2 Relations of samples C(2) and B between V_{1mA}/mm and tepmeratures of sintering and presintering and between α and temperatures of sintering and presintering.

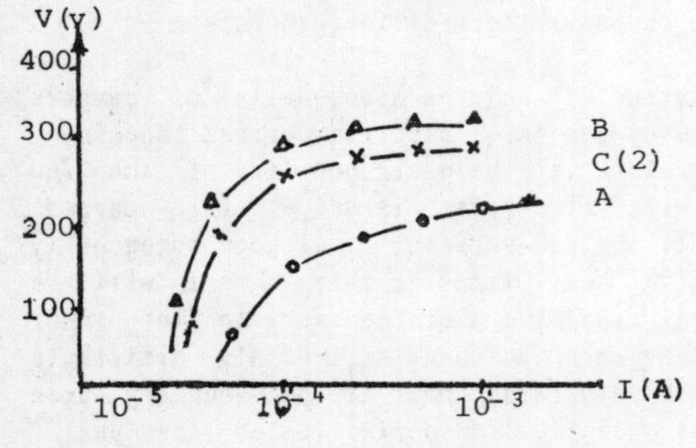

Fig. 3 V-I characteristics of samples of the ZnO varistor sintered at 1150℃

CONCLUSIONS

1. The homogeneity of crystal grain growth and the dispersity of the additives of the ZnO variator have been improved by the salt separation method and the sol precipitation method.

2. The density and the breakdown field of the ZnO varistor are influenced by different methods.

3. The bulk of the ZnO varistor can be prepared conveniently by the solution method, and the density of samples prepared by the sol-gel method can be improved by hot-pressing, thus obtaining high field varistors.

4. Thin and thick film varistors with excellent performance can be prepared by the sol-gel method.

REFERENCES

1. Li Biao Rong, The principle of electronic ceramic processing, HUST publisher (1986), P. 25.

2. LAUF, R. J., BOND, W. D., CERAMIC BULLETIN, Vol. 63, No. 2 (1984), P. 278.

CONTROLLING THE TEMPERATURE COEFFICIENT OF THE RESISTANCE OF THE METAL FILM RESISTOR

Li Huarui, Lu Yanxia
Beijing University of Science and Technology

Zhao Jianfa, Wu Yougun
Beijing First Radio Element Factory

Beijing, China

ABSTRACT

The relavant factors affecting the TCR and the method for controlling these factors were studied. The specimens studied included the Ni-Si, Cr-Si alloy films acquired by vacuum evaporation and the Ni-Cr-Si, Ni-Si alloy films acquired by sputtering technique.

The result of our study indicates that when the structure of alloy film consists of a mixture of crystalline phase and amorphous phase in proper proportion, the TCR approaches the value of zero.

INTRODUCTION

The metal film resistor, as one kind of electronic elements, has been widely applied in electro-mechanical and instrument-making industries. Due to their small volume, excellent high-frequency character and adequate stability, the production and application of the metal resistors has rapidly increased along the years. As the TCR (temperature coefficient of the resistance) is one of the most important characters of metal film resistor, the study of the factors affecting the TCR of the alloy film and the seeking of the methods for controlling these factors are of practical significance.

Usually metal films are prepared through the methods of vacuum evaporation or magnetic-controlled sputtering. The structure of the prepared films is amorphous or partially amorphous. In recent years, there have been many articles dealing with the electric properties of amorphous materials, especially on the electric properties of metallic glasses at low temperatures and considerable results have been achieved [1]. Thus for practical purpose, we laid our emphasis on the electrical properties of the amorphous alloy film at high temperatures.

As the temperature increases, the structure of the alloy film gradually changes from amorphous to partially crystalline, and their electric behavior calls for an explanation different from that of the purely amorphous structure.

* The Project Supported by National Natural Science Foundation of China.

EXPERIMENTAL PROCEDURE

1. The alloy films prepared by vacuum evaporation method were deposited on glass bases without being cooled. They are: $Ni_{90}Si_{10}$, $Ni_{80}Si_{20}$, $Ni_{70}Si_{30}$, $Cr_{90}Si_{10}$, $Cr_{80}Si_{20}$, $Cr_{70}Si_{30}$ respectively. The numeral after each symbol of the chemical element is its atomic percentage. The chemical analysis of these films was completed by an energy spectrum method on SEM with error ±10%. During the process of evaporation, the powders of various elements were mixed in different proportion. Then each powder mixture with alcohol was applied onto a tungsten wire which was used as the source of heating. However, the use of such a source resulted in the contamination of the prepared alloy film by about 2-3% tungsten.

2. Alloy films prepared by sputtering were also deposited on glass bases without cooling. Targets for sputtering were melted and cast into discs of dia. 65 x 6mm in vacuum. The chemical composition of the targets are $Ni_{80}Si_{20}$, $Ni_{70}Si_{30}$ and $Ni_{50}Cr_{15}Si_{35}$ respectively.

3. The electric resistance measurement and the heat treatment were performed under an argon atmosphere.

4. In this article, the TCR was defined by the following equation: $TCR=(R_{120^\circ C}-R_{20^\circ C})/R_{20^\circ C} \times 100$

RESULTS AND DISCUSSIONS

1. The temperature-resistance dependence of

sputtered Ni_{80} Si_{20} film in the temperature range of 20-700°C is shown in Fig. 1. It is a typical curve for the amorphous film. In the temperature range of 20°C-120°C, however, the resistance at starting gradually decreased with temperature and the TCR of this amorphous alloy film became eventually negative. Such a fact can be explained by the Ziman theory [2,3].

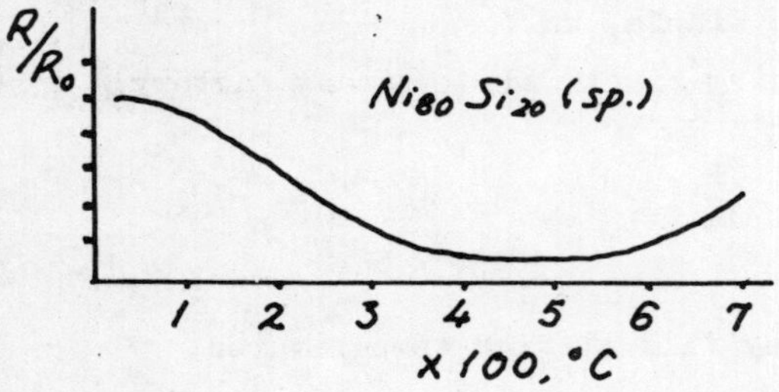

Fig. 1

Then the decrease of the resistance of this film soon quickened to minimum value as the temperature reached 460-538°C. The process of crystallization came to a stop as the temperature reached 538°C, when a Ni solid solution and a Ni_xSi_y are found to have formed. The result of the temperature-resistance dependence of the evaporated Ni-Si films is shown in Fig. 2.

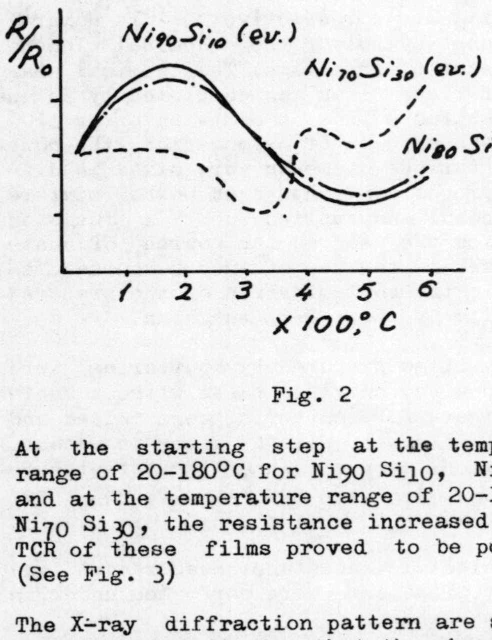

Fig. 2

At the starting step at the temperature range of 20-180°C for Ni_{90} Si_{10}, Ni_{80} Si_{20} and at the temperature range of 20-130°C for Ni_{70} Si_{30}, the resistance increased and the TCR of these films proved to be positive. (See Fig. 3)

The X-ray diffraction pattern are shown in Fig.4, which indicates that the phase structure of the sputtered film was amorphous and for evaporated film, partially amorphous and partially crystalline. Thus it can be inferred that the TCR of these alloy films directly depends upon the extent of crystallization.

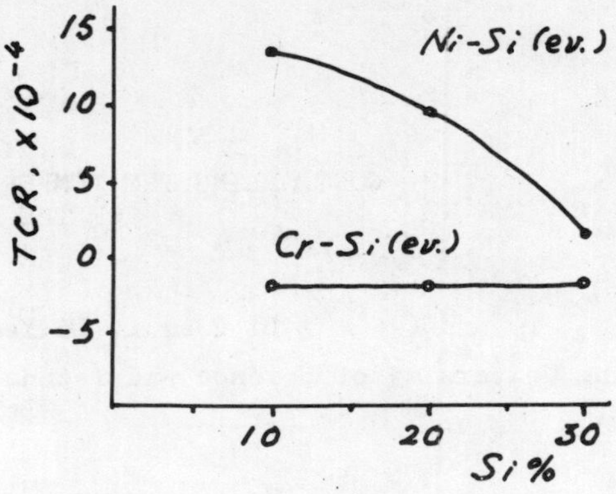

Fig. 3

What the above fact proved is that the structural state can be an important factor directly affecting the TCR of these thin alloy films.

2. The temperature-resistance dependence of evaporated Cr_{80} Si_{20} and Cr_{70} Si_{30} is illustrated by Fig. 5.

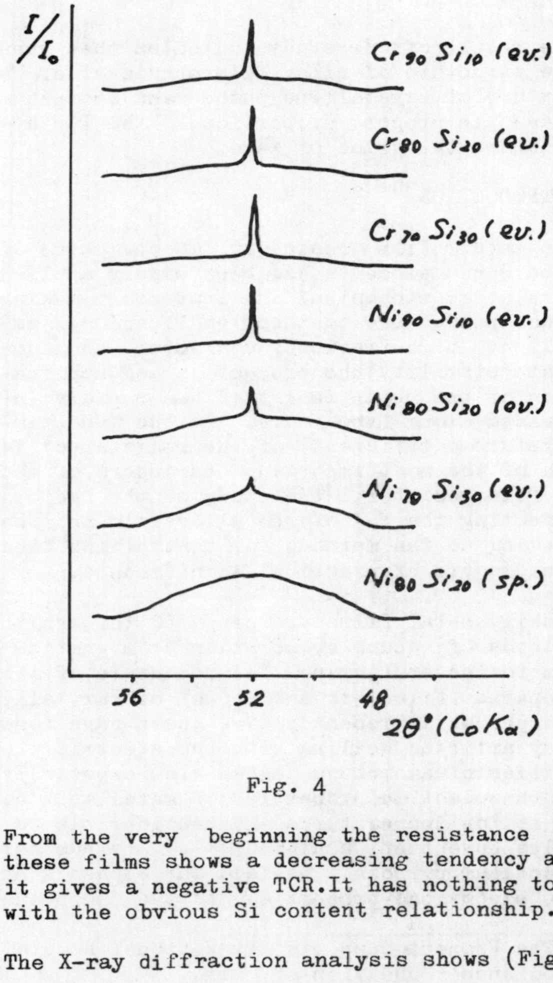

Fig. 4

From the very beginning the resistance of these films shows a decreasing tendency and it gives a negative TCR. It has nothing to do with the obvious Si content relationship.

The X-ray diffraction analysis shows (Fig.4)

that the phase structure of the evaporated Cr-Si alloy films is partially amorphous.

Compared with the phase structure of evaporated Ni-Si alloy films, bearing the same structural state, we found that the difference of chemical composition in alloy films proved to be another factor affecting the TCR of alloy films.

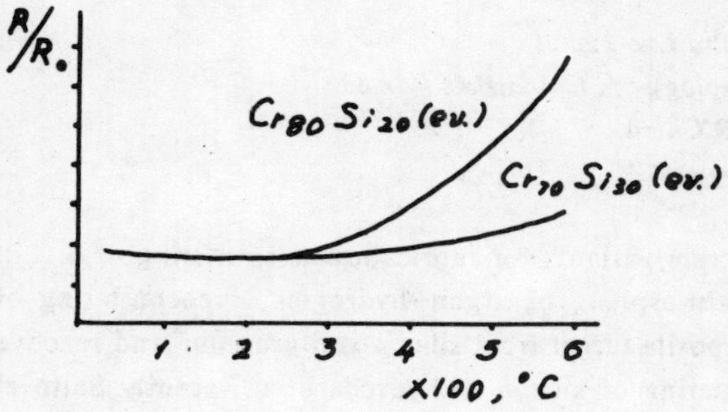

Fig. 5

3. The curve of resistance, changing with temperature for sputtered $Ni_{50}Cr_{15}Si_{35}$ alloy film is shown in Fig. 6.

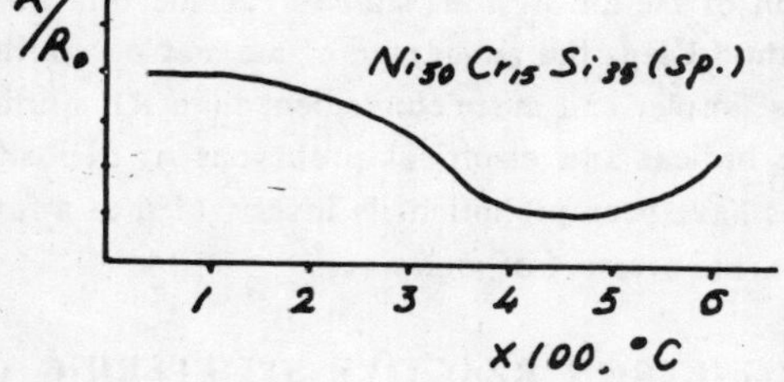

Fig. 6

The X-ray diffraction patterns of film specimens with various tempering temperature after sputtering is given in Fig. 7.

It is interesting to compare these X-ray patterns with Fig. 8, which shows a great dependence TCR upon the tempering temperature.

For the $Ni_{50}Cr_{15}Si_{35}$ alloy film, after low tempering temperature, when its structural state is amorphous, the TCR of this film is negative. And after high tempering temperature, when the structrual state of this film is entirely crystalline, the TCR shows a positive value. Once when the two phases of amorphous and crystalline are in proper proposition, the TCR will reach an approximately minimum value. Thus, through the proper control of the initial phase structure and of the temperature of the followed tempering of the alloy film the optimum TCR can be obtained.

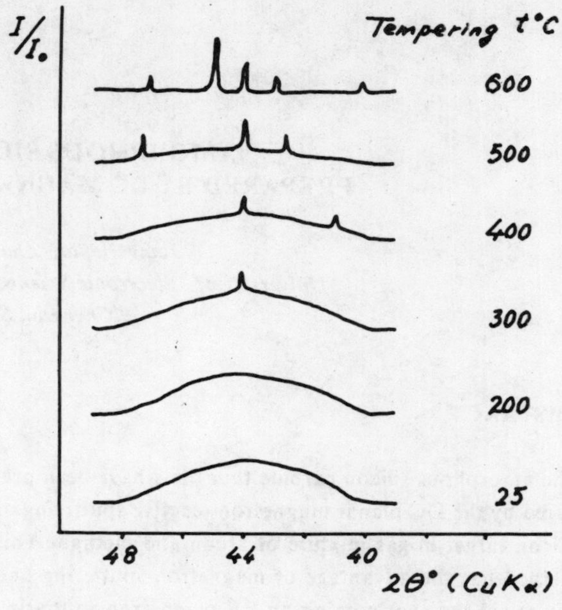

Fig. 7

ACKNOWLEDGEMENT

We thank Miss Gao Jia, Miss Li Danmei, Miss Huang Xiajun, Mr. Tan Li, Mr. Xu Guohong for their help in the X-ray diffraction experiments, resistance measurements and the preparation of specimens.

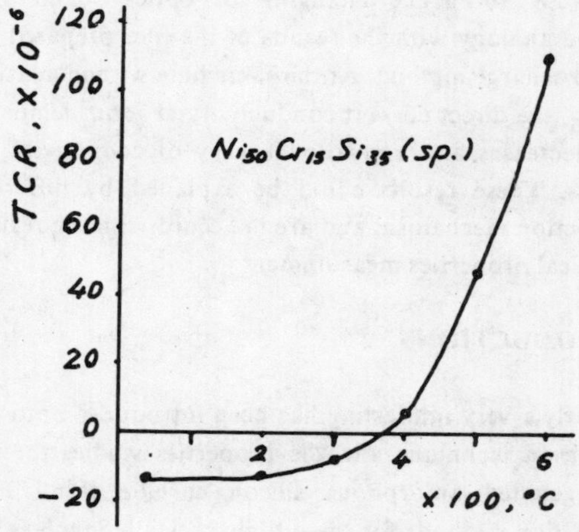

Fig. 8

REFERENCES

1. Meisel L. V. and Cote P. J., Phys, Rev., B 30 (1984), 1743.
2. Ziman J. M., Phil, Mag., 6(1961), 1013.
3. Cote P. J., Sol. St. Com., 19(1976),1311.

AMORPHOUS SILICON CARBIDE FILMS PREPARED BY DC MAGNETRON REACTIVE SPUTTERING

Jiang Jiejian, Zhu Yaojiang and Luo Jiahui
University of Electronic Science and Technology of China,P.O.Box 83
Chengdu, Sichuan, P.R.China

ABSTRACT

The amorphous silicon carbide thin films have been prepared by the DC planar magnetron reactive sputtering of silicon target in gas mixture of argon and methane.This methcd has the advantage of magnetron sputtering and can avoid the shortcoming by RF magnetron sputtering .The carbon content in the film was varied by changing the partial pressure of methane with respect to the total pressure .The optical and electrical properties of the thin films deposited by the different partial pressure of methane have been investigated. With increasing of partial pressure of methane, the optical band gap increases from 1.54eV at $P_{CH4}=0$ to 2.06cV at $P_{CH4}=2\times 10^{-2}$Pa. The optical band gap decreases again when $P_{CH4}>2\times 10^{2}$Pa.The changing of optical band gap showed analogy with the results of the film prepared by glow discharge method .At the same time with increasing of P_{CH4},the direct current conductivity at room temperature decreases and activation energy of coductivity increases. These results could be explained by different conduction mechanism and are in accord with the results of optical properties measurement.

INTRODUCTION

Recently,a very interesting has been introduced onto the deposition technique and the properties studies for the hydrogenated amorphous silicon carbide thin films. Deposition methods for these films basically involves the glow discharge decompostion and sputtering.In the glow discharge method gas mixture of silane and hydrocarbon was used as raw material, also alkylsilane was used[1].However,the sort of used hydrocarbon has significant effect on the structure and properties of films.For example, as the window material in a p–i–n a–Si solar cell the methane based a–SiC:H film is superior to other a–SiC:H films[2].In the sputtering method there are many concrete variants for fabrication,as sputtering of SiC in an atmosphere of argon–hydrogen[3],or cosputtering of composite target from silicon and graphite[4],and reactive sputtering of silicon in hydrocarbon[5].Recently Saito et al. reported that hydrogenated amorphous silicon carbide thin films were prepard by the magnetron sputtering of silicon in a gas mixture of methane and argon[6,7].However,all of these sputtering method belongs to the RF sputtering.

We first used the DC magnetron reactive sputtering for deposition of the amorphous silocon carbide thin films. This method keeps the advantage of magnetron sputtering,but is simpler and more convenient than RF sputtering .The optical and electrical properties of deposited thin films have been preliminarily investigated as a function of partial pressure of methane.

DC MAGNETRON REACTIVE SPUTTERING OF THE THIN FILMS

The amorphous silicon carbide thin films were deposited by DC diode planar magnetron sputtering apparatus.We had designed and fabricated magnetron target pole system[8].The outside diameter of target bay was 129mm. Within the target bay the magnetic bead and magmetic colum fabricated from different metallic constant magnetic material were constructed so that planar magnetic field was established on the surface of silicon target whose another suface was in tightly contact with the base of target polc .Within thc target bay input and output water–pipes were constructed and mobile water could cool the magnetron target pole system.In order to avoide the cffect from stray magnetic ficld without target pole system the magnetic screen was desposed. Moreever,also electric screen was installed to avoid bombardment of target bay edge by position ions.Because on surfac of target material presented certain strength magnetic field whose orientation was orthogonal with respect to that of electric field, during sputtering the electrons, which were emitted from target or produced by ionization, moved according to the cycloid.Thus,collision probability between the eletrons and gas molecules was increased by significant increasing in electron movement range, so that sputtering efficiency was raised and the voltage for sputtering was decreased to less value.

In the experiments diameter of the used single crystal silicon wafer was 114mm and its thickness was 3.5mm.The effective outside diameter of glow discharge cycle was about 87mm owing to the effect by electric screen and wall thickness of magnetic bead .The level component of magnetic induction on the surface of silicon wafer was 0.028 T. The distance between the surface of target material and substrate was 38mm.

The 99.99% pure argon was sputtering gas and the 99.99% pure methane was reactiving gas. The usage of methane as reativing gas not only could apply its carbon to fabricate silicon carbide thin film,but also could apply its hydrogen to appropriately hydrogenate the film.Before sputtering ,the chamber for sputtering was evaculated to pressure 2.2×10^{-3}Pa. Before deposition of the thin films the silicon target was always pre–sputtered in pure argon for 30min or more.During presputtering the argon pressure was 4 Pa and discharge power was the same as that for sputtering.The methane was leaked through needle valve into the chamber for sputtering.The thin films with different carbon cotent were deposited by changing partial pressure of methane with respect to the total pressure.During sputtering the total pressure was 4 Pa. It could be regulated and controlled by HY–1 type automatic pressure controller with the piezoelectric valve. The DC Voltgae for sputtering was 500~ 680V and the power density for sputtering was 1~ 1.2W / cm^2.The thickness of deposited film mostly was about 1.2μm. The deposition time was 2h and deposition rate was about 100Å / min. During sputtering the substrate was not specially heated.

OPTICAL PROPERTIES

Optical absorption coefficient of the samples prepered by reactive sputtering method in varying the partial pressure of methane with respect to the total gas pressure has been measured by the type of 7321 and 721 spectrophotometer.Generally,the relation of absorption coefficient α and photon energy $h\nu$ of amorphous semiconductors is $\alpha = B(h\nu - E_{opt})^2 / h\nu$ (E_{opt} is optical band gap of the materal and B is a constant).The optical band gap E_{opt} can be determind from the $(\alpha h\nu)^{1/2}$vs $h\nu$ plot. Fig 1 shows the results for the film prepared in varying partial pressuer of methane P_{CH_4}.The dependence of the optical band gap E_{opt} on P_{CH_4}obtained from fig.1 has been showed in fig.2.

The results of the fig.2 shows $E_{opt} = 1.54$ eV at $P_{CH_4} = 0$ and the E_{opt} increases with the $P_{CH_4}$when the pressure of methane P_{CH_4}being small,which could be owing to the bydrogenation of silicon and increase in the number of Si–C bond if introducing the methane. A maximum value of E_{opt} has been observed when increasing of P_{CH_4}further.The result is in according with that of the film prepered by glow discharge method.

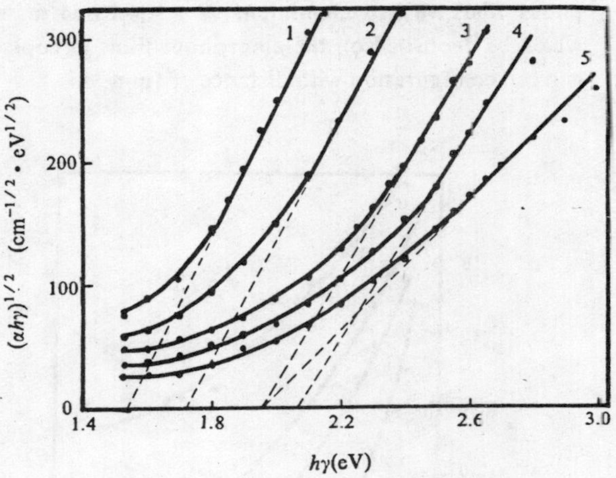

Fig.1 The dependence of $(\alpha h\nu)^{1/2}$and $h\nu$at different partial pressure of methane . 1–0; 2–0.8; 3–1.33; 4–2, 5–4$\times10^{-2}$Pa.

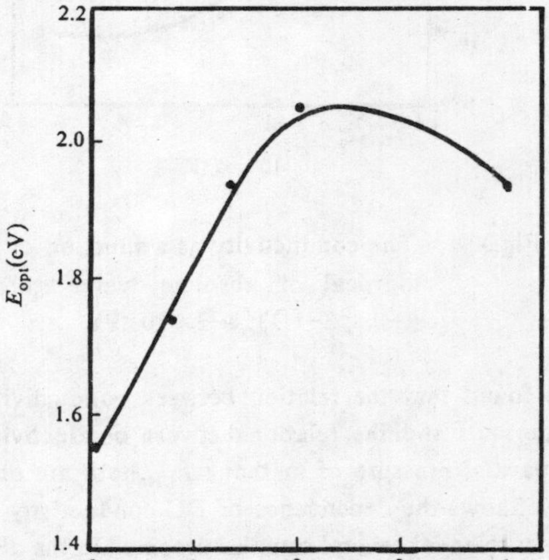

Fig.2 The dependence of optical band gap and the partial pressure.

Meanwhile, We also observed that every line is parallel each other in Fig.1 and B is a constant when $P_{CH4}$is low and the value of B decreases, When $P_{CH4}$is high. this is,propably,because of the formation of C−$H_n$bond and graphite−like structure.

ELECTRICAL PROPERTIES

The condutive properties of the film has been investigated by ZC−36 type higher resistance meter and RCJ−3 type isolation resistance instrument. The a−SiC film was depositid on the substrate of glass. In order to get a excellent ohmic contact between the electrod and amorphous films,we use aluminium as a electrode materal which be deposited on the amorphous films in coplanar gap cell configuration with distance of 1mm.

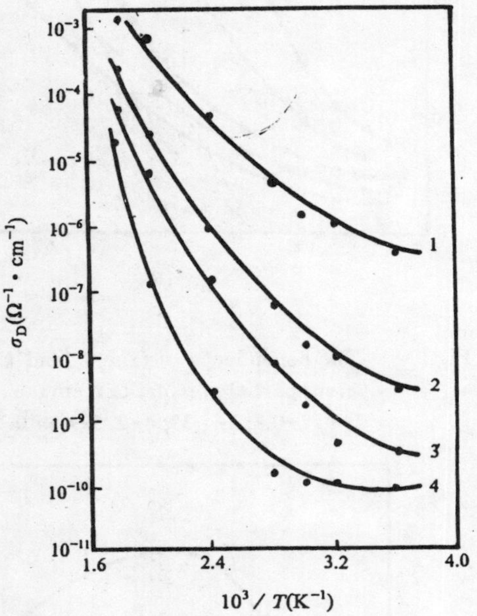

Fig.3 The conductivity as a function of the reciprocal of absolute temperature 1−0; 2−0.8; 3−1.33; 4−2 × 10^{-2}Pa.

It is found that the relation between conductivity and temperature,and the relation between conductivity and the paritial pressure of methane P_{CH4}both are obvious. Fig.3 Shows the dependence of DC conductivity $\sigma_D$and temperatuer for several samples prepered in the different partial pressure of P_{CH4}.The value of DC condutivity increases with the temperature.The dependence between the DC condtivity $\sigma_D$and the activation energy of carrier $E_o$is $\sigma_D \propto \exp(-E_o / kT)$ at higher temperatuer range.The hopping conduction is predominant at $P_{CH4}=0$. With the increase of P_{CH4}, the value of conductivity decreases rapidly and the type of conduction changes from the model of hopping to activating above room temperature. This phenmenon has been interpreted as due to the hydrogegnation of silicon and the formation of Si−C bond, so as to the density of localized states and the random degree of film structure decreases.

Activation energy $E_o$may be obtained from the slope of curve in Fig.3. $E_o$increases from 0.5 eV to 0.82eV, when P_{CH4}increases from o to 2 × 10^2Pa,the change tendency is in agreement with the change of E_{opt}.

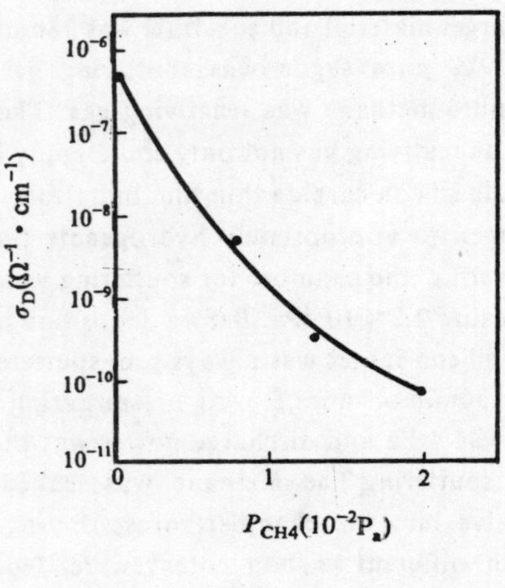

Fig.4 The room temperature conductivity of the film as a fuction of the partial pressure of methane.

The dependence of the conductivity at room temperature and the partial pressure of methane has been showed as Fig.4. For example,the room temperature conductivity decreases from 5.0 × $10^{-7}\Omega^{-1}$.$cm^{-1}$to 1.0 × $10^{-10}\Omega^{-1}$.cm^{-1}, when P_{CH4}increases from o to 2 × 10^{-2}Pa. It is indicated from these results above that the film conductivity may be changed continuously by the change of composition and it is discovered from our study also that the activation energy and DC conductivity will be decreased and increased respectively, if the partial pressure of methane increases further.The tendency of variety is in agreement with that between optical band gap E_{opt} and the partial pressure of methane.

CONCLUSION

We first prepared amorphous silicon carbide thin films by DC planar magnetron reactive sputtering method.This method can keep the advantage of magnetron sputtering and avoid shortcoming by RF sputtering. The dependence of cptical and clectric propcrties for prepared thin films on the partial pressure of methane has been perliminarily investigated. These results showed analogy with those of thin films fabricated by glow discharge method or RF sputtering method.

ACKNOWLEDGEMENTS

The authors are greatful to Professor Qu Xixin for his interest in and encouragement of this work. This work was supported by the department of Electronic Materials.

REFERENCES

1. J.Bullot,M.Gauthier, M.Schmidt,Y.Catherine and A.Zamouche,Philos.Mag., B49(1984),489.
2. Y.Tawada,H.Okomoto and Y.Hamakawa, Appl.Phys.Lett.,39(1981),237.
3. R.Dutta,P.K.Banerjee and S.S.Mitra, Solid State Commun., 42(1982), 219.
4. A.Sathyamoorthy and W.Wcisweiler,Thin Solid Films, 87(1982), 33.
5. M.Le Contellec,J.Richard, A.Guivarch, E.Ligeon and J.Fontenille, Thin SolidFilms, 58(1979), 407.
6. N.Saito, T.Yamada,T.Yamaguchi, I.Nakaaki and N.Tamaka, Philos.Mag., B52(1985),987.
7. N.Saito, Appl.Phys.Lett., 46(1985),61.
8. Jiang Jiejian and Zhu Yaojiang,J. Chengdu Inst. Radio Eng., 17(1988),296.

THREE-COMPONENT POLYCRYSTALLINE FILM MATERIALS $Zn_xCd_{1-x}Te$

Sun Chengyong, Wang Xusheng, Li Yawei
Xidian University
Xi'an, China

ABSTRACT

As known II – VI compounds are one of important research fields of electronic materials now. The physical properties of three component polycrystalline thin film materials are varied with proportion of these components. The photosensitive polycrystalline materials $Zn_xCd_{1-x}Te$ have high absorption and good electro-optical properties in their base absorbing zone. Sintering takes place at high temperature, after synthesis at low temperature, for preparation of the three-component polycrystalline thin film materials. The microstructrues of synthetic polycrystalline $Zn_xCd_{1-x}Te$ and of polycrystals after sintering have observed by electron microprobe. The effects of technological conditions were discussed in this paper. The X-ray diffraction patterns show that these polycrystals are pure compounds. We measured and analysed electrical and optical properties of these compounds. The results show that polycry-stalline $Zn_xCd_{1-x}Te$ are suitable materials for photosensitive thin film components.

I Technology for Preparation of Three-Compound Polycrystalline Materials $Zn_xCd_{1-x}Te$

1. Synthesis at Low Temperature

Technological process is shown in Fig. 1.
The polycrystals ZnTe, CdTe (or $Zn_xCd_{1-x}Te$ with various proportion) are synthesized by highly pure powder of Zn, Cd, Te, in which the principle of solid phase diffusion reaction at low temperature is adapted. In order to assure that in synthesis processes they are in solid state all the time, the adaptable temperature should be lower than the melting point of Zn, Cd and Te — 419.5℃, 320.9℃ and 450℃. So the synthesis temperatures of ZnTe, CdTe should be respectively 400℃ and 300℃.

Pure powders
↓
Ball-grinding, mixing, drying, mesh separating
↓
Tabletting
↓
Synthesis at low temperature
↓
Size degradation, mesh separating, tabletting
↓
Sintering at high temperature

Fig. 1 Technological process for preparation of polycrystalline materials

Before synthesis, raw powder materials which are fully mixed are treated to tablet shape by using dry press method. It can be considered that solid phase diffusion has limited source, which is equivalent to the diffusion of Zn atoms or Cd atoms on the surface of Te atom group. It can be supposed that diffusion takes place in X direction, so the concentration of oneside diffusion is,

$$C(X,\ t)=\frac{Q}{\sqrt{\pi Dt}}\quad \exp(-x^2/4Dt) \qquad (1)$$

in that equation, t is reaction time, Q is mass of diffusion material (Q=const), D is intrinsic diffusion coefficient. The concentration on the surface(x=0) is $Q/\sqrt{\pi Dt}$, so

$$C(x,\ t)/C(0,\ t)=\exp(-x^2/4Dt) \qquad (2)$$

It can be supposed that the center of Te atom group is at the point where x=0.5d, when there the concentration of oneside diffusion of Zn or Cd atom is equal to 50% surface concentration, the synthesis reaction finished essentially. That is to say, Zn or Cd atoms had diffused completely evenly into Te atom group and reacted with Te atoms to form ZnTe or CdTe, so the diffusion depth X1 is defined as the value of x, where the concentration is half of that on the surface, so X1=d/2.
The diffusion coefficient of Zn, Cd in Te is D, D =2φ(-8)mm/s, the diameter of raw material grain d is 0.04mm.

$$C(x_1,t)/C(C,t)=\exp(-x_1^2/4Dt)$$

so,

$$50\% = \exp(-x_1^2/4Dt) = \exp(-d^2/16Dt)$$

$$t = d^2/16D\ln_2 = 7213.5\,\text{sec} \approx 2\ \text{hr}$$

In order to react completely, synthesis reaction time is usually about 3 hr. It is shown by X-ray diffraction patterns of synthesized materials, that compounds ZnTe, CdTe are formed essentially.

Discussion:

(1) The smaller the raw material grains, the more completely the reaction carries out, the less the synthesis time is needed.

(2) The optimal pressure chosen in tabletting process is related to the character of raw material itself, the shape of grains, grain-size, size and thickness of tablets. For Zn, Cd, Te powder materials, when diameter of tablets is 15mm, thickness is 0.8-1.5mm, the suitable pressure put on should be 200-300 kg/cm^2

(3) The evaporation of elements can bring about deviation resulted from chemical measure ratio on the compound surface. The same materials should be used as cover plate and definite gas pressure of N_2 should be kept in tubular sintering furnace, then the evaporation of elements is controlled. According to semi-quantitative analysis of synthesized CdTe and ZnTe by using electron-microprobe, it can be shown that the atomic proportion of compound CdTe is Cd : Te = 58.76 : 41.24, and that of ZnTe is Zn : Te = 66.29 : 33.71. In blending the concentration of Te may be added suitably, in order to correspond with the concentration of each component of compounds mentioned above.

2. Sintering at High Temperature

Sintering is a recrystal process carring by mass transfer at high temperature, which is accompanied with the processes of crystalline grain's growing, crystal edge's moving and structure densification, all of that are physical changes mainly. Under the function of high temperature and because of propelling of surface energy the powder may be diffused. In the final stages of sintering, the process of structrue densification ends, crystalline grains grow gradually, gas holes are excluded, then densely-combined polycrystals can be formed.

After regrinding synthesized CdTe and ZnTe, must be sifting, mixing up according to the necessary proportion of components, pressing grains, then the materials can be sintered. Finally three-compound polycrystalline materials $Zn_xCd_{1-x}Te$ can be formed.

Because the melting points Tm of ZnTe and CdTe are 1098℃ (1371K) and 1238℃(1511K) respectively, and the beginning sintering temperature T_b is 0.57 Tm(K) (CdTe, 781K, ZnTe, 861K). The density of materials at optimal sintering temperature Ts reaches the maximum. If keeping at this temperatrue necessary time the expectant densified polycrystal can be gained. Because $T_b = 0.75$ Ts, so Ts(CdTe)=677℃, Ts(ZnTe)=784℃. So the sintering temperature zone of CdTe is chosen amorng 640—710℃ and of ZnTe is 740-820℃. For three-compound $Zn_xCd_{1-x}Te$, according to varied x that temperature is chosen among 680-760℃. The time of keeping temperature is chosen 2 hours. The microstrcture is shown in Fig. 2.

Fig. 2 Micrograph of $Zn_{0.1}Cd_{0.9}Te$(× 3200) after sintering(730℃, 2 hr)

By using X-ray diffraction patterns of CdTe, ZnTe and $Zn_xCd_{1-x}Te$ with varied proportion, we know that very pure compounds have been formed.

II. The electric Properties of Polycrystalline Thin Film Materials $Zn_xCd_{1-x}Te$

At varied sintering temperature and at varied time of keeping temperature, the dark resistivity and the light resistivity of the materials with three x value(0.05, 0.1, 0.2) are detected. The results are shown in Table 1 and Fig. 3.

Tab. 1 Resistivity($10^3\Omega \cdot cm$) of $Zn_xCd_{1-x}Te$

x	Sintering ℃	hr	Dark resistivity ρ	Light resistivity ρ 1*	ρ 1**
0.05	650	2	264	133	88.5
0.05	730	2	85.3	4.77	4.39
0.1	670	2	560	551	122
0.1	700	2	10.7	9.86	8.00

Tab.1 (continued)

x	Sintering ℃	hr	Dark resistivity ρ	Light resistivity ρ 1*	ρ 1**
0.1	730	1	148	50.1	37.8
0.1	730	2	9.95	5.99	4.26
0.1	730	2.5	36.8	22.7	15.1
0.1	730	4	10.8	9.59	8.52
0.2	730	2	8.12	6.12	4.73

* Light power density 12.8 mW/cm^2

** Light power density 35.8 mW/cm^2

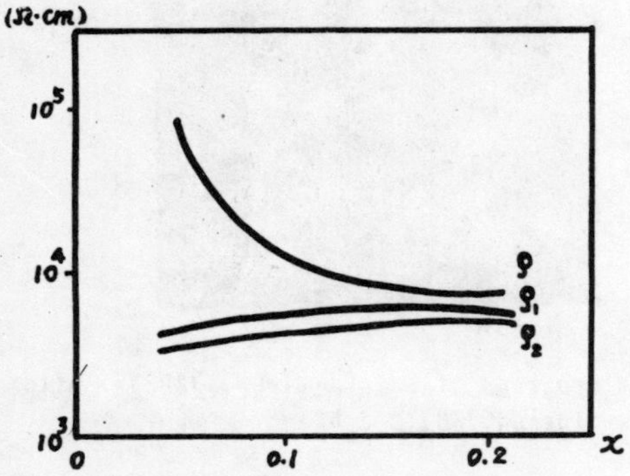

Fig.3 Resistivity of $Zn_xCd_{1-x}Te$(730℃ 2 hr) VS. compose x

Discussion:

1. The effect of sintering temperature (2 hr):
When sintering temperature is above 700℃, dark electric-resistivity changes slightly. It is shown that expectant densified polycrystal has been formed. For x=0. 1, the sintering temperature should be among 700-730℃.

2. The effect of keeping temperature time, when the time of keeping temperature reaches 2 hr, the difference between light and dark resistivities is maximum (photo-sensitivity is optimal), so the time should be two hours.

3. The effect of component proportion x (730℃, 2 hr):
When x=0. 05, photo-sensitivity is optimal and $\Delta\rho$ is maximum. As x increases, the optimal photo-responsive wave length moves towards short-wave direction. In order to make photo -sensitive sensors or solar cells which have excellent light response, x=0.2 is suitable.

Ⅲ. The Optical Properties of Polycrystalline Thin Film Materials $Zn_xCd_{1-x}Te$

1. The Relationship between Index of Refraction n and x on The Edge of Absorbing Znoe

The optical properties of polycrystalline thin film materials are indicated as index of refraction n, photoexcitation coefficient k and absorption coefficient $\alpha=4\pi k/\lambda$. Then the relationship between index of refraction n of $Zn_xCd_{1-x}Te$ on the edge of absorbing zone and x can be shown in Fig.4.

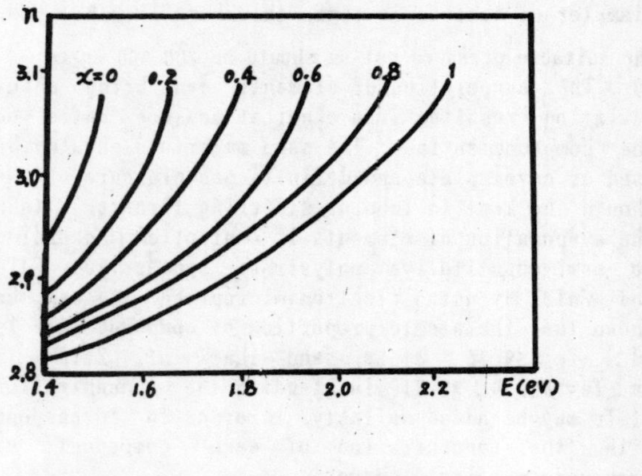

Fig.4 Index of refraction n of $Zn_xCd_{1-x}Te$

2. The Relationship between n, k and λ:

The reflective table R(λ) and transparency table T(λ) in the area of wave length of 0.4-2. 5μm are gained after measurement of the film thickness. By two-layer-film formula the relationship between the index of refraction n, photoexcitation coefficient k of films and wave length λ can be calculated. See Fig.5.

3. The Relationship Between α and λ

In the base absorbing zone the absorptive coefficient of $Zn_xCd_{1-x}Te$ reaches 10^3-10^5 cm^{-1}. See Fig.6.
It is shown that polycrystalline materials $Zn_xCd_{1-x}Te$ are very suitable thin film materials. From the Fig. 5 it can be also shown that the absorptive region is in visible region, so $Zn_xCd_{1-x}Te$ can be applicated to solar cells or photo-sensitive sensors.

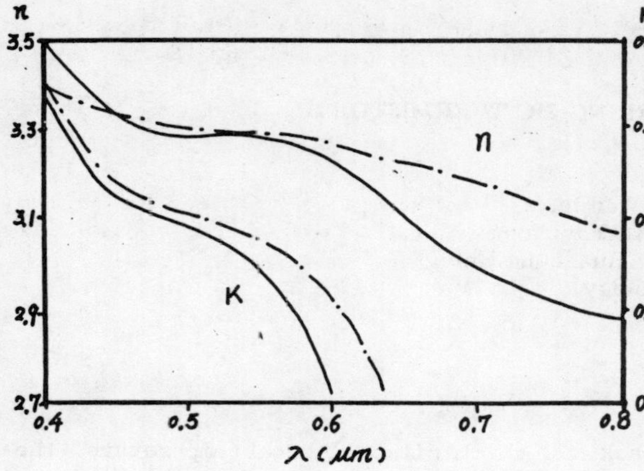

Fig. 5 Refractive index n and photoexcitation coefficient k of $Zn_xCd_{1-x}Te$ vs. wave length λ —— x=0.5, —·—·— x=0.7

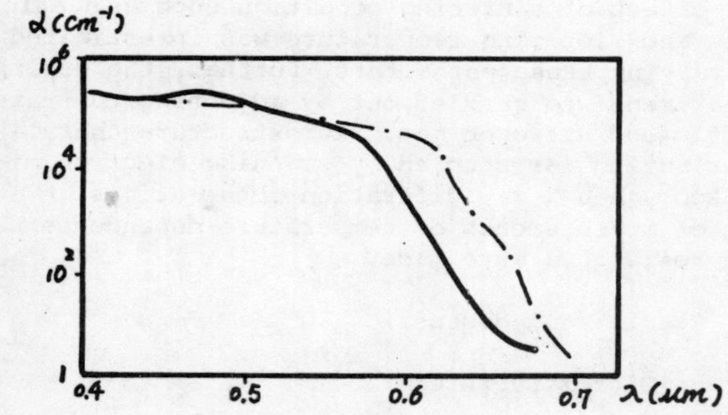

Fig.6 Absorbance α of $Zn_xCd_{1-x}Te$ vs. wave length λ —— x=0.5, —·—·— x=0.7

IV. Fabrication, Properties and Application of $Zn_xCd_{1-x}Te$ Thin Film

1. Thin-film Technology

The electron beam evaporation technique allows deposition of polycrystalline materials $Zn_xCd_{1-x}Te$. Average evaporation rate is 20 Å /S. High vacuum, $10^{-5}-10^{-6}$ torr, temperature of substrates, 220—250℃.

2. Properties of $Zn_xCd_{1-x}Te$ Thin Films

Use optical glass as reference plate, the heat treatment of thin films was necessary after evaporation at 300℃ about 30-40 min, to make thin film grow evenly. The electrical properties are shown in Tab.2 and optical properties are shown in Fig.7. The film thickness of samples is in the area of 1700-4600 Å .

Tab.2 Electrical properties of $Zn_xCd_{1-x}Te$ thin films

x	Conductivity ($\Omega^{-1}\cdot cm^{-1}$)	Migration rate (cm^2/v.s)
0.5	17.24	31.1
0.7	23.89	36.8
1.0	35.83	43.2

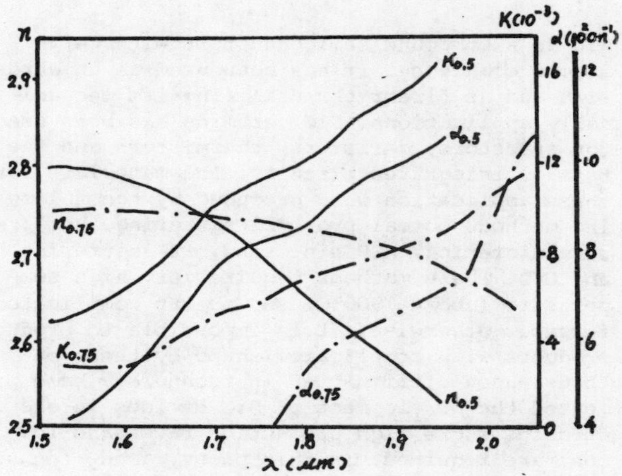

Fig.7 n, k and α of thin films $Zn_xCd_{1-x}Te$ —— x=0.5, —·—·— x=0.75

3. Application

In its base absorptive zone $Zn_xCd_{1-x}Te$ has very high absorption coefficient and its absorption limitation cover all visible light zone. So it is a kind of very suitable materials of solar cells. It has been tested that whole thin film solar cells have been made by thin film materials $Zn_xCd_{1-x}Te$ and $Zn_yCd_{1-y}S$ (x=0.1, y=0.2), certain photo-electric response can be gained. The follwing research work should relate to further improving the efficiency of photo-electric transfer.

REFERENCES

1. R. Pampuch, Ceramic Materials, 1976

2. S. J. Fonash, Solar Cell Device Physics, 1981

3. C.W.M kow etc. Thin Solid Films, 88(1982), 141

THE EXPERIMENT STUDY OF SINTERING SiC THERMISTOR

Qiu Sichou, Wu Fan, Lin Wenjiu,
Qiu Mochin, Hwang Hanrau, Xu Yanzhong
Dept. of Solid State Electronics of Huazhong Univ.
of Science and Technology
China

ABSTRACT

SiC is a compound semiconductors with very strong covalency. It has been used as an abrasive and as firebricks utilizing its mechanically applications. SiC ceramics has been used for arrestors, varistors, thermistors and the bass of integrated circuits. The materials for these application were produced by the following methods normal pressure sintering, hot pressure sintering. HIP sintering, rf-supttering and CVD. These methods require very high temperature (above 1900°C) and a very complicated furnace. Otherwise, it is impossible to produce products with complicated shape by these methods. Such difficulties in technology have prevented the development of SiC devices in electronics where high production rates and low cost are required. In this paper, a new technique to produce SiC thick film thermistors was developed. The SiC ceramics was made by using the polycarbosilanes as the sintering aid. This method did not require high temperature (1100°C-1300°C) and complicated furnace, It is possible to produce samples with complicated shape by this method.

I. Experiment

The procedure to produce SiC thick film thermistor is shown in Fig. 1. a feature of this method is that polycarbosilane as the binder is converted into fine particle inorganic β-SiC by heat treatment, which fills the grain boundary of the matrix SiC.

powdered pc→mixed pc, SiC→printed →sintering→making electrode.

Fig. 1 Procedure to produce SiC thick film thermistor.

Polycarbosilane was dissoved in the hexane, then, was mixed with SiC powder. The paste was formed into thick film by using printed technique. these were crosslinked by surface oxidation in air at 500°C in the tube furnace, and then further sintered at high temperature for one hour in an argon or nitrogen atmosphere to make black SiC thick film. At high temperature, the polycarbosilane pyrolyzed and acts as binder:

$$\left(\mathrm{Si}(CH_3)(H) - CH_2 \right)_n \dashrightarrow \beta\text{-}SiC + CH_3 + H_2$$

At last, the high temperature silver paste was printed on the SiC thick film and sintered as the electrode.

The effect of sintering condition such as heating rate and sintering temperature was investigated by varying these parameters. Further, the doping experiment was carried out by adjusting the rate of flow of nitrogen gas. Microstructure characterization was studied by scanning electron microscopy and X-ray diffration observations. A set of measurements of temperature-dependence of the resistance were made.

II. Results and discussion

1. The Microstructure

Scanning electron microscopy study reveal β-SiC pyrolyzed via polycarbosilane showing amorphous patterns and having definite grain strutures but on a very fine scale. The SEM photograph of SiC ceramics using the polycarbosilane as bined indicates that there are three phases in the ceramics: (i) α-SiC grains (about 10μm); (ii) intergranular amorphous and microcrystalline mixed material; and (iii) some porosity (about 0.1 to 1μm). Fig. 2 and Fig. 3 are SEM photographs of β-SiC pyrolyzed from pure polycarbosilane and SiC ceramics respectively. In the X-ray powder diffration pattern of SiC ceramics sintered at 1200°C, shown in Fig.4, both diffration line of α-SiC type and β-SiC type phase appearing.

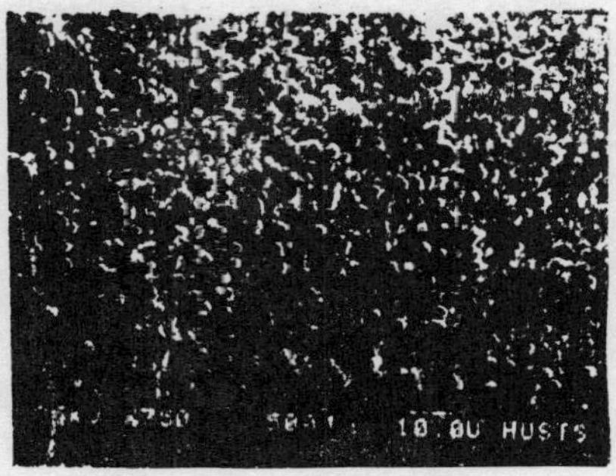

Fig.2 The SEM photograh of SiC pyrolyzed from pure polycarbosilane

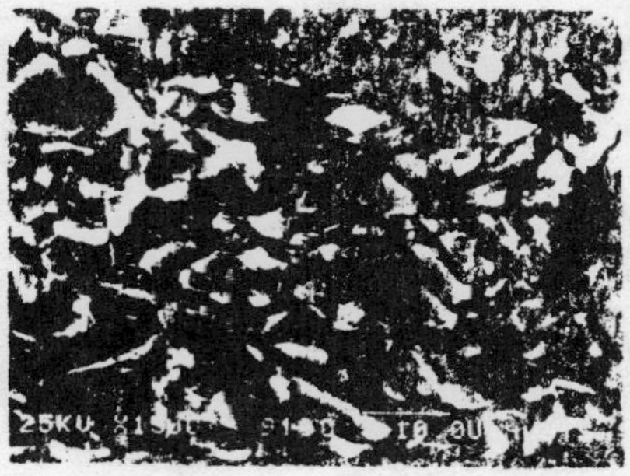

Fig.3 The SEM photograph of SiC ceramic

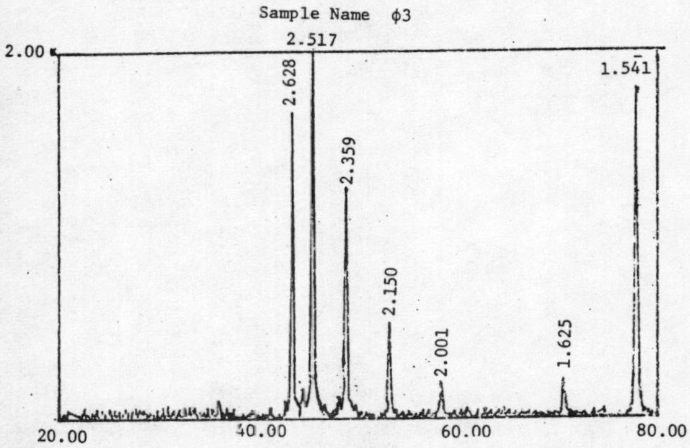

Fig.4 X-ray diffration pattern of SiC

2. Effect of doping

SiC starting powder is a n-type semiconductor. When the powder is sintered with polycarbosilane in Ar or N atmosphere, it is still a n-type semiconductor, however, when the powder is sintered with polycarbosilane and Fe addition, result of thermal electromotive force measurement show SiC become p-type semiconductor.

Another effect of doping is to change the resistivity of SiC ceramics. Table I shows the resistivities at room temperature of the samples sintered at argon and nitrogen atmosphere. It can be seen the value of the resistivities vary from $10^3\Omega$. cm to $10^8\Omega$·cm. At last line, the resistivity of the sample with rare-earth doping is given.

TABLE I The resistivities of the samples sintered at 1000°C

Sample No.	Sintering atmosphere (ml/min)	Resistivity (Ω. cm.)
1	Argon (100)	7×10^8
2	Nitrogen (20)	5×10^4
3	Nitrogen (100)	6×10^3
R4	Argon (100)	3×10^4

The binder, polycarbosilane, has the representive structural unit shown below, which is approximately consistent with the empirical formula $SiC_{1.83}O_{0.03}H_{4.16}$ derived from analysis(1). IR spectroscopy shows loss of Si-H bonds

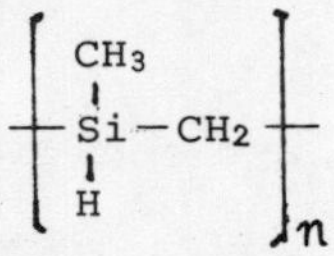

starts at 300 C and completes by 450 C. Loss of Si - CH and C - H complete by 800 C. In this process, Si atom and C atom bonds again. N atoms may diffuse into Si - C network and act as a donor in SiC grains. The excess electron derived from N atoms fill the defect states within the boundery region, lower the depletion layer barrier, thus, the resistance of the SiC thermistor decreases.

The resistivity of $10^7\Omega$. cm at 1000°C drops to $10^5\Omega$. cm at 1000°C. Difference between two samples probably reflects compositional difference between them.

By controlling the sintering condition, the resistance of the SiC thick film can be adjusted in a wide range (from several hundred ohms to several hundred kiloohms). It is possible to produce the thermistos with variant resistance and variant shape according to difference requirement.

3. Temperature dependence of the resistance

The resistance of SiC ceramics were measured in a temperature range from 273 to 1000°K. The following experimental equation which represents the dependence of the resistace R of the ceramics on temperature T was obtained:

$$R = R_{T_0}\exp(B(1/T - 1/T_0))$$

where R_{T_0} is the resistance of the sample at T_0 K;

R is the resistance of the sample at T°K;

B is the material constant determined experimentally.

The activation energy can be calculated from the data of the temperature dependence of the resistance and thermal stimulated current spectroscopy, a result of 0.23eV in n-type sample was obtained.

The temperature dependence of resistance of SiC ceramics is shown in Fig. 5.

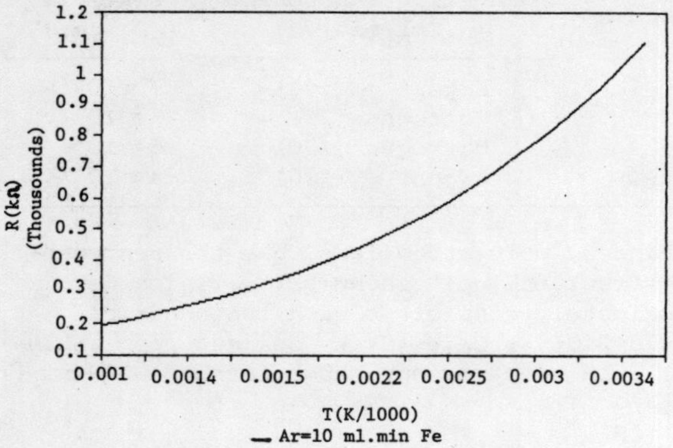

Fig.5 The temperature dependence of resistance of SiC ceramics

REFERENCE

S. Yajima, e al., Ceramic Bulletin, 56 (1977), 1061.

EFFECT OF ISOLATING TECHNIQUE ON PROPERTIES AND MICROSTRUCTURE OF $SrTiO_3$-BASED BLC

Longtu Li, Yujuo Jin, Xingjiang Zhou, Heping Zhou, Gangkai Zhao
Dept. of Mat. Sci. and Eng. Tsinghua University. Beijing, China

ABSTRACT

The isolating technique of vapor-phase diffusion with CuO in BLC had been studied. The isolating technique of vapor-phase diffusion with air and with CuO atmosphere had been compared. The effects of cooling conditions of second firing (isolating heat treatment) on the microstructure and properties of BLC have been also investigated. The BLC materials with $\varepsilon \geqslant 60000$, $\tan\delta \leqslant 0.02$, $\rho \geqslant 10^{11}\ \Omega$-cm, $\Delta C/C \leqslant 0.10$ (-25°C-+85°C) had been obtained. The microstructure of materials have been studied by means of SEM, TEM, EDAX. The relationship between the dielectric properties and microstructure of the grain boundary was discussed.

INTRODUCTION

The demand of the miniuturization of electronic circuits has increased for capacitors of small size and large capacitance. The strontium titanate-based grain Boundary Layer Capacitors (BLC) are suitable for miniuturization electronic circuits because of its high dielectric constant, the small temperature dependence of dielectric constant and the low dielectric loss. In grain boundary layer capacitors, semiconducting ceramic grains are surrounded by a thin dielectric layer. In general, the thin isolating layer is formed by the diffusion of metal oxides along a continuous liquid phase at the grain boundaries during the secondary heat treatment process. Recently, a new type of isolating technigue has been reported. This technique differ from normal procedure. It is vapor-diffusion with CuO, PbO into grain boundaries during second firing. Compared with normal metal oxides liquid phase diffusion, vapor-diffusion not only simplier, but also has the better dielectric properties than that of conventional liquid phase diffusion isolating technique.

The dielectric properties of BLC very dependence on the isolating layer. The cooling condition during second firing effect on microstructure and dielectric properties had been investigated. The results show that the quenching technique may be advantage to obtaine the good properties of BLC.

The dielectric constant and dielectric loss $\tan\delta$ were measured using LCR meter at 1 KHZ.

EXPERIMENTAL PROCEDURE

The TiO_2 of chemical reagent grade pure and $SrCO_3$, Nb_2O_5, GeO_2, CuO of analytically pure was mixed by wet milling for 24hrs. The composition formula is $SrTi_{0.994}Nb_{0.006}O_{3.003}+XGeO_2+YCuO$. The mixture was calcined at 1150°C for 3hrs. in air, and then ball mill for 48hrs. The powder was pressed into disk 10mm in diameter and 1mm thick at a pressure of 2000 kg/cm^2. The samples were sintered at 1450°C for 4hrs. in reducing atmosphere (80% N_2+20%H_2) cylinder gas with a flow rate of 100 mm^3/min and furnace cooled to room temperature. The atmosphere source of vapor diffusion was supplied by samples consisted of CuO:Al_2O_3=1:1. The condition of second firing is 1050°C-1250°C for 1 to 5hrs.

RESULTS AND DISCUSSIDN

1. Vapor Phase Diffusion Isolating

The dielectric properties as a function of heat treatment temperature are shown in Fig.1. It can be seen that the dielectric constant ε decreases, the resistivity ρ increases with the increasing temperature. When the heat treatment temperature are over 1200°C, the dielectric loss was increased greatly. These results show that the suitable temperature is about 1200°C.

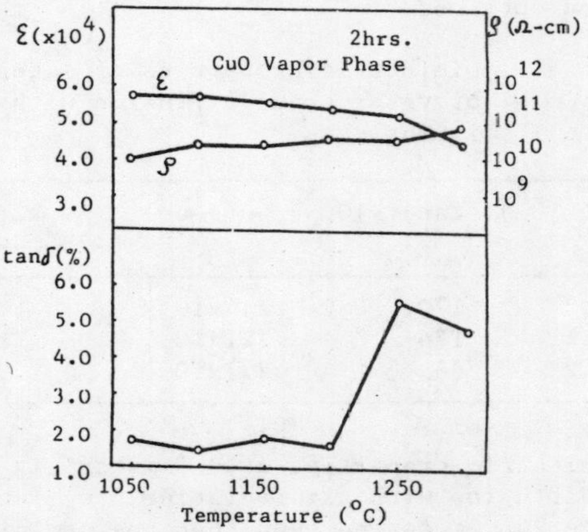

Fig.1 ε, ρ and $\tan\delta$ as a function of heat treatment temperature.

Fig.2 shows the dependence of ρ, $\tan\delta$, on the time of heat treament. It is suggested that the lower resistivity and the higher dielectric loss occur at the shorten soak of heat treatment. while prolonging the soak of heat treatment the resistivity increased and the dielectric loss decreased due to increasing the amount of CuO diffusion into grain boundary, the second phase layer and diffusion layer become wider. But the dielectric constant would be decreased when the soak of heat treatment was too long.

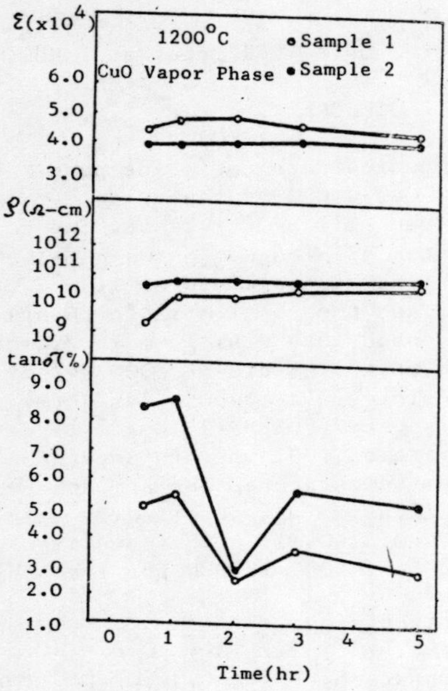

Fig.2 Dependence of ε, ρ, $\tan\delta$ on soak of second firing.

The properties of BLC materials of vapor phase diffusion with CuO isolating are listed in Table 1. It can be seen that the high dielectric constant $\varepsilon \geqslant 60000$, low dielectric loss $\tan\delta \leq 0.02$ and high resistivity $\rho \geqslant 10''\Omega$-cm could be obtained.

Table 1. Dielectric properties of materials of vapor phase diffusion with CuO isolating

	$\varepsilon(x10^4)$	$\tan\delta(x10^{-4})$	$\rho(\Omega$-cm$)$	$\Delta C/C$ (-25^oC-+85^oC)
1	6.2	170	$2.4x10^{11}$	10%
2	6.1	174	$1.1x10^{11}$	10%
3	6.2	88	$2.7x10^{11}$	10%

The dielectric properties of BLC materials vapor phase diffusion with air isolating are listed in Table 2. It can be seen that the capacitor with the higher dielectric constant and lower dielectric loss could be obtained, but the resistivity is about the two orders of magnitude lower than that of vapor phase diffusion with CuO. It is clear that the CuO plays an important role in maintaining the excellent properties of materials of vapor phase diffusion isolating, in particular for increasing resistivity.

Table 2. Dielectric properties of materials of vapor phase diffusion with air isolating

	$\varepsilon(x10^4)$	$\tan\delta(10^{-4})$	$\rho(\Omega$-cm$)$	$\Delta C/C$ (-25^oC-+85^oC)
1	6.8	58	$2.2x10^9$	10%
2	6.6	77	$5.5x10^9$	10%

The temperature variation of capacitance of vapor phase diffusion with CuO $\Delta C/C$ below 10% at -25^oC-+85^oC. The results show in Fig.3.

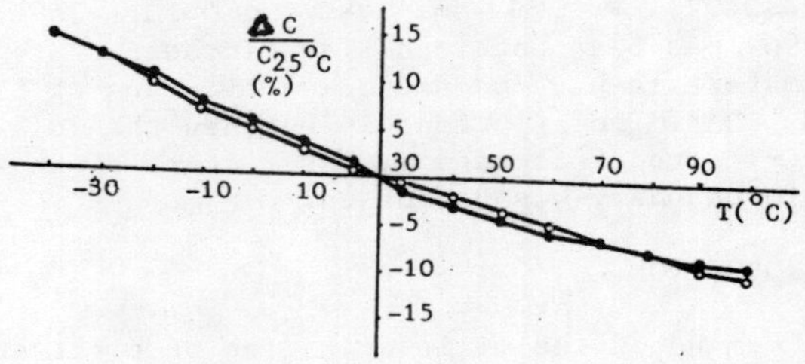

Fig.3 The Temperature Variation of Capacitance of vapor phase diffusion with CuO

As mentioned above, the BLC materials with high dielectric constant, low dielectric loss, high stability and uniform microstructure in vapor phase diffusion with CuO could be obtained.

2. Liquid Phase Diffusion Isolating Technique

The cooling condition during second firing (heat treatment) influence on dielectric properties of BLC had been investigated. Fig.4 shows that the ε, ρ and $\tan\delta$ as a function of composition in furnace cooling and quenching. Two kinds of isolating metal oxides of $PbO-Bi_2O_3-B_2O_3$ and $PbO-Bi_2O_3-CuO$ were used respectively. The second firing temperature and soak was 1150°C/30 min. It can be seen that the ε of quenching samples higher than that of furnace cooling and the $\tan\delta$ of the quenching samples lower than furnace cooling, the resistivity of both are compable.

Therefore, the quenching technique may be better than furnace cooling.

The difference of the properties between furnace cooling and quenching conditions related to the microstructure. The SEM micrograph of quenching and furnace cooling samples are shown in

Fig.5.

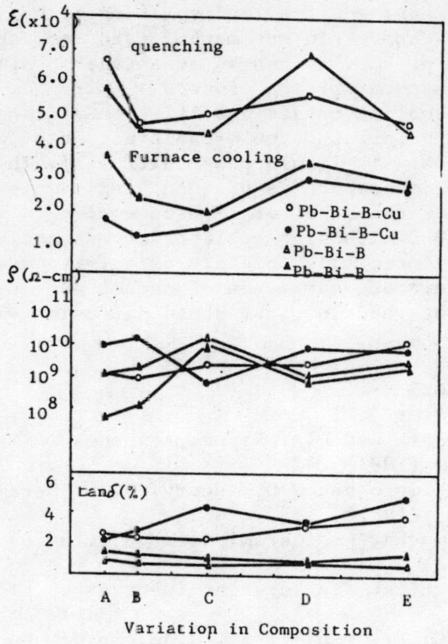

Fig.4 The properties of BLC as a function of composition at different cooling conditions.

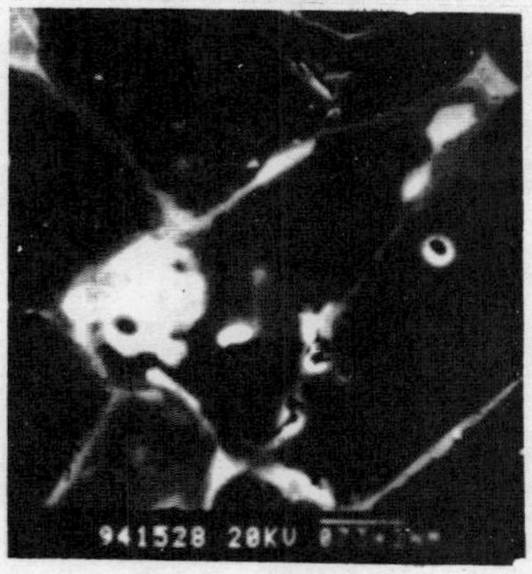

(a) quenching

(b) furnace cooling

Fig.5 SEM micrograpls of BLC materials

It is suggested that the quenching sample has thiner grain boundary. The wider diffusion layer were formed in furnace cooling sample. The composition of grain boundary had been analyzed by means of EDAX.

Fig.6,7 show the spectrum of grain boundary and triple intersection of quenching sample. It can be seen that only a small amount of Bi and Cu occur in the thin grain boundary, and a large amount of Bi and Cu segregate in triple intersection. It is suggested that the Bi_2O_3 and CuO concentrated in triple intersection principally in quenching sample. But in the case of furnace cooling is quite different.

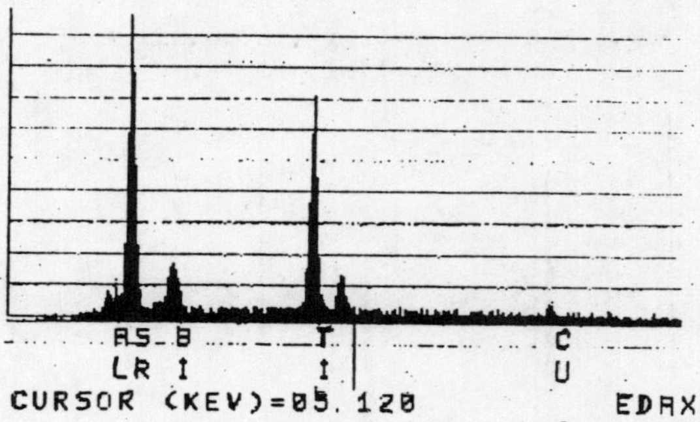

Fig.6 The EDAX specthcm of grain boundary in quenching sample.

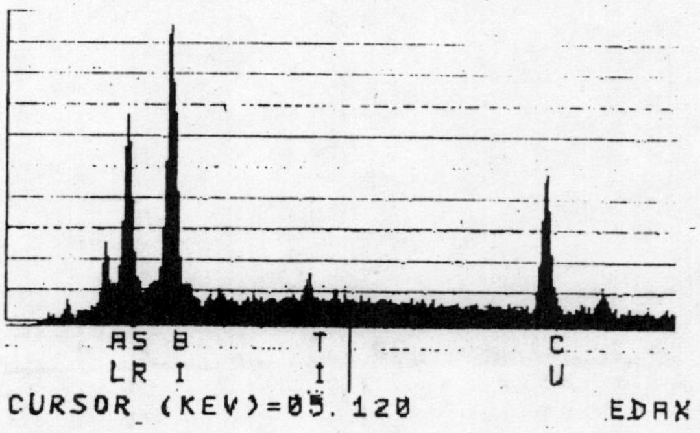

Fig.7 The EDAX spectrum of triple intersection in quenching sample.

Fig.8 shows the micrograph of grain boundary of furnace cooling sample. The EDAX spectrum of (a) up side, (b)left side, (c)right side of grain boundary are shown in Fig.9. (a)shows a small amount of Bi in grain boundary (b)shows a large amount of Cu in grain boundary and (c) shows a large amount of Bi, only a small amount of Cu occur at grain boundary.

It is suggested that the Bi_2O_3 and CuO had been separated during slow cooling. The results indicated that the amount of second phase of quenching samples are less and more uniform than that of furnace cooling samples. It is a possible reason of properties of quenching sample better than that of furnace cooling sample.

Fig.8 Micrograph of grain boundary of furnace cooling sample.

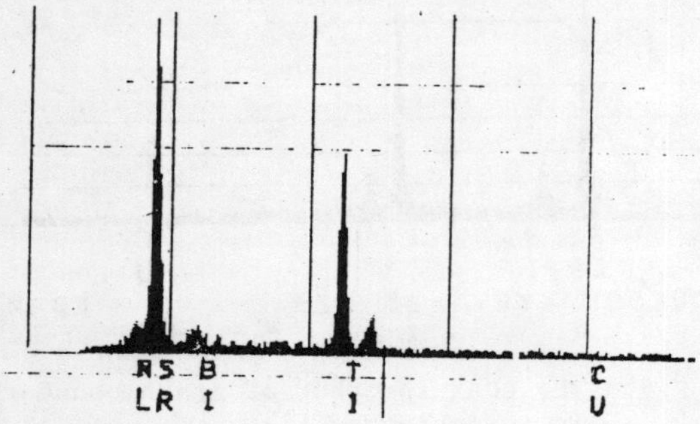

(a)

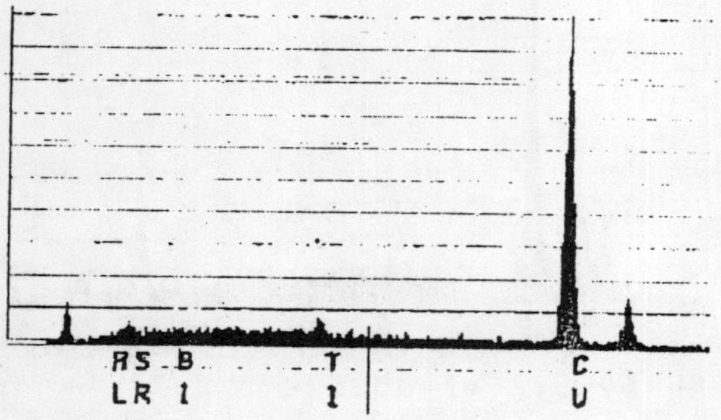

(b)

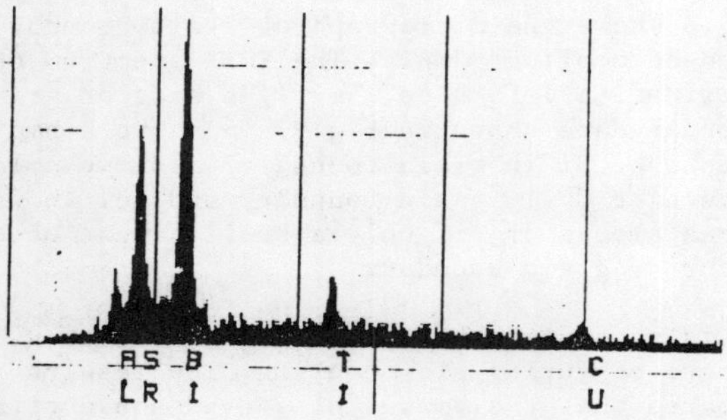

(c)

Fig.9 EDAX spectrum of triple intersection of furnace cooling sample.

CONCLUSIONS

1. Vapor phase diffusion isolating technique is a new and efficient method used for fablication of BLC. It has an advantage in simplify the second firing process.
2. The cooling conditions of second firing have strong influence on microstructure of grain boundary in liquid phase diffusion. The dielectric properties of quenching sample are better than that of furnace cooling samples.
3. The BLC with high dielectric constant. low dielectric loss, higher resistivity and low temperature variation of capacitance could be obtained in vapor phase diffusion with CuO.

REFERENCES

1. H.D.Park and D.A. Payne, Advances in Ceramics, Vol.1 (1981) 242
2. M.Fujimoto and W.D.Kingery, J.A,. Ceram. Soc. 68[4] (1985), 169
3. N.Yamaoka, Am. Ceram. Soc. Bull. 65(8) (1986) 1149
4. R.Wernicke, in Adrances in Ceramics, 1(1981) 242
5. P.E.C. Frenken et.al., J.Am, Ceram. Soc. 64 (1981), 687
6. W.Heywang, J.Amer.Ceram. Soc., 47[10],(1964), 484
7. J.Daniels et.al., Philips Res. Repts., 31 (1976), 489
8. I.Burn, S.Neirman, J.Mater. Sci., 17(1982), 3510
9. W.D.Kingery, J.Amer. Ceram. Soc., Vol. 57[1] (1974),1

ELECTRONIC STRUCTURE AT THE GRAIN BOUNDARIES IN ZNO VARISTORS WITH RARE EARTH METAL OXIDES

Kazuo Mukae and Koichi Tsuda
Fuji Electric Corporate Research and Development Ltd.
2-2-1 Nagasaka, Yokosuka, 240-01 Japan

ABSTRACT

Zinc oxide varistors which contain rare earth metal oxides rather than bismuth oxide and their electronic structures are reported. Varistors which contain Pr instead of Bi are applicable not only to low voltage circuits but also to high voltage electric power systems. Pr-type varistors have a different microstructure, i.e., two-phase structure, from Bi-type varistors. A new method of DLTS(Deep Level Transient Spectroscopy) was developed to investigate the interface states at the grain boundaries. The "zero-biased DLTS" method detected a new peak due to the interface states at higher temperature region. The depth of the interface states revealed to be 1eV below the conduction band edge. Those of bulk traps were obtained as 0.2eV and 0.3eV. ICTS(Isothermal Capacitance Transient Spectroscopy) gave similar results which suggest their higher suitability than DLTS because of the lower temperature measurement. The origin of the interface states were concluded to be chemisorbed oxygen ions at the grain boundaries.

INTRODUCTION

Zinc oxide varistors are electronic ceramic materials which show a highly nonlinear current-voltage relationship which deviates from the normal Ohm's law.

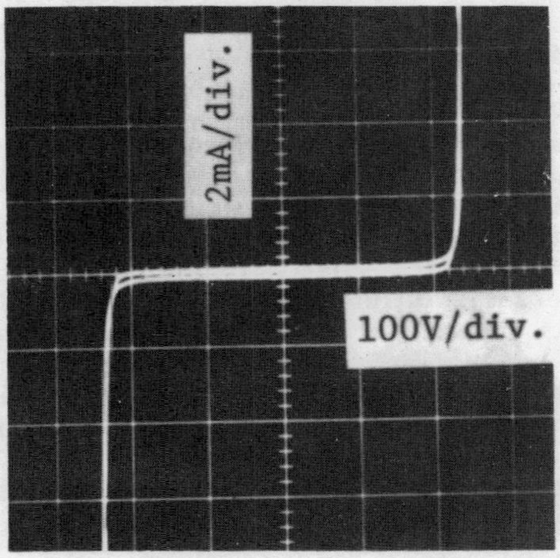

Fig.1 Typical current-voltage curve of ZnO varistors containing Pr.

Since sintered zinc oxide ceramics with a small amount of metal oxide additives were reported to have excellent nonlinearity,[1] they have been widely used for protection of electronic devices from transient overvoltage [2] and are recognized as the predominant type of varistors. In addition to this broad application, many reports have been published about the microstructure, the conduction mechanism, and the role of additive components. However most of these reports concern ZnO ceramics containing bismuth oxide. The present paper describes another type of zinc oxide varistors which contains praseodymium oxide rather than bismuth oxide as a principal additive component.[3] These varistors are applicable not only for low voltage devices but also for high voltage electric power systems. In order to evaluate the electronic structure which give a strong effect on the varistor characteristics, a new method will be discussed in relation to the role of praseodymium oxide on the electrical properties.

ZNO VARISTORS WITH PRASEODYMIUM

1. Nonlinear I-V Characteristics

Zinc oxide varistors with praseodymium oxide show highly nonlinear current-voltage relations. Figure 1 shows the typical I-V relation. The breakdown voltage which is the threshold for the steep current rise is proportional to the thickness of the element. In this case the breakdown voltage per unit thickness is 200V/mm. According to the empirical I-V relation expressed by the following equation, the nonlinear exponent exceeds 50.

$$I=KV^{\alpha} \quad \text{......................} \quad (1)$$

This nonlinearity is almost the same as those of varistors containing bismuth oxide.

2. Two-phase Microstructure

Figure 2 shows the microstructure of the element obtained by a scanning electron microscope(SEM). As can be seen in this figure, only two phase are present in the ceramic, namely, ZnO grains and an intergranular phase composed mainly of praseodymium oxide. This fact should be noted because bismuth-type varistors have been reported to contain a third phase in addition to ZnO grains and the intergranular phase. The third phase is a spinel phase of $Zn_7Sb_2O_{12}$[4] which plays no electrical role because of the high resistivity.

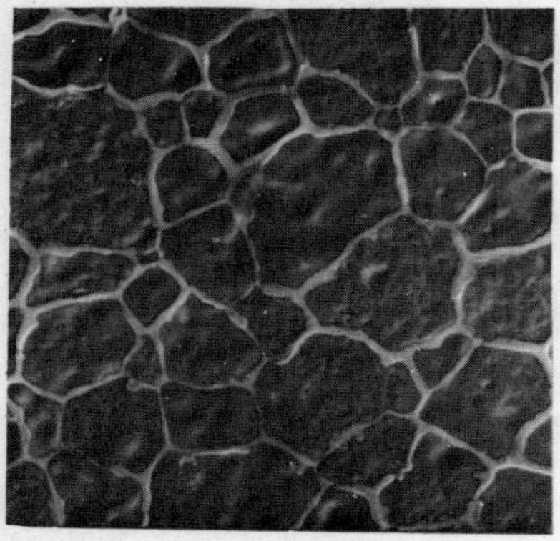

Fig.2 Scanning electron microphotograph of ZnO element. Two-phase structure can be observed.

Fig.4 Surge arrester installed at an actual electric power station. The system voltage is 154kV AC.

Figure 3 shows the simplified microstructure models of two-phase and three-phase microstructures. Since I-V nonlinearity originates from the grain boundary between ZnO grains, the absence of spinel particles in the two-phase microstructure increases the active grain boundary area through which electrical current flows. Therefore, the effective cross section area of the element is increased. This large effective cross section area resulted in high performances as surge-protection devices.

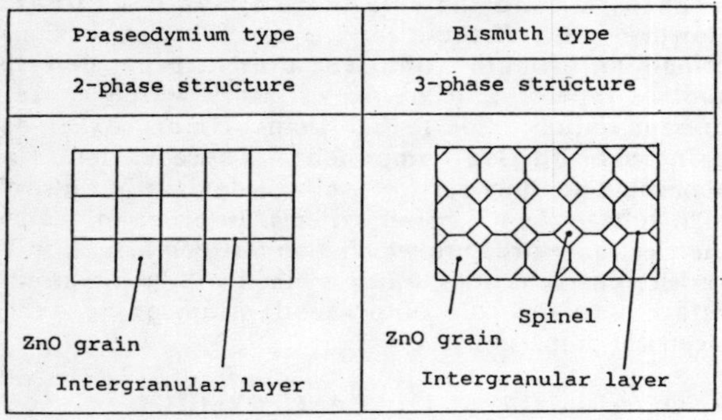

Fig.3 Microstructure model of ZnO element.

3. Application to High Voltage Power Systems.[5]

In addition to low voltage surge-protection devices, ZnO varistors with praseodymium oxide are also applied to surge arresters for high voltage electric power stations. Furthermore, the large cross section area produces high performance such as,

(1) High energy-absorption capability
(2) Excellent nonlinearity in high current
(3) Fast response time
(4) Long life.

Figure 4 is a photograph of the surge arrester actually installed at an electric power station.

ELECTRONIC STRUCTURE OF GRAIN BOUNDARIES

1. DLTS Method

When one discusses the conduction mechanism of ZnO varistors, it is important to know the electronic structure in the vicinity of the grain boundaries because the nonlinearity arises from the grain boundary phenomenon. Figure 5 shows the electronic structure of the grain boundary region. This structure is called a double Schottky barrier.[2] Since this barrier is formed by the electrostatic potential due to the electrons trapped at the interface states, characterization of the interface states in this area is considered as an important problem. Generally, Deep Level Transient Spectroscopy(DLTS) is a useful method to evaluate the deep traps in semiconductors.[6] In fact some investigators succeeded in detecting signals due to traps in ZnO varistors.[7-8] As DLTS spectra are obtained by measuring the capacitance transient of semiconductor junction after the voltage pulses. Therefore those of ZnO varistors give the information about the charge at the bulk traps and the interface states because the capacitance of the ZnO varistor is directly related to the double Schottky barrier.

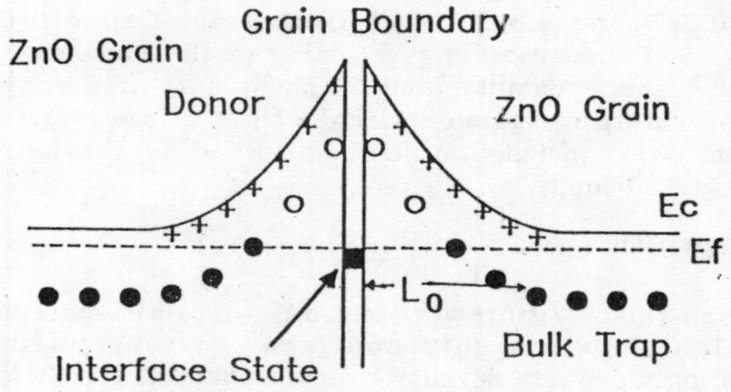

Fig.5 Double Schottky barrier at the grain boundary.

Figure 6 shows the typical spectra measured by the ordinary DLTS method at low temperature. Both Pr-type and Bi-type varistors were selected as samples. The bias voltage, the pulse height and the pulse width were -90V, 90V, and 1ms respectively. Two negative sharp peaks were observed at about -140°C and -80°C for both samples. Since the decrease in localized electrons in the grain boundary region causes the increase in the capacitance, the negative polarity of these peaks indicates that the electrons are emitted from the bulk traps which exist in the forbidden band of ZnO grains. The gate setting times t_1 and t_2 determine the emission rate e_{max} as following equation.

$$e_{max}=\ln(t_1/t_2)/(t_1-t_2) \quad \text{.......... (2)}$$

From the DLTS spectra, one can obtain a set of the temperature at the maximum of the DLTS peak and

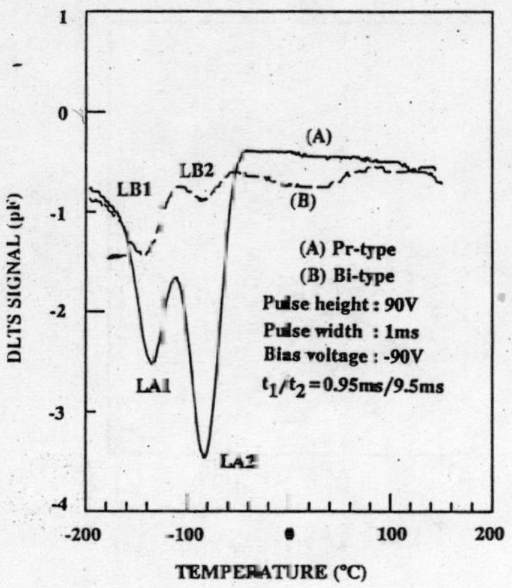

Fig.6
DLTS spectra of the bulk traps at low temperature.

calculated e_{max} from eq.(2). By repetition of similar measurements with different gate setting times, one can obtain an Arrhenius plot of the emission rate as shown in Fig.7. The resultant activation energy is considered as the depth of the trap below the conduction band edge. The calculated depth and density of the bulk traps from the Arrhenius plots are indicated in Table 1. The features of two bulk traps were almost the same although the Pr-type was quite different from the Bi-type in the additive components.

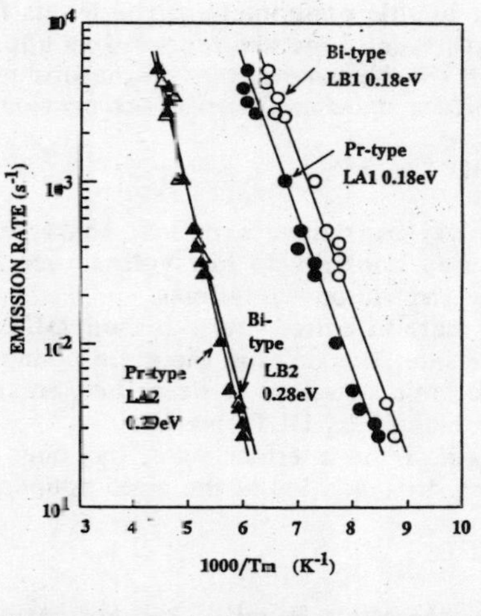

Fig.7 Arrhenius plots of the emission rate for the bulk traps.

Table 1 The level and density of the bulk traps measured by the ordinary DLTS at low temperature.

	L1		L2	
	Et(eV)	Nt(cm^{-3})	Et(eV)	Nt(cm^{-3})
Pr-type	0.18	2.2×10^{16}	0.29	3.0×10^{16}
Bi-type	0.18	1.1×10^{16}	0.28	0.7×10^{16}

2. Zero-biased DLTS Method

In the previous section, the ordinary DLTS method was shown as a useful method for detecting bulk traps. However bulk traps hardly relates to the I-V nonlinearity but to the degradation or the high speed response of varistor action, the detection of the interface states is required to understand the I-V characteristics. In order to detect these interface states, a new DLTS technique are developed. If DLTS measurements are performed without the bias voltage, DLTS signals explain the motion of electron trapping by the bulk traps which emitted electrons during the pulse and the motion of electron emission from the interface states after electron capture during the pulse. Since the amount of electrons captured by the interface states is extremely larger than that of emitted electrons from the bulk traps, the DLTS signal under zero-bias voltage is mainly composed of the electron motion of the interface states. However since the emission rate of the interface states is quite low, the peak due to the interface states should appear at higher temperature region. Figure 8 shows the comparison of zero-biased and ordinary DLTS spectra at higher temperature. Samples are the same Pr-type varistor as in Fig.6. A weak positive peak was observed by the ordinary DLTS method. The positive polarity indicates that this peak is due to the interface states because the bulk traps are negative as shown in Fig.6 . On the other hand, zero-biased DLTS shows a strong negative peak which explains the electron emission from the interface states.

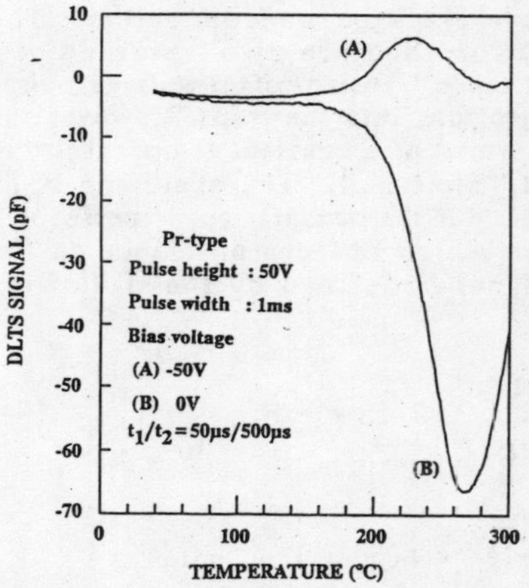

Fig.8
Comparison of the ordinary DLTS and zero-biased DLTS spectra of the interface states.

Fig.9 shows the observed DLTS spectra of two varistors. The peak of Pr-type was about 5 times larger than that of Bi-type. This difference reveals that the density of the interface states of Pr-type is higher than that of Bi-type because the peak height of DLTS signals is proportional to the density of the interface states. The Arrhenius plots of the emission rate gave 1.03eV and 0.94eV for the depth of the traps for Pr-type and Bi-type respectively as listed in Table 2. It is quite interesting that the interface

Table 2 The level and density of the interface states measured by zero-biased DLTS.

	Level(eV)	Density(cm^{-2})
Pr-type	1.03	4×10^{17}
Bi-type	0.94	7×10^{16}

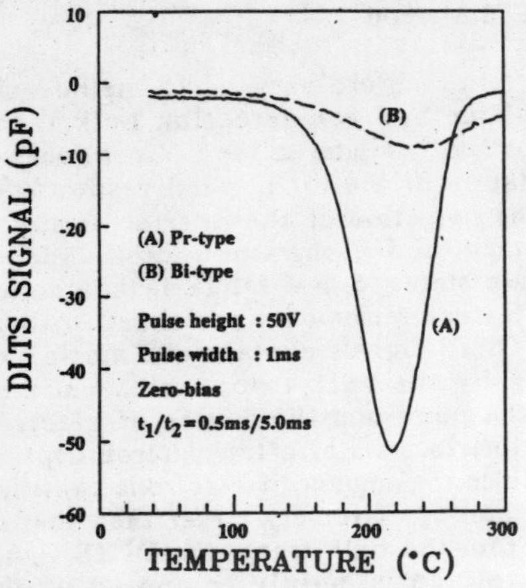

Fig.9 Zero-biased DLTS spectra of Pr-type and Bi-type aristors.

states for Pr- and Bi-type are almost the same although each varistor is entirely different in the additive components. This fact suggests that the same physical origin should generate the interface states in both varistors.

If the distribution of the interface states is assumed to be monoenergetic, the calculated DLTS spectrum fairly agreed with the observed spectrum, consequently the interface states is certainly monoenergetic.

3. ICTS Method

Similarly to the DLTS method, the Isothermal Capacitance Transient Spectroscopy(ICTS) method has been proposed to evaluate deep level traps in semiconductors.[9] The signals of the ICTS method are also obtained by measuring the capacitance transient after voltage pulses which inject the electrons into the traps. However the experiment is performed at a constant temperature in contrast to the DLTS method. Therefore the ICTS method is more suitable if the measurement is performed at low temperature which induces no effect on the traps.The ICTS signal is defined by the following equation.

$$S(t) = t\frac{d}{dt}\left[\frac{1}{C(t)} - \frac{1}{C_0}\right] \quad \text{.......... (3)}$$

where

$$\left[\frac{1}{C(t)} - \frac{1}{C_0}\right] \propto \exp(-t/\tau) \quad \text{.......... (4)}$$

τ is the time constant of the emission rate of the trap. Substitution of eq.(4) into eq.(3) reveals that S(t) becomes maximum at $t=\tau$. Figure 10 shows the observed ICTS spectra of a Pr-type varistor at various temperature. The Arrhenius plots of observed τ gave 0.66eV which was some what smaller than that obtained by the DLTS method. The reason for this discrepancy is still not clear.

ORIGIN OF THE INTERFACE STATES

In the previous chapter the interface states of Pr-type and Bi-type revealed to be similar to each other. Accordingly, it is natural to consider that the interface states should be generated from the common origin in both type. Experiments on other rare earth metal oxides as an additive component resulted in high nonlinear characteristics for Tb-type varistors. Comparison of Pr oxide and Tb oxide leads to the following conclusion that

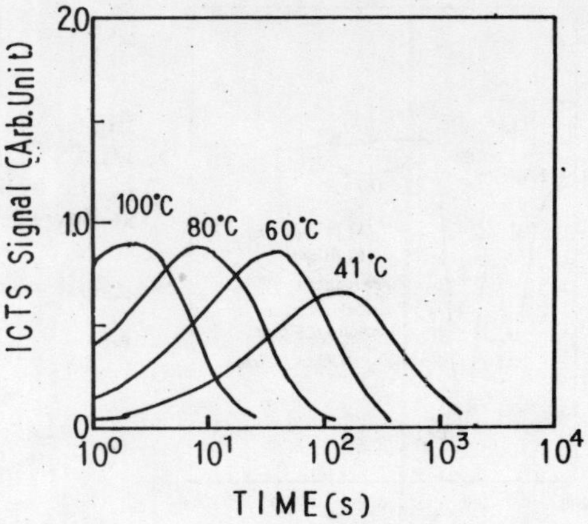

Fig.10 ICTS spectra of Pr-type varistor.

the origin of the interface states should be the chemisorbed oxygen ion because the oxygen is evolved by the following reactions during the sintering process of the varistors.[10]

$$Pr_6O_{11} = 3Pr_2O_3 + O_2 \quad \text{.......... (5)}$$

$$Tb_4O_7 = 2Tb_2O_3 + O_2/2 \quad \text{.......... (6)}$$

In addition to this phenomenon, the levels formed by chemisorbed oxygen ions are reported as approximately 1eV[11] which are the same as that of the observed results for the interface states as shown in the previous chapter.

CONCLUSION

(1) Pr-type varistors have excellent characteristics and can be applied not only to low voltage use but also to high voltage electric power systems.
(2) A new method called "zero-biased DLTS" method detected the interface states at the grain boundaries.
(3) The electronic structure is described on the basis of the results obtained by DLTS method.
(4) The origin of the interface states has been concluded as chemisorbed oxygen ion at the grain boundaries.

REFERENCES

1. M. Matsuoka et al., Suppl. J. Jpn. Soc. Appl., Phys., 39(1970), 94.
2. L.M. Levinson et al., Am. Cer. Soc. Bull., 65(1986), 639.
3. K. Mukae et al., Jpn. J. Appl. Phys., 16(1977), 1361.
4. M. Inada, Jpn. J. Appl. Phys., 19(1980), 409.
5. K. Mukae et al., IEEE Trans. Pow. Deliv., 3(1988), 591.
6. D.V. Lang, J. Appl. Phys., 45(1974), 3023.
7. N. Shohata et al., Jpn. J. Appl. Phys., 19(1980), 1793.
8. A. Nitayama et al., Jpn. J. Appl. Phys., 19(1980), L743.
9. H. Okushi et al., Jpn. J. Appl. Phys., 19(1980), L355.
10. K. Mukae et al., Advances in Ceramics, 1, Am. Cer. Soc.(1981), 331.
11. L. Ernst, Current Topics Materials Science, 7, North Holland Publishing Co.(1981), 364.

STUDY AND MANUFACTURE OF YSZ RF SPUTTER THIN-FILM

Zhang Zhisheng, Wang Wensheng, Liu Zhigang and Wei Ming
The Department of Electronic Engineering, Tianjin University
Tianjin, China

INTRODUCTION

Sensor technology has a close connection with some physical effect and discovery of some new materials. The solid electrolyte materials are sensitive materials developed of recent years. They are the solid with ionic conductivity. The technologists in studying sensors are devoted to study of solid electrolyte materials with great interest.

Within the scope of solid electrolytes, only ZrO_2 solid solution meets basically the requirements of sensor. YSZ (Yttria-Stabilized Zirconia) is a solid electrolyte of the ZrO_2 solid solution system.

The stabilized zirconia is a sensitive material for making oxygen sensor. At present, people basically adopt sintering to make oxygen sensor. Therefore the production processes fall behind. The development direction of the oxygen sensor is that it is made into thin-film. The stabilized ZrO_2 thin-film is the base of making thin-film oxygen sensor. We have manufactured ZrO_2 thin-film by rf sputtering.

First of all, YSZ rf sputter target is necessarily manufactured. Its dimension is that the diameter equals 80-100mm, thick equals 4mm. Its manufacture is very difficult. What is more, the sputter target must have a microstructure of cubic phase.[1]

EXPERIMENT

In order to make so big target, we must master the knowledge about crystal phase change of ZrO_2, especially about the relationship between crystal phase change and temperature. Furthermore, it is a very important point to select a proper stabilizer. We select Y_2O_3 as a stabilizer of ZrO_2, instead of selecting MgO or CaO which are more commonly used than Y_2O_3. This is because (a) their ionic lattice dimensions are: Zr^{4+} — 0.80Å; Y^{3+} — 0.96Å; Ca^{2+} — 1.04Å; Mg^{2+} — 0.66Å, Obviously Y^{3+} is closer to Zr^{4+} than Ca^{2+} and Mg^{2+}, (b) charge difference between Y^{3+} and Zr^{4+} is smaller, (c) ZrO_2 is linked up with Y_2O_3 by more covalent bonds and different polars.

The technological processes of YSZ rf sputter target mainly involve grinding, drying, sieving, making particles, forming, isopressing and sintering.

It is necessarily pointed out that in process of making particles we used polyethylene glycol 2000 as a bond line which is made into 3 per cent water solution. This is because the feature of ceramic material of ZrO_2 is dry, loose and hard.

Forming, isopressing and sintering processes are very key processes in which we must be very careful.[2] The proper sintering conditions are key of forming ceramic crystal. Our sintering conditions are that the lift temperature velocity is at 30°C/hr under 500°C, at 40°C/hr up 500°C. Because the lift temperature velocity is slow, it avoids disruption of the ceramic material. The highest temperature is 1750°C-1770°C at which the thermal retardation time is two hours. And the sintering atmosphere is oxidizing atmosphere so that ZrO_2 doesn't reduce oxidates with low valency.

By repeated experimental investigation we have eventually got a level, slick and compact YSZ target.

We adopt the following sputtering conditions in order to make stabilized ZrO_2 thin-film. The frequency of sputter voltage equals 13.5 MHz. When degree of vaccum gets at $1x10^{-4}$ torr, argon is aerified. When airstream dynamic equilibrium is established and the degree of vaccum is at $5x10^{-2}$ torr — $8x10^{-3}$ torr, we start sputtering ZrO_2. The sputter conditions are that the anode voltage equals 2500V, the anode current equals 0.12A-0.14A, the grid current equals 20mA. The sputter velocity is at 860Å/hr under the sputter conditions.

ANALYSIS AND DISCUSSION

By X-ray diffraction (XRD) we make phase analyses for the sputter target. The curve of XRD is shown in Fig.1.

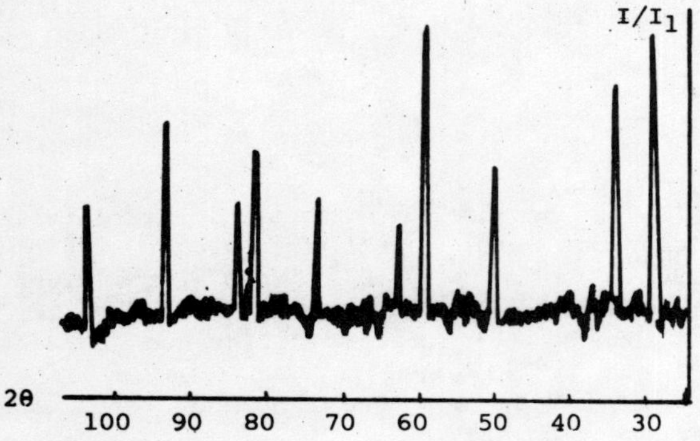

Fig.1 The curve of XRD of sputter target

There are ten peaks in Fig.1. We make measurement of angle for these peaks. We discover that the angles correlative with ten peaks and angles correlative with crystal plane of cubic phase ZrO_2 are basically correspondent. Therefore the sputter target have a microstructure of cubic phase. The data on angles corresponding to first four peaks are listed in Table 1.

Table 1 The data on angles of first four peaks

Order of peaks	1	2	3	4
Standard angles	30.60°	35.34°	50.42°	60.08°
Angles measured	30.30°	35.00°	50.30°	59.00°

We have made X-ray photoelectron spectroscopy (XPS) analyses for sputter thin-film. The analysis curves are shown in Fig.2, Fig.3 and Fig.4. It is proved that the thin-film is stabilized ZrO_2 one after the analysis curves are relative to standard curves of ZrO_2.

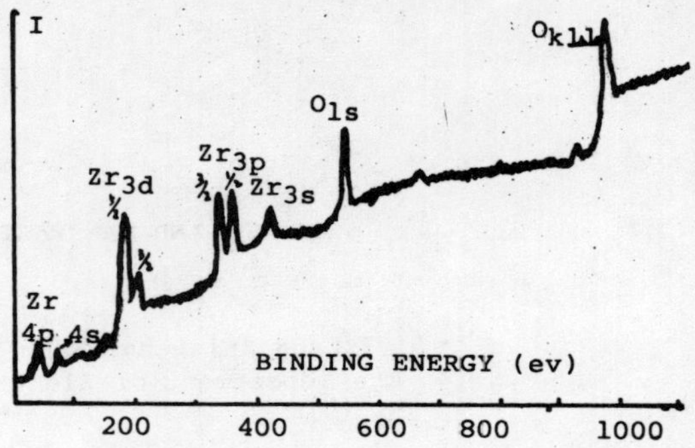

Fig.2 The whole spectrum of zirconia

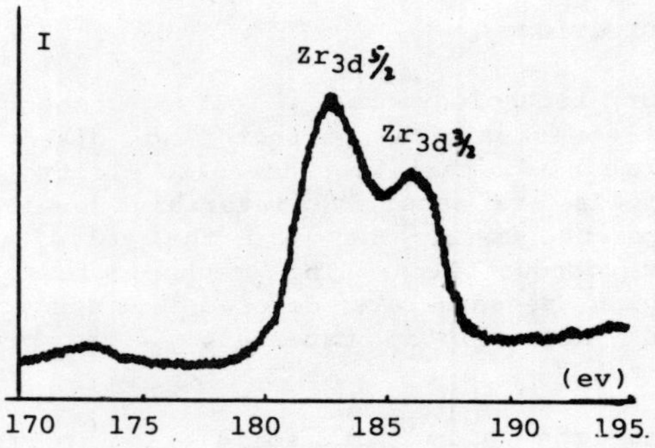

Fig.3 The narrow scan of zirconium

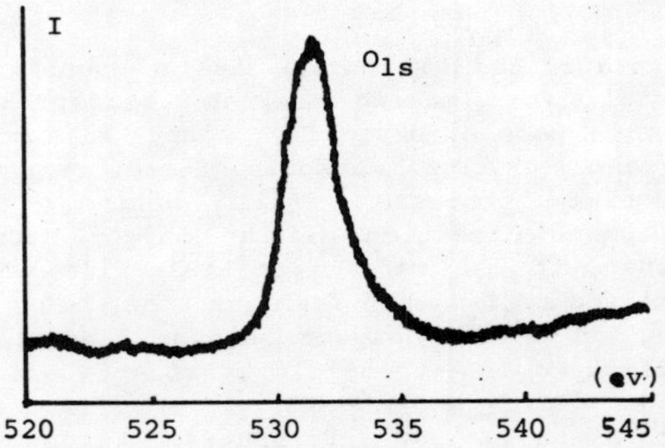

Fig.4 The narrow scan of oxygen

REFERENCES

1. Robert Ruh, K.S Mazdiyasni, P.G Valentine and H.O Bielstine, J. Am Ceram. Soc. [9] (1984), C-190-192.
2. Koji Tsukuma and Masahiko Shimada, Am. Ceram. Soc. Bull., 64, 2(1985), 310-13.

CRYSTAL STRUCTURE AND SENSITIVE CHARACTERISTIC OF La3-SYSTEM CERAMIC

Gu Xin*, Zhang zhong-tai**
*Beijing Glass Research Institute
**Department of Material science and Engineering Tsinghua University
Beijing, China

ABSTRACT

The structures of LaF_3-SrF_2 solid solution were investigated. It was found that the solid solutions despite containing same SrF_2 dopant content but had two kinds of crystallines—the tysonite-type and fluorite-type because of different sintered techniques, respectively. The electrical characteristics of LaF_3 and the solid solutions were studied. The doping SrF_2 up to 5 mole percent made the conductivity increase. While the doping of SrF_2 more than 5 mole percent let it decrease. The activation energies for conduction of fluorite-type were about 1.20ev, while those of tysonite-type were about 0.35ev, which reflected the two kinds of the mechanism of ion conduction. In this work, the electrical properties of solid-state sensing element using LaF_3 and the solid solutions electrolyte also had been investigated as a function of halogen containing gases.

INTRODUCTION

LaF_3 forms non-stoichiometric phases with bivalent metal fluorides(e.g. SrF_2), which crystallize in the fluorite or tysonite structure (1). In the solid solutions with flourite structure excess F^- ions is located in interstital sites, whereas there is excess F^--vacancies in phase of the tysonite structure. In each case these crystal defects should give rise to an enhancement of the ionic conductivity. Recently, Geiger and Schön (2) studied the single crystals of the nonstoichiometric tysonite phase $La_{1-x}Sr_xF_{3-x}$ ($0<x<0.14$). The polycrystalline of the tysonite phases was not as well investigated. In this work, we prepared the polycrystalline sinters of the solid solutions based on LaF_3 by two kinds of sintered techniques. And their conductivities as well their electrolytic properties were measured.
The fact that LaF_3 is a good F^- anion conductor with low electronic conductivity made it reasonable to investigate as an electrolyte for other anions. In this work, the electrical properties of the solid electrolyte based on LaF_3 and its solid solutions were studied as a gas sensing function of halogen.

EXPERIMENTAL

1. Preparation

In order to eliminate the ammonium fluoride included in the reagent, the LaF_3 (purity above 99%) preliminary heated at 400℃ in an argon stream which was dried with phosphorus pentoxide. The resultant powdery fluoride was mixed with strontium fluoride at a designed ratio and ground finely in an agate mortar. Then two kinds of techinques to sinter were adopted. Tech. I: The powder which filled in platinum crucible puted into a vacuum furance and were sintered at 900℃. Tech. II: The specimen were sintered under an argon environment at 1000℃.

2. X-ray diffraction and density measurements

The sintered fluorides were subjected to x-ray diffraction analysis to make sure of the formation of solid solutions and determine the lattice parameters. The theoretical densities of the speciments were calculated from the lattice parameters and compared with powder densities with the normal pycnometric method using kerosene at room temperature (about 25℃).

3. Conductivity measurements

The rectangle-shaped specimens were used for conductivity measurements. Platinum wire was bound tight on both ends of the specimen rod for electrodes, and it was put into the argon gas furnace. We could obtain the area and lenth of the specimen, and the current and voltage were measured. Then we could get the conductivity of specimen by calculation.

4. Sensitivity measurements

The sensor as illustratively(Fig. 1) shown was a symmetrical cylindrical configuration. The preferred embodiments were intended for the sensing of halogen

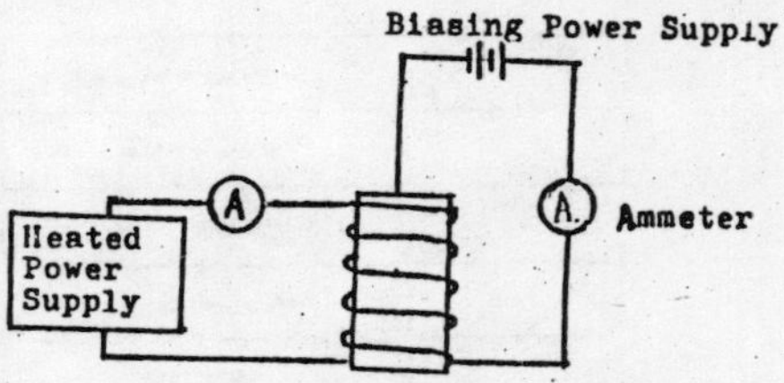

Fig. 1 Schematic of electrical circuit

containing gases, in which the sensing element was made from a ceramic material based on fluoride electrolyte. If a halogen containing gas such as Freon was present in atomsphere adjacent the sensing element, the current passing through the sensing element would be change as indicated on the ammeter.

RESULTS

1. Structure

Fig 2. shows the x-ray diffraction patterns of pure LaF_3 and SrF_2-doped specimens which were sintered by

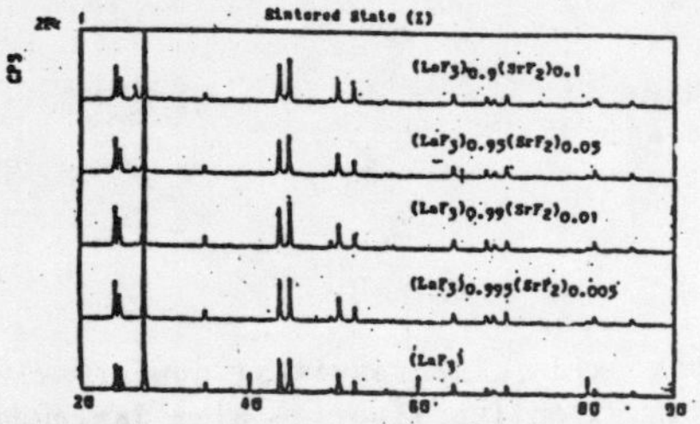

Fig. 2 The x-ray diffraction patterms of the specimens sintered by Tech. I

Tech. I . It was found there was tysonite phase in the specimen only when the dopant concentrations were less than or equal to 5mol%. However, in specimens doped with 10mol% SrF_2 the x-ray diffraction pattern gave small diffraction lines of SrF_2 which was fluorite phase. these results indicated that the solid solution formation ranged in these system up to, at least, 5mol% SrF_2. In Table I, the lattice parameters of the doped samples sintered by Tech. I were more than that of pure LaF_3. And the measured values were in good agreement with the values calculated based on the fluorous ion vacancies assaption, suggesting that the fluorous ion vacancies be formed by adding SrF_2 as a result of charge compensation. The x-ray diffraction patterns of $(LaF_3)_{1-x}(SrF_2)_x$ ($0.005 \leq x \leq 0.1$) sintered by Tech. II and pure SrF_2 are present in Fig. 3. The patterns showed there solid solutions were formed fluorite-type

Table I Lattice parameters and density of tysonite solid solutions

Specimen	Lattice a(Å)	parameter c(Å)	Calculated Density (g/cm3) Anion Vacancy	Calculated Density (g/cm3) Cation Interstitial	Measured Density (g/cm3)
LaF3(JCPD)	7.184	7.351			
LaF3 raw	7.183	7.352	3.13		3.14
(LaF3)0.995(SrF2)0.005	7.190	7.375	3.11	3.13	3.10
(LaF3)0.99(SrF2)0.01	7.200	7.377	3.10	3.12	3.08
(LaF3)0.95(SrF2)0.05	7.188	7.382	3.07	3.11	3.06
(LaF3)0.9(SrF2)0.1	7.198	7.385	3.01	3.09	3.01

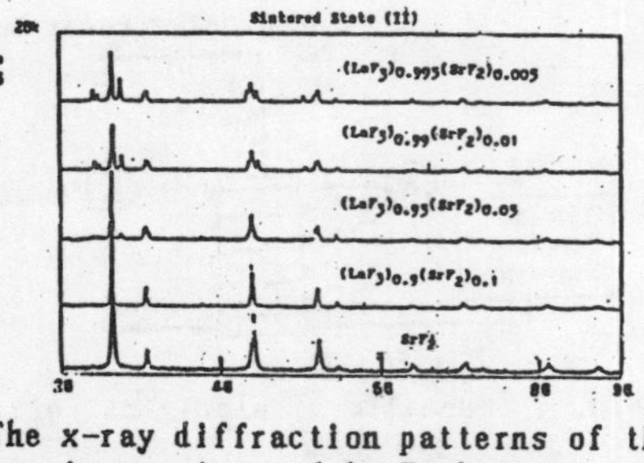

Fig. 3. The x-ray diffraction patterns of the specimens sintered by Tech. II

phase like pure SrF_2. And with increasing content of SrF_2 the fluorite phase increased. In this system, $(LaF_3)_{0.9}(SrF_2)_{0.1}$ is fluorite structure. As shown in Fig. 4, the specimen was tysonite-type phase before sintering, and it became fluorite-type phase after sintering. These results have not been reported and

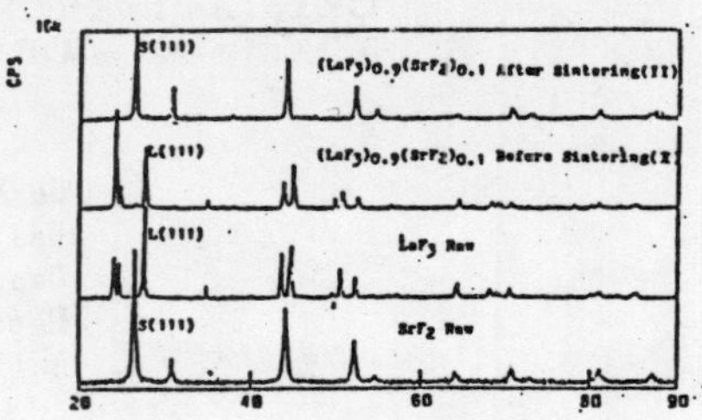

Fig. 4. The x-ray diffraction patterns of specimens after and before sintering.

did not square with the LaF_3-SrF_2 phase-graph (3). It has not been explained that why the solid solutions of $(LaF_3)_{1-x}(SrF_2)_x$ ($0.005 \leq x \leq 0.1$) sintered using Tech. II becamed the fluorite phase. With increasing content of SrF_2 the lattice parameters of the doped samples sintered by Tech. II were simlar to that of pure SrF_2 (see Table II), suggesting that these $(LaF_3)_{1-x}(SrF_2)_x$

Table II Lattice parameter of fluorite solid solutions

Specimen	SrF2 Dopant Content (%mol.)	Lattice Parameter a(Å)
SrF2(JCPD)		5.800
SrF2 raw	100	5.802
(LaF3)0.9(SrF2)0.1	10	5.799
(LaF3)0.95(SrF2)0.05	5	5.807
(LaF3)0.99(SrF2)0.01	1	5.814
(LaF3)0.995(SrF2)0.001	0.5	5.822

($0.005 < x < 0.1$) were the solid solutions with fluorite structure.

2. Conductivity

Fig. 5 shows the conductivity behaviour of pure LaF_3 and of $La_{1-x}Sr_xF_{3-x}$ sintered by Tech. I for the concentrations

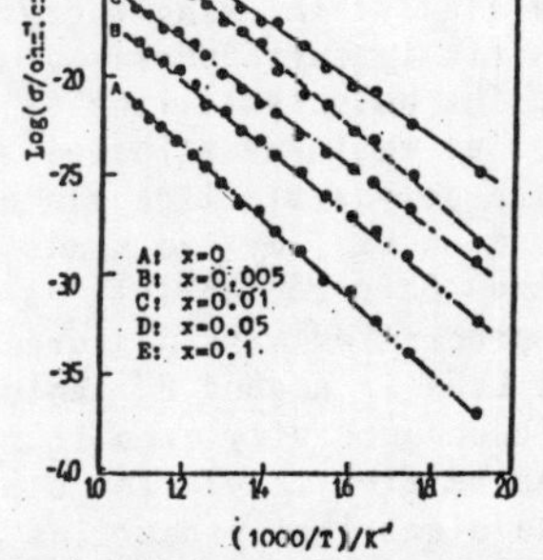

Fig. 5. The conductivity of pure LaF_3 and of $La_{1-x}Sr_xF_{3-x}$ sintered by Tech. I

with x=0.005, 0.01, 0.05, 0.1. The values for the ionic conductivity can be respesented by an Arrhenius equation. The conductivity of pure LaF_3 above 250° C was closed to that of the single crystal reported by Nagel et. al (4). The conductivities increased by adding SrF_2 to LaF_3 and maximum conductivity occurred at 5mol% SrF_2 compositions. The conductivity of this

fluoride was about one order of magnitude higher than that of pure LaF_3. But with a higher content of SrF_2, the dependance of conductivity on composition became ambiguous.
As shown in Fig. 6 Arrhenius plots of conductivity of pure LaF_3 and of the solid solutions sintered by Tech. II are present. It should be noticed that in this case

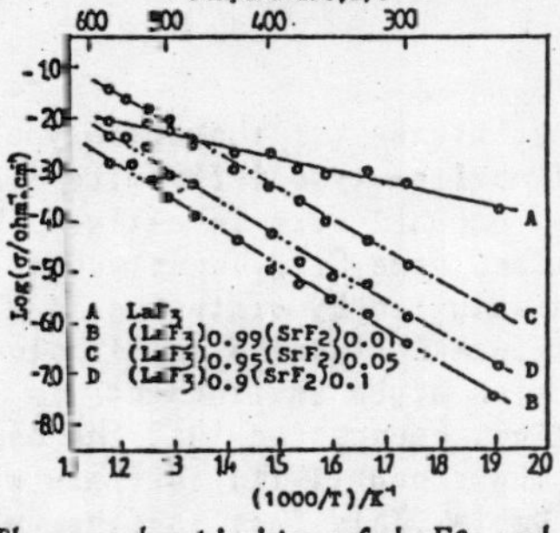

Fig. 6. The conductivity of LaF_3 and of $(LaF_3)_{1-x}(SrF_2)_x$ sintered by Tech. II

the conductivities of the $(LaF_3)_{1-x}(SrF_2)_x$ solid solutions were lower than that of pure LaF_3 below 450℃. And the slope of Arrhenius plot for pure LaF_3 is differed conspicuously from those for SrF_2-doped specimens, that is, the slops for fluorite-type solid solutions are rather steep compared to that for LaF_3. In this system, the conductivities also increased with increasing content of SrF_2 up to 5mol%.

3. Sensitivity

The sensor sensitivity S that determined the current passing through the element is given

$$S=(Ii-Io)/Io \qquad (1)$$

where Io is the background current in the absence of Freon. Ii is the monitored current in the given Freon concentration. Fig. 7 shows the sensitivity vs. parts per billion of Freon 12 for elements based on LaF_3 and $La_{1-x}Sr_xF_{3-x}$ solid solutions (tysonite phase). The $La_{1-x}Sr_xF_{3x-x}$ system element was more sensitive than

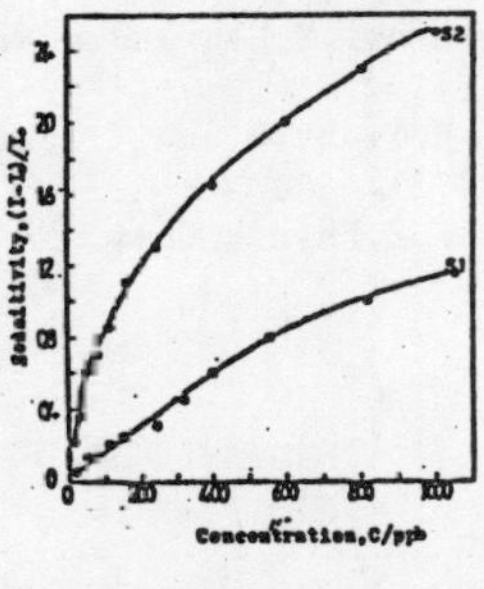

Fig. 7. Current sensitivity of Freon of elements S1: LaF_3-system, S2: $La_{1-x}Sr_xF_{3-x}$-system

the LaF_3-system one. Probably the vacancy concentration of $La_{1-x}Sr_xF_{3-x}$ was much more than that of LaF_3 so that the conductivities of the former S2 were larger than those of the later S1. The fact indicates that the sensitivity of element is own to the F^- ion conductivity in the solid electrolytes.

DISCUSSIONS

1. Conductivity of tysonite solid solution

In pure LaF_3 the intrinsic defects have been reported to be the Schottky type--F^- ion vacancies (5). The doping of SrF_2 sintered by Tech. I up to 5mol% makes the conductivity increase. A representative example for this can be seen in Fig. 8, in which the conductivities

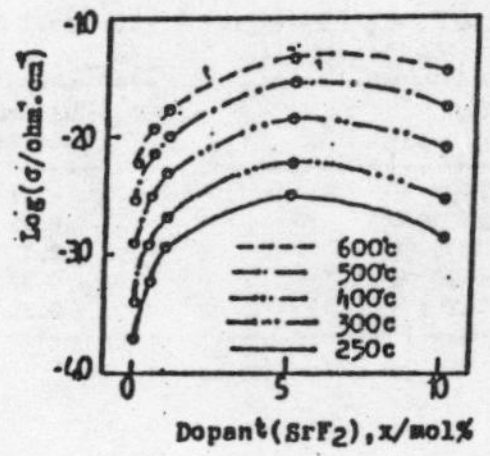

Fig. 8. Plots of logσ vs. SrF_2 content

were plotted against the dopant content for a SrF_2-doped sample. This fact suggests that the ionic current be carried by mobile fluorous ion vacancies, and that the vacancy concentration increase by doping SrF_2 as shown in the following equation

$$SrF_2(LaF_3) \longrightarrow Sr'_{La} + 2F_F^{x} + V_F^{\cdot} \qquad (2)$$

It is shown in Table I that the formation of fluorous ion vacancies by doping of SrF_2 was also supported by the result of density measurements. In fact, Takahashi et. al (6) observed a maximun in the composition dependence of the solid solution $Ce_{1-x}Ca_xF_{3-x}$ for x=0.05. Therefore, $Ce_{0.95}Ca_{0.05}F_{2.95}$ and the present solid solution $La_{0.95}Sr_{0.05}F_{2.95}$ are close to the composition for which optimum conductivities are expected. An association model, however, is inadequate to explain a maximum in the conductivity. Instead, ordering of oppositely charged defects can account for this type of composition dependence of the conductivity (7). As shown Fig. 4, when the SrF_2-doped concentration was more than 5mol%, the specimen gave the diffraction line of SrF_2, inducting that the second phase-SrF_2 phase was formed such as the following reaction

$$Sr'_{La} + V_F^{\cdot} + 2F_F^{x} = SrF_2(ppt) \qquad (3)$$

In this case the conductivity decreased.

2. Comparison tysonite with flourite in conductivity

Fig. 9 is conductivity as a function of temperature for

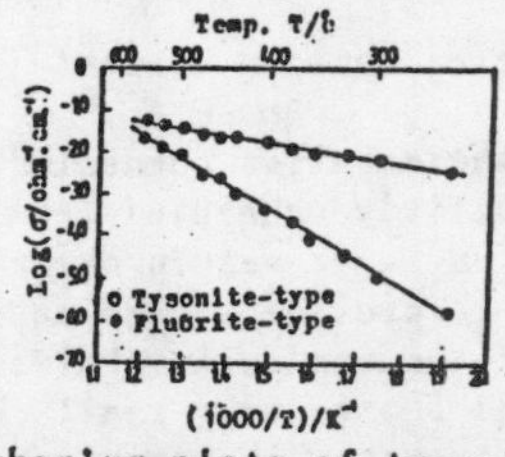

Fig. 9 Arrhenius plots of tysonite and fluorite solid solutions

$(LaF_3)_{0.95}(SrF_2)_{0.05}$ of tysonite-type and fluorite-type. The plot of tysonite differed conspicnously from that of fluorite. At low temperature the conductivity of tysonite was about three orders of magnitude higher than that of fluorite. It should be noticed that the slope of tysonite was much lower than that of fluorite, which meaned that the activation energy of tysonite for conduction was rather higher than that of fluorite. Table III summarizes the activation energies for conduction calculated from the slopes of Arrhenius plot.

Table III Activation nenrgy of tysonite and fluorite solid solutions

Specimen	Activation Energy (ev)	
	Tysonite	Fluorite
LaF_3	0.37	
$(LaF_3)_{0.995}(SrF_2)_{0.005}$	0.33	
$(LaF_3)_{0.99}(SrF_2)_{0.01}$	0.32	1.19
$(LaF_3)_{0.95}(SrF_2)_{0.05}$	0.30	1.19
$(LaF_3)_{0.9}(SrF_2)_{0.1}$	0.36	1.20

The activation eneryg for conduction of tysonite structure was about 0.35ev, and that of fluorite was about 1.20ev. The reason for this has not been fully explained. However, the values of the activation energy for conduction may be related to the defects, that is, the lower activation energies for ion vacancies motion and the higher for intersital ions movement. This suggests that the conduction mechanism of the two types of solid solutions should not be the same. The low activation energy for conduction in tysonite system is ascribed to the ion vacancies, whereas the high activation energy for motion in fluoride system may be owned to intersitial ions. However, quantitative discussions should not be made here, since the specimens examined were not single crystal but stintered polycrystalline samples.

3. Conduction mechanism of the sensing element

As shown in Fig.10, the method of operation of the sensing element was based on the fact that under a D.C. potential difference between points on the solid state ceramic sensing element, the positively charged ions

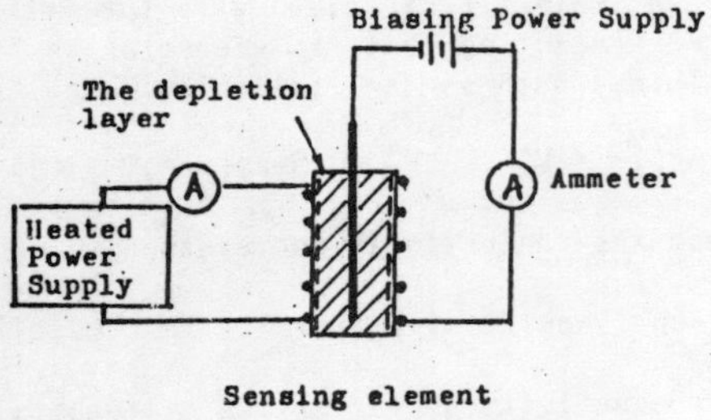

Fig. 10 Schematic of sensing element

F^- migrated from the negative connected element terminal to the positivly connected terminal. As a result, the depletion layer was formed at the surface of sensing element to provide a high resistance region at the surface, and the conduction between the terminals of element became very small. When a halogen containing gas was present, the negative ions of this gas would accumulate on the depletion layer of the sensing element. In order to maintain an electrical balance, some of the positively charged ions accumulated on the element centure terminal would migrate to the depletion lay, thus the current flow between the terminals would increase. Heating of the atmosphere by energizing the circuit from coil to heater power suppLty of coures aided in cracking any halogen containing gases in the atmosphere of the sensing element.

CONCLUSIONS

The sintered techniques for tysonite-type trifluoride LaF_3 and for fluori-te—type difluorite SrF_2 solid solutions based on LaF_3 were investiged. When the specimen contained same SrF_2 dopant content, it formed tysonite solid solution by sintering in a vacuum furance, but it formed fluoride solid solution by sintering under an argon environment. In the tysonite-type system it has been shown that the doping of SrF_2 to LaF_3 makes the conductivity increase with increasing dopants up to 5mol%. This fact together with the experimental result of density indicates that the tysonite structure specimens are fluorous ion conductors resulting from easy movement of fluorous ion vacancies. As to the fluorite solid solutions, in general, the conductivity is low and the activation energy for conduction is high. This suggests that the conduction mechanism of fluorite differ from that of tysonite, and it may be owned to intersital ion movement. It was taken advantage of the electrical properties of lauthanum fluoride electrolyte to prepare the sensing elements as a function of halogen containing gases. The sensing element having a depletion layer has high sensitivity.

REFERENCES

1. B.P.Sobolev and K.B.Seiranian, J.Solid State Chem., 39(1981), 337
2. H.Geiger, G.Schön and H.Strok, J.Solid State Ionics, 15 (1985), 155
3. R.H.Nafziger and N.Riazance, J.Am.Ceram.Soc., 55 (1972), 130
4. L.E.Nagel and M.O'Keffe, "Fast Ion Transport in Solids", Solid State Batteries and devices, W.van Gool, Editor, North Holland Publishing Co., Amsterdam (1973), 165
5. A.Sher, R.Solomon, K.Lee and M.W.Muller, Phys. Rev. 144 (1966), 593
6. T.Takahashi, H.Iwahara and T.Ishikawa, J.Electrochem soc., 124 (1977), 208
7. H.Schmalzried, Z.Physik.Chem. NF 105 (1977), 47

ROLE OF OXYGEN FOR VARISTOR AND PTC EFFECT

Satoru FUJITSU, Akira HASEGAWA, Kunihito KOUMOTO and Hiroaki YANAGIDA*
Department of Industrial Chemistry, Faculty of Engineering, the University of Tokyo
7-3-1 Hongo, Bunkyo-ku, Tokyo 113, JAPAN
*Research Center for Advanced Science and Technology, the University of Tokyo
4-6-1 Komaba, Meguro-ku, Tokyo 113, JAPAN

ABSTRACT

The role of oxygen in generating the energy barrier along the grain boundary in ZnO varistor and PTC thermistor has been examined from its analogy to the surface energy barrier formed by chemisorbed oxygen. It was proposed that the main chemical species to form the energy barrier in ZnO varistor and $BaTiO_3$ PTC thermistor is the excess ionized oxygen along the grain boundary.

INTRODUCTION

It is well known that ZnO based varistor and $BaTiO_3$ based PTC thermistor are the typical functional ceramics using the energy barrier along the grain boundary. The non-ohmic characteristic of the varistor is based on the energy barrier[1,2] and the resistivity change in PTC thermistor is based on the change in the barrier height which is inversely proportional to the dielectric constant[3]. These characteristics depend on the atmosphere, especially oxygen, during their fabrication process. When these materials were heated in the reducing ambient, the large non-ohmic characteristic in ZnO varistor and the large change in the resistivity in PTC thermistor cannot be observed, while these properties are restored by heating in air or oxygen.

The authors have pointed out that the acceptor state along the grain boundary to form such an energy barrier in these materials was the excess ionized oxygen along the grain boundary[4-7]. The excess ionized oxygen can be recognized as the chemisorbed oxygen on the surface, but such a state along the grain boundary has not been able to be detected. Then, they have examined the relation between the acceptor state along the grain boundary and the resulting Schottky type energy barrier, from the analogy to that between the chemisorbed oxygen and the resulting surface energy barrier by using the porous material.

In the present study, the electrical conductivity change in porous ZnO and porous semiconducting $BaTiO_3$ on the introduction of oxygen is examined from the viewpoint of the formation of the energy barrier at the neck region. The non-ohmic characteristic for ZnO and PTCR effect for doped $BaTiO_3$ will also be discussed from the view point of the effect of oxygen.

EXPERIMENTAL

Nominally pure ZnO powder was dispersed in an aqueous solution of $Al(NO_3)_3$, and dried. The resulting powder was calcined at 600 C for 4h in air, pressed into a rectangular bar and sintered at 800 C for 1h in air. Aluminum oxide addition was to maintain the ZnO body porous. When pure ZnO was sintered in the same manner, the relative density attained above 80%. The sample with surface area of $3.2m^2/g$ and 50% relative density was obtained. The average grain size of the porous sample was 0.3μm.

Pure $BaTiO(C_2O_4)_2 4H_2O$ and Sb_2O_3 were mixed in ethanol and dried. The mixed powder was calcined at 700 C for 1h in air, pressed into rectangular bar under the pressure below $1kg/cm^2$ and sintered at 1350 C for 1h in air. The sample with surface area of $0.7m^2/g$ and 73% relative density was obtained. The average grain size was 1μm.

As the starting condition, the sample(ZnO

or $BaTiO_3$) was heated in vacuum($<10^{-5}$ torr) at 600°C until the resistivity reached 0.05ohm m for ZnO or 5000ohm m for $BaTiO_3$. At prescribed temperature, oxygen was introduced into the holder, and changes in the pressure by chemisorption of oxygen and the electrical conductivity by four probe method were measured simultaneously.

After heated under various conditions of temperature, oxygen pressure and time, the sample was quenched in liquid nitrogen. The current-voltage (I-V) and capacitance-voltage (C-V) characteristics were measured at room temperature by curve tracer and LCZ meter at 100kHz, respectively. The temperature dependence of the electrical conductivity was measured by dc four probe method under the same atmosphere before quenching.

RESULTS AND DISCUSSION

(1) Chemisorption and the electrical conductivity

The electrical conductivity of the desorbed sample decreased with increasing amount of chemisorbed oxygen as shown in Fig.1(a) and (b) for ZnO and $BaTiO_3$, respectively.

The effect of oxidation and reduction in the bulk should be also considered. The dependence of electrical conductivity on oxygen partial pressure was larger than expected from the change in nonstoichiometric composition by atmosphere. And the conductivity did not change above the pressure to attain saturation of adsorption. From these experimental results it is judged that oxidation and reduction in the bulk would have a minor effect on changing the electrical conductivity in the present temperature and pressure region.

The assumption that the energy barrier whose height increases with increase in the amount of chemisorbed oxygen controls the total conductivity, could explain these results. Considering that the carrier mobility is constant throughout the sample, the electrical conductivity of a semiconducting material depends on the energy difference between the bottom of the conduction band (E_c) and the Fermi level (E_f) as $\sigma \propto \exp(-(E_c-E_f)/kT)$. The difference ($E_c-E_f$) is very small in vacuum. If the chemisorbed oxygen makes an acceptor type trap level on the surface, the resulting energy barrier to compensate the excess charge is generated. The height of such a surface energy barrier (ϕ_s) is estimated as[8]

$$\phi_s=(2eN_{ss}{}^2)/(\varepsilon\varepsilon_o N_d) \quad (1)$$

where N_{ss} is the trap state density which is proportional to the amount of chemisorbed oxygen, N_d the donor density and the other symbols have their usual meanings. When this barrier is formed into an interior of the narrow neck region, the total conductivity should be controlled by such necks as $\sigma \propto \exp(-(E_{cv}-E_f+\phi)/kT)$ where E_{cv} is the bottom

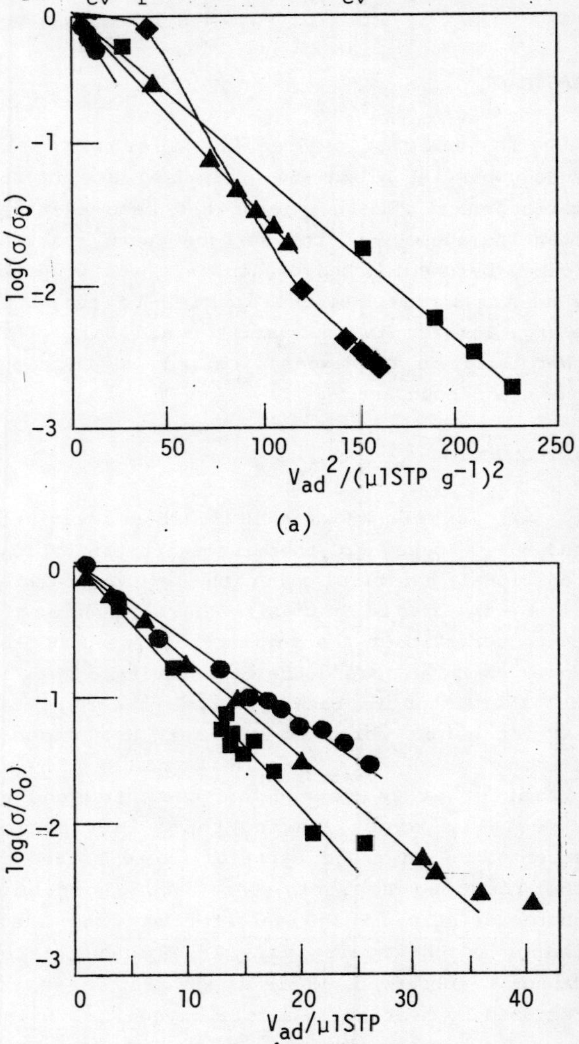

Fig.1 The change in electrical conductivity against the amount of chemisorbed oxygen for (a)porous ZnO:$\log(\sigma/\sigma)-V_{ad}^2$ and(b) porous $BaTiO_3$:$\log(\sigma/\sigma_o)-V_{ad}$: (●)300°C, (▲)400°C, (■)500°C ,(◆)600°C.

of the conduction band in vacuum and ϕ is the effective barrier height. Though the height of energy barrier at the surface is estimated by Eq.1, the effective height is determined by the lowest energy barrier at the center of the cross section of the neck.

If the conductivity measured in this experiment was controlled by the mechanism discussed above, $\log(\sigma/\sigma_o)$ in Fig.1 (a), (b) equal to ϕ/kT and the decrease in conductivity by two orders of magnitude corresponds to the generation of the energy barrier with an effective height to be 0.3eV at 500 C. Though the relation between $\phi/kT(=\log(\sigma/\sigma_o))$ and the amount of adsorved oxygen, $V_{ad}(\propto N_{ss})$ obeys to Eq.1 in porous ZnO at the temperature region 300 to 500°C, $\log(\sigma/\sigma_o)$ was proportional to V_{ad} in porous $BaTiO_3$ and porous ZnO at 600°C. Since when N_{ss} is large, the change in the height, $\Delta\phi_s$, is approximated to be proportional to the change in N_{ss}, it is considered that the chemisorbed oxygen or the other type of acceptor state such as a Ba vacancy remains in the sample.

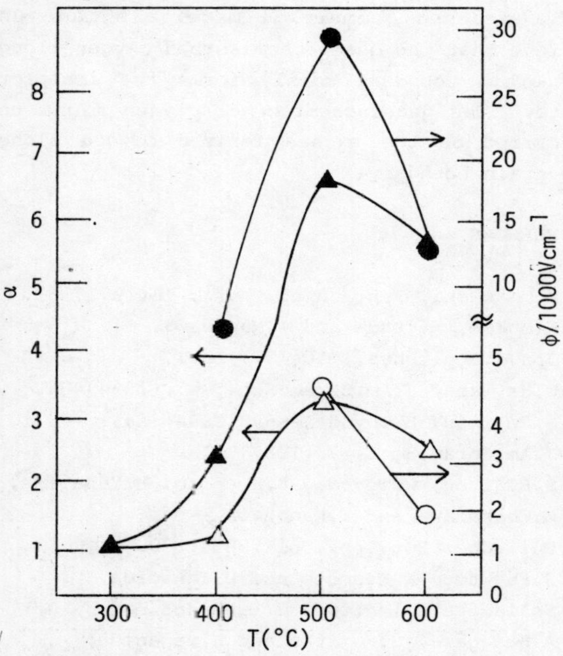

Fig.2 Annealing temperature dependence of the non-ohmic coefficient, α(△,▲), and the barrier height, ϕ(○,●), estimated by C-V relation. Open and closed marks show the non-doped and CoO-doped samples, respectively. The samples were evacuated at 600°C for 10h before annealing.

(2) The energy barrier showing non-ohmic characteristic and PTCR effect

As mentioned in the above section, the total conductivity is mainly controlled by the formed energy barrier at the neck region by chemisorbed oxygen. By this effect, the non-ohmic characteristic in porous ZnO was observed at room temperature when the sample was quenched. Figure 2 shows the non-ohmic coefficient, α, and the barrier height at room temperature for the samples quenched from the temperature indicated on the abscissa after attaining the saturated chemisorption. The non-ohmic coefficient is expressed as $I=SV^{\alpha}$ where I is the current, V the applied bias and S is a constant, and the barrier height was estimated by C-V relation where C is the capacitance. The temperature dependence of the non-ohmic coefficient and the barrier height corresponded to that of the saturated amount of the chemisorbed oxygen. The closed marks on Fig.2 show the results for the sample containing 0.5mol%CoO which is effective to modify the varistor characteristic[1]. When CoO was doped, the saturated amount increased. The relation between the barrier height at room temperature and the saturated amount of chemisorbed oxygen obeyed the rule expressed by Eq.1 in non-doped and CoO-doped porous samples[6,9]. on the contrary the barrier height and resulting non-ohmic coefficient decreased gradually by keeping the sample at room temperature. This decrease was due to the change in the oxidation state of the chemisorbed oxygen. The change in this state was confirmed by ESR and TPD(temperature programmed desorption) analyses[9]. When oxygen chemisorbs on the surface above 400°C, the oxidation state should be O^{2-}, while this state changes to O^- or O_2^- at lower temperature. Then, to form the effective surface energy barrier, the large amount of the high temperature type chemisorbed oxygen is demanded. Such a condition is attained by the addition of the dopants as CoO or MnO in porous ZnO. Such dopants are also effective to modify the varistor characteristic. This suggests that chemisorbed oxygen acts as the acceptor state to develop the energy barrier along the grain boundary in the commercial varistor.

In the case of porous semiconducting $BaTiO_3$[7], the change in the resistivity by PTC effect is larger than that of the sample with high density. This difference is explained in the following manner. The chemisorbed oxygen forms the acceptor state generating the energy barrier. The amount of chemisorbed oxygen is large in a porous material[10]. The resistivity change by PTC effect and the formed barrier height in a porous sample increased with increase in the amount of the chemisorbed oxygen as shown in Fig.3(a) and (b), respectively. In this case, the barrier height was proportional to the amount of chemisorbed oxygen as was expected from the result shown in Fig.1(b). A small resistivity change and the existence of barrier were observed in the sample without chemisorbed oxygen. This result indicates that the remaining chemisorbed oxygen or the other defect as a Ba vacancy forms the acceptor state as mentioned in the previous section. Since the resistivity change in the evacuated sample is 0.5 order of magnitude and the sample with the saturated amount of chemisorbed oxygen shows the resistivity change to be 5 orders of magnitude, it is judged that the main acceptor state to generate the energy barrier is the chemisorbed oxygen.

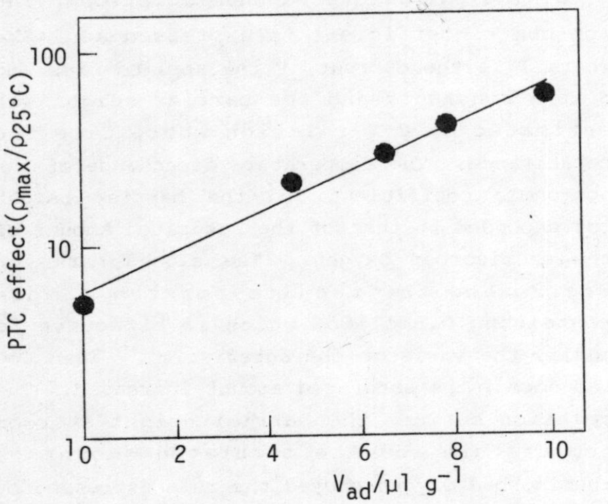

(a)

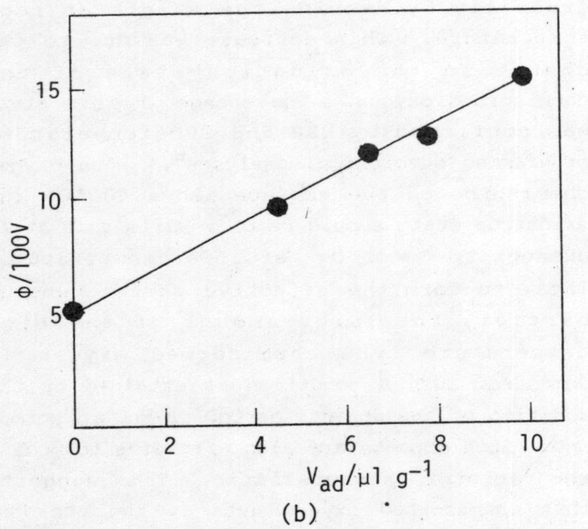

(b)

Fig.3 The relations (a)the magnitude of PTC effect and the amount of adsorbed oxygen and (b)the barrier height estimated by C-V characteristic and the amount of adsorbed oxygen.

(3) Grain boundary oxygen in varistor and PTC thermistor

The varistor characteristic and PTCR effect are produced by heating in oxidizing ambient. It is proposed that the chemical species to form the acceptor state is chemisorbed oxygen in the porous material, while that in the grain boundary has been not detected by any analytical method. The similarity of the change in the characteristics by the atmosphere between the porous materials and the dense commercial materials made us believe that the quasi-chemisorbed oxygen along the grain boundary mainly forms the acceptor state. The quasi-chemisorbed oxygen might be described as the excess ionized oxygen along the grain boundary.

REFERENCES

1. M.Matsuoka, Jpn.J.Appl.Phys., 16(1971)763.
2. K.Mukae, K.Tsuda and I.Nagasawa, Jpn.J.Appl.Phys., 16(1977)1361.
3. W.Heywang, J.Am.Ceram.Soc., 47,(1964)484.
4. S.Fujitsu, H.Toyoda and H.Yanagida, J.Am.Ceram.Soc., 70(1987)C71.
5. S.Fujitsu, H.Toyoda, K.Koumoto, H.Yanagida, M.Chikazawa and T.Kanazawa, Bull.Chem.Soc.Jpn., 61(1988)1979.
6. S.Fujitsu, K.Koumoto and H.Yanagida, Solid State Ionics, 32(1989)482.
7. A.Hasegawa, S.Fujitsu and H.Yanagida, J.Ceram.Soc.Jpn., 97(1989)549.
8. S.R.Morrison,Adv.Catal., 7(1955)259.
9. S.Fujitsu, unpublished data.
10. M.Kuwabara, J.Am.Ceram.Soc.,64(1981)639.

ELECTRONIC PROPERTIES OF B1 TYPE SOLID SOLUTION IN THE SYSTEMS OF MoN-MN (M= Nb, Zr)

G. C. Lai, M. Takahashi, K. Nobugai, and F. Kanamaru
ISIR, Osaka Univ., 8-1 Mihogaoka Ibaraki,
Osaka 567, Japan

ABSTRACT

The electrical properties and electronic structures for B1 type solid solution in the systems of $Mo_{1-x}M_xN$ (M= Nb, Zr) prepared by reactive sputtering have been investigated as a function of guest metal composition x, or the number of valence electrons (N_{VE}). The Mo-rich solid solution showed Tc below 7.6K, which was lower than the values predicted theoritically. The XPS measurement indicated the increasing in the density of states of d band for the Mo-rich samples, which was predicted from the theoritical calculation. The result of the measurement of Hall effect and stability of B1 structure in the Mo-rich samples, however, suggested that the d band seemed to split into two parts, bonding and antibonding parts and valence electrons began to fill the antibonding parts in the vicinity of N_{VE}=10.5. The relation between the electronic properties and N_{VE} is discussed.

INTRODUCTION

It is well known that transition metal carbides and nitrides possess a number of excellent properties combining high melting point and high hardness with metallic conductivity. Their crystal structures and properties have been extensively studied in relation to their distinctive chemical bonding and electronic structure. Recently, an attention has been paid to some B1 type (NaCl type) carbides and nitrides, particularly to NbN, as a good superconductor (Tc=17.3K[1]) with high critical current Jc, critical field Hc_2 and good resistance to neutron irradiation.[2,3] The electronic structure of the B1 type compounds is characterized by that the density of states in Fermi level [$N(E_F)$] increases with increase of the number of valence electrons (N_{VE}), and usually dealt approximatly with the Rigid Band Model. It is expected that the electronic structure and property of these compounds can be controlled by their chemical composition or N_{VE}. Some studies have been done on the B1 type nitrides with N_{VE} per formula less than 10, indicating that the $N(E_F)$ and Tc increased with increase of N_{VE}.[3,4] It has been predicted from the extrapolation of these results that MoN with N_{VE} of 11 should have more higher $N(E_F)$ and be also a superconductor with higher Tc as 25-30K,[4] and a similar prediction has been also made on the basis of band calculation.[5] However, such high Tc has not been realized for MoN so far, due to the metastability of B1 type MoN, and few information has been obtained in the relation between N_{VE} and electronic property for the B1 type compounds with N_{VE} more than 10.

In the present work, we prepared B1 type solid solution $Mo_{1-x}M_xN$ (M=Nb, Zr), in which N_{VE} varied from 9 to 11, and investigated the electrical and superconducting properties. A detail study was made for the dependence of the electronic properties on N_{VE}.

EXPERIMENTAL

The samples were deposited as films on fused silica and sapphire substrates by reactive sputtering in Ar/N_2 equimolar mixing gas at substrate temperature of 450°C and 550°C. The composite-targets composed of Mo (99.96%) and M (Nb, 99.95%; Zr, 99.7%) metals were used in the sputtering. The atomic ratio of the metals in the films was controlled by varying the area ratio of the metals on the targets. The deposition rate was about 10-20 nm/min, and the thickness of samples was controlled to 1.0-2.5 μm.

The identification of crystal phase was made by X-ray diffractometry, and the lattice parameter of the B1 type solid solutions was determined using the 200 and 400 reflections calibrated by silicon powder. The M metal contents, x, were determined by the intensity ratios of Mo-$L_{\alpha1}$ and M-$L_{\alpha1}$ radiations in the EPMA measurement. The electrical resistivities and their temperature dependences of the samples were measured using the standard four-point method, and Tco (the onset of Tc) was defined as the temperatures at 1% points of the resistivity change. The electronic state of the samples was measured by X-ray photoelectron spectroscope (XPS), using a spectrometer from Vacuum Generators Scientific Ltd. with unmonochromatized $Mg_{K\alpha}$ radiation before and after Ar ion beam etching. The vacuum in the analyzer chamber was below 3×10^{-7} Pa within the measurement, and the binding energy (BE) of the samples was calibrated against Ag $3d_{5/2}$ peak (368.3eV) of Ag sheet. The measurement of Hall effect was carried out in a magnetic field of 0.6 T at room temperature.

RESULTS AND DISCUSSION

1. MoN-NbN System

A. Crystal Structure

It was observed from the X-ray diffraction patterns that all as-sputtered samples have a single B1 type structure and a remarkable preferred orientation with (100) plane parallel to the samples plane. Fig. 1 illustrates the relation between the lattice parameter and chemical composition x in the solid solution prepared at 450°C. The lattice parameters of the both end-menbers MoN and NbN were consistent with the reported values, in which that for NbN was a experimental one (after JCPDS card 34-337) and that for MoN was evaluated from the effective radius of molybdenum and nitrogen atoms [4]. The lattice parameter of the solid solution increased continuously with increase of Nb, which has larger atomic radius than Mo. The dependence of the lattice parameter on x in the present system is in agreement with that of B1 type MoC-NbC system,[6] which was made by rapid quenching from the melt. It is considered, therefore, that a B1 type $Mo_{1-x}Nb_xN$ solid solution with near stoichiometric nitrogen composition has been obtained over the entire composition.

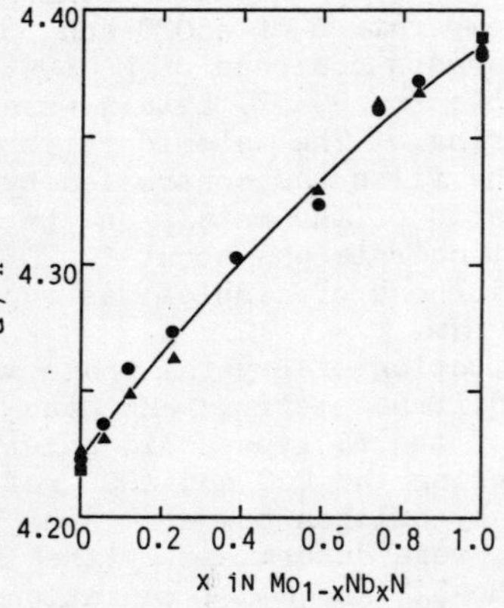

Fig. 1 Lattice parameters of the B1 type $Mo_{1-x}Nb_xN$ films versus x. ●, on fused silica; ▲, on sapphire; and ■, reported values.

B. Electrical Properties

Fig. 2 (a) shows the dependence of the electrical resistivity at 300 K on x in the solid solution. The resistivity of MoN (x=0) agreed roughly with the reported values for B1 type MoN films, [7,8] and NbN sample (x=1.0) showed a little higher resistivity than that of the NbN bulk.[9] The resistivity of the solid solution was higher than that of both end-menbers, and a maximum value was observed in the vicinity of x=0.4. Rudy et al.[10] studied the relation between electrical resistivity and chemical composition in many systems of IVa-VIa group transition metal carbides, and revealed that the resistivity of the solid solution was genenally larger than that of their end-menbers. The change of the residual resistivity ratio (R) of the samples, represented by $R=\rho(300K)/\rho(20K)$, as a function of x is shown in Fig. 2 (b). The solid solution showed smaller R than 1.0, and a minimum of R was observed near x=0.6.

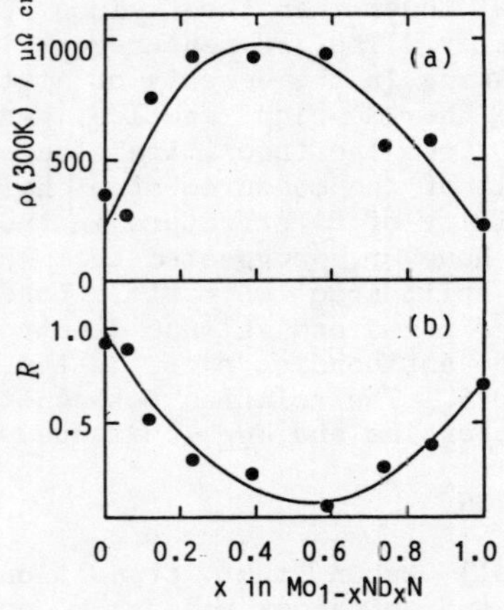

Fig. 2 (a)Electrical resistivity at 300K, and (b) residual resistivity ratio, $R=\rho(300K)/\rho(20K)$, of the $Mo_{1-x}Nb_xN$ samples.

Comparing with the thin nitride films of NbN-TiN and NbN-ZrN solid solutions having a very small temperature dependence of the resistivity,[12] the negative temperature coefficient of the resistivity (increasing resistivity with decreasing temperature) may be mainly associated with some inhomogeneity and disorder in the structure of the samples. Fig. 3 shows the measured T_{co} as a function of x in the solid solution. The calculated values by Papaconstantopoulos[5] are also shown in Fig. 3. NbN film had T_{co} of 15.4k, which was a little lower than the reported one (T_c=17.3K).[1] In NbN-rich samples ($x\geq0.74$), T_{co} decreased with decrease of x, and the tendency of the decrease of T_{co} was roughly

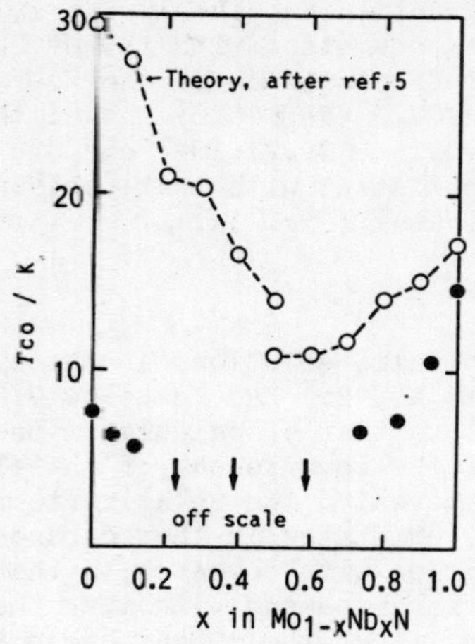

Fig. 3 Relation between Tco and x in the $Mo_{1-x}Nb_xN$ samples.

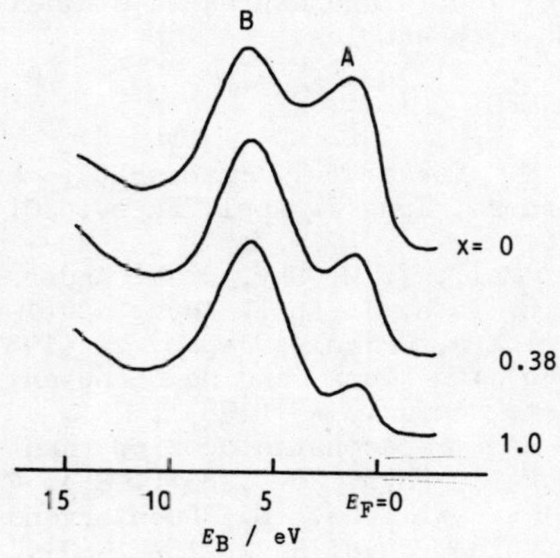

Fig. 4 XPS spectra of the valence band in several $Mo_{1-x}Nb_xN$ samples.

consistent with that of predicted values, although the observed Tco values were lower than that of the latter. Howerer, the Mo-rich samples showed Tc below 7.6K, which was much lower as compared with the calculated value. The superconductivity for B1 type MoN samples made by nonequilibrium reaction methods was reported by a lot of research groups, Tc were almost lower than 10K and the highest value was 12.8K.[11] The reasons for low Tc for the Mo-rich samples are thought to be due to (a) inhomogeneity and disorder in the samples, which was pointed out well so far, and (b) change of electronic state in Mo-rich samples with changing N_{VE} not as the theoritical calculation.

C. Electronic Structures

It was observed from the XPS measurement that the binding energy (BE) of 3d core levels in the both end-menbers, MoN and NbN, shifted to

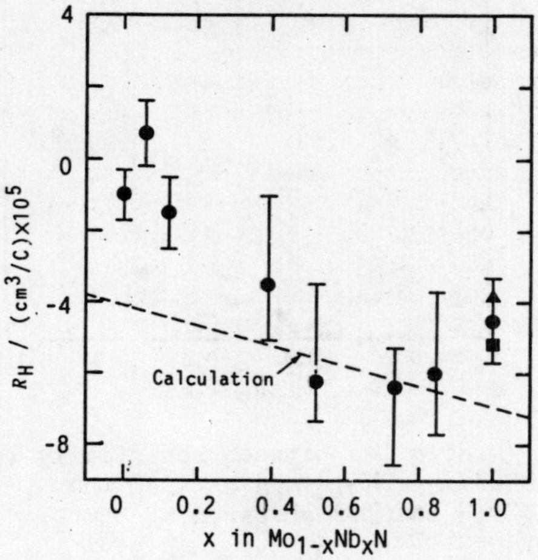

Fig. 5 Hall coefficients R_H of the $Mo_{1-x}Nb_xN$ samples as a function of x. ▲, after Piper (ref. 9); and ■, after Tonouchi (ref. 13).

the higher BE by 1.0 and 2.1 eV with compared to that of Mo and Nb metals, respectively. The BE shift of the metals in the solid solution was similar to that of both end-menbers. The N1s in all samples showed similar values and shifted to the lower BE side compared with that of nitrogen atom.[12] It is considered that the charge transfer occurred from metal atoms to nitrogen atoms in the solid solution. Fig. 4 shows the valence band region of the XPS spectra for three typical samples. Two bands named A and B were observed in these samples. Band A is mainly attributed to 4d state of the metal atoms, and band B to the p-d state of metal and nitrogen atoms. A relative increase can be seen in the band A (d band) for Mo-rich samples with compared to that of NbN (x=1.0). This suggests that the density of states near Fermi level increases in Mo-rich samples, and supports the theoritical calculation.[5]
In Fig. 5 the change of Hall coefficients R_H of the samples was plotted as a function of x. The calculated values of R_H was superimposed in Fig. 5 on the basis of the one-band approximation and the assumption of carriers being composed of only electrons. R_H of NbN is consistent with that reported by Piper et al.[9] and Tonouchi et al.[13] The R_H values of the Nb-rich samples with x>0.5 agreed roughly with the calculated ones. But for the Mo-rich samples, the measured R_H values are out of the range of calculated ones, and shift to the plus direction. This implies that the effect of holes on electrical conduction cannot be disregardful and a complex band structure, which does not agree with the one-band approximation, exists in the Mo-rich samples. Toth[9b] and Wijeyesekera[14] proposed that the decreased stability in B1 type transition carbides and

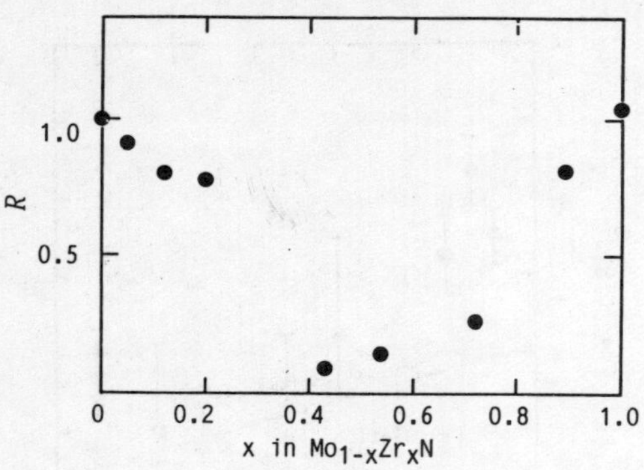

Fig. 6 Relation between resistivity ratio, R=ρ(300K)/ρ(20K), and x in the $Mo_{1-x}Zr_xN$ samples prapared at 550°C.

nitrides with high N_{VE} might be due to the existence of antibonding orbital. The occurrence of hole in the Mo-rich samples may be caused by that the d split into two parts, or bonding and antibonding parts, and valence electrons begin to fill the antibonding parts in the vicinity of x=0.5 or N_{VE} =10.5. The result seems to support the reason (b) for the low Tc in the Mo-rich samples mentioned above. In order to make it clear for the existence of the antibonding parts, we investigated the stability of the B1 structure in the solid solution by annealing them in ammonia gas at 500-900°C. The Mo-rich samples with x≦0.38 showed the phase transition from B1 type to WC (hexagonal) type, but no phase transition occurred in the Nb-rich samples with x ≧0.59.[15] The fact that the thermodynamic stability of B1 structure decreased with increasing N_{VE} in the Mo-rich samples also supported the discussion about the filling of valence electrons in the antibonding parts of the d band.

2. MoN-ZrN system

Single B1 type structure has been obtained in all samples prepared at 450°C and 550°C, respectively. But the samples prepared at 550°C showed sharp X-ray diffraction peaks and good linear relation between lattice parameter and composition x with compared to those prepared at 450°C. Fig. 6 indicates R values as a function of x in the samples made at 550°C. The samples with 0.43≦x≦0.73 showed relative lower R values than that of other samples. The uncontinuous change of R with x may be mainly due to the lattice strain or disorder caused by such factors as size effect. Tco of 5.0-7.6K was observed in the samples with x≦0.08, but no superconducting transition was observed in the other samples with x≧0.12 until 5K. The low Tc in Mo-rich region was consistent with the M= Nb system. The Hall coefficients R_H of the Mo-rich samples with x≦0.20 (N_{VE}≧10.6) showed the values of +0.6～-0.15 (±0.22)×10^{-5} cm^3/C. This is roughly consistent with that obtained from the M= Nb system (Fig. 5).

CONCLUSIONS

The Mo-rich solid solutions in the systems of $Mo_{1-x}M_xN$ (M= Nb, Zr) had Tc below 7.6K, which was lower than that of calculated ones. The results of the measurement of the electronic structure revealed the relative increase of the density of states in the d band for the Mo-rich samples with compared to that of NbN, which is in agreement with the theoritical calculation. But the d band seems to split into two parts, bonding and antibonding parts, and valence electrons begin to fill the antibonding parts of the d band in the vicinity of N_{VE}=10.5. This may be associated with the low Tc and decreased stability of B1 type Mo-rich samples.

REFERENCES

1. K. S. Keskar, T. Yamashita, and Y. Onodera, Jpn. J. Appl. Phys., 10(1971), 370.
2. H. Bell, Y. M. Shy, D. E. Anderson, and L. E. Toth, J. Appl. Phys., 39(1968), 2797.
3. H. Ihara, Ceramics Japan, 22 (1987), 559.
4. Zhao You-xiang and He Shou-an, Solid State Commun., 45 (1983), 281.
5. D. A. Papaconstantopoulos and W. E. Pickett, Phys. Rev., B31(1985), 7093.
6. R. H. Willens, E. Buehler and B. T. Matthias, Phys. Rev., 159(1967), 327.
7. G. Linker, R. Smithey and O. Meyer, J. Phys., F14(1984), L115.
8. H. Yamamoto, T. Miki and M. Tanaka, Advance Cryogenic enginn., 32(1986), 671.
9. (a) L. E. Toth, "Transition Metal Carbides and Nitrides", Academic press, new York, (1971), Chap. 6; (b) ibid. Chap. 8.
10. E. Rudy and F. benesovsky, Planseeber. Pulvermet., 8(1960), 72.
11. N. Terada, M. Naoe and Y. Hoshi, Advance Cryogenic Enginn., 32(1986), 663.
12. D. N. Hendrickson, J. M. Hollander, and W. L. Jolly, Inorg. Chem., 8(1969), 2642.
13. M. Tonouchi, T. Kobayashi, Y. Nakato, and H. Tsubomura, IEEE Trans. Mag., 23 (1987), 1007.
14. S. D. Wijeyesekara and R. Hoffmann, Organometallics, 3(1984), 949.
15. G. C. Lai, M. Takahashi, K. Nobugai, and F. Kanamaru, J. Solid State Chem., in press.

LIQUID SINTERING OF SEMICONDUCTIVE $SrTiO_3$

Zhao Gangkai Jin Yuzhuo Zhou Heping
Li Longtu Zhang Xiaowen
Department of Materials Science and Engineering
Tsinghua University
Beijing 100084, PRC.

ABSTRACT

The sintering process and microstructure of Nb-doped $SrTiO_3$ ceramics with CuO and GeO_2 additives were studied by High Temperature Microscope, SEM and TEM/EDAX. It was observed that there was liquid phase taking part in the grain growth during sintering. Inside some grains there existed precipitates around which were dislocations. There were also other dislocations such as loops and loops with stacking faults. Some interfaces exhibited dendritic protuberances. These phenomena were thought to be caused by rapid growth of grains from somewhat impure liquid.

INTRODUCTION

Sintering and grain growth rates in titanate ceramics vary strongly with cation stoichiometry and additives, usually due to the presence of a liquid phase.[1] In barium titanate, an excess of TiO_2 leads to rapid liquid phase sintering at temperatures above the Ti-rich eutectic (1320°C)[2]. In strontium titanate, the higher eutectic temperature (1440 ° C) necessitates additions of, e.g. , SiO_2 and Al_2O_3 in order to sinter materials such as barrier-layer capacitors at lowered temperatures.[3] The addition of different additives causes diverse grain growth rates. By addition of CuO or GeO_2 into Nb-doped $SrTiO_3$, the grain growth rate may be highly accelerated[4]. In this article, the sintering of Nb-doped $SrTiO_3$ with CuO and GeO_2 additives was observed by High Temperature Microscope and the microstructures of sintered ceramics were studied by SEM and TEM/EDAX.

EXPERIMENTAL

1. Sample Preparation

Powders of precisely determined chemical composition were made by wet milling process. These were then pressed into pellet form and sintered in a reducing atmosphere (N_2 90% + H_2 10%) at 1450 ° C for 4 hours. The compositions studied were $Sr(Ti_{0.994}+Nb_{0.006})O_3$ + xCuO + $yGeO_2$.

2. High Temperature Microscope

Pressed pellet was fired at 1100°C for 2 hours to have some strength. Then the pellet was polished and sintered in a HM-4SP type argon protected high temperature microscope to observe the changes of its microstructure during sintering process.

3. SEM

Polished pellets were etched using a solution made up of HNO_3 and HF. The samples were then cleaned ultrasonically in methanol and examined in the S-450 type SEM at 20kv.

4. TEM/EDAX

Samples were polished down to 20 um. This foil specimens were then prepared by ion beam thinning and examed in a H-800 electron microscope at 200kv.

RESULTS AND DISCUSSIONS

1. Observation of High Temperature Microscope

Microstructure changes of the sample during sintering were observed by high temperature microscope. When temperature was increased up to 1100°C, the spots of liquid phase started to appear. As the rising of temperature, these separate spots then contacted into lines and networks and particles inside the networks began to grow. Some small particles were dissolved into the liquid whereas some bigger ones grew from the liquid. In 1380°C, there were big grains formed. As the increasing of temperature, grains grew continuelly with the migrating and decreasing of liquid. During cooling procedure, some liquid abstracted into grain junctions and the boundaries became narrow and straight. After cooled to 1350°C, many fringes appeared in the surface of grains and there were precipitates along the interfaces which indicated the solidification of liquid phase. Finally, the sample had a microstructure of irregular grain and boundary shapes.

The surface of the sintered sample was analysed by ESCA, which found the existance of Cu, Ge, Sr, Ti and Nb. Then polished surface was analysed again by ESCA, but no Cu and Ge were discovered. These showed that the liquid phase in the surface of the sample contained Cu and Ge impurities, but the grain did not. So that CuO and GeO2 produced liquid phase during the sintering of the sample.

2. Observation of TEM/EDAX

Many tetragonal and hexahedral dislocation loops were observed by TEM, Fig. 1 is a tetragonal dislocation loop. Analysis of the compositions inside the loops by EDAX showed the precipitation of impurities, see Fig.2. Many loops with stacking faults were also found in some grains, see Fig. 3. Fringes inside the loops could be clearly seen, which was similar to that observed in metals.[5]

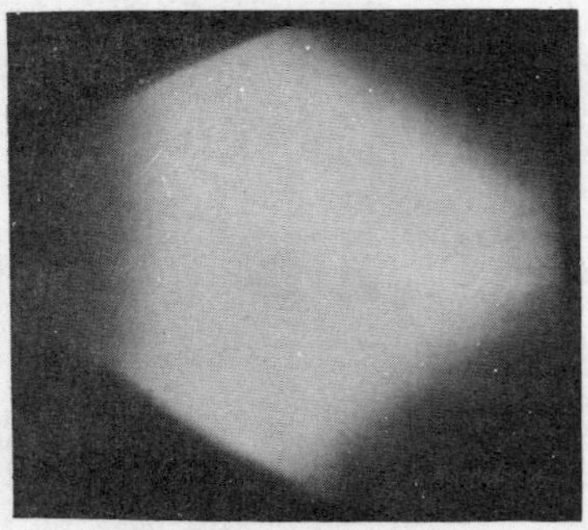

Fig. 1 Tetragonal dislocation loop observed by TEM.

The growth of grain from liquid phase gave rise to inhomogenity of grain compositions. During the cooling process, the impurities precipitated. When dislocation lines passed through these precipitates, the dislocation

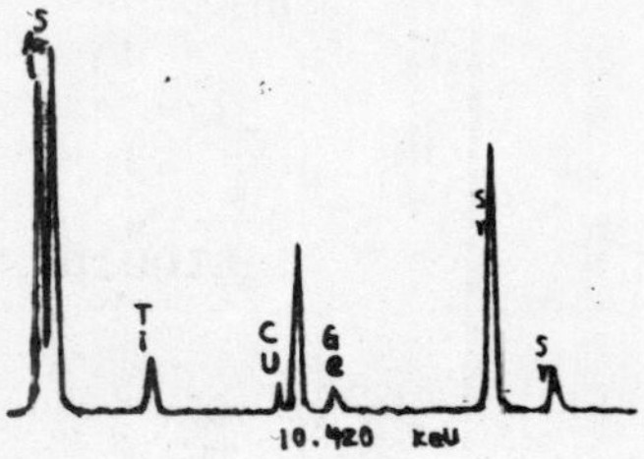

Fig. 2 EDAX spectrum of precipititate inside the tetragonal dislocation loop.

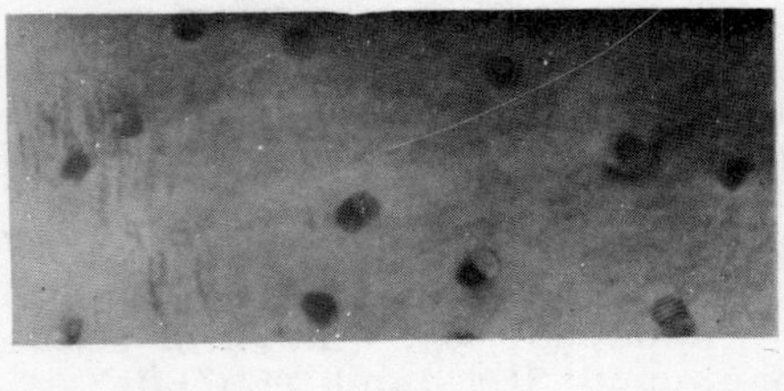

Fig. 3 Dislocation loops with stacking fault observed by TEM.

loops were formed around the precipitates. Fig. 4(a) is a schematic indication of this process[5]. This kind of dislocation loops produced by the interaction of dislocation line with precipitate was also observed in this experiment, see Fig. 4(b).

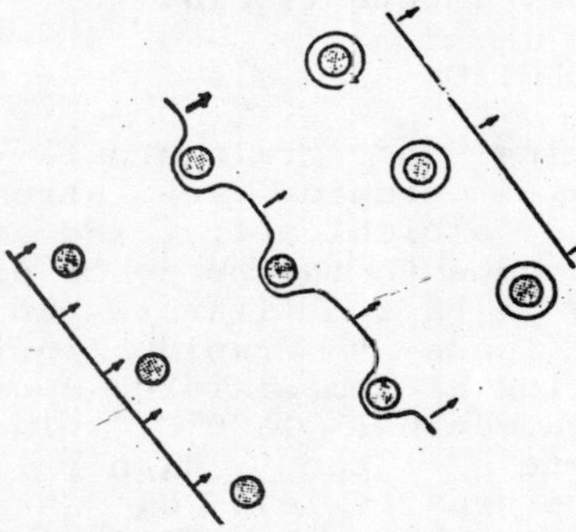

Fig.4(a) Schematic indicating of interaction between dislocation line and precipitates.

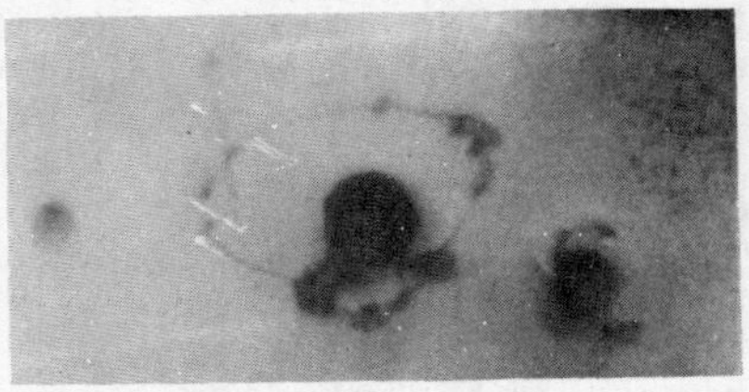

Fig.4(b) Precipitate and dislocation line around it observed by TEM.

3. Observation of SEM

Fig.5 is a second electron image of the sample by SEM. The dendritic protuberances of interfaces can be clearly seen.

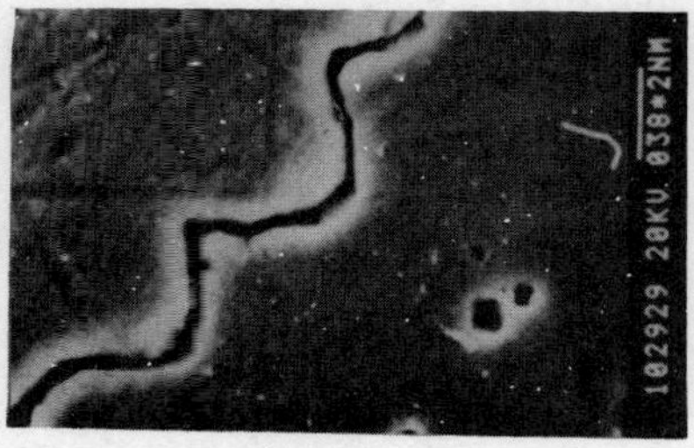

Fig.5 Dendritic protuberances of interface observed by SEM.

4. Discussions

The impurities of CuO and GeO2 in Nb-doped SrTiO3 powders can easily form liquid phase in lowered temperature because the melting points of CuO and GeO2 are 1026°C and 1150°C respectively. During the sintering process, GeO2 and SrO might possibly form eutectic liquid and CuO may change into Cu2O then form eutectic liquid with TiO2. These liquid phases were formed before the rapid growth of grains during sintering and surrounded the small particles just as observed in high temperature microscope. Then, by the medium of liquid phase, small particles were dissolved into the liquid whereas large particles grow from it. This is a reactive liquid phase sintering mechanism which accelerated the grain growth rate and caused great number of dislocations. ESCA analysis had confirmed the existance of CuO and GeO2 impurities segerated in the liquid on grain interfaces.

The driving force of the growth of grain from liquid is the difference of free energies between the two phases. Since the activation energy needed is very small, the grain growth rate is quite rapid. Now, let us consider a crystal which develops at the expense of a large volume of solution AB, of average concentration Co in the constituent B. With the x axis oriented away from the crystal towards the liquid growth takes place under the positive temperature gradient G=dT/dx, which moves with a positive velocity R (no supercooling).

The crystal formed has a concentration Cs of constituent B different from that C_L of the liquid in contact.

During growth, the excess (or lack) of the B atoms must evacuated towards the interior of the solution. This takes place by diffusion or by convection; it is not instantaneous. Thus growth of the crystal mades the solution locally richer (or poorer) in B, see Fig. 6. Since the liquid temperature T_L of a solution varies with its composition, the exact form of the interface depends now on the gradient, in its neighbourhood, not of its temperature T, but of the difference $T-T_L$. If, for example, $Cs<C_L$ and $dT_L/dC<0$, as is often the case in a solution not too rich in constituent B, see Fig. 7, there may be a small zone near the interface where $d(T-T_L)/dx$ is zero or even a little negative, although the temperature gradient G=dT/dx is positive.

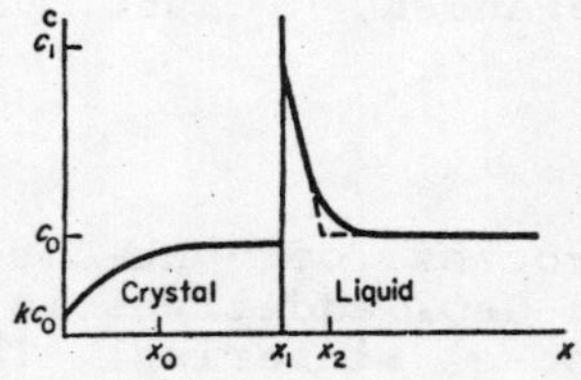

Fig. 6

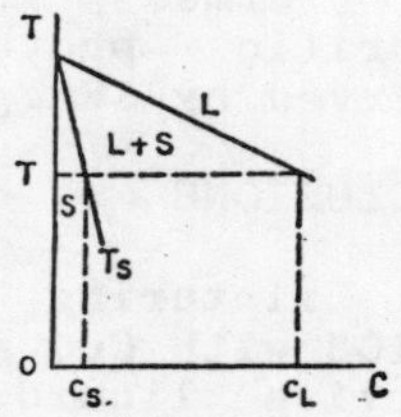

Fig. 7

Fig.6 Variation of the concentration of constituent B, with distance from the interface x, when a steady state is reached.

Fig.7 Phase diagram of an element as a function of its purity (L--Liquid, S--Solid).

This zone is then supercooled. One must expect this interface, which was initially plane, to become covered with small adjacent dendritic protuberances, with a height h of the order of the thickness of the zone. In grneral h is small, hence $d(T-T_L)/dx$ is not strongly negative: the protuberances must then take the form of pyramids limited approximately by faces of slow growth, if this is made possible by the lattice orientation of the interface. Finally the crystal which forms in the valleys separating the protuberances is in contact with a liquid richer in B than which forms the protuberances. The constituent B must then be concentrated, in the crystal, along the walls of a prismatic lattice very nearly parallel to the growth direction and which correspond to the successive positions of the valleys during growth, see Fig. 8.

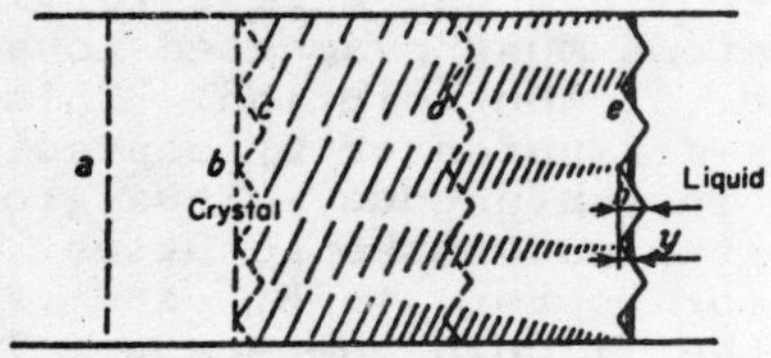

Fig.8 Possibility of local supercooling. Successive positions of the interface. a, b. planar interfaces; c, d,e. pyramidal projections.

A lattice of this type is indeed characteristic of the rather rapid growth of metals and of semiconductors starting from somewhat impure liquids.[6] In this experiment, CuO and GeO2 gave rise to impure liquid during sintering and grains grew from this impure liquid and formed the interfaces with dendritic protuberances, just as observed by SEM.

CONCLUSIONS

The sintering process of Nb-doped SrTiO3 with CuO and GeO2 additives is a reactive liquid phase sintering. The impurities CuO and GeO2 with some SrTiO3 powders formed eutectic liquid, i.e. impure liquid which took part in the growth of grains. During sintering, small particles were dissolved into liquid phase whereas large particles grew rapidly from it. This mechanism caused many dislocations and precipitation of impurities inside some grains and formed dendritic protuberances of interfaces.

REFERENCES

1. T. Negas, R. S. Roth, H. S. Parker and D. Minor, J. Sol. St. Chem., 9[3], (1974), 300.
2. A. Cocco and F. Massazza, Ann. Chim. 53 (1963), 892.
3. M. Fujimoto and W. D. Kingery, J. Am. Ceram. Soc., 68[4], (1985), 169.
4. N. Yamaoka and T. Matsu, Advances in Ceramics, Vol. 1 (1981), 232.
5. B.Henderson, Defects in Crystalline Solids, Edward Arnold Ltd., London, 1972.
6. J.. Friedel, Dislocations, Pergamon Press, 1964.

THE EFFECTS OF Co, Fe_2O_3 ADDITIVES ON $(V,Cr)_2O_3$ CERAMICS

Wu Baiyuan, Wu Shijun
South China University of Technology
Guangzhou, China

ABSTRACT

The effects of Co and Fe_2O_3 additives on the microstructure and electrical properties of PTC $(V,Cr)_2O_3$ ceramics have been studied using techniques such as SEM, AES, XRD and electrical resistivity measurements. The beneficial effects of introducing additives such as Co, which are mainly dispersed in the matrix in metallic form, on sample densities and electrical properties has been found to be significant.

INTRODUCTION

Since 1921, pure and doped V_2O_3 have been extensively studied[1,2] V_2O_3 exhibits there different phase in different temperature range.3 The transition from a paramagnetic metallic phase(PM) to a paramagnetic insulating phase(PI) occurs at about 525K, and it is this latter phase which shows the PTC characteristics. In 1981, R.S.Perkings et al made the posistor of $(V,Cr)_2O_3$ ceramics which shows a practically discontinuous change of resistivity from about $10^{-3}\Omega$.cm to about $10^{-1}\Omega$.cm between -100°C and 200°C. It is important for the effective additives to effect the sintering behaviour of ceramics and to improve its properties. It has been found that metallic additives like Fe, Co, Ni and Cu have a beneficial effect on sintering behaviour and electrical properties of $(V,Cr)_2O_3$. With SEM, EPMA, AES and XRD techniques, this paper studies the effects of Co, Fe_2O_3 additives on the microstructure and electrical properties of $(V,Cr)_2O_3$ ceramics, and its mechanism.

EXPERIMENT

In this experiment, the V_2O_5 was reduced in H_2 to obtain V_2O_3. The basic composition of the thermistor V_2O_3 material used was $(V_{1-x}Cr_x)_2O_3$ (x=0.0034), to which was added various amounts of Co or Fe_2O_3. $(V,Cr)_2O_3$ ceramics sample were prepared using standard electronic ceramics techniques. Mixtures of V_2O_5, Cr_2O_3 and Co or Fe_2O_3 (see Table 1) were ball-milled with alcohol for 24hr using steel balls. The dried powder mixtures were then calcined in H_2 for 3hr at 600°C, followed by 10hr at 1000°C. Pellets were diepressed using paraffin wax as a binder, and were subsequently sintered in H_2 at 1415°C for 10hr.

Table 1 Compositions prepared

Material code	A	B	C	D	E	F	G
Additive, wt%	0.1Co	1.0Co	2.5Co	4.0Co	7Co	10Co	3.5Fe_2O_3

The resistivity measurements on test samples were carried out using a four probe technique within the temperature range -30°C to 200°C at a heating rate of 2K/min. SEM, XRD, EPMA and AES were used to study the phases present and the microstructure of the materials produced.

RESULTS AND DISCUTION

The microstructure of pure $(V_{0.994}Cr_{0.006})_2O_3$ is shown at Fig.1. The effect of Co dopant on the electrical properties of $(V_{1-x},Cr_x)_2O_3$ is

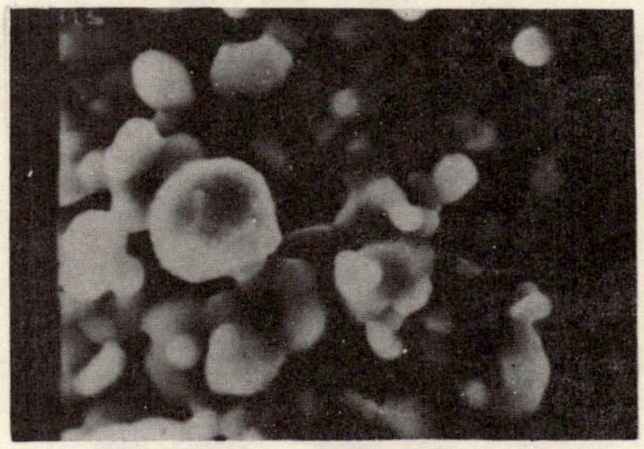

Fig.1 SEM image of fracture surface of the pure $(V_{0.994}Cr_{0.006})_2O_3$ ceramics

shown in Table 2. The minimum resistivity in the examined temperature range(-30 to 200°C) decreased with increasing Co content to reach a plateau above 2.5wt% addition. At 20°C, the resistivity was a minimum in the range 2.5-4wt% addition. The magnitude of the PTC effect, as denoted by ρ_{max}/ρ_{20} peaked at an addition of 2.5wt%. Their fracture surface are shown in Fig.2. Further examination of the effects of additives using XRD, SEM, EPMA and AES was made on materials with 3.5wt% Fe_2O_3(material G) and 4wt%

material (Fig.8) shows areas high in Co supporting the concept of metallic inclusions in the ceramics. Fig.9 displays the results of an analysis of a grain boundary region between two V_2O_3 grain, and indicates that there is only a low level of Fe present compared with the discrete Fe-containing areas. The auger electron spectroscopy (AES) has been used to examine the vertical (depth) distribution of elements in the Fe-doped material using argon ion surface stripping. The results of an analysis of a fracture are shown in Fig.10, which indicate that the Fe content decreases as the surface is ion-machined away. It is therefore believed that apart from forming a discrete second phase dispersed in the matrix, the dopant metals are also present at the surface of the grains. Generally speaking, the strength of a grain is much larger than that of grain boundary. When the sample is broken, it cracks along boundary. Therefore, the surface of grain on the

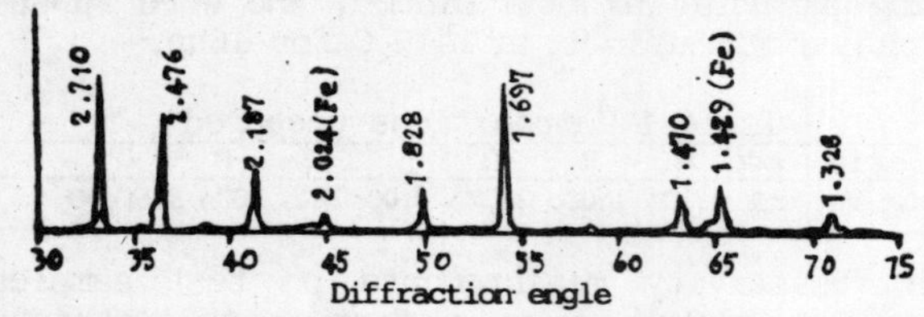

Fig.3 X-ray powder diffraction pattern for material doped with 3.5wt% Fe_2O_3. The peaks at 2.024 and 1.433A are due to metallic Fe.

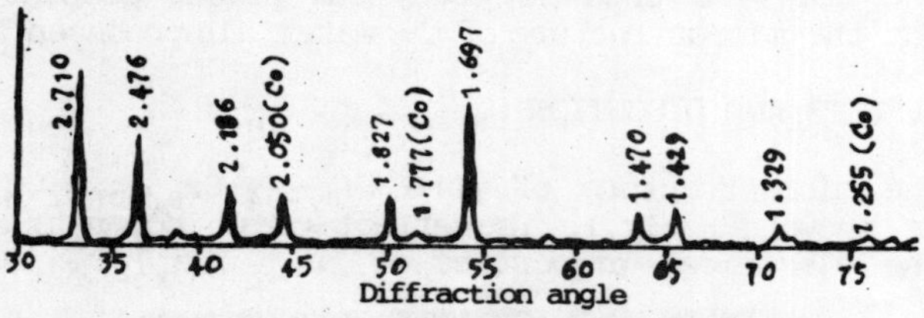

Fig.4 X-ray poeder diffraction pattern for material doped with 4wt% Co. The peaks at 2.050, 1.777 and 1.255A are due to metallic Co.

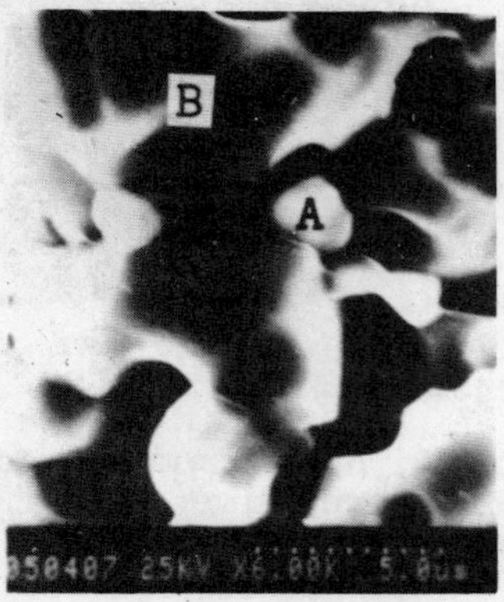

Fig.5 SEM image of fracture surface of material doped with 3.5wt% Fe_2O_3.

fracture is a part of grain boundary. The picture that reached from the above discussion is below: The Fe_2O_3 or Co additive is present in the form of metal in the compact after preparing. Therefore, their effects are the same. Up to high temperature, a small amout of metal melts to produce liquid phase and, at last, is dissoloved in the grain boundary of matrix. A large amount of metal remain as a discrete second phase in the ceramics. From the image of Fig.1, the grains of the pure $(V,Cr)_2O_3$ materials do not fully grow up, and there is linking pores. Even if the heating rate is about 200°C/hr, the pores are not put in the grains. Because the pores exhausted very slowly, there are many pores at the grain boundary. The hindering effect of the pore on the movement of the boundary decreases the boundary mobillity, and the pores are still at boundary partly and do not be exhausted fully after firing. The Fe or Co doped samples are different with the pure material. When the temperature rise up to the firing temperature, Fe or Co metal melt partly and produced a small quantity of liquid phase. The surface tension of the liquid phase makes the powders tight and close with each other. The material

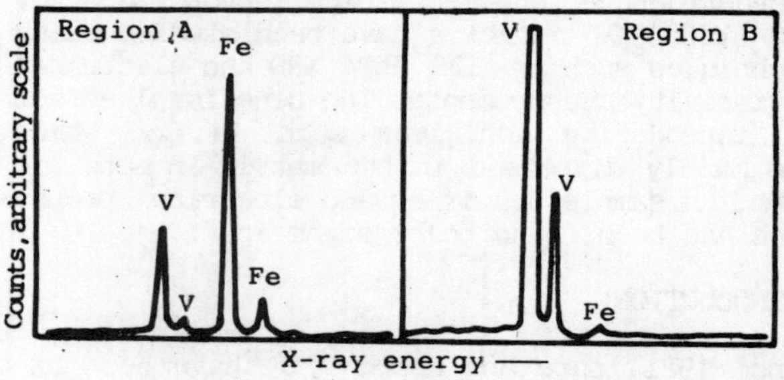

Fig.6 EPMA analyses of regions A (Fe-rich) and B (Fe-poor) at Fig.5

Fig.7 SEM image of fracture surface of material doped with 4wt% Co.

conveyance and the motion of the grain boundary are speeded up to promote densification. If the temperature rose overfast, the pore were pulled into the grain. It is thus clear that Fe_2O_3, Co dopants have the effection of promoting to fire the $(V,Cr)_2O_3$ ceramics. From the microstructure of material A--F (see Fig.2), the grains of the only Co 0.1wt% doped material A are larger and not well-distributed, and the grain boundary is not evident and there is the phenomenon of secondary crystallization in the ceramics. The grains of the Co 2.5wt% doped

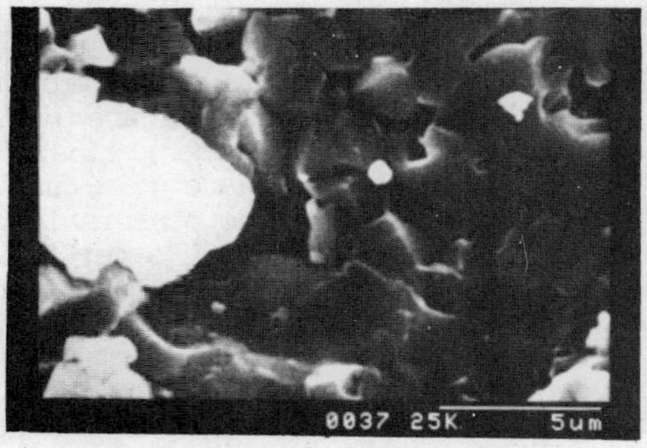

(a) Material A (SEM 6000X)

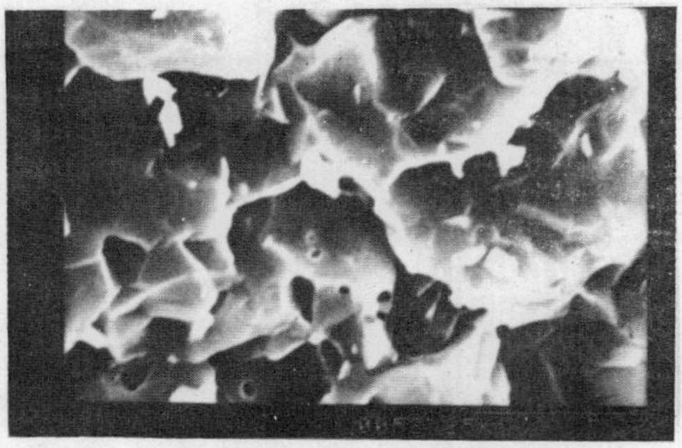

(b) Material B (SEM 6000X)

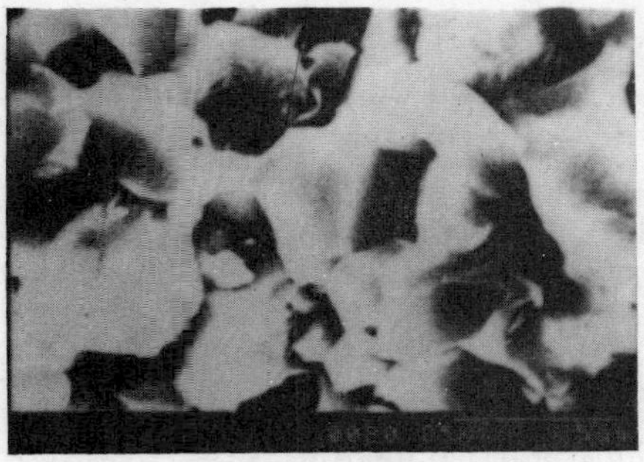

(c) Material C (SEM 6000X)

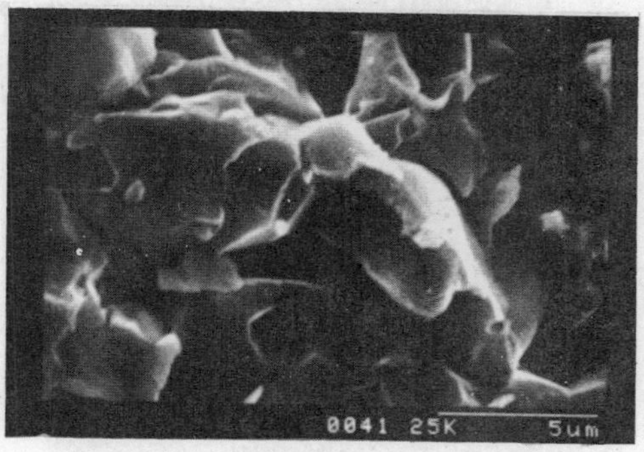

(d) Material D (SEM 6000X)

(e) Material E (SEM 6000X)

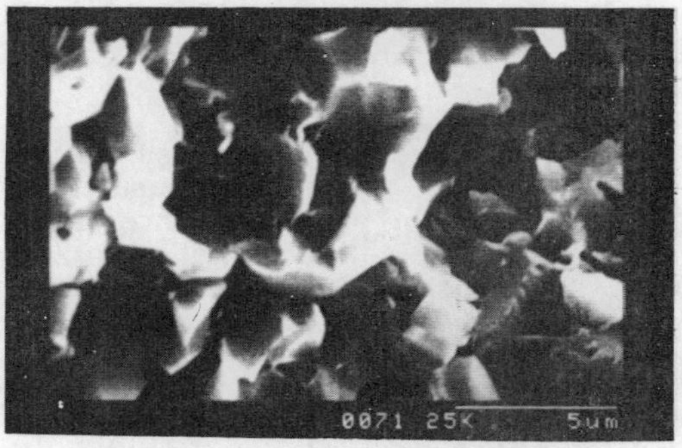

(f) Material F (SEM 6000X)

Fig.2 SEM image of fracture surface of the materials doped with different content of Co

Table 2 Effect of Co metal content on properties

Material	A	B	C	D	E	F
ρ_{min} (Ω.cm)	0.0046	0.0024	0.0013	0.0014	0.0016	0.0018
ρ_{20} (Ω.cm)	0.0143	0.0031	0.0017	0.0019	0.0028	0.0030
ρ_{max} (Ω.cm)	0.1291	0.0892	0.0660	0.0679	0.0421	0.0317
ρ_{max}/ρ_{20}	9	29	43	36	15	11
$T\rho_{max}$ (°C)	110	128	142	132	134	138

Co (material D). XRD revealed that the Fe-doped material contained a metallic Fe phase (Fig.3), the d-spacings of 2.024 and 1.433Å being very close to the values for pure iron, which is idential with what was reported but not confirmed in the reference.[5] In addition, Fe_2O_3 powder was reduced in the same condition as V_2O_3 powder and the reduced powder was analysed with TGA. Similarly, the Co-doped material showed peaks at 2.050, 1.777 and 1.255 Å, corresponding to metallic Cobalt (Fig. 5). The Fig. 6,8 show respectively the microstructure of 3.5wt% Fe_2O_3 doped material G and 4wt% Co doped material D . Fig.6 shows the results of EPMA analyses of the region of A,B in 3.5wt% Fe_2O_3 doped material G. Fig. 8 shows the results of EPMA analyses of the region of C,D in 4wt% Co doped material D. By comparing between the region A and B, it can be seen that region A incorporates a high level of Fe, while region B contain very little. Taken in conjunction with the XRD results, it is likely that the Fe-rich area is metallic Fe, with traces of V resulting from the penetration of the electron beam into the matrix phase. Therefore, Fe-doped material contains two kinds of grain phases. One is $(V,Cr)_2O_3$ grain phase, the other is Fe-grain phase. Similarly, examination of the Co-doped

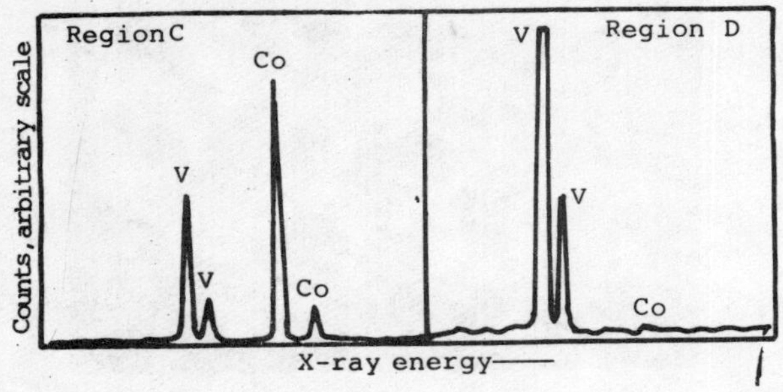

Fig.8 EPMA analyses of regions C (Co-rich) and D (Co-poor) at Fig.7.

(a)

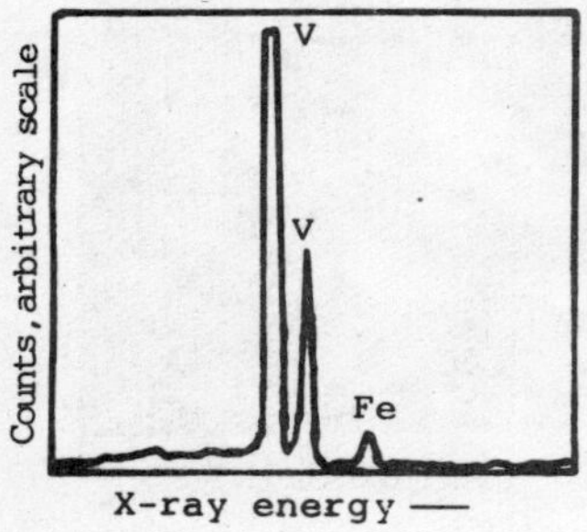

Fig.9 (a) SEM image of fracture surface Fe_2O_3 doped material indicating the grain boundary region subject to EPMA analysis shown in (b)

(b)

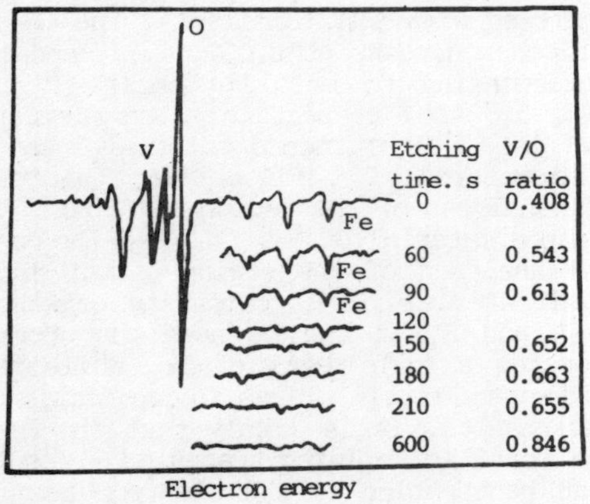

Fig.10 AES spectra of fracture of Fe_2O_3 doped material showing the reduction in Fe concentration and increase in V/O ratio with ion-etching time

material C are more unifprm and their size is 4-5 μ m. The small Co grains are observed in the material C with SEM. The grains of the Co 10wt% doped material F are more uniform, but their size is smaller. The observable grains of Co in the material F are larger than in the material C. Evidently, with the increase in the content of Co additive, the primary grain size decreases, and the size and shape of the grains become uniform, and the size and amount of the Co metal grains increase. This is because that Co additive will form a second phase among or between the grains of the $(V,Cr)_2O_3$ ceramics. The second phase at the boundary of the primary phase grains has the effect of pinning on grain boundaries, and reduce the mobility of the boundary to prevent excessive growth of the primary phase grains and to make the grains uniform and regular. From Table 2, it may be seen that the content of Co has an effect on the electrical properties of the $(V,Cr)_2O_3$ ceramics. The effects are interrelated with the effects of Co content on the microstructure. Since the PI-PM transition is related with the internal stress and defects of the materials. If the internal stress and defects increased, the transition would become more difficult and the hysteresis effect would be more strongthened.[4] The Co additive of just amount improve the microstructure, which decrease the internal stress and defects. Therefore, it makes the decreasing in ρ of the sample and the increasing in ρ / ρ with increasing in additive content. On the other hand, the metal in the ceramics play down the change of the resistivity of the $(V,Cr)_2O_3$ grain, so the excessive Co additive weakens certainly the apparent PTC effect of the materials.

CONCLUSION

With XRD, SEM, EPMA and AES techniques, it is comfirmed that the additives of Co, Fe_2O_3 are present in the form of metal in the $(V,Cr)_2O_3$ ceramics at last. The dopants appear to fulfil at least two rules during the sintering pocess: they form a small amount of liquid phase which promotes densification, whilst the majority remain as a dispersed second phase. This latter may then restrict grain growth by pinning grain boundary to improve the microstructure and control the electrical properties.

PRIMARY REFERENCES

[1] M.Foex, C.R.Acad, Sci. 223, (1946) 1126; 229, (1949) 880.
[2] L.J.Morin, Phys. Rev. Lett. 3, (1959) 34
[3] D.B.Mcwhan, A.Menth, J.P.Remeika, W. F. Brinkman and T.M.Rice, Phys. Rev. B7 (1973) 1920.
[4] R.S.Perkins, A.Ruegg and M.Fischer, Advances in Ceramics, Vol.7, 166.
[5] A.Ruegg, R.S.Perking, P.Streit, Sci. Ceramics, 11, (1981) 559.

Crystal Structure and Electrical Property of Mn-Co-Ni-(Fe) Oxide for Thermistor Materials in the Manufacture Process

T.Sasamoto*, T.Meguro**, T.Yokoyama**, Y.Abe***
and
N.Torikai**

* Dept. of Industrial Chem., Kanagawa Inst. Tech.,
** Dept. of Materials Chem., Yokohama National Univ.,
*** Technol Seven Co., Ltd.

Japan

Changes in crystal structure and thermistor constant in Mn-Co-Ni-Fe-oxide in firing, cooling, and annealing processes for making thermistor materials were investigated in defferent atmospheres. The starting oxide was prepared from each nitrate by calcining at 1000 C for 3 hs in air.

Phase changes in firing process were first investigated by the quenching method in defferent atmospheres in the temperature range from 1000 to 1400 C. A phase diagram of the Mn-Co-Ni-oxide was proposed, in which the temperature of crystal phase change was correlated with the concentration of oxygen. And second, changes in crystal structure and themistor constant, B, in cooling process below 1000 C after at 14 00 C have been studied. The B values changed with increasing heat-treatment temperature, and a remarkable depression of B value was observed between 550 and 650 C, which was associated with the change in c/a values in the tetragonal spinel structure observed in the same temperature range.

The samples, prepared by cooling to R.T. in the furnace after fired at 1400 C, was annealed in air, Ar, and pure O_2 at 310 and 370 C. A remarkable increase in axial ratio, c/a, in the tetragonal spinel structure was observed with increasing annealing time. This changes in axial ratio seems to be related to the electric conductivity change. An electric conduction mechanism and an effect of atmospheres in annealing processes on electric conductivity of Mn-Co-Ni-Fe oxide were investigated by measuring the electric conduction and Seebeck potential. The position of the acceptor level, the activation energy of the movility, and the jumping frequency were determined. These data can be explained satisfactorily in terms of a small-polaron hopping model.

PREPARATION OF MAGNETIC THIN FILMS FROM SOL-GEL PROCESS AND PROPERTIES

T. Tsuchiya, K. Yamashiro and J.D. Mackenzie*
Science University of Tokyo, Dept. of Mat. Sci. & Tech., Noda, Chiba, 278, Japan
* University of California, Los Angeles,Dept of Mat. Sci. & Eng., Los angeles, CA 90024, U.S.A

ABSTRACT
The preparation of the thin films and the powder of magnetoplumbite type ferrite ($MFe_{12}O_{19}$ M= Ba, Sr) was investigated by dip-coating from sol-gel process. The ferric nitrate and Ba, Sr-nitrate were used as raw materials. The solutions containing glycerol were prepared in stoichiometric ratio in the composition of the desired ferrite. A films were prepared by dipping a silica glass plate. Film thickness was about 0.08 μm for magnetoplumbite type ferrite. The film in the system of $15BaO.85Fe_2O_3$ showed the coercive force of the 3300 Oe value and the powder showed the 3050 Oe value. These films were shown to be much larger Hc than that the sintered powder. Effect of substitution of titanium and cobalt ion on barium hexaferrite films and powder were investigated. The coercive force (Hc) of films and powder were found to be controllable in a wide range without the reduction of magnetization by substitution of titanium and cobalt.

INTRODUCTION

Magnetoplumbite ferrite having general formula $MFe_{12}O_{19}$(M=Ba, Sr, Pb) are hard ferrite materials and are used as the permanent magnet. These excellent magnetic materials are required to be larger residual magnetization (σr) and coercive force (Hc). These magnetic powder are the particle having hexagonal sheet type and has been worthy of remark as the medium of vertical magnetic for high density memory. These powder are prepared by glass crystallized method 1), hydrothermally method 2), and hydrolysis of organometal 3). But these powder are used as the medium of painting type. Therefore the saturate magnetization is lower for containing organic binder of non-magnetic materials and are difficult to the preparation of thin films than that 1μm. Generally, the preparation method of the magnetic thin film are used to PVD, sputtering method and CVD. But these methods are difficult for the preparation of large sample and the compositional control. On the other hand, the dip-coating method from so-gel permits the preparation of thin films at relatively low temperatures and hense their compositions are more controllable. The dip-coating method from the solution containing metal salts is very useful for modifying the properties of substrate with a large surface area. But a films having magnetic properties by sol-gel process are not yet reported. This investigation is concerned with the preparation of thin films and powder containing magnetopulumbite ferrite ($BaFe_{12}O_{19}$, $SrFe_{12}O_{19}$) crystal. Moreover, the effect of substitution of titanium and cobalt ion on Ba or Sr-hexaferrite films and poder were investigated.

EXPERIMENTAL

1.Preparation of thin film

The composition studied were $yMO.(100-y)Fe_2O_3$, (M=Ba, Sr, y= 14.29, 15, 20, 25) and $MFe_{12-x}Co_xO_{19}$ (M= Ba, Sr),$yBaO.(100-y)$ [$(100-2x)Fe_2O_3$. $xCoOxTiO_2$] (y=14.29, 15, 20, 25, x= 0- 11.66) substituted by Ti and Co ions of Fe ion. Various metal salts were used as raw materials. Titanium oxyacetylacetonate was used as raw material for the composition containing titanium. Glycerol and folmamide were used as solvents. SiO_2 glass was used as substrate and the films were prepared by dip-coating method. After drying at 70°C for 2 hr, then the samples were heat-treated stepwise in an electric furnace up to 400°C and the ferrite thin films were prepared by heat-treatment at 800°C for 24 hr.

2. X-ray measurement and magnetic properties

Thin film was measured by X-ray diffractometer with thin film attachment. (Rigaku-CN 4148 type) Thin film thickness was measured by thin film step method. The coercive force (Hc), residual magnetization (σr) and saturate magnetization (σs) were measured by the eqipment of vibrating specimen magnetization (Riken electron Co Ltd). The surface of thin films were analyzed by ESCA. (JEOL- JPS-90SX type).

RESULT AND DISCUSSION

1. Preparation of thin film

It was completely dissolved in water or alcohol. After Ba-nitrate was dissolved in water, the homogeneous solution was obtained when glycerol was added as the solvent. But glycerol is the solvent having very high viscosity. Therefore the free crack films were easily prepared by adding water and holmamide. The thin film obtained was immersed in 3N-HCl and 3N-NaOH.

The film thickness was about 0.08 μm at one time dipping. Fig. 1 shows the film thickness versus number of dipping on Ba-hexaferrite heat-treated at 850°C for 48 hr. The film thickness increased with increasing number of dipping and was possibled to the control of film thickness.

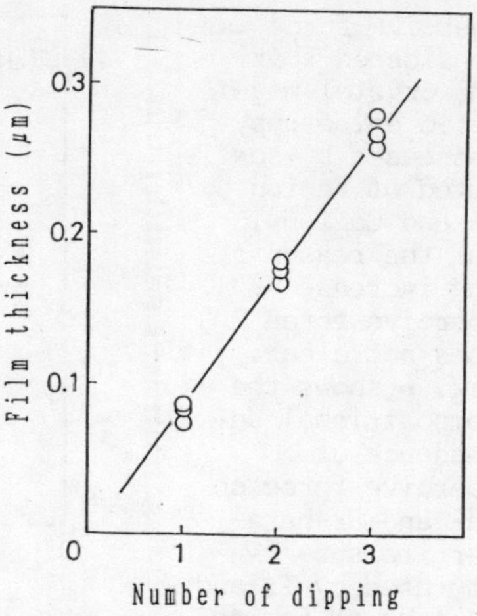

Fig. 1 Film thickness versus number of dipping for Ba-hexaferrite films

2. X-ray diffraction analysis

Fig. 2 shows the result by X-ray diffraction analysis of thin film in the composition of $15BaO.85Fe_2O_3$ and $20SrO.80Fe_2O_3$ heat-treated at 850°C for 48 hr. The peaks of Ba- and Sr-hexaferrite crystal were observed for all compositions. But the orientation of crystal was not observed.

3. Magnetic properties of ferrite

The coercive force (Hc), saturate magnetization (σs) and residual magnetization (σr) were measured from hystericis loop of thin film and powder at room temperature. Where the unit of magnetization (σ) is emu/cc for thin film and emu/g for powder. Table 1 shows the magnetic properties of thin film and powder heat-treated at 850°C for 48 hr. Hc of thin film showed the values of 3000-4700 Oe for Ba-hexaferrite and σs was obtained to 50-57 emu/g value. On the other hand, Hc showed the values of 4000-6700 Oe for Sr-hexaferrite and σs was obtained to 140-160 emu/g values. For the powder, Hc was 2400-4100 Oe and σs was 52-56 emu/g. Hc of the thin film and powder showed higher values than that of the reference data.

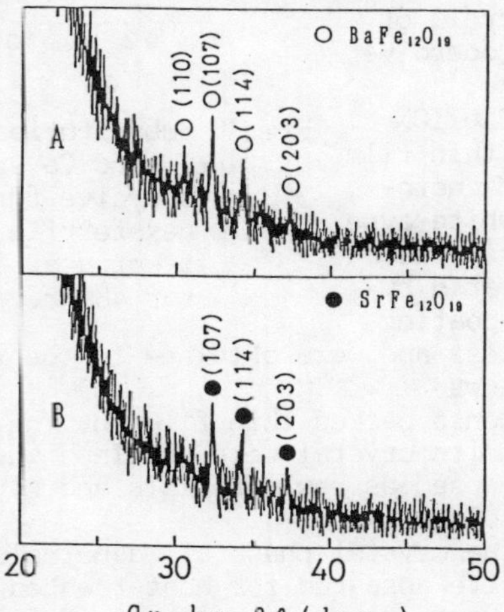

Fig. 2 X-ray diffraction patterns in the system of $15BaO.85Fe_2O_3$ film(A) and $20SrO.80Fe_2O_3$ film (B) heat-treated at850°C.

Moreover, Hc of thin film showed larger value than that of the powder. From the observation by SEM on the film and powder, the crystal particles having hexagonal sheet type was observed for the powder. But the crystal particle for the thin film could not observed. From the result above mentioned, the magnetic particles on the substrate for thin film were considered to the fine particles than that of the powder. Table 2 shows the magnetic properties of the thin film heat-treated at 950°C for 48 hr. The particles of the thin film were considered to the single domain particles because Hc increased with increasing heat-treatment temperature.

4. Analysis of thin film by ESCA

For analysis of the reaction on thin film and substrate, the surface and depth direction by ething were measured by ESCA. Fig. 3 shows the analysis of the depth direction by ESCA on $Ba3d_{5/2}$, $Fe2P_{3/2}$, Si2P and Cl $S_{1/2}$ of thin film in the system of $15BaO.85Fe_2O_3$ at 850°C for 48 hr. Peaks of Ba, Fe and C were observed in the surface of the film. But peak of carbon was found only on the surface. The peak of $Fe2P_{3/2}$ disappeared by ethching after 5min 30 minutes. The peak of Si2P appeared by ethching of 2 min 30 minutes and after 4 min the peak area was almost same. From these results, the interface between hexaferrite film and SiO_2 glass was considered to the appearance after about 3 min 30 minutes. On the other hand, the peak of Ba $3d_{5/2}$ appeared from the surface and showed the broad peak for ethching after 8 min. From these results, the diffusion of SiO_2 was confirmed in the ferrite and the diffusion in SiO_2 glass of Ba component are very easy than that of Fe

Table 1 Magnetic properties of the films and powder heat treated at 850°C for 48 hr.

Composition	Film			Powder		
	Hc (Oe)	σs(emu/cc)	σr(emu/cc)	Hc (Oe)	σs(emu/g)	σr(emu/g)
$14.29BaO \cdot 85.71Fe_2O_3$	3070	140	103	2920	51.0	29.2
$15BaO \cdot 85Fe_2O_3$	3310	133	110	3050	56.8	31.7
$20BaO \cdot 80Fe_2O_3$	4710	144	127	3600	53.8	31.2
$25BaO \cdot 85Fe_2O_3$	4650	140	111	3290	49.7	29.7
$14.29SrO \cdot 85.71Fe_2O_3$	4100	139	120	2350	53.4	31.7
$15SrO \cdot 85Fe_2O_3$	3970	156	90.5	3280	54.6	33.4
$20SrO \cdot 80Fe_2O_3$	6200	161	145	3670	56.1	33.6
$25SrO \cdot 75Fe_2O_3$	6720	162	144	4100	52.2	32.6

Table 2 Magnetic properties of the films heat-treated at 950°C for 48 hr.

Composition	Hc(Oe)	σs(emu/cc)	σr(emu/cc)
$14.29BaO \cdot 85.71Fe_2O_3$	3750	136	122
$15BaO \cdot 85Fe_2O_3$	3850	186	161
$20BaO \cdot 80Fe_2O_3$	4550	188	151
$25BaO \cdot 75Fe_2O_3$	4810	162	132
$14.29SrO \cdot 85.71Fe_2O_3$	4550	163	109
$15SrO \cdot 85Fe_2O_3$	4700	189	120
$20SrO \cdot 80Fe_2O_3$	6200	207	192
$25SrO \cdot 75Fe_2O_3$	6860	197	171

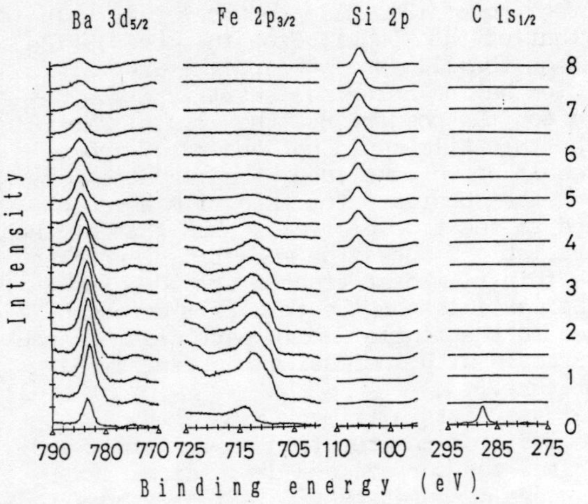

Fig. 3 Depth profiles of ESCA spectra in the system of $15BaO.85Fe_2O_3$ film heat-treated 850°C for 48 hr.

For the preparation of Sr-hexaferrite, the dense packed ceramics were obtained by adding 0.5 wt% SiO_2. Thse result shows that the liquid phase in the system of $SrO\text{-}Fe_2O_3\text{-}SiO_2$ are inhibited to the grain growth by inclusion between the particles 4). From the result above mentioned, the film and powder obtained from sol-gel process show very large Hc and has excellently properties as permanent magnetic materials.

5. Ba-hexaferrite film substituted by Ti and Co ions

Ba-hexaferrite has been worthy of remark as vertical magnetic memory media. But the thin film by sol-gel process are not suited for magnetic memory because the coercive force of the thin film has very large value. Therefore, we tried the improvement on the magnetic properties by substitution at Ti and Co ion of one part of Fe ion. Fig. 4 shows the change of coercive force on Ba hexaferrite thin film substituted by Ti and Co ions. Fig. 5 shows the change of σs. The coercive force decreased from 4000 to 200 Oe with increasing x substituted content. For y=15, 20 composition, the saturate magnetization increased with increasing x content and showed the maximum at x= 8.33. The decrease of

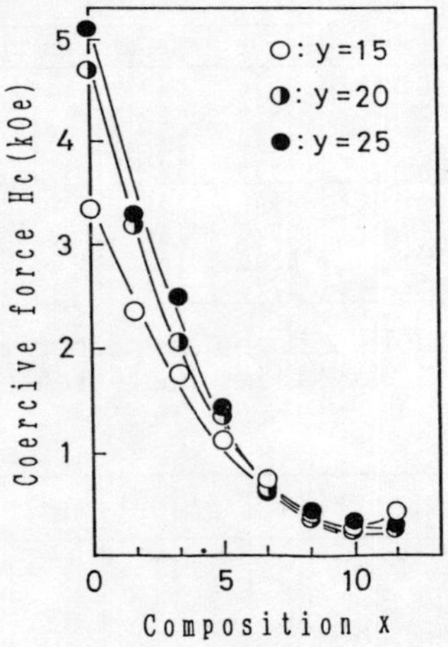

Fig. 4 Substitution effect of Ti and Co on coercive force in the system of $yBaO.(100-y)[(100-2x)Fe_2O_3.xTiO_2.xCoO]$ films.

coercive force was considered that the crystal magnetic anisotropy decreased by sustituted of Fe ion by Ti and Co ions. But the reason of the increase of coercive force does not clear. Fig. 6 shows the compositional dependence of coercive force on Ba- and Sr-hexaferrite substututed by Ti and Co ions of Fe ion. The coercive force of all compositions decreased with increasing substituted content. But the coercive force of the substituted by Ti and Co ions decreased the larger than that of substituted by Co ion. From the result above mentioned, the method of substitution by Ti and Co ions were excellented for the controll of the coercive force.

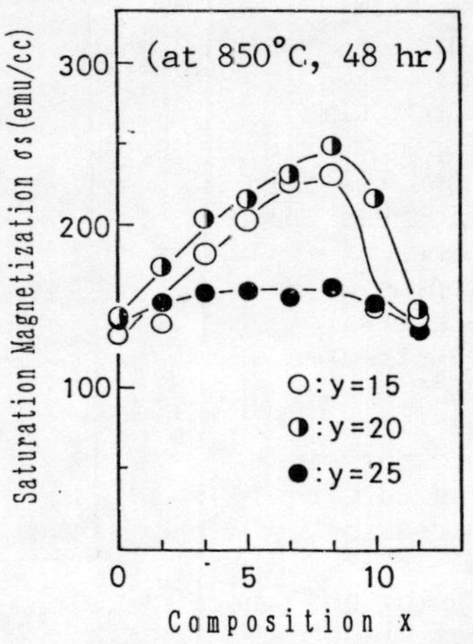

Fig. 5 Substitution effect of Ti and Co on saturate magnetization in the system of $yBaO.(100-x)[(100-2x)Fe_2O_3.xTiO_2.xCoO]$ films

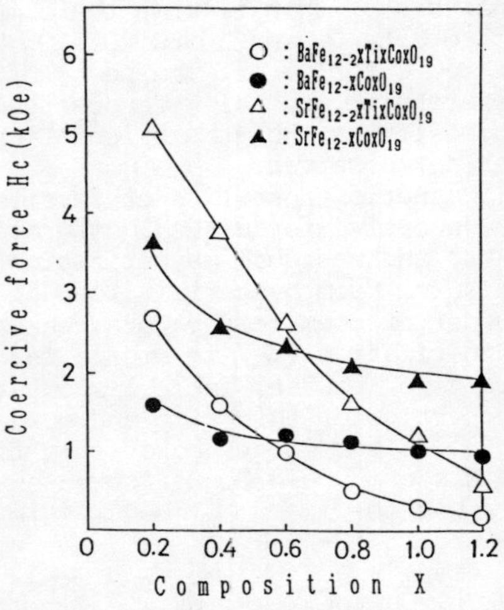

Fig. 6 Substitution effect of Ti and Co or Co ion on coercive force for Ba(Sr) hexaferrite films heat-treated at 850°C for for 48 hr.

CONCLUTION

The thin film of Magnetoplumbite type ferrite were prepared by dip-coating process and were obtained to the results as follows

1) Dense packed thin films having Magnetoplumbite crystal were obtained when each nitrate were used as raw materials and glycerol was used as solvent.

2) The crystal phase of Magnetoplumbite type ferrite appeared for heat-treated at 750°C and well grew crystal was obtained for heat-treated at 850°C.

3) The coercive force of thin film showed larger value than that of powder. This reason was considered to be due to the result that the magnetic particles of thin film has a single domain structure.

4) Coercive force of Ba-hexaferrite thin film was possibled to be the control from 4000 to 200 Oe by substitution Ti and Co ions.

REFERENCES

1. O.Kubo, T.Ido and H.Yokoyama, IEEE Trans. Mag. Mag-18, No 6, 1122 (1982)
2. H.Miura, K.Hashimoto, T.Kimura and K.Odan, Electric Comunication Soc. Tech. Report, 85 201 , 29(1985)
3. K.Higuchi, J.Naka and S.Hirano, 95, 186 (1987)
4. C.A.M.Broek and A.L.Stuijts, Philips. Tech. Rev., 37, 157(1977)

PREPARATION AND STUDIES OF HEXAGONAL FERRITES BY A NEW METHOD

*Feng Jie Mei Xue–Ming Guo Bi–Jun Zheng Zong–Yu*①
University of Electronic Science and Technology of China, Chengdu, P.R.C.

ABSTRACT

By means of a new methcd of coprecipitation combined with high temperature melting [1], we prepared and studied Ba(Sr) $Fe_{12-x}Al_xO_{19}$ ferrite powder with superhigh intrinsic coercivity and $(Zn_{1-x}Co_x)_2$–W. $Zn_{2-2x}(LiFe)_x$–W hexaferrites. We have got $BaFe_{12-x}Al_xO_{19}$ ferrite powder with intrinsic coercivity $H_{cJ} > 16000$ Oe and almost pure $(Zn_{1-x}Co_x)_2$–W, $Zn_{2-2x}(LiFe)_x$–W ferrites with $\sigma_s = 78.6$ emu / g.

INTRODUCTION

Among hexagonal ferrites, M–type ferrite has a higher anisotropy field H_A and W–type ferrite has a larger saturation magnetisation σ_s. For a long time, M–type ferrites were used as permanent magnets, such as Ba–ferrite and Sr–ferrite. The σ_s of them have reached to theoretical value, but the intrinsic coercivity H_{cJ} is lower than the theoretical value. C.D.Mee prepared Ba(Sr)$Fe_{12}O_{19}$ powder by chemical coprecipitation and got the intrinsic coercivity of unoriented assembly powder agreeable to the theoretical results of Stoner–Wohlfarth coherent rotation [2]. But if substituting Al^{3+} for Fe^{3+} in M–type Ba(Sr)–ferrite for the purpose of higher H_A, it is more difficult to obtain ferrite powder with H_{cJ} agreeable to S–W theory. Because the ferrite intrinsic coercivity is not only connected with the anisotropy field, it is also linked to the shape and size of the grain and to the crystalline of the grain which is sensitive to preparing process.

Because the σ_s of W–type ferrite is ~ 15% higher than that of M–type ferrite and its H_A comes nearly to that of M–type ferrite, W–type ferrite is also an ideal permanent magnet and medium of perpendicular magnetic recording.

By an opportune choice of the divalent cation Me^{2+}, we can adjust the value of magnetic anisotropy constant K_1 of W–type ferrite, so W–ferrite can also be used as microwave absorption material. But owing to the difficulties in preparation, W–type ferrite has not been used in commercial.

EXPERIMENT

We use related soluble chloride as raw materials, use mixture of NaOH and Na_2CO_3 as precipitant. Considering the deviation in preparing process, we select (Fe+Al) / Ba = 10.8 for M–type ferrite and select Fe / Ba = 15.4 for W–type ferrites in raw materials. The precipitate have PH value of 6.7 ~ 7.2. After coprecipitation, the soluble salt will remain in the slurry. The salt will act as fusion aid during the precipitate roasting. After roasting two hours for M–type, 0.5 hour for W–type in air, the reaction mixture was cooled and the soluble salt was washed out and leached out with deionized water and the desired ferrite powder was left.

For $BaFe_{12-x}Al_xO_{19}$ with superhigh H_{cJ}, the H_{cJ} values are measured by gate type magnetometer [3], the maximum magnetizing field is 23000 Oe. The principle of measurement is to magnetize the powder to saturation, then increase the demagnetizing field gradually, when the residual magnetization of the powder equals to zero, the demagnetizing field at this time is the H_{cJ} of this powder. For W–type ferrites powder, we use model 155 vibrating sample magnetometer made by EG & G Princeton Applied Research for measurement. The maximum magnetizing field is 10000 Oe.

The average diameter of powder was measured by Fisher method (a measurement by means of gas penetration).

RESULTS AND DISCUSSION

①Permanent address: Yi Bin P.O.Box25 Sichuan P.R.C.

1. M-type Ferrite With Al^{3+} Substitution

K.Haneda et. al. calculated the relations between Hc_J and Al^{3+} substitute in $BaFe_{12-x}Al_xO_{19}$ according to S-W theory, but their experimental results were much lower than theoretical results [4]. Ferrites powder prepared by our method avoid lenthy milling in ceramic method, so it is strain free. It is interesting to use our method to prepare and study the rule of H_{cJ} of $BaFe_{12-x}Al_xO_{19}$. We found that the H_{cJ} of $BaFe_{12-x}Al_xO_{19}$ is roasting temperature dependent. Fig 1 (a,b) shows the dependence of H_{cJ} of $BaFe_{12-x}Al_xO_{19}$ on roasting temperature.

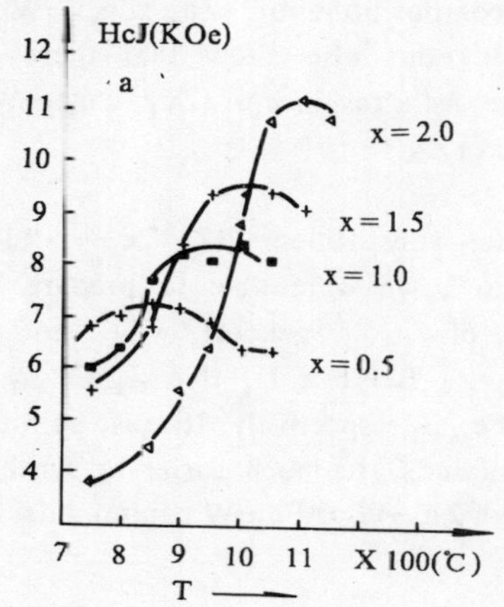

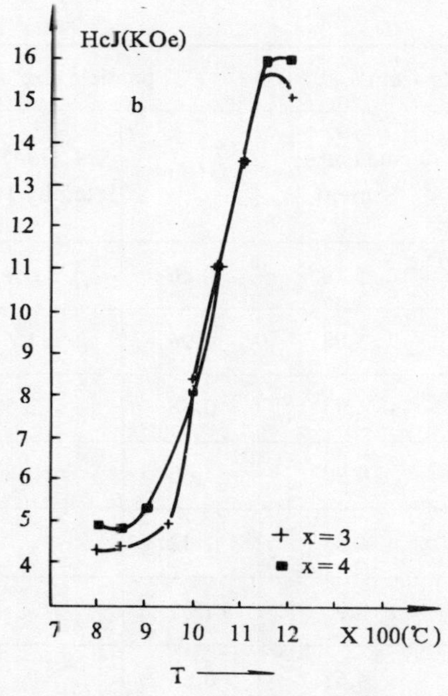

Fig. 1 (a, b) The dependence of H_{cJ} of $BaFe_{12-x}Al_xO_{19}$ on roasting temperatrue

From Fig 1(a,b) we can see that for different Al^{3+} substitution, with the change of roasting temperature the H_{cJ} has a peak respectively. We think in the roasting temperature with H_{cJ} at the peak, the grain of ferrite powder is fine, strain free and single domain.

Fig 2 is the plot of the H_{cJ} at the peak vs Al^{3+} substitute in $BaFe_{12-x}Al_xO_{19}$ in our experiment and the result of S-W theory calculated by K Haneda et al [4].

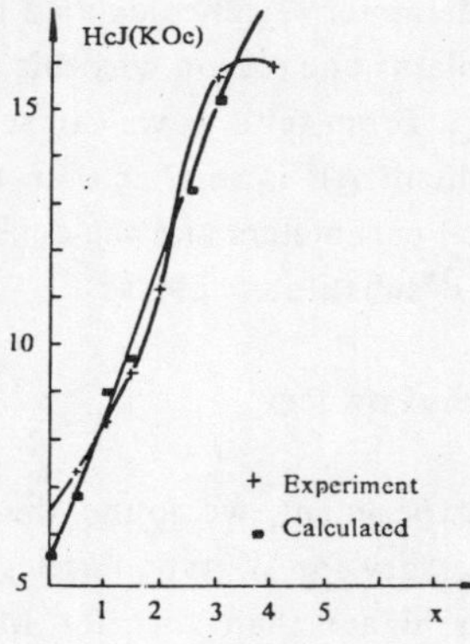

Fig. 2 Relations between the H_{cJ} and Al^{+++} substitute in $BaFe_{12-x}Al_xO_{19}$(experimental results and theoretical results)

We can see, when $x<3$ the results of our experiment are much close to the theoretical results. It means that we get M-type ferrites powder with fine crystalline in a wide range of Al^{3+} substitute. It is interesting to note that when $x=4$ the H_{cJ} is nearly the same as the H_{cJ} when $x=3$. This is probably because that the more the Al^{3+} substitute is, the higher the roasting temperature needed to make the ferrite crystalline fine is . But when roasting temperature increase some factors which prevent H_{cJ} from increase are caused before the crystalline of powder reaches perfect. So when $x=4$ H_{cJ} will begin to decrease before reaching the highest value.

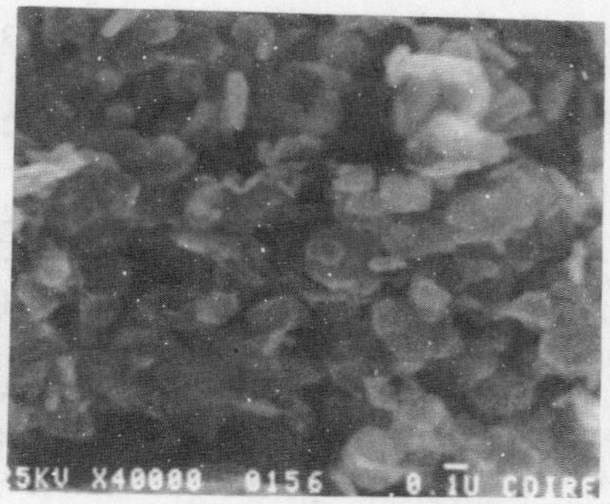

Fig 3. The SEM of $BaFe_{12}O_{19}$ ferrite powder

Fig 3 is the SEM of $BaFe_{12}O_{19}$ powder, the grain of

the powder have platelet shape which is suitable for pressing orientation.

Table 1 shows the properties of powder of $Ba(Sr)Fe_{12-x}Al_xO_{19}$. In table 1, $\bar{D}$ is average diameter of ferrites powder whose H_{cJ} is at the peak respectively. We find that the value of $\bar{D}$ is much smaller than the single domain critical diameter (2Rc) calculated by K. Haneda et. al. [4]. It explains one reason why this ferrite powder has such high H_{cJ}. From table 1, we can see, because the radius and weight of Al^{3+} is smaller than those of Fe^{3+}, the hexagonal cell parameters and the density of powder decrease while Al^{3+} substitute increase.

2. W-type Ferrites Powder.

From our experiment, we found that the roasting temperature necessary for W-type ferrites especially for Zn_2-W is much higher than that for M-type ferrite. Even if in our experiment the size of precipitate is less than 100Å and with NaCl as fusion aid, the roasting temperature for W-type ferrite is at least 1200℃ to avoid large scale spinel phase and M-type hexaferrite. For Zn_2-W, we have roasted it at 1300℃ for half an hour, but the reaction products are still mixture of W-type and spinel, M-type ferrite. As a result, σ_s of Zn_2-W ferrite powder is less than 70 emu / g. Substituting Zn^{2+} by some divalent cation can advance the formation of W-type ferrite and lower the rosting temperature. But, it is strange that we can not prepare almost pure $(Zn_{1-x}Cu_x)_2$-W ferrite by our process. We found that after thermal treatment, adding deionized water in reaction products the liquid has white precipitate. The white precipitate contains Zn^{2+} and it becomes heavy while roasting temperature changes from 1200℃ to 1250℃ and to 1300℃. We think the loss of Zn^{2+} leads to deviation of compositions of $(Zn_{1-x}Cu_x)_2$-W ferrite. X-ray diffractograms also show that there appears unknown phase. As a result, σ_s of $(Zn_{1-x}Cu_x)_2$-W ferrite is less than 70emu / g.

Using the substitution $2Zn^{2+} \leftrightarrow Li^+ + Fe^{3+}$, $Zn^{2+} \leftrightarrow Co^{2+}$, we were able to prepare W-type ferrites of general composition $BaZn_{2-2x}(LiFe)_xFe_{16}O_{27}$ and $Ba(Zn_{1-x}Co_x)_2Fe_{16}O_{27}$ respectively. It must be noted that all those compounds are much easier to synthesize as pure phases than Zn_2-W or Fe_2-W compounds. Fig. 4

Table 1 The properties of the particles of $Ba(Sr)\,Fe_{12-x}Al_xO_{19}$

Sample		lattice parameter		Density(g / cm³)		particle size	
Ferrite	x	a(Å)	c(Å)	X-ray calculation	measurement	$\bar{D}$ (μm)	2Rc(μm) calculated by Haneda
BaM	0	5.900	23.22	5.27	5.19	0.80	0.9
	1	5.864	23.10	5.22	5.08	0.96	1.6
	2	5.847	23.01	5.14	4.93	0.88	2.5
	3	5.814	22.91	5.07	4.90	1.00	
	4	5.812	22.87	4.94	4.66	1.02	
SrM	0	5.872	23.21	5.09	4.90	1.07	
	1	5.850	23.10	5.01	4.68	0.65	
	2	5.828	23.00	4.93	4.84	1.08	
	3	5.801	22.94	4.84	4.74	1.39	
	4	5.779	22.83	4.76	4.61	1.18	

are X–ray diffractograms of W–type ferrite powder.

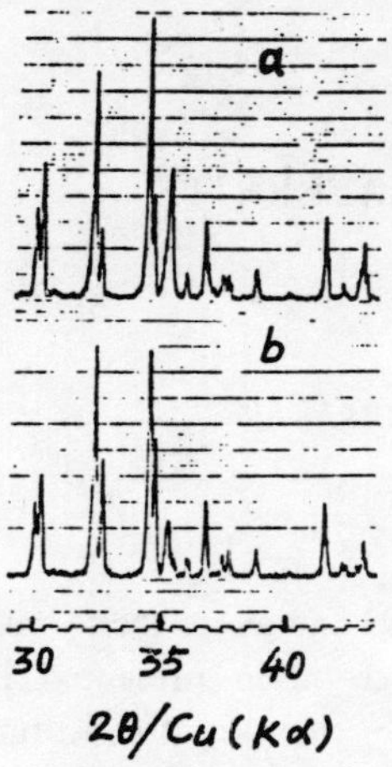

Fig. 4 X–ray diffractograms of W–type ferrite powder

a) (ZnCo)–W (1250℃ / 0.5h)

b) $Zn_{1.4}(LiFe)_{0.3}$–W (1250℃ / 0.5h)

From fig 4. we can see the phase of the powder is almost pure W–type phase. Fig 5. are relations of σ_s, H_{cJ} of $(Zn_{1-x}Co_x)_2$–W ferrite powder and x values

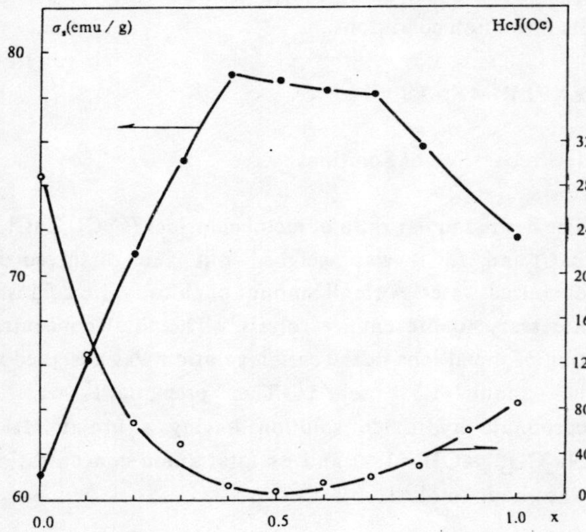

Fig. 5 Relations of σ_s, H_{cJ} and x value in $(Zn_{1-x}Co_x)_2$–W (1250℃ / 0.5h)

From Fig 5. we can see, Co^{2+} substitution advances the formation of W–phase. When x=0.4, the σ_s of $Zn_{1.2}Co_{0.8}$–W ferrite powder reaches 78.6 emu / g close to theoretical value. The H_{cJ} decreases with x. When x=0.5, the H_{cJ} reaches a minimum, after that the H_{cJ} begins to rise. This is a result of Co^{2+} substitution for Zn^{2+} so as to make magnetocrystalline anisotropy constant K_1 be compensated gradually. For that reason, we may produce absorption materials in a wide range frequency.

Fig 6 is the SEM of (ZnCo)–W ferrite powder

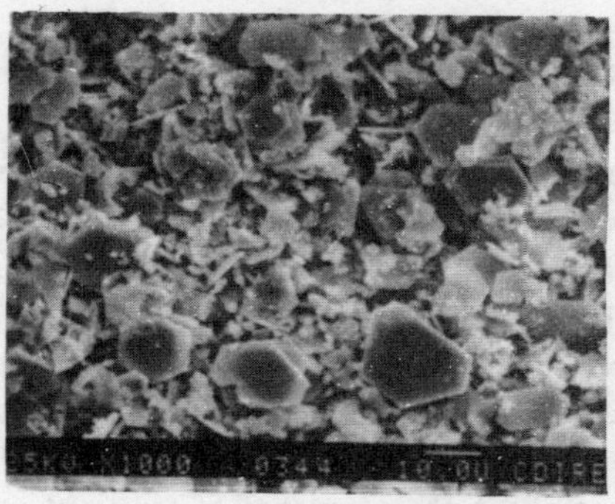

Fig. 6 The SEM of (ZnCo)–W (1250℃ / 0.5h)

Fig 6. Shows that because of such high roasting temperature, the size of W–type ferrite powder is bigger than that of M–type ferrite.

For W–type ferrites used as medium of perpendicular magnetic recording, it is necessary to decrease the size of particle and raise the H_{cJ} of powder.

ACKNOWLEDGMENTS

Discussion with prof. Zhang You–gang were very helpful.

REFERENCES

1. Zheng Zong Yu et. al. J. of M.M.M.78(1989) 73
2. C.D.Mee and J.C.Jeschke J.Appl. phys. 34 (1963) 1271
3. Zhang You–gang et. al. Rare Earth–Cobalt permanent Magnets And Their Applications (1983) 535
4. Koichi Haneda et. al. JAP. J Appl. phys. 12 (1973) 355

PREPARATION OF PURE LiFe–W HEXAFERRITE

Cai shou–Hui Rao Ming–Ming
University of electronic Science and Technology of China
Chengdu, Sichuan, P. R. China

ABSTRACT

This paper discribes a new method and conditions for the synthesis of the $BaZn_{2(1-x)}(LiFe)_xFe_{16}O_{27}(0<x<1)$. The new method that we used is "Coprecipitation combined with high–temperature melting". It is especially used for producing single crystal particles. According to the previously published papers, it is very difficult to obtain the pure W–phase ferrite. So special care had been given to the phase characteristics of the finished products and interesting effects had been revealed by studing the X–ray powder diffraction patterns.

It was found that pH of mixed slurry, sintering temperature, sintering period and cooling method will significantly influence the phase composition. Moreover, the W phase will decompose and an unknown X phase will form while the sintering temperature is higher than 1200℃. The content of X phase in the product will increase with increasing of the sintering period. The pure W phase, which was obtained by controlling pH of mixed slurry in the range of 8.0–8.5, sintering temperature at between 1250–1300℃ and sintering period for 20–30 min, has a saturation magnetization $\sigma_s = 76 Am^2/kg$ nearly as high as those of the previously known Zn_2–W or Fe_2–W ferrites.

INTRODUCTION

The W–type hexaferrite has a saturation magnetization about 10% higher than the widely used permanent–magnet material M–type hexaferrite and the anisotropy field is about equal. This opens up the possibility of using W as a permanent–magnet material with a higher remanence Br and energy product BHm [1]

Among the W–type hexaferrites, the compound $BaZn_2Fe_{16}O_{27}$ seems to be one of the most promising ones as permanent magnet material. It's main disadvantages are the relatively low uniaxial anisotropy and the high sintering temperature. In order to improve the characteristics of Zn_2–W and avoid the formation of spinel phase $ZnFe_2O_4$, which often presented in the products of W–type ferrite, the partial substitution of Zn^{2+} through the scheme $2Zn^{2+} \rightarrow Li^{+} + Fe^{3+}$ had been used and the composition of the substituted product was $BaZn_{2(1-x)}(LiFe)_xFe_{16}O_{27}(0<x<1)$. This kind of substitution was very successful by using standard ceramic technique according to the reference[2] in which the evolution of the magnetic specifications (magnetization σ_s, anisotropic field Ha and first anisotropic constant K_1) with composition x and temperature T had been researched. In the present paper, we mainly statement a new method for preparing the LiFe–W, and take the compound $BaZn(LiFe)_{0.5}Fe_{16}O_{27}$ as an example to discuss its formation conditions.

EXPERIMENTS

1. Preparation of Solutions

The desired molar ratio of metal chlorides ($FeCl_3$, $ZnCl_2$, $BaCl_2$ and LiCl) was weighed out and dissolved in deionized water. A small amount of chlorhydric acid was necessary to prevent hydrolysis. The total concentration of metal ions in the cation solution was designed to be about 1.5 mol/l. The precipitant was a carbonate–hydroxide solution having a molar ratio Na_2CO_3:NaOH = 1:40 and its total anion concentration was about 5mol/l.

2. Preparation of mixed Slurry

The metal ions in the cation solution was precipitated by adding the precipitant in it. The pH was monitored and when the desired pH was reached, the flow of precipitant was stopped, then stirred the mixed slurry for about 20 min to make it uniformity. The precipitate, which was not separated from mother liquor, was directly dried at between 160–200℃.

3. Stinter and Treatment of Product

The dried mixture was placed into Al_2O_3crucible and sintered in a batch furnace. The temperature was in the range of 1250–1300℃. During the high temperature, the NaCl in the mixture produced a molten reaction medium and the oxides formed by decomposition of metal carbonates and ydroxides reacted in the medium with each other to form ferrite. The sintered product was soaked with boilling water and after it perfectly loosed, washed with deionzed water until the acidified $AgNO_3$test was negative showing the absence of Cl^-. Then, the product was roasted at about 100℃.

PREPARING CONDITIONS AND PHASE COMPOSITIONS

We had carried out a series of experiments to make certain which factor had an influence on the phase composition of finished product and what influence it produced.We chose a given composition of the compound $BaZn_{2(1-x)}(LiFe)_xFe_{16}O_{27}$,taking x=0.5,so that we could conveniently research the other influence factors.The X-ray powder diffraction was used for determining the phase of the finished product.

1. pH of Mixed Slury and Phase Composition.

Took the mixed slurries pH=7, 7.5, 8, 8.5, 8.7 and 9, sintering temperature T=1250℃, sintering period t=30 min. The sintered product was cooled naturally in the furnace to room temperature.
Table 1 shows pH of mixed slurries and correspondent phase compositions

Table 1 pH of mixed slurries and phase compositions of the products

pH of slurries	7	7.5	8	8.5	8.7	9
Phase compositions	W+F**+X	W+F*+X	W+X	W+X	W+F*+X	W+F**+X

"W", "F"and "X" indicate W−type ferrite, $\alpha-Fe_2O_3$ and an unknown compound respectively. "*" indicates the relative content of a phase and the increase of the "*" indicates the increase of the phase content.

2. Sintering Temperature and Phase Composition

Took mixed slurries pH=8, sintering temperatures T=1150, 1200, 1250 and 1300℃, and sintering period t=30min. cooling method is the same as above.

Table 2 Sintering temperatures and phase compositions of the products

Sintering temperature(℃)	1150	1200	1250	1300
Phase compositions	M+S+X′	W+F+S+X″	W+X	W+X

"M", "S" and "X′, X″" indicate M−type ferrite, spinal ferrite and unknown oxides or carbonates respectively. The other symbolical meaning is the same as Table1.

3. Sintering Period and Phase Composition

Took mixed slurries pH=8, sintering temperature T=1250℃ sintering periods t=10, 20, 30, 40 and 105min. Cooling method was not changed. Table 3 shows the sintering period and correspondent phase compositions

Table 3 Sintering periods and the phase composition of the products

Sintering period(min)	10	20	30	40	105
Phase compositions	W+δF	W+X*	W+X**	W+X***	X+δW

"δ" indicate that the content of a phase is a very few.

4. Cooling Method and Phase Composition

Two kinds of cooling methods were designed: Ⅰ. the sintered product was cooled naturally in the furnace to the room temperature; Ⅱ. the sintered product was cooled coercively by opening the door of the furnace. Two sintering period were chosen t=20 and 30 min. The other conditions were the same as previous experiments: pH=8 and T=1250℃. Table 4 shows the result of the experiment.

Table 4 Cooling methods and phase composition of the products

Sintering periods (min)	20	20	30	30
Cooling methods	I	II	I	II
Phase compositions	W+X*	W	W+X**	W

DISCUSSIONS

1. Optimum pH

pH of mixed slurry influence significantly the phase composition of the finished product.The optimum pH is in the range of 8.0–8.5. $\alpha-Fe_2O_3$ will form during the suitable sintering temperature for LiFe–W hexaferrite out of the range even if 0.2–0.3. The range is not proper to all kinds of ferrites prepared by using our technique, it is 6.7–7.0 for M–type ferrite $BaFe_{12}O_{19}$for example.

But it is sure that the pH of mixed slurry can not be lower than 6.5 what ever kind of ferrite it is. It stands to reason that the CO_3^{2-}content will be reduced by evolution of $CO_2$when the acidity of the mixed slurry is raised, which will cause that the Ba^{2+}in the cation solution can not be precipitated in stoichiometry. The reaction is as shown as floolws:

$$CO_3^{2-} + H_2O \underset{OH^-}{\overset{H^+}{\rightleftharpoons}} HCO_3^- + OH^-$$

$$HCO_3^- + H_2O \underset{OH^-}{\overset{H^+}{\rightleftharpoons}} H_2CO_3 + OH^-$$

In addition to the phase composition is influenced sensitively by pH of mixed slurry, the saturation magnetization $\sigma_s$does also. The result is shown in Table 5.

Table 5 pH of slurries and saturation magnetizations $\sigma_s$of the products

pH of slurries	7	8.5	7	8.5
Sintering temperature(℃)	1250	1250	1300	1300
$\sigma_s(Am^2/kg$	60	73	62	76

2. Control of X Phase

It is seen from Table 3, that a unknown X phase will form while sintering period is t>20 min and the long the period is, the high the content of X phase. If the period is prolonged to t=105min, the main phase is the X (See Figure1).But at t=1150℃ there will be no the X phase even if sintering period is t=200min. We found that the formation of the X phase is at sintering temperature T>1200℃. The temperature 1200℃ is also the one of the beginning formation of W phase. Therefore, We can speculate that the formation of the X compound is in close relationship with the formation of W ferrite, in other words, with the decomposition of W, which is worthy of note only in melten salt sinter.

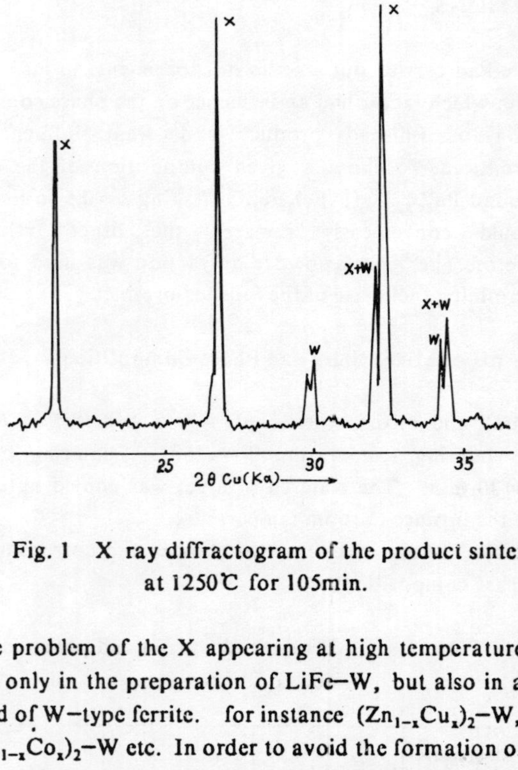

Fig. 1 X ray diffractogram of the product sintered at 1250℃ for 105min.

The problem of the X appearing at high temperature is not only in the preparation of LiFe–W, but also in any kind of W–type ferrite. for instance $(Zn_{1-x}Cu_x)_2$–W, $(Zn_{1-x}Co_x)_2$–W etc. In order to avoid the formation of X phase, it is necessary to control strictly the sintering period and use the coercive cooling method by opening the door of the furnace, which on the one hand can reduce the sintering period at any temperature during temperature drop after cutting off the power and on other hand can increase Po_2, avoiding the reduction of Fe^{3+}at high temperature.

The X compound is probably a magnetic matter. It was found by detecting in the presence of the X as main phase that the saturation magnetization of the product was not low, generally above 60Am^2/kg, but its σ_s reduced greatly if there was a little $\alpha-Fe_2O_3$ in the product. See Table 6.

Table 6 The phase compositions and saturation magnetization σ_s under different conditions

pH of slurries	8.5	8.5	8.5	8.5	7.5	7.5
Sintering temperature(℃)	1250	1300	1250	1300	1250	1300
Sintering period(min)	20	20	120	120	20	20
Phase compositions	W	W	X+δW	X+δW	W+δF	W+δF
σ_s(Am^2/kg)	72	75	65	68	59	61

Besides having magnetism, the X compound is soluble in chlorhydric acid. If the product which contains the X is soaked using 6 mol / l HCl for several hours, the X phase will disappear, while $\alpha-Fe^2O_3$ still remains. This shows that the X compound is easer to be dissolved by chlorhydric acid than $\alpha-Fe_2O_3$.

The pure $BaZnLi_{0.5}Fe_{16.5}O_{27}$W -type hexaferrite was obtained with a room temperature magnetization $\sigma=76Am^2/kg$(See Figure 2)and the value had not much change when the substitution amount x=0.1, 0.2, 0.3...0.9 and 1 for $BaZn_{2(1-x)}(LiFe)_xFe_{16}O_{27}$.

CONCLUSION

1. The sintering temperature larger than or equal to 1250℃ is necessary to obtain the pure W-type hexaferrite.
2. Control pH of mixed slurry in the range of 8.0-8.5 .
3. Control the sintering period at sintering temperature for 20-30 min and open the door of the furnace to cool the sintered product.
4. In comparison with the ceramic technique used by the author of reference[2], The coprecipitation combined with high-temperature has the advantage of saving electric power and time, in addition the grain of the product is more perfect because there is no injury caused by ball-milling.

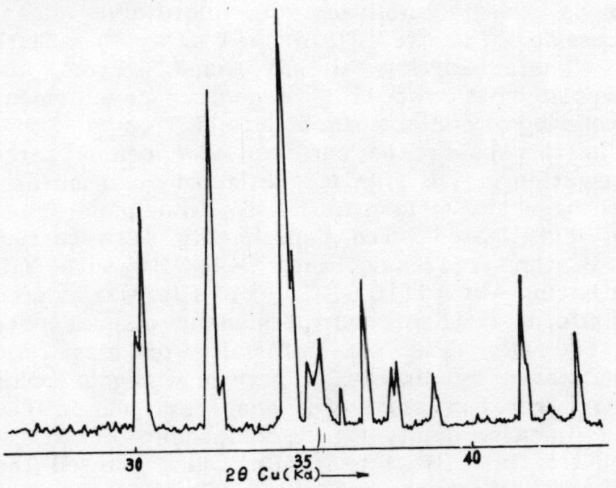

Fig. 2. X-ray diffractogram of the pure LiFe-W

REFERENCES

1. F.K.Lotgering, P.H.G.M.Vromans, and M.A.H.Huyberts, J.Appl.Phys. 51(11)(1980), 5913
2. J.P.Mignot, P.Wolfers and J.C.Joubert, J.Magn. Magn.Mat., 51(1985), 37

THE INFLUENCE OF PARABOLICAL NON-UNIFORMITY OF BIAS FIELD ON WAVE BEAM PATHS OF MSW

A.V. Vashkovsky, V.I. Zubkov, E.G. Lock, V.I. Shcheglov
IRE Acad Sci of the USSR
Moscow, 103907, K. Marx av 18, USSR

The use of the magnetostatic waves (MSW) expands the possibilities of microwave signal processing. The MSW-filters and delay lines with good characteristics already exist. From our viewpoint next step in MSW-devices development is the using of non-uniform bias field.

In this paper the surface MSW beams paths propagating in YIG film magnetized by non-uniform bias magnetic field parallel to film plane have been investigated. The experiments were carried out in the frequency range 3-4 GHz with YIG film having $4\pi M = 1810$ G. The YIG film was located in uniform field of the permanent magnet with H_{oz}=600 Oe. The non-uniformity of bias field was created by disposition between magnet poles of soft iron rod which was near film plane.. The probe measurements of z-component of static magnetic field (H_{oz}) near film plane showed the quadratic character of non-uniformity created by the iron rod:

$$H_z = H_{oz}\,[1+a(a(z_o-z)^2] - H_d \qquad (1)$$

Here z_o is coordinate of projection of the iron rod axis on film plane; a—constant; H_d—field displacement depending on the distance between the rod and the film. The quadratic law took place for $|z_o-z| < 30$ mm (see Fig.1)

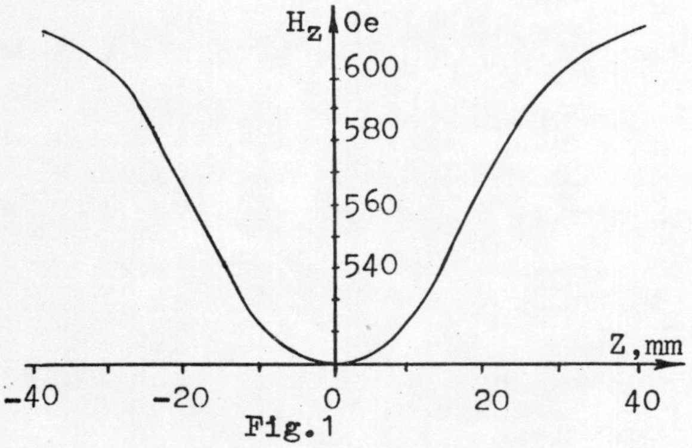

Fig.1

Exciting and receaving of the wave beam were realized by mobile antennas having length 4 mm and diameter 12μm. The antennas had the posibility to move independently one from another on the film surface and to orientate in any direction relatively H_{oz} (see Fig.2)

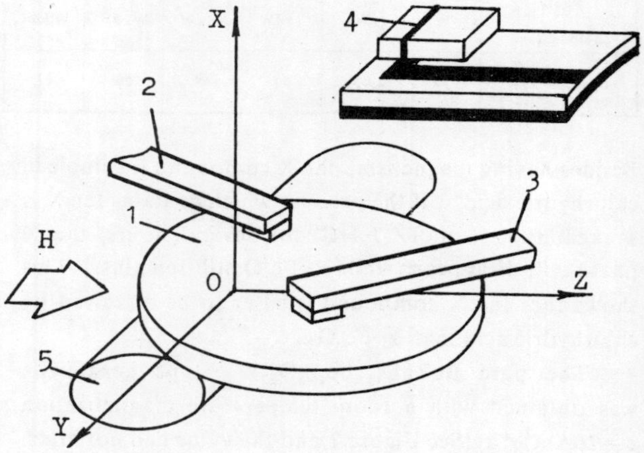

Fig.2 The elements of experimental setup: 1 - YIG film; 2,3 - mobile antennas; 4 - transducer of electromagnetic wave to MSW (see top of the Fig.2); 5 - magnetic-soft iron rod (parallel y-axis)

Two cases of a wave beam propagation were researched: 1) the exciting antenna was located in high magnetic field region, derecting the wave beam to less magnetic field region ($k\cdot\partial H_z/\partial z > 0$; 2) the exciting antenna was located in the region of magnetic field minimum. These cases are illustrated by Fig.3,4. The measured beam paths have been compared with paths calculated by Hamilton-Jackobyu method (points - experiment; lines - calculated).

In the first case (Fig.3), the experiment showed that wave beams paths with another frequency diverged in film plane quickly. The higher the frequency the more the beam deviation from perpendicular to antenna axis and the farther the beam penetrated into low bias field region. By antenna deviation at the angle =34° from z-direction the wave beams diverged at 25° by frequency change 300 MHz (1 - f=3.79, 2 - 3.89,

3 - 3.98, 4 - 4.07 GHz respectively). In this experiment the beams could not pass through the field minimum region, they were "reflected" from certain low magnetic field region and turned to the direction of bias magnetic field increasing.

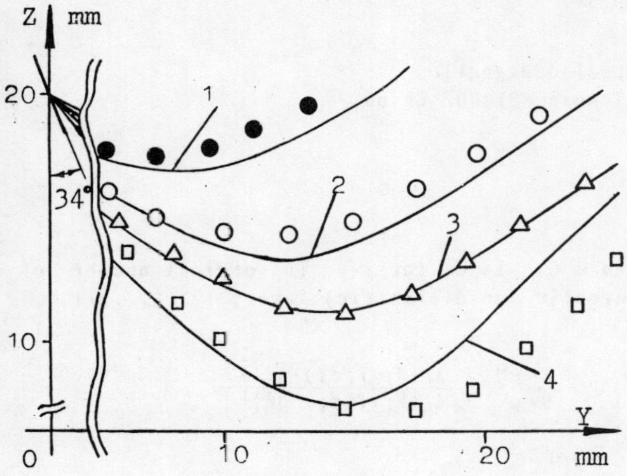

Fig.3

In the second case (Fig.4), the low deviation of antenna from z-direction was investigated (f=3.2 GHz, ϕ=-5°). The experiment showed the existence of the instability of surface MSW beam propagation along the region of bias field minimum. It was connected with the fact that half of the antenna rediated in the condition $k\cdot\partial H_z/\partial_z>0$ and another half - in condition $k\cdot\partial H_z/\partial_z<0$. The behaviour of diverging parts of the beam depends upon the signal frequency essentially. In both cases there is a good agreement between experimental and calculated data.

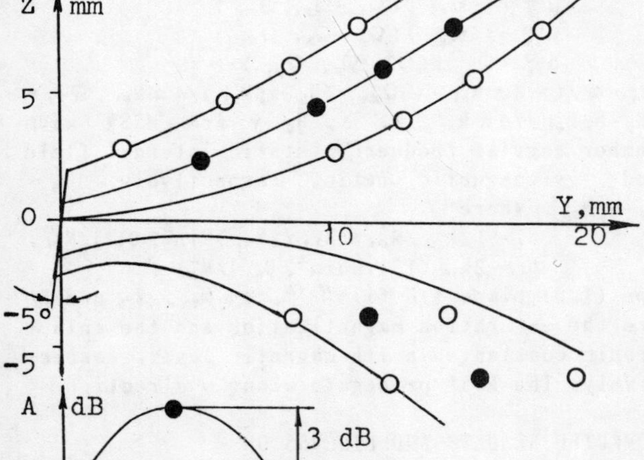

Fig.4

EFFECTS OF VARIOUS PARAMETERS ON MAGNETOSTATIC SURFACE WAVES (MSSW)

Hu Wenzhong
Southwest Institute of Applied Magnetics
P.O.BOX 105, Mianyang, Sichuan 621000, China

ABSTRACT

This paper has investigated the effects of the number of layers, thickness, magnetization and anisotropy for multilayer YIG film, and temperature on MSSW by the way of numerical computation. The results have shown that interactions of MSSW in all magnetic layers which the space between two layers is larger than the wave length are the root cause controlling MSSW dispersion by multilayer YIG film. Temperature not only changes MSSW operating frequency, but changes the delay time.

INTRODUCTION

Magnetostatic waves (MSW) are applied in fabricating new microwave and magnetic optical devices[1-4], because MSW have the natures of high operating frequency (2—20 GHz), electronic tunable, and interaction on optical wave. Except above described, magnetostatic surface waves (MSSW) have also the nature of nonreciprocal propagating at the surface of the sample[5,6].
Using a multilayer YIG film or/and a gradient field, the dispersion of MSW can be efficiently controlled[7-10]. But all dispersive equations for multilayer film are complication. The effects of various parameters on the dispersion of MSW are hardly seen through. This paper investigates the effects of the number of layers, thickness, magnetization and magnetic anisotropy for yttrium iron garnet (YIG) film, and temperature on MSSW dispersion by the way of numerical computation from dispersive equation, and shows computed results.

MSSW DISPERSIVE EQUATION

The geometry of the layered structure is shown in Fig.1. The dispersive equation[8] used in this paper is

$$\exp(2Nt_1)=G_1 \qquad (1)$$

where G_1 is G_i for i=1. The ordinal number of Magnetic (or dielectric) layers $i=1,2,\ldots,i_o$.

$$\begin{cases} G_i = \dfrac{N-\mu_{11}^{(i)} P_i A_i(R_i, P_i d_i)+\mu_{12}^{(i)} K}{N+\mu_{11}^{(i)} P_i A_i(R_i, P_i d_i)-\mu_{12}^{(i)} K} \\ G_{i_o+1} \equiv 1 \end{cases}$$

$$A_i(R_i, P_i d_i)=\frac{\exp(P_i d_i)-R_i \exp(-P_i d_i)}{\exp(P_i d_i)+R_i \exp(-P_i d_i)}$$

$$B_i(G_{i+1}, Nt_{i+1})=\frac{\exp(Nt_{i+1})-G_{i+1}\exp(-Nt_{i+1})}{\exp(Nt_{i+1})+G_{i+1}\exp(-Nt_{i+1})}$$

$$R_i=\frac{\mu_{11}^{(i)}P_i-\mu_{12}^{(i)}K-NB_i(G_{i+1}, Nt_{i+1})}{\mu_{11}^{(i)}P_i+\mu_{12}^{(i)}K+NB_i(G_{i+1}, Nt_{i+1})}$$

$$P_i^2=\left(\frac{n\pi}{W}\right)^2/\mu_{11}^{(i)}+K^2\mu_{22}^{(i)}/\mu_{11}^{(i)} \qquad \text{(in ferrite)}$$

$$N^2=\left(\frac{n\pi}{W}\right)^2+K^2 \qquad \text{(in dielectric)}$$

$\mu_{11}^{(i)}$, $\mu_{22}^{(i)}$, $\mu_{12}^{(i)}$ are the elements of permeability tensor for the ith magnetic layer,

$$\mu_{11}^{(i)}=1-\Omega_{yi}/(\Omega_i^2-\Omega_{xi}\Omega_{yi})$$
$$\mu_{22}^{(i)}=1-\Omega_{xi}/(\Omega_i^2-\Omega_{xi}\Omega_{yi})$$
$$\mu_{12}^{(i)}=\Omega_i/(\Omega_i^2-\Omega_{xi}\Omega_{yi})$$

$\Omega_i=\omega/(\gamma 4\pi M_{oi})$, $\Omega_{xi}=(H_o-a_i M_{oi})/4\pi M_{oi}$, $\Omega_{yi}=(H_o-b_i M_{oi})/4\pi M_{oi}$. k, ω, H_o, γ are MSSW wave number, angular frequency, static internal field and gyromagnetic ratio, respectively. $H_o'=H_o-a_i M_{oi}$. Where

$$a_i=-[2k_{1i}/M_{oi}+(K_{2i}/2M_{oi})\mathrm{Sin}^2 2\theta_i]/M_{oi},$$
$$b_i=-2K_{1i}(1-1.5\mathrm{Sin}^2 2\theta_i]/M_{oi}^2,$$

for (100) plane YIG film[11,12]. $4\pi M_{oi}$, K_{1i} and K_{2i} are the saturation magnetization and the anisotropic constants in ith magnetic layer, respectively. The MSSW propagate along y direction.

COMPUTED RESULTS AND DISCUSSION

In following computation, the (100) plane YIG film is considered, H_o'=500 Oe, $4\pi M_{oi}$=1750 G,

W=3 mm, the number of width modes n=2, $\theta_i=0$, K_{1i}/M_{0i}=-44 Oe, and K_{2i}/M_{0i}=-2 Oe .

1. Effect of the number of magnetic layers.

Let the number of magnetic layers i_0=1, 2, 3, respectively, the dispersive curves are computed from Eq.(1) in order to investigate the effect of the number of magnetic layers on MSSW dispersion. Note that the number of solutions for Eq.(1) equals i_0, because of satisfing excitation condition for MSSW in the all magnetic layers. The method, that solutions are found out one by one from f_1 $(=\gamma\,[H_0'(H_0'+4\pi M_{0i})]^{1/2})$ to f_3 $(=\gamma\,(H_0'+4\pi M_{0i}))$, is adopted in order to avoid losing the solution. Computed results and used parameters are shown in Fig.2. The M, D_i, and T_i in Fig.2. denote the dispersive curves for monolayer film, and for the ith layer magnetic film in double and three layered structures, respectively. The difference of M from D_1 and T_1 curves results from interaction of MSSW in other layers to MSSW in the first layer. This interaction is the root cause controlling the dispersion by the multilayer film. For K>600 1/cm in Fig.2(b), μ, D_1 and T_1 curves coincide with each other. This means that when the space between two magnetic layers is larger than the wave length, t_2=50 μm> λ =1/600 ≃ 17 μm, the interactions of MSSW in the all layers vanish.

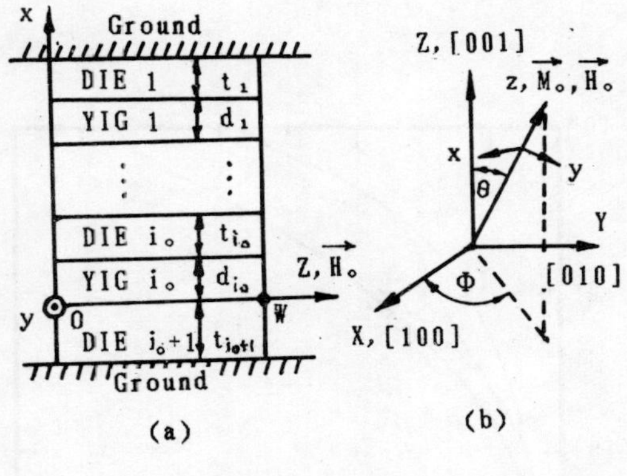

Fig.1. (a) The section schematic of layered structure.
(b) Dependence of two coordinate systems.

2. Effects of thickness of dielectric and magnetic layers.

Let i_0=3, the dispersion of MSSW in the first magnetic layer is only investigated and noted by the footnote 1 (shch as A_1). Varing the thickness t_i of dielectrics or d_i of magnetic layers, the dispersive curves are computed, respectively. The results are shown in Fig.3 and 4. {A} set of curves in Fig.3 is for condition t_{i_0+1}=0, and {B} set for condition t_{i_0+1}=∞. The condition t≥0.5 mm can be taken as t=∞ according to coincidence of A′ curve with B′. The effects of t_4, t_3, t_2 on the dispersion decrease gradually with increasing K, and the vanish one by one. This can be seen from that curves A_1 coincides with B_1, and A_1'' (B_1'', A_1''' and B_1'''), and A′ (B′, B) one by one. Coincidence of all curves with B for monolayer shows that interaction of MSSW in all layers vanish. Above montioned results demonstrate that the interaction of MSSW relates to the space between two magnetic layers and the wave length.
The curves for four sets of magnetic layer different thickness are shown in Fig.4. The results show that the effect of interaction of the MSSW in K≤400 1/cm region is equivalent to changed d_1 . For d_2, d_3 > d_1, it makes d_1 increase and for d_2, d_3 < d_1, it makes d_1 decrease. In sum, it makes d_1 change toward approaching d_2, d_3.
The effects of t_1 or t_{i_0+1}(t_4) on the dispersion are shown in Fig.5. The t_1 affects the dispersion in high K region, and the t_{i_0+1} the dispersion in low K region. The MSSW for t_1=0 called metallization mode have the largest exciting bandwidth ($\Delta f=f_3-f_1$). Comparing the curves (a) with (b) in Fig.2, it can be seen that t_{i_0+1} affects strongly the dispersion, and only affects the dispersion in the first magnetic layer.

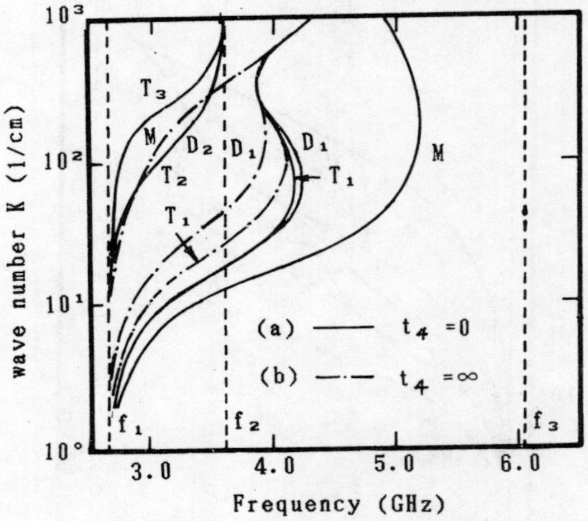

Fig.2 Dispersion curves for the different number of magnetic layers. $f_2=\gamma\,(H_0'+2\pi M_{0i})$. i_0=1: t_1=2 μm, d_1=5 μm. i_0=2: t_1=2 μm, d_1=5 μm, d_2=t_2=50 μm. i_0=3: t_1=2 μm, d_1=5 μm, d_2=t_2=d_3=t_3=50 μm.

3. Effects of saturation magnetization and magnetic anisotropy.

Let $\theta_1=\theta_3=0$ and only change θ_2, for two conditions of (a) $4\pi M_{0i}$=1750 G and (b) $4\pi M_{01}$=1750 G,

$4\pi M_{o2}$=1100 G , $4\pi M_{o3}$=600 G, the variation of the wave number versus frequency is computed, and its results are shown in Fig.6. It can be seen from Fig.6 that for same $4\pi M_{oi}$, the effect of YIG magnetic anisotropy on the dispersion in region of K< 300 1/cm is significant; for different $4\pi M_{oi}$, the effect of anisotropy is very slight. The anisotropy strongly affects the bandwidth exciting MSSW. The bandwidth for θ_2 =0 is the widest, and for θ_2 =45° the most narrow. But for different magnetizations, the anisotropy can not affect exciting bandwidth.

4. Effect of temperature

Let i_o=1, a monolayer (111) YIG film regarded as isotropic film[13] is considered. Assuming temperature coefficient of the magnetization to be -0.22 %/ C in the region of -20° ~40° C, the delay time curves for H_o=830 Oe at -20° , 10° ,40° C are computed from Eq.(1), respectively. Then the curves for H_o=863 Oe and 896 Oe at 40° C are computed, respectively. From the results shown in Fig.7, it can be seen that raising temperature, not only the curves move to lower frequency region, but also the delay time increases. This is the reason that a stable MSSW delay line for temperature can not be obtained only by the way of temperature compensation for magnetic field. Using electromagnet as bias field, a temperature test case, in which a MSSW delay line is packed, is put between the two poles. The dependences of delay time on frequency for different temperature or magnetic field are measured. The experimetal results are in good agreement with computation.

CONCLUSION

The interactions of MSSW in all magnetic layers which the space between two layers is larger than the wave length are the root cause controlling MSSW dispersion by multilayer YIG film. The magnetization and anisotropy can affect the exciting bandwidth. Temperature not only changes MSSW operating frequency, but changes the delay time.
Appropriately selecting the parameters, MSSW devices with good charateristic can be obtained.

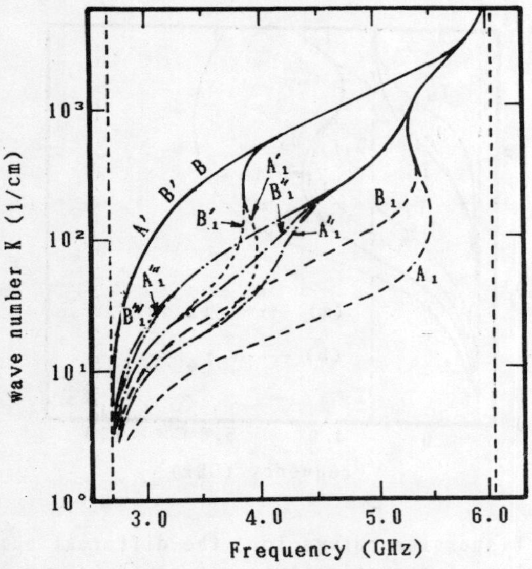

Fig.3. Dispersion curves in the structure with the different thickness of dielectric layers. d_1=5 μm, d_2=20 μm, d_3=50 μm, t_1=0.
Curves A', B': t_2=500 μm, t_3=5 μm
A_1''', B_1''': t_2=5 μm, t_3=500 μm
B : $d_2=d_3=t_2=t_3=0$
A_1'', B_1'': t_2=5 μm, t_3=100 μm
A_1', B_1': $t_2=t_3$=50 μm
A_1, B_1: $t_2=t_3$=5 μm

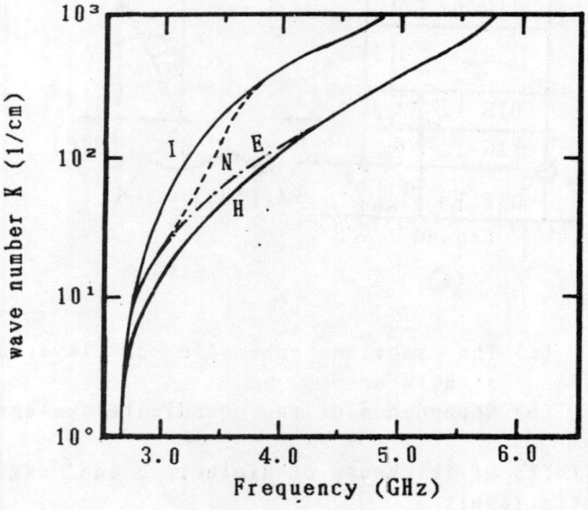

Fig.4. Dispersion curves in the structure with different thickness of magnetic layers. t_1=0, $t_2=t_3$=50 μm, $t_4=\infty$.
Curve H : $d_1=d_2=d_3$=15 μm
I : $d_1=d_2=d_3$=5 μm
N : d_1=5 μm, d_2=10 μm, d_3=15 μm
E : d_1=15 μm, d_2=10 μm, d_3=5 μm

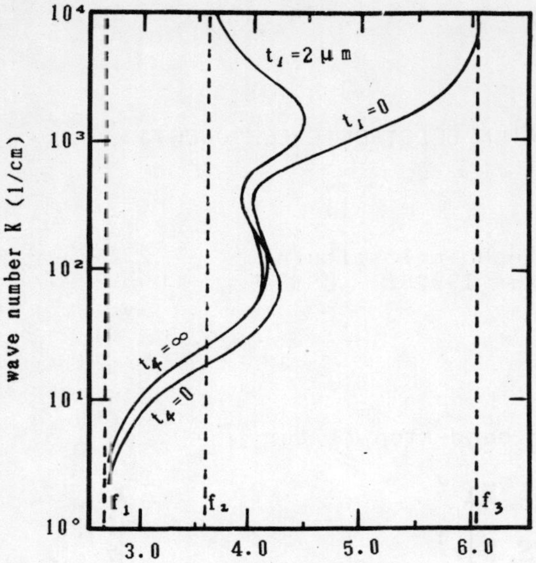

Fig.5. Dispersion curves of MSSW in the first magnetic layer for different t_i or $t_{i,+1}$. $t_2=t_3=50\,\mu m$, $d_1=5\,\mu m$, $d_2=d_3=50\,\mu m$.

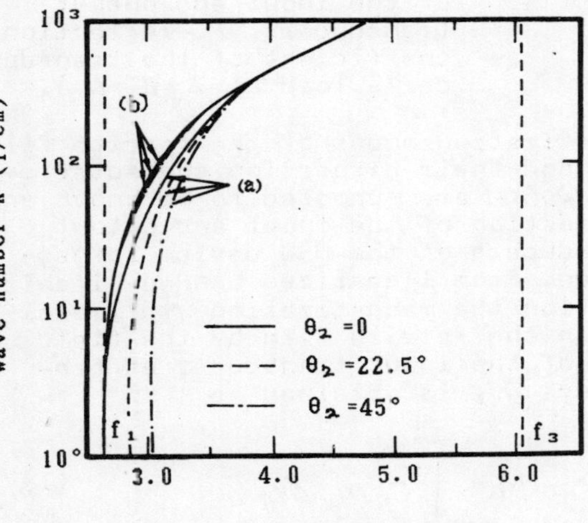

Fig.6. Effect of magnetic anisotropy and magnetization on dispersion for MSSW in the first magnetic layer.
curves (a): $4\pi M_{o1}=1750$ G
curves (b): $4\pi M_{o1}=1750$ G, $4\pi M_{o2}=1100$ G, $4\pi M_{o3}=600$ G.

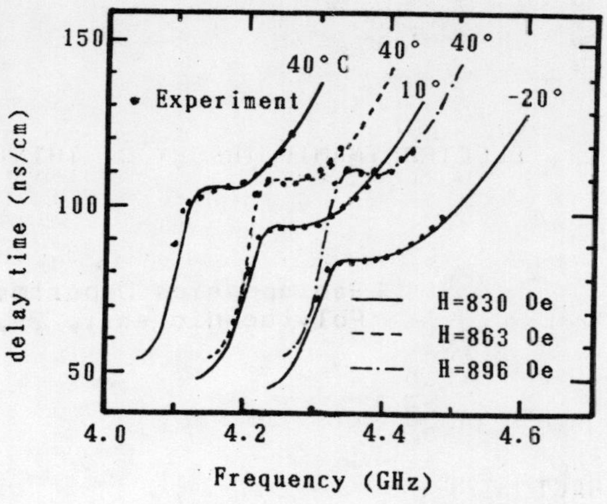

Fig.7. Effects of temperature and bias field on delay time and frequency of MSSW delay line. $d_1=15\,\mu m$, $t_1=300\,\mu m$, $t_2=\infty$.

REFERENCES

1. J.C. Sethares, J. Appl. Phys., 53(1982), 2646.
2. A.D. Fisher, E.S. Gaynor, and John N. Lee, 1983 IEEE Ultrasonics Symp., (1983), 226.
3. A.D. Fisher, John N. Lee, E.S. Gaynor, and A.B. Tveten, Appl. Phys. Lett., 41(1982), 779.
4. D. Young, and C.S. Tsai, Appl. Phys., Lett., 53 (1988), 1696.
5. W.L. Bongianni, J. Appl. Phys., 43(1972), 2541.
6. T.W. O'keeffe and R.W. Patterson, J. Appl. Phys., 49(1978), 4886.
7. N.S. Chang and Y. Matsuo, Proc. IEEE, 66 (1978), 1577.
8. Hu Wenzhong and Liu Farang, J. Magn. Magn. Mat., 35(1983), 153.
9. J.P. Parekh and K.W. Chang, IEEE. Trans. Magn. MAG-18(6), (1982), 1610.
10. L.R. Adkins, Circuits, Systems, and Signal Processing 4(1-2), (1985), 137.
11. C. Vittoria and N.D. Wilsey, J. Appl. Phys. 45 (1974), 414.
12. A.K. Ganguly and C. Vittoria, J. Appl. Phys. 45 (1974), 4665.
13. J.C. Sethares and T. Tsai, IEEE Trans. Magn. MAG-13(5), (1977), 1236.

ELECTRODYNAMIC THEORY OF THIN FERRITE FILM ELECTRONIC COMPONENTS

S. Zagriadski
Radiophysics Department, Polytechnic Institute
Polytechnic str., 29, Leningrad 195251 USSR

ABSTRACT

The present paper reports a theoretical study of microwave band-stop and band-pass filters (delay lines), using magnetostatic surface and volume waves in thin ferrite films.

INTRODUCTION

A new technology based on the propagation of magnetostatic waves (MSWs) in epitaxially grown thin ferrite films provides the physical basis for microwave planar electronic components and devices. MSW devices such as delay lines, filters and resonators are compatible with hybrid and monolithic microwave intergrated circuits and can provide signal processing at much higher frequencies than surface acoustic wave devices. Magnetic tuning of operating frequency is another advantage of MSW electronic components.

For excitation of MSWs in ferrite films different types of transducers are used: coplanar, microstrip and slotline (including multi-element gratings). Necessary dispersion characteristics are obtained by using the multiple ferrite films, or by varying the orientation of the bias magnetic field and the distance between the ferrite film and the ground plane.

In this paper the theory of excitation of MSWs in thin ferrite films by transducers of different types for arbitrary direction η of magnetizing field is discussed. The analysis is carried in electrodynamic and magnetostatic (dipole-exchange) approach.

INTEGRAL EQUATIONS

The geometry of the MSW filter (delay line) is shown in Fig. I. In the case of a band-stop filter $\Gamma = 0$.

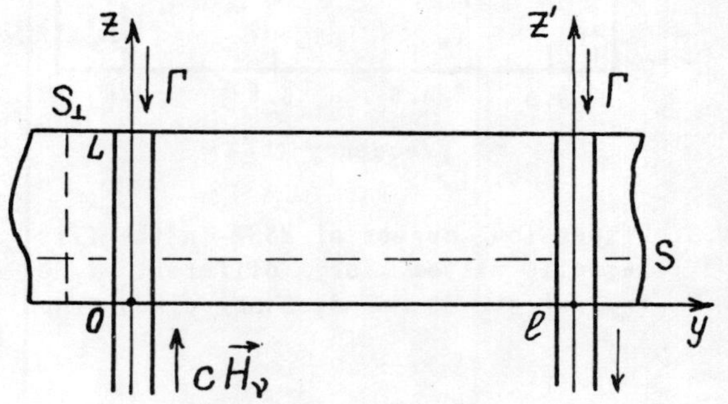

Fig. I Geometry of the MSW device (z, z'-longitudinal axes of the input and output transducers, Γ-reflection coefficient of the transducer's load at $z = z' = L$).

Magnetization modes of the ferrite film $\vec{m}_\mu$ and their dispersion characteristics $\omega_\mu(\kappa_\mu)$ are supposed to be known and interaction of the input and output transducers of the MSW device is neglected. From linearized Landau-Lifshits equation the magnetization $\vec{M}$, excited in the ferrite film by the field $\vec{h}$ of the input transducer at frequency ω, is obtained as

$$\vec{M} = \int_{-\infty}^{+\infty} c_\mu \vec{m}_\mu \, d\kappa_\mu , \qquad (I)$$

where

$$c_\mu = \varphi_\mu \int_V (\vec{h} \cdot \vec{m}_\mu^{*}) \, dV ,$$

$$\varphi_\mu = - \frac{i\omega_M(\omega + \omega_\mu)}{\Phi_\mu(\omega^2 - \omega_\mu^2 - 2iG_\mu\omega_\mu^2)} ,$$

$$\Phi_\mu = 2\pi \int\limits_{S_\perp} [\vec{m}_\mu^\circ \times \vec{m}_\mu^{\circ *}]_\eta \, dS_\perp ,$$

$$\vec{m}_\mu = \vec{m}_\mu^\circ (x,z) \cdot e^{-i\kappa_\mu y},$$

$$\omega_M = \gamma \mu_0 M_S ,$$

$S_\perp$, V - cross section and volume of the ferrite film, y - its longitudinal axis, G_μ - dissipation parameter of MSW. The magnetic field $\vec{h}$ is defined in dominant-mode approach by the field $c\vec{H}_\nu$ of the falling wave in the input line and the waves of the same mode ν, excited by the film

$$\vec{h} = (c + c_\nu)\vec{H}_\nu - \vec{M}_z + \left\{ c_{-\nu} + [c + c_\nu(L)]\Gamma e^{-2i\gamma L} \right\} \vec{H}_{-\nu} , \tag{2}$$

where

$$c_\nu(z) = -\frac{i\omega\mu_0}{N_\nu} \int_0^z dz \int\limits_S (\vec{M} \cdot \vec{H}_{-\nu}) \, dS ,$$

$$c_{-\nu}(z) = -\frac{i\omega\mu_0}{N_\nu} \int_z^L dz \int\limits_S (\vec{M} \cdot \vec{H}_\nu) \, dS ,$$

$$N_\nu = \int\limits_{S'} \left\{ [\vec{E}_{-\nu} \times \vec{H}_\nu] - [\vec{E}_\nu \times \vec{H}_{-\nu}] \right\}_z dS'$$

(the integration extends over the cross section of the input line S'), S, L - longitudinal section and width of the film, z - longitudinal axis of the transducer, Γ - reflection coefficient of the transducer's load at $z = L$. Here $\vec{E}_{\pm\nu}$, $\vec{H}_{\pm\nu}$, γ are modes and propagation constant for the structure in which the ferrite film is substituted by dielectric layer with permittivity ε the same of ferrite. The solution of the equations (I), (2) gives the amplitude coefficients for the waves in the input line $c_{-\nu}(0)$, $c_\nu(L)$ and for MSW c_μ. Similarly these equations (with z' instead of z) may be used for the cutput line of the device with the assumption

$$c = 0, \qquad \vec{M} = \vec{M}' + \int_{-\infty}^{+\infty} c_\mu \vec{m}_\mu \, d\kappa_\mu ,$$

where $\vec{M}'$ - the magnetization exciting the output transducer.

By solving the equations (I), (2) for both cases the characteristics of MSW filters and delay lines have been obtained.

The reflection coefficient of the magnetic field in the input line of the device at $z = 0$ is

$$\Gamma_1 = \{ c_{-\nu}(0) + \Gamma e^{-2i\gamma L} [c + c_\nu(L)] \} / c .$$

The transmission coefficient of the band-stop filter ($\Gamma = 0$) is obtained as

$$D = [c + c_\nu(L)]/c .$$

The transmission coefficient of the band-pass filter at $z' = 0$ is defined by the amplitude coefficients for the waves of λ-mode in the output line

$$K = [c_{-\lambda}(0) + c_\lambda(L)\Gamma e^{-2i\gamma L}]/c .$$

For instance, in the case of symmetrical filter ($\lambda \equiv \nu$) on surface MSWs and the assumption, that $\vec{m}_\mu^\circ$ is not a function of z and $\gamma L \ll 1$, this coefficient may be expressed as

$$K \approx \frac{i\pi\omega\omega_M\mu_0}{N_\nu v_\omega} \cdot \frac{|I_\nu(\kappa_\omega)|^2}{\Phi_\mu(\kappa_\omega)} \cdot e^{-(i\kappa_\omega + \kappa'')\ell} \cdot \frac{L^2(1+\Gamma)^2}{[1 + L^2(1+\Gamma)\,\mathrm{Re}\,F_\nu + iL^2(1+\Gamma)\,\mathrm{Im}\,F_\nu]^2} \tag{3}$$

where

$$F_\nu = \frac{i\omega\mu_0}{N_\nu} \int_{-\infty}^{+\infty} \varphi_\mu |I_\nu(\kappa_\mu)|^2 d\kappa_\mu ,$$

$$I_\nu(\kappa_\mu) = \int\limits_S (\vec{H}_{\nu 0} \cdot \vec{m}_\mu^*) \, dS ,$$

$\vec{H}_{\nu 0}$ - the magnetic field in the cross section of the line, κ'' - the dissipation parameter of MSW, v_ω - the group velocity of MSW. Index ω in (3) indicates that values are taken at $\omega_\mu = \omega$. Here we may draw the following conclusions.

A. The maximum efficiency of MSWs excitation is achieved by a shorted transducer ($\Gamma = I$).

B. In a band-pass filter there exists an optimum coupling of the transducers to the ferrite film, which provides the minimum loss in the filter.

C. In a band-stop filter the attenuation increases as the coupling is increased.

OPTIMUM DESIGN AND EXPERIMENTAL INVESTIGATION OF MAGNETOSTATIC SURFACE WAVE STEPPED GROUNDPLANE DELAY LINE

Huijin Xiang, Xiuchen Zhang, Huahui He.
(Huazhong University of Science and Technology, Wuhan, China)
Yaoxi Zhang
(State-Run Beijing Third Radio Appliances Factory, China)

ABSTRACT

By a stepped ground plane the inherent non-linear delay versus frequency characteristics of magnetostatic surface wave (MSSW) delay lines were controlled. The ground plane spacing and the length of each section were optimized so that the total group delay, i.e., the summation of the delay in each section, showed a minimum delay error from a ideal delay versus frequency characteristisc. Using this technique a linearly-chirped dispersive delay line with a 400 MHz bandwidth centered around 4 GHz and a non-dispersive. delay line with 200 MHz bandwidth around 4 GHz have been demonstrated.

INTRODUCTION

A magnetostatic wave (MSW) delay line constructed by a dielectric layered structure (i. e., a 3-layer propagation structure consisting of YIG-dielectric-ground plane) is called a simple delay line whose delay vs. frequency characteristic are dependent upon the saturation magnetization, bias fielb, YIG film thickness, and ground plane spacing. [1] Generally speaking delay vs. frequency variations of MSW delay lines are non-linear, which is the very major impediment of the MSW devices to meet with that most device applications require eather nondispersive (delay vs. frequency slope=0) or linearlydispersive(delay vs. frequency slope =constant) specifications. [2] Various approches to modify the intrinsic MSW dispersion have been investigated for a number of years. These studies have shown that a controlled variation of the ground plane spacing is most effective in the synthesis of linear delay with frequency relations. [3] Kok Wai Chang et al. [4] realized a laboratory version of linearly delay line operating in the MSSW mode. the method they used required a device geometry in which the ground plane spacing continuously varied as a function of distance in the direction of propagation. The resultant delay device showed a linear delay of 500 MHz bandwidth (between 2. 62 to 3.12 GHz) with a slope of 15.3 ns/100 MHz and a phase error from quadratic phase versus frequency characteristic of less than 13 degrees.
In this paper instead of continuously variable curve ground plane we employed a stepped ground plane[5] in order to facilitate fabrication. In effect this approach cascades a number of simple delay lines together to yield a net dispersion which can be properly adjusted. Our object was to obtain nondispersive or linearly dispersive MSSW delay line using this technique.

COMPUTER OPTIMIZING OF THE GROUNDPLANE SPACING AND THE LENGTH OF EACH SECTION

It was assumed that the MSSW delay line comprised several section(n) with different ground-plane spacing (t_n) and length (l_n) as shown in Fig.1.
Note that in our delay lines the input and output transducers were so arranged that the distance between them (that is, the summation of the length

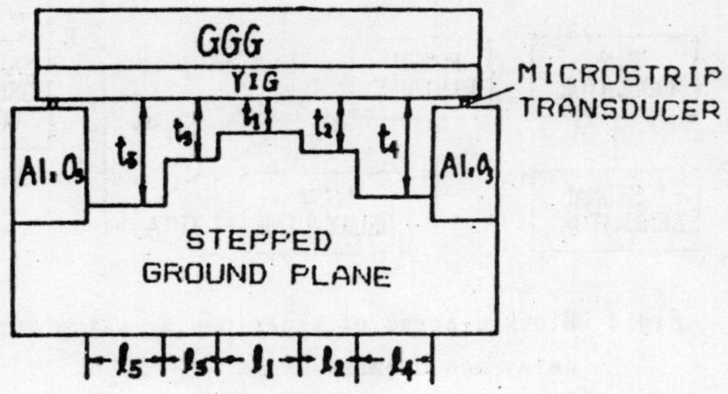

Fig.1 Schematic of stepped ground plane MSSW delay line

of n sections) was equated to a certain value(10mm). This means that of all n lengths only n-1 lengths were linearly independent. Therefore we designated $t1, \cdots, t_n, l_1, \cdots, l_{n-1}$ as the design variables, and in conventional practice devoted them by a vector x, that is,

$$x=[t_1,\cdots,t_n,\ l_1,\cdots,l_{n-1}]^T$$

In our programming problem T(X, f) was used to represent the total time delay, namely, the summation of the delay in each section, where f is frequency of the transmitted microwave signal. Moreover, the ideal delay characteristic was written symbolically as T(f), then

$$T(f)=C$$

for a non-dispersive delay line, and

$$T(f)=A+B\ f$$

for a linearly dispersive delay line, where A, B, and C are all constants independent of f. Let

$$(X)=(\int_{fL}^{fU}[T(X,\ f)-T(f)]^2df)/(f_U-f_L),$$

where the integrand indicates the square-error between the actual and the desired delay variations under a certain vector x, and thus DT(X) is the mean value of the square-error over frequency region $[f_L, f_U]$. DT(X) was referred to as the objective function to evaluate the actual delay characteristic. Some constraints were imposed on the design variables when minimizing DT(X). The reasons for doing it are the following. Ground current losses increase as the spacing decreases so the spacing of each section cannot be too small. Meanwhile, the spacing cannot be too large or else the ground plane will not useful for controlling of inherent dispersion. Moreover, for the sake of convenience machining the length of each section cannot be too short. Solving the above programming problem by SUMT penalty function method and Powell's method[6] we obtained optimal solutions as shown in Tables 1 and 2. Table 1 is for a linearly dispersive delay line with six sections and a YIG film of 25μm, and Table 2 for a nondispersive delay line sith three sections and a YIG film of 20μm. A separate program was used to calculated the delay vs. frequency variation of the two delay lines, calculated results are shown in Figures 2 and 3, respectively.

EXPERIMENTAL INVESTIGATION

The delay characteristics of the two devices were measured directly using the pulse delay technique[1] which allows the delay time of a transmitted microwave pulse to be measured.

Table 1. Ground plane spacings and lengths for a linearly dispersive delay line with six sections.

Section No., n	Ground plane spacing $t_n(\mu m)$	Section length $l_n(mm)$
1	75	4.19
2	119	0.84
3	151	0.97
4	195	1.23
5	251	1.20
6	369	1.57

Table 2. Ground plane spacings and lengths for a non-dispersive delay line with three sections.

Section No., n	Ground pland spacing tn(μm)	Section length $l_n(mm)$
1	30	5.56
2	171	1.13
3	227	3.31

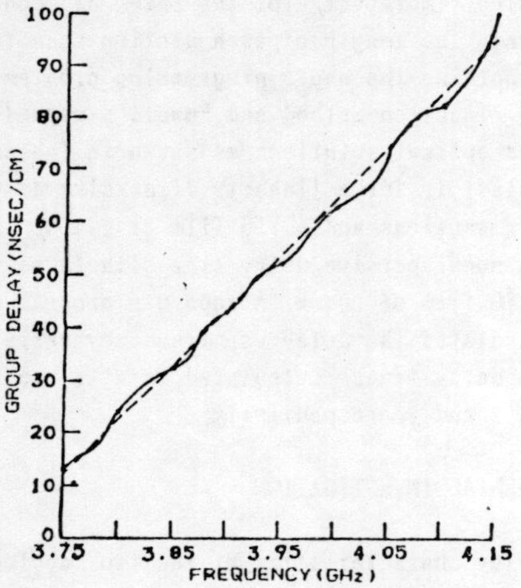

Fig. 2 Calculated result of group delay with frequency corresponding to Table 1.

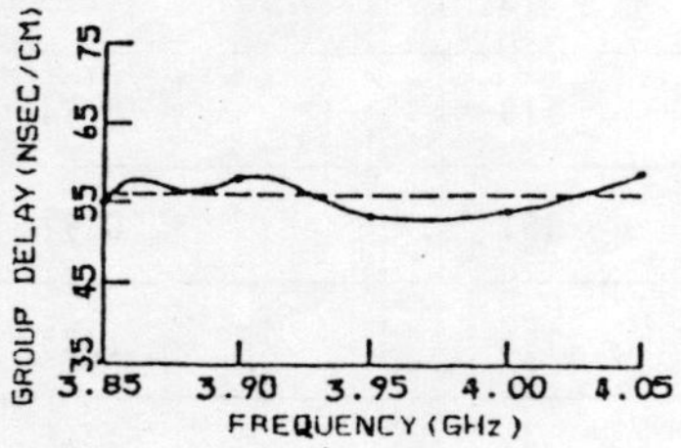

Fig. 3 Calculated result of group delay with frequency corresponding to Table 2.

Fig. 4 shows the experimental setup for pulse delay measurement. A cw microwave signal is modulated by a pulse signal to provide an rf pulse of proper width. This rf pulse is separated by a power divider into two branches, one serves as the input signal which is delayed in MSSW delay line whose output is then amplified and displayed on a sampling oscilloscope, the other is sent directly to the sampling oscilloscope as a comparison signal. Fixing the magnetic bias field H and varying the microwave signal frequency f, one can observe group delay times under different frequencies.

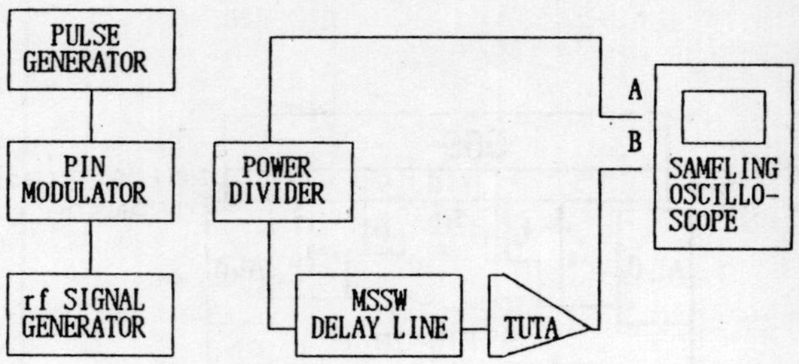

Fig. 4 Block diagram of experimental setup for pulse delay measurement.

Measurements of group delay as a function of frequency for both the linearly dispersive delay line and the non-dispersive delay line are shown in Figures 5 and 6 respectively. During measurement MSSW mode was controlled to propagate along +y direction and the frequency aperture was chosen to be 50 MHz. For each measurement point the input signal level was adjusted in order to show more clearly the delayed pulse. It was found that there was little pulse spreading and that EM direct leakage between input and output lines was negligible in comparison with the delayed pulse even for the shorter delay times. Comparing Figures 5 and 6 with Figures 2 and 3 one can see that the agreement with theory is within experimental accuracy for both delay lines.

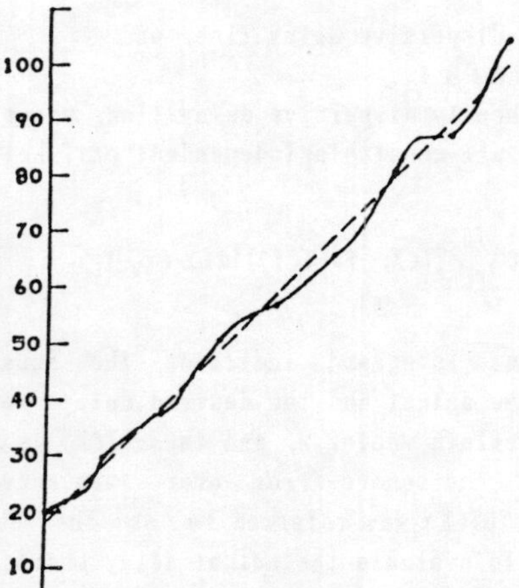

Fig. 5 Measured result of group delay with frepuency corresponding to Table 1.

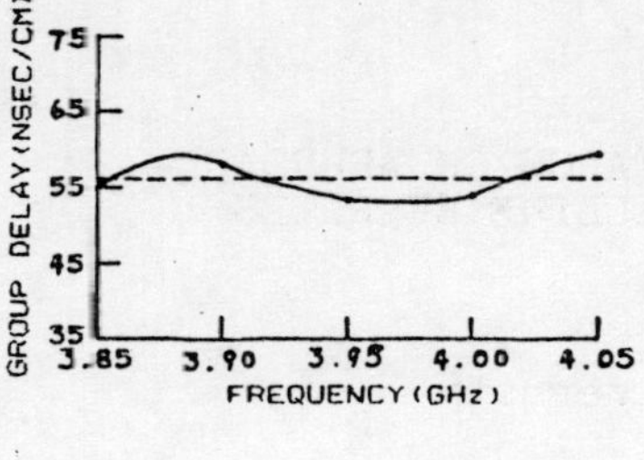

Fig.6 Measured result of group delay with frequency corresponding to Table 2.

REFERENCES

[1]. W.L.Bongianni, "Magnetostatic Propagation in a Dielectric Layered Structured," J.Appl. Phys., Vol.43, No 6, June 1972, P2541

[2]. L. R. Adkins, " Dispersion Control in Magnetostatic Wave Delay Line," Circuits Systems and Signal Processing, Vol. 4, No.1-2, 1985, P137

[3] . J.D.Adam, et al., " MSW Variable Time-Delay Technique, "RADC-TR-83-139, Final Techniqal Report, Sept., 1983

[4] . Kok Wai Chang, et al., " Linearly Dispersive Time-Delay Control of Magnetostatic Surface Wave by Variable Ground-Plane Spacing," Electronics Letters, 7th, July, 1983, Vol.19, No.14, P546

[5] . J.D.Adam, "Magnetostatic Volume Delay Lines with Stepped Ground Planes," 1984 IEEE MTT-S Digest, P87

[6] . D. M. Himmelblau, " Applied Non-Linear Programming," McGraw-Hill Book Company, 1972

COMPUTATIONAL ANALYSIS AND EXPERIMENTAL RESEARCH ON APPROACH-SATURATION CURVE FOR MICROWAVE POLYCRYSTALLINE FERRITES

Sheng-chuan Zhu and Fan-ping Wen
Department of Physics, Peking University
Beijing 100871, China

Yao-xi Zhang and Dao-zhi Zhang
Beijing 3rd Radio Appliances Factory
Beijing 100015, China

ABSTRACT

Computational analysis and experimental research on approach-saturation curve have been done for a series of microwave polycrystalline ferrites with various properties. The five coefficients describing approach-saturation curve can be determined simultaneously by computational method. The coefficients vary with different H interval and a suitable H interval can be obtained. A method for determining magnetic parameters is suggusted. The experimental results and their interepretations are given.

INTRODUCTION

The measurements of saturation magnetization and magnetocrystalline anisotropy for polycrystalline ferromagnetic materials are carried out usually on the basis of the approach to saturation.[1,2] We have already suggusted a method of determining simultaneously the five coefficents of the approach to saturation expression, and have found that the coefficients relate to the magnetic field region and the experimental point distribution.[3] In order to improve the measured precision of experiment and gain a clear idea of the variouson regularity of the coefficients, we have done a lot of experiment on microwave polycrystalline ferrites with various properties, and have performed the curve fitting and synthetical analysis.

THEORETICAL AND EXPERIMENTAL CONSIDERATION

In order to avoid the direct influence of porosity, we have measured the approach-saturation curves of the gram magnetic moment for polycrystalline ferrites, and considered that these curves can be expressed generally as follows,

$$\sigma = \sigma_s(1 - aH^{-1} - bH^{-2} - cH^{-3} - \ldots) + \chi_1 H \qquad (1)$$

where σ is the sample gram magnetic moment, σ_s is the saturation gram magnetic moment, H is the internal magnetic field strength, σ_s , a, b, c and are the coefficients determined by experiment. The five coefficients can be obtained simultaneously in different intervals of H by using the curve fitting method.

If we let $A = \chi_1$, $B = \sigma_s$, $C = -a\sigma_s$, $D = -b\sigma_s$, $E = -c\sigma_s$, Eq.(1) becomes

$$\sigma = AH + B + CH^{-1} + DH^{-2} + EH^{-3} \qquad (2)$$

According to the principle of least squares, the cofficients A , B , C, D, E should cause

$$\Delta = \sum_{j=1}^{n}(\sigma_j - AH_j - B - CH_j^{-1} - DH_j^{-2} - EH_j^{-3})^2$$

to have a minimum value, i.e., these coefficients can be determined by the following equations,

$$\sum_{j=1}^{n}H_j^2 A + \sum_{j=1}^{n}H_j B + nC + \sum_{j=1}^{n}H_j^{-1} D + \sum_{j=1}^{n}H_j^{-2} E = \sum_{j=1}^{n}\sigma_j H$$

$$\sum_{j=1}^{n}H_j A + nB + \sum_{j=1}^{n}H_j^{-1} C + \sum_{j=1}^{n}H_j^{-2} D + \sum_{j=1}^{n}H_j^{-3} E = \sum_{j=1}^{n}\sigma_j$$

$$nA + \sum_{j=1}^{n}H_j^{-1} B + \sum_{j=1}^{n}H_j^{-2} C + \sum_{j=1}^{n}H_j^{-3} D + \sum_{j=1}^{n}H_j^{-4} E = \sum_{j=1}^{n}\sigma_j H^{-1}$$

$$\sum_{j=1}^{n}H_j^{-1} A + \sum_{j=1}^{n}H_j^{-2} B + \sum_{j=1}^{n}H_j^{-3} C + \sum_{j=1}^{n}H_j^{-4} D + \sum_{j=1}^{n}H_j^{-5} E = \sum_{j=1}^{n}\sigma_j H^{-2}$$

$$\sum_{j=1}^{n}H_j^{-2} A + \sum_{j=1}^{n}H_j^{-3} B + \sum_{j=1}^{n}H_j^{-4} C + \sum_{j=1}^{n}H_j^{-5} D + \sum_{j=1}^{n}H_j^{-6} E = \sum_{j=1}^{n}\sigma_j H^{-3}$$

where $\sum_{j=1}^{n}$ is the sum to n terms and n is the number of experimental points. When $n > 5$, the equations have only one solution. Therefore, the values of σ_s, a, b, c, χ_1 obtained by this method are independent each other. The five coefficients in the approach-saturation expressisn can be determined simultaneously by a computational method.

Experimental research on polycrystalline ferrites with various chemical composition, saturation magnetization, magnetocrystalline anisotropy and porosity have been done. Gram magnetic moment measurements were carried out using a vibrating sample magnetometer with a precision 5×10^{-5} emu. The applied field H = 0 - 10000 Oe. The samples studied were spheres of fine spherecity. The effect of demagnetic field had been considered. All data was taken at room temperature.

EXPERIMENTAL RESULTS AND ANALYSIS

The approach to saturation σ - H curves for a series of polycrystalle ferrite samples were measured. There were about 100 experimental points on one curve. By curve fitting, the five coefficients can be obtained simultaneously in diffrent intervals and point-distribution of H.

1. Determination of the suitable H interval

In order to study the relation between the coefficient value and the magnetic field region, we have performed curve fitting over several regions of the approach-saturation curve for YIG sample. The results are shown in Table 1. Some conclusions can be obtained from Table 1:

A. The curve fitting can be performed successfully over the different regions for H, and the root-mean-square errors between the experimental and calculating curves for σ are small.

B. The coefficients vary with the H intervals. It can't be distinguished which coefficient plays an important role in which interval. Therefore, it is a mistaken method to obtained the coefficient A, B in a higer H interval and then the coefficient C, D, E in other lower H interval.

C. By choosing the suitable H-interval, the reasonable results can be obtained. For example, there are a>0, b>0, c<0 for the ferrites with $K_1<0$ (see Tab. 1, interval 1 and 2). It can be found that the optimum interval for H corresponds to the region σ/σ_m =0.94-0.99 (for H > $4\pi M_s$) and σ/σ_m =0.97-0.99 (for H < $4\pi M_s$), where σ_m is the maximim σ (see Fig. 1).

Table 1. The experimental results for YIG sample

Interval No.	1	2	3	4
H (Oe)	40-500	25-10000	500-10000	25-5000
σ_s (emu/g)	26.928	26.964	27.203	27.043
a	1.253	0.1875	17.781	4.010
b	15.56	28.58	-1.329×10^{4}	-3.370×10^{2}
c	-1399	-4.624	4.102×10^{6}	6.885×10^{3}
χ_1	1.355×10^{-3}	1.770×10^{-4}	4.526×10^{-5}	1.587×10^{-4}
rms er. (for σ)	0.016	0.047	0.015	0.015

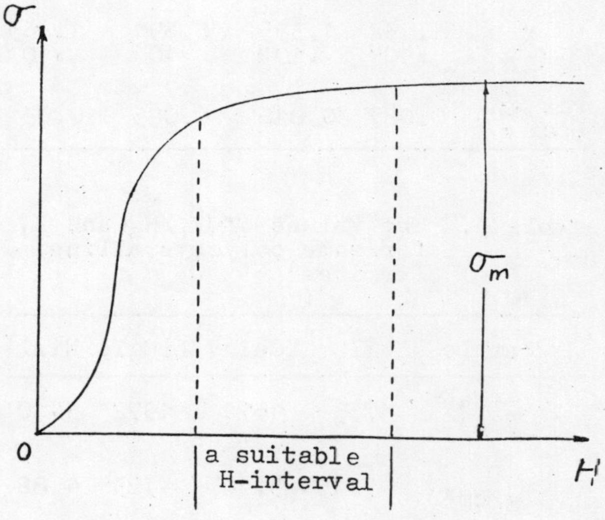

Fig. 1 A suitable H-interval in approach-saturation curve

D. The value of b relates to the interval for H. The ratio for two b values (interval 1 and 2) is about 0.5. It is in agreement with the results of Dionne.[1]

2. Determination of $|K_1/M_s|$ and K_1

The approach-saturation curves for YIG, YCaInV, LiMgTi, NiAl ferrites were measured. In a suitable interval (H = 50-500 Oe), the results are shown in Tab. 2.

A. The magnitude of H_a (the magneto-crystalline anisotropic field) can be seen from the values of b and c. The H_a value of YCaInV is the smallest, and that of NiAl ferrite is the biggest.

B. The coefficients for b and c can be expressed by $b = b_0(K_1/M_s)^2$, $c = c_0 \cdot (K_1/M_s)^3$. Since the value of $|K_1/M_s|$

Table 2. The experimental results for polycrystalline ferrites (H = 40 - 500 Oe)

sample	YCaInV	YIG	LiMgTi	NiAl
σ_s (emu/g)	24.273	26.928	37.093	23.292
a	0.6608	1.253	5.470	6.384
b	9.072	15.56	21.13	31.86
c	-824.8	-1399	-2817	-3255
γ_1	1.132×10^{-3}	1.335×10^{-3}	6.591×10^{-3}	1.263×10^{-2}
rms er. (for σ)	0.027	0.016	0.025	0.031

Table 3. The values of $|K_1/M_s|$ and K_1 for some polycrystalline ferrites

Sample	YIG	YCaInV	LiMgTi	NiAl
$4\pi M_s$ (Gs)	1753	1571	1972	1410
D (g/cm^3)	5.15	5.16	4.23	4.82
p	0.8%	0.5%	3%	6%
$|K_1/M_s|$ (Oe)				
from b		32	50	61
from c	(42.5)	36	53	56
average		34	51.5	58.5
K_1 (10^3 erg/cm^3)	-5.9	-4.3	-8.1	-6.6

for YIG single crystall equals to 42.5 Oe, so we have $b_0 = 8.615 \times 10^{-2}$ and $c_0 = 1.822 \times 10^{-2}$. By using these values of b_0 and c_0, the (K_1/M_s) and K_1 values for YCaInV, LiMgTi and NiAl ferrites can be obtained (see Tab. 3).

3. The determination steps for magnetic parameters

In order to eliminate the affection by the differences of H interval, fitting method and coefficient expression, and to improve the precision for determining magnetic parameters. We suggest the following steps:

A. The coefficients must be determined simultaneously by computational method.

B. The curve fitting must be performed in a suitable interval for H.

C. By using a standard specimen with given H_a, the unknown parameters in the coefficient expression can be obtained.

D. The values of H_a for studied samples can be determined, where $H_a = |K_1/M_s|$.

4. The Value of Coefficient a is Dependent on $p\sigma_s$

The porosity values (p) for YCaInV, YIG, LiMgTi and NiAl ferrite samples were about 0.5%, 0.8%, 3% and 6% espectively. It can be seen from Tab. 2 that the values of coefficient a increase with the increases of p . It is accordance with the theoretical result of Schloemann.[4]

CONCLUSION

The coefficients of approach-saturation expression can be determined simultaneously by computational method.

Because the values of coefficients depend on the H-interval, and the magnetic parameters relate to the expressions for b, c, only after the suitable H-interval, a standard specimen and the concrete expression for b and c have been obtained, can the values for H_a, K_1 be determined from the approach-saturation curves.

REFERENCES

1. D. F. Dionne, J. Appl. Phys., 40 (1969), 1839.
2. M. A. Atel'mashenko, Sov. Phys. soild state, 9(1967), 1137.
3. Yao-xi Zhang, Sheng-chuan Zhu and Fan-ping Wen, IEEE Trans. Magn., MAG-21(1985), 1803.
4. E. Schloemann, J. Appl. Phys., 38 (1967).

X-BAND WAVEGUIDE JUNCTION LATCHING FERRITE SWITCH

Shi Gu shan
Beijing Institute of Radio Measurement
P.O. Box 3923 Beijing, China

ABSTRACT

Waveguide Junction Latching Ferrite Switch is a fast switching device. It has the structure of internal magnetic circuit. It operates in the remanent magnetism state. When suitable current pulse (positive or negative) is fed through the three holes of the latching ferrite, its switching time is only several μs.

The switch is widely used in radar, in communication, remote sensing and space technology. It has such advantages as wide frequency, low loss, high isolation and light weight.

Its properties are as the following:

peak power	15 kw
average power	15 w
frequency width	10 %
reverse isolation	25 dB
loss	0.3 dB
VSWR	1.2
switching time	2 μs
switching energy	200 μJ
working temperature	-55—+85 ℃

The present paper discusses the following points:

1. working principle of the switch
2. how to widen the ranges of its frequency and working temperature
3. the driving form of the device

Further analyses show that the choice of proper working current and the adoption of magnetic flux feed back technology will greatly improve the properties of the switch.

WORKING PRINCIPLE OF THE DEVICE

Fig.1 shows the structure of the device. Excitation wire goes through the holes inside the ferrite. Strong pulse current flows along the wire, forming transversal magnetic field, and this makes an instant T type circulator.

Since the excitation wire has very few windings, the direction of the applied pulse current can be changed rapidly. So, the circulating direction of this instant T type circulator can be changed as rapidly.

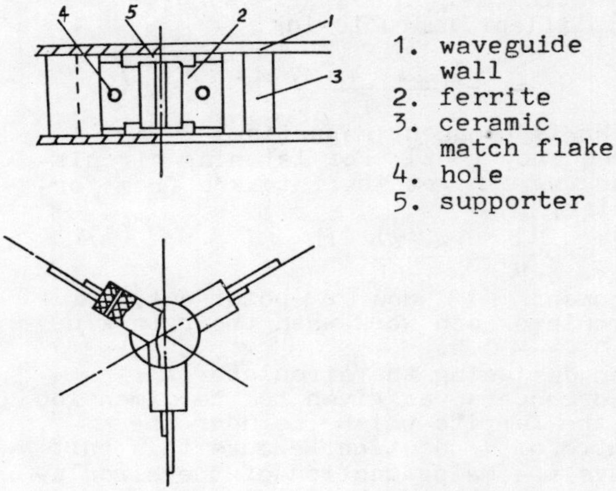

Fig.1 Structure of the device

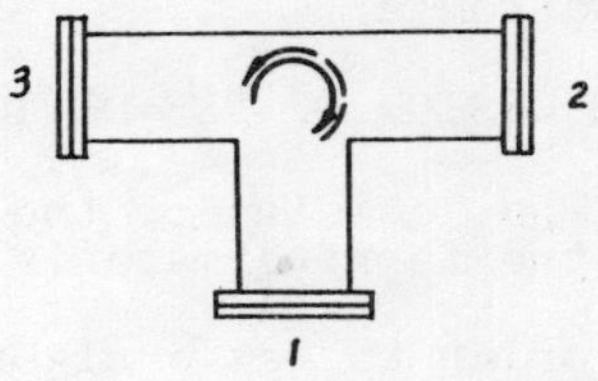

Fig.2

As Fig.2 shows, the microwave signal which is put in through port 2 will be put out from port 1. If the direction of the applied pulse current changes, the circulating direction will change accordingly, the microwave signal which is put in through port 2 will be put out from port 3 as a corresponding change. In this sence, this device has the function that can rapidly change the microwave passage.

If we connect a matched load at port 1 two states "on" and "off" will be formed between port2 and port 3. Hence, a rapid switch of microwave passage is formed. The similar theory of Latching

Circulator has been published before. One of the main results of the theory is that the function relationship between ferrite normalized radius and rate k/μ is deduced.
Here, k and μ refer to the components of Polder magnetoconductivity tensor. This normalized radius is decided by $K \cdot r$. Here "r" is the actual radius, component "K" is decided by formula:

$$K = \frac{2\pi}{\lambda_0}\sqrt{\varepsilon \mu_e} \qquad (1)$$

λ_0 refers to the free space wave length
ε refers to ferrite dielectric constant
μ_e refers to effective magnetoconductivity. When the device is working in a remanent magnetism state, the inner magnatic field H is zero, so μ_e can be simplified [3] as following:

$$\mu_e = 1 - \left[\frac{2.8 \times 4\pi M_r}{F}\right]^2 \qquad (2)$$

$4\pi M_r$ is remanent magnetism; F is frequency (GHz). For latching circulator, we can get the rate k/μ from formular:

$$\frac{k}{\mu} = \frac{2.8 \times 4\pi M_r}{F} \qquad (3)$$

Documents [1] and [2] point out, circulator can work when the rate k/μ is 0.2—0.5.
When designing the circulator, much concern was given to the dimension of the ferrite which is near the centre of Y junction. Because this part plays the main function of the circulator. The rate of the hole's inner area to the hole's outer area makes the key working point. This enables it to work in wide temperature range and wide band range.

THE DESIGN OF THE DEVICE

1. The desision of the properties of the ferrite material

In order to get a higher specification and to improve the utilization efficiency, we must take high saturant magnetization and high rate of remanent magnetism. Take the effect of natural resonance into consideration, the saturant magnetization $4\pi M_s$ must be limited to some extent. So, in the 3cm band, we think it proper to take 0.43 as the rate value k/u ($2.8 \times 4\pi M_r/F$). Meanwhile, to stand a certain average power, the ferrite must have a high Curie temperature. Since the device works in a remanent magnetism state, the ferrite must have high rectangle rate and a proper coercive force. On one hand, the switch must be easy to drive, on the other hand, the switch must have the ability to resist damage and have working stability. Hence, the the coercive force can neither be too large nor too small. We think it proper to take a compromized value 2—3 Oersted
The high power microwave Li-Mg-Cu ferrite [4] which we made meets the above needs perfectly.

2. The decision of the rate of S_{in} to S_{out}

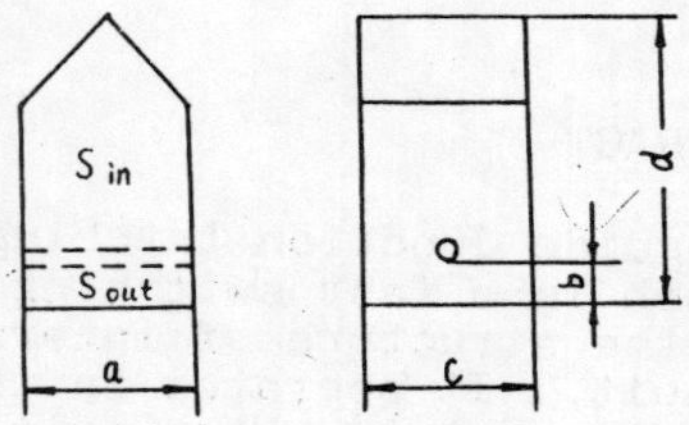

Fig.3
Cross section of ferrite

Fig.3 shows the section the the ferrite. In the published documents, S_{in} equals S_{out}. We found, to improve the property of the device, the areas of the two sections should be unequal.

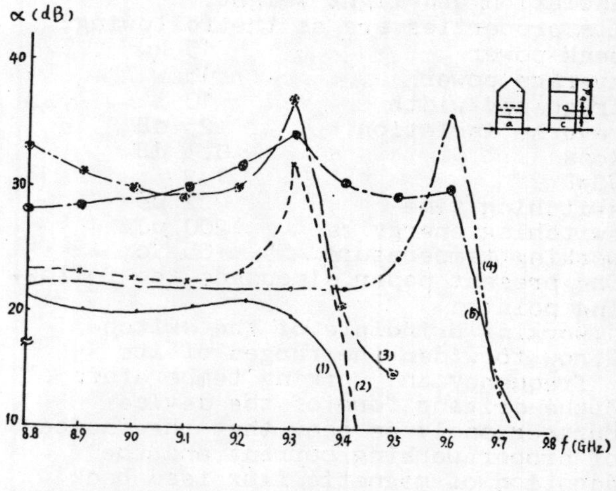

Fig.4 Effect of S_{in}/S_{out} on property

	(1)	(2)	(3)	(4)	(5)
S_{in}/S_{out}	1.28	1.326	1.388	1.41	1.48

Fig.4 shows the effect on the device due to the different rates of S_{in} to S_{out} As the fig. shows, when $S_{in}/S_{out}=1.48$ in 9% frequency width the isolations can reach about 30 dB. Further experiment proved that when $S_{in}/S_{out}=1.54$, in 10% frequency width above, the isolations are over 30 dB.

THE EFFECT OF MATCHING ON THE PROPERTIES OF THE DEVICE

Fig.5 shows the effect of the device's

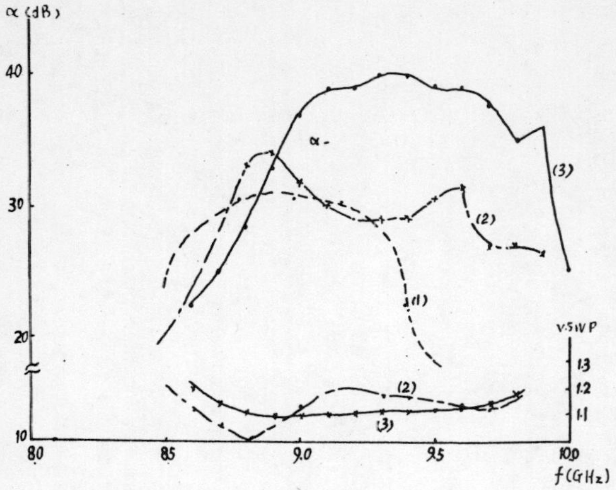

Fig.5 Effect of matching

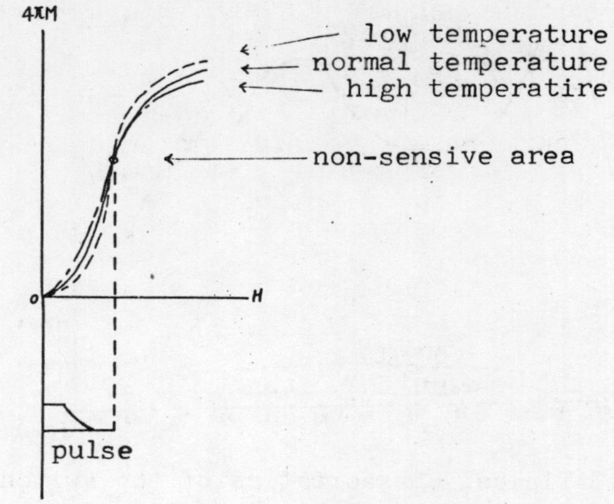

Fig.6 Magnetization curve of ferrite along with temperature

matching states on its properties. Compare curve (2) with curve(1), the dimension of the ferrite reduces by only 0.05mm, but the 30 dB belt width enlarges from 5% to 9.3%. The middle frequency moves 200 GHz towards higher frequency. This choice is just like the choice of ferrite radius r in the circulator.
The choice of proper ferrite dimension can meet the need of the best working condition.
Compare curve (2) with curve(3), curve (3) applies the technic of 2-step match Having improved the matching condition, the properties of the device are improved greatly.
Curve (3) shows that the 30 dB isolation belt width has reached 11.8%, among them, 6.4% belt width almost reaches 40 dB.

CHOICE OF WORKING POINT AND TEMPERATURE CHARACTERISTICS OF THE DEVICE

The ferrite of the device is made up of3 pieces of ferrite which are stuck together and shown in Fig.3. As the fig. shows,by changing the dimension of b, we can control the magnetic resistance in the magnetic circuit, hence to affect the applied field, that is to say to affect the applied controlling pluse current. In this way, we regulate the working point of the devive making the device work within a range where the device is not sensitive to the temperature. In other words, the ferrite is magnetized by applied field and tend to be saturated but just has not been saturated. See Eig 6.
If the choice of the working point is not exact, we use the temperature compensitation of the driver itself to regulate the temperature charicteristic of the device so that it can meet the need that the machine required to the device. Regulated this way, the characteristics of the device can be shown by Gig.7

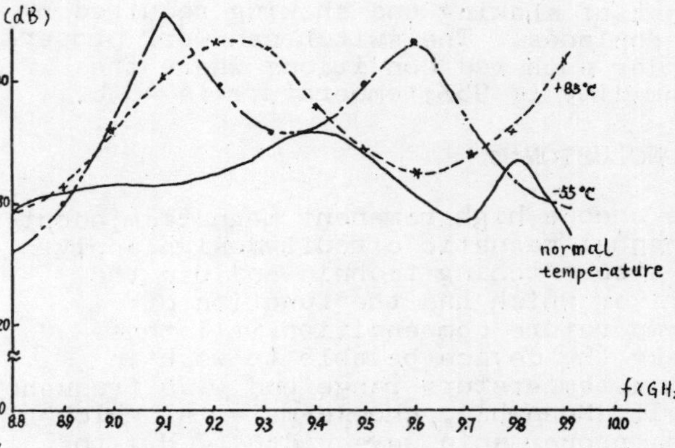

Fig.7 Temperature characteristics

You can see, the device, compensited by driver,the isolation (30 dB belt with; from -55 °c to +85 °c) almost does not reduce.

CHARACTERISTICS OF THE DEVICE

After the improvement, the distributed component in the driver have been replaced by flimsy circuit. So the size of the device has been reduced greatly. The size of the improved device is 43 x 43 x 44.8 (mm^3)
The tipical characteristics of the 3 cm latching ferrite switch is shown by Fig.8.

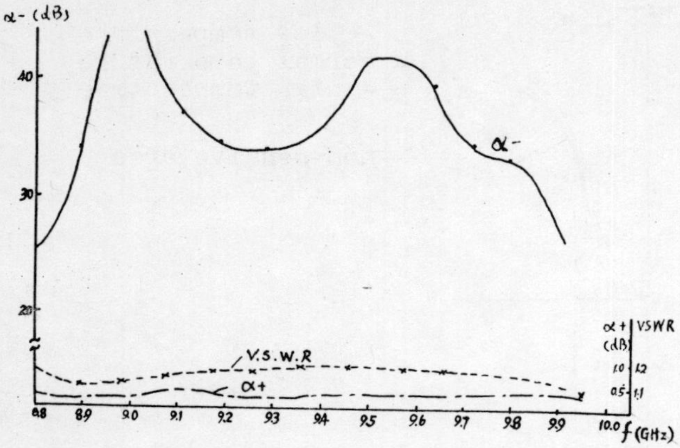

Fig.8 Tipical characristics of the switch

Within 10% belt width, all the isolations are over 30 dB; the insertion loss is not higher than 0.3 dB; the VSWR is lower than 1.20. When the peak power reaches 15 kw, the average power is 15 w, the insertion loss is has not obvious change The working temperature of the switch is -55 ℃—— +85 ℃. Switbhing time is less than 2 us. Switching energy is less than 200 uJ. This switch has taken the test of shaking and shcking required by aeroplanes. The switch can work properly under such bad conditions where the humidity is 95%; temperature is 40 ℃.

CONCLUSION

We choose high remanent magnetism; adopt unequal magnetic circuit design; apply 2-step matching technic and use the driver which has the function of temperature compensition. All these make the device be able to work in wide temperature range and wide frequency belt. Meanwhil, choose material which has proper spin wave width (ΔH_k) to make the device have a certain power capacity and also keep the rapidity of the device.

REFERENCES

1 Proc. IEEE (letters) Vol.55 P.407——408 1967
2 IEEE Trans Microwave Theory and Techniques, Vol MTT—16 P.177—— 183 1968
3 Lax and K.J. Button, Microwave Ferrites and Ferrimagnetics P.150 ——151,304 1962
4. Yanxin Su, Study of Practical Material of high-low power Microwave Li-Mg Ferrites, Beijing Institute of Radio Measurement, 1985,9

Recent Progress in Power IC Packaging Technology

by
Yehya Mohammed Kasem
The Semiconductor Group
Sprague Electric Company
115 Northeast Cutoff
Worcester, Massachusetts USA

ABSTRACT

As the rapid growth and increasing miniaturization of power and smart power integrated circuit devices continue, plastic packages are confronted with heat transfer problems and consequent thermo-mechanical stress effects, which impact device performance and reliability. Therefore, the advancement of power packaging technology is expected to have as much influence on the pace of power electronic systems progress through the 1990's as the progress achieved in the device technologies themselves.

Traditionally, insertion-mount plastic packages such as Dual-in-Line plastic packages (DIPs) have played a dominant role in high volume packaging of low cost IC devices. At present, Sprague Semiconductor Group, and other major power IC suppliers, are feeling the impact of Surface Mount Technology (SMT) on packaging trends. In response, a new generation of surface mount power packages has emerged. Thermal design of these packages is based on the same power tab and web concept employed in power DIPs.

This paper will present a state of-the-art review of progress in DIP technology, and surface mount packaging, for power integrated circuits. Thermal characteristics of representative, newly developed packages of each class will be discussed. Special leadframe design and thermal design guidelines will be reviewed. Results of 3-D finite element thermal modeling which provide explicit quantification for package temperature distribution and key thermal variables will be summarized.

INTRODUCTION

Power IC manufacturers are developing a wide range of power packages, since heat removal is among the major technical problems confronting power integrated circuit technologies. However, almost all low cost commercial power IC devices are still assembled in traditional insertion mount packages such as plastic DIPs with special leadframe design, which allows for more power dissipation than the standard DIPs. The first two classes of surface mount power packages have been developed with their leadframes designed similarly to the power DIP leadframe format, with central leads tied together and connected with the die pad to form the power tab.

Since power DIPs will be available for many years, while strong trends are now projecting toward SMT power packages, this paper will provide a brief overview of ongoing development activities in the two technologies. For example, the thermal performance of a recently developed 16-lead DIP is described, with 3 watt capability. Also, the superior thermal characteristics of a state-of-the-art 28 lead power DIP which is under development will be demonstrated. This package can dissipate up to twice the power of the standard power version. In addition, the thermal properties of Small Outline IC (SOIC) and Plastic Leaded Chip Carrier (PLCC) power packages will be presented. Finally, this study will describe critical thermal design factors which influence package power dissipation. The discussion will be illustrated by presenting procedures and results of 3-D finite element modeling (FEM), which has proven very effective as an alternative to building and testing prototype package designs.

PLASTIC POWER PACKAGES THERMAL DESIGN CONSIDERATIONS

The major task of power package thermal design is to increase the package power dissipation capability by optimizing the major thermal factors that are affecting the heat flow from the junction to the outside of the package. [1] PDIPs which are designed for power

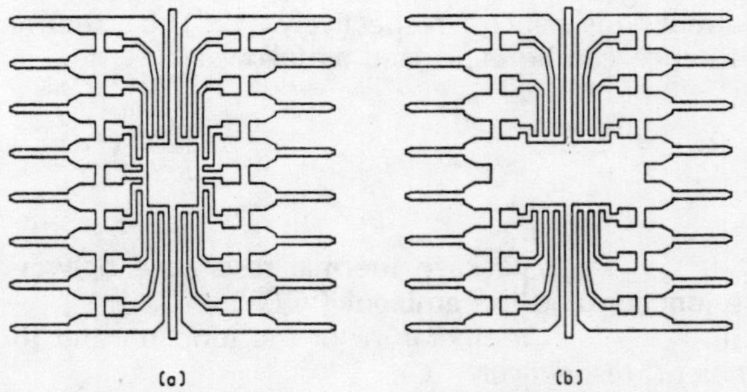

Fig. 1 LEADFRAME TYPES FOR PDIPs

integrated circuit devices are different from their low power versions, primarily in their internal leadframe design. The leadframe illustrated in Figure 1 utilizes the "power tab" concept, in which a number of internal leads on each side of the package are connected to the die paddle. In addition, a web can be formed between the central leads and connected to the paddle. The result is an increase of the effective heat transfer area and the creating of a short thermal conductive heat flow path from the junction to the printed circuit board. Further improvement can be attained by soldering a miniaturized heat sink to the web. A typical power DIP is shown schematically in Figure 2.

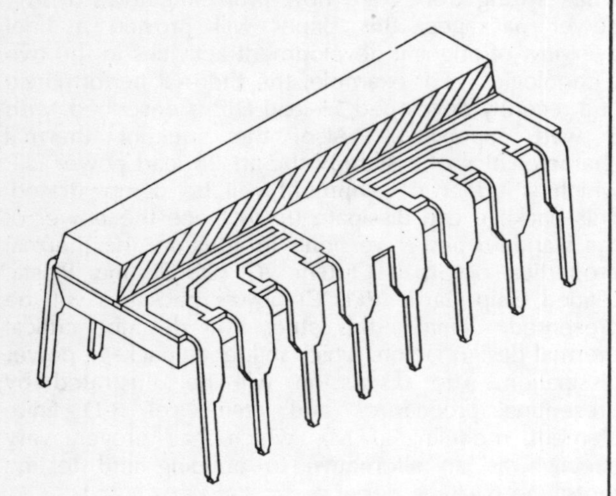

Fig. 2 16-LEAD POWER PDIP

The most frequent parameter used in package thermal design to assess package power dissipation capability is the package thermal resistance ($R_{\theta JA}$). This resistance is a measure of the ability of the package to conduct the generated heat from the junction to the environment, and represents the resistance to the heat flow from the chip to ambient. The concept of this thermal resistance is based upon an analogy between the electrical and thermal characteristics of components where the voltage, current and resistance are analogous to temperature, power dissipated and thermal resistance respectively. The thermal resistance can be expressed as follows:

$$R_{\theta JA} = \frac{T_J - T_A}{P_D} \qquad (1)$$

where

$R_{\theta JA}$ package thermal resistance between the junction and the ambient (°C/W)

T_J, T_A temperature of the junction and the ambient, respectively (°C).

P_D power dissipated (W).

Equation (1) implies that the amount of power that a package can dissipate is inversely proportional to its thermal resistance. The package thermal resistance must be minimized to achieve maximum power dissipation. This is accomplished by (i) improving the thermal conductivity of the major heat flow paths within the package and (ii) increasing the efficiency of the heat transfer mechanisms. To investigate the impact of these variables on the package thermal resistance for optimum design, several experimental measurements and computer modeling methods have been used.

EXPERIMENTAL THERMAL RESISTANCE MEASUREMENTS

The most common method used for thermal resistance measurements is known as the "Temperature-Sensitive Parameter" (TSP) method [2]. However, the accuracy of this method and the ability to interpret test data are affected by a number of factors. Some of these factors are, measurement delay time, calibration of substrate diode, uniformity of the heat generation on the die, and the board and ambient conditions.[3]

COMPUTER MODELING

Numerical simulation techniques such as finite difference and finite element methods have been recognized in electronic packaging as a useful means for expedited thermal analysis and design optimization. Nevertheless, it has been documented that the power integrated circuit packages are three dimensional structures. [4] Consequently, the heat flow and the temperature fields inside the package are orthogonally governed by the package structure. For this reason, the package is best described by three dimensional finite element modeling. Today, the method is routinely used as an in-house established technique for package thermal design and development. All package examples described in this article have been developed and designed with the aid of this method. A good agreement between the predicted performance and the experimental results has always been confirmed.

To demonstrate the effectiveness of the modeling scheme, an example of the modeling procedures and results will be given below. Though the discussion pertains to the thermal design of the new 16 lead power DIP whose thermal characteristics will be presented later, the modeling results and conclusions can serve as a general design guideline for power package development.

16-LPDIP REFERENCE MODEL

A reference model was first established for a typical

power 16-LDIP package, as seen in Figure 2. Due to symmetry, only half of the package was modeled. The model consists of 3032 nodes and 2270 elements. A steady state thermal analysis with free convection cooling is assumed. For half of the package, a 1.2 watt dissipated power was used to simulate a 150°C junction temperature. The power was assumed to be uniformly generated in a 0.025 mm thick active layer at the top of the silicon chip. For half of the chip (1.5 mm X 3.38 mm), the power was specified as heat generation per unit volume (9.49 watt/mm^3). The surfaces of the package and the external leads were assumed to have a convective heat transfer coefficient of 0.00001 W/mm^2°C.

PARAMETRIC STUDY

Following the verification of the reference model, a parametric study was performed to investigate the effects of different variables on the model thermal behavior. Major variables investigated in this study were: (1) leadframe material, (2) leadframe thickness, (3) tie bar size and layout, (4) lead lock hole size, (5) leadframe design, (6) die attach material, (7) die pad area.

The parametric study results are summarized in Figure 3. Several conclusions can be drawn from these results.

1 - In natural convection cooling for high conductivity leadframe material, $R_{\theta JA}$ has a minor dependency on the material thermal conductivity. An increase in thermal conductivity of 25% yields and 8% decrease in $R_{\theta JA}$. The leadframe thickness is of somewhat greater influence, yielding a 10% decrease in $R_{\theta JA}$ for only a 20% increase in thickness. Both $R_{\theta JC}$ and $R_{\theta CA}$ are decreased, due to the massive size of the thicker frame and its effect of reducing the package internal resistance and improving the convection mechanism.

2 - The tie bar is critical to the package thermal performance even in the presence of the power tab feature, because of its contribution in directing the heat flow throughout the package and disseminating heat to the package surfaces.

3 - Extending the chip pad outside the package to form a web between the central leads has a thermal contribution. A 1.6°C/W increase in $R_{\theta JA}$ was found when the web had been removed and the paddle was terminated inside the package.

4 - Lead lock holes of 0.2 mm^2 each have no effect on the package thermal performance if they are placed on all the leads except the four central leads.

5 - An improvement in $R_{\theta JA}$ of only 1.2°C/W was achieved by changing the die attach material from epoxy to solder, despite the large difference in their conductivities. This is attributed to the very small thickness of this layer.

6 - For a given chip, $R_{\theta JA}$ is insensitive to the increase in the die pad area beyond a critical dimension, since any increase in the paddle area in the longitudinal direction is accompanied by moving the lead tips away from the chip edge which results in increasing the plastic thickness between the chip and the leads, thus increasing lead resistance.

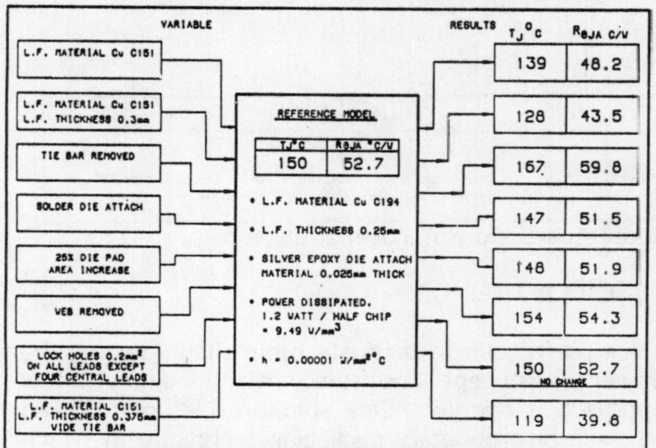

Fig. 3 NUMERICAL RESULTS OF 3-D FINITE ELEMENT ANALYSIS

PLASTIC POWER PACKAGING DEVELOPMENT UPDATE

INSERTION MOUNT TECHNOLOGY PROGRESS

In early 1970 the development of first power DIP was made possible by using copper as the leadframe material instead of Kovar. In a succeeding development, the central leads of the copper frame were tied together in a "web" shape and connected to the die pad, forming one piece called a "power tab," for a direct heat conduction path from the chip to the outside of the package. Since then, the major thrust has been the continuing development of cost effective, reliable plastic packages with higher power dissipation capability. DIPs are projected to maintain their popularity through the next decade. As a result, two state-of-the-art packages have been developed to satisfy the market demands for medium and high power DIPs. The first is 16-lead DIP (3 watts) which is now in production and the second is 28-lead DIP (6 watts) which currently is under development.

16-LEAD DIP

Before the introduction of this new package into the market this year there was no 16 pin DIP available which could dissipate 3 watts continuous operation in still air without being attached to a heat sink. This new package has eliminated this need and has proved its high thermal performance compared to other similar packages. Figure 4 compares the power dissipation capability of the package to five of its

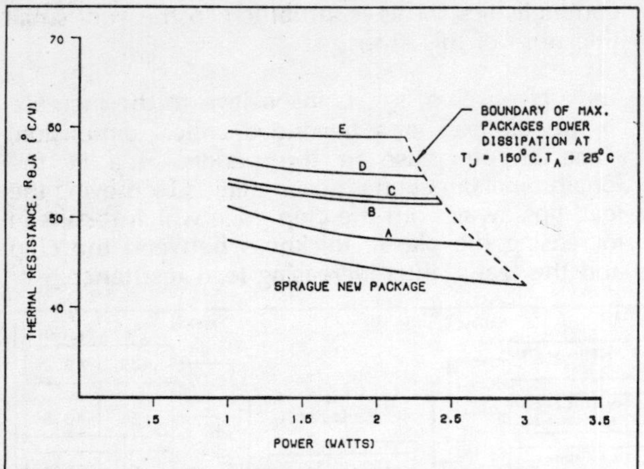

Fig. 4 THERMAL RESISTANCE vs POWER DISSIPATION CAPABILITY FOR VARIOUS PACKAGES

counterparts from major manufacturers.

28-LEAD POWER DIP

For a plastic DIP to dissipate more than 3 watts, the power tab concept is not sufficient and a different new approach is needed. One solution has been offered by using a proprietary packaging technique in which a copper heatspreader is molded on either the lower or upper side of the package, as illustrated in Figure 5. Because of this built-in copper heat sink feature, the power dissipation capability of the package is 100% higher than the standard plastic power 28-lead DIP. The package can dissipate 6 watts in still air for steady state operation. The concept has another manufacturing advantage, because the external dimensions of the package are identical to the existing standard packages, allowing the use of standard assembly, testing, forming, and shipping equipment. In addition, the package can effectively be further cooled by either soldering a heat sink to its copper top or can be inverted and directly soldered to the printed circuit board.

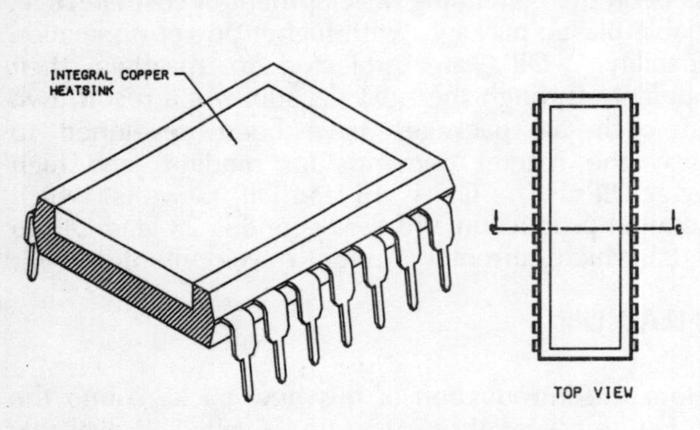

Fig. 5 28 LEAD PDIP - 3D CROSS SECTIONAL VIEW

STATUS OF SURFACE MOUNT POWER PACKAGING TECHNOLOGY

In general, the most important advantages of SMT are increased circuit density and improved electrical performance. There are other advantages which include reduced process cost, automated assembly and placement and easier rework and repair. Though the growth of this technology is rapidly taking shape for low power integrated circuit devices, its implementation into power and smart power packaging has been impeded. The main reason is that the advantage of increasing the pin counts and the packing density on the board has a thermal drawback. Due to the small size of surface mount packages, the heat is more concentrated and localized on the board, which can lead to increasing the thermal resistance of the overall system. Consequently, the thermal performance of the system is impacted. To counter these effects, and to enable the technology to further penetrate power packaging, some advances and innovative concepts have been made resulting in the evolution of the first generation of surface mount power packages. This development has been accomplished by adopting the same power tab and web concept of the copper leadframe that is employed in power DIP. Since 1986 Sprague has recognized the needs for SM power IC packages in two market segments. The first is in the medium power range(2 W or less). The second is high power (more than 2 W). Two package examples of these new classes are shown in Figures 6 and 7. They are 20-lead SOIC and 44-lead PLCC packages respectively.

20-LEAD POWER SOIC

The 20-lead power SOIC as shown in Figure 7 is manufactured with its four central leads on each side tied together and connected as an integral part with the paddle. The package meets the standard JEDEC outline and also allows for an external heat sink to be mounted on its web for further power dissipation.

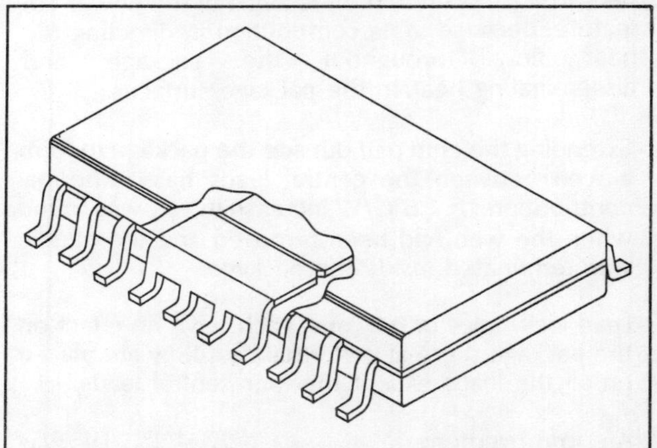

Fig. 6 20-LEAD POWER SOIC. (4 LEADS TIED TOGETHER ON EACH SIDE. 12+1)

However, from the thermal point of view, it is important to recognize the fact that in still air the power tab benefits are affected by the size of the package. For example, SOICs in general have smaller surfaces and their leads are very thin. Thus, their thermal resistance($R_{\theta JA}$) is higher than that of DIPs. Nevertheless, the present 20-lead SOIC can dissipate 2 watts in still air and is characterized by a 6°C/W junction to-tab thermal resistance, which is identical to that of the 16-lead power DIP. Therefore, if a heatsink is soldered to the web of the SOIC package it can be a direct replacement to the 16-lead DIP.

44-LEAD POWER PLCC

The two quadrant 44-lead power PLCC that is illustrated in Figure 7 is designed with 11 leads on each side of the package connected to the die pad to form the power tab, leaving the other 22 leads dedicated for signals. This leadframe format allows the package to be used for a range of smart power devices with the advantage of the small package size,

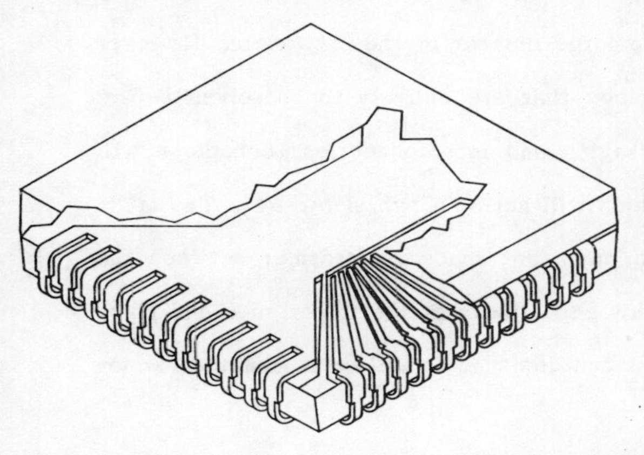

Fig. 7 44-LEAD POWER PLCC. (11 LEADS TIED TOGETHER ON EACH SIDE. 22+1)

high power capability and sufficient I/Os. In addition to the two quadrant configuration there are one or three quadrant options which can also be made without changing the manufacturing or handling equipment, and which conform to JEDEC outlines. In Figure 8 the thermal performance of the two-quadrant 44-lead PLCC is compared to that of a standard, low power PLCC. As can be seen, the heat tab concept has significantly increased the package power dissipation capability to 4.3 watts at 29.5°C/W junction-to-air thermal resistance in still air without external heat sinking. Since the amount of power that is dissipated by the package depends on the number of leadframe quadrants dedicated as a heat dissipator, the power dissipated in the package could vary from about 3.5 watts for a one-quadrant (33+11) package to 5 watts for a three-quadrant (11+33) package.

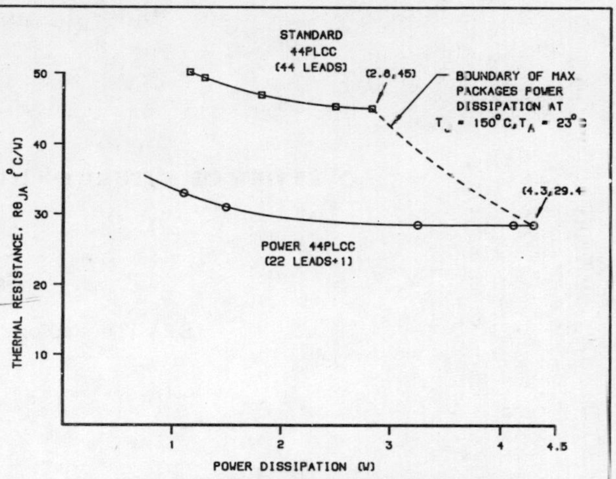

Fig. 8 THERMAL RESISTANCE vs POWER DISSIPATION CAPABILITY FOR 44-LEAD PLCCs

CONCLUSIONS

State-of-the-art thermal design has contributed to the advancement of plastic power packaging technology. In particular, today's DIPs are far superior to their previous generation with respect to high power dissipation capability and reliability. In addition, a new generation of surface mount power packages such as PLCCs and SOICs have emerged, promising higher power handling capability at low cost. At the same time they offer a sufficient number of leads for signal processing and smart power devices. This paper has presented a state-of-the-art review for both classes: insertion mount and surface mount technologies. Packages from each class have been described and their thermal performance discussed. In addition, an example of 3-D finite element modeling procedures has been given. The results of the modeling case provides a general thermal design guideline that is applicable to a host of power packages similar to those presented in this paper.

REFERENCES

[1] J.A. Andrews, et al., IEEE Transactions on Components, Hybrids, and Manufacturing Technology, Vol. CHMT-4, pp. 455-461, December 1981.

[2] D.L. Blackburn, Proceedings 13th Annual Reliability Physics Symposium, 1975, pp. 142-150.

[3] B.J. Shaker and R.T. Lambertson, Proceeding 36th Electronic Components Conference, 1986, pp. 161-168.

[4] Y.M. Kasem and L.G. Feinstein, Proceeding 39th Electronic Conference, 1989, pp. 719-728.

OVERVIEW OF FUTURE OF MULTICHIP MODULES IN ELECTRONICS

Nell Hartley

BPA (Technology & Management) Ltd., UK

Existing IC packaging and interconnection is creating a barrier for advances taking place in IC technology. Wafer scale integration (WSI) is being proposed as the next major step in electronics, bringing with it the removal of a large number of these barriers. However there are still major problems with WSI technology that are unlikely to be solved before the late nineties. In the meantime a new packaging and interconnection technology will be introduced - the multichip module (MCM), which will act as a hybrid for WSI. The MCM will not make the PWB obsolate, but act as an important space transformer between the IC die and the next level of interconnection. This paper looks at the importance and place of the MCM, predicting typical costs, applications and demand. It will be an updated version of that given at Nepoon West 1989.

HIGH-DENSITY MULTILAYER CHIP INTERCONNECTION MODULES

H.M.B. Bird
AT&T Bell Laboratories, Engineering Research Center
Princeton, N.J. 08540, U.S.A.

ABSTRACT

Over the past few years, researchers have developed a variety of high-density technologies for interconnection of VLSI integrated circuits. In these technologies, the basic substrate materials are different — they can be printed-circuit board (such as FR-4), ceramic (such as alumina) or silicon. But the fundamental approaches are similar — multilayer interconnection fabrics which consist of patterned metal layers separated by organic dielectrics. In some cases the dielectric materials are photodefinable. This presentation reviews a number of developments in AT&T that have been directed at achieving high-density, high-speed VLSI interconnection. Similar techniques that are directed at achieving printed optical interconnections and active optical devices are also described. Emphasis is on the POLYHIC technology which uses a photodefinable dielectric material to achieve multilayer chip connection on ceramic substrates, but other approaches are also discussed.

INTRODUCTION

The story of electronics for the past several decades is recognized as one of continuous miniaturization; technologists are able to put more functions into less space, and do so for less cost per function. The step from electromechanical and vacuum-tube technologies to solid-state devices is usually regarded as the major breakthrough. But this breakthrough brought with it a technological capability that permitted an even greater rate of progress — namely, printing large numbers of tiny interconnections in the same way that the active device features are "printed". This capability allowed the development and inexorable progress of integrated circuits (ICs) — as well as provided technology for connecting the circuits together. On any scale, mass-producing an interconnection fabric by photolithographic means is considerably less expensive and of higher quality than discrete wiring.

Printed-circuit boards (PCBs), on substrates such as FR-4, paper-phenolic, etc., are the first obvious embodiment of printed interconnections. Their use goes back to the latter days of vacuum tubes and early discrete transistors. As functional densities of integrated circuits increased by several orders of magnitude, Rent's Rule forced a corresponding increase in interconnection density on circuit boards. As ICs progress from through-hole mounting through surface-mount technology to bare chip-on-board, the demands for higher density on the circuit-boards intensify (Fig. 1).

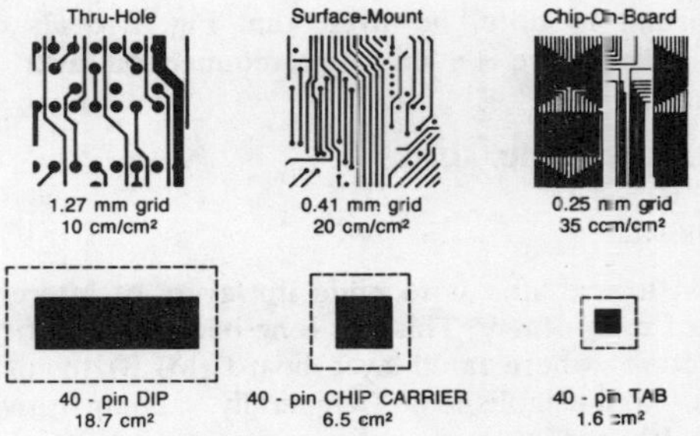

Figure 1. Increasing Interconnection Density

Steadily increasing functionality on ICs brings added complexity in terms of designing and testing devices and aggregations of devices. Modern ICs and systems require very sophisticated computer-driven design tools to partition the functions, lay out the circuits, generate the photomasks, generate the test vectors, and so on. The design of aggregations of ICs onto circuit-boards also requires extremely sophisticated design tools, which need to be coupled to the tools for IC design. At this point, clustering ICs into functional, testable "modules" or subsystems becomes attractive.

The most common manifestation of such "modules" is the well-known "hybrids" technology, in which IC's are attached to substrates (usually ceramic) that have a pre-printed thick-film or thin-film interconnection fabric. Resistors and capacitors can also be printed as part of the process, and laser-trimmed thin-film circuits are capable of very high-precision performance.

Finally, several workers around the world are beginning to use silicon as a substrate material. Typically, a very dense interconnection fabric is printed on the surface of a wafer, using common IC processing techniques, and then bare IC chips are mounted directly on the face of the wafer. This approach will be discussed in more detail later in this paper.

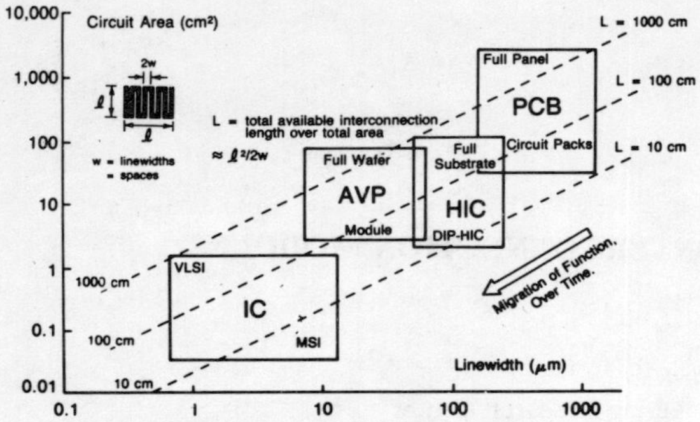

Figure 2. Interconnection Media

Fig. 2 is a log-log plot of interconnection density in which the total circuit area is represented against linewidth. The rectangular areas show the "regimes" occupied by typical media: printed-circuit boards (PCB), hybrid integrated circuits (HIC), silicon substrates (called "advanced VLSI packaging" — AVP) and integrated circuits (IC). The diagonal lines are contours of constant "total interconnection length" - the total length of interconnection that can be serpentinely compressed onto the surface of a circuit, given the total area and the minimum width of lines and spaces. It is important to note, however, that Fig. 2 deals only with circuits where there is a **single** interconnection layer.

MULTILAYER APPROACHES

1. Traditional

The next logical step is to build up layers of interconnection on top of each other. This has long been the practice in PCB manufacture, where multi-layer board (MLB) technology has become well-established. Generally, comparatively thin individual boards (innerlayers) are made separately, and then laminated together in a sandwich-like structure. Connections between various layers are made by plating drilled through-holes or by buried vias. Some suppliers are manufacturing MLBs with 40 and more layers.

HIC technology has begun to use a multi-layer approach, with ceramics. IBM's well-known Thermal Conduction Module[1] uses a structure of metallized "green" ceramic layers which are stacked and then fired together. And DuPont is offering its "Green Tape" technology — a low-temperature co-fired ceramic approach.

ICs themselves use multilayers to interconnect the thousands and hundreds of thousands of active devices. In this case, the metallization layers are built up one at a time, and separated by insulating or dielectric layers in a "grafted" structure. The dielectric or separation layers are generally inorganic materials such as silicon dioxide. This "grafted" approach to high-density interconnection, rather than the "stacked" approaches, is the subject of the rest of this paper.

2. Emerging

There are a number of ways in which grafted interconnection layers can be built up sequentially on a substrate[2]. Today, most workers use polymeric organic materials that can be applied in a number of ways. One advantage of a polymers is its comparatively low dielectric constant (typ. 3.0 to 4.0) which leads to higher-speed circuits. Dry films have been used, but in general these materials are applied in the liquid state — by spinning, spraying, screening, roller-coating or curtain-coating. They are subsequently cured by heat and/or ultraviolet light to their final dielectric state. Finally, the next layer of metal is applied (by plating, evaporation, sputtering, etc.) and photolithographically defined on top of the polymer. The process is repeated, building up a grafted multilayer structure.

Interconnections must be made between each metallization layer to one or more layers beneath it. Therefore, "vias" have to be opened in each dielectric layer to expose metal contact points beneath before the next layer of metal is applied. These vias are defined photolithographically in one of two ways. The first is to apply a photoresist, pattern it, use the pattern as a mask for the etching of the vias, and finally remove the remaining resist. The second way is to make the polymer itself photodefinable. This has the advantage of reducing the number of processing steps, thereby reducing costs and increasing quality.

It should be noted that "grafted" interconnection manufacturing requires very high process yields. With laminated structures, individual layers can be tested and rejected before final assembly. An unrepairable defect on the Nth layer of a grafted structure results in the rejection of the entire structure.

PRINTED-CIRCUIT BOARD SUBSTRATES

An example in AT&T of using a photodefinable dielectric to make a high-density interconnection fabric on PCBs[3] is called "sequential multilayer", or Grafted Multi-Layer (GML), shown in Fig. 3. It was motivated by a desire to make very small vias (microvias) that were smaller than typical drilled vias, thereby saving routing space. These vias could even be located right under mounting pads of surface-mounted devices (SMDs).

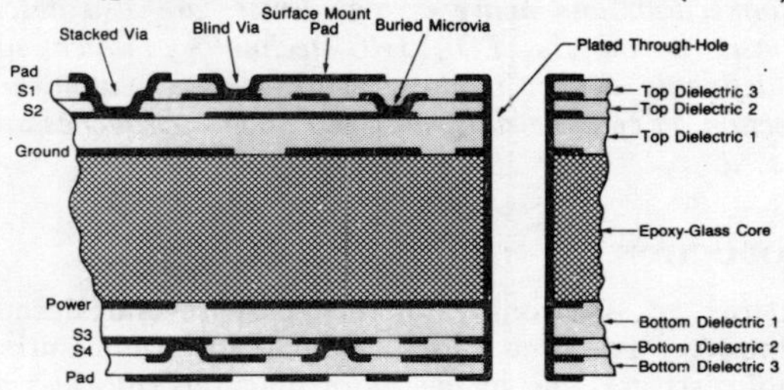

Figure 3. Four Signal-Layer GML Stack-Up

AT&T's GML uses a specially-developed dielectric material[4] that can have the vias defined photolithographically before the final curing process. This material is a mixture of an epoxy resin, rubber resin modifier, reactive monomer, photoinitiator, pigment and certain additives, and is blended to give the required properties. It is a negative-acting resist which is solvent-developable. Besides being liquid, photodefinable, and metallizable, it has to withstand all subsequent manufacturing operations (including soldering), and meet all the reliability criteria imposed on PCBs that are to be used in telecommunications systems in the field. Thermal stability, chemical resistance, adhesion and imageability are but a few of the properties that had to be optimized.

Fig. 3 is a sketch of a representative GML structure. Typical GML parameters are: thickness = 75-125μm, dielectric const. ~ 3.5, via diameter $\geq$ 150 μm, metal linewidth $\geq$ 100μm, max. number of metallization layers = 4/side, photosensitivity ~ 300 mJ/cm^2 @ λ = 365 nm, water absorption $\leq$ 2% min., Cu peel strength = 80 Kg/m, IR = 10^8-10^{10} Ω, min. T_g = 100°C, temp. cycling (-40°C to +150°C) > 500 cy.

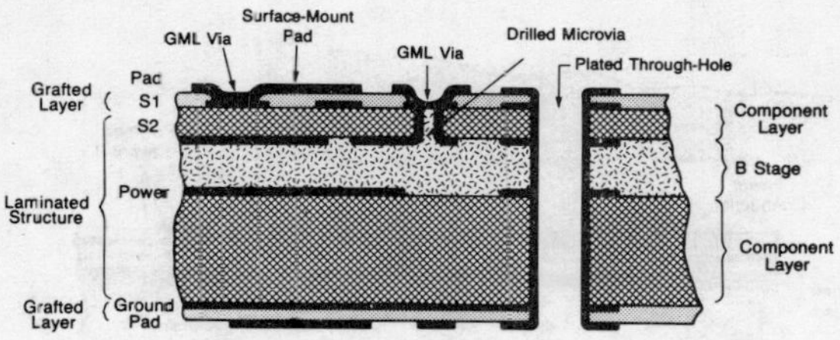

Figure 4. Composite Grafted Multi-Layer (CGML)

A variation of GML is called Composite Grafted Multi-Layer (CGML). In this scheme, multilayer boards are made by standard MLB technology. Then GML layers are applied on both sides of the structure, as in Fig. 4. The MLB layers carry power and ground, while the CGML layers are used as signal layers and mounting pads. An advantage of this approach is that it permits very accurate registration of the final signal layers and the mounting-pad layers, on both sides of the panel. Through-hole drilling and metallization can take place before or after the polymer layers are applied.

DIGRESSION: SOLDERMASKS

Once a robust photodefinable dielectric material was available, it was natural to extend its use to serve as a soldermask. Today it is used extensively in AT&T for this purpose. Again, the combination of liquid application, photodefinability, and aqueous development are extremely attractive. The specific formulation[5] used as a soldermask (called "PDD") has the further advantages of being flame-retardant and of being aqueously developable after UV exposure. Aqueous development is highly desirable from the points of view of environmental protection and industrial hygiene. The benefits that have been realized include low-cost material and process, ability to coat a range of thicknesses, and high surface planarity (for SMD assembly).

CERAMIC SUBSTRATES: POLYHIC

Once the advantages of grafted interconnection structures on PCB substrates were seen, the concepts were extended into the realm of thin-film HICs. Workers in a number of companies have published examples of their work where a dielectric material, often a polyimide, is used to separate layers of metallization on a ceramic substrate.

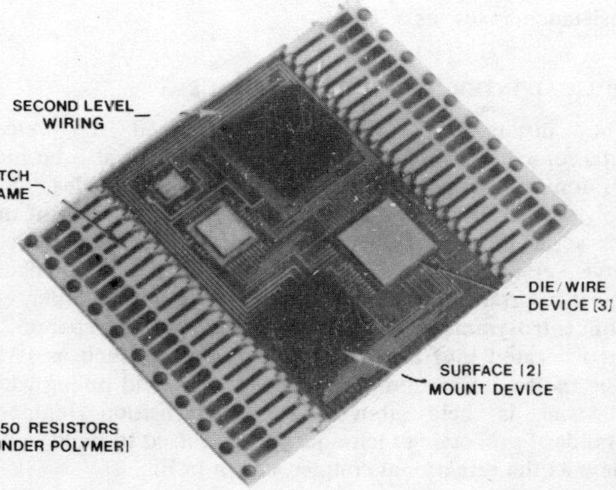

Figure 5. Completed POLYHIC Circuit

AT&T has developed a polymer-on-ceramic technology, called POLYHIC[6] [7] (Fig. 5). As with GML and CGML a special liquid, photodefinable (negative-acting), metallizable polymer is used. The HIC requirements were sufficiently different from those of GML that a new material had to be developed[8].

The POLYHIC material is triazine-based and gives important advantages to high-density high-speed hybrid modules. With a low effective dielectric constant (~ 2.8) and controlled line impedances on the order of 50 Ω, data rates of greater than 2 Gb/s are readily achievable. The triazine resin is a thermosetting material with high cross-link density and good solvent resistance after thermal curing. For toughness, adhesion and photosensitivity it is modified with an acrylated acrylonitrile-butadiene rubber, an acrylated epoxy resin, N-vinyl pyrrolidone, a photoinitiator and a photo-crosslinker. For a single polymer layer and two metallization layers, the operating temperature range is 0 °C to 120 °C, the operating voltage is up to 90V, and the structure can meet –40 °C to +130 °C temperature cycle requirements (500 cy). Other thermal tests such as 1000 hrs at 150 °C in air and THB (500 hrs at 85 °C, 85 %R.H., 2.5 KV/mm) show no physical or electrical degradation of the structures. The stability of underlying tantalum resistors is typically $\Delta R \leq 0.5\%$. through THB testing. Other properties of the polymer include T_g = 150 °C, dielectric strength = 35kV/mm, resistivity ~ 4×10^{12} Ω-cm, long-term thermal stability = 180 °C.

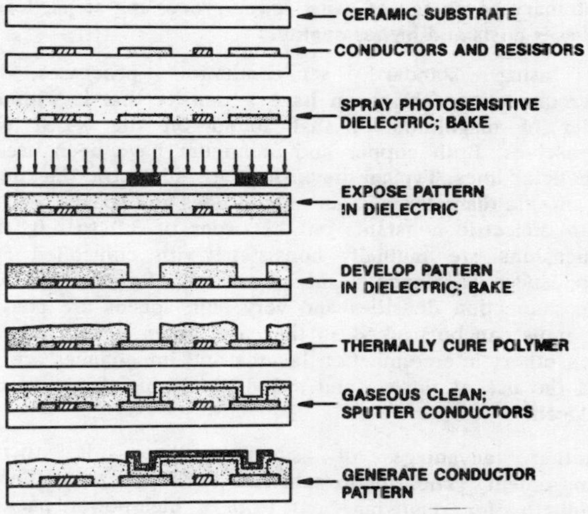

Figure 6. POLYHIC Fabrication Process Schematic

The polymer has very little effect on underlying thin-film precision resistors (0.1% stability over 20 yrs. @ 65 °C). The first polymer layer is typically 50 μm thick, after curing, and via diameters are on the order of 150 μm. Metal lines and spaces as small as 50 μm and 75 μm, respectively, are achieved. POLYHIC structures (Fig. 6) can be made on one or both sides of a ceramic substrate. The underlying tantalum nitride film resistors, tantalum film capacitors and first-layer conductor films are applied and patterned on the ceramic substrates by standard sputtering, photodefinition and laser-trimming processes. The liquid polymer is sprayed onto the surface, baked to remove solvent, photoexposed through a mask, and developed to provide vias to the underlying conductors. After thermal curing of the polymer, adhesion of the next layer of conductor is promoted by a plasma treatment, and the conductor film is applied by sputtering. Features in this conductor layer are defined by standard pattern-generation processes. The sequence can be repeated to build up subsequent layers of polymer and conductors. Active devices, packaged or unpackaged, can be attached by a variety of means in the same module; AT&T has used soldered chip carriers, beam-lead bonded chips, and wire-bonded chips on POLYHICs. The result is very high-density, high-speed functional modules.

POLYHIC modules have been manufactured by AT&T for a number of years. A common device is the Integrated Services Line Unit (ISLU), Fig. 5. So far applications have required only a single polymer layer (hence two metal layers) on each side of the substrate. But demand for very-high-density, very-high-bandwidth modules is growing and the first POLYHIC units with multiple polymer layers will appear soon.

MULTILAYER ON SILICON SUBSTRATES

The next step is to extend the grafted polymer multilayer concept down to the realm of silicon substrates. Many workers around the world[9] are pursuing a variety of "silicon-on-silicon" approaches. In all cases, a multilayer interconnection fabric is built up on a silicon substrate, and then bare IC chips are attached to form functional modules. While SiO_2 has been used, the inter-metal dielectric is usually a polymer, often a polyimide. Most workers have used non-photodefinable dielectrics, but recently suitable photodefinable substances have begun to be used. The latter have the customary advantage of using fewer processing steps, leading to lower costs and higher quality.

By using standard semiconductor processes, the interconnection fabric can have a density that is within an order of magnitude of that found on the VLSI chips themselves. Both copper and aluminum have been used as conductor lines. Typical dimensions are: dielectric thickness = 10 μm, via diameter = 25 μm, metal lines and spaces = 10 μm. With dielectric constants on the order of 3.0 to 4.0, these dimensions are mutually consistent with controlled 50 Ω impedances. Hence, modules with extremely high interconnection densities and very high speeds are possible. IC chips can be packed so they are within 0.5mm or so of each other; interconnection lengths and impedances are low, and the use of power- and speed-consuming drivers can be reduced.

Another advantage of silicon-on-silicon is thermal management. The silicon substrate itself is a very effective conductor for removing heat from a high-power package. And, since the ICs and the substrate are the same material (silicon) with the same coefficient of thermal expansion, thermal stresses are greatly reduced.

A silicon-on-silicon approach, like all multi-chip module (MCM) approaches, permits the separate fabrication and testing of active VLSI devices and the interconnection substrate — before final assembly. Thus, only good parts are committed to the assembly operation. In addition, the MCM approach permits the use of "mixed technology" (CMOS, ECL, III-V, etc.) on the same substrate.

AT&T's version of this technology[10] [11] [12] is called Advanced VLSI Packaging (AVP) and is shown in Fig. 7. It uses a non-photodefinable polyimide dielectric, plated copper metallization, and solder-bump technology for chip attachment.

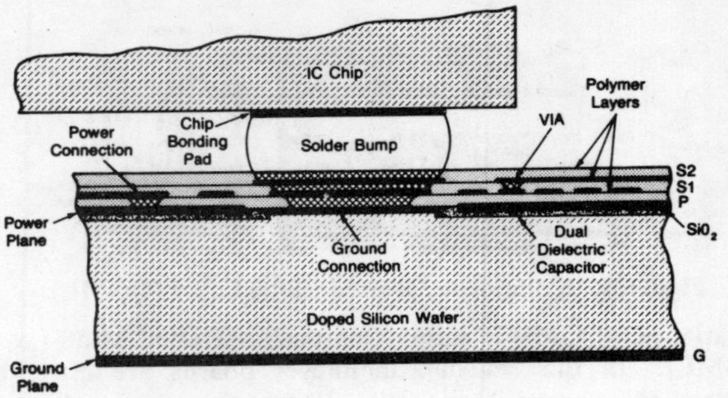

Figure 7. Advanced VLSI Packaging (AVP) Structure

A patterned oxide layer can be built onto the surface of the silicon substrate, thereby providing decoupling capacitors. Then a base metal layer is sputtered onto the surface. Photoresist is applied and patterned; it serves as a mask for electroplating the copper conductors to their full thickness. After resist stripping, the rest of the base layer of metal is chemically removed. Polyimide dielectric is spun onto the wafer, planarizing the surface. The polyimide is patterned for via holes which will connect to the first conductor layer. Then the vias are filled with electrolessly plated nickel. The process can be repeated to add successive layers of conductors and dielectric.

After the interconnection fabric has been made, IC devices are attached by flip-chip solder-bump technology. A layer of interface metallization is sputtered, photoresist is applied and patterned, solder bumps are electroplated, and the resist and base metal film are removed. VLSI chips are tacked into place accurately by high-precision robotic techniques and then the solder is reflowed to make the final connections.

The resulting structures have excellent electrical performance, with line capacitance ~ 1 pF/cm, and low-inductance connections between chips and the substrate. Power and ground lead inductance is less than 0.1 nH and signal-lead capacitance is less than 20 pF. The copper conductors (2 μm thick, 10 μm wide) have a resistance of 10 Ω/cm, and the maximum signal lead length on typical structures is less than 20 cm. Devices with more than 200 I/Os have been assembled with the solder-bump method. Operating speeds of several hundred MHz are achieved. The structure can be bonded to a metal heat sink, providing a chip-to-heat-sink thermal resistance as low as 5 °C/W.

OPTICAL INTERCONNECTION FABRICS

The discussion so far has focused on electrical interconnection modules, but mention should also be made of photonic interconnection and devices. The idea of using polymers as nonlinear photonic materials has been of interest for some time[13]. Polymeric materials can be prepared that have very desirable electro-optical properties, and figures of merit approaching those of $LiNbO_3$. For 2nd-order effects, noncentrosymmetric molecules have been prepared[14] and incorporated into a glassy polymer matrix (such as PMMA). The molecules can be aligned by electric-field poling while the material is held above its glass-transition temperature. Standard processing techniques can be used to build structures such as the directional coupler shown in Fig. 8.

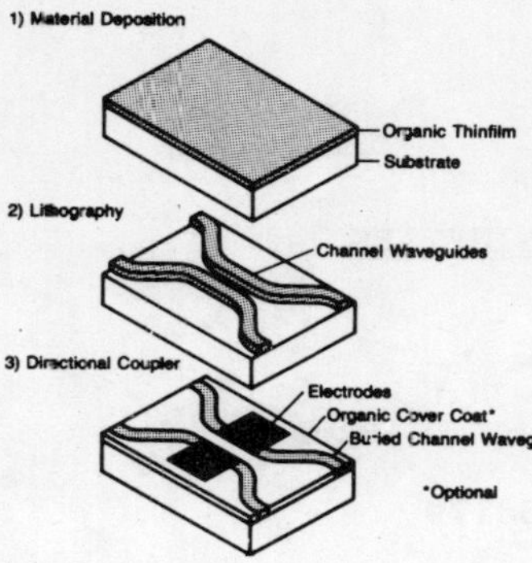

Figure 8. Polymeric Electro-Optical Structure

Workers at AT&T have used these materials to demonstrate a number of optical interconnection and device structures.[15] [16] A new copolymer containing methyl methyacrylate and a dicyanovinyl-terminated azo dye side chain can be coated onto a substrate (such as indium-tin-oxide glass), and orientational order imparted by electric-field poling with a corona discharge at elevated temperature. The material can then be patterned by standard lithographic techniques into structures. Substantial improvements have been seen in the magnitude and long-term stability of nonlinear susceptibilities when compared with guest-host systems ordered by electrode poling. Current work is directed at making practical demonstration devices with this new technology, with the expectation that it will permit the manufacture of low-cost, high-performance optical interconnections and device modules. It is another example of employing tailored organic materials to serve in interconnection fabrics.

SUMMARY

This paper has explored the use of tailored polymeric materials as essential ingredients in interconnection fabrics, both electronic and photonic. Current worldwide efforts are focused on high-density grafted multilayer structures for multichip modules. The technology has progressed from PC Boards through hybrids, and now silicon-on-silicon modules such as AVP attract much attention. Nonetheless, impressive performance characteristics can be obtained even at the Hybrids level, using structures such as POLYHIC. Moreover, the concepts can be extended into photonic interconnection, where the use of specially-developed molecules in a polymeric matrix offer the promise of low-cost, high-performance optical interconnections and devices.

ACKNOWLEDGEMENTS

This paper reviewed the efforts of many people at several Bell Laboratories and other AT&T locations, including Princeton, Whippany, Richmond, Murray Hill, Hawthorne, Indian Hill, Network Software Center, Allentown and Merrimack Valley. Space does not permit the listing of such a large number of people here, but their names appear throughout many of the references.

REFERENCES

1. A.J. Blodgett & D.R. Barbour, "Thermal Conduction Module: A High-Performance Multilayer Ceramic Package", IBM J. Res. Devel., **26**, pp30-36, Jan. 1982.
2. There are believed to be over 100 companies working in this area. An appreciation of the scope of the effort can be derived from perusing Sessions TS-11, TS-18, TS-25 & TS-32 of the Proc. of the Tech. Program, Natl. Electronic. Pkg. & Prodn. Conf., NEPCON-WEST '89, Anaheim, CA, March 1989.
3. C-y. Lin, G.B. Fefferman, J.F. D'Amico & J.A. Ors, "Multilayer Printed Wiring Boards Made by Sequential Layer Process", IPC 32nd Annual Meeting, Lake Buena Vista, FL, April 23-28, 1989.
4. G.B. Fefferman, C-y. Lin, J.F. D'Amico, J.A. Ors & F.R. Wight, "Material-Process Interactions in the Sequential Multilayer Process", IPC 32nd Annual Meeting, Lake Buena Vista, FL, April 23-28, 1989.
5. U.S. Patent Nos. 4,511,757 4,628,022 4,795,693 and other applied for.
6. D.B. Bucholtz, C.C. Faudskar, C.C. Shifflett, J.L. Markham & R.D. Small, "High-Density Multilayer Hybrid Circuits Made with Polymer Insulating Layers (POLYHICs)", Proc. of the 1986 Int'l. Symp. on Microelectronics (ISHM), pp481-486, Oct. 1986.
7. J.L. Brandner, C.C. Faudskar, M.E. Lindenmeyer, S.R. Hoffmann, D.B. Bucholtz & J.E. Ballantyne, "Electrical Characterization of POLYHIC, A High-Density, High-Speed Package Medium for Digital Circuits", Proc. 39th Electronic Components Conf., pp759-764, May 1989.
8. U.S. Patent No. 4,554,229.
9. A partial list of companies includes Hitachi, AT&T, Honeywell, NTT, Raychem-APS, IBM and Rockwell.
10. H.J. Levinstein, C.J. Bartlett & W.J. Bertram, Jr., "Multi-Chip Packaging Technology for VLSI-Based Systems", 1987 Int'l. Solid-State Ccts. Conf. (IEEE), New York, Feb. 1987.
11. C.J. Bartlett, J.M. Segelken & N.A. Teneketges, "Multichip Packaging Design for VLSI-Based Systems", IEEE Trans. on Comp. Hybrids & Mfg. Tech. **CMHT-10**, pp647-653, Dec. 1987.
12. W.J. Bertram, "High-Density, Large Scale Interconnection for Improved VLSI System Performance", Proc. IEDM'87, Sect. 5.4, pp100-103, 1987.
13. D.J. Williams, ed., "Nonlinear Optical Properties of Organic and Polymeric Materials, ACS Symp. Series #233, 1983.
14. H.E. Katz, K.D. Singer, J.E. Sohn, C.W. Dirk, L.A. King & H.M. Gordon, "Greatly Enhanced Second-Order Nonlinear Optical Susceptibilities in Donor-Acceptor Organic Molecules", J. Amer. Chem. Soc. **109**, pp6561-6563, 1987.
15. K.D. Singer, M.G. Kuzyk, W.R. Holland, J.E. Sohn, S.J. Lalama, R.B. Comizzoli, H.E. Katz & M.L. Schilling, "Electro-Optic Phase Modulation and Optical Second-Harmonic Generation in Corona-Poled Polymer Films", App. Phys. Lett **53** (19), pp1800-1802, 7 Nov. 1988.
16. K.D. Singer, et al., to be published in Proc. S.P.I.E., San Diego, 1989.

INTERCONNECTION TRENDS INTO THE NINETIES

Dr Maurice Sage
BPA (Technology & Management) Ltd.
Abinger House, Church Street,
DORKING, SURREY, UK
Tel: 0306 884522 Fax: 0306 888179

INTRODUCTION

Advances in integrated circuit technology has started to create major challenges in the area of interconnection and packaging. In a recent issue of the Hughes Aircraft publication 'Vectors' a lead article on interconnection and packaging refers to the technology as having 'a starring role for the future' and 'the pacing technology for electronics' race into the 21st Century.' Associated with this trend the separate technologies that make up the final product are becoming increasingly interdependent and will require a systems engineering perspective from the entire electronics industry.

This paper highlights the major drivers and innovations in interconnection and packaging that will be experienced in the nineties.

THE INTEGRATED CIRCUIT (IC)

IC cost and performance is rapidly outstripping available interconnection and packaging technology. The industry is developing 'electronic engines' that are supported in progressively obsolete 'electronic chassis'. Table 1 forecasts trends in IC technology which clearly reflect the challenges that are being created.

Table 1

IC TECHNOLOGY ADVANCES

Feature size halving every 6 years
Frequency doubling every 6 years
Die size doubling every 6 years
Pincount doubling every 3 years

ms 407

From Table 1 the trend towards higher pin-count requires new forms of IC packaging which result in ever larger footprints. This is illustrated in Table 2 overleaf.

Table 2

FOOTPRINTS OF VARIOUS IC PACKAGES

Type	Dimensions inches	Area inches	Pin Spacing
20 pin DIP	0.4 x 1.0	0.40	100 mil
64 pin DIP	1.0 x 3.0	3.00	100 mil
64 pin LCC	0.5 x 0.5	0.25	25 mil
132 pin LCC	1.22 x 1.22	1.50	25 mil

MS 402

From Table 2 the move from a 20 pin DIP to a 64 pin DIP increases the footprint by 7.5 times. This situation is eased by a factor of twelve by moving to the LCC (pin spacing 25 mil) but as soon as the pin-count goes to 132 pins there is an increase in the footprint of 6 times. The continuous growth in pin-count has created a move to surface mount technology (SMT) which is even now pointing towards the use of unpackaged ICs, for direct attach, e.g. chip on board, flip and TAB.

Fig. 1 below shows the BPA world forecast of demand for different IC packages through to 1997. Essentially there are three generic types i.e. packages for insertion, packages for SMT and direct attach.

Figure 1

WORLD IC PACKAGE TYPES 1987-1997

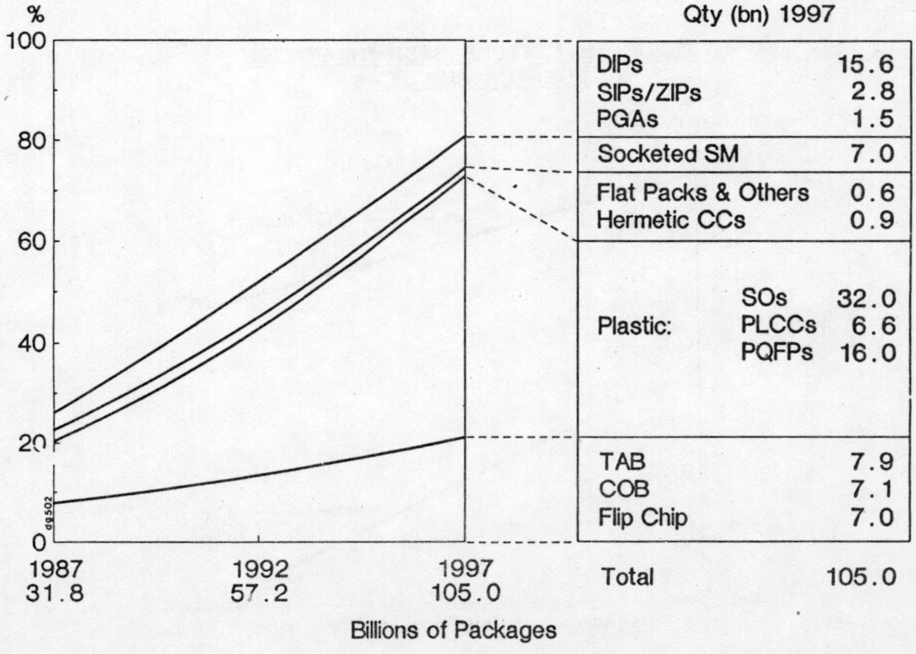

These three distinct sectors are plotted in Figure 2 with the lower curve indicating direct attach. This new and developing wave of IC attachment will require a number of new approaches to interconnection and packaging such as the multichip module (MCM).

Figure 2

IC PACKAGING SUBSTITUTION CURVE

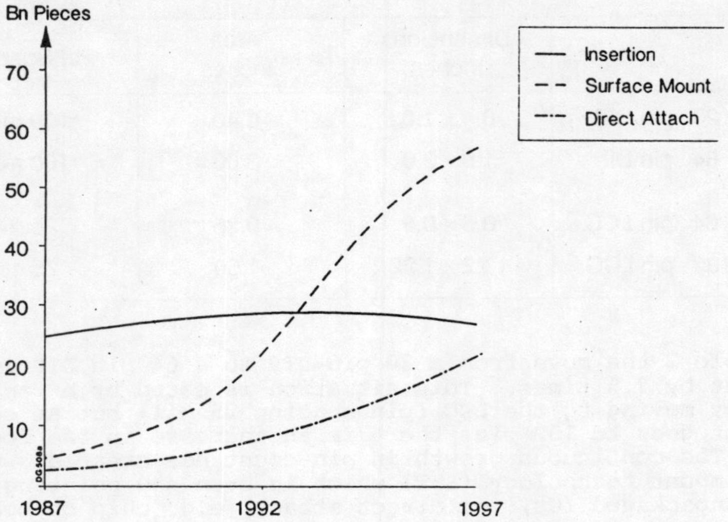

INTERCONNECTION DENSITY

As the density of IC interconnection continues to increase (feature sizes are expected to reach 0.25 microns during the nineties) the ability of PCB and thick film hybrid technology to complement this trend becomes increasingly difficult. The problems are particularly evident in escalating costs and limited interconnection density. Figure 3 illustrates the feature size 'gap' that is opening up between the IC and PCB technologies.

Figure 3

TREND IN FEATURE SIZE REDUCTION
FOR ICs AND PCBs

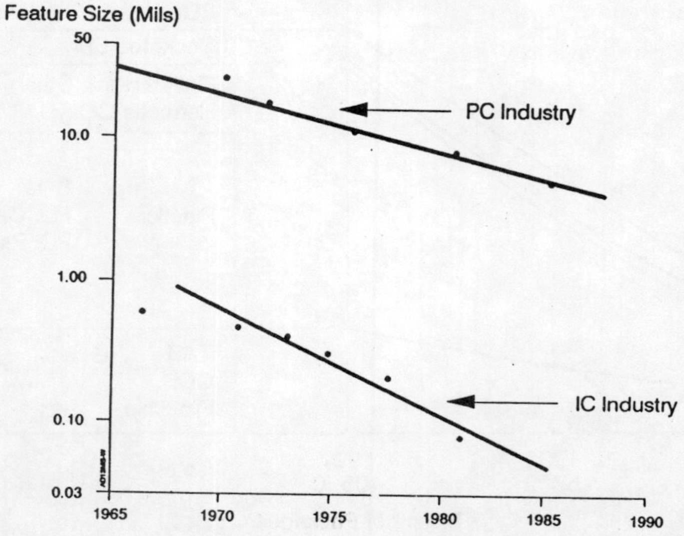

With larger pin-count there is also an increase in interconnection density. To take this into account a model has been developed by BPA over the past five years to assist in forecasting new interconnection and packaging technologies. The model plots cost against interconnection density for a wide range of technologies,e.g. PCBs, ICs. The algorithm used to normalise the horizontal axis, (in order to make comparative analysis) will not necessarily agree with those computed from other sources. The density measurement is given in inches/square inch, allowing for vias, ground and power planes, component packing densities etc. A calculation of 500 inches/square inch indicates a track width of 1 mil and a space of 1 mil (or 0.5 mil track 1.5 mil space) covering one square inch. Taken from a BPA study Figure 4 below shows the increasing price of PCBs against density.

Figure 4

PRICE/CIRCUIT DENSITY FOR PCBs
TOTAL OF USA, JAPAN AND EUROPE REGIONS

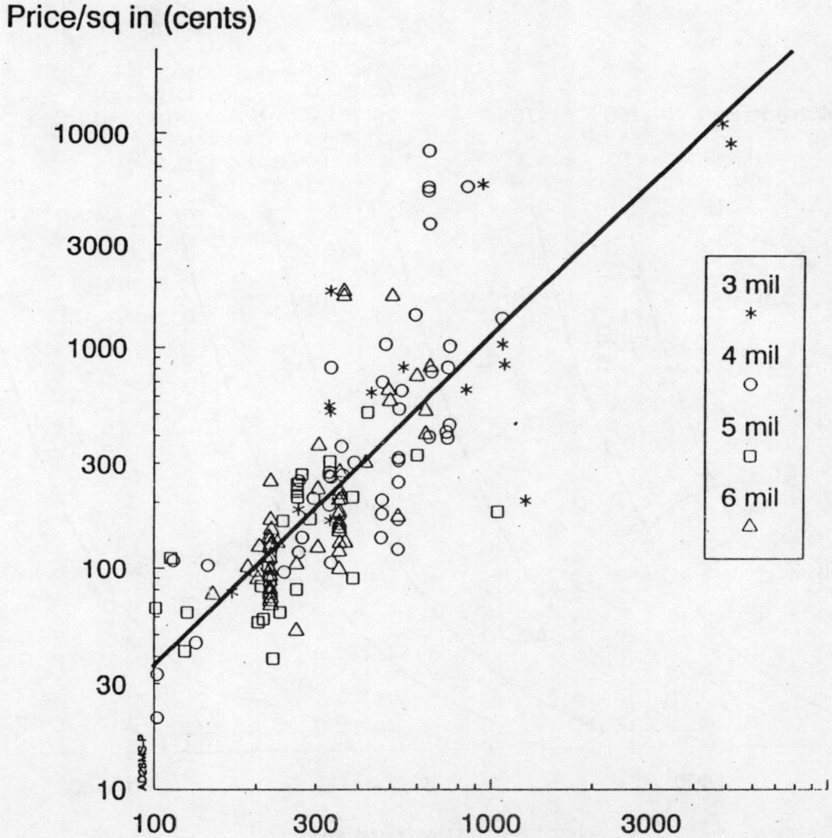

Another way of looking at this problem was also developed by BPA from its density modelling studies. It illustrates how feature size reflects into numbers of layers for particular densities. This is illustrated in Figure 5. which shows the layer count difference between the PCB (even with advanced boards) and that of the IC. It illustrates the 'gap' that is opening up between these technologies and the introduction of new ones such as the multichip module (MCM) (thin film multilayer interconnection) that is drawing much of its process technology from the IC industry.

To date BPA estimate that over $1bn have been spent developing MCM interconnection technology, with demand forecast to reach nearly $7bn by 1997 for circuitised substrates alone. This technology is already in use in Japan and the USA and is not confined to high performance systems. Like other interconnection technologies it will have across the board applications, including auto and consumer.

Figure 5

INTERCONNECTION DENSITY VERSUS LINE TECHNOLOGY

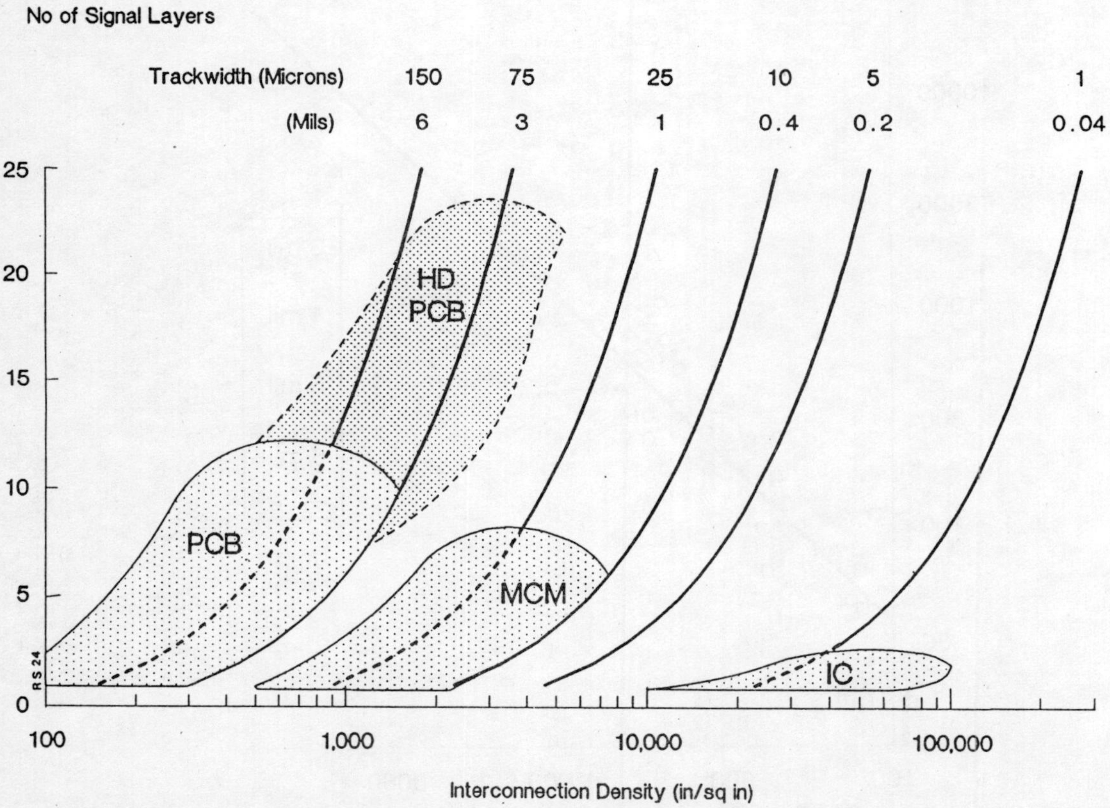

THE MOVE TO WAFER SCALE INTEGRATION (WSI)

From the packaging and interconnect challenges given so far why not move directly to wafer scale integration (WSI) ?. The idea of putting a complete system on a wafer has great appeal. However functional density in IC technology has been achieved more by reducing the feature size, with limited increases in die area. Figure 6 overleaf illustrates that between 1977 and 1983 the DRAM went from 16K to 256K bits but die size only increased from 20 to $40mm^2$ i.e. doubling every 6 years. The introduction of WSI on a significant scale is not forecast to happen until the late nineties. In the meantime and into the future the MCM (a hybrid for WSI) will be introduced.

Figure 6

PROGRESS IN CHIP SIZE VERSUS VLSI DEVICE DENSITY

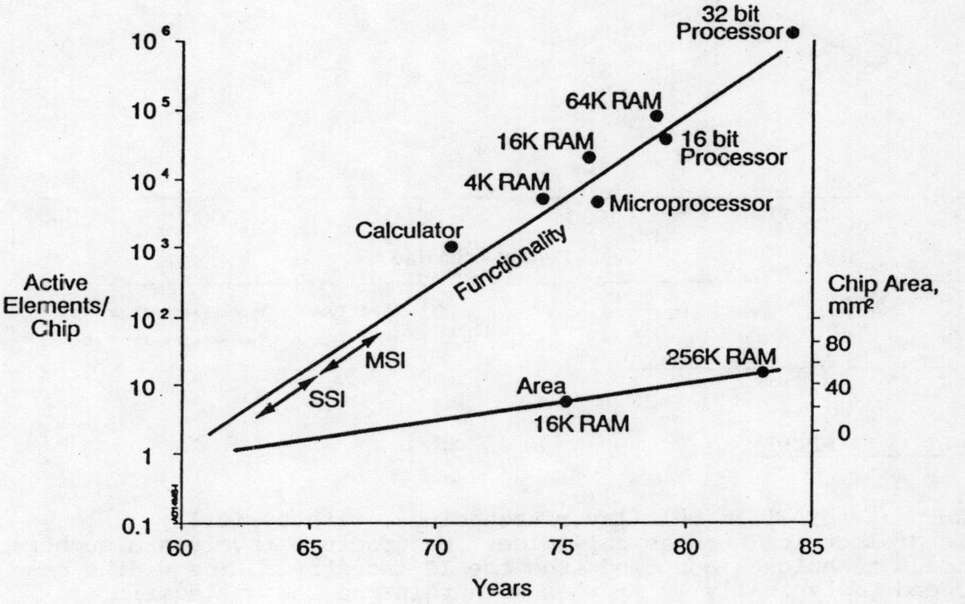

PRICE DENSITY MAP

The need to meet the increased density of interconnection that will be required between the advancing IC and the PCB will be met by technologies acting as space transformers. This developing situation is illustrated on the cost/density axis shown in Figure 7. with the data from Figure 4 replotted. The increasing density required from the PCB is giving rise to higher costs, but is still holding its own against the MCM type 1. However advantages of size, weight and reliability, even for the type 1 MCM, will ensure MCM viability in the longer term. The type 2 will have feature sizes down to 5 micron achieving very high levels of interconnection density.

Figure 7

PRICE DENSITY RELATIONSHIPS

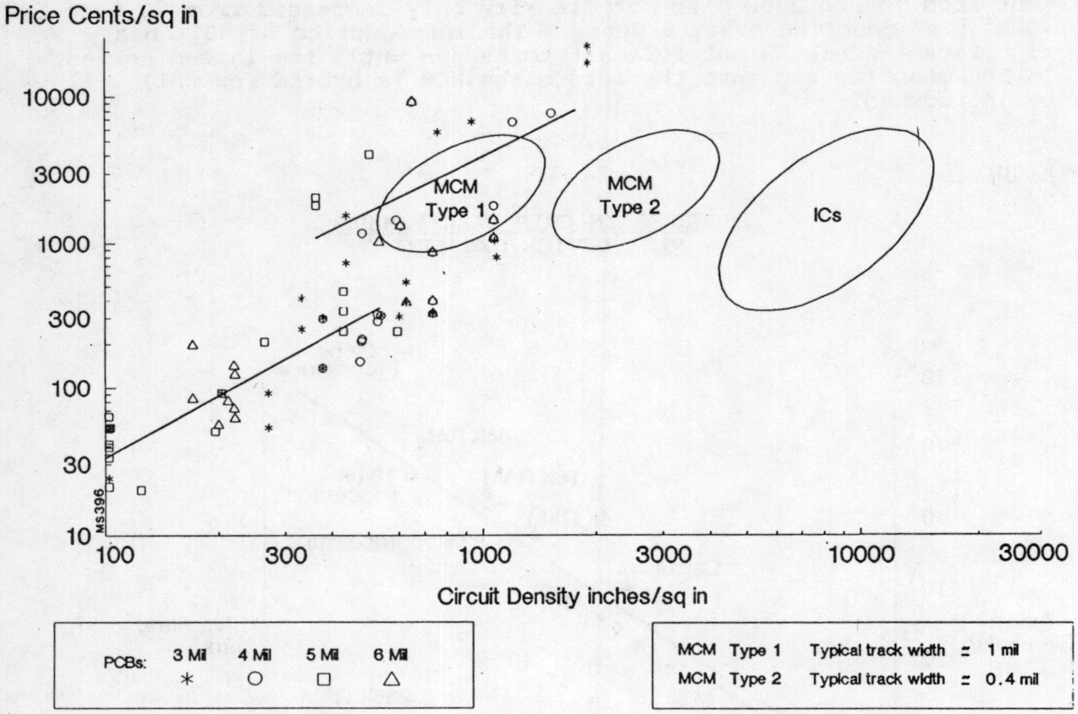

THE MULTICHIP MODULE

The MCM uses thin film multilayer technology, with typically organic dielectrics such as polyimide. Manufacture involves a number of process techniques borrowed from the IC industry. Line widths can be reduced to typically 10 microns with thin and low dielectric constant materials that enable the necessary controlled impedance value to be obtained, allowing increased density of interconnection with reduced crosstalk.

A figure of merit is derived for various technologies using the propagation speed in inches/p sec times the interconnection density; typical figures are given in Table 3. The figure of merit for VLSI will need to be partially adjusted (down) to take into account the line resistance and capacitance which forms a CR network. This effect is not significant in MCMs due to higher line conductivity, lower drive currents and lower output capacitance from the drivers.

Table 3

FIGURE OF MERIT SPEED - DENSITY

Technology	Figure of Merit (ins/p sec . density ins/sq in)
VLSI	28.0
MCM	14.6
PWB	2.2

MS401

MULTICHIP MODULE DEMAND

Computers - higher performance in CPUs and cache memory are prompting considerable development effort. Japanese computer companies are already using the technology in their products.

Telecomm/Comms - MCMs will find applications in Main Switches and Transmission Multiplexers. The demand is very dependent on the introduction of ISDN particularly the broadband services and this will vary throughout the world, with Japan and W Europe most possibly leading the way. Opportunities exist for MCMs in hand-held and pocket communication products.

Integrated Circuits - the yield problems of large die from probed wafers is creating a move to partitioning and rebuilding on silicon MCMs. Some companies have looked at restricting die size to 8mm/side with a view to adopting a flexible hybrid approach via the MCM.

Using a variety of forecasting approaches the MCM market for circuitised substrates in 1997 is estimated at $7bn - the PCB market alone at this time will be in excess of $40bn.

A NEW TECHNOLOGY

The MCM offers significant performance and cost advantages. However it is a new technology and its full introduction will take time, with take off forecast for early nineties. A list of advantages and cautions are given in Table 4 below. With all new technologies there will be plenty of 'good reasons' to downgrade it as many of the vacuum tube manufacturers were eventually to find to their cost in avoiding solid state.

Table 4

MCMs - ADVANTAGES AND CAUTIONS

Advantages	Cautions
Smaller size	A new technology
Reduced weight	Design and test problems
Increased interconnection density	Material and process problems
Lower heat dissipation	Not simply IC technology
Improved reliability	Connector problems
Ability to mix technologies	PCBs will compete for many applications
Repairability	Not an extension of thick film hybrids

CONCLUSIONS

The need to introduce new interconnection and packaging technologies is vital if the advantages in developing IC technology are to be realised. This trend does not mean that the PCB industry will disappear, far from it. However it must recognise the maturing nature of its industry and the considerable opportunities that will present themselves for diversification into complementary products during the nineties.

LOW COST TaAlN/TaN THIN FILM CHIP RESISTOR WITH CCNA NEW TYPE THIN FILM STRIP CONDUCTOR

Hu Nanshan, You Weimin
Chengdu Research Institute
of Telecommunication Technology, China

Yin Zhongyi
Huadong Research Institute
of Microelectronics Technology, China

ABSTRACT

This paper introduces a thin film chip resistor in use for microwave band. The low cost Cr-Cu-Ni-Au (CCNA) multilayer composite structure is used as thin film strip conductor. Copper is the main part of the conductor. The strip conductor is manufactured by a continue evaporation procedure. With this procedure only, the thickness (8 μm) of strip conductor with the sheet resistance of 0.003 ohms per square was obtained.

TaAlN/TaN thin film composite structure as the resistance film was obtained by R.F. reactive planar magnetron sputtering combined with both vacuum and atmosphere annealing. The temperature coefficient of resistance was measured as ± 20 ppm/°C.

INTRODUCTION

When the chip resistor is used at microwave band, the thin film structure is desirable at its higher frequency. But the thin film resistor with gold or platinum-gold precious metal as main part of conductor is costly. For this reason authors have developed a new type low cost CCNA conductor where copper is a main body and the thickness of gold layer is ~ 0.05 μm only. The CCNA structure is quite good at microwave band. Under the same applied condition, however, the thickness (8 μm) of gold is desirable for Cr-Au conductor. The sheet resistance of CCNA conductor is less than 0.003 ohms per square which is lower than the value (0.005 Ω/□) of Ti-Pt-Au conductor.

After a high temperature aging at 250°C, Cu-Au diffusion in CCNA was analysed by AES method. As compared with conventional Cu-Au structure, because the Ni barrier is contained in the CCNA conductor, it acts as a barrier to the interdiffusion of Cu and Au which improves the solderability and the high temperature property[1].

Using reactive sputtering to manufacture TaAlN resistance film, it is difficult to obtain a good film with constant component ratio and repeatability, unless to keep the ratio of N_2 and Ar, and gas flow accuracy and stable. Therefore the gas mixture set in a sputtering system was modified, and a pressure controller was used[2].

For resistance film a TaAlN/TaN composite thin film was used for the first time. With the composite film, TCR of the resistor was ± 20 ppm/°C in the higher sheet resistance range of 5 Ω/□ ~ 1kΩ/□.

The chip resistor have been fabricated in batch process and used for microwave hybrid integrated circuits and attenuators under 10 GHz.

EXPERIMENT

The substrates of the chip resistors are made of 97 % or 99 % Al_2O_3 . TaAlN/TaN thin film was deposited by R.F. reactive planar magnetron sputtering[3]. First the vacuum chamber was prepumped up to 6.65 x 10^{-4} Pa, then Ar and N_2 gas mixture was filled into the chamber (the ratio of Ar and N_2 was 100:1, the gas purity was 99.99 %). The resistance film was sputtered at pressure of (1~3) x 1.33 Pa, where R.F. input power applied on target was 1.2 KW and reflected power was 40 W. The magnetic field strength was 400 gauss (4 x 10^{-2} T). The diameter of the Ta-Al target (the atomic ratio of Ta and Al is 50% :50%) was 140 mm, and the thickness of the target was 8 mm. The R.F. planar magnetron sputtering system made in Shanghai Wire Factory possesses a substrate platform which acts as a planar planetary turning function.

CCNA strip conductor was made by evaporation with a shield foil method. Cr, Cu, Ni and Au were evaporated in series at the pressure of 6.6x10^{-4} Pa using the electron beam evaporation system with four electrods which was made in Nan-Guang Machinery Factory. A PHI 550 multifunction electron energy spectrum test system made in Perkin-Elmer Co., U.S.A., was used for AES analysis.

A SZ-82 numerical low resistance test meter with four probes made in Shanghai Electric Dynamic Tool Research Institute was used for measurement of sheet resistance of conductor.

EXPERIMENT RESULTS

1. Technique Properties of Chip Resistors

A. Resistance range: 5Ω ~ 10KΩ
B. Operation frequency: ⩽10GHz
C. TCR: ⩽±20 ppm/°C
D. Size of resistors:
    0.5x1.5x0.5 mm^3
    1x3x0.8 mm^3
    2x6x0.8 mm^3

2. Test for Chip Resistors

A. Sheet resistance of conductor ($R_□$)

By means of the experiment stipulated by ASTMF standard No. 390-70, U.S.A. the strip conductor samples were tested and the average value of the sheet resistance is: $R_□$ ⩽0.00295Ω/□ .

B. Temperature coefficient of resistance (TCR)

To arbitrarily extract 8 resistors as samples, and then to measure their TCR in the temperature range of 20°C ~ 120°C. The measured average value of TCR is: TCR= $\pm 7.935 \times 10^{-6}$ /°C.

C. Humidity testing

Eight samples by arbitrarily extracting were exposed to 85°C, 98% relative humidity (RH) for 24 hours and tested. The measured average value of resistance change was: $\Delta R/R = 5.793 \times 10^{-6}$.

D. High temperature storage testing

Seventeen chip resistor samples were exposed to 85°C for 5000 hours, and then the average value of resistance change was measured as: $\Delta R/R = 15.94 \times 10^{-6}$.

3. AES and ESCA Analysis

A. Surface Auger picture and AES analysis of CCNA strip conductor

Auger picture of Ni is shown in Fig. 1,

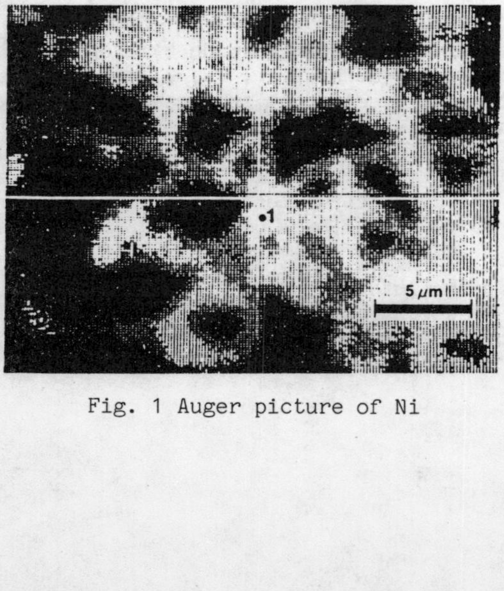

Fig. 1 Auger picture of Ni

which shows the surface fractional condensation in height. The Ni concentration area would be grain boundaries. The content of Ni and O_2 at point 1 in Fig. 1 is high as determined by automatic multipoint Auger analysis (see Fig.2).

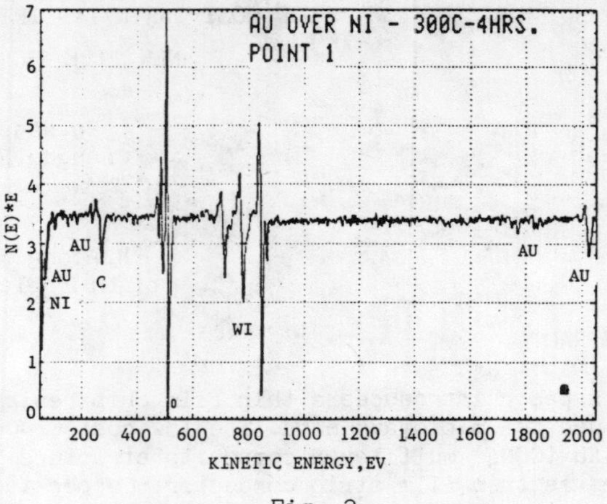

Fig. 2

Fig. 3 shows the relative space density of Au, Ni, and O_2 along with the horizontal line in Fig. 1 measured by Auger analysis.

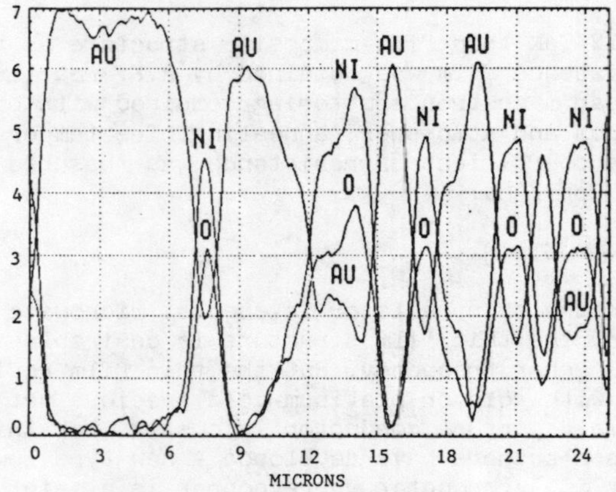

Fig. 3

In-depth profile of CCNA after 4 hours at 250°C using AES is shown in Fig. 4, and indicates that Ni barrier effectively prevents Cu and Au from interdiffusion.

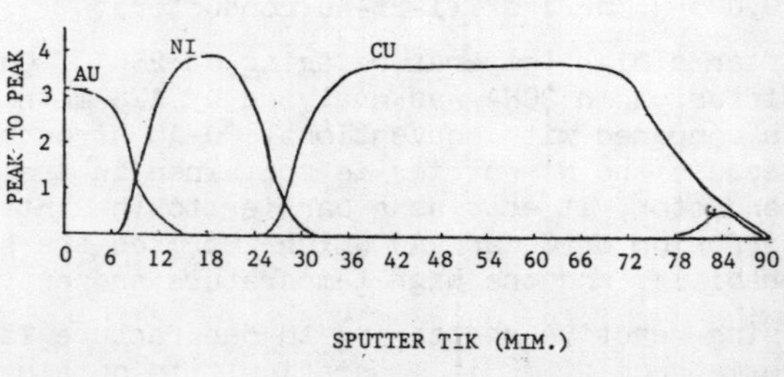

Fig. 4

B. ESCA analysis for TaAlN thin film[4]

ESCA analysis for TaAlN(Al:50 at. %) thin film indicates that basically the atomic ratio of Ta, Al, and Ni is kept the same within the whole film (see Fig. 5).

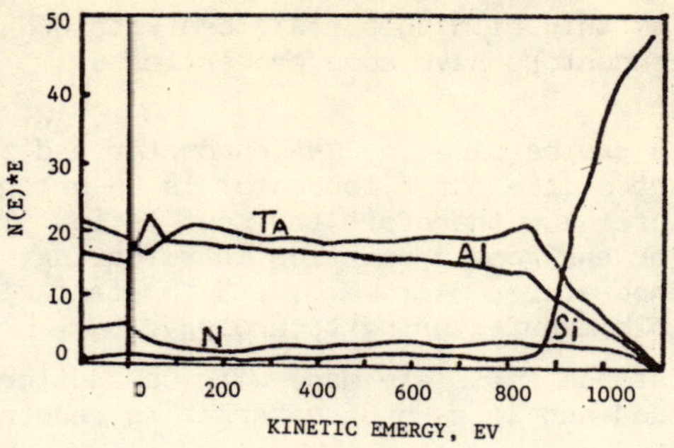

Fig. 5

DISCUSSION

1. To Improve the Film Property and Reduce Its Cost

Generally, Au and Pt-Au precious metals are used as the main part of strip conductors in thin film resistors at the microwave band, therefore the cost is expensive. Besides this, for Cr-Au thin film strip conductor, as gold is dissolved readily by Sn, and Cr does not match with Sn in the soldering process, the soldering characteristics to Cr-Au conductor are poor. For Cr-Cu-Au or FeCrAl-Cu-Au thin film strip conductor, the cost is low, but these stip conductor systems possess a poor high temperature property. In that conductor system interdiffusion between Cu and Au takes place after 250°C high temperature process. As copper diffuses into the gold surface, it will influence thermocompression bonding and sheet resistance for strip conductor. Pinnel, Bennett, and Borders et al.[5] have been indicated that grain boundary diffusion is the dominant diffusion mechanism for Cu and Au interdiffusion in the temperature range of 50~500°C. And interdiffusion takes place via a rapid saturation of the grain boundary in the gold film by copper to form $CuAu_3$ distributed uniformly along the gold surface, and a slower filling of the copper grain boundaries by gold to form Cu_3Au distributed at the Cu-Au interface[6].

To overcome Cu-Au interdiffusion, CCNA conductor was used for the thin film resistor where nickel plays a role as a diffusion barrier between Cu and Au layer. As compared with Pd, Pt, Rh barrier metals, the cost of Ni is lower and it is easy to process. Besides this the atomic radius of Ni is less than one of Cu, therefore the nickel was an effective barrier[7].

From Fig. 4 we are able to see that nickel is a diffusion barrier to Cu-Au interdiffusion after high temperature procedure. The contribution of authors is that the multicontinue evaporation procedure has been fulfilled, and CCNA strip conductor was made by an evaporation procedure only. The conductor manufactured by this way is much simple than other methods where Ti and Cu were evaporated followed by the plating of Cu, Ni, and Au[8]. And CCNA thin film strip conductor was used for the thin film chip resistor for the first time. Therefore it lays a foundation for manufacturing low cost thin film chip resistors in batch process.

The CCNA conductor possesses a superior soldering property as well and suits to a variety of soldering methods.

2. Reactive Magnetron Sputtering Ta-Al thin film by Doping with N_2

To change TCR and ρ of Ta-Al thin film, the doping method with reactive gas N_2 is commonly used for sputtering. The insulation compound (AlN) in atomic scale is formed in the thin film by doping with N_2, which makes ρ increase and TCR change[9]. But with excessive N_2 the anodization and long-term stability of film will become bad. Therefore at the time of reactive sputtering, the N_2 content must be controlled in a suitable range.

In our experiments the gas mixture set of sputtering system was modified, and a pressure controller was used. With this modified sputtering system the ratio of Ar and N_2 in the vacuum chamber and the gas flow can be controlled accurately[10]. The reactive sputtering stability was improved very much by this way. The TaAlN thin film with a constant ratio of Ta and Al, good uniformity and repeatability has been obtained.

3. Reduce the Manufacturing Cost of Target and Enhance the Uniformity Thickness of Film

A doped procedure with Al can enhance thermal oxidation-resistant property for uneasy melting metals. As Al content increase, the metal film will show a series of phase change. In our experiments amorphous films were obtained with 50 at. % Al. Under this condition the value of TCR is small, otherwise TCR will increase.

It is reported that Ta-Al target conventionally was made by Al foil inlaid into a Ta base target or by smelting Ta and Al. Generally the sputtering system with down-sputtering mechanism is commonly used. In our experiment we used a up-sputtering system where Al is used as a base target with several Ta foils on it to form a Ta-Al target. According to requirement it can be easily to change the proportion. Because of Al as a base target, the cost of Ta-Al target is low. To improve the thickness uniformity of films, a planetary turning platform for substrates fixed was used. In terms of hanging five small substrate plates (Ø 120 mm) on a big turnable plate (Ø 250 mm), uniform films with large loading were obtained, and it is useful to reduce the manufacturing cost of resistors.

4. Thermal Treatment on TaAlN/TaN Composite Film

TaAl thin film doped with N_2, and Ta thin film doped with N_2 have similar properties. Therefore according to the dual conductive model of metal conduct and tunnel effect, the thin film resistivity ρ_T is estimated as follows:

$$\rho_T = \frac{(m_1 T + c_1)[A \exp(\Delta E/kT)]\ t}{t_1 A \exp(\Delta E/kT) + t_2 (m_1 T + c_1)} \qquad (1)$$

where m_1, c_1, and A are constant, respectively; ΔE: activation energy; k: Boltzmann constant; T: absolute temperature; $(m_1 T+c_1)$: the contribution of metal to resistivity; $A \exp(\Delta E/kT)$: the contribution of tunnel effect to resistivity; t_1 and t_2 are proportion coefficient on metal conduct and tunnel effect conduct, respectively; $t=t_1+t_2$.

Usually the resistance film has two interface. To enhance the density of surface oxide layer, authors have specially designed the thin film resistance with TaAlN/TaN composite structure. The stability of the composite thin film resistance is influenced mainly by the change of oxygen gradient between the resistance film and its oxidation interface. Through the grain boundaries or other paths, oxygen is diffused into the resistance film, which makes the oxygen gradient at the interface gradually decrease. In this case the resistance change is confirmed to

$$\frac{\Delta R}{R} = A\dot{T}^{\frac{1}{4}} \qquad (2)$$

where A is constant, $\dot{T}$ is time.

To enhance the stability of the resistance film, the method with both vacuum thermal treatment and atmosphere thermal treatment are used[11]. By means of thermal treatment the internal stress of TaAlN/TaN will be eliminated, thus the crystal lattice will be rearranged. TCR of the two films can be compensated by means of selecting a suitable temperature as thermal treatment, and then the total TCR becomes low because the resistance film is composite structure. Furthermore it makes the surface of the resistor form a dual protected stable layer of Al_2O_3 and Ta_2O_5, which enhance the stability for resistors. Under the vacuum heat treatment as temperature is less than 750 °C, the change of TCR is small; as t>750°C, the change becomes larger; and at t=800°C, the TCR changes from negative to positive value. Besides these, TCR is related with the time of treatment, the proportion of Al, and the thickness of film. For example, after 2.5 hours treatment TCR is zero for atomic ratio of 56% Al in the target. In the case of 50% Al, however, the treatment time becomes short to make TCR zero. For the thicker film long treatment time is preferable for zero TCR because the time fo O_2 doping is long.

CONCLUSION

TaAlN/TaN thin film chip resistors with CCNA strip conductors have some properties as follows:

A. Ni is the barrier in CCNA conductor and the main part of the strip conductor is Cu which substitutes Au, therefore the cost is low. The conductor has good properties in soldering, high temperature, microwave, and together with its simple manufacturing technology.

B. Resistance films are made with up-sputtering method and Al as a base target to reduce the cost.

C. Resistance films made by reactive sputtering possess a stable component ratio and a good repeatability, because the flow and the ratio of Ar and N_2 can be controlled accurately.

D. In terms of composition with vacuum and atmosphere heat treatment the TCR was reduced and the resistance stability was enhanced.

ACKNOWLEDGMENT

The authors would like to thank Prof. Qu Xixin, Chengdu Telecommunication Engineering University, for his check and approve.

REFERENCES

1. You,W.M. and Hu, N.S. Microwave Journal (in Chinese), 2(1989).
2. Hu,N.S. Yao,G.R. Yu,Z.Z. and Qiao,S.Z. 35th E.C.C. Washington, DC. May (1985), 261.
3. Hu,N.S. Electronic Components, 3(1984).
4. Hu,N.S. Electronic Science. 2(1984),14.
5. J.A.Borders, Thin Solid Film, 19(1973),359.
6. J.M.Morabito, J.H.Thomas, IEEE, PHP11, 4(1975), 253.
7. Qu,X.X. Electronic Components and Materials, 4(1983),60.
8. M.J.Bill, Unpublished Bell Laboratories Data, June 1(1971).
9. Qu,X.X. Film Physics, Shanghai Science Technology Press, (1986).
10. Yu,Z.Z. Hu,N.S. and Cai,B.C. 4th IMC, Kobe. May (1986),165.
11. Hu,N.S. Electronic Components and Materials, 4(1983). 72.

MONOLITHIC INTEGRATED MIM CAPACITORS AND INDUCTORS

Feng Shen, Mao Jinghuai, Qu Weixuan, Lu Xiaobing
Nanjing Electronic Devices Research Institute
Nanjing, China

I. INTRODUCTION

Monolithic integrated capacitor and inductor are two important components in GaAs monolithic microwave integrated circuit. This paper describes the fabrication technology, design models and applications of monolithic integrated Metal-Insulator-Metal (MIM) capacitors and inductors with vertical broken-line and square spiral structures, which have been made in Nanjing Electronic Devices Research Institute (NEI), China.

II. FABRICATION TECHNOLOGY

Si_3N_4 dielectric film of MIM capacitor has been grown by magnetic-controlled RF sputter, in which a continuous two rotating sputter technology has been used. Compared with the Si_3N_4 and SiO_2 made by PECVD and Al_2O_3 made by anode oxidation, the film can realize a more accurate thichness, lower density of pin-hole, higher breakdown voltage, and is completely suitable for GaAs Monolithic Microwave Integrated Circuits (MMIC) technology. By means of AuGeNi/Au as the down electrode instead of Al, the fabrication process has been simplified and the capacitor yield has been increased. The upper electrode of the capacitor is Ti/Au structure, and is selectively plated to 2.5-3.0μm in order to reduce the loss. the bonding pad and the upper electrode are over-connected by a air-bridge. The capacitors have a capacitance of $200pF/mm^2$ and the breakdown voltage over 20 volts, and the yield of 11 pF and 2.6 pF capacitors have reached 90% and 95%, respectively. A photograph of the MIM capacitors has been shown in Fig.1(a).

The adhesion between the monolithic inductor and GaAs substrate is a critical parameter in a high yield processing technology. In our work, the AuGeNi/Au or Ti/Au has been used for the ground metallized layer, which has achieved a good adhesion and matched other processes of GaAs MMICs perfectely. Air-bridges have been used for the spiral lines over-connecting the leading line from the inner electrode to the bonding pad, which has reduced the parasitic couple, and the spiral lines have been selectively plated to 2.5-3.0μm. The 7nH vertical broken-line inductor and 1-2 nH square spiral inductors (4-types) have the yield over 97% and 95%, respectively. A photograph of the four type square spiral inductors has been shown in Fig.1(b).

III. DESIGN MODEL

Fig.2(a) and (b) have given design models for monolithic integrated MIM capacitor and square spiral inductor, respectively. In the design model of MIM capacitor, the capacitor has been considered a low impedance microstrip line with a open terminal, and the influences of the down electrode and the back grounding plate of GaAs chip have been also included. Based on the theory of coupling microstrip lines, the back grounding plate and coupling effect among the spiral lines have been considered in the model of monolithic inductors. Meanwhile, the loss of the dielectrics (Si_3N_4, GaAs) and metallization layers can be calculated in these models. GaAs monolithic integrated MIM capacitors and inductors have been analysed on IBM PC/XT computer with a MMIC lumped elements CAD program —— SLECAD, which has been developed in NEI. Some data have been given through DC, low frquences (1KHz and 1MHz) and microwave measurements. The experimental data of capacitance, inductance and Q-factor are in agreement with the modeling predictions.

IV. APPLICATION

1) 7 nH vertical broken-line inductor for 4-8 GHz monolithic integrated SPDT switch. A photograph of the switch has been shown in Fig.3(a). From 4.5 to 8.0 GHz, the SPDT switch has reached a insertion loss of 1 to 2 dB, isolation better than 21 dB, and 3 watts power-handling capability at 5 GHz. Experiments also verify the switching time being less than 0.6 nano-second.

2) 2.6 pF and 11 pF MIM capacitors for X-band and Ku-band MMIC low noise amplifiers and filters. A photograph of these amplifiers has been shown in Fig.3(b). A two stage X-band amplifier has 2.8dB noise figure and 17.5 dB associated gain at 9.5 GHz, and a two stage Ku-band ammplifier has 3.0 dB noise

figure and 14.0 dB associated gain at 12.4 GHz.

3) 1-2 nH square spiral inductors for integrated filters. High-pass and low-pass filters with 2.6 pF MIM capacitor and 1.2 nH square spiral inductor (spiral number=3.5) have been measured by 8510B network analyzer from 0.2 to 18.0 GHz. A further research has been conducting in NEI in order to reduce the insertion loss in the filters and to improve the design models.

REFERENCE:

<1> S. Feng, S.S.E. Research & Development, Vol. 8, No. 2.

<2> S. Feng, Semiconductor Magzine, No. 4, Aug. 1988.

<3> J. H. Mao, S. Feng, etal, S.S.E. Research & Development, Vol.9, NO.1.

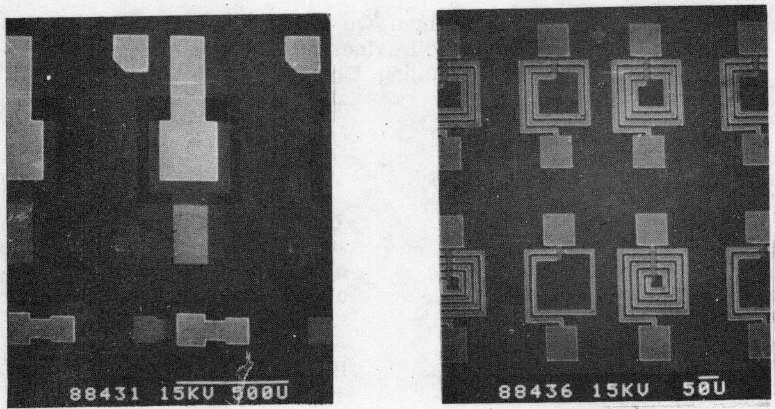

(a) (b)

Fig.1 Photographs of monolithic MIM capacitors(a),inductors(b)

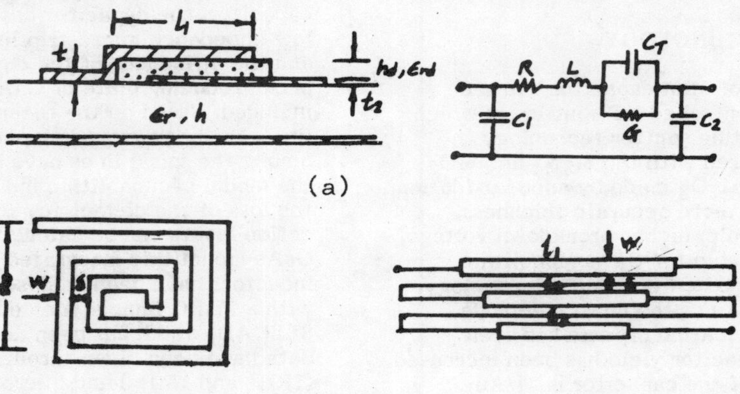

(a)

(b)

Fig.2 Design models of MIM capacitor(a) and inductor(b)

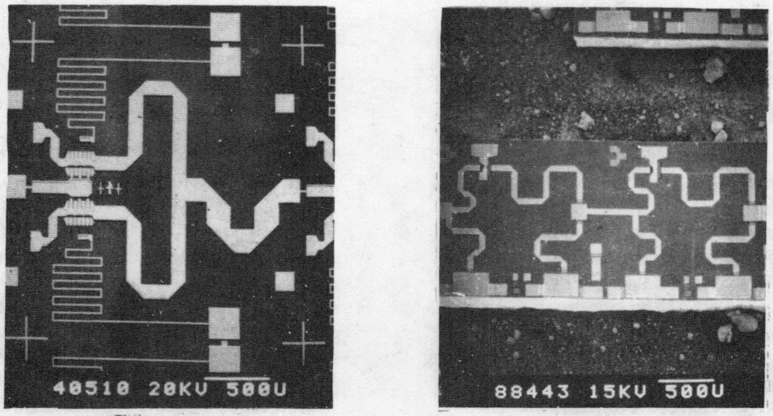

(a) (b)

Fig.3 GaAs monolithic SPDT switch(a) and X-band LNA(b)

Growth of $KNbO_3$ Single Crystal for Optical Nonlinear Application

K.Murata, H.Kumagai, N.Kito, Y.Tanabe, K.Sato and T.Fukuda*

Research Center, Asahi Glass Co.,Ltd.
1150 Hazawa-cho, Kanagawa-ku, Yokohama, Japan.
*Institute for Materials Research, Tohoku University.
2-1-1 Katahira, Aoba-ku, Sendai, Japan.

ABSTRACT

Phase diagram of K_2O-Nb_2O_5 system between 46 and 57 mol% K_2O is investigated with DTA. There is no sign of the metastable phase near 54.5 mol% K_2O which is reported by Flückiger. Suitable initial composition for top seeded solution growth (TSSG) is determined to be 52.5 mol%K_2O by the investigation of the phase diagram and growth experiments of slow cooling method.

$KNbO_3$ single crystals up to 30×30×17 mm^3 were grown by TSSG apparatus which has three zone heaters. Thickness of the crystals depend on vertical temperature profile of the apparatus and the crystal rotation rate.

Bundles of subgrain propagated from a seed crystal towards the center of crystal faces. The subgrain-free crystals are grown by increase of the crystal rotation rate up to 80 rpm.

Some crystals show colors, which depend on the solution composition and the soaking conditions. The blue crystals are not bleached with heat treatments.

Second Harmonic Generation (SHG) in the single domain crystals was performed with CW dye laser. The shortest phase matching wavelength of the crystal was 860nm at room temperature.

INTRODUCTION

$KNbO_3$ single crystal is one of the most promising nonlinear optical materials because it has the largest nonlinear figure of merit d^2/n^3 among ferroelectric oxides[1]. Recently, device oriented investigation on SHG and holographic storage has been reported by many authors[2,3,4]. However, $KNbO_3$ crystal is not widely used because it is very difficult to get good quality and large crystals.

Phase diagram of K_2O-Nb_2O_5 system is very important to crystal growth. But two different phase diagram were proposed.

One is first reported by Reisman et al., which shows the incongruent melting of $KNbO_3$[5]. The other is reported by Flückiger et al., who investigated between 46 and 57 mol% K_2O in detail[6]. They suggest that the $KNbO_3$ is incongruent composition and there is a metastable phase at 54.5 mol% K_2O.

If the metastable phase exists, possible initial composition of the solution for crystal growth should be limited to very narrow range.

We make a phase diagram between 46 and 57 mol%K_2O by DTA to confirm the existence of the metastable phase and determine the possible composition range for the crystal growth.

Optimum initial solution composition for TSSG is determined by the investigation of the phase diagram and growth experiments of the slow cooling method. $KNbO_3$ crystals are grown by TSSG method with the optimum initial composition.

We discuss defects of the crystals such as color and subgrain bundles. SHG in the crystal is performed with CW dye laser.

PHASE DIAGRAM AND INITIAL COMPOSITION FOR THE CRYSTAL GROWTH

K_2CO_3 and Nb_2O_5 powders of high purity (99.99%) were mixed into compositions, $(K_2CO_3/(K_2CO_3+Nb_2O_5))$ = 47.5 - 58 mol%. The mixture was placed in a 5mmϕ×3mmh platinum cell and DTA apparatus (Rigaku, DTA-TG8112BH). All measurements were carried out in the air.

The heating and cooling rate were 10°C/min. The reaction of K_2CO_3 and Nb_2O_5 was observed about 720°C by DTA peaks and TG curves. After the reaction and clarification at 1200°C, DTA was measured during another heating process.

Fig. 1 shows the DTA curves of 47.5, 52.5 and 55.5 mol%K_2O. The liquidus temperature of $2K_2O{\cdot}3Nb_2O_5$ was not determined because the endothermic peaks were small and uncertain.

Liquidus, peritectic and eutectic curves are shown in Fig. 2 . The peritectic temperature is 1055°C. The composition of the peritectic is estimated at between 50 - 52 mol% K_2O.

There is no sign of the metastable phase near 54.5 mol%K_2O which was reported by Flückiger. Initial composition between the peritectic and the eutectic are possible for the growth of $KNbO_3$ crystal.

Some crystals were grown by the slow cooling method from solutions containing 50.7, 52.5 and 54.0 mol% K_2O. The solutions in 50cc platinum crucible are maintained at 1200°C for 10hr and cooled at rates at 6, 40 and 60°C/hr for the growth, separately.

$KNbO_3$ crystals up to 1 cm^3 were obtained by this method. Some crystals show various colors such as pale reddish brown and dark blue depending on the growth conditions. The colors and the growth conditions are tabulated in Table 1.

When the initial composition is 52.5 mol% K_2O and cooling rate is below 6°C/hr, grown crystals are mostly transparent. When the cooling rate is faster than 6°C/hr or the starting composition is deviated from 52.5 mol%K_2O, the grown crystal shows blue or pale reddish brown.

The color of crystals is undesirable for optical applications. We suppose that 52.5 mol% K_2O is the most suitable initial composition for the transparent $KNbO_3$ crystal growth.

CRYSTAL GROWTH BY TSSG METHOD

K_2CO_3:Nb_2O_5 = 52.5:47.5 (mol ratio) is adopted as the initial composition for TSSG experiments. The TSSG furnace is shown in Fig.3. We developed the furnace with three zone heaters to change vertical temperature profile and control crystal nucleation and growth rate. 450g and 1500g of solution are

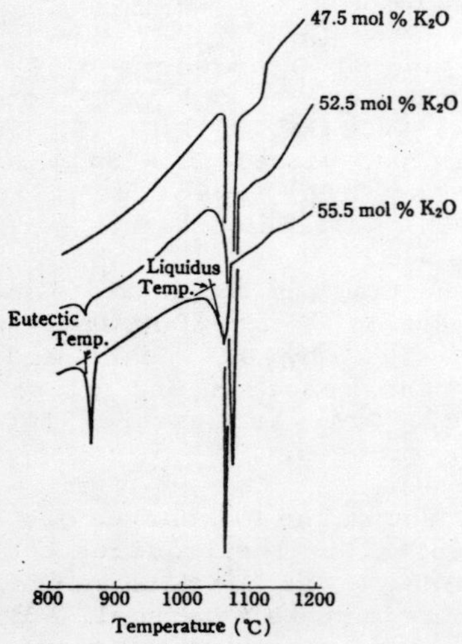

Fig. 1 DTA curves of 47.5, 52.5 and 55.5 mol%K_2O.

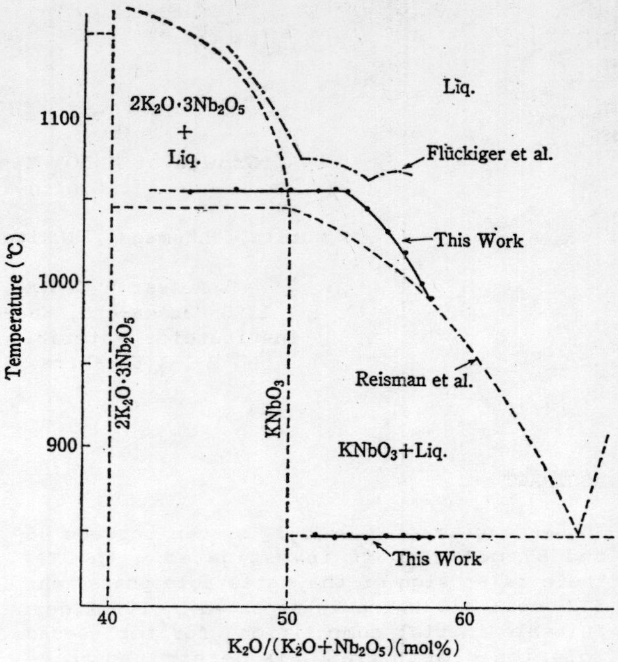

Fig. 2 Phase diagram of K_2O-Nb_2O_5 system.

contained in a 150cc and 550cc platinum crucible, respectively. The growth conditions are listed in Table 2. All experiments are carried out in the air.

The single crystals up to 30×30×17 mm^3 were grown. The crystals showed simple crystallographic large "{100}" faces. The thickness of the crystals depend on the temperature profiles and the crystals rotation rates.

Fig.4 shows subgrains of the crystals. In crystal grown under crystal rotation rate below 60rpm, bundles of subgrains are observed to propagate from seed juncture toward the center of crystal faces. The crystal orientation of the bundles are deviated up to 5 degrees from outer matrix crystals.

Table 1 Crystal colors and growth conditions of the slow cooling method.

Cooling Rate	K_2O concentration (K_2O mol%)		
	50.7	52.5	54.0
6°C/hr	Pale Reddish Brown	Colorless	Blue
40°C/hr	Pale Reddish Brown	Blue	Blue
60°C/hr	Pale Reddish Brown and Blue	Blue	Blue

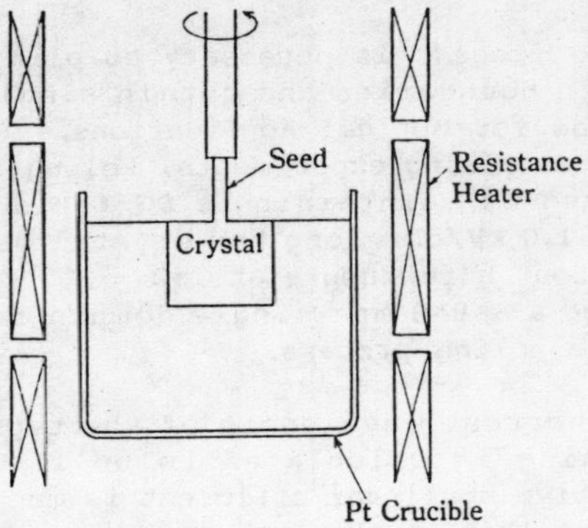

Fig. 3 Apparatus of top seeded solution growth method.

Table 2 Growth condition of TSSG method.

Starting composition	$K_2CO_3:Nb_2O_5$ =52.5:47.5 (mol)
Seed Orientation	‹100› (cubic)
Growth Cooling Rate	0.2 - 2.0 deg/hr
Crystal Rotation Rate	30 - 100 rpm

The subgrain bundles are often surrounded cloudy milky color which consists of complicated structure of ferroelectric domains and twin boundaries (Fig.5). The domains and the twins bound with the subgrains can not be removed by poling process.

Subgrain-free crystals are grown by increase of the crystal rotation rate above 80rpm.

We suppose that the increase of the crystal rotation promotes stirring of the solution and supply of solute to the crystal-liquid interface so that the subgrains are repressed.

Fig. 4 Subgrains of $KNbO_3$ crystal surface.

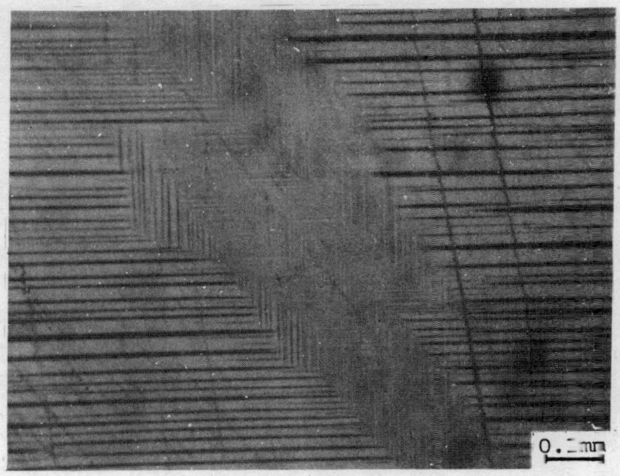

Fig. 5 Microphotograph of twins and multi-domain structure.

In some cases, blue thin layers are observed in the crystals (Fig.6), although the crystals are grown from the determined optimum composition 52.5mol% K_2O . The layers are perpendicular to growth direction and constitute growth striation of (100) faces.

More inclusions between 1 and 10 μm are distributed in the color layers rather than in colorless regions(Fig.7).

It is suggested that the discoloration as well as inclusion is an effect of unstable growth.

We found that high temperature and long time soaking (at 1250°C for 100hr) led to the complete colorless crystal growth, and low temperature and/or short time soaking (at 1100°C for 10hr ,at 1250°C for 30hr) led to the coloration.

Insufficient soaking which has an effect on the discoloration, is assumed to cause inhomogeneity of solution.

It is suggest that the coloration is attributed to the instability of growth rate and inhomogeneity of solution.

Fig. 6 Blue thin layers of $KNbO_3$ crystal.

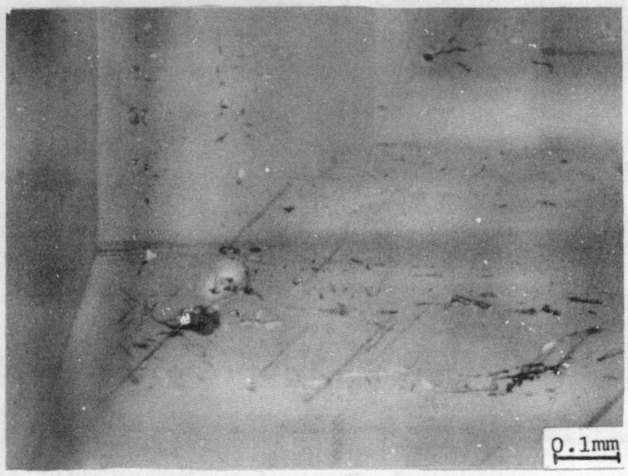

Fig. 7 Many inclusions, distributed in the color layers of Fig.6.

Effect of heat treatments on the crystals color was investigated. Two types of samples were prepared from crystal grown with the slow cooling method. One is colorless, and the other is blue in as-grown.

The heat treatment conditions are at 950°C ,10hours under reductive (H_2 gas flow)and oxidative (O_2 gas flow) atmosphere. The colorless crystals were colored to dark blue by the reductive heat treatment and reversibly bleached by the oxidative heat treatments. The blue crystals in as- grown were insensible on both heat treatments (Table 3).

These results suggest that there are at least two types blue color in $KNbO_3$ crystals. One is the discoloration by the reductive heat treatment that is attributed to oxygen vacancies. Another is the discoloration in as-grown crystals that is not simply attributed to oxygen vacancies but related to other defects.

POLING AND SHG

Since as-grown $KNbO_3$ crystals include many small ferroelectric domains and scatter incident light beams in the domain boundaries,

Table 3 Effects of heat treatment on crystals color.

As-Grown	H_2 Gas Flow, 950°C, 10hr	O_2 Gas Flow, 950°C, 10hr
Colorless →	Dark Blue →	Colorless
Blue →	Dark Blue →	Blue
Blue →	→	Blue

poling process is necessary to eliminate the domain boundaries and obtain single domain crystal for optical applications.

In our poling experiments, Poling is accomplished by maintaining a DC-field of about 0.5 - 1.0 kV/cm along the crystal polar axis for about a few hours at 150 - 200°C. We obtained a 8×8×8 mm^3 single domain crystal of by the poling process.

The shortest wavelength of the type 2 phase matching is calculated to be 1170 nm and effective nonlinear efficient is much smaller than in that of the type 1 phase matching.

In the type 1 phase matching, the shortest wavelength is calculated to be 843nm. In our experiments with a cw dye laser, 860nm was the shortest phase matching wavelength.

CONCLUSION

1. The K_2O-Nb_2O_5 system between 46 and 57 mol% K_2O phase diagram in detail was studied by DTA. there is no indication of the metastable phase near 54.5 mol% K_2O.

2. $KNbO_3$ single crystals of up to 30×30×17 mm^3 were grown from 52.5mol% K_2O by Top Seeded Solution Growth (TSSG) method. The crystal thickness are affected by the temperature profile of the furnace and the crystal rotation rate. The subgrain bundles are repressed by increase of the crystal rotation rate up to 80 rpm.

3. Some crystals show blue layers which are not bleached with heat treatments. The coloration as well as inclusion is an effect of unstable growth. And insufficient soaking which cause inhomogeneity of solution, has an effect on the discoloration.

4. The single domain crystals obtained by poling processes are used for Second Harmonic Generation (SHG) experiments. The shortest phase matching wavelength is 860nm.

REFERENCES

1. Uematsu,Y., Japan.J.Appl.Phys., 13 (1974) 1362
2. Uematsu,Y. and Fukuda,T., Japan.J.Appl. Phys., 12 (1973) 841.
3. Baumert,J.C., Günter,P. and Melchior,H., Optics Comm., 48 (1983) 215.
4. Krumins,A. and Günter,P., Appl.Phys., 19 (1979) 153.
5. Reisman,A. and Holtzberg,F., J.Am.Chem. Soc., 77 (1955) 2115.
6. Flückiger,U. and Arend,H., J.Cryst. Growth, 43 (1978) 406.

HIGH PRECISION SC-CUT CRYSTAL UNIT

Luo Fu Zhen
Chen Xing Electronic Component Factory
Beijing, China

ABSTRACT

We have produced in excess of 500 of the 5MHz 3rd overtone SC-cut crystal units and 100 of the 10MHz 3rd overtone SC-cut units to date . Each crystal unit produced is measured in an automatic test system which records crystal Q, ressitance , and frequency at multiple temperatures as well as the equivalent circuit parameters .

The frequency-temperature characteristics of the SC-cut is much better than the AT-cut . The paper also gives the frequency stability of the SC-cut precision crystal units , the aging rate is 8.6×10^{-12}/day for a 5MHz 3rd overtone cryatal unit and the short term stability is 3.6×10^{-13}/s .

INTRODUCTION

Since electrode stresses cause a large component of long term aging for high precision crystal units . It is necessary to find a new cut which have least stresses coefficient to improve the aging of crystal units . EerNisse was the first man to calculate the stresses coefficient of the C-mode thickness-shear resonators along the locus of the zero temperature coefficient . On the basis of theoretical calculations , EerNisse predicted that a doubly rotated quartz plate cut at the angles of ϕ =22.5° and θ =34.3° should be free of frequency change due to stresses . He called this new cut the SC-cut (stress compensated). In 1977, Kusters et al showed experimentally that the results predicted by EerNisse were obtained with a plate having the orientation angles ϕ =21.93° and θ =34.11° .

SC-cut has proved to have a number of advantages over the conventional AT-cut, among these are:[1]

1. Superior warm up characteristics due to reduction of the effects of stress associated with thermal grandients in the plate.
2. Reduced AF effect, oscillators can work at higher drive level.
3. Almost complete freedom from activity dips.
4. A flatter frequency-temperature curve, especially at the turnover points.
5. Reduced tipover effect.
6. Better short-term stability due to reduced response to thermal transients .
7. Reduced aging associated with relaxation of electrode stresses .
8. Relative immunity to the effects of radiation
9. Less aging due to changes in mounting stresses.

EXPERIMENT

We began to research the SC-cut resonators in 1978. The work reported here was mainly confined to the development of a 5MHz, 3rd overtone design which is suitable for high precision oven controlled oscillators. The orientation, Yxwl 21° 56′ /34° 08′ is determined to be the optimun cut. The crystal units are plano-convex, 15mm in diameter and polished on both major surfaces. The blanks are base-plated with gold film electrodes, 11mm in diameter. The base-plating are done by an oil free plating machine with an ion pump. Finally the resonators are sealed in a glass enclosure and pumped in an oil free vaccum system with 250℃ bake out.

1. Crystal Parameters

Up to date we have produced in excess of 500 of the 5MHz 3rd overtone SC-cut high precision crystal units. These units have been measured by an automatic test system (Saunders 2100) which records Q, frequency, resonance resistance, temperature coefficients, turnover points and equivalent circuit parameters of the crystal units. Table I shows a brief list of the parameters of the 5MHz 3rd overtone SC-cut crystal units .

Table I

Mode	C mode 3rd overtone
Frequency	5MHz
Resonance resistance	60-100 ohms
Q	2.4×10^{6}
Turnover point	Type A 55° - 65℃
	Type B 65° - 75℃
Shunt Capacitance	6.4 pF
Motional Capacitance	0.17 fF
Motional Inductance	6.1 Henry

From Table I we can see that the Q value of a SC-cut crystal units is much better than comparable AT-cut crystal units (1.6×10^{6} typically).

2. Frequency-Temperature Characteristics

A group of f-T curves of 5MHz, 3rd overtone crystal units are shown in Fig I. From Fig I we can see that the orientation angles is Yxwl 21° 56'/ 34° 08' for turnover point at 60℃. In the range of turnover point ±2℃ the frequency temperature coefficient can be no larger than 1×10^{-8}/℃, which is much less then that of AT-cut.

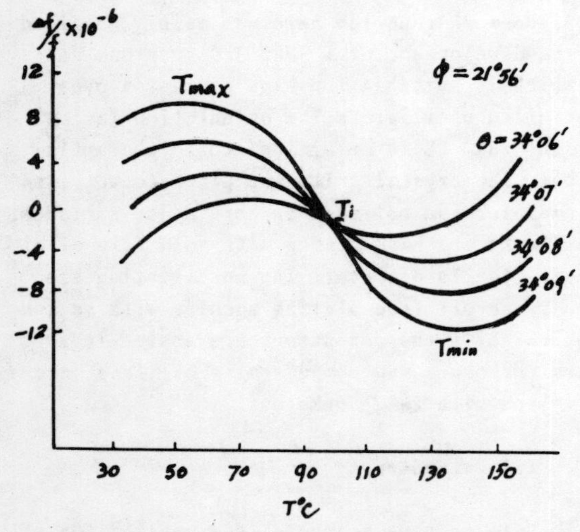

Fig. I f-T curves of SC-cut

3. Suppression of B-mode

It is well known that there are three modes in a thickness-vibration plate. They are the A-mode, B-mode and C-mode. The A-mode is a thickness extension mode, the B-mode is a fast thickness shear mode and the C-mode is a slow thickness shear mode. The C-mode is the wanted mode. For SC-cut the A-mode is far from the C-mode, we neednot consider it. But the B-mode is close to the C-mode, the frequency of the B-mode is only 9.4% higher than the frequency of the C-mode and it is very strong, even stronger than the C-mode. So it requires special attention in the design of oscillator and filter circuits in the case of employing the SC-cut crystal unit. In other hand it is important to suppress the B-mode in the design of the crystal units.

Three groups of the 5MHz, 3rd overtone SC-cut crystal units with different curvatures have been made. Table II shows the resonance resistance of the B-mode and the C-mode with different curvatures.

Table II

Radius of Curvature	100mm	120mm	150mm
Resonance Resistance of the B-mode Rr(B)	67.5Ω	63.8Ω	98.5Ω
Resonance Resistance of the C-mode Rr(C)	84.3Ω	70.9Ω	66.3Ω
Rr(B)/ /Rr(C)	0.80	0.90	1.38

It can be seen in Table II, the suppression of B-mode depends on the curvature of the plate. If Rr(B)/Rr(C) > 1, the B-mode can be suppressed when the radius of curvature is 150mm, the Rr(B)/Rr(C) is 1.38 the B-mode is well suppressed.

4. Frequency Stability

(1) Long term stability

The typical aging rate is 10^{-11}/day for 5MHz, 3rd overtone SC-cut high precision crystal units and the best is 8.6×10^{-12}/day.[2]

(2) Short term stability

Table III shows the short term stability of a 5MHz, 3rd overtone SC-cut high precision crystal unit used in an oven controlled oscillator.

Table III

Time domain (σ, τ)	
1ms	1.6×10E-10
10ms	1.7×10E-11
100ms	1.7×10E-12
1S	3.6×10E-13
Frequency domaim (phase noise spectrum density)	
10Hz	-133.5db/Hz
100Hz	-143.5db/Hz
1KHz	-150db/Hz
10KHz	-152db/Hz

From the data above we can see SC-cut crystal units have either excellent long term stability or superior short term stability.

5. 10MHZ 3rd Overtone SC-cut Crystal Units

In the end of this paper, we would like to introduce something about our research work in 10MHz 3rd overtone SC-cut crystal units simply.
The blanks are plano-convex, 14mm in diameter and polished on both major surfaces, plated with gold film electrodes 8.5mm in diameter. The manufacture process are same to the 5MHz 3rd overtone SC-cut crystal units. Table IV shows the parameters of the 10MHz, 3rd overtone SC-cut crystal units.

Table IV

Mode	C mode 3rd overtone
Frequency	10MHz
Resonance Resistance	60-80 ohms
Q	1.2×10^{6}
Shunt Capacitance	5.8 pF
Motional Capacitance	0.158 fF
Motional Inductance	1.6 Henry

The aging rate of the 10MHz 3rd overtone SC-cut high precision crystal units is 10^{-11}/day. The short term stability of a 10MHz, 3rd overtone SC-cut high precision crystal unit is shown in Table V.

Table V

Time domain (σ, τ)	
1ms	1.4×10E-10
10ms	1.1×10E-11
100ms	1.3×10E-12
1s	5.7×10E-13
Frequency domain (phase noise spectrum density)	
10Hz	-128db/Hz
100Hz	-138db/Hz
1kHz	-144db/Hz
10kHz	-146.5db/Hz

Up to date we have produced more than 100 of the 10MHz, 3rd overtone SC-cut crystal units.

CONCLUSION

Doubly rotated SC-cut crystal units have mamy advantage compare with the AT-cut, such as high Q, good short term stability, low aging rate and excellent frequency-temperature characteristics. Now the SC-cut crystal units have already come out of the laboratory and into the marketplace. We have produced high precision SC-cut 5MHz and 10MHz crystal units for several years and the products are widely used in our country.

REFERENCES

1. V.E.Bottom "Introduction to Quartz Crystal Units Design" Van Norstrand Reinhold company 1982.
2. Luo Fu Zhen "High Precision SC-cut Crystal Resonator" Journal of Hunan University Vol 15. No.3 .

A STUDY OF QUARTZ CRYSTAL ELECTRODIFFUSION

Gao Yuying, Luo Zhaoyuan and Fan Zhenqiang
Institute of Radio Metrology and Measurement
Beijing, China

ABSTRACT

This paper describes the experimental results of quartz crystal electrodiffusion (sweeping). Sweeping technique has been used to process natural and synthetic quartz. Its effects have been evaluated by ionic migration current measurement and infrared absorption spectra and X-ray diffraction spectra analyses. A set of AT-cut 5 MHz resonators with good performance have been fabricated from swept quartz. Measurements taken on an oscillator using one of the resonators have shown satisfactory results, namely, aging rate of 3.56×10^{-11}/day, frequency stability of 1.5×10^{-10}/1ms, 1.9×10^{-11}/10ms, 1.8×10^{-12}/100ms and 8.6×10^{-13}/1s, and phase noise of -137.5 db(10Hz), -148db(100Hz), -151db(1kHz) and -152db(10kHz). Frequency of swept and unswept resonators exposed to Co-60 gamma radiation have been measured. The results show that the frequency stability of swept quartz has been substantially improved.
Keywords, electrodiffusion(sweeping), swept quartz crystal, quartz resonator AT-cut, frequency stability, quartz oscillator.

INTRODUCTION

As early as in the 1950s, A. Kats discovered the migration phenomena of impurity ions in quartz crystal electrodiffusion experiments[1]. Since then, numerous authors held that sweeping could change the positive charge of compensating ions, stabilize defects due to Al^{3+} in quartz lattice, enhance the antiradiation behavior of quartz resonators and so on. But little study of quartz sweeping was made in china. The authors of this paper have used sweeping technique to process high Q quartz crystals in an attempt to obtain a set of high quality resonators. Effects of sweeping on natural and synthetic quartz have been evaluated by measuring the migration current of ions, infrared absorption spectra and X-ray diffraction spectra.

EXPERIMENTAL PROCEDURE

1. Sweeping and measurements

Z-cut 50 x 18 x 18mm plates have been prepared from natural and synthetic quartz crystals and have been each sputtered with thin platinum electrodes on both Z-faces to improve the electric contact. As shown in the sweeping apparatus schematic in Fig.1, it is placed in turn in a boron nitride container enclosed in a 50 mm diameter quartz tube which extends at each end about 30 cm outside a furnace with automatic temperature control. High d. c. voltages (applied across the plates) of 1000 V/cm heating temperature of 500°C and heating periods of 100 hours have been used. Migration currents have been measured at two temperature rise rates, i. e., 25°C/30min and 10°C/60min. Infrared spetra of natural and synthetic swept quartz crystals have been examined using infrared spectrophotometer at 25°C and 100° K. X-ray diffraction spectra of synthetic quartz before and after sweeping have been measured with a diffractometer.

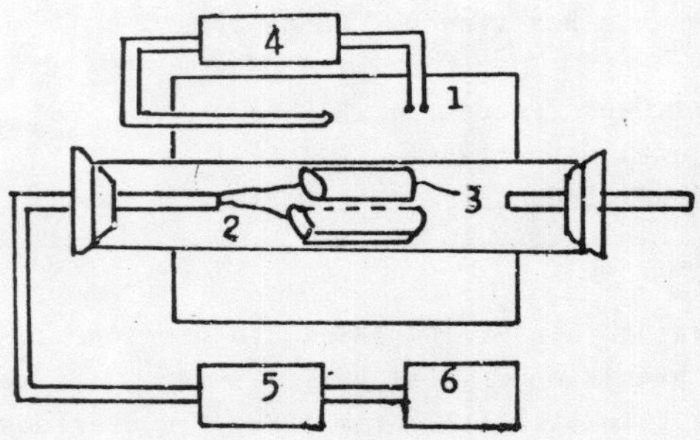

(1) furance (2) quartz (3) sample container
(4) temperature controller (5) micro ammeter
(6) voltage source

Fig.1 sweeping apparatus

2. Irradiation test

Swept resonators have been exposed to Co-60 gamma-ray of dose rate of 10^4 rads/hour for 28 hours. The frequencies and resistance of resonators have been measured during and after exposure.

3. Swept quartz resonators

Swept resonators have been fabricated by standard technology. Except five spoiled in fabrication, nine resonators have turned out to be satisfactory. Electric performance of swept resonators has been tested at room temperature.

RESULTS AND DISCUSSION

1. Migration current

Fig.2 gives the migration current as a function of time in sweeping run. The basic pattern of the current is the same as the results exhibited in reference (2), but we have found that the time for the current to reach peak value and then to fall to substate is related to the rate of temperature rise. The time of reaching peak value is 8 hours and 26 hours at the temperature rise rates of 25°C/30min and 10°C/60min respectively, whereas that of falling to substate is 13 and 68 hours at these temperature rise rates. The results show that choosing the best temperature rise rate is essential to improving the efficiency of sweeping.

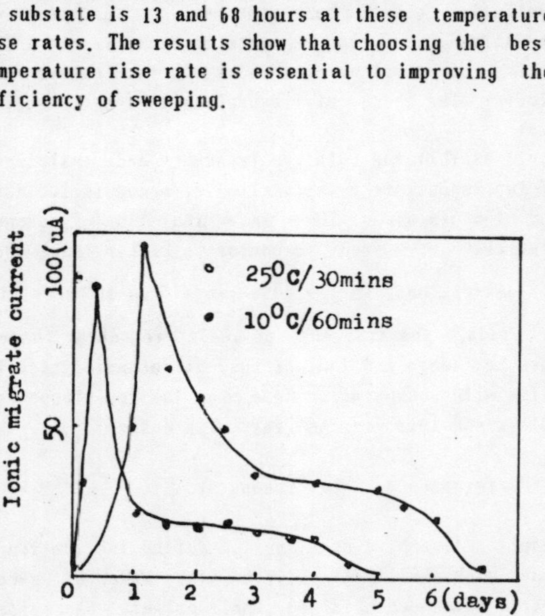

Fig.2 Ionic migration current

2. Infrared spectra of swept quartz

Infrared spectra of swept and unswept synthetic quartz at 25°C and 100° K are shown in Fig. :. Infrared spectra of swept and unswept natural quartz at 25°C and 100° K are given in Fig. 4(a) and Fig. 4(b) respectively. From these figures it is readily seen that infrared spectra of swept and unswept quartz are substantially different from each other between wave numbers of 2500cm^{-1} and 3000cm^{-1}. Infrared transmittance of swept natural quartz whether at room or low temperature, is higher than that of unswept one. As impurities in synthetic quartz are quite low, the difference in infrared transmittance of swept and unswept synthetic quartz is insignificant at room temperature, but it becomes very marked at low temperature. The results suggest that sweeping may decrease impurities in quartz. A quantitative analysis there of has yet to be made.

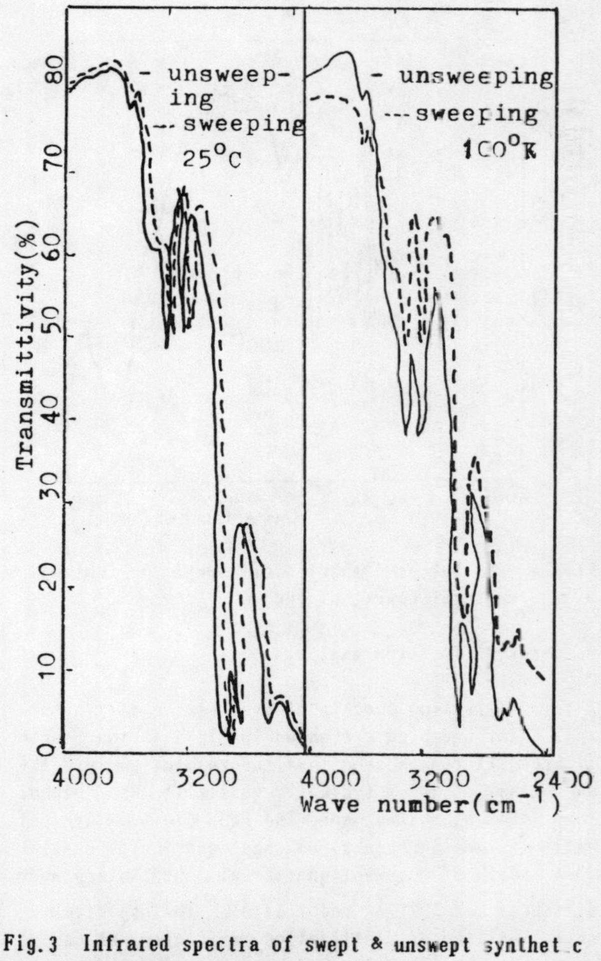

Fig.3 Infrared spectra of swept & unswept synthetic quartz at 25°C & 100° K

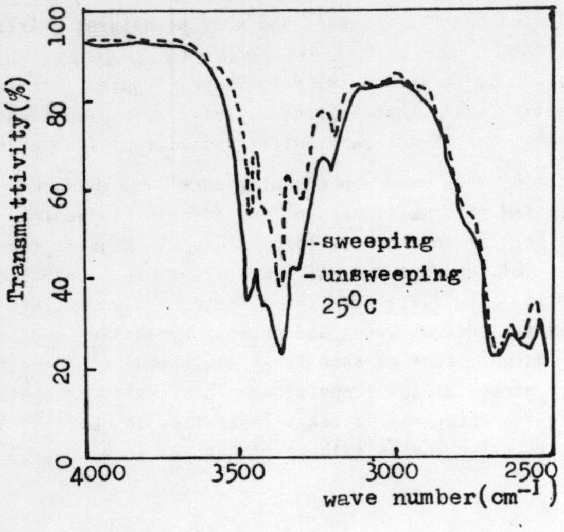

Fig. 4(a) Infrared spectra of swept & unswept natural quartz at 25°C

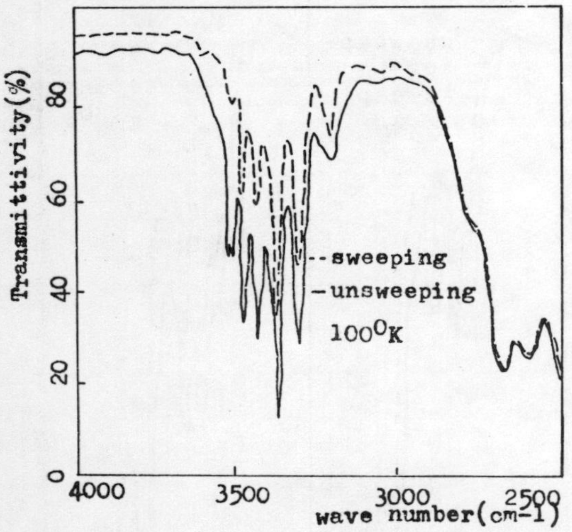

Fig. 4 (b) Infrared spectra of swept & unswept natural quartz at 100° K

3. X-ray diffraction analysis

X-ray diffraction spectra of synthetic quartz before and after sweeping are shown in Fig.5. From these figures it can be seen that the resolution of the Kα lines of swept synthetic quartz is much higher than that of unswept one. The FWHM (full width at half maximum Kα peaks) of swept quartz is smaller than that of unswept quartz at the interplanar distances of 1.817Å and 1.541Å . The spectra show that the crystallization perfection of swept quartz is significantly improved.

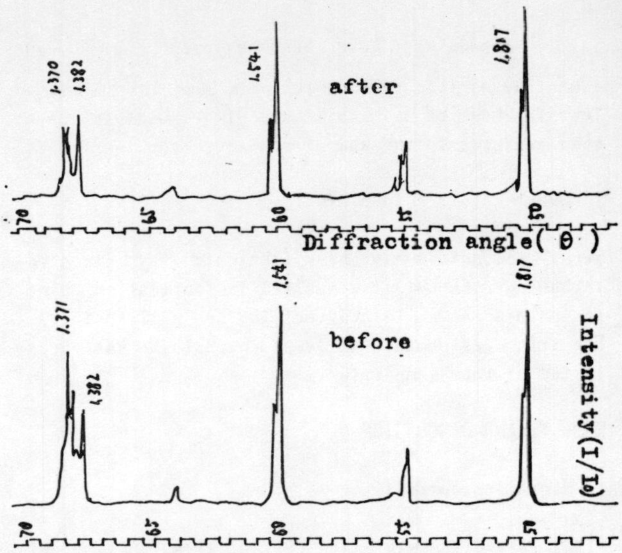

Fig.5 X-ray diffraction spectra of synthetic quartz before & after sweeping

4. Radiation effect on swept quartz

The relative frepuency drift of swept natural and synthetic quartz resonator as a function of radiation dose during exposure is shown in Fig.6 The curves of swept natural and synthetic quartz are similar. The change of frequency with dose is quite small.

Fig.7 exhibit the relative frequency and resistance of two resonators as a function of accumulation dose and temperature. They show that the frequency stability of a swept resonator is higher than that of unswept one. In the dose range from 10^4 rads to 10^6 rads, the frequency of swept resonator under dose no change and that of unswept resonator clearly falls with accumulation dose, but the resistance of both resonators does not vary with dose.

5. Performance of swept resonators

Despite five have been spoiled during fabrication, nine high quality AT-cut resonators made of swept quartz have been obtained. Their primary electrical parameters have been measured and listed in Table 1. Resonator No. ss84224 has been used in an oscillator and its primary electrical parameters are satisfactory as shown in Table 2. At the 4th National Quartz Oscillator Contest, its frequency stability and phase noise levels have won first places.

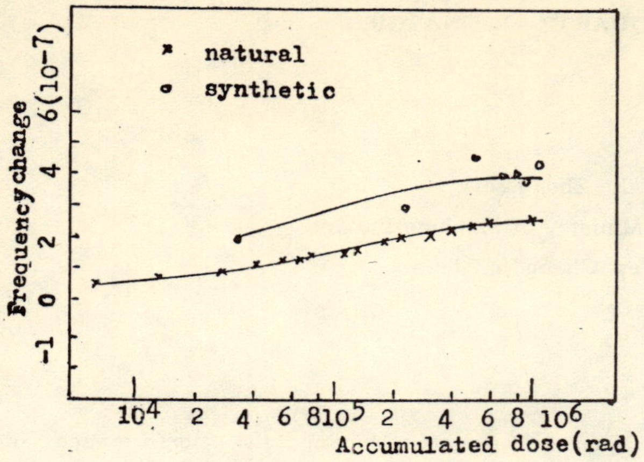

Fig.6 Radiation effect of swept natural & synthetic quartz resonators

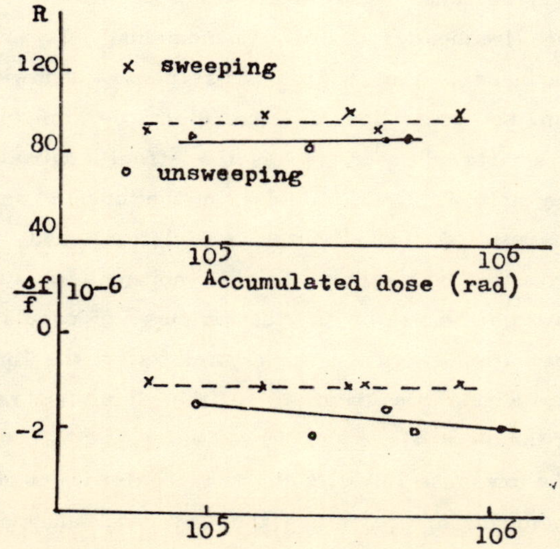

Fig. 7 Frequency & resistance of swept & unswept synthetic quartz resonators VS. accumulative dose

Table 1

No.	F(MHz)	f/F(10^{-6})	R(Ω)
ss84214	5	1.0	100
ss84215	5	1.0	95
ss84220	5	1.2	100
ss84221	5	1.7	100
ss84222	5	2.0	100
ss84223	5	1.2	100
ss84224	5	2.0	100
ss84226	5	1.0	95
ss84232	5	2.0	95

Table 2

parameters	test value			
Ageing rate	3.56×10^{-11}			
Frequency stability	1ms	10ms	100ms	1s
	1.5×10^{-10}	1.9×10^{-11}	1.8×10^{-12}	8.6×10^{-13}
Phase noise (db)	10Hz -137.5	100Hz -148	1kHz -151	10kHz -152

CONCLUSION

The above-mentioned results show that sweeping can enhance quartz purity and crystallization perfection and improve resonator performance.

ACKNOWLEDGMENT

The authors wish to thank all the participants in this study for their help.

REFERENCES .

1. Philips Journal of Research, No.2, Vol. 17, 1962.

2. Herbert G. Lipson, Ferdinand Euler & Alton F. Armington "Low temperature infrared absorption of impurities in high grade quartz" Proceeding of The 32th Annual Symposium on Frequency Control P11 1978.

APPLICATION OF ENERGY TRAPPING THEORY IN THE DEVELOPMENT AND DESIGN OF LOW NOISE QUARTZ RESONATOR

Chen Zhiyuan   Xu Zhongmei
The 10th Research Institute of Ministry of Machine Building and Electronics Industry, Chengdu, China

ABSTRACT

In this paper the results of energy trapping analysis by Shouckley, etc.(1) is applied to control noise in quartz resonator, the AT-cut quartz resonator wafer can be equivalent to a sound waveguide, from which the "response bandwidth" and the "inter modulation" can be obtained, its the source of noise in resonator, the "response bandwidth" and the "inter modulation"(2) can be controlled by adjusting the electrode thickness and diameter ratio of resonator. The low noise 100 MHz quartz resonator is designed by above method, some results are listed and discussed, it is shown that the design method is effective, the short period frequency stability over $2x10^{-11}/1$ ms and $5.5x10^{-12}$/sec are obtain through application of 100 MHz quartz resonator in oscillator.

INTRODUCTION

In order to measure and control accurately and reliably the rapidly flying target or transfer rapidly information, the crystal oscillator as a master frequency source in the radar and communication equipment is required to have the properties such as short period frequency stability and low noise, etc. For this reason it is also necessary to develop and design the quartz resonator having the performance of excellent "Squclch" besides perfect desgin of low noise oscillator circuits.

On the basis of developing the self noise in quartz resonator the study of behaviour mechanism on quartz crystal vibrator from inner or outer noise is emphasized in this article. According to the trapped energy mode of quartz vibrator and its analysis made by Shouckley etc. the quartz resonant wafer is equivalent to the sound waveguide for transmission of sound wave, and the "response bandwidth" which is drawn from the frequency cut-off phenomenon of sound waveguide, the width of response bandwidth can be controlled by the electrode thickness of quartz vibrator, it has the effect controlling the noises. The FM noise in the quartz wafer is resulted from "inter modulation" and the criterion controlling the FM noise is obtained through the use of distribution curve of caculated resonation energy in the crystal wafer, the "inter modulation" is relative to the diameter ratio of the electrode to the crystal wafer.

The low noise 100 MHz quartz resonator is designed by means of controlling the "response bandwidth" and "inter modulation" of crystal vibrator presented by the author in this article. In the article the measuring method(3) for performance of 100 MHz quartz resonator and its results are listed and discussed, it is shown that the design method is effecient, the short period frequency stability over $2x10^{-11}/1$ ms and $5.5x10^{-12}$/sec. are obtained through application of 100 MHz quartz resonator in the crystal oscillator.

The energy trapping analysis to the quartz vibrator made by Shouckley, etc. and obtained results have played an important role in the development and control of inharmonic parastics, and in the design of the monolithic crystal filter, moreover the author thinks that the method will have real significance in the devlopment and design of the low noise quartz resonator, too.

ENERGY TRAPPING THEORY AND LOW NOISE QUARTZ RESONATOR

1. Conclusions and valuable applications obtained by Shouckley, etc. developing "T" mode using "energy trap" theory.

In the analysis of "T" mode wave propagation characteristics of AT-cut crystal made by Shouckley, etc. using energy trap theory, its model simplified shown as Fig. 1 is proposed.

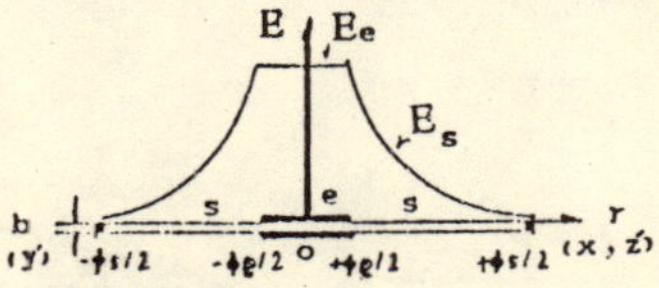

Fig. 1

In Fig. 1 the thickness b of crystal plate is along direction Y', the main plane is xz' surface, the diameter of plate is ϕ_s electrode diameter is ϕ_e, to simplify the analysis, let $\phi_s >> \phi_e$ and suppose that resonant energy in "e" region should propagate isotropically, in "s" region, and it is only relative to the distance r(x or Z') away from the centre o.

From analysis it is shown that there are cutoff angular frequencies Wec and Wsc in "e" and "s" regions, and there is a series of separative standing frequencies (energy trap wave) Wer, where Wec<Wer<Wsc, having studied the existence of energy trap mode and the propagation characteristics, it is concluded that:

a) the propagation of "energy trap" resonant mode in "e" region is a free vibration and is expressed in following fomular

$$a_e = A_e \cos\beta_e r, \qquad [r \leqslant \varphi_e / 2] \qquad (1)$$

where A_e amplitude,

β_e , Propagation factor,

its propagation in "e" region is free and lossless

b) Only a little of resonant energy of energy trap mode in "e" region is entered into "s" region, and attenuated exponentially with the increase of r in "s" region, it can be expressed in fomular as follows

$$a_s = A_s e^{-\gamma_s r}, \qquad [r > \varphi_e / 2] \qquad (2)$$

where A_s amplitude,

γ_s attenuation factor is "s" regioforn.

According to above conclusions, Shouckley, etc. have proposed important criterion for the control of inharmonical parasite nearby the main mode of quartz resonator, that is

$$\varphi_e / b \leqslant (Mn/p)[\Omega_0 / (1 - \Omega_0)]^{\frac{1}{2}} \qquad (3)$$

where p, the overtone times, p=1,3,5······

$$\Omega_0 = f_s / f_e \,, \qquad Mn = 2.4$$

They have also proposed the conditions of sound imdependance or couple realized with mult-electrodes on the monolithic plate, which has actual application value for the design of monolithic crystal filter.

2. Exploration of the development and design of low noise quartz resonator using"energy trap" theory.

a) "Response bandwidth" and its squelch behaviour of quartz resonator.

From the analysis made by Shouckley, etc. a AT-cut crystal plate with a part of electrode may be actually quivalent to a combined sound waveguide, shown as Fig. 2(a). Its "e" and "s" wavguide have different cutoff frequencies Wec and Wsc respectively, the sound waveguide made by "e" and "s" have formed a frequency response region shown in Fig. 2(b), namely "I" is the region where at W<Wec, frequency response is not produced in crystal plate and the vibrating energy is suppressed by the conbined waveguide. "II" is the region where at Wsc>W>Wec, the vibrating energy is propagated freely and lossless in "e" waveguide and attenuated rapidly in "s" waveguide, "III" is the region where at W>Wsc, the vibrating energy is propagated only in "s" waveguide.

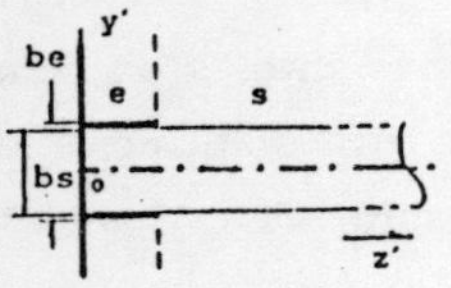

Fig. 2(a)

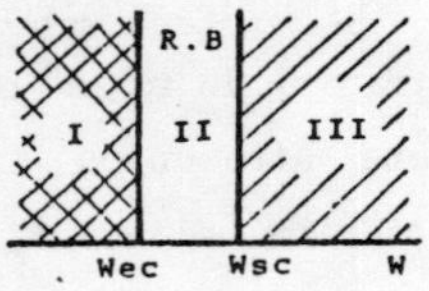

Fig. 2(b)

It is noted that the elastic wave in piezoelectric crystal plate is generally excited by "e" region, while the resonant frequency Wr in "e" region is only Wsc>Wr>Wec, the signal component over Wsc is not resonant and the energy is very low, so there is a "response bandwidth" ΔWse=Wsc-Wec in a "ES" combined waveguide, which can control the noise in a quartz resonator, namely squelch behaviour. It is considered that the control of response bandwidth can be made through the design of plate-back frequency amount of a crystal in actual application and that a good squelch effect can be obtained.

b) Additional noise generated with "inter modulation" in quartz resonator and its control method.

According to the conclusions drawn by Shouckley, etc. that is, a small part of resonant energy in "e" region is entered into "s" region and attenuated exponentially with the distance away from "e" region in "s" region. A calculated attenuation curve is shown in Fig. 3, this curve indicates the relation between the energy attenuation (dB) reached to crystal plate edge and ϕ_s/ϕ_e.

A small part of resonant energy in "e" region is always entered into the edge and a part of this energy will be returned back to "e" region from the edge, so the returned back vibrating waves and the vibrating wave in "e" region will generate two casses:

a) Form the overlapping of waves with a same frequency and a different phase: resonant mode a_c (We, Φ_e) in "e" region has a reflection wave $\varepsilon_1 a_c(We, \Phi_1)$, ε_1 is the reflection factor when propagated to the edge of crystal plate, and forms a overlapping wave with a same frequency and a differnt phase in "e" region.

$$\begin{aligned} a_{e+1} &= a_e(W_e, \varphi_e) + \varepsilon_1 a_e(W_e \cdot \varphi_1) \\ &= A_e Cos(W_e t + \varphi_e) + \varepsilon_1 A_e Cos(W_e t + \varphi_1) \\ &= A_{e+1} Cos(W_e t + \varphi_{e+1}) \end{aligned} \tag{4}$$

where

$$A_{e+1} = A_e[(Cos\varphi_e + \varepsilon_1 Cos\varphi_1)^2 +$$
$$(Sin\varphi_e + \varepsilon_1 sin\varphi_1)^2]^{\frac{1}{2}}$$
$$\varphi_{e+1} = tan^{-1} [(Sin\varphi_e + \varepsilon_1 sin\varphi_1)/$$
$$(Cos\varphi_e + \varepsilon_1 Cos\varphi_1)]$$

In reality reflection wave $\varepsilon_1 a_c$ (We, Φ_i) has some (i=1,2,3•••), its combined wave may be indicated with a vector diagram shown in Fig.4.

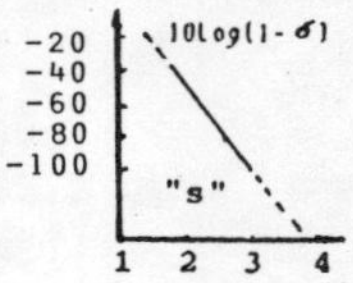

Fig. 3

Fig. 4

Seen from formula (4) and Fig. 4, if only there is a reflection wave in the edge of crystal plate, the change of main frequency in "e" region will be occured. Actually, as the edge is affected with a force (or heat), the amplitude and phase of the combined wave will be luctuating, which is one of phase noise sources generated in a crystal plate.

b)Form the overlapping of waves with a defferent frequency: in the above "response bandwidth" there always is a discrete or continuous wave spectrum generated by the outside injection or internal standing wave besides the main frequency, the main frequency, but the small signal nearby main frequency, but the small signal nearby the main frequency will also gerate the overlapping of waves with a different frequency to the main frequency. For a clutter W_r its combined wave with the main frequency wave can be expressed with a formula:

$$B_{o+1} = B_o Cos\omega_o t + B_1 Cos\omega_1$$
$$= B_{o+1} Cos(\omega_o t + \Phi_{o+1}) \quad (5)$$

where

$$B_{o+1} = [(B_o^2 + 2B_o B_1 Cos(\omega_1 - \omega_o \quad t + B_1^2]^{\frac{1}{2}}$$
$$\Phi_{o+1} = tan^{-1}\{ [B_1 Sin(\omega_1 - \omega_o)t] /$$
$$[B_o + B_1 Cos(\omega_1 - \omega_o)t] \}$$

Seeing from the above expression, the amplitude and phase of the conbined wave is periodically changed with a differential frequency (W_1-W_c). In the same way, for the overlapping of some frequencies a complicated cross modulation is generated in crystal plate.

Above two modulations, the sources of additional noise or short period frequency instability generated in a quartz resonator will be called the "inter modulation". To make a quartz resonator use in low phase noise application, we can obtain good squelch effect only to properly select the value of ϕ_s/ϕ_e according to Fig. 3, to control "response bandwidth" Δfse and excited amount or caused nonlinear degree.

DESIGN AND STUDY OF THE QUARTZ RESONATOR OF LOW PHASE NOISE

The piezoelectric crystal materials for making low noise quartz resonator should have high purity and high Q value, because of the properties of perfect crystal lattice and low internal loss, the materials have much lower self-noise that other electronic materials. However, from above analysis to obtain low noise, it is necessary to decrease the internal modulation and to narrow property of the reponse bandwidth. In the following paragraph we will present the design and study of the 100 MHz quartz resonator which has the L(f=1 KHz)<-160 dB/Hz

1. Design of crystal plate

Select AT-cut 5th overtone plane crystal plate of 100 MHz and design with a general method, then

a) Cut-angle ϕ_1=35°28'

b) Diameter ϕ_s=7 mm

c) Thickness b=0.083 mm

2. Design and test of electrode diameter and plate-back frequency Δfse.

a) Selection of electrode diameter ϕ_e

From the above conclusion, in order to reduce the addional noise caused by the inter modulation in crystal plate, the reflection in the edge should be reduced as far as possible, here the requirement of design method is to make the energy reflected into electrode region from the edge attenuate to a<-160 dB, which will be equivalent to the attenuation of -80 dB from electrode to edge, from Fig. 3. We know it is equivalent to ϕ_s/ϕ_e>3.5, then calculated ϕ_e<2 mm.

b) Selection of the plate-back frequency Δf_{sc}. Holding the high Q Value of quartz resonator and the suppression of non-resonant energy trap, the "response bandwidth" Δf_{sc} of the quartz resonator should be narrowed as far as possible, this will be designed by use of the formular (3) of Shouckley, etc.

When Mn<5 for the oscillating crystal in Table 1 are pressented the results of Δf_{sc} value in the selection of different electrode diameter

Table.1

M_n	Δf_{se}(KHZ)			
	$\varphi_e = 1_{mm}$	$\varphi_e = 1.5$	$\varphi_e = 2$	$\varphi_e = 3$
2.4	160	71	40	17
5	694	309	173	77

For holding high Q value, the selected ϕ_e<2 mm proper suppressing of parasite and realization in technology, we consider the Δf_{sc} 200 KHz could be selected.

c) Test and result analysis

In order to evaluate the above design, we select three type of designed crystal for test, the measuring block diagram is shown in Fig. 5 and the self-adjusting property of the system is presented in Fig. 6, it is indicated that -170 dB/Hz at f>1 KHz is generally in agreement with requirement of -160 dB/Hz.

The measured L(f) characteristic curves of No. 4, 11 and 23 crystal (or ϕ_e=1.4 mm, 3 mm and 3.5 mm) are shown in Fig. 7,8 and 9 repectively.

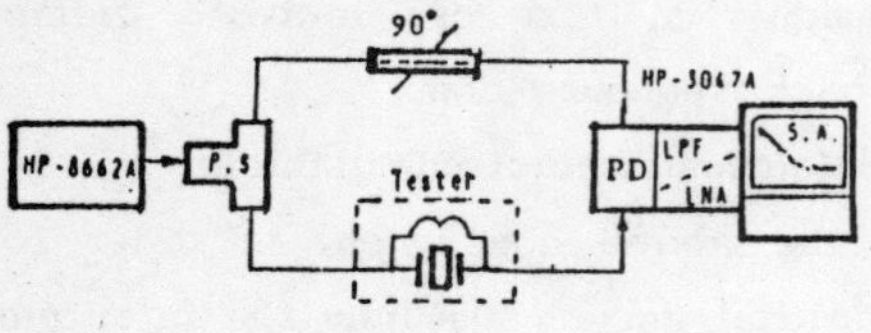

Fig. 5

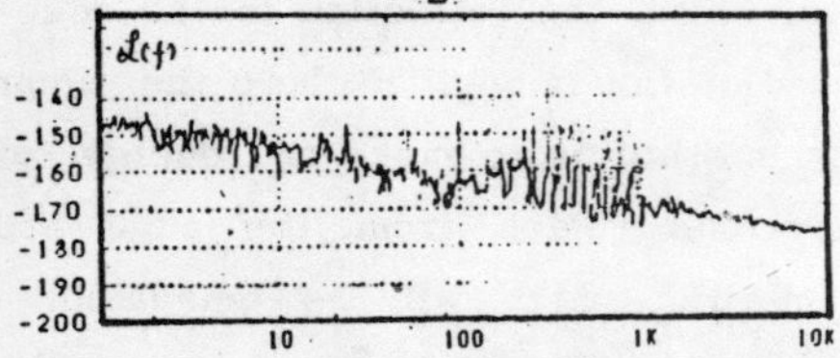

Fig. 6

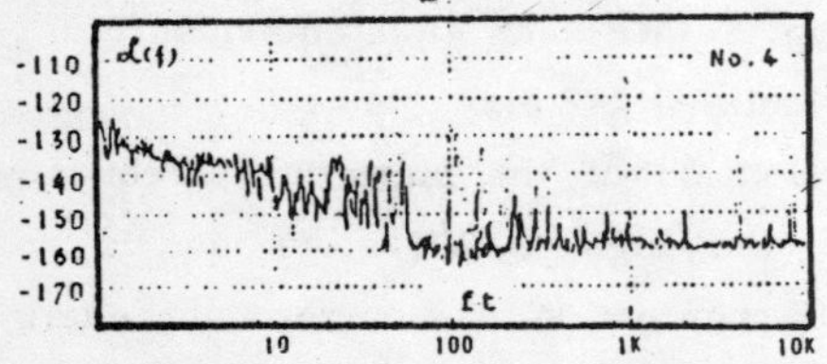

Fig. 7

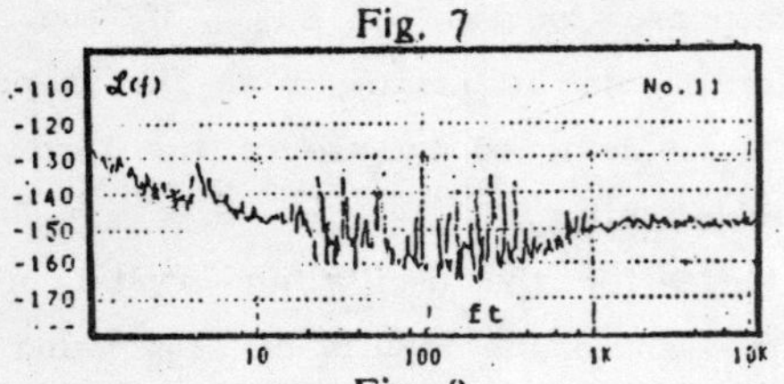

Fig. 8

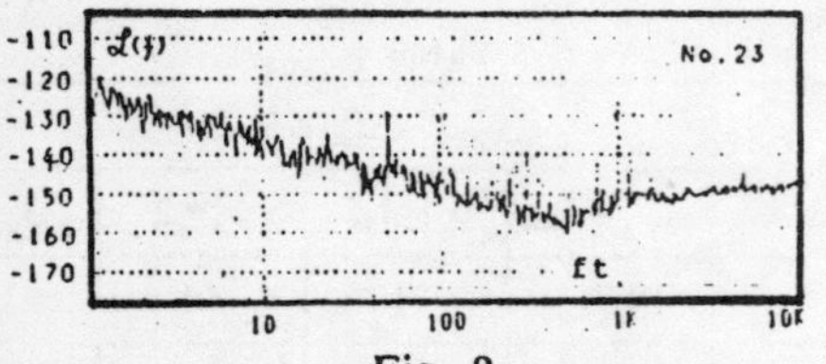

Fig. 9

From the analysis we have found the fact that proper decrease of ϕ_e would cause the decrease of tuning point frequency f_t, meanwhile, the white noise level is also decreased, it is shown that at a given Δf_{sc}, there is a optimum value of to obtain good squelch effect. Here we select Φ_e<2 mm for a parameter of design of 100 MHz resonator.

Seeing from the above measured results, we have found that the resonator has a low phase noise property in frequency f range of 100-1000 Hz. It can be predicated that resonator has an excellent Short period frequency stability in the application of crystal oscillation.

APPLICATION RESULT AND CONCLUSION OF 100 MHz LOW PHASE NOISE QUARTZ RESONATOR

Measured frequency domain property of the higher Short period stability 100 MHz quartz resonator which is made with above design method for application of crystal oscillator is shown in Fig. 10, its time domain property is shown in Table 2.

Table.2

NO. (oscillator)	$\sigma(\tau)\times{}^{-12}$			
	τ = 1sec	0.1	0.01	0.001
05	20	12	12	19
19	5.5	4.6	15	34
20	5.6	4.2	6	26

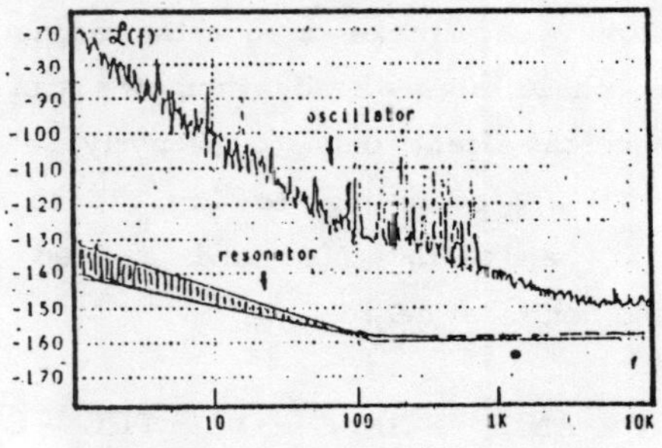

Fig. 10

In Fig. 10, we have found that if the circuit noise is further decreased, the performance will be improved more than 5 - 8 dB/Hz. From the application results, it is shown that the designed 100 MHz low noise quartz resonator has an excellent practice useful effect and the analysis and design presented here is perfect.

ACKNOWLEDGEMENT

The author would like to thank senior engineer

Lee Chunhai for assistance of providing application results.

REFERENCES

(1) W. Shouckley, D.R. Durran and D.J. Koneval "Energy Trapping and Related Studies of Multiple Electrode Filter Crystal" The 17h AFCS (1963).

(2) Chen Zhiyuan " ' Response Bandwidth' and 'Inter-modulation' of Quartz Resonator". The 3rd conference of CIE, China (1982).

(3) Stephen W. Hinch "High Performance Phase Noise Tester for Crystal" Microwave Journal, September, 1985.

DEVELOPMENT AND APPLICATIONS OF THE HIGH FUNDAMENTAL FREQUENCY QUARTZ RESONATOR ETCHED WITH ION BEAM

Wang Yong Bin
The Tenth Research Institute of Ministry of the
Machine–building and Electronics Industry,China

ABSTRACT

This paper presents firstly the performance of ion beam etching machine and selection of optimal operation parameters,and then describes electrode form,equivalent parameters measurement,vibration and shock test,store age experiment for quartz resonator. Results are listed in table. Applications of high fundamental frequency quartz resonator in FM crystal oscillator,low noise microwave oscillator and broadband crystal filters has achieved satisfactory results.

INTRODUCTION

High frequency quartz resonator usually uses AT–cut,SC–cut etc. thickness shear vibration quartz plate,which resonant frequency is inversely proportional to the thickness of quartz plate($f_n = nK_f / t$). If you want to raise frequency,it is necessary to reduce thickness of quartz plate or increase overtone number n. For a long time,the method of machinary polishing can obtanin quartz plate of thickness 55μm,fundamental frequency 30MHz. Large productive high–order overtone crystal can operate at 13th or 15th overtone,frequency is up to 350MHz. In recent ten years,chemical etching method and ion beam etching technology thin center part of quartz plate and greatly increase fundamental frequency. It was reported that chemical polishing could process quartz plate being of 400MHz fundamental frequency,3rd overtone up to 1200MHz[1],maximum fundamental frequency up to 1655MHz[2],thickness near 1μm. However,chemical etching is anisotropic to quartz and produces etched channels on the some given axes. It is more difficult to get a good flatness surface. Ion beam technology which thins quartz plate can obtain ideal etching surface and process quartz plate of 30~1000MHz fundamental frequency. The 3rd and 5th overtone can be up to 3GHz.

ION BEAM TECHNOLOGY

Ion beam etched technology is a procedure of entire physical property. Material selection is not rigorous. Etched figure has very good definition and high resolution. Therefore, ion beam technology has been widely applied to semiconductor microminiature processing, SAW device, magnetic bubble, the plane helical inductance, thinning quartz plate and so on.
Ion beam equipment chiefly consists of ion source, vacuum system and power supply. Ion source is designed by low pressure Kaufman. Cathode uses thorium tungsten filament, efficiency of transmission elcetron is very high. Argon gas enters from the top and vacuum is kept to (1 ~ 2)x10^{-4}Torr. Transmitted elcetron from the heated cathode does helical movement or oscillatory movement under the action of magnetic field. The transmitted electron produces elastic and non–elastic impact with argon atom and ionizes argon to form argon's plasma. Ion optical system extracts argon ion from plasma and completes focussing and acceleration, so as to form collimation ion beam. Surface etching processing is completed through the momentum exchanging of argon ion with quartz atom.

SURFACE QUALITY CONTROL

Good and bad of etched surface is closely related to quartz crystal material, preprocessing of quartz plate and operation parameters.
Quartz crystal material used as etching, its impurity concentration and density of defects must be very low, Q is above 2.4×10^6. The natural quartz or high–grade cultured quartz through the swept processing, surface injury after the etching is the least.
Before etching, quartz plate is polished to 25~ 30MHz and thickness 66~ 55μm by use of machinary way. Requirement of surface does not damage, flatness and parallelism must be well.
Selection of operation parameters is also a key getting a good etching surface. For Ar ion source,the anode volt-

age 50v is the optimal parameter. Ion beam energy can be adjusted continuously,generally selective value is 250 ~ 550cv,beam current 60~ 90mA,density of beam current below 1mA / cm^2,etching rate 0.8~ 3.0μm / hour. High energy and big beam current may achieve high etching rate,but quartz surface will be damaged; Low energy and small beam current may achieve better etching surface,but etching rate is not high. Not only the angle size(incidence angle)between ion beam and quartz plate normal has an effect upon etching rate but also upon surface quality. When ion beam is vertical to quartz surface,incidence angle is zero. At this time the stricken out atom from surace must change an angle,atom can ran out from surface. At the same time some of atom returns to the surface again and surface can not be etched uniformly,so etching rate and surface quality are reduced. When incidence angle becomes big slowly,etching rate increases and flatness of the surface gets well. But,if incidence angle is too large,beam current will be covered by mask. Thus,periphery etching is slower than the center,surface appears concave,Experiment has shown that proper etching rate and high—quality surface could be achieved when incidence angle is 30° ~35°.

DESIGN AND FABRICATION OF RESONATOR

Diameter of quartz plate is 7mm and 5.5mm. Using graphite mask, diameter of center etching area is 4mm , 3mm and 2.5mm.

Electrode film material uses high—net aluminium,Selection of electrode size and thickness is basing on trapped energy concept,in other word,not only equivalent resistance must be small,but also there is a good resonant response. Spurious mode suppression of fundamental frequency oscillation require to be more than 8dB,spurious mode suppression in overtone operation is greater than4dB. The higher fundamental frequency is,the smaller electrode is. For example,when fundamental frequency is 1000MHz,electrode size is not greater than 40 × 40μm^2. Therefore,in order to make two surface electrodes aim at,cross form is used,shown as Fig.1.

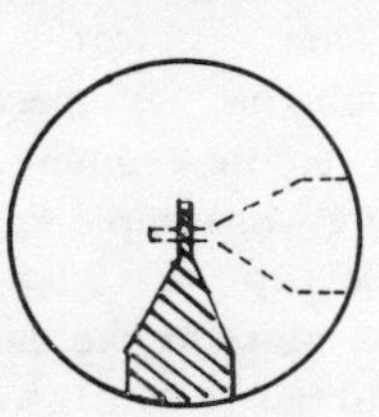

Fig. 1 Shape of electrode

It is quite difficult to precisely adjust frequency of higher fundamental frequency resonator,because slight change of film layer will cause bigger frequency change. For example,quartz resonator of 500MHz fundamental,film layer change 1A˚ ,frequency will vary 15KHz,namely 3×10^{-5}. The same,quartz resonator of 1000MHz fundamental frequency, film layer changes 1A˚ , frequency will vary 60KHz, namely 6×10^{-5}. Thus,it can be seen,it is rather difficult that accuracy of frequency wants to arrive at 10^{-5} order.

At present,for HF and UHF quartz resonators,only method of adjusting frequency is to use etching to increase frequency and to use anodic oxidation to reduce frequency.

Property of quartz resonator depends on packaged form in very degree. High fundamental frequency crystal can seal in a metallic case(tin seal,resistance seal),glass case and ceramics box,shown as Fig.2. Because the vibration of high fundamental frequency quartz plate is only limited in an extreme small center of etching area,so support has no any effect on the vibration of quartz plate.

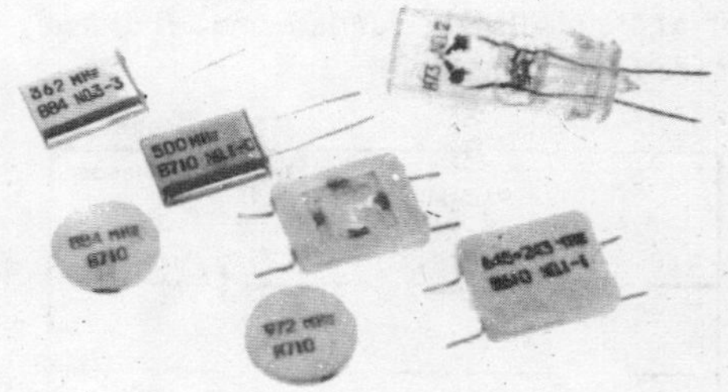

Fig.2 Packaged form of high fundamental frequency quartz resonator.

PERFORMACE TESTING

In recent years,increasing of quartz resonator frequency is very quick,but instrumentation and measurement method for precisely measuring equivalent parameters is not enough complete. Experiment below 300MHz uses measuring method of π—type network zero phase[3],experiment above 300MHz uses method of impedance or network analyzer .

Method of π—type network zero phase measures parameters V_{BO},V_B,$f_r$and f_p,computing from formula:

$$R_r = [(V_{BO} / V_B) - 1] \times 25$$
$$C_1 = 2Co[(f_p - f_r) / f_r]$$
$$L_1 = 1 / (4\pi^2 f_r^2 C_1) \qquad Q = 2\pi f_r L_1 / R_r \qquad (1)$$

Basing on the relation of resonator impedance and phase change with frequency in a small range near f_s ($\triangle f / f_s < \pm 2 \times 10^{-5}$),measured by impedance analyzer:

$$L_1 = \frac{1}{4\pi}\frac{dx}{df} \qquad C_1 = \frac{1}{\pi f_s^2 \frac{dx}{df}}$$

$$Q = \frac{f_s}{2}\frac{d\Phi}{df} \qquad R_1 = \frac{1}{2\pi f_s Q C_1} \qquad (2)$$

Measurement results are listed in Table 1.

Table 1. Equivalent parameter of fundamental frequency quartz resonator.

f_s (MHz)	C_0 (PF)	L_1 (mH)	C_1 (10^{-4}PF)	R_1 (Ω)	Q (10^4)
93.750	0.63	4.40	6.50	43	6.1
129.175	0.48	2.10	7.20	31	2.6
236.134	0.57	0.90	5.20	26	5.0
270.301	0.63	0.54	6.50	36	2.5
305.982	0.71	0.31	8.61	33	1.8
461.060	0.52	0.43	2.80	37	3.4

Resonant response is measured by sweeping frequency meter. Fig.3 shows resonant property of quartz resonator at 360.070MHz fundamental frequency and 3rd overtone 1079.609MHz.

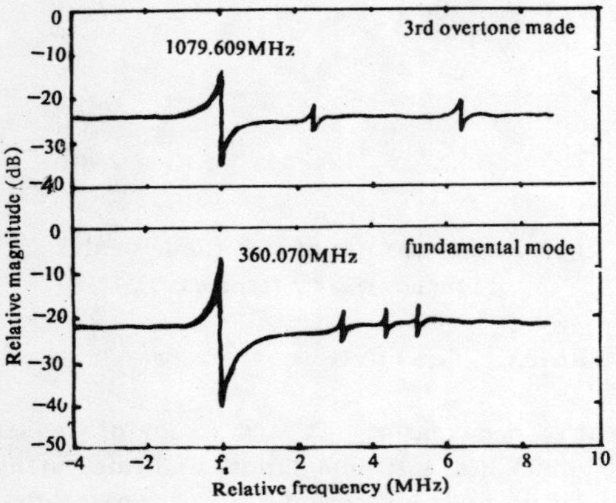

Fig.3 Resonant property of quartz resonator at fundamental frequency 360.070MHz and 3rd overtone 1079.609MHz.

Store age experiment was done,store temperature 80℃ ,store time 203 days.After store age all the crystals operate in normal state,results are listed in Table 2.

Table 2. Measurement results of store age

before store		after store		relative change	
f(MHz)	R_r(Ω)	f(MHz)	R_r(Ω)	Δf / f(10^{-6})	ΔR / R_r(%)
150.025370	40	150.035422	32	+67	−8
202.560900	36	202.553021	50	−39	+39
197.181503	38	197.192421	52	+55	+37
287.607855	65	287.609212	80	+4.7	+15

Vibration and shock experiment were done. Vibrationn conditions are: vibration frequency 50Hz,acceleration 6g,respectively vibrating 20 minutes in three mutual vertical directions; shock conditions are:acceleration 15g,100 times per minute,respectively shocking 1333 times in three mutual vertical directions. Results before and after the experiment are listed in Table 3. Seeing from the talbe,the relative change of frequency after vibration and shock is not greater than 10.5×10^{-6},the relative change of resistance is not greater than 13.6%. This is because oscillation of quartz plate is only limited in an extreme small center area,support influence is nearly zero. So capability of anti–vibration and anti–shock has been raised.

Talbe 3. Experiment results of vibration and shock.

before vibration and shock		after vibration		after shock	
f(MHz)	R_r(Ω)	Δf / f(10^{-6})	ΔR / R_r(%)	Δf / f(10^{-6})	ΔR / R(%)
269.444990	46.4	−3.6	3.0	1.8	4.6
290.337600	35.2	−2.0	13.6	0	−6.3
260.807000	26.0	−3.8	6.1	1.5	−6.5
290.703600	28.4	−5.5	1.7	2.7	1.4
283.821000	40.6	−10.5	7.6	−0.4	−1.1

APPLICATION

In recent years,application of high fundamental frequency quartz resonator in engineering has achieved satisfactory results. Ultra–thinned telemetry transmitter made by use of microstrip technique and quartz resonator of 362MHz and 545MHz fundamental frequency has applied to telemetric equipment in tactical missile. Design requirement is satisfied through the various routine expreiments. Stability is better than 5×10^{-5} under the operational ambient temperatre −40 ℃ ~ +50 ℃.

1GHz and 1.5GHz crystal oscillators which are developed by 3rd overtone have good short–term stability,low phase noise,small volume,light weight and low power consumption. They can apply to radar,communication,high speed frequency agility,telemetry and remote equipment. Experiment results show as following:1ms sample up to $1.1\times10^{-9}$and 3×10^{-9}, 1s sample up to $7.9\times10^{-10}$and 2.3×10^{-9}respectively;phase noise at the deviation oscillation frequency 1KHz up to −101dB / Hz and −93dB / Hz, at 10KHz up to −121dB / Hz and −113dB / Hz respectively.

471MHz FM crystal oscillator for coal–cutting telemetry transmitter has 150KHz deviation of FM, ambient temperature −10 ℃ ~ +45 ℃, frequency stability more than 3×10^{-5}.

Most of high frequency broadband crystal filters use also high fundamental frequency quartz resonator, such as

70MHz crystal filter, bandwidth $\triangle f_{3dB} = 70KHz$, $\triangle f_{40dB} = 194KHz$, group delay smaller than 8μs.

COMPARESION OF BULK ACOUSTIC WAVE, SAW AND DIELECTRIC RESONATOR

1. Applied frequency range

For a long time past, the fundamental frequency of quartz resonator is usually 30MHz or so, which high order overtone frequency is up to 350MHz. The covered range of frequency of SAW resonator is large,upper frequency limit in theoretic design can be up to 2~3GHz, but practical application frequency is 100~ 1500MHz. At this time, SAW resonator in VHF has widely applied. In recent ten years, chemical polishing method and ion beam etching technology have succeeded in thinning quartz plate, fundamental frequency raises to 500MHz in one leap, the most level up to 1665MHz, overtone frequency up to 3GHz. At present, the covered range of frequency of the two resonators is approximate to the same. Dielectric resonator is a novel microwave frequency source, which operation frequency is beyond 2GHz.

2. Quality factor Q

In 100~500MHz range, Q of quartz resonator is $(1\sim 6) \times 10^4$, and that of SAW resonator is about 1×10^4. Over the higher frequency ,Q of quartz resonator is slightly more than SAW resonator's Q, but practical Q of quartz resonator is smaller than limiting Q of material. It can also increase through the improving of technology and design. Q of SAW resonator has approached the limiting Q of material. Q of dielectric resonator is 8000~10000 at 2.5GHz.

3. Property of frequency to temperature

The relation between frequency change of AT-cut quartz resonator and temperature depends on third order temperature coeficient. Property of SAW resonator frequency to temperature depends on second order temperature coeficient. In the range of wide temperature the later is about everal times greater than the former. Temperature coeficient of dielectric resonator is adjustable, usually $Tf = (-1 \sim 8) \times 10^{-6} / ℃$.

4. Frequency adjustment and processing technology

Frequency adjustment of quartz resonator is more convenient, adjustable range is large. Frequency adjustment of SAW resonator is difficult, adjustable range is smaller. Processing of interdigital transducer of SAW resonator is very fine. Line width below 1μm is rather difficult by use of optical etching technology, and processing of quartz plate is easier relatively.

CONCLUSION

In general ,ion beam technology which thins quartz plate is successful, main characteristics as following:

1. Fundamental frequency is spreaded to 500~1000MHz, and overtone frequency can be up to 3GHz.

2. Vibration area of high fundamental frequency quartz plate is very small, thus volume can be reduced relatively. The plane assembly is able to implement.[4]

3. High fundamental frequency and low overtone resonators oscillate in the circuit easily, and circuit is simplified.

4. Sensibility of vibration, shock and acceleration is low. But, high fundamental frequency quartz resonator which uses ion beam etching has higher productive cost and lower efficiency. Therefore, it is impossible to produce on a large scale at present.

REFERENCE

1. W.P.Hanson, " Chemically polished high frequency resonators" , Proc. 37th Ann. Frequency control Symp., pp. 261–264, 1983.
2. R.C.Smythe, J.R.Hunt and M.D.Howard, "Chemically–milled VHF & UHF quartz resonator & monolithic filter", 1985 Ultrasonics Symposium Proceeding. New York: IEEE, PP.302–305, 1985.
3. Basic method for the measurement of resonance frequency and equivalent series resistance of quartz crystal units by zero phase technique in a π–network, IEC Recommendation Publication 444, 1973.
4. J.S.Wang, A.Kong, K.F.Lau and K.H.Yen, "Recent developments on membrane bulk–acoustic–wave resonators" , Proc. 39th Ann. Frequency Control Symp. , PP.356–360, 1985.

MEASUREMENT OF ALL THE ELECTRO-ELASTIC CONSTANTS OF LITHIUM TETRABORATE ($Li_2B_4O_7$) CRYSTAL

Wang Jin-feng Zhang Lei Qin Zi-kai
(Department of Physics, Shandong University, P.R.China)
Wang Hong
(Institute of Crystal Materials, Shandong University, P.R.China)
He Da-zhen Hu Shao-qin
(Institute of Piezoelectrical and Acousto-Optical Technology, Sichuan, P.R.C.)

ABSTRACT

A new approach to determine all the electro-elastic constants of the piezoelectric crystals in class 4mm has been put forward. As an example, all the constants of Lithium Tetraborate ($Li_2B_4O_7$) crystal have been measured by means of the approach. Only employing in measuring three crystal plates differently oriented is the outstanding advantage of the approach.

I. INTRODUCTION

Following $LiNbO_3$, $LiTaO_3$ and $AlPO_4$ crystals, another new piezoelectric crystal -$Li_2B_4O_7$ has been synthesized. Because of having higher electro-mechanical coupling coefficients and lower temperature coefficients, $Li_2B_4O_7$ has given rise to public attention. On the research of physical property, N.M. Shorrocks[2] had measured the electro-elastic constants of $Li_2B_4O_7$ crystal, afterwards M.Adachi[3] had newly measured these constants. From their results we can see that there are big deviations between the two set of data. M.Adachi had not mentioned detailed cut types of tested specimens. N.M.Shorrocks had used 9 cut types of tested specimens, including 3 cubic specimens which are very difficult to fabricate.

By means of the theory[4] on thickness vibration of piezoelectric plates, a new approach to determine the dielectric, piezoelectric and elastic constants of $Li_2B_4O_7$ crystal has been put forward. In the new approach only 3 simple specimens have been employed.

II. THEORY ANALYSES

Piezoelectric crystal in class 4mm possesses two independent dielectric, three independent piezoelectric and six independent elastic constants. Its matrices of dielectric, piezoelectric stress, and elastic stiffness constants, respectively, are shown in the following:

$$\begin{bmatrix} \varepsilon_{11} & 0 & 0 \\ 0 & \varepsilon_{11} & 0 \\ 0 & 0 & \varepsilon_{33} \end{bmatrix}, \begin{bmatrix} 0 & 0 & 0 & 0 & e_{15} & 0 \\ 0 & 0 & 0 & e_{15} & 0 & 0 \\ e_{31} & e_{31} & e_{33} & 0 & 0 & 0 \end{bmatrix},$$

$$\begin{bmatrix} c_{11} & c_{12} & c_{13} & 0 & 0 & 0 \\ c_{12} & c_{11} & c_{13} & 0 & 0 & 0 \\ c_{13} & c_{13} & c_{33} & 0 & 0 & 0 \\ 0 & 0 & 0 & c_{44} & 0 & 0 \\ 0 & 0 & 0 & 0 & c_{44} & 0 \\ 0 & 0 & 0 & 0 & 0 & c_{66} \end{bmatrix}$$

If its thickness is much smaller than its other dimensions, the piezoelectric plate may be considered as an infinite plate. In this case, the analyses about thickness vibration of piezoelectric plate is much simplified.

A. (YZL)45°-cut plate

The following equation can be got through piezoelectric constitutive relations:

$$\begin{aligned} T_2 &= c'_{22} S_2 - e_{31} E_1 , \\ T_4 &= c^E_{44} S_4 - e_{15} E_2 , \qquad (1) \\ T_6 &= c'_{66} S_6 , \end{aligned}$$

$$\begin{aligned} D_1 &= \varepsilon^S_{11} E_1 , \\ D_2 &= e_{15} S_4 + \varepsilon^S_{11} E_2 , \qquad (2) \\ D_3 &= e_{31} S_2 + \varepsilon^S_{33} E_3 . \end{aligned}$$

From Eq.(1) we can see that stress T_6 is not stimulated by piezoelectric effect, when only applying electric stimulating. Finding out, at first, the relation between E_2 and S_4 by $\nabla \cdot D=0$, and introducing the relation into

Eq.(1), then introducing Eq.(1) into dynamical equations of acoustics, one obtains

$$\begin{bmatrix} c'_{22}-c_y & 0 \\ 0 & \bar{c}_{44}-c_y \end{bmatrix} = 0 .$$

one root of the determinant equation is

$$c'_{y1} = c'_{22} = \tfrac{1}{2}(c^E_{11} + c^E_{12}) + c^E_{66} , \qquad (3)$$

another root is

$$c'_{y2} = \bar{c}_{44} = c^E_{44} + e^2_{15}/\varepsilon^S_{11} . \qquad (4)$$

In above equations c'_y is the effective elastic stiffness constant. We can turn Eq.(4) into

$$c^E_{44} = \bar{c}_{44}(1 - k^2_{24}) , \qquad (5)$$

where

$$k^2_{24} = e^2_{15}/\bar{c}_{44}\varepsilon^S_{11} . \qquad (6)$$

By way of the free mechanical boundary conditions, $T_2 = T_4 = T_6 = 0$, and the electrical boundary conditions, one gets

$$c'_{y1} = 4\rho(f^1_{sn}t/n)^2, \ n=1,3,\cdots \qquad (7)$$
$$c'_{y2} = 4\rho(f^2_{pn}t/n)^2, \ n=1,3,\cdots \qquad (8)$$

where ρ is the mass density of the crystal, t is the thickness of the piezoelectric plate, n is the overtone order, f_p is the parallel resonant frequency, and f_s is the series resonant frequency.

B. 0° Z-cut plate

By way of piezoelectric constitutive relations, one can get

$$\begin{aligned} T_3 &= c^E_{33}S_3 - e_{33}E_3 , \\ T_4 &= c^E_{44}S_4 - e_{15}E_2 , \qquad (9) \\ T_5 &= c^E_{44}S_5 - e_{15}E_1 , \\ D_1 &= e_{15}S_5 + \varepsilon^S_{11} E_1 , \\ D_2 &= e_{15}S_4 + \varepsilon^S_{11} E_2 , \qquad (10) \\ D_3 &= e_{33}S_3 + \varepsilon^S_{33} E_3 . \end{aligned}$$

Through analyses similar to (YZL)45°-cut plate, we get

$$\begin{bmatrix} c^E_{44}-c_z & 0 & 0 \\ 0 & c^E_{44}-c_z & 0 \\ 0 & 0 & \bar{c}_{33}-c_z \end{bmatrix} = 0 .$$

one root of the determinant equation is

$$c_{z1} = \bar{c}_{33} = c^E_{33} + e^2_{33}/\varepsilon^S_{33} , \qquad (11)$$

namely

$$c^E_{33} = \bar{c}_{33}(1 - k^2_{33}) , \qquad (12)$$

where

$$k^2_{33} = e^2_{33}/\bar{c}_{33}\varepsilon^S_{33} . \qquad (13)$$

By means of boundary conditions we obtain

$$c_{z1} = \bar{c}_{33} = 4\rho(f^1_{pn}t/n)^2 \qquad (14)$$

C. (ZYL)45°-cut plate

From piezoelectric constitutive relations we obtain

$$\begin{aligned} T_3 &= c'_{33}S_3 + c'_{35}S_5 - e'_{13}E_1 - e'_{33}E_3 , \\ T_4 &= c'_{44}S_4 - e'_{24}E_2 , \qquad (15) \\ T_5 &= c'_{35}S_3 + c'_{55}S_5 - e'_{15}E_1 - e'_{35}E_3 , \\ D_1 &= e'_{13}S_3 + e'_{15}S_5 + \tfrac{1}{2}(\varepsilon^S_{11} + \varepsilon^S_{33})E_1 , \\ D_2 &= e'_{24}S_4 + \varepsilon^S_{11} E_2 , \qquad (16) \\ D_3 &= e'_{33}S_3 + e'_{35}S_5 + \tfrac{1}{2}(\varepsilon^S_{11} + \varepsilon^S_{33})E \end{aligned}$$

When only applying parallel electric field E and E , one gets

$$\begin{bmatrix} c'_{55}-c'_z & 0 & c \\ 0 & c'_{44}-c'_z & 0 \\ c & 0 & c'_{33}-c'_z \end{bmatrix} = 0 .$$

one root of the determinant equation is

$$c'_{z1} = c'_{44} = \tfrac{1}{2}(c^E_{44} + c^E_{66}) . \qquad (17)$$

The sum and product of another two root are, respectively,

$$\Sigma'_z = c'_{z2} + c'_{z3} = c'_{33} + c'_{55} \qquad (18)$$
$$\Pi_z = c'_{z2}c'_{z3} = c'_{33}c'_{55} - (c'_{35})^2 , \qquad (19)$$

where

$$\begin{aligned} c'_{33} &= \tfrac{1}{4}(c^E_{11} + c^E_{33}) + \tfrac{1}{2}c^E_{13} + c^E_{44} , \\ c'_{55} &= \tfrac{1}{4}(c^E_{11} + c^E_{33}) - \tfrac{1}{2}c^E_{13} \\ c'_{35} &= \tfrac{1}{4}(c^E_{11} - c^E_{33}) . \end{aligned}$$

When only applying electric field E_2, one gets

$$c'_{z1} = 4\rho(f^1_{sn}t/n)^2 . \qquad (20)$$

When only applying electric field E_1, one gets

$$c'_{z2} = 4\rho(f^2_{sn}t/n)^2 , \qquad (21)$$
$$c'_{z3} = 4\rho(f^3_{sn}t/n)^2 . \qquad (22)$$

When only exerting perpendicular field E_3, one gets

$$\begin{bmatrix} \bar{c}'_{55}-\bar{c}'_z & 0 & \bar{c}'_{35} \\ 0 & c'_{44}-c'_z & 0 \\ \bar{c}'_{35} & 0 & \bar{c}'_{33}-\bar{c}'_z \end{bmatrix} = 0$$

One root of the determinant equation is the same as Eq.(11), the sum of another two roots is

$$\bar{\Sigma}'_z = \bar{c}'_{z2} + \bar{c}'_{z3} = \bar{c}'_{33} + \bar{c}'_{55} \quad , \qquad (23)$$

where

$$\bar{c}'_{33} = c'_{33} + 2(e'_{33})^2/(\varepsilon^S_{11}+\varepsilon^S_{33}) \quad ,$$

$$\bar{c}'_{55} = c'_{55} + 2(e'_{35})^2/(\varepsilon^S_{11}+\varepsilon^S_{33}) \quad ,$$

$$e'_{33} = \frac{\sqrt{2}}{4}(2e_{15}+e_{31}+e_{33}) \quad ,$$

$$e'_{35} = \frac{\sqrt{2}}{4}(e_{31}-e_{33}) \quad .$$

From boundary conditions, we get

$$\bar{c}'_{z2} = 4\rho(f^2_{pn}t/n)^2 \quad , \qquad (24)$$

$$\bar{c}'_{z3} = 4\rho(f^3_{pn}t/n)^2 \quad . \qquad (25)$$

Through Eq.(23) and Eq.(18), we obtain

$$\bar{\Sigma}'_z - \Sigma'_z = 2[(e'_{33})^2 + (e'_{35})^2]/(\varepsilon^S_{11}+\varepsilon^S_{33}) . \quad (26)$$

III. DETERMINATION PROCEDURE

1. (YZL)45°-cut plate .

A. At perpendicular field .
a. Measuring the capacitances of the specimen at 1 KHz and 50 MHz, respectively, we can obtain the free dielectric constant ε^T_{11} and the clamped dielectric constant ε^S_{11}.
b. Measuring the parallel resonant frequencies f_{pn} (or substituting f_{pn} with high order series resonant frequencies f_{sn} in the same overtone order), we can get c_{y2} by Eq.(8).
c. Measuring f_{s1}, f_{s3}, f_{s5}, . . ., we can get k_{24} by overtone ratio method[5], then we can get c^E_{44} and e_{15} by Eq.(5) and Eq(6), respectively.
B. At parallel field .
The electrode configuration for tested specimen is identical with that of reference 6. Measuring the series frequencies f_{sn} , we can get c'_{22} by Eqs.(3) and (7).

2. 0°Z-cut plate .

At perpendicular field .
a. Calculating out ε^T_{33} and ε^S_{33} through the measured capacitances of the specimen.
b. Finding out c_{z1} by Eq.(14).
c. Finding out k_{33} by overtone ratio method, then finding out c^E_{33} and e_{33} by Eqs.(12) and (13), respectively.

3. (ZYL)45°-cut plate .

A. At parallel field E_2 .
Calculating out c'_{z1} by Eq.(20), then finding out c^E_{66} by Eq.(17).
B. At parallel field E_1 .
Finding out c'_{z2} and c'_{z3} by Eq.(21) and Eq.(22), respectively, then finding out c^E_{11} by Eq.(18), c^E_{12} by Eq.(3) and c^E_{13} by Eq.(19) successively.
C. At perpendicular field E .
Finding out $\bar{c}'_{z2}$ and $\bar{c}'_{z3}$ by Eq.(24) and Eq.(25), respectively, then finding out e_{31} by Eq.(26).

IV. MEASUREMENT RESULTS AND DISCUSSION

The density of $Li_2B_4O_7$ crystal is 2439Kg/m³. In Table 1 the electro-elastic constants measured at room temperature for $Li_2B_4O_7$ have been listed. In testing, Hewlett-Packard HP-4191A, HP-4192A impedance analyzer and HP-3585A spectrum analyzer have been used.

Table 1 Electro-elastic constants of $Li_2B_4O_7$ crystal

constants	unit	ours	ref.2	ref.3
c^E_{11}		12.76	12.67	13.5
c^E_{12}		0.89	0.05	0.357
c^E_{13}		3.2	3.0	3.35
c^E_{33}		5.47	5.39	5.68
c^E_{44}		5.75	5.50	5.85
c^E_{66}	$(10^{10}\,N/m^2)$	4.82	4.60	4.67
c^D_{11}		12.84		
c^D_{12}		0.97		
c^D_{13}		3.5		
c^D_{33}		6.67		
c^D_{44}		5.94		
c^D_{66}		4.82		
s^E_{11}		9.26		
s^E_{12}		1.7		
s^E_{13}		-5.9		
s^E_{33}		25.2		
s^E_{44}		17.4		
s^E_{66}	$(10^{-12}\,m^2/N)$	20.7		
s^D_{11}		9.17		
s^D_{12}		1.6		
s^D_{13}		-5.2		
s^D_{33}		20.4		
s^D_{44}		16.8		
s^D_{66}		20.7		
e_{15}		0.39	0.36	0.472
e_{31}	(C/m^2)	0.24	0.19	0.290
e_{33}		0.93	0.89	0.928
g_{15}		0.0821		
g_{31}	(m^2/C)	-0.0320		
g_{33}		0.231		
h_{15}		4.95		
h_{31}	$(10^9\,N/C)$	3.34		
h_{33}		12.9		
ε^T_{11}		9.32		
ε^T_{33}		10.1		
ε^S_{11}	relative unit	8.91	8.97	8.90
ε^S_{33}		8.13	8.15	8.07

In measuring and calculating there are some aspects to be noteworthy:

1. The ration of width to thickness of the plate should be suitably large so as to approach the condition of a infinite plate.

2. The accuracy of the tested plate directed by x-ray is less ± 1′ , the deviation of the parallelism for the two main plates of the specimen with t = 1mm is less ± 1μm. Especially, the accuracy of (ZYL)45°-cut plate must be satisfied, otherwise big errors of c_{13}^E and e_{31} would be given rise to.

3. In order to test the measurement accuracy of c_{13}^E and e_{31}, we can take following steps: At first, measuring the constant d_{33} with 0°Z-cut plate by the quasistatic method, then measuring constant d_{31} with the bar taken from 0°Z-cut plate by the resonant and anti-resonant method, the values of c_{13}^E and e_{31} can be tested by following two equations

$$e_{33} = 2d_{31}c_{13}^E + c_{33}^E d_{33},$$

$$e_{31} = d_{31}(c_{11}^E + c_{12}^E) + c_{13}^E d_{33}.$$

In addition, and then combining following equation

$$\varepsilon_{33}^T - \varepsilon_{33}^S = 2d_{31}^2(c_{11}^E + c_{12}^E) + 4d_{31}d_{33}c_{13}^E + d_{33}^2 c_{33}^E,$$

we can multiply test the consistency of c_{13}^E , e_{31} or ε_{33}^S.

The specimens used are provided by Institute of Piezoelectrical and Acousto-Optical Technology, Sichuan, P.R.China. The Institute is the first unit in China which has synthesized $Li_2B_4O_7$ crystal.

The authors wish to thank Song-yang Zhou for her help in depositing the electrodes by vacuum evaporation.

REFERENCES

1. D.S.Robertson and I.M.Young, J.Mat. Sci., 17 (1982), 1729.
2. N.M.Shorrocks, R.W.Whatmore, F.W. Ainger and I.M.Young, Proc. IEEE Ultrasonics Symposium, (1981), 337 .
3. M.Adachi, T.Shiosaki, H.Kobayashi, O.Ohnishi and A.Kawabata, Proc. IEEE Ultrasonics Symposium, (1985), 228 .
4. H.F.Tiersten, J.Acoust.Soc.Am., 35 (1963), 53 .
5. M.Onoe, H.F.Tiersten and A.H.Meitzler J.AcoustSoc.Am., 35 (1963),36 .
6. A.W.Warner, M.Onoe and G.A.Coquin, J.Acoust.Soc.Am., 42 (1967), 1223 .

POROUS SiC CERAMICS FOR HIGH-TEMPERATURE THERMOELECTRIC ENERGY CONVERSION

Kunihito KOUMOTO, Chul Hoon PAI, Shunji TAKEDA and Hiroaki YANAGIDA*

Department of Industrial Chemistry, Faculty of Engineering, The University of Tokyo, 7-3-1 Hongo, Bunkyo-ku, Tokyo 113, Japan
*Research Center for Advanced Science and Technology, The University of Tokyo, 4-6-1 Komaba, Meguro-ku, Tokyo 153, Japan

ABSTRACT

Porous SiC ceramics were proposed to be promising materials for high-temperature thermoelectric energy conversion. Through the thermoelectric property measurements and microstructure observations on the porous β-SiC, it was experimentally clarified that elimination of stacking faults and twin boundaries by grain growth is effective to increase the Seebeck coefficient and better grain-to-grain connectivity gives rise to higher electrical conductivity.

INTRODUCTION

The development decades ago of thermoelectric meterials has recently been resurrected in a flurry of activity once again for high-temperature thermoelectric power generation. For effective utilization of solar heat, nuclear energy and waste heat from industries, materials possessing high thermoelectric energy conversion efficiencies and usable at high temperatures are intensely required.

Thermoelectric semiconductors composed of heavy elements, such as Bi_2Te_3, PbTe, etc., have already been commercially applied to both refrigeration and power generation [1]. Other materials as $FeSi_2$ [2], boron carbide [3], or BP [4] are also considered as potential candidates. However, the materials so far developed are easily oxidized, decomposed, or melt above 1000°C and are not satisfactorily suitable to high-temperature applications.

Since ceramic semiconductors in general have high thermal stability and corrosion resistance, it would be highly effective to apply these materials to high-temperature thermoelectric energy conversion. Porous ceramics of silicon carbide have recently been found by the present authors [5] to show high thermoelectric conversion efficiency at 700-1350K. Electrical conductivity, σ, of porous n-type SiC was comparable to or even higher than the reported values of single crystals, while thermal conductivity, κ, was kept as low as 1/10-1/30 of that for a dense ceramic. The striking phenomenon observed then was that the absolute value of Seebeck coefficient, α, for both n-type and p-type SiC always increases with increasing temperature. In this paper is reviewed how the microstructure of a porous SiC ceramic affects its thermoelectric property and are introduced some newly deduced guiding principles to improve the conversion efficiency.

THERMOELECTRIC MATERIALS DESIGN

The figure of merit, Z, for a thermoelectric semiconductor is usually used as a factor to express its energy conversion efficiency[6]:

$$Z = \sigma\alpha^2 / \kappa \qquad (1)$$

where σ is electrical conductivity, α Seebeck coefficient, and κ thermal conductivity. Conversion efficiency becomes larger with increasing Z. Guiding principles in designing thermoelectric materials possessing high figures of merit so far having been relied on can be summarized as :

(A) Selection of compound semiconductors composed of heavy elements
(B) Solid solution formation
(C) Amorphitization
(D) Microstructure control of porous semiconductors

Compound semiconductors already commercially available, such as Bi_2Te_3, PbTe, and their solid solutions, satisfy the principles (A) and (B). These materials often show high carrier mobility and control of carrier concentration gives rise to both high electrical conductivity and the Seebeck coefficient, while solid solution formation increases the number of phonon scattering centers and hence further decreases thermal conductivity [6].

Newly developed techniques for the preparation of thin films are being applied to the amorphitization of thermoelectric materials as $FeSi_2$ [7], $Si_{1-x}Ge_x$ [8], and boron carbide [9], which is thought of primarily as aiming at decreasing thermal conductivity through disordered structure. Amorphitization (C) is also an attractive method since the disturbance of the electronic system as well as the phonon system may lead to the appearance of the unexpected effect which is still inexplicable by a usual semiconductor theory. In fact, abnormally high Seebeck coefficient as high as 5-15mV/K has been reported in the case of amorphous iron disilicide, $FeSi_2$ [10].

Guiding principle (D) has been first proposed by the present authors [11]. Even in the case of porous SiC which never satisfies the principles (A)-(C), introduction of surfaces and interfaces into its microstructure enhacing the phonon scattering would become effective in realizing high energy conversion efficiency. In the thermoelectric power generation utilizing large temperature difference between the two ends of a material, the risk of fracture due to thermal stress should always be taken into consideration. Making materials porous would also be profitable in improving their thermal shock resistance.

PHONON-DRAG EFFECT

The temperature dependences of Seebeck coefficient found in our previous study [1,11] for porous SiC ceramics (Fig.1) cannot be explained by a simple semiconductor theory taking account only of the electron diffusion generating the thermoelectromotive force under the temperature gradient. Phonon contributions to the Seebeck effect, which is called a phonon-drag effect, have been found for metals

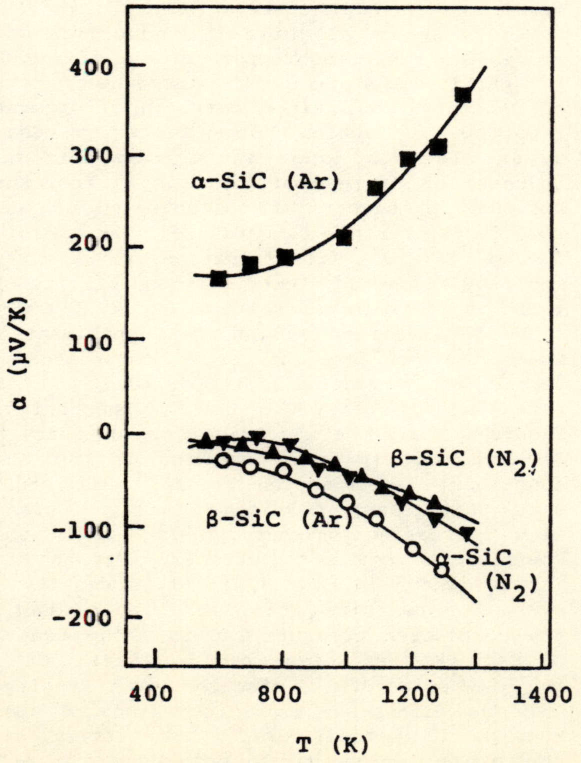

Fig.1 Temperature dependences of the Seebeck coefficient for porous SiC ceramics [1,11].

and compound semiconductors at low temperatures [6]. Phonon-drag effect is associated with the interaction between current carriers and phonons leading to an apparent increase in the absolute Seebeck coefficient. The temperature dependences of Seebeck coefficient shown in Fig.1 were then postulated to be indicating a phonon-drag effect taking place in porous SiC ceramics [11].

According to Parrott [12], the phonon-drag Seebeck coefficient can be expressed as follows:

$$\alpha = \alpha_0 + \frac{\tau_p}{\tau_{ep}} \cdot \frac{m^* s^2}{eT} \qquad (2)$$

$$\frac{1}{\tau_p} = \frac{1}{\tau_{pp}} + \frac{1}{\tau_{pb}} + \frac{1}{\tau_{pe}} + \cdots \qquad (3)$$

where α_o is the normal Seebeck coefficient without a phonon-drag effect, s the mean velocity of phonons, m* the effective mass of a charge carrier, e the electronic charge, T the absolute temperature and τ is the relaxation time, respectively. Subscripts for relaxation time τ represent the types of collision process and for example τ_{ep} means the relaxation time for electrons scattered by phonons. Subscripts e, p, and b denote electron, phonon, and boundary, respectively. If phonon scattering by boundaries (stacking faults, dislocations, etc.) is predominant, relaxation time for the responsible phonon scattering would be expressed as follows:

$$\frac{1}{\tau_p} = \frac{1}{\tau_{pb}} = \frac{s}{L^*} \qquad (4)$$

and this assumption was the only one which could explain the abnormal temperature dependences of Seebeck coefficient shown in Fig. 1. Then, α_p, can be written as follows, assuming current carriers (electrons or holes) are predominantly scattered by phonons :

$$\alpha_p = \frac{L^* s}{\mu_0 T} \qquad (5)$$

where μ_o is carrier mobility and its temperature dependence is expressed by the following equation.

$$\mu_0 = CT^{-n} \qquad (6)$$

where C is a constant and the value n depends on the phonon mode. Carrier scattering by acoustic and optical phonons would give n values of 1.5 and 2.5, respectively.

Using these equations combined with a usual semiconductor theory, the observed temperature dependencies of Seebeck coefficient were simulated by curve-fitting analysis. Solid lines in Fig. 1 represent the best fits obtained by a least squares method.

The L*s is required to be increased to increase the absolute value of Seebeck coefficient and hence the figure of merit for thermoelectric energy conversion. Since L* is the mean free path of acoustic phonons and is directly related to the microstructure, microstructural features should be taken into consideration for this purpose. In order to obtain high L*, average grain size should be increased and structural defects within the SiC grains should be eliminated. However, the microstructural variation would also alter both electrical conductivity and thermal conductivity, so that the optimum microstructure must be sought for to realize the highest thermoelectric conversion efficiency.

ROLE OF MICROSTRUCTURE

It was attempted to clarity the role of microstructure in thermoelectric properties of porous β-SiC ceramics [13]. Commercially available β-SiC powders (Central Glass Co., Ltd., average particle size ~0.4 μm) were isostatically pressed and sintered at 1750°-2200°C in nitrogen atmosphere. A rectangular bar was cut out from a sintered compact, polished and cleaned. Electrical conductivity and Seebeck coefficient were measured in Ar atmosphere using the technique previously reported [1]. Thermal conductivity wsa measured by the laser flash method. Scanning electron microscope (SEM) and high-resolution electron microscope (HREM) observations were carried out to examine the microstructure and stacking

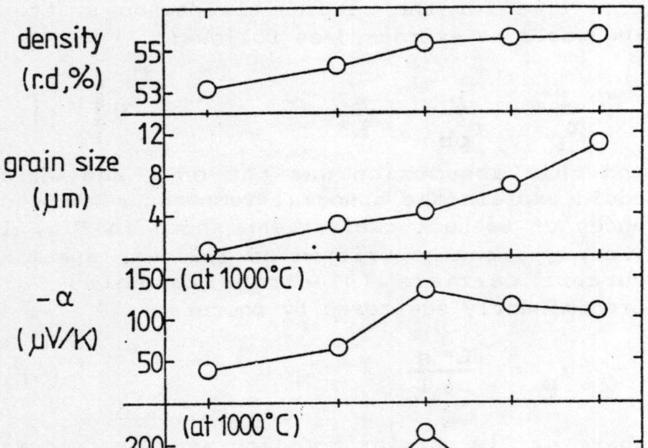

Fig.2 Thermoelectric properties vs. sintering temperature measured at 1000°C. Specimens were sintered for 3h in N_2 atmosphere (Compation pressure = 200kg/cm^2).

faults, respectively. XRD measurements were performed on the crushed powders using CuKα radiation with Ni filter and graphite monochromator.

Thermoelectric properties of β-SiC vary largely with sintering temperature and hence with microstructure variation. The measured σ, α, and κ at 1000°C are plotted against sintering temperature in Fig. 2. Both σ and α increase with increasing sintering temperature up to 2000°C and then decrease at higher temperatures, while κ practically remains unchanged. Similar tendencies were observed at other measuring temperatures.

Average size of β-SiC grains increased remarkably with increasing sintering temperature without noticeable densification as shown in Fig. 2. The authors have already reported [13,14] that the density of stacking faults, whose HREM structural images are shown as an example in Fig. 3, decreased with increasing sintering temperature. The fact that the absolute value of Seebeck coefficient increases with increasing grain size (Fig. 2) and hence decreasing stacking fault density is closely related to the phonon-drag effect explained in the previous section. Decrease in the stacking fault density accompanied by grain growth increased the phonon mean free path, L*, enhancing the phonon-drag effect leading to apparently larger Seebeck coefficient. That the Seebeck coefficient slightly decreased at above 2000°C (sintering temperature) must have been caused by phase transformation which was detected by both XRD measurement and SEM observation.

Increase in electrical conductivity as increasing sintering temperature up to 2000°C is partly ascribed to the increase in carrier concentration, since nitrogen incorporated into the SiC lattice during sintering behaves as an electron donor and its amount would increase as increasing sintering temperature. Decrease in electrical conductivity at above 2000°C (sintering temperature) was possibly due to the β → α transformation, since carrier mobility and hence electrical conductivity of α-SiC is lower in general compared to β-SiC [15].

We have carried out another experiment in which the pressure of powder compaction was varied while sintering temperature and time were fixed to be 2000°C and 3h, respectively. Figure 4 shows the electrical conductivity and Seebeck coefficient increased at 1000°C as functions of compaction pressure. It can be seen that electrical conductivity increases with increasing pressure, while the Seebeck coefficient remains practically unchanged. As is also shown in Fig. 4, relative density and average grain size are almost independent of the compaction pressure, so that the behavior of Seebeck coefficient can be explained from the view point of the stacking fault density. Only the difference among four kinds of specimens lied in their grain connectivity, i.e. the grain-to-grain connectivity (in other words the neck radius) became larger as increasing compaction pressure. Thus, better

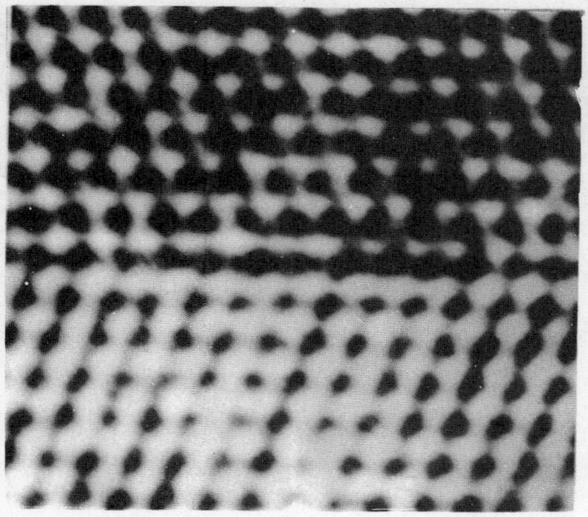

(a)

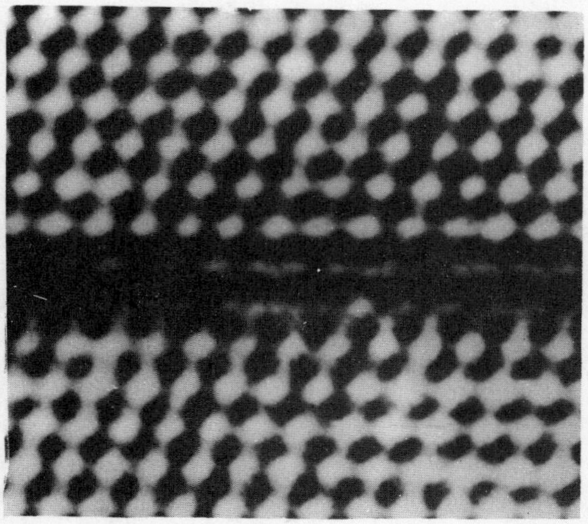

(b)

Fig.3 HREM structural images of typical stacking faults found in β-SiC ceramics sintered at 2000°C. (a) twin boundary, (b) stacking fault.

grain-to-grain connectivity must have played a role partly in increasing conductivity.

Finally, thermoelectric properties of the specimens prepared under the same compaction pressure (200kg/cm^2) and sintered for 0.5-3h at 2000°C were compared. Figure 5 shows that the grain size, Seebeck coefficient, and electrical conductivity all increase slightly with increasing sintering time, while relative density remains unchanged. These results are considered to be consistent with the above mentioned tendencies in that slight increase in grain size gives rise to lower stacking fault density and better grain-to-grain connectivity hence causing larger Seebeck coefficient and higher electrical conductivity, respectively.

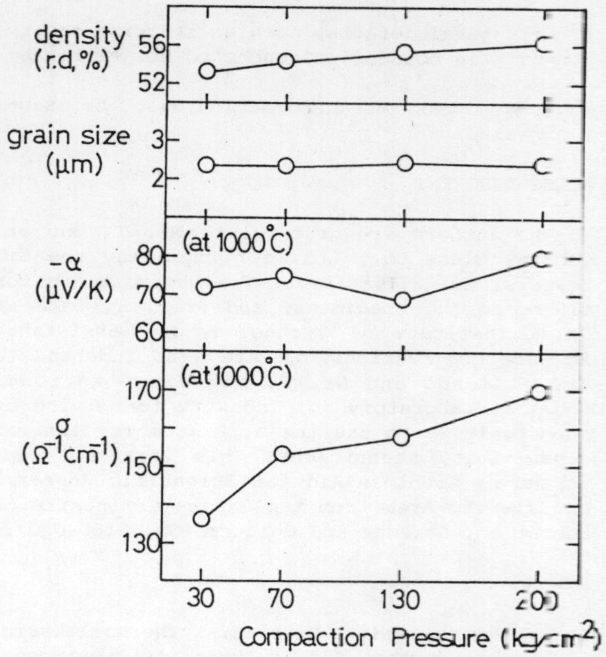

Fig.4 Thermoelectric properties vs. compaction pressure measured at 1000°C. Specimens were sintered at 2000°C for 3h in N_2 atmosphere.

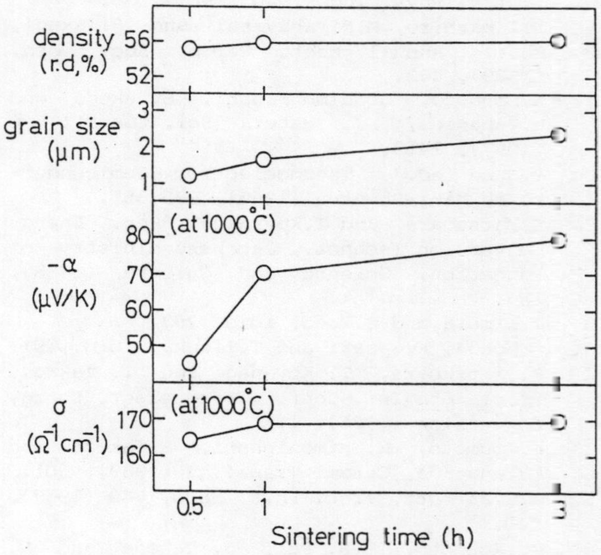

Fig.5 Thermoelectric properties vs. sintering time. Specimens were sintered at 2000°C in N_2 atmosphere (Compaction pressure = 200kg/cm^2).

CONCLUSION

Some of the guiding principles so far obtained to improve the thermoelectric conversion efficiency of n-type β-SiC can be summarized as follows:

(1) High grain-to-grain connectivity should be achieved in a porous body.

(2) Structural defects, such as stacking faults and twin boundaries, ought to be eliminated.

(3) β → α phase transformation must be suppressed.

ACKNOWLEDGMENT

The authors are grateful to Mr. T. Ono of Central Glass Co., Ltd. for supplying the SiC powders, Mr. H.Iwasaki of Toa Nenryo Kogyo for sintering the specimens, and Prof. T. Sato of Chiba Institute of Technology for assistance in HREM observations. We are also indebted to Dr. W. Kondo and Dr. S. Mizuda of National Chemical Laboratory for Industry for giving us opportunities to use the apparatus for thermal conductivity measurement. This work was supported by Grant-in-Aid for Scientific Research on Priority Area from the Japanese Ministry of Education, Science and Culture (No. 01603020).

REFERENCES

1. K.Uemura and I.Nishida, Thermoelectric Semiconductors and Their Applications, Nikkan Gogyo Shinbun (1988).
2. T.Kojima, M.Okamoto and I.Nishida, Proc. 5th Int. Conf. Thermoelect. Energy Conversion (1984), 56.
3. D.Emin, Phys. Today, 40 (1987), 55.
4. Y.Kumashiro, M.Hirabayashi and S.Takagi, Abst. Annual Meet. Ceram. Soc. Jpn. (1989), 463.
5. K.Koumoto, M.Shimohigoshi, S.Takeda and H.Yanagida, J. Mater. Sci. Lett., 6 (1987), 1453.
6. Y.Suge (ed.), Thermoelectric Semiconductors, Maki Shyoten (1966), 295-355.
7. K.Matsubara and T.Koyanagi, Rep. Energy Conversion Technol., Japanese Ministry of Education, Science and Culture (1988), 285.
8. T.Kimura and S.Yugo, ibid.,297.
9. S.Gonda, K.Sasaki and T.Iizuka, ibid.,291.
10. K. Matsubara, T. Koyanagi and T. Takagi, Proc. 6th Int. Conf. Thermoelect. Energy Conversion (1986), 1.
11. K.Koumoto, M.Shimohigoshi, S.Takeda and H.Yanagida, Ceram. Trans., 2 (1989), 501.
12. J.E.Parrott, Proc. Phys. Soc., B70 (1957), 540.
13. K. Koumoto, C.H. Pai, S. Takeda and H. Yanagida, Proc. 2nd Europe. Conf. Thermoelect., in press.
14. K.Koumoto, C.H.Pai, S.Takeda, T.Sato, and H.Yanagida, J. Am. Ceram. Soc., in press.
15. K. Koumoto, Sentan Gijutsu Highlight, 43 (1987), 1-5.

THE EFFECT OF CERAMIC SUBSTRATES ON TCR OF ELECTROLESS DEPOSITED NI-P FILM RESISTORS

Wang Wensheng and Hu Ming
Department of Electronic Engineering, Tianjin University
Tianjin, China

ABSTRACT

The effect of chemical composition and surface states of the ceramic substrates on TCR of electroless deposited Ni-P film resistors has been investigated by using SEM. The results indicate that if the grains on the surface of the ceramic substrates are tiny and uniform, the sizes of Pd crystalline grains of activation centre are small, or equivalently the density σ is large, then a compact perfect Ni-P film can therefore be deposited. In this case the tunnel effect dominates the conductivity. The TCR value is then small. In order to produce Ni-P film resistors with low TCR value, the ceramic substrates with tiny and uniformly distributed grains should be selected.

INTRODUCTION

The electroless deposited metal film resistors are made by depositing Ni-P thin film on activated surfaces of ceramic substrates. Since these kinds of resistors have low resistance and small value of temperature coefficient of resistance (TCR), they are usually used as the complement for the low resistance value serises of precision metal film resistors. During recent years, it has been noticed that the effect of the surface of ceramic substrates on TCR is quite large. If the ceramic grains on the surface of substrate are tiny and distributed uniformly, the value of TCR is low and is about±50ppm/°C. Contrarily, if they are large, the TCR value is about±100ppm/°C.

To probe into the causes, SEM has been used to analyze the topographic microstructures and chemical compositions of different ceramic substrates, and the measured TCR data has been studied by using the theories of surface chemical physics and thin film physics. The results show that if the ceramic grains are tiny, the sizes of the crystalline grains of activation centre Pd atoms are small and the density σ of activation centres is large. So the compact perfect Ni-P film can be deposited. In this case, as the tunnel effect dominates the conductivity, the TCR value is then small. Otherwise, if the tunnel effect is not the main cause of conductivity, the TCR value becomes large. Therefore, in order to produce precision metal film resistors with low TCR values, we ought to select the ceramic substrates with tiny grains which are uniformly-distributed on the surfaces of the substrates.

EXPERIMENTAL RESULTS

The measured data of TCR of the resistors made by depositing Ni-P film on different ceramic substrates under the same conditions are listed in Table 1. SEM images and analysis results of chemical composition for two kinds of substrates are shown in Fig.1, Fig.2 and Table 2, respectively.

Table 1 The measured TCR data (ppm/°C)

Substrate		A		B	
Resistors		¼w,15Ω	¼w,75Ω	¼w,4.7Ω	½w,8.2Ω
25°C	max	43.120	48.306	69.498	97.741
∫	min	40.062	41.418	52.416	43.630
+125°C	avg	41.374	45.114	61.052	66.370
25°C	max	30.562	24.583	70.979	88.451
∫	min	25.916	17.893	49.733	37.255
-55°C	avg	28.785	20.977	60.695	63.427

Table 2 Chemical composition analysis on the surface of two kinds of substrates (%)

	Al_2O_3	SiO_2	CaO	BaO	MgO	Fe_2O_3
Substrate A	72.09	23.11	0.86	3.67	0	0.26
Substrate B	81.71	10.19	0.52	0.38	6.70	0

Fig.1 SEM image of substrate A

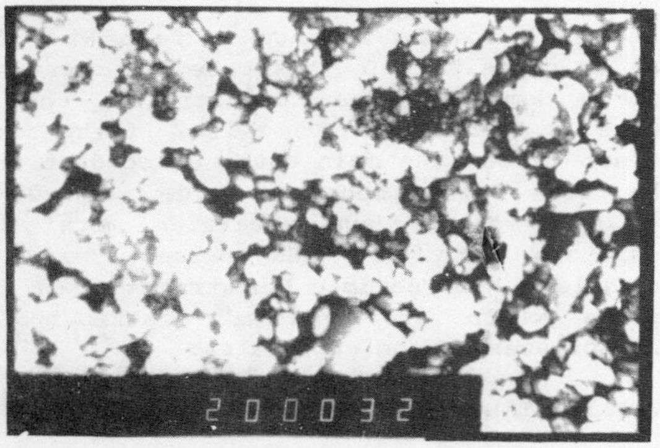

Fig.2 SEM image of substrate B

The experimental results above show the remarkable effect of the substrate surface on the TCR of electroless deposited metal resistors, especially in the low temperature range. From the results we see that the average value of TCR of substrate B is about two or three times that of A.

ANALYSIS AND DISCUSSION

Why the effect of substrate surface state on the TCR of electroless deposited metal film resistors is remarkable? This question can be answered by using the theories of thin film physics and surface chemical physics. It is known[1,2] that the temperature T affects crystal lattice scatter and tunnel effect. The resistivity that caused by crystal lattice scatter is in proportion to T, and that caused by tunnel effect is an exponential function of T. They may be expressed as

$$\rho_b = \frac{m v_o}{nq^2} \cdot \frac{1}{\lambda_b}, \qquad \lambda_b = \frac{A}{T}$$

$$\rho_a = \frac{4}{3} \cdot \frac{a}{d^2} \cdot q^2 \exp\left(\frac{+4\pi d}{h}\sqrt{2m\phi}\right) \cdot \exp[+\phi(x)/KT]$$

where ρ_b and ρ_a represent the resistivity of crystal lattice scatter and tunnel effect respectively. m is the electron mass, v_o is the moving speed of the electron, n is the electron concentration, q is the electric charge of electron, λ_b is the average free distance, A is a constant, a is the radius of conductive particle, d is the distance of tunnel gap, h is Planck's constant, ϕ is the work function of conductive material, K is Boltzmann's constant, $\phi(x)$ is the active energy, respectively.

According to the relationship between

$$R = \rho\frac{l}{w \cdot b}, \quad \text{and}$$

$$TCR = \frac{1}{R}\frac{dR}{dT} = \frac{1}{\rho}\frac{d\rho}{dT} + \frac{1}{l}\frac{dl}{dT} - \frac{1}{w}\frac{dw}{dT} - \frac{1}{b}\frac{db}{dT}$$

(where l, w, b are the length, width and thickness of the resistor, respectively). If we neglect the changes of geometric size of resistors with temperature T, TCR can be expressed

$$TCR \doteq \frac{1}{\rho}\frac{d\rho}{dT}$$

Substituting ρ_b and ρ_a into the expression of TCR, we obtain, for the crystal lattice scatter

$$TCR = \frac{1}{\rho_b}\frac{d\rho_b}{dT} = \frac{1}{T} > 0$$

for the tunnel effect

$$TCR = \frac{1}{\rho_a}\frac{d\rho_a}{dT} = \frac{-\phi(x)}{T^2} < 0$$

Comparing these two equations, it is clear that the tunnel effect results in a low TCR value.

In order to increase the tunnel effect on TCR, it is necessary to increase boundary contact areas of conductive particles. Suppose that the conductive particle is a cube with 10mm long on each side, its total contact area is then 600mm². If we cut this cube into 8 cubes each with 5mm long on each side, the total contact area is 900mm². The difference is 300mm². It can be easily shown that with the decrease of cubes volums, their total contact areas are increased. Similarly for the conductive particles in the film,if we reduce the size of the conductive particles, their boundary contact areas are increased. As a result, the conductivity due to tunnel effect becomes a main factor, and it reduce the TCR of electroless deposited film resistor.

Now why it also can result in small

crystalline grains of Ni-P film when the particles on the ceramic substrates are tiny? It is known from the theories of the chemical physics of surfaces[3,4] that substrates may be activated by immersion in solutions of $SnCl_2$ and $PdCl_2$ before eletroless Ni-P film deposition to form the Pd atom activation centres. Normally to form a large number of activation centres, it is required to make the dispersion D which is defined as the ratio of the number of atoms exposed to the number of atoms added equal to unity. Because a high dispersion means that the activation agent is being used effectively, so that the Ni-P film can be deposited uniformly over the surface. In contrast a low dispersion implies that it is possible to form large Pd atom particle bodies. As a result there are no activation atoms in some regions, so that the actions of activation centres are reduced.

Activation centre Pd atoms lie normally on the sites of the substrate surface where the chemical properties are active. These sites are usually at the boundaries, steps, displacements and defects. This can be observed by comparing Fig.1 with Fig.2 in which the surface of substrate A is uniform and with small particles. It is evident that the active chemical reaction points on A are much more than that of B. Surface states of substrate A approximately satisfy the condition of D to be unity. At the same time the sizes of the bodies of activation centre Pd atom are small, which is the favourable condition for growing the tiny structure films. In contrast with A, for the substrate B, the surface defects and Pd atom body are large in size, so the value of D is small. The particles of film deposited on substrate B are large and heterogeneous. The physical quantity , the number of activation centres can be expressed in terms of the activation centre density σ which is determined by the number of activation centres per unit area. Fig.3 shows the microstructure of deposits Ni-P film on glass substrates with different activation densities. It is clear that for large σ Ni-P film is formed in compact tiny crystallin structure; while for small σ Ni-P film is deposited in porous net structure.

From the dicussion above, after chemical activation treatment with $SnCl_2$-$PdCl_2$, the distribution of activation center Pd atoms (expressed as D), the number of activation centers per unit area(expressed as σ), the particle size of activation center (expressed as D) and so on are all related closely to the surface states of ceramic substrates, and have influence on structure of Ni-P film deposited, and therefore have influence on the electrical property (TCR) of Ni-P film.

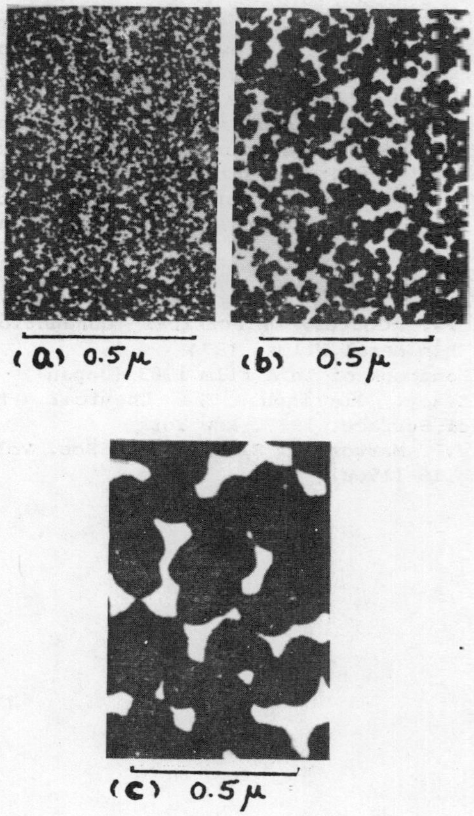

Fig.3 Transmission images of deposits on substrates with different activation densities:
(a) $10^4/\mu^2$; (b) $10^3/\mu^2$ (c) $10^2/\mu^2$
Note the dependence of film geometry on activation density

CONCLUSIONS

From the experimental research and theoretical analysis, the conclusions can be drawn as follows:

1. The effect of surface state of ceramic substrate on TCR of electroless deposited metal film resistor can be explained in two different conductive mechanism: crystal lattice scatter and tunnel effect.

2. To decrease the TCR value of the resistor, it is essential to increase the contribution from the tunnel effect.

3. When the surfaces of ceramic substrates are consist of very tiny particles which are uniformly distributed, the contribution from tunnel effect is large. This is caused by the small size of the crystalline grains of activation center formed by Pd atoms (dispersion D is nearly 1) and the large

density value σ of the activation centers.

4. The Pd atom crystalline grains of activation centers grow easily at the points where the chemical property is active, and most of the points are in the boundary among the crystalline grains. So if the surface of the substrate is formed by tiny particles which distribute uniformly, there will be large number of boundaries of particles and large D and σ values can then be obtained.

REFERENCES

1. T.J. Coutts, Electrical Conduction in Thin Metal Films. 1974
2. Handbook of Thin Film 1983 (Japan)
3. S.Roy. Morrison, The Chemical Physics of Surface. 1977. New York
4. J.P. Marton, J. Electrochem. Soc. Vol.115, p.16 (1968)

HEAT TREATMENT OF ELECTROLESS DEPOSITED Ni-P FILM RESISTOR

Liu Zonge
Electronic Engineering Department, Tianjin University, Tianjin, China
Zhu Jiaqi
Beijing Second Radio Appliance Factory, P.O. Box 554, Beijing, China

ABSTRACT

In this paper, the effects of heat treatment on temperature coefficient (TCR) and moisture coefficient of resistance (MCR) of metal film resistor made of electroless plating are studied. It has been found that the excellent properties of electroless Ni-P film can be obtained by heating for several hours at a certain tempereture. The film has been analyzed by means of TEM and SEM.

INTRODUCTION

It is reported that electroless Ni-P film has been used as metal film resistor since a comprehensive investigation program was started on it in 1959. To manufacture precision metal film resistor many researchers have exerted themselves to improve the unstable properties both in thermal and moist ambient condition.

It is known that the Ni-P film performance are not only related to the solution composition in which the substrate is plated electrolessly, its pH, plating temperature and the the thikness of film formed. In addition, it is possible to improve the TCR and MCR of this kind of resistors by means of heat treatment on the film.

The optimum conditions of heating treatment on Ni- P film have been found, with them the TCR and MCR of the resistor made of electroless plating tend to decrease obviously. The morphology and diffraction patterns of the film are also studied.

EXPERIMENT

1. Heat Treatment Technology

For manufacturing the electroless plated Ni-P film resistor samples, two kinds of rods are used: one comes from Japanese which is called YM simply and another from Qing Huang Dao in China which is called GHD. These samples are produced by Beijing second Radio Appliance Factory. They are treated by heating for certain time at several temperatures shown in Table 1.

Table 1 Heat treatment technologies

Rods	Temp (C°)	Time (hour)	The number of sample
YM	170	1.5 4 5	20
	220	1.5 4 5	
	240	1.5 4 5	
GHD	170	1.5 4 5	20
	220	1.5 4 5	
	240	1.5 4 5	

2. By means of TEM the diffraction patterns before and after treated by heating the Ni-P films deposited on the sample rods have been observed.

3. Those morphologies said above have also been obtained by SEM.

RESULTS AND DISCUSSIONS

Table 2 gives the average TCR and MCR values for the different samples made of various heat treatment technologies.

There are three average TCR_+ values-heating treatment temperature curves in Fig.1, They are treated for 1.5, 4 and 5 hours at the

Table 2 TCR and MCR for various heat treatment samples

Heat treatment technologies	YM			GHD		
	TCR_+	TCR_-	MCR	TCR_+	TCR_-	MCR
170°C 1.5 hurs	33.6	37.7	0.36	32.7	50.8	0.13
170°C 4 hurs	47.6	38.2	0.23	37.8	52.3	0.14
170°C 5 hurs	34.6	43.4	0.23	47.6	34.4	0.19
220°C 1.5 hurs	47.4	45.8	0.18	44.4	56.4	0.41
220°C 4 hurs	21.7	44.5	0.25	48.2	41.6	0.25
220°C 5 hurs	42.0	44.1	0.44	62.4	47.9	0.20
240°C 1.5 hurs	43.1	41.7	0.20	52.0	57.9	0.39
240°C 4 hurs	50.3	43.6	0.59	58.6	50.8	0.57
240°C 5 hurs	46.2	36.7	0.18	60.3	66.3	0.22

Note: TCR(PPM/°C) MCR(ΔR/R%)

maintaining temperatures respectively.

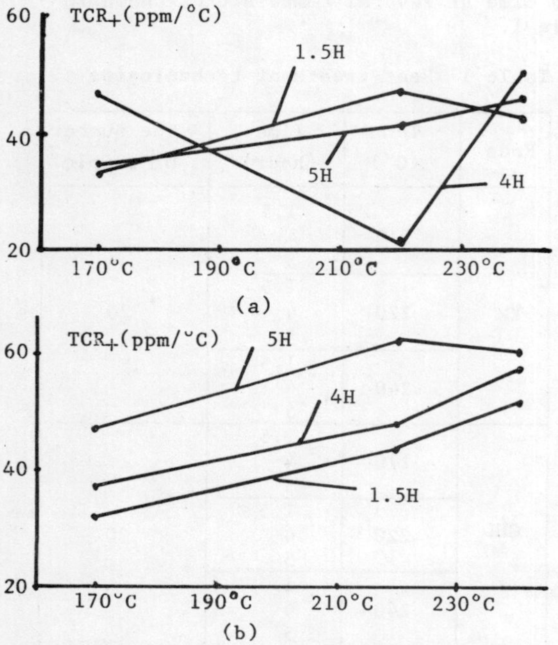

Fig.1 TCR_+ — heat treatment temperature curves of the resistors
(a) rod YM (b) rod GHD

The top one (a) in Fig.1 exhibits the case in which resistor rod is YM and under one (b) is for GHD. The variations of MCR with heat treatment temperature are shown in Fig.2 where (a) and (b) represent rod YM and CHD respectively.

It is known that Ni-P film formed by electroless plating process occurs always amorphous and its lattice is irregular. This kind of film isn't dense enough and the composition distribution in film is also not uniform. Therefore, the resistor consisted by this kind of film have a higher TCR and MCR. But with

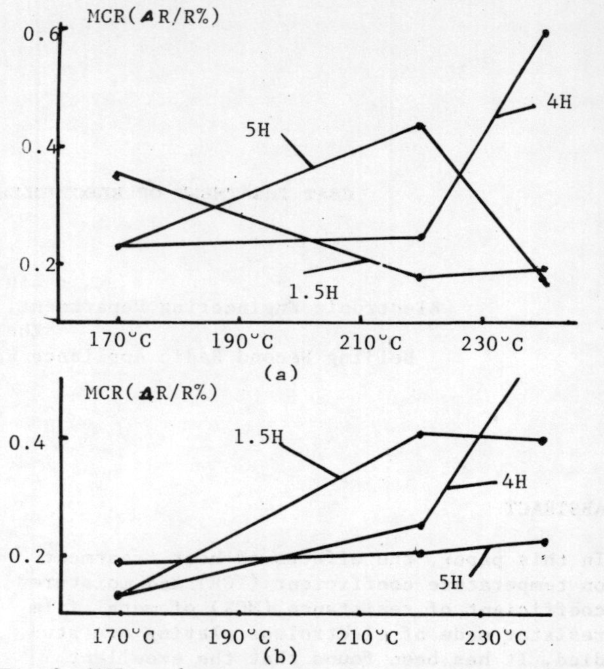

Fig.2 MCR — heat treatment temperature curves of the resistors
(a) rod YM (b) rod GHD

heat treatment the film properties will be improved obviously. The action of the heat treatment is as follows:

1. It can make the amorphous film crystallization;
2. It can eliminate the interior stress in Ni-P film;
3. It can decrease the surface defects. So long as the heat treatment is done properly, the sizes of microcrystal alloy particles will be more analogous and uniform. Film construction becomes more and more dense.

Since the atomic radii of Ni and P are quite close to each other, the P occupies most likely the lattice sites of Ni in the film solid solution. If the heat treatment process is proper, the crystals in sample will be fine and uniform. As a result its TCR and MCR are all improved. Figure 1 and 2 show that effects of heat treatment on the TCR and MCR of the samples are evident. The optimum condition for decreasing TCR of the sample is to treat the Ni-P film for 4 hours at 220°C when YM rods are used and for 1.5 hours at 170°C when the Ni-P film is deposited onto GHD rods. But optimum condition for decreasing MCR of the sample is to treat Ni-P film for 1.5 hours at 220°C when YM rods are used and for 1.5 hours at 170°C when the Ni-P film is deposited onto GHD rods.

By means of TEM the electron diffraction patterns for Ni-P film before and after treating are obtained. It is obvious that the presence of a scattering circle shown in figure 3 illustrates the film being not treated by heating is amorphous, but heating the film makes occuring corresponding regular diffraction points in the pattern in figure 4. This demonstrates that by heat treatment the crystaline is to grow step by step in the samples.

The Ni-P film morphologies for different heat treatment processes have been examined by SEM. The films having not been treated by heating have morphology patterns shown in figure 5 and 6 in which the substrates are YM and GHD respectively. Fig.7 and Fig.9 show morphology patterns of Ni-P films of the sample deposited on YM rods after heat treatment for 4 hours at 220°C and for 1.5 hours at 260°C respectively. Fig.8 shows morphology pattern of Ni-P film of the sample deposited on GHD substrate after treating for 5 hours at 170°C.

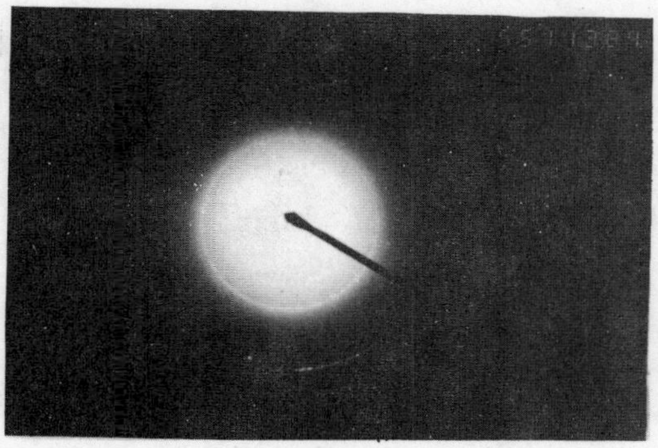

Fig.3 Transmission electron diffraction pattern of a typical Ni-B film on YM substrate before heat treatment

Fig.4 Transmission electron diffraction pattern of a typical Ni-P film on YM substrate after heat treatment for 4 hours at 240°C

They are illustrated in Fig.5,7 and 9 that with increasing treatment temperature crystal grain in Ni-P films deposited on YM substrate is becoming larger and larger. If the grain is growing too large, the large crystal grain makes the TCR increasing too high so that the

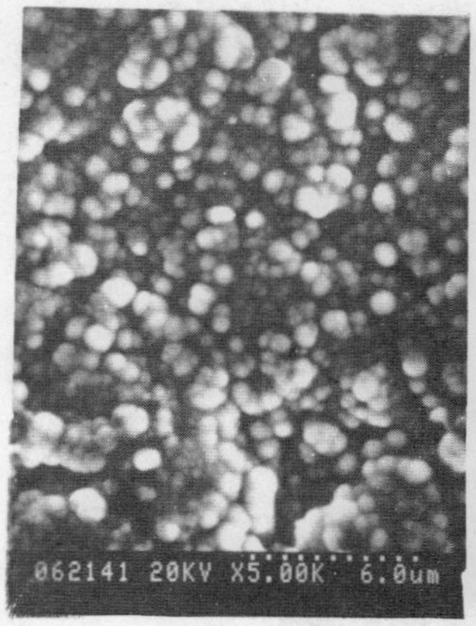

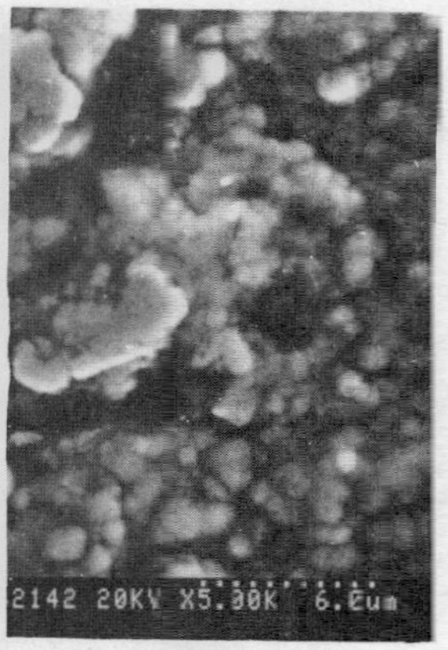

Fig.5 morphology pattern of a typical Ni-P film on YM substrate before heat treatment

Fig.6 morphology pattern of a typical Ni-P film on GHD substrate before heat treatment

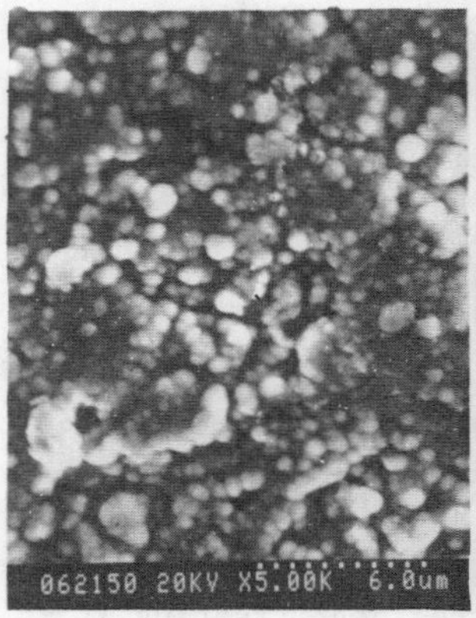

Fig.7 morphology pattern of a typical Ni-P film on YM substrate after heat treatment for 4 hours at 220°C

Fig.8 morphology pattern of a typical Ni-P film on GHD substrate after heat treatment for 5 hours at 170°C

Ni-P film property deteriorates, see Fig.9. But if the film is formed on a substrate with proper aspect and composition, for instance on GHD substrate and treated with suitable temperature and time, the grain can grow to become moderate in size and uniform. The data tested indicate that both of TCR and MCR are greatly decreased.

Structure and electronic properties of Ni-P films of the samples are not only related to heat treatment technology, but are also, even more important, related to composition, surface state and crystaline state of the substrate. Especially it is related to the contents in Ni-P films and plating technology. As des-

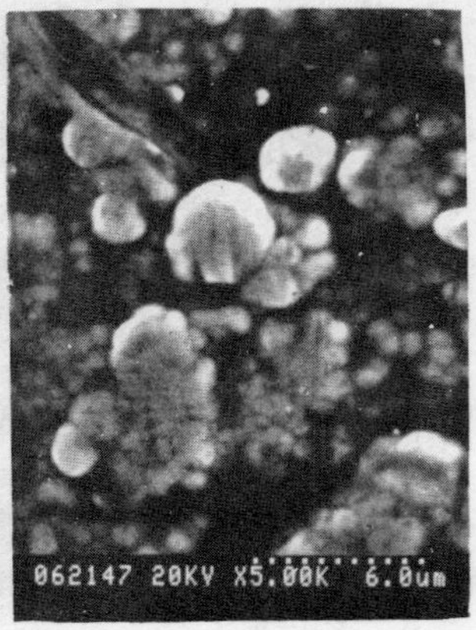

Fig.9 morphology pattern of a typical Ni-P film on YM substrate after heat treatment for 1.5 hours at 260°C

cribed above, this paper doesn't discuss them further more.

CONCLUSIONS

1. In order to improve electronic properties of the resistors and decrease temperature coefficient and moisture coefficient of resistance, there are different optimum heat treatment technologies for resistors deposited on different ceramic substrates using the same plating solution and technology.

2. Ni-P film structure of resistos is related to heat treatment technology and substrates if treated with same conditions above.

3. The Ni-P metal film occurs amorphous before heat treating, but after heat treating it becomes crystaline partially and sometimes perfectly.

REFERENCES

1. J.DEARDEN, Electroless plating-Its Applications in Resistor Technology. Electrocomponent Science and Technology 1976, Vol.3, PP.103-111

2. S.T. Pai and J.P. Marton, Annealing Effects on the Structure and Resistivity of Ni-P Films, J.Appl. Phys, Vol.43, No.2, February 1972.

THICK CERMET FILM FORMED BY DECALCOMANIA

Zhu Jiaqi and Pu Shoupeng
Beijing Second Radio Appliance Factory,
P.O. Box 554, Beijing, China

ABSTRACT

This paper lays stress on discussing the technical possibility of applying decal to thick film field. The drying and firing processes of cover coat-attached substrate have been analysed and the optimum conditions on which the pasty pattern configuration can be kept and the resulting cermet properties be the same as those obtained using the direct printing process have been worked out. On the basis of the technological processes established some resistor specimens have been made and their electrical performances studied in detail.

INTRODUCTION

Decalcomania is a kind of process for transferring frits to a substrate. It is different from the direct printing process. By decal the pasty ink containing powdered precious metals and frits is printed on a decal paper which is composed of backing paper covered with a layer of cover coat, such as polyvinyl butyral.[1] Separate the pattern-printed cover coat from the backing paper and attach it to a surface of a ceramic ware. Dry the cover coat-attached body by heating and bake it to form thick cermet pattern on substrate.

Decal is advantageous in its applicability to both flat and curve substrates. Pattern-attached decal paper may be sold as a semi-product. Its use allows many decal paper to be printed at a time thereby facilitating mass production of considerably precision printed substrates. In addition, if use the decal to make patterns in multilayers on a same substrate, the next one cannot be dirtied by pattern paste having printed already but not dried thoroughly before. There are many different constructions and varieties of decal[2,3,4] at present. Using decal some researchers have decorated metal film on ceramic wares[5,6] and studied its electric performances[7], and Kojima et al[8,9] have made great strides in applicating the decal to electronics. But due to presence of cover coat there are two problems to solve for transplanting decal to form cermet. They are: 1. Is pasty pattern transferred thoroughly? and 2. Is the stability of cermet product influenced by decal?

PROCESSES ANALYSIS

1. How to Make Paste Transfer Thoroughly

Difference between the substrates made of direct printing process (a) and that of decal (b) is shown in figure 1:

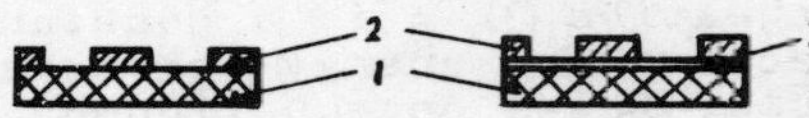

Fig. 1.
(a) direct printing substrate (b) decal composition
1. substrate, 2. pasty pattern dried, 3. cover coat attached to substrate

From figure 1 it is known that the substrate made of decal has only a cover coat layer more than that of direct printing. Cover coat known as organic polymer will adhere to substrate due to its melting or by means of some adhesives. At the time of drying it will frizzle, even shrink into itself if the heat is given improperly. If the suitable polymers of cover coat and/or adhesive are well chosen and the drying rate is got under, the glue produced by cover coat and/or adhesive will adhere the coat and substrate thereunder tightly. As a result, the adhesive force between both of them will be larger than the stress due to the heat shrinkage of the cover coat. And it is possible to decompose thermally the coat and/or adhesives everywhere during same time. Thus the paste that has been printed on the coat before will be transferred onto the substrate perfectly and kept its **original** configuration after firing.

2. How to Avoid Destruction of Pasty Pattern by Gas

Together with the temperature being elevated, if the carbon that is decomposed thermally from the cover coat and the adhesive is converted to gas suddenly, the gas such formed will disturb even break the pasty patterns. The reaction between C and O_2 is shown in equation (1).

$$C+O_2 \xrightarrow[\text{abundant } O_2]{\text{firing in oven}} CO_2 \qquad (1)$$

as time increased for dt at temp. T_0	1-dx	dx
Temperature rises from To to T_1 as time extending from t_0 to t_1	$1-x_1$	x_1

Its reaction rate V is illustrated as

$$V=d(O_2)/dt= -k(O_2) \qquad (2)$$

or

$$-d(O_2)/(O_2)= kdt \qquad (2a)$$

where the rate constant k is related with temperature T of the system:

$$k = A \exp (-E/RT) \qquad (3)$$

In equation (3), A is a constant and E is an activation energy for an active molecule. Based on equation (3), (2a) can be written as equation (4) when the elevated temperature is dT during the time increment dt

$$-d(O_2)/(O_2) = (k+dk)\ dt \qquad (4)$$

If the rate constant at temperature T_o is k_o, and so

$$-d(O_2)/(O_2) = (k_o+dk)\ dt \qquad (4a)$$

When the reaction time is passing by from t_0 to t_1 and temperature from T_0 to T_1 (the rate constant is k_1 at T_1), the (CO_2) in system increases from 0 to x_1 and (O_2) decreases from 1 to $(1-x_1)$. To calculate x_1, equation (4a) can be integrated against t and k as following:

$$\int_1^{1-x_1} \left[-d(O_2)/(O_2)\right] = \int_{t_0}^{t_1}\int_{k_0}^{k_1}(k_o+dk)dt$$

$$\text{i.e. } \ln(1/1-x_1) = k_1 \Delta t \qquad (5)$$

$$\text{or } \ln(1/1-x_1) = A \exp(-E/RT_1)\cdot\Delta t \qquad (5a)$$

Provided that there isn't any reaction at temperature T_0 and C begins to oxidize at the elevated T_1 point where T_1 has a certain value, from (5a) it is known that

1) If $\Delta t \to 0$, and so $x_1 \to 0$, that is, the fewer the time when C is oxidized into CO_2 at T_1 is spent, the lower the concentration of CO_2 produced;

2) When the temperature reaches T_2 from T_1 and goes on up, the higher the temperature rises, the more rapid the rate increases, and x_1 will increase with the elevated T_2;

3) At point T_1 the increase of x_1 is only dependent on Δt(prolong of time spent). Assuming that the system for the CO_2 going up to release* and the frits sinking down to adhere with substrate there-under doesn't be affected by any external forces, there is a relation according to the momentum conservation principle:

$$m_g \bar{V}_g = m_{CO_2} \bar{V}_{CO_2} \qquad (6)$$

where m and $\bar{V}$ are the substance mass and its average rate respectively and g represents frits. In the same time interval Δt the equation (6) becomes:

$$m_g \Delta S_g = m_{CO_2} \Delta S_{CO_2} \qquad (6a)$$

where ΔS_g and ΔS_{CO_2} are the travelling distance of frits and CO_2 molecules respectively. If $\Delta t \to 0$, and so $x_1 \to 0$, that is, $m_{CO_2} \to 0$. Due to the regular m_g, therefore $m_g \gg m_{CO_2}$, and

$$\Delta S_g/\Delta S_{CO_2} = m_{CO_2}/m_g \to 0 \qquad (7)$$

Equation (7) can be established only in the case which $\Delta S_g \to 0$ (but it isn't available) or $\Delta S_{CO_2} \to \infty$. It shows that in the shortest time interval the pasty powder (frits) cannot be almost considered to move, but a few of CO_2 molecules which have obtained higher energy can go glimmering from the system, they can never disturb the top and surrounding powders. But if we cannot get the elevated temperature under, because Δt and T_1 all have their fixed values, so the $x_1(m_{CO_2})$ will be constant (refer to(5a)), hence

$$\Delta S_g/\Delta S_{CO_2} \doteq \text{Const.}$$

*According to $\bar{V}$=8RT/$\pi\mu$ (μ is the molecular weight of CO_2) the average rate for CO_2 molecule at 723^{0}K is about 1.5 times more than at 298^{0}K, its kinetic energy is about 2.5 times more because of $E_k \propto \bar{V}^2$.

Provided that the system is cooled suddenly when the frits are being melted, blister will be contained in the resulting pattern.

3. How to Provide Cermet Pattern with Fully Satisfactory Properties

The tests show that if the pyrolysis of polymer isn't right in degree, some reductant, such as C and CO, can perhaps react with the relative contents in the frits so that their presence in system is going to produce the deleterious effect on the performance of the resulting cermet pattern. They can make the cermet resistance unstable.

1) Reduction of C and CO to Pb^{2+} and Bi^{3+}: When firing frits in which conductor phase consists of precious and/or base metal oxide, both the C content produced by decal paper and CO and CO_2 coming from carbon are more than that produced by the direct printing process. The standard electrodynamical potential between the two electrodes of the Voltaic cell is the sum of potentials both C or CO electrode and PbO or Bi_2O_3 electrode in pasty powder: For C, there are two cell reactions:

$$C+PbO+OH^-=Pb+HCO_2^-$$

$$\mathcal{E}^o=E^o_{PbO/Pb}-E^o_{HCO_2^-/C}=-0.02(v) \qquad (8)$$

$$3C+3BiOOH=2Bi+3HCO_2^-+Bi^{3+}$$

$$\mathcal{E}^o=E^o_{BiOOH/Bi}-E^o_{HCO_2^-/C}=+0.06(v) \qquad (9)$$

For CO, the reactions are respectively:

$$HCO_2^-+PbO+OH^-=Pb+CO_3^{2-}+H_2O$$

$$\mathcal{E}^o=E^o_{PbO/Pb}-E^o_{CO_3^{2-}/HCO_2^-}=+0.47(v) \qquad (10)$$

$$3HCO_2^-+3BiOOH=2Bi+3CO_3^{2-}+Bi^{3+}+3H_2O$$

$$\mathcal{E}^o=E^o_{BiOOH/Bi}-E^o_{CO_3^{2-}/HCO_2^-}=0.55(v) \qquad (11)$$

The data mentioned above show that if there is some water in system when pasty powder being melted, although C cannot reduce Pb^{2+} in alkali and can only be reluctant to reduce Bi^{3+}, yet CO will react with both Pb^{2+} and Bi^{3+}, and its reductibility is a little higher for Bi^{3+} than Pb^{2+} (alkali oxides Ag_2O and PdO decompose into the corresponding metals respectively in high temperature. The reduction of C and CO is in favour of forming Ag and Pa). Therefore, it is necessary to restrict the quantity CO approximately equal to zero in a short time interval at a temperature point.

2) Confine the quantity CO: When firing at a higher temperature, following reactions may take place:

$$2C+O_2 = 2CO + 52,780 \text{ cal} \qquad (12)$$

$$C + CO_2 = 2CO - 41,400 \text{ cal} \qquad (13)$$

$$2CO + O_2 = 2CO_2 + 135,280 \text{ cal} \qquad (14)$$

$$C + O_2 = CO_2 + 94,030 \text{ cal} \qquad (15)$$

$$C + H_2O = H_2 + CO - 31,380 \text{ cal} \qquad (16)$$

LeChatelier Law shows that in order to make the reaction towards to form CO_2, it is necessary: a. to conduct heat out from system, b. to confine the water content in system, c. to put fresh air through and increase the O_2 content, and/or fire in an oxidation atmosphere to avoid forming CO, and d. to restrict concentration or partial pressure of CO_2 (P_{CO_2}).

Reaction (13) is the source providing CO from CO_2. It can be considered to be a cell composed of the following two electrodes:

$$(-)\quad C + 1/2\ O_2 - 2e = CO$$
$$(+)\quad CO_2 + 2e = CO + 1/2\ O_2$$

The potential difference between the two electrodes is:

$$\mathcal{E}=E^o_{CO_2/CO}-E^o_{CO/C}+\frac{RT}{nF}(\ln P_{CO_2}/P^2_{CO}), \text{ i.e.}$$

$$\mathcal{E}=-0.626+\frac{RT}{2F}(\ln P_{CO_2}/P^2_{CO}) \qquad (17)$$

where R is a gas constant, T is absolute temperature, n is electron numbers, F is Faraday constant and P is the partial pressure of gas.

From equation (17) it is seen that as long as P_{CO_2} increases a little, it is possible to make $\mathcal{E}>0$, lead to form CO. Therefore, it is necessary that the CO_2 formed by other reactions per second in the system must be controlled so that the fewer the CO_2 forms, the better the result is.

3) Optimum firing conditions: The discussion above shows that in order to provide the resulting cermet with fully satisfactory properties the optimum firing conditions, in the final analysis, are:

a. The polymer for cover coat and/or adhesive must be chosen properly. Drying by heating must be controlled carefully;
b. Get the elevated rate in temperature well under when firing and stay properly at the carbonization and oxidization points;
c. Increase O_2 and/or air quantity;
d. Keep the firing furnace well ventilated and take out all of the organic compounds with lower molecular weight, water, carbon dioxide and heat probably produced in the system.

EXPERIMENT AND RESULTS

Using decal process according to the optimum firing conditions derived above the leadless chip resistor which two electrodes are extended to both back sides of substrate (see the table below),

Table: Sample (1/8 W 27kΩ ±5%) properties

Items	Tested results Max.	Min.
R(kΩ)	27.647	25.658
TCR(25°-+125°C), ppm/°C	172	31.4
TCR(25°--55°C), ppm/°C	324	20.5
Temp. cycling (3 times, -55°-+125°C), ΔR/R,%	-0.83	-0.01
Humidity(R.H.-93±3%, 40°±2°C,96 hrs.),ΔR/R,%	-2.41	0
Low pressure (7.5 Torr, 1 hr.), ΔR/R,%	-0.58	-0.05
Vibration(50±5Hz)ΔR/R,%	-0.32	0
Load (1.5 Pr, 40°±2°C), ΔR/R, %	-1.14	-0.15

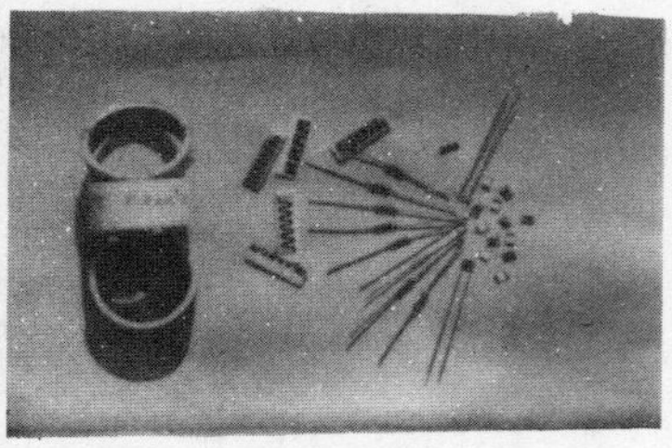

Photo 1 The various specimens

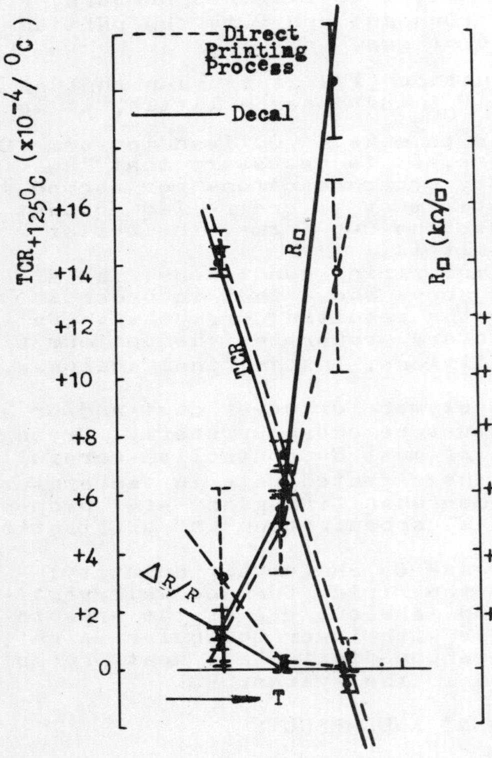

Fig. 2 Relationships of R□, TCR or ΔR/R with firing temperature (T)

the miniature cylindrical resistors (Ø1.6x6) and the resistors with helix on the interior ceramic wall (photo 1) have been made. The results tested for chip resistors comparing with the direct printing process are shown in figure 2.

CONCLUSION

1. From some physical and chemical principles this paper derives four necessary optimum conditions for transplanting decal to produce thick film components and devices, Based on them the thick film components such as chip resistors have been made and the performances tested don't be inferior to that of direct printing process, even better;
2. Using decal the pasty patterns can be deposited onto the surfaces of ceramic rods to form thick film conveniently. Because decal processes are successive when firing in a tunnel furnace, it is in favour of process mechanization and **automation** ;
3. Due to rapid developing of cermet itself, it is possible that the cylindrical cermet resistor by decal would perhaps take the place of carbon and/or metal film products to certain degree.

Based on the inclusions described it is known that decal is an effective means for depositing thick cermet film. Further investigating to apply decal in electronics field there is possible to exert a farreaching influence on production of discrete cermet components and hybrid circuits.

ACKNOLEDGMENT

We wish to express our appreciation to Mr. Shu Yugui (Central Arts and Crafts Institute) and Mr. Song Xinyi (Kuenmin Institute for Precious Metals) for their help.

REFERENCES

1. A group of editors in Jingdezhen, Handbook For Ceramics Industry For daily Necessities (graft, the internal issues), (1977), Paragraph 7 of Chapter 4.
2. USP. 3,616,176.
3. USP. 3,562,087.
4. USP. 4,044,181.
5. USP. 3,615,980.
6. USP. 4,012,552.
7. USP. 4,018,728.
8. USP. 3,981,761.
9. Fr. Pat. 2,173,196.

MODEL RI-8 HIGH VOLTAGE AND INDUCTIONLESS GLASS GLAZE RESISTOR FOR DISCHARGING APPLICATION

Wang Di-Yun and Chen Yan-Xian
Eastern China Microelectronic Circuits Research Institute P. O. Box 68, Hefei, Anhui, China

INTRODUCTION

The operational characteristics such as high resistance value, large overload capacity, high reliability, speedy discharge and low cost are critical requirements for discharge resistors which are used in power transmission network, lightning arrester and other applications. However, the requirements for tolerance and temperature coefficient of resistance(TCR) of this kind of resistors are not critical. In current domestic market, most of resistors are made of composition film, carbon film or metal film as resistive materials. Because the resistivities of such materials are low, the resistance value of resistors should be adjusted to high levels by engraving. The method of engraving adjustment will decrease the cross sectional area of resistor and produce inductive reactance. Therefore, the discharging speed of resistors will be reduced. In addition, the above mentioned resistive films are very thin and their withstanding temperature is low, therefore their overload capacity is weak.

In the world market, the resistors of high voltage and high resistance value are usually made of noble metals which have very low TCR and are used for potential-divider network and other applications. For instance, Japanese TAMA Electric. Co manufactured Model GS thick film resistors of high voltage and high resistance by means of engraving method. American Coddock Co. used inductionless printing method to make the high voltage resistors. However, the adoption of any above referred processing can cause the decrease of cross sectional area of resistive paths and also can weaken the reliability and overload capacity of resistors. In addition, the costs of noble metal materials are too expensive to be suited for manufacturing discharge resistors.

With the purpose of speedy discharge, the resistors should be designed free from inductance. To meet the requirements of the market, we have developed the new type high voltage resistors which can avoid inductive reactance caused by engraving and overcome the shortages of above mentioned materials. Thus, the purpose of this research work is to develop the materials suited to make this kind of resistors which have high resistivity, and are resistant to high electrical field intensity and high temperature, and also have cheaper price.

After all efforts more than one year, we have developed a new resistor material which is comprised of base metal semiconductive SnO_2 containing pyrochlore-related structure. The electrical properties of the resistors are as follows:

Sheet resistance, > 100MΩ/□
TCR, < +1500 ppm/℃
Voltage stress, > 400 V/mm

This material has already been used to fabricate Model RI-8 inductionless discharge resistors which have widely been applied to power transmission networks and lightning arresters. The properties of this resistors are in keeping with the requirements of users.

FABRICATING PROCESS

The flow chart of fabricating of high voltage resistor is as follows,

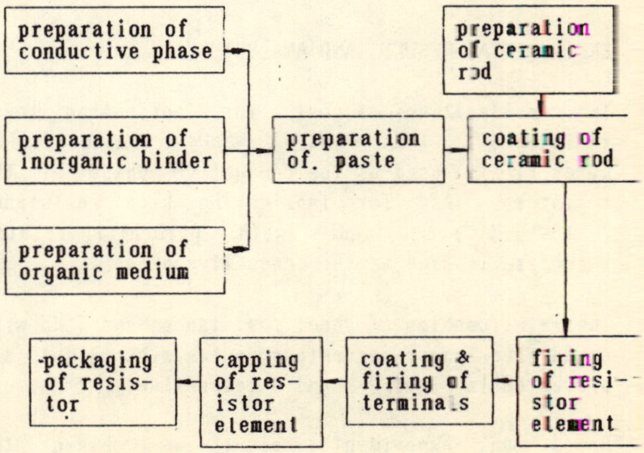

1. Conductive Phase Preparation

The preparation of such conductive phase of the resistor consists of melting tin and antimony together and treating the melted alloy (Sn, Sb) by acid and firing and oxidizing it at high temperature to form semiconductive SnO_2 doped with Sb. Thereafter, the pyrochlore-related compounds are prepared by firing the admixture of finely divided particales of semiconductive SnO_2, SnO and Ta_2O_5 in desired proportions at high temperature in a protective atmosphere and quenching, finely divided them again to form the final material of conductive phase.

2. Binder Phase Preparation

The inorganic binder is prepared by thoroughly mixing SiO_2, B_2O_3, BaO and a small quantity of CaO, MgO and ZrO_2 in desired proportions, and heating the admixture at 1400℃ to form a melt. The melt is then poured into cold water to subject to quenching. The crude frits separated from water are then treated by drying, ball milling and screening through a 400 mesh screen to remove any large particales.

3. Organic Medium Preparation

The organic medium is to serve as a vehicle for dispersion of finely divided solid in resistive paste. The vehicle is prepared by mixing with 8-5% wt. of ethyl cellulose and 92—95% wt. of resinol and then heated by water bath to solve them.

The paste of the resistor is prepared by thoroghly dispersing the powder mixture of conductive phase and glass phase in vehicle using a three-roll mill. The resistor substrates (ceramic rods) are then coated with the pastes, dried and fired at high temperature in an atmosphere condition. After firing, both ends of the resitor element are coated with silver paste and fired to form terminals. Finally, through capping and packaging, the resistors are then formed.

EXPERIMENTAL RESULTS AND ANALYSIS

In consideration of cost, for fabricating lower resistance (10KΩ～1MΩ), SnO_2 semiconductive material is used as the conductive phase of the resistors. And for fabricating high resistance (1～100MΩ) , SnO_2 with pyrochlore-related structure is used as the conductive phase.

The relationships of sheet resistances and TCRS with conductive phase percentage contents by weight are presented in Fig. 1, 2, 3 and 4 respectively.

Based on experimental result we obtain the parameters of the resistors shown as follows:

Operation temperature:	-55～+125℃
Sheet resistance:	10KΩ/□～100MΩ/□
TCR:	≤+1500 ppm/C
resistance to electrical field intensity:	400V/mm
sensitivity to processing:	≤1%/℃
stability:	≤3～5%
inductive reactance:	no

The experiments of firing process with respect to two different kinds of conductive phases are carried out. The experimental results are summarized and shown in Fig. 5～10. It is observable that all kinds of electrical properties are suitable, also the sensitivity to processing is low.

DISCUSSION

1. By alloy displacement method, the semiconductive SnO_2 doped with Sb is the basis of conductive phase. The reasons for using SnO_2 material as conductive phase of high voltage resistors are as follows: much more resistant to field intensity up to 400V/mm. large over load capacity, low sensitivity to processing and inexpensive firing condition. According to our experiments, 10% wt. of doped Sb is preferred. On this condition, the crystal cells are saturated, thus the resistivity and TCR of resistors are in the lowest degree. Because the sheet resisance of semiconductive SnO_2 material can only reach 5MΩ/□, it is unsuited to make the resistance much higher.

2. The conductive phase made by SnO_2 with pyro-chlore-related structure can increase resistivity and decrease TCR. Futhermore, in the wide ranges of resistance values of this type resistors, the variations of TCR are in a small degree. Also, it brings about fabricating high voltage, inductionless discharge resistor without adjustment by engraving, and meeting the specail needs of speedy discharge and high reliability.

The reason is that during firing period the solidoid reaction of Ta_2O_5, SnO and a small amount of SnO_2 takes place. As a result, it forms a pyrochlore related compound.
The formula is:

$$Sn^{2+}_{2-x}\ Ta_{y_2}\ Sn^{4+}_{y_1}\ O_{(7-x-y)/2}$$

where $x=0.55,\ y_2=0\sim2,\ y_1=0\sim0.5$
and $y_2+y_1=2$

The analysis data show that, there are no free states of Ta_2O_5 and SnO_2 in the material. This new conductive phase contains 20% wt. of pyrochlore and 80% wt. of free semiconductive SnO_2. Because the pyrochlore-related structure is cubiclattice cell, the lattic space can be easily occupied or replaced by varieties of metal ions. Thus, the resistivity of the materials is increased obviously. As an example, the higher sheet resistance above 100MΩ/□ was obtained. The TCR value of pyrochlore constituent of the conductive phase is positive, which can compensate for negative TCR of high resistance semiconductive SnO_2. Meanwhile, there are a large amout of semiconductive SnO_2 in the conductive phase, so the resistor element could be fired at atmosphere condition. However, to avoid the decrements of pyrochlore constituents which can cause decrease of resistivity, the firing atmosphere and constant temperature time should be strictly controlled.

3. The inorganic binder of the resistor paste is barium borosilicate system which does not contain lead, cadmium and bismuth. Because the oxides of Pb, Cd and Bi can form tinnate with SnO_2 at high temperature and cause isolation in resistors. For example (at high temperature):

$$2PbO + SnO_2 = Pb_2SnO_4$$

where, Pb_2SnO_4 is an isolator.

In addition, the change of glass viscosity with temperature strongly affects the sensitivity of processing. But for high resistance material the ratio of glass frits to conductive phase is large. To control the change of glass frit's viscosity with temperature, it is necessary, to blend different frits which have different softening points and expansion coefficients.

4. For fabrication of high voltage and inductionless discharge resistors, the resistivity of the material should be high and the particle size of paste should be small than 10 μm. Besides these requirements, the conductive material, glass phase and organic vehicle must be mixed uniformly. The coating layer of the resistor element also should be homogeneous to insure the electrical field at all points of conductive area uniform after firing. Otherwise at the high voltage some portions of resistor films will be broken down to cause irreversible change or drift of resistance value. Thus, the stability of resistance is effected strongly.

CONCLUSION

1. All fabricating processes of Model RI-8 high voltage resistor with base metal SnO_2 glaze including treatment of raw materials, techniques and construction are reasonable. Using Ta_2O_5, SnO and semiconductor SnO_2, the conductive phase containing pyrochlore-related structure can be formed under the condition of high temperature firing in protective atmosphere. This conductive phase can substantially increase resistivity and decrease TCR. It is suitable to make inductionless and high voltage resistor.

2. Because the resistors are inductionless, resistant to high temperature, high voltage and surge voltage, have large overload capacity and high reliability, they can meet the special requirements for speedy discharge.

3. The resistors with stable property and good quality have been used in practice and produced in quantity for power system, lightning arresters and other equipments. The resistors were used in quantity for manufactruing lightning arrester, which is able to block carrier waves and is patented by National Patent Office of China. In addition, this type of resistors has been used to make recently developed " Series spaced metal oxide lighting arrester" which is in keeping with the requirement for electrical power transmission protection. The work will continue to open up vast vistas.

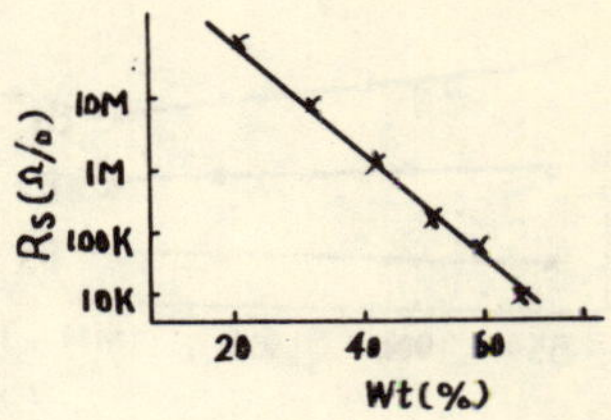

Fig. 1 Sheet resistance versus conductive phase contents of SnO_2 semiconductor.

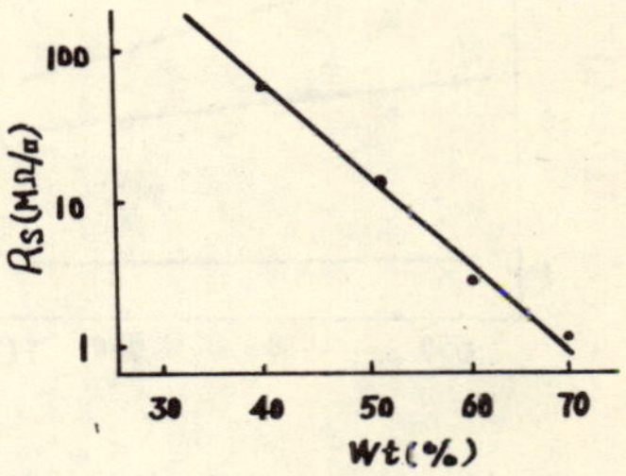

Fig. 2 Sheet resistance versus conductive phase contents of SnO_2 pyrochlore compound

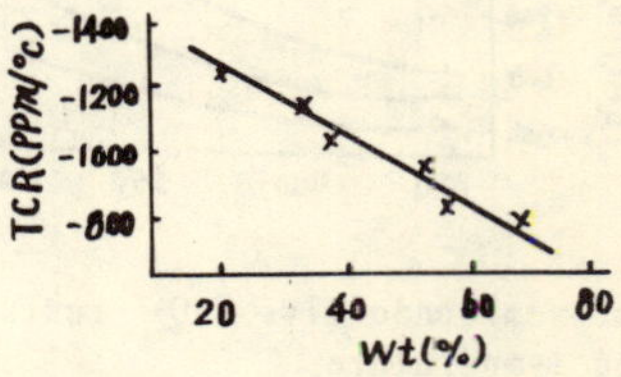

Fig. 3 TCR versus conductive phase contents of SnO_2 Semiconductor.

* B1, D1: for high resistance
B2, D2: for medium resistance
B3, D3: for low resistance

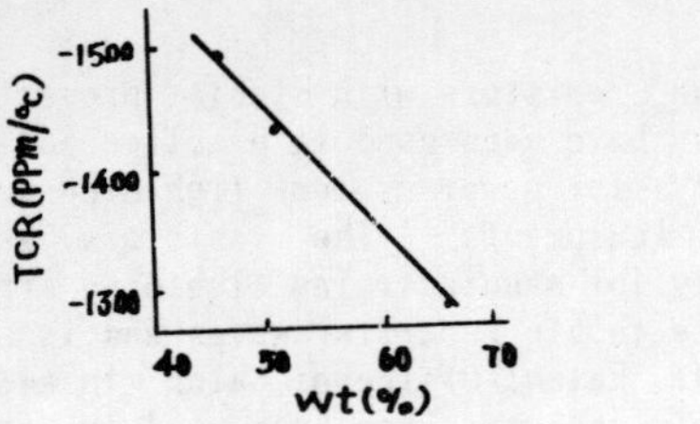

Fig. 4 TCR versus conductive phase contents of SnO_2 pyrochlore Compound.

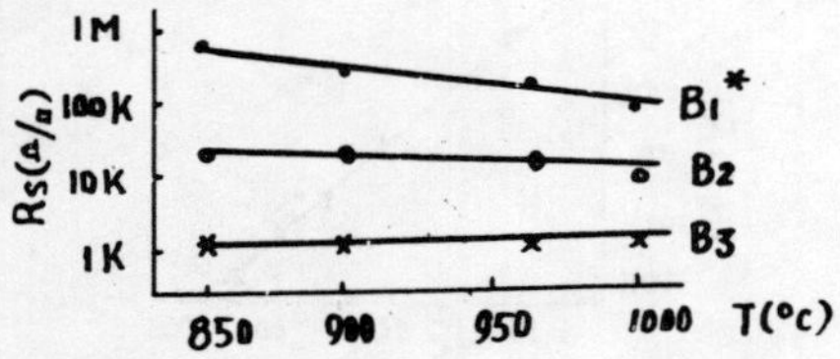

Fig. 5 Sheet resistance versus firing temperature of semiconductive SnO_2 resistor.

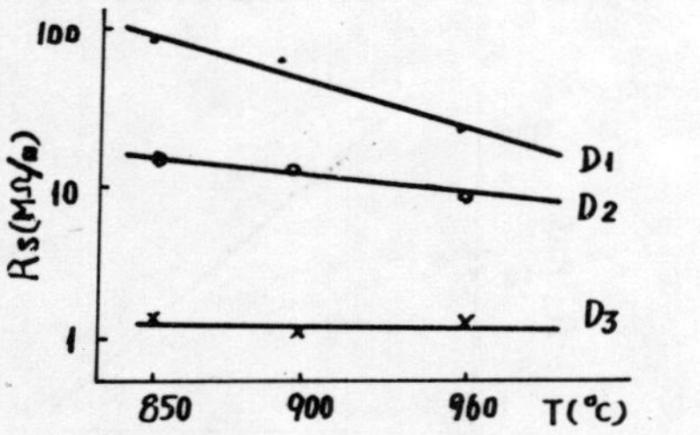

Fig. 6 Sheet resistance versus firing temperature of SnO_2 -pyrochlore compound resistor.

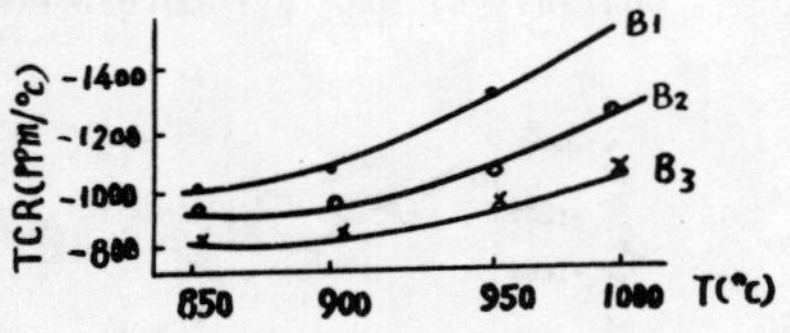

Fig. 7 TCR of semiconductive SnO_2 resistor versus firing temperature.

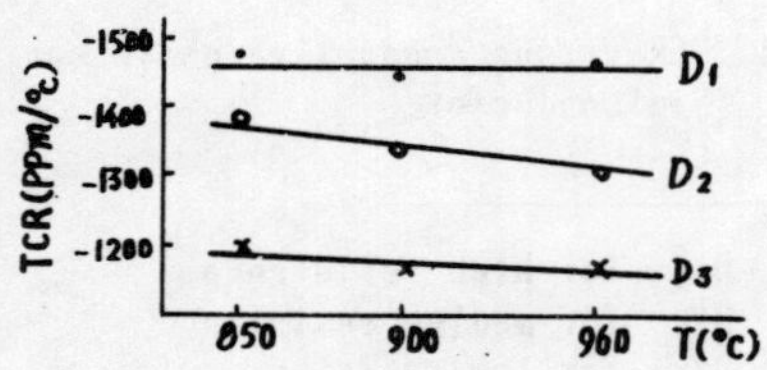

Fig. 8 TCR of SnO_2-pyrochlore compounds resistor vesus firing temperture.

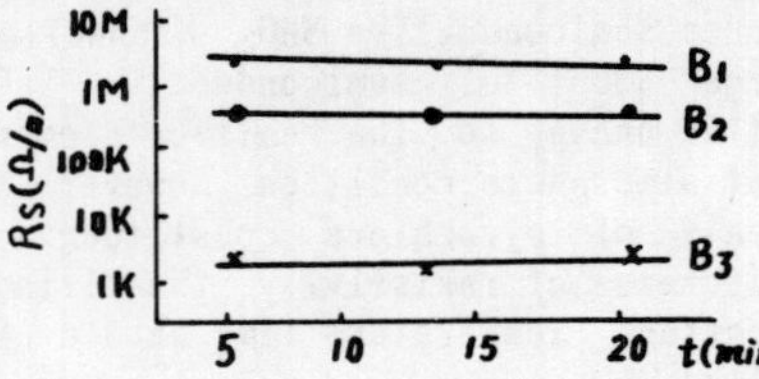

Fig. 9 Rs of semiconductive SnO_2 resistor versus retained time at firing temperature.

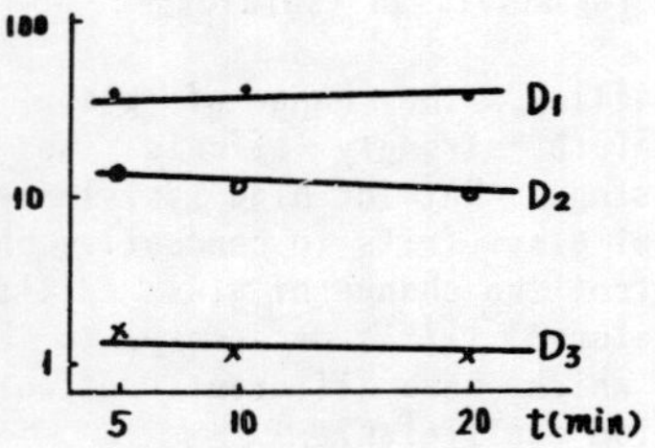

Fig. 10 Rs of SnO_2 -pyrochlore compound resistor versus retained time at firing temperature.

REFERENCES

1. US 4397915

2. GB 2017676

3. US 4065743

THE APPLICATION OF COMPUTER AIDED DESIGN ON PLATE SHAPE DESIGN FOR VARIABLE CAPACITOR

Pan Zheng Li
Shanghai Fudan Capacitors Works No.406 Gui Lin Road
Shanghai, China

Computer aided design of the plate shapes of variable capacitor (PSVC CAD for short) is a new technique of design, a technical improvement upon the traditional method of design.

1. Principles of PSVC CAD

The variable capacitor brings about its capacity change through the change of the covered area of the plates (rotor and stator). Besides an iso-radius semicircle taken on by the rotor of the variable capacitor under the straight-line law of capacity change, the rotor will invariably present a noncircular curve under other laws of capacity change.

In addition to its structural design, the shape design of the rotor is also a very important link. In general, the shape design of the rotor should consist of the following three aspects,

A. Design of parameters of curve points.

Several points on the curve of each rotor may be determined by C for the capacity of capacitor, N for the total number of plates, D for the distance between two plates, εr for dielectric constant and K for the variable coefficient of capacity.

B. Design of parameters for the fitting of circular arcs.

Make several circular arcs and make the neighbouring circular arcs tangential to one another through all the above points, thus forming a smooth curve to press on towards the non-circular curve.

C. Recalculate the effective area and capacity.

Recalculate the corresponding capacity of every angle according to the effective covered area enclosed and formed by the curve.
The principal mathematical models of PSVC CAD to be conducted on the basis of the above-mentioned three requirements are as follows,

(1) Adopt the capacity formula of parallel-plate capacitor and the average vector formula to determine the value of every point .

The capacity formula of parallel-plate air-dielectric capacitor is

$$C= \frac{0.00884 \cdot (N-1) \cdot S}{D} \qquad (1)$$

of which, N for the total number of plates, S for the covered area of plates and D for the distance between two plates.

In order to reduce the distance between effective plates, a dielectric with higher dielectric constant than air is inserted between the plates of PVC products. Then formula (1) will be evolved into

$$C= \frac{0.00884 \cdot (N-1) \cdot S}{Da + Dp / \varepsilon r} \qquad (2)$$

of which, Da stands for the air clearance between plates, Dp for the thickness of the dielectric material and εr for the dielectric coefficient. Then

$$S= \frac{C \cdot (Da + Dp/ \varepsilon r)}{0.00884 \cdot (N-1)} \qquad (3)$$

Since the curve of the rotor is a non-circular one, the radius vector of the rotor ρ (the distance from the centre of the rotor to the point on the curve) is not a constant. If we divide a 180° rotor into ten 18° sectors, it can almost be considered that there is no much change of radius vector in each sector. From applying the calculating formula for sector area

$$\Delta S = \frac{1}{2} (\rho^2 - \gamma_i^2) \cdot \Delta \theta ,$$

the formula of average vector can be derived,

$$\bar{\rho} = \sqrt{\gamma_i^2 + 2 \Delta S / \Delta \theta} \qquad (4)$$

rotating angle θ (°)	18	36	54	72	90	108	126	144	162	180
Variable coefficient of capacity K(%)	5.23	13.11	21.52	30.50	40.12	50.42	61.48	73.37	86.18	100.00

of which γ_0 represents the inside diameter of the notch of the stator, $\Delta\theta$ is 18°, i.e. 0.314(rad), and ΔS is derived from formula (3) according to the variable coefficient of capacity K.

Take the rotor of frequency modulation CBM-443BF produced by Shanghai Fudan Capacitors Works for an example. The curve of this frequency modulation is B curve with its law of change as the above table,

Suppose, C=20pf; Da=0.04mm; Dp=0.04mm; εr=2. 3; N=2+1=3

Apply formula (3) to get $S=64.92(mm)^2$

By calculating at 18°, $\Delta S_1=64.92\times5.23\%$
$=3.395(mm)^2$

By calculating at 90°, $\Delta S_5=64.92\times(40.12\%-30.50\%)$
$=6.245(mm)^2$

Let γ_0=2.9mm

Apply formula (4)to get

$$\bar\rho_1=\sqrt{2.9^2+2\times3.395/0.314}=5.480(mm)$$

$$\bar\rho_5=\sqrt{2.9^2+2\times6.245/0.314}=6.942(mm)$$

Obtain $\Delta S_1\sim\Delta S_{10}$ and $\bar\rho_1\sim\bar\rho_{10}$ according to the above methods. We can match the central angle value of each sector 9°, 27°, 45°…… 153°, 171° with the radius vector value $\bar\rho_1$, $\bar\rho_2$, $\bar\rho_3$, …… $\bar\rho_9$, $\bar\rho_{10}$ respectively. Each couple of angle and radius vector values is termed as a point value. Since the equilibrium point of PVC products is not 0° but 5.4°, ΔS_1, $\bar\rho_1$ and the angle of the first point should be specially dealt with.

(2)Insert a new point. Adopt the fitting parabola equation.

In order to solve the sound fitting of circular arcs, there must be the tangential condition at the starting end and terminal end of a curve. In general, such condition cannot be provided. We have to seek a new point value on condition that the starting end of the rotor is 5.4° ~ 18° while its terminal end is 171°~180°. Since the tabular curve fails to do so, we have to adopt the method of parabola equation to approximately work it out. We know that area is linear with capacity. Suppose the relation between capacity and angle is quadratic curve, then the relation between area and angle is also a quadratic curve.Take the variable coefficient of capacity of three points as the data for the purpose of obtaining the parabola of the three points.

Suppose the rotating angle is θ and the variable coefficient of capacity is K, we get the parabola equation $K=\alpha\theta^2+\beta\theta+\gamma$ (5) of which,

let $\alpha=(P-Q)/(\theta_3-\theta_1)$;

$\beta=\theta-\alpha(\theta_1+\theta_2)$;

$\gamma=K_1-\alpha\theta_1^2-\beta\theta_1$;

$P=(K_1-K_2)/(\theta_1-\theta_2)$;

$Q=(K_2-K_3)/(\theta_2-\theta_3)$.

Still, take the FM rotor plate of 443BF for example. Now, the new point 174.6° will be inserted, i.e., the corresponding value of radius vector towards 174.6° will be sought. According to the tabular curve, θ_1=144, θ_2=162, θ_3=180, K_1=73.37, K_2=86.18, K_3=100, Apply the above formulas, we get,
P =0.768; Q =0.712; α =0.00156; β =0.236; γ =7.007
In formula (5), $K=0.00156\theta^2+0.236\theta+7.007$.

In order to eventually find $\bar\rho$ at 174.6° angle, the variable coefficient of capacity at 169.2° angle must first be found.
Let θ =169.2° We get K=91.60
From 169.2° to 180°, the increment of the variable coefficient of capacity is 100-91.60=8.40

$\Delta S(169.2°\sim180°)=64.92\times8.40\%=5.453(mm)^2$
$\Delta\theta(169.2°\sim180°)=180°-169.2°=10.8°=0.1884(rad)$
$\bar\rho(174.6°)=\sqrt{2.9^2+2\times5.453/0.1884}=8.142(mm)$

For the same reason, arbitrary point may be inserted, This program may be inserted both by any arbitrary angle smaller than 18° angle at the starting end of a curve and by 174. 6° at the terminal end of the same curve according to the enduser's need.

(3)Adopt isosceles double circular arcs to make several mathematical models of circular arcs.

Suppose a non-circular curve is traced out from N number of points Pi(Xi,Yi) (i=1, 2, ……N). It is required to fit every neighbouring two points with two sections of tangent circular arcs in between. Moreover, the fitted tangent point should fall upon the vertical bisector of the chord $\overline{Pi\ Pi+1}$. Such two sections of circular arcs are termed as isosceles double circular arcs. In this program, eleven points are designed in all and there are ten pairs of isosceles double circular arcs in all.

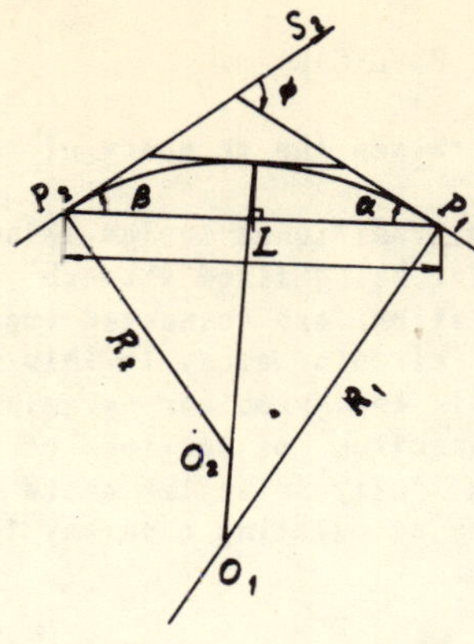

See the figure. Suppose P_1, P_2 are two neighbouring points, L is the length of chord $\overline{P_1P_2}$, S_1 stands for the tangent direction of the curve at the P_1 point, α is the correspouding chordal tangent angle; S_2 stands for the tangent direction of the curve at the P_2 point, B is the corresponding chordal tangent angle. ϕ is the intersecting angle in between two tangent directions, S_1 and S_2, Q point is the tangent point of two sections of circular arcs which falls upon the vertical bisector of $\overline{P_1P_2}$. Suppose R_1 is the radius of the circutar arc from the P_1 point while R_2 is the radius of the circular arc from the P_2 point. Through mathematical deduction we get the following equations,

$$\angle QP_1P_2=\angle QP_2P_1=\phi/4=(\alpha+\beta)/4 \qquad (6)$$

$$R_1=L/\left[4COS\frac{1}{4}(\alpha+\beta)\cdot SIN\frac{1}{4}(3\alpha-\beta)\right] \qquad (7)$$

$$R_2=L/\left[4COS\frac{1}{4}(\alpha+\beta)\cdot SIN\frac{1}{4}(3\beta-\alpha)\right] \qquad (8)$$

Take P_1 point as the centre of a circle, R_1 as the radius and intersect the vertical line of S_1 at O_1 point. Take P_2 point as the centre of a circle, R_2 as the radius and intersect the vertical line of S_2 at O_2 point. O_1 and O_2 are just the centres of circles of two sections of circular arcs on the left and right. Thus, the two sections of circular arcs of the two points can be fixed. By adopting the same method, the parameter of double circular arcs between other neighbouring two points can also be found.

According to cordinates of P_1, $P_2 \cdots\cdots P_{11}$ points and the tangent directions of S_1 and S_{11} (called boundary conditions) we can find S_2, S_3 …… S_{10} through 9-step linear equations, that is, we can get values of αi and βi.

(4) After setting up coordinates of centres of circles (a, b) and radius R, we can find the corresponding actual radius vector $\rho(\theta)$ of the angle θ and the covered effective area $S(\theta)$ from the starting end with mathematical methods.

Circular equation $(X-a)^2+(y-b)^2=R^2$

Substituted by $x=\rho(\theta)\cdot COS\theta$ and $y=\rho(\theta)\cdot SIN\theta$, we get $[\rho(\theta)\cdot COS\theta-a]^2+[\rho(\theta)\cdot SIN\theta-b]^2=R^2$

Solve this retated equation and use the positive value for $\rho(\theta)$

$$\rho(\theta)=aCOS\theta+bSIN\theta+\sqrt{R^2-(aSIN\theta-bCOS\theta)^2} \qquad (9)$$

$$S(\theta)=\frac{1}{2}\int_{\theta_1}^{\theta_2}[\rho^2(\theta)-\gamma_0^2]\,d\theta$$

$$=\frac{1}{2}(R^2-\gamma_0^2)\theta+\frac{1}{2}(aSIN\theta-bCOS\theta)\cdot\rho(\theta)$$

$$+\frac{R^2}{2}SIN^{-1}[(aSIN\theta-bCOS\theta)/R]\Big|_{\theta_1}^{\theta_2} \qquad (10)$$

Let $\theta_1=5.4°=0.0942(rad)$, θ_2 be the corresponding radian for 45°, 90°, 135°, 180° respectively. Then, the effective covered area from 5.4° to the above different angles can be worked out and the capacity can further be worked out . If the curve consists of several circular arcs, integration can be conducted by sectional function.

2. Introduction to the PSVC CAD program .

The PSVC CAD program contains three subprograms, design of common plate shapes, revision of the design of plate shapes and verification of the design of plate shapes. These three subprograms, which may be chosen at will by designers according to their needs, are respectively briefed as follows,

(1) Design of common plate shapes .

This subprogram applies to the design of plate shapes for standard curves. The so-called standard curves refer to the curves stipulated in the general technical specification of China's Film Dielectric Variable Capacitors (SJ2665-86). There have been 5 curves A. B. C. D. F in the memory of this subprogram. The required technical parameters to be inputted are, capacity C, total number of plates N (including rotor and stator), the inside diameter of the notch of the stator γ_0, thickness of dielectric material Dp, thickness of material of stator Ds, thickness of gaskets of rotor plates Dw, the dielectric constant of the dielectric $\varepsilon\gamma$, deviation volume Db and the angle of first point θ. Simultaneously, it is required to designate a curve of capacity change.

When a variable capacitor is designed, we always assume that the distance between the stator and rotor is a constant. However, it often happens in reality that the stator and rotor are not smooth or not parallel during assembly, or there exists actual deviation of plates, especially for PVC products. In general, the material thickness of stator is 0.03 ~ 0.07mm and, moreover, is of laminated type. Therefore, the phenomena of excessive assembled capacity often happens when the covered area calculated by theory is put into actual operation. It is for the solution of this problem that the physical concept of deviation volume is introduced, thus making the designed covered area of plate shapes meet practical requirements.

Owing to the use of deviation volume, the formula (3) will be evolved into

$$S=\frac{C}{0.00884(N-1)}\cdot\frac{(Da + Dp / \varepsilon\gamma)^2-Db^2}{Da + Dp / \varepsilon\gamma} \qquad (11)$$

of which, Db is deviation volume and the value range is from 0 to Da. Concrete value will be fixed by technological conditions. When Db=o, formula (11) is equal to formula (3), When Db=Da, the deviation volume will reach the maximum value.

In the course of calculating FM rotor plate of 443BF, input data are,
Capacity C=20(pf), total number of plates N=3, the inside diameter of the notch of the stator γo=2.9(mm), thickness of dielectric material Dp=0.04(mm), thickness of material of stator Ds=0.05(mm), thickness of gaskets of rotor Dw=0.21(mm), the dielectric constant of the dielectric material $\varepsilon\gamma$ = 2.3, deviation volume Db = 0.02713(mm); the angle of first point $\theta=9°$ and the chosen curve is curve B.

The air clearance between plates calculated by computer is ,

$$Da=(Dw-Ds)/2-Dp \qquad (12)$$

We get Da=0.04(mm). According to formula(11), we get S=50.40557(mm)^2

In this subprogram, there are four tables of output data.

Table 1, Table of technical parameters for designing rotor.

Table 2, Comprehensive analytic table of data for rotor .

Table 3, Table of data for curve points of rotor .

Table 4, Table of curve-processing data for rotor.

(2) Revision of the design of plate shapes.

This subprogram is specially designed for slight revision of some data for rotor, and is also applicable to the processing of data for rotor in the tabular drawings. The input technical parameters to be required are the number of points, the inside diameter of the notch of the stator , the data of points and boundary conditions (if any). There are also four output tables.

(3) Verification of the design of plate shapes.

This subprogram applies to the traditional design of plate shapes and the comprehensive analytic table of data for output rotor.

3.Features of PSVC CAD.

(1) PSVC CAD raises the accuracy of calculation .

According to traditional design method, all the points, after being fixed through a series of manual calculation, are connected together by 3～4 sections of circular arcs. In this case, there usually exist two problems—greater error and unsmooth connection of sections of circular arcs. PSVC CAD can fully solve the above two problems and ensure the calculating accuracy to reach the level of 10^{-4}mm.

(2) PSVC CAD enhances the speed of calculation .

The traditional design method usually takes a week or so for a plate shape and it calls for repeated material verifications. But it takes only a few minutes by PSVC CAD. Designers may make comparison and adjustment by contrast amongst a series of data output from computer. By doing so, not only can the design time be saved, but what is more significent is that much time and material consumption necessary from moulding, parts-machining up to material verification can also be saved, thus greatly reducing costs, shortening trial-production cycle of products and enhancing the dependability of products.

(3) Overall level of PSVC CAD.

All the specialized factories at home and abroad are developing the design of the plate shapes of variable capacitor in the direction of CAD. The technology-advanced countries combine CAD with CAM. The tasks undertaken by CAD are relatively simple. At present, the wire-cutting machine tool with computer in our country is still in the test stage. However, the automatic programing languages for the wire-cutting calculation (such as SKG. SAPT languages) have been widely popularized and used. Characteristic of CAM And the calculated results from PSVC CAD output are required to be suited to it. This PSVC CAD program comprises the point design of rotor, design of parameters for the fitting of circular arcs and recalculation of effective area, thus forming a closedcycle for the design of plate which should further be directly connected with the calculating program of wire-cutting. the application of the PSVC CAD program adopts the menu form , which being convenient and visual, has advancedness, completeness and practicality in technology.

CATHODE DEPOLARIZER OF ALUMINUM ELECTROLYTIC CAPACITOR

Zhang Jibao, Cao Wanzheng
Xi'an Jiaotong University
Xi'an, China

ABSTRACT

Cathode depolarizer is effective additive for improving the electrical performance and load life of Aluminum electrolytic capacitors. This paper analyzes the generaly machanism of depolarizing of it, and verify by experiments. By the oxidic efficiency of electrolyte and the related theories of Electrochemistry, an approximate calculation of the content of cathode depolarizer under certain conditions is provided.

INTRODUCTION

Cathode depolarizer (or hydrogen remover), which influence the characteristics of capacitor especially for the middle and high voltage ones, is a constituent part of the working electrolyte solution. Generally when the electrolytic capacitor is in store or work, the dielectric characteristics of the anodic oxide film will usually be deteriorated, while electrolyte repairs the wounded oxide film, hydrogen gas is produced at the cathode of it, thus the inner pressure will be increased. So the load life time will be shortened. But if depolarizer such as resocinol is added to electrolyte, the life characteristic of capacitor will be much improved, because it will react with the releasing hydrogen and gripe together.

Cathode depolarizer is an effective additive for improving the characteristics and load life of Aluminum electrolytic capacitor, it has been confirmed by numerous experiments and now widely used in capacitor manufacturing. But systematic study for it is not so common. This paper is to analyze the general machanism of depolarizing of the capacitors and verify by experiments. By the oxidic efficiency of electrolyte and the related theories of Electrochemistry, an approximate calculation of the content of depolarizer under certain conditions is provided.

BASIC THEORY

1. Basic Principles of Cathode Depolarizing

Generally, when depolarizer is added to electrolyte the influence on the properties of electrolyte is very small, but very obvious on load life characteristics of the capacitor. In electrolyte solution there is isomer with unsaturated bond in depolarizer because of action of hydrogen bond. So while electrolyte repairs the wounded oxide film, the releasing hydrogen gas at cathode is absorbed because those two will react each other.

2. Oxidic Efficiency of Electrolyte[1]

Under given current density and forming voltage, oxidic efficiency of electrolyte is generally defined as the ratio of theorical oxidic time and actual oxidic time, which is given by the following equation:

$\eta = t_r/t_g$ where $t_r = F \cdot D/N \cdot j \times 10^{-3}$, $D = \alpha \cdot U_f \cdot D_o$
η -Oxidic efficiency of electrolyte, F=96500 C
t_r-theorical oxidic time (second),
t_g-actual oxidic time (second)
D -weight of oxide film per unit area (g/cm^2)
D_o-density of oxide film (g/cm^3),
N -gram equivalent of oxide
j -forming current density (mA/cm^2),
U_f-forming voltage (v)
α -forming coefficient

For Aluminum electrolytic capacitior,
$D_o = 3.42\ g/cm^3$, $a = 1.2 \times 10^{-7}$ cm/v, N=16.99g
So we have

$$\eta = 2.33 U_f / j \cdot t_g \qquad (1)$$

3. Leakage Current of Capacitor

Under given temperature, the relation between leakage current and time at the begining will be attenuated exponentially and finally reach a stable value[2]. i.e.

$$I_L = i_o \cdot \exp(-bt) + i_R \qquad (2)$$

Where I_L-leakage current at t time, b-constant related to electrolyte, i_o-initual leakage current, i_R-final leakage current

For the same electrolytic capacitor which is made up with the same systems of electrolyte solution there are stable value for i_o and i_R.

4. Content of Cathode Depolarizer

Supposed that the oxidic efficiency of electrolyte equals to current efficiency of anode electrode under working condition. The effective electrical quantity which pass through anode within dt time is $dQ=I_L\cdot\ \ \cdot dt$. The total electrical quantity passing through anode at t time is

$$dQ=\int_o^t[i_o\exp(-bt)+i_R]\cdot\eta\cdot dt$$
$$=\eta\ i_o(1-\exp(-bt))/b+i_R\cdot T] \quad (3)$$

According to Faraday's law, we have

$$Q=n\cdot F \quad (4)$$

where n is gram equivalent count reacting at electrode.

By (3) and (4) we have

$$n=\eta\ [i_o(1-\exp(-bt))/b+i_R\cdot t]/F \quad (5)$$

Assume that gram equivalent of depolarizer is M. By electrochemical equivalent's law the content of cathode depolarizer may be expressed in following equation

$$w=n\cdot M=\eta\ [i_o(1-\exp(-bt))/b+i_R\cdot t]/F$$
$$=0.24\ U_f\cdot M[i_o(1-\exp(-bt))/b+i_R\cdot t]/(j\cdot t_g) \quad (6)$$

EXPERIMENTAL

Taking systems of ammonum hexandiate-ethylene glycol electrolyte solution as example. Under $j=2mA/cm^2$, $U_f=160v$ we got the oxidic time $t_g=324$ seconds, so $i_o=80$ uA, $i_R=3.6$ uA and at t=60 seconds $I_L=10.6$ uA. So we have b=0.02 by (2) equation. For 1000 hrs estimated load life time of capacitor, the theorical calculating content of depolarizer is w=0.1 g by (6) equation and the same as that w=0.12 g for the system of ammonum benzoate-ethylene glycol.

The systems of ammonum hexandiate-ethylene glycol 1,2 and 3 electrolyte solution and that of ammonum benzoate-ethylene glycol 4,5 and 6 are made up, in which the content of depolarizer is less than, equal to and more than theorical calculating quantity respectively and then put into experiments.

A. Under given current density the relation between oxidic efficiency and forming voltage of 1 6 electrolyte solution are given by the Fig.1.

B. The 160 v 33 uF 85 °C Aluminum electrolytic capacitors had been made up with 1 6 electrolyte solution, and the relation between time and leakage current of the capacitors were determined (Fig.2.)

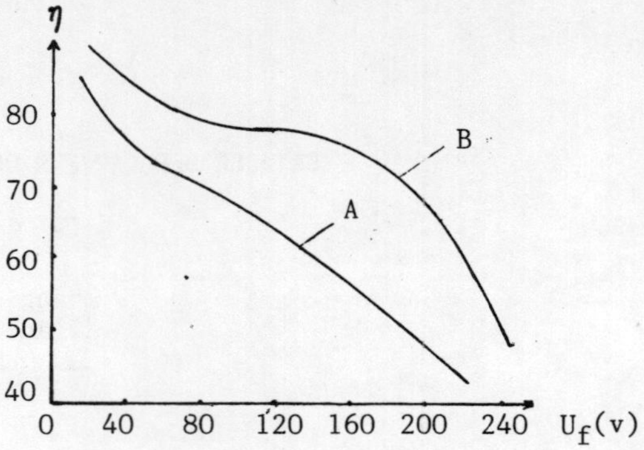

Fig. 1 Oxidic efficiency vs forming voltage for electrolyte solution
A. Systems of ammonum hexandiate-ethylene glycol.
B. Systems of ammonum benzoate-ethylene glycol.

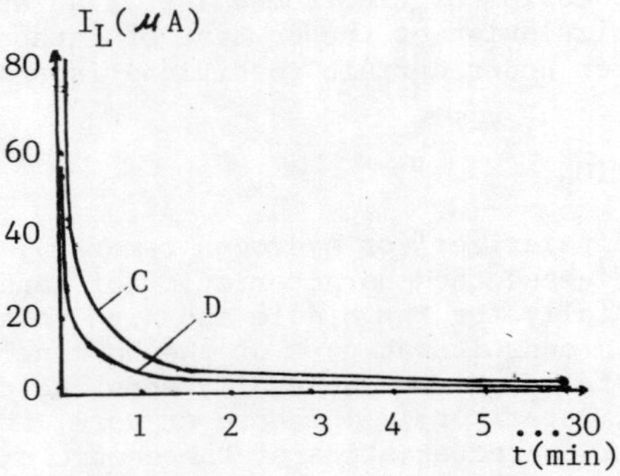

Fig. 2 Leakage current vs time for electrolytic capacitor.
C. Systems of ammonum hexandiate-ethylene glycol.
D. Systems of ammonum benzoate-ethylene glycol.

C. Under 85 °C and working voltage, the load life characteristics with time were tested (Fig.3 - Fig.8)

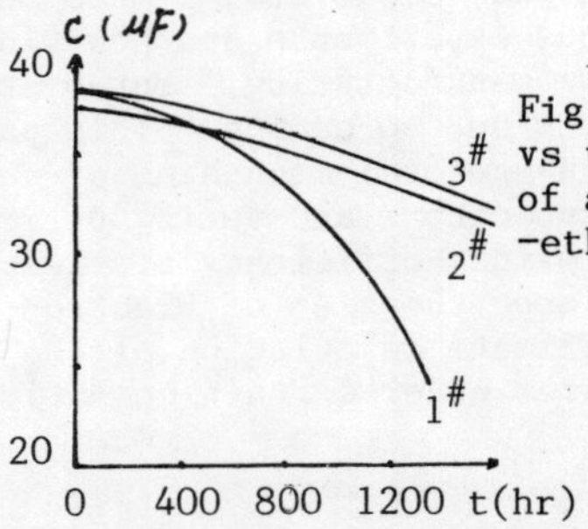

Fig. 3 Capacitance vs time for systems of ammonum hexandiate-ethylene glycol.

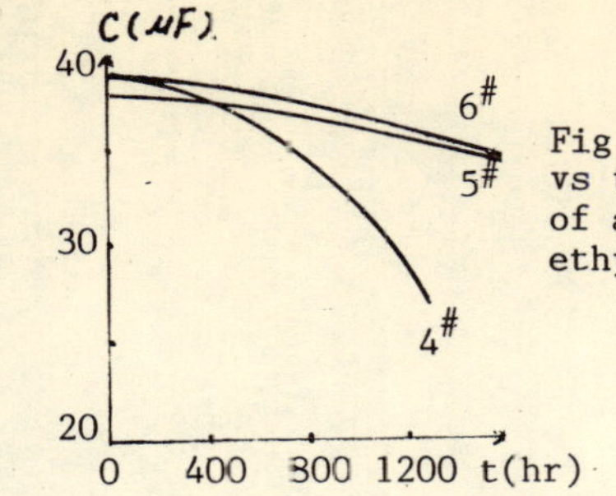

Fig. 4 Capacitance vs time for systems of ammonum benzoate-ethylene glycol.

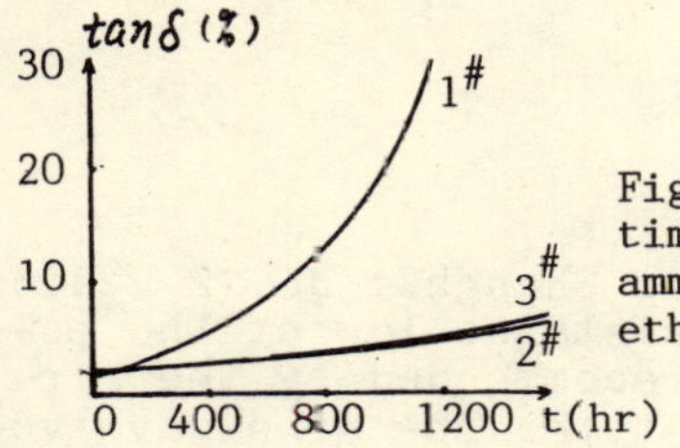

Fig. 5 Tangent vs time for systems of ammonum hexandiate-ethylene glycol.

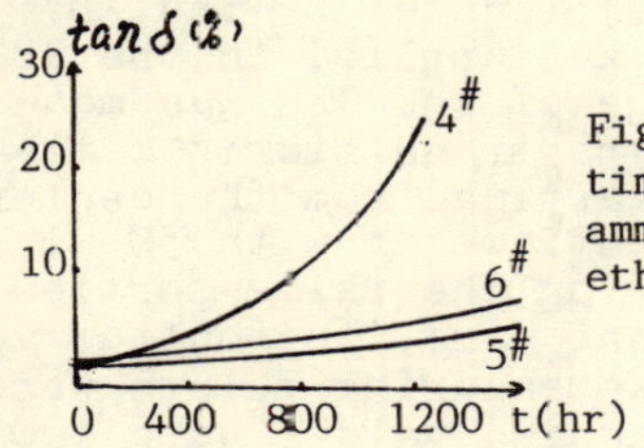

Fig. 6 Tangent vs time for systems of ammonum benzoate-ethylene glycol.

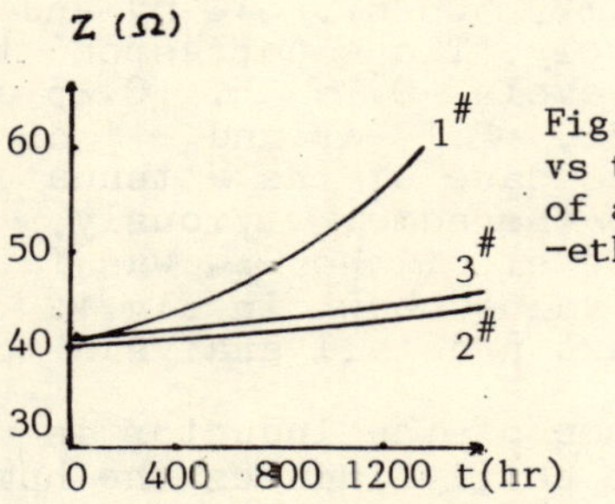

Fig. 7 Resistivity vs time for systems of ammonum hexandiate -ethylene glycol.

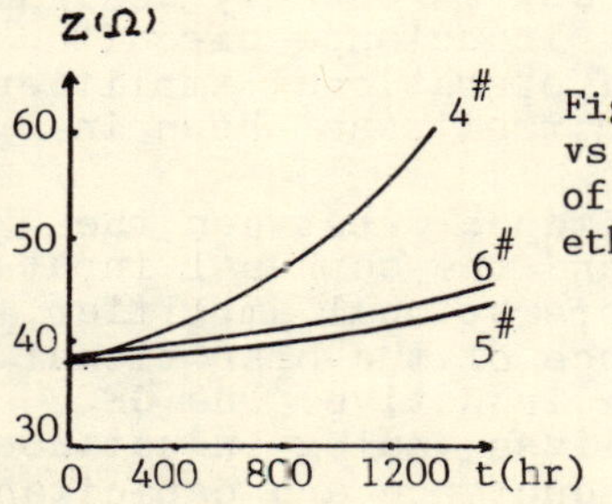

Fig. 8 Resistivity vs time for systems of ammonum benzoate-ethylene glycol.

RESULT ANALYSIS AND DISCUSSION

A). For Aluminum electrolytic capacitors the process of repairing wounded oxide film mainly rely on the properties of electrolyte and solvent, and not on the content of depolarizer. Under given voltage, for the same systems of electrolyte solution and with different content of depolarizer, their oxidic efficiency generally have a fixed value. For example, under 160 v the oxidic efficiency of systems of ammonum hexandiate-ethylene glycol is 57.5 percent and 75.8 percent for the systems of ammonum benzoate-ethylene glycol (Fig.1)

B). Under forming and working conditions of the capacitor, when there are flaws on dielectric film, the leakage current of it is decreased rapidly within the bebining tens second time, the decreasing rate will be decreased with time and finally come to a stable value. The content of depolarizer has no influence on leakage current of the capacitor. The influence is mainly on the load life characteristics of the capacitor (Fig. 2)[1]

C). Working under high temperature the content of depolarizer in electrolyte will be decreased with wounded oxide film being repaired. After sometimes the action of depolarizing will become much weak. So the inner pressure of it will be increased with the relesing of hydrogen gas at cathode and the volatilization rate of electrolyte solution will be increased, together with the lose rate of capacitance, resistivity and tangent with time. Finally the capacitor will not be effective in circuit.

For 1 and 4 electrolyte solution there isn t much content of depolarizer in it, so after 700 hrs load life time the characteristics of the capacitor begin to be bad and after 1000 hrs they have changed very greatly. But for 2,3 and 5,6 electrolyte solution, the characteristics of the capacitors varying with time are very small within load life time, i.e. when the content of depolarizer equals to and more than theorical quantity the capacitor are effective within estimated life time.

CONCLUSION

A. The content of depolarizer may effect the characteristics of electrolyte very faintly but very promptly to that of the electrolytic capacitor.

B. For the same systems of electrolyte solution the effect of the content of depolarizer is mainly reflected to load life characteristics.

C. From the results of identity between theorical calculating content of depolarizer and our experiments, we confirm that equation (5) conform to the actual demands.

REFERENCES

1. Chen Guoguang, Electrolytic Capacitors, Xi'an Jiaotong University Press, (1986). 115.
2. J. E. Gordon, The Organic Chemistry in Electrolytic Solution, (1975).

RESEARCH ON THE POTENTIOMETER FOR MULTI-BAND GRAPHIC EQUALIZER

Chang Shi
Shanghai No.12 Radio Factory
Shanghai, China

Graphic Equalizer(GEQ), as a kind of tone control circuit, is widely used in the Hi-Fi stereo audio equipment. GEQ is composed of multi-narrow-band sections. It is generally divided into five to thirty sections. Each band is controlled by a slide type potentiometer. Fig.1 is the diagram of the five-bands GEQ on the panel.

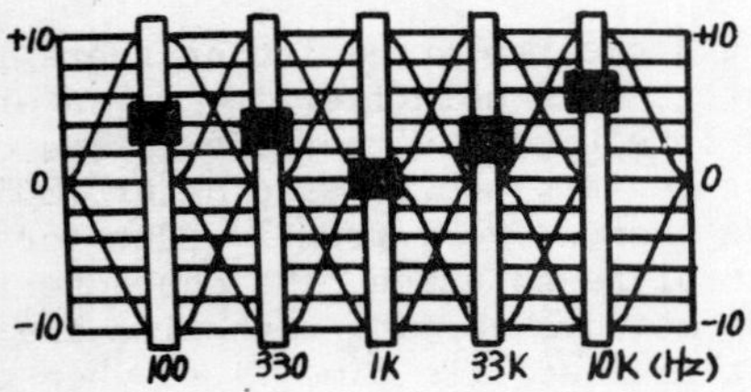

Fig.1 The panel of five-bands GEQ

According to the acoustics characteristic of the environment and the audio feeling of the programs, the users may raise or attenuate the levels of some frequency bands to modify the timbre of programs. On this condition, the standing wave will be minimized, the whistling sound of acoustic feedback will be attenuated in narrow-frequency band and the Hi-Fi will reappear. So people named it as "Audio Beauty Master". Besides, the diagram formed by the rods of slide potentiometers implies the comprehensive frequency response curve of the GEQ.
But after having made a survey of several kinds of market available tape recorder with GEQ circuit, I found that about center of the potentiometer travel man can hardly feel the change of the controlled level, but the change was obvious at the positions near both terminals.
For the purpose of illustration, the GEQ circuit of most popular SHARP GF-700 recorder was tested. The operational amplifier was LM324. The potentiometer was WH25-3(products of Shanghai No.12 Radio Factory). The resistance law of the potentiometer, as recommended by the circuit, was linear type. The sine-wave voltage(1KHz, 500mV)produced by the XD7 signal generator was applied to the GEQ. The output voltage of the GEQ was measured by DB-16 type transistorized millivolt meter. Starting from the center of the slide potentiometer with 30 mm travel and adjusting the raise portion with each increment 3 mm, the output voltages at the six measuring points were measured, the measured values were 490 mV, 515 mV, 530 mV, 565 mV, 640 mV and 1.77 V respectively. These correspond to the controlled levels -0.15 dB, +0.25 dB, +0.5 dB, +1.05 dB, +2.15 dB and +11 dB respectively. The case of the attenuate portion was also the same. Obviously, the controlled level of the GEQ was far from the actual value shown in Fig.1. Why so? Let's make a detail analysis to the GEQ.
Because the volume of the inductor is too large and it easily induces the hum, today multi-band GEQ is usually designed to use the analog inductance circuit which consists of operational amplifier. The typical GEQ circuit was shown in Fig.2.
Connecting capacitance C_1 between the output terminal and the cophasal input terminal of the operational amplifier, the input impedance of the operational amplifier appears inductive. The GEQ circuit thus realizes analog inductance. The equivalent inductance and capacitance C_2 form the series resonant circuit and the impedance of the series resonant circuit is as follows:

$$Z=R_1+j(\omega R_1 R_2 C_1-\frac{1}{\omega C_2})$$

Because IC_1 is a differential amplifier, we may simplify Fig.2 into Fig.3.
Z equals to the reactance network. Rab

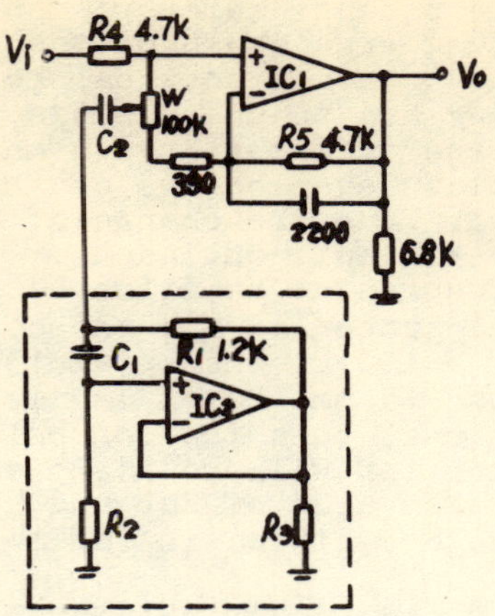

Fig.2

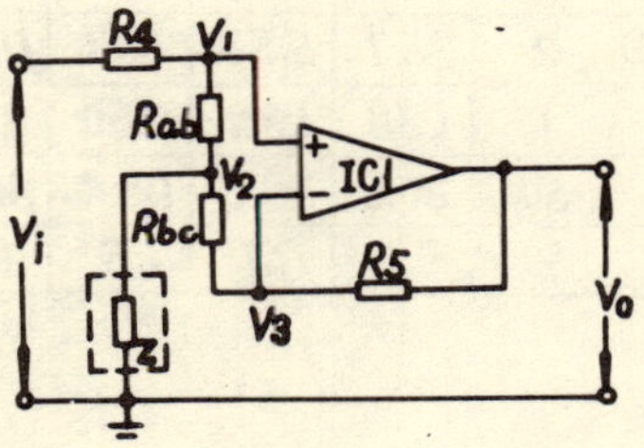

Fig.3

and Rbc are equivalent to potentiometer W. The circuit's transmission coefficient K can calculated by the node-pair method:

$$\begin{cases} \dfrac{Vi}{R_4}+\dfrac{V_2}{Rab}=\left(\dfrac{1}{R_4}+\dfrac{1}{Rab}\right)V_1 \\ \dfrac{Vi}{R_4+Rab}+\dfrac{Vo}{R_5+Rbc}=\left(\dfrac{1}{R_4+Rab}+\dfrac{1}{R_5+Rbc}+\dfrac{1}{Z}\right)V_2 \\ \dfrac{V_2}{Rab}+\dfrac{Vo}{R_5}=\left(\dfrac{1}{Rbc}+\dfrac{1}{R_5}\right)V_3 \end{cases}$$

In this circuit, we usually suppose $R_4=R_5$, so the above three formulas can be simplified into:

$$\begin{cases} V_1=\dfrac{RabVi+R_5V_2}{R_5+Rab} \\ V_2=\dfrac{Z(R_5+Rbc)Vi+Z(R_5+Rab)Vo}{(2R_5+W)Z+(R_5+Rab)(R_5+Rbc)} \\ V_3=\dfrac{R_5V_2+RbcVo}{R_5+Rbc} \end{cases}$$

If the open-loop voltage gain of the differential amplifier is A, the output voltage of the differential amplifier is:

$$Vo=A(V_1-V_3)$$

$$=A\left[\frac{Rab(R_5+Rbc)Vi-Rbc(R_5+Rab)Vo}{(R_5+Rab)(R_5+Rbc)}+\frac{R_5(Rbc-Rab)V_2}{(R_5+Rab)(R_5+Rbc)}\right]$$

Suppose

$$P=(2R_5+W)Z+(R_5+Rab)(R_5+Rbc)$$

Then

$$(R_5+Rab)(R_5+Rbc)Vo+ARbc(R_5+Rab)Vo-\frac{AZR_5(Rbc-Rab)(R_5+Rab)Vo}{P}=ARab(R_5+Rbc)Vi+\frac{AZR_5(Rbc-Rab)(R_5+Rbc)Vi}{P}$$

Transmission coefficient:

$$K=\frac{Vo}{Vi}$$

$$=\frac{\left[PRab+ZR_5(Rbc-Rab)\right](R_5+Rbc)}{\left[\dfrac{P(R_5+Rbc)}{A}+PRbc-ZR_5(Rbc-Rab)\right](R_5+Rab)}$$

Because the open-loop gain of the operational amplifier is higher than 10000, then

$$K=\frac{\left[RabP+ZR_5(Rbc-Rab)\right](R_5+Rbc)}{\left[RbcP-ZR_5(Rbc-Rab)\right](R_5+Rab)}$$

According to Fig.2, W=100 k, R =4.7 k and in resonance case Zo=R =1.2 k, then

$$K=\frac{\left[Rab(623.37+RabRbc)+5.64(Rbc-Rab)\right]}{\left[Rbc(623.37+RabRbc)-5.64(Rbc-Rab)\right]}\times\frac{(4.7+Rbc)}{(4.7+Rab)}$$

If the resistance law curve is symmetrically about the center point of the travel, the raising and attenuating of the controlled characteristic must be the same. Then, we may only discuss the raising portion. Taking linear resistance law and "S" resistance law(Fig.4) as examples, calculate respectively the trasmission coefficient K and the corresponding controlled level. The obtained values are listed in Table 1 and Table 2. From Table 1, we know the maximum raised magnitude of the GEQ is 13.83 dB. From 50% to 90% portion of total travel, the variation of the controlled level is

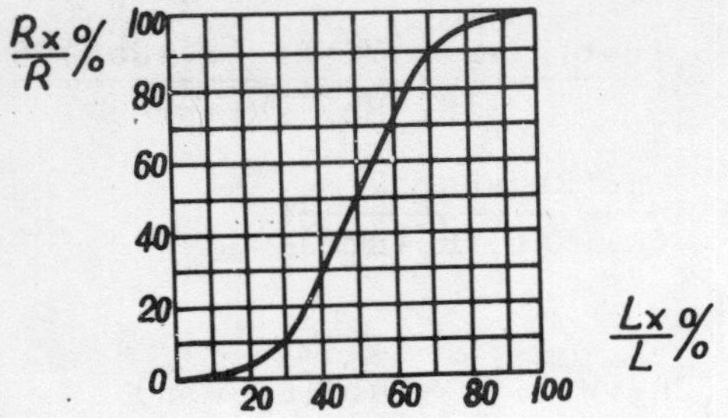

Fig.4 "S" type resistance law

Table 1 The GEQ characteristic obtained by appling potentiometer of linear type resistance law

Lx(mm)	0	3	6	9	12	15
Rab(k)	50	60	70	80	90	100
Rbc(k)	50	40	30	20	10	0
K	1	1.03	1.08	1.16	1.35	4.92
Level(dB)	0	0.30	0.67	1.26	2.62	13.83

Table 2 The GEQ characteristic obtained by appling potentiometer of "S" type resistance law

LX(mm)	0	3	6	9	12	15
Rab(k)	50	70	90	95	97.5	100
Rbc(k)	50	30	10	5	2.5	0
K	1	1.08	1.35	1.68	2.19	4.92
Level(dB)	0	0.67	2.62	4.53	6.81	13.83

only 2.62 dB. At the remainder portion (10%) of this travel the controlled level increases sharp from 2.62 dB to 13.83 dB. This case is not coincident with the degrees shown on the panel. In other words, the GEQ with linear resistance law potentiometer may not be referred to "Graphic" type. As shown in Tabble 2, except that the controlled level of the first 3 mm travel is 0.67 dB, the controlled levels of the subsequent 3 mm travels are about 2 dB respectively and the last 3 mm travel is still 7 dB. Apparently, "S" type resistance law is better than linear type resistance law in linearity of controlled level.
As a number of recorder manufacturers have not checked the conformity between the GEQ's panel degree and actual controlled characteristics, they have not paid enough attention to the contradiction rising from the potentiometer. Only a few manufacturers which produced professional GEQ requested the resistance law of the potentiometer to conform with the GEQ degree shown on the panel. Before they can get an ideal potentiometer, they had to use a slide switch to alter the different resistance in order to obtain even change of the controlled level. Thus the change of controlled level is not continuous.
Now we discuss the relation between the ideal resistance law and the controlled level. If W=100 k, then Kmax=13.83 dB. Suppose the 15 mm travel is divided into five portions, then the controlled level of each portion will be 13.83 dB ÷ 5 = 2.77 dB. The value of Rab and Rbc were calculated and listed in Table 3.

Table 3 The ideal resistance law

Lx(mm)	0	3	6	9	12	15
Level(dB)	0	2.77	5.53	8.30	11.06	13.83
K	1	1.38	1.89	2.60	3.57	4.92
Rab(k)	50	90.6	96.3	98.4	99.6	100
Rbc(k)	50	9.4	3.7	1.6	0.4	0

The ideal resistance law curve is shown in Fig.5.

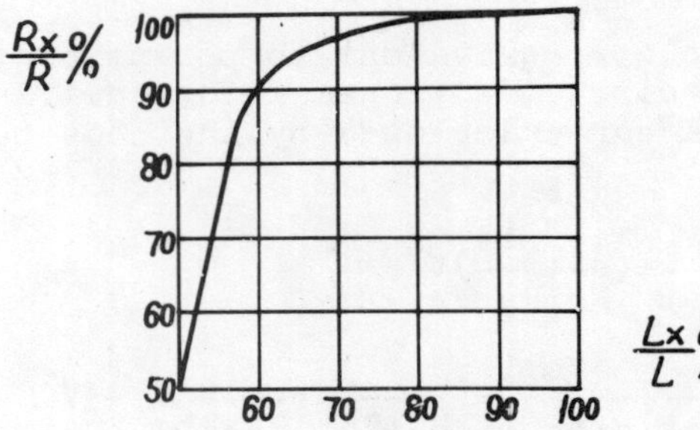

Fig.5 The ideal resistance law curve

It is similar to a logarithmic curve. The full travel (including raising travel and attenuating travel) forms a twin logarithmic curve. According to the above mentioned theory, we have produced potentiometers with "S_1" type resistance law. Shanghai Feilo Electroacoustic General Factory has applied the potentiometer to the GEQ of composite tuning equipment (TLZ-12 type____twelve bands and one-third octave JQ31-1 type____ thirty one bands). The controlled characteristic met the requirements. The test instruments included XFS-9B type signal generator. PTT-2 type frequency characteristic oscilloscope and 2307 type level recorder (producs of B.K & Co.Denmark). Fig.6 shows the controlled characteristic curve(at 1KHz).

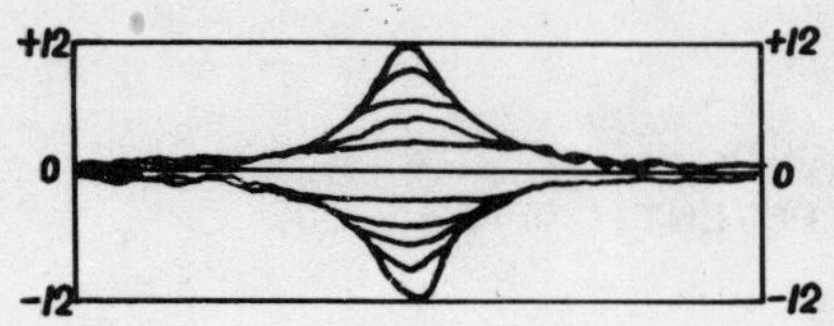

Fig.6

Conclusion:

The resistance law of the potentiometer applied to the multi-band GEQ should be twin logarithm type. It is useful to improve the uniformity of the controlled levels. Now the GEQS applied to the tape recorder exist in name only. Must pay all attention when stereo recorders and professional audio equipments are designed.

MICROCHIP BASE METAL GLAZE POTENTIOMETER

Zhang Han Zhao
The Hua Xing Electronic Components and Materials Factory
POB № 9 Xian Yang City Shaanxi Province P.R.China

ABSTRACT

During the past several years, in order to cope with the fresh development demand of the SMT(surface mounting technique) and the high-density, high-velocity, automation mounting, there have been spotlighting development of the micro-chip electronic elements. In the course of microminiaturization, reduction in thicknes and reduction in weight, a major step forward in miniaturization came from the micro-chip potentiometer. This article introduced major structure characteristics of micro-chip base metal glaze potentiometers, the methed of making a base metal glaze resistive paste and a ceramic substrate. the role of the adjustment of resistive paste composition in obtaining the required properties et cetera is also discussed.

I .INTRODUCTION OF PRODUCTS

At present, in the course of using micro-chip electionic elements widely, we have seen that the assemblage of hybrid -circuit with micro-chip electronic elements is easy to do, the availability serviceability, and reliability are more better than the performance of semiconductor integrated circuit. It copes with not only the problem of that the semiconductor integrated-circuit has bad dissipation performance of heat, but the problem of that the limiting performance of semiconductor integrated -circuit can not be adaptable to high accuracy, high power and high current, also can further develop in to the modularized functional block assembly system, it constitutes the functional block in the high-density assembly. When it is used in printed circuit, it advances the development of the program automatic assembly and welding effectively, accelerates the innovation of assembling technology. Now, along with the development of micro-cassette tape recorder, camera, portable VTR, electron-camera and insert-hearing aids etc, the new requirement for micro-chip electronic elements has been put forward, on the other hand the requirement encourages pots to develop to the microminiaturization. In past several years, in order to adapt the requirement of surface mounting technology, high-density, high-velocity and automatic assembly, the pots have been researched on microminiaturization, reduction in thickness and reduction in weight. Along with the appearance of micro-chip pots the process of miniaturization has taken a large step. We particularly introducte the structure and main parameters of the pots here:

① The dimensions of the pots mainly depend on the size of the resistive track substrate, we have chosen two types of dimensions (4x4.5x2.4 mm) and (6.6x8x3 mm), refering the general requirement, more over the pots are designed in two styles according to the different using:

1. Current adjust type (two electrode lead-out).
2. Voltage adjust type (three electrode lead-out).

The configuration and dimensions are shown in fig-1:

② Main parameters

Ambient Temperature : -55-+100℃ (max temp of full wattage +70℃).
Relative Humidity : 93±3% (40±2℃).
Barometric pressure : 44 mbar.
Vibration intensity : acceleration 5g.
Nominal resistance : 100Ω-1MΩ E6 series.
Tolerance of total resistance : ±20%.
Angle of effective electrical rotation: ≥120° ; ≥250° .
Starting torque : 0.2-2.4 Ncm.
Rated wattage : 0.05 W; 0.125 W.

Resistance law : linear law; logarithmic; exponential law.
Terminal resistance : $<20\Omega$ for $<1K\Omega$; $<\times 0.5\%$ of the nominal resistance for $>1K\Omega$.
Temp coefficient of resistance : ± 250 ppm/℃.
Rotational life : 50 cycles.

③ Micro-chip base metal glaze pots consist of few parts, the main parts are spindle, moving contact, grommet and resistive track substrate.

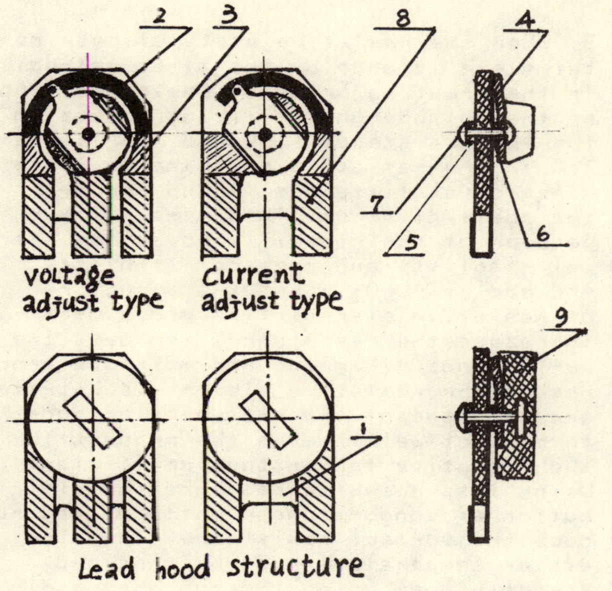

1. Substrate (alumina plate or temp-resistant plastic base)
2. Resistive element
3. Transition-layer
4. Moving contact
5. 6. Washer
7. Riveted spindle
8. Conducting layer
9. Spindle with hood

Fig-1: The sketch map of micro-chip potentiometer

II. MAIN PRODUCING TECHNIQUE

1. The schedule scheme of production of the micro-chip base metal glaze pots is shown in Fig-2:

2. The production of ceramic substrate: The major material is aluminium oxide, the material is rolled into tape and several tapes are arranged in laminated state, then punched and sliced on a large scale area, the schedule scheme is shown in fig.-3:

materials pretreatment → Resistive paste preparation
ceramic substrate preparation
↓
firing inner electrode
↓
firing intermediate electrode → preproduction
batch print ← plating electrode with tin
↓
sintering → resistance classification
electrothermal ageing
surface checking ← resistance classification ← measuring static noise
↓
assembling → electrical measuring → code
packing ← checking

Fig-2: The schedale scheme of micro-chip base metal glaze potentiometers

powder preparation → mixing → rough rolling
finish rolling
finishing ← sintering ← prebake ← blank punching and slicing
↓
classification → checking

Fig-3: The process flow of the ceramic substrate

3. The produce of base metal glass glaze resistive paste. It can be produced in different ways.

① The first method. To prepare the base metal glaze resistive paste, we mainly utilize molybdenum silicide and thallium oxide by mixing it with certain amount of glass powders according to the requirement of resistance values. The ethyl-cellulose and terpineol can be used as binder.

At first we grind the thallium oxide powder to fine, then pretreat the powder at 490℃ ± 15℃ to improve the crystallinity and the resistance stability. The glass is produced by mixing the lead oxide (PB_3O_4) boracacid, silicon oxide. Alumina, zirconia bismuth oxide, tin oxide, zinc oxide, barium carbonate and antimonous oxide, and then the mixture is processed as follows: firing, quenching, cleaning, ballmilling: sieving, checking etc.

② The second method. According to the different resistance range, we mix 60-90 percent by weight glass with about 10-40% by weight silicide admixtures which consist of magnesium silicide, molybdenum disilicide and tantalum silicide. We have added proper aluminium oxide and a little manganese disilicide to expend the resistance range and improve the stability of character. The diameter of the silicide admixture particle after grinding is 0.05-0.8 μm, the value of oxide alumina is 0.1-1 μm.

After mixing at different rate the resistance material blank is preformed, heated a half hour to five hours at 900°-1200℃. (prefered to prebeat and press the blank, in order that the silicide and the oxide can cover the glass frit, protecting the silicide from oxidation). After cooling, the material is grinded into grain which dimension distribution is 0.5-5 μm. Before using the resistive paste, the resistive powder as mentioned above is mixed up with terpineol containing ethyl-cellulose properly to meet the needs for screen printing.

4. Ceramic substrate must be washed with washings, then boiled about ten minutes, washed out repeatly, boiled in distilled water about ten minutes, taken out, dried at 120℃, then sintered 15 minutes again. After cooling it to ambient temperature, dust prevention must be taken for stock.

5. When we are making resistive track on ceramic substrate, firstly, inner electrode and then intermediate electrode, we first spray inner electrode and intermediate electrode with silverink and sinter them, then the resistive element and transition-layer are screen-printed and sintered. At last they are covered with a layer of tin as outer electrodes by immersing them in tin-bath.

6. Some other techniques such as mounting checking marking etc. are similar to the method of making the others pots.

Ⅲ. INTERPRETATIONS OF CHARACTERISTICS

1. The simple structure and few parts make micro-chip pots of base metal glass glaze reducing the time of riveting saving the materials, shortening the production, cycle period, reducing the cost and being suitable for mass production.

2. Because the pots have not any riveted terminal, the troubles that arise from the riveting are all eliminated. The accumulate of contact resistance is reduced that all make pots keeping their natural performance, reducing the stray capacitance and inductance in using, improving the reliability of the product and what is more, because the electrode of this kind of pots is soldered on the circuit board, so it can endure the various stress accordingly guarantee the circuit contact reliability.

3. When the resistive paste of base metal glaze is mode by the first method, in the great majority of cases, we choose the molybdenum disilicide to use in the low resistance (100Ω-3.3KΩ). The ZnO in the resistive paste makes glass glaze crystallize easly, and reduces the coefficient of linear thermal expansion of the glaze, improves the thermal stability and chemical stability. And adding AL_2O_3 makes the structure of resistive element fine and close, increases the resistance, improves the performance of ageing and moisture proof, makes the resistive element, substrate and the base of the siliver electrode to contact well, makes the product to show negative temperature coefficient. Using less graphite can make the distribution of conduct phase uniform, and reduce the surface resistance. Then the sinter temperature must be controed strictly when using the molybdtnum disilicide as the main material, the general sinter temperature is 790-800℃. If the sintering temperature is low, the temperature coefficient is bad or negative. If the sintering temperature is high, the ageing and moisture proof property are bad. The fig.4 and fig.5 show the relation of them.

Fig-4: Relation of AL_2O_3 content to resistance

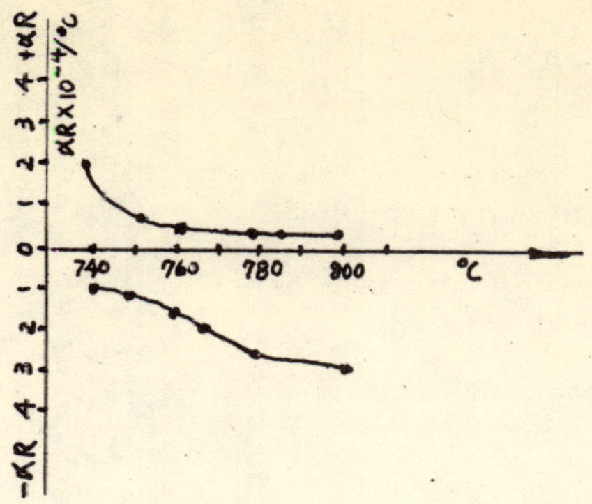

Fig-5: Relation of sintering temp to temp coefficient

The thallium oxide resistive paste mainly adopts the resistance section of 2.2KΩ-1MΩ. We heattreat the thallium oxide to increase the crystallinity of the oxide, increase the resistance stability, widen the sinter temperature range, when the sintering temperature is 540-580℃, the infiltration of the material is better. When the glaze was softening, the combining of the substrate and the thallium oxide is firm, avoiding volatilize and reforming the light conduct layer. If the sintering temperature is controled lower, the temperature coefficient and the performance of temperature cycling test are good, but if the temperature is too low, the struucture of resistance element will be loose and porous, and the property of endurance wiping and moisture proof will be bad. The suitable thickness of the layer is 5-10μm. If the layer thickness is thin, the temperature coefficient and the performance of temperature cycling test will be better, but if the layer thickness is too thin, the property of endurance wiping and moisture proof will be bad.

4. When the base metal glaze resistive paste is made by the above second method, you must notice the using quantity of silicide mixture. If the silicide is too few, the effect of the high resistivity is bad. And if the quantity of AL_2O_3 or glaze are few, the advantage of AL_2O_3 is not very obvious, or to make the stick action badly. But if the quantity of AL_2O_3 is more, the repeatability of resistivity is bad. In order to improve the performance of current noise and continuous electrical loading, the niobium oxide is added to glaze as addition.

Fig6 and Fig7 are the relation between the resistance and the temperature coefficient current noise of the pots produced by that resistive paste.

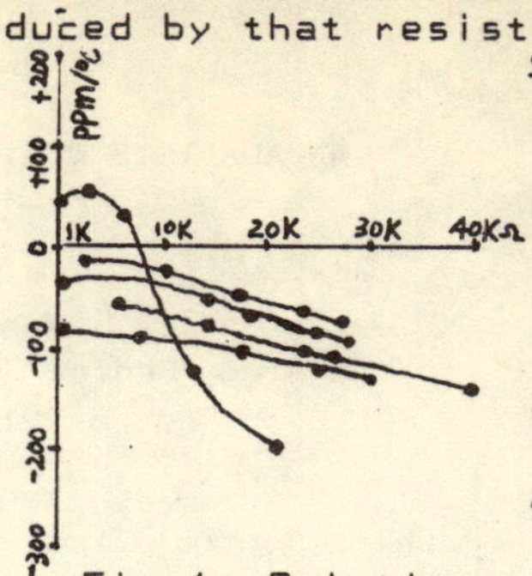

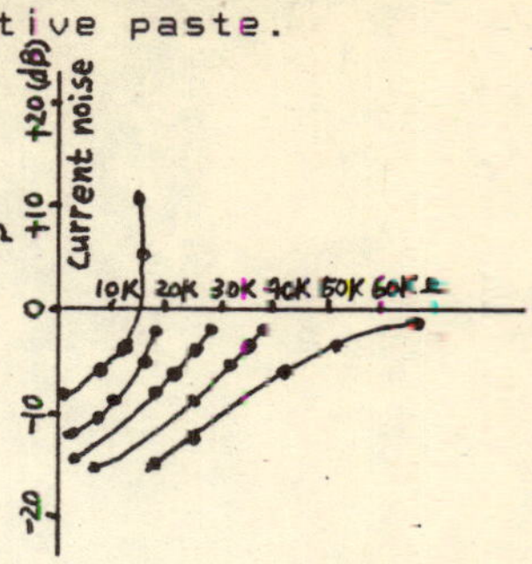

Fig-6: Relation of resistance with temperature coefficient.

Fig-7: Relation of resistance with current noise.

Ⅳ. APPLICATION AND DEVELOPING TREND

Because the volume of the micro-chip base metal glaze pots is small, the weight is light, the shape and measurement is going to standardize. This is the prerequisite of assembly of electronic machine in high density. It not only is assembled on the both side of the print circuit board, also suitable to make flat circuit, in order to auto-assemble in high speed, so it can improve the assembling efficiency to a considerable degree, and reduce the cost of the electronic machine. It is a essential prerequisite to make the electronic equipment microminiaturization, lamina and light duty. On the other hand, the resistance rang of this product is wide, the temperature coefficient can reach below ±100ppm/℃. In fact particularly, its surface is as smooth as a mirror and its resolution can reach below 0.01μm, the chemical stability, accuracy and endurance wiping life of the pots are better than the carbon composition or other film and wirewound potentiometer. It suits the mass production all the more, and the cost of the pots can be lower than the other kind of pots. Now, the pots are used in the micro-cassette recorder, lamina radio, television camera, stereo sound system, electron camera, videocorder etc. Before long, the pots will be used in the field of munition industry, aviation industry, automobile industry, agriculture machines sewing machine, medical appliance domestic, electronic product, and will be used further in the field of consumer goods, in the various localization adjust system, and in the micro computers to control the needle position. In the future, the requirement and applied range of the pots will be large quantity and wide.

AN ANALYSIS OF THE FAILURE OF CL80-TYPE FILM CAPACITORS

Wang Peixian
Beijing Radio Components Factory No. 2
Beijing, PRC

Wang Sheguan
China Central Iron and Steel Research Institute
Beijing, PRC

ABSTRACT

The article relates the reasons of the failure of the CL80-type high-voltage metallized polyester film capacitor. An Auger electronic microscope was used to make an argon-ion sputtering peel analysis. Measures have been proposed with a view to improvement of the product.

1. Preface

The CL80-type high-voltage metallized polyester film capacitor is a key component compatible with transformers used in color television. It is made of two round wicks in containers, each of which is formed by winding the aluminum metallized polyester film six times series in side. The pressure tolerance is C for 1,000 hours three defects appeared: a reduction of capacity; a complete failure of capacity; or an increase of the tangent value of the loss angle. Examination of the failed capacitors showed three abnormal changes: (a) the film had turned black; (b) the film had a mottled appearance; and (c) the aluminium film had almost completely fallen off.

2. Experimental Results and Analysis

An Auger electronic microscope was used to effect an argon-ion sputtering peel analysis of the four forms of film shown in Figures 1-4. The results are shown in Figures 1-4.

As a result of the Auger analysis, we found the following:

A. Chemical features. The film was composed of only three elements, namely, aluminum, oxygen, and carbon, not including hydrogen, as shown in Figures 1-4. Therefore the failure of the film could be caused only by the various contents and the corresponding metallurgical micrograph and microstructure of these three elements. Therefore the argon-ion sputtering peel was conducted. The results are shown in Figures 5-8.

B. Distribution features. The distribution features of the three elements as a function of the thickness of ordinary film are shown in Figure 5.

a. An almost pure aluminum region appears, as noted by the portion of the line denoted by A to B at about 300 Angstroms. It shows about 90% aluminum content and the lowest content of oxygen and carbon, the typical characteristics of almost pure aluminum film.

b. The oxygen content was low, about 10%, with a very narrow range of variation.

c. As the aluminum film was peeled, the content of carbon increased as a result of the sputtering of the polyester film by the argon-ion.

C. The causes of blackened film. Comparing Figure 6 with Figure 5 it can be seen that the distribution of the aluminum content of the blackened film has changed. The highly pure range (line A-B) has moved to the right a little; the aluminum content has decreased; and that of carbon and oxygen has increased in the 900 Angstrom range from the aluminum surface. That means that the surface had been contaminated by carbon and oxygen, which increased the thickness of the aluminum film.

During the depositing process contamination may come from these sources:

a. (a) The carbon contamination may come from the evaporation of the graphite crucible used during the process. It can be seen from Figure 5 (ordinary film) that carbon is deposited at any depth of the film. Weight loss and pitting of the graphite bed or crucible indicated that carbon was evaporated together with aluminum and co-deposited on the polyester. Obviously, when the aluminum was almost exhausted in the crucible the content of carbon will increase both on the vapor and the film, causing the black color. In order to prevent this condition, it is recommended that sufficient aluminum be kept in the crucible throughout the process; (b) carbon contamination may come from the oil returning from the diffusion pump; (c) as temperature in the vacuum chamber becomes excessivly high, the speed with which the film is coiled

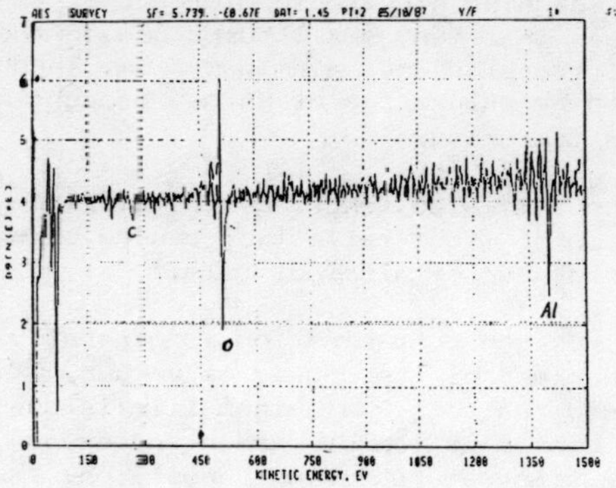

Fig.1 Auger spectrum of ordinary film

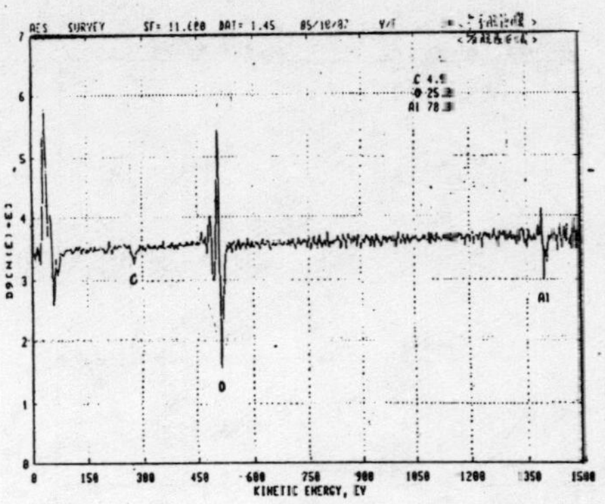

Fig.4 Auger spectrum of film showing loss of metal

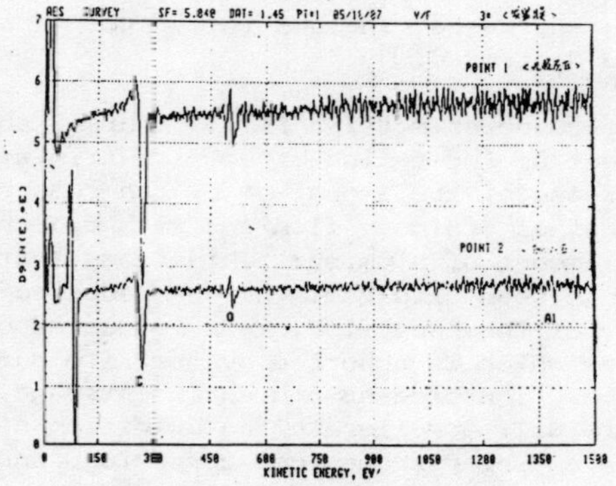

Fig.2 Auger spectrum of blackened film

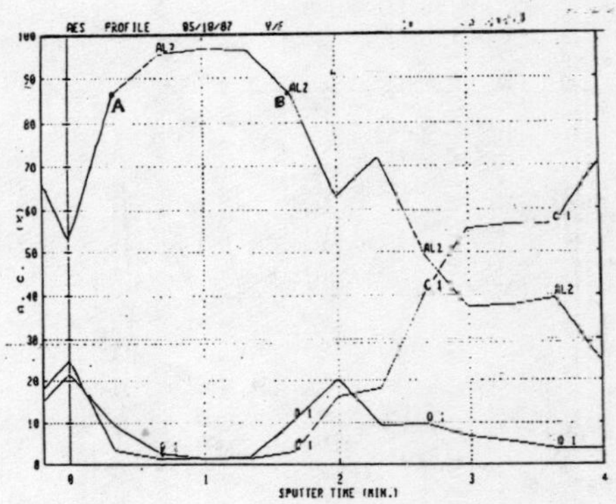

Fig.5 Distribution of aluminum, oxygen, and carbon as a function of film thickness: ordinary film

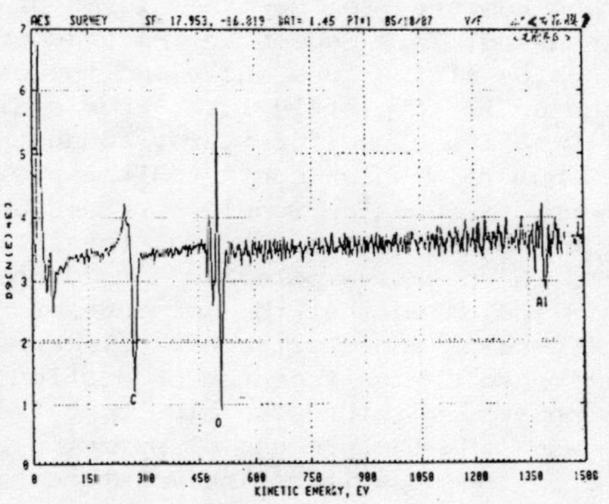

Fig.3 Auger spectrum of mottled film

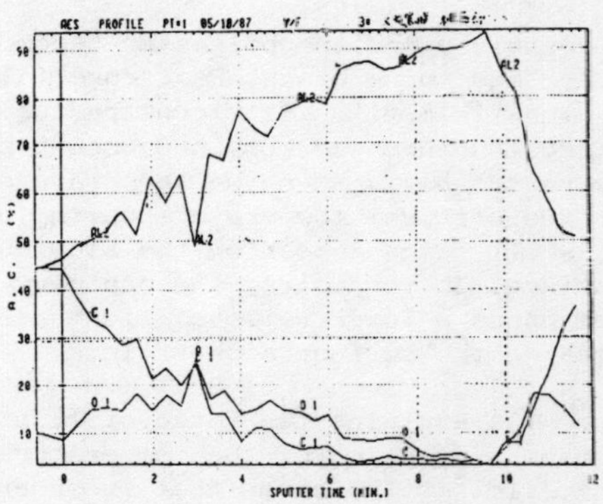

Fig.6 Distribution of aluminum, oxygen, and carbon as a function of film thickness: blackened film

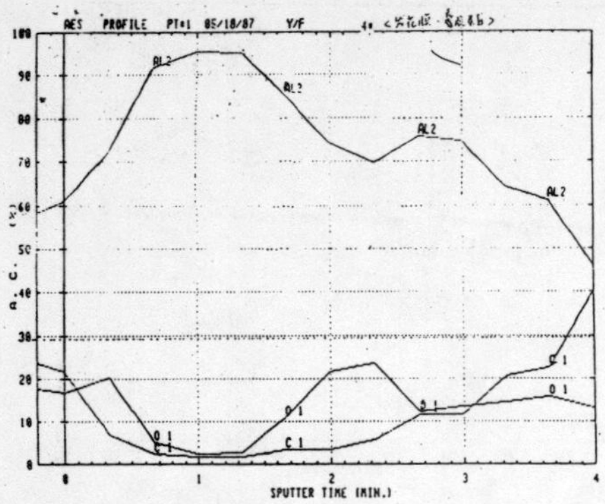

Fig.7 Distribution of aluminum, oxygen, and carbon as a function of film thickness: mottled film

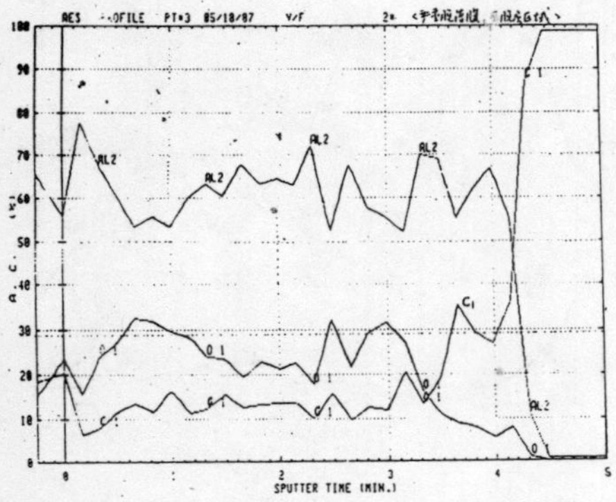

Fig.8 Distribution of aluminum, oxygen, and carbon as a function of film thickness: film showing loss of metal

slows excessivly and the polyester becomes overheated and turns black. Reference (1) explains blank film with its microstructure. This microstructure was obtained when temperatures were too high. When the temperature exceeds the critical temperature during deposition, small bumps appear on the aluminum film surface. These destroy the ordinary surface and cause a lower ratio of light reflection, making the surface appear black.

Reference (1) explains the ratio of light reflection as a function of the temperature of the base film. It indicates that in order to obtain a highly bright aluminum film it is important to lower the temperature of the base film. Older depositing equipment used water-cooling to reduce the temperature of the base film, but today more advanced equipment uses Freon cooling and has obtained good results.

Reference (1) explains a photograph obtained with a scanning electronic microscope. It can been seen that many small bumps are on the surface of the blackened aluminum film. These are produced by relaxation of stress brought about by high temperature.

Reference (1) also describes the decrease of the ratio of light reflection caused by excessive thickness of aluminum film.

b. Oxigen contamination is generally due to a decrease of the degree of vacuum. It is concluded that black aluminum film is due mainly to the contamination of carbon and oxygen, or to high temperature during deposition. Therefore the main measures to prevent blackened film are to control contamination by carbon and oxygen and to control excessive temperature during deposition. Otherwise, the conductivity of the metal film will be decreased and the value of the tangent of the loss angle of the capacitor will be increased. The quality of the capacitor will be destroyed.

D. The cause of mottled film. Figure 7 shows the spectrum Auger line of mottled film with an increase of the amount of oxygen compared with that of ordinary film and with a decrease of the amount of aluminum. This means that aluminum has been oxidized, thereby reducing the quality of the capacitor. This is caused by moisture which is absorbed by the film during the production process and during storage. Moisture will accellerate the oxidition of the Alumiam and electrochemical corrosion, and cause failure of the capacitor.

E. The cause of loss of aluminum from film. Figure 8 shows a sputtering spectrum of film which has lost most of its aluminum. A serious decrease of the aluminum content appears, and there is a greatly increased amount of oxygen and carbon compared to the other forms of film just discussed. This causes an increase of the tangent value of the loss angle and the capacity decreases or is completely lost. From a dissection of a failed capacitor it was found that the aluminum film which contacted the positive polar metal film was seriously corroded and much aluminum had fallen off. But the film which contacted the negative pole had not corroed at all. This indicated that the moisture promoted sparking erosion and electrochemical corrosion. In addition to coming from the production and storage phases, moisture can also come from metal powder and the process of spraying it onto the ends of the capacitor. The moisture appears as a film of water on the surface of the metal. This forms an electrochemical corrosive cell under high voltage and microvibration. This accellerated corrosive process will lead to the sprayed metal ends separating from the capacitor and complete failure.

3. Conclusion

A. The blackened film is due to excessively high temperature during deposition and to carbon and/or oxygen contamination. Therefore it is necessary to regulate the temperature carefully, to strictly control the degree of vacuum, to maintain sufficient aluminum in the crucible, to prevent oil from returning from the diffusion pump, and to clean the vacuum chamber regularly.

B. Mottled film and the film which lost most of its aluminum are caused by moisture. Therefore it is necessary to reduce the humidity of the air, to dry the metal powder in advance, and to filter and dry the compressed air during coiling and spraying processes and storage. it is worth exploring the possibility of coiling the capacitors in a vacuum.

Because of the results of this study, the measures mentioned above were carried out in Beijing Radio Components Factory No.2. Stable, high-quality capacitors were produced.

4. Reference

(1). Vacuum/volume 37/numbers 1 and 2/pages 183 to 186/1987.

5. Acknowledgement

The authors wish to express their thanks to Polly Trnavsky who assisted in proof the English version of this article.

FAILURE RATE DEMONSTRATION TEST OF THE ELECTRONIC COMPONENTS USING BAYESIAN METHOD

Su Deqing
Chinese Electronics Standardization Institute (CESI)
Beijing, China

Wang Lingling
Department of Statistics, East China Normal University
Shanghai, China

ABSTRACT

It has been realized that using the sampling plan given in the national standard GB1772 can be time consuming, and sometimes even impossible to determine the failure rate of electronic components. With the new Bayesian sampling method described below, by taking into account the historical data and the engineering experiences, one can reduce the test time and as a result, cut down the cost considerably.

1. Background

Our country, like the other countries in the world, developed a national standard in which a classical statistical theory was used to determine the failure rate for electronic components[1].

It will take a lot of time and expenditure to evaluate the reliability of a product with a very low failure rate because the classical statistical test method uses only the test result to evaluate the reliability of the product and does not involve other information in the process. In fact, there is still a great deal of untouched information that has accumulated in the previous process of production. Therefore a new evaluation method makes full use of the past information on as well as the present test results needed.

2. The essential points of the Bayesian method

Three types of information can be used in statistical inference:
A. Population information, such as the information contained in a certain type of life distribution;
B. Sample information, such as the information about the unknown parameters from the testing sample;
C. Prior information, the information acquired before the test of the product.

One of the essential points of Bayesian theory is to take the interested parameter such as the failure rate ($\wedge$) as a random variable with a certain distribution. The form of distribution is called prior distribution because it is based on what has been known before the sampling test about the failure rate of a product.

3. Determination of the prior distribution

The prior distribution of the item under test has to be found in order to develop a sampling plan using the Bayesian method.

Assume that the component life (t) obeys the exponential distribution:

$$F(t)=1-e^{-\lambda t},$$

where the unknown parameter $\lambda>0$, is the failure rate and is independent of t.

In the time-terminated test with replacement, the failure number r of n items will follow the Poisson distrbution on (0,t):

$$P(r \quad \lambda,T)=\frac{(\lambda t)^r}{r!}e^{-\lambda t}, \tag{1}$$

Where T=nt, the total test time; and, $\lambda \in (0,\infty)$.

From the above equation, it is natural to have an adjoint Gamma distribution G(λ;a,b) as the prior distribution, which is a biased distribution, with the density function as follows:

$$g(\lambda;a,b)=\frac{b^a\lambda^{a-1}}{\Gamma(a)}e^{-b\lambda} \quad (\lambda>0), \tag{2}$$

where a and b are parameters of shapes and scales respectively (see Fig.1), and both of them can be chosen for different products. Therefor, this distribution is widely applicable.

We have collected the data from all together 2009 lot time of qualification test for components, (see Fig.1), which represents the failure rate distribution of components produced in China; it also gives evidence that using

Gamma distribution is appropriate.

In order to determine parameters a and b, assume that product subject to Class K test meet the requirements of Class K-1, for example:

$$P_1=P(\lambda<10^{-K})=\int_0^{10^{-K}} g(\lambda;a,b)dt=0.30 \quad (3)$$

$$P_2=P(\lambda<10^{-(K-1)})=\int_0^{10^{-(K-1)}} g(\lambda;a,b)dt=0.90 \quad (4)$$

with the help of computer, we have obtained the following:

$$a=0.70877$$
$$b=0.1774\times10^{K} \quad (5)$$

4. Sampling plan for assessing failure rate

The LTPD plan is used in GB1772 to demonstrate the failure rate Class K. For the limit failure rate $\lambda_1=10^{-K}$, with the user's risk β given, acceptance probability is calculated:

$$L(\lambda_1)=P(r<c \quad \lambda_1)=\beta$$

Thus, the total test time T can be found for different acceptable failure number C. As an example, the demonstration test plan for Class 5 qualification (take $\beta=0.4$) is given in Table 1.

Table 1

c	0	1	2	3	4	5
T	91600	202000	311000	418000	524000	629000

The LTPD plan can also be used in Bayesian demonstration test. For Class K qualification test, $\lambda_1=10^{-K}$, and with the posterior user's risk β given, using the following equation,

$$P(\Lambda>\lambda_1 \quad r<c)=\beta^* \quad (6)$$

we can find the desired total test time T_B for the acceptable failure number c=0, 1, 2, ... If the prior distribution of failure rate Λ is Gamma distribution, $g(\lambda;a,b)$, and the failure number r in time-terminated life test obeys the Poisson distribution, the posterior user's risk can be derived from equation (1) and (2):

$$\beta^*=P(\Lambda>\lambda_1 \mid r<c)=\frac{P(\Lambda>\lambda_1, r<c)}{P(r<c)}$$
$$=\frac{1}{P(r<c)}\int_{\lambda_1}^{\infty}\sum_{r=0}^{c}\frac{(\lambda T)^r}{r!}e^{-\lambda t}\frac{b^a\lambda^{a-1}}{\Gamma(a)}e^{-b\lambda}d\lambda$$

$$=\frac{(b/(b+T))^a}{P(A)\Gamma(a)}\sum_{r=0}^{c}\frac{1}{r!}\left(\frac{T}{b+T}\right)^r[\Gamma(a+r)-\Gamma(a+r,(b+T)\lambda_1)] \quad (7)$$

where $\Gamma(a,x)$ is an incomplete Γ-function, and P(A) is the acceptance probability, i.e.,

$$P(A)=P(r<c)=\int_0^{\infty} P(r<c \mid \lambda)g(\lambda;a,b)d\lambda$$
$$=\frac{1}{\Gamma(a)}\left(\frac{b}{b+T}\right)^a\sum_{r=0}^{c}\frac{\Gamma(a+r)}{r!}\left(\frac{T}{b+T}\right)^r \quad (8)$$

For the posterior user's risk $\beta^*=0.4$ the Bayesian plan is given in Table 2.

Table 2

c	0	1	2	3	4	5
T	41100	79500	119500	160400	201800	243600
P(A)	0.427	0.473	0.487	0.493	0.496	0.498

The ratio of T_B, the total test time for Bayesian method, to T_C, for classical method, is shown in Table 3, demonstrating T_B smaller than T_C, and hence the time saving property of the Bayesian method.

Table 3

c	0	1	2	3	4	5
T_B/T_C	44.9%	39.4%	38.4%	38.4%	38.5%	38.7%

5. The eligibility appraisal procedure for Bayesian demonstration test plan

If the posterior distribution of the Class K-1 components satisfies equation (3), the parameter of the Class K prior distribution can then be decided by $b=0.1774\times10^K$. Therefor, it is necessary to verify the equation (3) before the Class K test is carried out. This process is called the eligibility appraisal. Solving equation (3), a set of T are found for different r, as shown in Table 4.

Table 4

Class \ T c	0	1	2	3	4	5
5	159700	852400	1653900	2496000	3360900	4240900
6	1597000	8524000	16539000	24960000	33609000	42409000

The historical data are mainly used for eligibility appraisal while some supplementary tests are still needed if the historical data are not sufficient. With the actual total failure number r from the historical data, a T_1 can be found in Table 4. The eligibility appraisal for Class K is deemed as passed if the actual test time is more than T_1. This means $\Gamma(a_K\ b_K)$ can be used as the prior distribution of Class K.

6. Maintenance and up-grading tests

After the Class K qualification test, the periodic maintenance tests are carried out to demonstrate the consistency of failure rate for products in continuing production. With the production in a controlled state, as shown by having passed the qualification test, a bigger posterior user's risk β^* can be chosen for the maintenance test. For instance, for the product having passed the Class 5 qualification test, the prior distribution of the maintenance test being the same as that of the qualification test, by taking $\beta^*=0.6$, we have the test plan and the acceptance probability as shown in Table 5.

Table 5

c	0	1	2	3	4	5
T	10500	26600	44100	61900	80000	98100
P(A)	0.720	0.744	0.749	0.750	0.750	0.750

An up-grading test may be taken at least four periods of maintenance after the qualification test with proof that the production process is stable and product reliability is perceived as higher.

After the qualification and maintenance tests, the posterior distribution of the failure rate is $G(\lambda; a+r, b+T)$, and the failure rate Λ must meet following requirements:

$$P(\Lambda \leqslant 10^{-(K+1)}) \geqslant P_1 \quad (9)$$

$$P(\Lambda \leqslant 10^{-K}) \geqslant P_2 \quad (10)$$

It has been proved that the expression (10) can be derived from expression (9) (see "A Bayesian qualification test method for assessing the failure rate of electronic components"[2]).From

$$\int_0^{10^{-(K+1)}} G(\lambda; a+r, b+T)\, d\lambda \geqslant P_1 \quad (11)$$

We can obtain the desired total test time T for every failure number r which can be used as a criterion of eligibility appraisal for Class K+1.

If $P_1=0.30$, $P_2=0.90$, T is found (see Table 4).

Prior distribution of λ, after the eligibility appraisal has been successfully passed, can be assumed as $\Gamma(a_{K+1}, b_{K+1})$. The up-grading test can then be conducted using the same procedure as that of the qualification test.

7. Sampling characteristic function

For comparison with the classical test plan, we can use the classical method to find the characteristic function of the Bayesian test plan, which will show us the properties of the Bayesian test plan when the failure rate Λ has been some given value λ. In the qualification and up-grading tests, for each (T,c) plan, the acceptance probability is:

$$L_1(\lambda) = P(r \leqslant c \mid \lambda) = \sum_{r=0}^{c} \frac{(\lambda T)^r}{r!} e^{-\lambda T} \quad (12)$$

For up-grading test, the eligibility appraisal is required. Let the probability of passing the appraisal be $P(\lambda)$, then, when $\Lambda=\lambda$, the acceptence probability is:

$$L(\lambda) = P(\lambda) L_1(\lambda) \quad (13)$$

Every plan in eligibility appraisal test with a certain r(r=0, 1, 2, ...) can be used to compute $P(\lambda)$, but for the qualification test, a definite (T,c) plan must be chosen before the test and will not be changed during the test; the total test time T_1 can be accumulated under the same test condition. From the equation in expression (9), we have,

$$P(\lambda) = e^{-\lambda T_1(0)} + \lambda T_1(0) e^{-\lambda T_1(1)}$$
$$+ \lambda^2 T_1(0)[T_1(1) - \frac{T_1(0)}{2}] e^{-\lambda T_1(2)}$$
$$+ \lambda^3 T_1(0)[T_1(1)T_1(2) - \frac{T_1(0)T_1(2)}{2} - \frac{T_1(1)}{2} + \frac{T_1^2(0)}{6}] e^{-\lambda T_1(3)}$$
$$+ \ldots \quad (14)$$

where $T_1(r)$, r=0, 1, 2, ..., represents the test time required by the eligibility appraisal for different r. The total test time T_1 in eliglibility appraisal for each test plan and the value of $P(\lambda)$ and $L(\lambda)$, when $\lambda=0.1, 0.3, 1, 2, 3, 5, 10 (\times 10^{-K})$, are given in the appendix.

REFERENCES

1. GB1772-79 "Test method for assessing the failure rate of electronic components"
2. "A Bayesian qualification test method for assessing the failure rate of electronic components" by Lingling Wang, Deqing Su, Heliang Fei and Bo Tao (to be published)

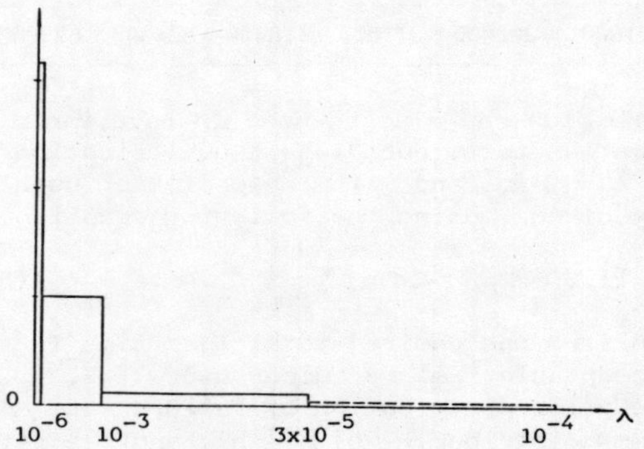

Figure 1

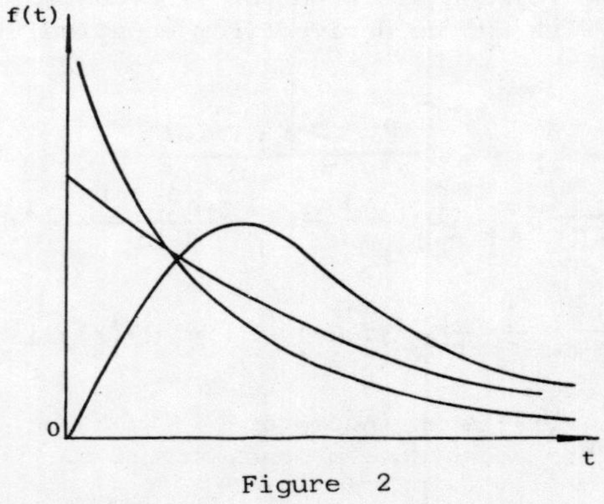

Figure 2

STUDY ON LIFE CHARACTERISTICS AND DETERIORATIVE MECHANISM OF ALUMINIUM ELECTROLYTIC CAPACITORS

Lin Xueqing
Xiamen Overseas Chinese Electronic Co. Ltd,
Xiamen, China

ABSTRACT

Life characteristics and deteriorative mechanism have been observed of miniature aluminium electrolytic capacitors with non-solid electrolyte which were tested for 10000 hours at high temperature with different loading. Results show that the capacitance decreased, and the impedance (Z) and the equivalent series resistance (ESR) increased with testing time which results in the deterioration of electrical characteristics of capacitors. Results also show that deterioration is caused mainly by degradation of dielectrical oxide film and exhuastion of electrolyte is related to capacitor's type, environmental condition, sealing rubber quality and sealing technology, etc. And relationships of $\Delta W/W$ with $\Delta C/C$, tgδ and Z were analysed gravimetrically.

INTRODUCTION

Aluminium electrolytic capacitors are widely used in electronic equipment and household appliances in virtue of their small size, large capacitance size retio and low price. The value $\Delta C/C$, however, may reduce, which results in the increase of dissipation factor (tgδ) and impedance (Z), after long operation because of diffusion of electrolyte and the deterioration of dielectrical oxide film. Therefore, aluminium electrolytic capacitors are components with "limited-life", which is the subject of this paper. The practical value of this discussion lies in the fact that to extend the life of electrolytic capacitors requires low vapour pressure and prevention of hydration of oxide film, etc. Also, life characteristics and deteriorative mechanism of miniature aluminium electrolytic capacitors at high temperature with different loading are discussed based on weight loss ratio, $\Delta W/W$.

TEST SAMPLES AND CONDITIONS

1. Test Samples

The type, case size and sample size of the capacitors are shown in Table 1. Electrolyte used in this study is ethylene glycol-dicarboxylic acid with water system, and sealed by EPT rubber.

Table 1. Details of tested samples

Item \ Group	1	2	3	4
Type	10uF 16v	220uF 16v	0.47uF 50v	47uF 50v
Case size (mm)	ϕ5x11	ϕ10x12	ϕ5x11	ϕ8x11.5
Maximum operation temperature (C)	85	85	85	85
Guaranteed life (hrs)	2000	2000	2000	2000
Sample size	20	20	20	20

2. Test Condition

Experiments were carried out at temperature of 85°C and under rated voltage application (high temperature loading) and non-voltage application (high temperature non-loading). The parameters of weight (W), capacitane(C), dissipation factor (tgδ), impedance (Z), leakage current (I_L) and equivalent series resistance (ESR) were measured at pre-specified time.
The data of high-temperature non-loading were achieved by following procedure: shelve the samples at room temperature for 2 hours after taking out of the baking oven, then apply rated voltage to it and leave it ageing for 0.5 hour.

RESULT AND DISCUSSION

1. Relationships of Weight Loss Ratio With Life

Changes of $\Delta W/W$ with life for different types of aluminium electrolytic capacitors with and without loading at high temperature are shown in Figures 1 and 2. The numbers 1, 2, 3 and 4 indicate the group numbers (cf. Table 1). In all figures, solid lines are for those with loading at high temperature and dotted lines for those with non-loading at high temperature. And each curve is drawn up by the arithmetic mean value of measured data for each sample.

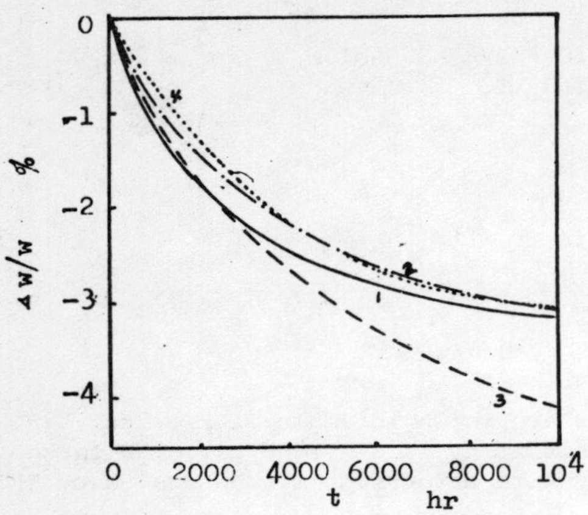

Fig. 1 Relationship of life with $\Delta W/W$ at high temperature of 85°C with loading.

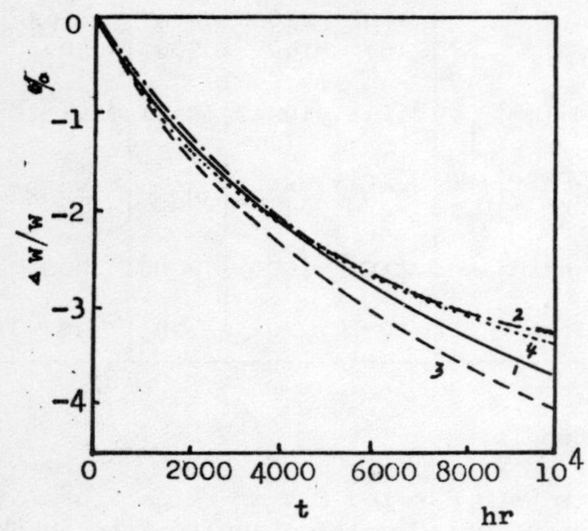

Fig. 2 Relationship of life with $\Delta W/W$ at high temperature of 85°C with non-loading.

It can be seen from the figures that regardless of rated or non-rated voltage, the change law of weight loss ratio follows a general form:

$$\text{Log } \Delta W/W = A \text{ Log } t \qquad (1)$$

Where A is a negative number.

The causes of weight loss of a capacitor is due to the ageing of the sealing rubber and PVC tube, and exhaustion of the electrolyte. The results of this test show that the latter is the main cause. Vapour pressure of electrolyte in the capacitor is about 17 kPa at 85°C. The vapour molecules slowly escape through the sealing rubber, lead hole on the dealing rubber and sealing edge, which results in gradual exhaustion of the electrolyte in the capacitor as time increase. And the speed of the exhaustion increases with temperature. According to diffusion theory, diffusion coefficient will change in accordance with following equation:

$$D = Do \exp (\Delta E/RT) \qquad (2)$$

where R - Boltzmann's constant,
ΔE - activation energy,
T - absolute temperature,
Do - a constant.

2. Relationship of $\Delta C/C$ $tg\delta$, I_L with t

For the sake of investigation into change law of $\Delta C/C$, $tg\delta$, I_L with life (t), electrolytic capacitors of 0.47uF 50V and 47uF 50V were tested. The results are given in Figures 3-6.

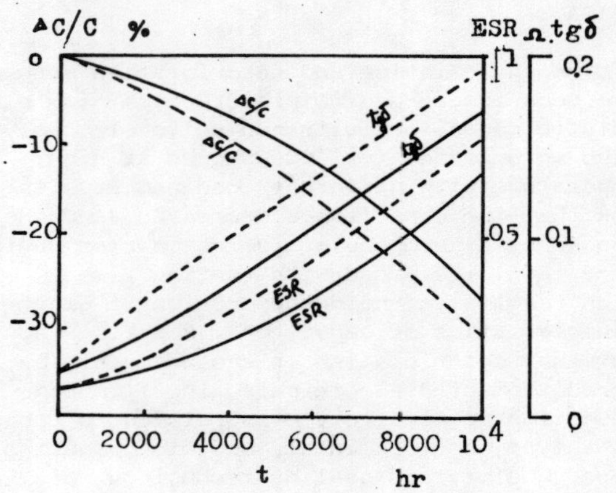

Fig. 3 Relationship of life with $\Delta C/C$, ESR and $tg\delta$ for 0.47uF 50V capacitors at °85 C.

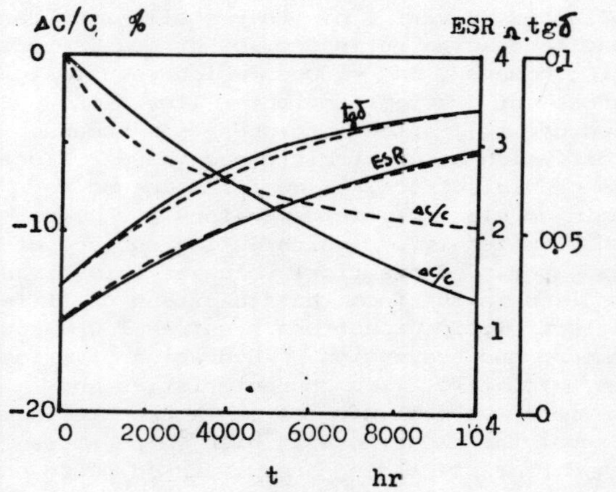

Fig. 4 Relationship of life with $\Delta C/C$, ESR and $tg\delta$ for 47uF 50V capacitors at 85°C.

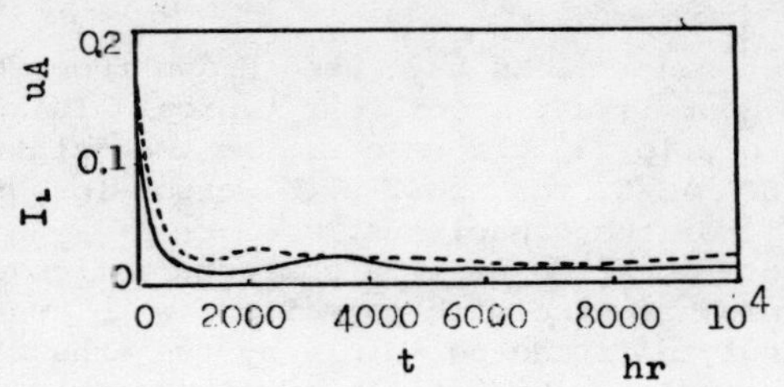

Fig. 5 Relationship of life with I_L for 0.47uF 50V capacitors at 85°C.

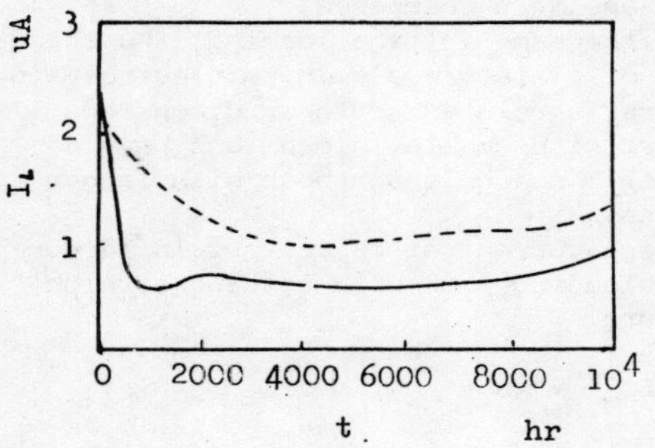

Fig. 6 Relationship of life with I_L for 47uF 50V capacitors at 85°C.

It can be seen from the figures that changes of $\Delta C/C$, tgδ, ESR and I_L with life of capacitors are complicated regardless of the type of capacitors. The complication is not only due to the exhaustion of electrolyte but also the corrosion and hydration of the dielectrical oxide film, which also gives rise to deterioration of electrical characteristics.

A. Investigation into the characteristics at high temperature with loading.

Dielectrical oxide film of an electrolytic capacitor every now and then can be self-repaired under electric field. But trace CL^-, SO_4^{--} present in its electrolytem spacer and sealing rubber, etc, may corrode anodic foil and release gas:

$$AL + CL^- \longrightarrow ALCL_3 \xrightarrow{H_2O} AL(OH)_3 + HCL$$

and increase tgδ and I_L, which results in increase of interior temperature and pressure of the capacitor that may even cause vent to open. Hydration, meanwhile, can also occur in the cathode, which builds up thickness of oxide film.

B. Investigation into the characteristics at high temperature without loading.

Characteristics of electrolytic capacitors are poorer under non-loading than under loading at high temperature, with a more evident tendency towards deterioration. Because of difficulty of defect repairation in oxide film on anodic surface at high temperature, higher grade r'-AL_2O_3 film with compact texture and better dielectric quality is degraded into a lower level. Also, water in electrolyte can diffuse into the oxide film and turn it into hydrated form $AL_2O_3nH_2O$ (n-1 to 3):

$$2AL+6H_2O \longrightarrow 2AL+6OH^-+6H^+ \longrightarrow 2AL^{3-}+6OH^- -3H_2\uparrow$$

$$\longrightarrow 2AL(OH)_3+3H_2\uparrow$$

$$AL_2O_3\cdot 3H_2O \text{ (Bayerite aluminium oxide)}$$

$$\longrightarrow 2ALO\cdot OH + 2H_2O + 3H_2\uparrow$$

$$AL_2O_3\cdot H_2O \text{ (Bohmite aluminium oxide)}$$

Same as in the anode, hydration can also occur on cathodic surface so that oxide film thickness increases and cathodic capacitance reduces. As a result of the joint effects mentioned above, withstanding voltage and capacitance of a capacitor decrease, and leakage current and tgδ increase.

3. Relationship of $\Delta C/C$, tgδ, Z with $\Delta W/W$

From the equivalent circuit analysis of electrolytic capacitors, it can be obtained that:

$$C_r = \frac{C\,[1 + (\omega C_p r_e)^2]}{1 + \omega^2 r_e^2 (C + C_p) C_p} \qquad (3)$$

in which C_p = capacitance with electrolytic spacer,
r_e = series resistance with electrolytic spacer,
C_r = equivalent capacitance.
when dielectrical loss of oxide film is negligible,

$$tg\delta = \frac{\omega C r_e}{1 + \omega^2 r_e^2 (C + C_p) C_p} \qquad (4)$$

From the above equations, it can be seen that because electrolyte in a capacitor diffuses outwards and gradually decreases in amount, series resistance of the capacitor (r_e) increases, resulting in the increase of tgδ and Z and the decrease of C_r. The relationship between those parameters achieved in those study are represented in Figures 7 to 9.

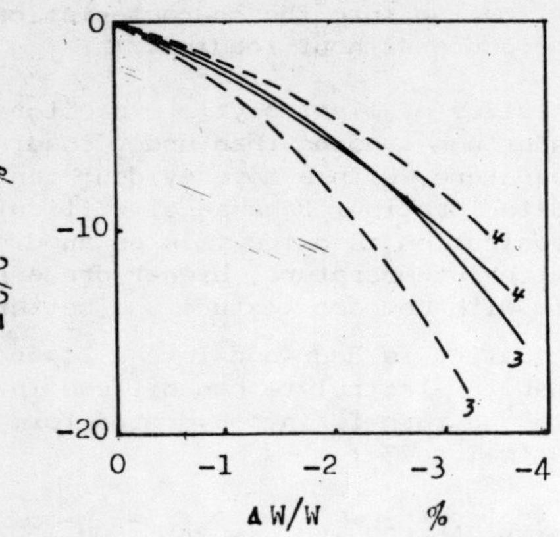

Fig. 7 Relationship of ΔW/W with ΔC/C at 85°C.

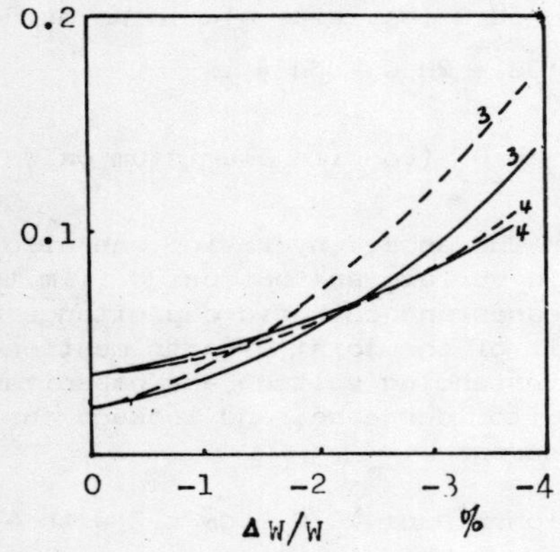

Fig. 8 Relationship of ΔW/W with tgδ at 85°C.

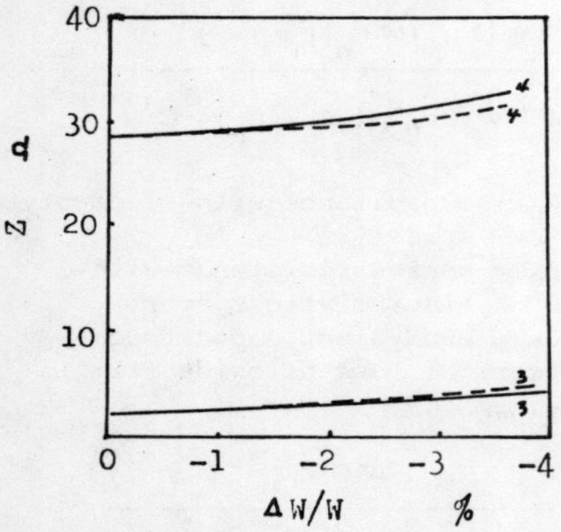

Fig. 9 Relationship of ΔW/W with Z at 85°C.

CONCLUSION

Several conclusions have been drawn from the 10000 hour life test of four typical aluminium electrolytic capacitors based on the relationship of ΔC/C, tgδ and Z with weight loss ratio (ΔW/W) of the capacitors:

1. Deterioration of life characteristics of aluminium electrolytic capacitors with non-solid electrolyte is caused mainly by the exhaustion of electrolyte, hydration and microcorrosion of dielectrical oxide film under operating conditions.
2. The weight loss ratio (ΔW/W) decreases exponentially with testing time increase.
3. Electrical characteristics of capacitors are closely related to the ratio of ΔW/W. ΔC/C decreases as ΔW/W decreases, and tgδ and Z increase as ΔW/W decreases.
4. With the same voltage loading, the change of ΔW/W of miniature aluminium electrolytic capacitors is related to their dimensions. Capacitors with smaller dimensions have a larger ΔW/W change than those with larger dimensions.
5. Leakage current of capacitors is larger with non-loading than with loading at high temperature.

RERERENCE

1. Izaya Nagata, Aluminium Electrolytic Capacitors with Liquid Electrolyte, (1983), (Japanese version).
2. G.E.Rhoades and A.W.H.Smith, 1984 Proceeding of 34th Electronic components conference, 159.
3. Hayama Izumoi et al., Report of Technology Research of Electronic Communication Society, R83-77 (1984), 53. (Japanese version).
4. K.Kiuchi et al., Report of Technology Research of Electronic Communication Society, PE85-20, 53. (Japanese version).

RELIABILITY TEST FOR LOW FREQUENCY
ELECTRONIC CONNECTORS

She Yufang, Su Deqing
Chinese Electronics Standardization Institute

ABSTRACT

This paper demonstrates the exponential law as the governing relationship between the inherent life of electronic connectors and temperature, as shown by the static load accelerated life test, and the life distribution may be represented by Weibull distribution with the shape parameter $m>1$. The early failure rate of connectors are very low because of their wearout life pattern. As in the case of p-type connectors under a normal evironmental condition(25°c) the mean failure rate of the first half year is less than $4x10^{-13}$/hour, while it is to go up to $1.5x10^{-6}$/hour after twenty years.

1. Introduction

Low frequency electronic connectors operate at frequencies lower than 3 MHz and include various types ranging from circular, rectangular, and concentric connectors PC connectors, and connectors for power supplies of electronic equipments.

Consistant stress accelerated life tests and the subsequent analysis have been carried out to investigate the static load life of electronic connectors.

2. Basic Hypothesis[1]

Hypothesis 1: The connectors life obeys or approximately obeys the Weibull distribution under the stress of So;

Hypothesis 2: The failure mechanism of the connectors remains same under the accelerated stress of S_i

Hypothesis 3: The characteristic life of the connectors under the applied temperature or electrical stress should satisfy Arrhenius equation or equation of inverse power, that is,

$$\eta = e^{a+b/T} \qquad (1)$$

$$\eta = 1/kv^{c} \qquad (2)$$

Where a.b.k and c are all constants.

The third hypothesis gives the functional relationship between the characteristic life and the stress. Much use of the relationship has been made for electric vacuum devices, capacitors and resistors. As for connectors, however, there has been little discussion both at home and abroad.

With mature or fully developed designs and processes, the static load life of connectors is usually affected mainly by the aging of insulation and resilient materials. For well-activated insulation and resilient materials a dynamic equilibrium should be maintained between the loss and the gain. Consequently, the main cause of failure (random failure not considered) is the breakdown of the equilibrium, in other words, greater loss and less gain indicate mainly the wear-out type failure. The reaction rate equation[2] shows as exponential characteristic of the relationship between physical and chemical process and the applied temperature under normal operating conditions as expressed by Arrhenius equation.

$$\frac{dM}{dt} = Ae^{-\frac{\Delta E}{kT}} \qquad (3)$$

Where $\frac{dM}{dt}$ denotes the rate of change of reactants

A is a constant
ΔE is activated energy (also diffused energy or evaporated energy)
T is absolute temperature
K is Baltzmanis constants

Integrating and taking the logarithmic transform of the equation(3), we have

$$\ln t = \ln a' + \frac{\Delta E}{K}\left(\frac{1}{T}\right)$$

supposing $a=\ln a'$, $b=\frac{\Delta E}{K}$, we have equation(1).

The effect of temperature on failure process can be seen from equation(3). Thus, a temperature rise speeds up the precess. Also, a continued rise in temperature will cause such factors as contact failure, insulation failure and mechanical connection failure to change their previous weights on the life of connectors. If dominated by any of these factors which leads to adverse breaking down of the dynamic equilibrium, the connector's failure rate will increase drastically.

The following tests have been conducted to verify this inference and understand temperature distribution in relation to the life of connectors:

(1) test of heat-resisting limit of insulation materials;
(2) accelerated life test at high temperature for low frequency electronic connectors;
(3) accelerated life test at high temperature for contact units.

3. Test and Analysis

3.1 Basic Tests

We have carried out step stress accelerated life tests in order to understand the heat-resisting limits for several insulation materials in common use for low frequency connectors. Table 1 shows the submitted samples for the tests. The consistant stress accelerated life tests were made for the insulation materials after that. Table 2 shows the arrangement for the test.

Table 1

Insulation materials	Code name	Colour	Quantity
Bakelite H1606-2	Shanghai Q/HGB-329-79	brown	6
Polysulfone S-140	Shanghai Q/HGB-257-79	blue	4
Polysulfone S-140	Shanghai Q/HGB-257-79	green	2
Glass fiber plastic FQBD-12	XA 50005-75	Yellow brown	3
Polysulfone S-140	Shanghai Q/HGB-257-79	gray brown	6
Reinforced nylon	1010	black	3
Reinforced nylon	1010	green	2
Reinforced nylon	1010	blue	2
Reinforced nylon	1010	white	1
Bolyacetylene Polycarbonate	1010	red	1

See Fig 1 for the methods and conditions for applying stress during the tests.

Table 2

Step	Stress T^oC	Parameters measured	Criteria for failure
1	175	insulation resistance	$< 500\ M\Omega$
2	185		
3	195	contact resistance	$> 0.005\Omega$
4	205		

See Table 3 for accelerated life test at high temperature for resilient materials.

Table 3

Step	Stress T^oC	Parameters measured	Criteria for failur
1	150	Separating force	< 0.09kg
2	170		
3	190		
4	216		

3.2 The Analysis of the Test Results

Fig 1 shows the life distribution curves for the connect units and insulation materials of low frequency connectors at temperature of 150°C, 170°C, 200°C and 210°C. The dots depict the life distribution for resilient materials while the data points x and straight lines for insulation materials. Life distribution and the distribution parameters for resilient and insulation materials and complete products can be thus obtained respectively.

Take resilient materials as an example. The Weibull distribution parameter and characteristic life can be found respectively with the shape parameter $\bar{m}$ being 5.1.

The dependence of the characteristic life on temperature, as at 150°C, 170°C and 210°C, is shown below:

$\eta_{150} = 1.76$ (hour)
$\eta_{170} = 57$ (hour)
$\eta_{210} = 13$ (hour)

Which brings forth the accelerated life curve for resilient materials (see Fig 2) and the acceleration equation:

$$\eta = 10^{a+b}/T \quad (4)$$

The parameters a and b in the above equation at different temperatures can also be found.

$$\hat{b} = (\lg\eta' - \lg\eta'')\frac{T'.T''}{T''+T'} \quad (5)$$

$$\hat{a} = \lg\eta - \frac{\hat{b}}{T} \quad (6)$$

The parameters a and b at the temperatures below 100°C can be derived respectively from equations (5) and (6) as:

$$\hat{b} = 3366 \qquad \hat{a} = -5.8$$

Using equation (4) can result in the characteristic life at any temperature. At 25°C or 40°C, for example, the characteristic life is

$$\eta_{25} = 312825 \text{(hour)}$$
$$\eta_{40} = 89984 \text{ (hour)}$$

In the same way, the reliability life t_R and the mean failure rate $\lambda(t)$ at any temperatures can be obtained as follows:

$$t_R = \eta\,(-2.30261gk)^{\frac{1}{\bar{m}}} \quad (7)$$

$$\lambda_{(t)} = t^{\bar{m}-1}/\eta^{\bar{m}} \quad (8)$$

Assume 25°C as a reference condition and we give the failure rate against time shown in Table 4 and Fig 3. Table 5 shows the basic failure rate as a dependent of temperature. The values λ_b are computed by using the acceleration factors in eqiuation (4) or Fig 2. (The suppose t in the fifth year is taken as the reference)

Table 4

Time_t (year)	Mean failure rate at 25°C, (t)
0.5	4.0×10^{-13}
1	7.0×10^{-12}
2	1.2×10^{-10}
3	6.3×10^{-10}
4	2.0×10^{-9}
5	5.1×10^{-9}
6	1.0×10^{-8}
7	2.0×10^{-8}
8	3.5×10^{-8}
9	5.7×10^{-8}
10	8.8×10^{-8}
15	4.6×10^{-7}
20	1.5×10^{-6}

Table 5

Temperature T°C	b (component/10^6 hour)
0	0.000016
10	0.000032
20	0.000066
30	0.00012
40	0.00021
50	0.00038
60	0.00069
70	0.0060
80	0.0015
90	0.0023
100	0.0035
110	0.0049
120	0.0068

The life distribution of insulation materials with respect to temperature can be determined in a similar way. It can be seen from Fig.1 that the life distribution parameter for insulation materials m is 3.3.

The physical meaning for the shape parameter m in Weibull distribution is to describe the extent to which the risk of terminating life for undamaged products in the use of them increases or decreases with time. The larger the m, the greater the risk.

Also from Fig.1 the characteristic life of insulation materials at 205°C can be determined as 27 hours and 105 hours respectingly, with life acceleration factor being τ 205~195=3.8 (times). Extrapolating to 150°C we have

$$\eta_{150} = 105 \times 3.8^4 = 21893 \text{(hours)}$$

From equation (8) the failure rates of insulation materials and resilient materials at 150°C in 100 hours are calculated with results as follows:

$$\lambda(100)\text{insulation} = 1.6 \times 10^{-8}$$
$$\lambda(100)\text{resilient} = 5.6 \times 10^{-4}$$

As analyzed in the second part of this paper the life-time of products is characterized by an equilibrium between the loss and the activation. Once the equilibrium is broken the wear-out failure will dominate. It can be seen from the test analysis that the life of insulation materials is not shorter than that of resilient materials, or in other words, the failure rate determined by insulation materials is lower than those by resilient materials. The conclusion through the above analysis agrees with the actual operation of connectors. The significance of this finding lies in that it leads to where the emphasis should be put to limprove the reliability of connectors.

4. Conclusions

A. (see Table 4) are basically concordant with those of the long-term storate life tests (see Table 6), providing further evidence for the validity of the above-made conclusions.

Table 6 Results of storage life tests

Connector type	Number of samples	Time (year)	Number of failure	Mean life	Mean failure rate
A_2^T-14	8	21	2	8.76×10^6	1.14×10^{-6}
$P20_J^k7_A^Q$	8	21	2	8.8×10^6	1.13×10^{-6}
$P40_J^k8_A^k$	7	21	3	5.825×10^5	1.7×10^{-6}

B. It can be shown from the life distribution for resilient materials and insulation materials that the connectors out of a validly designed and well-controlled manufacturing process have a very long static load life.

C. An important conclusion can be arrived at with the analysis of Fig.1 that low frequency connectors have a "guaranteed life" for a certain period of time (i.e. r=0). In Fig.1 r=28 at 150°C. The guaranteed life at 25°C is extrapolated to be 1000 hours (934 hours, for theoretical value) and the guaranteed life at 20°C is extrapolated to be 15000 hours (1492 hours for theoretical value).

D. The analysis shows that (random causes ignored) the basic failure rate of low frequency connectors will remain very low (less than 10^{-9} /hour) for a long period, four or five years after they are manufactured.

REFERENCES

1. "On Life Test Methods and Accelerated Life Test Methods - A Standard", Su Deqing, Standardization Report, CAS, No.3, 1980
2. Reliability Tests and Their Statistical Analysis, Dai Shusen, Su Deqing, Bai Hexiang, etc, Defence Industry Press, 1984, 1st edition, 2nd Vol. P487 501

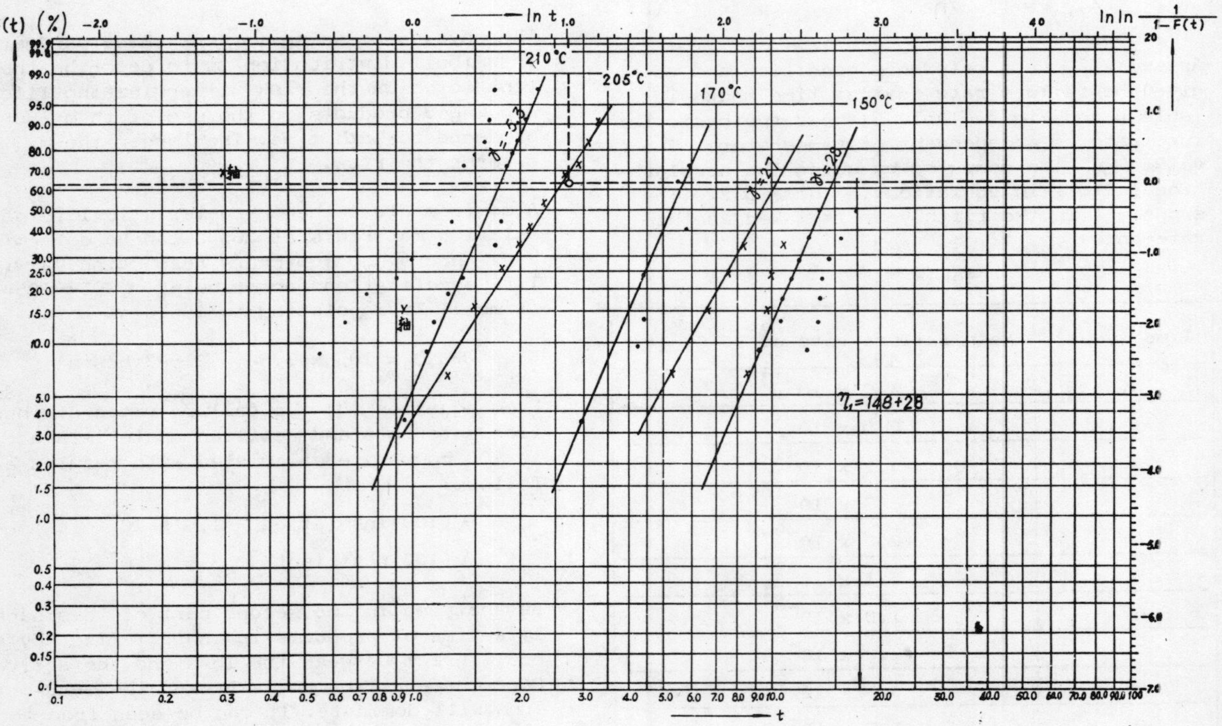

· insulation material
x resilient material

$F(t) = 1 - e^{-\frac{t^m}{t_0}}$ $\mu = t_0^{\frac{1}{m}} \Gamma(1+\frac{1}{m})$ $\eta = t_0^{\frac{1}{m}}$
$\sigma = t_0^{\frac{1}{m}} \{\Gamma(1+\frac{2}{m}) - \Gamma^2(1+\frac{1}{m})\}^{\frac{1}{2}}$

Weibul probability paper

Fig. 1

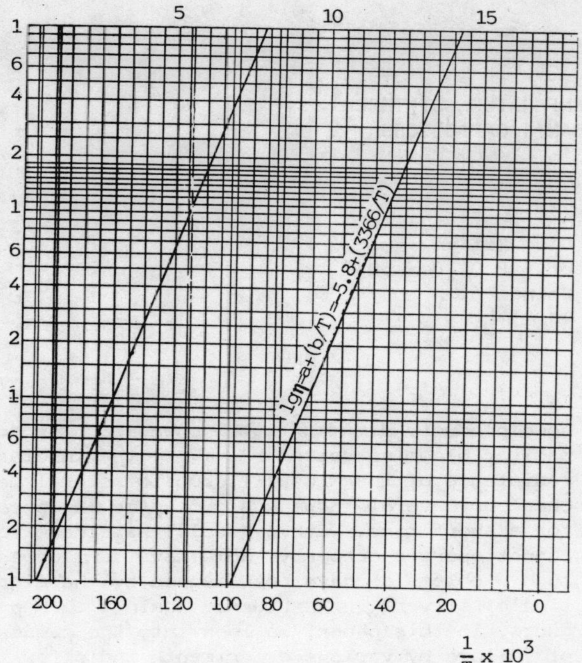
5
10
15
1
6
4
2
1
6
4
2
1
6
4
2
1
lgη=a+(b/T)=-5.8+(3366/T)
200
160
120
100
80
60
40
20
0
1/T x 10^3

Fig. 2

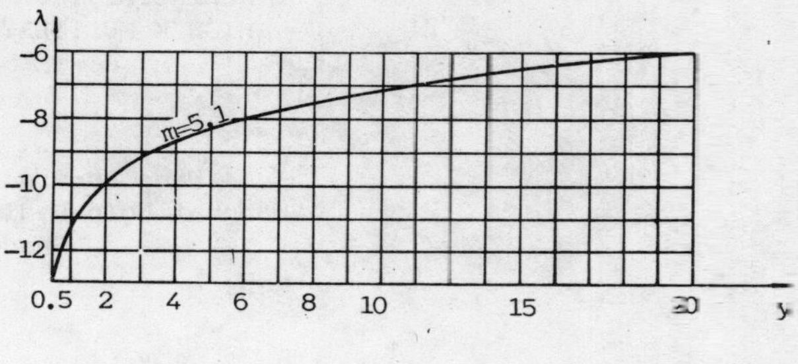
λ
-6
-8
-10
-12
m=5.1
0.5
2
4
6
8
10
15
20
y

Fig. 3

ON RELIABILITY OF LOW-TEMPERATURE SINTERED HIGH K MULTILAYER CERAMIC CAPACITORS

Wang Yuexiang and Ding Meifen
Shanghai No.6 Radio Factory, P.O. Box 5037
Shanghai, China

ABSTRACT

This paper describes the new achievements in improving the reliabity of low-temperature sintered high K multilayer ceramic capacitors(MLC) by researchers in China. The emphasis is placed on the mechanism analysis of capacitors'failure mode, i.e. "capacitor becomes resistor", and the suggestions for improvement are provided. The experimental results on these improved capacitors prove that their reliability has been improved significantly. This paper also describes in brief the accelerated life tests on MLCs, and obtained the acceleration factors for voltage and temperature, providing the theoretical basis for selecting various screening procedures to further improve the reliability of MLCs.

INTRODUCTION

With the progress in MLC technology, the sintering temperature of the dielectric materials in MLC is reduced increasingly. The high-temperature sintered $BaTiO_3$ dielectric materials fired at 1300°C and widely used in 1970s, are replaced by a new type of low-temperature sintered $BaTiO_3$, which is fired at 1100°C in 1980s. With the sintering temperature of dielectric materials decreasing, the electrode materials are changed from Pt--Au or Pd to 70%Ag-30%Pd, with a result that the costs of MLCs are reduced noticeably. In China, the development on low-temperature sintered high L multilayer ceramic capacitors began in the early 1970s, the dielectric materials used are composed of $PbMg_{\frac{1}{3}}Nb_{\frac{2}{3}}O_3$-$PbTiO_3$- ($Bi_2O_3$ or $PbCd_{\frac{1}{2}}W_{\frac{1}{2}}O_3$) - (glass powder or eutectic). The electric properties of such dielectric materials are by no means inferior in comparison with the traditional $BaTiO_3$ system materials but its dielectric constant is higher than that of $BaTiO_3$. The MLCs based on the above porcelain uses a 100% silver internal electrode fired at 920°C. Because in such capacitors no precious metals are used and they are sintered at a low-temperature, the energy con-sumption and the costs of the MLCs can be reduced greatly. Such capacitors have had a favourable market response due to their small size and low cost. But after they are used by customers for a time, some products demonstrate the phenomenon of "capacitor becoming resistor", i.e., the insulation resistance of the MLC will drop to several megaohms or several thousands ohm after it operates for a short time, with a result that the circuit cannot operate properly. Some have considered such problems to have been caused by the migration of silver ions and turned against such products, In this paper, we look into the causes of failure by various experiments and offer suggestions for improvement.

EXPERIMENTS AND RESULTS

The earlier type of dielectric materials used in MLCs were composed of $PbMg_{\frac{1}{3}}Pb_{\frac{2}{3}}O_3$-$PbTiO_3$-low-melting oxide. Many of the samples prepared by the above dielectric materials and packed with resin failed in operation due to the phenomenon of capacitor becoming resistor after the dry heat loading test and the humid heat loading test. The failure rate is especially high in the humid heat loading condition. Taking the sample of 0.47μF for example, if the 48V DC voltage is continously applied to the sample capacitor at the temperature of+85°C, its accumulated failure rate is 10-20% after 1000 hours elapse, and if the 40V DC voltage is continously applied to the same sample capacitor under conditions of-40°C and 95-98% RH, the accumulated failure rate will be as high as 40-50% after 1000 hours elapse.

We have made a series of mechanism analysis of the fact of "capacitor becoming resistor" which occured in early MLCs production and taken many measures for improvement. (For details, see "discussions and suggestions" which follows) The modified dielectric materials used in MLCs were composed of $xPbMg_{\frac{1}{3}}Nb_{\frac{2}{3}}O_3$-$yPbTiO_3$-$zPbcd_{\frac{1}{2}}W_{\frac{1}{2}}O_3$ ternary system[1], where x=0.6-0.9, y=0.05-0.5 and z=0.04-0.15. The thickness of dielectric sheet was 60-80μm and the sample was prepared with a typical process. The samples from 0.47μF capacitors was subjected to a life test under the same conditions as previously described. The experimental results showed a greater improvement in reliability of the capacitors. No failure occured after a 5000-hour humid heat loading testing and the measured electrical properties better. Under the humid heat loading condition, no sample capacitor packaged with the aluminum case, which had the same capacitance, failed after the 1000-hour testing. Their insulation resistance was not less than 4000MΩ.

We also put the imporved capacitor to an accelerated life test and obtained the acceleration factor values for voltage and temperature. All these factor values provide the theoretical basis for selecting various screening procedures to further improve the reliability of products. The experiments were performed for (2--6)x10^3 hours with three kinds of sample capacitors i.e. 0.015μF, 1μF and 4μF capacitors, under the stress conditions of V=63, 85, 110 and 150V, and t=125,150,165 and 185 C [2], [3].

It is well known that the life acceleration factor under the stresses of voltage and temperature can be expressed as:

$$A_L = \left(\frac{V_2}{V_1}\right)^C \cdot 2^{\frac{t_2 - t_1}{\theta t}}$$

where C and θt are acceleration factors for voltage and could not have temperature respectively. From this experiment we obtained the results as follows: C=1.71--2.04 and θt=7.03--8.89. The failure rate of the improved low-temperature sintered high K multilayer ceramic capacitors when subjected to a 25VDC external voltage (one half of the rating voltage) and examined at a temperature of 40°C and normal humidity (about 65%), with the 60 percent confidence intervals, can be reckoned based on the calculated values of C and θt, It should be 3 to 17 Fit.

The failure mode of the capacitor in the accelerated life test is that in case of voltage accelerating, most of the capacitors showed a decrease in insulation resistance and a few of them broke down, while in case of temperature accelerating, all of the capacitors became resistors due to reduction of IR. It can also be seen from the experimental results that the failure rate of the large value capacitors is evidently higher than that of the small value capacitors.

DISCUSSIONS AND SUGGESTIONS

The failure modes of the earlier produced sample capacitors after humid heat loading testing and dry heat loading test are similar, i.e. the capacitors become resistors due to the reduction of insulation resistance, but there exist some differences: in the former case, the IR of failure samples can be restored to a qualified value in normal temperature, and in the latter case, it cannot. The analysis for failure mechanisms in both cases mentioned above are as follows:

1. Based on the fact that the insulation resistance of the samples which failed in humid heat loading conditions can be restored to a normal value, we consider that the main causes for the failure of the sample MLCs are the less ideal packaging materials and poorly sintered compact of the capacitor porcelain, rather than the migration of silver ions. In other words, we consider the migration of silver ions too weak to cause the sample MLC to fail.
It is well known that the volume resistivity of water(H_2O)ρ_v =10^4-10^7Ω-cm, the radical of OH can migrate easily even under the effects of very small thermal energy and electric energy, and under the condition of humid heat loading, it will infiltrate into all gas cavities, laminations and cracks inside the sample capacitors with a result that the sample capacitors become resistors due to IR reduction.
2. Because the sample capacitors which failed under the dry heat loading condition and during the operation of the equipment, were operated under a drier condition and their failure occurred in early stage, the migration of silver ions could not have occurred due to humid in such a short time. The measurements on the failure sample MLCs have shown that the main causes of failure of the sample are the reduction of whole volume resistivity, i.e., the dielectric aging of the sample MLCs. If the failure sample capacitors are resintered, their resistance can be restored to a normal value. This experiment proves that the failure of the sample MLCs is caused by the dielectric aging instead of the migration of the silver ions. We have found that among a great many factors causing early dielectric aging, the following are more important:
(1) Different cooling temperatures have different effects on the dielectric aging of the capacitors[4]. For example, under the conditions of+85°C and 100VDC, 80% sample capacitors, which were cooled with the oven (slow cooling), became resistors and failed to pass the 1000-hour testing. While, under the same conditions as above, 100% capacitors, which were subject to the sharp quenching at 900°C, have passed the 1000-hour testing. The observations on sample capacitors mentioned above by electromicroscope showed that the capacitors sharp quenched at 900°C had a small grain size due to their quick cooling speed, and was prevented from forming coarse grain and devitrification while the slow cooling sample capacitors glowed a greater crystal to which attached a certain amount of smaller crystal (The addition of low-melting oxides will lead to a greater number of small crystals). The results of X-ray diffraction analysis showed that there existed no changes in phase characteristics of the main crystals of these two sample capacitors mentioned above. We may consider that the small crystals result from the crystalline segregation of the glass phases during slow cooling and the degradation of the insulation resistance is caused by the crystalline segregation.
Although the high reliable capacitors with high

IR can be manufactured by use of the sharp quenching process, some of the capacitors may become unstable and the glass phases of porcelain body will form an overcooling liquid in metastable state, if the sharp quenching process is controlled improperly. During the operation, it will be fired to transfer to the stable state under the effects of the thermal and electric energy, i.e., some of the crystallines will be separated out, and as a result, the insulation resistance will be reduced just as the results obtained in slow cooling.
(2) Different grades of niobium oxides (Nb_2O_5) have different effects on the dielectric aging of the capacitors. For example, after a 250-hour testing under conditions of 85°C and 100 VDC, 80% porcelain samples featuring high purity Nb_2O_5 failed and 40% porcelain samples containing 401 Nb_2O_5 failed to pass the test. But under the same testing conditions, a batch of porcelain samples containing 402 Nb_2O_5 have passed to the 1000-hour test without failure. It is because in case of using the first two Nb_2O_5 materials above mentioned, the underfiring of the porcelain will certainly lead to the increase of the amount of gas cavity, and as a result, the life of the MLCs will be shortened. By a further analysis, we know that the lower content of sinter-promoting impurity in high-purity Nb_2O_5 materials and the undue high content of sinter-impeding impurity in 401Nb_2O_5 materials are the factors which have an effect on the sintering temperature of the porcelains.

Based on the above analysis for failure products, it can be known that the poor moisture-resistance of the products, underfiring porcelain and the improper sharp quenching processes are what cause the capacitors to become resistors. Therefore, we propose the following measures to improve the products reliability:

1. Increase the moisture resistance of packaging materials and improve the packaging construction, As mentioned above, when the improved products use a non-hermetic (or semi-seal) aluminium case construction, it will have a good moisture-proof capability. It has been proved by the experiments that under the conditions of compact sintered porcelain body and better monolithic construction, the MLCs can also offer good moisture-resistance even when packaged by resin with good performance so as to give full play to the small-size features of MLCs.
2. Reduce the sintering temperature of the MLCs, raise the compactness and improve the moisture-proof capability by adding low-melting composite perovksite compound to porcelain formulation of the capacitors[5]. Of the earlier type of products, the low-melting oxides were used inproduction but with poor results. The improved formulation uses low-melting composite perovksite compound $PbCd_{\frac{1}{2}}W_{\frac{1}{2}}O_3$. The replacement of $(Mg_{\frac{1}{3}}Nb_{\frac{2}{3}})^{4+}$, $(Cd_{\frac{1}{2}}W_{\frac{1}{2}})^{4+}$ and Ti^{4+} is of the compensated ionic replacement. Because of its low melting and strong ionic activity, the presence of defect structures in the composite perovksite compound $PbCd_{\frac{1}{2}}W_{\frac{1}{2}}O_3$ after it forms the ternary system solid solution will lead to the increase of the vacancy concentrations with both positive and negative ions, thus speeding up the diffusion of ions and supporting the sintering. The composite perovksite compound $PbCd_{\frac{1}{2}}W_{\frac{1}{2}}O_3$ plays the role of sinter-promoting in early and intermediate stages of sintering and also that of modification in the latter stage of sintering in the resorption to the crystal lattice as the final phase, and thus, the insulation resistance of the capacitors can be improved, because the crystalline segregation will not always occur; it is not quite the same as that in addition of low-melting oxides. By adding excessive MgO and appropriate oxides, such as Cr_2O_3 , the vacancy concentration of O^{2-} may be increased. As a result, the sintering can be improved.
3. Create a technical specification on Nb_2O_5 applicable for low-temperature sintered multilayer ceramic capacitors. The practical applications show that only when the F content is controlled within a range of 0.15%, is the Si content less than 0.05% and the P content not greater than 0.02%, and can the formulation of Nb_2O_5 meet the requirements for applications. Meanwhile, it is necessary to lay out a process for calcinating the niobium hydroxide into the niobium oxide. The slow cooling method should be adopted,and the sintering temperature should be controlled to about 800°C. It is necessary to prevent Nb_2O_5 from phase-transformation at an excessively high temperature. Otherwise, the rate of gasification will be certainly increased.
4. Develop a scientific sharp quenching process, and strictly control the process. Because the sharp quenching process is finished in a very short time and performed by experience, it will certainly lead to a non-uniformity in products' performance. Hence, we suggest that a sharp quenching process be directed by a scientific method. Besides the prerequisite, i.e. ensure the products against crack or scorch, such requirements as cooling speed, temperature of cooling agent, cooling method and the necessary cooling equipment for meeting these requirements should be specified strictly. At the same time, skilled operators are required. In this way, the sharp quenching process can be controlled strictly with a result that the finished MLCs have a very high insulation resistance. Based on the relevant experiments, the sharp quenching temperature for common products should not be less than 750°C, or the performance of the products will degrade substantially.
5. Adopt a proper screening technique. In view of many factors affecting the reliability of MLCs, it is impossible to ensure the reliability of the products in mass production. The defective products of early failure degrading from capacitors to resistors can be rejected by using a proper screening technique. The basic method is as follows: work out various programs based on the valtage acceleration

factor and temperature acceleration factor; then select the screening conditions with short-time and good performance (which will not change the failure mechanism) from these programs by repeated experiments. Because the value of voltage acceleration factor is small in general, screening can be done by increasing the temperature. In addition, we have made an experiment on screening by use of low-voltage positive spike pulse based on the theory that a high electric field caused by the spike pulse will accelerate the dielectric ageing[6], and the experimental results show that the unqualified products due to the effect of capacitor becoming resistor can be effectively rejected by the method without damaging the good products.

Here we may draw the following conclusions:

A. The causes for the capacitors to become resistors demonstrated in low-temperature sintered MLCs as early failures are the less compact construction of the products, the underfiring of porcelain and the improper sharp quenching process, rather than the migration of silver ions.

B. The reliability of improved MLCs can meet the requirements of consumer electronic products. For special military products, the reliability may be further improved by using a screening technique.

REFERENCES

[1] Wang Yuexiang and Ding Meifen, "Development on $PbMg_{\frac{1}{3}}Nb_{\frac{2}{3}}O_3$ - $PbTiO_3$ - $PbCd_{\frac{1}{2}}W_{\frac{1}{2}}O_3$ ternary system porcelain" Electronic Ceramic Technology, No.2, 1976

[2] Wang Yuexiang and Ding Guoping, "Summary on accelerated life test of low frequency monolithic ceramic capacitors" (Restricted information) 1973--1980

[3] No.895 factory, "Summary on accelerated life test of monolithic capacitors" Reliability of the domestic electronic products and enviroment tests, No.2, 1977

[4] Unit 803, "Research on reliability of low-temperature sintered monolithic ferroelectric condensors" (restricted information) July, 1973

[5] Beijing Engineering University and No.798 Factory, "Mechanism of low-temperature sintering" Electronic Ceramics, No.1 1979

[6] Cheng Naixing "Research on failure mechanism of the low-temperature sintered high K monolithic ceramic capacitors" (Restricted information) Oct. 1987.

COPPER COMPATIBLE BASE-METAL THICK-FILM RESISTOR MATERIALS

DR. C. Y. KUO
TECHNICAL DIRECTOR
CTS CORPORATION
ELKHART, INDIANA 46514 USA

ABSTRACT

The materials of the early thick-film resistors were based on the Pd-PdO-Ag systems which were sensitive to process conditions, with a limited resistance range and inferior characteristics and performance. However, the Pd-PdO-Ag systems survived from the early 1960's to the beginning of the 1970's for more than ten years and established the foundation for thick film technology. For the 1970's and the 1980's, the ruthenium-based thick-film resistor systems have successfully and exclusively been used for communication, consumer, automotive, and industrial circuits because of their high reliability and excellent performance. All the early thick-film technology is based on the air-fired precious metals, used as both conductors and resistors.

In recent years, the copper thick-film technology has been advanced at a rapid rate because of its improvement in circuit density. Other advantages include high conductivity for high speed circuits and microware applications, resolderability, resistance to migration and leaching. The cost effectiveness of copper is also one of the main advantages for the copper circuit users. However, the copper circuits have to be fired under a neutral atmosphere to prevent the oxidation of copper. Therefore, the furnace and processes have to be redesigned to meet the firing of copper circuits.

There are also significant differences in materials between the air firing and the nitrogen firing conditions, including screening agents, glass frits, and the conductives. Firstly, most organics used for screening agents will be carbonized if they are fired under a non-oxygen atmosphere. The carbon residue in the fired film will dramatically decrease the adhesion, solderability and conductivity of the copper conductors. The carbon residue will also greatly affect the sheet resistance, TCR and stability for the base-metal resistors. Therefore, the evaporable carbon compound of the depolymerable screening agent with thixotropic rheology has to be employed for the nitrogen fired paste. Unfortunately, it is difficult to find good screening agents for nitrogen firing. Secondly, the glass frit fired under a nitrogen atmosphere should not be easily reducible, therefore, the low free energy materials including lead oxide and bismuth oxide which are used for most ruthenium-based resistors can not be used. Other requirements for glass include low soft point, moisture resistance, and preferable an absence of migrated alkali materials. Thirdly, it is difficult to employ one conductive to cover the whole seven-decade sheet resistance range and to maintain the excellent properties. For convenience, usually two conductives are used. The low end members are materials with metallic properties including allcys, intermetallics or a combination with metals. The high end members include semiconductors. Unfortunately, a blend of the two end member pastes will usually yield an anomalous results.

This paper discusses all these material problems, and illustrates two conductives which cover from 10 to 10^7 ohms per square with a TCR of within $\pm$100 ppm/degrees celsius, and excellent characteristics and stability. The TCR of 10^2 to 10^5 ohms per square can be further adjust to $\pm$50 ppm/degrees celsius for specific applications. The high end members cover from 10^3 to 10^7 ohms/sq using a pre-reacted tin oxide and the low end members from a few to 10^3 ohms/sq. using two titanium silicides, $TiSi_2$ and Ti_5Si_3.

Tin oxide has been used for thin and thick film resistors for many years. However, tin oxide resistors fired at 900 degrees celsius nitrogen atmosphere with a TCR less than $\pm$100 ppm/degrees celsius and excellent characteristics have not been reported. In the present approach, the metallo-organic coated tin oxide and glass powder, and tantula glass are the main materials in the paste. The deposited metal oxides on tin oxide from the metallo-organics can be either used in situ or further reduced under a low percent of hydrogen prior to making paste. Resistors made with compositions of various coated metallo-organics on tin oxide an/or glass powders, and various ratios of tantula glass frits are illustrated.

Resistors made from $TiSi_2$ powder show much higher resistance than those of Ti_5Si_3. $TiSi_2$ resistors also show a positive TCR while Ti_5Si_3 resistors exhibit a negative TCR. Alumina can be used for increasing resistances and decreasing TCR's. By combining different ratios of glass powders, $TiSi_2$, Ti_5Si_3, Al_2O_3, and selecting different particle sizes and surface areas of powders, the properties of titanium silicide resistors can be tailored to desired requirements.

Compared to other commercial base-metal materials, there are many advantages of titanium silicides when used for thick film resistors. Titanium silicides cost about one third of Ta, TaN, LaB_6, and In_2O_3. Resistance to oxidation and reduction, highest intrinsic resistivity and easy of adjusting certain properties are further advantages. Comparisons of other properties have been tabulated, and the compositions are described for various sheet resistances and TCR's.

The present compositions are less sensitive to processing conditions with resistance and TCR changes less than $\pm$10 percent. The various processing conditions tested include: firing at various temperatures from 880 to 920 degrees celsius; different firing times from 15 minutes up to one hour; different belt speeds from 4 to 8 inches per minute; and up to four firings. The resistor paints can be either fired in a conventional heater or IR belt furnace. The current noises and the termination resistances are also low. Based on long term environmental tests, the end life of these resistors is within ±0.5 or ±1 percent depending on specifications. For compatible conductors, we have developed copper paints which yield outstanding solderability, and superior adhesion. The characteristics and performance of these base-metal resistor system are comparable to those of the Ru-based counterparts.

THIN FILM METALLISATION ON AlN SUBSTRATES

Michael Feil and Andreas Drost
Fraunhofer Institute for Solid State Technology
Paul Gerhardt Allee 42, Munich, West - Germany

ABSTRACT

AlN has a good chance of becoming an important substrate material for high-quality hybrid circuits i.e. for HF high power modules because of its good thermal conductivity and its coefficient of thermal expansion, matching that of Si. Thin film metallisations (NiCr/Ni/Au, NiCr/TiW/Au and TiW/Au) with good adhesion properties could be realised without difficulties by sputtering. Surfaces of AlN ceramics in as-fired quality are to rough for high line resolution. Therefore, polished substrates had been used. On these substrates conducting stripes with 10 μm grid could be achieved. Pores in the surface of the ceramic decreased the number of good samples. NiCr resistors on AlN substrates did not differ from resistors on Al_2O_3. Changes in resistance and TCR were slightly smaller at AlN after long times of annealing at 300°C. The metallisation system NiCr/TiW/Au appeared to be more stable than the NiCr/Ni/Au system after annealing at 400°C at air.

INTRODUCTION

Apart from Si and SiC, AlN has a good chance of becoming a future substrate material for high-quality hybrid circuits because of its high thermal conductivity (substrates are available with up to 200 W/mK) and because of its coefficient of thermal expansion, matching to that of Si (TC(AlN) = 2.65 ppm/K, TC(Si) = 2.3 ppm/K at room temperature). Metallisation of AlN ceramics can be done by either thin film or screen printing. Whereas the use of paste systems, especially developed for AlN, is recommended with screen printing /1-3/, thin film processing apparently causes no serious problems. Reports about different metallisation systems such as Ta_2N/NiCr/Pd/Au, Ta_2N/NiCr/Au, Ta_2N/Ti/Pd/Au and Cr/Cu/Au /1,4-6/ have been published. Films with good adhesion properties were realised by evaporation or sputtering. The line resolution was limited to relatively high values (approx. 70 - 100 μm) because the AlN-substrates used in these experiments had an as-fired surface quality, i.e. a mean roughness of approx. 0.5 μm. This paper deals with the thin film metallisation systems of NiCr/Ni/Au and NiCr/TiW/Au used for the integration of resistors, and TiW/Au used for the wiring of connectors on polished AlN-substrates.

MATERIALS

1. Metallisation systems

The NiCr/Ni/Au metallisation is a system with integrated resistors, which is relatively easy to handle and widely used in thin film technology /7/. In this system NiCr acts as resistance and adhesion layer. In order to achieve a low temperature coefficient of resistance (TCR) the two components of NiCr each have 50 weight-% /8/. On the one hand Ni has to prevent diffusion of Au into the resistance film, on the other Ni has good solderable quality. Au has the function of a conductor metallisation, being corrosion resistant and bondable.
TiW is mainly used as a diffusion barrier in Si-technology, being very stable with regard to temperature and having a good adhesion to Si /9/. Therefore it is mainly Ti providing the adhesion, whereas W mainly prevents Au diffusion. The sputter target used in our experiments was composed of 10 weigth-% Ti and 90 weigth-% W. As will be shown later, the NiCr/TiW/Au system is more stable at high temperature tests compared to that of the Ni interlayer. Because of the good adhesion of TiW to both Al_2O_3 and AlN the TiW/Au system was tested for its application as a conduction metallisation. The disadvantage of TiW is its poor solderability; relatively expensive solders like AuSn or InPb must be used

2. Substrates

To achieve a high line resolution (≤ 50 μm is desirable in thin film technology) and a high edge resolution, polished AlN substrates were used, manufactured by Heraeus and Tokoyama Soda and sized 2" * 2". The mean roughness value was r_a = 35 nm and 10 nm respectively. The slightly different roughness can be seen in Fig 1. Moreover, we see that a defect appears in all AlN ceramics produced so far: the appearance of pores during the production process. AlN by Heraeus is dark gray, whereas Shapal - AlN by Tokuyama Soda is white and translucent. The purity is more than 98 % and 99.6 % respectively, according to the manufacturers data.

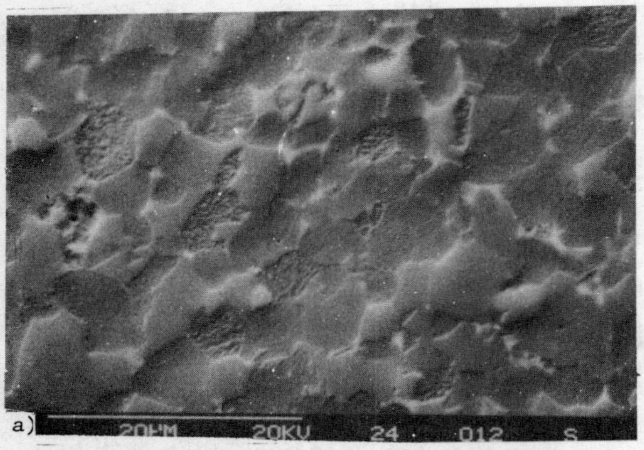

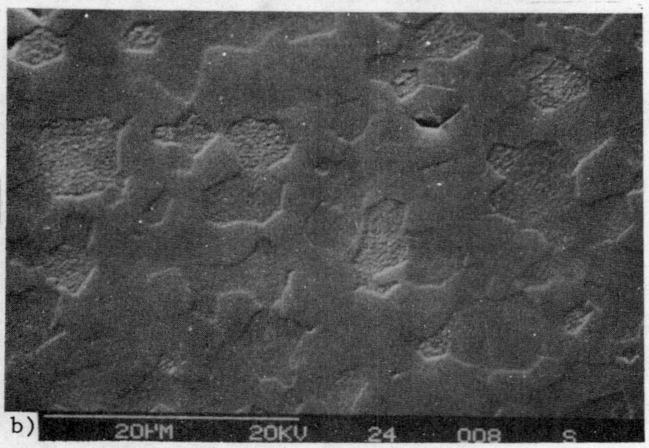

Fig. 1: SEM of polished AlN surfaces,
a) Heraeus, b) Shapal

For comparison purposes thin film Al_2O_3 substrates were included in the tests. Substrates manufactured by General Electric and Hoechst CeramTec had an as-fired quality with r_a - values of 140 nm and 75 nm respectively, substrates by Accumet had a roughness of r_a = 10 nm. The substrates used in our tests and their most important data are listed in Table 1.

PROCESSING METHOD

The different metallisations including a 100 - 200 nm Au film were realised by sputtering. The thickness of the NiCr film was determined by the desired value of 100 Ω/□ for the square resistance and was about 20 nm. The same sputter parameters were applied to all substrates, which caused different square resistances, depending on the roughness of the substrates. The increase in square resistance corresponded to the rise in roughness. The thickness of the Ni and TiW films was 30 - 40 nm.

Afterwards the structure of the conducting stripes was transferred on the substrates completely metallised by photolithography, and the Au was thickened in the resist canal by electroplating. After stripping the photoresist, the thin Au and the Ni or TiW film were removed by selective etching. The resistors were structured in a second lithographic process, and NiCr was removed by etching. The second photolithographic process was dropped for the TiW/Au metallisation.

RESULTS AND DISCUSSION

1. Fine line structures

In order to assess the maximum resolution, a high resolution stripe structure was used, consisting of 30 line groups per 15 line pairs each. The width of the grid was 1 mm to 1.6 μm. Fig. 2 shows that a grid of 10 μm could be achieved with a TiW/Au metallisation on polished Shapal AlN. Although this resolution is very good it could be improved

	AlN		Al_2O_3		
manufacturer	Heraeus	Tokuyama Soda	General Electric	Hoechst CeramTec	Accumet
name		Shapal	Alsimag 838	Rubalit 710	
short name	AlN(H)	AlN(S)	Al_2O_3(G)	Al_2O_3(R)	Al_2O_3(A)
purity	≥98%	99.6%	99.5%	99.5%	99.6%
surface	polished	polished	as fired	as fired	polished
r_a[nm]	35	10	140	75	10

Table 1: Substrates used in the tests

further by optimising the etching process. However, there is no point in this given the present state of art of ceramic because the surface pores mentioned above cause a narrowing or interruption of the metallisation (see Fig. 3).

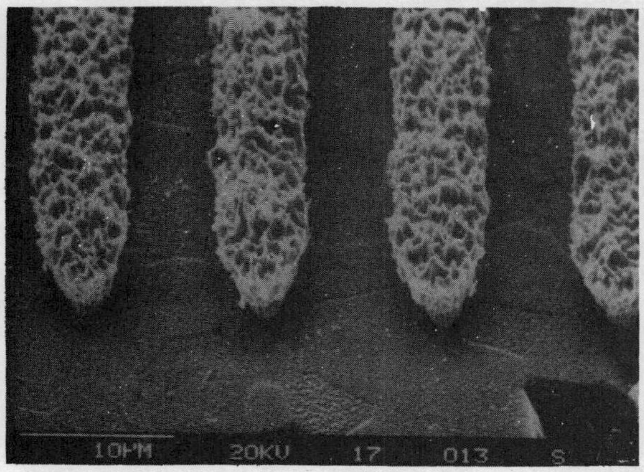

Fig. 2: SEM of fine line structures, grid 10 μm

Fig. 3: 50 μm wide NiCr resistor, narrowed by a pore

2. Properties of resistors (under normal test conditions)

For assessing resistors we used a test layout consisting of 27 resistors with L/W ratios from 1 to 500. The width of the resistors varied from 20 to 500 μm. As the desired square resistance was 100 $\Omega/\square$, this resulted in resistances from 100 Ω to 50 kΩ.
After the structuring process, the values for square resistances were 90 $\Omega/\square$ on substrates with very smooth surfaces (AlN(S), Al_2O_3(A)), 93 $\Omega/\square$ on AlN(H), and 105 $\Omega/\square$ and 111 $\Omega/\square$ on substrates with as-fired quality (Al_2O_3(R) and Al_2O_3(G)). The material of the substrate (AlN or Al_2O_3) did not influence the square resistance; the quality of the surface, however, turned out to be of importance. The different second metallisation Ni or TiW had no effect on the square resistance. The value of the square resistance differs by approx. ± 2 % within a batch and by approx. ± 5 % from batch to batch. The resistors with a width of 20 μm were often interrupted on AlN because of the pores.
The completely processed substrates were annealed at 300°C in an open oxidation furnace for stabilisation purposes. The relative

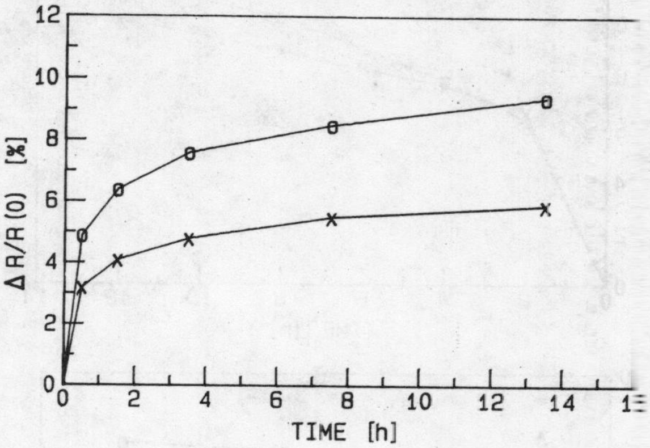

Fig. 4: Relative changes in resistance when annealing the NiCr/Ni/Au system at 300°C, x = AlN(H), o = Al_2O_3(G)

changes of resistance as a function of time are shown in Fig. 4 for the NiCr/Ni/Au system and in Fig. 5a for the NiCr/TiW/Au system. The curves are similiar, with the changes on AlN being slightly smaller than the changes on Al_2O_3. The greatest changes occur within the first two hours of annealing and so, therefore, an annealing stage of 2h at 300°C was taken as a standard for all subsequent tests.
The dependence of resistance on temperature was measured in a Heraeus-Vötsch temperature oven from -55°C to 125°C in steps of 20°C, and the temperature coefficient of resistance was calculated. Unless annealed the NiCr films show a hysteresis in their dependence on temperature which reflects instabilities of the film after the structuring process. After annealing, the hysteresis disappears. The TCR is independent of the substrate material, and it is usually sligthly negative, increasing to values < 30 ppm/K after annealing. Changes in the TCR as a function of annealing time are shown in Fig. 5b. Here we can also see that changes in the TCR are slight after the first 2h of annealing, and the TCR of NiCr resistors on AlN substrates is slightly smaller than that on Al_2O_3.
The noise index of the NiCr resistors was measured by means of a Quantech instrument which has a minimum noise index of about -40 dB. The noise index is a function of the roughness of the substrate surface. There is no dependence on the substrate material. After the structuring process the noise index

of AlN(S) and Al_2O_3(G) was -39 dB, of AlN(H) -37 dB and of Al_2O_3(G) -35 dB. It did not change after annealing on the polished substrates; the decrease on the unpolished substrates was 3 dB and the value became equal to that on polished substrates.

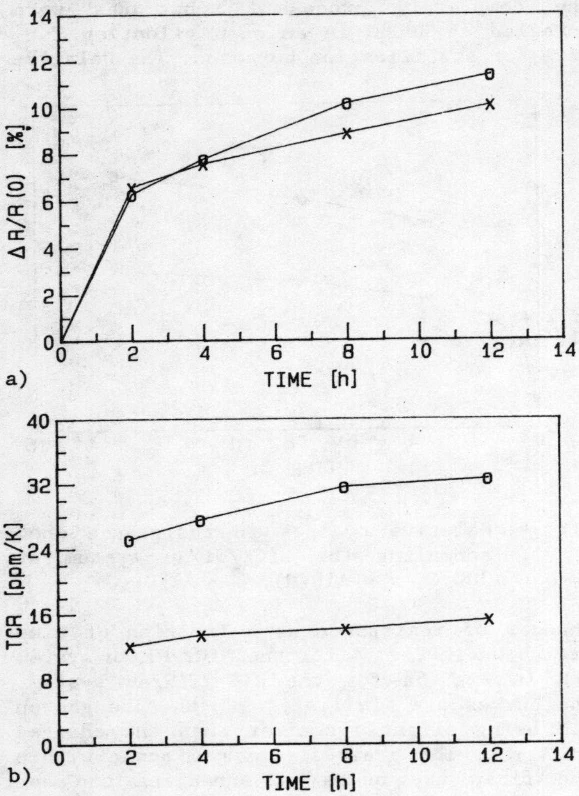

Fig. 5: a) Relative changes in resistance, b) TCR value when annealing the NiCr/TiW/Au system at 300°C, x = AlN(S), o = Al_2O_3(A)

3. Properties of resistors (under high temperature tests)

Because of their heat resistance, ceramics are a suitable substrate material for hybrid circuits in high temperature applications. Therefore we examined the properties of NiCr resistors during and after annealing at 400°C. For that purpose substrates metallised with NiCr/Ni/Au and NiCr/TiW/Au and then structured, and substrates completely metallised with NiCr were annealed at 400°C. This annealing process at 400°C was always preceded by a 2-hour annealing process at 300°C. The square resistance of the completely metallised substrates was determined by means of a VEECO four-point instrument (model AP150).

Fig. 6 shows the relative changes of the square resistance of the completely metallised substrates as a function of annealing time.

The correlation between the square resistance R□ and the time to failure during the annealing process at 400°C can be seen clearly. Fig. 7 shows AES depth profiles of three NiCr films on AlN(S) (marked with a,b,c in Fig. 6). Before annealing an even distribution of Ni and Cr over the cross section of the film can be observed. Every annealing process causes an oxidation of the NiCr film /10/. Because of the high affinity of Cr to oxygen, oxidation mainly results in the production of chromoxide. Oxidation also leads to a segregation of the film into two layers, one layer being at the surface of the film and rich in chromoxide, the other being at the substrate side and rich in Ni (see Fig. 7 b,c). After a long annealing period, oxygen can be found deeper in the film; oxidation of Ni also takes place.

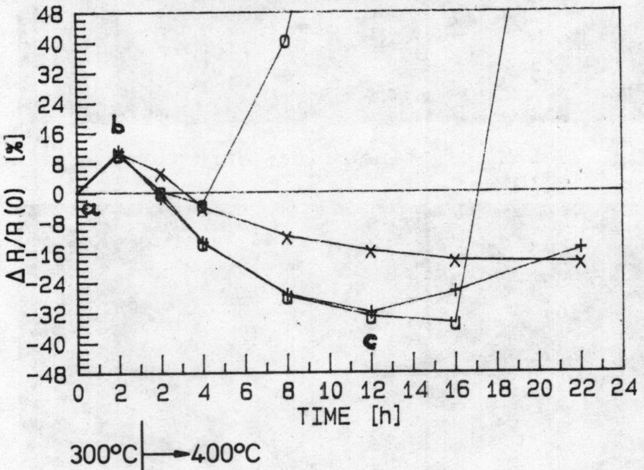

Fig. 6: Relative change in the square resistance of substrates completely metallised with NiCr
Square resistance R□ : o = Al_2O_3(R): 141 Ω/□, u = Al_2O_3(A): 113 Ω/□, + = AlN(S): 115 Ω/□, x = AlN(H): 121 Ω/□

The oxidation of the film is connected with an increase of resistance, leading to a failure of the resistance when the oxidation is complete. Failure occurs earlier at low values of the film thickness (r_a high, R□ high), because complete oxidation takes less time (see Fig. 6). The decrease in resistance in the substrates Al_2O_3(A), AlN(S) and AlN(H) can be explained by a crystallisation of Ni within the Ni-rich layer according to /11/. This crystallisation process occurs according to /11/ at temperatures > 325°C. The combination of oxidation and crystallisation leads to an increase or decrease in resistance according to the roughness of the substrate and the annealing temperature. Therefore the maxima and minima in Fig. 6 cannot be observed in the NiCr resistors in the 300°C test series (see Fig. 4 and Fig. 5a).

The assumption of crystallisation is supported by the results of the TCR measurement of the structured resistors (see Fig. 8b). The strong increase of TCR (starting at about 20 ppm/K) to values > 500 ppm/K points to an increasingly metallic character of the conducting Ni layer.

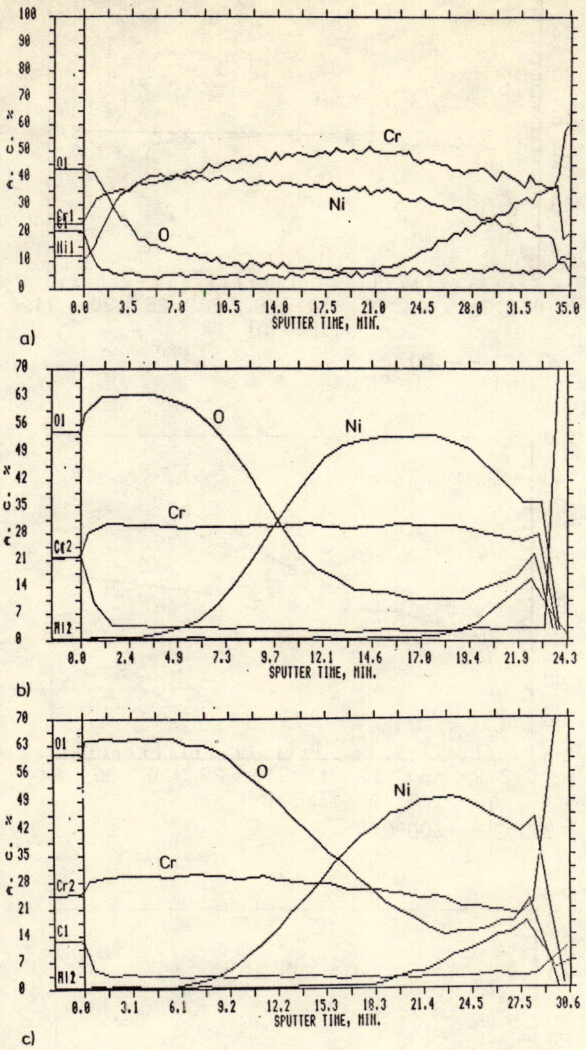

Fig 7: AES spectra of NiCr films on AlN(S)
a) before annealing, b) 2h, 300°C,
c) 12h, 400°C

In Figures 8 and 9 relative changes in resistance, TCR and noise index of structured NiCr resistors are shown as a function of annealing time for both the NiCr/Ni/Au (Fig. 8) and NiCr/TiW/Au (Fig. 9) systems. The structured resistors and the completely metallised substrates differ considerably in their changes in resistance. The clear decrease in resistance (Fig. 6) cannot be observed in the structured resistors.
First we consider the NiCr/Ni/Au system. After 4h of annealing at 400°C there is a small decrease in the resistance in the AlN(H) and AlN(S) substrates. This decrease is followed by a steep rise in the resistance after more than 8h of annealing, which is accompanied by resistor failures. The reason for this type of early failure lies in the complete oxidation of the NiCr resistors at one of its contact regions, the centre part of the resistor was still conducting in most cases. The region of failure is shown in Fig. 10.

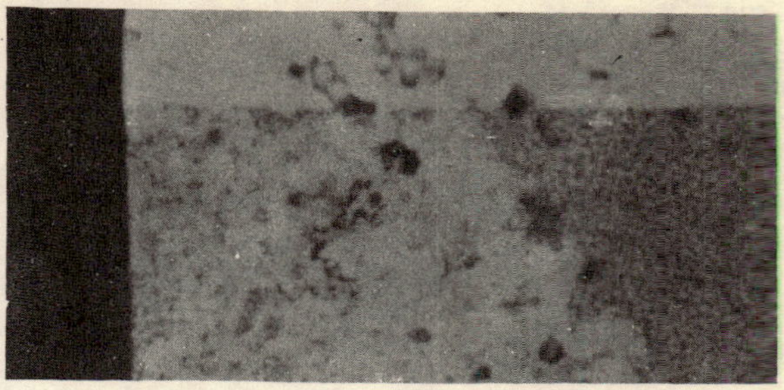

Fig. 10: Contact region of the NiCr resistor with the Au metallisation (made with transillumination), black region: Au metallisation, upper part: ceramic, lower part: NiCr resistor

Changes in the colour can be observed at the NiCr resistor, dark areas alternate with light areas. The light areas spread more often in the direction of the Au metallisation and they mark regions of complete oxidation. Further investigations will be carried out.
The increase of resistance clearly correlates with an increase of the noise index from an initial value of about -39 dB to > -25 dB (see Fig. 8c).

No failures could be observed in the NiCr/TiW/Au system up to annealing periods of 32h at 400°C. The resistance changes by about 15 % and is nearly constant after 4h of annealing. The TCR is sligthly higher (≃ 100 ppm/K) than that of the resistors having been annealed at 300°C. The noise index is as good as before (≃ 39 dB).
This behaviour which differs from that of the NiCr/Ni/Au system, is possibly due to a different stability of the diffusion layer Ni and TiW respectively during the annealing process at 400°C; an influence of the etching process on the quality of the NiCr film also seems possible. Differences in processing cannot be ruled out altogether.
It can be concluded that the stability of NiCr resistors is increased by a small roughness of the substrates used. An influence of the substrate material (Al_2O_3, AlN) could not be detected. Concerning metallisation, the NiCr/TiW/Au system seems to be more stable when annealed at 400°C. The reasons for this are still unknown.

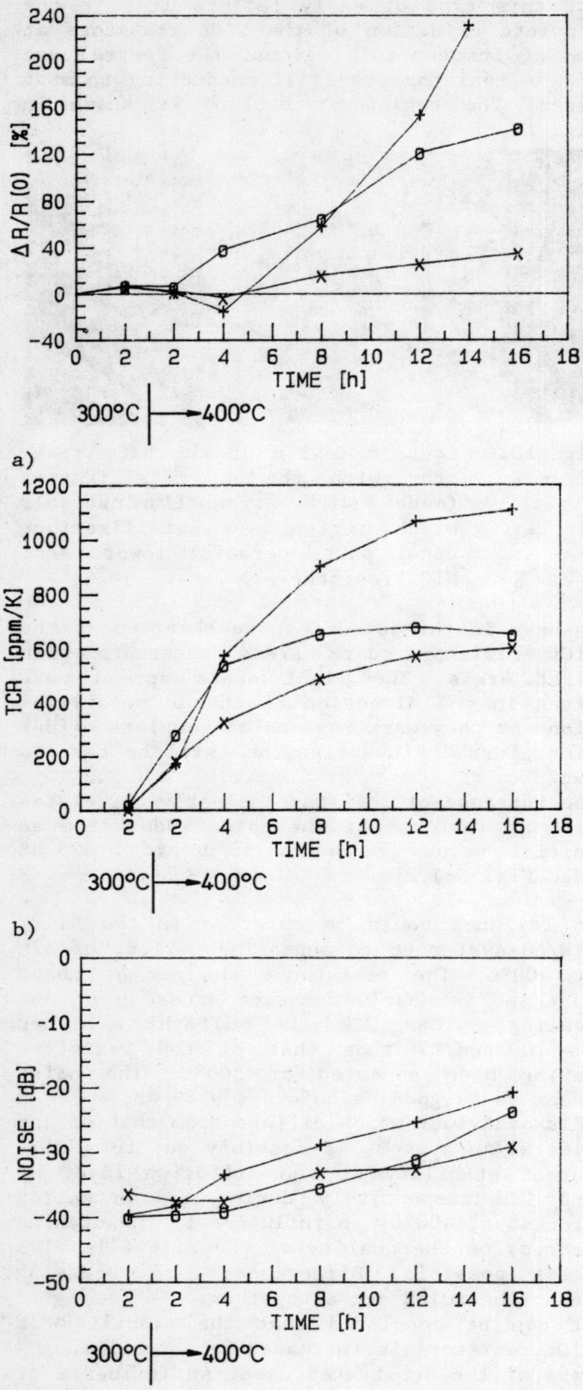

Fig. 8: a) relative changes in resistance, b) TCR, c) noise index for structured NiCr resisistors with the NiCr/Ni/Au system,
o = Al_2O_3(G), x = AlN(H), + = AlN(S)

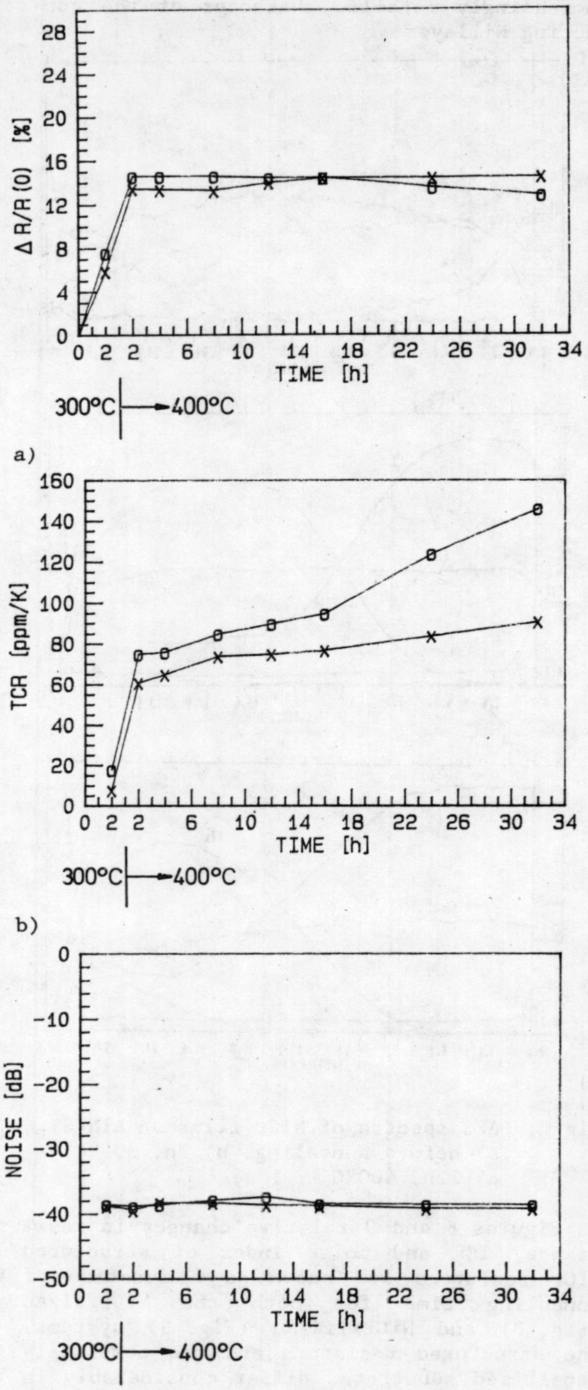

Fig. 9: a) relative changes in resistance, b) TCR, c) noise index for structured NiCr resisistors with the NiCr/TiW/Au system,
o = Al_2O_3(R), x = AlN(S)

4. Adhesive strength

For determing the adhesive strength, pieces of silver plated copper wire with an angle of 90° and a diameter of 0.8 mm were soldered by hand onto the pads with a size of 2 * 2 mm². The peel strength of wires was determined with a Dage Microtester model MCT 20. The testing wires were soldered on NiCr/Ni/Au with a Sn62Pb38Ag2 solder. The metallisations with TiW were first covered with an InPb solder, then the wires were fixed by means of the above solder.
The adhesion was good with all metallisations on all substrates. The values were generally higher than values known from screen printing (35 - 40 N). Different reasons for failure led to a high value for the mean variation. When the ceramic broke, the break value strongly depended on the way the ceramic was fixed. The results are summarised in Table 2. The Shapal AlN by Tokuyama Soda apparently had been polished; it had an initial thickness of 0.635 mm and a final thickness of only 0.5 mm. Therefore, the strength of this material was clearly reduced so that breaking of the ceramic occured in most cases (> 70 %) during the peel test. Concerning the TiW metallisation system the weakest point was the connection between wire and solder. Therefore the wire was often pulled out of the solder.

	adh. [N]	σ [N]	K	M	L
		TiW/Au			
AlN(S)	38	5.3	71	29	0
Al_2O_3(A)	60	16.1	5	26	69
Al_2O_3(R)	54	16.0	17	0	83
		NiCr/TiW/Au			
AlN(S)	41	9.2	58	0	42
Al_2O_3(G)	57	10.3	53	27	20
Al_2O_3(R)	46	6.7	40	50	10
		NiCr/Ni/Au			
AlN(H)	54	19.0	81	14	5
Al_2O_3(M)	74	19.0	15	35	50

Table 2: Overview over adhesion tests, Reasons for failure in %:
K = breaking of the ceramic,
M = detachment of the metallisation from the ceramic,
L = pulling of the wire out of the solder

5. Bondability

Bonding tests were done with ultrasonic wedge bonding, using an Al wire, and with thermosonic ball bonding using an Au wire. The thickness of the wires was always 25 μm. The length of the bond was about 1 mm. The adhesion of the bond connections was determined by peel test by means of the Dage Microtester mentioned above.
Bonding was good; no differences could be detected, neither in the different ceramics nor in the metallisation systems. This had to be expected because the upper layer was always a layer of electroplated Au with a thickness of 3 - 4 μm. The failure was caused in most cases by the wire tearing off above one of the two bonding sites.

CONCLUSION:

The tests showed that thin film technology can be done on AlN substrates and that there are no serious differences to thin film technology on Al_2O_3. A line resolution with 10 μm grid could be realised on polished surfaces. NiCr resistors on AlN appear to be slightly more stable than those on Al_2O_3. Concerning metallisation systems, the system using TiW as diffusion barrier seems to be more stable than the system with the Ni layer after annealing at 400°C at air. Adhesion of the metallisations is good on AlN substrates, so that ceramic broke at most of the tests. AlN as a substrate material is of interest in applications, where BeO is used up to now or/and where the coefficient of thermal expansion has to be matched to that of Si.

REFERENCES

/ 1/ E.S.Dettmer et. al.; "Hybrid Design and Processing Using Aluminum Nitride Substrates", Proc. of 1988 Symp. on Microelectronics, Seattle WA, Oct. 1988 545 - 553

/ 2/ K.K.Kachelrics, J.A.Olenick, S.G.Konsowski; "Evaluation of Thick Film Multilayers on AlN", Proc. of 1988 Symp. on Microelectronics, Seattle WA, Oct. 1988 170 - 173

/ 3/ M.Feil; "Experience with AlN Substrate" Hybrid Circuits, No 18, Jan. 1989 29 - 34

/ 4/ Y.Kurokawa; "AlN Substrates with High Thermal Conductivity", IEEE CHMT - 8, No. 2, 1985, 247 - 252

/ 5/ H.Arikawa et.al.; "Low Thermal Resistance Hybrid IC Package", Proc. of 1988 Symp. on Microelectronics, Seattle WA, Oct. 1988, 160 - 163

/ 6/ R.Chanchani; "Processability of Thin Film, Fine-Line Pattern on Aluminum Nitride Substrates", IEEE CHMT-11, No. 4, Dec. 1988, 427 - 432

/ 7/ A.P.Bhatt et al.; Proc. of Int. Symp. Microelectronics, Dallas Texas, Sept. 1984, 370

/ 8/ J.Rölcke, "Nichrome Thin Film Technology and its Application", Electrocomponent Science and Technology, Vol. 9, 1981, 51 - 57

/ 9/ C.Y.Ting, M.Wittner; "The Use of Titanium-Based Contact Barrier Layers in Silicon Technology", Thin Solid Films, 96, 1982, 327 - 345

/10/ J.Steffen, S.Hofmann; "Oxidation of NiCr and NiCrFe Alloys at Room Temperature", Surface and Interface Analysis, Vol.11, 1988, 617 - 626

/11/ W.Gawalek, "Resistance, Temperature Coefficient of Resistance and Long-Term Stability of Annealed Thin Ni-Cr-Si Films", Thin Solid Films 116, 1984, 205 - 210

ELECTRIC FIELD CONTROL OF INTERFACE PROPERTIES
OF CERAMIC SEMICONDUCTOR

Toshiyuki Shirakawa, Yoshinobu Nakamura, Masaru Miyayama
and Hiroaki Yanagida
Research Center for Advanced Science and Technology, University of Tokyo
4-6-1, Komaba, Meguro-ku, Tokyo, 153 Japan

ABSTRACT

The effect of the applied bias on the surface resistance of CuO/ZnO interface was investigated. The current-voltage characteristic of CuO/ZnO hetero-contact showed a rectifying character like a typical p-n junction diode. By applying reverse bias (CuO -, ZnO +) across the CuO/ZnO interface in the temperature range between 200°C and 400°C, the resistance of both CuO and ZnO surface at room temperature became larger than those without applying the reverse bias. Because the changes in resistance were always reversible, it was assumed to be due to the changes in the amount of adsorbed oxygen. Electric field control of surface electrical property was suggested to be possible.

INTRODUCTION

The electrical properties of the interface of semiconductor ceramic-ceramic contacts, such as p-n hetero-contact made by mechanical pressing of p-type and n-type ceramic semiconductors, are influenced by ambient combustible gases.[1,2] It is known that changes in electrical properties are due to the reaction of gases with adsorbed oxygen and to a subsequent decrease in the amount of adsorbed oxygen on the semiconductor surface. Recently, these changes influenced by ambient gases were applied to "gas sensors".[3,4] It was reported that a gas sensor of p-n hetero-contact made by CuO ceramics as a p-type semiconductor and ZnO ceramics as a n-type semiconductor was sensitive to CO gas selectively, and the sensitivity of this gas sensor was changed by the bias applied across the interface.[4]

In the present study, the effect of the applied bias on the surface resistance of CuO/ZnO interface was investigated and a possibility of the electric field control of the surface properties was discussed.

EXPERIMENTAL

Copper oxide(CuO) powder was obtained by calcining basic copper(II) carbonate ($CuCO_3.Cu(OH)_2.xH_2O$) powder (Kishida Kagaku Co.,Ltd.) at 600°C for 4h. The resulting powder was pressed into rectangular bars (15 X 1 X 5mm) under 157MPa and sintered at 820°C for 3h in air. Zinc oxide (ZnO) powder (99.99%pure, Kojundo Kagaku Kenkyu-jo Co.,Ltd.) was pressed into rectangular bars of the same size under 61.7MPa and sintered at 900°C for 3h in air. Surfaces of sintered bars were mechanically polished and cleaned in an acetone bath.

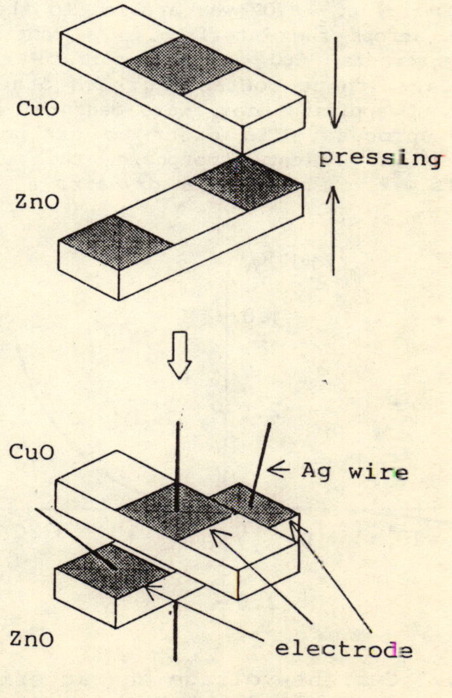

Fig.1 Sample for measurement.

As is shown in Fig.1, CuO sintered bar was mechanically contacted on ZnO sintered bar and silver wires were lead from the four electrodes, in order to measure surface electrical resistance of ZnO under applying voltage across the p-n contact interface. When surface resistance of CuO was measured, CuO sintered bar was replaced by ZnO sintered bar each other. As ohmic electrodes, usual Ag paste and ohmic Ag paste were applied to CuO and ZnO ceramics, respectively. Voltage-current characteristics of the contact were measured with a curve tracer (Kikusui Electronics Co., Ltd. Model 5802). Surface resistance and their temperature dependences of CuO and ZnO were measured with a d.c. digital electro-meter (ADVANTEST Co., Ltd. TR8652). Since the acceleration of adsorption and desorption rate of oxygen is expected above 200°C, the measurement of surface resistance was conducted in the following process.

(1) Room Temp. (R_1) → (2) T°C (R_2) —1hr.→ (3) T°C (R_3) —quenched→ (4) Room Temp. (R_4)

↑ (applying reverse bias or not) ↑

T=200, 250, 280, 300, 320, 350, 400°C

In the process, samples were heated from room temperature to T°C and kept at T°C for one hour and quenched to room temperature. When samples were kept at T°C and quenched to room temperature, the reverse bias (CuO -, ZnO +) of -20V was applied to the contact across the interface. As contrastive measurements, CuO and ZnO surface resistances, without contacting each other and without applying biaş, ware measured in the same process. In order to exclude the effect of ambient atmosphere, the measurements were carried out in dry air.

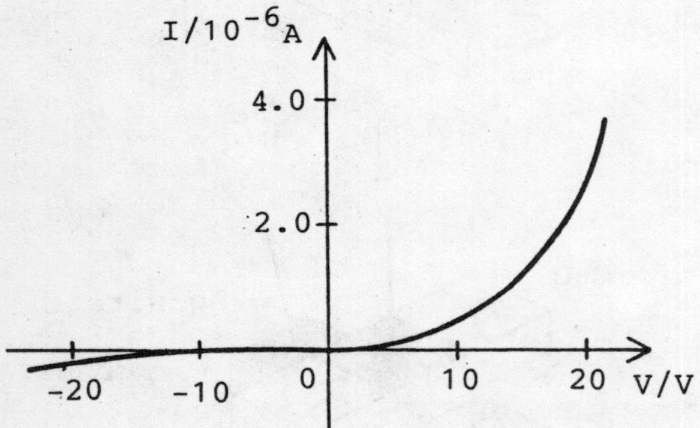

Fig.2 Current-voltage characteristic of CuO/ZnO hetero-contact.

RESULTS AND DISCUSSION

Fig.2 shows a V-I characteristic of CuO/ZnO ceramic hetero-contact at room temperature. A typical rectifying character like a p-n junction diode was observed, where the current flow was only under forward bias (CuO +,ZnO -). From the result of these V-I characteristic of CuO/ZnO, the reverse bias of -20V was chosen as a sufficient bias which caused the change in surface states.

Fig.3 shows temperature dependence of CuO and ZnO surface resistance without contacting each other. CuO surface resistance decreased as temperature increased. This temperature dependence of CuO surface resistance was a typical one for semiconductors. on the contrary, temperature dependence of ZnO surface resistance had a maximum at about 300°C. This temperature dependence of ZnO surface resistance is supposed to be due to a change in valence of adsorbed oxygen species on ZnO surface. That is to say, a transformation of the adsorbed oxygen from O^- to O^{2-} above 200°C makes the electron concentration of ZnO decrease,[5] resulting in a increase in surface resistance.

Table 1 shows a result of changes of CuO and ZnO surface resistances measured in the process described above, in which temperature(T) was 300°C. Numbers 1 and 2 are the results measured without contacting (case 1), numbers 3 and 4 are those measured without applying voltage (case 2), and numbers 5 and 6 are those measured with applying reverse bias of -20V (case 3). The values of the initial surface resistance (R_1) were somewhat different, depending on surface polish condition and contacting condition. The values of surface resistance at 300°C (R_2) of CuO were smaller than R_1, and those of ZnO were larger than R_1, as expected from Fig.2. The values of R_2 were

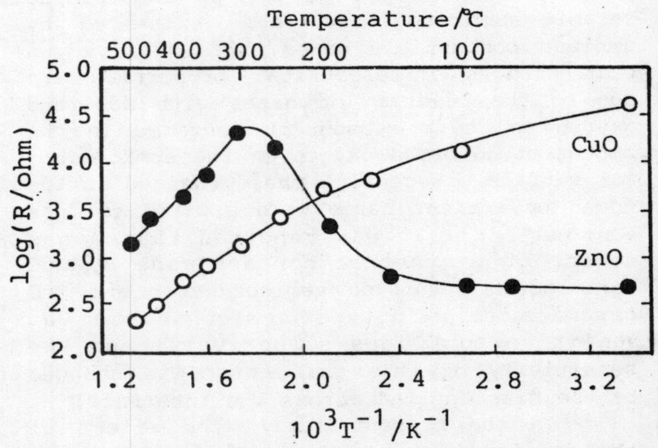

Fig.3 Temperature dependence of CuO and ZnO surface resistance.

Table.1 The change of CuO and ZnO surface resistances in which temperature(T) was 300℃.

No.			$R_1/k\Omega$	$R_2/k\Omega$	$R_3/k\Omega$	$R_4/k\Omega$	R_4/R_2
1.	Case1	CuO	355	2.83	2.62	472	167
2.		ZnO	0.58	33.2	32.3	0.53	1.60×10^{-3}
3.	Case2	CuO	24.9	1.22	1.22	26.2	21.5
4.		ZnO	0.53	38.7	37.9	0.47	1.21×10^{-3}
5.	Case3	CuO	32.1	1.19	1.15	40.6	34.1
6.		ZnO	0.58	51.8	45.3	0.56	1.08×10^{-3}

Case1:Without contacting
Case2:Without applying bias
Case3:Applying reverse bias of -20V

more uniform than those of R_1. In order to evaluate the effect of contact and applied bias on the changes of surface resistance, changes of surface resistance was denoted by the value of R_4/R_2, and the values of R_4/R_2 in three cases were compared with.

Fig.4 shows the comparison of R_4/R_2 in the three cases. It was found that both CuO and ZnO surface resistances in the cases 1 and 3 are larger than that in the case 2 in most cases. Both CuO and ZnO surface resistance after the heat and bias treatment (R_4) returned to the initial value (R_1) when the sample was kept at room temperature for about 12h. From this result, it is supposed that these changes of surface resistance were caused by the changes in adsorbed oxygen on CuO and ZnO surface.

An oxygen adsorbs chemically in the form of anions (like O^- and O^{2-}) on oxide semiconductor surface.[6] Therefore, when the amount of chemisorbed oxygen increase, concentration of electrons near the semiconductor surface decreases. As a result, the resistance of p-type semiconductor, whose major carrier is hole, decreases, but, on the contrary, the resistance of n-type semiconductor, whose major carrier is electron, increases. From this viewpoint, following phenomena are expected; the amount of chemisorbed oxygen on CuO surface is increased by contacting to ZnO and it is decreased by applying reverse bias of -20V. The amount of chemisorbed oxygen on ZnO surface, in contrast, decreased by contacting to CuO and it increased by applying reverse bias of -20V.

In order to be easy to understand a degree of the change of surface resistance by applying bias, the value of R_4/R_2 in the case 3 (applied bias) was normalized by dividing the value of R_4/R_2 in the case 2 (without bias). Fig.5 shows the value normalized in this way as a function of temperature(T). Increases in both CuO and ZnO

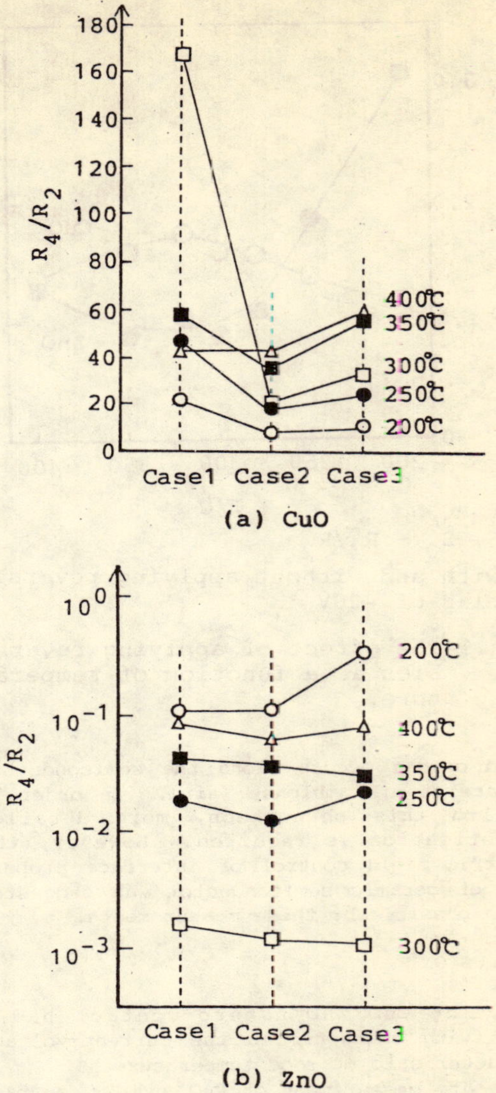

Case1:Without contacting
Case2:Without applying bias
Case3:Applying reverse bias of -20V

Fig.4 The effects of contact and applying reverse bias on the change in surface resistance.

surface resistance by applying reverse bias of -20V can be observed in Fig.5, although there are some differences in magnitude and temperature dependence.

From these results, it appears that chemisorbed oxygen of CuO/ZnO interface transferred from ZnO surface to CuO surface by contacting each other, but that it transferred from CuO surface to ZnO surface by applying reverse bias of -20V. It is not clear yet that this transference of chemis-

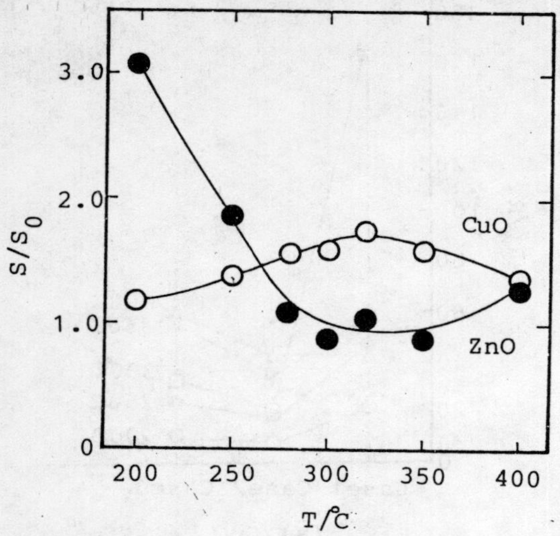

$S, S_0 = R_4/R_2$
with and without applying reverse bias of -20V

Fig.5 The effect of applying reverse bias as a function of temperature.

orbed oxygen occurs along the semiconductor surface or via ambience (air). In order to confirm this phenomenon, more detailed investigation is required. However, the electric field control of interface properties of ceramic semiconductor was suggested to be possible by these measurements.

CONCLUSION

The CuO/ZnO hetero-contact had a rectifying character in the current-voltage characteristic at room temperature.

The measurement of CuO and ZnO surface resistance, after heat treatment of 200°C to 400°C with and without reverse bias of -20V across the interface, showed following results. Both CuO and ZnO surface resistances were decreased by contacting each other but they were increases by applying reverse bias. These changes were always reversible. It was thought that these changes was due to the changes in the amount of chemisorbed oxygen of CuO/ZnO interface. From these results, a possibility of electric field control of interface properties of ceramic semiconductor was suggested.

REFERENCES

1. K.Kawakami and H.Yanagida, Yogyo-Kyokai-Shi, 87(1979), 112.
2. Y.Nakamura, M.Ikejiri, M.Miyayama, K.Koumoto, and H.Yanagida, Nihon-Kagaku-kai-Shi, (1985), 1154
3. N.Yamamoto, S.Tonomura, T.Matsuoka, and H.Tsubomura, Surface Science, 92(1980), 400
4. Y.Nakamura, A.Ando, T.Tsurutani, O.Okada, M.MIyayama, K.Koumoto, and H.Yanagida, Chemistry Letters, (1986), 413
5. M.Takata, D.Tsubone, and H.Yanagida, Journal of American Ceramic Society, 59(1976), 4
6. M.Takata and H.Yanagida, Yogyo-Kyokai-Shi, 87(1979),13

Unusual Property of a-Si:H in Vacuum and Humidity Atmosphere

Wu Dao-huai, Cheng Zhi-qiang
Shanghai Institute of Ceramics, Academia Sinica
Tan Song-sheng, Lin Cheng-lu, Huang Wei
Ion Bean Laboratory
Shanghai Institute of Metallurgy, Academia Sinica
Shanghai 200050, China

ABSTRACT

Unusual properties have been observed in the glow-discharge a-Si:H films with boron ion implantation and rapid thermal annealing. The experiments show that the film resistivity decreases when the sample is in a vacuum chamber, and it increases with the increasing of the moisture in the chamber. The Raman scattering spectroscopy shows that there are two broad peaks near 520 cm^{-1} and 480 cm^{-1}, and indicates that microcrystalline and amorphous state coexist in the films. The film sensitive mechanism and its potential application are discussed.

I. INTRODUCTION

Recent studies on μc-Si films have shown that μc-Si films are sensitive to temperature and pressure and their sensitivity can be enchanced by boron doped and rapid thermal annealing. Those properties have been used to make temperature sensor and pressure sensor[(1-2)]. In this study, we discovered that the glow-discharge a-Si:H films with boron ion implantation and rapid thermal annealing are sensitive to vacuum and humidity. To the auther's knowledge none of these phenomena has been reported.

2. Sample Preparation and Measurement

a-Si:H films were deposited by r.f. glow-discharge method. Quartz glass was used as substrate, and the deposited temperature was about 300°C. Then the a-Si:H films undergo boron ion implantation and rapid thermal annealing successively. Parallel electrods were prepared using painted graphite at film surface. The sample is placed in a chamber with vacuum pump, heater and water pot, then the film resistivity is measured in various temperature, pressure and humidity.

3. Experimental Results and Discussion

The film resistivity in atmosphere is about $9\times10^{7}\ \Omega\cdot cm$ measured at room temperature. The dependence of film resistivity on temperature are shown in Fig. 1. The result indicates that the film resistivity decreases with the rise of temperature, the same as usual a-Si:H film.

Novel properties have been observed in vacuum and humidity atmosphere compared with usual a-Si:H films. During the pressure in measurement chamber is pumped down, the film resistivity of the sample decreases and reaches a value of $2\times10^{6}\ \Omega\cdot cm$ decreasing by a factor of 45 at a pressure of 10^{-2} torr, as shown in Fig. 2. The resistivity restores to initial value when the measurement chamber returns to atmosphere. The film resistivity was also measured as a function of saturation vapor pressure of water. The experimental results show that the film resistivity increases with the rise of moisture in the measurement chamber. For example, it increases by a factor of 230 when the dew points

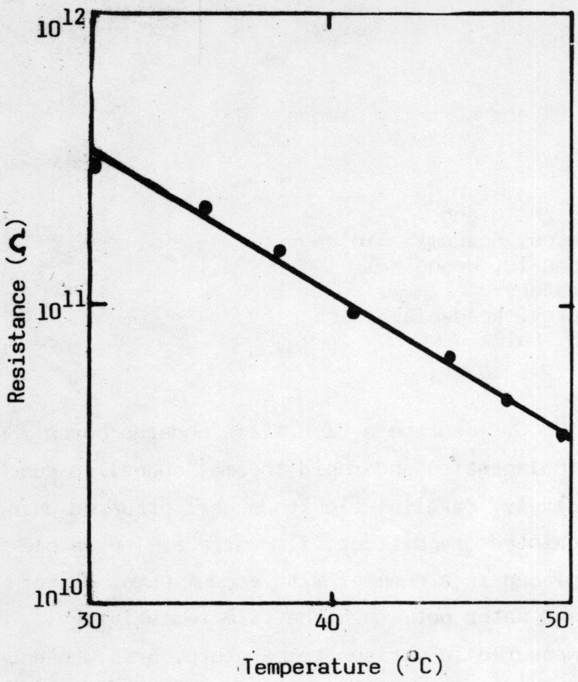

Fig. 1. The dependence of film resistivity on temperature

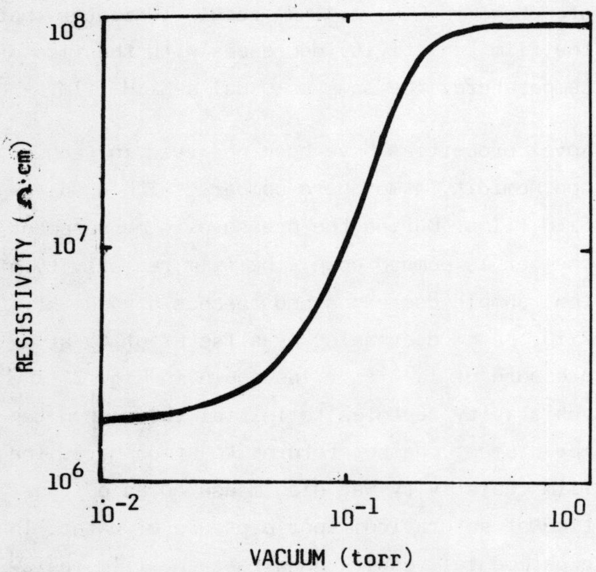

Fig. 2. The dependence of film resistivity on pressure

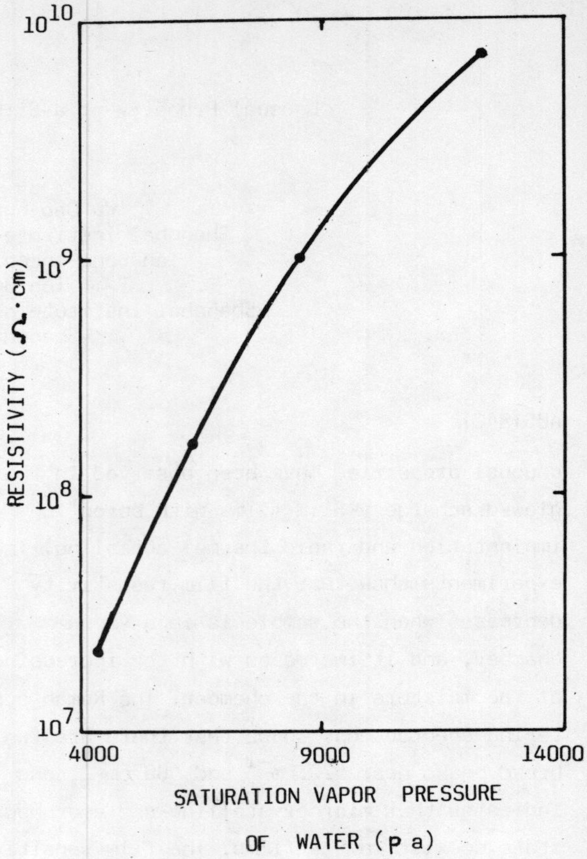

Fig. 3. The dependence of film resistivity on moisture

in chamber changing from 30°C to 50°C, as shown in Fig. 3. Then it restores to its initial value when the moisture recovere to normal. In all cases, the experiments are reversible and repeated within the measurement accuracy.

The Raman scattering spectroscopy has been determined to identify the film microstructure. Two broad peaks near 520cm^{-1} and 480cm^{-1} are observed as shown in Fig. 4, and indicate that microcrystalline and amorphous state coexist in this film.

Since the resistivity of a-Si:H don't change at the same way, the unusual film properties may be caused by the microcrystalline structure.

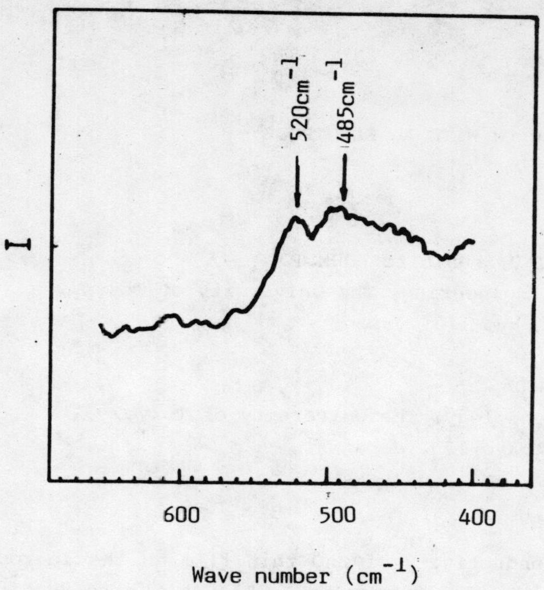

Fig. 4. Raman scattering diagram of film

More works are needed to be done in order to understand the sensitive mechanism of this film. The application of these novel properties to fabricating vacuum gauge and moisture sensors will be discussed in detail.

References

(1) W. Germer, Sensors and Actuators, 7, 2(1985) 135-142.

(2) Ernst Luder, Sensors and Actuators, 10, 1 & 2 (1986) 9-23.

ELECTRICAL PROPERTY OF ZnO THIN FILM WITH Ag ELECTRODE

Hirotaka.TAKAYAMA, Satoru FUJITSU, Kunihito KOUMOTO
Department of Industrial Chemistry, Faculty of Engineering, The University of Tokyo,
7-3-1 Hongo, Bunkyou-ku, Tokyo 113, Japan
and
Hiroaki YANAGIDA
Resaerch Center for Advanced Science and Technology, The University of Tokyo,
4-6-1 Komaba, Meguro-ku, Tokyo 153, Japan

ABSTRACT

A decrease in the conductance of ZnO thin film on the introduction of reducing gas was observed. This decrease could not be predicted from the properties of n-type semiconductor such as ZnO. This abnormal behavior was observed only when silver paste was used as an electrode. From the ionic polarization measurement, it was suggested that this abnormal behavior was due to the conduction of Ag ions penetrated from the electrode.

INTRODUCTION

Zinc oxide is well known to be an n-type semiconductor with excess Zn atoms at interstitial sites or oxygen vacancies acting as electron donors, while the conductivity of ZnO at lower temperature ($<600^{o}C$) depends on the amount and the nature of the chemisorbed oxygen[1-3]. The negatively chemisorbed oxygen is formed by trapping the conduction electrons from the bulk, so that the carrier density decreases and the electron depletion layers are generated on the surface simultaneously[4,5]. Hence porous or thin film ZnO shows low conductivity in air. When a reducing gas is introduced, it reacts with and thus removes the chemisorbed oxygen. Drastic increase in the conductivity of porous or thin film ZnO caused by the introduction of reducing gas has been utilized in gas sensing[6,7].

We have observed a decrease in the conductivity of ZnO thin film on the introduction of reducing gas. Though the conductivity of p-type semiconductor decreases on the introduction of reducing gas, it is not expected that the semiconduction type of ZnO changes from n to p even if Ag ions are dissolved into ZnO[8]. From the results of ionic polarization measurement, it was considered that a decrease in the conductivity of ZnO was due to a decrease in ionic conductance of Ag ions[9]. In this study, the changing process in the main carrier from electron to Ag^+ was examined.

EXPERIMENTAL

Zinc oxide films were prepared by RF sputtering using a zinc target (99.9%, from Mitsubishi Metal Co.,Ltd.) in a commercially available sputtering instrument (ANELVA SPF-210). Silica glass was employed as a substrate and maintained at 25°C. After pumping down to a pressure of 1.3×10^{-4}Pa, the gas

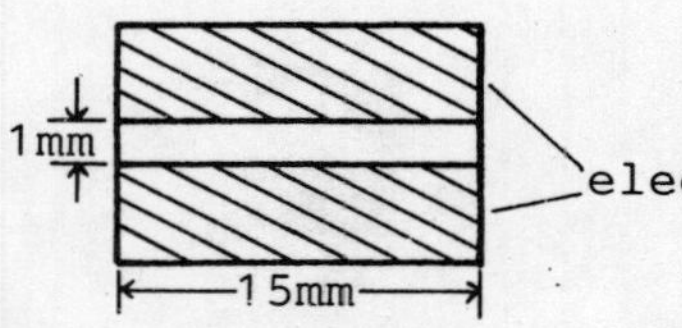

Fig.1. Electrode configuration.

containing 80%Ar-20%O_2 was introduced into the sputtering system, while the background pressure was maintained at 7Pa. The RF system frequency was 13.56MHz. The sputtering voltage and power were maintained at 1.5kV and 150W, respectively. Before sputtering, silver or platinum paste was painted on the substrate as electrodes as illustrated in Fig.1 and calcined at 760°C for 1 h. In this method, the deposition rate was about 1 μm/h and the film obtained was dense and polycrystalline with preferential c-axis orientation perpendicular to the substrate. The sputtering time was fixed to be 45 min.

Conductance was measured in both air and air containing 0.4% reducing gases (CO, H_2, C_3H_8 and n-C_4H_{10}) at 450°C under the constant voltage. The air containing reducing gas will be denoted as gas-air. Since it was difficult to divide the total conductance into surface, interface and bulk conductivity, the conductance measured with the electrode illustrated in Fig.1 was used in the present study. A gas flow system with a flow rate of 100ml/min was employed to obviate any change in gas concentration caused by the oxidation of the reducing gas.

RESULTS and DISCUSSION

As-sputtered sample showed a normal behavior, i.e. conductance increased when the reducing gas was introduced. The Ag-ZnO-Ag sample (electrodes on both sides were silver) showed an increase in conductance in air after annealed at 450°C and then showed an abnormal behavior, i.e. the conductance decreased when the reducing gas was introduced as shown in Fig.2. The ratio of the conductance in air to that in gas-air is shown in Table.1. Though the conductance did not change on the introduction of C_3H_8-air, the conductance decreased on the introduction of the other gases such as H_2, CO and n-C_4H_{10}. Such a decrease in conductance was also observed on the introduction of pure Ar or pure N_2. The conductance of the sample with the electrode other than Ag did not increase in air even by annealing for 4 days and the conductance change on the introduction of gas-air was of normal type as observed in the as-sputtered sample. Therefore, the Ag electrode may have played an

Table.1. The ratio of the conductance in air to that in gas-air measured 30 min after the introduction of gas-air.

CO	H_2	C_3H_8	n-C_4H_{10}
2.5	15.3	1.0	5.0

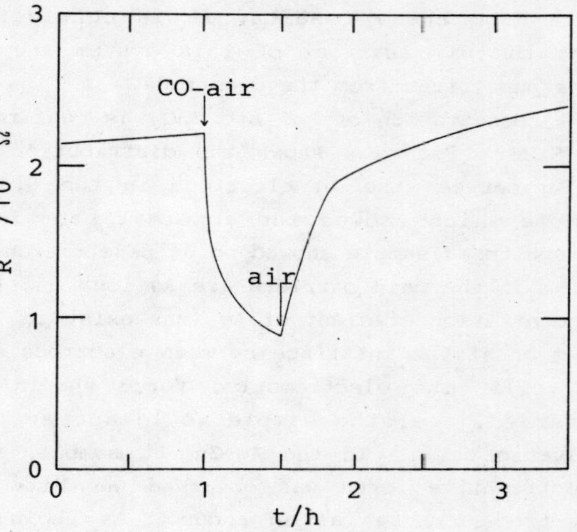

Fig.2. The conductance change in the Ag-ZnO-Ag sample on the introduction of CO-air.

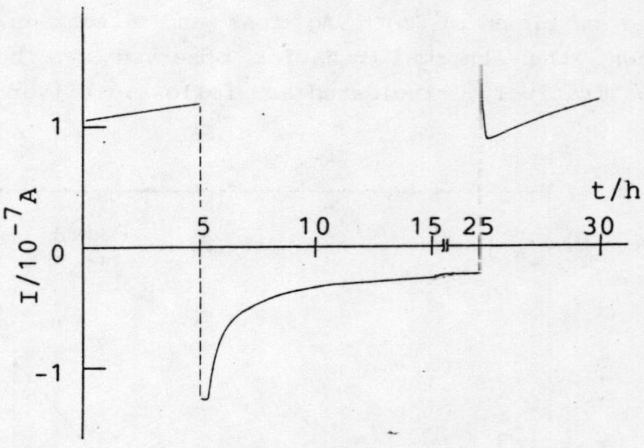

Fig.3. The current change in the Ag-ZnO-Pt sample with the change in the current direction. The current direction was changed from Ag→Pt to Pt→Ag at 5 h and revised at 25 h.

important role in giving rise to this abnormal behavior.

Such an abnormal behavior was also observed in the Ag-ZnO-Pt sample (one electrode was Ag and the other was Pt). This behavior was observed when the positive bias was applied to the Ag electrode. However, when the direction of current flow was turned from Ag→Pt to Pt→Ag, the conductance decreased. It was judged that this decrease in conductance was due to the ionic polarization. The current under the constant voltage (60mV) in air is shown in Fig.3.

From these results, it is considered that the main carrier of this system are Ag ions penetrated from the electrode.

Penetration of Ag into ZnO was confirmed by SIMS. Figure 4 shows the distribution of ^{107}Ag between the Ag electrode in the 0.5 μm sample which showed the abnormal behavior. As-sputtered sample showed no Ag penetration.

If the main carriers are Ag ions and the concentration gradient of Ag ions exist in ZnO film or at the interface between electrode and ZnO film, the electromotive force should be observed, i.e. the sample would act as the galvanic cell. In the Ag-ZnO-Pt sample, the electromotive force was observed when the Ag electrode was set as an anode. As shown in Fig.5, the short-circuit current of the Ag-ZnO-Pt sample decreased on the introduction of CO-air possibly due to a decrease in the ionic conductance.

Consequently, the sample showing an abnormal behavior is a mixed conductor having the carriers of both Ag ions and electrons. Then, the abnormal behavior observed can be qualitatively elucidated as follows: silver

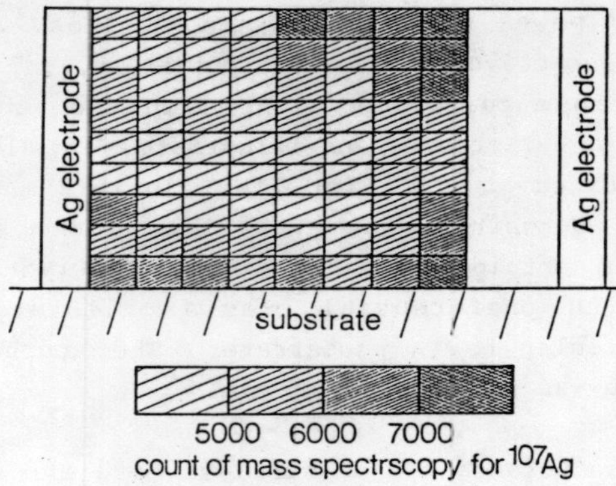

Fig.4. Distribution of ^{107}Ag in the 0.5 μm sample showing the abnormal behavior. The value is the count of mass spectroscopy for ^{107}Ag measured by SIMS.

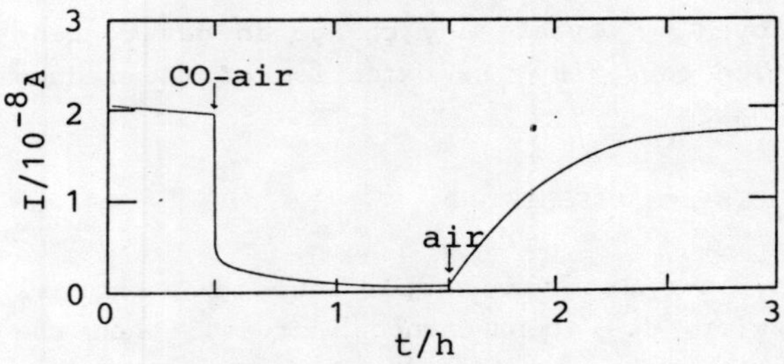

Fig.5. The short-circuiting current in Ag-ZnO-Pt sample on the introduction of CO-air.

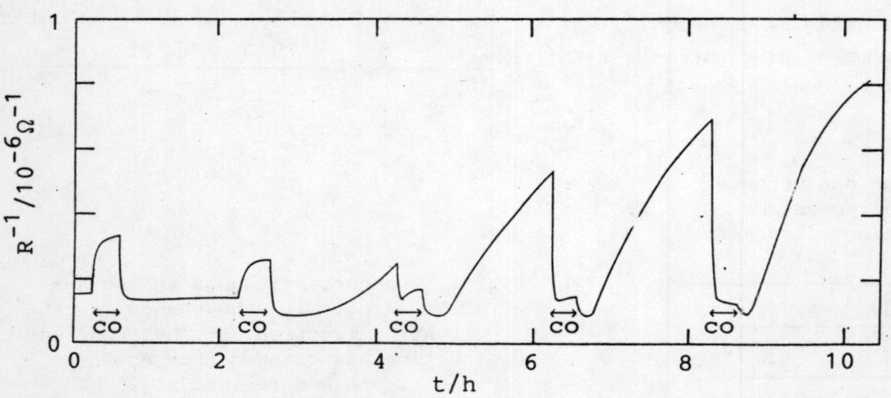

Fig.6. The conductance change in Ag-ZnO-Ag sample by annealing at 450°C in air. CO-air was introduced to check the property of the sample for 20 min at 2-hour intervals.

ionic conductance is high in air and decreases on the introduction of gas-air, while electronic conductance is low in air and increases on the introduction of gas-air.

The conductance change in the as-sputtered Ag-ZnO-Ag sample with annealing time at 450°C is shown in Fig.6. Carbon monoxide was introduced for 20 minutes at 2-hour intervals. On the first and second introduction of CO-air, the conductance in CO-air was higher than that in air, i.e. the n-type semiconduction in ZnO was not lost, and the ionic polarization was not observed in both air and CO-air. This result suggests that the contribution of ionic conduction was negligible. After annealing the sample at 450°C for 3h, the conductance increased gradually in air. On the third introduction of CO-air, the conductance decreased. In this stage, the ionic polarization was observed in air but was not observed in CO-air. The ratio of the conductance in air to that in CO-air increased with the increase in the ionic conductance in air and with the decrease of the electronic conductance in CO-air. The decrease in the electronic conductance by keeping at 450°C might be caused by the decrease in the conduction electron density through the incorporation of Ag^{+} into ZnO lattice.

The reason why Ag ionic conductance decreased in gas-air has not been clarified yet. Possible mechanism to explain this phenomenon is the coulombic effect between the Ag ions and the film. When the sample is kept in air, oxygen is chemisorbed on the surface and forms a positively charged electron depletion layer in the vicinity of the surface. The depletion layer might be eliminated by the introduction of reducing gas which reacts with and removes the chemisorbed oxygen. If the coulombic repulsion force between Ag ions and the depletion layer enhances the ionic conduction, high ionic conductivity should be observed in air, which is consistent with the experimentally observed behavior.

REFERENCES

1. M.H.Sukker and H.L.Tuller, Additives and Interfaces in Electronic Ceramics, eds. M.F.Yan and A.H.Heuer (The Am. Ceram. Soc., Columbus, OH, 1983) P.71.
2. K.Koumoto and H.Yanagida, Ceram. Jpn., 18 (1983) 927.
3. K.I.Hagemark and L.C.Chocka, J.Solid State Chem., 15 (1975) 261.
4. N.M.Beekmans, J.Chem.Soc.,Faraday Trans., 174 (1978) 31.
5. S.Fujitsu, H.Toyoda, K.Koumoto, H.Yanagida and T.Kanazawa, Bull.Chem.Soc.Jap., 61 (1988) 1979
6. S.Saito, M.Miyayama, K.Koumoto and H.Yanagida, J.Am.Ceram.Soc., 68 (1985) 40.
7. N.Ichinose and H.Okuma, Oyo Buturi, 45 (1976) 1183
8. G.Neumann, Current Topics in Materials Science, ed. E.Kalids, Vol.7 (North-Holland, Amsterdam, 1981) P.166.
9. H.Takayama, S.Fujitsu and H.Yanagida, Solid State Ionics, in press.

ZnO FILMS WITH PREFERENTIAL C-AXIS ORIENTATION PREPARED BY RF CYLINDRICAL MAGNETRON SPUTTERING

Xiao Yunxue and Ren Huaying
Sichuan Institute of Piezoelectric and
Acoustooptic Technology
Sichuan, China

ABSTRACT

Thin film ZnO is a multifunction electronic material with good properties and has found wide application. This paper presented a RF cylindrical magnetron sputtering technique and the equipment for sputtering films in our institute.
ZnO films with preferential c-axis orientation were prepared with the equipment. The film properties are examined by different analysis methods. The equipment with large uniform sputtering area can be used for mass production of ZnO films as well as for the sputtering deposition of other dielectric or metallic films.

INTRODUCTION

Thin film ZnO as a multifunction electronic material with good properties, has been drawn much attention. It not only has large electromechanical, optoelectronic, optoelastic coefficients and good temperatare stability, but also can be easily deposited on variety of substrates. Therefore, it has found wide application. ZnO films have been successfully applied to many kinds of SAW and BAW devices, as well as many sensing devices. It is also regarded as one of most promising materials for planar optic wave guide and acoustooptic interaction devices.

In recent years, the investigation of ZnO films for new applications are very active, and some new devices have been obtained(1-3). It is expected that ZnO films will play more important role in miniaturization and integration of electronic devices.

ZnO films and devices have been developed for more than ten years in china, using DC sputtering and RF planar magnetron sputtering techniques(4-6) . Meanwhile, TVIF SAW filter with ZnO/glass layer structure(7), microwave delay-line(8) and ultrasonic microscope(9) using ZnO thin films have also been produced successfully.

Moreover, SAW correlators/convolvers and gas sensors are being developed. Since more and more devices used ZnO thin films have been put into practice, RF planar magnetron sputtering technigue can not provide large quantity of good ZnO films and we have developed cylindrical magnetron sputtering system.

PRINCIPLLE OF RF CYLINDRICAL MAGNETRON SPUTTERING

Magnetron sputtering deposition is a high rate sputtering technique developed in 1970's. It is a particular form using magnetic field to enhance gas discharge in the vacuum chamber, which operation principle is shown in figure 1.

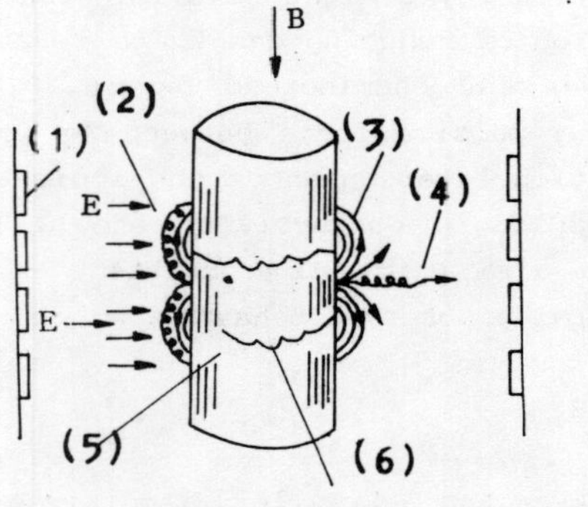

Fig. 1. The principle of magnetron sputtering. (1) substrate (2) low energy electrons (3) magnetic line (4) escaping electrons (5) circular plasma (6) first electrons.

In the sputtering system, a cylindrical target is placed in the center of vacuum chamber and the substrates are arranged surrounding the target. Axial and radial electric fields form a orthogonal electromagnetic field. Hence, the glory discharge generated in the vaccum chamber is activated by both electric and magnetic fields simultaneously. In the sputtering process, the electrons which generated from gas ions bombarding the target are acted by Lorentz force. Most of them are confined near target surface and move trochoidally. Both their movement path in magnetic circular gap and collision probability of gas molecules increase. Hence, a high density circular plasma is formed near the target and the sputtering rate increase greatly comparing with ordinary aputtering techniques.

FILM GROWTH

Fig. 2 shows a schematic diagram of the RF cylindrical magnetron sputering equipment. A internal magnetic cathode is used, because this kind of structure is simple, the magnetic field intensity can be easily adjusted according to requirements, so it is useful for depositing large area films.

The equipment consists of supply power system (D. C power, RF power and matching networks), electrode system et al. The limit vaccum is < 5×10^{-6} pa. RF output power is 4kw and D.C. output power is 5kw.

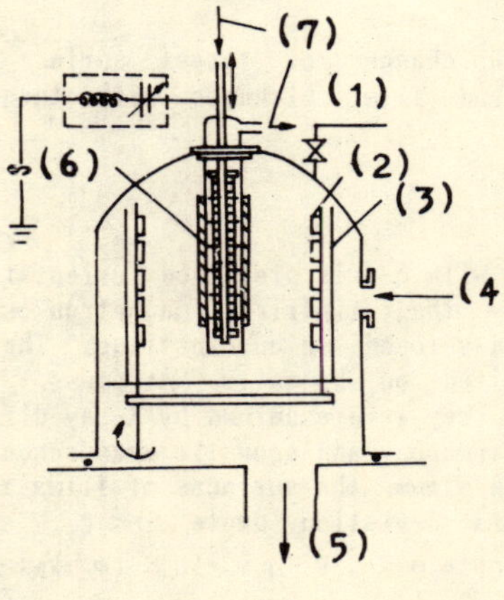

Fig. 2. Aschematic diagram of the RF cylindrical magnetron sputtering equipment. (1) airing system, (2) substrate, (3) heating system, (4) He-N laser monitor, (5) pumping system, (6) target, (7) cool water.

The substrate temperature can be freely controlled in a range of 20--350℃. The magnet can be recipracated up and down at a certain rate and the substrates move around the target in order to get uniform deposition. As a result, the utilization coefficient of target and the film thickness uniformity were increased. The sputtering system is equiped with a He-Ne laser monitoring system to detect the film thickness in sputtering process.

The target is a sintered ZnO cylinder with Li dopant and 60 mm in diameter, 300 mm in length. The typical sputtering conditions are as follows:

sputtering gas:	O : Ar = 1 : 1
gas pressure:	0.5-0.2pa
anode voltage:	1.5-2.2kv
substrate temp.:	150-300℃
distance from target to substrates:	90 mm
deposition rate:	0.3-0.8 μm/h

RESULTS AND DISCUSSION

ZnO thin film grown with orientation are a polycrystal, which electric, optic and acoustic properties strongly depend on its crystalline structure.

The film quanlity also has a close relation with sputtering conditions, thus, to select suitable conditions is very important for obtaining ZnO films with high reproducibility and fine properties.

The most important factor determing the quality of ZnO films is substrate temperature, which can change film structure and standard deviation angle in large degrees. The temperatrue can affect the mobility and reevaperation of absorbed atoms on the substrate surface and cause the change of crystalline structure of the film(10). The relationship between substrates and standard deviation angle are shown in Fig.3.

As many nucleating cores have been formed, minimum grains may be grown at higher deposition rate. In cylindrical magnetron sputtering process, the deposition rate increases as input power is increased. This is shown in Fig.4. However, when the power increases, the energy of electrons, second electrons and neutral particles also increases greatly. They bombard the substrates and harm the films and make the temperature of substrates higher.

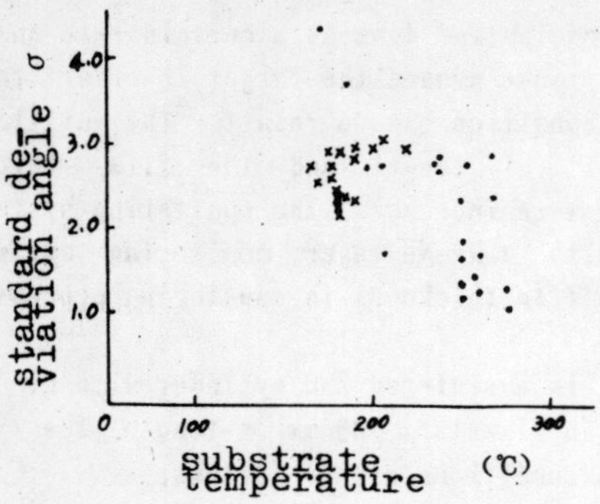

Fig. 3. The relation between substrate temperature and standard deviation.

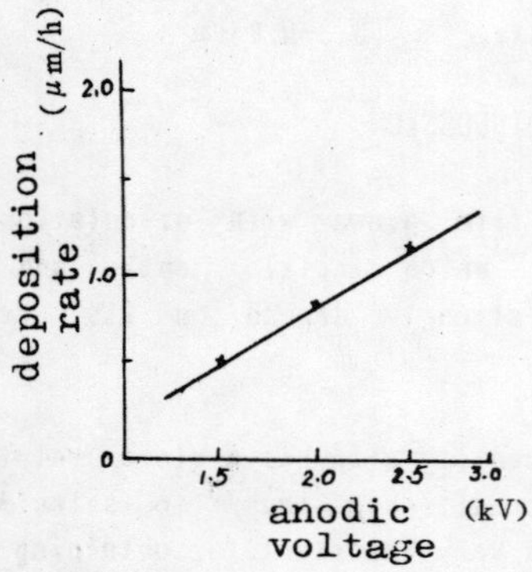

Fig. 4. The relation between deposition rate and input power.

The composition of gas is also a factor affected the quality of films. (002) prefferried orientation enhances as the content of O_2 in sputtering gas is increased, but the futher increase of O_2 content may make the deposition rate lower and the mean deviation angle m or standard deviation angle σ may also be affected under the lower deposition rate. The effect of O_2/Ar on ZnO film orientation is given in Table 1.

Tabke 1, The effect of O_2/Ar on ZnO film orientation

	O /Ar	deposition rate(μm/h)	normalized thickness (LPS／m)	orientation	m
C-142	1 : 1	0.54	505	3.4	0.1
C-144	3 : 2	0.43	523	3.9	0.8
C-143	1 : 0	0.35	820	4.1	1.0

From Table 1, we can see that O : Ar=1 : 1 is suitable for ZnO film orientation.

As described above, during cylindrical sputtering the electrons are acted by magnetic and electric fields simutaneously hence the distribution of intensity of magnetic field affects the deposition rate and the film thickness homogeneous.

Fig. 5, shows some results in our experiment. These results mean that the thickness homogeneous is very good, the thickness variation is less than ±3%.

On the other hand, we found that the distance from target to substrates, the types of substrates and state of the substrate surface also are the factors influenced the ZnO film quality.

It is valuable to notice that the interaction among above factors is existed. We must consider them overall to get the optimal technical conditions.

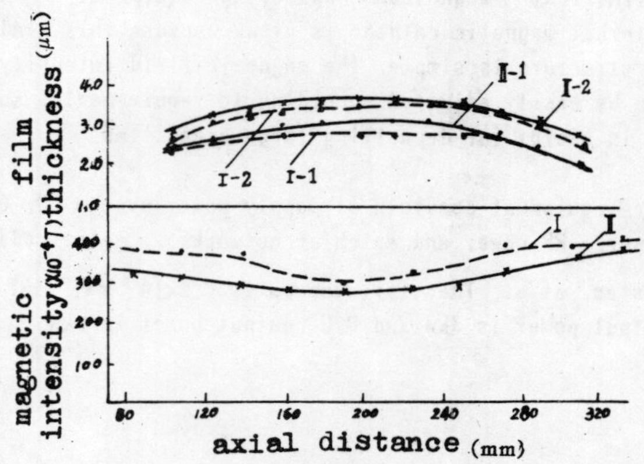

Fig. 5. The changes of target surface magnetic intensity and film thickness with target axial distance.

SUMMARY

ZnO films with c-axis preferred orientation were prepared by the cylindrical magnetron sputtering equipment developed in our institue. These films were deposited on k, galls, Al/glass, Au/glass substrates. They were examined by X-ray diffraction, electron microscopy and acoustic measurement. Their grains are minimum, the surfaces of films are smooth, the standard deviation angle $\sigma < 2.5°$, mean deviation angle $m < 0.8°$ $\rho v \gtrsim 10^8 \Omega$-cm, $V_S \approx 3200$m/s.

TV-IF devices produced used these films have good properties, IL< 25dB. The X-ray diffraction and RED patterns and the photographs of electronic microscopy of these films are shown in Fig. 6-8 respectively.

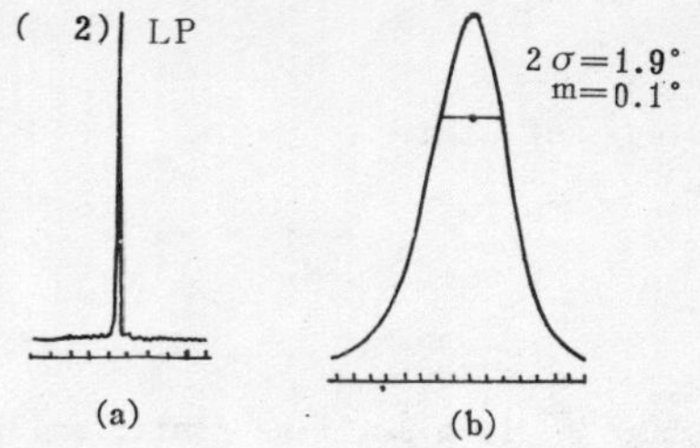

Fig. 6. ZnO film X-ray diffraction pattern and trochoidal curve.

Fig. 7. ZnO film reflection electronic diffraction pattern.

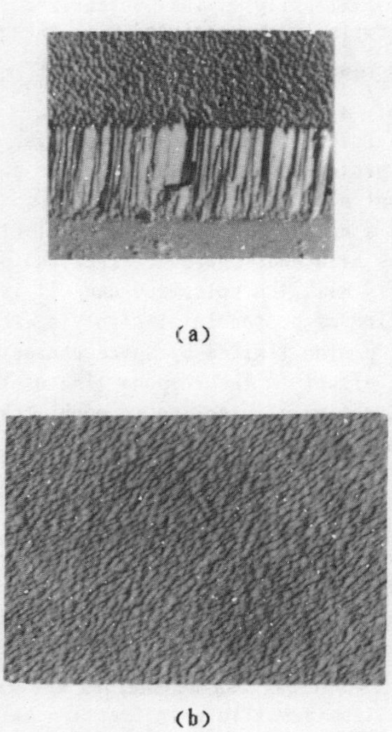

Fig. 8. The electronic microscopy photographs of ZnO films.

The results shows that the sputtering epuipment design is reasonable and has large production capacity. The equipment can be used for mass production of ZnO films and the sputtering deposition of other dielectric or metallic films.

REFERENCES

1. H.Fabricus et al. Appl. Opt., 1986, 25(6) 2764.
2. P.L.Palla et al. Proc. Ultrasonics symposium, 1985, p.495.
3. H.G.Horsthuis, Thin Solid Films, 1986,(137)1‡5.
4. Piezoelectrics & Acoustooptics, 1979, 1 (4) 10 (in chinese).
5. Piezoelectrics & Acoustoopics, 1981, 3 (3) 38 (in chinese).
6. Piezoelectrics & Acoustoopics, 1985, 7 (6) 20 (in chinese).
7. Piezoelectrics & Acoustoopics, 1985, 7 (1) 16 (in chinese).
8. The research report of Piezoelectrics & Acoustooptics Institute, 1986 (in chinese).
9. The research report of QingHua University, 1983 (in chinese).
10. F.S.Hickernell, Proc. Ultrasonics Symposium, 1980, p.785.

INVESTIGATION ON SOL-GEL ELECTROCHROMATIC FILM WO_3, V_2O_5

Chen Ai, Zhang Weimin × D.Hitchcock × ×
University of Electronic Science and Technology of China
Chengdu, Sichuan, SC 610054 P.R.C.

ABSTRACT

The noncrystalline sol-gel film WO_3 prepared by colloid electrophoresis method has been described in this paper. The film is dyed black-blue from transparent when the ions Li^+(or/and H^+) and electrons are double injected to it after some appropriate heat-treatment. The electrochromic properties are: contrast Ct>1:11,signal storage time at open circuit $\tau_s \approx 64$Hr, response time $t_{res} \approx 1$sec and cyclic lifetime $n \approx 8 \times 10^4$ cycles. The experiments are finished in a reversible battery constructed by carbon/2-methyl sulphoxide+$LiClO_4$ (Liquid electrolyte)/WO_3(V_2O_5)/ITO.
The effects of heat-treatment conditions, measurement voltage, film thickness etc. on structure of sol-gel film, water content, optical density exchange, cyclic lifetime, coloured reversibility and signal storage property are also discussed.
V_2O_5 film changes colour from orange to green, it is possible to make a multicolour display, but the cyclic lifetime and contrast is too low so far.

INTRODUCTION

Felectrochromatic effect is a reversible colour-(bleach-)process through the electrochemic action of materials. There are lot of materials have this effect, among which the transition metal elements oxides MxOy, such as WO_3, V_2O_5, Ir_2O_3, Ni_2O_5 etc. are most practical. Those materials will be coloured owing to the valance change of inos$^{2y/x}$ when ions and electrons are double injected to them simultaneously from two electrods.

For electrochromatic film WO_3, ions W^{6+}reduces to W^{5+} and a widely optical absorption band with a peak of 0.9—1.46 eV will be found when ions A^+(H^+, Li^+) and electrons double injected in it, thus WO_3 is coloured from transparent to dark-blue ($LixWO_3$, or $HxWO_3$). By "F-Center" model, Dexter[1] and Deb[2] has proposed that transport of ions H^+(or Li^+) in WO_3 as a fast ionic way, the ion H^+ reduce to H atom leading in electron, and than it reacted with WO_3 to $HxWO_3$,

$$W^{6+}O_3 + xH^+(Li^+) + xe^{-1} \rightleftarrows H_xW^{5+}O_3 \quad (1)$$
(transparent) (dark-blue)

For Hx(Lix)WO_3, W^{6+} is a positive center and it will be a quasi-hydrogen atom which is in another weak coupling excited state when the excited electron is injected in it. Some absorptine band on specific wave length is adjoint of this exchange process and colouring happens simultaneouslyl. The density of colour-center is in following equation by Smakula[2].

$$N.f = 8.7 \times 10^6 \frac{n}{(n^2+2)^2} (K_{max}L)W_{1/2} \quad (2)$$

where N is center amount per unit volume(cm^{-3}), n is refracting power of film, f is trasition intensity of colour center, that is allowable transition of diople for an electron, for WO_3, $f \approx 0.1$, α_{max} is peak absorption coefficient (cm^{-1}), $W_{1/2}$ is full breadth of half extreme for the absorption band (eV). For WO_3 film, it can be estimated at $N \approx 1.7 \times 10^{22}/cm^3$ substituting corresponding value above.

In addition, Shirmer[3] raises the" Optical transition model" yet. This model can explain qualitatively the shap, energy of absorpted band and viberative intensity. The colourized (charge injected) current is associated with the potential barrier in injective point of A^+ and e^{-1}, the mobility of A^+ in electrolyte and WO_3, and the potential barrier of counter electrode.The potential barrier in proton and electron injective point will play a controlling action if WO_3-ITO is in ohmic contact and a high ionic conductivity electrolyte is selected.The bleached (charge extraction) processing is unsymmetry with the colurized one, it is decided by transition of proton in electrolyte. The motion velocity of proton limited by space charge will be a determinant effect to the respone time of the device, due to the motion of electron is much faster than which of proton.

EXPERIMENT

The electrochromic film is obtained by sol-gel eletrophoresis method in this project. Except Ⅷ B family elements, almost all transitional metal oxide can form sol-gel and these are ideal ionic conductor of H^+(or Li^+). Sol-gel can be charged due to absorb, ionize, lattice substitute or fraction mainly. The

× Shanghai Institute of Metallurgy.
× × Visiting specialist of UESTC on leave from L.B.L, U.S.A

diffused electric double layer structrue is formed after it absorbs opposite ions which is in liquid electrolyte. The charged particles(with absorbed layer) move towards electrod with opposite charge selative to (liquid) dispersing medium under the applied electric field, this is called "electrophoresis phenomenon".

If the proportionally Na_2WO_4(0.1M) and Na_2MO_4(0.1M) pass through an ion exchange resin tower of strong acidity (IR-120, Dowex-50), the light yellow solution (ph≈4) can be obtained. Then it is treated in a hot bath for a right temperature and period, we can get the yellow green transclucent liquid WO_3. Tyndall effect is applied to identify whether it is sol-gel or not. Sol-gel V_2O_5, which is orange colour, is gained by similar processing from Na_2VO_4(0. 1M) Its chemical reaction is

$$2Na^+ + W^{6+}(V^{6+}) + 8(OH)^- + 2R^-SO_3H(Resin) \rightarrow$$

$$WO_3(V_2O_5)(Sol\text{-}gel) + 2R^-SO_3Na + 5H_2O \qquad (3)$$

The measuremental equipment is indicated as fig 1. In order to simplify the problem, film WO_3 is adopted in liquid electralyte $LiClO_4$+2-methy sulphoxide. The measurement signal of different wave shape and frequency is produced by electrochemical synthetic meter model DHZ-1A. Spectral analycis is carried out using spectrophotometer model 7321. The topographical analysis of specimen surface by SEM model S-450. Water content exchange is measrued by thermal-weight analyser model C-T160. The valencestate analysis of film before and after coloure by energy spectrumer model WE-W_4. Film thickness by interference microscope model 6J-A. X-ray spectrum by X-ray spectrumer model D/max-II A

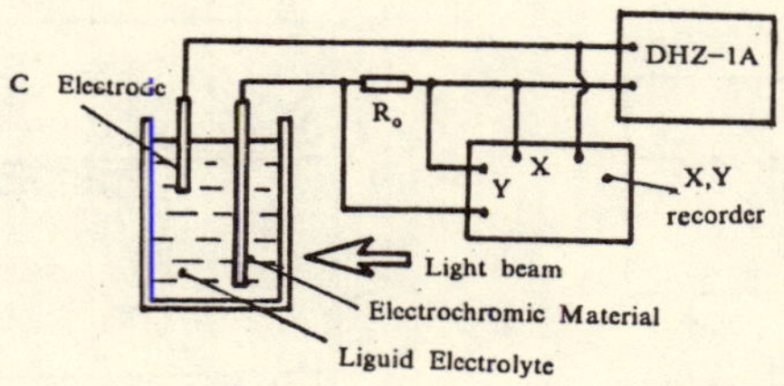

Fig.1 schematic diagram of specimen test facility

RESULTS ANS DISCUSSION

1. The effect of heat-treatment temperature on structure and water content of sol-gel film WO_3.

Table I Heat-treatment conditions of specimens

No	treatment temp (℃)	time for raising temp(hr)	time for keeping temp(hr)	structure form
# 1	untreatment			uncrystalline
# 2	180	15	1	uncrystalline
# 3	310	15	1	uncrystalline
# 4	500	15	1	crystalline

The untreated sol-gel film WO_3 is easy to solute in liquid electrolyte, and the large water content makes the layer loose, so the cycle lifetime is reduced, but the response time is raised due to ion H^+ removes in the film is more convenient. In order to explore the optimum relationship between film properties and water content, the experiments in different heat-treatment conditions have been tried here (see tablr I).

Water content is measured by temperature gravity analysis (TGA) and differential temp. gravity analysis (DTGA). TGA records directly the weigh loss information due to heat decomposition, thus giving the related data concerning reaction mechanism and its products [Fig.2]. DTGA indicates the derivative curve of weightlessness and gives the starting and ending point of this action clearly. There are three peaks appear in Fig 2, 42.7℃, 231.3℃, and 297.3℃. After some calculations its weightless rate is 5.71%, 2.01% and 1.95% respectively. So the whole process can be expressed as

$$WO_3 \cdot 2H_2O \xrightarrow[(1)]{42.7℃} WO_3 \cdot 1.15H_2O \xrightarrow[(2)]{231.3℃} WO_3 \cdot 0.8H_2O$$

$$\xrightarrow[(3)]{297.3℃} WO_3 \cdot 0.5H_2O \qquad (4)$$

WO_3 is a layer structrue constructed by common vertex octahedron[W—O]. Every ion W^{6+} in the layer contacts with one terminal oxygen Ot along derection axis b, and four bridge oxygens Ob along plane(a, c)[4]. The water molecular absorbing between layers is model I, and which joinding by short bond [W—O] along axis b is model II. The weak banding water molecular model I is lost at the first and second stage. The stronger banding water molecular model II is lost partly when anneal temp. up to third stage (T>297.3℃), which causes the lattice rearrengement and the film crystalline.

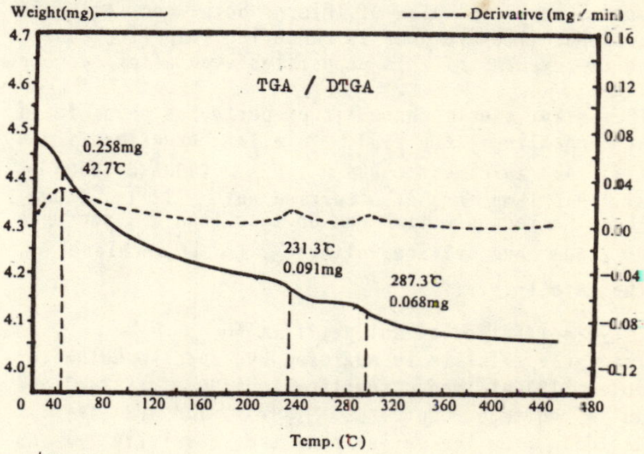

Fig.2

2. The effect of heat-treatment temp. on optical density variable.

The contrast of sol-gel film can be characterized by transformation of optical density as follow

$$C_T = I / I_0 = e^{-\bar{\alpha}\theta} = e^{-\Delta OD} \qquad (5)$$

I, I_0 is transmission indensity when film is

colored and bleached, $\triangle OD$ is induced change of optical density, $\bar{\alpha}$ is colouration efficiency, that is the induced optical density per charge density $[(mc \cdot cm^{-2})^{-1}]$, Q is charge density $[c \cdot cm^{-1}]$.

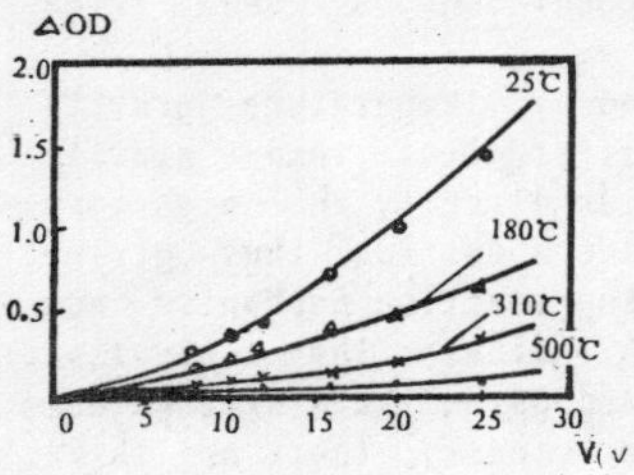

Fig.3 Relationship between optical density and voltage of the specimen at different heat-treatment temp.

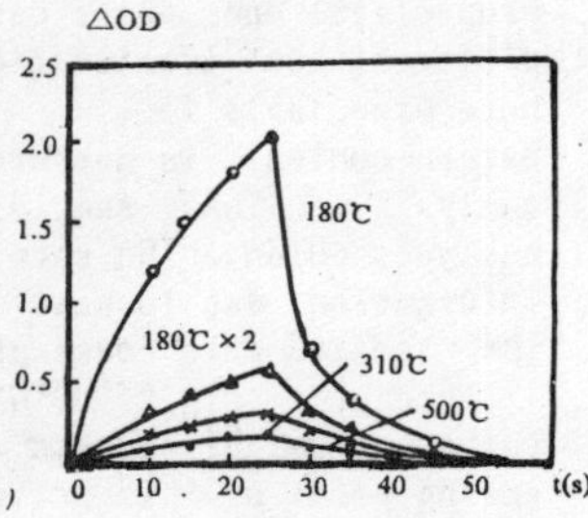

Fig.4 Relationship between optical density and time of the specimen at different heat-treatment temp.

The relationship curve between $\triangle OD$ and voltage is coming gentle, (Fig.3) and colourized effect is getting poor as heat-treatment temperture arise. Specimen #4 almost can't be coloued at this voltage.

The reason is that the injection of H^+ is difficult and the injection of electron is limitd when the film crystalline and the increment of density as (O/W) date goes up. Given suitable (O/W) ratio, such as it is treated in inert gas, the double injective effect is not degraded, also the film intensity increase. Therefore the result of coloration is improved. Fig. 4 indicates that the effect of heat-treatment temp. on the relationship between optical density and time at applied voltage 2. 8V. The maximum colourized darkness appears at the 25th sec. from Fig.4, and contrast of different specimen is $C_{T1} \approx 1:11$, $C_{T2} \approx 1:2.0$, $C_{T3} \approx 1:1.4$, $C_{T4} \approx 1:1.2$ separately, response time is $t_1 \approx 1$ sec, $t_2 \approx 8$ sec, $t_3 \approx 15$ sec and $\triangle OD < 0.3$ of specimen #4 in same condition. In view of this, both contrast and respone time are poor as annealing temp. increase, no improvement of this properties even after double anneals.

The better electrochromatic property bas been found at annealing temp 250℃. Similar experiments on different thickness (900Å, 1800Å, 2800Å, 5000Å) at this temp. are also carried out[5]. It is found that the response time can be less than 0.5 sec when d=2800Å and driving voltage $V_d > 3V$, it achieves the latest level[4].

3.Cycle lifetime of sol-gel film WO_3

The cycle lifetime is measured by spectrophotometer model 7320 at monochromatic light (λ=633nm) and pulse voltage (V_p=+1.5—-1.0V). The results is decided from the variable of transmissivity $\triangle T$ as Table II.

It is thus clear from tab. II that raising treated temp. suitably can increase the lifetime, but raising treated times has no effect clearly. The relationship between transmissivity T, circuit current i and time t for the specimens of different cycle times under same applied voltage (2V) is shown in Fig 5. The specimen annealing at 250℃ has better chromical property (Fig 5-b) and which of specimen treating in 310℃ (Fig 5-d) appears unusually increase of T, i when cyele over 10^4 time. It is considered that chromic effect of the film is influenced by two factors, the soluting action of liquid electrolyte drops down the colourized property and the water content improves it.

Table II Cycle lifetime and measurement condition of sol-gel film WO_3

Specimen	Cycle Voltage (V)	(△T) start	(△T) end	Cycle lifetime
#1(anneal at 180℃)	+1.5—-1.0	15	3	1.2×10
#2(double an. at 180℃)	+1.5—-1.0	27·	4	10
#3(310℃)	+1.5—-1.0	19	6	2.8×10
#4(500℃)	+1.5—-1.0	unclear		

4. Colour reversibility and storage charaacteristics of uncrystalline film WO_3.

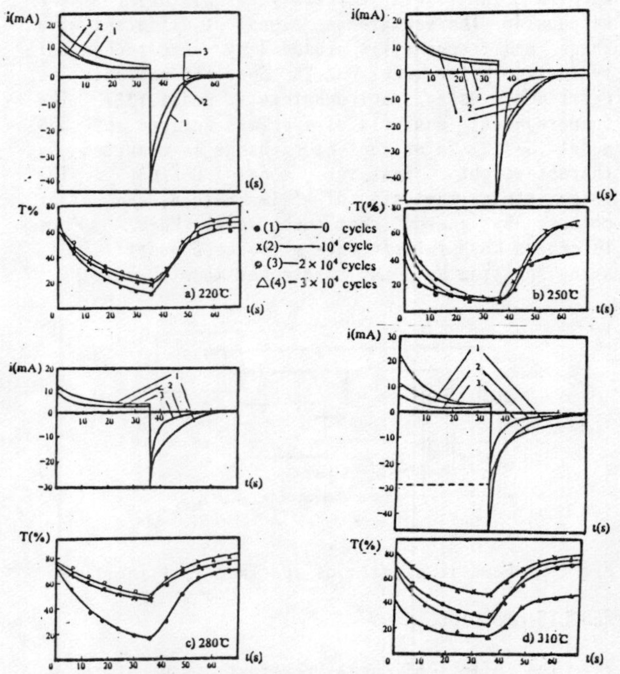

Fig.5 Relationship curve between current (transmissivity) and time of specimen at different treated temp.

Researching of X-ray scattering on small angle points that the atomic arrangement of WO_3 only has short range order, and its intrinsic energy is bigger. There are triangle and big size pentagon, hexagon tunnels except the quadriateral on structured by [W-O] octagon when it is crystallized. The tunnel transmitting ion is widened by oxygen vacancy and the additional unsymmetric structure distortion caused by dislocation and the maximum tunnel appears at(O/W)~2.5. The accumulation parttern is not change when Li^+ injects into it, ion Li only motions among

its interstials.
The reference electrode is mercury, counter electrode is carbon and the signal is triangular pulse (-2V—+1.5V) with scanning speed 100mV/sec adopted in this experiment for uncrystalline film WO^3. The colourized cycle curve is shown in Fig.6, it has more satisfired colour reversibility.

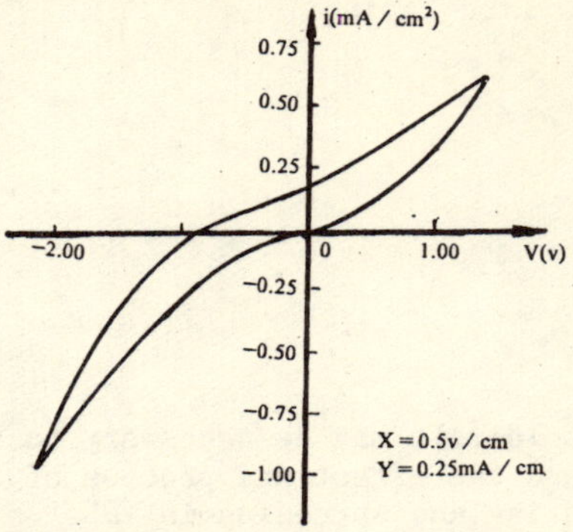

Fig.6 Relationship between V and i of uncrystalline sol-gel film V_2O_5 under triangular pulse

Uncrystalline sol-gel film WO_3 has good colour storage characteristics at open circuit. After deep colouring (transmissivity T from 70% to 7%). The storage time is $Ts \approx 64$hr. by $\triangle$OD drops to 0.3. Bleaching discharge is formed through electronic conductance of material mainly, so the storage characterictic would improve if electronic conductiuity is reduced further.

5. Preliminary investigation on chromic property of gol-gel film V_2O_5
It is found from X-ray diffraction that electropharesis sol-gel film α-V_2O_5 is noncrystalline state obviously when treatment temp. lower than 240℃, crystallized partly at 400℃, the crystallinity increases and the structure of film changes from nebular packing to leayer constituting of network fibre as temperature goes up continually. The direction of fibre layer is free and the distance between layers is decided by water content. One molecular V_2O_5 includes 1.6 H_2O moleculars after it dries in shade. The water loss is steped when heat-treated temperature increases.

$$V_2O_5 \cdot 1.6H_2O \xrightarrow[\text{wideendo-thermic peak}]{\sim 120^\circ C} V_2O_5 \cdot 0.5 \xrightarrow[\text{weak endo-thermic peak}]{\sim 210,270^\circ C} V_2O_5 \cdot 0.1H_2O \xrightarrow[\text{sharp-endothermic}]{\geq 300^\circ C} V_2O_5(\text{uncrystal}) \xrightarrow[\text{exothermic-peak}]{> 400^\circ C} V_2O_5(\text{crystal}) \quad (6)$$

formula (6) shows that mass reversible weak coupling water is released at ~120℃, the distance between layers d reduce to 8.75A° from 11.55A° . The crystal water which links by more strong H-bond in fibre network layer loses when temperature up to ~ 300℃. It caurses rearrangement of the layer structure and crystallize. After dehydration of the film which can expand again due to the water absorption. It is indicated that from X-ray diffraction and neutron scattering the distance between layers will increase steply take the unit almost 3A, which is just right a water molecular size. The increase of data d is continued when n>6. Compare with WO_3, film V_2O_5 is more stable but is not formed a bridge bond, so it has poor adhesive stength. When ion Li^+(or H^+) and electron double inject into V_2O_5, it will be colourized from orange to green $Li_{2x}(H_{2x})V_2O_5$. The possibility of multidisplay can be realized by V_2O_5. Three electrolyte solutions (acetonitrile+$LiClO_4$), (methanol+$LiClO_4$) and (2-nethy sulphoxide+$LiClO_4$) are used in this project, the first two have very poor lifetime, and that of third one equals to 5×10^2. It is possible to raise the lifetime if using solid electrolyte SiO_2 or SiO_2($\triangle T$)$_{\%}$, the average change of transmissivity before and after coloration, is about 25.18%, which is much lower than that of WO_3($\triangle Two \approx 53.50\%$). It will limit the practical application of this material.

CONCLUSION

1. Uncrystalline sol-gel film WO_3 is a new model electrochromatic material which has practical value of application. Its changement of vertical transmissivity $\triangle T$ is 55.3%, contrast C_T is over 1:11after colouration, and it has better celectivity of dynamic adjustable spectrum. The storage time τ_s is 64hr. at open cricuit, response time is almost 1 sec. cycle lifetime n is 8×10^4. The crystal sol-gel film WO_3 is not suitable to do the display although strength and adhesion is good, but the contrast is too low.
2. Uncrystalline sol-gel film WO_3 is a water contening material which has poprous and loose structure. Heat treatment is key technology to improve its chromical property, strength and adhesion. The experiments decided that treatment at 250℃ for one time has optimal effect. Increasing layer thickness is in favored of improvement of contrast and response time, but it is unfavorable to cycle lifetime and strength, the selection of optimal thickness is 3000A.
3. For uncrystalline sol-gel film WO_3, the response time can be reduced to almost 500ms, if the square resistance of transparent electrode ITO from 80Ω/□ to 10Ω/□, the driving voltage from 3V up to 4V and using the liquid electrolyte acrylic carbon ester or even solid electrolyte in stead of 2-methy sulphoside[6].
4. V_2O_5 also is a hopful sol-gel electrochromatic material, it can be coloured from orange to green and provides the possibility of mutlicolour display, but the cycle lifetime and contrast is pending further improvement.

REFERENCE

1. D.L.Dexter, "Solid state physics", vol 6, Ed. by F.Seits, at el. 353-
2. I.F.Chung, 《Nonemissive Electrooptic Displays》, New York(1975)115-96.
3. O.F.Shirmer et al., J.Electrochem. Soc. solid state Sci. Tech. 124(1977)749
4. B.W.Faughan, et al. Appl. phys. Lett. 27(1975)275.
5. W.M.Zhang, Master thesis, Univ. of Elec. Sci. and Tech. of China,(1989) Chengdu, P.R.C.(in Chinese)
6. A. Chemseddine, et al. Solid State Ionics, 9210(1983) 357-62.

CONDENSATION HEATING AND SOLDER SPREADING IN VAPOR PHASE REFLOW SOLDERING

Cui Dianheng, Wang Tingyue, Zhang Jianping, Ding Yaping
Nanjing research Institute of
Electronics Technology
P.O.B. 1315
Nanjing, P.R.C.

ABSTRACT

Condensation heating process in Vapor Phase Reflow Soldering is discussed.Temperature —time response curves of many kind of component were tested.It had been found that the posture of body had obvious effect on heating process. FC-70 resists spreading of rosin,so the wetting and spreading of solder on FC-70 coated surface will be affected.But the effect of FC-70 on spreading of solder does not change the basic nature of solderability.

INTRODUCTION

Vapor Phase Reflow Soldering(VPS)is a soldering method depended on phase transition heating.[1] VPS has its own feature----rapid and uniform heating,absence of hazard of overheating and charring,and others.For ten years ago,many type of VPS machine(batch or in-line type) and over ten kinds of working fluid has been developed. [2] VPS has become one of main reflow soldering technique utilized in SMT and HIC.Recently the development of mixed fluid(double vapor) and superheat technology may improve the temperature flexibility in VPS.[3]

The heat in VPS will flow as follows:

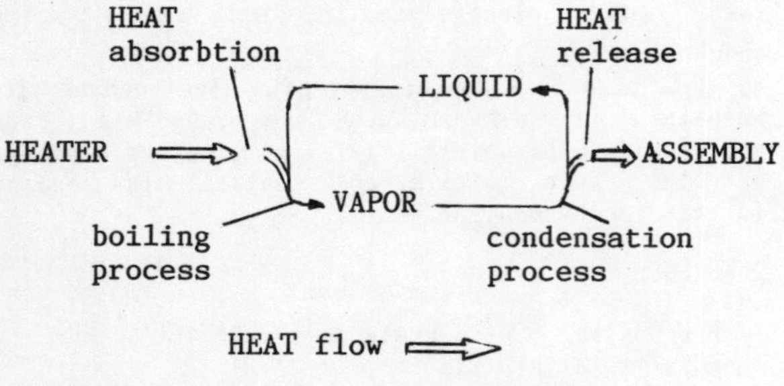

Vapor of FC-70 continuosly condenses on cold surface and releases its latent heat,and the liquid formed will drain off.Even if the temperature of surface approches the boiling point of FC-70 and the condensation-vaporization process of FC-70 is in dynamic equilibrium, the surface of body is still coated by FC-70 film.FC-70 is not only a heat transfer medium, but also has some effect on wetting and spreading of solder.It may be necessary to study and understand the effect and process of condensation heating and spreading in VPS.

BOILING AND CONDENSATION

The boiling style of FC-70 can be devided into four types: free-convection boiling, nucleate boiling,unstable film boiling and stable film boiling.In nucleate boiling regime a peak heat flux can be reached(6W/cm^2).This is main boiling style of FC-70 in VPS.In order to prevent heater from overheat,we would rather apply less heat flux(3W/cm^2).

Comparing with film condensation,drop condensation may have more heat transfer coefficient.But it is hardly to find any metal that can not be wetted by FC-70.

Condensation always occurs simultaneously on all surfaces contacted with vapor.Although the temperature in the saturate vapor zone is constant,the heat transfer coefficient of horizontal,vertical and inclinded surface is not the same.The main reason is the difference of draining away process of liquid coming from condensation.

Thermal resistance in condensation heating is mainly coursed by liquid film formed.It has to be membered that to keep good draining away of liquid and do not have any concave or blind holl on the assembly.This is useful to increase heat transfer and reduce drag out of fluid.

HEATING PROCESS IN VPS

Cup-type VPS microsystem used in lab is shown in fig.1.After heater is turned on,four temperature-time responce curves of point A,B,C and D are tested(fig.2).The temperature of liquid(point A) gradualy increases,sharp change of slope is not found. However we can find a sharp change of slope on curve B or C.Before top of vapor touches thermocouple at point B or C,the temperature of B or C still increases slowly by convection .Once touching,the temperature will suddenly increase to boiling point of FC-70.We can clearly find that temperature of point D,which is located above the lever of condensation coil or out off saturate vapor zone,very slowly increases to a lower temperature and can not reach boiling point of liquid at all.Such starting-up process stated above is special in VPS.

In vapor zone the temperature along hight or at same lever is very uniform except a little portion near the air-vapor interface.This feature is just belong to VPS too.

The heating process of multilayer ceramic hybrid circuit in VPS is shown in fig. 3.The hybrid module consists of 40 pieces of 16 I/O LCCC.The tip of three thermocuples was fixed in different position--on ceramic substrate,on gold-plated covar cap and in the cavity of LCCC. In result,the heating rate of point C in cavity is slightly slow.Comparing with point B,the time to rise to 183°C delays about 2-3 sec.This is coursed by gathering condensed liquid in cavity.Other heating method does not possess this rapid and uniform heating.

The temperature-time responce curves of solid and hollow triangular prism are shown in fig.4.It is clear that heating rate of hollow body is more quick than solid body,and thin wall body,bottom of which is face-down,possesses the most quick heating rate.All of this are according to theory of condensation heating.

From fig.5 and 4,it is clear that the heat rate generally is rather rapid in low temperature and gradually decreases as temperature increases.It is difficult to regulate heating process of condensation heating.

INFLENCE OF FC-70 FILM

Clean copper wire was heated in FC-70 vapor for different time(20,180 and 1800 sec),then was quickly put in water to cool them.The wetting force of copper wires was tested and the results are shown in table 1.As seen,the time to zero-Newton and wetting force at third and tenth sec is similar for copper wires heated for different time. The data are very similar to copper wire stored for 24 hs at room temperature,and slightly diferent with as cleaned wire.May sure that this difference is coursed by oxidation during cooling cycle.The deterioration is by no means due to oxidation of FC-70 vapor.

The greater part of clean and bright copper plate was inserted into saturate vapor and left 50 mm of it above condensation coil.After 10 min heating the portion of plate in vapor was still bright as before--no oxide.The portion at air-vapor interface gradually became red colour--with thin oxide film.The portion out off vapor was oxided to black or purplish red --with thick oxide film.So the inert characteristics of FC-70 is very obvious.

Some liquid,for example glycerine,has fluxaction.Not only it can be used as heating medium but also as flux.But the measurement of wetting force has indicated that(fig.6 a) neither pure FC-70 nor FC-70 saturated by rosin(0.12% rosin contain) does not help solder spreading.

In order to measure wetting curve of rosin on FC-70 coated and uncoated copple wire,rosin about 4 mm depth was melted on molten solder bath.The results tested with SKC-2 solderability tester are shown in fig.6.The wetting speed of rosin on copper is very slow at 200 C,but rosin can more rapidly wet at 215°C(fig.6 a,b).For copper wire coated by FC-70,although the surface tension reduces,but the wetting of rosin deteriorates.The main reason is resistant effect of FC-70 on wetting .

The wetting process of solder paste on FC-70 coated copper wire was imitated and tested with SKC-2 tester.The wire tested is 1 mm diameter, 20 mm length copper wire.FC-70 was applied on some wires precoated by rosin.Samples with different coating are shown in fig.5 and table 2. The resurts are shown in fig.7.

DISCUSSION

1.Condensation heating is a heat conduct process coursed by phase-transition.Heat is transfered from extenal heat power to cold assembly.This heating method possesses high heat transfer coefficient and does not have overheat hazard.It is clear that surface in different position has different heat transfer coefficient coursed by draining off style.

2.According to condensation heating theory, the effect of surface dimension on heating is rather weak.Horizontal downward surface is easier to heat than upward surface.Small part under large downward surface may be affected by condensation process and droplets.In this case tombstone or movement of small part may become a problem sometimes.

3.The surface of body heated in FC-70 vapor will be always coated by FC-70 film.Even if FC-70 is inert and does not have flux-action,it may have some effect on spreading of rosin(fig6). So FC-70 has some resistance effect on spreading of solder probably.This had been proved by measurement of wetting process(fig.7).

In rellow soldering applied solder paste,i.e. prespreading of rosin is necessary,FC-70 may resist spreading of solder(fig.7 d,e),and will produce a inference on apperence of solder joint. Although this effect is obvious,it is not always a bad thing.People does not always hope to produce very long flow of solder,and form overconcave solder fillet.Excessive concave of solder fillet may deteriorate low cycle creep-fatigue performance.If very good wetting of solder is requried,prefluxing on pad of component and substrate will be necessary.

4.When rosin is precoated,i.e.reflow soldering with prefluxing is used,it is very clear from fig.7 c that FC-70 will not have any effect on solder spreading.

5.Although FC-70 resists spreading of rosin and solder sometimes,but these effects are just on wetting speed and degree.The effect of FC-70 does not change the basic nature of solderability of solder.

REFERENCES

1. George M. Wenger, Roop L Mahajan
Condensation soldering fluids and heat transfer.
Insulation/Circuits Sep. 1979

2. Cui Dianheng
VPS technology
----Principle, Equipment, Fluid and application
Dianzi Gongyi Jishu No12 1986

3. D S L Slinn
Temperature fiexibility in vapor phase soldering
NEPCON 1987/EAST

Table 1. Wetting data of samples heated in vapor

	heating 20"	heating 180"	heating 1800"	heating No (as cleaned)	stored at room tem. for 24 hrs
Time to zero-Newton (second)	4.8	5.6	4.7	1.5	5.2
Wetting force at third secod (Newton)	$+2.1x10^{-4}$	$+2.1x10^{-4}$	$+2.1x10^{-4}$	$-4.7x10^{-4}$	$+3.0x10^{-4}$
Wetting force at tenth secod (Newton)	$-3.2x10^{-4}$	$-3.3x10^{-4}$	$-5.2x10^{-4}$	$-5.7x10^{-4}$	$-3.5x10^{-4}$

Table 2. Samples with different coating

No	1	2	3	4	5	6
Sample			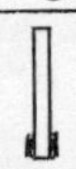			
Coating	Rosin 5 mm	Rosin 5 mm FC-70	Rosin 2 mm	rosin 2 mm FC-70	Paste 2 mm	Paste 2 mm FC-70
Dipping distance: 2 mm						

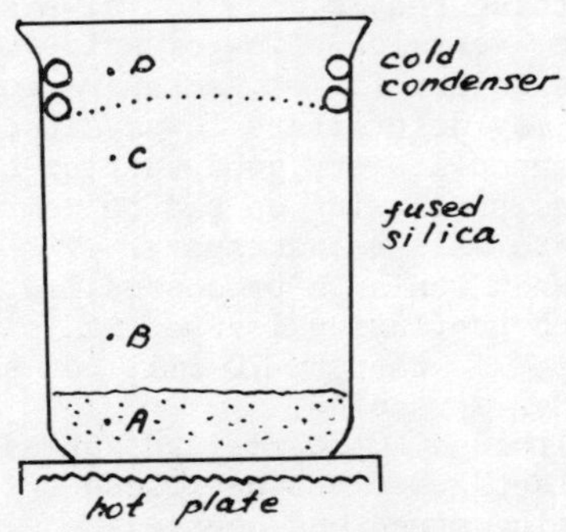

fig.1 Cup-type VPS microsystem

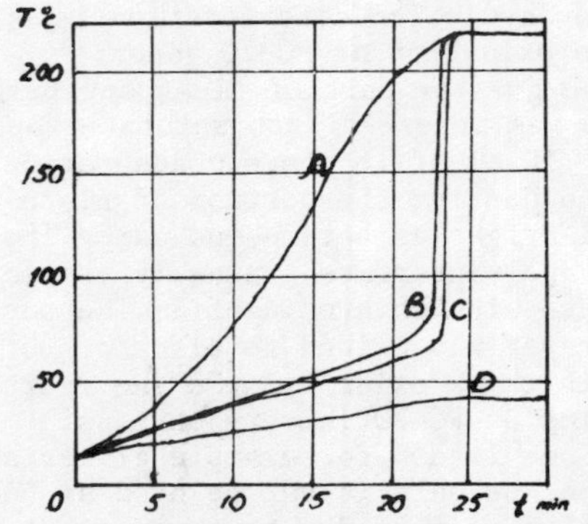

fig.2 Temperature-time curves of cup-type unit

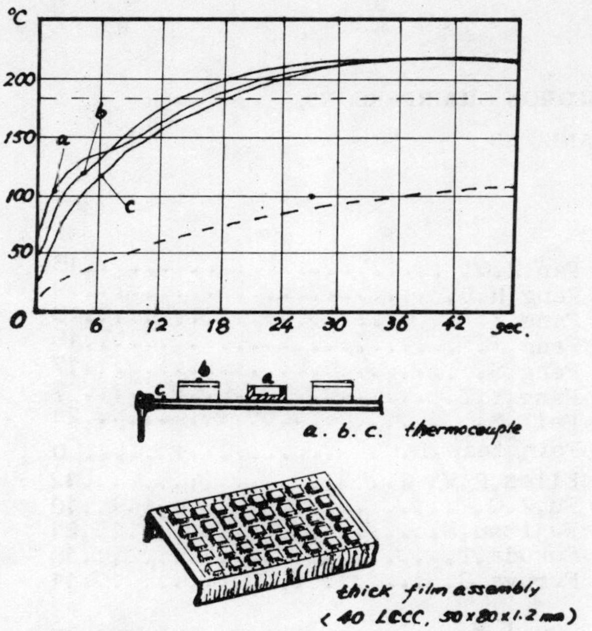

fig.3 Temperature-time response curves of multilayer ceramic hybrid module

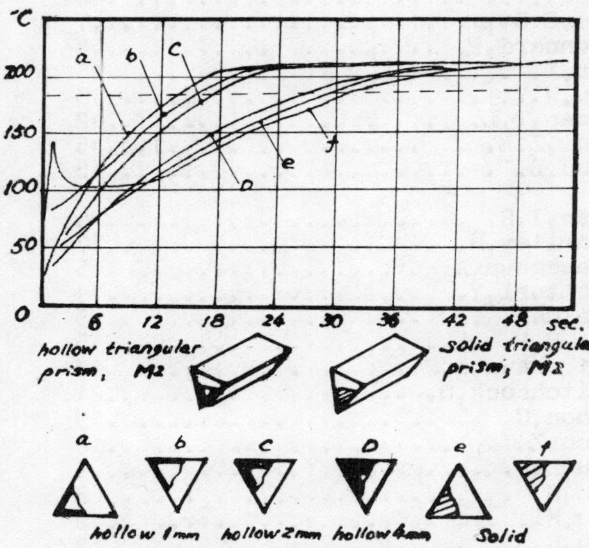

fig.4 Temperature-time response curves of solid and hollow triangular prism

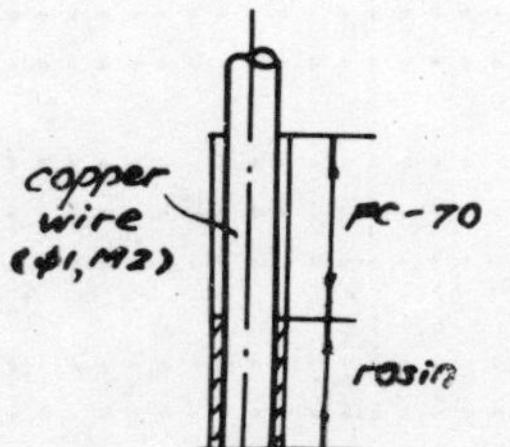

fig.5 Schematic drawing of sample

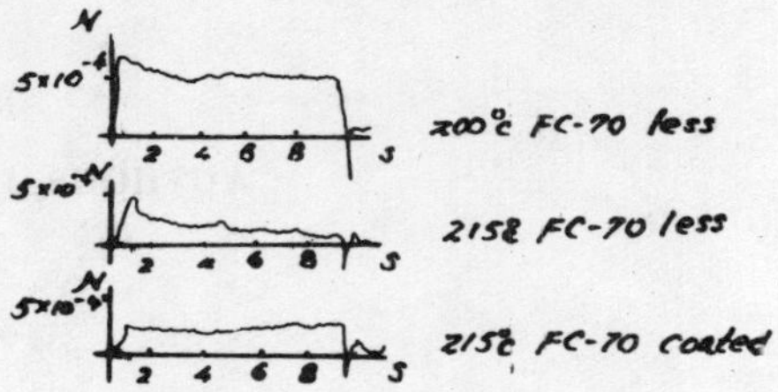

fig.6 Wetting curves of rosin on copper

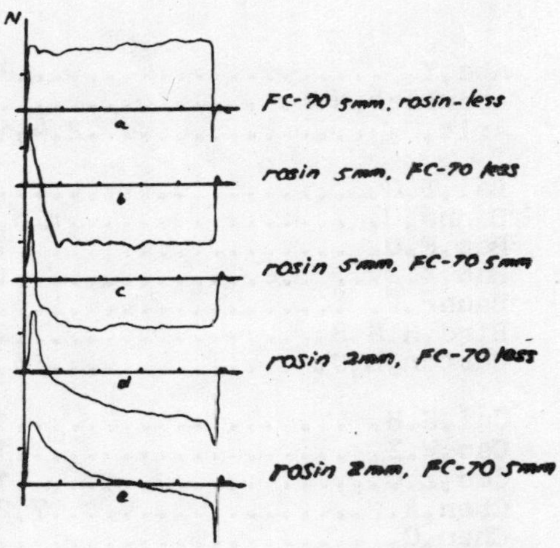

fig.7 Wetting curves of samples coated by rosin and FC-70 (T=215°C, dipping 2 mm, 63Sn/37Pb)

AUTHORS, PANELISTS, SESSION CHAIRPERSONS, AND ORGANIZERS